Catalysis and Electrocatalysis at Nanoparticle Surfaces

Catalysis and Electrocatalysis at Nanoparticle Surfaces

edited by

Andrzej Wieckowski
University of Illinois at Urbana-Champaign
Urbana, Illinois, USA

Elena R. Savinova
Boreskov Institute of Catalysis at the Russian Academy of Sciences
Novosibirsk, Russia

Constantinos G. Vayenas
University of Patras
Patras, Greece

CRC Press
Taylor & Francis Group
Boca Raton London New York

CRC Press is an imprint of the
Taylor & Francis Group, an **informa** business

CRC Press
Taylor & Francis Group
6000 Broken Sound Parkway NW, Suite 300
Boca Raton, FL 33487-2742

First issued in paperback 2019

© 2003 by Taylor & Francis Group, LLC
CRC Press is an imprint of Taylor & Francis Group, an Informa business

No claim to original U.S. Government works

ISBN-13: 978-0-8247-0879-5 (hbk)
ISBN-13: 978-0-367-39545-2 (pbk)

Visit the Taylor & Francis Web site at
http://www.taylorandfrancis.com

and the CRC Press Web site at
http://www.crcpress.com

Foreword

Catalysis and Electrocatalysis at Nanoparticle Surfaces reflects many of the new developments of catalysis, surface science, and electrochemistry. The first three chapters indicate the sophistication of the theory in simulating catalytic processes that occur at the solid–liquid and solid–gas interface in the presence of external potential. The first chapter, by Koper and colleagues, discusses the theory of modeling of catalytic and electrocatalytic reactions. This is followed by studies of simulations of reaction kinetics on nanometer-sized supported catalytic particles by Zhdanov and Kasemo. The final theoretical chapter, by Pacchioni and Illas, deals with the electronic structure and chemisorption properties of supported metal clusters.

Chapter 4, by Batzill and his coworkers, describes modern surface characterization techniques that include photoelectron diffraction and ion scattering as well as scanning probe microscopies. The chapter by Hayden discusses model hydrogen fuel cell electrocatalysts, and the chapter by Ertl and Schuster addresses the electrochemical nanostructuring of surfaces. Henry discusses adsorption and reactions on supported model catalysts, and Goodman and Santra describe size-dependent electronic structure and catalytic properties of metal clusters supported on ultra-thin oxide films. In Chapter 9, Marković and his coworkers discuss modern physical and electrochemical characterization of bimetallic nanoparticle electrocatalysts.

Bönnemann and Richards lead off the section on synthetic approaches with a discussion of nanomaterials as electrocatalysts to tailor structure and interfacial properties. Teranishi and Toshima as well as Simonov and Likholobov discuss preparation and characterization of supported monometallic and bimetallic nanoparticles.

Van der Klink and Tong cover NMR studies of heterogeneous and electrochemical catalysts, and X-ray absorption spectroscopy studies are the focus of the chapter by Mukerjec. In Chapter 15, Stimming and Collins discuss STM and infrared spectroscopy in studies of fuel cell model catalyst.

Lambert reviews the role of alkali additives on metal films and nanoparticles in electrochemical and chemical behavior modifications. Metal-support interactions is the subject of the chapter by Arico and coauthors for applications in low temperature fuel cell electrocatalysts, and Haruta and Tsubota look at the structure and size effect of supported noble metal catalysts in low temperature CO oxidation. Promotion of catalytic activity and the importance of spillover are discussed by Vayenas and coworkers in a very interesting chapter, followed by Verykios's examination of support effects and catalytic performance of nanoparticles. In situ infrared spectroscopy studies of platinum group metals at the electrode–electrolyte interface are reviewed by Sun. Watanabe discusses the design of electrocatalysts for fuel cells, and Coq and Figueras address the question of particle size and support effects on catalytic properties of metallic and bimetallic catalysts.

In Chapter 24, Duo and coworkers discuss metal oxide nanoparticle reactivity on synthetic boron–doped diamond surfaces. Lamy and Léger treat electrocatalysis with electron-conducting polymers in the presence of noble metal nanoparticles, and new nanostructure materials for electrocatalysis are the subject of the final chapter, by Alonso-Vante.

There is no doubt that this book will be a valuable addition to the library of surface scientists, electrochemists, and those working in interface sciences who are interested in research that is being carried out at the frontiers of solid–liquid and solid–gas interfaces as well as nanomaterials. The contributors are outstanding senior researchers and the volume gives a glimpse of the present and the future in this frontier area of physical chemistry and interface chemistry: molecular-level catalysis and electrochemistry, and their applications in the presence of nanoparticles. The editors should be commended for doing such an excellent job in collecting superb chapters for this outstanding volume.

Gabor Samorjai
University of California at Berkeley
Berkeley, California, U.S.A.

Preface

Catalysis and Electrocatalysis at Nanoparticle Surfaces is a modern, authoritative treatise that provides comprehensive coverage of recent advances in nanoscale catalytic and electrocatalytic reactivity. It is a new reference on catalytic and electrochemical nanotechnology, surface science, and theoretical modeling at the graduate level.

Heterogeneous catalysis and aqueous or solid-state electrochemistry have been treated traditionally as different branches of physical chemistry, yet similar concepts are used in these fields and, over the past three decades, similar surface science techniques and ab initio quantum mechanical approaches have been used to investigate their fundamental aspects at the molecular level. The growing technological interest in fuel cells—both high-temperature solid oxide fuel cells (SOFC) and low-temperature polymer electrolyte membrane (PEM) fuel cells—has drawn the attention of the electrochemical community to three-dimensional electrodes with finely dispersed active components (metals, alloys, semiconductors) anchored to appropriate conducting supports. This has raised the question of specific electrochemical properties of small-dimension systems. Utilization of nano-sized materials introduces a number of phenomena specific for nanoscale, including particle size effects, metal-support interactions, and spillover, which are common for heterogeneous catalysis and electrocatalysis, and calls for joint efforts from the experts in these fields. This has brought the catalytic and electrochemical communities closer, as the need for heterogeneous catalytic knowledge in designing and operating efficient fuel cell anodes and cathodes is being more widely recognized. The electrical potential dependence of the electrochemical (electrocatalytic, i.e., net charge transfer) reaction rate has traditionally been considered a characteristic feature that distinguished it from the catalytic (no net charge transfer) reaction rate. Yet recent advances have shown that this distinction is not so rigid, as the rates of catalytic processes at nanoparticles in contact with solid or aqueous electrolytes also depend profoundly on catalyst potential, similar to the electrochemical reaction rate, and that a classic electrochemical double-layer approach is also applicable to the kinetics of catalytic reactions.

The collection of 26 invited chapters by prominent scholars originating from both the electrochemical and the catalytic schools of thought portrays this growing and mutually enriching merge of the fields of heterogeneous catalysis and electrocatalysis to address new technological and fundamental challenges of catalytic and electrocatalytic reactivity at nanoparticle surfaces.

Comprehensiveness is the key qualifier for this book. Its unique feature is that it provides nearly complete coverage of the main topics in contemporary nanoscale catalysis and electrocatalysis, with coherent, state-of-the-art presentation of theory, experimentation, and applications. This is the first volume of its kind to bring together such a broad and diverse profile of surface science and electrochemistry from the perspective of unique nanoscale surface properties. Similarly, it emphasizes the emergence and progress of knowledge of nanoparticle surface properties, the newest subfield of surface science and electrochemistry, which has not been reviewed before except in selected proceedings volumes. Surface science and electrochemical surface science, jointly presented in this volume, have enabled scientists to develop atomic- and molecular-level perspectives of reactions occurring at gas–solid and liquid–solid interfaces. Such perspectives, both current and forecast for the future, are amply demonstrated.

The strength of the field lies in its employment of basic science for technological applications, in relation to materials, catalysis, and energy efforts. The main thrust of this book is that it presents and unifies two main interfaces for use in heterogeneous catalysis and electrocatalysis, for both basic science and applications. The contributors were asked to make their chapters accessible to advanced graduate students in surface science, catalysis, electrochemistry, and related areas. The tutorial feature of the volume adds to its strength and educational value.

The book is divided into six parts: theory of nanoparticle catalysis and electrocatalysis; model systems from single crystals to nanoparticles; synthetic approaches in nanoparticle catalysis and electrocatalysis; advanced experimental concepts; particle size, support, and promotional effects; and advanced electrocatalytic materials. This facilitates access to the general reader's interests. Each chapter begins with a summary and a table of contents to provide an overview of its scope.

Part I presents the state of the art of the theory of catalysis and electrocatalysis at clusters and nanoparticles. This section provides the current frame of modeling of the interaction of clusters with substrates as well as catalytic and electrocatalytic kinetics on clusters and nanoparticles, including ab initio quantum mechanical calculations, and epitomizes recent advances in understanding the relation between electronic structure and catalytic/electrocatalytic activity.

Part II addresses the key question of how electronic and geometric properties of catalytic surfaces change upon moving from single crystals to nanoparticles and how this influences their catalytic/electrocatalytic activity. The use of model catalysts and electrocatalysts in conjunction with powerful surface spectroscopic techniques provides new insight into the unique phenomena observed upon decreasing the size of catalyst particles down to nanoscale and builds a link between catalysis and electrocatalysis at perfect single crystals and nanoparticles. The chapters in this section scrutinize approaches to electrochemical nanostructuring of surfaces and provide insight into unique behavior of small, confined systems.

Part III presents the state of the art of the synthesis of nanoparticle catalysts and electrocatalysts, including bimetallic nanoparticles. Particular emphasis is given to carbon-supported nanoparticles due to their technological significance in the fabrication of electrodes for PEM fuel cells.

Part IV describes recent breakthroughs in the use of advanced experimental techniques for the in situ study of nanoparticle catalysts and electrocatalysts, including X-ray absorption spectroscopy, NMR, and STM.

Part V addresses the critical issues of particle size effects, alloying, spillover, classic and electrochemical promotion, and metal–support interactions on the catalytic and electrocatalytic performance of nanoparticles. Recent experimental evidence is presented on the functional similarities and operational differences of promotion, electrochemical promotion, and metal–support interactions.

Part VI discusses novel advanced electrocatalytic materials, including polymer-embedded nanoparticle electrodes for PEM fuel cells and synthetic diamond–supported electrocatalyst nanoparticles for toxic organic compound treatment.

This book will be an indispensable source of knowledge in laboratories or research centers that specialize in fundamental and practical aspects of heterogeneous catalysis, electrochemistry, and fuel cells. Its unique presentation of the key basic research on such topics in a rich interdisciplinary context will facilitate the researcher's task of improving catalytic materials, in particular for fuel cell applications, based on scientific logic rather than expensive Edisonian trial-and-error methods. The highlight of the volume is the rich and comprehensive coverage of experimental and theoretical aspects of nanoscale surface science and electrochemistry. We hope that readers will benefit from its numerous ready-to-use theoretical formalisms and experimental protocols of general scientific value and utility.

Andrzej Wieckowski
Elena R. Savinova
Constantinos G. Vayenas

Contents

Contributors

Nicolás Alonso-Vante   Université de Poitiers, Poitiers, France

P. L. Antonucci   University of Reggio Calabria, Reggio Calabria, Italy

V. Antonucci   Institute for Transformation and Storage of Energy, Messina, Italy

A. S. Arico   Institute for Transformation and Storage of Energy, Messina, Italy

Santanu Banerjee   University of Southern California, Los Angeles, California, U.S.A.

Matthias Batzill   University of Southern California, Los Angeles, California, U.S.A.

Helmut Bönnemann   Max Planck Institut für Kohlenforschung, Mülheim an der Ruhr, Germany

S. Brosda   University of Patras, Patras, Greece

J. A. Collins   Technical University of Munich, Munich, Germany

C. Comninellis   Swiss Federal Institute of Technology EPFL, Lausanne, Switzerland

Bernard Coq   ENSCM-CNRS, Montpellier, France

Achille De Battisti   Università di Ferrara, Ferrara, Italy

I. Duo   Swiss Federal Institute of Technology EPFL, Lausanne, Switzerland

Gerhard Ertl   Fritz-Haber-Institut der Max-Planck-Gesellschaft, Berlin, Germany

Sergio Ferro   Università di Ferrara, Ferrara, Italy

François Figueras   Institut de Recherche sur la Catalyse, Villeurbanne, France

D. W. Goodman   Texas A&M University, College Station, Texas, U.S.A.

Masatake Haruta   National Institute of Advanced Industrial Science and Technology, Tsukuba, Japan

Brian E. Hayden   The University of Southampton, Southampton, England

Claude R. Henry   CRMC2-CNRS, Marseille, France

Francesc Illas   Universitat de Barcelona, Barcelona, Spain

B. Kasemo   Chalmers University of Technology, Göteborg, Sweden

Bruce E. Koel   University of Southern California, Los Angeles, California, U.S.A.

Marc T. M. Koper   Eindhoven University of Technology, Eindhoven, The Netherlands

Richard M. Lambert   Cambridge University, Cambridge, England

C. Lamy   Université de Poitiers-CNRS, Poitiers, France

J.-M. Léger   Université de Poitiers-CNRS, Poitiers, France

V. A. Likholobov   Boreskov Institute of Catalysis at the Russian Academy of Sciences, Novosibirsk, Russia

N. M. Marković   University of California at Berkeley, Berkeley, California, U.S.A.

Sanjeev Mukerjee   Northeastern University, Boston, Massachusetts, U.S.A.

Matthew Neurock   University of Virginia, Charlottesville, Virginia, U.S.A.

Gianfranco Pacchioni   Università di Milano-Bicocca, Milan, Italy

C. Pliangos   University of Patras, Patras, Greece

V. Radmilovic   University of California at Berkeley, Berkeley, California, U.S.A.

Ryan Richards*   Max Planck Institut für Kohlenforschung, Mülheim an der Ruhr, Germany

P. N. Ross, Jr.   University of California at Berkeley, Berkeley, California, U.S.A.

A. K. Santra   Texas A&M University, College Station, Texas, U.S.A.

Rolf Schuster   Fritz-Haber-Institut der Max-Planck-Gesellschaft, Berlin, Germany

P. A. Simonov   Boreskov Institute of Catalysis at the Russian Academy of Sciences, Novosibirsk, Russia

U. Stimming   Technical University of Munich, Munich, Germany

Shi-Gang Sun   Xiamen University, Xiamen, China

Toshiharu Teranishi   Japan Advanced Institute of Science and Technology, Ishikawa, Japan

Y. Y. Tong   Georgetown University, Washington, D.C., U.S.A.

Naoki Toshima   Tokyo University of Science, Yamaguchi, Japan

---

* *Current affiliation*: International University of Bremen, Bremen, Germany

D. Tsiplakides   University of Patras, Patras, Greece

Susumu Tsubota   National Institute of Advanced Industrial Science and Technology, Ikeda, Japan

J. J. van der Klink   Ecole Polytechnique Federale de Lausanne, Lausanne, Switzerland

Rutger A. van Santen   Eindhoven University of Technology, Eindhoven, The Netherlands

Constantinos G. Vayenas   University of Patras, Patras, Greece

Xenophon E. Verykios   University of Patras, Patras, Greece

Masahiro Watanabe   University of Yamanashi, Kofu, Japan

V. P. Zhdanov   Chalmers University of Technology, Göteborg, Sweden, and Boreskov Institute of Catalysis at the Russian Academy of Sciences, Novosibirsk, Russia

1

Theory and Modeling of Catalytic and Electrocatalytic Reactions

MARC T.M. KOPER and RUTGER A. VAN SANTEN

Eindhoven University of Technology, Eindhoven, The Netherlands

MATTHEW NEUROCK

University of Virginia, Charlottesville, Virginia, U.S.A.

CHAPTER CONTENTS

SUMMARY

A brief overview is given of the main theoretical principles and computer modeling techniques to describe catalytic and electrocatalytic reactions. Our guiding principles in understanding the relationship between electronic structure and catalyst performance are the periodic table and the Sabatier principle. We discuss the different modeling tactics in describing quantum-mechanically adsorbate-surface interactions (i.e., cluster vs. slab models) and their accuracy in modeling (small) catalyst particles. The role of the solvent in electrocatalytic processes is described both from the point of view of quantum–chemical density–functional theory calculations, and from more classical Marcus theory-type considerations using molecular dynamics simulations. Furthermore, we discuss how kinetic Monte Carlo methods can be used to bridge the gap between microscopic reaction models and the catalyst's macroscopic performance in an accurate statistical and mechanical way.

1.1 INTRODUCTION

With the incessant boom in computational power of present-day computers, the role of theory and modeling in the understanding and development of catalytic processes is becoming increasingly important. Theory and computational modeling have already played a significant role in heterogeneous gas-phase catalysis for quite some years, but comparable developments in the electrocatalysis of reactions at metal–liquid interfaces have been less forthcoming. However, new theory developments in electrochemistry, the emergence of new computational possibilities with modern machines and codes, as well as the increasing overlap between electrochemistry and surface science promise a similarly prominent role of "computational electrochemistry" in the electrocatalysis community.

The aim of this chapter is to provide the reader with an overview of the potential of modern computational chemistry in studying catalytic and electrocatalytic reactions. This will take us from state-of-the-art electronic structure calculations of metal–adsorbate interactions, through (ab initio) molecular dynamics simulations of solvent effects in electrode reactions, to lattice-gas-based Monte Carlo simulations of surface reactions taking place on catalyst surfaces. Rather than extensively discussing all the different types of studies that have been carried out, we focus on what we believe to be a few representative examples. We also point out the more general "theory" principles to be drawn from these studies, as well as refer to some of the relevant experimental literature that supports these conclusions. Examples are primarily taken from our own work; other recent review papers, mainly focused on gas-phase catalysis, can be found in [1–3].

The next section gives a brief overview of the main computational techniques currently applied to catalytic problems. These techniques include ab initio electronic structure calculations, (ab initio) molecular dynamics, and Monte Carlo methods. The next three sections are devoted to particular applications of these techniques to catalytic and electrocatalytic issues. We focus on the interaction of CO and hydrogen with metal and alloy surfaces, both from quantum-chemical and statistical-mechanical points of view, as these processes play an important role in fuel-cell catalysis. We also demonstrate the role of the solvent in electrocatalytic bond-breaking reactions, using molecular dynamics simulations as well as extensive electronic structure and ab initio molecular dynamics calculations. Monte Carlo simulations illustrate the importance of lateral interactions, mixing, and surface diffusion in obtaining a correct kinetic description of catalytic processes. Finally, we summarize the main conclusions and give an outlook of the role of computational chemistry in catalysis and electrocatalysis.

1.2 OVERVIEW OF COMPUTATIONAL METHODS

1.2.1 Electronic Structure Calculations

Electronic structure calculations are generally concerned with solving the electronic ground state of a certain molecular ensemble based on first-principles quantum mechanics. The techniques are called *ab initio* if they require only the atomic numbers of atoms involved in the ensemble, and no adjustable parameters taken from experiment or other calculations. A good introduction into the various ab initio and other quantum-chemical methods is given in Jensen's book [4].

By far the most popular method currently used in the quantum-chemical modeling of adsorbate–substrate interactions is that based on the density functional theory (DFT). Contrary to the more "classical" quantum-chemical techniques, DFT is not directly based on electronic wave functions but rather on electronic density. The formal cornerstone of DFT is a theorem derived by Hohenberg and Kohn [5], which states that the ground-state electronic energy is a unique functional of the electronic density $n(r)$, with r the space coordinate. In other words, there exists a one-to-one correspondence between the r-dependent electronic density of the system and the energy. However, the exact functional giving the exact energy is not known, and in practice one must therefore resort to one of the many approximate expressions available. The quality of these functionals is now such that one may calculate overall energies to within an accuracy of about 5–10% of the exact result. For many purposes, this accuracy is quite sufficient.

In practical calculations, the electronic density is still calculated from wave functions, using a method originally devised by Kohn and Sham [6]. These so-called Kohn–Sham orbitals correspond to a "virtual" system of noninteracting electrons, similar to Hartree–Fock methods. Even though these Kohn–Sham orbitals have no clear physical meaning, they may still be quite useful in interpreting the results from a DFT calculation in terms of molecular orbital theory.

The great advantage of DFT over other, say Hartree–Fock-based, methods, is that it does not require too much computational effort to include the effects of electronic correlation in the calculation, which is known to be very important in obtaining reliable energies of chemical bonds. All it takes is an improved functional using the same set of Kohn–Sham orbitals, whereas the Hartree–Fock-based methods can only account for electronic correlation by considerably increasing the number of wave functions, leading to very long computation times for even the smallest system sizes. This makes DFT the method of choice for handling large systems, such as adsorbates interacting with surfaces.

The most popular series of functionals used in computational chemistry are those based on the generalized gradient approximation (GGA). These functionals include not only the electronic density, but also its spatial derivative (gradient), in order to account more accurately for longer-ranged electronic correlation effects (though very long-ranged electronic correlation effects, such as Van der Waals forces, are not accurately treated within the GGA). This type of calculation is often referred to as DFT-GGA. Various DFT-GGA functionals are available; some popular ones have been tested in [4,7,8]. As mentioned, the accuracy of these methods is roughly 5–10%.

Because extended surfaces and even nanoparticles are large systems from the atomistic point of view, the question still arises of what kind of molecular ensemble would constitute a reasonable model for the adsorbate–catalyst interaction. Here, essentially two different types of approaches have been taken. One approach is to model the surface as a small cluster of atoms, where a typical cluster size is 10–50 atoms. It is important to note that this size is still smaller than that of a typical catalytic nanoparticle, which consists of at least several hundreds of atoms (and usually more)[9]. A second approach that has become popular in the last 10 years is to model the surface as a periodic slab, in which a certain ensemble is periodically repeated in three dimensions by applying periodic boundary conditions in the computational setup. A surface, and a molecule-surface ensemble, is then modeled as a number of layers (typically 3–5 layers) of atoms in one part of the periodic cell, and a vacuum in the other part of the cell. The vacuum region is chosen large enough such that the different periodic images of the surface have a negligible interaction with each other. One final comment to be made about clusters and slabs concerns the different types of basis sets in which the Kohn–Sham orbitals are expanded. Cluster calculations are usually carried out with a standard quantum-chemical package employing so-called localized basis sets. These localized basis sets enable one to interpret the Kohn–Sham orbitals as molecular orbitals as introduced in quantum-chemical molecular orbital theory [10]. This may have great advantages in obtaining a more chemistry-based understanding of binding trends. Slab calculations, on the other hand, usually employ so-called plane waves as their basis set, because of the periodicity of the simulation cell [11,12]. Plane waves are not localized, and hence it is more difficult (though not impossible) to obtain molecular orbital-type information. However, it is now generally agreed that slab calculations give more reliable quantitative results when the binding energetics of the molecule-surface interaction is concerned. More local information, such as binding distances and vibrational properties, may still be calculated with good accuracy using cluster models. We discuss these issues in some more detail in Section 1.3.

1.2.2   Ab Initio Molecular Dynamics

One very prominent development in DFT has been the coupling of electronic structure calculations (which, when the ground state is concerned, apply to zero temperature) with finite-temperature molecular dynamics simulations. The founding paper in this field was published by Carr and Parrinello in 1985 [13]. Carr and Parrinello formulate effective equations of motion for the electrons to be solved simultaneously with the classical equations of motion for the ions. The forces on the ions are calculated from first principles by use of the Hellman–Feynman theorem. An alternative to the Carr–Parrinello method is to solve the electronic structure self-consistently at every ionic time step. Both methods are referred to as ab initio molecular dynamics (AIMD) [14].

AIMD is still a very time-consuming simulation method and has so far mainly been used to study the structure and dynamics of bulk water [14,15], as well as proton transfer [16] and simple S_N2 reactions in bulk water [17]. AIMD simulations are as yet limited to small system sizes and "real" simulation times of not more than a few picoseconds. However, some first applications of this technique to interfacial systems of interest to electrochemistry have appeared, such as the water–vapor interface [18] and the structure of the metal–water interface [19]. There is no doubt that AIMD simulations of electrochemical interfaces will become increasingly important in the future, and some first results are described in Section 1.4.2.

1.2.3 Molecular Dynamics

In classical molecular dynamics (MD) methods [20] the atoms and molecules in the system of interest interact through effective pair potentials, which may be obtained from previous ab initio electronic structure calculations. However, the quantum nature of the system is not explicitly taken into account, and the time evolution of the system is obtained by solving Newton's classical equations of motion. Averaging of the MD trajectories over a sufficiently long simulation period allows one to extract thermodynamic, dynamical, and other macroscopic properties. Depending on the system size (which can include up to several thousands of atoms) and the type of particles involved, typical "real" simulation times in classical MD studies can be as long as several nanoseconds. As we discuss in Section 1.4.1, MD simulations are of great value in investigating the role of solvent reorganization in electron-transfer reactions.

1.2.4 Monte Carlo

An alternative to the MD method in obtaining statistical averages is the Monte Carlo (MC) simulation method, which is based on an efficient sampling of low-energy configurations rather than on generating a dynamical trajectory [20]. A drawback of the standard MC algorithm is that it does not give any dynamical information, though for certain purposes it may be a more efficient way of calculating thermodynamic properties.

A method closely related to the standard "Metropolis"-type MC algorithm is a simulation technique known as kinetic Monte Carlo or dynamic Monte Carlo (DMC). This method is especially useful for studying processes on lattices, such as, for instance, catalytic reactions taking place on the reaction sites of a catalyst surface. One first defines the adsorption rates, desorption rates, reaction rates, and diffusion rates of the various reactions and processes assumed to take place on the surface, quantities that may in fact be estimated from first-principles electronic structure calculations. Next, the evolution of the entire system is obtained by solving the so-called master equation using an MC-type algorithm [21–24]. The algorithm is designed such that the exact time dependence is obtained, i.e., that the subsequent configurations generated satisfy the correct detailed balance. As Section 1.5 shows, this method is very useful in obtaining the correct dynamic behavior of a reactive system from the elementary processes. There is no need to resort to statistical-mechanical approximation schemes such as the mean-field approximation [9] in which all rates are expressed in terms of average coverages. Such approximation schemes may break down severely if the catalyst surface is very inhomogeneous, for instance due to poor mixing of the reactants on the catalyst surface or to a low concentration of catalytically active sites, or when strong interactions between adsorbates exist. DMC is the method of choice for the microkinetic modeling of catalytic reactions on surfaces when ordering, island formation, and slow surface mobility are deemed important. Lattice sizes in DMC can be as large as 1000×1000 sites (corresponding to 1 million surface sites), whereas the "real" computation times obviously depend on the (lowest) rate constants but can span time scales from a few milliseconds to several minutes.

1.3 REACTIVITY AND CATALYST ELECTRONIC STRUCTURE

1.3.1 Modeling Small Particles

The role of particle size in catalysis and electrocatalysis is a subject of longstanding interest. It is not our intention here to discuss in detail the available experimental and theoretical literature. Extensive reviews on particle-size effects in gas-phase catalysis and electrocatalysis can be found in the papers of Henry [25] and Kinoshita [26], respectively. Also, several monographs, reviews, and conference proceedings discuss particle-size effects from experimental, theoretical, and computational points of view [9,27,28].

It is well known that the electronic properties of small metal particles depend on the particle size. The work function or cluster ionization potential is a most dramatic example of this effect. Knickelbein et al. [29] have shown that for Ni clusters consisting of 3 to 90 atoms the cluster ionization potential continues to change and for the largest particle size is still about 0.2 eV higher than the work function of an extended surface. There is, however, a growing awareness that changes in reactivity and chemisorption properties with particle size reflect a much more local phenomenon. Smaller particles have more edges and kinks than larger particles, and these local sites usually possess a reactivity that is substantially different from those of the terraces. Surface-science studies in UHV and electrocatalysis have convincingly demonstrated the special reactivity of defects, showing that in certain cases the catalytic reaction takes place only on step and defect sites.

These facts obviously raise the question of what constitutes the best computational model of a small catalytic particle. As catalysis is often a local phenomenon, a cluster model of the reactive or chemisorption site may give quite a reasonable description of what happens at the "real" surface [1,3,30]. However, the cluster should still be large enough to eliminate cluster edge effects, and even then one must bear in mind that the cluster sizes employed in many computational studies are still much smaller than real catalytic particles (say 10–50 versus 50–1000 atoms, respectively). Hence, a slab model of a stepped surface may provide a much more realistic model of the active site of a catalytic nanoparticle. Hammer [31,32] has carried out quite extensive DFT-GGA slab calculations of N_2 and NO dissociation at stepped Ru and Pd surfaces, showing how the dissociation energy is significantly lower at the low-coordination step sites compared to terrace sites. The special reactivity of step sites for the dissociation of NO and N_2 has been demonstrated in several recent surface-science studies [33,34]. Also, the preferential adsorption of CO on step sites has been demonstrated in UHV [35], under electrochemical conditions [36], as well as by means of DFT-GGA slab calculations [37].

1.3.2 Catalytic Reactivity and the Periodic Table

Because a catalytic reaction generally consists of a series of elementary steps, one of the key questions in understanding catalytic reactivity is which of these steps is rate-limiting. The rate-limiting step determines not only the overall catalytic activity, but often also the selectivity of the reaction. In search of an optimal catalyst, the overall rate of the reaction is often studied as a function of a certain "reactivity parameter" of the catalyst material. A typical "reactivity parameter" is the heat of compound formation, e.g., surface oxide formation, or the enthalpy of adsorption of what is believed to be a key intermediate. Plots of catalyst reactivity versus the reactivity parameter usually yield the well-known "volcano plots," the rationalization for which was first given by Sabatier [38]. The Sabatier principle essentially states that an active catalyst should adsorb a key intermediate neither too weakly nor too strongly. A weakly adsorbed intermediate has a low reactivity because of its low concentration, and a strongly adsorbed intermediate leads to a low reactivity because of the difficulty with which it can desorb. An intermediate interaction strength often leads to the optimum activity and selectivity.

There are innumerable examples of volcano plots and the Sabatier principle in catalysis, as we have recently elaborated in relation to heterogeneous gas-phase catalysis [2]. An illustrative recent example of the Sabatier example in electrocatalysis is the ammonia oxidation in alkaline solution [39]. The selectivity of the ammonia electro-oxidation toward N_2 is determined by the stability and concentration of hydrogenated nitrogen adsorbates NH_x ($x = 1,2$), as the atomic nitrogen adsorbate is inactive in producing N_2 at room temperature. As a result, metals that adsorb N strongly, such as Ru, Rh, and Pd, show no selective NH_3 oxidation, as the dehydrogenation capacity of these metals is too high, leading to a fully inactive N_{ad}-covered surface. Coinage metals such as Cu, Ag, and Au show very little dehydrogenation capacity, and hence no activity for ammonia oxidation under electrochemical conditions. Only Pt and Ir, which show an intermediate dehydrogenation capacity, are able to oxidize ammonia to N_2 with a reasonable steady-state activity. The dehydrogenation capacity may be related to the chemisorption energy of atomic nitrogen: a high adsorption energy would imply a high dehydrogenation capacity.

Given the importance of the Sabatier principle in catalytic science, the question arises about whether the reactivity parameter, usually the chemisorption energy of some key intermediate, can be related to a more fundamental property of the catalyst surface. In particular, one would like to make a connection between the reactivity parameter and the properties of a metal or alloy surface as they can be deduced from the periodic table. An important development along these lines has been a model proposed by Hammer and Nørskov [3,40]. Their model singles out three surface properties contributing to the ability of the surface to make and break Adsorbate bonds: (1) the center ε_d of the d-band; (2) the degree of filling f_d of the d-band; and (3) the coupling matrix element V_{ad} between the adsorbate states and the metal d-states. These quantities have been calculated from extensive DFT calculations for a significant portion of the periodic table. The basic idea of the Hammer–Nørskov model is that trends in the interaction and reactivity are governed by the coupling of the adsorbate states with the metal d-states, since the coupling with the metal sp-states is essentially the same for the transition and noble metals.

Hammer and Nørskov have tested their model quite extensively [3] and have shown a good semiquantitative agreement between their model for the d contribution of the adsorption energy and the DFT-computed total adsorption energy of atomic adsorbates such as oxygen and hydrogen and molecular adsorbates such as carbon monoxide. Also, the activation energy, and hence the reactivity, of hydrogen dissociation onto various metal surfaces is described semiquantitatively by their model. Significantly, many of the observed variations among the different metals can be traced to variations in ε_d, and the hence total adsorption energy varies as $E_{ads} \sim \gamma\varepsilon_d$. Shifts in the (local) d-band energy, and hence variations in reactivity, can be brought about in a number of different ways. Variations in ε_d occur at steps and defects (see previous section), by alloying, or by making an overlayer system. We make use of this "d-band shift model" in the next section.

1.3.3 Hydrogen and Carbon Monoxide Adsorption on (Bi)metallic Surfaces

Alloy surfaces are of substantial importance in catalysis, such as in the hydrogenation of unsaturated hydrocarbons, Fisher–Tropsch synthesis, steam reforming of methane, and many other processes [41]. In electrocatalysis, they have recently received attention in relation to the development of CO-tolerant fuel-cell catalysts [42]. In many of these processes, atomic hydrogen and carbon monoxide are the most important intermediates or poisons; therefore, these two adsorbates have received a great deal of attention in theoretical and computational studies.

Pallassana and Neurock [43–44], for example, use periodic DFT calculations to examine the chemisorption of hydrogen over different Pd_xRe_{1-x} surfaces in order to understand the properties that control chemisorption over these alloys. The results for hydrogen adsorption over the bare Pd and Re surfaces along with pseudomorphic overlayers of Pd on Re(0001) and Re on Pd(111) are presented here in Figure 1 [43]. The binding energy for atomic hydrogen is significantly

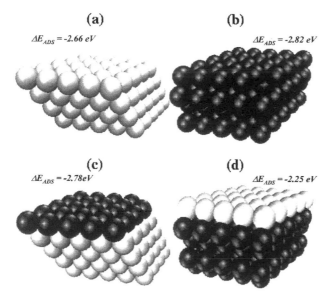

Figure 1 Adsorption of atomic hydrogen on Pd–Re model surfaces. (a) pure Pd(111), (b) pure Re(0001), (c) monolayer of Re on Pd(111), and (d) monolayer of Pd on Re(0001). (Adapted from Ref. [43].)

stronger on Re than it is on Pd. Re lies much further to the left in the periodic table and therefore has a much more open d-band. Re, therefore, contains more vacant states near the Fermi level that can accommodate electron transfer from the adsorbed hydrogen, thus leading to a stronger binding energy. The d-band center model proposed by Hammer and Nørskov [40] suggests that there should be an increase in orbital overlap since the center of the d-band shifts closer to the Fermi level. In addition, Pauli repulsion is reduced as we move from Pd to Re since Re has fewer filled states. Depositing an atomic layer of Re over Pd(111) or Pd over Re(0001) creates ideal pseudomorphic $Re_{ML}/Pd(111)$ and $Pd_{ML}/Re(0001)$ surfaces that have unique properties due to the degree of charge transfer between the two metals. The density of states projected onto the d-states at the top metal layer of Pd(111), Re(0001), $Pd_{ML}/Re(0001)$, and $Re_{ML}/Pd(111)$ were calculated in order to probe the electronic properties for each of these surfaces [43]. The results, which are shown in Figures 2 and 3, indicate that Pd and Re form a very strong bond at the Pd–Re interface of the Pd_{ML}–Re(0001) surface. An analysis of the density of states shows a significant shift of the d-band well below the Fermi energy. The density of states near the Fermi level looks quite similar to that of bulk gold. This would suggest that adsorbates should bind much more weakly on a the Pd_{ML}–Re(0001) surface than those on the pure Pd or pure Re surfaces. Indeed, the binding energy decreases from 2.77 to 2.35 eV as we move from $Pd_{ML}/Pd(111)$ to $Pd_{ML}/Re(0001)$. As we move to the Re surfaces, we find that there is little change in the hydrogen adsorption energies. The binding energy of hydrogen on Re is sufficiently strong that there is basically no change as we move from $Re_{ML}/Re(0001)$ to $Re_{ML}/Pd(111)$. The calculated binding energies for hydrogen on these surfaces correlated very well with the d-band center of the surface layer. The results are shown in Figure 3.

All the trends from these pseudomorphic studies readily fall out of the d-band center model. Pallassana and Neurock have shown that this model can be extended to predict what will happen over the alloyed surfaces as well [44]. In general, the hydrogen binding energy increases as the level of Re in the Pd surface layer is increased over the Pd(111) slab and decreases as the amount of Pd in the Re surface increases over the Re(0001) slab. Modeling the effect of the alloy, however, requires that we distinguish between Pd and Re adsorption sites since the energies on these metals are quite different. A weighted d-band model of the surface was therefore defined in order to account for differences between Pd and Re. The weighted d-band model is given in Eq. (1):

$$\in_{d-weighted} = \frac{(V_{Re}^2 \cdot \in_d^{Re} \cdot N^{Re} + V_{Pd}^2 \cdot \in_d^{Pd})}{(V_{Re}^2 \cdot N^{Re} + V_{Pd}^2 \cdot N^{Pd})} \tag{1}$$

V^2 here refers to the d-band coupling matrix element for the surface metal atoms, whereas N^{Re} and N^{Pd} refer to the number of Re–H and Pd–H bonds, respectively [44].

Hydrogen adsorption was examined on a series of alloyed surfaces including $Re_{ML}/Pd(111)$, $Pd_{33}Re_{66ML}/Pd(111)$, $Pd_{66}Re_{33ML}/Pd(111)$, $Pd_{ML}/Pd(111)$, $Pd_{ML}/Re(0001)$, $Pd_{66}Re_{33ML}/Re(0001)$, $Pd_{33}Re_{66ML}/Re(0001)$, and $Re_{ML}/Re(0001)$. The calculated hydrogen-binding energies on these surfaces are given in Figure 4. The weighted d-band model appears to capture most of the changes to the electronic structure. A plot of the weighted d-band against the DFT calculated binding energies

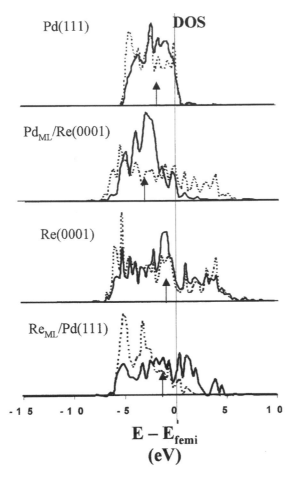

Figure 2   The density of states projected to the d-band of Pd–Re model surfaces. Solid line refers to the DOS projected to the d-band of the top layer. The dotted line refers to the second layer. (a) pure Pd(111), (b) $Pd_{ML}/Re(0001)$, (c) Re(0001), and (d) $Re_{ML}/Pd(111)$. The center of the d-band for each surface is depicted by the arrow. (Adapted from Ref. [43].)

on these alloyed surfaces indicates that this model works quite well for predicting binding energies over the PdRe alloys (Fig. 5). Some of the outlier points are systems where the most favored binding site changed due to a change in the surface composition. On the $Pd_{33}Re_{66}$ surface, for example, Pd acts to isolate smaller Re dimers structures. These dimers are unique. They behave like site-isolated Re clusters and therefore bind hydrogen more strongly. This would readily explain the deviations from the simple d-band center model.

The adsorption of carbon monoxide on metal surfaces can be qualitatively understood using a model originally formulated by Blyholder [45]. A simplified molecular orbital picture of the interaction of CO with a transition metal surface is given in Figure 6. The CO frontier orbitals 5σ and $2\pi^*$ interact with the localized d metal states by splitting into bonding and antibonding hybridized metal-

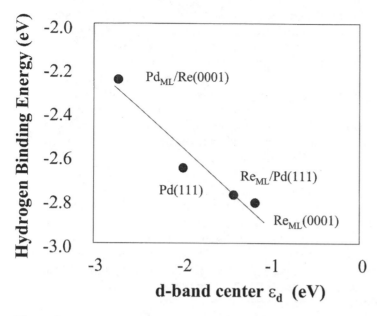

Figure 3   The binding energy of atomic hydrogen at the 3-fold fcc site on the Pd–Re surfaces as a function of the d-band center relative to the Fermi energy. (Adapted from Ref. [43].)

chemisorbate orbitals, which are in turn broadened by the interaction with the much more delocalized sp metal states.

Hammer et al. [46] show how this simple picture emphasizing the interaction of the CO frontier orbitals with the metal d-states can explain trends in the binding energies of CO to various (modified) metal surfaces. Generally, CO binds stronger to the lower transition metals (i.e., toward the left in the periodic table), mainly due to the center of the d-band ε_d moving up in energy, leading to a stronger back donation. The same Blyholder model may also be used to explain why on metals in the upper-right corner of the periodic table (Pd, Ni) CO prefers multifold coordination, whereas toward the lower left corner (Ru, Ir) CO preferentially adsorbs atop (although steric considerations also play an important role in the argument) [47].

Of special interest to fuel-cell catalysis is the adsorption of CO to Pt-based bimetallic surfaces, in particular Pt–Ru. These surfaces have long been known to be considerably more active in the electrochemical CO oxidation than pure Pt or Ru. They are also more CO-tolerant hydrogen oxidation catalysts and better methanol oxidation catalysts for use in low-temperature fuel cells. Watanabe and Motoo [48] suggest a bifunctional mechanism to explain the unique reactivity of Pt–Ru surfaces. In their mechanism, the oxygen species with which CO reacts, presumably chemisorbed OH formed from the dissociative water oxidation, is preferentially formed on the Ru sites. The active site is a Pt–Ru pair with CO on Pt reacting with OH on Ru. Dynamic Monte Carlo simulations of this mechanism, showing the importance of CO mobility [49], are discussed in Section 1.5.2. Although the higher oxophilicity of Ru compared to Pt renders the bifunctional mechanism a reasonable assertion, it does not take into account any changes in the CO binding properties to the Pt–Ru surface compared to pure Pt or Ru.

We have recently carried out extensive slab calculations of CO and OH interacting with Pt–Ru surfaces [50,51]. Results from both our groups show that on the surface of a bulk alloy the CO binding to the Pt site weakens whereas that to the Ru site gets stronger. For example, on the surface of a homogeneous $Pt_2Ru(111)$ alloy, the CO atop binding to Pt is weakened by about 0.2–0.3 eV and that to Ru strengthened by about 0.1 eV. However, real catalytic surfaces are not homogeneous but may show the tendency to surface segregate. Figure 7 shows the binding energy and vibrational properties of CO to an atop Pt site on a series of surfaces with a pure Pt top layer, but for which the bulk composition changes from pure Pt, to Pt:Ru 2:1, to Pt:Ru 1:2, to pure Ru. It is observed that a higher fraction of Ru in the bulk causes a weakening of the CO bond to the Pt overlayer. This electronic alloying effect can be understood on the basis of the Hammer–Nørskov d-band shift model. Alloying Pt with Ru causes a downshift of the local d-band center on the Pt sites because the flow of electrons from Pt to Ru lowers the local d-band on Pt in order to maintain local electroneutrality. Interestingly, Figure 7 also illustrates that there is no correlation between the binding energy and the C–O stretching frequency, even though such a correlation is often assumed in the literature. On the other hand, the Pt–C stretching mode correlates well with the binding energy: a stronger bond causes the expected increase in the Pt–C frequency.

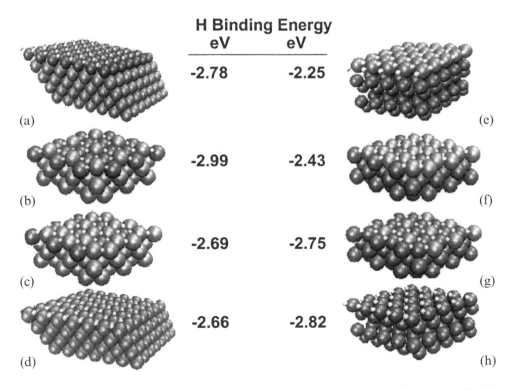

Figure 4 The binding energy of atomic hydrogen at the most favorable sites on Pd–Re alloyed surfaces: (a) $Re_{ML}/Pd(111)$, (b) $Pd_{33}Re_{66ML}/Pd(111)$, (c) $Pd_{66}Re_{33ML}/Pd(111)$, (d) Pd(111), (e) $Pd_{ML}/Re(0001)$, (f) $Pd_{66}Re_{33ML}/Re(0001)$, (g) $Pd_{33}Re_{66ML}/Re(0001)$, and (h) $Re_{ML}/Re(0001)$. (Adapted from Ref. [43].)

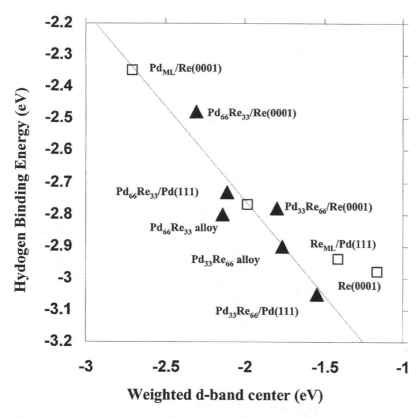

Figure 5 A comparison of the binding energy of atomic hydrogen at the most favorable adsorption sites on the Pd–Re alloyed surfaces and the d-band center relative to the Fermi energy. (Adapted from Ref. [44].)

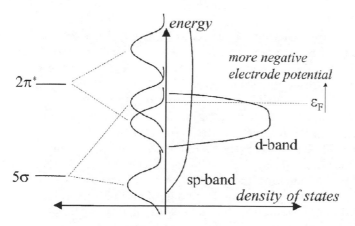

Figure 6 Model of the orbital interaction diagram for CO adsorbed on transition-metal surfaces. (Adapted from Ref. [47].)

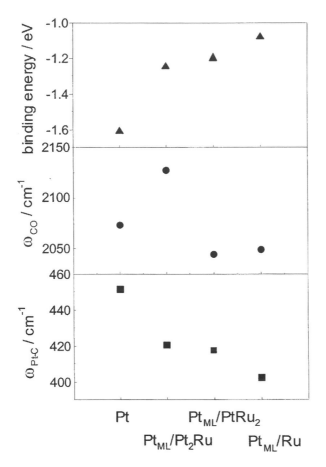

Figure 7 The binding energy, C–O stretching frequency ω_{C-O}, and Pt–C stretching frequency ω_{Pt-C} on four different surfaces, all with a surface layer of Pt, but different bulk compositions (Pt, Pt_2Ru, $PtRu_2$, Ru). (Adapted from Ref. [51].)

The opposite effect is observed when a surface with a pure Ru overlayer is considered, and the bulk is stepwise enriched with Pt. The CO binding to the Ru overlayer is strengthened with a higher content of Pt in the bulk. In fact, the Ru/Pt(111) surface is the one that shows the strongest CO binding. A similar effect is also observed when the top layer of a Pt(111) surface is mixed with Ru, leading to a surface alloy. As the Ru content of the surface layer increases, the binding to the Pt surface sites is weakened and to the Ru surface sites is strengthened.

The binding of OH to Pt–Ru surfaces shows trends that are quite similar to CO. In general, those surface sites that bind CO strongly (which are usually rich in Ru) also show a strong OH binding, and those surface sites that bind CO weakly (which are usually rich in Pt) also show a weak OH binding. This raises the question of whether Pt–Ru is a good example of a truly bifunctional catalyst with one surface component adsorbing one reactant and the other surface component the other reactant. In fact, on the basis of their extensive experimental investigations, the

Berkeley group has suggested that Pt–Sn and Pt–Mo alloys may constitute better examples of bifunctional catalysts [52]. Indeed, our most recent DFT-GGA slab calculations [53] show that on these surfaces CO may preferentially bind to Pt, whereas OH prefers to bind to Sn or Mo.

1.3.4 Electric Field Effects on CO Adsorption

One effect of special interest to electrochemists is the potential-dependent chemisorption of ions and molecules on electrode surfaces. A particularly well-studied example is the adsorption of CO on platinum single-crystal electrodes. In collaboration with the Weaver group at Purdue, we have recently undertaken detailed DFT calculations of the potential-dependent chemisorption of CO on platinum-group (111) surfaces [47,54,55], modeled as clusters, for comparison with the extensive vibrational characterization of these systems as carried out by the Purdue group [56,57]. The electrode potential in these studies is modeled as a variable external electric field applied across the cluster, an approach many others have taken in the past.

Two issues are of interest in relation to the potential-dependent bonding of CO to transition-metal surfaces: the potential-dependent binding energy and the resulting potential-dependent site preference; and the potential-dependent vibrational properties, in particular the internal C–O stretching frequency and the metal-chemisorbate Pt–CO stretching mode. Figure 8 shows the effect of the applied field on the binding energy of CO in a onefold (atop) and multifold (hollow) coordination on a 13-atom Pt cluster. What is particularly significant in this figure is the change in site preference predicted by these calculations, from atop coordination at positive

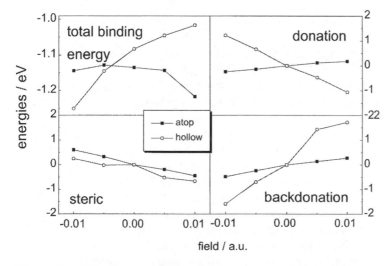

Figure 8 Field-dependent binding energy plots, and constituent steric, donation, and back donation components, for CO adsorbed atop and hollow on a 13-atom Pt (111)-type cluster. The orbital components are plotted with respect to their zero-field values. 0.01 field a.u. corresponds to 0.514 V/Å. (Adapted from Ref. [54].)

fields to threefold coordination at negative fields. Qualitatively, such a potential-dependent site switch is indeed observed experimentally [56,57]. By decomposing the binding energy in steric and orbital contributions, and deconvolving the contributions of the different Kohn–Sham orbitals according to the different symmetry groups, it can be appreciated that the main factor driving this preference for multifold coordination toward a negative field is the back donation contribution. This increasing importance of back donation with a more negative field can be understood on the basis of the classical Blyholder picture. A more negative field implies an upward shift of the metal electronic levels with respect to the chemisorbate $2\pi^*$ orbital, leading to an enhanced back donation of metal electrons. Since the interaction of the metal with the $2\pi^*$ orbital is a bonding interaction, and bonding interactions tend to favor multifold coordination, CO will favor multifold adsorption sites at negative fields.

The change in the intramolecular C–O stretching mode with electrode potential has also been a subject of longstanding theoretical interest and has become known as the interfacial Stark effect. Important theoretical contributions in this field have been made by Lambert, Bagus, Illas, Curulla, and Head-Gordon and Tully [58–62]. We have recently reexamined this issue in the context of a detailed comparison of DFT calculations with the spectroscopic data of CO and NO adsorbed on a variety of single-crystalline transition-metal electrodes [47]. These calculations have confirmed that NO generally possesses a higher Stark tuning slope (i.e., change in C–O stretching mode with potential) than CO and that multifold CO exhibits a higher Stark tuning slope than atop CO. In agreement with earlier calculations, a decomposition analysis shows that the lowering (blueshift) of the internal C–O stretching mode upon adsorption is mainly due to the back donation contribution, as metal electrons occupy the $2\pi^*$ antibonding orbital of CO [63,64]. This back donation contribution is offsetting the steric contribution, also known as the "wall effect," which by itself leads to a redshift upon chemisorption. Although the positive Stark tuning slope is also mainly the result of the back donation contribution, i.e., the negative field leading to enhanced back donation and hence a lowering of the C–O stretching mode, the different slopes found for the different metals and coordination geometries are generally found to be the result of a subtle interplay between back donation, donation, and steric effects [47].

The field dependence of the substrate-chemisorbate M–CO stretching mode has received much less attention as its low frequency makes it difficult to be observed by in-situ IR spectroscopy. However, recent advances in the preparation of transition-metal surfaces for surface-enhanced raman spectroscopy (SERS) have made it possible to study in some detail the potential dependence of this vibrational mode [65,66]. Our recent DFT calculations suggest that, if a sufficiently wide potential window is available, the M–CO mode on Pt, Pd, and Ir exhibits a maximum [55]. The M–CO Stark tuning slope is positive for (very) negative fields, and negative for positive fields. A charge and dynamic dipole analysis indicates that this behavior is at least partially related to the ionicity of the surface bond. Negatively charged adsorbates, such as chemisorbed Cl or CO at negative fields, exhibit a positive Stark tuning slope, whereas positively charged adsorbates, such as chemisorbed Na and CO at positive fields, exhibit a negative Stark tuning slope. Covalently bonded adsorbates, such as CO at intermediate electric fields, do not exhibit a strong dependence of surface bond vibration on the electric field, and in fact for these adsorbates the relationship between ionicity and Stark tuning slope is not one to one [67]. Hence, the potential dependence of the surface bond vibration may serve as a (semiquantitative) measure of the bond ionicity, provided the bond is indeed ionic enough. Experimentally, the Pt–CO vibration exhibits a negative Stark tuning slope, and the maximum predicted by the DFT calculations has yet to be observed experimentally.

1.4 THE SOLVENT IN ELECTRODE REACTIONS

1.4.1 Marcus Theory and Molecular Dynamics of Electrode Reactions

One major complication that distinguishes electrocatalytic reactions from catalytic reactions at metal–gas or metal–vacuum interfaces is the influence of the solvent. Modeling the role of the solvent in electrode reactions essentially started with the pioneering work of Marcus [68]. Originally these theories were formulated to describe relatively simple electron-transfer reactions, but more recently also ion-transfer reactions and bond-breaking reactions have been incorporated [69–71]. Moreover, extensive molecular dynamics simulations have been carried out to obtain a more molecular picture of the role of the solvent in charge-transfer processes, either in solution or at metal–solution interfaces.

In the Marcus theory, the electron-transfer act between a donor and an acceptor molecule is a radiationless event accommodated by a suitable non-equilibrium configuration of surrounding solvent molecules. Hence, the relevant reaction coordinate in electron-transfer reactions is a kind of collective solvent coordinate, more precisely defined as the electrostatic potential at the location of the accepting or donating redox species due to the prevailing configuration of polar solvent molecules. Clearly, the energy associated with creating these nonequilibrium configurations is a free energy, as a multitude of solvent configurations may correspond to one-and-the-same value of this electrostatic potential. The idea of this collective or generalized solvent coordinate is hence to map all different solvent configurations onto a single reaction coordinate.

More explicitly, we can consider the prototype of a simple electron-transfer redox reaction, the ferri/ferro couple:

$$Fe^{3+} + e^- \leftrightarrow Fe^{2+} \tag{2}$$

at some inert electrode material such as gold. Figure 9 shows a typical free-energy surface. In the Marcus theory, the free-energy surface consists of two parabolic curves with minima at the equilibrium solvent configuration(s). At the equilibrium electrode potential, the free energies of the two minima are equal. The solvent fluctuates around the minimum, and when it reaches the intersection point of the two curves, an electron may be exchanged with the metal in a radiationless fashion. This is the electron-transfer equivalent of the Franck–Condon principle, as during the electron exchange at the crossing point the nuclei do not change their positions. Whether the electron transfer really occurs is determined by the adiabaticity of the process. Usually, if the reaction takes place sufficiently close to the electrode, the electron transfer will always take place and the reaction is adiabatic.

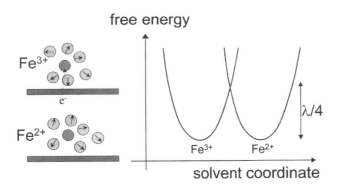

Figure 9   Free-energy curves for simple "outer-sphere" electron-transfer reactions. $\lambda/4$ is the activation energy at equilibrium, λ being the solvent reorganization energy.

The activation energy for electron transfer is thus determined by the free energy at the intersection point (i.e., the transition state), which in turn depends on the curvature of the parabolic curves. This curvature, traditionally denoted by the symbol λ (see Figure 9), is known as the solvent reorganization energy. In the formula originally given by Marcus, the solvent reorganization energy depends on the static and optical dielectric properties of the solvent, ε_{opt} and ε_s, and the radius a of the reactant(s). For a single ion reacting at an electrode, the Marcus formula reads

$$\lambda = \frac{e_0^2}{a}\left(\frac{1}{\varepsilon_{opt}} - \frac{1}{\varepsilon_s}\right) \tag{3}$$

where we have neglected the image interaction of the ion with the metal electrode. This expression assumes a linear response between the charge on the ion and the polarization induced in the surrounding solvent, resulting in the prediction that the reorganization energy depends only on the size of the ion and not on its charge. However, recent detailed MD simulations [72] have shown that this is not an accurate description, and the reorganization energy of multivalent ions is lower than that of uncharged species or ions of low valency with the same size. The main reason for this is the dielectric saturation occurring near strongly charged ions, which may be interpreted in an effective lowering of the static dielectric constant ε_s near the ion, leading to a lowering of λ according to Eq. (3). Strongly nonlinear solvent responses also occur in the transition from a neutral species to a singly charged species, as the ion–dipole interactions lower the effective radius a in Eq. (3). The effect is particularly strong for a negatively charged species as the water molecule can approach closer to a negatively charged species than to a positively charged species [72].

Many technologically important electrochemical processes are more complicated than the simple electron transfer in the Fe^{3+}/Fe^{2+} couple, in which neither of the reactants is supposed to adsorb onto the electrode surface. Typically, however, reactants or products are adsorbed onto the electrode surface, or the electron-transfer act induces the breaking of a bond in the reacting molecule. It is quite clear that in these more complex situations the Marcus theory is expected to break down and a more detailed description of the solvent–reactant–metal interaction is required.

It is in the understanding of these processes that molecular dynamics (MD) currently play a vital role.

The electron exchange between the relevant energy level on the reactant and the metal levels can be treated by a so-called Anderson–Newns model [73]. The interaction with the water solvent is treated by extensive MD simulations. In a sense, this represents a kind of highly simplified tight-binding MD. The energy barrier for electron transfer is calculated by so-called umbrella sampling techniques, which is a systematic way of sampling trajectories away from the equilibrium configurations [20]. The electron transfer itself is treated adiabatically.

We have applied this formalism to the simplest type of bond-breaking electron-transfer reaction at a metal electrode [71,74]. The type of reaction we have in mind is the reductive cleavage of an R–X molecule:

$$R\text{-}X + e^- \rightarrow R^\bullet + X^- \tag{4}$$

where typically $R^\bullet$ is some carbon-centered radical fragment and X^- a halide ion. The R–X molecule is separated from the metal electrode surface by at least one layer of water molecules, in agreement with the experimental deductions from the Savéant group for the methylchloride reduction [75].

Assuming a simple Marcus-type model for the interaction with the solvent, one can derive an analytical expression for the potential (free) energy surface of the bond-breaking electron-transfer reaction as a function of the collective solvent coordinate q and the distance r between the fragments R and X [71,75]. The activation energy of the reaction can also be calculated explicitly:

$$\Delta G_{act} = \frac{(\lambda + D_e + \eta)^2}{4(\lambda + D_e)} \tag{5}$$

where λ is the solvent reorganization energy, D_e the bond dissociation energy, and η the free energy of the reaction, i.e., the distance from the equilibrium potential ("overpotential"). Savéant, who originally derived Eq. (5), carried out extensive experiments to test its qualitative validity [75].

In order to test the (in)correctness of the Marcus solvent model, we have carried out extensive MD simulations of a bond-breaking electron-transfer reaction in water at a platinum electrode. Figure 10a shows the computer simulated potential energy surface obtained by a two dimensional umbrella sampling technique. Analysis of the results in Figure 10a brings to light two important effects of the solvent the Marcus model does not account for.

First, the presence of an additional minimum at $r = 3\,\text{Å}$ along the reaction path is clearly discernible in the computer-simulated contour plot. The small energy barrier between this minimum and the minimum at $r = 6\,\text{Å}$ is due to the molecularity of the solvent: there is a free-energy cost associated with the making of the hole between the two fragments in order to fit in a water molecule.

Second, we have analyzed the curvature of the computer-simulated potential energy surface by a parabolic fitting of the energy surface along the q solvent coordinate at various separation distances r, in order to investigate any possible r-dependence of the effective solvent reorganization energy. The result is shown in Figure 10b and clearly shows that the "local" solvent reorganization energy depends both on the well (i.e., R–X or $R^\bullet$ and X^-) and the interfragment distance r. The

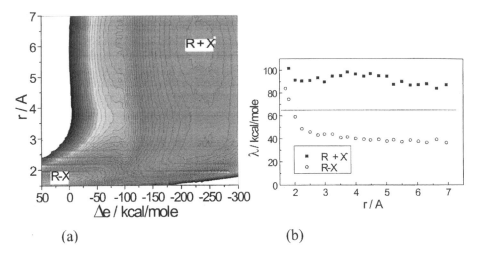

(a)   (b)

Figure 10 (a) Computer-simulated contour plot of the free-energy surface obtained by molecular dynamics for the methylchloride reduction on Pt(111) electrode. (b) The interfragment distance-dependent solvent reorganization energy extrapolated from the molecular dynamics simulations. The dashed line gives the Marcus solvent reorganization energy obtained by a local harmonic fit of the reactant well. (Adapted from Ref. [74].)

dashed line in Figure 10b shows the solvent reorganization energy as determined from a local harmonic fit to the reactant well. Hence, the comparison of the analytical theory [71] and the MD computer simulations [74] clearly demonstrates the important role of solvent molecularity and nonlinearity in the breaking of bonds at electrode surfaces. Both features are not incorporated in the traditional Marcus theory.

1.4.2 Effect of Water on Catalytic Reactions from DFT and AIMD Calculations

The presence of a protic solvent at the surface of a metal significantly alters the local environment at the surface. This can impact both chemisorption as well as surface reactivity. Modeling the effects of solution, however, presents a major challenge for first-principle methods. A variety of semiempirical models exist that use continuum methods to describe solvation effects directly in the Hamiltonian. Amovilli et al. [76] provide an elegant review of the recent advances of polarizable continuum models. These methods lead to the most efficient analyses of solvent effects. This approach has been used extensively in the area of modeling enzymes and proteins. The difficulty with this approach, however, lies in the empirical description of the solvent cavity. A second method for treating solvent effects involves the introduction of one or two explicit solvent molecules into the calculations. This approach has proven to be quite valuable in the area of modeling homogeneous catalytic systems and reactions in zeolites. Catalysis on surfaces, however, presents a more difficult challenge.

Desai et al. [77] extend this idea by adding between 8–26 explicit water molecules into a unit cell of first-principles DFT slab calculations. The vacuum layer essentially is now completely filled with solvent molecules. The layer thickness between the slabs, the unit cell size, and the number of explicit water molecules added are manipulated to maintain the appropriate density of liquid water at the surface. The water molecules that fill the vacuum region effectively form an extensive hydrogen-bonding network. The adsorption of water on Pd(111) in the presence of liquid water was examined to show the effect of solution on the chemisorption properties. The calculated structure and energetics are shown in Figure 11. Although there is only a small change in the structure of water molecules adsorbed at the surface, there is a much more dramatic change in heat of adsorption. In the gas phase, water adsorbs to Pd through the lone pair of electrons on oxygen. Water sits at an angle of 17% tilted from the surface normal. This is consistent with known experimental results for water adsorbed on Pd(100). Water molecules that adsorb directly on the Pd surface lie flat along the surface layer. The surface layer is subsequently stabilized by the formation of a hydrogen-bonding network that organizes in the water multilayers above the surface. Water rapidly begins to lose its orientation and structure (i.e., crystallinity) with respect to the metal in the layers of molecules that are farther removed from the surface. The changes in structure of water are shown in Figure 11. The resulting adsorption configuration found here is quite similar to the "bilayer" model proposed by Doering and Madey [78].

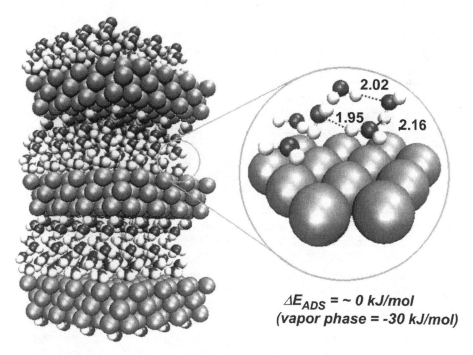

$$\Delta E_{ADS} = \sim 0 \ kJ/mol$$
$$(vapor \ phase = -30 \ kJ/mol)$$

Figure 11 The adsorption of liquid water on Pd(111). The binding energy for water on Pd(111) in the vapor phase is 30 kJ/mol. The binding energy for liquid water is −2.5 kJ/mol. (Adapted from Ref. [77].)

The formation of a hydrogen-bonded network weakens the metal–oxygen bonds at the expense of forming hydrogen bonds with the neighboring water molecules in solution. The adsorption energy of water on Pd(111) is reduced from $-30\,kJ/mol$ to $-2.5\,kJ/mol$ as we move from the vapor phase to solution. Experimental results show a similar trend. The experimental binding energy for water on Pd(100) is estimated to be $-43\,kJ/mol$ in the vapor phase [79]. The experimental TPD results for multilayers of water on Ag(110) by Bange et al. [80] suggest that the strength of the multilayer water–water interaction is nearly the same as water–metal interaction. This would indicate a net binding energy of about zero on Ag(110) in the presence of water multilayers.

The presence of solution can dramatically affect dissociative chemisorption. In the vapor phase, most metal-catalyzed reactions are *homolyticlike*, whereby the intermediates that form are stabilized by interactions with the surface. Protic solvents, on the other hand, can more effectively stabilize charge-separated states and therefore aid in heterolytic activation routes. Heterolytic paths can lead to the formation of surface anions and cations that migrate into solution. This is directly relevant to methanol oxidation over PtRu in the methanol fuel cell. The metal-catalyzed route in the vapor phase would involve the dissociation of methanol into methoxy or hydroxy methyl and hydrogen surface intermediates. Subsequent dehydrogenation eventually leads to formation of CO and hydrogen. In the presence of an aqueous media, however, methanol will more likely decompose heterolytically into hydroxy methyl (-1) and H^+ intermediates.

We used periodic DFT and ab initio (DFT) MD simulations to examine the activation of acetic acid over Pd(111) [77]. The results from both of these methods agreed quite well with one another. In the vapor phase, acetic acid dissociates to form a surface acetate intermediate and chemisorbed surface hydrogen. If carried out in the vapor phase over Pd(111), the overall reaction energy for this step was calculated to be $+28\,kJ/mol$. The calculated charges on the acetate and hydrogen surface intermediates compared much more favorably with those calculated from a simple gas-phase *free-radical* reaction than with a gas-phase *ionic* reaction. This indicates that the reaction is more *homolytic*. The adsorbed acetate intermediate carries a charge of 0.20 units greater than the acetate free radical, while the hydrogen atom gains a charge of 0.14 units on adsorption.

In the presence of a water solution, the reaction now becomes *heterolytic* [77]. Acetate anions along with protons are formed. At a low temperature the acetate anions remain chemisorbed to the metal surface, whereas the protons migrate into solution one solvation shell away from the surface, thus forming $H_5O_2^+$ intermediates. The structure of the chemisorbed product state is shown in Figure 12. We clearly see the formation of the well-known double layer. Acetate anions form a layer of negatively charged layer at the surface, which is stabilized by the layer of $H_5O_2^+$ that sits directly above it. A Mulliken charge analysis indicates an excess negative charge of 0.55 units on the adsorbed oxygen atoms of the acetate groups as compared to the acetate species formed by the dissociation of acetic acid over Pd(111) in the vapor phase. In addition, the atomic hydrogen formed in the presence of water has a net positive charge of 0.26 units as compared to the hydrogen intermediate formed in the presence of water by the dissociation of acetic acid in the vapor phase. This suggests that the dissociation of acetic acid over Pd(111) is much more heterolytic in the presence of solvent molecules than in the vapor phase.

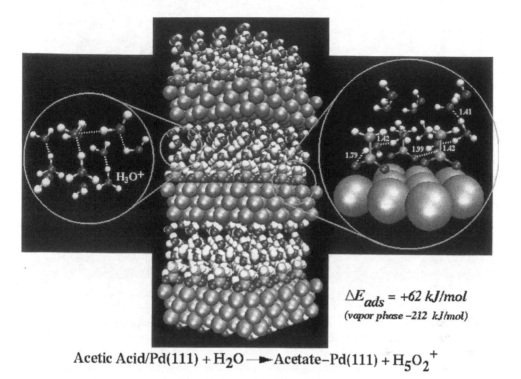

$$\Delta E_{ads} = +62 \ kJ/mol$$
$$(vapor \ phase \ -212 \ kJ/mol)$$

Acetic Acid/Pd(111) + H$_2$O $\longrightarrow$ Acetate–Pd(111) + H$_5$O$_2^+$

Figure 12 The adsorption of acetic acid on Pd(111) in the presence of a water solution. The binding energy for the acetate anion on Pd(111) in the vapor phase is -212 kJ/mol. The binding energy for acetate in the presence of a water solution is estimated to be $+62$ kJ/mol. (Adapted from Ref. [77].)

The dissociative adsorption of acetic acid therefore leads to

$$(CH_3COOH)_{ads/aq} \rightarrow (CH_3COO)^-_{\ ads/aq} + H^+_{\ ads/aq} \qquad (6)$$

Interestingly, it was found that the acetate anion formed in this reaction was more stable in solution rather than on the surface. The anion, however, forms quite readily on the surface and appears to be rather stable.

Ab initio MD studies were carried out to help understand the elementary processes that occur at the metal–solution interface [77]. At temperatures less than 300 K, acetic acid decomposed on Pd, leaving an adsorbed acetate intermediate along with a proton in solution. Above 300 K, however, surface acetate recombines with a proton in solution to form acetic acid. Acetic acid is then displaced from the surface by water. Once the acetic acid finds its way into solution, it redissociates to form acetate and protons in solution that are now more efficiently stabilized by water. A series of snapshots that portray some of the images from the simulation is shown in Figure 13. These results were further corroborated with a more conventional transition-state search approach, which showed that desorption of acetate from the surface was an activated process. The process is quite complicated, involving the simultaneous breaking of the acetate–metal bond, the formation of an

(a)   (b)   (c)

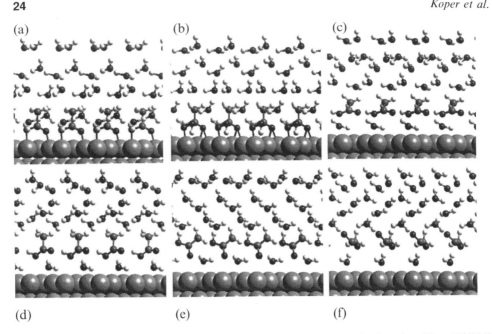

(d)   (e)   (f)

Figure 13   Snapshots from an ab initio MD simulation ($T = 300$ K) of acetic acid on Pd(111) in the presence of water. (a) Acetate forms at the surface along with an H_5O_2+ intermediate adjacent to the acetate layer; (b) acetate species and protons react at the surface to form acetic acid; (c) water displaces acetic acid from the surface; (d) water adsorbs to the surface; (e) acetic acid rotates in solution toward the water layer; (f) acetic acid dissociates in solution to form acetate anions and protons in solution. (Adapted from Ref. [77].)

O–H bond, and the displacement of acetic acid by water. This is followed by the subsequent redissociation of acetic acid in solution phase. This is quite different than the gas-phase metal-surface reaction, which typically involves a nonactivated desorption process. Activated desorption processes in electrochemical systems are well known [69,70].

As was just outlined, the presence of a solvent can dramatically affect metal-catalyzed reaction chemistry. It is well established that polar solvents can enhance reactions that have a greater degree of charge separation in the transition state than in their reactant state. The solvent acts to stabilize the transition state over the reactant state. This effectively lowers the activation barrier. Solution effects are, therefore, of critical importance to electrocatalytic systems.

Desai and Neurock [81] use period DFT calculations along with ab initio (DFT) MD simulations in order to establish the effects of protic solvents on the hydrogenation of formaldehyde to methoxy and hydroxy methyl intermediates over Pd(111). Although this work was carried out in order to gain insight into solvent effects for selective hydrogenation reactions, it also has direct relevance to the oxidation of methanol since these two steps are simply the microscopic reverse reactions for the decomposition of hydroxy methyl and methoxy.

The activation barrier for the addition of hydrogen to either the carbon or the oxygen end of formaldehyde leads to the formation of methoxy and hydroxy methyl intermediates, respectively. The activation barriers for methoxy and hydroxy methyl formation in the vapor phase over Pd(111) are 76 and 87 kJ/mol, respectively. In the presence of water, however, the activation barriers are reduced to 59 and 65 kJ/mol. Water stabilizes the negative charge that forms on the CH_3O and CH_2OH surface intermediates as well as the positive charge localized on the protons that form. The reaction path to the hydroxy methyl is slightly less favored.

Interestingly, the presence of a protic water solvent can actually play a direct role in mediating this chemistry. Desai and Neurock found that the lowest energy path was mediated through solution rather than over the surface. In the presence of water, atomic hydrogen can donate an electron to the surface to form a proton, which subsequently migrates through solution. The barrier is lowered to 42 kJ/mol. Water provides the medium for the formation and subsequent shuttling of the proton. The primary roles of Pd here are to anchor formaldehyde and also to dissociate H_2. This path is quite similar to those proposed for electrocatalytic systems, which involve the formation and transfer of protons.

Similar calculations that examine the effect of solution on the chemistry at the anode for both the hydrogen and the direct methanol fuel cells are currently begin carried out. While detailed studies on the effect of the potential dependence and solution effects have been studied, no one has begun to couple the two studies. It is clear that this will be very important for future efforts.

1.5 MONTE CARLO SIMULATIONS OF CATALYTIC REACTIONS

Dynamic or kinetic Monte Carlo methods have been used to simulate the catalytic surface chemistry for various different reaction systems. The vapor-phase oxidation of CO to form CO_2, however, has been the most widely studied due to its simplicity as well as its general applicability. Pioneering work by Ziff [82] and Zhdanov [83] shows the formations of interesting phase transitions as a function of the kinetics and lateral interactions. Many subsequent studies by various other groups extend the basic models to cover more general features.

1.5.1 Ab Initio-based MC Simulations of Ethylene Hydrogenation

Ethylene hydrogenation serves as a model system for the hydrogenation of other olefin as well as aromatics. In addition, it is directly relevant to the selective hydrogenation of acetylenic intermediates from ethylene feedstocks. Hansen and Neurock [84,85] developed an ab initio-based dynamic Monte Carlo simulation to follow the pathways and associated kinetics over Pd and PdAu surfaces in order to understand the microscopic features that govern this chemistry. A comprehensive set of first-principles density functional theoretical calculations was performed to determine the binding energies for ethylene and all reaction intermediates, including ethyl, ethylidene, ethylidyne, along with the CH_x decomposition products [86]. The activation barriers for the critical steps for ethylene and ethyl hydrogenation were calculated by performing an detailed set of transition-state searches of the potential energy surface. A plot of the energetics associated with the elementary reaction steps is shown in Figure 14. These energies were all calculated at the zero coverage limit.

DFT calculations were also used to calculate lateral interactions between coadsorbed intermediates. These results cover only the interactions for specific arrangements of adsorbates examined. The number of conceivable reaction

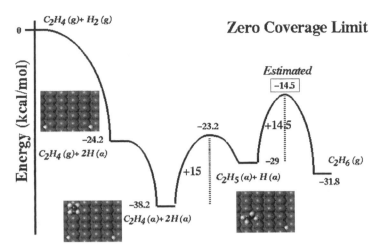

Figure 14 The DFT calculated potential energy profile for ethylene hydrogenation over Pd in *the zero coverage limit*. The sequence involves the dissociative adsorption of H2, the adsorption of ethylene, the addition of hydrogen to form ethyl, and the addition of hydrogen to ethyl to form ethane. (Adapted from Ref. [84].)

environments that can form in the simulation, however, is on the order of hundreds of thousands. Therefore, while the ab initio calculations provide a start, they cannot calculate all the conceivable states. Instead, the ab initio results can be used to establish simpler models that can be called internally within the simulation. The DFT results calculated for the ethylene system were used to regress bond-order conservation and force-field models that were subsequently called "on the fly" in the simulation to calculate the through-space and through-surface interactions between adsorbates.

The simulation uses a variable-time step method in order to simulate the kinetics over different Pd and PdAu surfaces [85]. The temporal behavior of all intermediates is explicitly tracked throughout the simulation. All atop, bridge, and 3- and 4-fold hollow sites are specifically followed as a function of time. The simulation follows all lateral and through-space interactions between coadsorbed intermediates within a cut off of two nearest-nearest neighbors.

The simulation was used as a "virtual experiment" in order to monitor the surface coverage, the surface binding energies of all intermediates, along with the reaction rates. The simulation enables us to back out overall turnover frequencies as well as individual elementary step turnover frequencies (for hydrogen adsorption, ethylene hydrogenation, ethyl hydrogenation, and the desorption of products). The lateral interactions between coadsorbed intermediates proved to be quite important in dictating both the surface coverage as well as the reaction rate. Most of the interactions in this system were found to be repulsive. Repulsive interactions weaken the binding energies and act to lower hydrogenation barriers. The barriers for both ethylene and ethyl hydrogenation dropped from 15 kcal/mol for a surface coverage of 0 to about 9 kcal/mol for surface coverages that are greater than 0.3 ML. The calculated barriers here are in very good agreement with experimental results reported by Davis and Boudart [86]. Our simulation results were also taken at different partial pressures and used to establish the reaction orders for both ethylene and atomic hydrogen. These results are in remarkable agreement with those from experiment as is seen in Eqs. (7) and (8), especially considering that the simulation results were derived solely from first principles.

$$R_{\text{Ethane}}^{\text{Simulation}} = 10^{5.4 \pm 0.07} \exp\left(-\frac{9.5 \pm 2.5 \,\text{kcal/mol}}{RT}\right) P_{\text{H}_2}^{0.65-1.0} P_{\text{C}_2\text{H}_4}^{-0.4-0.0} \tag{7}$$

$$R_{\text{Ethane}}^{\text{Experiment}} = 10^{6.3 \pm 0.07} \exp\left(-\frac{8.5 \pm 2.5 \,\text{kcal/mol}}{RT}\right) P_{\text{H}_2}^{0.5-1.0} P_{\text{C}_2\text{H}_4}^{-0.3-0.0} \tag{8}$$

The simulations provide a full disclosure of how the atomic structure of the adsorbed layer changes with changes in processing conditions.

The direct accounting for atomic structure within the Monte Carlo algorithm enables one to explore how changes in the atomic arrangement of metal atoms at the surface impact kinetics and ultimately the slate of products produced. Mei et al. [87] extend their results on Pd to PdAu in order to examine the effects of alloying. They found that alloying has little impact on the overall rate of reaction. The turnover frequency on a per-palladium atom basis remained constant over various compositions of Pd and Au and various ensemble sizes. The results for various alloys are shown in Figure 15. These results are consistent with experimental results by Davis and Boudart [86], who show very little change in the turnover frequency as Au is alloyed with Pd.

The simulation results by Mei et al. [87] indicate that the invariance in turnover frequency is due to both geometric as well as electronic effects. A snapshot taken from the simulation of ethylene hydrogenation over $Pd_{93.5\%}Au_{6.25\%}$ is shown in Figure 16. The addition of Au into the surface lowers the number of sites for hydrogen activation. This leads to a lower surface coverage of hydrogen. This is purely a geometric whereby gold shuts down sites, which slows down the overall rate. This decrease in the rate, however, is offset by a relative increase in the rate due to the weaker interaction of ethylene and hydrogen on the alloyed surfaces. This weakening of the metal–adsorbate bond by alloying is purely electronic. The two effects balance one another out, whereby the overall turnover frequency relative to the number of Pd sites remains constant. Although the relative activity is insensitive to the alloy, there is a much more dramatic effect on the selectivity. Gold acts to shut down the larger surface ensembles that are necessary for the formation of the decomposition intermediates such as ethylidyne, CH, and carbon.

1.5.2　Electrochemical CO Oxidation on Pt–Ru Alloy Surfaces

Dynamic Monte Carlo simulations have been employed to study the effect of the bimetallic catalyst structure and CO mobility in a simple model for the electrochemical oxidation of CO on Pt–Ru alloy electrodes. The Pt–Ru surface was modeled as a square lattice of surface sites, which can either be covered by CO or OH, or be empty. The important reactions taken into account in the model reflect the generally accepted bifunctional model, in which the OH with which CO is

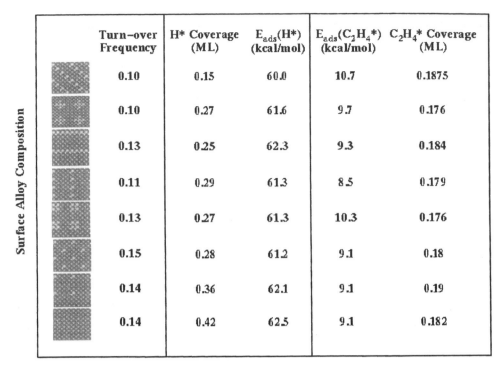

Surface Alloy Composition	Turn-over Frequency	H* Coverage (ML)	$E_{ads}(H*)$ (kcal/mol)	$E_{ads}(C_2H_4*)$ (kcal/mol)	C_2H_4* Coverage (ML)
	0.10	0.15	60.0	10.7	0.1875
	0.10	0.27	61.6	9.7	0.176
	0.13	0.25	62.3	9.3	0.184
	0.11	0.29	61.3	8.5	0.179
	0.13	0.27	61.3	10.3	0.176
	0.15	0.28	61.2	9.1	0.18
	0.14	0.36	62.1	9.1	0.19
	0.14	0.42	62.5	9.1	0.182

Figure 15 The effect of alloying Au with Pd on ethylene hydrogenation. The addition of gold shows little change with respect to the overall turnover frequency on a per-palladium site basis. Au reduces the hydrogen coverage but also weakens the adsorption energy of hydrogen.

supposed to react is preferentially formed on the Ru surface sites:

$$H_2O + \overset{*}{_{Ru}} \leftrightarrow OH_{Ru} + H^+ + e^- \tag{9}$$

$$CO_{Ru} + OH_{Ru} \leftrightarrow CO_2 + H^+ + 2\overset{*}{_{Ru}} + e^- \tag{10}$$

$$CO_{Pt} + OH_{Ru} \leftrightarrow CO_2 + H^+ + \overset{*}{_{Pt}} + \overset{*}{_{Ru}} + e^- \tag{11}$$

The DMC simulations explicitly take into account the (finite) mobility of CO, by specifying the rate at which an adsorbed CO can exchange places with an empty site (or, more realistically, with a physisorbed water molecule):

$$CO_{Pt,Ru} + \overset{*}{_{Pt,Ru}} \leftrightarrow \overset{*}{_{Pt,Ru}} + CO_{Pt,Ru} \tag{12}$$

The rate of this reaction is proportional to the diffusion coefficient D.

The simulations were designed to mimic as closely as possible the experiments carried by Gasteiger et al. [88]. These authors prepared well-characterized Pt–Ru alloys in UHV before their transfer to the electrochemical cell. The surface consisted of a random mixture of Pt and Ru sites, onto which a saturated monolayer of CO was adsorbed. The monolayer was oxidized in a CO-free electrolyte by stripping voltammetry, and the catalytic activity of the Pt–Ru alloy was studied as a function

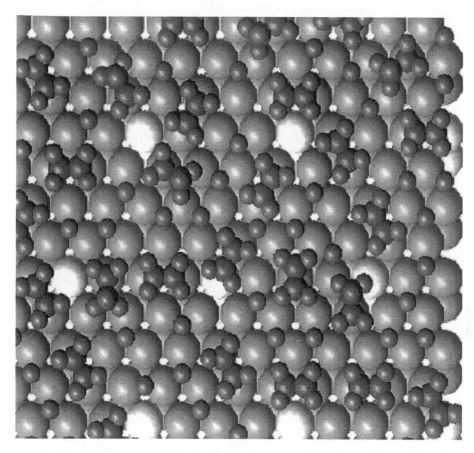

Figure 16 A single snapshot from the simulation of ethylene hydrogenation over the well-dispersed $Pd_{93.5\%}Au_{6.25\%}$ alloyed surface. Ethylene adsorbs on both Pd and Au sites. Atomic hydrogen however prefers only the threefold *fcc* sites of Pd.

of the fraction of Ru sites on the surface. The lower the potential needed to oxidize the CO layer, the higher the catalytic activity of the surface.

In our DMC simulations, the Pt–Ru lattice was initially filled with CO up to 99%, with no preference for either site. The CO adlayer was oxidized at a series on randomly mixed Pt–Ru surfaces with varying Ru content. The resulting stripping voltammetry for high diffusion rates D is shown in Figure 17, and the dependence of the peak potential, at which the oxidation current reaches a maximum, as a function of Ru fraction and diffusion rate in Figure 18. (The motivation for the choice for the exact numerical values of the other rate constants can be found in the original paper.) It is clearly observed from Figure 18 that a relatively fast diffusion rate is necessary for the surface to show a significant catalytic enhancement. For fast diffusion, the optimum catalytic activity is observed at about 0.5 Ru fraction, in agreement with experiment. This optimum can be understood as being the result of a

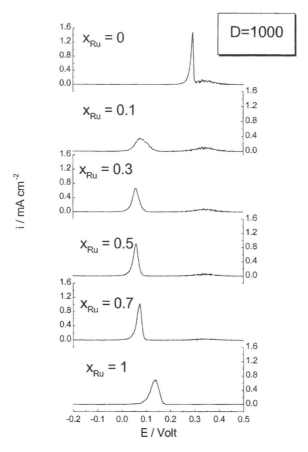

Figure 17 CO stripping voltammetry from Dynamic Monte Carlo simulations, for pure Pt ($x_{Ru} = 0$), various Pt–Ru alloy surfaces, and pure Ru. Details of the kinetic rate constants can be found in the original publication. CO surface diffusion is very fast, hopping rate D from site to site of $1000\,s^{-1}$. (Adapted from Ref. [49].)

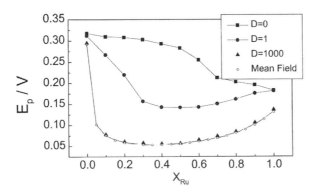

Figure 18 CO stripping peak potential E_p from dynamic Monte Carlo simulations as a function of the Ru fraction on the Pt–Ru model surface, for three different surface diffusion rates $D = 0$, 1, and $1000\,s^{-1}$ and for a mean-field model. (Adapted from Ref. [49].)

maximization of the number of Pt–Ru neighbors. More importantly, the DMC simulations for fast diffusion show that only a small fraction of Ru is needed to observe a major negative shift (about 200 mV) in peak potential. This fact is in excellent agreement with the stripping experiments of Gasteiger et al. Moreover, the shape of the stripping voltammetry shown in Figure 17, from relatively broad at low x_{Ru}, narrower at intermediate x_{Ru}, and broader again at the highest x_{Ru}, agrees very well with the experimental results. These results clearly suggest the importance of CO surface mobility in explaining the electrocatalytic activity of Pt–Ru alloys.

1.6 CONCLUSIONS

In this chapter we have described some recent applications of various computational methods to understanding some basic principles of complex catalytic and electrocatalytic processes. These methods rely on either quantum-mechanical or statistical-mechanical principles, or a combination of both, and obviously the level of detail and the kind of insight into a certain catalytic problem will depend on the chosen method.

Quantum-chemical electronic structure calculations, in practice usually DFT-GGA calculations, allow one to calculate binding energies and activation barriers of processes taking place on well-defined catalyst surfaces. From many detailed calculations, it can be concluded that many adsorption processes and surface reactions are controlled by the energy level of the d-band at the site where the process is taking place. We have illustrated how this model (or modifications thereof), suggested by Hammer and Nørskov, explains the basic binding-energy trends in the adsorption of hydrogen and carbon monoxide on PdRe and PtRu alloy surfaces. The field-dependent binding of carbon monoxide can also be described by a model in which the d-band shifts with respect to the carbon monoxide frontier orbitals as a result of the applied electric field. The DFT calculations also show that, unfortunately, the field-dependent or substrate-dependent vibrational properties of adsorbed CO are usually not good indicators of changes in the CO binding strength.

Finite-temperature molecular dynamics simulations are required to model the influence of the solvent on catalytic reactions taking place at electrified metal–liquid interfaces. In general, the presence of the solvent leads to charge transfer or charge separation across the interface, due to the charge-stabilizing properties of the polar solvent. The standard model for electron-transfer reactions is the classical Marcus model. Molecular dynamics simulations clarify the molecular role of the solvent reorganization accompanying charge-transfer reactions. In general, the solvent reorganization responds in a nonlinear fashion to changes in the charge on the reacting species, so that the Marcus continuum models are often not accurate. Effects of solvent molecularity, dielectric saturation, and electrostriction are important, especially when both neutral and charged species are involved in the reaction. Ab initio MD simulations also underscore the importance of charge stabilization in catalytic reactions at the metal–water interface. For the dissociation of acetic acid, it was found that at the metal–gas interface this reaction is homolytic because the interaction with the surface dominates, whereas it is rendered heterolytic at the metal–water interface due to the strong interactions with the protic solvent. Methanol decomposition on a Pd surface was also found to be strongly influenced by the water solvent. The solvent may lower the energy barrier of reactions that involve charged transition states or intermediates and may thereby change the lowest energy path when several possible pathways exist.

Finally, dynamic Monte Carlo simulations are very useful in assessing the overall reactivity of a catalytic surface, which must include the effects of lateral interactions between adsorbates and the mobility of adsorbates on the surface in reaching the active sites. The importance of treating lateral interactions was demonstrated in detailed ab initio-based dynamic Monte Carlo simulations of ethylene hydrogenation on palladium and PdAu alloys. Surface diffusion of CO on PtRu alloy surfaces was shown to be essential to explain the qualititative features of the experimental CO stripping voltammetry. Without adsorbate mobility, these bifunctional surfaces do not show any catalytic enhancement with respect to the pure metals.

REFERENCES

1. R.A. van Santen, M. Neurock, Catal. Rev. Sci. Eng. 37, 357, 1995.
2. R.A. van Santen, M. Neurock, in Encyclopedia of Catalysis, New York, Wiley, 2001, in press.
3. B. Hammer, J.K. Nørskov, Adv. Catal. 45, 71, 2001.
4. F. Jensen, Introduction to Computational Chemistry, Chicester, Wiley, 1999.
5. P. Hohenberg, W. Kohn, Phys. Rev. 136, B864, 1964.
6. W. Kohn, L.J. Sham, Phys. Rev. 140, A1133, 1965.
7. J.P. Perdew, J.A. Chevary, S.H. Vosko, K.A. Jackson, M.R. Pederson, D.J. Singh, C. Fiolhais, Phys. Rev. B46, 6671, 1992.
8. B. Hammer, L.B. Hansen, J.K. Nørskov, Phys. Rev. B59, 7413, 1999.
9. J.M. Thomas, W.J. Thomas, Principles and Practice of Heterogeneous Catalysis, Weinheim, VCH, 1997.
10. J.P. Lowe, Quantum Chemistry, San Diego, Academic Press, 1993.
11. M.C. Payne, M.P. Teter, D.C. Allen, T.C. Arias, J.D. Joannopoulos, Rev. Mod. Phys. 64, 1045, 1992.
12. G. Kresse, J. Furthmüller, Comput. Mater. Sci. 6, 15, 1996.
13. R. Carr, M. Parrinello, Phys. Rev. Lett. 55, 2471, 1985.
14. M. Sprik, J. Hutter, M. Parrinello, J. Chem. Phys. 105, 1142, 1997.
15. P.L. Silvestrelli, M. Parrinello, J. Chem. Phys. 111, 3572, 1999.
16. D. Marx, M.E. Tuckerman, J. Hutter, M. Parrinello, Nature 397, 601, 1999.
17. B. Ensing, E.J. Meijer, P.E. Blöchl, E.J. Baerends, J. Phys. Chem. A 105, 3300, 2001.
18. P. Vassilev, C. Hartnig, M.T.M. Koper, F. Frechard, R.A. van Santen, submitted.
19. S. Izvekov, A. Mazzolo, K. VanOpdorp, G.A. Voth, J. Chem. Phys. 114, 3248, 2001.
20. B. Smit, D. Frenkel, Understanding Molecular Simulation, Boston, Academic Press, 1996.
21. D.T. Gillespie, J. Comp. Phys. 22, 403, 1976.
22. R.M. Nieminen, A.P.J. Jansen, Appl. Catal. A 160, 99, 1997.
23. J.J. Lukkien, J.P.L. Segers, P.A.J. Hilbers, R.J. Gelten, A.P.J. Jansen, Phys. Rev. E58, 2598, 1998.
24. M.T.M. Koper, A.P.J. Jansen, R.A. van Santen, J.J. Lukkien, P.A.J. Hilbers, J. Chem. Phys. 109, 6051, 1998.
25. C. Henry, Surf. Sci. Rep. 31, 231, 1998.
26. K. Kinoshita, in J.O'M. Bockris, B.E. Conway, R.E. White, eds., Modern Aspects of Electrochemistry, Vol. 14, New York, Plenum Press, 1982, pp 557–630.

27. G. Pacchioni, P.S. Bagus, F. Parmigiani, eds., Cluster Models for Surface and Bulk Phenomena, NATO ASI Series B283, New York, Plenum Press, 1992.
28. R.M. Lambert, G. Pacchioni, eds., Chemisorption and Reactivity on Supported Clusters and Thin Films, NATO ASI Series E331, Dordrecht, Kluwer, 1997.
29. M.B. Knickelbein, S. Yang, S.J. Riley, J. Chem. Phys. 93, 94, 1990.
30. J.L. Whitten, H. Yang, Surf. Sci. Rep. 218, 55, 1996.
31. B. Hammer, Phys. Rev. Lett. 83, 3681, 1999.
32. B. Hammer, J. Catal. 199, 171, 2001.
33. T. Zambelli, J. Wintterlin, J. Trost, G. Ertl, Science 273, 1688, 1996.
34. S. Dahl, A. Logadottir, R.C. Egeberg, J.H. Larsen, I. Chorkendorff, E. Törnqvist, J.K. Nørskov, Phys. Rev. Lett. 83, 1814, 1999.
35. J. Xu, J.T. Yates Jr., J. Chem. Phys. 99, 725, 1993.
36. N.P. Lebedeva, M.T.M. Koper, E. Herrero, J.M. Feliu, R.A. van Santen, J. Electroanal. Chem. 487, 37, 2000.
37. B. Hammer, O.H. Nielsen, J.K. Nørskov, Catal. Lett. 46, 31, 1997.
38. R.A. van Santen, J.W. Niemantsverdriet, Chemical Kinetics and Catalysis, New York, Plenum Press, 1995.
39. A.C.A. de Vooys, M.T.M. Koper, R.A. van Santen, J.A.R. van Veen, J. Electroanal. Chem. 506, 127, 2001.
40. B. Hammer, J.K. Nørskov, Surf. Sci. 343, 211, 1995.
41. J.H. Sinfelt, Bimetallic Catalysts: Discoveries, Concepts, and Applications, New York, Wiley, 1983.
42. P.N. Ross, in J. Lipkowski, P.N. Ross, eds., Electrocatalysis, New York, Wiley-VCH, pp 43–73.
43. V. Pallassana, M. Neurock, L.B. Hansen, B. Hammer, J.K. Nørskov, Phys. Rev. B 60, 6146, 1999.
44. V. Pallassana, M. Neurock, L.B. Hansen, J.K. Nørskov, J. Chem. Phys. 112, 5435, 2000.
45. G. Blyholder, J. Phys. Chem. 68, 2772, 1964.
46. B. Hammer, Y. Morikawa, J.K. Nørskov, Phys. Rev. Lett. 76, 2141, 1996.
47. M.T.M. Koper, R.A. van Santen, S.A. Wasileski, M.J. Weaver, J. Chem. Phys. 113, 4392, 2000.
48. M. Watanabe, S. Motoo, J. Electroanal. Chem. 60, 275, 1975.
49. M.T.M. Koper, J.J. Lukkien, A.P.J. Jansen, R.A. van Santen, J. Phys. Chem. B 103, 5522, 1999.
50. Q. Ge, S. Desai, M. Neurock, K. Kourtakis, J. Phys. Chem. B, in press.
51. M.T.M. Koper, T.E. Shubina, R.A. van Santen, submitted.
52. N.M. Markovic, P.N. Ross, Electrochim. Acta, 45, 4101, 2000.
53. T.E. Shubina, M.T.M. Koper, in preparation.
54. M.T.M. Koper, R.A. van Santen, J. Electroanal. Chem. 476, 64, 1999.
55. S.A. Wasileski, M.T.M. Koper, M.J. Weaver, J. Phys. Chem. B 105, 3518, 2001.
56. M.J. Weaver, S. Zou, C. Tang, J. Chem. Phys. 111, 368, 1999.
57. M.J. Weaver, Surf. Sci. 437, 215, 1999.
58. D.K. Lambert, Electrochim. Acta 41, 623, 1996.
59. P.S. Bagus, G. Pacchioni, Electrochim. Acta 36, 1669, 1991.
60. F. Illas, F. Mele, D. Curulla, A. Clotet, J. Ricart, Electrochim. Acta 44, 1213, 1998.
61. D. Curulla, A. Clotet, J. Ricart, F. Illas, Electrochim. Acta 45, 639, 1999.
62. M. Head-Gordon, J.C. Tully, Chem. Phys. 175, 37, 1993.
63. P.S. Bagus, G. Pacchioni, Surf. Sci. 236, 233, 1990.
64. F. Illas, S. Zurita, J. Rubio, A.M. Márquez, Phys. Rev. B 52, 12372, 1995.
65. P. Gao, M.J. Weaver, J. Phys. Chem. 90, 4057, 1986.
66. M.F. Mrozek, M.J. Weaver, J. Am. Chem. Soc. 122, 150, 2000.
67. S.A. Wasileski, M.T.M. Koper, M.J. Weaver, in preparation.

68. R.A. Marcus, J. Chem. Phys. 43, 679, 1965.
69. W. Schmickler, Chem. Phys. Lett. 237, 152, 1995.
70. M.T.M. Koper, W. Schmickler, in J. Lipkowski, P.N. Ross, eds., Electrocatalysis, New York, Wiley-VCH, pp 291–322.
71. M.T.M. Koper, G.A. Voth, Chem. Phys. Lett. 100, 282, 1998.
72. C. Hartnig, M.T.M. Koper, J. Chem. Phys., in press.
73. W. Schmickler, J. Electroanal. Chem. 204, 31, 1986.
74. A. Calhoun, M.T.M. Koper, G.A. Voth, J. Phys. Chem. B 103, 3442, 1999.
75. J.M. Savéant, Acc. Chem. Res. 26, 455, 1993.
76. C. Amovilli, V. Barone, R. Cammi, E. Cances, M. Cossi, B. Mennucci, C.S. Pomelli, J. Tomasi, Adv. in Quant. Chem. 32, 227, 1999.
77. S.K. Desai, P. Venkataraman, and M. Neurock, J. Phys. Chem. 2001, in press.
78. D. Doering, T.E. Madey, Surf. Sci. 123, 305, 1982.
79. E.M. Stuve, S.W. Jorgensen, R.J. Madix, Surf. Sci. 146, 179, 1984.
80. K. Bange, T.E. Madey, J.K. Sass, E. Stuve, Surf. Sci. 183, 334, 1987.
81. S.K. Desai, M. Neurock, in preparation.
82. R.M. Ziff, E. Gulari, Y. Barshad, Phys. Rev. Lett. 56, 2553, 1986.
83. M. Neurock, E.W. Hansen, Comp. in Chem. Eng. 22, 1045, 1998.
84. E.W. Hansen, M. Neurock, J. Catal. 196, 241, 2000.
85. M. Neurock, R.A. van Santen, J. Phys. Chem. B 104, 11127, 2000.
86. R.J. Davis, M. Boudart, Catal. Sci & Technol. 1, 129, 1991.
87. D. Mei, E.W. Hansen, M. Neurock, J. Phys. Chem. B, 2001, in press.
88. H.A. Gasteiger, N.M. Markovic, P.N. Ross, E.J. Cairns, J. Phys. Chem. 98, 617, 1994.

2

Simulations of the Reaction Kinetics on Nanometer-Sized Supported Catalyst Particles

V.P. ZHDANOV

Chalmers University of Technology, Göteborg, Sweden, and
Boreskov Institute of Catalysis at the Russian Academy of Sciences, Novosibirsk, Russia

B. KASEMO

Chalmers University of Technology, Göteborg, Sweden

CHAPTER CONTENTS

SUMMARY

The reaction kinetics on supported nm-sized catalyst particles may be quite different compared to those observed on macroscopic poly- or single-crystal surfaces, because the very function of the catalyst is often affected by decreasing the particle size, due to inherent factors connected with the properties of small particles alone, or as a result of new kinetic effects arising on the nm scale. These proven or suspected differences between supported catalysts and macroscopic surfaces have long been recognized to be a central part of the so-called structure-gap and pressure-gap problems in catalysis. To bridge these gaps and to form a conceptual basis for the understanding of reactions occurring on supported catalysts, we summarize in the present review the results of simulations scrutinizing qualitatively new effects in the reaction kinetics on the nm scale. Attention is paid to such factors as reactant supply via the support, interplay of the reaction kinetics on different facets, adsorbate-induced reshaping of catalyst particles, selectivity on the nm scale, and oscillatory and chaotic kinetics on nm catalyst particles. The kinetics of the growth of nm particles is briefly discussed as well.

2.1   INTRODUCTION

Understanding of the kinetics of heterogeneous catalytic reactions (HCR) is of high practical importance because heterogeneous catalysis constitutes cornerstones for the chemical industry and environmental technologies [1]. Physically and chemically, the kinetics of HCR are of interest due to their richness and complexity related to such factors as adsorbate–substrate and adsorbate–adsorbate lateral interactions, surface heterogeneity, and/or spontaneous and adsorbate-induced surface restructuring [2–4] and manifested in such phenomena as chemical waves, kinetic oscillations, and chaos [2–9]. In basic academic studies, HCR are usually explored on macroscopic poly- or single-crystal surfaces. In practice, however, HCR often run on very small ($\approx 10\,\mathrm{nm}$) crystalline particles, deposited on the walls of pores of a more or less inactive support. The specific catalytic activity (i.e., the activity of an adsorption site) of such particles may be quite different compared to that of macroscopic samples. The collective set of such differences and the challenge to understand and explain them is referred to as the *structure (or materials) gap* in catalysis. An additional factor complicating bridging academic and applied studies is the *pressure gap*, expressing that practical conditions involve pressures of 1 atm or higher, while many of the most detailed academic studies have been performed at vacuum conditions, typically at 10^{-9}–10^{-4} Torr.

   The physics behind the structure gap is usually believed to be related with unique electronic properties of nm metal particles, contribution of the facet edges or other nonideal sites to the reaction rate, and/or so-called metal-support interaction [10]. These factors may also play a role for the pressure gap, which in addition is affected by the difference in population of adsorption sites at high and low pressures. The gaps may also be connected with the purely kinetic effects such as, e.g., the interplay of the reaction kinetics occurring on different facets [11] and spillover effects. All these effects are complicated by the fact that the adsorbate coverages at practical conditions are often appreciably higher than those inherent for surface-science-based studies. Extrapolation of kinetic data from low- or moderate-coverage regimes to those with high coverages (for relevant discussion, see [12] and [13]) is not straightforward due to the nonideality of the HCR kinetics.

Despite the longstanding interest, understanding the factors behind the structure and pressure gaps is far from complete. This state of development is connected first of all with limited experimental information on both the structure and reactant populations of nm catalyst particles under reaction conditions. This problem can be experimentally addressed by applying more sophisticated physical methods of investigation to real porous catalysts and/or using various modern techniques aimed at preparation of model-supported catalysts consisting of well-controlled arrays of catalyst particles deposited on planar surfaces [11]. For very small particles, these methods are, e.g., carefully controlled evaporation and annealing of condensed particles, or deposition of clusters by cluster beams. The deposits can be characterized by scanning-probe techniques or electron microscopy even under reaction conditions. Such model systems were recently reviewed comprehensively by Henry [14]. For somewhat larger particles ($\geqslant 10\,nm$), electron-beam lithography can be employed [15,16] to make nearly perfect arrays of supported particles. These experimental preparation techniques have demonstrated that it is possible to fabricate particle arrays of fairly uniform size and shape distributions, which can be varied systematically. Recent examples of using such arrays in catalytic studies can be found in [17–19].

In parallel with the experimental efforts, the pressure- and structure-gap problems can and should be addressed theoretically. In particular, using as an input the structural and kinetic data supplied by surface science, one can construct kinetic models of HCR running on nm particles. At present, the applicability of such models to specific real systems is usually limited, because the input data are, as a rule, incomplete. Nevertheless, this approach makes it possible (1) to clarify the conceptual basis of our understanding of the kinetics of HCR occurring on supported catalysts and (2) to use this information for planning and analysis of related experimental research. With the latter two points as primary goals, we have executed a series of simulations [20–30] aimed at identifying and quantifying novel, purely kinetic effects inherent to HCR on the nm scale. A detailed review of our simulations up to 1999 was recently published in *Surface Science Reports* [11]. The present paper briefly outlines the results reviewed earlier [11]. The simulations performed after [11] are discussed in more detail. [For a recent review of the mean-field (MF) and Monte Carlo (MC) treatments of the conventional (no nm specifics) kinetics of HCR, see [13] and [31], respectively.]

2.2 REACTANT SUPPLY VIA THE SUPPORT

Kinetics of HCR can be affected by adsorption of reactants on the support followed by diffusion to catalyst particles, and vice versa (spillover). In the case of CO oxidation, for example, the first experimental reports indicating that the CO supply via the support may be important for model nm catalysts, obtained by evaporating Pd onto mica, Al_2O_3, SiO_2, and MgO(100), were published in the 1980s (see the reviews [14,32]). On the noble metals, this reaction runs via the standard Langmuir–

Hinshelwood mechanism,

$$CO_{gas} \rightleftharpoons CO_{ads} \tag{1}$$
$$(O_2)_{gas} \rightarrow 2O_{ads} \tag{2}$$
$$CO_{ads} + O_{ads} \rightarrow (CO_2)_{gas} \tag{3}$$

Additional steps related to CO adsorption on the support are especially significant when the relative CO pressure is low and the reaction occurs far from the CO adsorption–desorption equilibrium. In this limit, the contribution of the support-mediated steps to the reaction rate was analyzed qualitatively by Boudart and co-workers [33] by using the "collection zone" concept and quantitatively by Henry [34] by employing the MF reaction-diffusion (RD) equations. In the latter treatment, CO diffusion jumps from the support to the metal were considered to be rapid and irreversible (mathematically, this means that the CO coverage on the support in the vicinity of the catalyst boundaries was assumed to be zero). A more general MF analysis [20] of the problem includes treatment of all the elementary steps with self-consistent boundary conditions. The catalyst particles are considered to be regularly distributed on the support. Typical reaction kinetics obtained in the case when the CO diffusion (collection) zones around different catalytic particles are not overlapping (i.e., the particles are sufficiently far apart) are shown in Figure 1 together with the conventional kinetics when there is no CO supply via the support. Due to the support-mediated CO adsorption, the position of the maximum reaction rate is seen to be shifted to a lower value of the relative CO pressure, $P_A/(P_A + P_{B2})$. In addition, the dependence of the reaction rate on the reactant pressure (for the regime where P_{CO} ($P_{CO} \equiv P_A$) is small and the surface is predominantly covered by oxygen) is changed considerably (it becomes almost linear) compared to the case without CO diffusion from the support. With increasing CO pressure, the system exhibits a transition from a regime where the reaction rate is almost completely controlled by CO supply from the support to a regime where diffusion from the support is negligible.

Note that the CO supply from the support—from a practical point of view—has a negative effect. It makes the self-poisoning transition to a CO-covered surface to occur more easily than without the support channel. A situation where the support would instead supply oxygen would have a positive effect. The latter occurs on ceria, which is an example of practical importance of the reactant supply via the support. The Pt-on-ceria system may operate in car exhaust catalysts in the so-called oxygen-storage regime when a metal oxide (e.g., ceria) provides an oxygen uptake/release function for reactions occurring on the noble metal catalyst [35,36]. An equally important and similar process is temporary NO_x storage in Pt–BaO structures where NO is oxidized to NO_2 on Pt (in O_2 excess) and stored as Ba–NO_x species, until the reverse process of NO_2 release from Ba–NO_x occurs, where NO_2 is reduced to N_2, during transient reducing conditions [37]. Understanding such processes is yet far from complete (see, e.g., the discussion in [38]) but can in principle be treated along the same line as the CO-spillover case. The difficulty is that at present the details of the oxygen (on ceria) and NO_2 (on BaO) transport and storage and the associated kinetic constants are too uncertain to allow detailed kinetic simulations.

2.3 INTERPLAY OF REACTION KINETICS ON DIFFERENT FACETS

The equilibrium geometric shape of nm catalyst particles is determined by the Wulff rule [39], stating that the shape is a consequence of minimizing the total surface free energy (for application of this rule during the adsorption–desorption equilibrium, see, e.g., [40]). Practically, this means that catalyst particles of fcc metals contain primarily the (111) and (100) facets. The total rate of reaction occurring on such particles is often believed to be a sum of the reaction rates corresponding to independent facets. During catalytic reactions, adjacent facets can, however, communicate with each other by reactant diffusion. Physically, it is clear that this communication is of minor importance if reaction runs near the adsorption–desorption equilibrium, because in this case the net diffusion flux will be negligibly low. For reactions occurring far from the adsorption–desorption equilibrium, in contrast, the diffusion-mediated facet–facet communication may dramatically change reaction kinetics. This was explicitly demonstrated in our MC simulations [21,25] of CO oxidation on nm catalyst particles. (For related MC simulations of CO oxidation on a field emitter tip, see [41,42].)

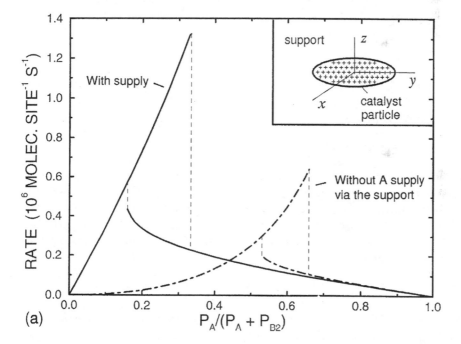

Figure 1 (a) Reaction rate and (b) reactant coverages for a rapid $2A + B_2 \rightarrow 2AB$ reaction under steady-state conditions. The solid and dashed lines respectively show the kinetics with and without A supply via the support. In both cases, the kinetics are bistable. During the highly reactive regime (at relatively low P_A), the surface is covered primarily by B, i.e. $\theta_A \ll \theta_B$. The regime with lower reaction rate (at relatively high P_A) occurs with CO domination on the surface ($\theta_B \ll \theta_A$). The results have been obtained for $P_A + P_{B2} = 0.01$ bar and $T = 450$ K with the kinetic parameters typical for CO oxidation on Pt, Pd, or Rh (for the details of calculations, see [20]). The insert on panel (a) displays the model used in calculations. (Redrawn from Ref. [20].)

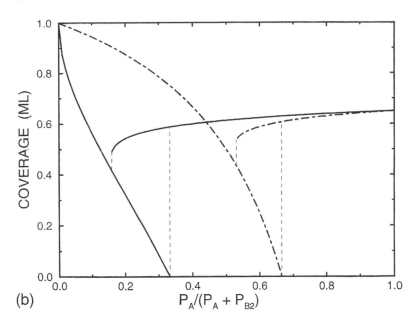

(b)

Figure 1   Continued.

Below we illustrate [43] the importance of facet–facet communication by analyzing a kinetic model of oxidation of saturated hydrocarbons. As a specific example, we treat propane oxidation,

$$C_3H_8 + 5O_2 \rightarrow 3CO_2 + 4H_2O \tag{4}$$

on Pt under lean-burn conditions. This reaction is of high current interest because hydrocarbons are one of the constituents of exhaust gases. Of special interest is the reaction kinetics during oxygen excess, in view of the legislation-driven trend toward lean combustion. In this case, the reaction is considered to be limited by C_3H_8 chemisorption accompanied by breaking one of the C–H bonds formed by the central carbon atom [44],

$$(CH_3\!-\!CH_2\!-\!CH_3)_{gas} + O_{ad} \rightarrow (CH_3\!-\!CH\!-\!CH_3)_{ads} + (OH)_{ad} \tag{5}$$

Concerted CH_3–CH_2–CH_3 chemisorption with the formation of $(OH)_{ad}$ seems to be more plausible compared to chemisorption resulting in the formation of H_{ad}, because the O–H bond is probably much stronger than the H–Pt bond. Subsequent reaction steps (after breaking the first C–H bond) are assumed to be rapid. Thus, adsorbed atomic oxygen is the dominant species under reaction conditions.

Surface-science-based studies indicate (see the discussion in [45]) that (1) the lateral interactions between oxygen atoms adsorbed on Pt are strong (≈ 2–3 kcal/mol) and (2) the oxygen sticking coefficient rapidly decreases with increasing oxygen coverage. Although this information is not sufficient to simulate the kinetics of C_3H_8 oxidation with no fitting parameters, it allows an estimate of the magnitude of the effects modifying the Langmuir kinetics of O_2 and C_3H_8 adsorption at relatively high oxygen coverages inherent for the reaction under lean conditions. In particular, the model [45] taking into account adsorbate–adsorbate lateral interactions makes it possible to obtain natural understanding and explanation of the apparent reaction orders with respect to C_3H_8 and O_2. The simulations [45] were executed for the simplest case when the catalyst surface is uniform, i.e., the specifics of nm particles were ignored. In our present treatment, the main findings obtained earlier [45] are used as inputs of a generic MF model focused on the interplay of the reaction kinetics on different facets of an nm catalyst particle.

The particle shape is considered to be a truncated pyramid (Fig. 2), with top and bottom (100) facets and (111) side facets, with the largest (100) facet attached to the support (such particles are often observed in experiments [39]). The reaction is assumed to occur primarily on the facets. The contribution of the facet edges is neglected.

If C_3H_8 oxidation is complete (CO_2 and H_2O are the final products) and oxygen desorption is negligible [the latter is the case at temperatures of practical interest (below 700–800 K)], the steady-state balance of adsorbed species is described as

$$W_{O_2} = 5W_{C_3H_8} \tag{6}$$

where W_{O_2} and $W_{C_3H_8}$ are the adsorption rates of O_2 and C_3H_8, respectively. The factor of 5 on the right-hand part of this equation takes into account that one needs five O_2 molecules in order to convert one C_3H_8 molecule to CO_2 and H_2O.

Equation (6) is applicable both to single-crystal surfaces and nm particles. In both cases, one should first calculate the O_2 and C_3H_8 adsorption rates corresponding to the (111) and (100) surfaces. In our model, these surfaces are represented by triangular and square lattices, respectively. To describe the reactant adsorption kinetics on these lattices, we take into account that under the lean-burn

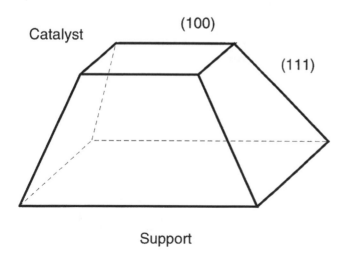

Figure 2 Pyramidal particle.

conditions Pt is covered primarily by atomic oxygen. The coverages of other species (including the propyl fragment) are considered to be negligibly low. With these conditions, the adsorption rates depend on the arrangement of adsorbed oxygen atoms, which in turn depends on oxygen–oxygen lateral interactions. In general, the interactions are rather complex [46]. For our goals, it is sufficient to take into account only nearest-neighbor repulsive interactions (Fig. 3), because these interactions dominate at the relatively high temperatures ($\approx 600\,K$) typical for catalytic C_3H_8 oxidation. At such temperatures, the statistics of adparticles can be described by employing the quasi-chemical (QC) approximation. In particular, the rate of O_2 adsorption is given by [45,47]

$$W_{O_2} = k_{O_2} P_{O_2} \mathscr{P}_{00} S^{2z-2} \tag{7}$$

where k_{O_2} is the adsorption rate constant on the clean surface, P_{O_2} is the oxygen pressure, $\mathscr{P}_{00}$ is the QC probability that two nn sites are vacant, and

$$S = \frac{\mathscr{P}_{00} + 0.5\mathscr{P}_{A0}\exp(-\epsilon_1^*/k_B T)}{\mathscr{P}_{00} + 0.5\mathscr{P}_{A0}} \tag{8}$$

is the factor taking into account the nonideality of adsorption ($\mathscr{P}_{A0}$ is the QC

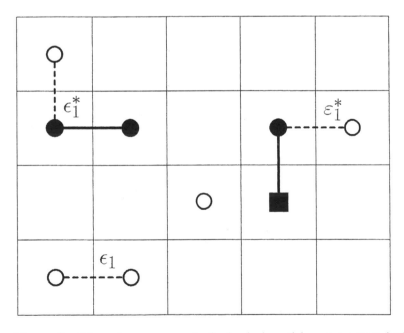

Figure 3 Schematic arrangement of adsorbed particles on a square lattice. Open circles show oxygen atoms. The dimer composed of filled circles represents the activated complex for O_2 adsorption. The dimer, consisting of a filled square and circle, represents the activated complex $[O–(C_3H_8)]^*$ for C_3H_8 adsorption [the circle and square represent O^* and $(C_3H_8)^*$, respectively]. The dashed lines indicate lateral adsorbate–adsorbate interactions. (Redrawn from Ref. [45].)

probability that a pair of nn sites is occupied by one oxygen atom, ϵ^*_1 is the nearest-neighbor interaction in the activated state for O_2 dissociation, and $z = 6$ or 4 is the number of nearest-neighbor sites for adsorption on triangular and square lattices, respectively). The probabilities $\mathscr{P}_{00}$ and $\mathscr{P}_{A0}$, defined by Eqs. (3.3.22)–(3.3.24) in [47], depend on oxygen coverage and the nearest-neighbor O–O interaction in the ground state, ϵ_1. The power $2z - 2$ in Eq. (7) is related to the number of sites adjacent to the two sites occupied by the activated complex formed during O_2 adsorption.

Assuming C_3H_8 adsorption to occur in a concerted way on a vacant site near an adsorbed oxygen atom so that the activated state includes this atom and taking into account only nn O–O lateral interactions in the ground and activated state, we have [45,47]

$$W_{C_3H_8} = k_{C_3H_8} P_{C_3H_8} 0.5 \mathscr{P}_{A0} V^{z-1} \tag{9}$$

where $k_{C_3H_8}$ is the coverage-independent rate constant, $P_{C_3H_8}$ the C_3H_8 pressure, and

$$V = \frac{\mathscr{P}_{AA} \exp[-(\varepsilon^*_1 - \varepsilon_1)/k_B T] + 0.5 \mathscr{P}_{A0}}{\mathscr{P}_{AA} + 0.5 \mathscr{P}_{A0}} \tag{10}$$

the factor describing the nonideality of adsorption (ε^*_1 is the O–O interaction in the activated state for C_3H_8 dissociation).

In the equations above, we have six parameters, k_{O_2}, $k_{C_3H_8}$, ϵ_1, ϵ^*_1, ε^*_1, and z, for each surface. Not to obscure the main message, it makes sense to minimize the number of free parameters. Following this line, we employ the same rate constants k_{O_2} and $k_{C_3H_8}$ and interactions $\epsilon_1 = 3$ kcal/mol and $\epsilon^*_1 = 2$ kcal/mol for both surfaces. For ε^*_1, we use 2 and 0 kcal/mol for the (111) and (100) faces, respectively. With these parameters, the O_2 and C_3H_8 adsorption rates are higher (Fig. 4) for the (100) face, because z and ε^*_1 are lower in this case. As we will see, these differences among the facets are sufficient to produce unique new kinetics, compared to the individual facets.

Solving Eq. (6) with explicit expressions for the O_2 and C_3H_8 adsorption rates makes it possible to calculate the reaction rate, which is identified below with the O_2 adsorption rate per site, i.e., $W_r \equiv W_{O_2}$. In particular, Figure 5 shows the reaction rates for the (111) and (100) surfaces. Using these rates and assuming the (111) and (100) facets to operate independently, one can obtain the reaction rate for a catalyst particle with noncommunicating facets. This approach yields (Fig. 6)

$$W_r = \alpha^{(1)} W_r^{(1)} + \alpha^{(0)} W_r^{(0)} \tag{11}$$

where $\alpha^{(1)}$, $\alpha^{(0)}$ ($\alpha^{(0)} \equiv 1 - \alpha^{(1)}$), $W_r^{(1)}$, and $W_r^{(0)}$ are the fractions of adsorption sites and reaction rates (Fig. 4) corresponding to the (111) and (100) surfaces [the subscripts (1) and (0) refer to (111) and (100), respectively].

In reality, the (111) and (100) facets communicate via oxygen diffusion. In this case, Eq. (6) can be read as

$$\alpha^{(1)} W_{O_2}^{(1)} + \alpha^{(0)} W_{O_2}^{(0)} = 5\alpha^{(1)} W_{C_3H_8}^{(1)} + 5\alpha^{(0)} W_{C_3H_8}^{(0)} \tag{12}$$

where $W_{O_2}^{(1)}$, $W_{C_3H_8}^{(1)}$, $W_{O_2}^{(0)}$, and $W_{C_3H_8}^{(0)}$ are the reactant adsorption rates per site on the

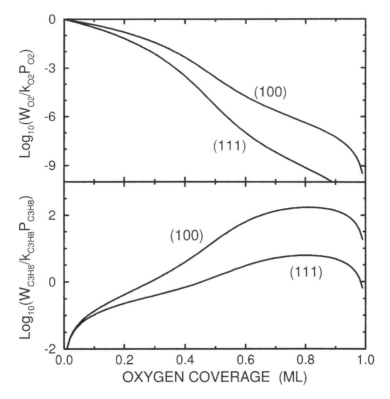

Figure 4 O_2 (top) and C_3H_8 (bottom) adsorption rates (normalized to $k_{O2}P_{O2}$ and $k_{C3H8}P_{C3H8}$) on the (111) and (100) surfaces as a function of oxygen coverage at $T = 600$ K [according to Eqs. (7) and (9), respectively].

(111) and (100) facets. These rates, given by Eqs. (7) and (9), depend respectively on the oxygen coverages, $\theta_O^{(1)}$ and $\theta_O^{(0)}$, of the (111) and (100) facets. To solve Eq. (12), we need a relation between these coverages.

At relatively high temperatures typical for C_3H_8 oxidation, oxygen diffusion is rapid compared to other steps (for the Arrhenius parameters for this process, see [21]). In this case, the relation between $\theta_O^{(1)}$ and $\theta_O^{(0)}$ is given by the grand canonical distribution, i.e., one should have

$$\mu^{(1)}(\theta_O^{(1)}) = \mu^{(0)}(\theta_O^{(0)}) \tag{13}$$

where $\mu^{(1)}$ and $\mu^{(0)}$ are the chemical potentials of oxygen on the (111) and (100) facets.

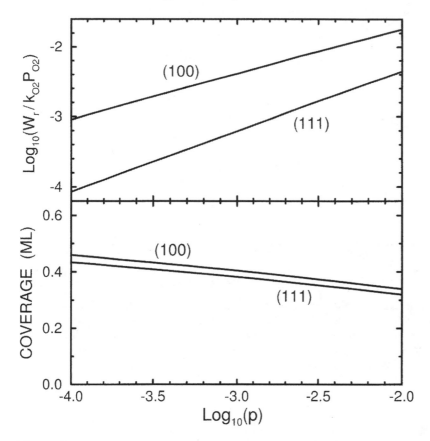

Figure 5 Reaction rate (normalized to $k_{O2}P_{O2}$) and oxygen coverage as a function of p ($p \equiv 5k_{C3H8}P_{C3H8}/k_{O2}P_{O2}$) for the (111) and (100) surfaces at $T = 600$ K.

For these potentials, the QC approximation yields [47]

$$\mu^{(1)}(\theta_O^{(1)}) = k_B T \ln\left[\frac{\theta_O^{(1)} F^6(\theta_O^{(1)})}{1 - \theta_O^{(1)}}\right] \tag{14}$$

$$\mu^{(0)}(\theta_O^{(0)}) = \Delta E + k_B T \ln\left[\frac{\theta_O^{(0)} F^4(\theta_O^{(0)})}{1 - \theta_O^{(0)}}\right] \tag{15}$$

where ΔE is the adsorption energy difference between the facets at low coverages, and

$$F(\theta_O) = \frac{\mathscr{P}_{AA}\exp(\epsilon_1/k_B T) + 0.5\mathscr{P}_{A0}}{\mathscr{P}_{AA} + 0.5\mathscr{P}_{A0}} \tag{16}$$

is the factor taking into account lateral interaction.

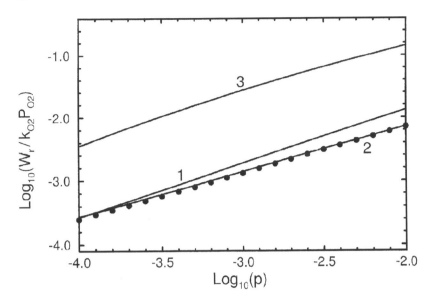

Figure 6 Reaction rate (normalized to $k_{O2}P_{O2}$) for an nm catalyst particle with $\alpha^{(1)} = 0.8$ at $T = 600\,K$. Solid lines 1, 2, and 3 show the results obtained by taking into account the interplay of the reaction kinetics on the (111) and (100) facets [Eqs. (12) and (13) with $\Delta E = 10$, 0, and $-10\,kcal/mol$, respectively]. The filled circles correspond to the case when the (111) and (100) facets operate independently [Eq. (11)].

To use Eq. (15), we need ΔE. This parameter can be estimated from the O_2 temperature-programmed desorption spectra. Scrutinizing the spectra collected in [48], we obtain $\Delta E \simeq -10\,kcal/mol$ [the negative sign of ΔE means that oxygen adsorption is more favorable on the (100) face].

Employing Eqs. (12) and (13), we have calculated (Fig. 6) the specific rate of C_3H_8 oxidation on an nm catalyst particle for $\Delta E = 0$ and $\pm 10\,kcal/mol$. For $\Delta E = 0$ and $+10\,kcal/mol$, the reaction rate is found to be nearly equal to the rate calculated by assuming the (111) and (100) facets to operate independently. In the most important case when $\Delta E = -10\,kcal/mol$, the reaction rate is *much higher* than the former ones. To rationalize this interesting finding, we show the contribution of the (100) facet to the total rates of O_2 and C_3H_8 adsorption and also oxygen coverages, $\theta_O^{(1)}$ and $\theta_O^{(0)}$, of the (111) and (100) facets for $\Delta E = -10\,kcal/mol$. In this case, oxygen adsorption is thermodynamically favorable on the (100) face and accordingly, due to oxygen supply from the (111) facet, the oxygen coverage of the (100) facet is higher than in the case of the independent (100) surface (cf. Figures 5 and 7). The oxygen coverage of the (111) facet is accordingly lower than in the case of the independent (111) surface. Under such circumstances inherent for an nm particle, the contribution of the (100) facet to the total rate of O_2 adsorption is nearly negligible, i.e., O_2 adsorption occurs primarily on the (111) facets. In contrast, C_3H_8 adsorption takes place almost exclusively on the (100) facet. The total rates of both these processes are higher than the corresponding rates for independent (111) and (100) surfaces, because the oxygen coverages of the (111) and (100) facets are respectively lower and higher than those of the independent surfaces. For these reasons, the

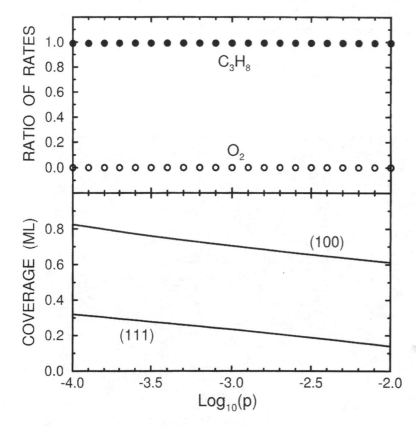

Figure 7 Ratio of the rate of O_2 adsorption on the (100) facet to the total rate of O_2 adsorption (open circles), similar ratio of the rates of C_3H_8 adsorption (filled circles), and oxygen coverages of the (111) and (100) facets of an nm particle as a function of p for $\Delta E = -10\,\text{kcal/mol}$.

reaction rate on an nm particle is higher than that calculated by ignoring communication between the facets.

In summary, our analysis of the kinetics of C_3H_8 oxidation on nm-supported Pt particles shows that, due to the purely kinetic factors related to the interplay of the reaction kinetics on different facets, the activity of a catalyst particle may be appreciably higher than that calculated by assuming that the facets operate independently. This important effect, found to apply for a wide range of reactant pressures, may of course occur in many other catalytic reactions as well, since the underlying mechanism in the present analysis is generic.

2.4 PHASE SEPARATION ON THE NM SCALE

Phase separation or, more broadly, island formation in HCR is possible due to attractive adsorbate–adsorbate lateral interactions (thermodynamic mechanism) and/or limited mobility of adsorbed species (kinetic mechanism). Under transient conditions, island formation has been experimentally observed (by using STM) in plenty of reactions [49]. Experimental data on island formation under steady-state regimes are, however, scarce. Theoretically, phase separation occurring due to attractive lateral interactions during the simplest $A + B$ reaction under steady-state conditions was studied in [50–52]. The bulk of simulations was executed for the case

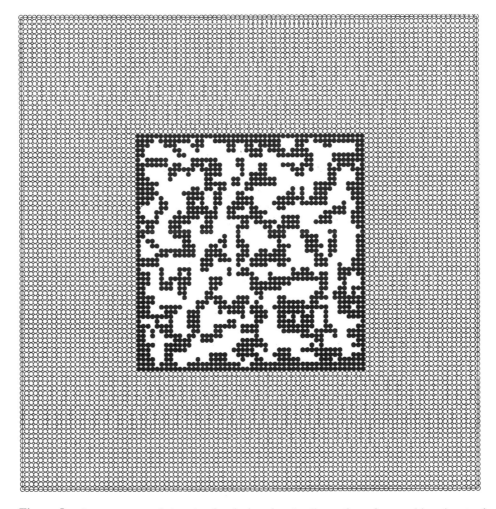

Figure 8   Arrangement of A molecules during the $A + B$ reaction after reaching the steady-state regime. The reaction runs on a 100×100 lattice mimicking a pyramidal supported catalyst particle (Fig. 4). In the simulations, the top and side facets of the pyramid are represented by the central and peripheral sublattices of the lattice (A molecules, located on these sublattices, are shown by filled and open circles, respectively). Reaction occurs only on the central sublattice via the Eley–Rideal mechanism, including A adsorption on vacant sites and A consumption in collisions with gas-phase B particles. The peripheral sublattice is able to adsorb A. Attractive nn lateral interaction, introduced only for A molecules located on the central sublattice, is chosen so that $T = 0.5T_c$. A diffusion on and between the sublattices is rapid compared to reaction. (Redrawn from Ref. [51].)

when reaction runs on a single-crystal surface. The case of supported catalyst particles was briefly treated [51] as well. For example, Figure 8 shows adsorbate distribution on a catalyst particle under the chemically reactive conditions at $T = 0.5T_c$. Referring to an Ising model, one could expect that in this case all the facets were either free of or completely occupied by adsorbed particles. In contrast, the model predicts small islands formed on one of the facets. The island size depends on the interplay of reaction and phase separation.

2.5 OSCILLATIONS AND CHAOS

During the past two decades, regular and chaotic oscillations were found in about 30 reactions on practically all types of catalysts including single crystals, poly-crystalline samples (foils, ribbons, and wires), and supported catalysts over a pressure range from 10^{-12} bar to atmospheric pressure [2,5,6,8]. The experience accumulated indicates that oscillations are often observed in systems where a rapid bistable catalytic cycle is combined with a relatively slow "side" process, e.g., with oxide formation [53], carbon deposition [54], or adsorbate-induced surface restructuring [2,6]. Despite the fact that in many cases a mechanism of oscillations is considered to be established, the understanding of this intriguing phenomenon is still limited especially in the situations when the experiments are executed on supported catalysts (see, e.g., [55,56]). A few aspects of the latter problem were analyzed in our recent MC simulations [11,27–29].

The kinetics of CO oxidation on a pyramidal supported catalyst particle (Fig. 4) was simulated [11,27] by assuming CO adsorption to cause restructuring of the top (100) facet. The restructuring was described on the basis of the lattice-gas model, predicting phase separation in the overlayer. CO diffusion was much faster compared to other steps. Oscillatory and chaotic kinetic regimes were found in the simulations. One of the reasons of irregular oscillatory kinetics was demonstrated to be the interplay of the reactions on the (100) and (111) facets.

Oscillations connected with adsorbate-induced surface restructuring were studied also in [29]. The model used was aimed at mimicking oscillations in NO reduction by H_2 on a mesoscopic Pt particle containing two catalytically active (100) areas connected by an inactive (111) area that only adsorbed NO reversibly. NO diffusion on and between facets was much faster than other steps. The results obtained show that the coupling of the catalytically active sublattices may synchronize nearly harmonic oscillations observed on these sublattices and also may result in the appearance of aperiodic partially synchronized oscillations. The spatiotemporal patterns corresponding to these regimes are nontrivial. In particular, the model predicts that, due to phase separation, the reaction may be accompanied by the formation of narrow NO-covered zones on the (100) sublattices near the (100)–(111) boundaries. These zones partly prevent NO supply from the (111) sublattice to the (100) sublattices.

In the simulations [11,27,29] the size of a lattice representing a catalyst particle was relatively large [typically (100 × 100)]. The effect of the lattice size on oscillatory kinetics was demonstrated [28] in simulations of CO oxidation accompanied by oxide formation. To mimic nm catalyst particles, the lattice size was varied from 50 × 50 to 3 × 3. With rapid CO diffusion, more or less regular oscillations were found (Fig. 9) for sizes down to 15 × 15.

The simulations discussed above are focused on the behavior of single catalytic oscillators at fixed reactant pressures. In the full-scale analysis of reactions on nm-supported particles, the reactant pressures should be calculated self-consistently with the reaction kinetics. At present, due to computational limitations, the self-consistent treatment can, however, be done only by using the MF equations (see, e.g., recent simulations [57] of oscillations in CO oxidation in a continuously stirred tank reactor). The MF approach does not, however, make it possible to scrutinize the reaction kinetics on the nm scale. Under such circumstances, the MC and MF treatments are complementary. In particular, the MC results may be employed in order to understand the limits of applicability of the MF approximation.

2.6 SELECTIVITY ON THE nm SCALE

HCR involving complex polyatomic molecules usually occur via a large number of steps and result in the parallel formation of several products. In this case, the reaction rate and selectivity may easily be affected by the geometric details of nm-sized catalyst particles. To simulate such reactions, one can use (with proper modifications) general approaches developed [58] to describe adsorption of complex

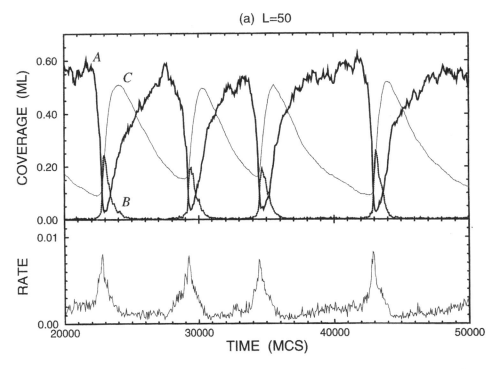

Figure 9   CO (*A*), O (*B*), and oxide (*C*) coverages and reaction rate (CO_2 molec. per site per MCS) as a function of time for $L = 50$ (a), 30 (b), 20 (c), 10 (d), 5 (e), and 3 (f). With decreasing lattice size, kinetic oscillations become more irregular. For the smallest size ($L = 3$), oscillations rapidly disappear due to complete poisoning of the lattice by oxygen (this is possible because O_2 adsorption is considered to occur on nn vacant sites). (Redrawn from Ref. [28].)

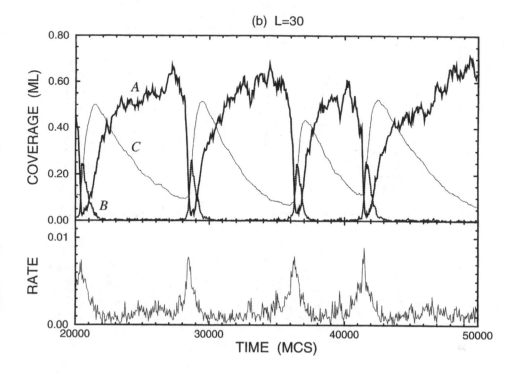

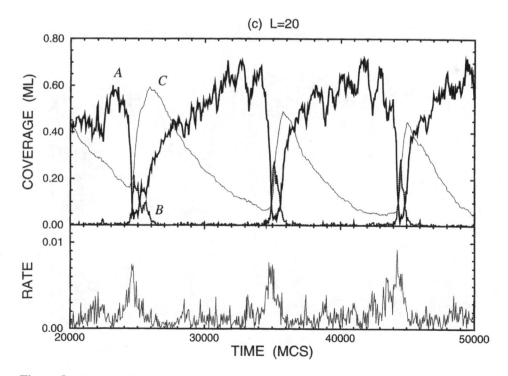

Figure 9　Continued.

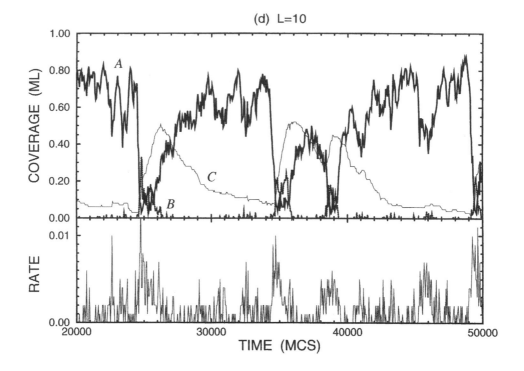

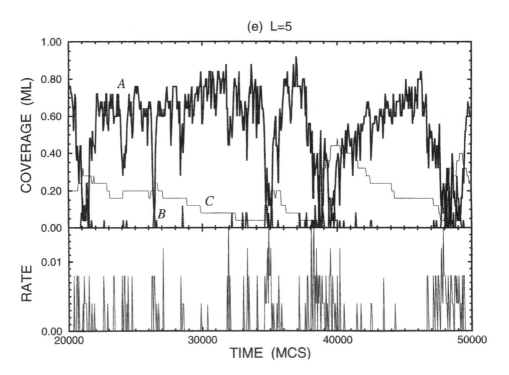

Figure 9 Continued.

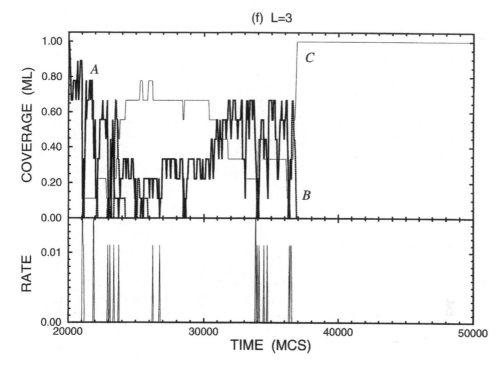

Figure 9 Continued.

molecules. The first works in this direction were published by McLeod and Gladden [59,60] (see also our review [11]).

2.7 ADSORBATE-INDUCED RESHAPING OF CRYSTALLITES

Crystallite shape transformations, due to adsorbed reactants, may affect the steady-state kinetics of catalytic reactions. Such effects can be studied phenomenologically by employing the Wulff rule [39] in order to find the optimum crystallite shape. The use of this rule implies that the crystallite reshaping time scale is shorter than the experimental time scale. The latter can be very long in practical systems.

An interesting example of application of the Wulff rule is given by Ovesen et al. [61]. They have analyzed the kinetics of methanol synthesis on nm Cu particles supported by ZnO. The generalized surface tension for the particle–substrate interface was assumed to be dependent on the reduction potential of the gas phase. The latter resulted in the dependence of the areas of the (111), (100), and (110) facets on the gas-phase concentrations (such changes were observed by using EXAFS). The total reaction rate, represented as a sum of the reaction rates on different facets, was found to be affected by the changes in particle morphology.

In our work [22], the Wulff rule was employed to analyze adsorbate-induced reshaping of crystallites during the $A + B_2$ reaction mimicking CO oxidation.

Myshlyavtsev and co-workers [62] recently tried to describe explicitly adsorbate-induced changes in the shape of catalyst particles by using the solid-on-solid (SOS) model. The results obtained are, however, somewhat artificial from the physical point of view, because the shape of particles predicted on the basis of this model has little in common with crystallites.

Generally, the possibility of a shape change of a catalyst particle, as the gas-phase composition, and consequently the coverages of different species change, is both very interesting and a complicating factor—when it occurs—for interpretation of kinetic data. In an experimental situation, the largest change in adsorbate coverage (at constant temperature) occurs when the gas mixture is changed so that the system passes the rate maximum or passes over a kinetic phase transition. In both cases, there is a change in the dominant surface species. For example, for the $A + B_2$ reaction we have discussed above, there is a change from dominant B coverage to dominant A coverage as the gas-mixture ratio $P_A/(P_A + P_{B2})$ is varied from P_A below to above the rate maximum. The crystallite shapes, predicted by the Wulff rule in these situations, may be different. (An interesting open question concerns the limits of applicability of the Wulff rule in the chemically reactive systems. Originally, this rule was derived for thermodynamic equilibrium, and accordingly one could expect that it would be applicable at adsorption–desorption equilibrium. Often, the latter condition is, however, not necessary in order to use the Wulff rule.)

2.8 REACTIONS ON ULTRASMALL METAL CLUSTERS

In the previous sections, we have discussed the kinetics of HCR occurring on nm-sized crystallites. The situations when the catalytic metal particles are ultrasmall are of practical importance as well. In zeolites, for example, metal particles often contain only a few atoms. Such particles called clusters can also be obtained on more conventional supports (see, e.g., recent studies of CO adsorption on Rh, Pd, and Ir clusters formed on alumina [63] and acetylene cyclotrimerization on Pd clusters on MgO(100) [64]). Reaction kinetics on clusters should be described by explicitly taking into account all possible configurations of adsorbed particles. For example, we refer to recent simulations [65] of the kinetics of a rapid $A + B_2$ reaction occurring via the Langmuir–Hinshelwood mechanism (1)–(3) on ultrasmall clusters containing a few metal atoms (see, e.g., Figure 10 showing a four-site cluster). For the infinite lattice, this reaction is bistable and accordingly may exhibit hystresis. If reaction occurs on a small cluster, the bistability and hysteresis disappear (Fig. 11). Physically, this difference in the reaction behavior results from the fact that the fluctuations of the number of adsorbed particles on a small cluster are much higher than those on the large lattice.

In general, the concepts of bistability and hysteresis are related not only to the reaction kinetics but also to the time scale of the experiment. For this reason, the lattice size corresponding to the boundary between "small" and "large" lattices depends on the time scale of the experiment as well. MC simulations [66] indicate that often the boundary size is about 5×5. For somewhat larger lattices, one can observe the fluctuation-driven transitions between the high- and low-reactive regimes [42].

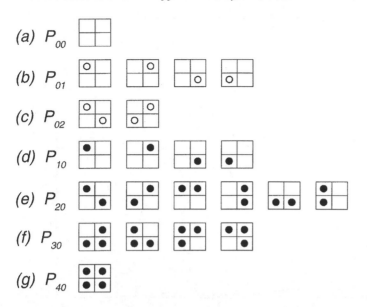

Figure 10 Seven possible types of configurations of A and B particles (filled and open circles) occurring during rapid $2A + B_2 \to 2AB$ reaction on a four-site cluster. The model used implies that A and B_2 adsorption are competitive, i.e., each site can be either vacant or occupied by A or B particles. The Langmuir–Hinshelwood step is assumed to be so fast compared to A and B_2 adsorption that there are no configurations with simultaneous adsorption of A and B particles. B particles are immobile, and A diffusion is rapid compared to the LH step. In addition, the model takes into account that strong repulsive lateral interactions between nn B particles prevent B_2 adsorption on nn sites. Specifically, B_2 adsorption is considered to occur on vacant pairs of next-nearest-neighbor sites provided that adjacent sites are not occupied by B (this detail is significant, because it prevents poisoning of the surface by B). (Redrawn from Ref. [65].)

2.9 DIFFUSION LIMITATIONS

The reaction kinetics in porous catalysts are kinetically controlled only if the reaction is sufficiently slow, while the rate of rapid reactions becomes limited by reactant diffusion via pores. Quantitatively, the role of diffusion is usually scrutinized by employing the phenomenological reaction-diffusion equations [1]. For example, in the case of the simplest first-order reaction, one has

$$\partial n/\partial t = D_{\mathrm{ef}}\Delta n - k_{\mathrm{ef}}n \tag{17}$$

where n is the reactant concentration, D_{ef} the effective diffusion coefficient, and $k_{\mathrm{ef}} = k_0 N_c$ the effective reaction rate constant (k_0 is the rate constant corresponding to a single catalyst particle, and N_c is the concentration of catalyst particles). This approach is based on the assumption that the rate constant k_{ef} (or k_0) characterizes the true reaction kinetics. In other words, this constant is usually assumed to be independent of the rate of diffusion, i.e., the interplay of reaction and diffusion inside single pores is not treated explicitly. The latter aspect of the problem was

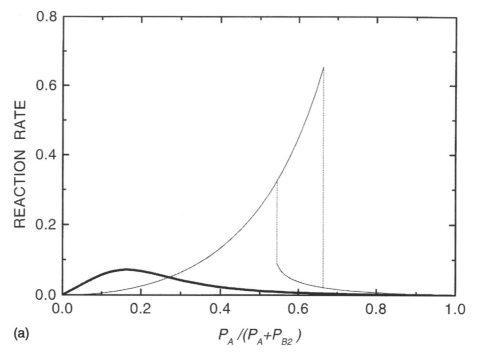

(a)

Figure 11 (a) Rate of the $A + B_2$ reaction (10^6 AB molec. site^{-1} S^{-1}) and (b) A and B coverages as a function of the relative A pressure for the infinite lattice (thin lines) and four-site cluster (thick lines). The results for the infinite lattice were obtained by using the conventional MF equations. The kinetics corresponding to a four-site cluster were calculated by employing the master equations, taking into account all possible configurations of adsorbed particles (as shown in the previous figure). The model parameters used in the calculations are typical for CO oxidation on noble metals at $P_{CO} + P_{O2} = 0.01$ bar and $T = 500$ K. (Redrawn from Ref. [65].)

analyzed in detail in [67,68]. Specifically, the first-order reaction was assumed to occur on the catalytically active walls of pores (the results obtained [67] justify application of the phenomenological approach). For supported catalysts, this model makes sense if the pores are so large that a multitude of catalytic particles is inside a single pore. In mesoscopic pores, this is often not the case. If, for example, the size of catalyst particles is comparable with the pore radius, a single pore will, as a rule, contain no or only one catalytic particle. Under such circumstances, the reaction rate on a single catalyst particle can be limited by diffusion inside the pore where this particle is located (see, e.g., Figure 12 reproduced from [30]). Diffusion limitations may also be significant in reactions occurring on model planar-supported catalysts. The latter case was treated in [23].

2.10   HEAT DISSIPATION ON CATALYST PARTICLES

General equations for estimation of the time scales characterizing dissipation of heat released in exothermic reactions occurring on nm-supported catalyst particles have

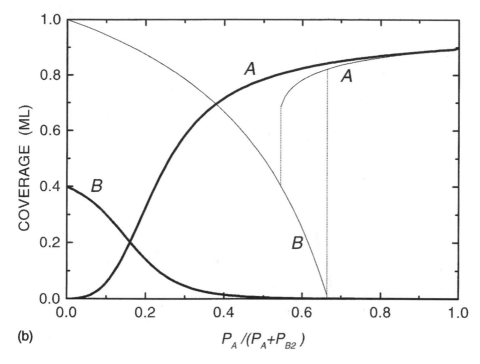

(b) $P_A/(P_A+P_{B2})$

Figure 11 Continued.

been derived in [69,70]. The results obtained indicate that overheating a single nm particle is usually negligible. One cannot, however, exclude local overheating of a porous catalyst on the larger scale. The necessary conditions for observation of the latter phenomenon are still open for discussion.

2.11 GROWTH OF SUPPORTED nm CRYSTALLITES

Although the main subject of this review is the kinetics of HCR on nm-sized supported crystallites, it is instructive to discuss briefly the kinetics of growth (sintering) of such particles. In practical conditions, this process is usually not desirable, because it results in a decrease of the active catalyst area (one type of catalyst aging). Still it is a common phenomenon during the initial and long-term life of real catalysts. Experimental studies of this phenomenon are numerous [71–73], but the relative importances of the many factors affecting the growth are still not quite clear. In vacuum, larger crystallites usually grow at the expense of smaller ones via the Ostwald ripening scenario including 2D evaporation of metal atoms, diffusion along the support, and condensation [diffusion and collisions of crystallites may also be important but only if their size is small ($\approx 1–2$ nm)]. In atmosphere or under the chemically reactive conditions, the sintering often occurs faster. While for H_2 or N_2 this effect is usually relatively minor, oxygen-containing atmospheres may result not only in rapid sintering, occurring presumably with participation of volatile or adsorbed PtO_2 particles, but also in redispersion of crystallites.

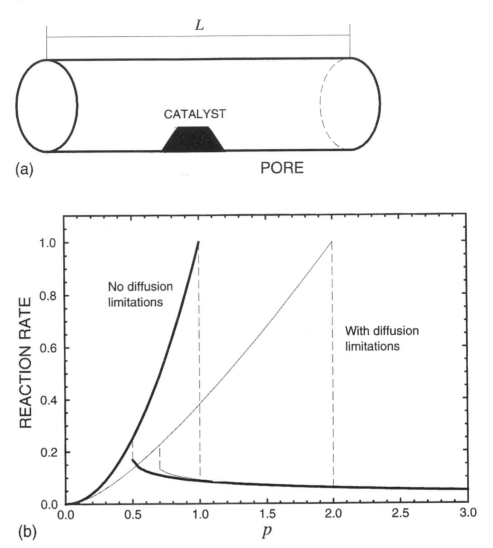

Figure 12 (a) Schematic arrangement of a catalyst particle inside the pore of length L. (b) Normalized rate of the $A + B_2$ reaction as a function of the ratio of the impingement rates of A and B_2 for the cases when A diffusion limitations inside the pore are negligible (thick lines) and significant (thin lines). The model parameters employed are typical for CO oxidation on noble metals at $P_{CO} + P_{O2} = 0.01$ bar and $T = 500$ K. (Redrawn from Ref. [30].)

Phenomenologically, the sintering kinetics are usually described by employing a power law for the average linear crystallite size, $l(t) \simeq \mathscr{A} + \mathscr{B}t^x$, where x is the growth exponent, $\mathscr{A}$ is the constant introduced to take into account that in the beginning the growth is far from the asymptotic regime, and $\mathscr{B}$ is the constant corresponding to the asymptotic growth. Often, the exponent x is replaced by $1/n$, i.e.,

$$l(t) \simeq \mathscr{A} + \mathscr{B}t^{1/n} \tag{18}$$

Experimentally, the sintering of supported nm metal particles has been studied during several decades [71–73] by using x-ray diffraction, transmission electron microscopy, and temperature-programmed desorption (the former two techniques measure the average size of particles, the latter measures surface area). Numerous data obtained in inert and reactive conditions indicate that the growth of nm catalyst particles is usually described by Eq. (18) with $n = 5$–11. With increasing temperature, n often decreases.

Using the conventional Lifshitz–Slyozov arguments [74,75] based on the Kelvin equation, one might expect [73,76] that the 3D crystallite growth, occurring via 2D diffusion, should follow Eq. (18) with $n = 4$. This value, however, is much lower than observed in experiments. The appreciable difference between the theory and experiment is actually not surprising, because the applicability of the Lifshitz–Slyozov model to nm crystallites is far from obvious (e.g., the curvature of such crystallites is an ill-defined quantity).

To tackle the problem under consideration by employing MC simulations, we have adopted [77] the restricted-solid-on-solid model (RSOSM) used earlier to explore surface roughening [78,79] (this model is somewhat more realistic compared to the conventional SOS model). Applying this model to supported catalyst particles, we determine the latter as 2D arrays of columns on a square support lattice. For the nearest-neighbor columns, the heights are allowed to differ by at most 1. Lateral interaction between nearest-neighbor metal atoms forming columns is considered to be attractive, $\epsilon_1 < 0$. The binding energy of a single atom on the support is assumed to be lower compared to that on the top of a crystallite (this corresponds to the nonwetting condition). The initial disordered state was formed by successively depositing one-half monolayer of metal atoms on the support (a deposition attempt on a randomly chosen site was accepted if an arriving atom did not violate the RSOSM constraint). The algorithm for describing the system at $t > 0$ consisted of attempts of diffusion jumps (we used the simplest Metropolis dynamics) to nearest-neighbor and next-nearest-neighbor sites.

Typical snapshots illustrating evolution of the distribution of crystallites are shown in Figure 13. At early stages (see, e.g., Fig. 13a for $t = 10^2$ MCS), the crystallites look like bilayer islands. With increasing time, the height of crystallites becomes larger (Fig. 13b). At the latest stages, the shape of the crystallites is pyramidal (Fig. 13c).

Analyzing quantitatively the crystallite growth at temperatures of practical interest (below the roughening temperature), we have found (see, e.g., Figure 14) that it can be described by Eq. (18) with $n = 7$–8 (n is lower at higher temperatures). Despite the simplicity of the model, the results obtained are in much better agreement with numerous experimental data compared to those predicted by the Lifshitz–Slyozov theory. Thus, our simulations explicitly show that in general the latter theory is not applicable to describing the growth of nm crystallites. (Additional simulations treating this problem on the basis of more realistic models are, of course, desirable.)

(a) $t=10^2$ MCS

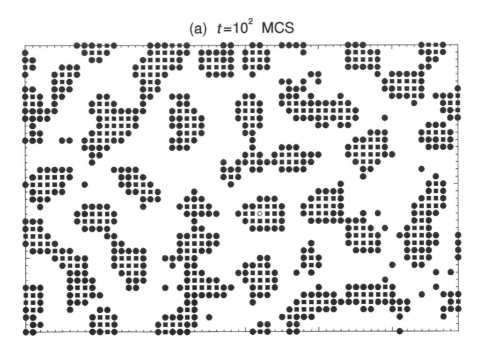

(b) $t=10^4$ MCS

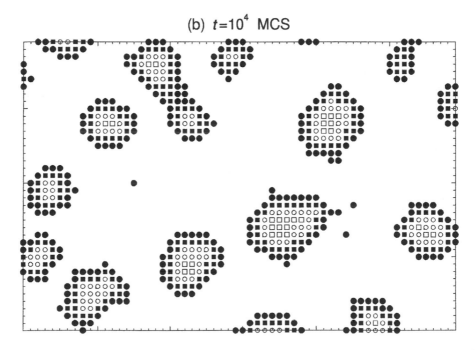

Figure 13 A 40 × 60 fragment of the 200 × 200 lattice after (a) 10^2, (b) 10^4, and (c) 10^6 MC steps (MCS) at $T = 0.5|\epsilon_{MM}|/k_B$. Columns with the heights $h = 1, 2, 3, 4, 5$, and 6 are indicated by filled circles and squares, and open circles and squares, and plus and cross signs, respectively. (Redrawn from Ref. [77].)

(c) $t=10^6$ MCS

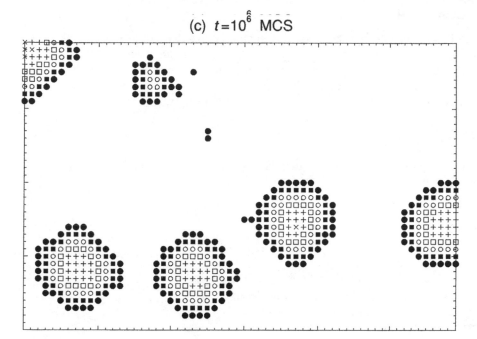

Figure 13 Continued.

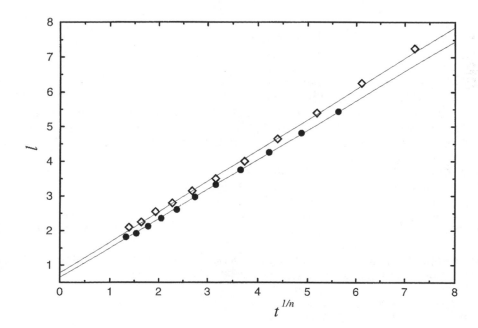

Figure 14 Average crystallite size versus $t^{1/n}$ for $T = 0.5|\epsilon_1|/k_B$ with $n = 8$ (filled circles) and $T = 0.7|\epsilon_1|/k_B$ with $n = 7$ (open diamonds). (Redrawn from Ref. [77].)

2.12   CONCLUSION

The theoretical results presented in this review show that, due to purely kinetic factors, the kinetics of catalytic reactions occurring on nm-sized metal particles, exposing different crystalline facets, may be unique compared to those observed on poly- or single-crystal surfaces. Experimental studies focused on such factors are still rare but certainly will attract more attention in the near future.

ACKNOWLEDGMENTS

Financial support for this work has been obtained from TFR and from the Competence Center for Catalysis at Chalmers.

REFERENCES

1.   J.M. Thomas, W.J. Thomas, Principles and Practice of Heterogeneous Catalysis, VCH, Weinheim, 1997.
2.   M. Gruyters, D.A. King, Faraday Trans. 93 (1997) 2947.
3.   H.C. Kang, W.H. Weinberg, Chem. Rev. 95 (1995) 667.
4.   V.P. Zhdanov, Surf. Sci. 500 (2002), 966.
5.   G. Ertl, Adv. Catal. 45 (2000) 1.
6.   R. Imbihl, G. Ertl, Chem. Rev. 95 (1995) 697.
7.   H.H. Rotermund, Surf. Sci. Rep. 29 (1997) 265.
8.   F. Schüth, B.E. Henry, L.D. Schmidt, Adv. Catal. 39 (1993) 51.
9.   V.P. Zhdanov, B. Kasemo, Surf. Sci. Rep. 20 (1994) 111.
10. M. Che, C.O. Bennett, Adv. Catal. 36 (1989) 55.
11. V.P. Zhdanov, B. Kasemo, Surf. Sci. Rep. 39 (2000) 25.
12. V.P. Zhdanov, B. Kasemo, Surf. Sci. Rep. 29 (1997) 90.
13. P. Stoltze, Progr. Surf. Sci. 65 (2000) 65.
14. C.R. Henry, Surf. Sci. Rep. 31 (1998) 231.
15. P.L.J. Gunter, J.W. Niemantsverdriet, F.H. Ribeiro, G.A. Somorjai, Catal. Rev. Sci. Eng. 39 (1997) 77.
16. K. Wong, S. Johansson, B. Kasemo, Faraday Disc. 105 (1997) 237.
17. J. Zhu, G.A. Somorjai, Nano Lett. 1 (2001) 8.
18. J. Libuda, I. Meusel, J. Hartmann, L. Piccolo, C.R. Henry, H.-J. Freund, J. Chem. Phys. 114 (2001) 4669.
19. S. Johansson, L. Österlund, B. Kasemo, J. Catal. 201 (2001) 275.
20. V.P. Zhdanov, B. Kasemo, J. Catal. 170 (1997) 377.
21. V.P. Zhdanov, B. Kasemo, Surf. Sci. 405 (1998) 27.
22. V.P. Zhdanov, B. Kasemo, Phys. Rev. Lett. 81 (1998) 2482.
23. V.P. Zhdanov, B. Kasemo, Catal. Lett. 50 (1998) 131.
24. S. Johansson, K. Wong, V.P. Zhdanov, B. Kasemo, J. Vac. Sci. Techn. A 17 (1999) 297.
25. H. Persson, P. Thormählen, V.P. Zhdanov, B. Kasemo, J. Vac. Sci. Techn. A 17 (1999) 1721.
26. H. Persson, P. Thormählen, V.P. Zhdanov, B. Kasemo, Catal. Today 53 (1999) 273.
27. V.P. Zhdanov, B. Kasemo, Phys. Rev. E 61 (2000) R2184.
28. V.P. Zhdanov, Catal. Lett. 69 (2000) 21.
29. V.P. Zhdanov, B. Kasemo, Chaos 11 (2001) 335.
30. V.P. Zhdanov, B. Kasemo, Catal. Lett. 72 (2001) 7.
31. E.V. Albano, in M. Borowko, ed., Computational Methods in Surface and Colloid Science, Marcel Dekker, New York, 2000, p. 387.

32. I. Stara, V. Matolin, Surf. Sci. Lett. 4 (1997) 1353.
33. F. Rumpf, H. Poppa, M. Boudart, Langmuir 4 (1988) 722.
34. C.R. Henry, Surf. Sci. 223 (1989) 519.
35. M. Shelef, G.W. Graham, Catal. Rev. Sci. Eng. 36 (1994) 433.
36. H. Muraki, G. Zhang, Catal. Today. 63 (2000) 337.
37. E. Fridell, M. Skoglundh, S. Johansson, B. Westerberg, A. Toncrona, G. Smedler, Stud. Surf. Sci. Catal. 116 (1998) 537.
38. V.P. Zhdanov, B. Kasemo, Appl. Surf. Sci. 135 (1998) 297.
39. L.D. Marks, Rep. Prog. Phys. 57 (1994) 603.
40. H. Graoui, S. Giorgio, C.R. Henry, Surf. Sci. 417 (1998) 350.
41. Yu. Suchorski, J. Beben, E.W. James, J.W. Evans, R. Imbihl, Phys. Rev. Lett. 82 (1999) 1907.
42. Yu. Suchorski, J. Beben, R. Imbihl, E.W. James, D.-J. Liu, J.W. Evans, Phys. Rev. B 63 (2001) 165417.
43. V.P. Zhdanov, Phys. Rev. B 64 (2001) 193406.
44. R. Burch, T.C. Watling, J. Catal. 169 (1997) 45.
45. V.P. Zhdanov, B. Kasemo, J. Catal. 195 (2000) 46.
46. V.P. Zhdanov, B. Kasemo, Surf. Sci. 415 (1998) 403.
47. V.P. Zhdanov, Elementary Physicochemical Processes on Solid Surfaces, Plenum, New York, 1991.
48. M.A. Morisson, M. Bowker, and D.A. King, Compr. Chem. Kin. 19 (1984) 1.
49. J. Wintterlin, Adv. Catal. 45 (2000) 131.
50. V.P. Zhdanov, Surf. Sci. 392 (1997) 185.
51. V.P. Zhdanov, Langmuir 17 (2001) 1793.
52. M. Hildebrand, A.S. Mikhailov, G. Ertl, Phys. Rev. Lett. 81 (1998) 2602.
53. B.C. Sales, J.E. Turner, M.B. Maple, Surf. Sci. 114 (1982) 381.
54. N.A. Collins, S. Sundaresan, Y.J. Chabal, Surf. Sci. 180 (1987) 136.
55. M.A. Liauw, P.J. Plath, N.I. Jaeger, J. Chem. Phys. 104 (1996) 6375.
56. J. Lauterbach, G. Bonilla, T.D. Pletcher, Chem. Eng. Sci. 54 (1999) 4501.
57. M.M. Slinko, E.S. Kurkina, M.A. Liauw, N.I. Jaeger, J. Chem. Phys. 111 (1999) 8105.
58. A. Milchev, in M. Borowko, ed., Computational Methods in Surface and Colloid Science, Marcel Dekker, New York, 2000, p. 509.
59. A.S. McLeod, L.F. Gladden, J. Catal. 173 (1998) 43.
60. A.S. McLeod, Catal. Today 53 (1999) 289.
61. C.V. Ovesen, B.S. Clausen, J. Schiotz, P. Stoltze, H. Topsoe, J.K. Norskov, J. Catal. 168 (1997) 133.
62. E.V. Kovalyov, E.D. Resnyanskii, V.I. Elokhin, B.S. Balzhinimaev, A.V. Myshlyavtsev, in L. Petrov et al., eds., Heterogeneous Catalysis; Proc. of the 9th Intern. Symp, Institute of Catalysis, Varna, 2000, p. 199.
63. M. Frank, R. Kühnemuth, M. Bäumer, H.-J. Freund, Surf. Sci. 454 (2000) 968.
64. S. Abbled, A. Sanchez, U. Heiz, W.-D. Schneider, A.M. Ferrari, G. Pacchioni, N. Rösh, Surf. Sci. 454 (2000) 984.
65. A.V. Zhdanova, Phys. Rev. B 63 (2001) 153410.
66. V.P. Zhdanov, B. Kasemo, Surf. Sci. 496 (2002) 251.
67. L. Zhang, N.A. Seaton, Chem. Eng. Sci. 49 (1994) 41.
68. J.S. Andrade, D.A. Street, Y. Shibusa, S. Halvin, H.E. Stanley, Phys. Rev. E. 55 (1997) 772.
69. C. Steinbrüchel, L.D. Schmidt, Surf. Sci. 40 (1973) 693.
70. V.P. Zhdanov, B. Kasemo, Catal. Lett. 75 (2001) 61.
71. R. Hughes, Deactivation of Catalysts, Academic Press, New York, 1984.
72. C.H. Bartholemew, Catalysis 10 (1993) 41.
73. P.J.F. Harris, Intl. Mater. Rev. 40 (1995) 97.

74. I.M. Lifshitz, V.V. Slyozov, J. Phys. Chem. Solids 19 (1961) 35.
75. E.M. Lifshitz, L.P. Pitaevskii, Physical Kinetics, Pergamon, Oxford, 1981.
76. K. Shorlin, S. Krylov, M. Zinke-Allmang, Physica A 261 (1998) 248.
77. V.P. Zhdanov, B. Kasemo, Surf. Sci. 437 (1999) 307.
78. K. Rommelse, M. den Nijs, Phys. Rev. Lett. 59 (1987) 2578.
79. V.P. Zhdanov, B. Kasemo, J. Chem. Phys. 108 (1998) 4582.

3

Electronic Structure and Chemisorption Properties of Supported Metal Clusters: Model Calculations

GIANFRANCO PACCHIONI

Università di Milano-Bicocca, Milan, Italy

FRANCESC ILLAS

Universitat de Barcelona, Barcelona, Spain

CHAPTER CONTENTS

SUMMARY

In this chapter we review the field of electronic structure calculations on metal clusters and nano aggregates deposited on oxide surfaces. This topic can be addressed theoretically either with periodic calculations or with embedded cluster models. The two techniques are presented and discussed underlying the advantages and limitations of each approach. Once the model to represent the system is defined (periodic slab or finite cluster), possible ways of solving the Schrödinger equation are discussed. In particular, wave function based methods making use of explicit inclusion of correlation effects are compared to methods based on functionals of the electron density (DFT). The second part then describes a series of applications, mostly based on DFT cluster model approaches. We start with a systematic presentation of the features of isolated metal atoms on regular and defect sites of simple binary oxides like MgO and SiO_2: the first can be considered as a prototype of an ionic oxide, while the second is a typical oxide with strong covalent character of the bond. The analysis is then extended to TiO_2 as an example of transition metal oxide. The role of point defects in the stabilization of supported metal clusters and in the activation of very small clusters or even metal atoms is discussed in the last part of this chapter.

3.1 INTRODUCTION

Metal–ceramic interaction is relevant in several areas of modern science such as corrosion, adhesion, microelectronic devices, photovoltaic cells, protective coating of metals, etc. [1–7]. In addition, highly dispersed metal particles supported on oxides make up an important class of heterogeneous catalysts. Oxide surfaces serve not only as inert support for the active component, but they are actually able to stabilize metal particles of a particular dispersion grade or to alter electronic and chemisorption properties of these species. Experimental findings of significant changes in the catalytic performance of supported transition metal species have stimulated a particularly high interest in the question of how the support affects the chemisorption and the catalytic properties of active sites. The complicated phenomenon of metal–support interaction comprises a variety of mechanisms, among them the direct local electronic effect of oxide surfaces on supported metal particles. Despite this considerable technological importance, very little is known about the microscopic nature of the interface between the surface of an oxide substrate, like MgO, SiO_2, Al_2O_3, TiO_2, etc., and the contact metal atoms of a supported particle or of a metallic overlayer. This lack of information is both structural and theoretical. Most of the technologically interesting materials are based on amorphous components, but there are open structural questions even for single-crystal experiments, like the possibility of surface reconstruction, the presence of defects and dislocations, local disorder, etc. For these reasons the theoretical description of the metal–oxide interaction is especially challenging.

More recently, considerable experimental effort has been spent to better characterize the formation of metallic overlayers, emphasizing the very first stages of metal deposition [8–17]. This intense experimental activity is complemented by a rather limited number of "first-principles" theoretical studies dealing with the general problem of metal–ceramic interaction [18–28]. For adsorbed metal atoms, two types of interactions are usually assumed, chemical bonds (mainly with the surface oxygen atoms) on the one hand, or van der Waals interaction and/or weak polarization bonds with no metallization of the surface [29] on the other hand. In the initial step of forming a metallic film, metal atoms impinge on the substrate. These atoms can be reflected from the surface or they may stick to the surface, diffuse on it, and eventually re-evaporate. Condensation can occur if the flux of adsorbed atoms is larger than the flux of re-evaporated atoms, and it is clear that the strength of the bond with the surface plays an essential role in this process.

In this chapter we review the results of calculations on single metal atoms and on nano-clusters interacting with oxide substrates, in particular MgO, SiO$_2$, and TiO$_2$, and also provide a minimum theoretical background to understand the capabilities and limitations of these model calculations. We discuss the interaction of single, isolated, transition metal atoms with the sites of the MgO surface to gain a more systematic understanding of the interface bond [30,31]; we consider small metal clusters interacting with the terrace sites of MgO to better understand the early stages of metal deposition and cluster growth [32–34]. Since an important aspect of the metal–oxide interface is related to point defects where the metal growth usually occurs, some attention is given to the main defects present on the MgO surface and to their interaction with adsorbed metal atoms. We then consider the interaction of isolated metal atoms with simple models of the nondefective and of the defective silica surface [35], and we discuss the structure of small Cu clusters on this surface [36,37]. The third substrate considered is TiO$_2$, and we describe the adsorption properties of alkali metal atoms as well as of Cu, Ag, Pd, and Au atoms on this surface [38,39]. Finally, we consider the extent of the perturbation induced by the substrate on the electronic properties of the supported metal cluster and the changes induced by the metal–oxide interface bond on the reactivity of the supported species.

3.2 SURFACE MODELS

3.2.1 Periodic Models

The electronic structure of solids and surfaces is usually described in terms of band structure. To this end, a unit cell containing a given number of atoms is periodically repeated in three dimensions to account for the "infinite" nature of the crystalline solid, and the Schrödinger equation is solved for the atoms in the unit cell subject to periodic boundary conditions [40]. This approach can also be extended to the study of adsorbates on surfaces or of bulk defects by means of the supercell approach in which an artificial periodic structure is created where the adsorbate is translationally reproduced in correspondence to a given superlattice of the host. This procedure allows the use of efficient computer programs designed for the treatment of periodic systems and has indeed been followed by several authors to study defects using either density functional theory (DFT) and plane waves approaches [41–43] or Hartree–Fock-based (HF) methods with localized atomic orbitals [44,45].

The presence of the adsorbate in the surface unit cell, however, results in a periodic repetition of the ad-atom or molecule in the two directions of space, hence modeling high coverage. The only way to reduce the adsorbate concentration is to increase the size of the unit cell, a solution that implies a very large computational cost. Periodic calculations for supercells containing several tens of atoms are routinely done today. Even for large supercells containing $\approx$ 100 atoms, however, the coverage may be too large. The supercell approach is therefore based on the assumption that the adsorbates do not interact appreciably except when they are very close to each other, so that rapid convergence is achieved with increasing size of the supercell. With charged adsorbates (e.g., ions adsorbed at electrochemical cells) the supercell approach is feasible but less reliable because of the long-range Coulomb interaction between the adsorbed ions. Methods to include correction terms to account for these spurious interactions have been proposed [46].

3.2.2 Cluster Models

An alternative approach to the periodic band structure methods to study solids is the cluster approach [47–50]. Here one explicitly considers only a finite number of atoms to describe a part of the surface while the rest is treated in a more or less simplified way (embedding). The main conceptual difference is that in the cluster approach one uses molecular orbitals, MO, instead of delocalized bands. The description of the electronic properties is thus done in terms of local orbitals, allowing one to treat problems in solids with the typical language of chemistry, the language of orbitals. This is particularly useful when dealing with surface problems and with the reactivity of a solid surface. In fact, the interaction of gas-phase molecules with a solid surface can be described in exactly the same way as the interaction of two molecules. Of course, the cluster model is also not free from limitations. The most serious one is that the effect of the surrounding is often taken into account in a more or less approximate way, thus leading to some uncertainties in the absolute values of the computed quantities. It is also possible that some properties are described differently depending on the size of the cluster used. It is therefore necessary to check the results versus cluster size and shape. The advantages, besides a smaller computational cost, are (1) that in describing adsorbates a very low coverage is considered so that no mutual adsorbate–adsorbate interaction is present in the model and (2) that theoretical methods derived from quantum chemistry can be applied. The latter is an important advantage and should not be underestimated. In fact, in this way it is possible (1) to explicitly include correlation effects in the calculations through, for instance, a configuration interaction (CI) procedure (see below) and (2) to treat exactly the nonlocal exchange as in the Hartree–Fock formalism; in DFT, in fact, the exchange is taken into account in an approximate way through the exchange-correlation functional. This second aspect can be particularly important for the description of magnetic molecules or radical species.

Therefore, cluster calculations represent an alternative way of describing localized bonds at surfaces as well as defects in ionic crystals. The problem is to introduce in a reasonable way the effect of the rest of the crystal. Completely different strategies can be adopted to "embed" clusters of largely covalent oxides, like SiO_2, or of very ionic oxides, like MgO. In SiO_2 and related materials the cluster dangling bonds are usually saturated by H atoms [48,50]. The saturation of the dangling bonds with H atoms is an important aspect of the embedding, but not the only one. In fact, in this way one neglects the crystalline Madelung field. While this term is less important in more covalent materials like silica, it is crucial in the description of solid surfaces with more pronounced ionic character, like that of MgO.

3.2.3 Embedding Schemes

The very ionic nature of MgO implies that the Madelung potential is explicitly included. Indeed, several properties of MgO are incorrectly described if the long-range Coulomb interactions are not taken into account [51]. A simple approach is to surround the cluster of Mg and O ions by a large array of point charges (PC) of value ± 2 to reproduce the Madelung field of the host at the central region of the cluster

[52]. However, the PCs polarize the oxide anions at the cluster border and cause an incorrect behavior of the electrostatic potential [53]. The problem can be eliminated by placing at the position of the ± 2 PCs around the cluster an effective core potential, ECP, representing the finite size of the Mg^{2+} core [54]. No basis functions are associated to the ECP [55], which accounts for the Pauli or exchange repulsion of the O^{2-} valence electrons with the surrounding. This is a simplified approach to the more rigorous ab initio model potential (AIMP) method [56,57] but is computationally simple and reliable. In the AIMP approach the grid of bare charges is replaced by a grid of AIMPs that account not only for the long-range Coulomb interaction but also for the quantum mechanical short-range requirements of exchange and orthogonality without explicitly introducing extra electrons in the model.

The addition of the ECPs to the cluster results in a better representation of the electrostatic potential and hence of the electrostatic contribution to the surface bonding. Still missing from this simplified approach is the polarization of the host crystal induced by an adsorbed species. This effect can be particularly important for charged adsorbates.

The polarization, E_{pol}, induced by a charge on the surrounding lattice can be estimated by means of the classical Born formula [58]:

$$E_{pol} = -(1 - 1/\varepsilon)q^2/2R \tag{1}$$

where ε is the dielectric constant of MgO, q is the absolute value of the charge, and R is the radius of the spherical cavity where the charge is distributed. Since a certain degree of ambiguity remains in the definition of R, this correction is only qualitative. A more refined approach that has been used for the study of the ground state of oxygen vacancies (F centers) in MgO [59] makes use of the ICECAP program [60]. In this approach instead of PCs the cluster is surrounded by polarizable ions described according to the shell model [61,62]; in this way the polarization response of the host is taken into account self-consistently up to infinite distance. In the shell model an ion is represented by a point core and a shell connected by a spring to simulate its dipole polarizability. A similar method has been applied recently to the study of energy states of defect sites at the MgO surface [63]. A quantum cluster has been embedded in a finite array of PCs placed at the lattice sites. The part of the ions closest to the quantum cluster has been treated by the shell model in such a way that they interact among themselves and the quantum cluster via specific interatomic potentials. The positions of the cores and shells of the ions are optimized in response to the changes in charge density distribution within the quantum cluster to minimize the total energy of the system [63].

An alternative, more rigorous approach has been developed in recent years by Pisani and co-workers [64–66]. It is named the *perturbed cluster method* and is based on the EMBED computer program [67]. With this approach the properties of adsorbates at the surface of MgO have been studied at the HF and MP2 levels. The method relies on the knowledge of the one-electron Green function G^f for the unperturbed host crystal, which is obtained by means of the periodic program CRYSTAL [68,69]. A cluster (C) containing the adsorbate is defined with respect to the rest of the host (H). The molecular solution for the cluster C in the field of H is corrected self-consistently by exploiting the information contained in G^f in order to allow a proper coupling of the local wave function to that of the outer region.

3.3 ELECTRONIC STRUCTURE METHODS

In this section we provide a unified point of view of the different theoretical methods used in the study of electronic structure. This includes two rather different families of methods that nevertheless arise from the principles of quantum mechanics. On the one hand, one has the ab initio methods of computation of electronic wave functions and, on the other hand, one has the methods based in the modern density functional theory. In the forthcoming discussion we attempt to focus mainly on the physical significance rather than on mathematical foundation and technical aspects of computer implementation. Details of the methods outlined in this section can be found in specialized references, monographs [70], and textbooks [71].

Finally, one of the main goals of the methods of quantum chemistry is to explore potential energy surfaces and, thus, determine geometries of stable molecules or cluster models as well as of intermediates, transition-state structures, energy barriers, and thermochemical properties. This means that one does not need to compute only accurate energies but also energy gradients and second derivatives with respect to nuclear displacements. Energy derivatives are not trivial, and some methods offer special technical advantages when gradients or higher-order derivatives are to be computed. In addition, the proper interpretation of electronic spectra requires simultaneously handling several electronic states, and not only the ground state. When choosing a particular computational method, one must contemplate the problem to be solved, and in any case the choice a compromise between accuracy and feasibility.

3.3.1 Wave Function-Based Methods in Quantum Chemistry

The best attainable approximation to the wave function and energy of a system of N electrons is given by the full configuration interaction (FCI) approach. In the FCI method the wave function is written as an expansion of Slater determinants with the electrons distributed in the orbitals in all possible ways and the expansion coefficients given by the corresponding secular equation. Indeed, this is the exact solution (in a given finite orbital subspace) and it is independent on the N-electron basis used provided the different basis considered expand the same subspace. This is a very important property because it means that the total energy and the final wave function are independent of whether atomic or molecular orbitals are used as one-electron basis to construct the Slater determinants. The use of atomic orbitals leads to valence bond (VB) theory [72], whereas use of molecular orbitals leads, of course, to the molecular orbital configuration interaction (MO-CI) theory [71]. In the FCI approach both theories are exactly equivalent and provide the same exact wave function although expressed in a different basis. The VB wave functions are relatively easy to interpret [73] because the different Slater determinants can be represented as resonant forms and usually only the valence space is considered within a minimal basis description. The reason for the use of such a limited space is that these VB wave functions are difficult to compute because of the use of a nonorthogonal basis set [74]. On the other hand, the MOs are usually taken as orthonormal, a choice that permits us to carry very large CI expansions and extended basis sets [75–77]. Notice that once a finite set of "$2m$" atomic/molecular spin orbitals is given, the dimension of the N-electron space and, hence, of the N-

electron subspace, is also given. This is because for N electrons and "m" spin orbitals, $m > N$, the number of Slater determinants that can be constructed is

$$\text{dim FCI} = \binom{2m}{N} \qquad \text{or, more precisely,} \qquad \text{dim FCI} = \binom{m}{N_\alpha}\binom{m}{N_\beta} \qquad (2)$$

if the system contains N_α and N_β electrons with alpha and beta spin, respectively. The dimension of the FCI problem, i.e., the dimension of the secular equation or dim FCI, grows so fast that practical computations can be carried out for systems with a small number of electrons. Therefore, the FCI method is often used to calibrate more approximate methods [78].

The simplest N-electron wave function that can be imagined is a single Slater determinant, and the energy is computed as an expectation value. Of course, constraining the wave function to just one Slater determinant largely reduces the variational degrees of freedom of the wave function, and the energy is uniquely defined by the one-electron basis used to construct this particular Slater determinant.

$$|\Psi_0\rangle = \frac{1}{\sqrt{N!}}\det|\Phi_i\Phi_j \ldots \Phi_N\rangle \qquad (3)$$

The only variational degree of freedom concerns the orbital set, which is therefore chosen to minimize the energy expectation value with the constraint that the orbitals remain orthonormal. This leads to a set of Euler equations that in turn lead to the Hartree–Fock equations, finally giving the ϕ_k set although in an iterative way because the Hartree–Fock equations depend on the orbitals themselves. This dependency arises from the fact that the HF equations are effective one-electron eigenvalue equations

$$\hat{f}\phi_i = \varepsilon_i\phi_i \qquad (4)$$

where $\hat{f}$ is the well-known one-electron Fock operator, sum of the kinetic energy, nuclear attraction energy, and the Coulomb and exchange effective potential operators. These effective potentials average interaction with the rest of electrons, which, of course, is given by the orbitals themselves. The final optimum orbitals are therefore those for which the effective average potential used to construct the Fock operator is *exactly* the same that will be obtained using the solutions of (4) and the effective potential is self-consistent. The optimum orbitals are then named self-consistent and HF is synonymous with self-consistent method. Solving Eq. (4) is not simple, especially for molecules, and in practice (4) is transformed to a matrix form by expanding the orbitals through the well-known MO-LCAO method originally designed by Roothaan [79], which leads to

$$\mathbf{FA} = \mathbf{SAE} \qquad (5)$$

where $\mathbf{A}$ is the matrix grouping the coefficients entering in the LCAO and $\mathbf{S}$ is the overlap matrix appearing because the orbitals used in the LCAO are centered in different nuclei and, hence, are not orthogonal. The matrix Eq. (5) is also solved iteratively, and the whole procedure is termed the HF-SCF-LCAO method. An important remark here is that, because a variational approach is used, the HF scheme is aimed to approximate the ground-state (of a given symmetry) wave function only.

The resolution of the Hartree–Fock equations leads to "$2m$" spin orbitals but only N, i.e., the occupied orbitals, are needed to construct the HF determinant (3). The remaining spin orbitals, unoccupied in HF or virtual orbitals, can be used to construct additional Slater determinants. A systematic way to do it is by substituting 1, 2,..., N, occupied spin orbitals by virtual orbitals leading to Slater determinants with 1, 2,..., N substitutions with respect to the HF determinant. The determinants thus constructed are usually referred to as single-, double-..., N excitations and, including all possible excitations, lead to the FCI wave function. Clearly, the FCI wave function is invariant with respect to the orbital set chosen to construct the Slater determinants. However, using the Hartree–Fock orbitals has technical advantages because, at least for the ground-state wave function, the Hartree–Fock determinant contribution to the FCI wave function is by far the dominant term. The fact that the electronic Hamiltonian includes up to two-electron interactions suggests that double excitations would carry the most important weight in the FCI wave function; this is indeed found to be the case. Therefore, one may design an approximate wave function in which only the reference Hartree–Fock determinant plus the double excited determinants are included. The result is called the doubly excited CI (DCI) method and is routinely used in ab initio calculations. The practical computational details involved in DCI are not simple and are not described here. Adding single excitations is important to describe some properties such as the dipole moment of CO [71]; this leads to the SDCI method. Extensions of SDCI by adding triple or quadruple excitations, SDTQCI, are also currently used although the dimension of the problem grows very rapidly.

The truncated CI methods described above are variational, and finding the energy expectation values requires the diagonalization of very large matrices. An alternative approach is to estimate the contribution of the excited determinants by using the Rayleigh–Schrödinger perturbation theory up to a given order. This is the basis of the widely used MP2, MP3, MP4,..., methods that use a particular partition, the Møller–Plesset one, of the electronic Hamiltonian and a HF wave function as zero-order starting point [80,81]. A disadvantage of perturbation theory is that the perturbation series may converge very slowly or even diverge. However, the MP(n) methods have a special advantage over the truncated CI expansions. In the DCI and related methods the relative weight of the different excitations differs from the one in the exact FCI wave function because of the normalization of the DCI wave function. This normalization effect introduces spurious terms and, as a result, the energy of N interacting molecules does not grow as N. This is the so-called size consistency problem and is inherent to all truncated CIs. On the other hand, the MP series is size-consistent order by order. Successful attempts to render truncated CI expansions size consistent have been reported recently [82]. However, the resulting methods are strongly related to the family of methods based on the cluster expansion of the wave function [83]. The coupled-cluster (CC) form of the wave function can be derived from the FCI one as in the case of the DCI method although here the terms included are not selected by the degree of excitation with respect to the HF determinant only. The additional condition is that the different terms fulfill the so-called linked cluster theorem [83]. The resulting system of equations is rather complicated and is not usually solved by diagonalization but rather by means of nonlinear techniques and are not variational [84–86].

The truncated CI and CC methods perform rather well when used to approximate the ground-state wave function. This is because the HF determinant provides an adequate zero-order approach. However, this is not necessarily the case, especially when several excited states are to be studied. The logical extension of the truncated CI expansion is the so-called multireference CI (MRCI) approach where excitations, usually single and double, for a set of reference determinants are explicitly considered [87,88]; the method is referred to as the MR(SD)CI method. Energies and MRCI wave functions are obtained by solving the corresponding secular equation. Again the concept is quite simple, but solving the corresponding eigenvalue problem is not a simple task and the different computational approaches involve very smart ideas and specialized codes coupled to vector, parallel, or vector-parallel processors. For problems of chemical interest the dimension of the MRCI problem is so large that often a small block of Hs in the matrix form of the secular equation is diagonalized and the effect of the rest is taken up to second order by means of perturbation theory in different partitions. The reference space can be constructed by selecting either important determinants or important orbitals. The first idea is used in the CIPSI [89–91] method, whereas the second one is the basis of the CASMP2 [92] and CASPT2 [93–95] methods, where CAS stands for complete active space, the active space defined once a subset of orbitals is chosen and it is complete because a FCI is performed within this orbital space.

Except for the simplest Hartree–Fock approach, the logic of the methods that we have discussed is based on solving the secular problem in a finite subspace defined by the one-electron, orbital, basis chosen or in finding suitable approximations. In all cases the orbital set is fixed and, usually, obtained from a previous HF calculation. Then, the contribution of the different Slater determinants in the CI expansion or the cluster amplitudes in the CC methods is obtained either variationally, i.e., the different CI methods, through perturbation theory, i.e., the MP series and related methods, or by mixed approaches, i.e., the CIPSI method. Nothing prevents one from using the variational method to optimize the orbital set and the configuration contribution at the same time. This is the basis of the multiconfigurational self-consistent field (MCSCF) methods, which are the logical extension of HF-SCF to a trial wave function made as a linear combination of Slater determinants [96]. The mathematical problem is conceptually very similar to that of the HF-SCF approach, namely finding an extreme of a function (the energy expectation value) with some constraints (orbital orthonormality). The technical problems encountered in MCSCF calculations were much more difficult to solve than those of the single-determinant particular case. One of the problems faced by the earlier MCSCF methods was the poor convergence of the numerical process and the criteria to select the Slater determinants entering into the MCSCF wave function. The first problem was solved by introducing quadratically convergent methods [97,98] and the second one by substituting the determinant selection by an orbital selection and constructing the MCSCF wave function using the resulting CAS. The resulting MCSCF approach is known as CASSCF and has turned out to be a highly efficient method [99–101]. The CASSCF wave function is always precisely the zero-order wave function in the CASMP2 and CASPT2 methods and in CIPSI if desired. The CASSCF wave function has some special features worth mentioning. It is invariant with respect to

written as a combination of terms, all of which depend on the one-electron density only:

$$E[\rho] = T_s[\rho] + V_{ext}[r] + V_{coulomb}[\rho] + V_{XC}[\rho] \qquad (7)$$

the first term is the kinetic energy of the noninteracting electrons, the second term accounts for the contribution of the external potential, the third corresponds to the classical coulomb interaction of noninteracting electrons and, finally, the fourth term accounts for all the remaining effects, namely the contribution to the kinetic energy due to the fact that electrons are interacting, the exchange part due to the Fermi character of electrons, and the correlation contribution due to the fact that electron densities are correlated. Obviously, the success of DFT is strongly related to the ability to approximate E_{XC} in a sufficiently accurate way. Now, Eq. (7) plus the Hohenberg–Kohn variational theorem permit us to vary the density by varying the orbitals, which are now referred to as Kohn–Sham orbitals. In addition, the Kohn–Sham orbitals can be expressed in a given basis set as in the usual LCAO approach. When a CGTO basis is used, one has the LCGTO-DF methods [107,108]. The orbital variation must preserve the orthonormality of the orbitals in the Kohn–Sham reference system to hence maintain the number of electrons. The overall procedure is then very similar to Hartree–Fock, and the orbitals minimizing (7) while preserving orthonormality are those satisfying a one-electron eigenvalue problem such as (4), but here the one-electron operator contains the exchange and correlation effective potentials as well and indeed as local one-electron operators. Mathematically the exchange and correlation potentials are the functional derivative of the corresponding energy contributions in Eq. (7). Once $E_{XC}[\rho]$ (or more precisely $E_X[\rho]$ and $E_C[\rho]$) is known, the effective potentials are known and solving the Kohn–Sham equations is similar to solving the Hartree–Fock equation with the important difference that here one may find the exact solution if the $E_{XC}[\rho]$ is the exact one. Notice that there is no guarantee that the final electron density arises from a proper wave function of the corresponding Hilbert space through Eq. (6). This is the famous representability problem; it does not affect the practical use of DFT and is not discussed further here. The interested reader is addressed to the more specialized literature [109].

Several approaches to $E_{XC}[\rho]$ have been proposed in the last several years with increasing accuracy and predictive power. However, in the primitive version of DFT the correlation functional was ignored and the exchange part approximated following Slater's $\rho^{1/3}$ proposal, the method was known as $X\alpha$. In 1980 Vosko et al. [110] succeeded in solving the electron correlation for a homogeneous electron gas and establishing the corresponding correlation potential. The resulting method, including also the exchange part, is known today as the local density approximation (LDA) and has been successful in the description of metals, bulks, and surfaces, although it has experienced more difficulties in the description of molecules and ionic systems; for instance, LDA incorrectly predicts NiO to be a metal [111,112]. The Kohn–Sham equations were initially proposed for systems with a closed-shell electronic structure and hence suitable to study singlets. The study of open-shell systems can be carried out using a spin unrestricted formalism. In this case different spatial orbitals are used for alpha and beta spin orbitals. In the case of LDA the resulting formalism is known as the local spin density approximation (LSDA), or simply local spin density (LSD). This is similar to the well-known unrestricted

Hartree–Fock (UHF) formalism and suffers from the same drawbacks when dealing with open-shell systems [71]. This is especially important when attempting to study magnetic systems with many open shells. The unrestricted Kohn–Sham determinant corresponds necessarily to a mixture of the different possible multiplets in the electronic configuration represented by the Kohn–Sham determinant.

In spite of the inherent simplifications, LDA (and LSDA) predictions on molecular geometries and vibrational frequencies are surprisingly good. However, bonding energies are much less accurate and require going beyond this level of theory. This is also the case when dealing with more difficult systems such as biradicals or more delicate properties such as magnetic coupling in binuclear complexes or ionic solids [113–115]. The DF methods that go beyond the LDA can be grossly classified in gradient corrected (GC) and hybrid methods. In the first set the explicit calculation of the $E_{XC}[\rho]$ contributions involves not only the density, ρ, but also its gradient, $\nabla\rho$. The number of GC is steadily increasing but among the ones widely used we quote the Becke exchange functional [116] (B) and the Perdew–Wang [117,118] (PW) exchange correlation functional. The latter is usually referred to as the generalized gradient approximation (GGA) and is particularly used in condensed matter and material science. Another popular gradient corrected correlation functional is the one proposed by Lee-Yang-Parr [119] based on the work of Colle and Salvetti on the correlation factor [120]. In some cases one uses B for the exchange and PW for the correlation part—this is usually referred to as BP—or B for the exchange and LYP for the correlation part, giving rise to the BLYP method. The term *hybrid functionals* is used to denote a family of methods based on an idea of Becke [121]. This approach mixes DF and Fock exchange and local and GC correlation functional in a proportion that is obtained from a fit to experimental heats of adsorption for a wide set of molecules. The most popular hybrid method is B3LYP, where the number indicates that three parameters are fit to experiment. For a more detailed description about GC and hybrid methods, the reader is referred to general textbooks [122] or to more specialized literature [108,109].

Aside from the great advances in the development of new exchange-correlation functionals, the question of which functional provides the best chemical accuracy is still under discussion. In the wave function-based methods it is possible to check the accuracy of a given level of theory by systematic improvement of the basis set and of the level of treatment of electron correlation. Unfortunately, this is not completely feasible in the framework of DFT. One can improve the basis set, but there is no way to systematically improve the $E_{XC}[\rho]$. Therefore, one needs to establish the accuracy of the chosen approach by comparing several choices for $E_{XC}[\rho]$. For many systems this choice is not critical, and the use of several functionals permits one to add error bars to the computed quantities. However, in the systems of interest in the present work, namely the description of the transition metal–oxide interface, the choice of the functional has been shown [123] to be crucial. An extreme case is that of Cu adsorption on MgO: the reported adsorption energies range from a practically unbound Cu atom at the Hartree–Fock level to a moderate adsorption, 0.35–0.90 eV, at gradient corrected DF level, to strong adsorption, about 1.5 eV, using the local density approximation, LDA. Recently Ranney et al. [124] extrapolated the adhesion energy of a single copper atom on MgO from their microcalometric measures of the heat of adsorption and found a value of 0.7 eV. The comparison of this value with the computed adsorption energies of [123] indicates that, among the currently used approximations of the exchange-correlation functional, the pure DF ones seem to provide the best answer, while hybrid functionals slightly underestimate the adsorption energy.

3.4 THE MgO SUBSTRATE

3.4.1 Transition Metal Atoms on MgO(001)

Understanding the bonding mechanism of individual metal atoms with the surface of a simple oxide like MgO is an essential prerequisite for the theoretical study of more complex systems, like clusters or organometallic fragments, supported on oxide substrates. In the following we review the interaction of various transition metal (TM) atoms, Cr, Mo, W, Ni, Pd, Pt, Cu, Ag, and Au, with the oxygen anions of the MgO(001) surface. We restrict the analysis to these sites as it is known from other theoretical investigations that the Mg cations of the surface are rather unreactive. The cluster used to simulate the MgO(001) surface is O_9Mg_9 (see Figure 1 for an example of an MgO cluster), and the calculations have been performed at the DFT level using the BLYP method; i.e. Becke's exchange functional [116] and the Lee-Yang-Parr correlation functional [119]. For further details see [30].

Cu, Ag, Au

The presence of a filled d shell and a singly occupied s orbital prevents the Cu, Ag, and Au atoms ($d^{10}s^1$ atomic ground state) from easily changing their configuration during bond formation. The lowest-lying excited state, d^9s^2, is in fact about 1.7 eV higher than the ground state for Cu and Au, and about 4 eV higher for Ag. On the other hand, Cu, Ag, and Au could, in principle, form relatively strong bonds with oxygen atoms of not completely reduced oxide surfaces by partial transfer of their outer electron to the substrate. We will see that this is the case for TiO_2; see Section 3.6.1. The ground state of the surface cluster plus adsorbate is $^2A_1$ for all atoms of

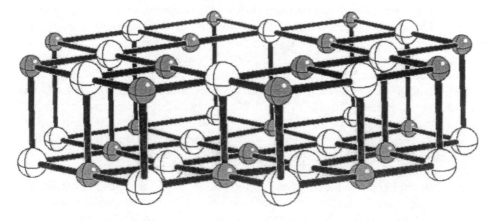

Figure 1 $Mg_{21}O_{20}$ cluster model of the MgO surface. The central atom in the first layer has been removed to represent an oxygen vacancy or F center. The cluster is embedded in a large array of point charges (not shown).

the triad and the adsorbed atoms retain their $nd^{10}(n+1)s^1$ atomic configuration. However, a very small mixing of the metal σ orbitals with the O $2p_\sigma$ orbital also occurs. All atoms are weakly bound to the surface. The dissociation energy goes in fact of 0.2–0.3 eV for Cu, Ag, and Au. The weak bonding is shown also by the very low values of the frustrated translation, perpendicular to the surface, which is characterized by a vibrational frequency of $100\,\mathrm{cm}^{-1}$ and smaller.

The analysis of the bonding mechanism excludes the occurrence of a charge transfer from the metal atom to the substrate. The bonding can thus be described as mainly due to the polarization of the metal electrons by the surface electric field, accompanied by a relatively modest chemical interaction. This result suggests that the oxide anions of the MgO surface are highly charged, with little tendency to ionize the adsorbed coinage atoms. The relatively weak bonding found for Cu, Ag, and Au is therefore the consequence of the large ad-atom size (due to the singly occupied valence s orbital), which determines the long surface–adsorbate distance. Ag behaves somewhat differently from Cu and Au: the bond distance is even longer than for the other two members of the triad. The smaller propensity for a metal s–d hybridization and for a mixing with the oxygen levels is most likely the reason for the weaker bonding of Ag.

Ni, Pd, Pt

The interaction of Ni, Pd, and Pt atoms with MgO is considerably more complex than that of the coinage metal atoms because of the interplay between the d^{10}, the d^9s^1 and, at least for Ni, the d^8s^2 configurations of the atoms. The interaction of free atoms with the MgO cluster gives rise to several states; the lowest triplet and singlet states, $^3B_2$ and $^1A_1$, correlate at infinite ad-atom–surface separation with $MgO + M(d^9s^1)$ and $MgO + M(d^{10})$, respectively. The magnetic state, $^3B_2$, features the lowest energy for long surface-metal distances. At shorter distances, however, a crossing of the $^3B_2$ with the $^1A_1$ state occurs. At short distances the $^1A_1$ state becomes lower in energy, although the detailed nature of the ground state is not completely established [125]; the potential energy curve exhibits a deeper minimum and a rather high adsorption energy of about 1 eV or more, depending on the metal. The curve crossing implies that a magnetic quenching accompanies the formation of the bond. The key mechanism for the bonding is the formation of s–d hybrid orbitals, which can conveniently mix with the O $2p_\sigma$ orbitals; in fact, in the $^1A_1$ configuration the metal s, d_σ, and the O $2p_\sigma$ orbitals are strongly hybridized.

The bonding has a somewhat different character in Ni than for Pd and Pt. Ni exhibits a nonnegligible polarity of the bond, while Pd and Pt form a more covalent bond. This is consistent with the fact that Ni is easier to ionize. However, by no means can the bonding be viewed as a charge transfer from the metal atom to the oxide surface. The orbital analysis clearly shows a substantial mixing (hybridization) of metal nd and O 2p orbitals. The fact that all three atoms in the group form relatively strong bonds with the surface, compared for instance to the coinage metals, can be explained by the more pronounced tendency of the group 10 atoms to form s–d hybrid orbitals and a more direct involvement of their valence d orbitals in the bond with the oxygen. This tendency can be generally related to the s→d (or d→s) transition energies, which are considerably smaller for Ni, Pd, and Pt than for Cu, Ag, and Au.

Cr, Mo, W

The interplay between high-spin and low-spin states is of great importance in the interaction of Cr, Mo, and W atoms with MgO. These atoms have high-spin atomic ground-state configurations, $nd^5 (n+1)s^1$ ($^7A_1$ in C_{4v} symmetry); their valence d shell is therefore half-filled, and a comparison of their interaction with that of the late TM atoms is particularly instructive. For Cr the ground state of the M/MgO cluster, $^7A_1$, correctly dissociates into Cr and MgO ground states and exhibits a very shallow potential energy curve, with a minimum around 2.8 Å and a binding energy of 0.34 eV. Also for Mo the $^7A_1$ state is lowest in energy, with an equilibrium distance at large separation from the surface, close to 3 Å, and a binding energy of 0.33 eV. At adsorption heights below 2.2–2.3 Å, the low-spin $^5A_1$ state becomes energetically preferred, but again with a local energy minimum slightly higher than for the $^7A_1$ state. The two atoms, Cr and Mo, also have similar bonding properties. The bonding is largely due to polarization and dispersion, with little, if any, chemical mixing of the metal orbitals with the surface electronic states. A completely different bonding arises from the interaction of W with MgO. The interaction of W in the high-spin d^5s^1 configuration results in a flat curve, very similar to those computed for Cr and Mo. This state, however, is not the lowest one, even for long surface-adsorbate distances. The low-spin $^5A_1$ state exhibits a deep minimum near 2.15 Å. In this minimum, W is bound to MgO by 0.72 eV, a chemical bond of similar strength than that calculated for the Ni triad. A population analysis has shown that in this state the d^4s^2 atomic configuration is mixed not only with the d^5s^1 state, but also with the O 2p orbitals. This is consistent with the fact that the s–d transition energy is smallest in W where the s^1d^5 and the s^2d^4 states are almost isoenergetic, in contrast to the situation for Cr and Mo. The facile s–d hybridization is actually the reason for the strong bonding of W. As for adsorbed Ni, the bond with W exhibits a considerable polar character.

In summary, the Ni triad is the only one for which strong interface bonds are formed. The adhesion energy of these atoms on top of surface oxygens is about 1 eV/ atom or more. The only other metal atom considered here that exhibits a tendency to form strong bonds with the surface is W. For W, in fact, the bond strength is comparable to that of the Ni triad. The tendency to form strong bonds is connected to the fact that metal s and d orbitals hybridize and that these hybrid orbitals mix with the p orbitals of the surface oxygen. Pd is somewhat special in the Ni triad as it has the smallest binding energy. This reflects the general tendency toward a nonmonotonous behavior in many chemical properties as one moves down a TM group; in particular, second-row TM atoms often exhibit weaker interactions than the isovalent first- and third-row atoms. To some extent, this trend is observed also for the other two triads considered. The bonding of Cr, Mo, Cu, Ag, and Au to a MgO substrate, however, can be classified as weak, arising mainly from polarization and dispersion effects with only minor orbital mixing with the surface oxygen orbitals. This explains the very long bond distances found in some cases and the flat potential energy curves that result in very small force constants.

Thus, the TM atoms considered can be classified into two groups: atoms that tend to form relatively strong chemical bonds with the surface oxygen anions of MgO (Ni, Pd, Pt, and W), and atoms that interact very weakly with the surface, with adsorption energies of the order of one third of an eV or less (Cr, Mo, Cu, Ag, and

Au). The interaction *does not imply a significant charge transfer from the metal to the surface*. This is an important conclusion, connected to the highly ionic nature of the MgO surface where the surface oxygen atoms have their valence almost saturated. To a first-order approximation MgO can be described in terms of classical ionic model, $Mg^{2+}-O^{2-}$. This means that the oxygen centres at the *regular surface sites of MgO(001) are almost completely reduced and are not able to oxidize adsorbed metal atoms*. This conclusion is valid for the regular surface sites, but of course adsorption phenomena at oxide surfaces can be substantially different when occurring at defect sites. In the following section we consider some of these sites and their interaction with deposited metal atoms.

3.4.2 Pd Atoms at MgO Defect Sites

Often the most important properties of materials are directly or indirectly connected to the presence of defects and in particular of point defects [126,127]. These centers determine the optical, electronic, and transport properties of the material and usually dominate the chemistry of its surface. A detailed understanding and a control at the atomistic level of the nature (and concentration) of point defects in oxides are therefore of fundamental importance also to understand the nature of the metal–oxide interface. The accurate theoretical description of the electronic structure of point defects in oxides is essential for understanding their structure-properties relationship but also for a correct description of the metal–oxide interface and of the early stages of metal deposition on oxide substrates.

MgO is a particularly well-studied oxide; the structure of the (100) single-crystal surface is extremely flat, clean, and stoichiometric. Recent grazing incident x-ray scattering experiments have shown that both relaxation, $-0.56 \pm 0.4\%$, and rumpling, $1.07 \pm 0.5\%$, are extremely small [128]. However, no real crystal surface consists of only idealized terraces. A great effort has been undertaken in recent years to better characterize the MgO surface, in particular for polycrystalline or thin-film forms, which in some cases exhibit a heterogeneous surface, due to the presence of various sites. All these sites can be considered as defects. The identification and classification of the defects are of fundamental importance. In fact, the presence of appreciable concentrations of defects can change completely the chemical behavior of the surface.

The most important defects present at the surface of MgO have been recently reviewed [129,130] (Table 1 and Figure 2). These are (1) low-coordinated Mg^{2+} ions (Mg^{2+}_{4c}, Mg^{2+}_{3c}, etc.) with a number of neighbors lower than on the flat (100) terraces; (2) low-coordinated anions O^{2-} sites that exhibit a completely different chemistry than the corresponding five-coordinated terraces, O^{2-}_{5c}, sites [131]; (3) hydroxyl groups [132,133]; (4) oxygen vacancies (the so-called color centers or F centers); these can have different formal charges, F or F^+ and can be located at terrace, step, and corner sites [52,134–137]; (5) cation vacancies, often classified as V_s centers [138]; (6) divacancies, created by removing a neutral MgO unit from the surface [139]; (7) impurity atoms like substitutional Ni ions, as in MgO–NiO solid solutions, or monovalent dopants like Li^+; (8) oxygen radical anions, O^-, which can be formed by various means on the surface, like doping the material with alkali metals ions; (9) neutral electron traps at the MgO surface, like corner Mg ions, inverse kink, etc. [63]; (10) (111) microfacets, inverse kinks and other morphological irregularities. The

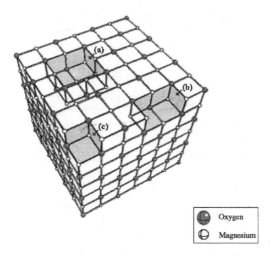

Figure 2 Schematic representation of oxygen vacancies (F centers) formed at the terrace (a), edge (b), and corner (c) sites of MgO.

complexity of the problem is increased by the fact that the point defects can be located at various sites, terraces, edges, steps, and kinks [140] and that they can be isolated, occur in pairs, or even in "clusters." Furthermore, the concentration of the defects is usually low, making their detection by integral surface-sensitive spectroscopies very difficult. A microscopic view of the metal–oxide interface and a detailed analysis of the sites where the deposited metal atoms or clusters are bound become essential in order to rationalize the observed phenomena and to design new materials with known concentrations of a given type of defects.

Table 1 Summary of Most Important Surface Defects in MgO

Defect	Symbol	Schematic description
Low-coordinated cation	Mg_{nc}^{2+} $(n = 3,4)$	Coordinatively unsaturated cation
Low-coordinated anion	O_{nc}^{2-} $(n = 3,4)$	Coordinatively unsaturated anion
Hydroxyl group	(OH)	Proton attached to O^{2-}
Anion vacancy	F_{nc}^{m+} $(m = 0,1,2; n = 3,4,5)$	Missing oxygen with trapped electrons
Cation vacancy	V_{nc}^{m-} $(m = 0,1,2; n = 3,4,5)$	Missing cation with holes at O neighbors
Divacancy	$V_{Mg} V_O$	Cation and anion vacancy
Impurity atoms	M^{n-}/O^{n-}; Mg^{2-}/X^{2-}	Substitutional cation (M) or anion (X)
Oxygen radical	O_{nc}^{-} $(n = 3,4,5)$	Hole trapped at O anion
Shallow electron traps	None	Morphological sites with positive electron affinity (inverse kink, etc.)
(111) Microfacets	None	Small ensemble of Mg_{3c} or O_{3c} ions

At least for the case of isolated Pd atoms, a number of these sites have been explored and a classification of the defects in terms of their adhesion properties is possible [141]. We first consider the case of Pd interacting with anion sites, O_{5c}, O_{4c}, or O_{3c}. The binding energy of a Pd atom with these sites increases monotonically from $\approx 1\,eV$ (O_{5c}) to $\approx 1.5\,eV$ (O_{3c}), and consequently the distance of the Pd atoms from the surface decreases (Table 2). This is connected to the tendency of low-coordinated anions on the MgO surface to behave as stronger basic sites as the coordination number decreases [131]. However, the difference in binding energy between an O^{2-} at a terrace site or at a corner site, 0.5 eV, is not too large. Much larger binding energies have been found for other defect sites like the oxygen vacancies or F centers.

On these sites the bonding of Pd is about three times stronger than on the O anions [141–143]. There is no large difference in Pd atom adsorption energy when the F center is located at a terrace, at a step, or at a corner site; see Table 2 ($E_b \approx 3.5 \pm 0.1\,eV$). Notice that the binding energy of Pd to a F center turns out to be the same with different computational methods. In fact, cluster model studies [141] and periodic plane wave calculations [143] give almost exactly the same energy and geometry for Pd on an F_{5c} center. Thus, the Pd atoms are likely to diffuse on the surface until they become trapped at defect sites like the F centers, where the bonding is so strong that only annealing at high temperatures will induce further mobility.

The last group of defect sites considered here is that of the paramagnetic F^{+} centers. These consist of a single electron trapped at the vacancy; their electronic structure has been studied in detail in polycrystalline MgO samples by EPR spectroscopy [144]. Formally these centers are positively charged as a single electron replaces an O^{2-} ion in the lattice. On these sites the binding of Pd is between that of the O_{nc} sites and of the F centers (Table 2). For instance, on F_{5c}^{+}, Pd is bound by 2.10 eV, while E_b for Pd on O_{5c} is 0.96 eV and that for F_{5c} is 3.42 eV (Table 2).

Recently, the adsorption of a Pd atom on a model of a surface OH group on MgO has shown that the binding energy at this site, 2.6 eV [145], is in between that found for neutral and charged F centers, suggesting that hydroxyl groups at the oxide surface are good candidates for metal nucleation and growth [132].

Therefore, although this list is far from being complete, it provides strong evidence that in the initial phases of metal deposition the defect centers play a crucial role in stabilizing the metal atoms and in favoring nucleation and growth.

3.4.3 Metal Clusters on MgO(001)

The interaction of TM metal clusters with the surface of MgO has been studied with both cluster [32–34] and slab [146] models, but only very small clusters have been considered, containing up to 4 to 5 atoms. The metals considered are Co, Ni, Cu, Pd, and Ag, and the various isomers studied are shown in Figure 3. They can be classified into three main groups: planar or nearly planar structures with the cluster plane "parallel" to the MgO surface (Fig. 3 a–c), planar structures with the cluster plane "normal" to the surface (Fig. 3 d–f), and tetrahedral or distorted tetrahedral structures (Fig. 3 g–i). The results show that microclusters adsorbed on the regular MgO(001) surface do not necessarily tend to adhere to the surface with the largest

Table 2 Adsorption properties of Pd on O Anions, F and F^+ Centers Located at Terrace (5c), Edge (4c), and Corner (3c) Sites of the MgO Surface

	Terrace			Edge			Corner		
	O_{5c}	F_{5c}	F_{5c}^+	O_{4c}	F_{4c}	F_{4c}^+	O_{3c}	F_{3c}	F_{3c}^+
$z(Pd),^{(a)}$ Å	2.210	1.524	1.497	1.981	1.092	1.124	1.760	0.694	0.693
$E_b(Pd),^{(b)}$ eV	0.96	3.42	2.10	1.35	3.64	2.41	1.51	3.66	2.35

(a) Vertical distance between the Pd atom and the surface plane.
(b) Computed as $E(MgO) + E(Pd) - E(MgO/Pd)$.

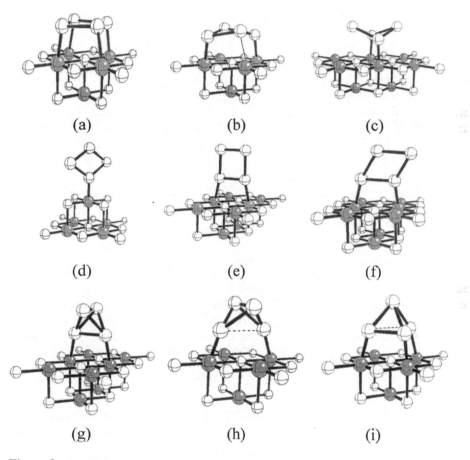

Figure 3 Structure of various isomers of Co, Ni, Cu, Ag, and Pd clusters deposited on the regular MgO(100) surface.

possible number of metal atoms (surface wetting), but rather they keep some bond "directionality." This results from the balance of various terms, the energy gain due to the bond formation with the O anions, the Pauli repulsion with the surface, and the loss of metal–metal bonding within the cluster due to distortions of the metal frame. When the metal–oxygen bond is sufficiently strong, e.g., as for Ni and Pd, the formation of new interface bonds may compensate for the loss of metal–metal bonds and the cluster distorts from its gas-phase geometry (Fig. 3i). For weaker metal–oxide bonds, as found for Co, Cu, and Ag, the metal–metal interactions tend to prevail over the metal–MgO ones so that the cluster maintains, to a large extent, its electronic and geometric structure (Fig. 3f). Ni_4 and Pd_4, which form relatively strong bonds with MgO, tend to adhere with more metal atoms with the surface leading to a distorted tetrahedron while in the gas-phase they both assume a tetrahedral shape (Fig. 3i). Ag_4 and Cu_4 clusters, which interact weakly with MgO, assume a rhombic form with the cluster plane normal to the surface (Fig. 3f); their structure on the surface is not too different from what they have in the gas phase. Co_4, which forms bonds of intermediate strength between Ag and Cu on one side, and Ni and Pd on the other, exhibits several surface isomers with different structures but similar stabilities. The magnetization on the magnetic Ni and Co clusters is largely unchanged also in the supported species. In some cases, however, we observe a partial quenching of the magnetic moment, which is generally restricted to the metal atoms in direct contact with the oxide anions [34]. Thus, despite relatively strong MgO/M_4 bonds (Co_4 is bound on MgO by 2.0 eV, Ni_4 by 2.4 eV), the electronic structure of supported transition metal moieties is only moderately perturbed. These conclusions are valid only for an ideal defect free surface; investigations of the interaction of metal clusters with surface defects may lead to quite different conclusions.

3.5   THE SiO$_2$ SUBSTRATE

3.5.1   Metal Atoms on Nondefective SiO$_2$

Despite the importance of silica as support for metal catalysts, few experimental studies have been dedicated to the interaction of metal atoms with the surface of silica. Cu, Pd, and Cs are among the few metals for which experimental data exist [12,147–150]. Because many catalysts consist of small particles on high surface-area powders of SiO_2 or Al_2O_3, understanding at a microscopic level the metal–oxide interaction, in particular at the defect sites, is important for understanding the catalytic activity. The interaction of the isolated Cu, Pd, and Cs atoms with the regular surface sites of silica has been modeled by the $(HO)_3Si–O–Si(OH)_3$ cluster [35] using the DFT approach with the B3LYP hybrid functional described earlier [119,121]. All three metals interact very weakly with the two-coordinate oxygen sites of silica, with interaction energies of the order of 0.1–0.2 eV [35]. The potential energy curves are flat and the equilibrium distances are rather long. Given the weak nature of the interaction, adsorption geometries where the adsorbate interacts with more than a surface oxygen seem to be preferred. Thus, the regular sites of the silica surface are very unreactive toward the metal atoms. It is worth noting that the adsorption of Pd on the O atoms of silica shows a bond strength that is even lower than that of the rather inert MgO surface. This means that atoms deposited on silica from the gas phase will not be trapped at the regular sites but instead will diffuse on the surface, remain trapped at a defect, or re-evaporate. The role of defects to understand the mechanism of metal deposition and cluster growth is therefore even more important than for MgO.

3.5.2 Metal Atoms on Defective SiO_2

Here we consider some of the most common defect sites at the surface of dehydroxylated SiO_2. These are (1) E′ defect centers corresponding to a Si singly occupied sp^3 dangling bond, $\equiv Si^\bullet$; (2) nonbridging oxygen (NBO), $\equiv Si-O^\bullet$, centers; (3) neutral oxygen vacancies, $\equiv Si-Si \equiv$ or V_O. While the E′ and NBO defects are paramagnetic and detectable by EPR spectroscopy, the neutral oxygen vacancy can only be detected with other techniques. All these defects have attracted a great interest in the past 20 years because of their role in the degradation of Si–SiO_2 interfaces in microelectronics devices or in the absorption of light in optical fibers [127]. Very similar defects are present in the bulk of amorphous silica and α-quartz [151] and on the surface of mechanically activated silica [152,153], of SiO_2 thin films [154], or of UHV-cleaved α-quartz single crystals [155,156]. The fingerprints of the presence of these centers on the surface are typical EPR signals and absorption bands in the optical spectra of the material [157]. It has been suggested that these defect centers are the primary cause of the interface bond formation.

Surface E′$_s$ Centers

Si sp^3 dangling bonds, $\equiv Si^\bullet$, are present on the (0001) and (10$\bar{1}$0) surfaces of α-quartz, on mechanically activated silica, as well as on UHV-grown thin SiO_2 films. The presence of these defects on the SiO_2 surface is shown by a hyperfine coupling constant of ≈ 470 G with the ^{29}Si nuclide [152,153] and by an intense optical transition centered around 6 eV similar to that observed in bulk silica [154–156]. The $\equiv Si^\bullet$ surface radical is rather reactive toward molecular species but also toward metal atoms. However, very different bonding interactions occur with different metals. Cu and Pd, in fact, form rather strong bonds with the defect site, ≈ 2.3 eV. The formation of such a strong bond is reflected in the rather short Si–M distance, ≈ 2.2 Å. The bonding can be described as largely covalent, and the polarization of the bonding electrons is toward the metal. For the case of Pd adsorption, the resulting spin density is almost entirely located on the metal atom.

Despite the similar structural characteristics, the bonding mechanism of Cu and Pd with a $\equiv Si^\bullet$ center is somewhat different. In Cu it arises from the coupling of the Si sp^3 singly occupied orbital and the metal 4s open-shell orbital with formation of a σ bonding level. This orbital appears as an impurity level in the band gap of the material. Pd, on the other hand, forms a bond through a direct involvement of the 4d orbitals with the SiO_2 surface state. This implies a configuration change that increases the $4d^9 5s^1$ character of the metal atom.

In principle, one could expect for Cs a similar bonding mechanism as for Cu. Given the large size of the Cs atom, the Si–M bond distance is very large, about 3.5 Å, and the bond strength is of 0.9 eV. However, the nature of the interaction is different from the Cu case. The adsorbed Cs, in fact, becomes positively charged, indicating the formation of a Si–Cs covalent bond strongly polarized toward Si.

Nonbridging Oxygen

Nonbridging oxygens at the surface of silica, $\equiv Si-O^\bullet$, represent probably the most important defects for the reactivity of mechanically activated SiO_2[152]. These broken bonds have been proposed as the centers where impinging Cu atoms are trapped from metastable impact electron spectroscopy experiments (MIES) [150]. The fingerprint of their existence is given by a characteristic doublet in the EPR signal and by a hyperfine splitting with the ^{17}O nuclide as well as by an optical absorption band at about 2 eV [157]; this absorption band, however, is rather weak in glassy silica and may be difficult to detect at the SiO_2 surface.

   Cu, Pd, and Cs form strong bonds with the NBO centers. The computed D_e is of 3.8, 2.9, and 3.5 for Cu, Pd, and Cs, respectively. The metal atoms directly bound to the surface oxygen become partially oxidized, although to a different extent. Cs donates almost an entire electron to the SiO_2 substrate and becomes Cs^+; Cu and Pd form bonds with more covalent character with the $\equiv Si-O^\bullet$ group, but with an important polarization toward oxygen (partial charge transfer). A net residual positive charge also forms on adsorbed Cu and Pd atoms. The fact that the metal atoms become partially (Cu and Pd) or fully (Cs) ionized leads to important interactions with the neighboring surface O atoms. The adsorbate interacts, mostly electrostatically, with the exposed two-coordinated O atoms of the surface, which become effectively three-coordinated. This results in the formation of rings where the metal atom binds to two or even three surface oxygens. The formation of six-member rings has recently been demonstrated from XAFS measurements on the deposition of Ni^{2+} ions on amorphous silica [158].

Neutral Oxygen Vacancy

Neutral oxygen vacancies, V_O, are quite abundant in glassy silica or in irradiated α-quartz. The defect arises from the recombination of two $\equiv Si^\bullet$ dangling bonds derived from the removal of a neutral oxygen, $\equiv Si-Si \equiv$. As a consequence, the Si–Si distance, which in nondefective silica is of 3.06 Å, in a V_o center becomes ≈ 2.5 Å [159]. This defect is diamagnetic, and its presence can be detected only from a strong optical absorption around 7.6 eV [159]. The presence of the V_O centers at the (1010) surface of α-quartz or in SiO_2 thin films is clearly shown by the EELS spectra exhibiting a prominent feature at 7.2–7.5 eV [154–156].

   The behavior of the three metal atoms with this defect center is very different. Pd and Cu are bound by 2.2 and 0.6 eV, respectively; Cs is very weakly bound, 0.1 eV, and the bond arises entirely from dispersion forces. Thus, while the TM atoms form strong bonds with the Si atoms of the surface, Cs does not react at all with this defect. The main reason for the different behavior lies, besides the different electronic structure, in the atomic size. In fact, the bonding interaction of the $\equiv Si-Si \equiv$ defect with the metal atom results in a $\equiv Si-M-Si \equiv$ structure, where M replaces the missing O atom of the nondefective surface. This leads to a strong geometrical relaxation and, in particular, to a considerable increase of the Si–Si distance, which goes from ≈ 2.5 Å to ≈ 4 Å (Cu) and ≈ 3.5 Å (Pd). The presence of a singly occupied diffuse 5s orbital is the reason for the larger distortion of the lattice and of the weaker bonding in the Cu case. The inclusion of Cs in the Si–Si bond would probably require an even higher relaxation, but this would result in a very strained structure. As a consequence, only a very shallow minimum is found for Cs at about

4.3 Å from the Si atoms, which retain the Si–Si distance of the free-surface defect, 2.49 Å.

There is a clear indication that the two-coordinated bridging oxygens of the surface, $\equiv$Si–O–Si$\equiv$, are very unreactive toward metal atoms. Only weak interactions occur between the metal and these sites. This means that the atoms adsorbed from the gas phase will rapidly diffuse on the surface without being trapped except at defect sites. Therefore, these are the sites where clustering and metal aggregation processes are likely to begin. These results are consistent with measurements of the sticking coefficient of Cu on SiO_2. Zhou et al. found that at 300 K only one third of the initially incident Cu atoms stick to the surface [12]; Xu and Goodman found that the sticking depends markedly on the temperature, varying from 0.6 at 90 K to 0.1 at 400 K [147]. Both studies agree with the fact that the bonding of Cu with the clean surface is weak and that sticking occurs only at the defect sites. The very weak interaction with the substrate is probably the reason for the Volmer–Weber growth mode of metal overlayers on silica with formation of 3D clusters and small metal particles [160,161]. It has not been possible so far to identify experimentally which defects are responsible for the sticking. Among the three major defect sites considered here, the nonbridging oxygen, $\equiv$Si–O$^\bullet$, is the most reactive followed by the surface E'_S centers, $\equiv$Si$^\bullet$, and by the neutral oxygen vacancies, $\equiv$Si–Si$\equiv$.

3.5.3 Cu Clusters on SiO_2

In this section we consider the interaction mode of small Cu clusters, from Cu_2 to Cu_5, with the E' and NBO centers as a representative example of point defect at the silica surface [36,37]. At the Cu_n–SiO_2 interface, the clusters are bound to the E' and NBO centers with rather different distances, $r(Si–Cu) \approx 2.34 \pm 0.06$ in E' and $r(O–Cu) \approx 1.95 \pm 0.07$ Å in NBO. However, all clusters are anchored to the surface through more than one Cu atom (see Figure 4). This is a general characteristic; in fact, besides the direct O–Cu or Si–Cu covalent bonds, weaker interactions occur between the metal cluster and the bridging oxygens of the surface. In NBO the partial charge transfer from Cu to SiO_2 leads to a depletion of electronic charge from the metal cluster. The interaction with the substrate induces substantial geometrical changes within the metal unit. The addition of a second atom to the $\equiv$Si–Cu complex results in the formation of a supported Cu dimer. The Cu–Cu distance in the supported molecule, 2.34 Å, is about 0.1 Å longer than in gas phase, 2.26 Å. Notice that the same molecule adsorbed on an NBO center shows an elongation of almost 0.2 Å, consistent with a stronger bond with this surface defect. Gas-phase Cu_3 is bent, C_{2v}, with an internal angle of 75.7° and Cu–Cu distances of 2.326 Å, while Cu_3^+ is a closed-shell equilateral triangle with Cu–Cu distances of 2.394 Å. The addition of a Cu atom (doublet) to the $\equiv$Si–Cu_2 surface complex (doublet) results in the closed-shell $\equiv$Si–Cu_3 system. The distances within supported Cu_3 are considerably elongated with respect to the gas-phase unit, but the cluster retains the C_{2v} structure, with two long and one short Cu–Cu distances. On an NBO center, on the contrary, two Cu atoms of Cu_3 interact with the NBO and Cu_3 assumes an almost perfect equilateral triangular geometry with internal angles of $60 \pm 1°$ and Cu–Cu distances of about 2.4 Å. Thus, a quite different structure is found for the

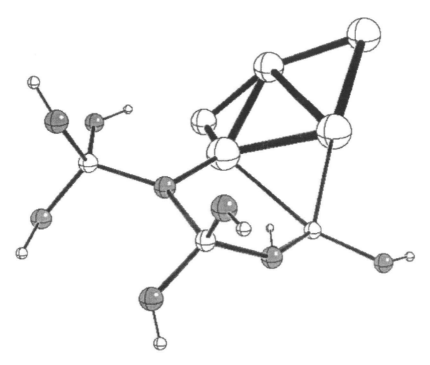

Figure 4 Structure of a Cu_5 cluster interacting with an Si E′ defect center (an Si dangling bond) at the surface of SiO_2.

same cluster interacting with the two defect centers, showing the nonnegligible role of the substrate in modifying the properties of nano-clusters.

Free Cu_4 has a planar rhombic structure with $r(Cu–Cu) = 2.455$ Å. When deposited on an E′ center of SiO_2, Cu_4 remains nearly planar, but one of the Cu–Cu distances is strongly elongated. Also in this case the structure is different from that of an NBO center, where Cu_4 assumes a pseudo-tetrahedral shape; the structure on the NBO can be better described as that of a bent rhombus (butterfly). Free Cu_5 has a planar trapezoidal structure obtained by adding a Cu atom in the plane containing the rhombic Cu_4; the internal angles are close to 60°, and the Cu–Cu distances go from 2.404 to 2.475 Å. The structure of supported Cu_5 is that of a flat pentamer that resembles free Cu_5. This structure is not too different from that found on an NBO center where Cu_5 is bound with two Cu atoms to the surface, with the Si atom of the E′ center and with a bridging oxygen (Fig. 4). In average, the Cu–Cu distances of supported Cu_5, 2.45 Å, are practically coincident with those of the free cluster, 2.44 Å, and very similar to those of the $\equiv Si–O–Cu_4$ surface complex (NBO). This suggests that the geometrical distortions within the metal cluster due to the bonding with different substrate defects disappear quite rapidly as the cluster size increases.

Small Cu clusters, more polarizable than a single Cu atom, interact more strongly with the nondefective SiO_2 surface. Still, the role of defects for the diffusion, adhesion, and nucleation processes is crucial. All Cu clusters interact with the E′

center with adhesion energies that go from 1.7 to 3.3 eV. These values are always smaller than for the same clusters interacting with NBO. The energy required to atomize the cluster increases with cluster size and converges to the cohesive energy of the bulk metal for very large metallic aggregates. For a supported cluster, the atomization energy provides a measure of the additional stability of the cluster due to the bond at the interface. An additional stabilization of the supported compared to the free cluster is present and is more or less constant for all clusters. Even for a supported pentamer there is a nonnegligible contribution from the bond at the interface to the overall stability of the cluster toward atomization.

An important quantity determining the mechanism of cluster growth is the nucleation energy, E_{nuc}, defined as the energy gain due to the addition of an isolated Cu atom to a supported Cu_n cluster ($E_{nuc} = -[E(Cu_n/SiO_2) - E(Cu) - E(Cu_{n-1}/SiO_2)]$). Recently, accurate microcalorimetric measurements of the heat of adsorption of a metal atom to a metal cluster supported on an oxide surface have been reported [2,124]. Therefore, the nucleation energy is a quantity becoming available also through experimental studies, even for small aggregates. On oxide surfaces nucleation is believed to occur through diffusion of isolated atoms or eventually dimers; therefore, it is useful to compare the nucleation energy for free and supported clusters. The addition of an extra Cu atom leads to a stabilization that is larger for the supported than for the free Cu clusters with the exception of the dimer (the energy gain for the process $Cu + Cu \rightarrow Cu_2$ is obviously larger than for the $\equiv Si-Cu + Cu \rightarrow \equiv Si-Cu_2$ one because of the closed-shell nature of the Si-Cu bond). This important conclusion shows the role of the substrate in the growth process of a supported particle. In fact, since isolated Cu atoms are weakly bound to the regular SiO_2 surface, they will diffuse with low activation barriers. The diffusion process will stop only at defects or at sites where the nucleation has already started. It should be noted that the nucleation energy seems to increase with the cluster size for the supported clusters, going from about 1 eV for the formation of the dimer to about 2.7–3.0 eV for the case of the pentamer. In other words, for very small aggregates the energy gain of the process $SiO_2-Cu_n + Cu \rightarrow SiO_2-Cu_{n+1}$ seems to increase for larger n. This is in part due to the fact that as the cluster becomes larger, the cohesive energy increases due to the increase of the coordination number of the atoms in the cluster; another effect, however, is that larger clusters are more polarizable and the interactions with the bridging oxygens also increase. Of course, by further increasing the cluster size the nucleation energy will tend first to that of the corresponding isolated Cu particles and then, for larger crystallites, to the cohesive energy of the bulk metal. In other words, the perturbation induced by the strong bond with the surface defect is rapidly screened by the conduction band electrons of the metal cluster, and the electronic modifications induced by the bond with the surface defect disappear for aggregates of a few tens of atoms.

To summarize, Cu clusters containing from 2 to 5 metal atoms form strong bonds with both E' centers and nonbridging oxygens, but the adhesion energy with these latter centers is higher [36]. The interaction arises in part from the formation of a covalent polar bond between the metal cluster and the paramagnetic center, either E' or NBO, and in part from the polarization interaction of the cluster electron density, which leads to direct, although weaker, bonds with a two-coordinated oxygen of silica. As a result, due to the interface bond the fragmentation energy of the supported cluster increases compared to the free, gas-phase counterparts. The shape of the supported clusters can differ substantially from that of the gas-phase units. This is true in particular for the Cu clusters interacting with the NBO centers, where the interaction is stronger. By growing larger Cu particles these will assume nearly spherical three-dimensional structures since the metal–metal bonds will dominate over the weak electrostatic Cu–SiO$_2$ interactions. In this respect point defects on the SiO$_2$ surface act as strong anchoring sites for the entire cluster, limiting the diffusion process and favoring the nucleation. These results are consistent with a Volmer–Weber growth mode of Cu overlayers on silica with formation of 3D particles [162].

3.6   THE TiO$_2$(110) SUBSTRATE

3.6.1   Cu, Ag, and Au Atoms on TiO$_2$(110)

TiO$_2$ and metal-promoted TiO$_2$ in rutile or anatase forms play a very important role in photolysis and photo-oxidation reactions, electrocatalysis, gas sensors, and catalysis by supported metals [1,163–165]. The technological importance of this material has stimulated an intense activity in the area of surface science to better characterize the TiO$_2$ surface and adsorbates on TiO$_2$. The face studied in greater detail is the rutile (110) one; in this case the defectivity of the surface has also been subject to experimental investigations [166]. Metal deposition on rutile is another topic receiving increasing interest. Several metal atoms and clusters have been deposited on the TiO$_2$ (110) surface, alkali metals [167–178], simple metals like Al [179], as well as transition metals [163,180–188]. However, despite the intense experimental activity, a detailed understanding of the general rules governing the bond at the metal–TiO$_2$ interface is still missing. For instance, while K is known to chemically reduce the surface by transfer of the 4s valence electron to the Ti 3d states [189], Pd does not show a similar tendency and forms more covalent polar bonds with no net charge transfer [38]. Also, the preferred adsorption site of metal atoms on TiO$_2$ is a matter of discussion. While alkali ions have been shown to adsorb on the triangular faces formed by the bridging and the basal oxygen atoms of the surface [176], it has been suggested that the Pd atoms prefer to adsorb on the five-coordinated Ti sites from the analysis of STM images [183]. These are questions where quantum-mechanical calculations can in principle provide an answer. However, the complexity of the metal–TiO$_2$ system is one reason for the limited amount of theoretical work done on this subject [189–194]. In fact, not only is the number of theoretical studies on metals deposited on TiO$_2$ scarce, but almost all studies deal with adsorbed alkali metals [38,178,190–192].

As an example of the metal–TiO$_2$ interaction we consider here the adsorption of isolated Cu, Ag, and Au atoms. Recently several experimental studies on the deposition of clusters and films of these metals have been reported [182,188,195–199]. The rutile surface has been represented by finite clusters, Ti$_{13}$O$_{26}$ and Ti$_{15}$O$_{30}$ embedded in PCs and ECPs or by periodic supercells using the VASP code [200–202]. The calculations [39] have been performed at the DFT level using the B3LYP [119,121] and the GGA–Perdew and Wang [118] exchange and correlation functionals.

Cu/TiO$_2$

Four possible sites can be considered for Cu adsorption on TiO$_2$(110): on top of the protruding O$_{2c}$ atoms, bridging two O$_{2c}$ atoms, on top of a Ti$_{5c}$ cation, and bridging two Ti$_{5c}$ cations (also called fourfold hollow). The nature of the bonding has been analyzed by computing properties like the dynamic dipole moment for the vertical motion of the atom [203], by means of the net Mulliken charges, or of the spin density and spin population. These two latter quantities in particular are very useful for the characterization of the bond at the interface. In fact, Cu, Ag, and Au atoms in their ground state have the $(n+1)$s level singly occupied. In case of full charge transfer, this level is empty and the unpaired electron must reside on some level state of the substrate oxide. This is the situation observed for K on TiO$_2$, where the K 4s level is empty and filled Ti 3d states appear in the gap [189,190]. The strongest adsorption is found for Cu bridging two O$_{2c}$ atoms, with $D_e = 2.9$ eV. The adsorption on top of O$_{2c}$ is only slightly less favorable. On the other hand, the bonding of Cu on top of Ti$_{5c}$ or in the "open" sites along the basal plane is much smaller, from 0.5 to 0.8 eV. Therefore, there is a clear preference for Cu to bind at the O sites, at variance with Pd which forms bonds of similar strength with the Ti and O sites, about 1.2– 1.4 eV [38]. The metal–oxygen distance for adsorption on the O$_{2c}$ sites is that expected for strong Cu–O bonds; on the contrary, on the Ti sites the distances are long, consistent with a weak interaction.

The analysis of the bond at the interface shows a net transfer of one electron from Cu to TiO$_2$. A direct measure of the occurrence of the charge transfer is given by the dipole moment curve as a function of the displacement of the metal atom along the normal to the surface, $\mu(r)$. In general, a linear curve and a large slope are indicative of the presence of a charged adsorbate [203,204]. For instance, the interaction of a single K atom on TiO$_2$, characterized by a net charge transfer, results in linear dipole moment curve with large slope, $d\mu/dr \approx +1.1$ au [189]. The linearity is measured by the second derivative of the curve, $d^2\mu/dr^2 \approx 0$. For Cu atoms on TiO$_2$ the dipole moment curve is linear with a large slope, $d\mu/dr = +1.0$ au, confirming the charge transfer nature of the interaction. The situation of Cu is thus reminiscent of that of K, despite the different ionization potentials of the two metal atoms, IP(K) = 4.34 eV, IP(Cu) = 7.73 eV. Therefore, while Cu atoms interact with the MgO surface with a weak bonding dominated by polarization effects, a completely different interaction occurs on TiO$_2$, thus reflecting the different nature of the oxide anions in the two materials.

Completely different is the interaction of a Cu atom with the Ti rows. Here, in fact, not only is the bonding weak, but the spin density is also partially on Cu and partially on the surface atoms where the metal is adsorbed. This is due to the fact that on these sites the bonding is mainly due to the metal polarization induced by the surface electric field with little covalent bond at the interface. Again, there is a difference with respect to the Pd case, where a mixing of the 4d levels on Pd with the 3d empty states of the Ti cations leads to the formation of a relatively strong bond with a covalent polar character [38].

These results, obtained with cluster models embedded in point charges and ECPs, have been confirmed by slab calculations with periodic boundary conditions [39]. Even a periodic calculation, however, is not necessarily conclusive. In fact, the cluster model provides a simplified description of very low coverage, i.e., of a single adsorbate in absence of any adsorbate–adsorbate repulsion. In a periodic calculation, unless extremely large supercells are used, the calculation refers to a finite coverage, usually half or a quarter of a monolayer. This can be particularly critical in the case of charge transfer interactions. For alkali metals on metal surfaces it is known that the bonding nature changes as function of the coverage, going from fully ionic at very low coverage ($\theta < 0.2$) to metallic for higher coverages [205,206]. This is the direct consequence of the coulomb repulsion between the adsorbed K^+ ions when the density of adsorbates becomes too large: the adsorbate–adsorbate repulsion increases to the point where the charge transfer is no longer energetically favorable [207]. In these conditions a metallic K layer forms in contact with the metal substrate. Therefore, a supercell calculation of Cu/TiO_2 at low coverage ($\theta = 0.25$) can in principle give two answers: if a charge transfer is found, the cluster results are confirmed; if no charge transfer occurs, it is possible that the result is due to a coverage effect and a larger supercell, representing a lower coverage, is required. Periodic calculations on K/TiO_2 have shown that at $\theta = 0.25$ the charge on K is $\approx +1\,|e|$, but that at $\theta = 0.5$ it is reduced to $\approx +0.5\,|e|$ [189]. The results of the slab calculations provide a picture that is qualitatively similar to that obtained with the cluster models. In fact, the most stable adsorption sites are the O_{2c} atoms. They also have shown the great importance of surface relaxation for the strength of the bond at the interface and the role of adsorbate–adsorbate repulsion, which considerably reduces the adsorption energy. Also, the periodic calculations show that a net charge transfer occurs when the Cu atom is adsorbed on the O_{2c} sites. The density of states (Fig. 5) shows that the Cu 4s states are above the Fermi level in the Cu/TiO_2 system while the Ti 3d states are crossed by the Fermi level and are spin-polarized, indicating the localization of the unpaired electron in these levels. Consistent with the cluster results, no charge transfer is observed when Cu sits along the Ti rows.

To summarize this section, isolated Cu atoms deposited on TiO_2 show the tendency to transfer their valence electron to the surface in more or less the same way observed for K/TiO_2. The charge transfer occurs only when Cu is interacting with the bridging O_{2c} sites with formation of a strong bond, while the interaction with the Ti sites is weak and of the polarization type. As a consequence, the surface relaxation plays a very important role when Cu is adsorbed on the O sites and plays much less a role on the Ti sites. Finally, the strength of the interaction depends strongly on the coverage, in particular when adsorption on the O site is considered. This is connected to the ionic nature of the adsorbate and to the long-range coulomb repulsion in the adsorbed layer.

Ag/TiO₂ and Au/TiO₂

An Ag atom adsorbed on TiO_2 exhibits essentially the same characteristics of a Cu atom. The bonding is much stronger on the O_{2c} atoms of the surface than on the basal Ti rows, and the bridge adsorption on O_{2c} is preferred over the on-top O_{2c} case. The binding energy, 2.3 eV in the O_{2c} bridge position, is about 0.5 eV smaller than for Cu. Very weak bonds are formed with the Ti_{5c} atoms or in the hollow site. The Ag–Ti or Ag–O distances on the basal plane are of the order of 3 Å and indicate the absence of a covalent interaction. The most important difference between O_{2c} and Ti_{5c} sites is that in the former case Ag becomes Ag^+ with electron transfer, while on Ti the unpaired electron remains largely on the Ag 5s orbital. The charge transfer on O_{2c} sites is shown by the zero spin density on the adsorbed Ag atom. The analysis of

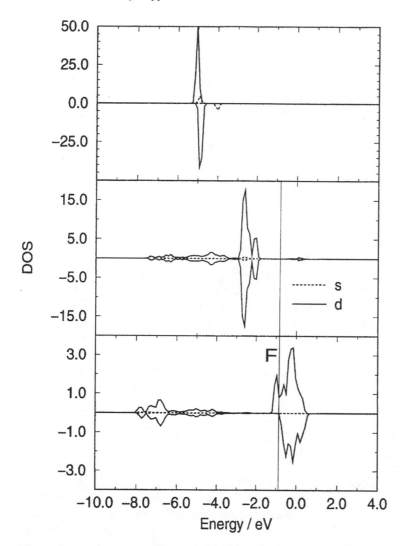

Figure 5 Majority and minority spin components of DOS curves for Cu adsorbed on a TiO$_2$ slab (Cu atoms in bridge position over O(2c) sites). (top) isolated Cu atom; (center) Cu 4s and 3d contribution to the total DOS of Cu/TiO$_2$. (bottom) Ti 3d contribution to the total DOS of Cu/TiO$_2$.

the dipole moment curve for Ag–Ti$_4$O$_8$ gives a linear curve and a slope $d\mu/dr = +0.60$. This latter value, while consistent with a large charge transfer, is smaller than in case of Cu and suggests that the bonding has a partial covalent character. The occurrence of charge transfer for Ag on TiO$_2$ is not surprising if we consider that Cu and Ag have very similar IPs [actually, IP(Ag) is about 0.1 eV smaller than IP(Cu)].

The case of Au is quite different from that of Cu and Ag. First of all, the interaction energy of an Au atom with the O$_{2c}$ sites is much smaller than for the other members of the series—of the order of 1 eV. This is connected with the fact that the charge transfer does not occur, as shown (1) by the residual spin density on the Au atom and (2) by the dipole moment curve, which exhibits a small slope, $d\mu/dr = 0.16$, and a significant curvature, $d^2\mu/dr^2 = -0.13$, compared to that for Cu–TiO$_2$. On the Ti rows the Au atoms are only very weakly bound by dispersion forces. One can conclude that the chemistry of Au atoms on TiO$_2$ will be rather different from that of the other atoms of the group, Cu and Ag. This difference is largely related to atomic properties, particularly the higher IP of Au compared to Cu and Ag.

In summary, Cu and Ag atoms give rise to strong interactions with the bridging oxygens of the surface, O$_{2c}$, with formation of Cu$^+$ and Ag$^+$ ions and full (for Cu) or nearly full (for Ag) transfer of the metal valence electron to the Ti 3d states, in complete analogy with the K–TiO$_2$ interaction. The spin density is largely localized on the Ti 3d levels. Even at a coverage of a quarter of a monolayer we observe the occurrence of the charge transfer. This may look surprising since the IP of Cu and Ag, 7.6–7.7 eV, is more than 3 eV larger than that of K, 4.34 eV. Clearly, the cost of the ionization is compensated by the gain connected to the electrostatic interaction between the positive ion and the "reduced" surface. The charge transfer, however, does not have a purely electrostatic origin. In fact, it occurs only when Cu and Ag are absorbed on the O$_{2c}$ sites and not when the atom is on top of the Ti$_{5c}$ ions of the surface or in the fourfold hollow positions along the Ti(5c) rows. In these cases, in fact, the interaction is weak and the bonding is of the polarization type. This reflects the partial oxidizing character of the TiO$_2$ surface, at variance with other substrates like MgO and SiO$_2$, where strong chemical bonds are formed only in correspondence of surface defects.

The picture of the Au–TiO$_2$ bonding is quite different. One reason is that the Au IP is high, 9.23 eV; the second one is that relativistic effects usually contribute to make Au different from Cu and Ag [208]. As a result, the preferred interaction sites are still the O$_{2c}$ rows, but with only minor charge transfer. The bonding is better described as covalent polar, and the polarization is toward the surface O atoms. Also for Au no bonding is formed with the Ti atoms of the basal rows.

The tendency to form only weak interactions with the Ti atoms contrasts with the results obtained on Pd–TiO$_2$, where a similar bond strength was found for the adsorption on the O$_{2c}$ or the Ti$_{5c}$ atoms of the oxide surface [38]. The explanation lies in the fact that Pd can mix (hybridize) the 4d and the 5s atomic levels to form strong covalent bonds with the Ti 3d levels. Cu, Ag, and Au have no chance to form such hybrid levels because they have completely filled nd shells and a partially occupied $(n+1)$s level.

3.7 THE METAL–SUPPORT INTERACTION: ROLE OF OXIDE DEFECTS

3.7.1 Reactivity of Size-Selected Pd Clusters on MgO

We have seen that a great variety of defect centers can form at the surface of an oxide like MgO (Table 1). Each surface defect has a direct and characteristic effect on the properties of absorbed species. This becomes particularly important in the analysis of the chemical reactivity of supported metal atoms and clusters [1–3]. The defects not only act as nucleation centers in the growth of metal islands or clusters [4,209], but also can modify the catalytic activity of the deposited metal by affecting the bonding at the interface [210–212]. The role of point defects at the surface of MgO in promoting or modifying the catalytic activity of isolated metal atoms or clusters deposited on this substrate has been recently investigated by considering the cyclization reaction of acetylene to form benzene, $3C_2H_2 \rightarrow C_6H_6$ [211,212]. This process has been widely studied on single-crystal surfaces from UHV conditions $(10^{-12}–10^{-8}\,atm)$ to atmospheric pressure $(10^{-1}–1\,atm)$ [213–216]. A Pd(111) surface is the most reactive one [217,218], and it has been shown unambiguously that the reaction proceeds through the formation of a stable C_4H_4 intermediate, resulting from addition of two acetylene molecules [219]. This intermediate has been characterized experimentally [220–222] and theoretically [223] and its structure is now well established. Once the C_4H_4 intermediate is formed, it can add a third acetylene molecule to form benzene, which then desorbs from the surface at a temperature of about 500 K [213,214]. The same reaction has been studied on size-selected Pd clusters deposited on an MgO surface.

A new experimental setup has recently been designed to study the chemical properties of size-selected metal clusters deposited on oxide substrates [210,211]. Pd clusters have been produced by a laser evaporation source, ionized, then guided by ion optics through differentially pumped vacuum chambers and size-selected by a quadrupole mass spectrometer [210–212]. The monodispersed clusters have been deposited with low kinetic energy (0.1–2 eV) onto an MgO thin-film surface. The clusters-assembled materials obtained in this way exhibit peculiar activity and selectivity in the polymerization of acetylene to form benzene and aliphatic hydrocarbons [224]. Figure 6 shows the temperature-programmed reaction (TPR) spectra for the cyclotrimerization of acetylene on supported Pd_n ($1 \leqslant n \leqslant 30$) clusters. Pd_1, Pd_2, and Pd_3 exclusively form benzene at temperatures around 300 K, while for cluster sizes up to Pd_8 a broad feature between 400 K and 700 K is observed in the TPR (Fig. 1) [211]. For Pd_7 an additional desorption peak of benzene is clearly observed at about 430 K. For Pd_8 this feature becomes as important as the peak at 300 K, while for Pd_{30} benzene mainly desorbs around 430 K. On a clean MgO(100) surface no benzene is produced at the same experimental conditions. This strongly size-dependent behavior is related to the distinct electronic and geometric properties of the metal clusters, making this new class of materials extremely interesting for understanding the structure-property relationship. Interestingly, according to these results even a single Pd atom can produce benzene. This is an important result that has allowed us to investigate the activity of the Pd atoms as a function of the support where it is deposited or of the sites where it is bound.

3.7.2 Mechanism of Benzene Formation on Pd/MgO

In the following we briefly review the most important aspects of the cyclization reaction over supported Pd atoms, looking in particular at the role of the various types of morphological defects present on the MgO surface, and performing an accurate analysis of the electronic effects involved in the metal–support interaction [211,212]. The review is largely based on DFT cluster model calculations to interpret the experimental data, either thermal desorption spectra (TDS) or infrared spectra (IR). The results provide an indication of the importance of defects like low-coordinated oxygen anions, neutral and charged oxygen vacancies, etc. in changing the reactivity of a supported Pd atom.

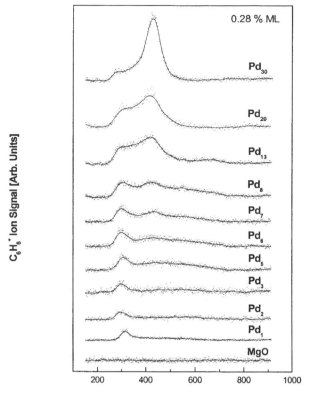

Figure 6   Temperature-programmed reaction spectra of C_6H_6 formation on Pd clusters of various size deposited on a MgO thin film. The bottom spectrum shows that for clean MgO(100) films no benzene is formed.

The possibility to catalyze the acetylene trimerization depends critically on the ability of the metal center to coordinate and activate two C_2H_2 molecules and then bind the C_4H_4 intermediate according to reaction:

$$Pd + 2C_2H_2 \rightarrow Pd(C_2H_2)_2 \rightarrow Pd(C_4H_4) \tag{8}$$

The activation of the acetylene molecules is easily monitored, for instance, by the deviation from linearity of the HCC angle due to a change of hybridization of the C atom from sp to sp^2, or by the elongation of the C–C distance, d(C–C), as a consequence of the charge transfer from the metal 4d orbitals to the empty π^* orbital of acetylene. This process can be followed by optimizing the geometry of one and two acetylene molecules coordinated to an isolated Pd atom, and of the corresponding $Pd(C_4H_4)$ complex (Fig. 7). One acetylene is strongly bound to Pd, by almost 2 eV (the results refer to the BP functional); the second is bound by 1 eV. Also, the C_4H_4 intermediate is quite strongly bound to the Pd atom and is more stable than two adsorbed acetylene molecules. A similar result has been obtained for the Pd(111) surface [223]. The level of activation of acetylene is clearly larger when a

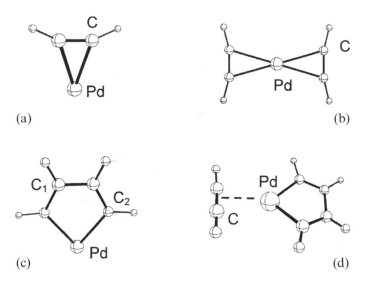

Figure 7 Structure of (a) $Pd(C_2H_2)$, (b) $Pd(C_2H_2)_2$, (c) $Pd(C_4H_4)$, and (d) $Pd(C_4H_4)$ (C_2H_2) complexes. In $Pd(C_4H_4)(C_2H_2)$ the third acetylene molecule is weakly bound and is not activated (linear shape as in the gas phase).

single C_2H_2 is adsorbed, consistent with the idea that the electron density on the metal is essential for promoting the molecular activation. However, it should be kept in mind that a stronger bonding may imply a larger barrier for conversion of two C_2H_2 units into C_4H_4. The following step, $C_4H_4 + C_2H_2 \rightarrow C_6H_6$, requires the capability of the metal center to coordinate and activate a third acetylene molecule according to reaction:

$$Pd(C_4H_4) + C_2H_2 \rightarrow Pd(C_4H_4)(C_2H_2) \rightarrow Pd(C_6H_6) \tag{9}$$

However, the third C_2H_2 molecule interacts very weakly with $Pd(C_4H_4)$ with a binding of $<0.3\,eV$. The structure of the third molecule is not deformed compared to the gas phase (Fig. 7). The molecule is bound to the $Pd(C_4H_4)$ complex by dispersion forces only, and its distance from the Pd atom is therefore very long, $\approx 2.6\,\text{Å}$. This result indicates that *an isolated Pd atom is not* a catalyst for the cyclization process, at variance with the experimental observation for Pd_1–MgO.

The role the support plays therefore becomes a critical aspect of the interaction. As we have seen earlier in section 3.4.1, the Pd–MgO bonding is not characterized by a pronounced charge transfer and is better described as covalent polar. However, only regular adsorption sites have been considered and the bonding mode at defect or low-coordinated sites has attracted less attention so far [225,226]. In this respect one can formulate two hypotheses. One states that the surface oxygen anions of the MgO surface still have some oxidizing power to deplete charge from the metal atom. The other possibility is that the O anions are fully reduced, O^{2-}, and donate charge to the metal atom, thus leading to electron-enriched species (the surface acts as a Lewis base). In the two cases one would end up with positively or negatively charged supported Pd atoms, respectively. In principle, both situations

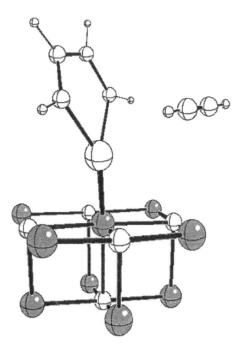

Figure 8   Optimal structure of a Pd(C$_4$H$_4$)(C$_2$H$_2$) complex formed on a five-coordinated O anion at the MgO(100) surface. As for a gas-phase Pd atom (Fig. 7), the third acetylene molecule is weakly bound and not activated.

can lead to an activation of the supported metal. In fact, the bonding of unsaturated hydrocarbons to metal complexes and metal surfaces is classically described in terms of σ donation from a filled bonding level on acetylene to empty states on the metal and π back donation from the occupied d orbitals on the metal to empty antibonding π^* orbitals of the ligand [227,228]. Both mechanisms result in a weakening of the C–C bond and a distortion of the molecule. A simple model study where the charge on Pd has been artificially augmented has shown that the presence of a charge on Pd, either positive or negative, reinforces the bonding of acetylene to the Pd(C$_4$H$_4$) complex; however, only for negatively charged Pd atoms (electron-rich) is a substantial activation of the third acetylene molecule observed [211]. This result shows that the increase of the electron density on Pd is a key mechanism to augment the catalytic properties of the metal atom. The role of the substrate is just that of increasing the "basic" character of the Pd atoms (or clusters).

When Pd is deposited on MgO, the activation of C$_2$H$_2$ is much more efficient than for an isolated Pd atom. The structural distortion of adsorbed C$_2$H$_2$ follows the trend corner > edge > terrace > free atom [212]. The contribution of the charge transfer from acetylene to Pd or Pd$_1$–MgO, the σ donation in the Dewar–Chatt–Duncanson model [227,228], is similar for the four cases and is much smaller than the charge transfer in the other direction, i.e., from Pd or Pd$_1$–MgO acetylene. This confirms that the interaction with the MgO substrate increases the basic character of the Pd atom. The donor capability of Pd increases as a function of the adsorption site on MgO in the order terrace $<$ edge $<$ corner, the same order found for the deformation of the acetylene molecule. Therefore, one can conclude that (1) the substrate plays a direct role in the modification of the properties of the supported Pd atom and (2) low-coordinated sites are more active than the regular terraces. This is consistent with the idea that the degree of surface basicity in MgO is larger for the low-coordinated anions because of the lower Madelung potential at these sites [131].

In principle several defect sites can be active in promoting the catalytic activity of a deposited Pd atom. The surface of polycrystalline MgO presents, in fact, a great variety of irregularities [130] like morphological defects, anion and cation vacancies [133,144], divacancies [139], impurity atoms, etc. (Table 1). Recent studies have shown that even MgO thin films deposited on a conducting substrate are not completely defect free, as shown by the different adsorption properties of CO on MgO thin films and single-crystal surfaces [229]. It is therefore difficult to answer the question of which defect sites of the MgO surface are more likely involved in the acetylene trimerization. To contribute to a clarification of the role of defects, the structure and stability of Pd atoms, $Pd(C_2H_2)$, $Pd(C_2H_2)_2$, $Pd(C_4H_4)$, and $Pd(C_2H_2)(C_4H_4)$, complexes formed on various regular and defect sites at the MgO surface have been considered. In particular the formation of Pd–hydrocarbon complexes has been considered on O sites, neutral oxygen vacancies (the F centers), and charged oxygen vacancies (the F^+ centers).

Let us consider the reactivity of the supported Pd atoms. A Pd atom bound at terrace sites is able to add and activate one or two acetylene molecules and to form a stable O_{5C}–$Pd(C_4H_4)$ complex. However, the addition of the third acetylene molecule does not occur. The situation is reminiscent of that of an isolated Pd atom where the third acetylene is weakly bound by dispersion forces (Fig. 7). Therefore, the complex is unstable and acetylene does not bind to O_{5C}–$Pd_X(C_4H_4)$ (Fig. 8). The situation is only slightly better on the low-coordinated O sites. The formation of $Pd(C_4H_4)$ on four- and three-coordinated oxygens and the consequent addition of the third acetylene molecule result in a surface complex, O_{4c}–$Pd(C_2H_2)(C_4H_4)$ or O_{3c}–$Pd(C_2H_2)(C_4H_4)$, that is very weakly bound or even unbound. Thus, the O anions at low-coordinated defects are also not good candidates for increasing the catalytic activity of supported Pd atoms.

Things are different on the F and F^+ centers located on various sites. They all show a similar activity, with formation of a stable $Pd(C_4H_4)$ complex and substantial stabilization of the third acetylene molecule; the strength of the bonding of C_2H_2 to the supported $Pd(C_4H_4)$ complex is larger than for the morphological defects and becomes quite large, $>2\,eV$, for a neutral F_{3c} center. The degree of activation is comparable in the three cases, with C–C distances of about 1.31 Å and HCC angles of about 140°. Thus, F centers are promoting the catalytic activity of the deposited Pd atoms.

The study of the cyclization reaction of acetylene to benzene on Pd atoms supported on MgO allows one to clarify some of the experimental aspects and to draw some general conclusions about the role of defects on oxide surfaces. The results show that *only in the presence of surface defects does a single Pd atom become an active catalyst for the reaction*; in fact, an isolated Pd atom is not capable of adding and activating three acetylene molecules, an essential step for the process. The change in the electronic structure of supported Pd is connected to the electron donor ability of the substrate, which does not simply act as an inert substrate. The oxide surface acts in pretty much the same way as a ligand in coordination chemistry and provides an additional source of electron density, which increases the capability of the Pd atom to back-donate charge to the adsorbed hydrocarbon. In this respect it is remarkable that on transition metal complexes the acetylene trimerization reaction follows a very similar mechanism as on heterogeneous supported catalysts [230]. The oxide anions of the MgO(001) surface, located either on terraces, steps, or corners, do not significantly change the Pd catalytic activity. The Lewis basicity of these sites in fact is too low and in any case not sufficient to significantly increase the density on the metal. Things are completely different on oxygen vacancies, the F or F^+ centers. Due to the electron(s) trapped in the cavity left by the missing oxygen, these centers are good basic sites and promote the activity of an adsorbed Pd atom. Indeed, it has been shown recently that F centers on the surface of polycrystalline MgO are able to reduce very inert molecules like N_2, leading to metastable N_2^- radicals [136].

3.8 CONCLUSIONS

The study of the metal–oxide interface and of the mechanisms that govern the nucleation and growth of small metal particles with ab initio methods is still in its infancy. However, the interest in this topic is increasing very rapidly, as shown by the number of papers dedicated to this problem every year. Only five years ago, just a few studies of this type were available in the literature. From the work done so far, it is possible to draw some general conclusions that form the basis for further, more detailed studies.

Not surprisingly, the bonding of metal atoms with oxide surfaces differs substantially from oxide to oxide and from metal to metal. On simple binary oxides of nontransition metals like MgO and SiO_2, where the metal cation has no d orbitals available for bonding, the interaction occurs preferentially with the oxide anions; much weaker bonds, mostly of electrostatic nature, are found on the metal cations. The oxygen sites of both MgO—a fully ionic substrate—and SiO_2—a rather covalent oxide—give rise to relatively weak interactions with metal atoms. Transition metal atoms with partially filled d shells form stronger bonds, mainly through the formation of covalent bonds at the interface. In fact, a general characteristic common to both MgO and SiO_2 is that little charge transfer occurs between the oxide and the metal. This result has the important consequence that the regular sites of these two substrates are not good nucleation centers and that the metal atoms, once deposited, diffuse quite easily on the surface even at low temperatures. When a transition metal oxide like TiO_2 is considered, with a more complex electronic structure, the interaction of the metal atoms is much more complex. Even on the nondefective surface in fact the possibility for the Ti cations to change their oxidation state from +IV to +III leads to a tendency to favor the charge transfer from low-ionization potential metal atoms to the substrate. This is the case for alkali metals but also for coinage metals like Cu and Ag, which adsorb preferentially on the bridging oxygen rows and transform in full cations by transferring one electron to the 3d states of the Ti ions. The other difference is that on TiO_2 transition metal atoms like Pd can form relatively strong bonds also with the metal cations so that a competition exists between the two bonding sites.

The unusual nature of the metal–oxide bonding mechanism is also the reason why its description with modern computational tools may depend significantly on the choice of the method used. In particular, within DFT schemes quite different bond strengths can be obtained depending on which exchange-correlation functional is used. For this reason, good benchmark calculations with explicitly correlated wave functions can be of considerable importance to assess the reliability of the various DFT approaches.

The other consequence of the weak metal–oxide interactions on the nondefective surfaces is that small metal clusters once deposited on the substrate tend to keep the same structure as they have in the gas phase. Of course, the actual structure of the deposited cluster is a delicate balance between the strength of the metal–metal bond within the cluster and the metal–oxide interface bond. Also in this case, however, it is likely that the small clusters will diffuse on the surface until they become stabilized at some specific defect site.

The topic of defect sites at oxide surfaces therefore becomes crucial in order to fully understand the metal–oxide bonding. This subject has been addressed theoretically only recently. In this review we have shown how defect sites at both MgO and SiO_2 surfaces play a fundamental role in both stabilization and nucleation, but also that they modify the cluster electronic properties. In particular, some defect centers that act as electron traps like the oxygen vacancies at the MgO surface are extremely efficient in increasing the electron density on the deposited metal atoms or clusters, thus augmenting their chemical activity toward other adsorbed molecules. Understanding the metal–oxide interface and the properties of deposited metal clusters also needs a deeper knowledge of nature, concentration and mechanisms of formation, and conversion of the defect sites of the oxide surface.

ACKNOWLEDGMENTS

We are indebted to our co-workers and collaborators Thomas Bredow, Cristiana Di Valentin, Anna Maria Ferrari, Livia Giordano, and Nuria Lopez, who have contributed over the last years to the study of metals deposited on oxide surfaces. We also thank for the fruitful discussions and ongoing collaboration Stephane Abbet, Elio Giamello, Ueli Heiz, Javier Fernandez-Sanz, Jacek Goniakovski, Konstantin Neyman, and Notker Rösch. This work has been supported by the Italian Istituto Nazionale per la Fisica della Materia through the PRA-ISADORA project.

REFERENCES

1. H.-J. Freund, Angew. Chem. Int. Ed. Engl. 36 (1997) 452.
2. C.T. Campbell, Surf. Sci. Rep. 27 (1997) 1.
3. R.M. Lambert, G. Pacchioni (eds.), Chemisorption and Reactivity of Supported Clusters and Thin Films, NATO ASI Series E, Vol. 331, Kluwer, Dordrecht, 1997.
4. C. Henry, Surf. Sci. Rep. 31 (1998) 231.
5. G. Renaud, Surf. Sci. Rep. 32 (1998) 1.
6. S.A. Chambers, Surf. Sci. Rep. 39 (2000) 105.
7. D.P. Woodruff (ed.), The Chemical Physics of Solid Surfaces—Oxide Surfaces, Vol. 9, Elsevier, Amsterdam, 2001.
8. J.W. He, P.J. Møller, Surf. Sci. 180 (1987) 411.
9. J.W. He, P.J. Møller, Surf. Sci. 178 (1986) 934.
10. J.W. He, P.J. Møller, Chem. Phys. Lett. 129 (1986) 13.
11. I. Alstrup, P.J. Møller, P.J. Appl. Surf. Sci. 33/34 (1988) 143.

12. J.B. Zhou, H.C. Lu, T. Gustafsson, E. Garfunkel, Surf. Sci. 293 (1993) L887.
13. M.C. Wu, W.S. Oh, D.W. Goodman, Surf. Sci. 330 (1995) 61.
14. F. Didier, J. Jupille, Surf. Sci. 307–309 (1994) 587.
15. M. Meunier, C.R. Henry, Surf. Sci. 307–309 (1994) 514.
16. C.R. Henry, M. Meunier, S. Morel, J. Crystal Growth 129 (1993) 416.
17. T. Kizuka, T. Kachi, N. Tanaka, Z. Phys. D 26 (1993) S58.
18. C. Noguera, G. Bordier, J. Phys. III France 4 (1994) 1851.
19. U. Schönberger, O.K. Andersen, M. Methfessel, Acta Metall. Mater. 40 (1992) S1.
20. C. Li, A.J. Freeman, Phys. Rev. B 43 (1991) 780.
21. C. Li, R. Wu, A.J. Freeman, C.L. Fu, Phys. Rev. B 48 (1993) 8317.
22. J.R. Smith, T. Hong, D.J. Srolovitz, Phys. Rev. Lett. 72 (1994) 25.
23. N.C. Bacalis, A.B. Kunz, Phys. Rev. B 32 (1985) 4857.
24. A.B. Kunz, Phylosoph. Magazine B 51 (1985) 209.
25. Y. Li, D.C. Langreth, M.R. Pederson, Phys. Rev. B 52 (1995) 6067.
26. G. Pacchioni, N. Rösch, Surf. Sci. 306 (1994) 208.
27. G. Pacchioni, N. Rösch, J. Chem. Phys. 104 (1996) 7329.
28. A.M. Ferrari, G. Pacchioni, J. Phys. Chem. 100 (1996) 9032.
29. M. Gautier, J.P. Duraud, J. Phys. III France 4 (1994) 1779.
30. I. Yudanov, G. Pacchioni, K.M. Neyman, N Rösch, J. Phys. Chem. 101 (1997) 2786.
31. L. Giordano, J. Goniakovski, G. Pacchioni, Phys. Rev. B, 64 (2001) 075417.
32. G. Pacchioni, N. Rösch, J. Chem. Phys. 104 (1996) 7329.
33. A.M. Ferrari, C. Xiao, K.M. Neyman, G. Pacchioni, N. Rösch, Phys. Chem. Chem. Phys. 1 (1999) 4655.
34. L. Giordano, G. Pacchioni, A.M. Ferrari, F. Illas, N. Rösch, Surf. Sci. 473 (2001) 213.
35. N. Lopez, F. Illas, G. Pacchioni, J. Am. Chem. Soc. 121 (1999) 813.
36. N. Lopez, F. Illas, G. Pacchioni, J. Phys. Chem. 103 (1999) 1712.
37. G. Pacchioni, N. Lopez, F. Illas, Faraday Discus. 114 (1999) 209.
38. T. Bredow, G. Pacchioni, Surf. Sci. 426 (1999) 106.
39. L. Giordano, G. Pacchioni, T. Bredow, J. Fernandez-Sanz, Surf. Sci. 471 (2001) 21.
40. P.A. Cox, The Electronic Structure and Chemistry of Solids, Oxford Science Publications, Oxford, 1987.
41. A. Gibson, R. Haydock, J.P. LaFemina, Appl. Surf. Sci. 72 (1993) 285.
42. A. Gibson, R. Haydock, J.P. LaFemina, Phys. Rev. 50 (1994) 2582.
43. L.N. Kantorovich, J.M. Holender, M.J. Gillan, Surf. Sci. 343 (1995) 221.
44. E. Castanier, C. Noguera, Surf. Sci. 364 (1996) 1.
45. R. Orlando, R. Millini, G. Perego, R. Dovesi, J. Molec. Catal. A 119 (1997) 253.
46. M. Leslie, M.J. Gillan, J. Phys. C 18 (1985) 973.
47. G. Pacchioni, P.S. Bagus, F. Parmigiani (eds.), Cluster Models for Surface and Bulk Phenomena, NATO ASI Series B, Vol. 283, Plenum Press, New York, 1992.
48. J. Sauer, P. Ugliengo, E. Garrone, V.R. Saunders, Chem. Rev. 94 (1994) 2095.
49. G. Pacchioni, Heter. Chem. Rev. 2 (1995) 213.
50. P.S. Bagus, F. Illas, The Surface Chemical Bond in Encyclopedia of Computational Chemistry, in P.V. Schleyer, N.L. Allinger, T. Clark, J. Gasteiger, P.A. Kollman, H.F. Schaefer III, P.R. Schreiner (eds.), Vol. 4, p. 2870, John Willey & Sons, Chichester, UK, 1998.
51. G. Pacchioni, A.M. Ferrari, A.M. Marquez, F. Illas, J. Comp. Chem. 18 (1997) 617.
52. A.M. Ferrari, G. Pacchioni, J. Phys. Chem. 99 (1995) 17010.
53. A.M. Ferrari, G. Pacchioni, Int. J. Quant. Chem. 58 (1996) 241.
54. M.A. Nygren, L.G.M. Pettersson, Z. Barandiaran, L. Seijo, J. Chem. Phys. 100 (1994) 2010.
55. W.J. Stevens, H. Basch, M. Krauss, J. Chem. Phys. 81 (1984) 6026.
56. Z. Barandiaran, L. Seijo, J. Chem. Phys. 89 (1988) 5739.

57. V. Luaña, L. Pueyo, Phys. Rev. B 39 (1989) 11093.
58. M. Born, Z. Physik 1 (1920) 45.
59. R. Pandey, J.M. Vail, J. Phys.: Condens. Matter 1 (1989) 2801.
60. J.H. Harding, A.H. Harker, P.B. Keegstra, R. Pandey, J.M. Vail, C. Woodward, Physica B & C 131 (1985) 151.
61. B.G. Dick, A.W. Overhauser, Phys. Rev. 112 (1958) 90.
62. C.R.A. Catlow, M. Dixon, W.C. Mackrodt, in Computer Simulation of Solids, C.R.A. Catlow, (ed.), Springer, Berlin, 1982, p. 130.
63. P.V. Susko, A.L. Shluger, C.R.A. Catlow, Surf. Sci. 450 (2000) 153.
64. C. Pisani, J. Mol. Catal. 82 (1993) 229.
65. C. Pisani, F. Corà, R. Nada, R. Orlando, Comput. Phys. Commun. 82 (1994) 139.
66. C. Pisani, U. Birkenheuer, Comput. Phys. Commun. 96 (152) 1996.
67. C. Pisani, U. Birkenheuer, F. Corà, R. Nada, S. Casassa, EMBED96 User's Manual, Università di Torino, Torino, 1996.
68. C. Pisani, R. Dovesi, C. Roetti, in Hartree-Fock Ab-initio Treatment of Crystalline Systems, Lecture Notes in Chemistry, Vol. 48, Springer, Heidelberg, 1988.
69. V.R. Saunders, R. Dovesi, C. Roetti, M. Causà, N.M. Harrison, R. Orlando, C.M. Zicovich-Wilson, CRYSTAL98 User's Manual, Università di Torino, Torino, 1998.
70. P.V. Schleyer, N.L. Allinger, T. Clark, J. Gasteiger, P.A. Kollman, H.F. Schaefer III, P.R. Schreiner, (eds.), Encyclopedia of Computational Chemistry, John Willey & Sons, Chichester, UK, 1998.
71. A. Szabo, N.S. Ostlund, Modern Quantum Chemistry: Introduction to Advanced Electronic Structure Theory, Macmillan Publishing Co., Inc., New York, 1982.
72. R. Mc Weeny, Methods of Molecular Quantum Mechanics, Academic Press, London, 1992.
73. G. Sini, G. Ohanessian, P.C. Hiberty, S.S. Shaik, J. Amer. Chem. Soc. 112 (1990) 1407.
74. R. Broer, W.C. Nieuwpoort, Theor. Chim. Acta 73 (1988) 405.
75. P.J. Knowles, N.C. Handy, J. Chem. Phys. 91 (1989) 2396.
76. H.-J. Werner, P.J. Knowles, J. Chem. Phys. 89 (1988) 5803.
77. P.J. Knowles, H.-J. Werner, Theor. Chem. Acta 84 (1992) 95.
78. P.R. Taylor, Accurate Calculations and Calibration, in Lecture Notes in Quantum Chemistry, Roos, B.O. (ed.), Vol. 58, p. 325, Springer-Verlag, Berlin, Heidelberg, 1992.
79. C.C.J. Roothaan, Rev. Mod. Phys. 23 (1951) 69.
80. J.A. Pople, J.S. Binkley, R. Seeger, Int. J. Quantum Chem. Symp. 10 (1976) 1.
81. K. Raghavachari, J.A. Pople, E.S Replogle, M. Head-Gordon, J. Phys. Chem. 94 (1990) 5579.
82. J.P. Malrieu, J.L. Heully, A. Zaitsevskii, Theor. Chim. Acta 90 (1995) 167.
83. J. Paldus, Coupled Cluster Theory in Methods of Computational Molecular Physics, S. Wilson, G.H.F. Diercksen, (eds.), NATO ASI Series, Series B: Methods in Computational Molecular Physics, Vol. 293, p. 99, Plenum Press, New York, 1992.
84. G. Peris, J. Planelles, J. Paldus, Int. J. Quantum Chem. 62 (1997) 137.
85. R.J. Bartlett, G.D. Purvis, Int. J. Quantum Chem. 14 (1978) 516.
86. G.D. Purvis, R.J. Bartlett, J. Chem. Phys. 76 (1982) 1910.
87. R.J. Buenker, S.D. Peyerimhoff, W. Butscher, Mol. Phys. 35 (1978) 771.
88. S.D. Peyerimhoff, R.J. Buenker, Chem. Phys. Lett. 16 (1972) 235.
89. B. Huron, P. Rancurel, J.P. Malrieu, J. Chem. Phys. 75 (1973) 5745.
90. S. Evangelisti, J.P. Daudey, J.P. Malrieu, Chem. Phys. 75 (1983) 91.
91. F. Illas, J. Rubio, J.M. Ricart, P.S. Bagus, J. Chem. Phys. 95 (1991) 1877.
92. S.R. Langhoff, E.R. Davidson, Int. J. Quantum Chem. 7 (1973) 999.
93. K. Andersson, P.-Å. Malmqvist, B.O. Roos, A.J. Sadlej, K. Wolinski, J. Phys. Chem., 94 (1990) 5483.
94. K. Andersson, P.-Å. Malmqvist, B.O. Roos, J. Chem. Phys. 96 (1992) 1218.

95.   L. Serrano-Andrés, M. Merchán, I. Nebot-Gil, B.O. Roos, M. Fülscher, J. Chem. Phys. 98 (1993) 3151.
96.   J. Hinze, J. Chem. Phys. 59 (1973) 6424.
97.   R. Sheppard, I. Shavitt, J. Simmons, J. Chem. Phys. 76 (1982) 543.
98.   E. Dalgaard, P. Jorgensen, J. Chem. Phys. 69 (1978) 3833.
99.   B.O. Roos, P. Taylor, P. Siegbahn, Chem. Phys. 48 (1980) 157.
100. P. Siegbahn, A. Heiberg, B.O. Roos, B. Levy, Phys. Scr. 21 (1980), 323.
101. User's manual for MOLCAS 4.0 program.
102. J. Gerratt, Adv. At. Mol. Phys. 7 (1971) 141.
103. D.L. Cooper, J. Gerratt, M. Raimondi, Int. Rev. Phys. Chem. 7 (1988) 59.
104. D.L. Cooper, J. Gerratt, M. Raimondi, Chem. Rev. 91 (1991) 929.
105. R. Pauncz, Spin Eigenfunctions, Plenum Press, New York, 1979.
106. P. Hohenberg, W. Kohn, Phys. Rev. B 136 (1964) 864.
107. B.I. Dunlap, N. Rösch, Adv. Quantum Chem. 21 (1990) 317.
108. N. Rösch, S. Krüger, M. Mayer, V.A. Nasluzov, in Recent Developments and Applications of Modern Density Functional Theory. Theoretical and Computational Chemistry, Vol. 4, p. 497, J.M. Seminario (ed.), Elsevier, Amsterdam, 1996.
109. R.G. Parr, W. Yang, Density Functional Theory of Atoms and Molecules, Oxford Science, Oxford University Press, London, 1989.
110. S.H. Vosko, L. Wilk, M. Nusair, Can J. Phys. 58 (1980) 1200.
111. K. Terakura, T. Oguchi, A.R. Williams, J. Klüber, Phys. Rev. B 30 (1984) 4734.
112. Z.-X. Shen, R.S. List, D.S. Dessau, B.O. Wells, O. Jepsen, A.J. Arko, R. Barttlet, C.K. Shih, F. Parmigiani, J.C. Huang, P.A.P. Lindberg, Phys. Rev. B 44 (1991) 3604.
113. F. Illas, I de P.R. Moreira, C. de Graaf, V. Barone, Theoret. Chem. Acc. 104 (2000) 265.
114. R.L. Martin, F. Illas, Phys. Rev. Lett. 79 (1997) 1539.
115. C. de Graaf, F. Illas, Phys. Rev. B 63 (2001) 014404.
116. A.D. Becke, Phys. Rev. A 38 (1988) 3098
117. J.P. Perdew, Y. Wang, Phys. Rev. B 45 (1992) 13244.
118. J.P. Perdew, J.A. Chevary, S.H. Vosko, K.A. Jackson, M.R. Pederson, D.J. Singh, C. Fiolhais, Phys. Rev. B 46 (1992) 6671.
119. C. Lee, W. Yang, R.G. Parr, Phys. Rev. B 37 (1988) 785.
120. R. Colle, O. Salvetti, J. Chem. Phys. 79 (1983) 1404.
121. A.D. Becke, J. Chem. Phys. 98 (1993) 5648.
122. I.N. Levine, Quantum Chemistry, 5th ed., Prentice-Hall, New Jersey, 2000.
123. N. Lopez, F. Illas, N. Rösch, G. Pacchioni, J. Chem. Phys. 110 (1999) 4873.
124. J.T. Ranney, D.E. Starr, J.E. Musgrove, D.J. Bald, C.T. Campbell, C.T. Faraday, Discuss. 114 (1999) 195.
125. A. Markovits, M.K. Skalli, C. Minot, G. Pacchioni, N. López, F. Illas, J. Chem. Phys., in press (2001).
126. G. Pacchioni, Solid State Sci. 2 (2000) 161.
127. G. Pacchioni, L. Skuja, D.L. Griscom (eds.), Defects in SiO_2 and Related Dielectrics: Science and Technology, NATO Science Series II, Vol. 2, Kluwer, Dordrecht, 2000.
128. O. Robach, G. Renaud, A. Barbier, Surf. Sci. 401 (1998) 227.
129. G. Pacchioni, in The Chemical Physics of Solid Surfaces—Oxide Surfaces, P. Woodruff, (ed.), Vol. 9, pp. 94–135, Elsevier, Amsterdam, 2001.
130. G. Pacchioni, Surf. Rev. and Lett. 7 (2000) 277.
131. G. Pacchioni, J.M. Ricart, F. Illas, J. Am. Chem. Soc. 116 (1994) 10152.
132. A. Bogicevic, D.R. Jennison, Surf. Sci. 437 (1999) L741.
133. F. Giamello, M.C. Paganini, F.M. Murphy, A.M. Ferrari, G. Pacchioni, J. Phys. Chem. B 101 (1997) 971.
134. E. Scorza, U. Birkenheuer, C. Pisani, J. Chem. Phys. 107 (1997) 9645.

135. G. Pacchioni, A.M. Ferrari, E. Giamello, Chem. Phys. Lett. 255 (1996) 58.
136. E. Giamello, M.C. Paganini, M. Chiesa, D.M. Murphy, G. Pacchioni, R. Soave, A. Rockenbauer, J. Phys. Chem. B 104 (2000) 1887.
137. A. D'Ercole, C. Pisani, J. Chem. Phys. 111 (1999) 9743.
138. M. Sterrer, O. Diwald, E.J. Knözinger, J. Phys. Chem. B 104 (2000) 3601.
139. L. Ojamäe, C. Pisani, J. Chem. Phys. 109 (1998) 10984.
140. G. Pacchioni, P. Pescarmona, Surf. Sci. 412/413 (1998) 657.
141. S. Abbet, E. Riedo, H. Brune, U. Heiz, A.M. Ferrari, L. Giordano, G. Pacchioni, J. Am. Chem. Soc. 123 (2001) 6172.
142. I. Yudanov, S. Vent, K. Neyman, G. Pacchioni, N. Rösch, Chem. Phys. Lett. 275 (1997) 245.
143. J. Goniakovski, Phys. Rev. B 58 (1998) 1189.
144. M.C. Paganini, M. Chiesa, E. Giamello, S. Coluccia, G. Martra, D.M. Murphy, G. Pacchioni, Surf. Sci. 421 (1999) 246.
145. S. Abbet, U. Heiz, A.M. Ferrari, L. Giordano, C. Di Valentin, G. Pacchioni, Thin Sol. Films, in press (2001).
146. V. Musolino, A. Selloni, R. Car, J. Chem. Phys. 108 (1998) 5044.
147. X. Xu, D.W. Goodman, Appl. Phys. Lett. 61 (1992) 1799.
148. X. Xu, W. He, D.W. Goodman, Surf. Sci. 284 (1993) 103.
149. J.B. Zhou, T. Gustafsson, E. Garfunkel, Surf. Sci. 372 (1997) 21.
150. M. Brause, D. Ochs, J. Günster, T. Mayer, B. Braun, V. Puchin, W. Maus-Friedrichs, V. Kempter, Surf. Sci. 383 (1997) 216.
151. J. Arndt, R. Devine, A. Revesz (eds.), The Physics and Technology of Amorphous SiO_2, Plenum, New York, 1988.
152. V.A. Radtsig, Chem. Phys. Rep. 14 (1995) 1206.
153. G. Hochstrasser, J.F. Antonini, Surf. Sci. 32 (1972) 644.
154. X. Xu, D.W. Goodman, Appl. Phys. Lett. 61 (1992) 774.
155. F. Bart, M. Gautier, F. Jollet, J.P. Durand, Surf. Sci. 306 (1994) 342.
156. F. Bart, M. Gautier, J.P. Durand, M. Henriot, Surf. Sci. 274 (1992) 317.
157. G. Pacchioni, G. Ieranò, Phys. Rev. B 57 (1998) 818.
158. J.Y. Carriat, M. Che, M. Kermarec, M. Verdaguer, A. Michalowicz, J. Am. Chem. Soc. 120 (1998) 2059.
159. G. Pacchioni, G. Ieranò, Phys. Rev. Lett. 79 (1997) 753.
160. B. Bellamy, S. Mechken, A. Masson, Z. Phys. D 26 (1993) 61.
161. X. Xu, D.W. Goodman, J. Phys. Chem. 97 (1993) 683.
162. X. Xu, S. Vesecky, D.W. Goodman, Science 258 (1992) 788.
163. U. Diebold, J.-M. Pan, T.E. Madey, Surf. Sci. 331 (1995) 845.
164. See Faraday Discuss. 114, 1999, and references therein.
165. V.E. Henrich, P.A. Cox, The Surface Science of Metal Oxides, Cambridge University Press, Cambridge, 1994.
166. J. Purton, D.W. Bullett, P.M. Oliver, S.C. Parker, Surf. Sci. 336 (1995) 166.
167. K. Prabhakaran, D. Purdie, R. Casanova, C.A. Muryn, P.J. Hardman, P.L. Wincott, G. Thornton, Phys. Rev. B 45 (1992) 6969.
168. P.J. Hardman, R. Casanova, K. Prabhakaran, C.A. Muryn, P.L. Wincott, G. Thornton, Surf. Sci. 269/270 (1992) 677.
169. R. Casanova, K. Prabhakaran, G. Thornton, J. Phys.: Condensed Matter 3 (1991) S91.
170. A.G. Thomas, P.J. Hardman, C.A. Muryn, H.S. Dhariwai, A.F. Prime, G. Thornton, E. Roman, J.L. de Segovia, J. Chem. Soc. Faraday Trans. 91 (1995) 3569.
171. R.J. Lad, L.S. Dake, Mat. Res. Soc. Symp. Proc. Vol. 228 (Mat. Res. Soc. 1992) p. 823.
172. R. Heise, R. Courths, Surf. Sci. 331–333 (1995) 1460.
173. R. Heise, R. Courths, Surf. Sci. and Lett. 2 (1995) 147.
174. A.W. Grant, C.T. Campbell, Phys. Rev. B 55 (1997) 1844.

175. H. Onishi, T. Aruga, C. Egawa, Y. Iwasawa, Surf. Sci. 199 (1988) 597.

176. J. Nerlov, S.V. Christensen, S. Wichel, E.H. Pedersen, P.J. Møller, Surf. Sci. 371 (1997) 321.

177. R. Souda, W. Hayami, T. Aizawa, Y. Ishizawa, Surf. Sci. 285 (1993) 265.

178. J. Muscat, N.M. Harrison, G. Thornton, Phys. Rev. B 59 (1999) 15457.

179. L.S. Dake, R.J. Lad, J. Vac. Sci. Technol. A 13 (1995) 122.

180. K.D. Schierbaum, S. Fisher, M.C. Torquemada, J.L. de Segovia, E. Roman, J.A. Martin-Gago, Surf. Sci. 345 (1996) 261.

181. Y. Gao, Y. Liang, S.A. Chambers, Surf. Sci. 365 (1996) 638.

182. C. Xu, W.S. Oh, G. Liu, D.Y. Kim, D.W. Goodman, J. Vac. Sci. Technol. A 15 (1997) 1261.

183. C. Xu, X. Lai, G.W. Zajac, D.W. Goodman, Phys. Rev. B 56 (1997) 13464.

184. D. Abriou, D. Gagnot, J. Jupille, F. Creuzet, Surf. Rev. and Lett. 5 (1998) 387.

185. P.J. Møller, M.C. Wu, Surf. Sci. 224 (1989) 265.

186. J.M. Pan, B.L. Maschhoff, U. Diebold, T.E. Madey, Surf. Sci. 291 (1993) 381.

187. T.E. Madey, in The Chemical Physics of Solid Surfaces and Heterogeneous Catalysis, Vol. 8, D.A. King and D.P. Woodruff (eds.), Elsevier, Amsterdam, 1997.

188. X. Lai, T.P. St. Clair, D.W. Goodman, Faraday Discuss. 114 (1999) 279.

189. T. Bredow, E. Aprà, M. Catti, G. Pacchioni, Surf. Sci. 418 (1998) 150.

190. J.F. Sanz, C.M. Zicovich-Wilson, Chem. Phys. Letters 303 (1999) 111.

191. C.J. Calzado, J. Oviedo, M.A. San Miguel, J.F. Sanz, J. Mol. Catal. A 119 (1997) 135.

192. M.A. San Miguel, C.J. Calzado, J.F. Sanz, Surf. Sci. 409 (1998) 92.

193. X. Wei-Xing, K.D. Schierbaum, W. Goepel, J. Solid State Chem. 119 (1995) 237.

194. L. Thien-Nga, A.T. Paxton, Phys. Rev. B 58 (1998) 13233.

195. L. Zhang, F. Cosandey, R. Persaud, T.E. Madey, Surf. Sci. 439 (1999) 73.

196. M. Valden, D.W. Goodman, Israel J. Chem. 38 (1998) 285.

197. L. Zhang, R. Persaud, T.E. Madey, Phys. Rev. B 56 (1997) 10549.

198. S.C. Parker, A.W. Grant, V.A. Bondzie, C.T. Campbell, Surf. Sci. 441 (1999) 10.

199. X. Lai, T.P. St Clair, M. Valden, D.W. Goodman, Prog. Surf. Sci. 59 (1998) 25.

200. G. Kresse, J. Hafner, Phys. Rev. B 47 (1993) RC558.

201. G. Kresse, J. Furthmüller, Comput. Mater. Sci 6 (1996) 15.

202. G. Kresse, J. Furthmüller, Phys. Rev. B 54 (1996) 11169.

203. P.S. Bagus, G. Pacchioni, M.R. Philpott, J. Chem. Phys. 90 (1989) 4287.

204. G. Pacchioni, P.S. Bagus, Surf. Sci. 286 (1993) 317.

205. G.A. Benesh and D.A. King, Chem. Phys. Lett. 191 (1992) 315.

206. G. Pacchioni, P.S. Bagus, Surf. Sci. 286 (1993) 317.

207. P.S. Bagus, G. Pacchioni, J. Chem. Phys. 102 (1995) 879.

208. M. Mayer, G. Pacchioni, N. Rösch, Surf. Sci. 412/413 (1998) 616.

209. G. Haas, A. Menck, H. Brune, J.V. Barth, J.A. Venables, K. Kern, Phys. Rev. B 61 (2000) 11105.

210. A. Sanchez, S. Abbet, U. Heiz, W.D. Schneider, H. Häkkinen, R.N. Barnett, U. Landmann, J. Phys. Chem. A 103 (1999) 9573.

211. A. Sanchez, S. Abbet, U. Heiz, W.D. Schneider, A.M. Ferrari, G. Pacchioni, N. Rösch, J. Am. Chem. Soc. 122 (2000) 3453.

212. A. Sanchez, S. Abbet, U. Heiz, W.D. Schneider, A.M. Ferrari, G. Pacchioni, N. Rösch, Surf. Sci. 454/456 (2000) 984.

213. W.T. Tysoe, G.L. Nyberg, R.M. Lambert, J. Chem. Soc., Chem. Commun. (1983) 623.

214. W. Sesselmann, B. Woratschek, G. Ertl, J. Kuppers, H. Haberland, Surf. Sci. 130 (1983) 245.

215. P.M. Holmblad, D.R. Rainer, D.W. Goodman, J. Phys. Chem. B 101 (1997) 8883.

216. I.M. Abdelrehim, K. Pelhos, T.E. Madey, J. Eng, J.G. Chen, J. Mol. Catal. A 131 (1998) 107.

217. T.M. Gentle, E.L. Muetterties, J. Phys. Chem. 87 (1983) 2469.
218. T.G. Rucker, M.A. Logan, T.M. Gentle, E.L. Muetterties, G.A. Somorjai, J. Phys. Chem. 90 (1986) 2703.
219. H. Hoffmann, F. Zaera, R.M. Ormerod, R.M. Lambert, J.M. Yao, D.K. Saldin, L.P. Wang, D.W. Bennett, W.T. Tysoe, Surf. Sci. 268 (1992) 1.
220. C.H. Patterson, R.M. Lambert, J. Am. Chem. Soc. 110 (1988) 6871.
221. R.M. Ormerod, R.M. Lambert, J. Phys. Chem. 96 (1992) 8111.
222. R.M. Ormerod, R.M. Lambert, H. Hoffmann, F. Zaera, J.M. Yao, D.K. Saldin, L.P. Wang, D.W. Bennet, W.T. Tysoe, Surf. Sci. 295 (1993) 277.
223. G. Pacchioni, R.M. Lambert, Surf. Sci. 304 (1994) 208.
224. S. Abbet, A. Sanchez, U. Heiz, W.D. Schneider, J. Catal. 198 (2001) 122.
225. A.V. Matveev, K.M. Neyman, I.V. Yudanov, N. Rösch, Surf. Sci. 426 (1999) 123.
226. J.F. Goellner, K.M. Neyman, M. Mayer, F. Nörtemann, B. Gates, N. Rösch, Langmuir 16 (2000) 2736.
227. M.J.S. Dewar, Bull. Soc. Chem. Fr. 18 (1951) C71.
228. J. Chatt, L.A. Duncanson, J. Chem. Soc. (1953) 2939.
229. R. Wichtendahl, M. Rodriguez-Rodrigo, U. Härtel, H. Kuhlenbeck, H.J. Freund, Surf. Sci. 423 (1999) 90.
230. J.H. Hardesty, J.B. Koerner, T.A. Albright, G.Y. Le, J. Am. Chem. Soc. 121 (1999) 6055.

4

State-of-the-Art Characterization of Single-Crystal Surfaces: A View of Nanostructures

MATTHIAS BATZILL, SANTANU BANERJEE, and BRUCE E. KOEL

University of Southern California, Los Angeles, California, U.S.A.

CHAPTER CONTENTS

SUMMARY

Atomic level characterization of solid surfaces is important for understanding and tailoring properties of nanoparticles. In the past 30 years, numerous electron and ion spectroscopic techniques, in addition to microscopic or imaging techniques, have been established to provide this information. In this chapter, we briefly describe several techniques that provide state-of-the-art characterization of the structure and morphology of single-crystal surfaces. Such surfaces serve as models to understand and predict the behavior of nanoparticles or are directly relevant as supports (substrates) for nanoparticles. The basis for X-ray photoelectron diffraction (XPD) and holography and low energy ion scattering (LEIS) is discussed, along with several examples of applications in structure determination including adsorbate structure and adsorbate-induced restructuring at metal surfaces, ultrathin metal films and bimetallic surfaces, and oxide surfaces. Scanning probe microscopy (SPM), including scanning tunneling microscopy (STM), atomic force microscopy (AFM), and near-field scanning optical microscopy (NSOM) is discussed, and examples of results obtained by utilizing these techniques are given that illustrate their applications. These include ultrahigh vacuum experiments with STM, high-pressure STM studies, and STM investigations in electrolyte solutions.

4.1 INTRODUCTION

Advances in the characterization of solid surfaces have an important role to play in understanding and tailoring catalysis and electrocatalysis at nanoparticle surfaces. This is because of the strong "surface"–"nanoparticle" connection. There are two aspects to this: (1) surface *properties of* nanoparticles, and (2) nanoparticles *on* surfaces. By necessity, nanometer scale objects have a large part of their material present at the surface. Also, nanoparticles and nanometer-scale devices and structures are often deposited or constructed at a solid interface, and thus the surface properties and chemistry of that interface can control the construction, stability, and properties of the particle or device. This is important, because 30 years of surface science research has taught us that surface behavior is *not* that in the bulk. New thermodynamic variables for the surface must be specified. Abrupt termination of the bulk phase at the surface leads to changes in coordination (nearest neighbors) and charge distribution, and the existence of "dangling bonds," electronic surface states, and other properties that are different from the bulk. Relaxation and reconstruction of the atoms in the top layer(s) occurs, causing new lattice structures (e.g., pseudo-morphic crystalline films templated by the substrate structure) and altered equilibrium shapes of nanoclusters due to the importance of the interface energy. Nanocluster geometries and properties in the gas phase are not the same if those clusters are deposited on a surface. Equilibrium bulk compositions and phase diagrams are not predictive and must be reevaluated, e.g., two metals that are immiscible in the bulk may alloy in the topmost layer to relieve strain at the interface. Surfaces generally are less stable than the bulk and, for example, have lower melting temperatures, higher vapor pressures, and greater reactivity.

It is also important to remember that all surfaces in ambient environments are covered with at least a monolayer of adsorbed water or other material. This means that experiments performed in vacuum or ultrahigh vacuum (UHV) conditions study different surface interactions. The structure of a surface is strongly influenced by adsorbed layers (surfactants), which can be used to control the surface structures present or the morphology of film growth. Even in UHV, the presence of gaseous adsorbates can cause large changes in alloy composition at the surface, because of chemisorption-induced segregation, and in the 2D and 3-D structure at the surface because of adsorbate-induced reconstruction, faceting, or massive step-bunching.

Thus, characterization of surfaces is important to the field of catalysis and electrocatalysis of nanoparticles. In the past 30 years, numerous electron and ion spectroscopic techniques, in addition to microscopic or imaging techniques, have been established to provide this information. Figure 1 provides high-resolution transmission electron microscopy (TEM) images of a practical (real), high-surface-area, Au/anatase–TiO_2 heterogeneous catalyst that show the small Au nanoparticles

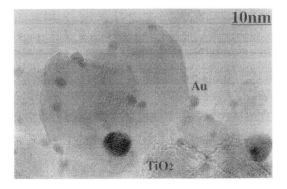

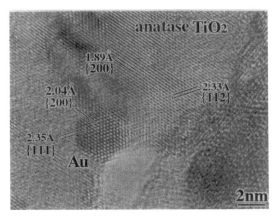

Figure 1 High-resolution transmission electron microscopy (TEM) images of an Au/anatase-TiO₂ catalyst. (From T. Akita, ONRI, Japan.)

that are thought to be responsible for the remarkable catalytic activity of these supported Au catalysts [1]. These images also illustrate the preferred orientational relationship between the crystalline Au nanoparticles and crystals of the TiO₂ support.

In this chapter, we briefly describe several techniques that provide state-of-the-art characterization of the structure and morphology of single-crystal surfaces. Such surfaces serve as models to understand and predict the behavior of nanoparticles or are directly relevant as supports (substrates) for nanoparticles. It is beyond the scope of this chapter to provide a comprehensive review of work in this field, but rather we provide a number of examples of results obtained by utilizing these surface characterization techniques which illustrates their applications.

4.2 PHOTOELECTRON DIFFRACTION AND ION SCATTERING

4.2.1 Introduction

Numerous spectroscopic techniques, such as x-ray photoelectron spectroscopy (XPS) [2], Auger electron spectroscopy (AES) [3], ultraviolet photoelectron spectroscopy (UPS), soft x-ray absorption spectroscopy (SXA) [4], high-resolution electron energy loss spectroscopy (HREELS) [5], and Fourier transform infrared spectroscopy (FTIR) [6] can be used to characterize surfaces and adsorbed films. These techniques probe different aspects of surfaces, and a multitechnique approach is usually necessary to get a complete picture.

The most powerful techniques commonly used to determine structure at single-crystal surfaces include low-energy electron diffraction (LEED) [7], x-ray photo-electron diffraction (XPD) [8–10], and low-energy ion scattering (LEIS) [11,12]. Herein, we focus on the latter two. These are element-specific tools for determining the local geometric structure around a probed atom at the surface. XPD and LEIS are highly surface-sensitive and can provide a direct determination of surface structure. In this section, we briefly describe some essential features of XPD, photoelectron holography, and LEIS, as they relate to determining the geometric structure of ordered, single-crystal surfaces. We then describe some recent examples of applications of these techniques to investigate adsorbate structure and adsorbate-induced restructuring at metal surfaces, thin films, and bimetallic surfaces, and oxide surfaces. This list is obviously not comprehensive but was chosen to illustrate the usefulness of these techniques and indicate the state-of-the-art in experimental and theoretical developments related to these methods.

4.2.2 X-ray Photoelectron Diffraction (XPD) and Holography

XPS is a quantitative, element-specific analytical probe that provides information on the chemical nature of atoms in the near-surface region of a solid. In XPS, x-rays photoionize atoms in a target, producing photoelectrons, and the kinetic energy of electrons emitted from the surface is analyzed. The focus is typically on photoelectrons produced by photoionization of deep, atomiclike core levels, because core-level binding energies are characteristic of each element and photoabsorption cross sections for these levels are independent of the chemical environment. Furthermore, these have relatively narrow linewidths and chemical shifts in the core-level binding energies can be related to changes in valence electronic state (i.e., oxidation state) of the probed atom.

For single-crystal substrates, there is additional information about the geometric structure of the surface contained in the intensity angular distribution (IAD) of the photoemitted electrons. This information is obtained in XPD measurements that often directly indicate bond directions at surfaces. A Fourier transform of the XPD data can also be used to determine element-specific surface structure in a method known as *photoelectron holography*. This approach can be used to directly determine surface structures without any starting model.

X-Ray Photoelectron Diffraction (XPD)

X-ray photoelectron diffraction is the coherent superposition of a directly photo-emitted electron wave with the elastically scattered waves from near-neighboring atoms. This gives element-specific structural information about the near surface atoms in a single crystal [8–10]. The short inelastic mean free path of the electron waves at the kinetic energies of interest (15 to 1000 eV) leads to surface sensitivity and determination of the atomic geometry of the emitting atom. The known energies of narrow XPS core-level peaks lead to element specificity. The resolution of surface peaks and chemical shifts may even sometimes lead to a chemical state-specific structure determination.

One classification scheme for XPD uses the kinetic energy regime of the photo-emitted electrons. At sufficiently high kinetic energies (above 500 eV), the scattering factor is highly forward-peaked, which leads to a direct interpretation of bond angles from the angular intensity oscillations using single scattering theory. The interpretation of the data is particularly simple because there are substantial intensity enhancements along interatomic directions. At lower kinetic energies, the scattering factor is more isotropic, which leads to more complex XPD patterns because of increased multiple scattering. Therefore, at low energies, extensive numerical simulations including multiple scattering calculations have to be used to determine the structure.

Experimentally, XPD data are obtained by one of two approaches that are distinguished by the scanning parameter: angle-scanned mode and energy-scanned mode. Angle-scanned XPD is more common because it can be performed using any available, lab-based x-ray source typically used for XPS, and collecting the diffracted intensity over a wide range of angles. Energy-scanned XPD is performed using a synchrotron light source and an appropriate monochromator so that the photon energy can be scanned while keeping the photoelectron kinetic energy fixed for a few high-symmetry orientations of the crystal.

A schematic of the XPD experiment is shown in Figure 2. X-rays incident on the sample create photoelectrons detected by an electron-energy analyzer. Angular intensity modulations are created by elastic scattering of the photo-emitted electrons from the neighboring atoms. In angle-scanned XPD, these angular distributions are obtained by rotating either the sample or the analyzer to scan different exit angles for the photo-emitted electrons and then used to determine the surface structure.

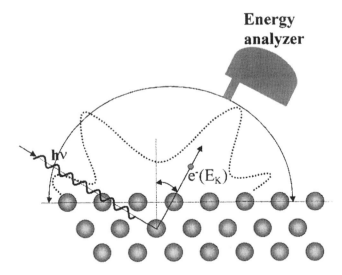

Figure 2 Schematic for photoelectron diffraction. Photo-emitted electrons are energy-analyzed at different exit angles. Direct photo-emitted electron waves interfere with the scattered waves to give angular intensity variations. These angular variations contain element-specific information about the near-surface structure.

Figure 3 illustrates the use of angle-scanned XPD data to determine the near-surface structure of a single-component metal, single-crystal sample. Figure 3a shows an XPD pattern for a face-centered cubic (*fcc*) Pt(111) single crystal using the Pt 4*f* core-level peak (at 1183 eV kinetic energy using Mg K_α excitation). The diffraction pattern is plotted with the center corresponding to the surface-normal direction and the edges corresponding to grazing exit angles for photoelectrons. The radial distance is proportional to the polar angle, as measured with respect to the surface normal. This diffraction pattern is representative of any *fcc*(111) surface. The angular intensity enhancements can be understood according to a simple forward-scattering model, denoted by Egelhoff [13] as the "searchlight effect," by looking at the schematic drawing in Figure 3b. This shows a vertical cut along a high-symmetry azimuth of the *fcc*(111) crystal. The angular position of each high-intensity feature in the Pt(111) XPD pattern corresponds to a direction between a pair of near-surface atoms. For example, if we take a line scan along the $[1\bar{2}1]$ azimuth in the XPD pattern (i.e., a polar angle scan), the main feature is a strong peak near 35° that corresponds to the [101] direction. This angular peak originates from an emitter atom located one layer below the scattering atom. Similarly, in the radially opposite azimuth $[\bar{1}2\bar{1}]$ three peaks correspond to the [112], [114], and [001] directions from emitters located at the 3rd, 4th, and 2nd layers, respectively, considering that the scattering atom is in the 1st layer. The intensity at the center corresponds to the normal direction [111] and originates from 4th-layer emitters.

Figure 3c shows an XPD pattern for the *fcc* Ni(100) surface using the Ni LMM Auger peak (at 841-eV kinetic energy) in XPS. This pattern is representative of *fcc*(100) single crystals. Figure 3d shows a vertical cut through the *fcc*(100) crystal along the [100]-like azimuths corresponding to a vertical or horizontal line in the XPD pattern. The three main features along this azimuth are in the [001], [103], and [101] directions, originating from emitters in the 3rd, 4th, and 3rd layers, respectively, considering the scattering atoms to be in the 1st layer. Similarly, the main features along the [110]-like azimuth in the XPD pattern, i.e., along a line oriented 45° from the horizontal or vertical direction in the XPD pattern, can be explained by viewing the schematic in Figure 3e of a vertical cut along the [110] azimuth. The [114], [112], and [111] directions correspond to emitters in the 5th, 3rd, and 3rd layers if the scattering atom is in the 1st layer. In general, in these diffraction patterns the intensity enhancements are inversely proportional to the interatomic distances involved, i.e., angular intensity modulations from scattering of a 4th layer emitter by a 1st-layer scatterer is weaker than that for a 2nd layer emitters (see, for example, [101] versus [111] in Figure 3a).

In a theoretical description for the angle-dependent photoelectron-intensity variation, one considers the wave nature of the photoelectrons. The photon-emitted electron is scattered by the surrounding atoms. The interference of the photoelectron wave with its scattered waves results in an intensity modulation that depends on the geometrical arrangement of the scatterers (lattice atoms) and the atomic scattering factor. This ultimately is responsible for the angle dependency of the photoelectron intensity and is therefore directly related to the structure of the surface layers. For a

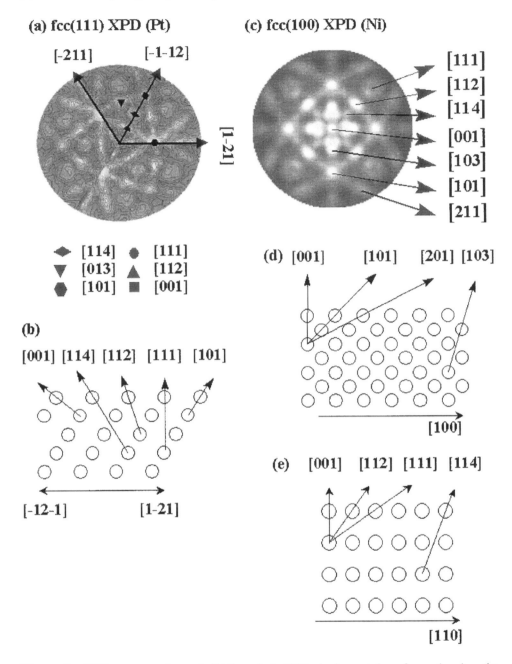

Figure 3 XPD patterns from *fcc*(111) and *fcc*(001) single-crystal surfaces showing the correspondence between enhanced intensity and near-neighbor directions. The center of the pattern corresponds to the direction normal to the surface and the radial distance is proportional to the polar angle with respect to the surface normal. (a) XPD pattern for *fcc* Pt(111). (b) Vertical cross section of an *fcc*(111) crystal along a high-symmetry direction. (c) XPD pattern for *fcc* Ni(100). (d) Vertical cut of the *fcc*(001) crystal along the [100] azimuth. (e) Vertical cut of the *fcc*(001) crystal along the [110] azimuth.

plane incident wave ϕ_0 (photoelectron), the scattered wave ϕ_j can be written as

$$\phi_j = \phi_0 f_j(\theta_j) \frac{e^{ikr_j}}{r_j} \tag{1}$$

where $f_j(\theta_j)$ is the atomic scattering factor (complex number), θ_j is the scattering angle, k is the electron wave number, and r_j is the emitter–scatterer distance. In the small atom approximation, the scattering factor can be calculated using the partial wave method:

$$f_j(\theta_j) = \frac{1}{k} \sum_l (2l+1) e^{i\delta_l^j} \sin \delta_l^j P_l(\cos \theta_j) \tag{2}$$

where $\{\delta_l^j\}$ is a set of phase shifts for the jth scatterer, and $\{P_l(\cos \theta_j)\}$ are Legendre polynomial functions. This approximation is accurate in general for all but nearest-neighbor distances of $\leqslant 5\,\text{Å}$ from the emitter. At smaller distances where the distance between the emitter and the scatterer is not large compared to the size of the scatterer, or the wave number of the electron is small, corrections have to be implemented [14–17] for the curved wavefront. When the emitted electron wave is other than a simple s wave, the atomic scattering factor can be calculated using a high-energy approximation [14]. The total diffracted intensity is then given by the interference of the photoelectron wave with the scattered waves. This intensity can be expressed by the square of the sum of the waves. [9]:

$$I(\mathbf{k}) = \left| \phi(\mathbf{k}) + \sum_j \left(f(\mathbf{k}, \mathbf{r}_j) \frac{e^{i(kr_j - \mathbf{k} \cdot \mathbf{r}_j)}}{r_j} \phi(r_j) \right) \right|^2 \tag{3}$$

The variation with the electron wave vector $\mathbf{k}$ is associated with an intensity variation in the experimentally observed polar and azimuth angles. (In order to include vibrational attenuation of interference effects, each scattered wave has to be multiplied by the temperature dependant Debye–Waller factor.)

At relatively high kinetic energies (above a few hundred eV), single scattering theory gives the position of diffraction features quite accurately. However, it overestimates the intensity of diffraction features [18] along low-index directions, because the theory does not consider the defocusing effect along chains of atoms. For lower energies, the errors are much more severe. This can be seen from the shape of the scattering factor. At high energies, the scattering factor is highly forward-peaked and most of the diffraction intensity lies along low-index internuclear directions. At low kinetic energies, the scattering factor is much more isotropic and, hence, multiple scattering becomes important in all directions. Hence, multiple scattering effects must be included in the theory and analysis to accurately interpret the intensity modulations.

X-Ray Photoelectron Holography

Gabor [19] first proposed holography in 1948 as a means to overcome barriers in doing "lensless" electron microscopy and avoiding inherent spherical aberrations. He proposed that if a known wave is allowed to interfere with an unknown wave, the resulting interference pattern can be stored as a "hologram," which contains most of the information needed to restore the original unknown wave. In his two-step model, a known spherical wave called the "reference" wave first interferes with a wave scattered by the object, called the "object" wave, and the interference modulations are recorded on film. Then, the reconstruction is done by illuminating the film by a similar reference wave to generate a three-dimensional holographic image of the object.

The large field of optical holography [20] emerged with the advent of laser sources. This technique is illustrated in Figure 4. A laser source, required for its large coherence length, which has to be of the order of the object to be imaged, is used as a reference wave. A beam splitter splits the laser beam in order to illuminate the object and allow for a direct wave to propagate to the film as the reference wave. Beams reflected from the object back to the film are the object waves, and the film records the interference patterns between the reference and object waves. Subsequently, a laser beam incident on the film gives a three-dimensional reconstructed image.

In 1986 Szöke [21] suggested that a photoelectron diffraction pattern from a single crystal may be treated as a hologram. In photoelectron holography the direct

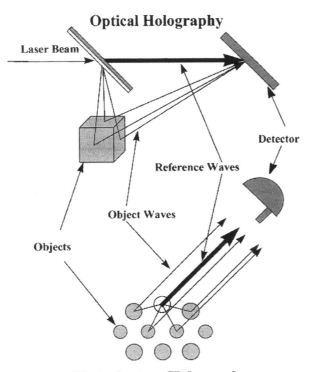

Figure 4 Analogy between the principles of optical and photoelectron holography. In optical holography (top), the incident laser beam is the reference wave, the reflected waves are the scattered waves, and a film is used as a detector. In photoelectron holography (bottom), the direct, photo-emitted waves are reference waves that interfere with the scattered waves from neighboring atoms, and a hemispherical, electron energy analyzer is used as a detector.

wave can be considered as the reference wave. The scattered wave is similar to the object wave, and the photoelectron diffraction pattern can be used for holographic reconstruction. In 1988 Barton [22] proposed a computer algorithm based on the Helmholtz–Kirchoff principle of optics that directly inverts the XPD data to obtain a three-dimensional image of the local environment around the emitter atom. In order to suppress multiple scattering and twin images, an algorithm using holograms taken at different energies has been proposed [23].

A holographic transformation was first used successfully by Harp *et al.* [24] on high-energy, single x-ray, photoelectron patterns. In the case of the backscattering (low energy), which is needed to study adsorbate or thin layers, Zharnikov *et al.* [25] demonstrated the validity of the method. Here we show, using the backscattering geometry, that a simple and direct transformation of x-ray photoelectron data, without any corrections requiring a previous knowledge of the structure, can lead to a determination of the surface structure.

Barton [26] suggested that one may use a Fourier transform formula to convert the intensity modulation into an image function, $A(\mathbf{r})$:

$$A(\mathbf{r}) = \int \int I(\mathbf{k}) e^{-i\mathbf{k} \cdot \mathbf{r}} d\mathbf{k}_{\parallel} \tag{4}$$

where $\mathbf{k} = (k_x, k_y, k_z)$ is a wave vector, $\mathbf{k}_{\parallel} = (k_x, k_y)$ is its component parallel to the surface, and k_z is given by $k_z = \sqrt{k^2 - k_x^2 - k_y^2}$. The intensity $I(\mathbf{k})$ in single scattering is given by Eq. (3).

By expanding Eq. (3) and inserting it in Eq. (4), we get four terms. Of these four terms, the first term ($|\phi(k)|^2$) is the direct term, the cross terms are the image term and a twin image term, and the last term is the self-interference term. Assuming that the first term $|\phi(k)|^2$ does not vary very much with $\mathbf{k}$, i.e., $f(\mathbf{k}, \mathbf{r})$ is a smoothly varying function of $\mathbf{k}$, and assuming that the self-interference term is negligible (true for backscattering geometries), it can be seen that the $A(\mathbf{r})$ has a maximum for $\mathbf{r} = \mathbf{r}_j$ due to the image term and at $\mathbf{r} = -\mathbf{r}_j$ due to the twin image term. Thus the intensity of $A(\mathbf{r})$ directly gives the emitter–scatterer distance and orientation and therefore the crystallographic structure.

Including multiple energy diffraction patterns, Barton's algorithm can be rewritten as

$$A(\mathbf{r}) = \int \int \int_{k,k_x,k_y} \chi(k_x, k_y, k) \exp(i\mathbf{k} \cdot \mathbf{r} - ikr) dk_x dk_y dk \tag{5}$$

where $\mathbf{k}$ is the wave vector *inside* the crystal, k_x and k_y its components parallel to the surface directions, and k its modulus (the z-axis is normal to the surface, z is positive out of the surface). The advantage of including energy integration is that it removes the twin image completely. This improves the interpretation of the results because the twin-image intensities are located at positions different from atomic positions.

4.2.3   Low-Energy Ion Scattering (LEIS)

Ion scattering spectroscopy (ISS) was introduced by Smith [25] using noble gas ions and has become a powerful tool for surface analysis. In low-energy ion scattering (LEIS), a monoenergetic beam of low-energy ions in the 0.2–5-keV range is directed toward a surface and the backscattered incident ions are energy-analyzed at a known scattering angle [11,12]. Inert gas ions (e.g., ^{4}He and ^{20}Ne) are often used in typical applications. Mean free paths of ions in solids at low energies are extremely short such that only the topmost layer composition is probed normally in LEIS (in contrast to XPS and AES).

The process of ion scattering is illustrated schematically in Figure 5. Because collision times are very short (10^{-15} to 10^{-16} s), the interactions can be approximated as elastic binary collisions [28] between the incident ion and a single surface atom (i.e., with an effective mass equal to the atomic mass). Diffraction effects are negligible. The basic equation in ISS, using energy and momentum conservation, is

$$\frac{E_1}{E_0} = \left(\frac{\cos\theta \pm \sqrt{m_2^2/m_1^2 - \sin^2\theta}}{1 + m_2/m_1} \right)^2 \tag{6}$$

where E_0 and E_1 are the initial and final energies of the incident ion of mass m_1 scattered through an angle θ by a target atom of mass m_2. Since the final energy E_1 depends only on the mass ratio m_2/m_1 for a fixed scattering angle θ, the energy spectrum gives a direct picture of the surface composition. Only the plus sign applies for those cases in which the target atom is heavier than the incident ion ($m_2/m_1 > 1$) and each target mass can be identified by a single peak in the spectrum. If $m_2/m_1 < 1$, both signs apply and each target mass gives rise to two peaks at different energies [Detection of forward-scattered, recoil ions in elastic recoil spectrometry (ERS) has been particularly important for analysis of surface hydrogen.]

In addition to elemental analysis, LEIS can be used to provide information about the local surface structure at the probed atom. This application of LEIS was improved by using alkali ions (e.g., ^{7}Li and ^{23}Na) and large scattering angles (near 180°) in so-called impact-collision ISS (ICISS) [29] Alkali ions have strong trajectory-dependent neutralization cross sections and give relatively intense

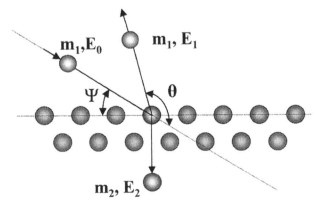

Figure 5 Schematic of a binary, elastic collision in LEIS. An incident ion of mass m_1 and energy E_0 gets scattered by a stationary target atom of mass m_2 in a crystal. The final energy E_1 of the scattered, incident ion only depends on the mass ratio m_2/m_1 for a fixed geometry.

angular-dependent signals in alkali ion scattering (ALISS) studies of surface atomistic structure. Most of the results discussed in this section are based on ALISS experiments in which the scattered alkali ions are energy-analyzed for a range of incidence angles while monitoring the scattered intensities.

For structural analysis, the interaction potential between the incident and target atoms must be considered to calculate the trajectory of the ions during the scattering process. The interaction between two charged particles (here, the incident ion and target nuclei) is given by the Coulomb potential. At low energies (~ 1 keV), screening due to the electrons has to be considered also. Thus, a screened Coulomb potential is used to describe the scattering

$$V(r) = \frac{Z_1 Z_2 e^2}{r} \phi\left(\frac{r}{a}\right) \tag{7}$$

where Z_1 and Z_2 are the atomic numbers of the incident ion and target atom, respectively, and r is their instantaneous separation. An analytical approximation, called the Thomas–Fermi–Moliere [30] (TFM) screening function of the form

$$\phi(x) = 0.35e^{-0.3x} + 0.55e^{-1.2x} + 0.1e^{-6.0x} \tag{8}$$

is the most commonly used screening function. The screening length a suggested by Firsov [31] is

$$a = Ca_F = \frac{C(0.8853)a_B}{\left(Z_1^{1/2} + Z_2^{1/2}\right)^{-2/3}} \tag{9}$$

where $a_B = 0.529$ Å, the Bohr atomic radius. The adjustable parameter C was introduced to improve agreement with experimental results.

The concept of a "shadow cone," i.e., a region behind the target atom where no ion can penetrate, is useful in surface-structure determination using ISS. This region is created by the repulsive potential between the incident ion and target atom. Figure 6a illustrates the shadow cone region behind a target atom. When a parallel beam of mono-energetic ions interact with an atom at varying impact parameters, the envelope of the ion trajectories creates a region behind the target atom that is inaccessible to any of the incident ions. A universal, empirical relation for this shadow cone at a distance l from the target atom was given by Oen [32] as

$$\frac{r(l)}{2\sqrt{bl}} = 1 - 0.12\alpha + 0.01\alpha^2 \qquad \text{for } 0 \leqslant \alpha \leqslant 4.5 \tag{10}$$

$$\frac{r(l)}{2\sqrt{bl}} = 0.924 - 0.182\ln\alpha + 0.0008\alpha \qquad \text{for } 4.5 \leqslant \alpha \leqslant 100 \tag{11}$$

where $b = Z_1 Z_2 e^2 / E_0$, $\alpha = 2\sqrt{bl}/a$, and a is the screening length. The shadow cone can also be calculated by solving the scattering integrals numerically with an appropriate scattering potential. Figure 6b illustrates the determination of local surface structure using the shadow cone concept. At a sufficiently low angle of incidence ψ (polar angle) of the ion beam relative to the surface, all the surface atoms are hidden in the shadow cones of the preceding atom in the "chain" of scatterers. As the angle of incidence is increased above a critical angle ψ_c adjacent, top-layer atoms

(a)

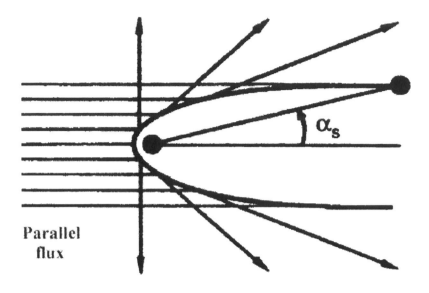

α_s : shadowing critical angle

Figure 6 (a) Trajectories of a parallel flux of ions impinging on an atom with varying impact parameters. The envelope of these trajectories creates a region behind the target atom which is inaccessible to the incoming ions. This region is called the shadow cone. (b) With increasing incidence angle, a critical angle ψ_c is reached where nearest-neighbor atoms emerge from the shadow cone of the preceding atom. This causes a sharp increase in the ion scattering intensity, and thus a measurement of ψ_c can be used to determine unknown surface structures.

emerge out of these shadow cones and there is a sharp increase in the ion scattering signal at the detector. Simple geometry relates ψ_c to the shadow cone radius r at a distance l from its apex and at a distance d between two neighboring atoms:

$$\left. \begin{array}{l} r = d \sin(\psi_c) \\ l = d \cos(\psi_c) \end{array} \right\} \tag{12}$$

These relationships can be used to obtain the shadow cone radius and interaction potential for known distances (structures) as a calibration. Then, unknown distances and structures can be solved using the known shadow cone radius.

Figure 7 illustrates the appearance of critical angles in LEIS data. As described above, the locations of the critical angles are directly related to the atomic geometry and can be utilized in solving surface structures. Figure 7a gives a polar scan in ALISS using 1-keV Na^+ ions incident on the *fcc* Pt(111) surface along the [$\bar{2}11$] direction. At incident angles below ψ_1, all atoms are in the shadow cone of their preceding atoms and hence no ALISS signal can be observed. As the angle reaches

(b)

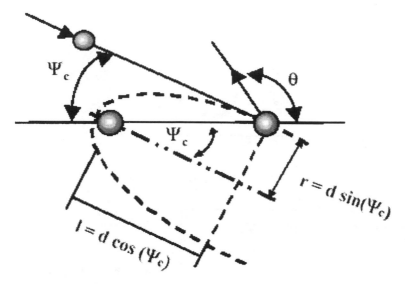

Figure 6   Continued.

ψ_1, the atoms emerge out of the shadow cone of the preceding atom, leading to a sharp rise in intensity corresponding to the ψ_1 peak in the ALISS scan. Similarly, as the polar scan is increased further to the ψ_2, the second-layer atoms emerge out of the shadow cone of the first-layer atoms, giving rise to the ψ_2 peak. Figure 7b shows an azimuthal ALISS scan for 5-keV Na^+ ions incident on a *fcc* Ni(111) surface at a grazing polar angle of 12°. When the angle of incidence is along a low-index azimuth direction, like [110] or [112], the shadow cone of the preceding atom prevents the incident ions from reaching the target atoms for polar angles below the critical angle. This causes the dips in ion scattering intensity along the [110] and [112] azimuthal directions. Such azimuthal scans can provide the structure, orientations, and symmetry of the neighboring atoms for a surface-layer target atom.

4.2.4   Applications to Structure Determination at Single-Crystal Surfaces

Studies of single-crystal surfaces under UHV conditions have allowed us to quantify fundamental interactions at surfaces, and the majority of surface-science studies have been conducted in this manner. Utilization of XPD and LEIS techniques require the studies to be conducted under high vacuum, and studies of clean surfaces or precisely controlled adsorbate layers require UHV conditions. Here we discuss a few examples of the use of these two techniques in studies of single-crystal surfaces, illustrating their power and limitations. The surfaces discussed are metal surfaces that contain controlled amounts of adsorbates, ultrathin metal films, two-component metal alloy surfaces, and oxide surfaces.

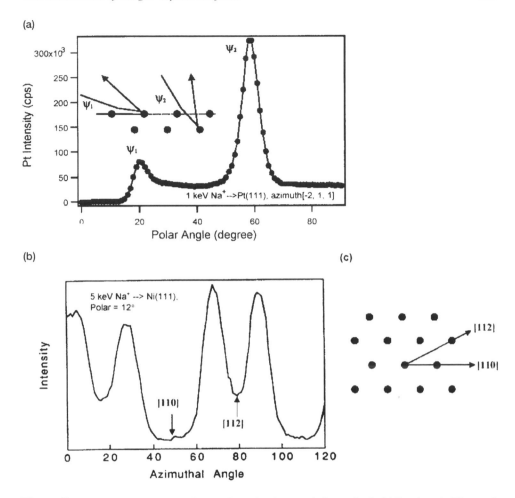

Figure 7 (a) ALISS polar scan for an *fcc* Pt(111) crystal along the [−211] azimuth. The peak at ~20° originates from scatterers in the surface layer. The peak at ~60° originates from second-layer scatterers. (b) ALISS azimuthal scan at a low polar angle for an *fcc* Ni(111) surface. At low polar angles, there are dips in the [110] and [112] directions because of the shadow cone effect.

Adsorbate Structure and Adsorbate-Induced Restructuring at Metal Surfaces

Formation of a surface always requires energy. However, the surface free energy can often be minimized by interplanar relaxation or, more severe by reconstruction of atoms at the surface to positions that deviate greatly from those expected from an ideal termination of the bulk lattice. The presence of adsorbates can alter or induce such reconstructions, and adsorbates themselves can form a variety of structures in the adlayer as well. In the past, LEED and other surface-science techniques have been employed to characterize the structure of these reconstructed surfaces and ordered adsorbate layers. The chemical specificity of XPD and extreme surface sensitivity of LEIS are particularly useful in this regard.

Adsorbate-Induced Surface Reconstruction of Ag(110)–(2 × 1)–O

Atomic and molecular adsorbates that form chemical bonds to surface atoms may cause a restructuring of the surface in order to minimize the total energy of the system. Such adsorbate-induced surface reconstructions have been observed for many reactive adsorbates, e.g., H, C, N, O, and S adatoms. In the particular case of O/Ag(110), LEED studies on the Ag(110)–(2 × 1)–O surface concluded that O adatoms were adsorbed on long-bridge sites, but reconstruction of the Ag(110) surface was not considered. Later, LEIS and STM studies indicated a "missing-row" reconstruction for the Ag(110)–(2 × 1)–O structure in which alternating rows of Ag were removed. Scanned-energy mode XPD was used to unambiguously confirm this missing-row structure [33].

Figure 8 shows the environment of the O atoms at the surface of the (2 × 1) reconstruction. This result was obtained using the "projection method" to get the approximate atomic geometry without a starting model. The vertical and horizontal cuts near the surface establish the position of O atoms to be at the long bridge sites.

Figure 9 shows the comparison between theory and experiment for the missing-row model of the Ag(110)–(2 × 1)–O system. After consideration of all possible (2 × 1) reconstructions, the missing-row model has the lowest (best) R-factor, i.e., a one-parameter expression for the quality of fit between two curves.

The locations of the O atoms deduced from XPD are consistent with previous studies on this system using LEIS, STM, and surface-enhanced x-ray absorption fine-structure spectroscopy (SEXAFS). Similar (2 × 1)–O phases on Cu(110) and Ni(110) also have missing-row reconstructions.

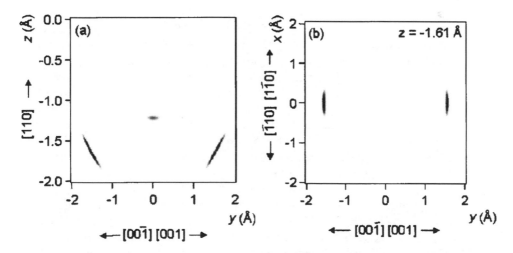

Figure 8 Intensity plot for the projection-method calculation. (a) Vertical cut through the [−110] azimuth. (b) Horizontal cut 1.61 Å below the oxygen atom emitter, which is at (0,0,0). (From Ref. 33.)

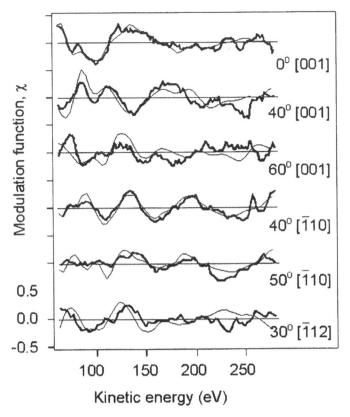

Figure 9 Comparison of results from theoretical simulations (thin lines) with energy-scanned, O 1s XPD data (thick lines) for the Ag(110)–(2 × 1)–O system. (From Ref. 33.)

Orientation of Adsorbed C_{60} Molecules Determined by XPD

XPD has been used to determine the structure of C_{60} molecules chemisorbed on single crystal metallic substrates [34]. The origin of the XPD pattern from a chemisorbed C_{60} molecule is schematically illustrated in Figure 10a. All 60 carbon atoms act as photo-emitters. The photo-emitted electrons are scattered by neighboring carbon atoms, and this gives enhanced intensity along the C–C bond directions, as shown in Figure 10b due to "forward focusing." Analyzing the position and symmetry of the high-intensity spots therefore gives a direct determination of the relative locations of the carbon atoms. Figure 10c shows a calculated diffraction pattern for the orientation of the C_{60} molecule, as shown in Figure 10a. The dark spots correspond to interatomic directions.

Figure 11 shows the C 1s diffraction patterns observed in experiments on C_{60} monolayers on Cu(111), Al(111), Cu(110), and Al(001) surfaces. Figures 11a and b have sixfold symmetry, which shows that a six-carbon ring is "facing" the Cu(111) and Al(111) surfaces. The diffraction patterns between the two are quite similar except for a 30° azimuthal rotation. This shows that the C_{60} molecules are rotated by 30°. None of the groups has fivefold symmetry, suggesting that the 5-ring is not

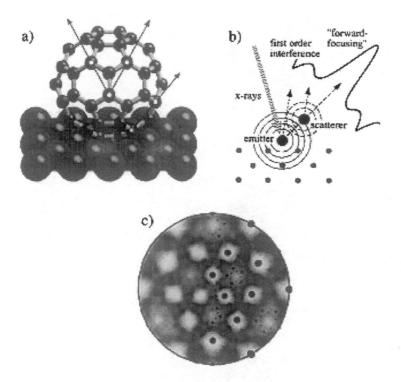

Figure 10   (a) Schematic depiction of a C_{60} molecule adsorbed on a substrate showing the relative orientation of the 60 carbon atoms. (b) Schematic drawing of the enhanced intensity in XPD along interatomic directions caused by forward focusing. (c) Calculated XPD pattern for a C_{60} molecule adsorbed on a substrate with a 6-ring directed toward the substrate. The dark spots correspond to C–C bond directions with the size of the spots inversely proportional to the interatomic distance. (From Ref. 34.)

facing the surface in any of the cases. The orientation of the C_{60} molecules adsorbed on different substrates is illustrated schematically in Figure 12. The atoms closest to the surface are shown in black. In each of the four cases, theoretical single scattering cluster (SSC) simulations of the rigid C_{60} cage structure closely reproduce the corresponding diffraction patterns in Figure 11 after domain averaging. The twofold symmetry in Figure 11c can be reproduced well by considering that the two C atoms from a 6-ring and 5-ring (5–6 bond) face the surface. The fourfold symmetry in Figure 11d can result from a single edge atom adsorption (between two 6-rings and one 5-ring).

Ultrathin Metal Films and Bimetallic Surfaces

The interest in the morphology of ultrathin (monolayer, bilayer, etc.) metal films on metal substrates has been fueled partly by the possibility of growing novel materials with unique chemical and physical properties. For example, giant magneto-resistance (GMR) structures are built by alternating thin nonmagnetic and ferromagnetic layers, and the intermixing in the interface region has a large effect

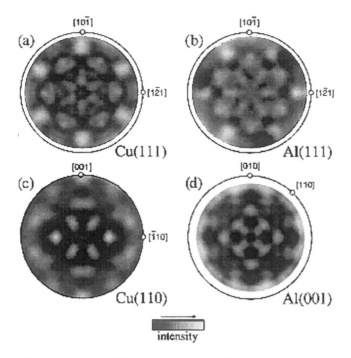

Figure 11 C 1s XPD pattern, using a Mg K_α source, from monolayer films of C_{60} adsorbed on (a) Cu(111), (b) Al(111), (c) Cu(110), and (d) Al(001). The patterns have been azimuthally averaged by considering the appropriate rotational symmetry. (From Ref. 34.)

on the strength of the GMR effect. XPD and LEIS are both well suited to studying such structures since both of these techniques can directly probe aspects of the composition and structure of individual layers at and near the interface.

Bimetallic alloy surfaces are also of great importance. Most metallic materials used commercially, and in particular metal-based, heterogenous catalysts, have multicomponent alloy phases. Despite their obvious importance, alloy single-crystal surfaces have not been studied so extensively in the past. A first step in understanding the chemistry of these surfaces is a thorough characterization of the structure of such surfaces. These attempts are part of efforts to overcome the "material gap" between commercial catalysts and surface-science studies.

Fe/Ni(001) Studied by XPD

The *fcc* phase of Fe (γ-phase) at low temperature is of interest to understand the effects of magnetic properties on atomic volume. While the γ-phase in bulk, solid Fe is only stable at high temperatures (above 910°C), epitaxial growth of Fe thin films on suitable substrates can stabilize the γ-phase at room temperature. Thin films of Fe on Ni(001) are of interest because of lattice matching and the possibility of novel magnetic phases that may arise from the influence of the structure and magnetism of the substrate.

The growth mode of Fe on Ni(001) was studied using forward-scattering photoelectron diffraction [35]. Figure 13 shows the IAD along the [110] azimuth for

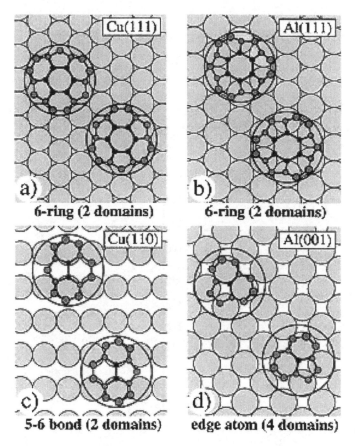

Figure 12   Molecular orientations of C_{60} in monolayer films on (a) Cu(111), (b) Al(111), (c) Cu(110), and (d) Al(001) as determined from the XPD patterns in Figure 11. For clarity, only the lower carbon atoms of the molecules are shown. The atoms closest to the surface are shown as black dots. (From Ref. 34.)

the Fe/Ni(001) surface with Fe coverages of 0–14 ML. The presence of forward-scattering features from Fe emitters at 0.5-ML and 1-ML coverage shows that the Fe film growth mode is not "layer by layer." The position of the forward-scattering features of Fe([112] shifted by $-1.4°$ from clean Ni(001)) and the corresponding Ni $3p$ IAD (not shown here) strongly suggest that the growth mode is one of island formation, rather than intermixing (alloying). The shift in the Fe[112] peak position from 0–2 ML can be explained by elastic strain. The Fe[112] peak at 1 ML coverage shifts by $-1.4°$, which compares to $-1.3°$ expected from the vertical expansion required to retain the atomic volume of bulk Fe. As the Fe film thickness increases, the [112] peak shifts to even lower angles, reaching a final value of $31.7°$, which far exceeds values from elastic strain calculations. The IAD features for the thicker films can be explained by a strain relief transition of Fe to a body-centered cubic (*bcc*) (110) phase (with a $bcc\langle111\rangle \parallel fcc\langle001\rangle$ in-plane orientation) between 2- and 3-ML coverage. The in-plane lattice constant between the *bcc*(110) and *fcc*(001) surface unit cell is within 0.1%. Along the *bcc*[111] azimuth, the forward-scattering features

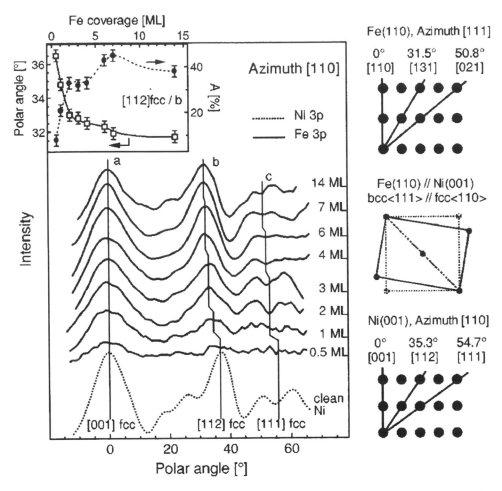

Figure 13 IAD curve from XPD data for 0–14-ML Fe on Ni(001). Angular scans from the Fe 3p core level are shown and compared with those from the Ni 3p core level. The scans were taken along the [110] azimuth of the Ni substrate. The inset shows the position of the [112] peak and the anisotropy function. Cross sections of the *bcc*(110) azimuth [111] and *fcc*(001) azimuth [110] are shown at the right, along with an in-plane schematic diagram of the *fcc*(100) and *bcc*(110) surface cells. (From Ref. 33.)

are expected to be at 31.5° and 50.8°, which match quite well with the features denoted as "*b*" (31.7°) and "*c*" (51°) for the 14-ML Fe coverage. The in-plane angular rotation between the *bcc*(110) and *fcc*(001) surface can be explained by including the sum of the four possible (110) domains leading to the same symmetry as the (001) surface.

Table 1 shows the results of fitting quantitative multiple scattering calculations to the XPD data for clean Ni(001) and Fe coverages of 3 and 7 ML. The calculations for 8-ML *fcc* Ni(001) give a "best fit" for a Ni lattice constant of 1.75 Å and lattice parameter of 2.50 Å, which agrees well with the known, bulk values of 1.76 and

Table 1 Best-Fit Structural Parameters for Ni(001), 3-ML Fe/Ni(001), and 7-ML Fe/Ni(001)

Experimental structural	Ni(001)	3-ML Fe/Ni(001)	7-ML Fe/Ni(001)
Best-fit structure	8-ML *fcc* ($R = 0.050$)	2-ML *bcc*/3-ML *fcc* ($R = 0.038$)	4-ML *bcc*/3-ML *fcc* ($R = 0.030$)
Best-fit	$d_{1-2,2-3} = 1.75 \pm 0.01$	$d_{1-2} = 2.05 \pm 0.06$	$d_{1-2,2-3,3-4} = 2.04 \pm 0.04$
Parameters (Å)	$a_{1,2,3} = 2.50 \pm 0.01$	$d_{2-3} = 2.01 \pm 0.03$	$d_{4-5} = 2.01 \pm 0.03$
		$d_{3-4,4-5} = 1.85 \pm 0.03$	$d_{5-6,6-7} = 1.85 \pm 0.03$
		$a = 2.49 \pm 0.02$	$a_{1,2,3,4} = 2.47 \pm 0.02$

Source: Ref. [35].

2.49 Å, respectively. Best fits for the 3-ML Fe coverage data indicate a structure of 2-ML *bcc*-ML *fcc* Fe. The spacing corresponds well with the expected spacing of 2.04 Å for the *bcc* phase and 1.84 Å for the tetragonally expanded *fcc* phase. The thickness indicated by the best fit in the calculation is greater than the nominal coverage (5 ML versus 3 ML), and this is consistent with an island-formation growth mode. Calculations for the 7-ML Fe film show a best fit for a 4-ML *bcc*/3-ML *fcc* Fe structure. The Fe lattice parameter was relatively unchanged in both Fe films. The consistently low *R*-factor values for all three structures show a very reliable goodness of the fit for the *fcc* to *bcc* phase transition above 3-ML Fe thickness on Ni(001).

c(2 × 2)–Mn/Ni(001)

MnNi and MnCu binary alloys are found to have ordered surface reconstructions. Theoretical arguments based on total energy calculations suggest that the magnetic properties of these surfaces are the reason for the existence of the unusual surface structures. Indeed, novel magnetic properties of these alloys have been found using X-ray absorption spectroscopy (XAS) and X-ray magnetic circular dichroism (XMCD), showing an enhanced magnetic moment of Mn of $4 \mu_B$.

LEED studies showed an outward corrugation or "buckling" of Mn atoms out of the surface. Photoelectron holography and quantitative XPD [36] was used to unambiguously determine the element-specific structure of these surface alloys. Photoelectron holography was used to get an initial estimate of the structure. Figures 14a and b compare a model structure with the 3-D perspective view of the holographic reconstruction. The reconstruction was performed using Barton's algorithm [Eq. (5)] on 14 full-hemisphere diffraction patterns equispaced in momentum space. In the model structure, the Mn emitter atom is at the origin and the neighboring atoms are all chosen to be Ni from the observed $c(2 \times 2)$ symmetry of the surface. The Mn emitter is located at the origin also in the holographic reconstruction. This determines the structures to be a surface alloy from the relative position of the holographic intensities along the substrate crystallographic directions. The difference in height between the top-layer holographic intensities from the origin is assigned to a large, outward buckling of Mn atoms.

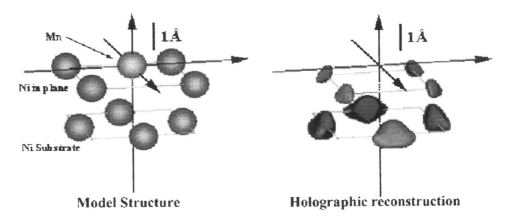

Model Structure   **Holographic reconstruction**

Figure 14   Comparison of a 3D projection for (a) a model structure and (b) a holographic reconstruction for the MnNi surface alloy produced using XPD data. The model structure is a substitutional alloy with Mn atoms buckled out of the surface plane by 0.4 Å. All of the nearest neighbors around Mn atoms can be seen in the holographic transformation, and these are located close to the position of atoms expected from the model structure. A large buckling (0.5 Å) for Mn is also observed in the holographic reconstruction.

To get better quantitative accuracy, simulations were then performed using multiple-energy diffraction patterns and calculating R-factors over four different kinetic energies. The kinetic energies selected were 60, 66, 80, and 94 eV, where the backscattering intensities were strong and the experimental patterns varied considerably. Comparisons between experiment and theory for the Mn diffraction patterns are shown in Figure 15. The simulations reproduce the main features in each of the distinct diffraction patterns.

A two-dimensional R-factor is shown in Figure 16 with the top-layer Mn-height d_{Mn} varied from 1.6 Å to 2.4 Å, while the Ni height d_{Ni} is varied from 1.4 Å to 2.2 Å. The R-factor is plotted in a reverse intensity scale where the maximum intensity corresponds to the minimum in R-factor. We observe the presence of a global minimum of the R-factor corresponding to $d_{Mn} = 2.1$ Å and $d_{Ni} = 1.7$ Å. This shows the structure where the Mn atoms are buckled out of the surface by 0.4 Å with respect to the top-layer Ni atoms. This is also accompanied be an inward buckling of the Ni atoms by 0.06 Å.

Buckling in Bimetallic Alloys of Pt Determined by ALISS

Bimetallic Pt–Sn catalysts are useful commercially, e.g., for hydrocarbon conversion reactions. In many catalysts, Pt–Sn alloys are formed and play an important role in the catalysis. This is particularly true in recent reports of highly selective oxidative dehydrogenation of alkanes [37]. In addition, Pt–Sn alloys have been investigated as electrocatalysts for fuel cells and may have applications as gas sensors. Characterization of the composition and geometric structure of single-crystal Pt–Sn alloy surfaces is important for developing improved correlations of structure with activity and/or selectivity of Pt–Sn catalysts and electrocatalysts.

While bulk, single-crystal samples of alloys or intermetallic compounds can sometimes be obtained, another approach is to anneal films prepared by depositing

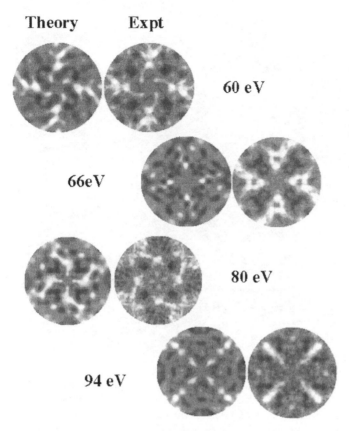

Theory **Expt**

60 eV

66eV

80 eV

94 eV

Figure 15 Comparison between theoretical (left) and experimental (right) results for diffraction patterns obtained at four different kinetic energies: 60, 66, 80, and 94 eV. The diffraction patterns are from Mn emitters in a $c(2 \times 2)$ MnNi surface alloy. The model structure used for the theoretical results was a substitutional alloy with the Mn atoms buckled out of the surface layer.

one metal on a second metal, single-crystal substrate. It is particularly important to know whether the deposited metal forms an adlayer or is incorporated into the surface layers to form an alloy structure. The $c(2 \times 2)$–Sn/Pt(100) structure provides a good example of this situation, as shown in Figures 17a and b. Ordered Sn adlayers and intermixed, alloy-surface layer models can both account for this structure. Using most surface science techniques, it is quite difficult to distinguish between such structures. However, the large difference in the critical angles for Sn scattering in ALISS polar scans for the two structures can immediately determine the actual structure. At low-incidence angles, Sn ad-atoms are shadowed by other Sn adatoms that are located at relatively large distances compared to the situation for incorporated Sn atoms, which are shadowed by closer-in, neighboring Pt atoms. Figure 17c shows that incorporation of Sn into the surface layer increases ψ_c by about 6°.

Sn has been shown to form ordered $(\sqrt{3} \times \sqrt{3})$ R30° surface alloys at the (111) surfaces of several late-transition metals upon annealing. Because of the lattice

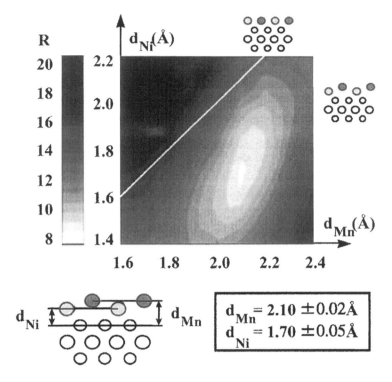

Figure 16 Two-dimensional, R-factor calculation for variations in the height of Mn and Ni atoms above the first subsurface layer in a $c(2 \times 2)$ MnNi surface alloy. The R-factor was calculated as the square of the difference between experiment and theory, summed over all points in the diffraction patterns that were obtained at four different energies. The bar on left shows the intensity scaling of the R-factor, and the plot shows that a global minimum is present.

mismatch between Sn and substrate atoms (the atomic diameters of Sn, Ni, Cu, Rh, and Pt are 2.81 Å, 2.49 Å, 2.56 Å, 2.69 Å, and 2.77 Å, respectively), Sn atoms "buckle," i.e., are displaced outward from the surface plane, to relieve the strain. ALISS has been used to determine the geometric location of Sn atoms in these bimetallic alloys for the Sn/Ni(111) [38] Sn/Cu(111) [38] Sn/Rh(111) [39] Sn/Pt(111) [38]. Figure 18 shows that a linear relationship exists between the buckling of the Sn atoms and the lattice mismatch with the substrate atoms.

Oxide Surfaces

Although oxides are of increasing industrial importance for a large number of applications, structural studies on oxide surfaces are still relatively rare compared to those on metals and semiconductors. Oxides have a variety of chemical compositions and structures, and a wide range of properties. For instance, perovskites range from insulating to superconducting, range from transparent to opaque, and exhibit dielectric constants between 30 to 30,000 within a relatively small variation of composition. In chemical applications, oxides are used as gas sensors or support

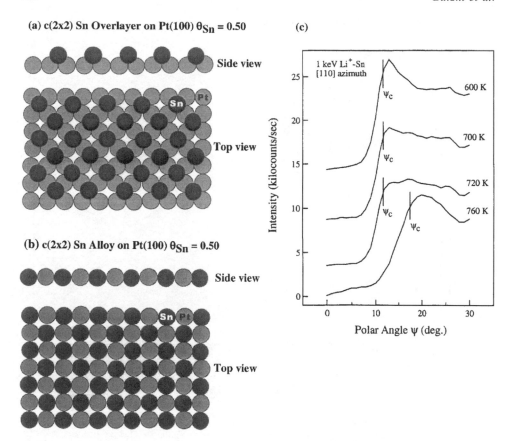

(a) c(2x2) Sn Overlayer on Pt(100) θ_Sn = 0.50

Side view

Top view

Sn   Pt

(b) c(2x2) Sn Alloy on Pt(100) θ_Sn = 0.50

Side view

Top view

Sn Pt

(c)

1 keV Li$^+$-Sn
[110] azimuth

600 K
700 K
720 K
760 K

ψ_c

Intensity (kilocounts/sec)

Polar Angle ψ (deg.)

Figure 17   Two possible structures for the $c(2 \times 2)$ Sn–Pt(100) surface. (a) Overlayer model with the Sn atoms located above the Pt(111) surface plane in threefold hollow sites. (b) Surface alloy model with the Sn atoms replacing every second Pt atom in the surface plane. (c) ALISS is ideally suited to distinguish between these two structures with high accuracy, as indicated by the shift in the critical angle for Sn-scattering upon alloying between 720–760 K. (From Ref. 73.)

materials for nanodispersed catalysts. XPD and LEIS can be used to study oxide surfaces, even those that are insulating, and problems due to charging at insulating surfaces can be overcome. These are powerful structural tools because XPD and LEIS determine the element-specific real space structures, which break down the structure of complex oxide surfaces into simpler building blocks.

Anatase TiO$_2$ Surface Using XPD

Rutile and anatase are two phases of TiO$_2$. High-quality rutile single crystals can be found in nature as it is the more stable polymorph. Naturally occurring anatase minerals usually contain impurities. Single-crystal anatase phase with few impurities can be grown on SrTiO$_3$(001) single crystals due to its close lattice match with the anatase phase. XPD has been used to characterize the anatase phase grown on SrTiO$_3$(001) and check the ordering of O and Ti in the crystal [40].

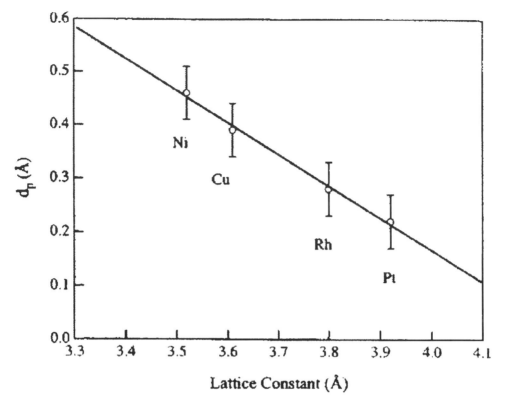

Figure 18 A linear correlation exists between the Sn-buckling distance d_p and the lattice constants of late-transition metal, $fcc(111)$ substrates. (From Refs. 38 and 39.)

Figures 19a and b show the experimental XPD patterns of O $1s$ and Ti $2p$ core levels from anatase TiO_2 grown on an $SrTiO_3(001)$ single crystal. Figures 19c and d and Figures 19e and f show the corresponding single-scattering cluster (SSC) simulations for the anatase and rutile phase, respectively. We can see that the experimental results match very closely to the anatase phase, whereas it is quite different from the rutile phase. Hence, we can conclude that the TiO_2 phase grown on $SrTiO_3$ is a high-quality anatase single crystal.

Rutile TiO_2 Surface Using ALISS

Studying the titanium dioxide surfaces are of fundamental importance in understanding heterogeneous catalysis. High-quality rutile phase TiO_2 are easy to obtain and prepare. The $TiO_2(110)$ surface has been studied extensively, as it has the highest thermodynamic stability, but widely different structural models have been proposed using experiment and theory.

Recently ALISS experiments and simple classical theory have been used to directly get the surface structure of $TiO_2(110)$ [41]. Figure 20 shows an unrelaxed stoichiometric TiO_2 surface with bridging oxygen rows.

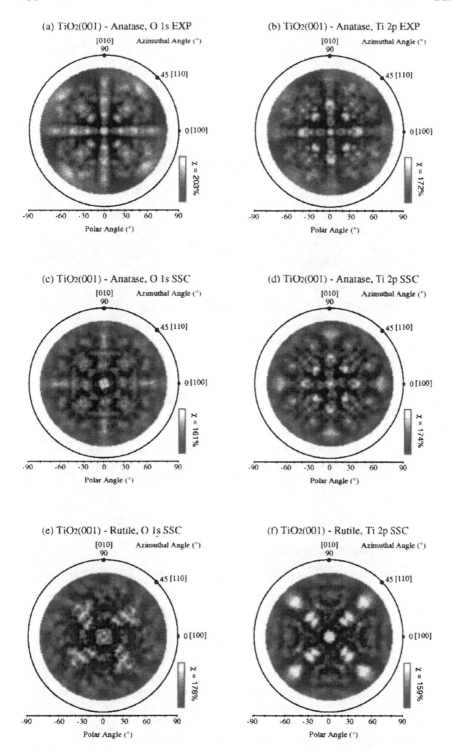

Figure 19 Experimental and theoretical results for XPD patterns from TiO$_2$ surfaces. (From Ref. 40.)

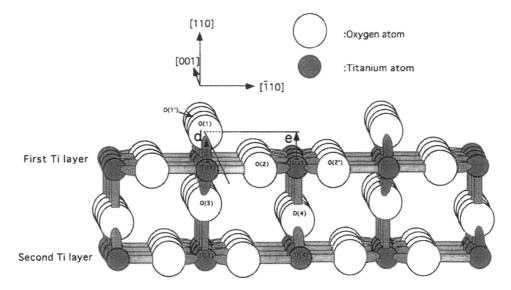

Figure 20 Unrelaxed structure of the TiO$_2$(110) surface with bridging oxygen rows. (From Ref. 41.)

Figure 21 shows the Li$^+$ ALISS polar scans obtained along the [001] and [$^-$110] azimuths of the TiO$_2$(110)–p(1 × 1) surface and the corresponding theoretical simulations. Peaks I through **VI** have been reproduced in the theoretical simulations, and the difference with bulk structure is illustrated in Figure 21b. The structure determined from the critical angles shows that the bridging oxygens are located at 1.2 ± 0.1 Å above the sixfold titanium atoms at the surface. A large relaxation of about −18 ± 4% (−0.6 ± 0.1 Å) was observed between the first and second titanium layers.

4.3 SCANNING PROBE MICROSCOPY

4.3.1 Introduction

The invention of the scanning tunneling microscope (STM) by Binnig and Rohrer at the IBM Research Laboratory in Zürich in 1981 has revolutionized the science of surface imaging and characterization. STM laid the basis for numerous scanning probe microscopy (SPM) techniques now successfully employed in many surface-science experiments. Today modern surface science is difficult to imagine without SPM techniques to provide structural information that is either complimentary to other techniques or unique observations of local defect structures on an atomic scale. In this section we briefly review the physical principles of some of the most important SPM techniques. Then we discuss the use of STM. We describe results obtained for single-crystal surface characterization on metal and metal oxide surfaces under UHV conditions in some detail. The reader interested in the enormous number of STM studies on semiconductor surfaces is referred elsewhere [42]. Although most single-

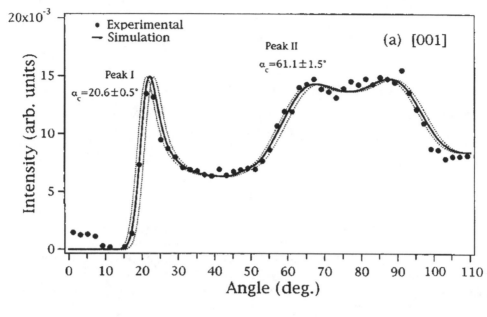

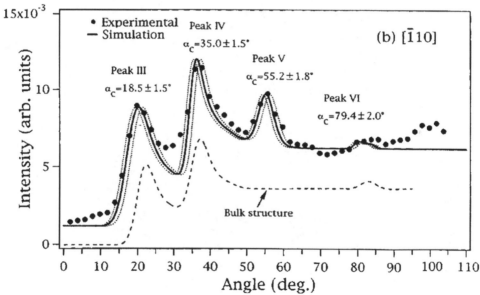

Figure 21 ALISS polar scans of a TiO$_2$(110)–p(1 × 1) surface taken along the (a) [001] direction and (b) [−110] direction using 1-keV Li$^+$ ions backscattered at 160°. The solid circles are experimental data points and the solid curve is the result of theoretical calculations. Dotted lines demonstrate simulations for an error in critical angle by ±1.0°. The dashed line shows the simulations for the bulk structure shown in Figure 20. (From Ref. 41.)

crystal characterization is done under UHV conditions, SPM techniques do not rely on those conditions for their operation. Thus, we also describe results obtained on surface structures in other controlled environments, in particular under high-gas-pressure conditions and in electrolyte solutions.

4.3.2 Scanning Probe Microscopy (SPM)

All SPM techniques have in common that a sharp probe is raster-scanned across the surface, utilizing piezoelectric transducers to control the position of the probe relative to the surface with sub-Ångstrøm precision. The various SPM techniques exploit different interactions between the probe and the surface to obtain locally resolved information about the surface. This information can be presented as a two-dimensional map (image) of the properties of the probed surface. Depending on the technique, various physical and chemical properties of the surface can be imaged. To illustrate the versatility of scanning probe microscopes, we briefly describe three commonly used SPM techniques utilizing different probe-surface interactions to obtain surface-sensitive information: STM, atomic force microscopy (AFM), and near-field scanning optical microscopy (NSOM). There are numerous variations of these techniques and other techniques utilizing different physical phenomena to image the surface [43–50].

Scanning Tunneling Microscopy (STM)

The operation of an STM and the resulting resolution are very simple conceptually. A schematic diagram illustrating this is shown in Figure 22. An atomically sharp metal tip (commonly W, Pt, PtIr, or Au) is brought in close proximity to a conducting sample, only separated by a few Ångstroms. A bias voltage applied

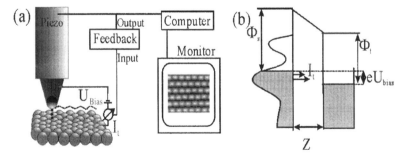

Figure 22 (a) Schematic diagram of a scanning tunneling microscope. An applied bias voltage between the tip and the surface causes a tunneling current to flow, which can be measured and used as an input signal for a feedback loop. During raster-scanning the tip across the surface, the tunneling current is kept constant by changing the z-position of the tip by applying a voltage (feedback output) to a piezoelectric transducer. An image of the surface is generated by monitoring the feedback signal at different positions of the tip. (b) Tunneling mechanism under the influence of an applied bias voltage between tip and sample. Electrons tunnel from occupied states in the tip to empty states in the sample. The tunneling barrier is defined by the separation between tip and sample and workfunctions of the tip Φ_t and sample Φ_s.

between the sample and the tip can cause electrons to tunnel through this gap, which acts as a barrier, from occupied electronic states in the tip to empty states in the sample, or vice versa, depending on the polarity of the applied bias voltage. This tunneling current depends exponentially on the width of the barrier (i.e., the distance between the tip and the surface) and directly on the electronic structure of the surface. Thus the change in the tunneling current caused by scanning the tip across the surface gives an image that is a convolution of the topography and the electronic structure of the surface. In most STM experiments the tunneling current is kept constant by means of a feedback loop. A voltage applied to a piezoelectric transducer adjusts the tip–sample separation to maintain a constant, preset tunneling current.

A more detailed description of the tunneling current between the tip and the surface can be derived from Bardeen's tunneling-current formalism [51] and expressed as

$$I = \int_0^{eV} \rho_s(r, E)\rho_i(r, E - eV)T(r, E, eV)dE \tag{13}$$

where $\rho_s(r, E)$ and $\rho_t(r, E)$ are the electronic density of states of the sample and the tip at location r and energy E, respectively, V is the applied bias voltage, and T is the tunneling transmission probability. This is given by

$$T(E, eV) = \exp\left(-\frac{2z\sqrt{2m_e}}{\hbar}\sqrt{\frac{\varphi_s + \varphi_t}{2} + \frac{eV}{2} - E}\right) \tag{14}$$

where φ_s and φ_t are the workfunctions of the sample and tip, respectively, and z is the tip–sample separation. Thus in "constant-current" mode, the STM tip follows a complex contour line dictated by the surface density of states and the transmission probability, which critically depends on the workfunction of the sample.

Each atom is sensed locally in STM, and in contrast to most surface-science techniques, data are not generated by an average over ensembles of many atoms. The electrons involved in STM have energies of a few electron volts, often smaller than chemical bond energies, and this allows nondestructive, atomic-resolution imaging.

Atomic-Force Microscopy (AFM)

STM relies on a conducting sample for its operation. This restriction inspired the invention of a new scanning probe microscope, the atomic-force microscope (AFM). In contrast to the STM, which senses tunneling current, the AFM probes the force between the tip and sample. To sense the force over a small area, a sharp tip with a radius of curvature of a few nanometers is mounted at the end of a fine cantilever micromachined out of silicon. These tips are usually etched from Si or SiN_3. Carbon nanotubes have been either attached or grown at the end of a tip, in order to create tips with even-higher aspect ratios, and various tips have been functionalized to add chemical sensitivity to the AFM. Bending of the cantilever is proportional to the applied force (Hook's law) and can be monitored and used as a feedback signal to keep the force on the tip constant.

The AFM can be utilized in various operational modes, with the most popular ones being contact, noncontact, and tapping modes.

In contact mode the tip is "touching" the surface and is hindered from penetrating the surface by repulsive forces. These repulsive forces are mainly Coulomb repulsion due to incomplete shielding of nuclear charges of atoms in the tip and the surface and the Pauli exclusion principle for the electrons. Scanning is a dynamic process, and because the tip is in mechanical contact with the sample, frictional or lateral forces can act on the tip in addition to the surface-normal forces. These lateral forces result in a twisting of the cantilever. By separating the twisting and bending of the cantilever one can simultaneously obtain information about normal and lateral forces (lateral force microscopy).

In noncontact mode, attractive forces or force gradients between the tip and surface are sensed. These attractive forces mainly arise from van der Waals interactions between the tip and the surface. Since these interactions ($\sim 10^{-11}$ N) are smaller than the forces encountered in contact-mode AFM, they must be detected by some resonance-enhancement technique. This is commonly done by vibrating the cantilever just above its mechanical resonance frequency. Force gradients encountered by the tip in close proximity to the surface shift the resonance frequency of the cantilever. This causes in a change in the amplitude and/or a change in the time lag (phase shift) between the driving oscillation and vibration of the cantilever. These shifts can be monitored and used as feedback signals to keep the tip–sample separation constant. Thus an image of the topography of the surface can be obtained. Samples with ferromagnetic or electrostatic components can be probed by special tips that are sensitive to those interactions and used to image magnetic domains or electrostatic charges. Because the interaction forces in noncontact-mode AFM are considerable smaller than for contact-mode operation, the risks of "accidentally" manipulating the surface by the measurement are reduced. This allows for imaging soft (e.g., biological) materials by using noncontact AFM that would be otherwise altered by contact-mode AFM.

Some operational limitations of noncontact-mode AFM, in particular trapping of the tip in a water layer that is omnipresent on samples under ambient conditions, can be overcome by utilizing a tapping mode of operation that retains a low impact rate of the tip on the surface. In this mode the tip is vibrated at higher amplitude than in noncontact-mode AFM. The tip makes contact with (i.e., taps on) the sample, and the change in amplitude of the oscillating cantilever due to this contact is measured. The force exerted by the tip on the sample can be very small because small changes in the amplitude can be measured. Furthermore, lateral (shear) forces on the sample are virtually zero. This allows delicate samples to be examined without altering their surfaces.

Near-Field Scanning Optical Microscopy (NSOM)

The resolution in conventional optical microscopy is limited to ($\lambda/2$) by the diffraction limit. This limitation can be avoided by using near-field microscopy. A small light source brought close to the sample (<10 nm) interacts with the sample in the "near field" of the light. Imaging the surface is accomplished by measuring the interaction of the light with the surface, by reflection, refraction, or by absorption and fluorescence mechanisms, and by detection of the light in the far field.

The light source in this subwavelength optical probe is usually a nanofabri-cated optical fiber or a metal-coated micropipette. Raster-scanning of the probe in close proximity to the surface is usually controlled by combining the NSOM with an AFM feedback loop. This means that the contact of the probe with the sample is controlled in the same way as in an AFM, but the image is generated from the detected light.

4.3.3   UHV Experiments with STM

As mentioned before, the bulk of surface-science studies have been conducted under UHV conditions. Most atomic-scale STM studies have been performed also in UHV, even though STM itself does not require vacuum conditions, because it is necessary to prepare and keep clean a well-defined surface (often highly reactive) in order to obtain meaningful, reproducible results. Also, STM is used under UHV conditions so that complementary electron- or ion-based analytical techniques can be used to characterize the same surface.

In this section we discuss a few illustrative examples of the use of STM in studies of single-crystal surfaces. Given the large and increasing number of STM studies in recent years we have made no attempt at completeness. The contributions and limitations of STM for the characterization of different single-crystal surfaces are demonstrated. The surfaces discussed are low Miller-index, clean metal surfaces, two-component metal alloy surfaces, metal surfaces that contain controlled amounts of adsorbates, ultrathin films that have been epitaxially grown on single-crystal substrates, oxide surfaces, and finally nanoclusters on single-crystal supports.

STM of One-Component, Single-Crystal Metal Surfaces

As discussed above, STM is primarily sensitive to the electronic structure of surfaces. Thus, on an atomic scale, protrusions may not necessarily correspond to the positions of atomic nuclei, even if the measured periodicity corresponds to the anticipated atomic lattice. However, in contrast to semiconductor surfaces, studies on clean metal surfaces have shown that this caution can generally be ignored and protrusions can be assigned to atom-nuclei positions. Furthermore, while large corrugations are observed on semiconductor surfaces due to the presence of dangling bonds, the atomic corrugation on metal surfaces is smaller by a factor of 50 to 100 than on semiconductor surfaces. Thus a higher resolution is needed for imaging metal surfaces with atomic corrugation compared to semiconductors.

An unknown factor in the imaging of surfaces is usually the state of the tip. It is generally believed that a single-atom tip, i.e., a tip with a single atom protruding farther than all the others, is required in order to achieve a well-resolved, atomic corrugation image of the surface. Since no reliable, reproducible way exists to product such a tip and the shape and even chemical state of the tip frequently change during the acquisition of an STM image, the resolution and even the measured electronic structure of the "surface" can change with the tip. Even though this appears at first to be a great disadvantage, different tip geometries may allow additional fitting parameters for calculated, theoretical STM images to compare to experimental results. Usually only a few feasible tip geometries have to be considered for the different appearances of the experimental STM images. Thus, the change of the appearance of the surface with different tip geometries allows a more accurate picture of the surface to be obtained. Such procedures have been successfully applied for more complex surfaces where the electronic structure of the surface did not allow the simple assignment of protrusions to atomic sites [52].

Surface Reconstructions

"Broken" bonds at the surface give surface atoms a lower coordination number compared to bulk atoms, and this causes the position of surface atoms to be altered. In some cases this results in a relaxation of the interatomic spacing between the atoms in the surface layer and the second layer compared to the bulk lattice spacing. In other cases the surface reconstructs, i.e., surface atoms occupy completely different positions, giving rise to a larger surface unit cell than that for an ideal termination of the bulk at the surface. In principle, STM can be used to determine the atomic arrangement of the surface layer. However, although STM has been successful in elucidating surface structures of reconstructed surfaces and discriminating between different surface structures proposed by using other experimental techniques and theoretical calculations, STM has rarely been used alone to identify the structure of a reconstructed surface. Other techniques (e.g., I-V LEED, XPD, and LEIS) still have the edge over STM for quantitatively determining exact atomic positions within the surface layer. Nevertheless, "real-space" images of surface reconstructions provided by STM uniquely provide information on domain sizes, defects, and interactions of reconstructions with other surface irregularities, e.g., step edges.

Au(110) Surface Reconstruction

Examples of simple reconstructions with a small unit cell are the "missing-row" (2×1) reconstructions of Pt, Ir, and Au(110) surfaces. The Au(110) reconstruction was one of the first structures studied by STM. The STM results confirmed the (2×1) reconstruction determined earlier by other techniques and allowed further studies of domain sizes and the order–disorder phase transition that LEED found to occur at $\sim 700 \, \mathrm{K}$. Annealing the sample within $10 \, \mathrm{K}$ of this transition temperature and then suddenly quenching the sample caused the half-order LEED spots to almost disappear. STM revealed that (2×1) domains still existed, but with small domain sizes of the order of 2–4 nm, far less than the usual coherence length required for sharp spots in electron diffraction. Furthermore, STM showed that the domains were separated by (1×3) and (1×4) missing-row structures, in addition to steps. This illustrates how STM can provide important information about "local" atomic structure, even when the ordered domains are too small (smaller than the coherence length) to provide a signal in diffraction techniques.

Pt(100)–Hex Reconstruction

More complex reconstructions are formed on Ir, Au, and Pt(100) surfaces. These surfaces form quasi-hexagonal overlayers on the square substrate lattice. This results in a large surface-unit cell. On the Pt(100) surface, two reconstructions are known to exist. There is a metastable Pt(100)-hex reconstruction that forms by annealing the crystal to $\sim 1000 \, \mathrm{K}$, and a stable reconstruction that is rotated by $0.7°$, i.e., the Pt(100)–hexR$0.7°$, that forms upon annealing to above $1200 \, \mathrm{K}$. The exact structure of the surface-unit cell of the reconstruction is still disputed. LEED studies suggest a unit cell of $\left| \begin{smallmatrix} 14 & 1 \\ 1 & 5 \end{smallmatrix} \right|$, which would give a fivefold symmetry in approximately the [011] direction. However, more sensitive He-scattering experiments show a splitting of the one-fifth-order diffraction peak and thus indicates a much larger until cell of $\left\lfloor \begin{smallmatrix} 13 & 1 \\ 18 & 43 \end{smallmatrix} \right\rfloor$. High-resolution STM images exhibit a clear modulation of the surface atoms with a periodicity of 30 atoms in $\sim[01\bar{1}]$ and 6 atoms in the $\sim[011]$ direction, as shown in Figure 23. This corresponds to 29 and 5 atoms on the (1×1) lattice, and thus a (29×5) until cell. However, Figure 23 shows that there is also a less well-pronounced, long-range modulation visible along the [011] direction. While rows 3 and 5 appear equally bright in cell A in Figure 23, the same rows of the "unit" cell in cell B are different in appearance. This implies that the surface is not described by a (29×5) unit cell but by a very large unit cell in the [011] direction, or it may even be incommensurate in this direction. A thorough inspection of the modulation in such STM images suggests that the whole period of the unit cell consists of 26 (29×5)-cells in the [011] direction, i.e., 156 surface atoms or 129 substrate atoms. This result agrees closely with the He-scattering data.

Standing Electron Waves

Apart from the determination of topography and atomic structure of surfaces, the STM also has a unique capability to image the local electronic structure of surfaces. Delocalized surface states of electrons, so called Shockley states, behave very much like a two-dimensional free-electron gas. These electrons are scattered at step edges and point defects at the surface. Reflected electron waves from step edges can interfere with incident waves, leading to an oscillation in the local density of states. It is possible to image these surface states by STM [53]. The formation of standing waves by the scattering of surface states at defect sites was first observed on a

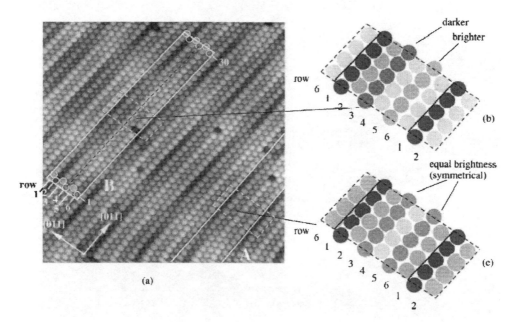

Figure 23 High-resolution STM image of a reconstructed Pt(100) surface. A (29×5) cell with (30×6) surface atoms is indicated. The differences between cells A and B are highlighted in the schematic drawings at the right. These inequalities between the (29×5) cells indicate a larger unit cell for the reconstruction. (From Ref. 62.)

Cu(111) surface at low temperatures. The constant-current images clearly show standing electron waves at step edges and around point defects in Figure 24. It was found that the oscillation wavelength is energy-dependent. The wavelength increases as the energy is lowered with respect to the Fermi energy. Similar surface-state oscillations have been observed on other metal surfaces such as Au(111) and Ag(111) that exhibit Shockley states.

Adsorbate-Induced Surface Restructuring

STM can be used to study this restructuring on an atom-by-atom basis, but it also allows studies of the dynamics of this process, i.e., the nucleation and growth of reconstructed domains and the mass transport that is involved.

Oxygen on Cu(110)

The oxygen-induced reconstruction of a Cu(110) surface has been studied by recording consecutive STM images of the same surface area during oxygen exposure. Some of these images are reproduced in Figure 25. It was found that the nucleation and growth of the (2×1) reconstruction proceed via the release of Cu atoms from step edges to combine with adsorbed oxygen atoms, forming Cu–O "added rows" on top of terraces running in the $\langle 001 \rangle$ directions [54]. These added rows tend to

Figure 24 Constant-current STM image of a Cu(111) surface measured at 4 K ($V_t = 0.1$ V and $I_t = 1.0$ nA). Spatial oscillations with a periodicity of 15 Å are clearly emanating from monatomic step edges and point defects. (From Ref. 53.)

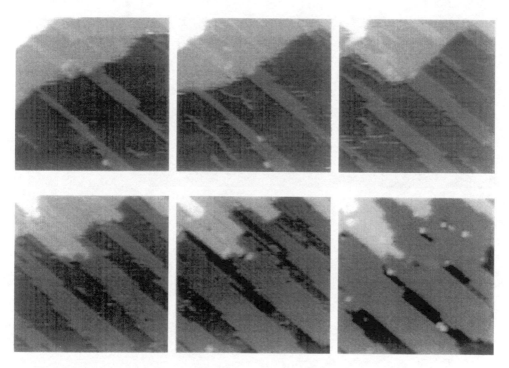

Figure 25 Series of STM images recorded while a Cu(110) surface was exposed to an oxygen (O$_2$) background pressure of $\sim 10^{-8}$ Torr. The imaged area is 235 × 256 Å^2. Cu atoms are removed from step edges and the added Cu–O– rows nucleate and grow on the terraces along the ⟨001⟩ direction. (From Ref. 54.)

agglomerate with increasing oxygen exposure to form (2 × 1) islands. Once most of the step edges of the Cu surface are pinned by (2 × 1) islands, another reaction channel becomes competitive. Cu atoms also can be expelled from flat terraces to form rectangular troughs of missing first-layer Cu atoms. If the amount of oxygen exceeds the saturation coverage for the (2 × 1) reconstruction (0.5 ML), a more complex reconstruction evolves and a c(6 × 2) structure is formed. STM images reveal that this structure preferentially nucleates and grows from step edges and appears to form on top of the coexisting (2 × 1) structure. STM shows protrusions situated in short bridge sites of the underlying Cu–(1 × 1) lattice forming the c(6 × 2) superstructure. These protrusions are associated with Cu atoms sitting on top of the (2 × 1) Cu–O rows. A structural model for the c(6 × 2) surface is depicted in Figure 26.

Lifting of the Pt (100)–Hex Reconstruction by CO Adsorption

Adsorption of various molecules (CO, NO, O$_2$, and C$_2$H$_4$) onto the reconstructed Pt(100) surface causes a lifting of the surface reconstruction and the formation of a (1 × 1) Pt surface covered by an adsorbed layer. The driving force for lifting the reconstruction is the higher heat of adsorption of the molecules on an

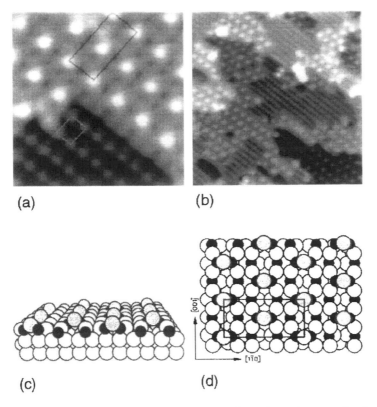

Figure 26 STM images of Cu(110) showing (a) coexistence of a $(2 \times 1)O$ structure with a $c(6 \times 2)O$ structure $(5.1 \times 5.0 \, nm^2$ scan), and (b) heterogenous nucleation of a $c(6 \times 2)$ structure at step edges and its subsequent anisotropic growth $(15.7 \times 17.1 \, nm^2$ scan). Panels (c) and (d) show perspective and top-view atomic models for the $c(6 \times 2)$ structure, respectively. (From Ref. 54.)

unreconstructed surface layer compared to that on a quasi-hexagonal, reconstructed overlayer.

The hexagonal, platinum top layer of the reconstructed surface is more densely packed than the square (1×1) Pt(100) lattice. As a consequence, lifting the reconstruction causes platinum atoms to be expelled from the surface layer. These platinum adatoms rapidly nucleate to form Pt islands at 300 K. Lifting of the reconstruction and formation of Pt islands can be directly observed by STM at room temperature. Figure 27 shows a series of STM images that were taken as the Pt (100)–hex surface was exposed to CO gas. With increased CO adsorption, the areas covered with Pt islands increase until the reconstruction has been removed from the whole surface. The nucleation and growth of the Pt islands appear to be rather anisotropic. Once islands have formed, these regions grow rather than nucleating new areas. The islands have an elongated shape in the long-periodicity direction of

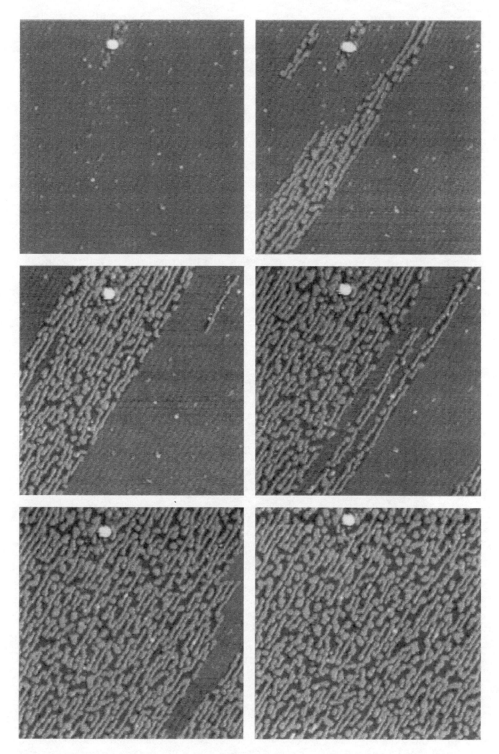

Figure 27 Series of STM images of a Pt(100) surface at 300 K exposed to CO gas. Lifting of the hex-reconstruction of the clean surface and formation of Pt ad-islands can be observed.

the hex-reconstruction, indicating that adatom diffusion and/or lifting of the reconstruction proceeds preferentially along this direction. Annealing such a surface to ~450 K causes the islands to assume a compact, square shape (Figure 28), as one would expect for an equilibrium island shape. The absorbed CO molecules on this surface can also be imaged by STM. The CO molecules form $c(2 \times 2)$ domains (Figure 28), both on top of the islands as well as on the region between the islands. The $c(2 \times 2)$ structure is also observed by LEED.

Metal Alloy Surfaces

In order for STM to be a useful tool to characterize surfaces with multiple (here we limit ourselves to two) components, it has to be able to discriminate between the different elements present at the surface. Such a sensitivity of the STM to different chemical elements is not obvious a priori and cannot be expected generally. Such a discrimination, however, has been observed for many systems. The mechanism for such "chemical contrast," a discrimination between two different elements, is not yet completely understood. In a few cases the state of the tip appears to be critical, and it was proposed that "trapping" of an adsorbate atom at the end of the tip may be critical. This tip-adsorbate atom may then form chemical bonds preferentially with one element in the surface. Such a precursor of a chemical bond may increase the

Figure 28 Annealing the surface probed in Figure 27 to 450 K causes the Pt ad-islands to assume a compact, square shape. Adsorbed CO forms a $c(2 \times 2)$ adlayer on top of the unreconstructed Pt(100) islands and surface regions between the islands. (Inset $10 \times 9.3 \, \text{nm}^2$.)

local density of states (LDOS) between the adsorbate and the surface and/or move the adsorbate closer to the surface. Both mechanisms result in higher tunneling probability and thus increased contrast. In such a case, frequent "tip changes" are usually observed that cause alterations of the chemical contrast as atoms are picked up and dropped by the tip. A more controlled contrast has been observed on surfaces that exhibit large variations in the LDOS between the two elements in the surface. This is the case for PtRh alloys, for instance. In this case, the density of states near the Fermi level is significantly larger above Rh than above Pt atoms. This results in a larger corrugation of the Rh compared to Pt atoms in the STM images.

    Chemical contrast on alloy surfaces is a direct way of determining the composition of the surface layer. For example, the image shown in Figure 29 of a

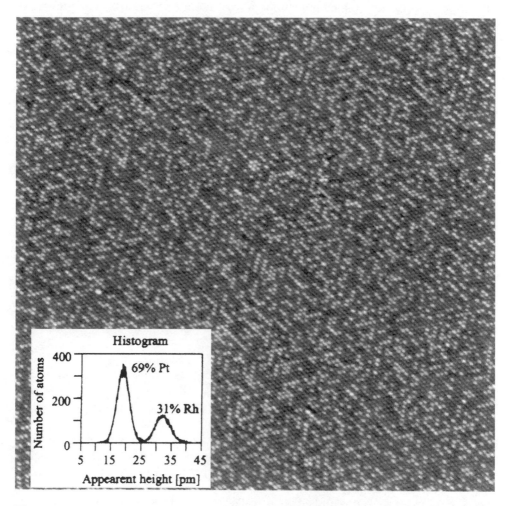

Figure 29   STM image of a $Pt_{25}Rh_{75}$ (111) surface showing chemical contrast ($30 \times 30 \, nm^2$). A histogram of the centers of the atoms results in two separate peaks and the concentration of the components can be determined by a Gaussian fit. (From Ref. 55.)

$Pt_{75}Rh_{25}(111)$ crystal surface shows 69% Pt and 31% Rh atoms in the surface layer [55]. Furthermore, information on the short-range order of the atoms can be extracted from atomically resolved images. From Figure 29, a small but significant increase in the number of hetero-neighbor atoms compared to a random distribution was deduced.

STM images of alloyed surfaces may also show a contrast difference between the same element with different numbers of hetero-neighbor atoms due to the effects of alloying, like charge transfer between atoms and rehybridization. This arises from changes in the LDOS around the Fermi level. Such an effect was observed for Au atoms alloyed in Ni(111) (Figure 30, top) and also for Sn/Pt(111) surface alloys (Figure 30, middle). Two ordered surface alloys can be formed for Sn alloyed into the top layers of a Pt(111) crystal, i.e., the (2×2) and the $(\sqrt{3} \times \sqrt{3})R30°$ structures. The two structures have two and three Sn-neighbor atoms per Pt atom, respectively. On a surface that exhibited domains of both surface structures, it was found that Pt ensembles with fewer Sn-neighbor atoms per Pt atom appeared brighter than Pt ensembles with more Sn-neighbor atoms (Figure 30, middle and bottom). This effect was explained by a depletion of the DOS of the Pt atoms near the Fermi level due the effect of alloying with Sn.

Metal-on-Metal Epitaxy

STM has greatly contributed to obtaining a better understanding of the nucleation and growth of metals deposited on metal substrates. Traditionally three growth modes are distinguished: 2D layer-by-layer growth (Frank–van der Merwe), 3-D island growth (Volmer–Weber), and 3-D island growth on top of a 2D wetting layer (Stranski–Krastanov). These growth modes are only valid in the limit of thermodynamic equilibrium. Depending on the temperature and concentration (e.g., supersaturation) of metal adatoms at the surface, growth modes deviate from the thermodynamic equilibrium and kinetic effects play an important role. The dynamic range of the STM to image surfaces from an atomic scale to micron-sized areas allows direct information to be gathered on atomic processes governing the growth kinetics, such as the size of the critical nucleus, motion of individual adatoms and clusters, and shape and branch thickness of individual dendritic islands. Information on island densities, morphological defects, and step densities can be obtained as well.

Diffusion-Controlled Island Morphology—Ag on Pt(111)

Information on island morphologies can provide details about atomistic diffusion processes. For the growth of Ag on Pt(111), island growth proceeds via anisotropic branching after reaching a critical size. Two types of ramified island structures were observed as shown in Figures 31a and b. A fractal structure, i.e., a randomly ramified island, results when ad-atoms "hit and stick" to the growing island, and a dendritic structure, i.e., symmetrically branched islands, results if the growth rate is lowered. Anisotropic ad-atom diffusion around the perimeter of the islands has been identified as an important process for the formation of dendritic islands. There are two types of close-packed island edges, so-called "A" and "B" steps, on an *fcc*(111) surface. It is apparent from simple geometric reasoning that diffusion from a corner site to the A step can occur via an *hcp*-hollow site while diffusion to the B step has to

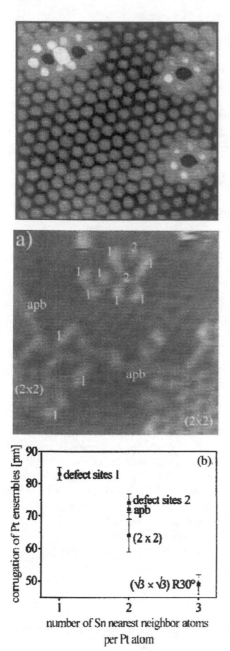

Figure 30 (top) Au atoms alloyed randomly into the Ni(111) surface appear as "holes." Ni atoms surrounding an alloyed Au atom appear brighter than Ni atoms without Au-neighbor atoms. This may be explained by a change in the electronic structure of the Ni atoms due to the effect of alloying. (From Ref. [63].) (middle and bottom) Mixed domain surface of the ordered (2 × 2) and ($\sqrt{3} \times \sqrt{3}$)R30° Sn/Pt(111) surface alloys. The brightness of Pt ensembles depends on the number of Sn-neighbor atoms. This is interpreted as increasing depletion of the Pt electronic states close to the Fermi level with increasing number of Sn neighbors. This results in a lower tunneling probability and thus a decreased contrast in STM. (From Ref. 64.)

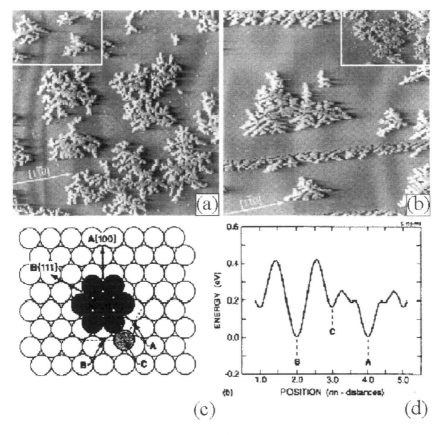

Figure 31 STM images ($120 \times 120\,\text{nm}^2$) showing (a) fractal (randomly ramified) and (b) dendritic Ag aggregates grown on Pt(111) at rates of $1.6 \times 10\text{–}5\,\text{ML/s}$ in (a) and $1.1 \times 10\text{–}3\,\text{ML/s}$ in (b). For both cases, the total Ag coverage was $\Theta = 0.12\,\text{ML}$. (From Ref. [65].) (c) Schematic diagram illustrating the A and B directions for the diffusion of an Ag ad-atom along the edges of an Ag heptamer on Pt(111). (d) Calculated total energy of an Ag ad-atom diffusing along the edges of an Ag heptamer. The diffusion path with the lowest barrier from a corner site (C) to the A step site is via the *hcp* site close to it. Diffusion to a B step site from the C-corner site is hindered by a larger diffusion barrier associated with diffusion over an atop site. (From Ref. 66.)

occur via an on-top site. This results in different diffusion barriers and thus anisotropic diffusion rates as illustrated in Figures 31c and d.

Surfactant-Influenced Growth—Homoepitaxial Growth of Pt

Ad-atom diffusion, in particular diffusion across step edges, can be influenced by the presence of "impurity" atoms on the surface, or so-called surfactants. The positive influence that surfactants can have on crystal growth has long been known and exploited. However, new microscopic insight was obtained from recent STM studies. Figure 32 shows how the presence of oxygen ad-atoms decisively influences the homoepitaxial growth of Pt on Pt(111) surfaces, producing a "flatter" surface.

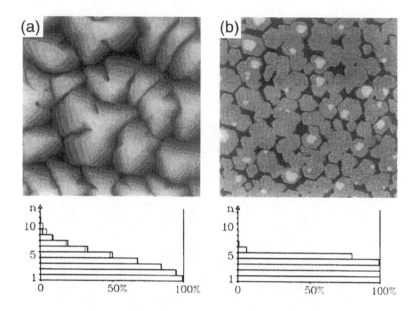

Figure 32 STM topographs of 5-ML Pt deposited on a (a) clean and (b) oxygen precovered Pt(111) surface at 400 K. The imaged area is (220 × 220 nm^2) for both images. The percentage of completion of the deposited layers versus the layer number n is plotted as histograms below each topograph. For the oxygen precovered surface, a completion of layers is favored before new layers nucleate, indicating a reduced interlayer diffusion barrier. (From Ref. 67.)

Oxygen reduces the diffusion barrier for interlayer diffusion of Pt and a more perfect, layer-by-layer growth is achieved.

Strain Relief due to Misfit Dislocations—Cu on Ru(0001)

Lattice mismatch between the adlayer and substrate material in heteroepitaxy can have pronounced effects on the morphology of the grown film. Pseudomorphic films, in which the adlayer adopts the substrate lattice spacing, can be significantly strained. There are several ways for this stress to be relieved. One way is to form islands instead of forming a continuous film. This can be done by forming either compact or elongated island structures or by forming islands on top of a strained wetting layer (Stranski–Krastanov growth). STM is an ideal tool for characterizing island density, shape, and size. Another strain-relief mechanism available for the film is the formation of dislocation networks. Ultrathin Cu films grown on Ru(0001) [56], for instance, grow pseudomorphically in the first monolayer, but form domains with *fcc* and hcp atom-stacking, similar to the Au(111) surface reconstruction, if a second layer forms in order to uniaxially relieve the stress in the film. These domains are separated by misfit dislocations where the Cu atoms occupy slightly higher, bridge sites. Since the domains are anisotropic, with the misfit dislocations running in ⟨120⟩ directions, the Cu atoms are still in registry with the substrate along this direction and consequently under tensile stress. This anisotropy is lifted for the third Cu layer, where the misfit dislocations are arranged in a pseudoisotropic triangular pattern. For a fourth Cu layer, the in-plane stress is relieved by rotation of the Cu overlayer with respect to the Ru(0001) substrate by $\sim 1°$, forming a Moiré pattern. The small differences in the topography in the dislocation lines or in the Moire pattern can be imaged by STM (Figure 33), and this allows the most direct determination of the surface morphology.

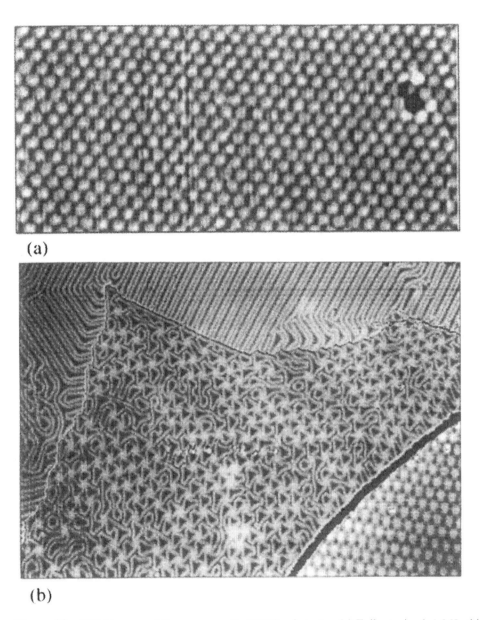

(a)

(b)

Figure 33   STM images of Cu grown on Ru(0001) substrates. (a) Fully strained, 1-ML-thick pseudomorphic Cu layer in registry with the Ru substrate. (b) Relaxation of the lattice strain for different layer thicknesses: 2 ML (top), 3 ML (middle), and 4 ML (bottom right). The film in (b) had a nominal film thickness of 3 ML. (From Ref. 56.)

Oxide Surfaces

For STM and many other surface-science techniques, conductive samples are needed. However, many oxides with perfect stoichiometry are insulators. This problem has been addressed by studying thin oxide films grown on metal single-crystal substrates. Epitaxial, thin oxide films can be grown for the right choice of metal substrates, which exhibit surfaces with structures similar to those for bulk oxide samples.

Thin Oxide Films on Metal Supports—Al_2O_3 on NiAl(110)

Al_2O_3, one of the most important support materials for dispersed metal catalysts, can be grown in continuous films that are only a few Ångstrøms thick by oxidizing a NiAl alloy. Figure 34 shows an example of an Al_2O_3 film on NiAl(110) [57]. The film is atomically flat but exhibits line defects (antiphase and reflection domain boundaries). The line defects are a consequence of strain in the oxide film resulting from the lattice mismatch between the substrate and oxide film. The formation of these defect lines can partially relieve the strain in the oxide layer. Although these oxide films are very thin, they can possess properties similar to bulk oxide surfaces, and thus they are often used as substitutes for bulk oxide samples. These films allow for the use of STM and other surface-science techniques that rely on conducting samples. As one example, such oxide films have been used as supports to grow metal nanoclusters, which can then be used as model systems for nanodispersed metal, heterogeneous catalysts.

Bulk Oxide Crystal Surfaces—TiO_2(110)

Bulk oxides can also be studied by STM if the sample has a sufficiently high electrical conductivity. This may be achieved for some oxides by simply annealing the sample in vacuum in order to create oxygen defects. An oxygen deficiency of 0.1% alters the

Figure 34 STM image ($45 \times 45\,nm^2$) of an Al_2O_3 film formed on NiAl(110) by oxidation. A step edge (S), reflection-domain boundary (R), as well as an antiphase boundary (A) are indicated. (From Ref. 57.)

electrical conductivity of $SrTiO_3$ from $<10^{-8}$ $(\Omega cm)^{-1}$ to 10^{-2} $(\Omega cm)^{-1}$. However, such annealing procedures may change the surface stoichiometry, because oxide surfaces accommodate variations in composition relatively easily. Consequently, a large variety of surface compositions and structures are expected and observed. Furthermore, transition metal cations have multiple valence states and concomitant variations in local bonding geometries. Thus, various surface-science techniques have been used to show numerous stable surface structures of oxides and the sensitivity of the surface structure to the thermal and chemical history of the sample.

Ambiguities arise in the interpretation of STM images of oxide surfaces from the convolution of electronic and topographic information in the STM data. Electronic effects at oxide surfaces are much more pronounced than on metal surfaces. Oxygen-deficient oxides are n-type semiconductors, i.e., the Fermi level is at the top of the band gap. Therefore, tunneling should probe the d-derived states of the cation. STM images should show cation sites as bright protrusions and oxygen sites as dark depressions. However, this assignment of cation and oxygen sites is disputed, and opposite assignments can be found in the literature.

The rutile $TiO_2(110)$ surface has been extensively studied by STM, and Figure 35 shows one image from the (1×1) $TiO_2(110)$ surface. There is now accumulated evidence from both theoretical calculations and studies of adsorbed molecules that

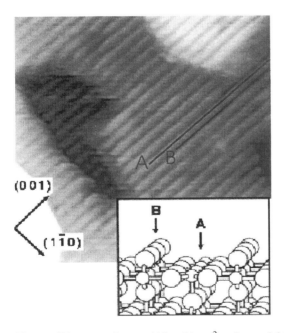

Figure 35 STM image $(14 \times 14\,nm^2)$ of a stoichiometric 1×1 rutile $TiO_2(110)$ surface. Dark rows on the terraces correspond to bridging oxygen rows, while the bright rows are due to titanium rows. The inset shows a ball-and-stick model of the rutile $TiO_2(110)$–(1×1) surface. Large balls represent oxygen atoms, and small balls represent titanium atoms. (From Ref. 68.)

are assumed to bind to the Ti sites at the surface that the bright rows in the STM image correspond to Ti rows along the $\langle 001 \rangle$ direction and the dark rows correspond to bridging oxygen rows.

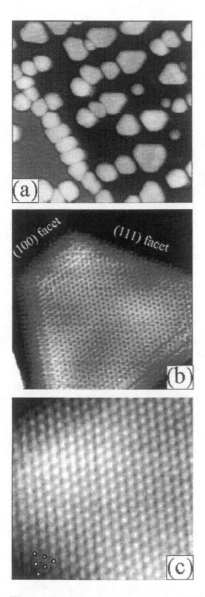

Figure 36 STM topographs of Pd clusters grown on an Al_2O_3 thin film on a NiAl(110) substrate. (a) Image ($65 \times 65 \, nm^2$) recorded after deposition of ~ 2-ML Pd at room temperature. Pd clusters have preferentially nucleated at a step and along domain boundaries. (b) and (c) Atomic-resolution images of nanosize, crystalline Pd clusters. The top of the clusters are (111)-terminated. The side facets are (100)- or (111)-terminated. (From Ref. 57.)

Nanoclusters on Single-Crystal Supports

STM is ideally suited to characterize the morphology of nanostructures grown on single-crystal substrates. The self-organization of nanoclusters with a preferential size distribution on semiconductor surfaces is being exploited to form quantum dots, and a huge number of studies in this technologically important field have been conducted. Here, however, we provide two examples that relate to catalysis and electrocatalysis.

Metal Clusters on Oxide Supports—Pd on Al_2O_3

Metal clusters on single-crystal oxide substrates have been studied with an aim to better understand metal/oxide interfaces and to create model systems for supported, nanodispersed metal catalysts for surface-science studies. Growth of metal deposits on oxide crystal surfaces proceeds via nucleation and ad-atom incorporation into existing clusters. Initial nucleation can occur by trapping of an ad-atom at a defect site (heterogenous nucleation) or by fluctuations that form a critical cluster size or nucleus during recombination of ad-atoms (homogenous nucleation). For homogenous nucleation, the density of nuclei is dependent on the ad-atom diffusion coefficient and the deposition flux of ad-atoms. Both parameters can be varied easily experimentally and thus one can control the density of nanoclusters and their sizes. For heterogenous nucleation, the saturation cluster density is independent of the incident atom flux and surface diffusion coefficient. In this case, only a change in the density of defect sites (nucleation sites) on the substrate allows for an adjustment of the nanocluster distribution. STM is a powerful probe of the cluster-size distribution for different preparation conditions.

Pd deposits on thin Al_2O_3 films grown on an NiAl(110) substrate have been studied by STM. Since Al_2O_3 is an insulator with a wide band gap of $\approx 8\,eV$, tunneling at low-bias voltages takes place from, or to, states in the underlying metal substrate. Determining the real height of metal clusters on the surface requires a subtraction of the oxide film thickness from the apparent height measured with STM at low bias. Islands preferentially nucleate along line defects like surface steps and antiphase or reflection boundaries of the Al_2O_3 thin film as shown in Figure 36a. Atomic-resolution STM images of these Pd nanoclusters were achieved for clusters with a width larger than $\sim 40\,Å$ (Figures 36b and c). The top faces of the Pd islands were found to be (111)-terminated. The facets on the side of the islands were also predominantly (111) faces and only small (100) facets were exposed. Quantitative information about the work of adhesion could be derived from the observed shape of the islands and a Wulff construction based on calculated surface energies [57].

MoS_2 Islands on Au(111)

Atomic-scale, structural information has also been obtained on MoS_2 nanoclusters grown on a Au(111) substrate [58] STM results for these single-layer, MoS_2 nanoclusters showed that the S-terminated step edges are preferred to Mo-terminated step edges. This results in a triangular-shaped island. A reconstruction of the sulfur atoms along the step edges was also observed. This can be seen by using the grid superimposed on the STM image in Figure 37a. S atoms at the edges are shifted by onehalf of a lattice constant along the edge compared to the hexagonal lattice of the basal plane. Exposing the surface to atomic hydrogen "strips off" sulfur atoms from the step edges due to chemical reactions and forms defect sites with

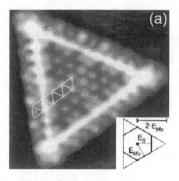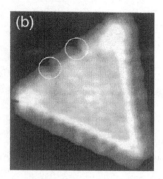

Figure 37 Atom-resolved STM images of MoS_2 nanoclusters on Au(111). (a) A superimposed grid shows the registry of edge atoms relative to those in the basal plane of the MoS_2 triangle. The inset shows the Wulff construction for the MoS_2 crystal. E_{Mo} and E_S denote the free energy for Mo and S edges, respectively. (b) An MoS_2 nanocluster after it was exposed to atomic hydrogen at 600 K. S vacancies were formed at the edges (indicated by circles). (From Ref. 58.)

undercoordinated Mo atoms, as shown in Figure 37b. It is believed that such sites are catalytically active for hydrodesulfurization reactions. This approach makes it possible to directly image active sites for catalysis with STM.

4.3.4 High-Pressure STM Studies

Most surface-science techniques use electrons or ions to probe surfaces and thus require vacuum conditions for their operation. STM does not involve free electrons. This allows STM to be used in various ambient environments, such as air, liquid, or a pressurized cell with a well-defined background gas. The latter situation is described in this section. High-pressure STM studies are designed to close the "pressure gap" between the conditions for surface-science studies and those encountered in industrial catalytic processes. Studies of adsorption of molecules on surfaces under UHV conditions are necessarily carried out at low temperatures (often below 300 K) to create high coverages of molecules similar to those that might form at high pressures and elevated temperatures, i.e., those conditions under which most catalytic processes take place. However, adsorbate structures formed at low temperatures are not necessarily in equilibrium with the gas phase and thus may correspond to kinetically trapped structures that have little resemblance to the thermodynamically stable structures formed at high pressures and temperatures. The use of low temperatures also severely limits observations of surface reactions that occur via activated processes.

High-pressure STM apparatuses are now available in a few laboratories. This allows one to scrutinize the low-temperature results, identify new, ordered adsorbate structures, and study activated processes such as adsorbate-induced, surface reconstruction, and reactions that only occur in the presence of adsorbates at high temperatures. These studies may shed new light on mechanisms of catalytic reactions under industrial conditions.

Although these measurements are taken at elevated pressures, an UHV chamber is still normally used to prepare a clean, well-defined surface prior to exposing the surface to gases. In practice this is often achieved by having a two-chamber design with sample transfer between the chambers. One chamber includes sample preparation and standard surface-science analysis techniques, while the second chamber contains the STM and can be pressurized conveniently.

Molecular Adsorbates at High Pressures—CO on Pt(111)

An example of an adsorbate structure identified at high pressure that is different from those observed under UHV conditions comes from studies of CO adsorption on Pt(111). An STM image of a Pt(111) surface in a background pressure of 150 Torr CO and 50 Torr O_2 is shown in Figure 38a. The surface exhibits an ordered, hexagonal superstructure with a uniform spacing of 12 ± 1 Å. A comparison with $\langle 110 \rangle$-type step edges of the Pt(111) surface indicates that the close packed rows of the hexagonal superstructure are aligned along this crystallographic direction. Thus, the hexagonal structure shows only one rotational domain, in contrast to the CO structures observed under UHV conditions. The structure observed in STM at high pressure is a Moiré-type structure, due to an ordered, close-packed layer of CO that is incommensurate with the Pt(111) substrate. The observed orientation of the Moiré structure indicates that rows of CO molecules are formed that are parallel with rows of the Pt lattice. The 12 ± 1 Å Moiré structure is most likely explained by a model for the CO overlayer in which CO molecules are separated by 3.7 Å, with three CO molecules spanning approximately four Pt atoms, as shown in Figure 38b. Such a structure corresponds to a coverage of 0.60 ML.

This surface structure is different from those formed at high-CO coverage under UHV conditions and stabilized by low temperatures. Under UHV conditions,

 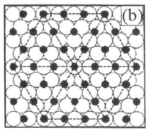

Figure 38 (a) STM image (56×36 nm^2) obtained in 150-Torr CO and 50-Torr O_2 after annealing to 183°C. The STM image (160×200 nm^2) in the inset indicates the [110] direction of the step edges of a triangular hole and mesa of Pt atoms. The close-packed rows of the Moiré pattern align parallel to the step edges of the Pt structures in the inset. (b) Proposed model for the high-pressure, CO overlayer on Pt(111). Small black circles correspond to CO molecules and large open circles correspond to the Pt(111) substrate. (From Ref. 69.)

these adlayers always exhibit several domains due to different angular epitaxies with the Pt substrate. Thus, even though the detailed model for the CO overlayer structure can be disputed, it is apparent that CO molecules adopt a different surface order at high pressures than under low-temperature, UHV conditions. It is likely that this structure is relevant for catalytic processes occurring at elevated pressures and temperatures.

Surface Reconstruction—Pt(110) under High Pressures of H_2, O_2, and CO

Topographical changes due to massive surface reconstructions may occur if surfaces are exposed to high pressures. This can be illustrated by the example of a Pt(110) surface. In one experiment, a clean Pt(110) surface was exposed to hydrogen, oxygen, and CO at pressures of above 1 bar and annealed to 425 K for several hours. The surface was imaged in situ by using a high-pressure STM before, during, and after gas exposure.

Hydrogen exposure causes rows running in the $\langle 110 \rangle$ direction with separations varying by multiples of the lattice spacing (Figure 39a). Corrugation of the rows increased with their separation. It was concluded from these observations that the

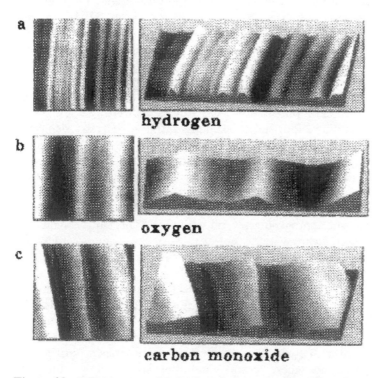

Figure 39 STM images of Pt(110) surfaces exposed to different gases at high pressures. (a) Pt(110) surface in 1.6-atm H_2 after heating to 425 K for 5 h, showing a randomly nested, $(n \times 1)$ missing-row reconstruction. Image size: $73 \times 70\ nm^2$. (b) Pt(110) surface in 1-atm O_2 after heating to 425 K for 5 h, showing a facetted "hill and valley" surface structure. Image size: $90 \times 78\ nm^2$. (c) Pt(110) surface in 1-atm CO after heating to 425 K for 4 h, showing flat terraces separated by multiple-height steps. Image size: $77 \times 74\ nm^2$. (From Ref. 70.)

surface consists of missing rows with ($n \times 1$) unit cells, where n is an integer between 2 and 5. This surface reconstruction was unchanged by annealing to 425 K or evacuating the STM chamber.

Oxygen exposure causes a different surface morphology. Instead of the moderately corrugated, missing-row reconstruction, the surface was dominated by 10–30 nm structures as shown in Figure 39b. These features were identified as (111) microfacets resulting in a larger "hill and valley" surface topography compared to the same surface exposed to hydrogen.

CO exposure causes yet another surface morphology. The surface appears to be atomically flat with no missing-row reconstructions on a small length scale. This is in agreement with UHV studies that showed that CO lifts the missing-row reconstruction. On a larger length scale, steps of multiple-layer heights can be identified in Figure 39c. Thus, this surface consists of flat terraces separated by multiple-height steps.

4.3.5 STM in Electrolyte Solutions

STM is one of only a few techniques that can be used to obtain detailed structural information at the solid/liquid interface. Surface x-ray scattering (SXS) can be used and probes the local order of the surface with higher resolution than STM, but the information is averaged over an extended area of the surface. STM has the advantage that it can image lighter atoms that do not scatter x-rays well, but it is fairly insensitive for distinguishing between atomic or molecular species. Thus, as usual, a combination of the two techniques can provide a more detailed description of the solid–liquid interface.

The potential of the surface (which acts as an electrode) has to be controlled independently of the tunneling tip potential in STM studies in electrolyte solutions. This is commonly done by using a four-electrode configuration such as that depicted in Figure 40. The potential of the substrate and tunneling tip relative to a reference electrode can be controlled independently by using a bipotentiostat. Electrochemical current flowing through the substrate and counter electrode can be monitored and the tunneling current I_t can be measured by using another amplifier. The tunneling wire has to be insulated because it is immersed in solution and could generate a Faradaic background current approaching several milliamps. This can be achieved by completely coating the tunneling wire, except at its very end, using a variety of coating materials, e.g., soft glass, polyethylene and other polymers, Apiezon® wax, or even nail polish.

Preparing a clean surface is often a prerequisite for surface-science studies. UHV-based methods of sample preparation and characterization are established, and these may be exploited for studies of surfaces immersed in solution by interfacing an electrochemical cell with an UHV chamber. Samples can then be transferred from UHV and immersed into electrolyte solution under a purified-Ar atmosphere. However, even under these "clean" conditions, some metals oxidize or get contaminated prior to immersion. Other techniques for the preparation of clean surfaces that do not require UHV techniques are available for some metals. For example, flame annealing and quenching have been successfully used, but this procedure is probably limited to Au, Pt, Rh, Pd, Ir, and Ag substrates. In this technique, substrates are annealed in an oxygen flame and quenched in pure water.

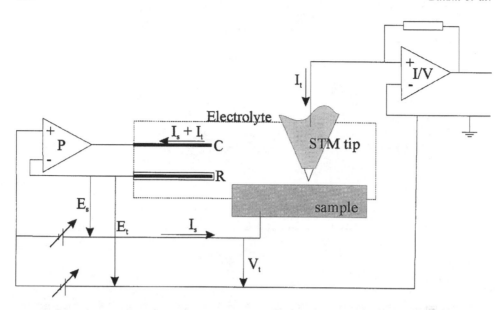

Figure 40 Schematic diagram of STM in an electrochemical cell with a four-electrode configuration using a potentiostat P with current-carrying counter electrode C and currentless reference electrode R for controlling the potentials E_s and E_t with adjustable voltage sources. (From Ref. 71.)

Iodine adlayers can provide a protective layer against oxidation and contamination while single-crystal surfaces are handled in ambient atmospheres. Also, the adsorbed iodine can be replaced by CO, which in turn can be electrochemically oxidized in solution to yield a clean metal surface. Anodic dissolution of various metals occurs at step edges under carefully adjusted electrochemical conditions, and this is a promising method for in-situ preparation of atomically flat terraces. Such layer-by-layer dissolution has been demonstrated for Ni, Ag, Co, and iodine-modified Pd and Cu surfaces.

Below, we describe two examples that illustrate electrochemical applications of STM. Other topics that can be studied by solution-phase STM include surface reconstructions of single crystals, adsorption of molecules, and anodic dissolution [59,60].

Adsorbed Anions—Iodide (I⁻) on Au(111)

The adsorption and structure of anions such as bromide, cyanide, sulfate/bisulfate, and iodide on metal electrodes have been extensively studied by in-situ STM in electrolyte solutions. Figure 41a displays a cyclic voltammogram for an Au(111) electrode in 1-mM KI solution. The anodic/cathodic peaks below 0 V versus Ag/AgI are associated with adsorption/desorption of iodine at the surface. The smaller peaks at ~0.5 V are due to a phase transition in the adsorbed iodine layer, as can be observed by STM images taken at various electrode potentials. STM images shown in Figure 41b taken at a potential of −0.2 V show a periodic structure with perfect

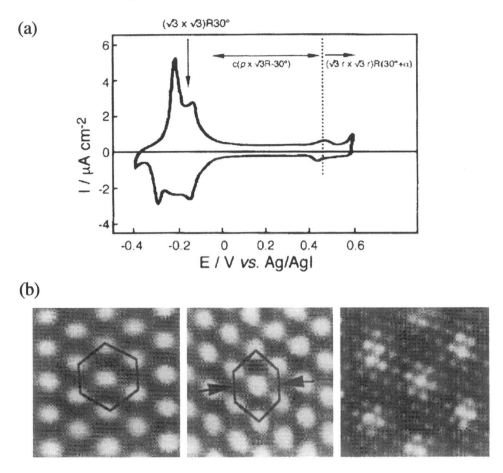

Figure 41 (a) Cyclic voltammogram of Au(111) in a 1-mM KI solution at a scan rate of 50mV/s. (From Ref. [74].) (b) In-situ STM images at potentials of −0.2, 0.3, and 0.5 V versus Ag/AgI, respectively. (From Ref. 72.)

threefold symmetry, and a 0.50-nm periodicity. This corresponds to a $(\sqrt{3} \times \sqrt{3})R30°$-iodine adlayer. A shift of the electrodes more positive potential results in a more densely packed ad-atom arrangement with an uniaxial compression of the unit cell, as shown in Figure 41b. At an electrode potential of 0.5 V, the surface exhibits a true sixfold symmetry. In this structure, the iodine interatomic spacing is smaller than in the $(\sqrt{3} \times \sqrt{3})R30°$-iodine adlayer, and the lattice appears to be rotated by several degrees with respect to the $(\sqrt{3} \times \sqrt{3})R30°$ structure. This type of adlattice has been denoted as a "rot–hex" structure. The additional periodic features that can be clearly identified in the STM images are associated with a Moiré pattern due to the mismatch between the iodine adlattice and the Au(111) substrate lattice.

In addition, ex-situ LEED measurements were also performed. Samples immersed at a potential of −0.2 V exhibited a $(\sqrt{3} \times \sqrt{3})R30°$ structure. Samples immersed at a more positive potential exhibited a more complex LEED pattern that can be associated with a rectangular $c(n \times \sqrt{3}R30°)$ structure, with n a rational number less than 3 (for $n = 3$ the structure is $(\sqrt{3} \times \sqrt{3})R30°$). This describes the uniaxially compressed unit cell that was observed in STM. However, samples immersed at potentials more positive than 0.5 V still exhibited a $c(n \times \sqrt{3})R30°$ structure in the ex-situ LEED studies, even though a rotated hexagonal structure was observed in the STM data. This discrepancy was interpreted as arising from the instability of the rot–hex structure under UHV conditions, and thus this serves as a demonstration of the limitations of ex-situ techniques.

Underpotential Deposition—Cu on Au(111)

Cyclic voltammograms of an Au(111) electrode in 0.05-M H_2SO_4 and 1-mM $CuSO_4$ solutions, as shown in Figure 42, show an anodic peak at 1.25 V and a cathodic peak

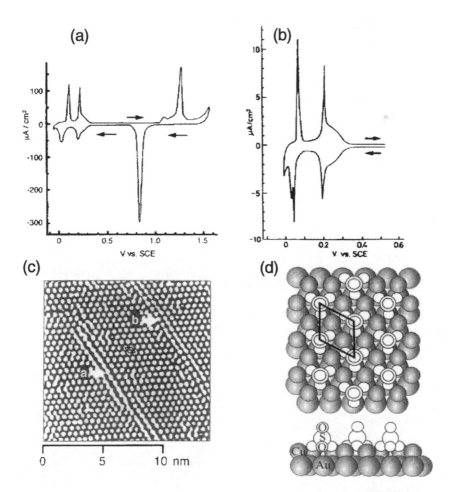

Figure 42   (a) and (b) Cyclic voltammograms for an Au(111) electrode in a solution of 0.05-M H_2SO_4 + 1-mM $CuSO_4$. Scan rates were (a) 20 mV/s and (b) 1 mV/s. (c) STM image of the Cu adlayer on Au(111). (From Ref. 75.) (d) Top and side views of a model for the coadsorbed Cu_2^+ and SO_2^+ adlayer on Au(111). (From Ref. 61.)

at 0.82 V due to oxidation and reduction of the Au(111) surface, respectively. In the potential region between 0 and 0.35 V versus SCE, two clear Cu underpotential deposition waves were observed. High-resolution STM images of the surface under underpotential deposition conditions revealed $(\sqrt{3} \times \sqrt{3})R30°$ domains covering the surface almost entirely. Coulometric curves obtained simultaneously indicate that the ratio of charges consumed during the first and second underpotential deposition waves was roughly 2:1. This implies a surface Cu coverage of about 2/3 after the first underpotential deposition peak. Therefore, the observed $(\sqrt{3} \times \sqrt{3})R30°$ structure, which corresponds to a surface coverage of 1/3 for a primitive unit cell, cannot be explained by Cu ad-atoms only. SXS examinations of this surface [61] lead to a model where Cu ad-atoms located at threefold-hollow sites of the substrate form a "honeycomb" lattice. Sulfate ions are adsorbed in the honeycomb centers, with the three oxygen atoms of the sulfate ion bound to Cu atoms. Thus, the sulfate ions form a $(\sqrt{3} \times \sqrt{3})R30°$ structure, and it is likely that this is the structure observed in STM images. STM is not very sensitive for distinguishing between different chemical species, and this is a serious limitation. Once again, this illustrates the importance of a multitechnique approach to nanostructure determination.

ACKNOWLEDGMENTS

This work was partially supported by the Analytical and Surface Chemistry Program in the Division of Chemistry, National Science Foundation. The authors thank Jooho Kim for a critical reading of the manuscript, and Jooho Kim and Dr. Take Matsumoto for help with preparing some of the figures.

REFERENCES

1. M. Haruta, Catalysis Today 36 (1997) 153.
2. D. Briggs, M.P. Seah, Practical Surface Analysis by Auger and X-ray Photoelectron Spectroscopy, John Wiley and Sons, Chichester (1983).
3. D. Briggs, M.P. Seah, Practical Surface Analysis by Auger and X-ray Photoelectron Spectroscopy, John Wiley and Sons, Chichester (1983).
4. Joachim Stöhr, NEXAFS Spectroscopy, Springer-Verlag, New York (1992).
5. John T. Yates, Jr, Theodore E. Madey, Vibrational Spectroscopy of Molecules on Surfaces, Plenum Press, New York and London (1987).
6. John T. Yates, Jr, Theodore E. Madey, Vibrational Spectroscopy of Molecules on Surfaces, Plenum Press, New York and London (1987).
7. M.A. van Hove, W.H. Weinberg, C.-M. Chan, Low-Energy Electron Diffraction: Experiment, Theory and Surface Structure Determination, Springer-Verlag, New York (1986).
8. C.S. Fadley, Prog. Surf. Sci. 16 (1984) 275.
9. C.S. Fadley, in *Synchrotron Radiation Research: Advances in Surface Science*, Plenum, New York (1990).
10. S.A. Chambers, Surface Sci. Reports 16 (1992) 261.
11. E. Taglauer, in Ion Spectroscopies for Surface Analysis, A.W. Czanderna and D.M. Hercules, eds., Plenum Press, New York and London (1991).
12. J.W. Rabalias, Low Energy Ion-Surface Interactions, John Wiley, Chichester (1994).

13. W.F. Egelhoff, Jr, in Ultrathin Magnetic Structures I. An Introduction to Electronic, Magnetic, and Structural Properties, J.A.C. Bland and B. Heinrich, eds., Springer-Verlag, Berlin (1994).

14. J.J. Rehr, R.C. Albers, C.R. Natoli, E.A. Stern, Phys. Rev. B 34 (1986) 4350.

15. J. Mustre de Leon, J.J. Rehr, C.R. Natoli, C.S. Fadley, J. Osterwalder, Phys. Rev. B 39 (1989) 5632.

16. J.J. Rehr, R.C. Albers, Phys. Rev. B 41 (1990) 8139.

17. A.G. Mckale, G.S. Knapp, S.K. Chan, Phys. Rev. B 33 (1986) 841.

18. M.L. Xu, J.J. Barton, M.A. Van Hove, Phys. Rev. B 39 (1989) 215.

19. D. Gabor, Nature 161 (1948) 777.

20. P.J. van Heerden, Appl. Opt. 2 (1963) 393.

21. A. Szöke, in Short Wavelength Coherent Radiation: Generation and Applications, D.J. Attwood and J. Booker, AIP Conf. Proc. 147, AIP, New York (1986).

22. J.J. Barton, M.L. Xu, M.A. van Hove, Phys. Rev. B 37 (1988) 10475.

23. J.J. Barton, Phys. Rev. Lett. 67 (1991) 3106.

24. G.R. Harp, D.K. Saldin, B.P. Tonner, Phys. Rev. Lett. 65 (1990) 1012.

25. M. Zharnikov, M. Weinelt, P. Zebisch, M. Stichler, H.P. Steinrück, Phys. Rev. Lett. 73 (1994) 3548.

26. L.J. Terminello, J.J. Barton, D.A. Lapiano-Smith, Phys. Rev. Lett. 70 (1993) 599.

27. D.P. Smith, J. Appl. Phys. 38 (1967) 340.

28. H. Goldstein, Classical Mechanics, Addison-Wesley, Reading, MA (1965).

29. M. Aono, R. Souda, Jpn. J. Appl. Phys. 24 (1985) 1249.

30. G. Molière, Z. Naturforsch. 2a (1947) 133.

31. O.B. Firsov, Sov. Phys. JETP, 9 (1959) 1517.

32. O.S. Oen, Surf. Sci. 131 (1983) L407.

33. M. Pascal, C.L.A. Lamont, P. Baumgärtel, R. Terborg, J.T. Hoeft, O. Schaff, M. Polcik, A.M. Bradshaw, R.L. Toomes, D.P. Woodruff, Surf. Sci. 464 (2000) 83.

34. R. Fasel, P. Aebi, R.G. Agostino, D. Naumovic, J. Osterwalder, A. Santaniello, L. Schlapbach, Phys. Rev. Lett. 76 (1996) 4733.

35. G.C. Gazzadi, P. Luches, A. di Bona, L. Marassi, L. Pasquali, S. Valeri, S. Nannarone, Phys. Rev. B 61 (2000) 2246.

36. S. Banerjee, S. Ravy, J. Denlinger, X. Chen, D.K. Saldin, B.P. Tonner, ICSOS-5 Proceedings, Surf. Rev. and Lett. 4 (1997) 1131.

37. A.S. Bodke, D.A. Olschki, L.D. Schmidt, L.D. Science 285 (1999) 712.

38. S.H. Overbury, Y.-S. Ku, Phys. Rev. B 49 (1992) 7868.

39. Y. Li, M.R. Voss, N. Swami, Y.-L. Tsai, B.E. Koel, Phys. Rev. B 56 (1997) 15982.

40. G. Herman, Y. Gao, T. Tran, J. Osterwalder, Surf. Sci. 447 (2000) 201.

41. E. Asari, T. Suzuki, H. Kawanowa, J. Ahn, W. Hayami, T. Aizawa, R. Souda, Phys. Rev. B61 (2000) 5679.

42. J.A. Kubby, J.J. Boland, Surf. Sci. Rep. 26 (1996) 61.

43. C. Bai, Scanning Tunneling Microscopy and Its Applications, 2nd ed., Springer, Berlin (2000).

44. H.-J. Güntherodt, R. Wiesendanger, Scanning Tunneling Microscopy I: General Principles and Applications to Clean and Adsorbate-Covered Surfaces, Springer, Berlin (1994).

45. H.-J. Güntherodt, R. Wiesendanger, Scanning Tunneling Microscopy II: Further Applications and Related Scanning Techniques, Springer, Berlin (1995).

46. H.-J. Güntherodt, R. Wiesendanger, Scanning Tunneling Microscopy III: Theory of STM and Related Scanning Probe Methods, 2nd ed., Springer, Berlin (1996).

47. D.A. Bohnell, Scanning Tunneling Microscopy and Spectroscopy: Theory, Techniques, and Applications, VCH Publishers (1993).

48. C.J. Chen, Introduction to Scanning Tunneling Microscopy, Oxford University Press, New York (1993).

49. D. Sarid, Scanning Force Microscopy: With Applications to Electric, Magnetic, and Atomic Forces, Oxford University Press, New York (1991).

50. R. Wiesendanger, Scanning Probe Microscopy, Springer, Berlin, Heidelberg (1998).

51. J. Bardeen, Phys. Rev. Lett. 6 (1961) 57.

52. M.A. van Hove, J. Cerda, P. Sautet, M.-L. Bocquet, M. Salmeron, Prog. Surf. Sci. 54 (1997) 315.

53. M.F. Crommie, C.P. Lutz, D.M. Eigler, Nature 363 (1993) 524.

54. F. Besenbacher, Rep. Prog. Phys. 59 (1996) 1737.

55. E.L.D. Hebenstreit, W. Hebenstreit, M. Schmid, P. Varga, Surf. Sci. 441 (1999) 441.

56. C. Günther, J. Vrijmoeth, R.Q. Hwang, R.J. Behm, Phys. Rev. Lett. 74 (1995) 754.

57. K. Højrup Hansen, T. Worren, S. Stempel, E. Lægsgaard, M. Bäumer, H.-J. Freund, F. Besenbacher, I. Stensgaard, Phys. Rev. Lett. 83 (1999) 4120.

58. S. Helveg, J.V. Lauritsen, E. Lægsgaard, J.K. Nørskov, B.S. Clausen, H. Topsøe, F. Besenbacher, Phys. Rev. Lett. 84 (2000) 951.

59. K. Itaya, Prog. Surf. Sci. 58 (1998) 121.

60. A.A. Gewirth, B.K. Niece, Chem. Rev. 97 (1997) 1129.

61. M.F. Toney, J.N. Howard, J. Richter, G.L. Borges, J.G. Gordon, O.R. Melroy, D. Yee, L.B. Sorensen, Phys. Rev. Lett. 75 (1995) 4472.

62. G. Ritz, M. Schmid, P. Varga, A. Borg, M. Rønning, Phys. Rev. B 56 (1997) 10518.

63. F. Besenbacher, I. Chorkendorff, B.S. Clausen, B. Hammer, A. M. Molenbroek, J.K. Norskov, I. Stensgaard, Science 279 (1998) 1913.

64. M. Batzill, D.E. Beck, B.E. Koel, Surf. Sci. 466 (2000) L821.

65. H. Brune, H. Röder, C. Romainczyk, K. Kern, Appl. Phys. A 60 (1995) 167.

66. H. Brune, K. Bromann, K. Kern, J. Jacobsen, P. Stoltze, K. Jacobsen, J.K. Nørskov, Surf. Sci. 349 (1996) L115.

67. S. Esch, M. Hohage, T. Micheley, G. Comsa, Phys. Rev. Lett. 72 (1994) 518.

68. U. Diebold, J.F. Anderson, K.-O. Ng, D. Vanderbilt, Phys. Rev. Lett. 77 (1996) 1322.

69. J.A. Jensen, K.B. Rider, M. Salmeron, G.A. Somorjai, Phys. Rev. Lett. 80 (1998) 1228.

70. B.J. McIntyre, M. Salmeron, G.A. Somorjai, J. Vac. Sci. Tech. A 11 (1993) 1964.

71. H. Siegenthaler, R. Christoph, in Scanning Tunneling Microscopy and Related Methods, R.J. Behm, N. Garcia, and H. Rohrer, eds., Kluwer Academic Publishers, Dodrecht, Netherlands (1990).

72. T. Yamada, N. Batina, K. Itaya, J. Phys. Chem. 99 (1995) 8817.

73. Y. Li, B.E. Koel, Surface Sci. 330 (1995) 193.

74. K. Itaya, Prog. Surface Sci. 58 (1998) 121.

75. T. Hachiya, H. Honbo, K. Itaya, J. Electroanal. Chem. 315 (1991) 275.

5

Single-Crystal Surfaces as Model Platinum-Based Hydrogen Fuel Cell Electrocatalysts

BRIAN E. HAYDEN

The University of Southampton, Southampton, England

CHAPTER CONTENTS

SUMMARY

The electro-catalytic oxidation of hydrogen, and reduction of oxygen, at carbon supported platinum based catalysts remain essential surface processes on which the hydrogen PEM fuel cell relies. The particle size (surface structure) and promoting component (as adsorbate or alloy phases) influence the activity and tolerance of the catalyst. The surface chemical behavior of platinum for hydrogen, oxygen, and CO adsorption is considered, in particular with respect to the influence of metal adsorbate and alloy components on close packed and stepped (defect) platinum surfaces. Dynamical measurements (employing supersonic molecular beams) of the reactions of hydrogen and oxygen on platinum are described, highlighting the importance of molecular precursor states and the role of defects and alloy components in molecular dissociation. The use of ex-situ methodology in transferring is well characterized, and metal modified platinum single crystals to the electrochemical environment is also described. The results of such experiments to study CO electro-oxidation on platinum surfaces modified by ordered overlayers of bismuth, clusters of ruthenium, and alloyed ruthenium are also summarized. The results are discussed in the light of the surface chemistry of the CO, and possible mechanisms of providing CO tolerant anode catalysts.

5.1 INTRODUCTION

Metal surfaces provide a means of catalyzing a wide variety of reactions, and heterogeneous catalysis at metal surfaces is carried out extensively at both the liquid–solid and gas–solid interface. The reactivity and selectivity of the surface are determined by its electronic and geometric structure, and understanding this structure/activity relationship provides one of the greatest challenges in the design of heterogeneous catalysts. Surface science has provided an important insight into the structure, thermodynamics, kinetics, and dynamics of adsorption and reaction at surfaces as a function of surface structure and composition. In this way it has provided a means of experimentally modeling heterogeneous catalytic systems. The main limitations of this approach are twofold: the catalytic situation is complex, with metal catalysts generally in the form of oxide or carbon supported nanoparticles, modified by promoters or inhibitors. In addition, the structure and composition of the surface may be quite different under the often harsh condition of reaction than those achieved in ultrahigh vacuum systems. In recognition of this, surface scientists have increasingly sought to extend their understanding of adsorption and reaction at the extended single-crystal metal surface to more complex systems or more realistic conditions. For example, oxide and oxide-supported metal surfaces have been the subject of increasing attention. Also, a number of high-pressure cells have been designed and used specifically to study reactions at higher pressures at single-crystal surfaces.

In the case of electrocatalysis, the same methodology of experimentally modeling the catalytic structure and assessing reactivity is applicable. This approach has perhaps been most advanced by the demonstration that well characterized single-crystal surfaces can be maintained and studied at the solid–electrolyte interface. In this case, however, the limitations derive from the differences in the double-layer environment and surface structure at the solid–electrolyte interface compared to the vacuum–solid environment, and the additional problems of modeling complex (supported) structures. The first of these limitations is exemplified by the modeling of the solid–electrolyte interface by studying hydrated structures in UHV: the full gambit of surface structural and spectroscopic techniques is available, but the interface is not that between the solid and liquid phase and is certainly not under electrochemical control. In-situ methods have been extensively developed to study the single-crystal–electrolyte interface, but the environment inevitably limits the flexibility and scope of the technique. For example, one of the most influential techniques, that of scanning tunneling microscopy (STM), can be applied at both the solid–gas and solid–electrolyte interfaces. However, the limitations of Faradaic contributions to the measured tip current introduce limitations to the effective range of tunneling potential in the electrochemical environment. This, in turn, limits the range of states into which, and out of which, one may tunnel and excludes the possibility of spectroscopic measurement. A third important experimental approach is the transfer of well-characterized surfaces between the UHV and the electrochemical environment, with a correlation of surface structure with electro-chemical reactivity. All these approaches have inherent strengths and weaknesses but taken together have contributed to a better understanding of the reactivity–structure relationship in electrocatalysis through studies of single-crystal, model catalyst surfaces.

Anode and cathode electrocatalysts in polymer electrolyte membrane (PEM) fuel cells are predominantly carbon-supported, platinum-based alloy nanoparticles, with particle sizes typically in the range of 1–3 nm. The high dispersion provides the advantage of high specific surface areas and utilization of the precious metal. The results are particle morphologies with high concentrations of low-coordinate atoms at facet edges, a distribution of facet structures, and the possibility that the electronic structure of the smallest particles may deviate from that of the extended bulk metal. Binary, and increasingly ternary, alloys are employed to enhance the reactivity, stability, and tolerance of the catalyst. The surface concentration of the alloy components will generally not be the same as that of the bulk, the distribution of components may not be homogeneous among particles, and the electrochemical reaction or environment may modify the alloy composition and structure through, for example, surface segregation or selective oxidative leaching.

Presented here are the results and analysis of a series of UHV/electrochemical transfer experiments carried out on clean and modified single-crystal platinum surfaces, with compositions and structures chosen in order to model platinum-based PEM fuel-cell catalysts. The results are presented together with information obtained through the other methodologies outlined above, and a comparison is made with results from supported fuel-cell catalysts. A brief outline of what is known about the purely chemical interaction of some of the poisons and reactants with platinum surfaces, such as CO, hydrogen, and oxygen is also summarized. Such interactions are important when trying to understand the overall electrochemical surface process, which takes place during the catalytic promotion of, for example, hydrogen electro-oxidation, or reformate poisoning at platinum-based PEM anodes, and oxygen electroreduction at the cathode.

5.2 THE PREPARATION, CHARACTERIZATION AND ELECTROCHEMICAL TRANSFER OF PROMOTED SINGLE-CRYSTAL SURFACES

Structural, spectroscopic, and dynamic measurements of adsorption at single-crystal surfaces form the basis of modern surface science, and more extensive reviews of these methodologies and the information that can be provided concerning adsorbates and surface reactions can be found elsewhere as well as within this volume. Most experiments on metal surfaces are carried out on single crystals of about 10-mm diameter which are cut and polished to within 0.5° of the desired crystal plane and cleaned and ordered in the UHV environment using ion bombardment, chemical reaction, and annealing. Single-crystal surfaces prepared in this way can be transferred cleanly to an electrochemical cell in order to make contact to an electrolyte under electrochemical control and carry out potentiostatic electrochemical measurements such as cyclic voltammetry and polarography. Before transfer, the surface can be modified by adsorption of a reactant molecule or a promoting metal atom. The structure and composition of the clean or modified surface can clearly be characterized by the combination of surface techniques available in the UHV chamber. Nevertheless, there is always the uncertainty that in contact with the electrolyte, and through the change in potential of what becomes a working electrode, the surface structure and composition may change. It is the application of in-situ techniques such as STM (scanning tunneling microscopy) and FT-RAIRS (Fourier transform reflection absorption infrared spectroscopy), and recharacterization following transfer back to the UHV chamber that can provide confidence that the structures are the same, or sufficiently similar to make meaningful conclusions concerning the structural/reactivity relationship. Cyclic voltammetry also provides a connecting characterization of adsorbed states and reaction overpotentials for the model electrocatalyst surface prepared using various routes, such as MVD (metal vapor deposition) under UHV, and deposition from solution under open-circuit conditions or under electrochemical control. In the case of platinum surfaces, the additional fingerprint of the reversible hydrogen adsorption–desorption region from cyclic voltammetry provides a valuable structure-sensitive characterization for both the clean single-crystal surfaces and those modified by metal adsorbates. Indeed a bead single crystal of Pt(533), orientated, polished, and annealed according to the recipes of Clavillier et al. [1], exhibits the same hydrogen adsorption structure in cyclic voltammetry as a single crystal cleaned and transferred from a UHV chamber [2].

The combination of surface characterization techniques used in this and other laboratories include XPS (x-ray photoelectron spectroscopy), AES (Auger electron spectroscopy), LEISS (low-energy ion scattering spectroscopy), LEED (low-energy electron diffraction), and TPD (temperature-programmed desorption). These provide a range of surface structural and compositional characterizations of the clean and metal-modified surface. It not usual to find more complete molecular structural and spectroscopic characterization of adsorbates provided by techniques such as FT-RAIRS, HREELS (high-resolution electron energy loss spectroscopy) or SEXAFS (surface extended x-ray absorption fine structure) to have been carried out before transfer, but TPD provides a good characterization of the adsorbate which may have been investigated more fully elsewhere. STM, either in the UHV or in the electrochemical environment, is also not usually incorporated in transfer experiments, probably as a result of the different requirements for sample mounting and manipulation for the STM and transfer itself.

XPS or AES is extensively used not only to indicate the cleanliness of the sample before transfer, but also to indicate the presence of adsorbates and their oxidation states following electrochemical experiments and transfer back into the UHV environment. In the case of model platinum-based electrocatalysts, the electron spectroscopies have been used to estimate the coverage of the adsorbate metal atoms or the alloy composition. In the case of alloys, or the nucleation and growth of metal adsorbate structures, the techniques give only the mean concentrations averaged over a depth determined by the inelastic mean free path of the emitted electrons. Adsorption and reaction at surfaces often depend on the composition of the uppermost layer, and AES or XPS is often complemented by the technique of LEISS. RHEED has also been used in combination with AES to characterize metal adsorption structures [3,4,93]. A system used for transfer experiments, incorporating a hemispherical electron analyzer and twin anode X-ray source for XPS, an ion gun for LEISS, optics for LEED, and quadruple mass spectrometer facilities for TPD, is shown in Figure 1. Multiple ports also provide facilities for gas dosing and MVD deposition of metals. A number of experimental requirements are necessary to achieve clean transfer of the single crystal on a manipulator from the UHV chamber into the electrochemical ante–chamber. Central to these is the need to bring the electrochemical chamber to atmospheric pressure using a pure (scrubbed) inert gas in order that the electrochemical cell can be brought into contact with the surface. The electrochemical cell requires, in addition to counter and reference electrodes, a second working electrode in order to maintain potential control while the single crystal (the working electrode) is making contact with the electrolyte [2]: the second working electrode is subsequently disconnected.

5.3 THE KINETICS AND DYNAMICS OF THE CHEMICAL AND ELECTROCHEMICAL REACTIONS

The activity of platinum surfaces for the oxidation and reduction of a range of molecules at the gas–solid interface is well known. The surface chemistry of the reactants, intermediates, and products have also been studied extensively. Many of the reactions, not surprisingly, have close counterparts at the electrochemical interface, mainly because it is the surface reactivity of platinum, which is the basis of the catalytic processes in both cases. The relationship of the chemical interactions and reaction steps at the electrochemical interface—and their relationship with the overall electrochemical reaction—provides one of the most challenging aspects of electrocatalysis. Of course, it has been widely speculated, and in some cases convincingly shown, that the presence of the electrolyte at a particular potential strongly modifies the surface reactivity, and the overall reaction path may not involve intermediates recognized as characteristic of the gas–solid reaction. A prime example of this is the adsorption and reaction of methanol on platinum for which the surface steps and intermediates important at the electrochemical interface are suggested to be quite different than those observed in UHV experiments [5]. On the other hand, CO and hydrogen adsorption at the two interfaces has much more in common. In order to understand the effects of the co-adsorbates (water and specifically adsorbed ions) and the local field in the inner Helmholtz layer induced by the electrode potential on adsorbed species at the electrochemical interface, it is important to be familiar with the short-range chemical interactions of adsorbates.

Some key adsorbates and reaction intermediates relevant to fuel-cell anodes are H_2 as the fuel, CO and CO_2 as poisons in hydrogen reformate feeds, and water as a co-adsorbate and potential oxidant. In the case of the cathode, oxygen is clearly the most important reactant. In the case of a number of these molecules, such as H_2, O_2, and H_2O, not only is the molecular adsorption important on platinum (or promoted platinum catalysts), but the dissociative adsorption of the molecules is important as well. With this in mind, some details concerning the dynamics of adsorption of these molecules, the associated dissociation barriers, molecular degrees of freedom, and energy partition are important to the overall catalytic processes. In addition to the

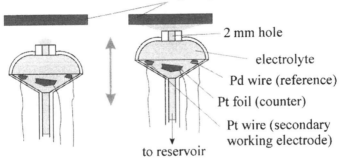

UHV   ELECTROCHEMISTRY

preparation
modification   cyclic voltammetry
characterization   electrochemical modification

XPS
LEED
TPD
LEISS
MVD

sorption pump
turbo pump

ELECTROCHEMICAL CELL

single crystal (working electrode)

2 mm hole
electrolyte
Pd wire (reference)
Pt foil (counter)
Pt wire (secondary
working electrode)

to reservoir

Figure 1 Schematic of the experimental UHV/electrochemical transfer system used for studies on modified platinum single-crystal surfaces. (From Ref. 26.) The UHV system has facilities for X-ray photoelectron spectroscopy (XPS), low-energy ion scattering spectroscopy (LEISS), low-energy electron diffraction (LEED), and temperature–programmed desorption (TPS). The electrochemical chamber allows the electrochemical cell, 0 with integral counter, reference, and secondary working electrode, to be brought to the surface allowing contact of the electrolyte with the transferred surface.

complication of particle size and support in trying to understand the catalysis of nanoparticles, the added complexity of surface defects and promotion on the overall surface reactivity must also be considered. To this end, the chemistry of hydrogen and oxygen on stepped surfaces of platinum, and alloy surfaces, has been considered in some detail.

5.3.1 Hydrogen Adsorption and Dissociation

Hydrogen dissociation and adsorption are key steps in the overall electro-oxidation of hydrogen at platinum anode catalysts, although the relationship of the adsorbed hydrogen species involved in the oxidation process (or hydrogen evolution) and the adsorbed UPD (under potential deposition) states of hydrogen are less clear. Adsorbed states of hydrogen have likewise been identified at the gas–solid interface, as have the dynamical channels to hydrogen dissociation. Most recently, the role of steps and the dynamical channels to dissociation on platinum have received some attention, as has the effect of surface alloy formation and blocking of dissociation pathways by, e.g., CO, one of the most pernicious poisons of reformate fuels.

Evidence that steps or defects are important in the dissociation of hydrogen through a molecular precursor on platinum surfaces had been made on the basis of both dissociative sticking measurements on Pt(111) [6–8] and the H_2/D_2 exchange reaction on Pt(111) and Pt(332) [9–13]. The dynamical measurements employing molecular beams can be used to determine the effect of the incident kinetic energy (E_i) of the reacting molecule on the dissociation process. The results can be used to distinguish direct (single-impact) and indirect (involving an adsorbed molecular intermediate) dissociation channels. Indeed, the suggestion that a molecular mobile precursor was responsible for the defect-mediated channel on Pt(111) [7] resulted in the calculations made by Muller [14,15], which indicated that at $E_i < 200$ meV vibrational zero-point energy release to parallel translation motion could result in a long-lived mobile molecular precursor state at the surface when combined with an inelastic process. More recently, supersonic molecular beam measurements [16,17] of the dissociation of D_2 on Pt(111) revealed only a direct channel to dissociation over the entire range of E_i investigated. On Pt(110)–(1 × 2) a nonactivated channel associated with the valley or summit sites, in addition to the direct activated channel on the (111) microfacets, was suggested as responsible for the decay of S_0 with E_i at low gas temperatures [18]. We have carried out a series of H_2 and D_2 sticking measurements using supersonic molecular beams on a Pt(533) surface [19]. This surface exposes 4-atom-wide terraces of (111) structure separated by (100) steps. Comparison with the results of similar sticking measurements on the Pt(111) surface [16,17] was used to establish the role of the step sites in the overall dissociation dynamics and to determine the dynamic characteristics of the channel.

Temperature-programmed desorption measurements (Figure 2) indicated that associative desorption of H_2 from (100) step sites on Pt(533) is observed from lowest exposures at 375 K and assigned the β_3 state. Saturation of this peak takes place at $\Theta_H = 0.14$ and corresponds to the filling of half of the available fourfold sites at the (100) step edge. Additional associative desorption takes place at higher coverages in a broad peak below 300 K. This is associated with desorption from the (111) terraces being similar to that observed (and assigned β_1 and β_2) for desorption from the Pt(111) surface. Note the similarity to the hydrogen UPD structure associated with the atomic states of hydrogen, the more strongly bound at the step site (0.28 V_{RHE}) and the weakly bound at the terrace site (0.13 V_{RHE}) on Pt(533) (Figure 16).

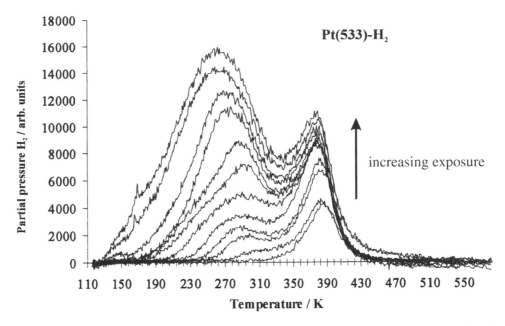

Figure 2 TPD of H_2 from Pt(533). (From Ref. 19.) The high-temperature peak is associated with associative desorption from step sites, and the low-temperature peak from the (111) terraces.

The initial dissociative sticking probability (S_0) of H_2 and D_2 as a function of incident energy E_i on Pt(533) is shown in Figure 3. (S_0) first decreases over the range $0 < E_i \text{(meV)} < 150$ (low-energy component) and subsequently increases (high-energy component). Comparison with D_2 dissociation on Pt(111), where S_0 increases linearly with E_i, leads to the conclusion that the step sites are responsible for the low-energy component to dissociation on Pt(533). The high-energy component is associated with a direct dissociation channel on (111) terraces of the Pt(533) surface. The surface-temperature dependence, coverage dependence, and incident angle dependence of dissociative sticking were also found to be consistent with the presence of two dissociation channels (indirect and direct) on Pt(533).

The results indicated that there are two contributions to sticking below 150 meV, where the step-mediated indirect channel dominates. At very low energy, where trapping into the physisorbed precursor is possible, a conventional precursor-mediated channel to dissociation on Pt(533) was proposed. A second indirect channel contributes over a much wider energy range, indeed right up to 150 meV, and was found to be much less dependent on surface temperature. The dynamic characteristics of the low-energy channel induced by the steps on Pt(533) are similar to those for the same component observed on Ni(997) [20] using a thermal beam source. They are also similar to a number of other surfaces exhibiting low-activation barriers, particularly clean and adsorbate-modified or alloyed W(100) [21–23], where it was suggested that steps play an important role in the dissociation process at low energies. An important observation regarding the alloy systems is that the low-energy indirect channel to dissociation, mediated by the step sites and defects,

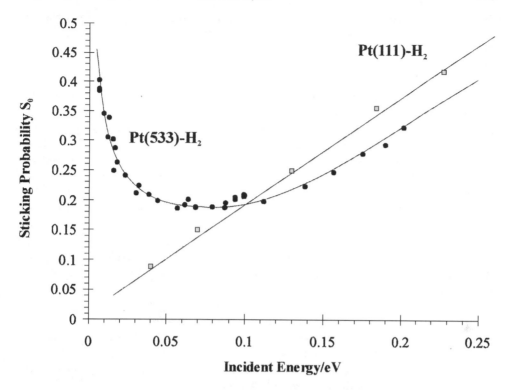

Figure 3 The initial sticking probability S_0 for H_2 and D_2 (not distinguished, see text) on Pt(533) (circles) in various seeding gases as a function of the incident kinetic energy E_i. (From Ref. 19.) The beam was incident normal to the (533) plane and at a surface temperature $T_s = 300\,K$. The initial sticking probability (S_0) of D_2 on Pt(111) (squares) (From Ref. 17.) is also shown, with the beam incident normal to the (111) plane and $T_s = 300\,K$.

remains intact even when the adsorption and/or dissociation of hydrogen is inhibited at the alloyed platinum terraces [24]. It is interesting in this regard to note that the effect of modifying platinum with metals as adsorbate or alloy has little influence on the rate of hydrogen electro-oxidation [25,26].

The importance of this low-energy (indirect) channel to dissociation to the electrocatalytic process becomes evident when considering the dominance of low-energy collisions in a Maxwellian distribution [19]. Keeping dissociation sites responsible for the low-energy channel will be crucial for catalysis involving hydrogen dissociation. Conversely, blocking these sites will also have a dramatic influence on the reactions leading to effective poisoning. We have recently shown [27] that the low-energy channel for hydrogen dissociation on Pt(533) [19] is switched off when CO (or oxygen) decorates the step sites. Both CO and oxygen preferentially adsorb at the steps. Because of the fast kinetics of the hydrogen electro-oxidation reaction, assessing the role of defects in the overall electrochemical reaction or the details of the effects of alloying or poisoning species is not straightforward. Nevertheless, experimental studies of hydrogen electro-oxidation kinetics on well-characterized stepped, poisoned, and metal-modified platinum surfaces, taken together with surface dynamic studies, provide a clearer indication of the importance of Tafel and Volmer processes [28] on modified and poisoned PEM anode catalysts.

5.3.2 Oxygen Adsorption and Dissociation

Full utilization and improvements in the activity of platinum-based alloy catalysts for electroreduction of oxygen at PEM fuel-cell cathodes continue to be the focus of significant effort, not least because this reaction is becoming rate-limiting under load in cells run with commercial membrane electrode assemblies. It is interesting to note that the extended close-packed platinum surface appears to be the most active catalyst per unit area [29], and the effects of reducing particle size and alloying, when simultaneous increases in effective surface area are accounted for, are to generally reduce the specific activity. The overall mechanism of the oxygen electroreduction reaction is complex and difficult to elucidate unambiguously from kinetic or spectroscopic measurements. Nevertheless, the role of the platinum in catalyzing the reaction is important, and one may therefore expect the surface chemical interactions involving adsorption and dissociation to influence the electroreduction kinetics and mechanism. It is not the purpose here to review the state of mechanistic understanding of oxygen electroreduction at the platinum surface. However, despite elegant adsorption and kinetic studies at single-crystal platinum electrodes [30–32], the absence of direct spectroscopic evidence for adsorbate intermediates makes mechanistic interpretation difficult. Most broadly, mechanisms of electroreduction of oxygen on platinum include the sequential hydrogenation of an adsorbed molecular species and a direct mechanism involving the reduction of Pt–O, i.e., reduction through the dissociative state. The enhanced oxygen-reduction activity afforded by alloys has been rationalized very generally on the basis of "the interplay between the electronic and the geometric factors on one hand, and their effect on the chemisorption behaviour of OH species from the electrolyte" [33].

It is shown below that a number of molecular species are important in the interaction of oxygen with platinum surfaces, and these are precursors to the dissociative state. The interception of these adsorbed states in the electrochemical process are important in determining the catalytic mechanism of electroreduction, and we also show below that surface defects play an important role in their reactivity and surface lifetimes.

The kinetic and dynamic aspects of the adsorption and desorption of O_2 on Pt(111), and the spectroscopic characterization of a number of molecular and atomic states, have been the subject of considerable attention. Temperature-programmed desorption (TPD) [34–37], valance-band [34,36,38–41] and core-level [41–43] photoelectron spectroscopy, high-resolution electron energy loss spectroscopy (HREELS) [34,39,40,44], and X-ray absorption spectroscopy (XAS) [41,45–47] have been successful in identifying and characterizing a molecular physisorbed state, one or two molecular chemisorbed states, and an atomic state of oxygen on Pt(111). The physisorbed state is present below 25 K and identified spectroscopically [43] as the precursor to a molecular chemisorbed state. In the temperature range of 90–135 K, two molecular chemisorbed states (seperoxo- and peroxolike configurations) are adsorbed. These two states are identified spectroscopically [41] as precursors to the dissociative or atomic state, which forms a (2×2) ordered overlayer with oxygen in threefold sites [39,41,48]. Dynamic studies on Pt(111) have shown that dissociation takes place through two dynamic channels [49–51]. At low incident energies, the fall in initial sticking probability S_0 with energy was interpreted as a result of dissociation mediated by trapping in a chemisorbed precursor [49–51]. A subsequent rise in $S_0(E)$ was interpreted as direct access to the dissociative state. It was subsequently shown [52] that both low- and high-energy channels to dissociation were indirect: at low energy, trapping in the physisorbed state takes place with partition between desorption and conversion to the chemisorbed molecular state at a surface temperature $T_s > 35$ K. At incident translational energies above about 200 meV, direct activated adsorption into the chemisorbed state takes place. The scattering distributions at higher energies are influenced by the molecular chemisorption potential and modified by the significant proportion of molecules that are trapped and go on to dissociate [53]. The molecular chemisorbed state on Pt(111) can subsequently partition following equilibration at T_s between desorption and dissociation. Direct access to the dissociative state is a minority dynamic channel at incident energies below 1 eV. The direct activated access to the chemisorbed state has recently been confirmed spectroscopically [54,55] and is identified as a peroxotype species with an activation barrier to dissociation of 0.29 eV.

The possibility that defects play a role in the precursor-mediated dissociation of oxygen is supported by the differences observed for the molecular and atomic adsorption states between terrace and step. The increased binding energy reflected in TPD [56] for both atomic and molecularly chemisorbed states, and the promotion of charge transfer in the molecular state at step sites, may lead to a concomitant reduction in activation barriers between states. Indeed differences between the dissociative sticking probability between Pt(112) and Pt(111) were rationalized by a reduced activation barrier between the molecular chemisorbed state and the dissociative state at the step site [56]. Since the molecular states of oxygen appear to be precursors to the dissociative state, the introduction of steps on a plane vicinal to Pt(111) provides a system well suited to the study of the role of steps in oxygen dissociation on platinum. Supersonic molecular beam experiments on Pt(533), i.e., Pt{4(111) × (100)}, provided direct evidence [57] of the crucial role of step defects on the dissociation dynamics by comparing the results with similar data obtained on Pt(111). The difference in the binding energy on Pt(533) of the atomic states at terraces and step sites is evidenced in the desorption temperatures (Figure 4) of the two species (β_1 and β_2, respectively). The low-temperature peak (α_3) results from desorption from a step-stabilized molecular species (the terrace molecular species are not stable at the adsorption temperature).

The dependence S_0 (E_i) for oxygen on Pt(533) at $T_s = 200$ in the energy range $52 < E_i (\text{meV}) < 1400$ is shown together with results for Pt(111) [49] in Figure 5. The sticking on Pt(533) exhibits similar E_i-dependent features to those observed on Pt(111)—i.e., an initial decrease and a subsequent increase in S_0 with increasing E_i. This suggests that the sequential precursor route to dissociation proposed for O_2 on Pt(111) [52,54,55] is also present on the Pt(533) surface. Although the functional form is the same for S_0 (E_i) on Pt(111) and Pt(533), the magnitude of S_0 for O_2 sticking on the Pt(533) surface is significantly greater than that reported on the Pt(111) surface at all energies. This difference is associated with the presence of steps on the Pt(533) surface. It was concluded from incident-angle, surface-coverage, and surface-temperature dependencies that at higher incident energies, dissociation takes place on Pt(533) primarily through the activated adsorption of the chemisorbed

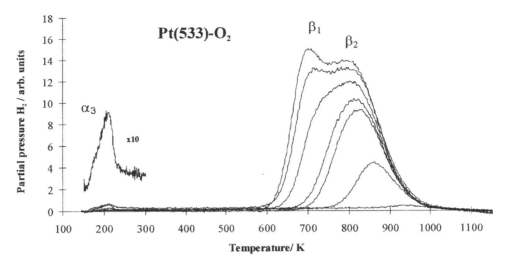

Figure 4 TPD of O_2 from the Pt(533) surface at 0.0019, 0.034, 0.089, 0.11, 0.16, 0.21, 0.24 ML of atomic oxygen, dosed at 150 K. (From Ref. 57.) The incident beam energy $E_i = 52$ meV, heating rate 0.6 K s^{-1}. The high-temperature peaks are associated with recombinative desorption from step and terrace sites. The insert shows a magnification of desorption below 300 K at 0.24 ML associated with the chemisorbed molecular state.

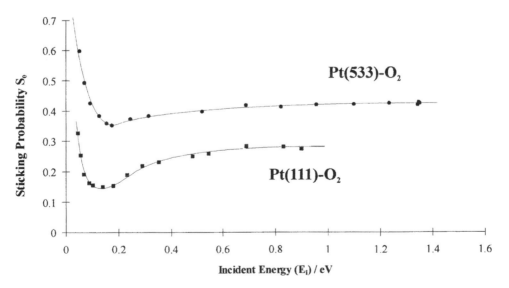

Figure 5 Initial sticking probability of oxygen on Pt(533) seeded in various gases (circles). (From Ref. 57.) The beam was incident normal to the (533) plane and at a surface temperature $T_s = 200$ K. For comparison the data measured on Pt(111) surface (squares) [49] under identical conditions have been included.

precursor on the (111) terrace and its subsequent partition between desorption and dissociation. Values for the ratio of the pre-exponential factor $v_d/v_{ca} = 0.27$ and activation energy $\Delta E = 71\,\text{meV}$ were obtained for the partition of the chemisorbed precursor on the (111) terraces of Pt(533). This compares with values obtained using the same analysis for this channel on Pt(111) of $v_d/v_{ca} = 3.5$ and $\Delta E = 86\,\text{meV}$ [52], and $v_d/v_{ca} = 2.2$, $\Delta E = 130\,\text{meV}$ [54,55]. The similarity in the values for ΔE on the Pt(111) and Pt(533) surfaces, but the large difference in the pre-exponential ratio v_d/v_{ca}, can be understood if defect or step sites dominate the dissociation of chemisorbed oxygen on both surfaces. This conclusion is consistent with the absence of any direct route to dissociation for O_2 adsorption on Pt(111) even at high incident energies (1.1 eV) [52] and the high barrier for dissociation (0.86–0.9 eV) of the chemisorbed precursor predicted theoretically [58].

The rapid decrease in $S_0(E_i)$ observed below 0.15 eV on Pt(533) has also been observed on the Pt(111) surface [49] and is consistent with a trapping mechanism where the need to dissipate energy limits the probability of adsorption, and subsequent dissociation, of the physisorbed precursor. We have estimated the sticking and subsequent dissociation probability S_0 at $T_s = 200\,\text{K}$ using the hard cube model. In so doing we show that in addition to dissociation through the physisorbed precursor, a significant contribution to dissociation ($\approx 50\%$) from partition of the chemisorbed precursor also contributes at $E_i = 0.05\,\text{eV}$, significantly higher than on Pt(111) ($\approx 10\%$) [52]. Once this "direct" contribution is subtracted, the dependence $S_0(T_s)$ can be analysed using a similar kinetic model to that applied to the chemisorption precursor. This gives values for Pt(533) of $\Delta E = 120\,\text{meV}$ and $v_d/v_{ca} = 80$, compared to $v_d/v_{ca} = 1000$ and $\Delta E = 60\,\text{meV}$ for Pt(111) [52]. *We conclude that steps dominate the conversion of the physisorbed to chemisorbed precursors on both surfaces, and the effective barrier to conversion at the (100) step on Pt(533) is effectively zero.* The barrier on the (111) terrace to interconversion is substantially higher, and the defect sites, or adsorbate sites [59] at finite coverages, are responsible for dissociation.

5.3.3 Co Adsorption and Oxidation

The overall process of the electro-oxidation of CO can be represented by

$$CO + H_2O \rightarrow CO_2 + 2e^- + 2H^+$$

The thermodynamic potential for this process lies close to $0\,V_{\text{RHE}}$, however, CO electro-oxidation is characterized by a significant overpotential on platinum of about $0.8\,V_{\text{RHE}}$. This overpotential is associated with the activation of the water (2) to produce an adsorbed oxidizing species that subsequently oxidized adsorbed CO through a Langmuir–Hinshelwood mechanism:

$$CO \rightarrow CO_a \tag{1}$$
$$H_2O \rightarrow OH_a + H^+ + e^- \tag{2}$$
$$CO_a + OH_a \rightarrow CO_2 + H^+ + e^- \tag{3}$$

Once the overpotential results in the activation of the water, the subsequent reaction kinetics are determined largely by (3). The overpotential and kinetics of CO electro-oxidation are sensitive to the CO coverage and packing density, the anions in the electrolyte, the CO concentration in the electrolyte, the structure of the surface, and the presence of adsorbed promoters. Only steps (1) and (3) have comparative counterparts at the gas–solid interface, and data obtained at this interface can be used to rationalize the kinetics associated with steps (1) and (3) [60,61]. For example, CO oxidation by adsorbed oxygen on platinum takes place at the edge of CO islands [62,63], and therefore it is perhaps not unsurprising that the CO packing density and coverage or CO in the electrolyte can influence the kinetics of electro-oxidation through step (3). With respect to the poisoning of reactions by CO, it is of interest to note the preferential adsorption of CO at step and defect sites on platinum [64,65], sites that are particularly active (Sections 5.3.1 and 5.3.2) in molecular dissociation. The nearest counterpart to step (2) would be the dissociative adsorption of water to produce an adsorbed hydroxyl species and an adsorbed hydrogen atom. The activation barrier to this proces has never been measured and may deviate from the barrier associated with (2) as a result of the stabilization of a hydrated proton. Nevertheless, one may qualitatively expect a reduction in the activation barrier at metal prompters that dissociate water more easily, and predict a lower overpotential for step (2).

5.4 THE PROMOTION OF ELECTRO-OXIDATION BY METAL OVERLAYERS

An adsorbed or alloyed metal promoter may influence electro-oxidation reactions involving CO in one of three ways [66]. First, the promoter can preferentially block sites for CO adsorption or active sites resulting in reactions (from, for example, methanol) producing CO. Second, it may alternatively modify the CO interaction with the surface in such a way as to enhance its oxidative removal. Third, the promoter may enhance the availability of reactive oxygen at the surface, enhancing oxidative removal of CO. In order to investigate these possibilities, there are a significant number of studies of CO electro-oxidation by platinum single-crystal surfaces modified by metal overlayers. The vast majority of these are for overlayers that have been deposited under electrochemical control or through dipping at open potential. The oxidation state, extent of alloying, coverage, and surface morphology of metal submonolayers and clusters produced in this way are not always known, which results in difficulties in interpreting the promotional or poisoning effects of the metal overlayer. When metal-modified surfaces are more fully characterized, however, a clearer understanding of the modifying effect of the co-adsorbed or alloyed metal is accessible. In-situ STM measurements or UHV transfer techniques, when combined with cyclic voltammetry of the metal overlayer, provide powerful tools for such characterization. An alternative to deposition of the metal modifier from the electrolyte is to use MVD deposition in UHV to synthesize structures in UHV transfer experiments. In some cases, it can be shown, as in the case of ruthenium on Pt(111) shown ahead, that the same electrochemical behavior can be produced by solution-phase and MVD deposited structures on platinum.

5.4.1 Bismuth on Platinum: A Model Adsorbate Metal Modifier

Bismuth has traditionally been considered a purely third body-(ensemble) type modifier [67] that exerts little influence on the platinum electronic structure, whereas metals such as ruthenium and tin are believed to promote reactions by the third mechanism outlined above [68]. It is also noteworthy that bismuth has been shown to promote formic acid oxidation but not methanol oxidation [69]. Bismuth adsorbed from solution was shown to promote the electro-oxidation of carbon monoxide on Pt(111) [70,71], while an apparent poisoning of CO electro-oxidation was found for MVD layers of bismuth on Pt(110) [72]. It is instructive to compare and reconcile the latter apparently contradictory findings. The bismuth/platinum system has the advantage of forming a series of well-ordered overlayers on both Pt(110)–(1 × 2) [72,73] and Pt(111) [74,75]; some examples of these are shown in Figure 6. Mixed adlayers of CO and bismuth are also formed over a wide coverage range with little perturbation of the bonding of the CO with the platinum, and a simple linear blocking of sites for CO adsorption is observed with bismuth coverage [72]. Figure 7 shows the effect of adsorbed bismuth on the TPD spectra of CO desorbing from Pt(110) [72,73]. There is very little change in desorption temperature of CO, illustrating the small perturbation in the CO binding energy. The inset in Figure 7 shows the linear, dependent blocking effect of an inert diluent, with complete blocking at saturation coverage of bismuth. In addition to the CO coverages estimated from the TPD of CO, CO coverages have also been estimated for a number of bismuth-covered surfaces for the CO stripped coulometrically following transfer [72]. The correspondence leads to confidence that the blocking effect of bismuth on CO is also operating at the electrochemical interface.

The bismuth/platinum system also has an advantage in model studies in the electrochemical environment because the modifier (bismuth) exhibits well-characterized oxidative stripping characteristics. Figure 8 shows the anodic stripping of

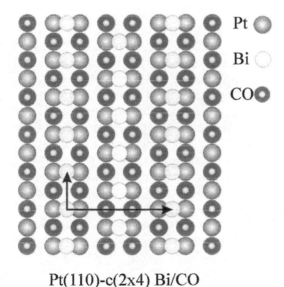

Pt ⬤   Bi ◯   CO⬤

Pt(110)-c(2x4) Bi/CO

Figure 6 A mixed ordered adsorbate layer of bismuth and CO on Pt(110) suggested on the basis of LEED and TPD measurements. (From Ref. 73.)

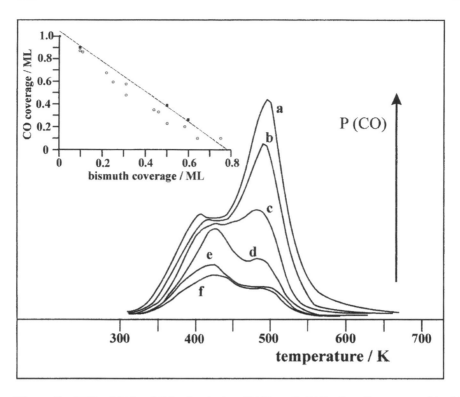

Figure 7   TPD of 50 L of CO adsorbed at 300 K on Pt(110)–(1 × 2) precovered by bismuth at (a) 0.0 ML; (b) 0.11 ML; (c) 0.31 ML; (d) 0.46 ML; (e) 0.58 ML; (f) 0.75 ML. (From Ref. 73.)

bismuth layers adsorbed on Pt(110)–(1 × 2) by MVD (Figure 8a) and through underpotential deposition (UPD) from solution (Figure 8b) [72]. Electrochemical characterization of the Pt(110)–Bi(3 × 1) structure formed through MVD of 0.7 ML of bismuth on Pt(110)–(1 × 2) MVD is shown in Figure 8a. The anodic features at 0.77 V and 0.88 V clearly resolved in the first cycle are associated with the oxidation of the bismuth. Continuous cycling to 1.2 V results in the disappearance of the bismuth-derived peaks, with the peak at 0.77 V reduced most quickly, and the restoration of the clean Pt(110)–(1 × 2) CV signature. Cycling under these conditions appears to remove the bismuth overlayer completely, with the gradual restoration of the Pt(110)–(1 × 2) oxidation signature and the recovery of the hydrogen UPD structure. Indeed, complete oxidative stripping of the bismuth layer was confirmed by XPS following transfer of the cycled electrode back to the UHV chamber. Transfer of a Pt(110)–c(2 × 2)Bi 0.5-ML overlayer, and several bismuth overlayers at slightly lower and higher coverages, indicate that the onset of the first oxidation peak at 0.77 V corresponds to the onset of the compressed bismuth structures observed in UHV [72,73] at coverages >0.5. The results are in agreement with previous studies on bismuth layers deposited on Pt(110) from solution [76]. Similar results to those observed for the transferred MVD layers of bismuth are

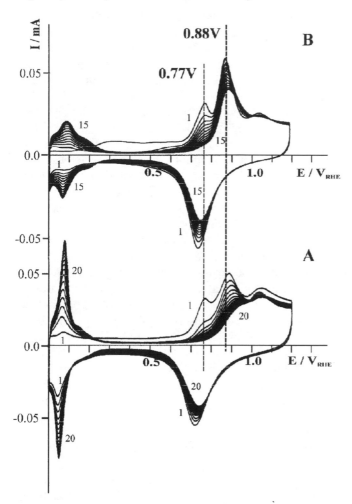

Figure 8 Cyclic voltammetry of a Pt(110)–Bi(3 × 1) overlayer (0.7 ML) deposited by (A) MVD and (B) UPD deposition (1 × 10⁻³ M Bi in 0.5 M H_2SO_4, 0.4 V for 120 s). (From Ref. 26.) The first and last cycles are indicated in each case. The peak at 0.77 V_{RHE} is associated with stripping of bismuth at coverages forming the Pt(110)–c(2 × 2)Bi structure and above. The working electrode area was about 50 mm².

observed for overlayers deposited on the Pt(110)–(1 × 2) surface by UPD or by the dipping technique (Figure 8b) with two bismuth oxidation peaks at 0.77 V and 0.88 V. Transfer of a saturated UPD overlayer back into the vacuum chamber showed similar LEED patterns to those produced by MVD, and XPS showed that only metallic bismuth was observed after washing the electrode, irrespective of the break potential [72]. This result indicated that the window of stability for any intermediate oxidized UPD bismuth layer prior to dissolution is very small on Pt(110). Higher oxidation states of bismuth were observed only if residual sulphate anions from the bulk electrolyte are transferred as a result of incomplete rinsing [72].

One may conclude that metallic bismuth overlayers deposited by MVD in the UHV environment or through UPD/dipping in the electrochemical environment are very similar in structure. A similar correspondence between the structures of ruthenium deposited by the two methods on Pt(111) is demonstrated below, but only following the gas-phase reduction of the solution-deposited ruthenium phase with hydrogen.

The electro-oxidation of CO on the Pt(110)–(1 × 2) surface and the Pt(110)–$c(2 × 2)$Bi surface [72] is shown in Figure 9. The first cycle of the CV showing CO electro-oxidation on the Pt(110)–$c(2 × 2)$Bi at a 0.5-ML surface differs from subsequent cycles in the observation of two distinct and sharp anodic peaks, the first at 0.83 V_{RHE} with a shoulder at a slightly higher potential. The second peak occurs at 1.05 V_{RHE}, which lies well into the region for oxidation of the platinum substrate, has a non-Gaussian shape with a sharp onset, and follows a region that shows an anodic current even below that of the Pt(110)–(1 × 2) surface itself. The first peak and shoulder are associated with the electro-oxidation of the CO, at a potential shifted from the clean surface value of 0.74 V_{RHE}. At a bismuth coverage of 0.5 ML, the co-adsorbed CO layer produces an intermixed phase with retention of the $c(2 × 2)$ ordered structure characteristic of the bismuth atoms alone [72,73]. Electro-oxidation in this closely packed mixed phase at a higher potential (0.83 V_{RHE}) overlaps the normal oxidation potential of bismuth in the absence of CO (shoulder at higher potential). CO oxidation is mediated by bismuth-mediated oxygen transfer. Oxidation mediated by the Pt(110)–(1 × 2) surface is hindered by the close-packed bismuth layer, resulting in an onset for CO electro-oxidation that is clearly higher than Pt(110)–(1 × 2). Once the CO has been oxidized and the product desorbs, the resulting bismuth layer is thermodynamically unstable as a result of the oxygen depletion. The result is a kinetically limited oxidation of the bismuth layer resulting in the second anodic peak at 1.05 V_{RHE}, which corresponds to the delayed oxidation of the bismuth. Subsequent cycles show a return to the normal oxidative stripping behavior of bismuth on Pt(110)–(1 × 2) expected at these bismuth coverages (Figure 8).

Figure 9 also shows results obtained for CO stripping under similar conditions to the Pt(110) data (H_2SO_4 electrolyte, scan rate 50 mV s^{-1}, CO adsorbed from solution and oxidized in CO-free electrolyte) on Pt(111) modified by bismuth adsorbed from solution [70,71]. This allows a direct comparison of the apparent promotional effect of bismuth observed on a Pt(111) surface (a reduction in overpotential of about 0.1 V) and the poisoning effect on Pt(110) (an increase in overpotential of about 0.1 V). The results are readily reconciled [72,77] in the light of the different overpotential associated with the two clean platinum surfaces and the very similar effect of bismuth in mediating the activation of water on the modified surfaces. The overpotential for CO electro-oxidation on the two modified surfaces is very similar (Figure 9).

What, if any, relevance do such results have when predicting the influence of adsorbed bismuth on the CO of supported platinum nanoparticle catalysts? In order to test the transferrability of results obtained on single crystals to practical fuel-cell anode catalysts, a series of experiments was performed [77] on a gas diffusion electrode of carbon-supported platinum (0.22 mg cm^{-2}) catalyst (Johnson Matthey). Figure 10 shows the results of polarization measurements for hydrogen oxidation at clean and bismuth-modified (0.65-ML) catalysts. In order to establish the CO tolerance of the electrodes, in addition to experiments involving pure H_2,

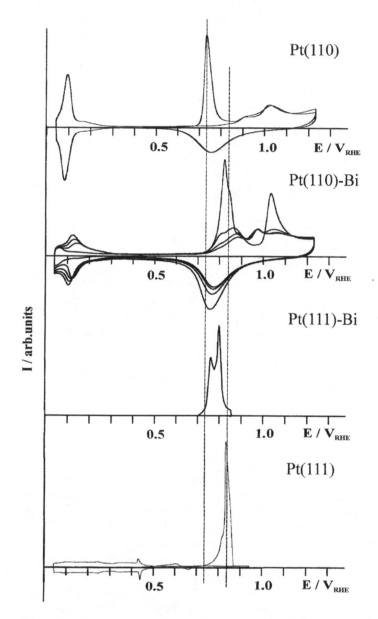

Figure 9 Stripping voltammetry of saturated CO overlayers on Pt(110) [26], Pt(110)–Bi [26], Pt(111)–Bi [70,71] and Pt(111) [98] in H_2SO_4 electrolyte at similar scan rates (50–100 m Vs^{-1}). The potentials marked by the dotted line correspond to the clean surface stripping potentials.

polarization curves of the hydrogen oxidation were carried out [77] after exposure to hydrogen feeds at various times containing 100 ppm CO in H_2. Figure 10b shows results for surfaces run only in clean fuel and run following exposure for 15 minutes to the poison containing fuel. It is interesting to note for the nonpoisoned electrodes the linear polarization response for both the clean and bismuth modified electrodes

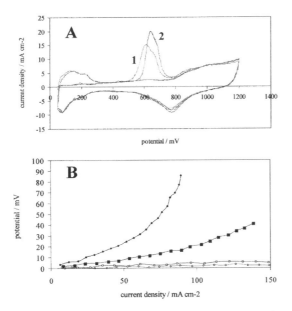

Figure 10 (A) Stripping voltammetry ($30\,\mathrm{mVs^{-1}}$ in $1\,\mathrm{M}$ H_2SO_4 at $80\,°C$) of saturated CO layers on (1) a pure platinum supported catalyst and (2) a bismuth-modified supported platinum catalyst (B) Steady-state polarization curves of hydrogen electro-oxidation at $80\,°C$ on the supported platinum catalyst (squares) and the bismuth-modified catalyst (circles). The measurements are made in pure hydrogen (unfilled points), and after running 15 minutes in synthetic reformate (100 ppm CO) (filled points). (From Ref. 77.)

up to the highest current densities investigated, with identical (within experimental error) and very small overpotential required to obtain high currents. Despite the high bismuth coverage, therefore, *the sites active in the oxidation reaction appear to remain unaffected*. Clearly because of the fast kinetics of the oxidation process, one must be careful not to overinterpret such a result, but it does appear that the site involved in determining the reaction rate on the platinum is not modified significantly by adsorbed bismuth. The active site in the reaction may or may not be associated with the hydrogen adsorption site, which we have noted is not blocked at the same rate on Pt(110)–(1 × 2) by bismuth as CO [72]. Previous studies on Pt(100) [78] have indicated that bismuth does lower the rate of the hydrogen evolution reaction, no strong electronic effect is exerted on the reaction by the bismuth, but the UPD hydrogen is not involved as a reaction intermediate. This question, however, underlines the fundamental question concerning the reaction kinetics and dynamics responsible for a surface process involving molecular dissociation at metals and alloys where several channels may be operating (Section 5.3). CO poisoning of the unmodified catalyst results in the increasing overpotential required to obtain the same current density as the nonpoisoned surface. A modification of the catalyst that resulted in increased CO tolerance would be reflected by a reduction in the overpotential required to obtain the same current density for the same exposure to CO. The same condition of CO exposure on the bismuth-modified surface results in an increase in the overpotential required to maintain oxidation currents, *indicating a marked decrease in CO tolerance.*

This result is consistent with the observed effective poisoning of the CO oxidation reaction as reflected in the increased potential induced by bismuth in the cyclic voltammetry on the supported platinum electrodes (Figure 10a). The voltammetry of CO stripping on the supported catalysts indicates a similar behavior to that found on Pt(110) in that bismuth results in a higher overpotential for CO oxidation. One must conclude that the morphology of the supported platinum catalyst results in facets more akin to the more open-packed Pt(110) surface than the Pt(111) surface, a conclusion supported by comparison of the bismuth redox chemistry on the supported catalyst and the single-crystal surfaces [77].

A similar correlation between the reduction in overpotential by an alloy component for the electro-oxidation of CO and the efficiency of the alloy in providing CO tolerance has been described for a platinum/molybdenum-supported catalyst. Figure 11A shows the CO stripping voltammetry on carbon-supported platinum and platinum/molybdenum alloy catalysts [79]. A significant reduction in the overpotential on the alloy catalyst over the pure-platinum catalyst is observed. The same supported catalysts were tested for CO tolerance in fuel-cell RDE and half-cell polarization measurements [79]; the latter is shown in Figure 11B. The requirement for a significant overpotential for hydrogen electro-oxidation in the presence of CO on pure platinum is significantly reduced on the platinum/molybdenum alloy catalyst.

5.4.2 Ruthenium on Pt(111)

Wieckowski et al. have investigated the formation of ultrathin films on low-index, platinum single-crystal surfaces up to coverages of 0.4 ML by the spontaneous deposition technique [80]. They have also investigated the oxidation of methanol on these modified surfaces and found a significant degree of surface-structure dependence in the oxidation currents obtained [81]. The spontaneous deposition of ruthenium on Pt(111) has been shown by STM to lead to nanometer-sized islands at these low coverages, mostly monatomic but with a tendency to nucleate in a bilayer configuration to about 10% of the total ruthenium coverage [82]. The open-circuit potential obtained for these deposition experiments on Pt(111) was 0.88 V, at which point the ruthenium species is likely to be considerably oxidized. XPS studies of electrodeposited Ru formed at these higher potentials have been demonstrated to form a complex mixture of oxides and hydroxides [83,84]. Hence an alternative approach has been to prepare platinum single-crystal surfaces modified by the vapor deposition of metallic ruthenium, which has the advantage that all the ruthenium is in the initial oxidation state Ru^0 [85–87]. Jarvi et al. have observed the voltammetry of Pt(111) modified with 0.38-ML ruthenium (coverage calculated by AES) in 0.1-M perchloric acid and have observed hydrogen adsorption features associated with Ru at 0.12 and 0.14 V [87]. These features disappear on annealing of the surface to 600 K in vacuum due to migration of the ruthenium into near-surface layers.

A number of electro-oxidation studies on ruthenium-modified, platinum single-crystal surfaces have recently appeared. Ruthenium has most commonly been deposited from solution on single-crystal platinum surfaces through spontaneous deposition or under electrochemical control [80–82,88–93]. In some cases MVD

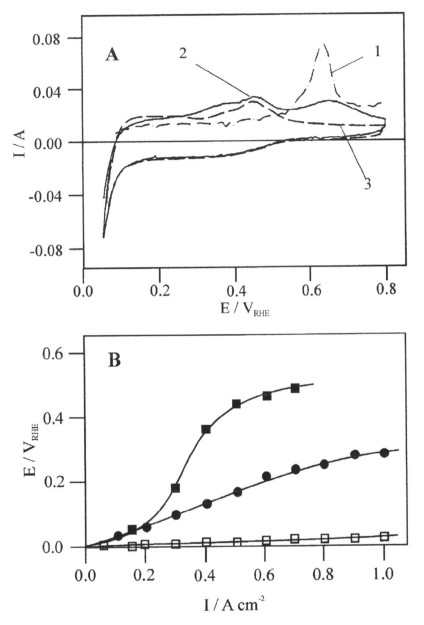

Figure 11 (A) Stripping voltammetry (20 m Vs^{-1} at 55 °C) of CO layers on humidified PEM fuel-cell anodes; (1) platinum catalyst; (2) platinum/molybdenum catalyst. Voltammetry in the absence of adsorbed CO on the platinum/molybdenum catalyst is shown in (3). Molybdenum-mediated electro-oxidation of adsorbed CO takes place on the alloy catalyst in the peak at 0.45 V and at lower overpotentials [79]. (B) Steady-state polarization curves of PEM fuel-cell anode at 85 °C for platinum (squares) and platinum/molybdenum catalysts in the presence of 100 ppm CO (filled points) and pure H$_2$ (unfilled points). (From Ref. 79.)

ruthenium-modified, platinum single-crystals have been investigated for their electrochemical activity [85–87,94]. Ruthenium layers electrodeposited on Pt(111) have been characterized by electron diffraction and AES [93] and form monatomic commensurate layers at low coverages, with three-dimensional clusters forming at higher coverages. The effect of the ruthenium clusters is to reduce the overpotential for CO electro-oxidation in $HClO_4$ from 0.71 V for unmodified Pt(111) to ≈ 0.50 –

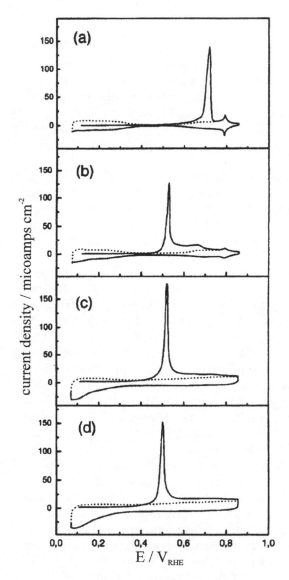

Figure 12   Stripping voltammetry of saturated CO layers (1 M $HClO_4$, 10 mVs^{-1}) from (a) Pt(111), and ruthenium-modified Pt(111) at ruthenium coverages (b) 0.2; (c) 0.6; (d) 0.75. (From Ref. 93.) The dotted lines show the voltammetry of the CO-free surfaces in each case.

0.52 V for coverages of ruthenium between 0.20 and 0.75 ML. The results of the stripping experiments [93] are shown in Figure 12. Note that a single stripping peak characterizes not only the clean Pt(111) surface, but also the ruthenium-modified surfaces. They conclude that the oxidation of CO takes place preferentially at the edge of the Ru islands, with significant mobility of CO on Pt to the active sites. Spontaneous deposition of ruthenium at open-circuit potential on low-index, platinum single crystals [80–82,88,90] results in low coverages of ruthenium, probably in the form of mixed oxide and hydroxide phases [83,84]. The oxidation of methanol on these surfaces exhibits a significant degree of surface-structure dependence in the oxidation currents obtained [81,90]. Spontaneous deposition on Pt(111) has been shown by STM to lead to nanometer-sized islands at these low coverages, mostly monatomic, but with a tendency to form bilayer structures for about 10% of the total ruthenium clusters [82]. Pt(110) [85,86] and Pt(111) [87,94] single-crystal surfaces modified by metallic ruthenium have been prepared and characterized in UHV, and the modified surfaces subsequently transferred for electrochemical characterization [85–87]. Ruthenium adsorption on the open-packed Pt(110)–(1 × 2) surface resulted in facile incorporation and alloying of the phases; the results of these experiments are described in more detail below when considering the alloy systems. In contrast, MVD of ruthenium submonolayers on Pt(111) results in nucleation and cluster growth.

As in the case of Pt(110) [85,86], a combination of XPS and LEISS was used to characterize the adsorbed ruthenium overlayers on Pt(111) [94]. Figure 13b shows the effect of flash-annealing an overlayer of ruthenium to increasing temperatures on the ruthenium coverages calculated from XPS and LEISS data. The ruthenium was deposited at 300 K, and the XPS and LEISS measurements were made following the cooling of the annealed surface to 300 K. For the "as-deposited" layer, there is a large difference between $\Theta_{XPS} = 0.60$ ML and the coverage obtained from LEISS of $\Theta_{LEISS} = 0.18$. This difference is much larger than that exhibited by ruthenium deposited on Pt(110)–(1 × 2) under the same conditions [85,86] and was ascribed to the clustering of the ruthenium ad-atoms at 300 K on the close-packed Pt(111) surface. This conclusion was supported by LEED measurements that indicate that the (1 × 1) pattern of the clean surface remained sharp with the presence of $\Theta_{XPS} = 0.60$ ML of ruthenium on the surface. It was also found consistent with the hydrogen UPD characteristics of surfaces, which remained similar to the clean Pt(111) surface without the long-range order structure. Only after annealing to intermediate temperatures, where incorporation of ruthenium takes place, were the LEED (1 × 1) spots slightly diffuse. This tendency to cluster is similar to the behavior found for the electrodeposition of ruthenium on the same surface in the presence of a weakly adsorbed anion [93]. XPS and LEISS were used to estimate the nucleation density and average cluster size of ruthenium deposited on the Pt(111) surface at 300 K. The XPS results are assumed proportional to the total amount of ruthenium in the clusters at submonolayers coverages. The scattering probabilities obtained in LEISS used to estimate the coverage in the top surface layer are proportional to the footprint of the cluster on the Pt(111). The experimental values Θ_{XPS} and Θ_{LEISS} are plotted in Figure 14. The fitted curve corresponds to a nucleation density of 1:90 platinum sites in the nucleation and growth of pyramidal clusters of ruthenium. The result does not change significantly for the nucleation and growth of truncated pyramid clusters.

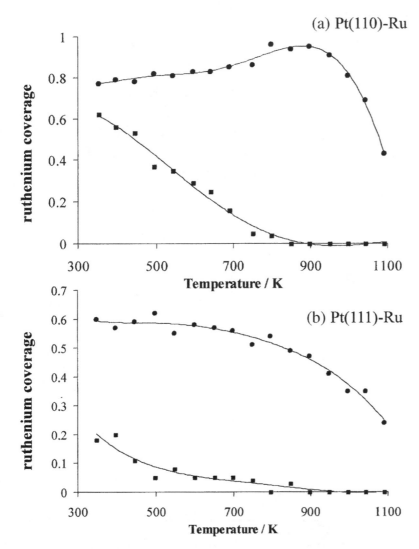

Figure 13 Ruthenium coverages on (a) Pt(110) and (b) Pt(111) estimated from LEISS (circles) and XPS (squares). The temperature corresponds to the temperature at which the surfaces have been annealed following ruthenium deposition by MVD. (From Refs. 86 and 94.)

Figure 15 shows a series of voltammograms for the oxidative stripping of CO adsorbed on the clean Pt(111) surface and shows the same surface modified with increasing coverages of ruthenium deposited at 300 K. CO was adsorbed from solution to saturation coverage at 0.05 V and the CO was then stripped by cycling in a fresh, CO-free 0.5-M H_2SO_4 solution. On the clean Pt(111) surface a single, sharp oxidation peak is observed at ≈ 0.80 V with a small prepeak at approximately 0.6. On stripping the CO monolayer, all the features associated with the clean Pt(111) are

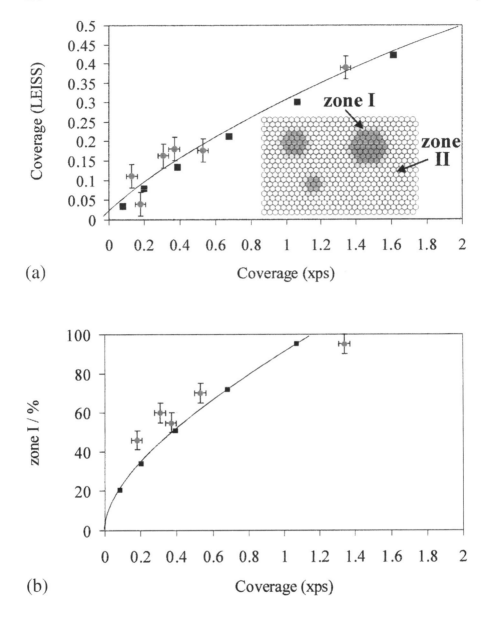

Figure 14 (a) Nucleation and growth of MVD ruthenium clusters on Pt(111) (inset). The square data points and curve correspond to a calculation of the ruthenium coverage measured by LEISS (a measure of the island footprint) as a function of the ruthenium coverage measured by XPS (total ruthenium coverage). A nucleation and growth model of hemispherical ruthenium clusters with a nucleation density of 1:90 has been assumed. The circles represent the measured LEISS and XPS data for the ruthenium-modified Pt(111) surfaces for deposition at 300 K. (From Ref. 94.) (b) Using the nucleation and growth model fit to the experimental data in (a), the percentage of platinum sites neighboring the ruthenium islands (Zone I, inset) has been calculated (square data points and curve). The experimental points (circles) correspond to the percentage of the charge associated with the first of the two CO oxidative stripping peaks on the ruthenium-modified Pt(111) surfaces (Fig. 15).

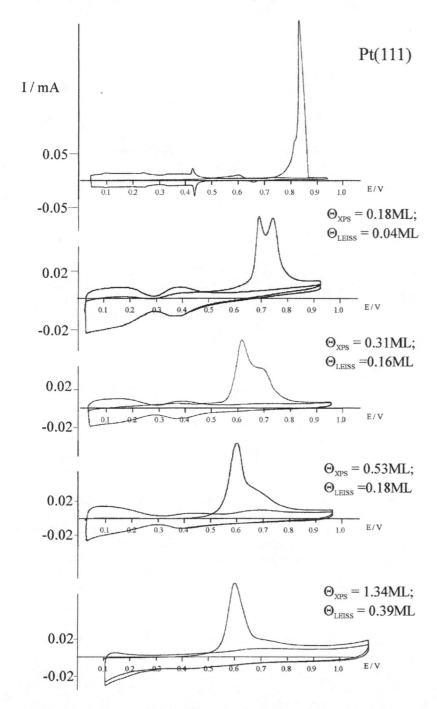

Figure 15 Stripping voltammetry (0.5 M H_2SO_4, 100 mVs^{-1}) of saturated CO layers on clean and MVD ruthenium-covered Pt(111) surfaces. The ruthenium coverages measured using XPS (Θ_{XPS}) and LEISS (Θ_{LEISS}) before transfer to the electrochemical cell are indicated. CO was dosed at 0.025 V vs. RHE and then oxidatively removed in fresh CO-free electrolyte. (From Ref. 94.) The working electrode area was about 50 mm^2.

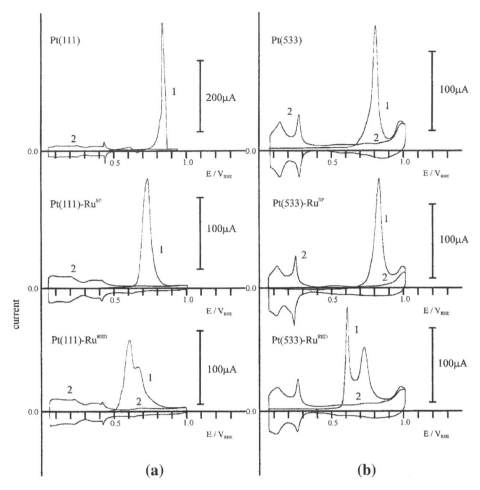

Figure 16 Stripping voltammetry (first cycle) $(0.5\,M\ H_2SO_4,\ 100\,m\,Vs^{-1})$ of saturated CO layers on clean and ruthenium-modified Pt(111) and Pt(533) surfaces. The ruthenium has been deposited by dipping, and following rinsing CO was adsorbed and stripped in CO-free electrolyte (Pt(111)^{SP} and Pt(533)^{SP}). The same surfaces were subsequently reduced in a flow of 10% H_2 in Ar, CO adsorbed and stripping voltammetry carried out (Pt(111)^{RED} and Pt(533)^{RED}). The second cycle in each case corresponds to the CO-free surface. (From Ref. 98.) The working electrode area was about 20 mm^2.

observed in the second cycle, indicating a total removal of the CO overlayer without loss of surface order. The overpotential for CO electro-oxidation on a ruthenium-modified surface with $\Theta_{XPS} = 0.18\,ML$ and $\Theta_{LEISS} = 0.04\,ML$ is significantly reduced. Oxidation of the CO takes place in two overlapping peaks, at 0.65 V and 0.75 V, with similar charges associated with each peak. Increasing the ruthenium coverage to $\Theta_{XPS} = 0.31\,ML$ and $\Theta_{LEISS} = 0.16\,ML$ leads to a slight downward shift of both peaks to 0.62 V and 0.72 V, with a decrease in the proportional charge from the upper oxidation peak. Further increases in ruthenium coverage do not lead to any further reduction in the overpotential for CO oxidation, with the two peaks remaining at the same potential. However, the contribution of charge associated with the high potential peak continues to decrease with increasing ruthenium coverage. At $\Theta_{XPS} = 1.34$-ML $\Theta_{LEISS} = 0.39$ ML, the upper peak is only present as a slight shoulder to the main feature at 0.62 V.

Two possible origins to the promotional behavior of the ruthenium in CO electro-oxidation were considered [94]. The first model involves the electronic perturbation of the CO by ruthenium, resulting in its susceptibility to oxidation and a concomitant increase in the rate of the surface reaction between adsorbed oxidant and CO. This was consider an unlikely explanation for the observed promotional effect on Pt(111) because of the significant unperturbed regions of the Pt(111) surface at these sub-monolayer coverages of ruthenium because of extensive nucleation. A significant overall promotion of CO electro-oxidation is apparent under these conditions (Figure 15). This is consistent with results obtained on the Pt(110)–Ru alloy surfaces [86], where it was shown that there is no correlation between the small reduction in CO binding energy by ruthenium in Pt(110) and the promotion of the electro-oxidation. *It was therefore proposed, rather, that the promotional effect of ruthenium clusters on the electro-oxidation of CO on Pt(111) is associated with the promoted activation of water to produce the surface-oxidizing species.* The subsequent oxidation takes place either at the edge of the ruthenium clusters (Zone I) or in regions of Pt(111) atomically more distant from the clusters (Zone II) (inset of Figure 14). This corresponds to a bi-functional mechanism initially proposed by Watanabe and Motoo [95,96]. The two peaks observed in the oxidative stripping voltammetry (Figure 15) on the ruthenium-modified surfaces correspond to CO oxidation taking place with the fastest Langmuir–Hinshelwood (L–H) kinetics in Zone I (the peak observed at lowest potential), and with slower L–H kinetics in Zone II (the peak observed at higher potential). These two zones are likely to be the atomically nearest-neighbor platinum sites surrounding the ruthenium clusters (Zone I) and sites one or more platinum atoms from the cluster (Zone II) and are shown schematically in Figure 14. An increase in the ruthenium coverage changes the proportion of Zone I and Zone II sites for CO adsorption and is responsible for the change in the proportion of charge associated with the two oxidation peaks. In order to substantiate this interpretation of the results, the XPS and LEISS data used to estimate the nucleation density and average cluster size of the ruthenium on the Pt(111) surface (Figure 14a) were used to establish the percentage of exposed platinum atoms in Zone I, and this is also shown in Figure 14b. Together with this curve are plotted the experimental points corresponding to the proportion of CO oxidized in Zone I and Zone II estimated from the integrated charges of the corresponding two peaks in the stripping voltammetry (from Figure 15). The close correspondence between the experiment and the calculated ensemble concentration indicates that the results on Pt(111)–Ru were consistent with the proposed bi-functional model.

This model was also used to explain the promotion of CO electro-oxidation by electrodeposited ruthenium islands on Pt(111) [93], with the surface oxidant being provided by the activating ruthenium atoms. However, in HClO$_4$ electrolyte only a single stripping peak is observed [93] at a sweep rate of $10 \, \text{mV s}^{-1}$ (Figure 12). This apparent difference with the Pt(111)–Ru stripping results obtained in sulphuric acid electrolyte (Figure 15) was tentatively attributed to the presence of a weakly (perchlorate) rather than a strongly (sulphate) absorbing anion and its effects on the

L–H kinetics, which follow water activation. It had been concluded that in the presence of perchlorate anions that the mobility of CO was high with respect to the rate of the L–H oxidation reaction [93]. It was demonstrated using partial stripping voltammetry that the more strongly adsorbing sulphate anion inhibits the mobility of the CO, resulting in the separation of the two kinetic regimes in sulphuric acid electrolyte [94]. However, the relatively immobile CO is oxidized at a lower overpotential in Zone II than the clean Pt(111) surface (Figure 15). This was attributed to the continued activation of water by the ruthenium, but with the additional requirement of spillover of the oxidant from the ruthenium cluster and its diffusion to Zone II for the L–H oxidation of Zone II CO to take place. Recent lattice gas simulations of CO electro-oxidation kinetics based on the bi-functional mechanism on ruthenium-modified platinum surfaces [97] have shown that either a single peak or two peaks may be observed in the CO stripping voltammetry depending on the ruthenium cluster size.

5.4.3 Ruthenium on Pt(533)

In order to demonstrate the importance of a local ensemble in the promotion by ruthenium of the L–H oxidation of CO, a number of experiments were carried out on stepped platinum surfaces [98]. The results of these experiments also provide an interesting comparison between surfaces modified by MVD ruthenium and through deposition from solution. Experiments were carried out [98] on Pt(111), Pt(533), and Pt(311) single-crystal surfaces. Ruthenium was dosed from an aged solution of $5 \times 10^{-4}\,M$ $RuCl_3$ in $0.5\,M$ H_2SO_4 (believed to contain the complex $\{RuO(H_2O)]^{2+}\}$ by spontaneous deposition at open-circuit potential) [80]. Experiments were carried out on the clean surfaces, following the spontaneously deposition of ruthenium, and on surfaces where the deposited ruthenium was reduced in a 10% H_2 in Ar gas mixture. CO was adsorbed on the variously prepared surfaces from solution and stripped in CO-free H_2SO_4 electrolyte.

Figure 16a shows CO stripping on Pt(111), Pt(111)–Ru[SP] (following spontaneous deposition), and Pt(111)–Ru[RED] (where the spontaneously deposited ruthenium has been reduced in hydrogen). Only a very small reduction in overpotential for CO electro-oxidation is observed for Pt(111)–Ru[SP]. The overpotential for CO electro-oxidation on the Pt(111) surface has been reduced, however, on the Pt(111)–Ru[RED] surface, and the latter exhibits a doublet structure. This CO stripping result on Pt(111)–Ru[RED] is nearly identical to that found on the Pt(111)–Ru surface where the ruthenium was MVD deposited (Figure 15): It was concluded that Pt(111)–Ru[RED] was decorated islands of Ru[0].

Figure 16b shows CO stripping on Pt(533), Pt(533)–Ru[SP], and Pt(533)–Ru[RED]. The inset of Figure 16 is a schematic of Pt(533) that consists of 4 atom terraces of Pt(111) with (100) steps. The overpotential for CO electro-oxidation on Pt(533)–Ru[SP] exhibits nearly no change from that of the clean surface. On Pt(533)–Ru[RED], however, behavior similar to that observed on Pt(111)–Ru[RED] is observed, i.e., a significant reduction in overpotential and a doublet stripping peak. Close comparison of CO stripping from Pt(533)–Ru[RED] and Pt(111)–Ru[RED] reveals that the second stripping peak for the two surfaces is at a similar potential. The first peak on Pt(533)–Ru[RED] is, however, considerably sharper than the first peak on Pt(111)–Ru[RED]. The result on Pt(533)–Ru[RED] was interpreted in the following manner. When reduced to Ru^0, the ruthenium decorates the inside step edges of Pt(533). This results in linear arrays of the Pt–Ru ensemble, highly active in CO electro-oxidation. Oxidation at the platinum site in the linear array of the ensemble results in the first sharp stripping peak (Figure 16). The second broader peak is analogous to the oxidation of CO on terrace sites, away from the Pt–Ru step-edge ensemble.

In order to substantiate this interpretation, experiments were carried out [98] on Pt(311)–Ru[RED]. This surface has a very high density of (100) steps. CO stripping experiments (not shown) on the Pt(311)–Ru[RED] surface were characterized by a dominance of anodic charge in the first of the two stripping peaks, i.e., that associated with the linear array of Pt–Ru ensembles. The charge associated with oxidation of CO adsorbed at terrace sites was very low, as one would expect for the low concentration of terrace sites on Pt(311).

5.5 ELECTROCATALYSIS AT SINGLE-CRYSTAL PLATINUM ALLOY SURFACES

While many studies have been performed for the oxidation of methanol and carbon monoxide on supported catalyst systems [66,99–103] and Pt–Ru bulk alloys [61,104–107], relatively few studies have been initiated on single-crystal platinum surfaces modified with ruthenium. Of those performed these have largely involved the investigation of platinum single crystals modified by ruthenium dosed electrochemically [92,93] or spontaneously [80–82,90,91] from aqueous chloride solutions. This approach is discussed in Section 5.4.

MVD of ruthenium on Pt(110) has been shown to provide an ideal system for the study of the promotion of electrocatalytic reactions on a well-characterized Pt–Ru alloy surface [85,86]. In transfer studies, XPS, LEISS, and LEED have been used to characterize the Pt(110)–Ru alloy system, and TPD and stripping voltammetry used to investigate the chemisorption behavior of CO, and the promotion of CO electro-oxidation as a function of incorporated ruthenium. The facile incorporation of ruthenium in the relatively open-packed Pt(110)–(1 × 2) surface provided an ideal model for the alloy system. It is also interesting to note also that the clean Pt(110) surface exhibits the highest hydrogen oxidation currents of the three basal planes of platinum [108].

The adsorption of 0.5 ML of ruthenium on Pt(110) at 300 K lifts the (1 × 2) reconstruction, with evidence of only a small amount of clustering. XPS and LEISS measurements demonstrate that flash-annealing submonolayer coverages of ruthenium leads to incorporation of ruthenium into the top surface layers and eventual bulk dissolution at temperatures above 1000 K. This behavior contrasts sharply with that found for ruthenium deposition on the close-packed Pt(111) surface (Section 5.4), and the differences are demonstrated by the changes in coverage measured by XPS and LEISS as a function of surface temperature shown in Figure 13a. The large differences in coverages estimated by the two techniques on Pt(111) are a result of clustering of the ruthenium overlayer, while the similar coverages on Pt(110) are a result of the wetting of the more open-packed surface. The incorporation of ruthenium onto Pt(110) results in the decrease in the LEISS coverage in the temperature range $300 < T_s/K < 900$, with little change in the average coverage in the uppermost three layers probed by XPS at these energies. The result of this behavior is that surfaces can be prepared: (A) where ruthenium is present as an adsorbed metal (following deposition at 300 K); (B) incorporated in the top layer as an alloy (following annealing to ≈ 500 K); (C) incorporated in the second or third atomic layer on the Pt(110) surface with no ruthenium in the top layer (following annealing to about 900 K); or (D) with ruthenium incorporating more deeply in the bulk of the crystal (annealing to about 1100 K). The reactivity of the surface phases of ruthenium in surface types A–D were investigated [85,86]. The modification of the chemisorption behavior of the CO was investigated by TPD in order to establish any correlation between electro-oxidation activity and CO bonding.

TPD from the clean and ruthenium-modified surfaces types A–D are shown in Figure 17. TPD of CO from the unmodified Pt(110) surface led to two desorption peaks at 415 K and 515 K, in good agreement with the literature [109–111]. Both peaks are known to be due to linearly bound CO as demonstrated by vibrational spectroscopy [110,111]. The low-temperature peak is observed at higher coverages as a result of the strong repulsive lateral interactions between the CO molecules at coverages greater than 0.5 ML [110]. On a type A surface (freshly deposited ruthenium corresponding to coverages $\Theta_{XPS} = 0.29$ and $\Theta_{LEISS} = 0.18$), TPD of CO results in two less distinct peaks, shifted down in temperature to ≈ 375 K and 500 K, are observed.

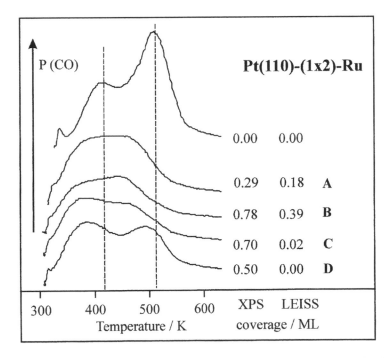

Figure 17 TPD of saturated CO layers from clean and MVD ruthenium-modified Pt(110)–(1 × 2). The ruthenium coverages are estimated by XPS and LEISS. Modified surface A corresponds to an unannealed surface, and B–D have been annealed to various temperatures [86].

Annealing a surface on which a higher coverage of ruthenium was adsorbed to 680 K produces a type B Pt(110)–Ru surface with a slightly higher top-layer concentration of ruthenium, but incorporated as an alloy ($\Theta_{XPS} = 0.78$ and $\Theta_{LEISS} = 0.39$). Further pre-annealing to 890 K results in a type C Pt(110)–Ru surface, with nearly no ruthenium in the top layer, but significant concentrations in the second and third layers ($\Theta_{XPS} = 0.70$ and $\Theta_{LEISS} = 0.02$). TPD of CO on types B and C surfaces are similar to type A, and with desorption temperatures similar to those reported by Iwasita et al. for the TPD of CO from bulk Pt–Ru alloys, where two peaks at 390 and 445 K were obtained [112]. *There is clearly a significant perturbation of CO adsorbed on the top-layer Pt atoms with ruthenium in the second and third layers, and this perturbation is similar to that found for surfaces modified by ruthenium in the top layer incorporated as either an alloy or an adsorbate.* After flash-annealing to 1100 K, the TPD features associated with CO adsorbed on Pt(110)–(1×2) are beginning to be regained as one may expect for the type D surface, where ruthenium is dissolving deeper into the bulk of the crystal (Figure 13a).

In contrast, the effect of the adsorbed and alloyed ruthenium on the electro-oxidation of CO has shown that promotion of the reaction is only evident if ruthenium is present in the top surface layer. Hence the mediation of the oxidizing species by top-layer ruthenium in the provision of Pt–Ru ensembles, rather than the modification of CO adsorption by ruthenium, promotes the electro-oxidation reaction [85,86].

Figure 18 show a series of CO stripping voltammograms for Pt(110)–Ru surfaces prepared by MVD and characterized by XPS and LEISS before clean transfer to the electrochemical cell. Surface A was prepared by MVD of Ru at 300 K with a coverage calculated from XPS (Θ_{XPS}) of 0.28, which gave a corresponding coverage of 0.28 in LEISS (Θ_{LEISS}). Surfaces B–D were prepared from an initial coverage of Ru of 1 ML followed by flash-annealing to a range of temperatures, as indicated above. The CO has been dosed from solution under potential control at $0.05\,V_{RHE}$. The initial cycle for each surface has been limited to $0.85\,V_{RHE}$ in order to minimize any irreversible oxidation or dissolution of the ruthenium overlayers. Subsequently CO was re-adsorbed from solution under potential control at $0.05\,V_{RHE}$ on the same surface that had been previously cycled to $1.25\,V_{RHE}$. Note that CO adsorption completely poisons the adsorption of hydrogen before it is stripped, as one may expect for a saturated CO overlayer.

Surface A, produced through deposition of ruthenium at 300 K to coverages corresponding to $\Theta_{XPS} = 0.28$ and $\Theta_{LEISS} = 0.28$, produces a surface on which the ruthenium is adsorbed but not yet incorporated into the Pt(110) surface. This adsorbed-ruthenium phase on Pt(110) displays similar voltammetry to that observed for pure ruthenium [85]. CO adsorbed on this surface exhibits a broad electro-oxidation peak centered at $0.72\,V_{RHE}$ (Figure 18a). The peak is not significantly shifted from the overpotential for CO oxidation on the Pt(110) surface (Figure 9), although the broadening is largely to lower potential. This contrasts sharply with the results obtained for surface B. This surface has been prepared in such a way as to exhibit a similar top-layer coverage of ruthenium, but under conditions in which the ruthenium is incorporated into the first three layers of the Pt(110) surface to form an alloy through annealing an initially larger coverage to 625 K. CO electro-oxidation on such a surface (Figure 18b) is strongly promoted in a single broad peak centered at $0.6\,V_{RHE}$ V_{RHE}. This promotion could be associated either with the incorporated

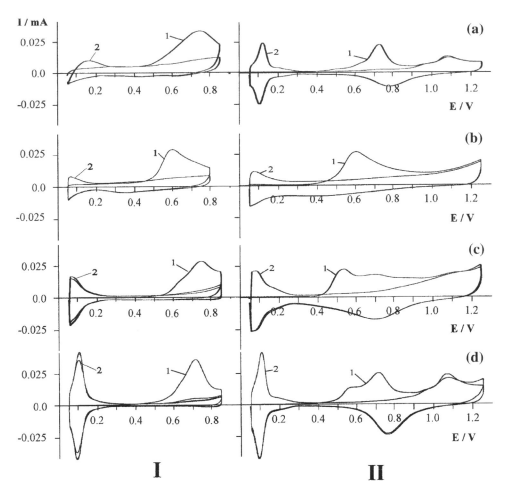

Figure 18 Stripping voltammetry for saturated CO layers on Pt(110) for modified surface types (a) directly after deposition at 300 K ($\theta_{XPS} = 0.31$ $\theta_{LEISS} = 0.28$); (b) annealing at 625 K ($\theta_{XPS} = 0.90$ $\theta_{LEISS} = 0.20$); (c) annealing at 750 K ($\theta_{XPS} = 0.90$ $\theta_{LEISS} = 0.05$); (d) annealing at 1050 K ($\theta_{XPS} = 0.50$ $\theta_{LEISS} = 0.01$). CO has been adsorbed from solution following transfer at $0.05\,V_{RHE}$. The second cycle corresponds to the CO-free surface. Following transfer, the first cycles are restricted to $0.8\,V_{RHE}$, and are shown in series I. The surfaces have subsequently been oxidatively cycled to $1.25\,V_{RHE}$ (ten cycles) and the CO adsorbed and stripped again in series II [85,86]. The working electrode area was about $50\,mm^2$.

ruthenium in the top Pt layer ($\Theta_{LEISS} = 0.2$), or with the underlying concentration of ruthenium in the second/third layers ($\Theta_{XPS} = 0.9$). It will become evident that ruthenium in the top layer is responsible for the promotion of CO electro-oxidation.

Reducing the top-layer concentration to $\Theta_{LEISS} = 0.05$ (Figure 18c), and keeping the average in the top three layers the same ($\Theta_{XPS} = 0.9$), is accompanied by a concomitant reduction in the size of the low potential peak: this promoted oxidation is observed as a small shoulder on the now major peak at $0.72\,V_{RHE}$

(Figure 18c), the overpotential characteristic of Pt(110). Further reduction of the top-layer concentration to $\Theta_{LEISS} = 0.01$ and reduction of the second/third-layer concentration to $\Theta_{XPS} = 0.5$ are accompanied (Figure 18d) by a reversion to the Pt(110) CO oxidation behavior. We can conclude from the measurements shown in Figure 18 that *CO electro-oxidation is promoted primarily by ruthenium incorporated in the top layer of Pt(110)*, by adsorbed ruthenium to a lesser extent, and not by Pt atoms modified by ruthenium in the second/third atomic layers. This does not correlate with the perturbation observed for CO adsorption by ruthenium in the TPD. Ruthenium in the second/third atomic layers had the most significant effect on CO adsorption, reducing the desorption temperature and lowering the saturation coverage. It is, therefore, not primarily the modification of CO adsorption behavior by ruthenium that promotes the CO electro-oxidation. Rather it is the effect of ruthenium in the top layer to activate water dissociation that is responsible for the promotional effect. The difference between the adsorbed and incorporated ruthenium (Figures 18a and 16b) is likely to be a result of a better distribution of reactive ensembles in the surface alloy. In the case of adsorbed ruthenium, reactive ensembles may be available only at the edge of adsorbate islands. These ensembles are similar to the highly promoting ensembles at the edges of ruthenium clusters on Pt(111) or in the linear arrays of ensembles on the ruthenium-modified step surfaces (Section 5.4). Alternatively, incorporated ruthenium may be a more effective promoter of water activation than the adsorbed state because of differences in electronic structure.

Cycling surface A to $1.25\,V_{RHE}$ and subsequently carrying out CO stripping voltammetry (Figure 18) results in the oxidative dissolution of a significant proportion of the adsorbed ruthenium. This is evidenced by the recovery of the hydrogen adsorption structure associated with Pt(110)–(1 × 2) and CO electro-oxidation, which takes place mainly in a sharper peak at $0.72\,V_{RHE}$, more characteristic of the clean Pt(110) surface. In contrast, CO electro-oxidation on the Pt(110)-ruthenium surface with a similar top-layer coverage (surface B) but on which the ruthenium is incorporated in an alloy exhibits no change on cycling the surface to $1.25\,V_{RHE}$ when compared to 0.8 V. The Pt(110)-ruthenium alloy phase is more resilient to oxidative modification, at least in so far as the removal by oxidative dissolution of the incorporated ruthenium phase. Comparison of CO stripping on the alloy surfaces prepared with low concentrations of ruthenium in the top layer (Figures 18c and 16d) and cycled only to $0.8\,V_{RHE}$ with the same surfaces that have previously been cycled to $1.2\,V_{RHE}$ reveal substantial differences. The result of cycling to $1.25\,V_{RHE}$ is to produce a surface that gives rise to an additional CO oxidation peak at lower potential to the peak at 0.7 V associated with Pt(110) sites, which dominates before oxidative modification. This new peak is observed at $0.52\,V_{RHE}$ on C and $0.56\,V_{RHE}$ on D. A concomitant reduction in charge associated with the 0.7-V peak is observed. Ruthenium is clearly promoting CO electro-oxidation on the oxidatively modified surfaces. A clue as to what is taking place by cycling to $1.25\,V_{RHE}$ to induce the promotional phase is given by the observation of a small increase in the redox currents in the Pt(110) double-layer region following the oxidative cycling procedure [85]. This is associated with ruthenium incorporated in the topmost layer of the Pt(110) surface. We must conclude that the ruthenium, which was present in the second/third layers, is the source of the ruthenium that segregates to the top layer. Note that the initial (before oxidative cycling) surface coverages were $\Theta_{LEISS} = 0.05$ and $\Theta_{LEISS} = 0.01$. The voltammograms obtained after cycling (Figures 18c and 16d) would indicate that a substantial proportion (50% and 35% in the case of Figure 5c and 6d, respectively) of the surface is in the promoting phase following oxidative cycling.

5.6 CONCLUSION

The activity, stability, and tolerance of supported platinum-based anode and cathode electrocatalysts in PEM fuel cells clearly depend on a large number of parameters including particle-size distribution, morphology, composition, operating potential, and temperature. Combining what is known of the surface chemical reactivity of reactants, products, and intermediates at well-characterized surfaces with studies correlating electrochemical behavior of simple and modified platinum and platinum alloy surfaces can lead to a better understanding of the electrocatalysis. Steps, defects, and alloyed components clearly influence reactivity at both gas–solid and gas–liquid interfaces and will understandably influence the electrocatalytic activity.

The surface structure has been shown to clearly influence the ability of platinum to dissociate hydrogen and chemisorb and dissociate oxygen. These processes are related to key steps in hydrogen electro-oxidation (Tafel) and oxygen electroreduction (through either sequential or direct pathways). In the case of hydrogen dissociation, an indirect channel to dissociation mediated at defect sites will dominate the reaction on platinum for thermal sources such as those found in fuel cells. Adsorbates such as CO and O preferentially adsorb at such sites and block this dissociation channel. The introduction of an alloy component in a metal does not block the defect-mediated indirect channel to hydrogen dissociation. Similarly, metal adsorbates and alloying components in platinum appear not to hinder hydrogen electrooxidation. Oxygen molecular chemisorption and dissociation also take place through indirect channels, and surface defects provide sites exhibiting low-energy barriers for the sequential conversion of the physisorbed to chemisorbed and dissociative states. It is therefore possible that the surface concentration of such active sites may determine the propensity of electroreduction of oxygen to take place through a sequential or a direct pathway. The higher specific activity of the extended surface over small particles in oxygen reduction is direct evidence of a structure-dependent reaction. Further studies of electro-oxidation and reduction kinetics at well-characterized stepped and modified single-crystal platinum surfaces are required to establish the details of the influence of surface structure on the overall reaction mechanism.

There is clear evidence that adsorbate and alloyed metal atoms on platinum surface promote CO electro-oxidation. The reduced overpotential is primarily a result of the promotion of the activation of water. The subsequent kinetics are determined by the details of a Langmuir–Hinshelwood reaction between the adsorbed oxidant (OH) and adsorbed CO. Evidence is also presented that relates this promotion (or poisoning) of CO electro-oxidation to tolerate CO in hydrogen feeds in the hydrogen electro-oxidation reaction. An alternative mechanism that may operate at low potentials [79,113] may be that the reduction in CO adsorption energy on platinum induced by Ru [86,113,114] results in a higher equilibrium concentration of nonpoisoned sites. The relative importance of these mechanisms is a function of the effective overpotential of the hydrogen electro-oxidation reaction of the PEM anode, and this itself is a function of CO coverage if reasonable kinetics are to be maintained. Studies of the hydrogen electro-oxidation kinetics on well-characterized promoted platinum surfaces [60] will provide a clearer picture of the mechanism responsible for the promotion of CO-tolerant PEM anodes.

REFERENCES

1. J. Clavilier, R. Faure, G. Guinet, R. Durand, J. Electroanal. Chem. 107 (1980), 205.
2. D.J. Pegg, in *Chemistry*, University of Southampton, Southampton, 1996, p. 294.
3. D.M. Kolb, G. Lehmpfuhl, M.S. Zei, in *Interfacial Electrochemistry*, C. Gutierrez and C. Melendres, eds., Kluwer Academic, The Netherlands, 1999, p. 361.
4. G. Lehmpfuhl, Y. Uchida, M.S. Zei, D.M. Kolb, in *Frontiers of Electrochemistry*, vol. 5, J. Lipkowski and P.N. Ross, eds., Wiley-VCH, New York, 1999, p. 57.
5. K. Franaszczuk, E. Herrero, P. Zelenay, A. Wieckowski, J. Wang, R.I. Masel, J. Phys. Chem. 96 (1992), 8509.
6. K. Christmann, G. Ertl, Surface Science 60 (1976), 365.
7. B. Poelsema, L.K. Verheij, G. Comsa, Surface Science 152 (1985), 496.
8. L.K. Verheij, M.B. Hugenschmidt, B. Poelsema, G. Comsa, Catalysis Letters 9 (1991), 195.
9. S.L. Bernasek, G.A. Somorjai, J. Chem. Phys. 62 (1975), 3149.
10. R.J. Gale, M. Salmeron, G.A. Somorjai, Phys. Rev. Lett. 38 (1977), 1027.
11. M. Salmeron, R.J. Gale, G.A. Somorjai, J. Chem. Phys. 67 (1977), 5324.
12. M. Salmeron, R.J. Gale, G.A. Somorjai, J. Chem. Phys. 70 (1979), 2807.
13. L.K. Verheij, M.B. Hugenschmidt, A.B. Anton, B. Poelsema, G. Comsa, Surface Science 210 (1989), 1.
14. J.E. Muller, Phys. Rev. Lett. 59 (1987), 2943.
15. J.E. Muller, Applied Physics A—Solids and Surfaces, 49 (1989), 681.
16. A.C. Luntz, J.K. Brown, M.D. Williams, J. Chemical Phys. 93 (1990), 5240.
17. P. Samson, A. Nesbitt, B.E. Koel, A. Hodgson, J. Chem. Phys. 109 (1998), 3255.
18. G. Anger, H.F. Berger, M. Luger, S. Feistritzer, A. Winkler, K.D. Rendulic, Surface Science 219 (1989), L 583.
19. A.T. Gee, B.E. Hayden, C. Mormiche, T.S. Nunney, J. Chem. Phys. 112 (2000).
20. H.P. Steinruck, M. Luger, A. Winkler, K.D. Rendulic, Phys. Rev. B, Condens. Matter 32 (1985), 5032.
21. D.A. Butler, B.E. Hayden, J.D. Jones, Chem. Phys. Lett. 217 (1994), 423.
22. D.A. Butler, B.E. Hayden, Chem. Phys. Lett. 232 (1995), 542.
23. D.A. Butler, B.E. Hayden, Surface Science 337 (1995), 67.
24. B.E. Hayden, A. Hodgson, J. Phys. Condens. Matter 11 (1999), 8397.
25. S. Mukerjee, J. McBreen, J. Electrochem. Soc. 143 (1996), 2285.
26. J.C. Davies, B.Sc. project report (1994).
27. B.E. Hayden, C. Mormiche, T.S. Nunney, J. Phys. Chem. (to be submitted).
28. S.J. Lee, S. Mukerjee, E.A. Ticianelli, J. McBreen, Electrochim. Acta 44 (1999), 3283.
29. A. Kabbabi, F. Gloaguen, F. Andolfatto, R. Durand, J. Electroanal. Chem. 373 (1994), 251.
30. M. Kita, H.-W. Lei, Y. Gao, J. Electroanal. Chem. 379 (1994), 407.
31. H. Kita, Y. Gao, K. Ohnishi, Chem. Lett. (1994), 73.
32. R.J. Nichols, in *Adsorption of Molecules on Metal Electrodes*, J. Lipkowski and P.N. Ross, eds., VCH, New York, 1993.
33. S. Mukerjee, S. Srinivasan, M.P. Soriaga, J. McBreen, J. Electrochem. Society 142 (1995), 1409.

34. J.L. Gland, B.A. Sexton, G.B. Fischer, Surf. Sci. 95 (1980), 587.

35. D.M. Collins, W.E. Spicer, Surf. Sci. 69 (1977), 85.

36. J. Grimblot, A.C. Luntz, D.E. Fowler, J. Electron Spectroscopy and Related Phenomena 52 (1990), 161.

37. N.R. Avery, Chem. Phys. Lett. 96 (1977), 371.

38. D.M. Collins, W.E. Spicer, Surf. Sci. 69 (1977), 114.

39. H. Steininger, S. Lehwald, H. Ibach, Surf. Sci. 123 (1982), 1.

40. S. Lehwald, H. Ibach, H. Steininger, Surf. Sci. 117 (1982), 342.

41. C. Puglia, A. Nilsson, B. Hemnas, O. Karis, P. Bennich, N. Martensson, Surf. Sci. 342 (1995), 119.

42. J.L. Gland, Surf. Sci. 93 (1980), 487.

43. A.C. Luntz, J. Grimblot, D.E. Fowler, Phys. Rev. B 39 (1989), 12903.

44. A. Cudok, H. Froitzheim, G. Hess, Surf. Sci. 309 (1994), 761.

45. J. Stohr, J.L. Gland, W. Eberhardt, D. Outka, R.J. Madix, F. Sette, R.J. Koestner, U. Doebler, Phys. Rev. Lett. 51 (1983), 2414.

46. D.A. Outka, J. Stohr, W. Jark, P. Stevens, J. Solomon, R.J. Madix, Phys. Rev. B. 35 (1987), 4119.

47. W. Wurth, J. Stohr, P. Feulner, X. Pan, K.R. Bauchspiess, Y. Baba, E. Hudel, G. Rocker, D. Menzel, Phys. Rev. Lett. 65 (1990), 2426.

48. O. Bjomeholm, A. Nilsson, H. Tilborg, P. Bennich, A. Sandell, B. Hemnas, C. Puglia, N. Martensson, Surf. Sci. 315 (1994), L983.

49. A.C. Luntz, M.D. Williams, D.S. Bethune, J. Chem. Phys. 89 (1988), 4381.

50. B. Williams, A.C. Luntz, J. Chem. Phys. 88 (1988), 2843.

51. M.D. Williams, D.S. Bethune, A.C. Luntz, J. Vacuum Sci. & Tech. Vacuum Surfaces and Films 6 (1988), 788.

52. C.T. Rettner, C.B. Mullins, J. Chem. Phys. 94 (1991), 1626.

53. A.E. Wiskerke, F.H. Geuzebroek, A.W. Kleyn, B.E. Hayden, Surf. Sci. 272 (1992), 256.

54. P.D. Nolan, B.R. Lutz, P.L. Tanaka, J.E. Davis, C.B. Mullins, Phys. Rev. Lett. 81 (1998), 3179.

55. P.D. Nolan, B.R. Lutz, P.L. Tanaka, J.E. Davis, C.B. Mullins, J. Chem. Phys. 111 (1999), 3696.

56. A. Winkler, X. Guo, H.R. Siddiqui, P.L. Hagans, J.T. Yates, Surf. Sci. 201 (1988), 419.

57. A.T. Gee, B.E. Hayden, J. Chem. Phys. 113 (2000), 10333.

58. A. Eichler, J. Hafner, Phys. Rev. Lett. 79 (1997), 4481.

59. T. Zambelli, J.V. Barth, J. Wintterlin, G. Ertl, Lett. to Nature 390 (1997), 495.

60. H.A. Gasteiger, N.M. Markovic, P.N. Ross, J. Phys. Chem. 99 (1995), 16757.

61. H.A. Gasteiger, N. Markovic, P.N. Ross, E.J. Caims, J. Phys. Chem. 98 (1994), 617.

62. J.L. Gland, E.B. Kollin, J. Chem. Phys. 78 (1983), 963.

63. J.L. Gland, R.J. Madix, R.W. McCabe, C. DeMaggio, Surf. Sci. 143 (1984), 46.

64. B.E. Hayden, K. Kretzschmar, A.M. Bradshaw, R.G. Greenler, Surf. Sci. 149 (1985), 394.

65. R.G. Greenler, K.B. Burch, K. Kretzschmar, R. Klauser, A.M. Bradshaw, B.E. Hayden, Surf. Sci. 152 (1985), 338.

66. B. Beden, F. Kadrigan, C. Lamy, J.M. Leger, J. Electroanal. Chem. 127 (1981), 75.

67. R.R. Adzic, in *Advances in Electrochemistry and Electrochemical Engineering*, Vol. 13, H. Gerischer, ed., Wiley, New York, 1984, p. 159.

68. M. Watanabe, S. Motoo, J. Electroanal. Chem. 60 (1975), 259.

69. S.A. Campbell, R. Parsons, J. Chem. Soc. Faraday Trans. 88 (1992), 833.

70. F. Herrero, J.M. Feliu, A. Aldaz, J. Catal. 152 (1995), 264.

71. E. Herrero, A. Rodes, J.M. Perez, J.M. Feliu, A. Aldaz, J. Electroanal. Chem. 393 (1995), 87.

72. B.E. Hayden, A.J. Murray, R. Parsons, D.J. Pegg, J. Electroanal. Chem. 409 (1996), 51.

73. D.C. Godfrey, B.E. Hayden, A.J. Murray, R. Parsons, D.J. Pegg, Surf. Sci. 294 (1993), 33.
74. M.T. Paffett, C.T. Campbell, T.N. Taylor, J. Vacuum Sci. & Tech. A—Vacuum Surfaces and Films 3 (1985), 812.
75. M.T. Paffett, C.T. Campbell, J. Chem. Phys. 85 (1986), 6176.
76. R.W. Evans, G.A. Attard, J. Electroanal. Chem. 345 (1993), 337.
77. B.E. Hayden, Catalysis Today. 38 (1997), 473.
78. R. Gomez, A. Fernandezvega, J.M. Feliu, A. Aldaz, J. Phys. Chem. 97 (1993), 4769.
79. S. Mukerjee, S.J. Lee, E.A. Ticianelli, J. McBreen, B.N. Grgur, N.M. Markovic, P.N. Ross, J.R. Giallombardo, E.S. De Castro, Electrochem. and Solid-State Lett. 2 (1999), 12.
80. W. Chrzanowski, A. Wieckowski, Langmuir 13 (1997), 5974.
81. W. Chrzanowski, A. Wieckowski, Langmuir 14 (1998), 1967.
82. E. Herrero, J.M. Feliu, A. Wieckowski, Langmuir 15 (1999), 4944.
83. S. Szabo, I. Bakos, J. Electroanal. Chem. 230 (1987), 233.
84. M. Vukovic, T. Valla, M. Milun, J. Electroanal. Chem. 356 (1993), 81.
85. J.C. Davies, B.E. Hayden, D.J. Pegg, Electrochimica Acta 44 (1998), 1181.
86. J.C. Davies, B.E. Hayden, D.J. Pegg, Surf. Sci. 467 (2000), 118.
87. T.D. Jarvi, T.H. Madden, E.M. Stuve, Electrochem. and Solid-State Lett. 2 (1999), 224.
88. E. Herrero, K. Franaszczuk, A. Wieckowski, J. Electroanal. Chem. 361 (1993), 269.
89. K.A. Friedrich, K.-P. Geyzers, U. Linle, U. Stimming, J. Stumper, J. Electroanal. Chem. 402 (1996), 123.
90. W. Chrzanowski, H. Kim, A. Wieckowski, Catalysis Lett. 50 (1998), 69.
91. G. Tremiliosi-Filho, H. Kim, W. Chrzanowski, A. Wieckowski, B. Grzybowska, P. Kulesza, J. Electroanal. Chem. 467 (1999), 143.
92. K.A. Friedrich, K.-P. Geyzers, A. Marmann, U. Stimming, R. Vogel, Zeitschrift fur Physikalische Chemie 208 (1999), 137.
93. W.F. Lin, M.S. Zei, M. Eiswirth, G. Ertl, T. Iwasita, W. Vielstich, J. Phys. Chem. B. 103 (1999), 6968.
94. J.C. Davies, B.E. Hayden, D.J. Pegg, M.E. Rendall, Surf Sci. in press (2001).
95. M. Watanabe, S. Motoo, J. Electroanal. Chem. 60 (1975), 267.
96. M. Watanabe, Y.M. Zhu, H. Igarashi, H. Uchida, Electrochem. 68 (2000), 244.
97. M.T.M. Koper, J.J. Lukkien, A.P.J. Jansen, R.A. vanSanten, J. Phys. Chem. B 103 (1999), 5522.
98. J.C. Davies, B.E. Hayden, M.E. Rendall, O. South, J. Electroanal. Chem. in preparation (2001).
99. R. Ianniello, V.M. Schmidt, Ber. Bunsenges. Phys. Chem. 99 (1995), 83.
100. M. Krausa, W. Vielstich, J. Electroanal. Chem. 379 (1994), 307.
101. J. Munk, P.A. Christensen, A. Hamnett, E. Skou, J. Electroanal. Chem. 401 (1996), 215.
102. C.E. Lee, S.H. Bergens, J. Phys. Chem. 102 (1998), 193.
103. P.S. Kauranen, E. Skou, J. Munk, J. Electroanal. Chem. 404 (1996), 1.
104. H.A. Gasteiger, N. Markovic, P.N. Ross, E.J. Cairns, J. Phys. Chem. 97 (1993), 12020.
105. H.A. Gasteiger, N. Markovic, P.N. Ross, E.J. Cairns, J. Electrochem. Soc. 141 (1994), 1795.
106. H.A. Gasteiger, N. Markovic, P.N. Ross, E.J. Cairns, Electrochimica Acta 39 (1994), 1825.
107. H.A. Gasteiger, N.M. Markovic, P.N. Ross, J. Phys. Chem. 99 (1995), 8290.
108. H. Kita, Y. Gao, S. Ye, K. Shimazu, Bull. Chem. Soc. Jpn. 66 (1993), 2877.
109. H.P. Bonzel, R. Ku, J. Chem. Phys. 58 (1973), 4617.
110. B.E. Hayden, A.W. Robinson, P.M. Tucker, Surf. Sci. 192 (1987), 163.
111. S.R. Bare, P. Hofmann, D.A. King, Surf. Sci. 144 (1984), 347.

112. T. Iwasita, R. Dalbeck, E. Pastor, X. Xia, Electrochim. Acta 39 (1994), 1817.
113. F. Buatier de Mongeot, M. Scherer, B. Gleich, E. Kopatzki, R.J. Behm, Surf. Sci. 411 (1998), 249.
114. E. Christofferson, P. Liu, A. Ruban, H.L. Skriver, N.J.K., J. Catal. 199 (2001), 123.

6

Electrochemical Nanostructuring of Surfaces

ROLF SCHUSTER and GERHARD ERTL

Fritz-Haber-Institut der Max-Planck-Gesellschaft, Berlin, Germany

CHAPTER CONTENTS

SUMMARY

Two different approaches for nanostructuring of surfaces are discussed, where electrochemical methods were proven particularly successful. First, ordering processes following electrochemically driven phase transitions are investigated. Depending on the driving force of the ordering process (i.e., the excess free energy of the system), the phase separation may take place either on the basis of nucleation and growth or through spinodal decay. The formation of structures ranging from compact islands up to labyrinthine island patterns on the nanometer scale was in-situ observed by electrochemical scanning tunneling microscopy. In the second approach, the application of ultrashort voltage pulses of only few nanoseconds duration to tiny tool electrodes allows for localized electrochemical dissolution or deposition of material. This was demonstrated for the formation of nanometer holes and clusters on Au(111) surfaces as well as for the three-dimensional machining of metal sheets on the micrometer scale.

6.1 INTRODUCTION

Many physical and chemical properties of materials are determined not only by the materials themselves, but also by their geometric dimensions. As discussed in detail in this book, the size of nanoparticles influences the catalytic activity, not only due to the enhanced surface area, but also due to particular electronic properties, different from those of bulk material. Confining electrons in small metal particles considerably changes the electronic band structure and therefore the chemical bonding of adsorbates [1–3]. In semiconductor nanocrystals, because of the lower charge carrier densities, such effects are even more pronounced, and the electron confinement leads to discrete energy levels rather than to electronic bands [4]. Tailoring the size and shape of metal or semiconductor structures allows for adjusting the electronic and chemical properties according to the actual application. For example, it has recently been shown by Maroun et al. that the electrochemical/ catalytic activity of PdAu(111) electrodes for the CO oxidation or hydrogen adsorption is dependent on the actual size of the Pd islands embedded in the Au(111) surface [5]. However, not only the size of the structures but also their geometrical arrangement may be of importance. In optoelectronic devices as quantum dot lasers, the lasing action depends crucially on the lateral coupling between the single quantum dots and, therefore, on their well-defined arrangement [6,7].

These two examples already imply two completely different approaches for the fabrication of nanoparticle arrays on surfaces. One strategy is the part-by-part manufacturing of the structural elements to obtain well-defined arrangements. The second approach is to exploit the "self-organization" of the system, e.g., upon an ordering process following a phase transition [8] or the pattern formation in dissipative systems [9]. The second strategy has the advantage of patterning large surface areas in parallel. A drawback is the rather predetermined structure, given in general by the thermodynamic or kinetic properties of the system.

How can such ordering processes be influenced and steered into a particular direction? Electrochemistry is particularly useful in this respect, since the free energy of the surface system is directly correlated with the electrochemical potential. A simple variation of the electrochemical potential changes the state of the system and may eventually drive a transition into a different surface phase. The electrochemical potential can in general be varied very rapidly, just limited by the time constant of the electrochemical cell, which is given by the capacity of the electrodes' electrochemical double layer and the electrolyte resistance [10].

In Section 6.2 we discuss the evolution of structures occurring upon such electrochemically driven phase transitions. The resulting surface morphology depends crucially on the driving force for the phase transition, i.e., the starting point in the phase diagram of the system. At low driving forces nucleation and growth processes will prevail, whereas at high enough deviations from the phase equilibrium the instantaneous decay of an unstable system can be observed. Nucleation and growth processes will be demonstrated for phase transitions in a Cu underpotentially deposited (UPD) layer on Au(111), which is in-situ observed by electrochemical scanning tunneling microscopy (STM) [11].

Reducing the distance between the electrodes into the micro-to nanometer range, e.g., by employing the tip of an electrochemical STM as a local counter electrode, decreases the time constant of the electrochemical cell well into the nanosecond range. This allows for the investigation of very fast ordering processes, usually inaccessible upon varying other thermodynamic variables of state as temperature or particle concentration. Since electrochemical reaction rates are exponentially dependent on the potential drop in the double layer, reactions can be driven on a time scale barely achievable with other methods. This offers new possibilities for the formation of kinetically determined structures upon the equilibration of the system: in Section 6.2.2 we present data on the ordering process of a thermodynamically unstable Au ad-atom gas [12]. Half a monolayer of Au atoms from the topmost Au(111) surface was removed within a 20-μs pulse to the tip of an electrochemical STM. The remaining Au ad-atom gas decayed into a labyrinthine pattern of Au islands, typical for the spinodal decomposition of a two-phase system. This is completely different from morphologies, expected upon nucleation and growth of a new phase, which usually governs "slow" deposition of material.

Also, for the part-by-part manufacturing of nano- and microstructures, electrochemical methods were proven to be very successful. Particularly in combination with scanning probe methods several approaches were employed. For instance, Kolb and co-workers fabricated arrays of thousands of Cu nanoclusters on an Au(111) surface by mechanical detachment from an STM tip, which was in-situ electrochemically plated with Cu [13]. Another example is the scanning electrochemical microscope (SECM), where an ultramicroelectrode (UME) is positioned in close proximity to the surface [14,15]. Surface modifications are achieved either by direct oxidation or reduction of the surface or by local production of reactants at the UME, which then diffuse toward the surface. The size of the produced structures is determined by the geometry of the electrode arrangement and the subsequent current distribution in the electrolyte or the diffusion length of the reactive species and, in general, lies in the range of several micrometers, even with very tiny electrodes. All these scanning probe methods have, however, to cope with a general problem, which prevents the localization of electrochemical reactions: electrochemical reactions are governed by the potential drop within a very narrow region at the electrode surface, i.e., the electrochemical double layer [10]. Upon application of an external voltage the ions in the solution redistribute and the potential varies essentially across the double layer, with the electrolyte providing an effective shortcut in between the double layers on the two electrodes. Therefore, spatial resolution in conventional electrochemistry rather stems from secondary effects as diffusion length of the reactive species or current density variations due to spatially varying electrolyte thickness. Even with such tiny electrodes as the tip of an electrochemical STM, only limited spatial confinement of conventional electrochemical reactions has to be expected.

To overcome this constraint, we performed experiments where we applied very short voltage pulses of only nanoseconds' duration to an STM tip in electrochemical environment. The limited diffusion of the ions during such short voltage pulses prevented the ion distribution in the electrolyte from equilibration and indeed led to confinement of the electrochemical reactions on a nanometer scale. This method is demonstrated for the formation of holes of about 5-nm diameter and up to 3 monolayers (ML) depth on an Au(111) surface by local anodic dissolution of Au and for the local deposition of Cu clusters of similar dimensions by reduction of Cu ions from a $CuSO_4$ electrolyte [16,17]. These experiments are reviewed and compared with other electrochemical approaches in Section 6.3.1. The application of short voltage pulses to the electrodes offers an additional means for conducting local electrochemistry. Besides the diffusion of ions, also their migration proceeds with finite velocity [12,18,19]. This is the reason for the finite time constant of potential changes in an electrochemical cell, where upon application of an external field ions are forced to migrate and charge the respective double layers at the electrodes, finally canceling the external field inside the electrolyte. The migration velocity is inversely proportional to the applied field, i.e., the ratio between applied voltage and electrode separation. Therefore, upon application of short voltage pulses, the electrodes' double layers can only be charged in locations where the electrodes are in close enough proximity. Since ion migration is enforced by an external field, it is in general much faster than thermal diffusion, and with ns pulses and concentrated electrolytes sufficient polarization of the double layers and subsequent electrochemical reaction occur only for electrode distances in the micro-rather than in the nanometer range. This method and recent experimental results are discussed in Section 6.3.2.

6.2 STRUCTURE FORMATION UPON ORDERING PROCESSES

Ordering processes upon first-order phase transitions usually involve long-range mass transport leading to the formation of intermediate structures in the course of the equilibration of the system [8]. For example, the small droplets nucleating upon water condensation in a cloud are the precursors of the complete phase separation into bulk water and saturated water vapor [20]. Similarly, the lamellar structure occurring upon quenching of an Fe/Al alloy is the consequence of the uncompleted demixing into Fe- and Al-rich phases [21]. Although at a first glance the morphologies of such unequilibrated systems differ considerably, their major structural features depend in principle only on one single parameter, the driving force for the ordering process, i.e., the magnitude of the deviation from the equilibrium situation, initiating the phase transition. This is discussed with the help of Figure 1, which shows a general phase diagram of a two-phase system, representative, e.g., for the vapor–liquid phase coexistence in water or the phase separation in a binary mixture of oil and water. Temperature is depicted versus an order parameter, which in the case of water might be the density. Inside the coexistence region, the simultaneous existence of two phases is energetically favorable; i.e., a homogeneous system driven into that region of the phase diagram will separate into the two phases at the right and the left side of that region to reach thermal equilibrium [(2) and (3) in Fig. 1]. If the driving force is weak, at positions close to the border of the two-phase region, the homogeneous system is in a metastable state [22] [(1) in Fig. 1]. In other words, the formation of the new phase requires surmounting an energy barrier. For example, upon the formation of water droplets from supersaturated vapor, due to the unfavorable surface-volume ratio of small droplets and their high surface tension, the addition of water molecules to small droplets costs energy rather than gaining energy by the formation of the stable condensed phase [20]. Only above a certain droplet size does the transfer of water molecules from the supersaturated gas phase to the condensed phase gain energy. Such a phase separation process is called nucleation and growth, and the droplet size above which the droplets' growth is energetically favorable is called the critical radius. Since the driving force for such phase transitions is rather low, and the

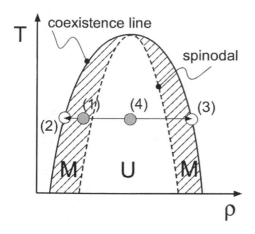

Figure 1 Sketch of the phase diagram of a two-phase system. Temperature is plotted versus an order parameter, which might be the density. Inside the coexistence region the spinodal (dotted line) separates metastable regions (M) from the unstable regime (U). The decay of a metastable system (1) into the two stable phases (2) and (3) at the coexistence line proceeds via surmounting a nucleation barrier and subsequent growth. Unstable systems (4) instantaneously separate into (2) and (3) via spinodal decomposition.

formation of the critical nucleus occurs via thermal fluctuations, nucleation and growth processes are slow and the resulting morphology of the system exhibits compact and round shaped islands, i.e., droplets.

However, if the driving force becomes large enough, the energy barrier hindering the phase separation will vanish, and the homogeneous system becomes unstable [8,22,23]. The decay into the two stable phases can now proceed instantaneously. In the framework of classical near-equilibrium thermodynamics, this instability occurs at the so-called spinodal (Figure 1), and the decay process is called spinodal decomposition. It has been described theoretically by Cahn and Hilliard for the decomposition of metal alloys and is completely different from the nucleation and growth process described above. Since the system is unstable, no critical nucleus has to be formed upon the reorganization. The system is ordering by instantaneous exponential growth of thermal density fluctuations of a particular wavelength, which finally leads to an interwoven, labyrinthine pattern of the two phases. In crystalline materials the overall structure can introduce anisotropy, which, e.g., leads to the lamellar structure of some magnetic materials [24]. It is noteworthy that the concept of the spinodal is derived from equilibrium thermodynamics, which is in principle not applicable during the beginning of the decomposition with locally strongly varying order parameter. Detailed theoretical investigations recently refined this picture, but without any consequence for the qualitative discussion here [8,23].

Since it is often difficult to achieve the unstable situation inside the spinodal, phase-ordering processes following the nucleation and growth scheme are much more common from daily experience. Upon changing the thermodynamic variables of state into the coexistence region, the system is in general crossing the metastable region of the phase diagram. If this change of the thermodynamic state is too slow, nucleation and growth processes will occur before rendering the system unstable. Moreover, the nucleation barrier gradually vanishes when proceeding toward the spinodal, i.e., with increasing supersaturation. In surface systems most observed phase-ordering processes therefore involve nucleation and growth processes. For example, in the deposition of thin metal films, the variable of state, i.e., the metal ad-atom surface concentration, is changed so slowly compared with the diffusion of the ad-atoms that nucleation events and island formation occur long before the final surface coverage is achieved [25]. Additionally, deposited metal atoms directly nucleate at the existing islands and the supersaturation of the metal ad-atom phase remains rather low.

The only possibility for entering the spinodal region without crossing the metastable region is via a temperature quench through the critical point of the system. This was achieved, e.g., in alloy systems, requiring, however, careful adjustment and control of the system parameters [8,23].

6.2.1 Nucleation and Growth Processes in Electrochemistry

Also in electrochemical systems, nucleation and growth processes play a prominent role in the formation of a new phase. Such processes were observed indirectly via analyzing current transients in potential step experiments or directly via microscopic observations, e.g., of electrochemical metal deposition. The dependence of the nucleation density of Ag crystallites on the overpotential was originally studied by Budewski and co-workers [26], initiating various experiments on the electrochemical growth of metal crystallites with controlled size distributions. Recently, the coevolution of hydrogen and its influence on the mass transport have been proven to be particularly successful for obtaining monodisperse metal cluster distributions [27]. In this section we present in-situ STM images of electrochemically driven phase transitions in Cu UPD layers on Au(111) surfaces with special respect toward the nucleation and growth of the domain pattern. Cu UPD on Au(111) in the presence of various anions has been studied intensively during the last two decades and can be considered as a model system for UPD metal deposition (see [28,29] and references therein). A $(\sqrt{3} \times \sqrt{3})R30°$ phase is formed by a honeycomb lattice of 2/3 of a monolayer (ML) of Cu with 1/3 ML of sulfate anions adsorbed in the honeycomb centers [30,31]. Lowering the potential further toward the bulk Cu deposition leads to the formation of a (1×1) structure of a full ML of Cu, also covered by sulfate anions [32–34].

The experiments were conducted in a conventional electrochemical STM [11,35]. Au films with a thickness of 250 nm on a Cr-coated (2-nm) glass substrate were used as samples. After repeated cycles of rinsing with triply distilled water and subsequent flame annealing, these films exhibited (111) terraces up to 100 nm wide. High-resolution images of these surfaces show the $(\sqrt{3} \times 22)$ reconstruction of the clean Au(111) surface in accordance with [36].

In-situ STM images of the evolution of the morphology upon the phase transition from the (1×1) phase into the $(\sqrt{3} \times \sqrt{3})R30°$ phase, driven by a potential step, are shown in Figure 2. Experimental details are described in [11]. Before recording Figure 2a, the surface was completely covered by the (1×1) UPD phase by holding the electrochemical potential only a few mV positive from the bulk Cu deposition. After a positive potential step of 200 mV at the beginning of recording

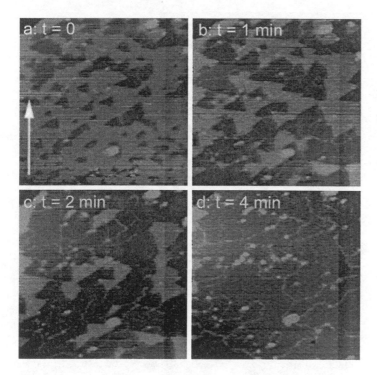

Figure 2 Series of STM images (100 nm × 100 nm; recording time 1 min per image) of the phase transition between the (1×1) and the $(\sqrt{3} \times \sqrt{3})R30°$ phase in a Cu UPD layer on Au(111) in 1 M CuSO$_4$ + 0.5 M H$_2$SO$_4$ solution. Directly before recording Figure 2a, the potential was stepped to 0.2 V versus Cu|Cu^{2+}, initiating the homogeneous nucleation of dark triangular domains of the $(\sqrt{3} \times \sqrt{3})R30°$ phase. The slow scan direction is indicated by an arrow. (From Ref. 11.)

Figure 2a, small dark triangular islands of the $(\sqrt{3} \times \sqrt{3})R30°$ structure homogeneously nucleate on the formerly gray (1×1) covered terrace. (The slow scan direction of the STM is directed upward in that image. Recording one image takes approximately one minute.) The domains of the $(\sqrt{3} \times \sqrt{3})R30°$ grow until they intercept each other. The $(\sqrt{3} \times \sqrt{3})R30°$ structure exhibits three translational domains on the Au(111) surface. Because all three domains are equidistributed among the nucleating islands, by chance the intercepting domains are out of phase and form domain walls, finally resulting in the domain wall pattern of Figure 2d (faint gray lines). The domain wall pattern is random. Repeating the experiment leads to a completely different pattern. In particular, Au clusters, which remained on the surface from the lifting of the $(\sqrt{3} \times 22)$ reconstruction of the Au(111) surface upon the UPD layer formation (bright white dots in the STM images), do not influence the domain pattern.

Also the reverse transition, the formation of a (1×1) structure on the domain network of the $(\sqrt{3} \times \sqrt{3})R30°$ phase can be followed in-situ by STM. Figure 3 shows the evolution of the surface structure upon a potential step from the $(\sqrt{3} \times \sqrt{3})R30°$ into the (1×1) Cu UPD phase. Starting from the domain

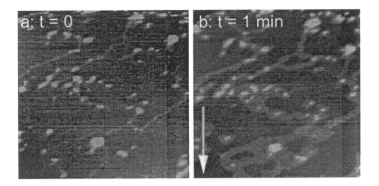

Figure 3 STM images (100 nm × 100 nm; recording time 1 min per image) of the heterogeneous nucleation of the (1×1) Cu UPD phase at the domain boundaries of the $(\sqrt{3} \times \sqrt{3})R30°$ structure. At the beginning of the recording of Figure 3b, the potential was switched to about 0 V versus Cu|Cu^{2+}. (From Ref. 11.)

configuration of the $(\sqrt{3} \times \sqrt{3})R30°$ in Figure 3a, upon switching the potential the (1×1) phase starts to grow instantaneously (Figure 3b, slow scan direction downward). The existing domain walls broaden until the (1×1) phase covers the whole surface (data not shown). In this case, the new phase nucleates heterogeneously at the defects of the $(\sqrt{3} \times \sqrt{3})R30°$ structure, i.e., the domain walls. Whether the domain walls in the $(\sqrt{3} \times \sqrt{3})R30°$ structure already exhibited the (1×1) phase or whether they merely supported the nucleation of the new phase cannot be unequivocally decided. In any case, the formation of the (1×1) phase at the existing defects/domain walls is kinetically favored over homogeneous nucleation, randomly distributed over the surface.

In these two examples of electrochemically driven growth processes, subsequent mass transport, i.e., the increase of the Cu coverage upon the (1×1) or the adsorption of sulfate during the $(\sqrt{3} \times \sqrt{3})R30°$ formation, occurs on a rather slow time scale. Therefore, the deviation from the thermodynamic equilibrium situation remains small during the reorganization of the system, and the evolution of the surface morphology is governed by nucleation and growth.

6.2.2 Spinodal Structures upon Fast Electrochemical Ordering Processes

Electrochemical phase transitions can in principle be driven very fast. The reaction rate of the electrochemical processes is exponentially dependent on the overpotential. Therefore, in the absence of mass-transport limitations, the only restriction is the finite charging time of the double layer, which is given by the cell geometry, i.e., the time constant of the RC network, consisting of the electrolyte resistance and the double-layer capacity. In conventional cells the length of the current path d between the counter and working electrodes is in the range of centimeters. With typical values for the specific electrolyte resistance $\rho \approx 100 \, \Omega \, \text{cm}$ and the double-layer capacity $c_{\text{DL}} \approx 10 \, \mu\text{F}/\text{cm}^2$, this results in a charging time constant $\tau = \rho d c_{\text{DL}} \approx 100 \, \Omega \, \text{cm} \cdot 1 \, \text{cm} \times 10 \, \mu\text{F}/\text{cm}^2 = 1 \, \text{ms}$. This is still not very fast, compared with typical jump rates of metal ad-atoms on metal surfaces. For example, from the data derived by Bott et al. [37] for the diffusion of individual Pt ad-atoms on a Pt(111) surface in vacuum at room temperature a jump rate of about 10^8 1/s is estimated.

However, since the charging time constant of the double layer linearly depends on the electrolyte resistance and, therefore, on the distance between counter and working electrode, the double-layer charging time constant can be significantly lowered simply by changing the size of the effective cell. Using the tip of an electrochemical STM as local counter electrode in a few nanometers' distance to the surface reduces the time constant well into the nanosecond range, even with diluted electrolytes. Moreover, the evolution of the surface morphology can be observed in situ by STM, directly after inducing the structural changes. In the following, we describe experiments where Au atoms of the topmost layer of an Au(111) surface were electrochemically oxidized and dissolved by microsecond-long voltage pulses [12]. The electrolyte was 1-M KCl, which ensures that the oxidized Au atoms are readily complexed and easily dissolved in the electrolyte. Therefore, the reaction occurs without mass-transport limitations, and solely the overpotential, i.e., the pulse height, determines the dissolution rate of Au. The reordering of the disturbed topmost layer is subsequently monitored by STM.

The experimental setup for such fast electrochemical experiments is sketched in Figure 4, employing a modified electrochemical STM previously described in [16,38]. The sample potential was potentiostatically controlled versus a reference electrode. For conducting local electrochemistry, a high-frequency pulse generator, which supplied voltage pulses as low as 10-ns duration and voltages up to ± 4 V, was switched to the tip for a few milliseconds. The current voltage converter of the STM was protected against the high-voltage pulses by a low-pass filter. Since the time

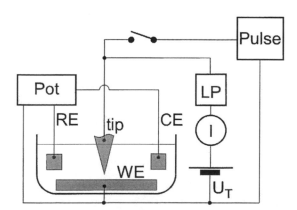

Figure 4 Sketch of the electrochemical STM for short-pulse surface modifications. The potential of the working electrode (WE) is controlled by a low-frequency potentiostat (Pot) versus the reference electrode (RE) via the counterelectrode (CE). The tunneling voltage (U_T) is supplied via the I/U converter of the STM (I_T) and a low-pass filter (LP). To apply the short pulses to the STM tip, a high-frequency pulse generator (Pulse) is switched onto the tip for a few milliseconds.

constant of the STM preamplifier and the controller is much longer than the duration of the short pulses, the vertical position of the tip remains unchanged during the application of the pulses, even if large charging and Faradaic currents flow for a few ns. Gating of the feedback loop is therefore optional, although in most cases the feedback loop of the STM was switched off for a couple ms, to allow for the recovery of the preamplifier, before stable tunneling conditions were reestablished. A mechanically ground and polished Ir wire was used as the STM tip in order to withstand the relatively high-voltage pulses during the electrochemical surface modification without degradation of the imaging quality. The side of the tip was coated with Apiezon wax W 100 to reduce parasitic Faradaic currents.

By proper choice of the electrolyte Au atoms can be dissolved in the electrolyte during short pulses. Figure 5 shows the result of a -4-V, 20-μs pulse to the STM tip, which was applied in the center of the image while slowly scanning downward in 4-M KCl (Figure 5a). In order to allow for large-scale reaction and unconstrained transport of ions in the electrolyte, the tip was retracted by about 2 nm during the application of the pulse. After the pulse was applied, an interconnected, labyrinthine pattern of monoatomically high Au islands formed on the surface. Comparison with previously recorded images revealed that only Au atoms from the topmost terraces were dissolved during the pulse without changing the terrace structure of the surface.

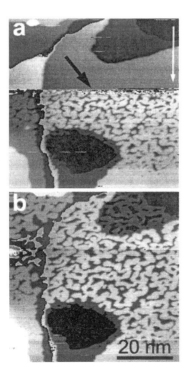

Figure 5 (a) During scanning of an Au(111) surface in 4 M KCl, the application of a 20-μs, -4-V pulse to the STM tip led to the formation of a labyrinthine pattern of Au islands. (b) The pattern coarsened slightly in the consecutive STM image, recorded 1 min later.

Therefore, during the short voltage pulse, about 1/2 ML of Au atoms from the topmost surface layer has been removed, leading to the ordering of the remaining 1/2 ML into the labyrinthine pattern.

The topography of the island pattern is completely different from that expected upon vacuum evaporation of material or slow electrochemical deposition or dissolution, as discussed in Section 6.2.1. In the latter case, the situation is governed by nucleation and growth, mostly leading to relatively compact, round-shaped islands. The labyrinthine patterns observed here rather point to the spinodal decay of an unstable Au ad-atom gas, which decomposed instantaneously. It is conceivable that upon the application of the high-potential step Au atoms from the topmost surface layer were statistically removed, leaving an Au ad-atom gas with a density of 1/2 ML behind. Since the dissolution of the material occurred about 6 orders of magnitude faster than the change of the ad-atom concentration with conventional vacuum or electrochemical methods, the metastable region of the phase diagram was surpassed before nucleation and growth of compact Au islands could start. Indeed, assuming that the particle interactions in the Au atom gas are dominated by nearest-neighbor interactions, a density of 1/2 ML equals the critical density [39], i.e., the ad-atom gas with 1/2 ML density is situated in the middle of the coexistence region, where it is definitely unstable (see Figure 1).

The labyrinthine pattern slowly coarsens on a time scale of minutes (Figure 5b). Eventually, after about 30 min, the pattern breaks up into single, compact islands, and Ostwald ripening sets in where big islands grow on the expense of smaller ones (images not shown here). It is noteworthy that the size of the reacted surface area in Figure 5 amounts to only about 60 nm × 60 nm, whereas the rest of the surface remains unaffected by the pulse. Apparently, due to the short duration of the pulse, electrochemical reactions were confined near the very end of the STM tip. In the following section, we discuss how short voltage pulses, applied to properly shaped tool electrodes, can indeed be employed to the part-by-part manufacturing of small surface structures.

6.3 ELECTROCHEMICAL PART-BY-PART MACHINING OF NANO- AND MICROSTRUCTURES

In conventional electrochemistry, where the current density distribution in the solution reached its steady state, the shape and size of the counter electrode are of only secondary influence on the lateral extension of the modifications. Even upon employing very small electrodes such as the tip of an electrochemical STM as a local counter electrode, the modified surface area far exceeds the length scale of the atomic structures resolved in the STM image. This is demonstrated in Figure 6a, which shows the result of the application of a −3-V, 10-μs-long pulse to the tunneling tip, while imaging an Au(111) surface in 10-mM H_2SO_4 + 1-mM $CuSO_4$. The experimental setup is identical to that of Section 6.2.2. After the pulse the previously atomically flat Au(111) surface exhibits the typical roughened morphology of an oxidized Au(111) surface with monolayer high islands (white) and monolayer deep holes (black) [40–42]. In contrast to the experiments in Section 6.2.2 the absence of Cl⁻ prevented the dissolution of Au, and the oxidation process led to the formation

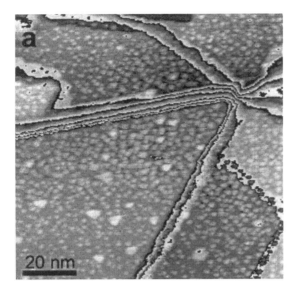

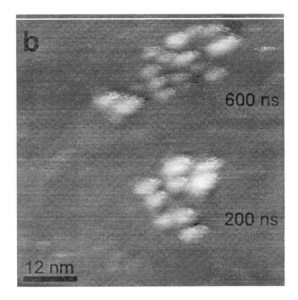

Figure 6 (a) Application of a 10-μs, −4-V pulse to the STM tip during imaging an Au(111) surface in 0.5 M CuSO₄ resulted in large-scale oxidation of the surface. (b) Reducing the pulse duration to 600 ns or 200 ns in 0.5 M CuSO₄ led to the spatial confinement of the electrochemical reaction. The oxidized region still exhibits the fingerprint of the electrochemical Au oxidation, i.e., the characteristic island and hole pattern.

of insoluble Au hydroxides and oxides on the topmost Au surface. Obviously, already within 10 μs the double layer on the Au surface was sufficiently charged over the whole imaged region, to induce the large-scale electrochemical oxidation. Because the recording time for an STM image is about 1 min, the STM tip did in effect not move during the short voltage pulse, inducing the modification. However, no hint on the actual position of the STM tip during the µs voltage pulse is discernible in the image.

It is noteworthy that several 100 nm away from the imaged surface region in Figure 6a the surface remains unchanged for two reasons. First, secondary effects as current density and voltage drop in the electrolyte become more important with larger distance between the electrodes, and, therefore, the reaction rate at more distant areas decreases. The second reason is that the charging of the double layer requires considerable current to flow through the electrolyte. Hence, with increasing distance between the electrodes and a longer current path through the electrolyte, the charging time of the double layer increases. Eventually the potential drop across the double layer, reached within the short pulse, will be insufficient to initiate the electrochemical reaction. However, according to the above approximation for the double-layer charging time constant, with 10-µs pulses this effect becomes efficient only at length scales of several 10 µm [18]. In Section 6.3.2 we discuss how this charging behavior of the double layer can be exploited to confine electrochemical reactions down to submicrometer precision. Even-higher spatial resolution, however, can be obtained by applying ns pulses to an STM tip, employing the confined geometry and limited electrolyte reservoir in the tip-surface arrangement of an STM, which is demonstrated first.

6.3.1 Nanoscale Surface Modifications with the Electrochemical STM

Figure 6a demonstrates that there is an intrinsic limitation of the achievable spatial resolution in (quasi) steady-state electrochemistry. To overcome this limit, which is determined by the equilibrium ion and current distributions in the electrolyte, we further reduced the duration of the voltage pulses. Figure 6b shows the result of two single pulses, 200 ns and 600 ns long. The locations where the two pulses were applied are clearly visible. They exhibit the typical fingerprint of an oxidized Au surface; however, the reactions are now sharply confined to small areas, only several 10 nm in diameter. Further reduction of the pulse duration to 50 ns even leads to the formation of well-defined holes on the Au surface with about 5-nm diameter and a depth between 1 and 3 ML (Figure 7). The STM image shows the in-situ formation of a single hole exactly at the tunneling position of the tip where the pulse was applied (indicated by an arrow). Since the slow scan direction of the STM image was directed upward, only the upper half of the hole is imaged. The three holes in the upper part of the image were previously produced by three single pulses.

To check the influence of the electrolyte on the local confinement of the reaction, we conducted experiments in various electrolytes such as H_2SO_4, HCl, KCl, $NaSO_4$, $CuSO_4$, and some of their mixtures. It turned out that the most important ingredient for the local surface modification was the use of highly concentrated electrolytes. In pure water, hole formation was impossible with ultrashort voltage pulses, which rules out a purely field-induced surface modification, due to the high field strength at the tip apex, and further supports the electrochemical nature of the process.

For an ultimate proof of the electrochemical nature of the process, however, it has to be checked whether upon inversion of the pulse polarity also the opposite reaction of the hole formation, i.e., the reduction of ions from the solution, becomes

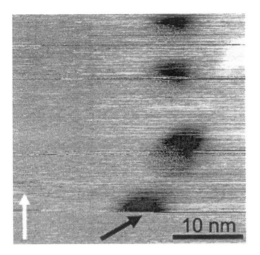

Figure 7 In-situ formation of a single hole on Au(111) by a 50-ns, -2-V pulse in 1 M $CuSO_4$ + 0.5 M H_2SO_4. The hole is formed at the location of the tip during the application of the pulse (black arrow). The slow scan direction is directed upward. The three holes in the upper part of the image were previously produced.

possible. This is shown in Figure 8a on an Au(111) surface immersed in 0.5-M $CuSO_4$. Three short, positive voltage pulses (50 ns, $+4$ V) led to the formation of three Cu clusters with about 5-nm diameter and monoatomic height. The clusters are located exactly at the position where the tip was tunneling at the moment when the pulse was applied. These clusters dissolved completely after increasing the sample potential to about 400 mV versus $Cu|Cu^{2+}$, substantiating that the precipitates indeed consisted of copper and did not constitute contaminations disposed off the tip. In this respect it should also be mentioned that the potential of the tip was at about 400 mV versus $Cu|Cu^{2+}$ prior to the application of the pulse, which excludes underpotential deposition of Cu on the tip. Therefore, the deposited Cu had to stem directly from reduction of ions of the electrolyte.

The respective Cu deposition experiment upon long voltage pulses is shown in Figure 8b. As expected, a single, long, positive pulse (20 μs, $+1.3$ V) led to the large-scale deposition of Cu clusters on the surface, which is the well-known growth mode for conventional electrochemical Cu deposition in the absence of surface-active substances [43–45].

When discussing the mechanism of the local nanostructuring, it should be kept in mind that the STM tip is a rather macroscopic object. Our tips are mechanically ground wires with a typical tip radius of about 1 μm. During tunneling the very apex of the tip resides at approximately 1-nm distance before mechanical contact with the Au surface occurs. This was deduced from tunneling-current versus tip-surface distance measurements. Therefore, the actual geometry in the vicinity of the reacted surface area rather resembles an extended gap with a slightly varying gap width, which increases from about 1 nm at the tip apex to about 10 nm at several 100-nm lateral distance. Due to the relatively small distance between the electrodes, the

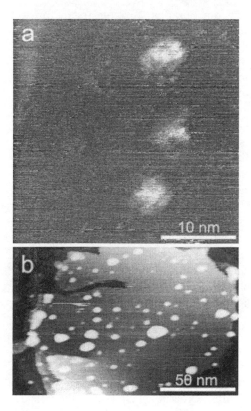

Figure 8   (a) Three single Cu clusters on Au(111) formed by three 50-ns, 3-V pulses to the STM tip in 1 M CuSO$_4$ + 0.5 M H$_2$SO$_4$. (b) Application of a 20-μs, 1.9-V pulse in 0.5 M CuSO$_4$ led to the large-scale deposition of Cu clusters.

migration of the ions in the applied field during the pulse is extremely fast. With an estimated mobility of 10^{-4} cm^2/Vs and a typically applied external field of 1 V/10 nm, the redistribution of the ions in this gap is finished after the first few nanoseconds of the pulse. However, these ions are incorporated in the double layer, whose charging requires an amount of ions, equivalent to about 0.1 ML for a 1-V change of the double-layer potential. Considering that even in highly concentrated electrolytes of about 1-M concentration, as employed here, only every 50th particle comprises an ion of the respective charge, the charging of the double-layer capacity instantaneously consumes about all the ions in the gap. The electrolyte in the gap is hence significantly depleted already at the very beginning of the pulse.

In contrast, refilling the gap with ions from the bulk of the solution happens on a much slower time scale because it is driven solely by the concentration gradient. With a typical diffusion constant of 10^{-5} cm^2/s the diffusion length within 100 ns for a delta-shaped concentration profile is only about 10 nm, which corresponds to about one tenth of the estimated lateral extension of the gap. Therefore, there is no significant refilling of the gap during the short pulse, leaving the electrolyte there highly depleted. The depleted solvent layer in the gap can hence be reasoned as local insulation of the macroscopic tip, which prevents Faradaic currents from flowing and therefore, suppresses the electrochemical reaction over most of the gap region. Only at the very apex of the tip, where the electrodes' distance is merely a few diameters of a solvent molecule, is this simplified picture of double layers at the electrodes and an electrolyte in between no longer applicable. Here, the distance between the ions in the double layers is of the same order of magnitude as the distance between the electrodes, and the notion of an electrolyte resistance is meaningless. It is expected that direct transitions of ions from one electrode to the counter electrode and vice versa become possible in the local electric field. Only there the electrochemical reaction proceeds and the Au surface is locally dissolved.

The importance of depletion and subsequent refilling of the electrolyte in the tip-surface gap for the confinement of the reaction is substantiated by Figure 9. This image was recorded after a 2-μs, −3-V pulse was applied to the tip while it resided at the middle of a 150-nm × 150-nm terrace. This terrace was part of a staircaselike morphology, steeply descending to the right and ascending to the left. Therefore, the tip-surface gap became narrower toward the left-hand side and opened to the right, which eases the refilling of the depleted electrolyte from that side. In agreement with the explanation above, the reaction was limited to regions near the descending step where the ion concentration increased considerably during the relatively long pulse, and the resulting finite electrolyte resistance allowed the large-scale electrochemical reaction.

Although the electrochemical nature of the nanostructuring process is convincingly demonstrated, the role of the externally applied electric field should be once more considered. Field-induced modifications of a surface on an atomic scale have been reported from various vacuum experiments [46–49]. The fields required for such manipulations are much larger than those applied here and usually necessitate very small distances between tip and surface, close to mechanical contact

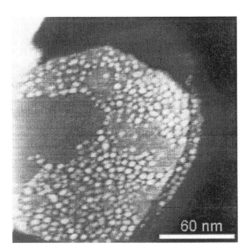

Figure 9   Application of a 2-μs, −3-V pulse in the middle of an Au(111) terrace in a stepped region led to the large-scale oxidation of the surface, proceeding from the descending (right) step edge (0.5 M $CuSO_4$).

[48]. Upon tunneling in an electrochemical environment the typical tip-surface distance is much larger, and the resulting external field during the voltage pulse is more than a factor of 5 lower than those applied for the vacuum modifications. However, it is conceivable that the external field can compensate for insufficient charging of the double layer, which might result from a deficiency of ions in the gap. This might provide a reason why hole formation on the Au(111) surface worked with almost every tip in concentrated electrolytes if the external voltage was raised to about 3 V. The external field adds to the electric field in the double layer and accelerates the electrochemical reaction. Such effects are, of course, only relevant for small distances at the very apex of the tip. In contrast to this, upon Cu cluster deposition by reduction of ions from solution, sufficient copper ions have to be present in the gap to form a microscopic, stable cluster. Because the number of ions is dependent on the specific shape of the gap, the success of the Cu cluster deposition depends much more on an appropriate tip-surface geometry than on the hole formation.

The local electrochemistry by ultrashort voltage pulses differs significantly from other electrochemical approaches for producing small structures with the help of an STM. One of these involved the electrochemical plating of the tunneling tip with Cu and the successive detachment of Cu clusters on the surface by mechanical contact [13] as mentioned above. Although the reproducibility of this method could be enhanced up to 10^4 almost identical clusters in a row, it is still limited to systems where the adhesion between the material to be deposited and the surface is high enough to allow detachment from the tip. In a recent experiment by Hoffmann et al. millisecond voltage pulses have been employed to redissolve material from the tip (in this case Co) where it had previously been deposited [50]. This led to a local, transient enhancement of the metal-ion concentration and, therefore, to local variations of the Nernst potential for its bulk deposition on the surface. Although this process is in principle determined by the diffusion properties of the species, the achieved resolution is about 20 nm with good reproducibility. Other experiments concern the structuring of surfaces in humid air by use of an STM tip. However, in such experiments on Au(111), voltage pulses to the STM tip longer than 100 ms were necessary for hole formation with reasonably low voltages [51,52]. This is directly conceivable from the low-ion concentration in pure water (10^{-7} M), which condensed in the tip-surface gap and forms the electrolyte in those experiments. Indeed, in experiments with bulk water, we found no reaction with ns pulses and large-scale reaction with μs pulse duration, during which substantial ion transport from the water bulk into the tip-surface gap could occur. Therefore, in experiments conducted in humid air the local confinement of the reaction has to be attributed to the formation of a small water neck in between the tip and the surface, due to capillary forces, as proposed in recent experiments on the oxidation of Si and Ti [53,54].

A different approach for producing small metal clusters on a surface was sucessfully demonstrated in Penner's group. Small Ag clusters were deposited on a graphite surface from an Ag^+ solution by applying positive voltage pulses of 50-μs duration to the STM tip [55,56]. The local metal deposition started a few microseconds after the beginning of the pulse and was attributed to nucleation of an Ag cluster in a small hole on the graphite surface formed instantaneously during the application of the pulse. Note that on graphite the hole formation proceeds independently of the polarity of the pulse, which points to a field effect rather than to electrochemical oxidation of the surface. Analogously, holes on the Au surface produced by ultrashort voltage pulses also can serve as local nucleation sites for Cu bulk deposition. Figure 10a shows three holes of nm dimensions on an Au(111) surface in 0.5-M CuSO$_4$ at a sample potential of 100 mV versus Cu|Cu^{2+}. Upon reducing the potential toward -60 mV, the Cu bulk deposition started in the holes (Figure 10b). Contrary to this, on the bare surface next to the holes the Cu bulk deposition was kinetically suppressed because of a substantial nucleation barrier, due to the surface energy of the Cu clusters (see, e.g., [57]). As discussed in more detail in [17] the height of these clusters is dependent on the overpotential rather than on the polarization time, reflecting the energy balance between the surface energy of the cluster and the electrochemical excess energy of the Cu reduction.

In conclusion, carrying out electrochemical reactions far from equilibrium conditions, i.e., far from the equilibration of the ion distribution in the electrolyte, can indeed lead to strongly localized surface modifications on the nm scale. Technological applications comparable to those of e-beam lithography might be possible but still suffer, however, from relatively slow processing speed, due to the sequential nature of the modification process. Confining electrochemical reactions on a surface by the application of short voltage pulses adds a new versatility to the variety of scanning probe modification techniques, which were recently reviewed, e.g., in [58–60].

6.3.2 Electrochemical Micromachining with Ultrashort Voltage Pulses

The preceding method (Section 6.3.1) relies completely on the limited equilibration of the ion distribution in solution, due to the hindered diffusion. Additionally, the charging time of the double layer can be exploited for the local machining of surfaces. As mentioned in Section 6.2.2, the time constant for the double-layer charging is given by the product of solution resistance and double-layer capacity. In the above experiments employing the tip of an STM, which is a few nms' distance to the surface, as local counter electrode this time constant is well below nanoseconds, even for diluted electrolytes. Nonetheless, for electrode separations in the

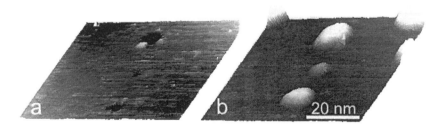

Figure 10   (a) Three holes on an Au(111) terrace produced by 75-ns, -2.4-V pulses in 1 M CuSO$_4$, imaged at a surface potential of 100 mV versus Cu|Cu^{2+}. (b) Reducing the electrochemical potential to -60 mV versus Cu|Cu^{2+} led to deposition of Cu clusters of up to 3-ML height in the holes. On the terrace no nucleation of Cu occurred at such small overpotentials.

micrometer range, this time constant amounts to about 10 ns (with typical values for the specific electrolyte resistance and the double-layer capacity, [18]). In other words, upon application of voltage pulses of 10-ns duration, the electrodes' double layers can only be charged if the electrode separation is below about 1 μm; if it is larger, no significant charging will occur during the pulse and the electrochemical reactions are effectively suppressed. This argument can be directly transferred to irregularly shaped electrodes in close proximity (Figure 11). The locally varying length of the current path through the electrolyte leads to locally varying electrolyte resistances (R_{wide}, R_{close} in Figure 11a) for the charging of the double-layer capacity C_{DL}. Upon application of a short voltage pulse to the electrodes, the double layer at electrodes' regions in close proximity is charged more strongly than at those farther apart (Figure 11b). Although the charging-time constant and, therefore, the degree of charging vary about linearly with the local electrode separation, electrochemical reaction rates are exponentially dependent on the voltage drop in the double layer, leading to an overall exponential dependence of the electrochemical reaction rate on the local electrode separation. Therefore, upon application of ultrashort voltage pulses, the electrochemical reactions can be strongly confined to electrode regions in very close proximity. Properly shaped tool electrodes can, for example, be locally etched into a workpiece without affecting surface regions farther apart. The pulse duration directly controls the maximum gap width, i.e., spatial resolution of this method.

A sketch of the experimental implementation is shown in Figure 12a. The workpiece is mounted in an electrochemical cell, attached to a piezo stage for three-dimensional manipulation with submicrometer precision. The average potentials of the workpiece as well as the baseline potential of the tool electrode are bipotentiostatically controlled. For local surface modifications, high-frequency pulses are supplied to the tool electrode. A typical current transient upon machining a Cu surface with a cylindrical Ø50-μm Pt tool in 0.01-M $HClO_4$ + 0.1-M $CuSO_4$ upon the application of 50-ns pulses of 1.6-V amplitude is depicted in Figure 12b. The rectangular background of the current peak corresponds to the insignificant charging of the more distant regions of the tool and workpiece electrodes: no increase of the double-layer voltage and hence no decline of the current were achieved there during the short pulses. To the contrary, electrodes regions that are in very close proximity, i.e., the front face of the cylindrical tool and the corresponding Cu surface regions, are strongly charged, signaled by the steeply decreasing current peak superimposed on the rectangular background. Those charged regions are quickly discharged after the end of the voltage pulse, indicated by the negative current peak. As expected, both the maximum current and the slope of the charging and discharging peaks are dependent on the tool workpiece distance (data not shown here, [18]).

The result of such a microstructuring experiment where the tool was etched into a Cu surface upon application of a sequence of 10^9 50-ns pulses of 1.6-V amplitude, which in total lasted about 10 min, is shown in Figure 13. The tool was first etched vertically into the surface by 15 μm and then moved along a rectangular path, to machine the trough around the small Cu tongue with only 2.5-μm thickness. The diameter of the tool was 10 μm, and the width of the trough was 16 μm, which signals a working gap width between tool and workpiece of about 3 μm. This value corresponds well with that expected from the double-layer charging time constant

a)

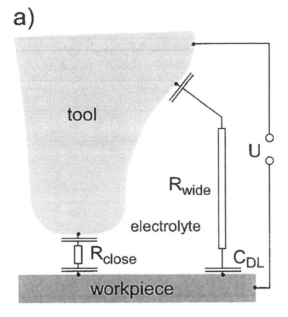

b)

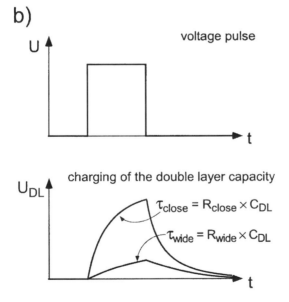

Figure 11 (a) Principle of the electrochemical micromachining with ultrashort voltage pulses. Due to the position-dependent electrolyte resistances (R_{wide} and R_{close}) the double-layer capacity (C_{DL}) is charged with varying time constants. (b) Upon application of a voltage pulse the double layers, where the electrodes are in close proximity, are charged more strongly than at larger distances.

a)

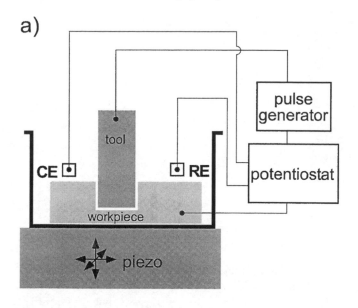

b)

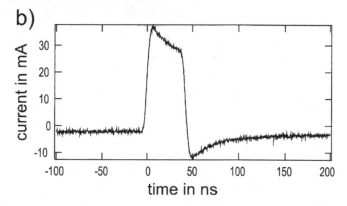

Figure 12 (a) Experimental setup for electrochemical micromachining. (See text for description.) (b) Measured current transient at close electrode distances. (Tool: Ø50-μm Pt cylinder, 1.6-V pulse amplitude, 0.01 M $HClO_4$ + 0.1 M $CuSO_4$.)

[18]. A detailed calculation, taking geometry and pulse amplitude into account, produced a value for the working gap width of about $2\,\mu m$, again in excellent agreement with the experiment. Because the charging time constant and local electrode separation are linearly related, the working gap width is expected to decrease linearly with the pulse duration. This was indeed experimentally confirmed for Cu dissolution down to pulse durations of 20 ns. Also, the specific electrolyte resistance, i.e., the concentration, linearly influences the spatial resolution. Therefore, simply by diluting the electrolyte or decreasing the pulse duration, one can significantly enhance the spatial resolution of the method. The only constraint to

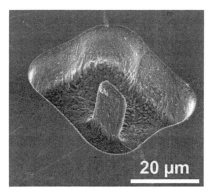

Figure 13 SEM image of a Cu tongue, machined into a Cu sheet with a 2-MHz train of 50-ns pulses, 1.6-V amplitude applied to a Ø10-μm tool in 0.01 M HClO$_4$ + 0.1 M CuSO$_4$.

meet is that the prerequisites for the simple picture sketched in Figure 11 must be fulfilled, i.e., that the charging of the double layer can be represented by an RC network. Apart from kinetic effects, as relaxation dynamics of electrolyte and double layer, which might come into play at such short time scales touching the picosecond range, the major condition is that the ions for the charging can be provided without significantly depleting the electrolyte in the working gap. Additionally, handling short pulses and feeding them to the tool electrode become considerably difficult for pulse durations reaching the lower nanosecond range. However, assuming pulse durations of 100 ps and fairly concentrated electrolytes of 0.3 M should result in a working gap width in the range of 10 nm. As is conceivable from Figure 13, the working precision is, in general, much higher than the working gap width, reflected by the smooth surface of the Cu tongue with a roughness well in the nanometer range. The electrochemical micromachining with ultrashort voltage pulses can in principle be applied to all electrochemical active materials. Pure metals such as Cu, Au, W, Ni, and Co, semiconductors such as p-Si, and alloys (nickel silver and stainless steel) were successfully machined. Also, the inverse process—the local deposition of materials—was demonstrated for Cu deposits on Au surfaces.

The electrochemistry of the above-mentioned materials differs considerably. While electrochemical Cu dissolution is very fast and extremely reversible, the electrochemical dissolution of stainless steel is considerably hindered by the formation of a passivating oxide layer. Because this passivation layer has to be chemically removed, the chemical composition of the electrolyte becomes very important in the latter case. A concentrated mixture of hydrofluoric and hydrochloric acids has been proven to be particularly successful. Figure 14 shows an SEM image of a stainless steel cube, etched directly out of a 1.4301 stainless steel sheet with a Ø50-μm cylindrical Pt tool. The microcrystalline structure of the stainless steel sheet becomes exposed in 3D, and the different grains are anisotropically etched. This directly reveals one of the strengths of the electrochemical method: the material removal proceeds without inducing thermal or mechanical strain or deformations. This is what usually hinders traditional mechanical methods such as milling or drilling for the fabrication of small high-aspect ratio structures.

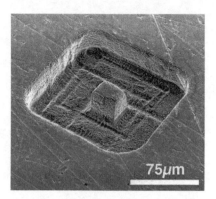

Figure 14 SEM image of a microcube, machined into a 1.4301 stainless steel sheet (100-ns pulses, 2-V amplitude, 3 M HCl + 6 M HF, Ø50-μm tool). The grain structure of the alloy is laid free.

Even undercuts can be produced with properly shaped tools. Figure 15a shows a freestanding cantilever directly machined into a stainless steel sheet with a small loop made out of a Ø10-μm Pt wire. The loop was first etched vertically into the surface, then moved horizontally by about 50 μm, before it was removed following the cut in the reverse direction. In order to probe the oscillation behavior of the cantilever, it was electrostatically excited with an auxiliary electrode brought into μm distance to the free end of the cantilever. The mechanical oscillations following a voltage step from 200 V down to 10 V were measured by the capacitive current flowing through the cantilever/auxiliary electrode capacitor (Figure 15b). Due to the small mechanical dimensions, the cantilever exhibits an oscillation frequency of 1.1 MHz. Such cantilevers provide sensitive microbalances. The resonance frequencies of a cantilever of the dimensions in Figure 15a are estimated to change by about 4 Hz upon the adsorption of one monolayer of a reactive gas as CO.

The electrochemical microstructuring method has to be compared with other micro- and nanostructuring methods such as lithography or ion beam milling, rather than with the scanning probe methods mentioned in Section 6.3.1. Its versatility concerning the materials as well as the possibility to machine three-dimensional structures on the micrometer to submicrometer scale fills the gap between the conventional micromachining methods such as micromechanical milling or electrical discharge machining, on the one hand side, and lithographic methods or ion beam milling, on the other. While mechanical milling and electrical discharge machining have been proven to be applicable down to structure sizes in the 10-μm range with a wide range of materials, the machining of high-aspect ratio structures becomes increasingly difficult for smaller structures [61]. Moreover, both methods suffer from considerable wearoff of the tools, which further restricts their applicability for the machining of small structures. Recently, there was significant progress in the development of laser ablation techniques, particularly due to the application of short pulses [62].

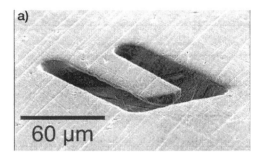

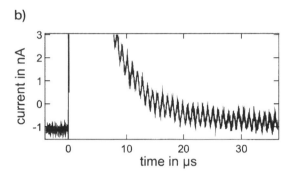

Figure 15   (a) SEM image of a freestanding cantilever, cut into a stainless steel sheet. A Ø10-μm wire loop was used as tool (100-ns, 2-V pulses, 3 M HCl + 6 M HF). (b) Electrostatic deflection of the cantilever with an auxiliary electrode yielded mechanical vibrations with 1.1-MHz frequency, which can be perceived from the capacitive current transients through the auxiliary-electrode/cantilever capacitor.

Nonetheless, such techniques suffer in general from the optical diffraction limit, which restricts the possibility of manufacturing high-aspect ratio structures with submicrometer precision. Lithographic methods achieved the highest resolution so far [63]; however, they are limited to the fabrication of essentially 2D structures, with only a few exceptions [64,65]. For the formation of thick, 3D structures, anisotropic etching has to be employed, which restrains both the available structures and materials. Polycrystalline metals are machinable with only limited precision [66,67]. In the so-called LIGA process the lithographic structuring of a polymer mask is followed by electrochemical metal deposition, which combines the high precision of lithographic methods with the versatility of metallic materials. However, it is still limited essentially to 2D structures with a fixed thickness in the third dimension [61]. The real 3D machining of metals on the nanometer scale is possible with ion beam milling and ion beam deposition, which suffer from high cost and slow processing speed [68]. Electrochemical pulse machining might widen that inventory of methods with special respect toward the versatile 3D machining of conductive, electrochemically active surfaces. In principle, the method can be paralleled similarly to lithographic techniques: with complex tools, exhibiting the negative of the desired structures, multiple structures can be stamped directly into the workpiece. However, the field for improvements is still fairly wide, and the feasibility of the ultimate goal, reaching machining precisions in the 10-nm range, comparable to today's best lithographic methods, still has to be proven experimentally.

6.4 CONCLUSIONS

In this chapter we present several methods for electrochemical surface modifications. This discussion focuses on the various possibilities offered by electrochemistry rather than on the actual applicability to catalytically active nanostructured surfaces. We believe that, due to the ease of their application, the low cost, and their potential for further development, electrochemical methods will gain importance for nanostructuring surfaces. The first signs of such a development can be perceived, e.g., with the fabrication of chip interconnects by electrochemical Cu deposition. The deposition from solution proved to be the only method to gain sufficient filling of such delicate structures as the 500-nm-wide trenches in a lithographically produced mask [69]. Another example is the in-situ formation and investigation of small structures by scanning probe techniques: recently, the electrochemical activity of a single Pt cluster was studied with STM after in-situ preparation of the cluster [70]. Also from the purely academic point of view, e.g., concerning the ultrafast change of the thermodynamic state of the system and the in-situ observation of the ordering behavior, electrochemical methods offer unprecedented opportunities to gain knowledge about such general phenomena as the ordering behavior during thermodynamic equilibration.

ACKNOWLEDGMENTS

This chapter presents results accumulated in collaboration with several co-workers and visiting scientists at our institute for several years. We want to explicitly acknowledge the contributions of Dr. X. Xia, who participated in the STM experiments, as well as of Dr. P. Allongue, whose contributions to the electrochemical microstructuring experiments are indispensable. We also want to heartily thank Dr. L. Cagnon, V. Kirchner, and M. Kock for their steady support and patience while conducting the experiments. The help of our machine shop and electronics laboratory on the implementation of the experimental setup is gratefully acknowledged.

REFERENCES

1. M. Boudart, G. Djéga-Mariadassou, Kinetics of Heterogeneous Catalytic Reaction, Princeton University Press: Princeton, NJ, 1984.
2. M. Haruta, Size and support-dependency in the catalysis of gold. Catal. Today 36, 153–166, 1997.
3. M. Valden, X, Lai, D.W. Goodman, Onset of catalytic activity of gold clusters on titania with the appearance of nonmetallic properties. Science 281, 1647–1650, 1998.
4. M.A. Kastner. Artificial atoms. Physics Today 46, 24–31, 1993.
5. F. Maroun, F. Ozanam, O.M. Magnussen, R.J. Behm, The role of atomic ensembles in the reactivity of bimetallic electrocatalysts. Science 293, 1811–1814, 2001.
6. T. Takaghara, Quantum dot lattice and enhanced excitonic optical nonlinearity. Surf. Sci. 267, 310–314, 1992.

7. C.R. Kagan, C.B. Murray, M. Nirmal, M.M.G. Bawendi, Electronic energy transfer in CdSe quantum dot solids. Phys. Rev. Lett. 76, 1517–1520, 1996.

8. J.D. Gunton, M.S. Miguel, P.S. Sahni, The dynamics of first-order phase transitions. In Phase Transitions and Critical Phenomena, C. Domb, J.L. Lebowitz, eds. Academic Press: London, 1983; Vol. 8; pp 267–466.

9. M.C. Cross, P.C. Hohenberg, Pattern formation outside of equilibrium. Rev. Mod. Phys. 65, 851–1091, 1993.

10. J.O.M. Bockris, A.K.N. Reddy, Modern Electrochemistry, Plenum Press: New York, 1970, Vol. II.

11. X.H. Xia, L. Nagle, R. Schuster, O.M. Magnussen, R.J. Behm, The kinetic of phase transitions in underpotentially deposited Cu adlayers on Au(111). Phys. Chem. Chem. Phys. 2, 4387–4392, 2000.

12. V. Kirchner, X. Xia, R. Schuster, Electrochemical nanostructuring with ultrashort voltage pulses. Acc. Chem. Res. 34, 371–377, 2001.

13. D.M. Kolb, R. Ullmann, T. Will, Nanofabrication of small copper clusters on gold(111) electrodes by a scanning tunneling microscope. Science 275, 1097–1099, 1997.

14. A.J. Bard, G. Denuault, C. Lee, D. Mandler, D.O. Wipf, Scanning electrochemical microscopy: A new technique for the characterization and modification of surfaces. Acc. Chem. Res. 23, 357–363, 1990.

15. A.J. Bard, M.V. Mirkin, eds. Scanning Electrochemical Microscopy, Marcel Dekker, Inc.: New York, 2001.

16. R. Schuster, V. Kirchner, X. Xia, A.M. Bittner, G. Ertl, Nanoscale electrochemistry. Phys. Rev. Lett. 80, 5599–5602, 1998.

17. X.H. Xia, R. Schuster, V. Kirchner, G. Ertl, The growth of size-determined Cu clusters in nanometer holes on Au(111) due to a balance between surface and electrochemical energy. J. Electroanal. Chem. 461, 102–109, 1999.

18. R. Schuster, V. Kirchner, P. Allongue, G. Ertl, Electrochemical micromachining. Science 289, 98–101, 2000.

19. V. Kirchner, L. Cagnon, R. Schuster, G. Ertl, Electrochemical machining of stainless steel microelements with ultrashort voltage pulses. Appl. Phys. Lett. 79, 1721–1723, 2001.

20. P.G. Debenedetti, Metastable Liquids: Concepts and Principles, Princeton University Press: Princeton, NJ 1996.

21. K. Oki, H. Sagana, T. Eguchi, Separation and domain structure of $\alpha + B_2$ phase in Fe-Al alloys. J. Phys. (Paris) C 7, 414, 1977.

22. H.C. Callen, Thermodynamics and an Introduction to Thermostatistics, Wiley: New York, 1985.

23. K. Binder, Theory of first-order phase transitions. Rep. Prog. Phys. 50, 783–859, 1987.

24. K.R. Rundman, Spinodal structures. In Metals Handbook, T. Lyman, ed. American Society for Metals: Metals Park, Vol. 8, 1973.

25. H. Brune, Microscopic view of epitaxial metal growth: Nucleation and aggregation. Surf. Sci. Rep. 31, 121–230, 1998.

26. E. Budewski, V. Bostanov, T. Vitanov, S. Stoynov, A. Kotzewa, R. Kaischew, Electrochim. Acta 11, 1697, 1966.

27. M.P. Zach, R.M. Penner, Nanocrystalline nickel nanoparticles. Adv. Mater. 12, 878–883, 2000.

28. M.H. Hölzle, U. Retter, D.M. Kolb, The kinetics of structural changes in Cu adlayers on Au(111). J. Electroanal. Chem. 371, 101–109, 1994.

29. M.A. Schneeweiss, D.M. Kolb, The initial stages of copper deposition on bare and chemically modified gold electrodes. Phys. Stat. Sol. (a) 153, 51–71, 1999.

30. Z. Shi, S. Wu, J. Lipkowski, Coadsorption of metal atoms and anions: Cu UPD in the presence of SO_4^{2-}, Cl^- and Br^-. Electrochim. Acta 40, 9–15, 1995.

31. M.E. Toney, J.N. Howard, J. Richer, G.L. Borges, J.G. Gordon, O.R. Melroy, D. Yee, L.B. Sorensen, Electrochemical deposition of copper on a gold electrode in sulfuric acid: Resolution of the interfacial structure. Phys. Rev. Lett. 75, 4472–4475, 1995.
32. L. Blum, H.D. Abruña, J. White, J.G. Gordon, G.L. Borges, M. Samant, O.R. Melroy, Study of underpotentially deposited copper on gold by fluorescence detected surface EXAFS. J. Chem. Phys. 85, 6732–6738, 1986.
33. O.R. Melroy, M.G. Samant, G.L. Borges, J.G. Gordon, L. Blum, J.H. White, M.J. Albarelli, M. McMillan, H.D. Abruña, In-plane structure of underpotentially deposited copper on gold(111) determined by surface EXAFS. Langmuir 4, 728–732, 1988.
34. A. Tadjeddine, D. Guay, M. Ladouceur, G. Tourillon, Electronic and structural characterization of underpotentially deposited submonolayers and monolayer of copper on gold (111) studied by in situ X-ray-absorption spectroscopy. Phys. Rev. Lett. 66, 2235–2238, 1991.
35. A.M. Bittner, J. Wintterlin, G. Ertl, Strain relief during metal-on-metal electrodeposition: A scanning tunneling microscopy study of copper growth on Pt(100). Surf. Sci. 376, 267–278, 1996.
36. W. Haiss, D. Lackey, J.K. Sass, K.H. Besocke, Atomic resolution scanning tunneling microscopy images of Au(111) surfaces in air and polar organic solvents. J. Chem. Phys. 95, 2193–2196, 1991.
37. M. Bott, M. Hohage, M. Morgenstern, T. Michely, G. Comsa, New approach for determination of diffusion parameters of adatoms. Phys. Rev. Lett. 76, 1304–1307, 1996.
38. A.M. Bittner, J. Wintterlin, G. Ertl, Effects of iodine coating and desorption on the reconstruction of a Pt(110) electrode: A scanning tunneling microscopy study. J. Electroanal. Chem. 388, 225–231, 1995.
39. G.M. Bell, D.A. Lavis, Statistical Mechanics of Lattice Models, Ellis Horwood Ltd.: Chichester, U.K. 1989.
40. H. Honbo, S. Sugawara, K. Itaya, Detailed in situ scanning tunneling microscopy of single crystal planes of gold in aqueous solutions. Anal. Chem. 62, 2424–2429, 1990.
41. X. Gao, M.J. Weaver, Nanoscale structural changes upon electro-oxidation of Au(111) as probed by potentiodynamic scanning tunneling microscopy. J. Electroanal. Chem. 367, 259–264, 1994.
42. C.M. Vitus, A.J. Davenport, In situ scanning tunneling microscopy studies of the formation and reduction of a gold oxide monolayer on Au(111). J. Electrochem. Soc. 141, 1291–1298, 1994.
43. N. Batina, T. Will, D.M. Kolb, Study of the initial stages of copper deposition by in situ scanning tunneling microscopy. Faraday Discuss. 94, 93–106, 1992.
44. R.J. Nichols, D.M. Kolb, R.J. Behm, STM observation of the initial stages of copper deposition on gold single-crystal electrodes. J. Electroanal. Chem. 313, 109–119, 1991.
45. R.J. Nichols, E. Bunge, H. Meyer, H. Baumgärtel, Classification of growth behaviour for copper on various substrates with in-situ scanning probe microscopy. Surf. Sci. 335, 110–119, 1995.
46. D.M. Eigler, E.K. Schweizer, Positioning single atoms with a scanning tunnelling microscope. Nature 344, 524, 1990.
47. J.A. Stroscio, D.M. Eigler, Atomic and molecular manipulation with the scanning tunneling microscope. Science 254, 1319–1326, 1991.
48. P. Avouris, I-W. Lyo, Probing the chemistry and manipulating surfaces at the atomic scale with the STM. Appl. Surf. Sci. 60/61, 426–436, 1992.
49. G. Meyer, L. Bartels, S. Zöphel, E. Henze, K-H. Rieder, Controlled atom by atom restructuring of a metal surface with the scanning tunneling microscope. Phys. Rev. Lett. 78, 1512–1515, 1997.
50. D. Hoffmann, W. Schindler, J. Kirschner. Electrodeposition of nanoscale magnetic structures. Appl. Phys. Lett. 73, 3279–3281, 1998.

51. C. Lebreton, Z.Z. Wang, Nanowriting on an atomically flat gold surface with scanning tunneling microscope. Scan. Microsc. 8, 441–448, 1994.
52. C. Lebreton, Z.Z. Wang, Critical humidity for removal of atoms from the gold surface with scanning tunneling microscopy. J. Vac. Sci. Technol. B14, 1356–1359, 1996.
53. P. Avouris, T. Hertel, R. Martel, AFM tip-induced local oxidation of silicon: Kinetics, mechanism, and nanofabrication. Appl. Phys. Lett. 71, 285–287, 1997.
54. H. Sugimura, T. Uchida, N. Kitamura, H. Masuhara, Scanning tunneling microscope tip-induced anodization for nanofabrication of titanium. J. Phys. Chem. 98, 4352–4357, 1994.
55. W. Li, J.A. Virtanen, R.M. Penner, Nanometer-scale electrochemical deposition of silver on graphite using a scanning tunneling microscope. Appl. Phys. Lett. 60, 1181–1183, 1992.
56. W. Li, G.S. Hsiao, D. Harris, R.M. Nyffenegger, J.A. Virtanen, R.M. Penner, Mechanistic study of silver nanoparticle deposition directed with the tip of a scanning tunneling microscope in an electrolytic environment. J. Phys. Chem. 100, 20103–20113, 1996.
57. E.B. Budevsky, G.T. Staikov, W.J. Lorenz, Electrochemical Phase Formation and Growth: An Introduction to the Initial Stages of Metal Deposition, VCH: Weinheim, 1996.
58. U. Staufer, In Scanning Tunneling Microscopy II, H.-J. Güntherodt, R. Wiesendanger, eds. Springer: Berlin, 1992.
59. R. Wiesendanger, Contributions of scanning probe microscopy and spectroscopy to the investigation and fabrication of nanometer-scale structures. J. Vac. Sci. Technol. B 12, 515–529, 1994.
60. R.M. Nyffenegger, R.M. Penner, Nanometer-scale surface modification using the scanning probe microscope: Progress since 1991. Chem. Rev. 97, 1195–1230, 1997.
61. C.R. Friedrich, R. Warrington, W. Bacher, W. Bauer, P.J. Coane, J. Göttert, T. Hannemann, J. Haußelt, M. Heckele, R. Knitter, J. Mohr, V. Piotter, H.-J. Ritzhaupt-Kleissl, R. Ruprecht, High aspect ratio processing. In Handbook of Microlithography, Micromachining, & Microfabrication, P. Rai-Choudhury, ed. SPIE: Bellingham, WA, 1997; Vol. 2, pp 299–377.
62. S. Preuss, A. Demchuk, M. Stuke, Sub-picosecond UV laser ablation of metals. Appl. Phys. A 61, 33–37, 1995.
63. P. Rai-Choudhury, ed. Handbook of Microlithography, Micromachining, & Microfabrication, SPIE Optical Engineering Press: Bellingham, WA, 1997, Vol. 1, 2.
64. R.J. Jackman, S.T. Brittain, A. Adams, H. Wu, M.G. Prentiss, S. Whitesides, G.M. Whitesides, Three-dimensional metallic microstructures fabricated by soft lithography and microelectrodeposition. Langmuir 15, 826–836, 1999.
65. S. Kawata, H.-B. Sun, T. Tanaka, K. Takada, Finer features for functional microdevices. Nature 412, 697–698, 2001.
66. M. Datta, L.T. Romankiw, Application of chemical and electrochemical micromachining in the electronics industry. J. Electrochem. Soc. 136, 285C–292C, 1989.
67. M. Datta, Microfabrication by electrochemical metal removal. IBM J. Res. Develop. 42, 655–669, 1998.
68. D.K. Stewart, J.D. Casey, Focused ion beams for micromachining and microchemistry. In Handbook of Microlithography, Micromachining, & Microfabrication, P. Rai-Choudhury, ed. SPIE: Bellingham, WA, 1997, Vol. 2, pp. 153–195.
69. P.C. Andriacacos, Copper on-chip interconnections. Electrochem. Soc. Interface 8, 32–37, 1999.
70. J. Meier, K.A. Friedrich, U. Stimming, Novel method for the investigation of single nanoparticle reactivity. Faraday Discuss. 121, 365–372, 2002.

7

Adsorption and Reaction at Supported Model Catalysts

CLAUDE R. HENRY

CRMC2-CNRS, Marseille, France

CHAPTER CONTENTS

SUMMARY

The knowledge of basic mechanisms in heterogeneous catalysis has tremendously increased in the last ten years thanks to the combination of surface science techniques with ab-initio calculations, however, a complete understanding of the effect of the intrinsic heterogeneity of real catalyst is still missing. This heterogeneity is mainly due to the finite size of catalytic particles, their morphology, and the presence of the support. The way to bridge this so-called "material gap" is to use supported model catalysts prepared by epitaxial growth on oxide single crystals that can be properly studied by surface sciences techniques combined with various microscopy techniques. In this chapter we will review the structure and morphology of nanometer-sized metal clusters, focusing on Au, Ag, and Pd metals. The electronic structure, as well as the chemical properties of clusters containing less than about 25 atoms, depend on the exact number of atoms contained in the clusters. For particles containing at least 100 atoms, the electronic properties evolve smoothly with particle size. The adsorption behavior for CO, O_2, and NO on Rh and Pd nanoparticles is reviewed. True catalytic reactions like CO oxidation and NO reduction by CO have been recently investigated by molecular beam techniques. While the reaction mechanism is basically the same as on metal extended surfaces, the reaction kinetics on supported particles is different due to the heterogeneity of these systems. The precise knowledge of this heterogeneity allows a quantitative comparison between these two types of model catalysts.

7.1 INTRODUCTION

The basic understanding of heterogeneous catalysis has rapidly increased in the last 20 years thanks to the thorough experimental investigations carried out on metal single crystals via the huge number of available surface-science techniques [1]. More recently, partly due to the development of new theoretical tools (like density functional theory) and partly due to the fast increasing power of computers, it has become possible to correlate the electronic structure of the surface atoms with their catalytic activity [2]. Thus, step by step, a general view of heterogeneous catalysis is building up. Even in some cases, this new theoretical frame combined with specially designed experiments is able to give new ways to improve or even find new catalysts [3]. However, investigations on metal single-crystal surfaces are not able to take into account some aspects of real catalysts. These differences, which are often identified as "material gaps," are primarily the finite size of the metal particles (often a few nm) and the presence of a support (most often an oxide powder). To bridge the material gap, a new type of model catalyst has been developed, mainly in the last 10 years, called the *supported model catalyst*. It is prepared by growing (generally by vapor deposition) metal clusters on an extended oxide surface. Due to the UHV environment during its preparation, it can also be studied by surface-science techniques. Several well-documented reviews on the supported model catalysts are now available [4–8]. In this chapter I focus on some aspects that have yet to be reviewed in detail, like the morphology of nanoparticles and results on true catalytic reactions (CO oxidation and NO reduction by CO) that have been recently obtained by molecular beam experiments. From these examples we will see the importance of the effects of particle size and particle morphology as well the support on the reaction kinetics.

7.2 PREPARATION OF SUPPORTED MODEL CATALYSTS

Supported model catalysts are prepared by deposition of atoms, under UHV, on a clean, well-ordered oxide surface. We do not review other preparation methods based on wet impregnation [5] and decomposition of metal-organic complexes [9], which are close to industrial methods but less subject to investigation by surface-science techniques. Oxide supports are single crystals or thin films. Bulk oxide single crystals like MgO [10], ZnO [11], and NiO [12] can be cleaved under UHV and provide clean and very well-ordered surfaces as checked by the diffraction of helium atoms [10]. However, most of the oxides cannot be cleaved and they have to be cut, polished, and prepared by standard techniques of ion sputtering and annealing under UHV or in oxygen partial pressure. This procedure is used for α-alumina [13], rutile [14], ceria [15], zirconia [16], and also sometimes for MgO [17] and ZnO [18]. Now, it is also possible to prepare well-ordered ultrathin films of oxide by oxidation of a metal or a metallic alloy. An advantage of this method is that STM and IRAS (infrared absorption spectroscopy in reflection mode) are possible. It is the case for alumina on NiAl(110) [19] or (111) [20], Ta(110) [21]; MgO on Mo(100) [22]; NiO(111) on Au(111) [23], and TiO$_2$ on Mo(100) [24]. Very recently, well-ordered silica thin films have been also obtained on Mo(112) [25]. Generally the thin films have more surface defects than cleaved single crystals. As we will see, if surface-point defects are nucleation centers for the growth of metal clusters, then one expects a much larger density of clusters on thin films. For HRTEM work powder oxides synthesized in situ and exposing well-defined facets (MgO cubes, ZnO prisms, etc.) are also used [6]. In some cases layered compounds are used as support for model catalysts; they are very easy to cleave and present very smooth and well-ordered surfaces. It is the case for graphite [26], mica [27], and molybdenite [28].

A calibrated beam of atoms is generated by a Knudsen cell or more simply by a tungsten filament on which a metal wire has been wrapped. The nucleation and growth are controlled by the beam intensity, the exposure time, and the substrate temperature [29]. On oxide surfaces the nucleation occurs mainly on surface defects; then the preparation of the substrate surface is very important [6]. Generally, the density of defects (i.e., of clusters) increases when we go from a UHV-cleaved surface to an air-cleaved surface and then to a thin film. Qualitatively, the number density of clusters increases with exposure time and reaches a plateau: the saturation density and, at the late stage of growth, clusters coalesce. The saturation density decreases when temperature increases, probably because the point defects are not perfect sinks for metal ad-atoms [30]. Another way to increase the density of clusters is to hydroxyl the oxide surface [31]. Along the plateau of saturation density the clusters grow uniformly. The mean cluster diameter generally follows a power law of the deposition time [32]. For most of the systems used as supported model catalysts, the exponent is between 1/4 and 1/2. Thus, it is relatively easy to independently control the cluster density and the mean cluster size. The nucleation and growth of the clusters can be studied in situ by several techniques, including SPA-LEED [33], He diffraction [34], STM [8], and AFM [35], but the most commonly used technique is TEM, which, despite some rare cases [36], is an ex-situ technique [29].

Bimetallic clusters can be prepared by the same way in using two metal beams. However, controlling the composition of the clusters is not an easy task, because it depends strongly on the deposition parameters: substrate temperature, beam fluxes, and (more annoyingly) the deposition time [37,38].

Another way to prepare model catalysts (under UHV conditions) is to grow the clusters in gas phase and depose them on the substrate. However, to avoid implantation, fragmentation, and dynamic coalescence, it is necessary to soft-land the clusters. A first possibility is to decrease the kinetic energy of the clusters to less than 0.2 eV/atom. This method is used by several groups without size selection [39,40] and in few cases with size selection [41,42]. The most convenient method to prepare gas-phase clusters is to use a laser vaporization source [43]; with this method bimetallic clusters with a constant composition are easily prepared [39]. The other possibility is to deposit the clusters, at a very low temperature, on a rare gas film condensed on the substrate and to evaporate the buffer layer by gently raising the temperature. It has been proven recently by STM that by this method it is possible to soft-land Ag clusters on Pt(111) [42]. Finally, it is also possible to use high-energy clusters pined on a smooth substrate like graphite [44].

However, except for the extensive work of Heiz and colleagues in Lausanne [45], the preparation of model catalysts by cluster deposition is scarce. Instead of using deposition of (true) size-selected clusters, which is a very heavy method that is, in fact, limited to small sizes (below 30 atoms), several groups now try to grow clusters with a narrow size distribution by atom deposition. As nucleation occurs randomly on the surface and during a finite time, the clusters grow nonuniformly. Alternatively, on a regular lattice of point defects clusters will nucleate rapidly (if the deposition rate is large) and grow at a constant rate; then a very sharp size distribution is expected. Such a result has already been obtained by creating a dislocation network, by growing a thin layer (2 ML) of Ag on a Pt(111) surface, which is due to the misfit between the two lattices [46]. On this dislocation network 2D silver clusters grow with a size dispersion ($\sigma = 0.20$) much smaller than for the growth on the clean Pt surface ($\sigma = 0.59$). However, these structures are not stable upon annealing, because they form alloys. Very recently it has been shown that ultrathin alumina thin films obtained by oxidation of NiAl(111) form a periodic lattice with a parameter of 4.5 nm that can be used as a template for the growth of a regular array of metal clusters having a narrow size distribution [47].

However, these lattices formed by misfit dislocations or surface reconstruction have a well-defined lattice parameter that fixes the density of clusters (and the maximum cluster size). Alternative ways to grow arrays of clusters using new nanolithography techniques are now being investigated. Several groups have already made model catalysts by using electron beam lithography [48,49]; however, this method is limited to particle sizes of about 10 nm and more severely by the slowness of this serial process. The speed of the process could be dramatically increased by the new technique called *nano-inprint* [50]. In this method a stamp is fabricated by electron beam nanolithography and pressed against a wafer covered by a resist; then the conventional lift-off process is applied. In that way the slow process of e-beam sensitization of the resist is made only once during the preparation of the stamp, which can be reused several times. The size limit of the metal dots, however, is the same as for e-beam lithography. Nanofabrication by STM [51] of AFM [52] offers very good size resolution but is still a serial, very slow process. A new cheap and fast process of nanolithography is colloidal lithography. In this method a drop of colloid composed by polystyrene spheres (100 nm to 1 μm) is deposited on a flat surface. During evaporation of the solvent the spheres become spontaneously self-organized in a compact, hexagonal array. This array is then used as a mask for the evaporation of a metal. After removing the colloidal spheres, a hexagonal array of triangular metal particles is obtained [53]. This method is presently limited to particle sizes of about 20 nm, but it will certainly decrease very rapidly when smaller colloids become available.

7.3 STRUCTURE AND MORPHOLOGY

7.3.1 Introduction

One of the key issues of supported model catalysts is to prepare collection of metal particles having a well-defined morphology. Indeed, if a catalytic reaction is structure-sensitive [54], it will depend on the nature of the facets present on the particles. Moreover, the presence of edges, the proportion of which is increasing rapidly below about 5 nm, can affect the reactivity by their intrinsic low coordination and also by their role as boundary between the different facets. In this section I first discuss the theoretical predictions of the shape of small particles and clusters, then I briefly describe the available experimental techniques to study the morphology, and finally I discuss from selected examples how it is possible to understand and control the morphology of supported model catalysts.

7.3.2 Theoretical Point of View
Structure of Small Metal Clusters

Taking a cluster containing a small number of atoms, what structure will it choose? From a geometrical point of view it has been known for a very long time that a compact arrangement of spheres gives rise to crystalline FCC and CC structures but also to a noncrystalline icosahedron structure [55]. To minimize the energy of a cluster, we have to maximize the bonding (i.e., highest compactness) and minimize the surface energy (i.e., to be close to a sphere). The icosahedron structure, which is the most compact and not far from a sphere (the truncated icosahedron, which is the shape of the C_{60} cluster, is the closest polyhedron to a sphere), is expected to have the lowest energy, but it cannot grow indefinitely because it has no translation symmetry and then strain is stored in the structure. At a given size the icosahedron must transform to a crystalline structure. Figure 1 shows some possible structures of metal clusters. Early simulations using Lennard–Jones-type pair potentials have shown that for small sizes, the icosahedral structure is more stable [56]. A 13-atom cluster is a perfect icosahedron, because it can grow layer by layer, forming the series of Mackay icosahedra [57]. Closed shells are obtained for the series of magic numbers of atoms: 13, 55, 147, 309, etc. These magic numbers have been clearly seen in the

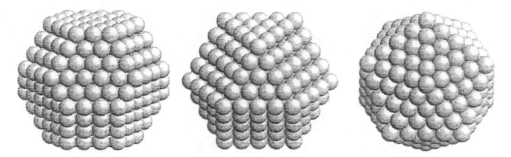

Figure 1 Ball model of possible structures for metal clusters: truncated octahedron (left), cuboctahedron (middle), icosahedron (right).

mass spectra of rare gas clusters prepared by supersonic expansion [58]. However, for metals it is now well established that pair potentials are not well adapted because of the long-range interaction of the metallic bond. In order to simulate the many-body character of the metallic bond, empirical many-body potentials have been introduced. These potentials (glue model [59], embedded atom method (EAM) [60], effective medium theory (EMT) [61], tight binding in the second moment approximation (TB-SMA) [62]) have to be parameterized on some bulk properties of the considered metal. All the energetic calculations (at 0 K) predict that the stable structure of a 13-atom metal cluster is the icosahedron.

However, the critical size, where the cluster changes from the icosahedral to the crystalline structure, often depends on the type of potential used. The energy as a function of the size of Pd clusters has been calculated with TB-SMA potential for icosahedron, cuboctahedron, and truncated octahedron (Wulff polyhedron for the FCC structure) [63]. Indeed, at very low sizes the icosahedron is more stable, but near 20 atoms the FCC structure becomes more stable. It is worth noting that the cuboctahedron is less stable than the truncated octahedron. Often only cuboctahedra are compared to icosahedra in the calculations because they have the same magic numbers (closed-shell structures). The same type of calculation has shown that the transition toward the FCC structure occurs at about 2300 atoms for Cu, around 300 atoms for Ag, and below 100 atoms for Au [63]. For Cu other calculations predict 1500 atoms by TB-SMA [64] and nearly 2500 atoms by EMT [65]. For Ni, EAM simulations [66] predict that the icosahedron is the most stable up to 2300 atoms, then the Marks-decahedron (decahedron with additional truncations that has been predicted by Marks [67]; see Figure 2) and the transition to the FCC structure occur only near 17,000 atoms. For gold the largest discrepancies are observed between the different calculations. With TB-SMA potential calculations have given a transition toward the FCC structure for clusters larger than 13 atoms [68] or less than 100 atoms [63], but EAM calculations predict that at about 30 atoms the icosahedron

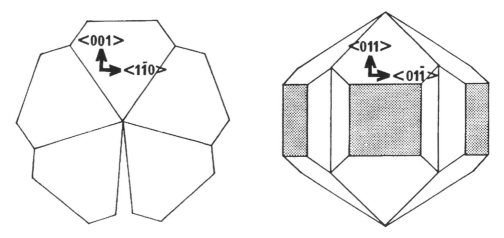

Figure 2   Schematic representation of a Marks-decahedron: ⟨111⟩ (left) and ⟨100⟩ (right) projections. (From L.D. Marks [67].)

transforms to a decahedron and around 1000 atoms only it becomes FCC [69]. At this stage, however, several remarks can be made. First, the energy between different configurations can be very small, especially for small sizes [70]. Second, between two closed shells different structures can be stabilized depending on the number of atoms. In fact, finite-temperature molecular dynamics simulations have shown that for Ag clusters, if the icosahedron is the most stable near 150 atoms, it is in competition with the decahedron during growth [71]. For gold at low temperatures, Marks decahedra are favored for 75 and 146 atoms, while the Wulff polyhedron is favored at 459 atoms; at high temperatures the icosahedron is favored, and it seems to be the precursor to fusion [72]. Kinetic and dynamic effects are very important especially during growth. In the case of Al clusters, zero K calculations (with EMT potential) show that the icosahedron is more stable below 2000 atoms, while the Wulff polyhedron becomes stable beyond [73]. However, during growth simulations at finite temperature, kinetic shapes (close to the octahedron) different from the equilibrium ones appear [73].

Equilibrium Shape of Crystals

The problem of the equilibrium shape of crystals was addressed 100 years ago by Wulff [74]. The problem was to find the shape that minimizes the surface energy for a given number of atoms. For an isotropic system the answer is obvious: it corresponds to the shape that minimized the surface—that is, a sphere. For a real crystal, which presents an anisotropy of the surface energy, the problem is less trivial. Assuming that the equilibrium shape must be a polyhedron, Wulff showed that the minimum energy is obtained when the surface energy of a given face (σ_i) divided by the central distance to this face (h_i) is a constant:

$$\sigma_i/h_i = \text{const.} \tag{1}$$

For an FCC structure one gets a truncated octahedron (see Fig. 1) that we generally call a Wulff polyhedron. It contains only (111) and (100) faces. At 0 K and for a simple broken-bond model, the ratio between the surface energies $\sigma(100)/\sigma(111)$ is $2/\sqrt{3}$, which fixes the relative extensions of these facets. However, when the temperature increases, the surface-energy anisotropy decreases, disappearing at the melting point.

Numerous TEM observations, mainly for Au and Ag small particles (about 3 to 10 nm), have shown shapes with fivefold symmetry like icosahedra and decahedra [75]. In fact, these particles are multiple twins; they are formed by 10 or 20 slightly deformed FCC tetrahedra for the decahedron and the icosahedron, respectively. Marks has modified the Wulff construction by introducing the twin boundary energy [67]. He showed that the decahedron is stabilized by the presence of small re-entrant (111) facets forming notches at the twin boundaries (see Figure 2) that decreases the total surface energy. Then the Marks decahedron can be more stable than the icosahedron at intermediate sizes. If the cluster size increases, the icosahedron becomes less stable than the Marks-icosahedron, which further becomes less stable than the Wulff polyhedron. For very small sizes the energy difference between these structures are very low; by thermal fluctuations the clusters change from one shape to other ones. This phenomenon, called *quasi-melting*, has been experimentally observed by HRTEM on gold clusters [76]. On the basis of these energetic

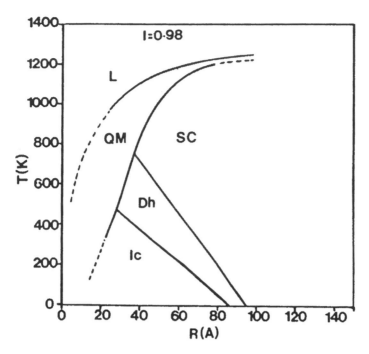

Figure 3 Phase diagram of gold clusters as a function of their size. **Ic**: icosahedron MTP, **Dh**: decahedron MTP, **SC**: single crystal, **QM**: quasimelt, **L**: liquid. (From P.M. Ajayan and L.D. Marks [77].)

calculations Marks [77] established a phase diagram for gold clusters, presented here in Figure 3. From this diagram, at RT quasi-melting occurs below a diameter of 4 nm, the transition between icosahedron and decahedron is predicted to occur at a diameter of about 8 nm, and the transition between decahedron and Wulff polyhedron occurs near 14 nm. At 500 K the icosahedron disappears, and at 800 K one finds only liquid, quasi-melting, and FCC particles.

We have seen in the previous paragraph that during growth the shape of the particles can be far from the equilibrium one. Meanwhile, by annealing one expects to reach the equilibrium. The kinetics toward equilibrium shape have been addressed from a macroscopic thermodynamic point of view by Herring [78] and Mullins [79]. The mechanism of the shape evolution is surface diffusion and the driving force is the difference of curvature. Calculations assume an isotropic surface energy (that is, in fact, true only close to the melting point). The relaxation time toward equilibrium is proportional to the particle size at the fourth power. More recently, Mullins and Rohrer [80] have shown that at low temperatures it is no longer the surface diffusion that is the rate-limiting step in the process but the nucleation of a critical nucleus to grow a new layer on a facet. They have estimated the nucleation barrier. The important result of this study is that this barrier becomes prohibitively large for facets larger than a limiting size (if no dislocation is present). Then for particles having facets larger than this size, it is virtually impossible to reach the equilibrium shape. This result explains why in annealing experiments of rough Pd particles supported on MgO, the shape is blocked (at 400 °C) on a faceted shape that is far from the equilibrium one [81]. It is interesting to notice that at the same temperature the equilibrium shape is easily obtained during growth. This is easily understood because with growth the supersaturation is very high and the size of the critical nucleus drops. Simulations of the relaxation toward the equilibrium shape of nanometer particles have recently been undertaken using pair potential in 2D [82] and 3D cases [83]. The simulations find the fourth-power dependence of the particle size at high temperatures, as predicted by Mullins. At low temperatures they show that the shape evolves noncontinuously with temporary blocking on some faceted shapes, but the relaxation time does not show a simple power-law dependence of the particle size.

Effect of the Support

Following the Wulff approach of the equilibrium shape of crystals, some authors have tried to include the presence of a substrate. The solution of this problem has been given by Kaishew [84] and Winterbottom [85]. It is known as the Wulff–Kaishew theorem. The crystal is now truncated at the interface by an amount Δh_s, which is related to the adhesion energy of the crystal on the substrate (β) and to the surface energy of the facet parallel to the interface (σ_s) by the following relation:

$$\frac{\beta}{\Delta h_s} = \frac{\sigma_s}{h_s} \tag{2}$$

h_s is the central distance to the facet parallel to the interface. From relation (2) one sees that when the energy of adhesion increases (i.e. the crystal–support interaction increases) the crystal becomes flatter. When the adhesion energy is two times larger than the surface energy (σ_s), $\Delta h_s = 2h_s$, it is like the crystal has sunk in the substrate. In fact, that means that the crystal is a 2D layer on the substrate; it corresponds to perfect wetting. However, this approach implicitly assumes that the deposited crystal has the same lattice constant as the substrate. But, in general, there is a misfit between the two lattices that will result in strain in the crystal (and also in the support), and the strain energy has to be taken into account in minimizing the total energy of the system. This problem has been addressed theoretically only recently. Markov [86] has given an atomistic formulation of the Wulff–Kaishew theorem in assuming that deposited crystal was uniformly strained (no relaxation). At the same time Kern and Müller [87] have followed the same approach but separated the surface and the volume elasticity. In both cases the equilibrium shape depends on the misfit between the two lattices but not on the crystal size: the equilibrium shape was still self-similar. More recently, Kern and Müller [88,89] have calculated the equilibrium shape of a parallelepiped crystal supported on a substrate with a variable misfit. The crystal and the (crystalline) support were relaxed in using the elasticity of continuous media. Now the strain is inhomogeneous, and the important result is that the equilibrium shape is no longer self-similar—it depends on the size of the deposited crystal. Figure 4 shows the main results of this study. From Figure 4a one sees that when the misfit is nonzero, the aspect ratio of the deposited crystal increases, i.e., the crystal tries to grow in height to minimize the interfacial strain. However, it is obvious that the crystal cannot grow indefinitely in height; at a critical

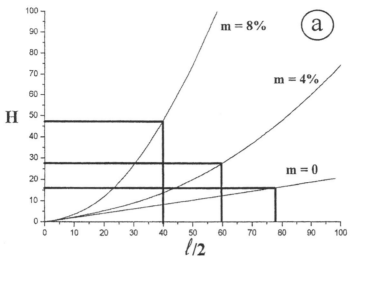

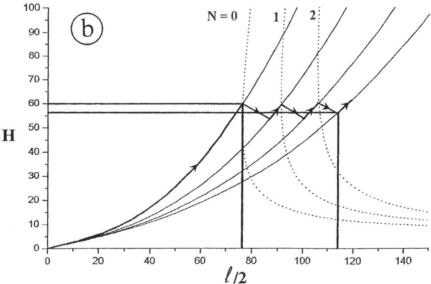

Figure 4 Effect of strain on the equilibrium shape of supported crystals. (a) Equilibrium shape of a box-shaped crystal of height H and length l as a function of the misfit m. The continuous curves are the trajectories of the edges. (b) Effect of the dislocation entrance on the equilibrium shape (N is the order of the dislocation entrance). The continuous curves represent the edge trajectories for decreasing misfits (4%, 3.4%, 2.9%, 2.6%) due to the dislocation entrance. The dotted curves correspond to the thermodynamic criterion for dislocation entrance. The arrowed curve is the trajectory followed by the crystal during growth. (From P. Müller and R. Kern [89].)

size it is cheaper in energy to create dislocations. From Figure 4b one can see that the aspect ratio increases with crystal size and that at a given size the aspect ratio decreases. This corresponds to the introduction of the first dislocation. Later the aspect ratio increases again up to the introduction of the second dislocation, and so on. The interesting feature is that upon introduction of dislocations the height of the crystal oscillates around a constant value. The same authors have extended their model to pyramidal shapes, which are more realistic crystal shapes, but the results are qualitatively similar [89].

Another theoretical approach of the equilibrium shape of supported crystals is molecular dynamics. The first approach has used the empirical interatomic potential [90–93]. For Au atoms deposited on MgO(100) it has been shown that surface defects (vacancies and steps) are nucleation sites [90]. In the case of Pd/MgO(100) the parameters of the potentials have been obtained from ab initio calculations [91]. Starting from a truncated pyramid containing 127 atoms it was shown that the cluster keeps this shape but that some relaxation occurs and the atoms at the corners disappear. The Pd cluster is dilated by 3.8% in average in the direction perpendicular to the interface, but the strain is nonhomogeneous. In the planes parallel to the interface the dilatation is much less pronounced (0.77% in the average). For the same system calculations have been made using different potentials [92]. Starting from a cube the Pd cluster (64 atoms) transforms to a half-octahedron, and this structure remains stable up to 1000 K. A similar approach has been made using for the Pd–Pd interaction a TB-SMA potential and for the Pd/MgO interaction a semi-empirical many-body potential issued from ab initio calculations [93]. By quenched molecular dynamics they found that the more stable shape of the Pd clusters (50 to 9054 atoms) is a truncated pyramid with the interfacial layer forming a re-entrant angle with the MgO surface. They found a dilatation of the Pd lattice parallel to the interface and a contraction in the perpendicular direction. However, the deformation is inhomogeneous and most of the dilatation is localized at the interface. The adhesion energy converges to $0.83 \, J/m^2$ for Pd clusters of 7 nm [93].

However, these molecular dynamics calculations suffer some limitations: the empirical nature of the potential (especially for the metal–support interaction) and the arbitrary separation between the metal–metal and metal–support interactions (the metal–metal potential is probably perturbed near the interface). Indeed, according to the type of potential used, very different results are obtained. In the case of Pd/MgO, a mean dilatation [91] or contraction [92] is observed. For finite-temperature molecular dynamics, the calculations are limited to very short times and it is not sure that the equilibrium shape is reached. As we have seen in the last section the cluster shape can be blocked for a long time on facetted metastable shapes.

Fully ab initio calculations could solve the problem of the metal–support interaction, but it is limited to very small systems or to infinite layers. However, these calculations have given very important information concerning the adsorption sites and the diffusion of atoms and very small clusters. Numerous DFT calculations on the adsorption of isolated atoms or pseudomorphic monolayers have been made in the last few years for metals on MgO [94–100], TiO_2 [101–103], and Al_2O_3 [104,105]. Almost all the calculations agree that the oxygen anion is the adsorption site, for titania calculations suggest that it is the twofold coordinated oxygen. Only for gold on $TiO_2(110)$ the fivefold coordinated Ti seems to be favored regarding oxygen sites [103].

However, it is now well admitted that surface defects play an important role in the nucleation of metals. Indeed, calculations show that metal atoms are much more strongly bound on surface defects (like oxygen vacancies or metal atom vacancies) than on the regular sites [96c,99]. Another important issue is the effect of hydroxyl ions on the nucleation. Indeed, it was shown experimentally that after hydroxylation of an alumina surface the density of metal clusters increased significantly [31]. Calculations have shown that the binding energy of a Pt atom is a little increased on (or next to) an isolated hydroxyl anion on an MgO surface, and it is almost doubled next to a neutral hydroxyl [99]. Thus, hydroxyls seem to be nucleation promoters. However, recent calculations have shown that the interaction of a metal (Cu and Pd) with a hydroxylated α-alumina surface is very weak [105]. To reconcile this result with the experimental observations [31], it is suggested that the hydroxylation process creates surface defects that promote nucleation [105], as already pointed out to explain the increased cluster density of metal clusters on an air-cleaved (and subsequently dehydroxylated) MgO surface [6].

Effect of the Adsorption

It is known that adsorption decreases the surface energy [78]. Because the adsorption energy often depends on the type of facet, one expects a change of the equilibrium shape. Shi and Masel [106,107] have calculated the equilibrium shape of a crystal at 0 K in a presence of gases. The calculations show drastic changes of the equilibrium shape already at coverages around 0.1 ML. Then during a catalytic reaction the particle shape can evolve, which can affect the reaction kinetics, as shown recently by Monte Carlo simulations [108].

7.3.3 Experimental Techniques

In this section we only briefly describe the techniques used to characterize the structure and the morphology of the supported clusters, because they are well detailed in several review papers [4–8,109]. The more commonly used technique to study the morphology and the structure of supported clusters is transmission electron microscopy (TEM) [110,111]. Often the difficult part for TEM observation is the sample preparation. For oxide supports an easy preparation method is to cover the sample by a thin (a few tenths' nm) layer of amorphous carbon. Then the carbon film containing the metal clusters is separated from the support by surface dissolution of the oxide support in an acidic solution. This method works with MgO [112] and alumina [113]. It is also possible to pre-thin a 3-mm disc of the oxide support by ion milling and to clamp this disc on the main sample substrate that will be studied by surface-science techniques; thus both supports will be treated in the same way [114]. Layered samples can be easily prepared for TEM observation by cleavage after the growth of the clusters [115]. Size distributions are obtained by computer analysis of the TEM images. To have a stable size distribution, it is necessary to count at least 2000 particles. If we are only interested in the mean particle size, a good approximation is already obtained in counting 100 particles. The particle shape can be also obtained by TEM. Top-view images provide only partial information on the 3D shape of the particles. Then it is possible to combine images in two perpendicular directions, and by taking into account the crystal symmetry (for facetted particles) the morphology can be deduced. Cross sections can be obtained by special sample preparation of thin sections obtained with an ultra-microtome [116]. More simply, cross sections can be obtained by folding the thin TEM sample; in this case it is necessary to properly align the sample relative to the electron beam by using, for example, electron diffraction [110]. For particles larger than 6 nm it is also possible to use the weak beam dark field imaging [115]. An image of the particle is obtained by selecting a single reflection to be in the dark field condition. Then the sample is tilted a few degrees off the Bragg condition, and thickness fringes appear in the image of the particles (see Figure 5). From the width of the fringes (*w*) one deduces directly the slope of the facets by the following relation:

$$w = (G_{hkl}\Delta\theta\tan\phi)^{-1} \tag{3}$$

G_{nkl} is the modulus of the vector of the reciprocal lattice associated to the (*hkl*) reflection, $\Delta\theta$ is the tilt angle from the Bragg condition, and ϕ is the angle between the lateral facets and the basal plane (if re-entrant angles are present, the width of the fringes is smaller [110]). If the epitaxial relationships are known, the nature of the facets is directly deduced from the ϕ angles [115]. The structure of individual

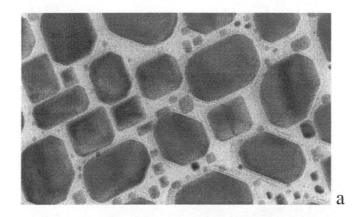

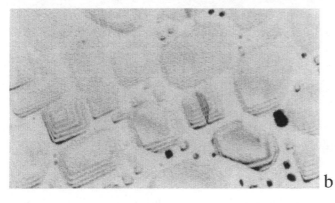

Figure 5 Bright field (a) and weak beam dark field (b) TEM pictures ($345 \times 212\,\text{nm}^2$) of large Pd particles supported on MgO (100).

particles is obtained by selected area electron diffraction [110]. If all the particles are in perfect epitaxy, the structural information can be obtained from diffraction on a large area. More details on the internal structure of the particles can be gained by high-resolution TEM (HRTEM) from the direct visualization of the lattice planes [110]. The morphology of very small particles (down to 1 nm) as well as the lattice deformations near the interface, due to the lattice mismatch, can be accurately measured [117]. Figure 6 shows an example of an HRTEM profile image of a gold cluster supported on MgO.

The main limitation of electron microscopy is its ex-situ character (except for the very rare true UHV electron microscopes). For direct in-situ observations (under UHV or eventually under a special atmosphere) STM and AFM scanning probe microscopies are well suited. STM can be used without any special preparation on semiconducting oxides (like TiO_2, ZnO) or on ultrathin films of insulating bulk oxide deposited on metals [7,8]. For bulk insulating oxide only AFM can be used. These two techniques are complementary to TEM. In TEM the lateral size of the particles is very accurately measured but the height is more difficult to measure (except in cross section). By scanning probe techniques the height of the particles is measured with a precision of a fraction of an Angstrom, while the width is always overestimated because of the convolution with tip shape [118]. The internal atomic structure of supported clusters can be obtained by HRTEM [110], while the surface structure can be obtained (not routinely) by STM [119,120] and also by AFM [121].

Figure 6 HRTEM picture, in a $\langle 110 \rangle$ projection, of a 4-nm gold cluster supported on MgO (100). (Courtesy of S. Giorgio.)

X-ray diffraction now appears also to be a powerful technique for in-situ study of the structure and morphology of supported clusters [122]. Grazing incidence X-rays scattering (GIXS) gives the epitaxy relationships and the lattice parameter of the metal particles, while grazing incidence small angle X-rays scattering (GISAXS) gives the particle shape [122]. This technique allows us to study in situ and in real time the growth of metal clusters, but it gives information averaged on a large number of particles. Wide-angle X-rays scattering (WAXS) has also been used to study the structure of small clusters down to 1 nm [123].

7.3.4 Some Experimental Results

We will not review all the experimental work published on the structure and morphology of supported clusters; rather we show the information gained recently by various techniques and by the comparison with theoretical results by looking mainly at three different metals: noble metals (Au and Ag), which have recently shown (for gold) very interesting catalytic properties [124], and Pd as a classical catalytic metal.

Gold Clusters

Au clusters embedded in mylar (that is, assumed to have a very weak interaction with the metal clusters) have been studied carefully by EXAFS [125,126]. The clusters show a lattice contraction due to the surface stress [127], which is proportional to the reciprocal diameter. The structure is FCC down to 50 atoms, and the morphology is a truncated octahedron (Wulff polyhedron). This result contrasts with the HRTEM observation of quasi-melting of gold clusters [76]. However, it is not excluded that the mylar substrate stabilizes the FCC structure. In fact, an interesting observation has been made in HRTEM on small (a few nm) gold clusters supported on MgO [128,129]. Gold particles showing quasi-melting (rapid fluctuations between different crystalline and noncrystalline structures) were very weakly bound to the support, while particles in epitaxy on the MgO support were stable and had a truncated octahedron shape. It was concluded [128] that strong interaction with the support prevents quasi-melting. In other words, particles in epitaxy are on a deep energy minimum. This result is in agreement with molecular dynamics simulations. In-situ TEM observations on larger epitaxial gold particles ($\geqslant 10$ nm) on MgO(100) have shown truncated pyramid shapes [130]. The amount of truncation by (100) facets increased with cluster size.

Another, more surprising, observation was that at a given size the gold particle grew mainly laterally, giving rise to flatter shapes. Very recently, in-situ HRTEM observations of the growth of gold clusters on MgO(100) have shown that at the beginning the cluster's habit was a half-octahedron and at a given size (100) truncations appeared [131]. At a later stage the perfect pyramid reappears and again becomes truncated. Up to 5 such shape oscillations have been observed, they disappear at a size corresponding to about 1400 atoms. Free gold clusters (around 2 nm in size) have been generated by a laser vaporization source and deposited on MgO(100) surfaces [40]. The Au clusters do not show any evidence of coalescence. HRTEM observations show that the deposited clusters have not an isotropic shape as expected in gas phase, but they have the shape of a truncated octahedron as in the case of epitaxial growth on the MgO surface. HRTEM pictures show that the gold lattice is accommodated on the MgO lattice at the interface. These interesting observations confirm that the stable state of a gold cluster on the MgO(100) surface is a truncated octahedron in (100) epitaxy.

The growth of gold on TiO_2(110) has recently been reviewed [132]. From TEM observations in cross section, a truncated octahedron shape was observed for large particles (≈ 200 nm). A value of 0.35 was obtained for $\Delta h/h$ [see Eq. (2)], equivalent to a contact angle of 122°. Assuming a surface energy of 1.045 J/m² for Au(111), an adhesion energy of 0.9 to 1.0 J/m² was deduced [132].

Recently, we have investigated the growth of gold particles on mica (100), in situ by AFM and ex situ by TEM [121,133]. Figure 7 shows a TEM picture of a 20-ML gold deposit. Three kinds of particles can be seen: small, more or less round particles; triangle particles; and large, quasi-hexagonal particles. Electron diffraction combined with dark field imaging has shown that the particles are in (111) epitaxy but that round ones are azimuthally disoriented. In a bright field (see Figure 7) the round particles show typical contrasts for icosahedron and decahedron MTP particles [134,135]. The origin of these MTP particles is controversial. They are attributed either to the growth from icosahedron nuclei preserving the fivefold symmetry [136] or to the coalescence of FCC tetrahedra due to their mobility [135]. We have observed that at the early stages of growth the proportion of MTP particles

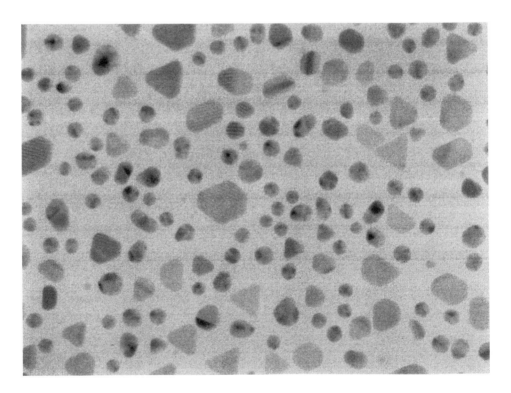

Figure 7 TEM picture (362×270 nm²) of gold particles supported on mica. (From S. Ferrero, A. Piednoir, and C.R. Henry [133].)

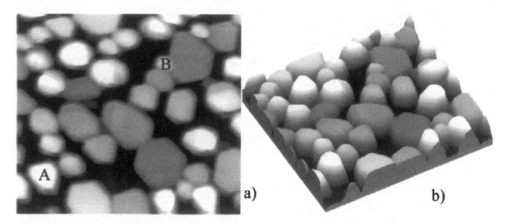

Figure 8 AFM pictures in 2D (a) and 3D (b) representations of gold particles on mica (same sample as in Fig. 7). (From S. Ferrero, A. Piednoir, and C.R. Henry [121].)

strongly drops while the proportion of triangular particles increases. These observations are only compatible with the postnucleation origin of the MTP particles. AFM gives additional information on the particle shapes. Figure 8 shows an AFM picture (in 2D and 3D representations) of the same sample corresponding to Figure 7. The same outline shapes as in TEM are observed. By AFM we have an accurate measurement of the particle heights, but the widths are enlarged by about 25% from the convolution with the tip shape. From the AFM pictures we deduced

Figure 9 Atomically resolved (111) top facet on a 30-nm Au particle in epitaxy on mica obtained by AFM in contact mode under UHV. (From S. Ferrero, A. Piednoir, and C.R. Henry [121].)

that the triangular particles are tetrahedra or top-truncated tetrahedra. The large hexagonal particles are flat, highly truncated tetrahedra. On these large particles it is possible to image the top facets at the atomic scale [121]. In Figure 9 we clearly see the atomic arrangement of a (111) Au plane. These facets are flat, but occasionally atomic steps are observed [121]. In some cases (100) lateral facets have been also identified by visualization of the ⟨110⟩ atomic rows [121]. In Figure 10 we have plotted the particle height as a function of the diameter measured by AFM on individual particles [133]. The height of the particles increases almost linearly, then reaches a maximum, decreases, and stays approximately constant. The same behavior has been also observed by STM on Pd particles on molybdenite [137] and on graphite [138]. Presently it is not possible to know if this behavior results from a kinetic effect or a thermodynamic effect. Indeed, if the particles have reached their equilibrium shape, because of the relaxation of strain (induced by the misfit between the particle and the substrate lattices), through the generation of dislocations, theoretical calculations [88,89] predict that the height of the particles should oscillate around a constant value. Conversely, if the particles have not reached their equilibrium shape, it is possible that the top facets stop their growth because of the

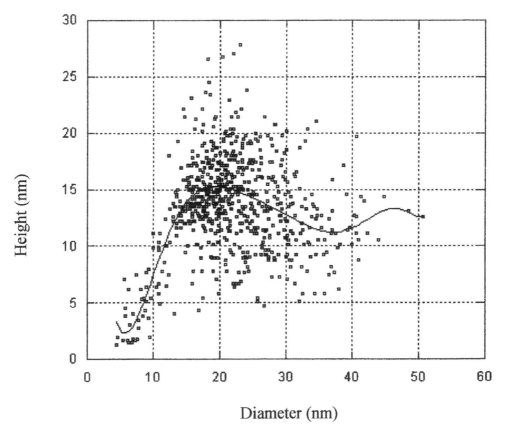

Figure 10   Height as a function of diameter measured by AFM under UHV on gold particles supported on mica. The continuous line is a polynomial fit of the data. (From S. Ferrero, A. Piednoir and C.R. Henry [133].)

difficulty to nucleate a new layer. This would imply that at small sizes the downhill diffusion was hindered presumably by a Schwoebel barrier [4].

Silver Clusters

The structure of free silver clusters has been studied for the first time by in-situ electron diffraction [139]. Besides the inherent difficulty of the method due to the mixture of different structures, it is clear that both icosahedra and decahedra coexist with FCC clusters up to sizes of 10 nm. By increasing the source temperature the proportion of icosahedra decreases, proving the metastable structure of the icosahedra. Molecular dynamics simulations have confirmed the metastability of the large silver icosahedra and that they can appear spontaneously during growth at low temperatures [140].

Early in-situ TEM work [141] has shown that the growth of silver clusters on MgO(100) followed the same trend as gold; they grow as square pyramids and then become truncated by (100) facets, and at a critical size they grow mainly laterally. From the shape of large particles (assumed to be at equilibrium) an adhesion energy of $0.45 \, J/m^2$ has been deduced [142]. More recently the adhesion energy of silver on MgO thin films was measured by microcalorimetry, which also results in a weak adhesion: $0.3 \, J/m^2$ [143]. An extensive study of the growth of silver particles on MgO(100) has been made by Renaud and co-workers using x-ray diffraction [144–146]. By GIXS they have previously shown that silver starts to grow at RT as perfectly registered 2D islands up to 0.4 ML [144]. Beyond 0.4 ML, the Ag islands are flat and coherent with the MgO lattice but relaxed at steps. At 1 ML the islands thicken, and they relax by generation of dislocations after 4 ML. At a thickness of about 10 nm, a dislocation network with the Burgers vector parallel to the $\langle 110 \rangle$ direction appears [145]. However, more recent measurements [146a] were interpreted by a 3D growth already at 0.2 ML and up to 1 ML, with relaxed islands, and between 1 ML and 4 ML islands should be coherent and partially relaxed at the edges. These discrepancies illustrate the difficulty, at low coverage (below 1 ML), to separate in the diffraction signal the different contributions coming from the coherent part and the relaxed part, which are averaged on many islands. By GISAXS it has been possible to follow in situ in real time the growth of the islands at RT [146]. Figure 11 shows the evolution of the average island size, island height, and mean distance between neighboring islands as a function of the silver coverage. The aspect ratio is nearly constant and close to 0.36 [146b].

Palladium Clusters

Pd grows in the (100) epitaxy on MgO(100) down to RT [36,112]. At RT the clusters are rough and relatively flat [81]. At 150 °C they grow as truncated pyramids with an aspect ratio of 0.22 [112]. At 400 °C they keep the same shape (below about 8 nm), but the aspect ratio is higher (0.40) [112]. The equilibrium shape of large particles (at least 10 nm) has been obtained by growth and annealing at 400 °C for 5 hr under UHV [147]. The shape is a truncated octahedron (see Figure 12) with re-entrant angles; the aspect ratio is 0.68. From this shape using the Wulff theorem a surface-energy anisotropy factor $\sigma_{100}/\sigma_{111} = 1.16$ was measured [147] close to that expected for an FCC crystal at low temperatures (1.15). From the amplitude of the truncation at the interface and applying the Wulff–Kaishew theorem [Eq. (2)] an adhesion energy of $0.91 \, J/m^2$ was obtained (assuming $\sigma_{100} = 1.641 \, J/m^2$ [148]). This value is in

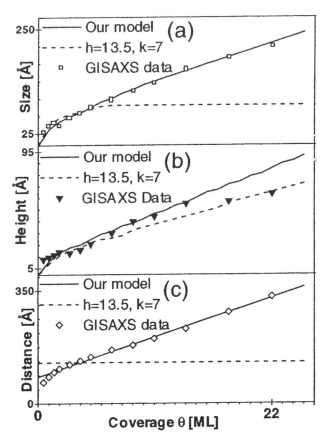

Figure 11 In-situ growth of Ag particles on MgO(100) studied by GISAXS. (a) Particle size, (b) particle height, (c) particle interdistance as a function of Ag coverage. The continuous curves represent the best fit with a growth model assuming a truncated pyramid shape. (From A. Barbier, G. Renaud, and J. Jupille [146b].)

good agreement with the contact angle measurement (115°) of a micron-sized liquid Pd droplet on MgO, which gives $\beta = 0.95\,\mathrm{J/m^2}$ [149]. This value is in reasonably good agreement with recent molecular dynamics calculations, which give $0.83\,\mathrm{J/m^2}$ [93]. HRTEM has shown that these large ($\geq 10\,\mathrm{nm}$) Pd clusters are relaxed and that dislocations are present at the interface [150]. At small sizes ($\leq 2\,\mathrm{nm}$) the equilibrium shape of Pd clusters is a truncated pyramid (the truncation at the top disappears below 1.5 nm) and their lattice (in the planes parallel to the interface) is dilated by 8% to become accommodated to the MgO [117]. In the size range of 3 to 7 nm, the Pd clusters have the shape of a truncated pyramid, but now only the first Pd layer is accommodated to the MgO lattice and the Pd lattice progressively relaxes to the bulk ones after 3 to 4 layers [150]. The different equilibrium shapes of the Pd clusters as a function of their size are represented schematically in Figure 12. The fact that the equilibrium shape depends on the particle size is due to the presence of strain. The large particles are relaxed and the equilibrium shape is that predicted from thermodynamics (Wulff–Kaishew). It is remarkable that down to 10 nm macroscopic

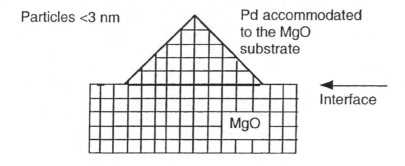

Particles <3 nm

Pd accommodated
to the MgO
substrate

Interface

MgO

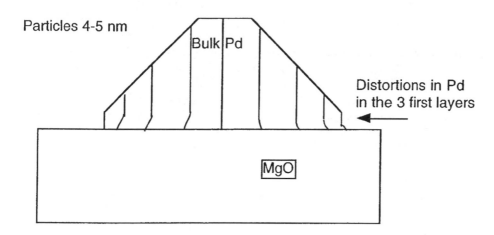

Particles 4-5 nm

Bulk Pd

Distortions in Pd
in the 3 first layers

MgO

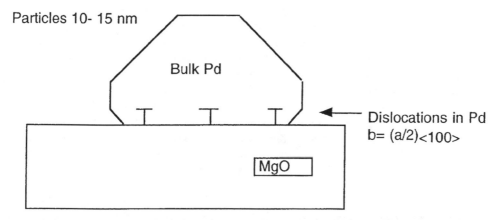

Particles 10- 15 nm

Bulk Pd

Dislocations in Pd
b= (a/2)<100>

MgO

Figure 12 Schematic representation of the equilibrium shape of Pd particles supported on MgO(100) as a function of their size. (From H. Graoui, S. Giorgio, and C.R. Henry [150].)

concepts are still valid. For smaller sizes the equilibrium shape is different, but in this case the lattice of the Pd is strained because of the 8% misfit between the Pd and MgO lattices. Unfortunately, these observations cannot be directly compared with the calculations from Kern and Müller [88,89] because the shapes used in the calculations are different. However, some qualitative features are in agreement. Calculations predict that when the strain increases the aspect ratio increases. Indeed, at very small sizes the Pd clusters are full pyramid and when the lattice of the Pd is relaxed, except at the interface, and (100) truncations appear at the interface. However, it is not clear why the re-entrant angles appear only at large sizes when the Pd lattice is fully relaxed.

On other supports with a threefold symmetry axis, like alumina [120], graphite [115], mica [151], or molybdenite [137], the Pd particles are in the (111) epitaxy and their shape evolves from a tetrahedron toward a truncated tetrahedron having a hexagonal outline. Contrary to the case of the (100) epitaxy, the top truncation is (obviously) a (111) plane and the lateral facets are (111) planes. Therefore, in general, in the (111) epitaxy the metal particles exhibit more (111) facets than in the (100) epitaxy when the particles are truncated. Meanwhile, in the early stage of growth in the both epitaxies they exhibit only (111) planes. However, it is always difficult to ascertain that the metal particles exhibit equilibrium shapes except in some cases where the supersaturation was very low and the particles annealed for a long time at high temperatures. For Pd particles (about 10 nm) on alumina, assuming that the equilibrium shape was reached, an adhesion energy around $3 \, J/m^2$ has been deduced from STM observations [120].

The effect of oxygen adsorption on the equilibrium shape of 10-nm Pd particles supported on MgO(100) has recently been investigated [147]. In these experiments Pd particles having the truncated octahedron shape under UHV have been annealed for 3 hr at high temperatures: 450–550 °C (to have a significant mobility of the Pd surface atoms) and under high oxygen pressure: 10^{-5}–10^{-3} Torr (to have a nonnegligible equilibrium oxygen coverage). After equilibration the Pd particles were rapidly quenched and covered by a carbon film before being observed ex situ by TEM [147]. After annealing under oxygen the surface anisotropy becomes inverted, which corresponds to an increase of the (100) facets and a shrinkage of the (111) facets. The experimental values are in rather good agreement with calculations of the decrease of surface energies using kinetic data from extended Pd surfaces and assuming no diffusion between the facets [147]. As an example, for annealing at 550 °C under 1×10^{-3} Torr of oxygen, the aspect ratio decreases from 0.68 to 0.49. This flattening of the particle is not only due to the large extension of the (100) facets but also results from the decrease of the adhesion energy measured by the truncation at the interface. The drop of the adhesion energy could be due to diffusion of oxygen at the interface between the Pd particles and the MgO surface.

7.4 ELECTRONIC PROPERTIES

7.4.1 Introduction

The knowledge of the electronic structure of metal clusters is of key importance to understand the origin of their catalytic activity. The electronic structure of bulk solids and of solid surfaces is now well understood in terms of band theory. On another side the electronic structure of atoms and molecules is also well developed on the basis of a succession of discrete energy levels. Building a cluster atom by atom up to macroscopic sizes, how does the electronic structure evolve from the discrete energy level scheme to the continuous band scheme? We will see that, in fact, it is possible to separate clusters in small ones having an electronic structure close to a molecule and large ones resembling bulk metals. Finally, we briefly describe a recent heuristic approach of the relationship between electronic structure and reactivity.

7.4.2 Small Clusters

One of the most beautiful results of the physics of small clusters is the electronic shell model first introduced by Knight and colleagues [152]. The electronic shell structure was discovered from the presence of intense peaks in the mass spectra of alkali metal clusters on well-defined masses (8, 20, 40, 58, etc.) called magic numbers. These numbers correspond to the closing of electronic shells made by the valence electrons (1s in the case of alkali metals) of the atoms in the cluster. This was the first direct evidence of the discreteness of the electronic structure of small metal clusters. For free-transition metal clusters, photoemission laser spectroscopy has shown that the electronic structure of small metal clusters depends on the exact number of atoms they contain. Figure 13a shows a series of photoemission spectra of Cu anionic clusters from 1 to 36 atoms obtained by Smalley and co-workers [153]. It is clear that around 20 atoms the embryo of the 3d and 4s bands appears. In Figure 13b the evolution of the onset of these bands is drawn as a function of the number of atoms. Two important features can be drawn from this figure. First, some minima are reminiscent of the electronic shell structure. Second, for clusters larger than 30 to 40 atoms, the evolution of the electronic structure is monotonous and no more dependent on the exact number of atoms they contain.

For supported small clusters little is actually known about their electronic structure [154].

7.4.3 Large Clusters

For large, free clusters containing more than 100 atoms, few experiments have been undertaken to study their electronic structure. Extrapolating the experimental results from smaller clusters, one expects a continuous evolution of a band structure. Indeed, theoretical calculations show that by increasing the cluster size the valence band shifts away from the Fermi level and its width broadens [63b,155]. Other calculations have shown that core levels shift by the same amount as the valence bands [156]. If the electronic structure of large clusters does not depend on the exact number of atoms they contain, the local electronic structure depends significantly on the local arrangement of the atoms. In general, the shift of the valence band and its sharpening increase when the coordination of the considered atom decreases [155].

Large supported clusters are generally studied experimentally by photoemission spectroscopy. From the experimental spectra the broadening of the valence band, when the cluster size increases, has been clearly evidenced [157]. However, the core levels and the valence band generally show a shift toward the Fermi level when the cluster size increases, but the amount of the shift depends on the nature of the support [158]. Meanwhile, the shift observed in photoemission results in a complex

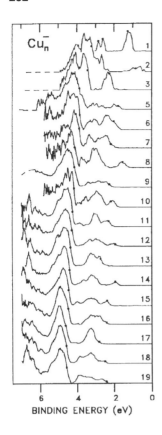

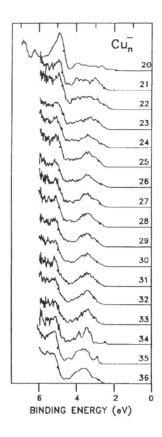

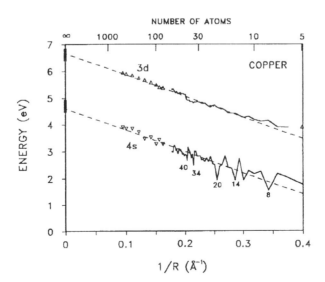

Figure 13   Photoemission spectra of free Cu$_n^-$ clusters for $n = 1$–36 (top) and onset of the 4s and 3d bands as a function of the reciprocal cluster radius (down). (From K.J. Taylor et al. [153].)

way from different contributions: an initial state shift, which represents the shift of the electronic levels, and a final state contribution, which corresponds to an increase of the binding energy that is due to the nonperfect screening of the hole created during the photoemission process (in the case of insulator substrates it is also possible to have static charges, giving rise to additional energy shifts). Thus, it is generally impossible to access the initial state shift by photoemission except in certain cases [159,6]. The electronic structure of supported clusters has been also investigated recently by STS (scanning tunneling spectroscopy). Pd, Au, and Ag clusters supported on TiO_2 have been studied by STS by Goodman [160]. For clusters smaller than 3 nm the opening of a gap has been observed. It is not yet clear if this gap is really the signature of a metal/insulator transition, which is expected from calculations to smaller sizes, or an effect of the quantification of the conductance in very thin, flat clusters [154].

7.4.4 Relationship Between Electronic Structure and Reactivity

From recent ab initio combined with tight binding calculations performed by Norskov and co-workers [2,161] a new comprehensive picture of the relationship between electronic structure and reactivity of metal surfaces has emerged. The key feature in the electronic structure of late-transition metals is the position of the d-band. When the d-band shifts up toward the Fermi level, the chemisorption energy (O, H, N, CO, NO, etc.) increases and the dissociation barrier (H_2, O_2, CO, NO, N_2, etc.) decreases. This simple picture is still valid for hydrocarbons, as recently shown [162]. Then, knowing the evolution of the position of the d-band, it becomes possible to know the reactivity trends. Thus we can understand why, when the coordination of surface atoms decreases, the adsorption energy and the dissociation rate increase. Besides coordination, two other factors can affect the position of the valence band (and then the reactivity); they are the strain [163] and the alloying [164]. From these new concepts, combined with experimental measurements, it has been possible to entirely design a new catalyst for steam reforming [165]. This new scheme of the relationship between electronic structure and reactivity seems applicable to predict the reactivity trends in many different systems and in particular to metal clusters that are sufficiently large to develop a band structure.

7.5 REACTIVITY

7.5.1 Introduction

The reactivity of supported model catalysts has already been reviewed in detail in 1998 [6]. Here I give a general overview of the specificity of supported nanometer-sized clusters as model catalysts compared to small clusters (less than 20 atoms) and single-crystal surfaces. Based on two catalytic reactions recently studied in detail—the CO oxidation and the NO reduction by CO—we will see the respective role of the support, the cluster size, and the cluster morphology.

7.5.2 Small Clusters

In the last three years extensive work has been performed by Heiz et al. on the reactivity of supported clusters containing from 1 to 30 atoms [41,45,166,167]. The clusters were prepared by laser vaporization, size selected, and soft-landed on an MgO(100) thin film (see Section 7.2). The striking result of these studies is that the reactivity depends on the exact number of atoms they contain. We see in Figure 14 the production of CO_2 during a TPR (temperature-programmed reaction) from preadsorbed CO and oxygen on Pt clusters supported on MgO as a function of the number of atoms they contain [41]. We see that the reactivity per Pt atom varies in a discrete manner with the number of atoms contained in the Pt clusters. This fact is due to the specific electronic structure of these small clusters, which changes abruptly with the number of atoms (see Section 7.4.2). In that respect, the reactivity of these small clusters cannot be extrapolated to large clusters that are used in catalysis which contain more than 30 atoms and exhibit a smooth variation with the size of their electronic structure (see Section 7.4.3). However, the great advantage of such studies is that it is possible to do exact ab initio calculations for these small clusters. A good example is given by CO oxidation on gold. The reactivity is zero below 8 atoms and then increases abruptly and varies as a function of the number of gold atoms [166]. Calculations performed by Landmann and co-workers show that the onset of reactivity at eight atoms was due to the particular structure of this cluster but also to the support. This cluster is only active if it sits on a oxygen vacancy, which results in a charge transfer from the MgO surface to the gold cluster [166]. The question we can ask is if these small clusters are good catalysts. It is not possible to answer from these experiments on size-selected, deposited clusters because the reactivity

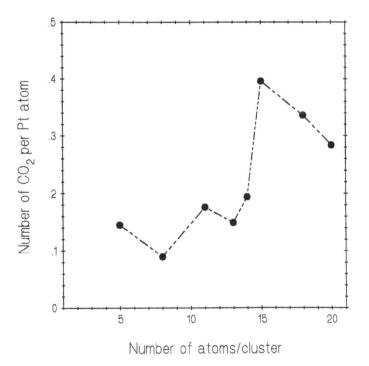

Figure 14   CO oxidation on size-selected Pt_n clusters soft-landed on MgO. CO_2 production per Pt atom as a function of n. (From U. Heiz et al. [41].)

measurements are not representative of a catalytic cycle; thus the TON (turnover number) cannot be extrapolated from the reactivity measurements. However, Gates has directly addressed this question in preparing Ir_4 and Ir_6 clusters on high-area supports (alumina, magnesia, zeolithes) [168]. The catalysts were prepared from metal carbonyl clusters by gently removing the CO ligands. It has been verified by EXAFS that the clusters kept their tetrahedral (Ir_4) and octahedral (Ir_6) structure after deposition [168a]. These model catalysts have been tested for the toluene hydrogenation reaction and have shown a weaker catalytic activity than conventional Ir catalysts corresponding to particle mean sizes from 1 to 4 nm [168b,169].

Why are these small clusters less active? From the concept of the relationship between the decrease of the coordination of the metal atoms and the increase of the reactivity (see Section 7.4.4), one expects an amplification of reactivity for tiny clusters. But it is not clear down to what size this concept developed for extended surfaces is still valid. An explanation could be the presence of the support modifies the electronic properties of these small clusters. This idea is supported by the fact that Ir_4 clusters on alumina are much less reactive than on magnesia [168b]. However, the presence of a support does not necessarily have a negative effect. For Au_8 the MgO support has a positive effect on the reactivity because bulk gold is not active at all. In this case the support effect is due to the presence of a surface defect (F center) that leads to a charge transfer toward the gold cluster that becomes (partially) negatively charged. This charge transfer is probably a key factor in the reactivity of this tiny gold clusters. This idea is further supported by the fact that free anionic gold clusters are able to adsorb oxygen while gold cations do not [170].

7.5.3 Large Clusters
Introduction

To illustrate the prominent features in the reactivity of supported nanosized metal clusters, we focus on two reactions that have recently been studied in detail mainly by molecular beam techniques: the CO oxidation and the NO reduction by CO on supported Pd clusters. First we discuss the adsorption of the reactants and then the catalytic reaction itself.

CO Adsorption

It is now well known that at very low coverage the binding energy of CO increases when the cluster size decreases below 5 nm [6,171]. This result has been explained by a stronger adsorption on low coordinated atoms (edges and corners), the proportion of which increases when cluster size drops [6]. This is understandable in the framework of the Hammer–Norskov theory [161] that predicts an increase of the CO binding energy when the d-band center shifts up toward the Fermi level. In fact, electronic structure calculations for a (5.7-nm) Pd cluster [155] have shown that the d-band center shifts in the right direction when the coordination of the surface atoms decreases, and the predicted maximum increase of CO adsorption energy (8.1 kcal/mol) is in agreement with the experiment [6]. On the contrary, at high coverage the binding energy of CO decreases with cluster size. This result has been established on Pd/alumina from the increasing proportion of the low-temperature tail in the TPD spectra when the cluster size decreases [8]. It is also in agreement with the decrease of the equilibrium coverage at low temperatures, when the size of the Pd clusters supported on MgO decreases [172].

The support plays an important effect in the adsorption kinetics of CO on supported clusters. Indeed CO physisorbed on the support is captured by surface diffusion on the periphery of the metal clusters where it becomes chemisorbed. The role of a precursor state played by CO adsorbed on the support is a rather general phenomenon. It has been observed first on Pd/mica [173] then on Pd/alumina [174,175], on Pd/MgO [176], on Pd/silica [177], and on Rh/alumina [178]. This effect has been theoretically modeled assuming the clusters are distributed on a regular lattice [179] and more recently on a random distribution of clusters [180]. The basic features of this phenomenon are the following. One can define around each cluster a capture zone of width X_s, where X_s is the mean diffusion length of a CO molecule on the support. Each molecule physisorbed in the capture zone will be chemisorbed (via surface diffusion) on the metal cluster. When the temperature decreases, X_s increases, then the capture zone increases to the point where the capture zones overlap. Thus the adsorption rate increases when temperature decreases before the overlap of the capture zones that occurs earlier when the density of clusters increases. Another interesting feature is that the adsorption flux increases when cluster size decreases. It is worth mentioning that this effect (often called *reverse spillover*) can increase the adsorption rate by a factor of 10. We later see the consequences for catalytic reactions.

CO does not dissociate on Pd clusters that have well-defined shapes with smooth facets (as on extended Pd surfaces); however, on ill-defined clusters (and maybe on very small clusters) CO dissociation can occur [6]. On Rh and Ir particles CO dissociation occurs very easily (much more than on extended surfaces) [181–183]. CO dissociation on Rh presents a very interesting size effect. We see in Figure 15 that the dissociation rate of CO on Rh clusters supported on alumina increases up to about 200 atoms and then decreases slowly [183]. The explanation of this behavior is related to the morphology of the Rh clusters. At the beginning of growth the Rh clusters are well-ordered 2D islands, and then the clusters become thicker, probably by nucleation of new layers parallel to the substrate surface, and then present steps where CO is dissociated. At a later stage of growth the Rh particles become completely facetted and the corresponding number of steps is low. The maximum dissociation would occur for a size corresponding to very rough and numerous clusters. Interestingly, when the growth temperature increases, the decline of dissociation starts earlier (see Figure 15) because the particle become smooth at an earlier stage due to the increased mobility of surface atoms.

NO Adsorption

As for CO [171,172,176], the adsorption of NO as been studied on Pd supported on MgO(100) using molecular beam techniques [184]. The scattering of NO on the clean MgO surface shows that 56% of the NO beam becomes physisorbed while 44% of the beam is quasi-elastically reflected [184]. When Pd clusters are present on the substrate, NO becomes becomes chemisorbed on them and, as in the case of CO (see former paragraph), part of the NO physisorbed on the MgO is caught by the Pd clusters and becomes chemisorbed [184]. The lifetime of NO molecules chemisorbed on the Pd clusters has been measured at various temperatures in modulating the beam by a square wave. At low coverage, the lifetime as a function of temperature

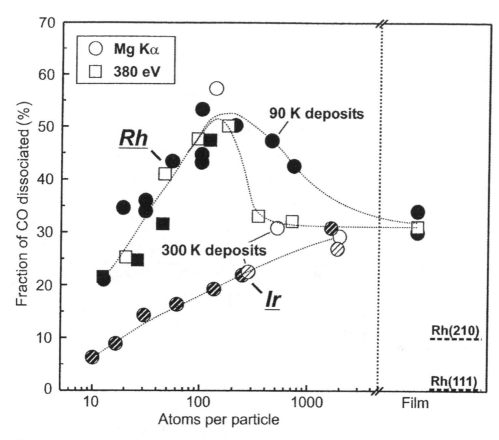

Figure 15 Dissociation rate of CO on Rh (grown at 90 and 300 K) and Ir (grown at 300 K) clusters supported on ultrathin alumina thin films, measured by XPS. The level of dissociation on extended Rh (111) and (210) surfaces is indicated. (From M. Frank and M. Bäumer [183].)

shows an Arrhenius behavior. Then one can deduce the adsorption energy of a NO molecule, which for large clusters (at least 10 nm) is around 32 kcal/mol [184]. This value is higher than for CO (29–30 kcal/mol [171]), and it explains why NO is able to displace adsorbed CO. However, unlike CO, NO can be dissociated on Pd clusters. Starting from a fresh sample (never exposed to NO) the dissociation is fairly high during the first pulse, then from pulse to pulse it decreases, and after the third pulse the dissociation rate is constant. The dissociation is directly proven by the emission of N_2 during the NO pulse. However, no oxygen is produced during the NO dosing. In fact, dissociated oxygen cannot desorb at 450 °C (the highest temperature used in this study). In additional experiments we have dosed by CO the amount of oxygen adsorbed on the Pd particles, and we have been able to prove that some adsorbed oxygen diffuses inside the lattice of the Pd particles [185]. The dissociation of NO on the fresh samples increases when particle size decreases in agreement with other observations, and it is attributed to the increasing fraction of low coordinated sites (edges and corners) when cluster size decreases [186,187]. In Figure 16 we have

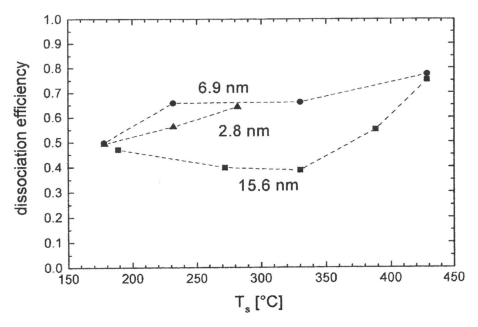

Figure 16   Steady-state dissociation rate of NO and Pd clusters of different sizes supported on MgO(100). (From L. Piccolo and C.R. Henry [216].)

plotted the NO dissociation rate, in steady-state conditions, i.e. after the three first pulses, as a function of temperature for three different cluster sizes. We see that the dissociation rate increases with temperature. The effect of particle size is not obvious; the lowest dissociation is observed for the largest particles, while the highest dissociation is observed for the medium-sized particles. This behavior is again due to the morphology of the Pd clusters, which we explain ahead.

Oxygen Adsorption

The dissociative adsorption of oxygen on supported Pd, Pt, and Rh clusters has already been studied by several groups [188–190]. It has been shown by molecular beam techniques that oxygen can be incorporated in the volume of the Pd clusters already at 300 K [188,190]. It has also been observed that the total coverage of oxygen incorporated in the metal lattice increases when the particle size decreases [189]. The incorporation of oxygen is higher for Rh than for Pd and Pt [189]. For Pd on silica it was suggested that oxygen atoms adsorbed on the metal clusters can diffuse on the support [177]. Meanwhile, it seems that this phenomenon could also be explained by diffusion of oxygen from the inside of the particle to their outer surface.

From studies on extended surfaces the oxygen adsorbed on Pd desorbs around 800 K [191], meaning that during CO oxidation and NO reduction by CO, in usual conditions (below 725 K), adsorbed oxygen will not desorb from the Pd. It has recently been shown by several techniques (mainly by XPS) that adsorbed oxygen can diffuse below the surface for Pd(110) [192] and Pd(111) [193].

CO Oxidation

The CO oxidation is certainly the most studied catalytic reaction on supported clusters since the pioneering work of Ladas, Poppa, and Boudart, 20 years ago, on Pd/alumina [194]. A detailed review on the work on CO oxidation on supported model catalysts until 1997 is presented in [6]. Since 1998 still a large number of papers have appeared for Pd/alumina [190,195,203], Pd/magnesia [199,201,202], Rh/alumina [198], and Au/titania [196,197,200]. Here we limit our attention to Pd clusters. Another very interesting metal is gold, for which the reaction mechanism seems different than for platinum-type metals [124,132] and is reviewed in the chapter by Wayne Goodman. From the very large number of studies on CO oxidation on Pd up to 1997 (see [6]), several important differences appear between extended surfaces and supported clusters. The first one and probably the more quantitatively important in the reaction kinetics is due to the reverse spillover of CO from the support to the metal particles. In the regime where the rate-limiting step for the reaction is the adsorption of CO (at high temperatures) the reaction rate is proportional to the CO pressure. The capture, by the metal clusters, of CO physisorbed on the support is equivalent to an increase of the CO pressure and the reaction rate increases. The model developed to take into account this reverse spillover effect [179] has allowed us to quantitatively explain the very careful kinetic measurements by Rumpf, Poppa, and Boudart on Pd/sapphire [204]. Particularly, the reaction rate increases when cluster size decreases in the temperature range where the capture zones are nonnegligible and do not overlap. Then if we are interested in a pure size effect for this reaction, it is necessary to correct the data from this effect. For Pd/MgO(100) several particles sizes (between 2.8 and 13 nm) have been investigated [205]. After correction of the reverse spillover effect, for the particles of 13 nm the steady-state TOR (turnover rate) was the same as on an extended Pd surface. However, for particles smaller than 7 nm, the reaction rate near the temperature corresponding to the maximum of TOR was higher (about a factor of 2) than on large particles. At very high temperatures the difference was reduced, and at low temperatures no significant difference was observed. This is clear evidence of a size effect. The proposed interpretation was the increase of CO adsorption at low coverage due to the large proportion of low coordinated sites on small particles. It has to be noticed that in the conditions for the maximum reaction rate the CO coverage is very small [6]. For a long time CO oxidation has been considered as structure-insensitive, but recent experiments have shown that the reactivity of (110) or stepped surfaces is higher than on (111) or (110) surfaces, but the structure sensitivity remains weak [206].

It is important to figure out that the structure sensitivity is only important in the low-CO coverage limit (high temperature, low-CO pressure), and in the high-CO coverage limit the reaction is truly structure-insensitive, as observed by Goodman's group [207]. Another peculiar feature of this reaction on supported Pd clusters was discovered in the transient regime. Taking the CO pressure constant and sending to the sample a pulse of CO, the CO_2 production shows, at low temperature, a transient pulse before reaching the steady state but, more surprising, at the end of the pulse (when the CO beam is shut down) the reaction rate decreases and suddenly increases to form a second pulse of CO_2 [208]. The area of this secondary peak (relative to the steady-state intensity) was found to increase when cluster size decreased. This surprising effect (the occurrence of the second peak) was explained by CO strongly bound on the edges, which should stay adsorbed on the Pd clusters a long time after shutting off the CO beam and could diffuse toward the facets, where it readily reacts with adsorbed oxygen. A microkinetic model based on this mechanism was able to reproduce the experimental data, but it was necessary to assume a very large barrier for a CO molecule to diffuse from an edge to a facet (27 kcal/mol) [201]; that is about 3 times larger than the diffusion energy of CO on a smooth, extended Pd surface. Recently, a new molecular beam experimental setup has been built in Berlin in the group of Freund; it is based on three independent molecular beams with simultaneous IRAS measurements [190,203]. With this experiment it has been possible to reproduce the preceding results [203]. In Figure 17a we see a series of CO_2 transients during a CO pulse and a constant O_2 beam for increasing ratio (x_{co}) of the CO pressure to the sum of the CO and O_2 equivalent pressures, at 440 K. We clearly see that the second peak of CO_2 appears for $x_{co} = 0.47$ when the steady-state reaction rate starts to decrease. In fact, the decrease of activity corresponds to the beginning of the domain of high-CO coverage where the adsorption of oxygen starts to be inhibited by CO. The IR spectra obtained, every second, during the reaction show only the presence of linear and bridge-bonded CO, which continuously decrease in intensity even when the CO beam is turned off [203]. These observations prove two different things: (1) there is no sharp variation of the populations of CO adsorbed on the Pd when the CO beam is turned off; (2) the CO responsible for the secondary peak has the same IR signature as that at steady state.

These observations are not compatible with the edge/facet model [201]. We have tried to compare the model with a previously developed microkinetic model based on a single type of site and a strong inhibiting effect for oxygen adsorption [209]. As we can see in Figure 17b, this model fits the experimental spectra very well. However, this model, based on one type of site, cannot explain all the experimental observations. First, the reaction window (reaction rate versus x_{co}) is not correctly reproduced [203], and the dip observed after the closing of the CO beam is more intense in the experiments [203,208] (see inset of Figure 17b), but the more important disagreement is the size dependence observed in the experiments [208,209]. In fact, the kinetic model is oversimplified; for example, we have not taken into account the coverage dependence on the adsorption energies and of the reaction barrier. The origin of the size effect can arise from two reasons: the CO reverse spillover (that has not been taken into account in the simulations) and the increased diffusion of oxygen under the surface, when cluster size decreases, that can affect the CO adsorption and then the blocking of the oxygen adsorption [192a] that is at the origin of this second CO_2 peak.

NO Reduction by CO

The NO reduction by CO is a very important reaction taking place in the car exhaust catalytic converters. It has received considerable attention on Rh single crystals [210] and much less on Pd [211,212]. On supported model catalysts many fewer experiments have been performed than for CO oxidation, and only in recent years [212–216]. From the work from Goodman's group it appears that the reactivity of Pd single crystals decreases when the surfaces become more open, and on supported clusters it decreases with cluster size [212]. We have studied this reaction by molecular beam techniques [209]. The steady-state reaction rate as a function of

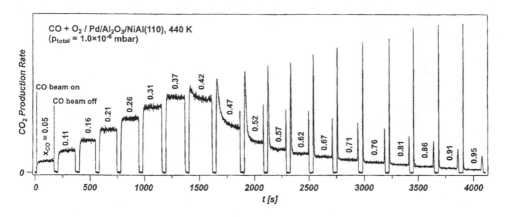

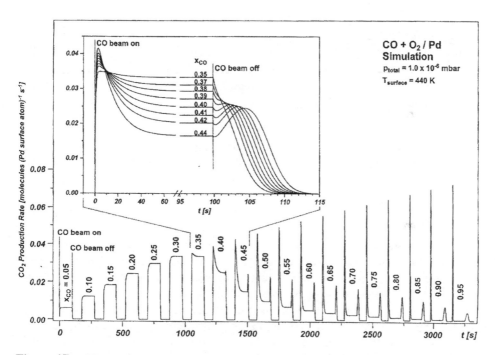

Figure 17 CO oxidation on Pd/alumina model catalysts. Top: CO_2 production versus time from a pulsed CO beam and a constant O_2 beam. X_{co} is the equivalent CO pressure divided by the sum of the CO and O_2 pressures. Down: Simulation of the experimental data (on top) with a kinetic model. The inset is an enlargement of part of the figure showing the presence of a small dip after closing of the CO beam. (From J. Libuda et al. [203].)

temperature shows a bell-shaped curve (see Figure 18) [215]. At low temperatures (high coverage) the rate-limiting step for the reaction is the dissociation of NO, and at high temperatures (low coverage) it is the adsorption of CO [216].

In Figure 18 we can see the TOR as a function of sample temperature for different cluster sizes. Apparently it increases when the particle size decreases.

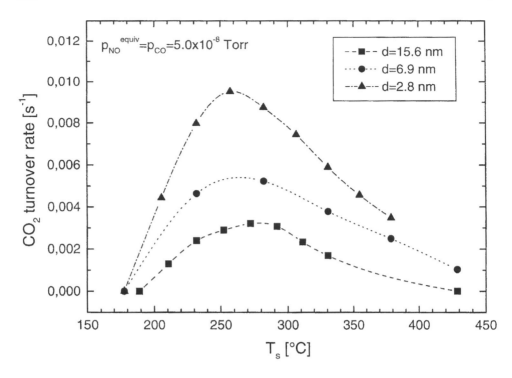

Figure 18 NO reduction by CO on a Pd/MgO(100) model catalyst. Steady-state TOR of CO_2 as a function of sample temperature and for different cluster sizes. (From L. Piccolo and C.R. Henry [215].)

However, in these curves the TOR is not corrected from the reverse spillover effect of NO (nor for CO), which is important, and as it depends on cluster size the direct measurement of the TOR can be misleading if we are looking for a size effect. Thus to study the size dependence on the reactivity it is more suited to calculate the reaction probability, which takes into account the actual flux of reactant molecules that joins the particles either by direct impingement from the gas phase or by surface diffusion on the substrate. In a molecular beam experiment these two contributions can be directly measured [184,215]. In figure 19 the reaction probability of NO is represented for the different particle sizes studied [216]. Now we see that the size effect is less pronounced than on the TOR curves (Fig. 18). Below 350 °C, where the maximum reactivity occurs, we see that the larger particles are the least active and the medium-sized ones are the most active. In fact, in this temperature range the NO dissociation limits the reaction and the reaction probability follows the trends of the dissociation rate of NO (compare with Figure 16). Meanwhile, we have to understand why there is no continuous decrease of the reactivity with cluster size as observed by Goodman (at a higher pressure) [212]. In fact, from the TEM characterization of the Pd particles, the large particles have coalesced and exposed mainly (100) facets while the other (smaller) particles had a shape close to equilibrium and exposed mainly (111) facets. If we recall the fact that the more open

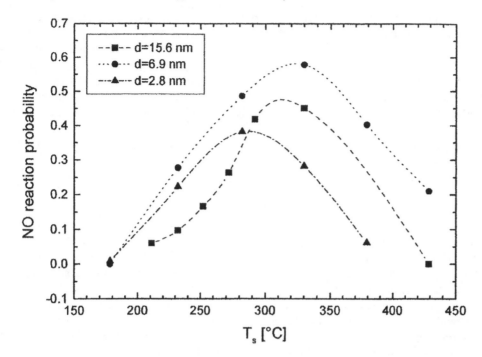

Figure 19 NO reduction by CO on a Pd/MgO(100) model catalyst. NO reaction probability versus temperature for various particle sizes (same samples as for Fig. 18). (From L. Piccolo and C.R. Henry [216].)

a Pd surface is, the less active it is for this reaction [211], we will understand why the large particles are less active. The decrease of reactivity with size for the other particles is probably due to the increased fraction of low coordinated edge sites, which are poisoned by strongly bound nitrogen that cannot recombine to form dinitrogen [212]. Therefore, for the NO reduction by CO we have seen both an effect of the particle morphology and an effect of the particle size.

7.6 CONCLUSION AND FUTURE PROSPECTS

In this chapter we have seen that the material gap in heterogeneous catalysis can be addressed in an efficient way by theoretically and experimentally studying metal clusters supported on flat, well-ordered surfaces. However, the bottleneck of these studies is the preparation of a uniform collections of particles with a well-defined shape. This goal can be reached by epitaxially growing the metal clusters on single crystals at high temperatures (and low supersaturation). The high-temperature morphology of nanometer-sized metal clusters depends on the surface energy (which can be modified by adsorption) and on adhesion energy (which depends on the presence of strain due to the misfit between the metal and the substrate). The morphology can be characterized using complementary techniques: STM and AFM for morphology related to the surface structure and TEM for the morphology related to the internal structure. Molecular dynamics with semi-empirical potentials are now able to reach the range of size relevant for catalysis, while purely ab initio calculations are still restricted to very small clusters and extended surfaces. At low pressures it has been shown that the reactivity on supported clusters differs from extended surfaces by the intrinsic heterogeneities related to the supported model catalysts. They are primarily the finite size of the particles, the presence of different facets (i.e., morphology), and the presence of the support. The size effects are mainly due to the nonnegligible proportion of low coordinated sites (edges and corners); the morphology effect takes place for the structure-sensitive reaction but also through the diffusion of reactants between the different facets. The effect of the support (beside the effect on the morphology and electronic structure for the very small clusters) appears through the diffusion of the reactants on the support.

In the future the main point to be addressed will be the pressure gap. Surface-science studies of catalytic reactions are generally performed at very low pressures (less than 10^{-5} Torr) that are about 8 to 10 orders of magnitude below the conditions used in industrial catalysis. In the last two years this severe problem has been, for the first time, approached by using new experimental techniques. The surface structure during adsorption and reaction can be studied in situ by STM [217]. The adsorbed species can be identified and quantified in situ by sum frequency generation (SFG) [218]. It will become possible to follow in situ the evolution of the internal structure of the catalysts during a reaction by the newly developed environmental HRTEM [219]. These techniques, combined with theoretical calculations, will probably solve the main problems in heterogeneous catalysis and also will help not only to improve actual catalysts but also to design new generations of powerful catalysts.

REFERENCES

1. G.A. Somorjai, Introduction to Surface Chemistry and Catalysis, Wiley, New York (1994).
2. B. Hammer, J.K. Norskov, Adv. Catalysis 45 (2000) 71.
3. F. Besenbacher, I. Chorkendorff, B.S. Clausen, B. Hammer, A.M. Molenbroek, J.K. Norskov, I. Stensgaard, Nature 279 (1998) 1913.
4. C.T. Campbell, Surf. Sci. Rep. 27 (1997) 1.
5. P.L.J. Gunter, J.W.H. Niemantsverdriet, F.H. Ribeiro, G.A. Somorjai, Catal. Rev. Sci. Eng. 39 (1997) 77.
6. C.R. Henry, Surf. Sci. Rep. 31 (1998) 231.
7. D.R. Rainer, C. Xu, D.W. Goodman, J. Mol. Catal. A 119 (1997) 307.
8. M. Bäumer, H.J. Freund, Prog. Surf. Sci. 61 (1999) 127.
9. S. Giorgio, C. Chapon, C.R. Henry, Langmuir 13 (1997) 2279.
10. C. Duriez, C. Chapon, C.R. Henry, J.M. Rickard, Surf. Sci. 230 (1990) 123.
11. E.F. Wassermann, K.A. Polacek, Surf. Sci. 28 (1971) 77.
12. F. Winkelmann, S. Wohlrab, J. Libuda, M. Bäumer, D. Kappus, M. Menges, K. Al-Shamery, H. Kuhlenbeck, H.J. Freund, Surf. Sci. 307–309 (1994) 1148.
13. G. Renaud, B. Villette, I. Vilfan, A. Bourret, Phys. Rev. Lett. 73 (1994) 1825.
14. U. Diebold, J. Lehman, T. Mahmoud, M. Kuhn, G. Leonardelli, W. Hebenstreit, M. Schmid, P. Varga, Surf. Sci. 411 (1998) 137.
15. J. Stubenrauch, J.M. Vohs, J. Catal. 159 (1996) 50.
16. G.S. Zafiris, R.J. Gorte, J. Catal. 132 (1991) 275.
17. O. Robach, G. Renaud, A. Barbier, Surf. Sci. 401 (1998) 227.

18. J. Yoshihara, J.M. Campbell, C.T. Campbell, Surf. Sci. 406 (1998) 235.
19. R.M. Jaeger, H. Kuhlenbeck, H.J. Freund, M. Wuttig, W. Hoffmann, R. Franchy, H. Ibach, Surf. Sci. 259 (1991) 235.
20. A. Rosenhahn, J. Schneider, C. Becker, K. Wandelt, J. Vac. Sci. Technol. B (2001).
21. D.R. Rainer, M.C. Wu, D.I. Mahon, D.W. Goodman, J. Vac. Sci. Technol. A (1996) 1184.
22. J.S. Corneille, J.W. He, D.W. Goodman, Surf. Sci. 338 (1995) 211.
23. C.A. Ventrice, T. Bertrams, H. Hannemann, A. Brodde, H. Neddermeyer, Phys. Rev. B 49 (1994) 5773.
24. C. Xu, D.W. Goodman, Chem. Phys. Lett. 263 (1996) 13.
25. T. Schröder, M. Adelt, B. Richter, N. Naschitzki, M. Bäumer, H.J. Freund, Surf. Rev. Lett. 7 (2000) 7.
26. J.J. Métois, J.C. Heyraud, Y. Takeda. Thin Solid Films 51 (1978) 105.
27. H. Poppa, A.G. Elliot, Surf. Sci. 24 (1971) 149.
28. H. Poppa, Z. Naturforschung 19 A(1964) 835.
29. H. Poppa, Catal. Rev. Sci. Eng. 35 (1993) 359.
30. K.R. Heim, S.T. Coyle, G.G. Hembree, J.A. Venables, M.R. Scheinfein, J. Appl. Phys. 80 (1996) 1161.
31. M. Heemeier, M. Frank, J. Iibuda, K. Wolter, H. Kuhlenbeck, M. Bäumer, H.J. Freund, Catal. Lett. 68 (2000) 19.
32. C.R. Henry, M. Meunier, Vacuum 50 (1998) 157.
33. M. Bäumer, J. Libuda, A. Sandell, H.J. Freund, G. Graw, T. Bertrams, H. Neddermeyer, Ber. Bunsenges. Phys. Chem. 99 (1995) 1381.
34. C.R. Henry, M. Meunier, S. Morel, J. Cryst. Growth 129 (1993) 416.
35. G. Haas, A. Menck, H. Brune, J.V. Barth, J.A. Venables, K. Kern, Phys. Rev. B 61 (2000) 11105.
36. K. Heinemann, T. Osaka, H. Poppa, M. Avalos-Borja, J. Catal. 83 (1983) 61.
37. A. Schmidt, V. Schünemann, R. Anton, Phys. Rev. B 41 (1990) 11875.
38. F. Gimenez, C. Chapon, C.R. Henry, New J. Chemistry 22 (1998) 1289.
39. (a) J.L. Rousset, A.M. Cadrot, F.J. Cadete Santos Aires, A. Renouprez, P. Mélinon, A. Perez, M. Pellarin, J.L. Vialle, M. Broyer, J. Chem. Phys. 102 (1995) 8574.
    (b) J.L. Rousset, L. Stievano, F.J. Cadete Santos Aires, C. Geantet, A. Renouprez, M. Pellarin, J. Catal. 197 (2001) 335.
40. B. Pauwels, G. van Tendeloo, W. Bouwen, L. Theil Kuhn, P. Lievens, H. Lei, M. Hou, Phys. Rev. B 62 (2000) 10383.
41. U. Heiz, A. Sanchez, S. Abbet, W.D. Schneider, J. Am. Chem. Soc. 121 (1999) 3214.
42. R. Schaub, Ph.D. thesis, EPFL, Lausanne (2000).
43. W. De Heer, Rev. Mod. Phys. 65 (1993) 611.
44. S.J. Carrol, S.G. Hall, R.E. Palmer, R. Smith, Phys. Rev. Lett. 81 (1998) 3715.
45. U. Heiz, W.D. Schneider, J. Phys. D 33 (2000) R85.
46. H. Brune, Surf. Sci. Rep. 31 (1998) 121.
47. C. Becker, A. Rosenhahn, K. von Bergmann, A. Wiltner, S. Degen, T. Mangen, K. Wandelt, Presented at the ASEVA Summer School "Growth and Behaviour of Metal-Oxide Interfaces," Avila (Spain) 2001 (to be published).
48. A.S. Eppler, G. Rupprechter, G.A. Somorjai. J. Phys. Chem. B 101 (1997) 9973.
49. K. Wong, S. Johansson, B. Kasemo, Faraday Disc. 105 (1996) 237.
50. S.Y. Chou, P.R. Krauss, W. Zhang, L. Guo, L. Zhuang, J. Vac. Sci. Techn. B (1997) 2897.
51. G.E. Engelmann, J.C. Ziegler, D.M. Kolb, Surf. Sci. 401 (1998) L420.
52. H. Dai, N. Franklin, J. Han, Appl. Phys. Lett. 73 (1998) 1508.
53. F. Burmeister, C. Schäfle, B. Keilhofer, C. Brechinger, J. Boneberg, P. Leiderer, Adv. Mater. 10 (1998) 495.

54. M. Boudart, Adv. Catal. 20 (1969) 153.
55. J.J. Burton, Catal. Rev. Eng. 9 (1974) 209.
56. M.R. Hoare, P. Pal, Adv. Chem. Phys. 20 (1971) 161.
57. A.L. Mackay, Acta Cryst. 15 (1962) 916.
58. O. Echt, K. Sattler, E. Recknagel, Phys. Rev. Lett. 47 (1981) 1121.
59. F. Ercolessi, E. Tosatti, M. Parrinello, Phys. Rev. Lett. 57 (1986) 719.
60. S.M. Foiles, M.I. Baskes, M.D. Daw, Phys. Rev. B 33 (1986) 7983.
61. K.W. Jacobsen, J.K. Norskov, M.J. Puska, Phys. Rev. B 35 (1987) 7423.
62. V. Rosato, M. Guillopé, B. Legrand, Phil. Mag. A 59 (1984) 321.
63. (a) C. Mottet, G. Tréglia, B. Legrand, Surf. Sci. 383 (1997) L719.
    (b) C. Mottet, Ph.D. thesis, University of Aix-Marseille II (1997).
64. G. D'Agostino, Phil. Mag. 68 (1993) 903.
65. S. Valkealahti, M. Manninen, Phys. Rev. B 45 (1992) 9459.
66. C.L. Cleveland, U. Landmann, J. Chem. Phys. 94 (1991) 7376.
67. L.D. Marks, Phil. Mag. A 49 (1984) 81.
68. G.D'Agostino, A. Pinto, S. Mobilio, Phys. Rev. B 48 (1993) 14447.
69. C.L. Cleveland, U. Landman, M.N. Shafigullin, P.W. Stephens, R.L. Whetten, Z. Phys. D 40 (1997) 503.
70. K. Michaelian, N. Rendon, I.L. Garzon, Phys. Rev. B 60 (1999) 2000.
71. F. Baletto, C. Mottet, R. Ferrando, Phys. Rev. Lett. 84 (2000) 496.
72. C.L. Cleveland, W.D. Luedke, U. Landman, Phys. Rev. B 60 (1999) 5065.
73. S. Valkealahti, M. Manninen, Z. Phys. D 40 (1997) 496.
74. G. Wulff, Z. Kristallogr, 34 (1901) 449.
75. L.D. Marks, Rep. Prog. Phys. 57 (1994) 603.
76. S. Iijima, T. Ichihashi, Phys. Rev. Lett. 56 (1986) 616.
77. P.M. Ajayan, L.D. Marks, Phys. Rev. Lett. 60 (1988) 585.
78. C. Herring, in Structure and Properties of Solid Surfaces, R. Gomer, C.S. Smith, eds. Univ. Press, Chicago, 1952, p. 581.
79. W.W. Mullins, J. Appl. Phys. 28 (1957) 333.
80. W.W. Mullins, G.B. Rorher, J. Am. Cerac. Soc. 83 (2000) 214.
81. C.R. Henry, M. Meunier, Materials Sci. Eng. A 217 (1996) 239.
82. P. Jensen, N. Combe, H. Harralde, J.L. Barrat, C. Misbah, A. Pimpinelli, Eur. Phys. J. B 11 (1999) 497.
83. N. Combe, P. Jensen, A. Pimpinelli, Phys. Rev. Lett. 85 (2000) 110.
84. R. Kaishew, Arbeitstagung Ferstkörper Physik, Dresden, 1952, p. 81.
85. W.L. Winterbottom, Acta Met. 15 (1967) 303.
86. I.V. Markov, In Crystal Growth for Beginners, World Scientific, Singapore, 1995, p. 75.
87. R. Kern, P. Müller, J. Cryst. Growth 146 (1995) 330.
88. P. Müller, R. Kern, J. Cryst. Growth 193 (1998) 257.
89. P. Müller, R. Kern, Surf. Sci. 457 (2000) 229.
90. M. Kubo, A. Stirling, R. Miura, R. Yamauchi, A. Miyamoto, Catal. Today 36 (1997) 143.
91. R. Yamauchi, M. Kubo, A. Miyamoto, J. Phys. Chem. B 102 (1998) 795.
92. J. Oviedo, J.F. Sanz, N. Lopez, F. Illas, J. Phys. Chem. B 4 (2000) 4342.
93. W. Vervisch, C. Mottet, J. Goniakowski, Phys. Rev. B65 (2002) 245411.
94. U. Schönberger, O. Andersen, M. Methfessel, Acta Met. 40 (1992) 51.
95. R. Wu, A.J. Freeman, Phys. Rev. B 51 (1995) 5408.
96. (a) G. Pacchioni, N. Rösch, J. Chem. Phys. 104 (1996) 7329.
    (b) I.V. Yudanov, G. Pacchioni, K.M. Neyman, N. Rösch, J. Phys. Chem. 101 (1997) 2786.
    (c) A.V. Matveev, K.M. Neyman, Y.V. Yudanov, N. Rösch, Surf. Sci. 426 (1999) 123.
97. V. Musolino, A. Selloni, R. Car, Surf. Sci. 402–404 (1998) 413.

98. (a) J. Goniakowski, Phys. Rev. B 58 (1998) 1189.
    (b) J. Goniakowski, Phys. Rev. B 59 (1999) 11047.
99. A. Bogicevic, D.R. Jennison, Surf. Sci. 437 (1999) L741.
100. Y.F. Zhukoskii, E.A. Kotomin, P.W. Jacobs, A.M. Stoneham, J.H. Harris, Surf. Sci. 441 (1999) 373.
101. X. Wei Xing, K.D. Schierbaum, W. Goepel, J. Solid State Chem. 119 (1995) 237.
102. T. Bredow, G. Pacchioni, Surf. Sci. 426 (1999) 106.
103. Z. Yang, R. Wu, D.W. Goodman, Phys. Rev. B 61 (2000) 14066.
104. C. Verdozzi, D.R. Jennison, A.A. Schutz, M.P. Sears, Phys. Rev. Lett. 82 (1999) 799.
105. Z. Lozdania, J.K. Norskov (submitted).
106. A. Shi, Phys. Rev. B 36 (1987) 9068.
107. A. Shi, R.I. Masel, J. Catal, 120 (1989) 421.
108. V.P. Zhdanov, B. Kasemo, Phys. Rev. Lett. 81 (1998) 2482.
109. R. Persaud, T.E. Madey, In The Chemical Physics of Solid Surfaces and Heterogeneous Catalysis, Vol. 8, D.A. King, D.P. Woodruff, eds. Elsevier, Amsterdam, 1997.
110. S. Giorgio, H. Graoui, C. Chapon, C.R. Henry, In Metal Clusters in Chemistry, P. Braunstein, ed. VCH-Wiley, 1999, p. 1194.
111. Z.L. Wang, J. Phys. Chem. B 104 (2000) 1153.
112. C.R. Henry, C. Chapon, C. Duriez, S. Giorgio, Surf. Sci. 253 (1991) 177.
113. E. Gillet, V. Matolin, Z. Phys. D 19 (1991) 361.
114. H. Poppa, Ultramicroscopy 11 (1983) 105.
115. C. Chapon, S. Granjeaud, A. Humbert, C.R. Henry, Eur. Phys. J AP 13 (2001) 23.
116. A.D. Polli, T. Wagner, T. Gemming, M. Rühle, Surf. Sci. 448 (2000) 279.
117. S. Giorgio, C. Chapon, C.R. Henry, G. Nihoul, Phil. Mag. B 67 (1993) 773.
118. E. Perrot, A. Humbert, A. Piednoir, C. Chapon, C.R. Henry, Surf. Sci. 445 (2000) 407.
119. A. Piednoir, E. Perrot, S. Granjeaud, A. Humbert, C. Chapon, C.R. Henry, Surf. Sci. 391 (1997) 19.
120. K.H. Hansen, T. Worren, S. Stampel, E. Laegsgaard, M. Bäumer, H.J. Freund, F. Besenbacher, I. Stensgaard, Phys. Rev. Lett. 83 (1999) 4120.
121. S. Ferrero, A. Piednoir, C.R. Henry, Nanoletters 1 (2001) 227.
122. G. Renaud, Surf. Sci. Rep. 32 (1998) 1.
123. M.J. Casanove, P. Lecante, E. Snoeck, A. Mosset, C. Roucau, J. Phys. III France 7 (1997) 505.
124. M. Haruta. Catal. Today 26 (1997) 153.
125. A. Balerna, E. Bernieri, P. Picozzi, A. Reale, S. Santucci, E. Burattini, S. Mobilio, Phys. Rev. B 31 (1985) 5058.
126. A. Pinto, A.R. Pennisi, G. Faraci, G.D'Agostino, S. Mobilio, F. Boscherini, Phys. Rev. B 51 (1995) 5315.
127. C.R. Henry, J. Cryst. Res. Techn. 33 (1998) 1119.
128. P.M. Ajayan, L.D. Marks, Phys. Rev. Lett. 63 (1989) 139.
129. S. Giorgio, C.R. Henry, C. Chapon, G. Nihoul, J.M. Penisson, Ultramicroscopy 38 (1991) 1.
130. H. Sato, S. Shinozaki, I.J. Cicotte, J. Vac. Sci. Techn. 6 (1969) 62.
131. T. Kizuka, N. Tanaka, Phys. Rev. B 56 (1997) R10079.
132. F. Cosandey, T.E. Madey, Surf. Rev. Lett. 8 (2001) 73.
133. S. Ferrero, A. Piednoir, C.R. Henry (submitted).
134. J.G. Allpress, J.V. Sanders, Surf. Sci. 7 (1967) 1.
135. K. Yagi, K. Takayanagi, K. Kobayashi, G. Honjo, J. Cryst. Growth 28 (1975) 117.
136. A. Howie, L.D. Marks, Phil. Mag. A 49 (1984) 95.
137. E. Perrot, Ph.D. thesis, University of Marseille II (1996).
138. S. Granjeaud, A. Humbert, C. Chapon, C.R. Henry (unpublished).
139. D. Reinhard, B.D. Hall, D. Ugarte, R. Monot, Phys. Rev. B 55 (1997) 7868.

140. F. Baletto, R. Ferrando, A. Fortunelli, F. Montalenti, C. Nottet, J. Chem. Phys. 116 (2002) 3856.

141. H. Sato, S. Shinozaki, J. Vac. Sci. Technol. 8 (1971) 159.

142. A. Trampert, F. Ernst, C.P. Flynn, H.H. Fischmeister, M. Rühle, Acta Met. 40 (1992) S227.

143. J.H. Larsen, J.T. Ranney, D.E. Starr, J.E. Musgrove, C.T. Campbell, Phys. Rev. B 63 (2001) 195410.

144. P. Guenard, G. Renaud, B. Villette, Physica B 221 (1996) 205.

145. G. Renaud, P. Guénard, A. Barbier, Phys. Rev. B 58 (1998) 7310.

146. (a) O. Robach, G. Renaud, A. Barbier, Phys. Rev. B 60 (1999) 5858.
     (b) A. Barbier, G. Renaud, J. Jupille, Surf. Sci. 454–456 (2000) 979.

147. H. Graoui, S. Giorgio, C.R. Henry, Surf. Sci. 417 (1998) 350.

148. C.L. Liu, J.M. Cohen, J.B. Adams, A.F. Voter, Surf. Sci. 253 (1991) 334.

149. A.F. Moodie, C.E. Warble, Phil. Mag. 35 (1977) 201.

150. H. Graoui, S. Giorgio, C.R. Henry, Phil. Mag. B 81 (2001) 1649.

151. R. Koch, H. Poppa, J. Vac. Sci. Techn. A 5 (1987) 1845.

152. W.D. Knight, K. Clemenger, W.A. De Heer, W.A. Saunders, M.Y. Chou, M.L. Cohen, Phys. Rev. Lett. 52 (1984) 2141.

153. K.J. Taylor, C.L. Pettiette-Hall, O. Cheshnovsky, R.E. Smalley, J. Chem. Phys. 96 (1992) 3319.

154. K.H. Meiwes Broer, In Metal Clusters at Surfaces, K.H. Meiwes Broer, ed. Springer, Berlin, 2000, p. 151.

155. C. Mottet, G. Tréglia, B. Legrand, Surf. Sci. 352–354 (1996) 675.

156. (a) B. Hammer, Y. Morikawa, J.K. Norskov, Phys. Rev. Lett. 76 (1996) 2141.
     (b) S. Sawaya, J. Goniakowski, C. Mottet, A. Saul, G. Tréglia, Phys. Rev. B 56 (1997) 12161.

157. C. Kuhrt, M. Harsdorff, Surf. Sci. 245 (1991) 173.

158. S. Kohiki, S. Ikeda, Phys. Rev. B 34 (1986) 3786.

159. Y. Wu, E. Garfunkel, T.E. Madey, J. Vac. Sci. Techn. A 14 (1996) 1662.

160. (a) C. Xu, W.S. Oh, G. Liu, D.Y. Kim, D.W. Goodman, J. Vac, Sci. Techn. A 15 (1997) 1261.
     (b) C. Xu, X. Lai, G.W. Zajac, D.W. Goodman, Phys. Rev. B 56 (1997) 13464.

161. B. Hammer, J.K. Norskov, In R.M. Lambert, G. Pacchioni, eds. Chemisorption, Reactivity of Clusters and Thin Films, NATO ASI Series E, Kluwer, Dordrecht, 1987, p. 285.

162. C. Pallassana, M. Neurock, J. Catal. 191 (2000) 301.

163. B. Mavrikakis, B. Hammer, J.K. Norskov, Phys. Rev. Lett. 81 (1998) 2819.

164. A. Ruban, B. Hammer, P. Stoltze, H.L. Skriver, J.K. Norskov, J. Mol. Cat. A 115 (1997) 421.

165. F. Besenbacher, I. Chorkendorff, B.S. Clausen, B. Hammer, A.M. Molenbroek, J.K. Norskov, I. Stensgaard, Science 279 (1998), 1913.

166. A. Sanchez, S. Abbet, U. Heiz, W.D. Schneider, H. Häkkinen, R.N. Barnett, U. Landman, J. Phys. Chem. A 103 (1999) 9573.

167. S. Abbet, U. Heiz, W.D. Schneider, J. Catal. 198 (2001) 122.

168. (a) B.C. Gates, Chem. Rev. 95 (1995) 511.
     (b) O. Alexeev, B.C. Gates, J. Catal. 176 (1998) 310.

169. C.R. Henry, Appl. Surf. Sci. 164 (2000) 252.

170. D.M. Cox, R. Brickman, K. Creegan, A. Kaldor, Z. Phys. D 19 (1991) 353.

171. C.R. Henry, C. Chapon, C. Goyhenex, R. Monot, Surf. Sci. 272 (1992) 283.

172. C. Duriez, C.R. Henry, C. Chapon, Surf. Sci. 253 (1991) 190.

173. V. Matolin, E. Gillet, Surf. Sci. 166 (1986) L115.

174. F. Rumpf, H. Poppa, M. Boudart, Langmuir 4 (1988) 722.

175. I. Jungwirthova, I. Stara, V. Matolin, Surf. Sci. 377–379 (1997) 644.
176. C.R. Henry, C. Chapon, C. Duriez, J. Chem. Phys. 95 (1991) 700.
177. M. Eriksson, L.G. Pettersson, Surf. Sci. 311 (1994) 139.
178. V. Nehasil, T. Hrncir, S. Zafeiratos, S. Ladas, V. Matolin, Surf. Sci. 454–456 (2000) 289.
179. C.R. Henry, Surf. Sci. 223 (1989) 519.
180. V.P. Zhdanov, B. Kasemo, J. Catal. 170 (1997) 377.
181. V. Nehasil, I. Stara, V. Matolin, Surf. Sci. 331–333 (1995) 105.
182. S. Andersson, M. Frank, A. Sandell, J. Libuda, A. Giertz, B. Brena, P.A. Brühwiler, M. Bäumer, N. Martensson, H.J. Freund, J. Chem. Phys. 108 (1998) 2967.
183. M. Frank, M. Bäumer, Phys. Chem. Chem. Phys. 2 (2000) 3723.
184. L. Piccolo, C.R. Henry, Surf. Sci. 452 (2000) 198.
185. G. Prevot, O. Meerson, L. Piccolo, C.R. Henry, J. Phys. Condens. Matters 14 (2001) 4251.
186. X. Xu, D.W. Goodman, Catal. Lett. 24 (1994) 31.
187. H. Cordatos, T. Bunluesin, R.J. Gorte, Surf. Sci. 323 (1995) 219.
188. I. Stara, V. Nehasil, V. Matolin, Surf. Sci. 365 (1996) 69.
189. E.S. Putna, J.M. Vohs, R.J. Gorte, Surf. Sci. 391 (1997) L1178.
190. I. Meusel, J. Hoffman, J. Hartmann, M. Heemeir, M. Bäumer, J. Libuda, H.J. Freund, Catal. Lett. 71 (2001) 5.
191. K. Yagi, D. Sekiba, H. Fukutani, Surf. Sci. 442 (1999) 307.
192. (a) S. Ladas, R. Imbihl, G. Ertl, Surf. Sci. 280 (1993) 14.
     (b) I.Z. Jones, R.A. Bennett, M. Bowker, Surf. Sci. 439 (1999) 235.
193. F.P. Leisenberger, G. Koller, M. Sock, S. Surnev, M.G. Ramsey, F.P. Netzer, B. Klötzer, K. Hayek, Surf. Sci. 445 (2000) 380.
194. S. Ladas, H. Poppa, M. Boudart, Surf. Sci. 102 (1981) 460.
195. V. Matolin, I. Stara, Surf. Sci. 398 (1998) 117.
196. M. Valden, X. Lai, D.W. Goodman, Science 281 (1998) 1647.
197. M. Valden, S. Park, X. Lai, D.W. Goodman, Catal. Lett. 56 (1998) 7.
198. V. Nehasil, S. Zafeiratos, S. Ladas, V. Matolin, Surf. Sci. 433–435 (1999) 2153.
199. H. Fornander, L.G. Ekedahl, H. Dannetun, Surf. Sci. 441 (1999) 479.
200. V.A. Bondzie, S.C. Parker, C.T. Campbell, Catal. Lett. 63 (1999) 143.
201. L. Piccolo, C. Becker, C.R. Henry, Europ. Phys. J. D 9 (1999) 415.
202. L. Piccolo, C. Becker, C.R. Henry, Appl. Surf. Sci. 164 (2000) 156.
203. J. Libuda, I. Meusel, J. Hoffmann, J. Hartmann, L. Piccolo, C.R. Henry, H.J. Freund, J. Chem. Phys. 114 (2001) 4669.
204. F. Rumpf, H. Poppa, M. Boudart, Langmuir 4 (1987) 722.
205. C. Becker, C.R. Henry, Surf. Sci. 352–354 (1996) 457.
206. (a) H. Uetsuka, K. Watanabe, H. Ohnuma, K. Kunimori, Surf. Rev. Lett. 4 (1997) 1359.
     (b) K. Watanabe, H. Ohnuma, H. Kimpara, H. Uetsuka, K. Kunimori, Surf. Sci. 402–404 (1998) 100.
207. X. Su, D.W. Goodman, J. Phys. Chem. 97 (1993) 7711.
208. C. Becker, C.R. Henry, Catal. Lett. 43 (1997) 55.
209. L. Piccolo, Ph.D. thesis, University of Marseille II, 1999.
210. V.P. Zhdanoy, B. Kasemo, Surf. Sci. Rep. 29 (1997) 31.
211. S.M. Vesecky, D.R. Rainer, D.W. Goodman, J. Vac. Sci. Techn. A 13 (1995) 1539.
212. D.R. Rainer, S.M. Vesecky, M. Koranne, W.S. Oh, D.W. Goodman, J. Catal. 167 (1997) 234.
213. M.C. Wu, D.W. Goodman, J. Phys. Chem. 98 (1994) 9874.
214. X. Xu, D.W. Goodman, Catal. Lett. 24 (1994) 31.
215. L. Piccolo, C.R. Henry, Appl. Surf. Sci. 162–163 (2000) 670.

216.   L. Piccolo, C.R. Henry, J. Mol. Cat. A 167 (2001) 181.

217.   L. Osterlund, P.B. Rasmussen, P. Thostrup, E. Laegsgaard, I. Stensgaard, F. Besenbacher, Phys. Rev. Lett. 86 (2001) 460.

218.   (a) G.A. Somorjai, G. Rupprechter, J. Phys. Chem B 103 (1999) 1623.
       (b) T. Dellwig, G. Rupprechter, H. Unterhalt, H.J. Freund, Phys. Rev. Lett. 85 (2000) 776.

219.   P.L. Hansen, J.B. Wagner, EUREM 12 Proceedings (Brno, Czech Republic) 2000, Vol. II, P. 537.

8

Size-Dependent Electronic, Structural, and Catalytic Properties of Metal Clusters Supported on Ultrathin Oxide Films

A.K. SANTRA and D.W. GOODMAN

Texas A&M University, College Station, Texas, U.S.A.

CHAPTER CONTENTS

SUMMARY

The electronic structure, morphology, and chemical reactivity of metal nanoclusters have attracted considerable attention due to their extensive technological importance. Chemical reactions and their catalytic relevance have been investigated on a variety of well-characterized, supported model catalysts prepared by vapor deposition of catalytically relevant metals onto ultrathin oxide films in ultrahigh vacuum conditions. Such ultrathin film supports are usually prepared by vaporizing a parent metal onto a refractory metal substrate in an oxygen atmosphere at a high temperature. These unique model systems are particularly well suited for surface-science studies and facilitate the study of many issues crucial to industrial catalysts such as the intrinsic effect of cluster size, role of the support, and metal–support interaction. Chemical and spectroscopic/microscopic techniques have demonstrated that these ultrathin oxide films are excellent models for the corresponding bulk oxide, yet are electrically conductive (via defects and tunneling to the underlying metal substrate). This conductivity makes them suitable for various spectroscopic and microscopic studies such as STM and AFM. Measurements on metal clusters using these ultrathin oxide model catalysts have revealed a metal to nonmetal transition, changes in the electronic structures (including band-width, band-splitting and core-level binding energy shifts), and that changes in catalytic activities do occur with variation in cluster size.

8.1 INTRODUCTION

The size-dependent electronic structure of metal clusters supported on solid substrates has been studied extensively for the last two decades [1–20]. This interest is due to their technological importance particularly in the area of heterogeneous catalysis, information storage, photographic and electronic imaging, microcircuitry, and the production of high-quality thin films [9–11,13,21–32]. Changes in energy and peak width are observed for the core levels and valance bands of metal clusters as a function of cluster size. These changes have been discussed in terms of initial- and/or final-state effects because of the differences in charge density and core-hole screening among bulk metal atoms and atoms in small clusters. For small clusters the metallic properties diminish and the electronic structure approaches that of an isolated atom [3,16,20,33–42]. X-ray photoelectron spectroscopy (XPS), ultraviolet photoelectron spectroscopy (UPS), Auger electron spectroscopy (AES), temperature-programmed desorption (TPD), and scanning tunneling microscopy/spectroscopy (STM/STS) have been used to follow this size-dependent electronic structure. STM-STS data have shown that a metal–nonmetal transition occurs with a decrease in cluster size [19].

In the bulk form, Au is known to be chemically inert compared to the other Pt group metals. However, it has recently been shown that Au clusters, deposited as finely dispersed, small clusters (<5-nm diameter) on reducible metal oxides like TiO_2, Fe_2O_3, and Co_3O_4, exhibit catalytic activity with respect to a number of industrially important reactions; i.e., CO oxidation, hydrogenation, partial oxidation of hydrocarbons, and selective oxidation of higher alkenes [13,19,25–28,43–45]. This unusual catalytic activity has been shown to be a function of metal cluster size.

Since the bulk metal oxides of typical catalytic supports are insulators, charging makes them unsuitable for electron spectroscopic and STM measurements. In our laboratory, these difficulties have been circumvented by synthesizing well-ordered, thin oxide films on refractory metal substrates [46–55]. These films are thin enough (~ 2–5 nm) to prevent charging, yet thick enough to exhibit electronic and chemical properties comparable to the corresponding bulk oxide. Since the preparation of a thin oxide film on a conducting support is critical to the synthesis of our model catalysts, the preparation of the thin oxide is discussed in detail. The effect of cluster size on acetylene cyclotrimerization, the CO + NO reaction, and CO oxidation is also discussed. The ultimate goal of these studies is to establish a definitive correlation among cluster size, cluster electronic properties, and cluster chemical/catalytic properties.

8.2 EXPERIMENTAL

$TiO_2(110)$ single crystals (Commercial Crystal Laboratories) become sufficiently conductive for STM and electron spectroscopic measurements after cycles of Ar^+ sputtering and annealing to 700–1000 K. Because the preparation of well-ordered, ultrathin films of TiO_2, SiO_2, and Al_2O_3 on refractory metals is crucial to these studies, particularly for electron spectroscopy and STM, these thin-film preparations are discussed separately. Deposition of the metal was typically carried out by resistive evaporation of high-purity metal wire wrapped around a W or Ta filament in vacuum. Such dosers are a clean, stable metal source after thorough outgassing. By controlling the filament current, the doser-to-substrate distance, and the substrate temperature, fine control can be achieved with respect to cluster size and density. Three different UHV chambers equipped with XPS, AES, high-resolution electron energy loss spectroscopy (HREELS), low-energy electron diffraction (LEED), STM, and ion scattering spectroscopy (ISS) were used for these studies.

8.2.1 Preparation and Characterization of Thin Oxide Films

The most commonly used methods for the preparation of ultrathin oxide films are (1) direct oxidation of the parent metal surface, (2) preferential oxidation of one metal of choice from a suitable binary alloy, and (3) simultaneous deposition and oxidation of a metal on a refractory metal substrate. The detailed procedures for (1) and (2) are discussed elsewhere [7,56,57]; procedure (3) is discussed here in detail. Preparation of a model thin-film oxide on a refractory metal substrate (such as Mo, Re, or Ta) is usually carried out by vapor-depositing the parent metal in an oxygen environment. These substrate refractory metals are typically cleaned by repeated cycles of Ar^+ sputtering followed by high-temperature annealing and oxygen treatment. The choice of substrate is critical because film stoichiometry and crystallinity depend on lattice mismatch and other interfacial properties. Thin films of several oxides have been prepared in our laboratories and are discussed below.

Alumina

Well-ordered Al_2O_3 films have been prepared using an Al deposition rate of $\sim 0.5\,MLE.min^{-1}$ (equivalent monolayer, calibrated on an Mo(110) substrate using AES and TPD), at a background O_2 pressure of 7×10^{-7} Torr, and an Mo(110) substrate temperature at 300 K [58]. Following deposition, the films were annealed to 1200 K in O_2 to improve the film order. Figure 1 shows representative AES spectra of alumina films at two different thicknesses [58]. An AES spectrum of a 2.0-nm film has a predominant Al^{3+} (LVV) transition at $\sim 45\,eV$ and an O (KLL) transition at $\sim 500\,eV$. The absence of a peak at 68 eV, characteristic of Al°, indicates that the film is fully oxidized. Features attributable to the Mo(110) substrate can be seen in the 100–250-eV region. The thickness of the oxide films was calculated from the attenuation of the AES intensity ratio of the Al^{3+} (LVV) feature relative to the Mo(MNN) feature.

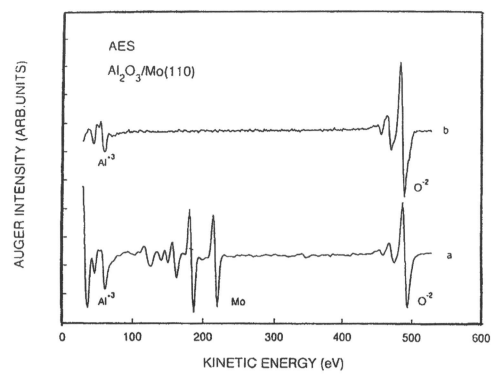

Figure 1   AES of Al_2O_3 films on Mo(110) of thickness (a) 0.44 nm and (b) 2.0 nm. The films were annealed to 1200 K in oxygen ambient to improve their crystalline quality. The spectra were acquired at $E_p = 2.0$ keV and a sample current of 5 μA. (From Ref. 58.)

Detailed investigations using low-energy electron diffraction (LEED) and ion scattering spectroscopy (ISS) of Al_2O_3 thin films have revealed that the films grow epitaxially with long-range order on Ta(110) and Mo(110) [55,59]. The LEED patterns are indicative of a slightly distorted ($\beta = 117.9°$) hexagonal lattice characteristic of close-packed O^{2-} planes. Initially the films grow two-dimensionally (2D), with a transition from 2D to three-dimensional (3D) growth occurring for films thicker than 1.5 nm. The initial stages of alumina film growth are governed by strain relief mechanisms that lead to a Kurdjumov–Sachs (KS) orientational relationship where the epitaxial overlayer exhibits coherence along one of the close-packed rows of the bcc(110) substrate [55,59].

Additional support for assignment of stoichiometric Al_2O_3 to the films was found by carrying out high-resolution electron energy loss (HREELS). Figure 2 shows HREELS data for two Al_2O_3 films: (a) 0.44 nm and (b) 2.0 nm thick, respectively [58]. The fundamental modes of the surface optical phonons are located below 1000 cm^{-1}. The thick films are characterized by three fundamental modes at 370, 640, and 875 cm^{-1}, while the thin films are characterized by two fundamental modes at 590 and 860 cm^{-1}. The energies of the three modes for the thick films agree well with the previously reported vibrational data for Al_2O_3 [60–63], while the data for a 0.44 nm film are consistent with those reported by Frederick et al. for Al_2O_3 [64,65].

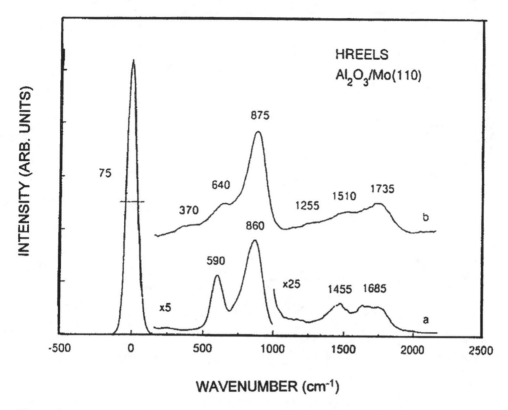

Figure 2 HREELS spectra of Al_2O_3 films grown on Mo(110) of thickness (a) 0.44 nm and (b) 2.0 nm. The spectra were acquired at $E_p = 4.0$ eV and at the specularly reflected beam direction. (From Ref. 58.)

Silica

Silica thin films were prepared on Mo(110) and Mo(100) substrates at various temperatures [66–68] where the deposition conditions strongly affect the film stoichiometry. AES can distinguish silicon and silicon oxide species based on their AES transition energies and characteristic line shapes. Elemental Si has a strong LVV feature at 91 eV while SiO_2 has three LVV transitions at 76, 63, and 59 eV, respectively. Figure 3 presents AES spectra for silica thin films deposited at different O_2 deposition pressures. Increasing the oxygen pressure to 1×10^{-6} Torr during deposition results in a two-phase film that contained both elemental Si and a new Si(LVV) feature at 76 eV attributed to SiO_2. However, at a deposition pressure of 4×10^{-6} Torr, the AES data clearly indicate a film composition consistent with SiO_2.

A detailed XPS investigation [66] has shown that a stoichiometric SiO_2 film can be obtained with a Si deposition rate of ~ 0.12 nm/min at an oxygen pressure of 2×10^{-5} Torr at 300 K, followed by an anneal of the film to ~ 1300 K. The representative XP spectra are shown in Figure 4 at three different annealing temperatures. The absence of any feature near 99.4 eV is indicative of a

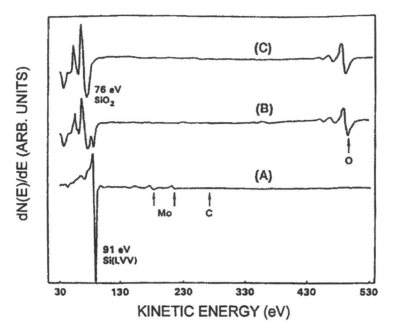

Figure 3 AES of Si-oxide thin film as a function of O_2 adsorption pressure: (A) no background O_2, (B) 1×10^{-6} Torr, and (C) 4×10^{-6} Torr. (From Ref. 66.)

stoichiometric SiO_2 film (with an Si2p binding energy of 103.4 eV) that contains no elemental Si.

Infrared reflection absorption spectroscopy (IRAS), electron energy loss spectroscopy (EELS), and LEED studies indicate that SiO_2 films deposited at low temperatures are disordered and likely consist of short-range networks of [SiO_4] tetrahedral units [66–68]. These units become more ordered and convert to a structure that consists of long-range [SiO_4] networks upon annealing to ~ 1300 K.

Titania

Well-ordered TiO_2 films were prepared by depositing Ti in 2×10^{-7} Torr of O_2 onto Mo(110) or Mo(100) at 600–700 K followed by an anneal to 800 K in $\sim 10^{-7}$ Torr O_2. Films prepared in this method [48] showed a very sharp 1×1 LEED pattern, and careful analysis confirms that the films are primarily TiO_2(100). The Ti LMM and LMV AES fine structure can also be used to distinguish titanium oxides. For TiO_2, for instance, the Ti LMM Auger transition exhibits a shoulder in the region 410–415 eV. Figure 5 shows a typical AES spectrum obtained from a ~ 30-MLE oxide film. The feature near 413 eV matches very well with that found for single-crystal TiO_2.

Titanium oxides of various types exhibit different Ti 2p core-level binding energies in XPS depending on the formal oxidation state of the Ti cation. In general, a higher titanium oxidation state exhibits a higher binding energy. In Figure 6 the as-prepared oxide film clearly shows a Ti $2p_{3/2}$ binding energy of 459.1 eV with a spin-orbit splitting of 5.7 eV, characteristic of TiO_2. However, the high-temperature

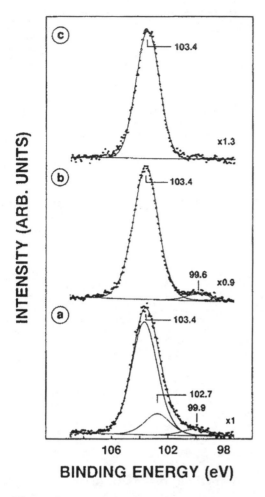

Figure 4   The Si2p XP spectra of silicon oxide films (3.6 nm) on a Mo(100) surface. The films were prepared by evaporating Si in an O_2 atmosphere of 2×10^{-5} Torr with a dosing rate of ~ 0.12 nm/min. The substrate temperature during the film preparation was ~ 350 K. The film was then annealed to 393 K (a), 1023 K (b), and 1373 K (c). (From Ref. 66.)

vacuum annealed film in this figure exhibits Ti oxidation states indicative of reduced TiO_2.

Figure 7 shows the HREELS spectrum acquired at room temperature and at an incidence angle of 60° from the surface normal for a ~ 30 MLE titanium oxide film prepared as described above. Fundamental phonons of 54 meV(v_2) as well as overtone losses and combinations are seen. The loss at 149 meV is due to $v_1 + v_2$, and the loss at 245 meV is due to $v_1 + 2v_2$. That TiO_2 films prepared in this manner are anatase and can be ruled out because the anatase phase exhibits fundamental surface phonons at 44 and 98 meV for the $\langle 100 \rangle$ face and 48 and 92 meV for the $\langle 001 \rangle$ face. The combined use of LEED, AES, XPS, and HREELS has revealed that the thin TiO_2 films prepared via the above procedure are rutile with a surface structural orientation primarily $\langle 100 \rangle$.

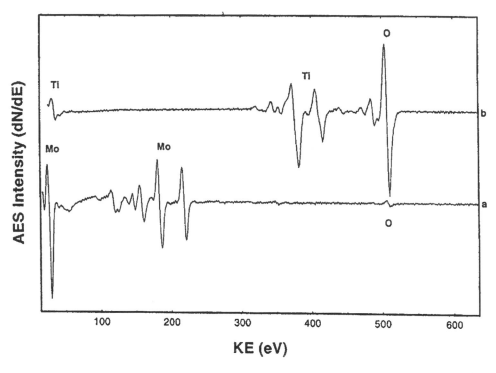

Figure 5 AE spectra (a) Mo(110) surface prior to film growth; (b) $\sim 30\,\text{ML}$ titanium oxide film on Mo(110). The primary energy was $3\,\text{keV}$; the spectra were collected at $100\,\text{K}$. (From Ref. 48.)

8.3 ELECTRONIC STRUCTURE AND GROWTH OF SUPPORTED METAL CLUSTERS

8.3.1 Transmission Electron Microscopy (TEM) and TPD

Several simple models for the growth morphologies have been proposed for the noble and transition metal clusters deposited on various solid substrates [69,70]. These models generally assume that the substrate contains randomly distributed nucleation sites and that metal cluster size depends on the distribution of these fixed nucleation sites and the amount of metal deposited [41,71]. For amorphous carbon substrates the number of nucleation sites is large. Egelhoff et al. [72] have proposed that on substrates with large numbers of nucleation sites, deposition of a small quantity of metal will produce isolated ad-atoms initially. These isolated ad-atoms then diffuse randomly on the substrate until sticking occurs at another ad-atom site [72,73]. Because the nucleation sites are randomly distributed and the diffusion is also random, a distribution of cluster sizes at every coverage results. For example, typical TEM images of Au clusters on amorphous graphite substrate at various coverages are shown in Figure 8. It can be seen that with an increase in Au coverage, there is a distinct increase in the cluster size [74]. However, growth of metal clusters on oxide surfaces could be quite different, it is discussed in Section 8.3.3.

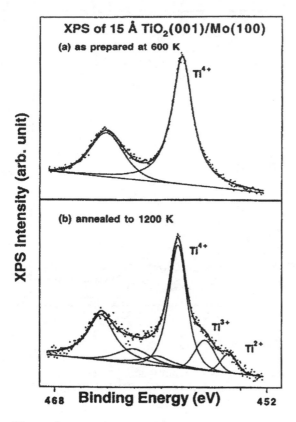

Figure 6 Ti 2p XP spectra for 1.5-nm TiO$_2$/Mo(100). (a) As prepared at 600 K; (b) annealed to 1200 K. The spectra were collected at 300 K. (From Ref. 48.)

TPD of the deposited metal clusters is very informative in evaluating the interaction strength of the metal with the substrate. Figures 9a and b show families of TPD spectra for Cu/Al$_2$O$_3$ [58] and Au/SiO$_2$ [75] systems. Generally, we see that the leading edge of the spectra increases continuously with the metal coverage. The width of the TPD curve is also a measure of the distribution of cluster diameter and height. An increase in the width of the desorption curve with an increase in metal coverage is evident, indicating an increase in the cluster size distribution, as seen earlier (Figure 8) via TEM. The lower desorption temperature at low coverage means that the substrate–metal interaction is less than the metal–metal interaction in the bulk metal. The heat of sublimation, E_{sub}, is shown in the inset as a function of metal coverage (cluster size). For higher coverages the heat of sublimation reaches the value of the bulk metal, suggesting that the metal–metal interaction is dominant at the large cluster limit.

8.3.2 Electron Spectroscopy

Electron spectroscopy, particularly photoemission, has been used successfully to study the electronic structure of small metal clusters deposited on well-characterized model systems [3–5,8,16,20,34–42,66,71,72,76–86]. The core-level binding energy

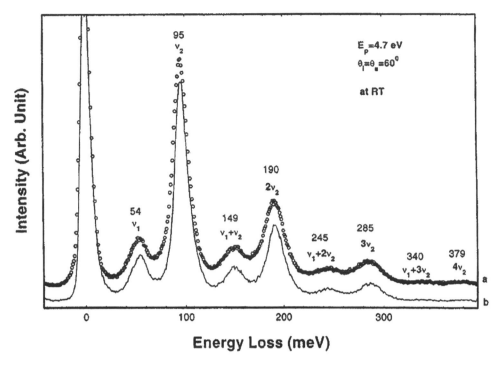

Figure 7 HREELS spectra at room temperature: (a) titanium oxide film with ∼30 MLE thickness on Mo(110); (b) TiO_2(110) single crystal. $E_p = 4.7$ eV. (From Ref. 48.)

generally decreases, and the splitting or width of the valance band increases with an increase in cluster size, ultimately reaching the bulk value. These general characteristics are well accepted; however, the physical origins of these spectral changes remain in dispute. Core-level binding energy is best expressed as $E_c = E_{\text{final}}$ (after photoemission) $- E_{\text{initial}}$ (before photoemission) and therefore is largely determined by the differences in the electronic charge densities of the initial and final states of the system. Final-state effects, or changes in E_{final}, can arise due to variabilities in the screening of the core hole created after photoemission. For a solid, it is convenient to separate the total screening into contributions from intra-atomic and extra-atomic screening. Intra-atomic screening mainly depends on the element; however, extra-atomic screening depends on the surrounding environment, i.e. coordination number, ligand, and substrate. For example, for a cluster, changes in the extra-atomic screening become very important; a reduction in the extra-atomic screening compared to that in the bulk metal can lead to an increase in the core-level binding energy. Initial-state contributions are due to differences in charge density before photoemission and may arise from a variety of factors such as (1) changes in the electronic structure of the clusters with respect to the bulk metal, (2) changes in chemical environment at the interface, (3) surface core-level shift, and (4) defect-induced metal-substrate interactions.

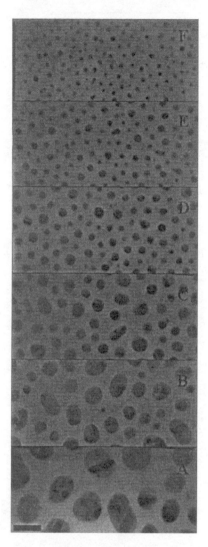

Figure 8 TEM micrographs of Au deposited on amorphous graphite at varying coverages. The bar indicates 20 nm. (From Ref. 74.)

Figures 10a and b show changes in the $Au4f_{7/2}$ core-level as a function of Au coverage (measure of cluster size) on $TiO_2(110)$ and $SiO_2/Mo(110)$ [75]. A similar plot on amorphous graphite is shown in Figure 11. The $Au4f_{7/2}$ binding energy with respect to the bulk metal (84.0 eV) decreases with an increase in cluster size, finally converging to the bulk value at large cluster sizes. It should be noted that the largest shifts observed on three different substrates are 0.55, 0.8, and 1.6 eV for graphite, $TiO_2(110)$, and SiO_2, respectively. Such differences have been explained as due to the relative abilities of these substrates to screen the core hole after photoemission, i.e., a final-state contribution. As graphite is the most conductive among the three screens, it thereby yields the least shift in the core-level binding energy. In general, there is an

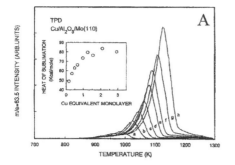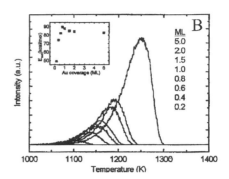

Figure 9 (A) A family of TPD spectra of Cu deposited as a function of equivalent monolayers, θ_{Cu}: (a) 0.16, (b) 0.33, (c) 0.50, (d) 0.67, (e) 0.98, (f) 1.25, (g) 1.55, and (h) 2.09. The inset shows the heat of sublimation, derived from the leading edge analysis of the spectra, as a function of Cu coverage in equivalent monolayers. (From Ref. 58.) (B) A set of TPD spectra of Au ($m/e = 197$) on a 0.25-nm-thick SiO_2 thin film on Mo(100) at Au cluster coverages ranging from 0.2 to 5.0 MLE. The inset shows a plot of E_{sub} determined from the leading edge analysis. (From Ref. 75.)

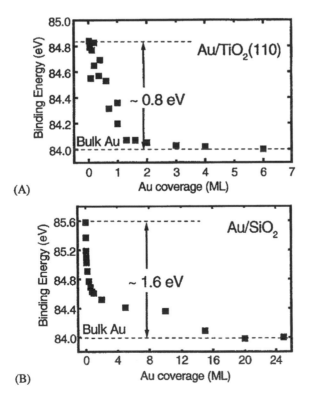

Figure 10 Plots of the $Au4f_{7/2}$ core-level binding energy as a function of Au cluster coverage (ranging from 0.02 MLE to bulk) on $TiO_2(110)$(A) and SiO_2(B) surfaces. (From Ref. 160.)

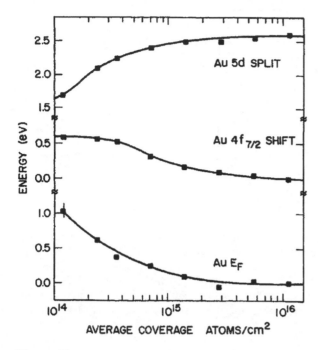

Figure 11   Coverage dependence plots of the Au4f binding energy, position of the Au Fermi edge, and Au5d splitting. (From Ref. 87.)

increase in the FWHM of the core-level XP spectra with a decrease in the cluster size and likely relates to the unavoidable distribution of cluster sizes.

Figure 11 shows how the Au 5d splitting, $Au4f_{7/2}$ binding energy, and Au E_F change with a change in cluster size on an amorphous graphite substrate. Wertheim et al. [87] have explained such changes as due to the poor conductivity of the substrate leading to an unscreened positive charge after photoemission. However, an increase in the Au E_F could also be due to a metal–nonmetal transition as the cluster size decreases; this possibility is discussed further in the next section. A decrease in the Au 5d splitting with a decrease in cluster size could be due to a decrease in the coordination number as shown by Mason et al. [16] using the Au-Cd alloy system. DiCenzo et al. [3] have shown how the binding energy of the $Au4f_{7/2}$ level changes with the mean coordination number of small clusters (Figure 12). If one assumes that the unscreened charge of Werthcim et al. [87] is solely responsible for the changes in binding energy, then the dashed line of Figure 12 follows. The dashed line considers only the effect of the Coulombic term, $e^2/2R$, where R is the radius of the cluster [3].

Ultraviolet photoemission valance-band spectra with increasing Pd coverage on $Al_2O_3/Re(0001)$ have been measured [33] and are shown in Figure 13. A gradual evolution of metallic valance bands with increasing cluster size is a manifestation of the increase in density of states near the Fermi level and the appearance of dispersing bands parallel and perpendicular to the substrate. The latter has been attributed to the formation of crystallites with a preferred orientation. The appearance of dispersion may also be used to define the boundary between a metallic and

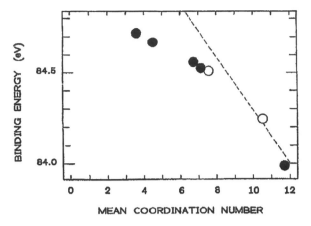

Figure 12 Au4f$_{7/2}$ binding energy versus coordination number plot for mass-selected Au$_n$ clusters deposited on amorphous carbon. The open circles are data for the clusters obtained by vapor deposition on amorphous carbon; the average coordination numbers for these were determined by correlating their corresponding Au5d splitting with those obtained for mass-selected clusters. The dashed line shows the binding energies expected for supported Au clusters allowing only for the Coulomb energy of the positively charged clusters in the photoemission. (From Ref. 3.)

nonmetallic state of the clusters. In the data presented, this transition appears to take place at an average cluster size of ~ 2.5 nm.

8.3.3 Scanning Tunneling Microscopy

The growth of metal clusters on TiO$_2$ is quite different from that seen on amorphous graphite (Figure 8, Section 8.3.1), due primarily to the presence of a relatively low density of nucleation sites in the former. The constant-current STM micrographs in Figure 14 show [88] how the size of the clusters changes with respect to the quantity of Au deposited. At relatively low coverages (0.1 MLE) of Au, hemispherical 3D clusters with diameters of 2–3 nm and heights of 1–1.5 nm are observed to grow mainly along the step edges. Well-dispersed quasi-2D clusters, having a height of 0.3–0.6 nm and a diameter of 0.5–1.5 nm, can be seen on the terraces. With increasing Au coverage, the clusters grow larger with little increase in cluster density. However, even at 4.0 MLE, some portions of the TiO$_2$ substrate are still visible. In contrast, on amorphous graphite substrate, at low coverages (Figure 8) a greater density of smaller-size clusters without any preferential spatial orientation is apparent.

A constant-current STM micrograph of 0.25 ML Au deposited onto single-crystal TiO$_2$(110)–(1 × 1) is shown in Figure 15a [15,19]. The deposition was carried out at 300 K, followed by annealing to 850 K for 2 min to stabilize the clusters. In the image we only see the Ti cations; the O^{2-} anions are not visible. The interatomic distance between the [001] rows is ~ 0.65 nm. Three-dimensional Au clusters, imaged as bright protrusions, have average diameters of ~ 2.6 nm and heights of ~ 0.7 nm (corresponding to 2–3 atoms thick) and are preferentially nucleated on the step edges. Quasi-2D clusters are characterized by heights of 1–2 atomic layers. Previous

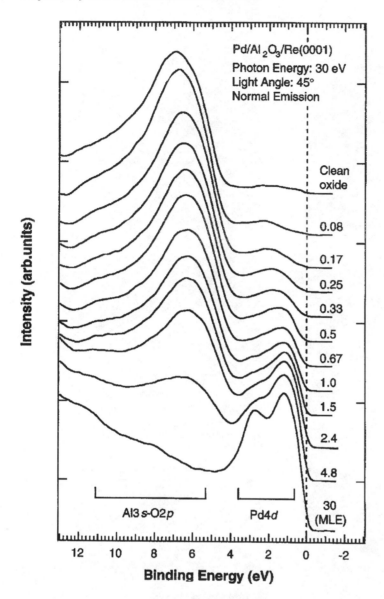

Figure 13 Normal emission spectra of the valance bands taken at 30 eV photon energy showing an overview of the growth of the Pd clusters for coverages ranging from 0.08 to 30 MLE. The binding energy referred to the Fermi edge of the Re substrate, which is identical to that of the thick metallic Pd film at a coverage of 30 MLE. (From Ref. 33.)

annealing studies show that the Au clusters form large microcrystals with well-defined hexagonal shapes.

Figure 15b shows STS acquired at various clusters on the surface, where the tunneling current (I) as a function of bias voltage (V) is measured. The length of the observed plateau at zero tunneling current is a measure of the band gap (along

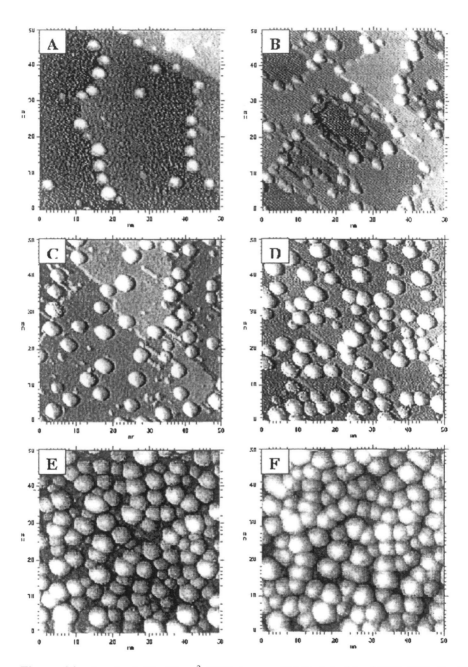

Figure 14 A set of 50×50 nm^2 STM images (2.0 V, 1.0 nA) of TiO$_2$(110)–(1 × 1) with different Au coverages: (A) 0.10 MLE, (B) 0.25 MLE, (C) 0.50 MLE, (D) 1.0 MLE, (E) 2.0 MLE, and (F) 4.0 MLE. With increasing coverage, Au clusters grow and gradually cover the surface. (From Ref. 88.)

the bias voltage axis) of electrons tunneling between the valance and conduction band of the cluster and tip. The electronic character of these clusters varies between that of a metal and a nonmetal, depending on their size. With an increase in size, clusters gradually adopt the metallic character with an enhanced density of states at the Fermi level, consistent with the UPS data presented in the previous section. Note that the cluster 2.5 mm × 0.7 nm in size has a larger band gap than that of a

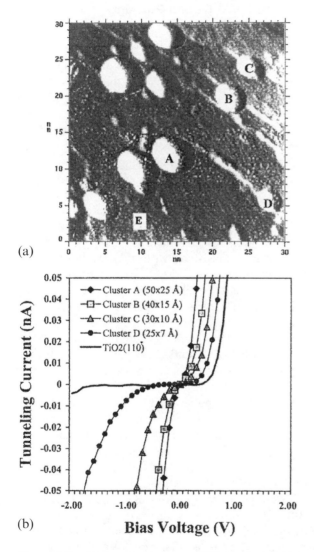

(a)

(b)

Figure 15 (a) A constant-current STM image of a 0.25 MLE Au deposited onto $TiO_2(110)$–(1×1) prepared just prior to a $CO:O_2$ reaction. The sample had been annealed to 850 K for 2 min; (b) STS data acquired for Au clusters of varying sizes on the $TiO_2(110)$–(1×1). An STS of the TiO_2 substrate, having a wider band gap than the Au cluster, is also shown as a point of reference. (From Ref. 19.)

5.0 nm × 2.5 nm cluster. Smaller clusters have a nonmetallic character, resulting in a significant band gap and a reduced density of states near the Fermi level. A similar metal–nonmetal transition with respect to cluster size has also been observed for Fe clusters deposited on GaAs(110) [89]. That the very small clusters are nonmetallic leads to changes in the electronic and chemical properties, as discussed below.

8.4 SIZE-DEPENDENT REACTIVITY OF SUPPORTED METAL CLUSTERS

8.4.1 Acetylene Cyclotrimerization on Pd/Al$_2$O$_3$ and Pd$_n$/MgO

The cyclotrimerization of acetylene to benzene on low-index single-crystal Pd surfaces has received considerable attention since its discovery by Tysoe et al. [90–98]. This reaction has been shown to be structure-sensitive with the Pd(111) facet being most active; i.e., the yield of benzene is 10 times larger than that for Pd(100); essentially no benzene production is observed on Pd(110) [99]. To understand the cluster-size effect on this reaction, supported Pd clusters of various sizes have been prepared on model Al$_2$O$_3$/Mo(110) thin films [100]. Figure 16 shows the TPD corresponding to benzene desorption ($m/e = 78$) as a function of Pd cluster size. These TPD data are very similar to the analogous low-index single-crystal results and reveal distinct cluster-size effects. For the smallest clusters ($\sim$1.5–2.0 nm), benzene desorption primarily occurs above 500 K, with very little low-temperature benzene evolution. As the cluster size increases, the high-temperature peak attenuates, while the desorption peak at 230 K increases in intensity. Both the high- and low-temperature benzene desorption features have been reproduced by adsorbing benzene on the Pd(111) surface, consistent with the reaction being desorption rate-limiting. The benzene desorption near 370 K has not been observed for single crystals and therefore is likely associated with interfacial defect sites. The lower activity for the smaller clusters has been explained as due to their enhanced curvatures and the absence of large ⟨111⟩ terraces, rather than a critical cluster-size effect. The high curvature and absence of ⟨111⟩ structures preclude the formation of an ordered, compressed overlayer of weakly bound benzene in the tilted configuration and favor the more strongly bound benzene, bonded parallel to the surface.

According to the "template" effect, the hexagonal Pd(111) face can correctly orient both C$_2$H$_2$ and intermediate C$_2$H$_4$ species to form a strongly bound benzene bonded parallel to the surface. This idea is consistent with benzene synthesis, requiring the participation of seven contiguous, flat Pd atoms. Indeed, recent studies using mass-selected Pd clusters deposited on MgO thin film have demonstrated a minimum requirement of a seven-atom cluster for the synthesis of a strongly bound flat-laying benzene desorbing at higher temperatures ($T_d \sim$ 430 K). Figure 17 shows TPD data for benzene from Pd/MgO surfaces with varying cluster size (Pd$_n$, $1 \leqslant n \leqslant 30$) [101]. In each case, the total coverage of Pd is constant at 0.28%. It is noteworthy that the production of benzene is observed even for a single palladium, though producing only weakly adsorbed ($T_d \sim$ 300 K) benzene in the tilted configuration. Density functional calculations show that a single Pd atom can be activated for benzene production by charge transfer when bonded to an oxygen vacancy site of the MgO substrate.

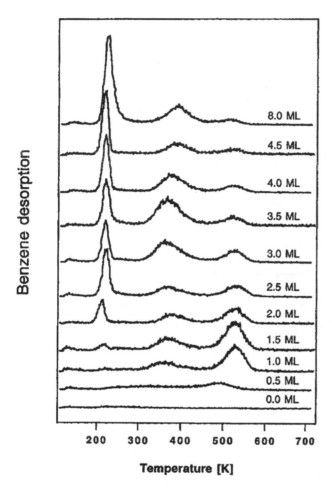

Figure 16 Desorption of benzene ($m/e = 78$) following 70 Langmuir acetylene at 150 K for various Pd coverages as indicated. (From Ref. 100.)

8.4.2 NO and CO Reactions on Pd and Cu/Al₂O₃

Recently there has been considerable interest in identifying alternatives to the expensive Pt/Rh catalyst used for three-way automobile exhaust gas conversion [102–106]. CO and NO reaction studies on Pd and Cu single crystals have shown promising results in terms of catalytic activity. Evidence for structure sensitivity has also been demonstrated [102–105] in high-pressure CO + NO reactions over Pd(111) and Pd(100), implying a cluster-size effect. In this section the results obtained for the CO + NO reaction with respect to cluster size on model Pd and Cu/Al₂O₃ catalysts are discussed. [15]NO adsorption experiments were carried out on a Pd/Al₂O₃/Ta(110) model catalyst [32] with several different Pd loadings. Clusters were exposed to a background pressure of 1×10^{-7} Torr [15]NO for 5 min at 550 K and then cooled to

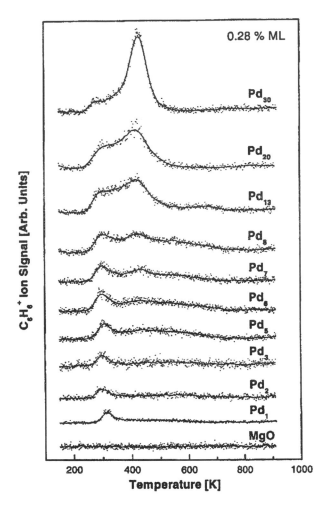

Figure 17 Catalytic benzene formation for different Pd cluster sizes obtained from TPR experiments. The bottom spectrum shows that for a clean MgO(100) film no benzene is formed. Cluster coverage is 0.28% for a monolayer for all cluster sizes, where one monolayer corresponds to 2.25×10^{15} atoms/cm^2. (From Ref. 101.)

350 K prior to ^{15}NO removal. The desorption spectra of ^{15}N$_2$ and ^{15}N$_2$O over several different Pd loadings are presented in Figures 18a and b, respectively. The behavior is very similar to that found for single crystals. For recombinative ^{15}N$_2$ desorption (Figure 18a), a low-temperature feature is observed at $\sim$520 K, along with a high-temperature feature above 600 K. The desorption temperature of the low-temperature feature is $\sim$70 K higher than for single crystals, consistent with the clusters' having an increased concentration of strongly binding step/edge defect sites.

Studies have shown the following steps to be critical:

$$^{15}NO_{(a)} \rightarrow {}^{15}NO$$
$$^{15}NO_{(a)} \rightarrow {}^{15}N_{(a)} + O_{(a)}$$
$$^{15}NO_{(a)} \, {}^{15}N_{(a)} \rightarrow {}^{15}N_2O$$
$$^{15}N_{(a)} + {}^{15}N_{(a)} \rightarrow {}^{15}N_2$$

The spectra in Figure 18a show enhancement of the low-temperature N_a (that recombines to N_2 or N_2O) with increasing cluster size. The peak maximum in each set of spectra shifts to lower temperature with an increase in cluster size. This trend is consistent with smaller clusters with higher surface defect sites stabilizing N_a to a greater extent than the large clusters. Figure 18b shows that with an increase in the cluster size, the N_2O desorption feature at ~ 530 K appears. Enhancement of the N_2O with increasing cluster size can be understood by comparing the single-crystal data for the $CO + NO$ reaction over (111), (110), and (100) faces of Pd. For these three Pd orientations, the N_2 selectivity follows the order $(110) > (100) > (111)$. This order is consistent with the fact that the surface concentration of N_a is enhanced on the more open $\langle 100 \rangle$ surface.

Arrhenius plots are shown in Figure 19 for CO_2 production from a $CO + NO$ reaction mixture for model oxide-supported Pd, Pd single crystals, and Pd/Al_2O_3 powder catalysts. For the powder and model catalysts, a pronounced increase in activity is seen with an increase in cluster size or loading. The larger clusters display the characteristics of the less open $\langle 111 \rangle$ plane.

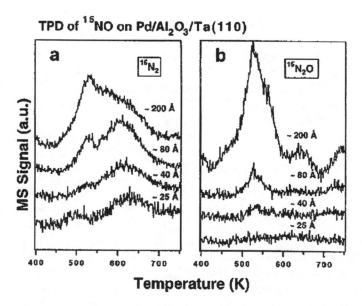

Figure 18 TPD monitoring (a) $^{15}N_2$ desorption and (b) $^{15}N_2O$ desorption over several different Pd coverages with the indicated average cluster sizes in a $Pd/Al_2O_3/Ta(110)$ catalyst after exposure to ^{15}NO at 550 K. (From Ref. 32.)

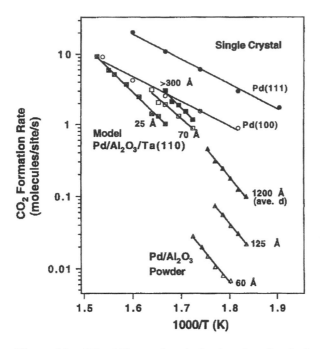

Figure 19   CO + NO reaction Arrhenius plots for single-crystal, model planner-supported, and Pd/Al$_2$O$_3$ powder catalysts. The powder catalyst data were taken in the flow reaction mode (4.4/5.2 CO/NO ratio, steady state), and the model catalyst and single-crystal data were acquired for a batch reaction mode in 1 Torr of each reactant. (From Ref. 32.)

The reaction of ^{15}NO with well-characterized Cu clusters deposited on highly ordered Al$_2$O$_3$ films has also been studied using TPD and HREELS [58]. Figure 20 shows that the ^{15}N$_2$ formed is sensitive to the cluster size in that no ^{15}N$_2$ is detected until $\theta_{Cu} > 0.13$ (~ 3.5 nm clusters). Desorption of ^{15}N$_2$ was observed at about 130 and 770 K; the 770 K feature is understood to be due to recombination of atomic ^{15}N$_{(a)}$. The reaction of CO + NO has also been examined for this catalytic system. The results show, in addition to ^{15}N$_2$ and ^{15}N$_2$O, small amounts of CO$_2$ are produced via reaction between adsorbed CO and adsorbed oxygen.

8.4.3   CO Oxidation on Pd/SiO$_2$/Mo(100) and Au/TiO$_2$

Oxidation of CO on transition metals has been a subject of interest for many years due to its technological importance in the area of automobile exhaust catalyst and the preparation of CO$_x$-free hydrogen for fuel cells [107–119]. Very interesting cluster-size effects with respect to the catalytic activity have been observed for certain catalysts. For example, CO oxidation has been studied over Pd/SiO$_2$/Mo(100) model catalysts [31,120]. The reaction conditions for the catalysts were 10 Torr CO, 5.0 Torr O$_2$, and reaction temperatures in the range 540–625 K. The conversions were maintained at less than 50% and were monitored using the pressure decrease in a static reactor of known volume (750 cm^3). Figure 21 shows Arrhenius plots of CO oxidation over three different model Pd/SiO$_2$ catalysts and a 5% loading of Pd on

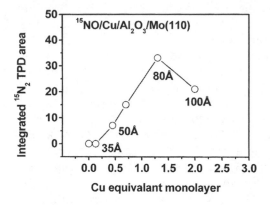

Figure 20 The integrated $^{15}N_2$ TPD area (β_2 state) plotted as a function of the Cu coverage in equivalent monolayers. The average sizes of the Cu clusters are also indicated. (From Ref. 58.)

powdered SiO_2. The average cluster sizes shown in Figure 21 were determined by CO, O_2 TPD, and ex-situ STM/AFM. The specific reaction rates were somewhat higher for the model catalysts than the high-surface-area catalysts; however, the activation energies are remarkably similar. This is because for a high-surface-area catalyst, reaction rates are calculated under the steady-state conditions, whereas initial reaction rates are considered in the case of a model catalyst, and generally the

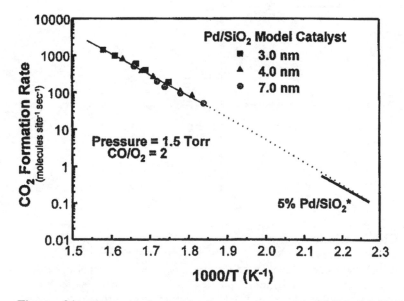

Figure 21 CO oxidation with O_2 over a model $Pd/SiO_2/Mo(100)$ catalyst and a conventional 5% Pd/SiO_2 catalyst. Reaction conditions were $P_{Torr} = 0.5$ Torr and $CO/O_2 = 0.2$. (From Ref. 31.)

steady-state rates are comparably lower due to poisoning effects. There was no noticeable dependence of the CO_2 formation rate on the Pd cluster size, indicating that CO oxidation over Pd/SiO_2 is structure-insensitive.

Interestingly, a significant correlation between the cluster size and catalytic activity has been observed for CO oxidation over an Au/TiO_2 system [15,19,45,121,122]. Investigations have been carried out on $Au/TiO2/Mo(100)$ as well as on $Au/TiO2(110)–(1 \times 1)$ for comparison. Figures 22a and b show a plot of CO oxidation activity [TOF = (product molecule/total Au atom) $\times$ sec] at 350 K as a function of Au cluster size supported on $TiO_2(110)–(1 \times 1)$ and $TiO_2/Mo(100)$ substrates. These results show similarities in the structure sensitivity of CO oxidation with a maximum activity evident at ~ 3 nm Au cluster size on both TiO_2 supports. For each catalyst, the activity and the selectivity of the supported Au clusters are markedly size-dependent. Although the TiO_2-supported Au catalysts exhibit a high activity for the low-temperature CO oxidation, the catalysts are often rapidly deactivated. Figure 23 shows a plot of TOF versus time for CO oxidation at 300 K on 0.25 ML Au on $TiO_2/Mo(100)$. The model catalyst, which exhibited a high initial activity, deactivated after a $CO + O_2$ (1:5) reaction of ~ 120 min at 40 Torr. This deactivation is due to agglomeration of the Au clusters with reaction time and has been documented by detailed STM measurements [15,19,45,121,122]. The STM data clearly demonstrate that under reaction condition the Au cluster Ostwald ripens; i.e., large clusters grow at the expense of small ones. Such a ripening depends on the strength of the cluster–support interaction as well as the gas pressure. The TOF for CO oxidation reaches a maximum with respect to cluster size and correlates with a metal to nonmetal transition that occurs at cluster size of ~ 3 nm (discussed in Section 8.3.3).

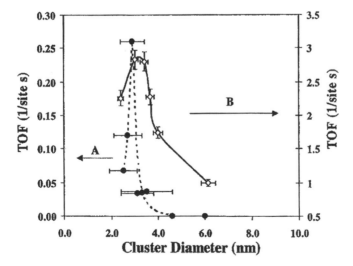

Figure 22 CO oxidation TOFs as a function of the Au cluster size supported on TiO_2. (A) The Au/TiO_2 catalysts were prepared by a precipitation method, and the average cluster size was measured by TEM at 300 K. (B) The Au/TiO_2 catalysts were prepared by vapor deposition of Au on planar TiO_2 films on Mo(100). The CO/O_2 mixture was 1:5 at a total pressure of 40 Torr and 350 K. (From Ref. 19.)

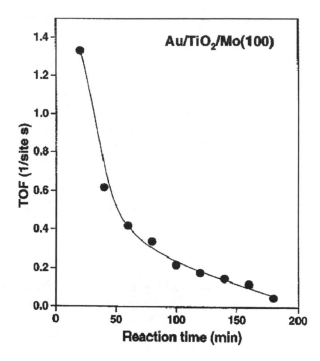

Figure 23 The specific activity for CO conversion as a function of reaction time at 300 K on a model $Au/TiO_2/Mo(100)$ catalyst. The Au coverage was 0.25 MLE, corresponding to an average cluster size of ~ 2.4 nm. (From Ref. 123.)

8.5 IMPORTANT FACTORS GOVERNING SIZE-DEPENDENT CATALYTIC PROPERTIES

The preceding sections have described how the structural, electronic, and catalytic properties are altered with a change in cluster size. The TOF of CO oxidation over Au/TiO_2 goes through a maximum near the onset of a metal to nonmetal transition, whereas for Pd/SiO_2, the same reaction shows little cluster-size dependence. For acetylene cyclotrimerization, a template effect is apparent whereas the TOF increases monotonically with cluster size. Therefore, the dependence of TOF on cluster size is very specific to the catalyst system and chemical reaction. A monotonic increase in the reaction rate with a decrease in cluster size can relate simply to the fact that the surface-to-volume ratio increases substantially. The fact that there is a regular lattice contraction with respect to a decrease in cluster size could amplify the effect of the increase in surface area. However, those cases where the bulk metals show no catalytic activity clearly cannot be understood by a simple surface-to-volume change as a function of cluster size. At the small cluster-size limit, electronic, geometric, and metal–support interaction effects obviously must be important.

8.6 CONCLUSIONS AND FUTURE PERSPECTIVE

The utility of planar-supported model catalysts, consisting of metal clusters deposited on metal oxide thin films, as models for high-surface-area industrial catalysts has been demonstrated. Indeed the use of thin metal oxide films as supports makes them suitable for ultrahigh vacuum measurements. Electronic structural changes shown by core-level binding energy, band splitting, position of the Fermi level, valance-band dispersion, etc. with respect to the cluster size are understood by considering both initial- and final-state effects. A metal to nonmetal transition with respect to the cluster size has been observed by STS and valance-band spectra. Size-dependent catalytic reactivity has been observed for a variety of industrially important systems. A critical comparison of acetylene cyclotrimerization on a single crystal of Pd, Pd/Al_2O_3, and mass-selected Pd clusters reveal a minimum requirement of seven Pd atoms for the production of strongly bound, flat-laying benzene. The $CO + NO$ reaction has been found to be size-dependent in terms of CO_2, N_2, and N_2O production. Very interestingly, low-temperature CO oxidation by O_2 has been found to have a dramatic size dependence for the Au/TiO_2 system, the maximum rate occurring at the onset of the metal–nonmetal transition.

ACKNOWLEDGMENTS

We acknowledge with pleasure the support of this work by the Department of Energy, Office of Basic Energy Sciences, and Division of Chemical Sciences and the Robert A. Welch Foundation.

REFERENCES

1. M. Baumer, H.J. Freund, Progress in Surf. Science 1999, 61(7–8), 127–198.
2. C.T. Campbell, Surf. Science Reports 1997, 27(1–3), 1–111.
3. S.B. Dicenzo, S.D. Berry, E.H. Hartford, Phys. Rev. B—Condensed Matter 1988, 38(12), 8465–8468.
4. U. Diebold, J.M. Pan, T.E. Madey, Surf. Science 1995, 333, 845–854.
5. J.W.F. Egelhoff, Surf. Science Reports 1987, 6, 253–415.
6. H.J. Freund, Physica Status Solidi B-Basic Research 1995, 192(2), 407–440.
7. H.J. Freund, H. Kuhlenbeck, V. Staemmler, Reports on Progress in Phys. 1996, 59(3), 283–347.
8. L.J. Gerenser, M.G. Mason, Bull. of the American Physical Society 1977, 22(3), 431.
9. D.W. Goodman, Abstracts of Papers of the ACS 1993, 205, 166–COLL.
10. D.W. Goodman, Surf. Science 1994, 300(1–3), 837–848.
11. D.W. Goodman, J. Phys. Chem. 1996, 100(31), 13090–13102.
12. D.W. Goodman, Abstracts of Papers of the ACS 2000, 219, 69-COLL.
13. M. Haruta, Catal. Today 1997, 36(1), 153–166.
14. C.R. Henry, Surf. Science Reports 1998, 31(7–8), 235–325.
15. X. Lai, T.P. St. Clair, M. Valden, D.W. Goodman, Progress in Surf. Science 1998, 59(1–4), 25–52.
16. M.G. Mason, Phys. Rev. B 1983, 27(2), 748–762.
17. H. Poppa, Catal. Rev.—Science and Eng. 1993, 35(3), 359–398.
18. W.D. Schneider, Analysis 1993, 21(8), M20–M22.
19. M. Valden, X. Lai, D.W. Goodman, Science 1998, 281(5383), 1647–1650.
20. G.K. Wertheim, Zeitschrift Fur Physik D-Atoms Molecules and Clusters 1989, 12(1–4), 319–326.
21. C.T. Campbell, Current Opinion in Solid State and Materials Science 1998, 3(5), 439–445.
22. M. Frank, R. Kuhnemuth, M. Baumer, H.J. Freund, Surf. Science 1999, 428, 288–293.

23. H.J. Freund, M. Baumer, H. Kuhlenbeck, Advances in Catal., 2000, 45, 333–384.
24. D.W. Goodman, Abstracts of Papers of the ACS 2000, 219, 10-CATL.
25. M. Haruta, T. Kobayashi, H. Sano, N. Yamada, Chem. Lett. 1987, (2), 405–408.
26. M. Haruta, 3rd World Congress on Oxidation Catal. 1997, 110, 123–134.
27. M. Haruta, B.S. Uphade, S. Tsubota, A. Miyamoto, Res. on Chemical Intermediates 1998, 24(3), 329–336.
28. M. Haruta, B.S. Uphade, K. Ruth, S. Tsubota, Abstracts of Papers of the ACS 2000, 219, 61-PETR.
29. C.R. Henry, C. Chapon, C. Goyhenex, R. Monot, Surf. Science 1992, 272(1–3), 283–288.
30. L. Piccolo, C.R. Henry, Applied Surf. Science 2000, 162, 670–678.
31. D.R. Rainer, M.C. Wu, D.L. Mahon, D.W. Goodman, J. Vacuum Science and Techn. A—Vacuum Surfaces and Films 1996, 14(3), 1184–1188.
32. D.R. Rainer, S.M. Vesecky, M. Koranne, W.S. Oh, D.W. Goodman, J. Catal. 1997, 167(1), 234–241.
33. Y.Q. Cai, A.M. Bradshaw, Q. Guo, D.W. Goodman, Surf. Science 1998, 399(2–3), L357–L363.
34. S.B. Dicenzo, G.K. Wertheim, J. Electron Spectroscopy and Related Phenomena 1987, 43(3–4), C7–C12.
35. M.G. Mason, L.J. Gerenser, S.T. Lee, Phys. Rev. Lett. 1977, 39(5), 288–291.
36. M.G. Mason, S.T. Lee, G. Apai, Chem. Phys. Lett. 1980, 76(1), 51–53.
37. M.G. Mason, S.T. Lee, G. Apai, R.F. Davis, D.A. Shirley, A. Franciosi, J.H. Weaver, Phys. Rev. Lett. 1981, 47(10), 730–733.
38. G.K. Wertheim, S.B. Dicenzo, D.N.E. Buchanan, Phys. Rev. B 1986, 33(8), 5384–5390.
39. G.K. Wertheim, Phys. Rev. B 1987, 36(18), 9559–9562.
40. G.K. Wertheim, Zeitschrift Fur Physik B—Condensed Matter 1987, 66(1), 53–63.
41. G.K. Wertheim, S.B. Dicenzo, Phys. Rev. B 1988, 37(2), 844–847.
42. G.K. Wertheim, Phase Trans. 1990, 24–6(1), 203–214.
43. M. Haruta, S. Tsubota, T. Kobayashi, A. Ueda, H. Sakurai, M. Ando, Studies in Surf. Science and Catal. 1993, 75, 2657–2660.
44. M. Haruta, Y. Souma, Catal. Today 1997, 36(1), 1.
45. M. Valden, S. Pak, X. Lai, D.W. Goodman, Catal. Lett. 1998, 56(1), 7–10.
46. J.S. Corneille, J.W. He, D.W. Goodman, Surf. Science 1994, 306(3), 269–278.
47. D.W. Goodman, J. of Vacuum Science and Tech. A—Vacuum Surf. and Films 1996, 14(3), 1526–1531.
48. Q. Guo, W.S. Oh, D.W. Goodman, Surf. Science 1999, 437(1–2), 49–60.
49. Q. Guo, D.Y. Kim, S.C. Street, D.W. Goodman, J. of Vacuum Science and Techn. A—Vacuum Surf. and Films 1999, 17(4), 1887–1892.
50. Q. Guo, S. Lee, D.W. Goodman, Surf. Science 1999, 437(1–2), 38–48.
51. X. Lai, C.C. Chusuei, K. Luo, Q. Guo, D.W. Goodman, Chem. Phys. Lett. 2000, 330(3–4), 226–230.
52. W.S. Oh, C. Xu, D.Y. Kim, D.W. Goodman, J. of Vacuum Science and Tech. A—Vacuum Surf. and Films 1997, 15(3), 1710–1716.
53. M.C. Wu, J.S. Corneille, C.A. Estrada, J.W. He, D.W. Goodman, Chem. Phys. Lett. 1991, 182(5), 472–478.
54. M.C. Wu, J.S. Corneille, J.W. He, C.A. Estrada, D.W. Goodman, J. of Vacuum Science and Tech. A—Vacuum Surf. and Films 1992, 10(4), 1467–1471.
55. Y.T. Wu, E. Garfunkel, T.E. Madey, J. of Vacuum Science and Tech. A—Vacuum Surf. and Films 1996, 14(4), 2554–2563.
56. S.A. Chambers, Surf. Science Reports 2000, 39(5–6), 105–180.
57. R. Franchy, Surf. Science Reports 2000, 38(6–8), 199–294.
58. M.C. Wu, D.W. Goodman, J. of Phys. Chem. 1994, 98(39), 9874–9881.

59. P.J. Chen, D.W. Goodman, Surf. Science 1994, 312(3), L767–L773.

60. P.J. Chen, M.L. Colaianni, J.T. Yates, Phys. Rev. B 1990, 41(12), 8025–8032.

61. J.E. Crowell, J.G. Chen, J.T. Yates, Surf. Science 1986, 165(1), 37–64.

62. J.L. Erskine, R.L. Strong, Phys. Rev. B 1982, 25(8), 5547–5550.

63. R.L. Strong, B. Firey, F.W. Dewette, J.L. Erskine, Phys. Rev. B 1982, 26(6), 3483–3486.

64. B.G. Frederick, G. Apai, T.N. Rhodin, J. of Electron Spectroscopy and Related Phenomena 1990, 54, 415–424.

65. B.G. Frederick, G. Apai, T.N. Rhodin, Phys. Rev. B 1991, 44(4), 1880–1890.

66. J.W. He, X. Xu, J.S. Corneille, D.W. Goodman, Surf. Science 1992, 279(1–2), 119–126.

67. X.P. Xu, D.W. Goodman, Surf. Science 1993, 282(3), 323–332.

68. X.P. Xu, D.W. Goodman, Appl. Phys. Lett. 1992, 61(7), 774–776.

69. R. Ludeke, Surf. Science 1983, 132(1–3), 143–168.

70. A. Franciosi, D.J. Peterman, J.H. Weaver, V.L. Moruzzi, Phys. Rev. B 1982, 25(8), 4981–4993.

71. M.G. Mason, S.T. Lee, G. Apai, Chem. Phys. Lett. 1980, 76(1), 51–53.

72. W.F. Egelhoff, G.G. Tibbetts, Phys. Rev. B 1979, 19(10), 5028–5035.

73. J.R. Arthur, A.Y. Cho, Surf. Science 1973, 36(2), 641–660.

74. D. Dalacu, J.E. Klernberg-Sapieha, L. Martinu, Surf. Science 2001, 472(1–2), 33–40.

75. K. Luo, D.Y. Kim, D.W. Goodman, J. of Molec. Catal. A—Chem. 2001, 167(1–2), 191–198.

76. G. Apai, S.T. Lee, M.G. Mason, Solid State Comm. 1981, 37(3), 213–217.

77. S.B. Dicenzo, G.K. Wertheim, Phys. Rev. B 1989, 39(10), 6792–6796.

78. U. Diebold, J.M. Pan, T.E. Madey, Abstracts of Papers of the ACS 1991, 202, 191-Coll.

79. U. Diebold, J.M. Pan, T.E. Madey, Surf. Science 1993, 287, 896–900.

80. U. Diebold, J.M. Pan, T.E. Madey, Phys. Rev. B—Condensed Matter 1993, 47(7), 3868–3876.

81. J.W. He, X. Xu, J.S. Corneille, D.W. Goodman, Surf. Science 1992, 279(1–2), 119–126.

82. H. Hovel, B. Grimm, M. Pollmann, B. Reihl, Phys. Rev. Lett. 1998, 81(21), 4608–4611.

83. H. Hovel, B. Grimm, M. Pollmann, B. Reihl, Eur. Phys. J. D 1999, 9(1–4), 595–599.

84. S.T. Lee, G. Apai, M.G. Mason, R. Benbow, Z. Hurych, Phys. Rev. B 1981, 23(2), 505–508.

85. W.D. Schneider, Appl. Phys. A—Mater. Science and Proc. 1994, 59(5), 463–467.

86. R.L. Whetten, D.M. Cox, D.J. Trevor, A. Kaldor, Phys. Rev. Lett. 1985, 54(14), 1494–1497.

87. G.K. Wertheim, S.B. Dicenzo, S.E. Youngquist, Phys. Rev. Lett. 1983, 51(25), 2310–2313.

88. X. Lai, T.P. St. Clair, M. Valden, D.W. Goodman, Progress in Surf. Science 1998, 59(1–4), 25–52.

89. P.N. First, J.A. Stroscio, R.A. Dragoset, D.T. Pierce, R.J. Celotta, Phys. Rev. Lett. 1989, 63(13), 1416–1419.

90. R.M. Ormerod, R.M. Lambert, H. Hoffmann, F. Zaera, J.M. Yao, D.K. Saldin, L.P. Wang, D.W. Bennett, W.T. Tysoe, Surf. Science 1993, 295(3), 277–286.

91. R.M. Ormerod, R.M. Lambert, H. Hoffmann, F. Zaera, L.P. Wang, D.W. Bennett, W.T. Tysoe, J. of Phys. Chem. 1994, 98(8), 2134–2138.

92. R.M. Ormerod, R.M. Lambert, D.W. Bennett, W.T. Tysoe, Surf. Science 1995, 330(1), 1–10.

93. W.T. Tysoe, G.L. Nyberg, R.M. Lambert, Surf. Science 1983, 135(1–3), 128–146.

94. W.T. Tysoe, G.L. Nyberg, R.M. Lambert, J. of the Chem. Soc. Chemical Comm. 1983, (11), 623–625.

95. W.T. Tysoe, G.L. Nyberg, R.M. Lambert, J. of Phys. Chem. 1986, 90(14), 3188–3192.

96. W.T. Tysoe, R.M. Ormerod, R.M. Lambert, G. Zgrablich, A. Ramirezcuesta, J. of Phys. Chem. 1993, 97(13), 3365–3370.

97. W.T. Tysoe, Abstracts of Papers of the ACS 1995, 209, 231–Coll.
98. W.T. Tysoe, Israel J. of Chem. 1998, 38(4), 313–320.
99. J. Yoshinobu, T. Sekitani, M. Onchi, M. Nishijima, J. of Phys. Chem. 1990, 94(10), 4269–4275.
100. P.M. Holmblad, D.R. Rainer, D.W. Goodman, J. of Phys. Chem. B 1997, 101(44), 8883–8886.
101. S. Abbet, A. Sanchez, U. Heiz, W.D. Schneider, A.M. Ferrari, G. Pacchioni, N. Rosch, Surf. Science 2000, 454, 984–989.
102. S.M. Vesecky, D.R. Rainer, D.W. Goodman, J. of Vacuum Science and Tech. A—Vacuum Surfaces and Films 1996, 14(3), 1457–1463.
103. S.M. Vesecky, D.R. Rainer, M.M. Koranne, D.W. Goodman, Abstracts of Papers of the 1996, 211, 1-HYS.
104. S.M. Vesecky, P.J. Chen, X.P. Xu, D.W. Goodman, J. of Vacuum Science and Tech. A—Vacuum Surfaces and Films 1995, 13(3), 1539–1543.
105. X.P. Xu, P.J. Chen, D.W. Goodman, J. of Phys. Chem. 1994, 98(37), 9242–9246.
106. T. Fink, J.P. Dath, R. Imbihl, G. Ertl, Surf. Science 1991, 251, 985–989.
107. J. Szanyi, D.W. Goodman, J. of Phys. Chem. 1994, 98(11), 2972–2977.
108. J. Szanyi, W.K. Kuhn, D.W. Goodman, J. of Phys. Chem. 1994, 98(11), 2978–2981.
109. D.W. Goodman, C.H.F. Peden, G.B. Fisher, S.H. Oh, Catal. Lett. 1993, 22(3), 271–274.
110. J. Szanyi, D.W. Goodman, Catal. Lett. 1993, 21(1–2), 165–174.
111. C.H.F. Peden, D.W. Goodman, M.D. Weisel, F.M. Hoffmann, Surf. Science 1991, 253(1–3), 44–58.
112. J.W. He, K. Kuhn, L.W. Leung, J. Szanyi, D.W. Goodman, Abstracts of Papers of the ACS 1991, 201, 59-COLL.
113. P.J. Berlowitz, C.H.F. Peden, D.W. Goodman, J. of Phys. Chem. 1988, 92(18), 5213–5221.
114. C.H.F. Peden, D.W. Goodman, D.S. Blair, P.J. Berlowitz, G.B. Fisher, S.H. Oh, J. of Phys. Chem. 1988, 92(6), 1563–1567.
115. D.W. Goodman, C.H.F. Peden, J. of Phys. Chem. 1986, 90(20), 4839–4843.
116. D.W. Goodman, C.H.F. Peden, Abstracts of Papers of the ACS 1985, 190(Sep), 197-COLL.
117. C.H.F. Peden, D.W. Goodman, J. of Vacuum Science and Tech. A—Vacuum Surfaces and Films 1985, 3(3), 1558–1559.
118. M. Eiswirth, P. Moller, K. Wetzl, R. Imbihl, G. Ertl, J. of Chem. Phys. 1989, 90(1), 510–521.
119. R. Imbihl, S. Ladas, G. Ertl, J. of Vacuum Science and Tech. A—Vacuum Surfaces and Films 1988, 6(3), 877–878.
120. D.R. Rainer, M. Koranne, S.M. Vesecky, D.W. Goodman, J. of Phys. Chem. B 1997, 101(50), 10769–10774.
121. Y. Iizuka, H. Fujiki, N. Yamauchi, T. Chijiiwa, S. Arai, S. Tsubota, M. Haruta, Catal. Today 1997, 36(1), 115–123.
122. M. Valden, D.W. Goodman, Israel J. of Chem. 1998, 38(4), 285–292.
123. X.F. Lai, D.W. Goodman, J. of Molecular Catal. A—Chemical 2000, 162(1–2), 33–50.

9

Physical and Electrochemical Characterization of Bimetallic Nanoparticle Electrocatalysts

N. M. MARKOVIĆ, V. RADMILOVIC, and P. N. ROSS, Jr.

University of California at Berkeley, Berkeley, California, U.S.A.

CHAPTERS CONTENTS

SUMMARY

In this chapter we review studies, primarily from our laboratory, of Pt and Pt-bimetallic nanoparticle electrocatalysts for the oxygen reduction reaction (ORR) and the electrochemical oxidation of H_2 (HOR) and H_2/CO mixtures in aqueous electrolytes at 274–333 K. We focus on the study of both the structure sensitivity of the reactions as gleaned from studies of the bulk (bi) metallic surfaces and the resultant crystallite size effect expected or observed when the catalyst is of nanoscale dimension. Physical characterization of the nanoparticles by high-resolution transmission electron microscopy (HRTEM) techniques is shown to be an essential tool for these studies. Comparison with well-characterized model surfaces have revealed only a few "nanoparticle anomalies," although the number of bimetallics studied in this manner is still relatively small. Two anomalous or unexpected results in this chapter: (1) Pt-Co, Ni, and Ru nanoparticles when prepared using certain (low-temperature) methods appear to have surface compositions very close to that of the overall stoichiometry of the metals, even though the bulk crystals show strong surface enrichment in Pt; (2) Pt-Mo nanoparticles when prepared using similar (low-temperature) methods appear to exhibit an anomalous surface enrichment in Mo versus surface enrichment in Pt observed with the bulk alloy. It may be that surface segregation thermodynamics unique to nanoparticles can explain both anomalies in a consistent manner. At the 2–10-nm dimensionality used in these studies, no crystallite size effect was observed for the ORR in nonadsorbing electrolytes like those employed in commercial fuel cells, e.g., perfluoroalkylsulfonic acid polymer and potassium hydroxide. The H_2 and H_2/CO oxidation reactions are structure-sensitive reactions that are expected to exhibit a significant crystallite size effect, particularly with Pt-Mo. Further experimental studies are needed to confirm this expectation.

9.1 INTRODUCTION

Decades before the prefix "nano" appeared so prominently in the scientific lexicon, catalytic chemists had already realized that unique catalytic properties were obtained by intermixing two metals at the nanometric scale, i.e., bimetallic particles 1–10 nm in characteristic dimension [1]. The terminology used to describe these types of materials has changed over the decades, and researchers from different fields working with the same catalyst may use a different terminology. Sinfelt and co-workers employed the term "supported bimetallic cluster" to describe the catalysts developed in his group at Exxon in the late 1960s [2]. More generally in inorganic chemistry, metallic clusters are aggregates of 2 to perhaps 100 or so metal atoms freestanding in the gas phase or isolated in a matrix. The characteristic dimension of metallic clusters is thus generally on the order of a subnanometer. In this scale of dimension, the aggregates of metal atoms have electronic properties distinct from the bulk material, e.g., the ionization potentials [3], and are thus distinct states of matter. It is now known, both experimentally and theoretically, that the electronic properties of metals converge to the bulk properties at characteristic dimensions above about 1 nm [4], albeit with much greater contributions from surface electronic states in the 1–10-nm (nanoscale) range than in bulk crystals. Metallic aggregates in this size range contain from 10^3–10^5 atoms. Because the electronic properties of metal aggregates in this size range are not unique, catalyst research with such materials has focused on the variation in the surface structure of the aggregates with characteristic dimension, i.e., how the distribution of different types of adsorption sites varies with dimension for a given structure. Boronin and co-workers [5] in particular pioneered this approach, which they termed "mitohedry," to understanding the "crystallite size effect" in heterogeneous catalysis. The crystallite size effect is, as the term implies, the variation in reaction rate or selectivity with the characteristic dimension of a metallic catalyst. Structure sensitivity, a companion property, refers to the dependence of reaction rate or selectivity on the geometry of the surface [6]. Hence a structure-sensitive reaction is expected to exhibit a crystallite size effect. In this chapter we use this same terminology for bimetallic electrocatalysts and focus on the study of both the structure sensitivity of the reaction and the resultant crystallite size effect expected or observed when the catalyst is of nanoscale dimension. For convenience in this chapter, we refer to all metallic and bimetallic aggregates of 1–10-nm characteristic dimension as *nanoparticles*.

Pt and Pt-bimetallic nanoparticle catalysts were employed in commercial prototype phosphoric acid fuel cells even in the mid-1970s [7], so in fact the concept of nanoparticle electrocatalysts is not new. Like many industrial catalysts, however, the catalysts are put into use long before their structure and properties are clearly understood, and that was certainly the case, for example, for the Pt–Co–Cr air cathode catalysts used at United Technologies [8]. In this chapter we review studies, primarily from our laboratory, of Pt and Pt-bimetallic nanoparticle electrocatalysts for the oxygen reduction reaction (ORR) and the electrochemical oxidation of H_2 (HOR) and H_2/CO mixtures.

9.2 EXPERIMENTAL METHODS

9.2.1 Methods of Nanoparticle Preparation

Because carbon black is the preferred support material for electrocatalysts, the methods of preparation of (bi)metallic nanoparticles are somewhat more restricted than with the oxide supports widely used in gas-phase heterogeneous catalysis. A further requirement imposed by the reduced mass-transport rates of the reactant molecules in the liquid phase versus the gas phase is that the metal loadings on the carbon support must be very high, e.g., at least 10 wt.% versus 0.1–1 wt.% typically used in gas-phase catalysts. The relatively inert character of the carbon black surface plus the high metal loading means that widely practiced methods such as ion exchange [9] are not effective. The preferred methods are based on preparation of colloidal precursors, which are adsorbed onto the carbon black surface and then thermally decomposed or hydrogen-reduced to the (bi)metallic state. This method was pioneered by Petrow and Allen [10], and in the period from about 1970–1995 various colloidal methods are described essentially only in the patent literature. A useful survey of methods described in this literature can be found in the review by Stonehart [11]. Since about 1995, there has been more disclosure of colloidal methods in research journals, such as the papers by Boennemann and co-workers [12].

9.2.2 Physical Characterization

The primary method for physical characterization of nanoparticle catalysts is transmission electron microscopy (TEM). Because the support is electronically conducting, the materials are free of nuisances like charging, and the samples are easily prepared for microscopy, e.g., dispersion in alcohol and adsorption onto a carbon grid full of holes. The large difference in electron scattering cross section between carbon and the metals used in electrocatalysts makes the metallic phases easy to distinguish from the carbon support. Essential properties like particle size and number density are relatively easily obtained and do not require a sophisticated microscope. However, the nanosize of these particles presents challenges to their microstructural characterization. Lattice structure characterization can be obtained by high-resolution electron microscopy (HREM), which can reveal the presence of defects, such as dislocations and/or twins. The advantages of the application of TEM in catalyst characterization are described in several review articles [13–15]. More specifically, HREM has been extensively used to determine faceting planes, geometric shapes, the presence of surface steps, as well as the size and distribution of supported metallic catalysts [16–18].

Conventional image formation in the transmission electron microscope is achieved by magnifying either the forward-scattered beam, to form a bright field (BF) image, or one of the Bragg-scattered beams, to form a dark field (DF) image. Weak beam dark field imaging (WBDF) [15–18] cannot be used for true nanoparticles because of its well-known limitations if the particle size is less than 10 nm. The mechanisms responsible for contrast in conventional image formation, called, *amplitude contrast*, are not limited by instrument resolution and are well understood. However, high-resolution electron microscopy in the imaging mode derives information about specimen by *phase contrast* methods [19–23]. These methods are based on the theory of microscopic image formation in which an electron beam of coherent radiation falls on a periodic object, such as a crystal structure, and passes through an imaging lens, producing a diffraction pattern in the back focal plane of the objective lens. It actually represents a mapping of the Fourier transform of the specimen. Maximum information about the specimen, i.e., at highest resolution, can be obtained by an inverse Fourier transform wherein all the Fourier coefficients are retained. This can be achieved by including higher-order reflections in the imaging aperture, which means that high-resolution detail is provided in the image by allowing the interaction of these higher-order reflections arising from planes of smaller spacing in the crystal and appearing at higher angles to the microscope optical axis. Phase contrast is derived from the variations in phase induced in the incident electron wave by both the specimen and the imaging system, because the TEM objective lens is imperfect and phase shifts are introduced from spherical aberration. The relative phase of interacting beams is an important factor in determining the resultant amplitude. By proper control of phase shifts the exit wave amplitude can be changed, yielding wide variations in contrast from the specimen. In order to get interpretable contrast, phase shifts must have a known relationship to the structure of the specimen. The simplest case is a very thin sample that behaves as "weak phase object," whereby high-resolution image contrast is linearly related to the phase shift. Image interpretation of sample structure in terms of a projected atomic potential is then rather straightforward. If the sample does not behave as a "weak phase object," it implies multiple scattering and attenuation processes that destroy the linear relationship between phase shifting of the incident wave and image contrast, and image interpretation is rather difficult. It requires a full dynamical interpretation of the exit wave function, described in detail elsewhere [12–22].

Radmilovic et al. [23,24] have used HRTEM image simulation to explore the changes in images of metallic nanoparticles under various imaging conditions, i.e., to relate the "real" particle size to its "apparent" size as derived from the HRTEM image. They demonstrate that even under optimum focusing conditions the spot intensity distribution does not correspond to the positions of the atoms in the particle due to the strong influence of Fresnel fringes near the edge of the particle. As expected, the deviations of the peaks representing atom positions that affect d-spacing are more pronounced in regions close to the edges of the small particles, where the potential drop is sharper, producing larger apparent d-spacing, i.e., apparent surface relaxation. Fresnel effects also produce "ghost" atoms outside the particle, producing an error in particle-size measurement that can be as large as ~20%. The results of this work [23,24] suggest that measurements of size and "relaxation" of outer planes in small particles must be accompanied by image formation simulation knowing all imaging parameters and taking into account the strong Fresnel effect.

Special electron microscopes, typically called analytical electron microscopes, equipped with spectrometers for elemental analysis can also provide information on the chemical composition of bimetallic nanoparticles. The two most prevalent spectrometers for this are electron energy loss (EELS) and x-ray emission spectroscopy. In principle, such analytical microscopes can give the composition of a single nanoparticle. Quantitation, however, requires careful calibration using samples of known compositions in a physical form that is as close as possible to that of the sample being analyzed [25]. This can be a problem in the case of supported catalysts. Another technique also available only with special microscopes is microdiffraction. As the name implies, this is the observation of an electron diffraction pattern from a single nanoparticle. Microdiffraction is especially useful for identifying twinning, particles with different crystal structures, e.g., phases, and ordering in bimetallic nanoparticles. In this chapter we present examples of all these possibilities in bimetallic electrocatalysts of interest.

9.2.3 Electrochemical Characterization

By design, nanoparticle electrocatalysts have a very high metallic surface area, e.g., $>10^5 \, cm^2$ per g of metal, producing a very high concentration of active sites per unit volume of electrode. This presents a special set of challenges in obtaining true kinetic measurements of the rates of electrochemical reactions. This challenge was recognized some time ago by Stonehart and co-workers [26], who were among the first to pursue rigorous measurements of electrochemical kinetics with nanoparticle catalysts. Ross and Stonehart [27] use the effectiveness factor method of analysis to demonstrate the difficulty in obtaining kinetic-only rates with these types of catalysts. Briefly, the catalyst layer must be very thin, e.g., 1–10 μm, for the effectiveness factor to approach unity. As the name implies, the effectiveness factor is a measure of the "effectiveness" of mass transport within the catalyst layer, and an effectiveness factor of unity means that all active sites in the layer see the same concentration of reactant. An effectiveness factor approaching unity is absolutely essential for analyzing the "crystallite size effect" in electrocatalysis. As the crystallite size changes, so do the metallic surface area and the concentration of active sites within the catalyst layer. Using a fixed catalyst layer thickness can produce an effectiveness factor that decreases with decreasing particle size, resulting in an apparent "crystallite size effect," which is nothing more than an an "effectiveness factor effect." It should be noted that the effectiveness factor effect produces an apparent crystallite size effect in which the apparent catalytic activity *decreases* with decreasing particle size, which as we see later in this chapter is the most frequently reported type of crystallite size effect in electrocatalysis. In light of this, all studies of a crystallite size effect in which mass transport within the catalyst layer is not carefully considered are suspect.

A breakthrough in methodology occurred with the advent of solubilized Nafion and the ability to recast catalyst-loaded Nafion films close to the thickness required, e.g., 1–10 μm [28]. A further refinement by Schmidt et al. [29], wherein 10–20 μg of supported catalyst is "glued" to a glassy carbon rotating-disk electrode by a 1–2 μm layer of Nafion, is now the most reliable method for kinetic measurements with nanoparticle catalysts. The total metal area in electrodes of this type is about 1 cm^2, or about 5 times the geometric area. Still, there are finite mass-transfer resistances even at infinite rotation rates, and these appear to come onto play at current densities above about 0.5 mA/cm^2 (geo) [30]. There is no simple way to correct for this resistance in this region of the polarization curve.

9.3 ELECTROCATALYSIS OF THE ORR ON PT AND PT-BIMETALLIC SURFACES

9.3.1 Structure Sensitivity of the ORR and Implications for a Crystallite Size Effect

There is no simple ideal structure that will model all the aspects of nanoparticle catalysts, particularly in the exact configuration they are used in electrolytic cells. Well-characterized single-crystal surfaces are reasonable models for some types of nanoparticles, such as highly facetted, crystalline metallic nanoparticles. This is illustrated in Figure 1. The equilibrium shape of an fcc metal, e.g., Pt, is predicted from fundamental thermodynamic considerations [31] to be a cubo-octahedral structure (Figure 1a), consisting of (111) and (100) facets bounded by edge atom rows that are like the topmost rows in the (110) surface. The surface atoms having the lowest coordination number are the corner atoms at the vertices of the crystal surface. HRTEM analysis of Pt nanoparticles close to the edge of carbon support clearly indicate this structure, as shown by the example in Figure 1c. Simple geometric considerations indicate that the relative concentration of surface atoms in edge and corner positions changes dramatically as the crystallite size decreases below 10 nm, i.e., in the 1–10-nm nanoparticle regime (Figure 1b). If these low-coordination sites are especially active (or inactive), then one should see a dramatic change in catalytic activity (per unit surface atom or turnover number) in the nanoparticle regime. This would produce what is now widely known as a crystallite size effect.

In his pioneering work, Somorjai [32] proposes that one could model the catalytic properties of the low-coordination edge and corner atoms using Pt single crystals with a step-terrace structure. However, for electrocatalysis, there has been and continues to be a concern about the stability of step-terrace surface structures in the electrochemical environment. Even if single-crystal surfaces cannot be used as specific one-to-one models of the ideal nanoparticles, studies employing well-controlled surface geometries can identify the structure sensitivity of the reaction on that particular catalyst. *A structure-sensitive reaction might then be expected to show a crystallite size effect with nanoparticle catalysts.*

The effect of Pt crystallite size on the kinetics of the oxygen reduction reaction is a long-standing problem in electrocatalysis. An excellent review of experimental work with carbon-supported Pt was presented by Kinoshita [33] for work prior to about 1990. Kinoshita concluded that the change in fraction of surface atoms on the

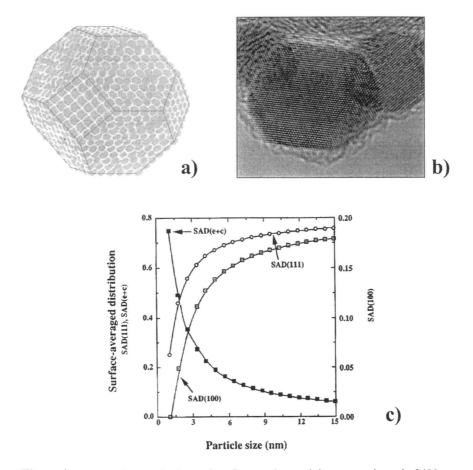

Figure 1 (a) A cubo-octahedron of an fcc metal containing approximately 2400 atoms with an equivalent sphere diameter of about 3 nm, (b) HRTEM image of Pt nanoparticles sitting on the surface of a carbon black primary particle, (c) distribution of surface atoms in (111) and (100) facets, and in edge (e) and corner (c) sites of an fcc cubo-octahedron as a function of equivalent sphere diameter. (From Ref. 33.)

(100) facets of Pt nanoparticles, which were assumed to be cubo-octahedral, could be correlated to the specific activity, the activity per unit area of Pt. There have not been a great number of studies with supported Pt catalysts since 1990. Fundamental studies have focused on finding the correlating structure sensitivity implied by this conclusion, i.e., that the (100) surface is much more active than either the (111) surface or any stepped-terrace structure. Early work (before 1990) on the kinetics of the ORR with Pt single crystals had been suspect due to concerns about the stability of the low-index surfaces in the ORR potential region, especially the Pt(111) surface [34]. In the last few years, however, considerable advances have been made in the establishment of the kinetics of the ORR on well-characterized Pt single-crystal surfaces, largely as a result of improvements in experimental techniques. With in-situ surface x-ray diffraction, it has been possible to determine unambiguously the

stability of specific Pt surface structures in electrolyte *under* ORR reaction conditions
[35]. In addition, a new adaptation of the classical rotating ring-disk electrode
(RRDE) technique has enabled the kinetics of the ORR on Pt(*hkl*) surfaces to be
examined under well-defined mass-transport conditions [36]. As summarized below,
these new techniques have produced new insight into the structure sensitivity of the
ORR on Pt(*hkl*) surfaces and further fundamental understanding of possible
crystallite size effects.

Figures 2–4   show a comparison of the ring-disk electrode for the oxygen
reduction kinetic data along with the base voltammetry in oxygen-free solutions for
each Pt(*hkl*) surface. Clearly, the kinetics of the ORR on Pt(*hkl*) surfaces vary with
crystal face in a different manner depending on the solution. In perchloric acid
solution, Figure 2, the variation in activity at 0.8–0.9 V is relatively small between the
three low-index faces, with the activity increasing in the order $(100) < (110) \approx (111)$.
A similar structural sensitivity is observed in KOH, Figure 3, with the activity

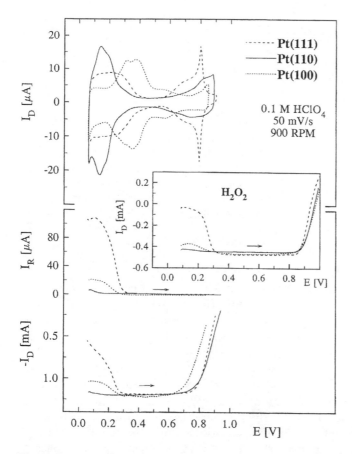

Figure 2   Top: Cyclic voltammetry of Pt(*hkl*) in oxygen-free 0.1-M HClO$_4$ electrolyte in the
RRDE assembly. Bottom: Disk (I_D) and ring (I_R) currents during oxygen reduction on Pt(*hkl*)
(ring potential = 1.15 V). Insert: Reduction of $1.2 \cdot 10^{-3}$ M H$_2$O$_2$ on Pt(*hkl*) mounted in the
RRDE assembly. Sweep rate: 50 mV/s. Rotation rate: 900 rpm. *E* versus RHE.

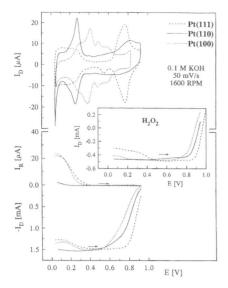

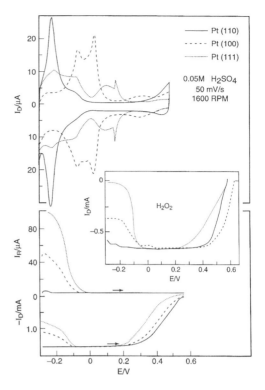

Figure 3 Top: Cyclic voltammetry of Pt(*hkl*) in oxygen-free 0.1-M KOH electrolyte in the RRDE assembly. Bottom: Disk (I_D) and ring (I_R) currents during oxygen reduction on Pt(*hkl*) (ring potential $= -1.15$ V). Insert: Reduction of $1.2 \cdot 10^{-3}$ M H_2O_2 on Pt(*hkl*) mounted in the RRDE assembly. Sweep rate: 50 mV/s. Rotation rate: 900 rpm. E versus RHE.

Figure 4 Top: Cyclic voltammetry of Pt(*hkl*) in oxygen-free 0.05-M H_2SO_4 electrolyte in the RRDE assembly. Bottom: Disk (I_D) and ring (I_R) currents during oxygen reduction on Pt(*hkl*) (ring potential $= 1.15$ V). Insert: Reduction of $1.2 \cdot 10^{-3}$ M H_2O_2 on Pt(*hkl*) mounted in the RRDE assembly. Sweep rate: 50 mV/s. Rotation rate: 900 rpm. Note: E versus SCE (saturated calomel electrode).

increasing in the order $(100) < (110) < (111)$, but with larger differences. In sulfuric acid solution, Figure 4, the variations in activity with crystal face are much larger, with the difference between the most active and the least active surface being about two orders of magnitude, and increase in the opposite order, i.e. $(111) << (100) < (110)$. To understand the structure sensitivity more clearly, it is useful to make comparison of the kinetics of the ORR on the same platinum low-index single-crystal plane but in different solutions, as summarized in Figure 5. The Pt(111) surface shows the largest variation in activity among the three electrodes, with the activity in sulfuric acid being nearly three orders of magnitude lower than in either of the other two solutions. This is consistent with the relatively strong adsorption of bisulfate anions on this surface [37,38]. The (100) surface has the least variation (factor of 2 to 3) in activity with electrolyte, while (110) is similar to (100) with somewhat larger differences with $HClO_4$ and H_2SO_4 solutions. A summary of the kinetic parameters for the ORR on Pt(*hkl*) in aqueous solutions is given in

Table 1. Results for other anions are not discussed here, but the interested reader is referred to detailed studies for the effects of Cl_{ad} [39] and Br_{ad} [40] on the ORR kinetics.

The Pt(hkl)/KOH system is the most direct probe of the effects of OH_{ad} on the rate of the electrode reaction, since no other anions are co-adsorbed with hydroxyl species. Figure 3 shows that in the potential range where O_2 reduction is under combined kinetic-diffusion control ($E > 0.75$ V), the order of activity of Pt(hkl) in 0.1-M KOH increases in the sequence (100) < (110) < (111). As shown in the insert in Figure 3, the same order is observed for the reduction of HO_2^-. The shift in half-wave potential between the least active and the most active surfaces was about 110 mV, which is more than an order of magnitude difference in the kinetic rate. These differences can be attributed to the structure sensitivity of hydroxyl anion adsorption on Pt(hkl) surfaces and its inhibiting (site-blocking) effect on both the oxygen and HO_2^- kinetics. While we do not contend that the two effects can be distinguished under all conditions on all three surfaces, the weight of experimental evidence favors the conclusion that the reversible and irreversible forms of OH_{ad} affect the ORR in different ways: the reversible form has little or no measurable effect on the reaction mechanism, while the irreversible form ("oxide") affects both the reaction rate and the reaction pathway. In perchloric acid, the variation in structural sensitivity is also

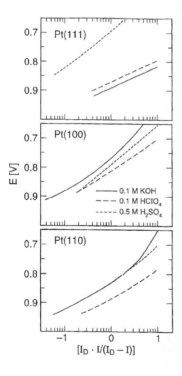

Figure 5   Tafel plots for Pt(hkl) electrodes corrected for diffusional resistance. Tafel plots are derived from the polarization curves shown in Figures 2–5.

Table 1 Some Kinetic Parameters for the ORR of Pt(hkl) in Aqueous Solutions at 298°K

(hkl)	0.1 M HClO$_4$				0.05 M H$_2$SO$_4$				0.1 M KOH		
	n $0.3 > E/V > 0.3$	m	$\partial E/\partial \log i$ mV/dec	$k \times 10^2$ cm s^{-1}	n $0.3 > E/V > 0.3$	$E/\partial \log i$ mV/dec	$k \times 10^2$ cm s^{-1}	$\Delta H^{0\#}$ kJ mol^{-1}	n $0.3 > E/V > 0.3$	$\partial E/\partial \log i$ mV/dec	$k \times 10^2$ cm s^{-1}
111 (1 × 1)	4 to 2	1	77	2.45	4 to 2	120	0.10	42	4 to 3	75	3.8
100 (1 × 1)	4 to 2.5	1	lcd 100 hcd 120	0.85	4 to 2.5	lcd 65 hcd 120	2.0	42	4 to 3.2	lcd 86 hcd 168	0.64
110 (1 × 2)	4	1	lcd 110 hcd 82	2.5	4	lcd 80 hcd 120	2.5	42	4	lcd 89 hcd 269	0.76

lcd low current density; hcd high current density; n, number of electrons; $\partial E/\partial \log i$, Tafel slope: lcd and hcd stand for high current density and low current density; m, reaction order; k, rate constant obtained from Levich–Koutecky plots at $E = 0.8$ V; and $\Delta H^{0\#}$, apparent activation energy.

relatively small between the three low-index faces, with activity increasing in the order $(100) < (111) < (110)$; see Figure 2. The structure sensitivity of the ORR in the potential region of OH adsorption in $HClO_4$ has the same origin as that in KOH, namely the structure-sensitive adsorption of the hydroxyl species in the supporting solution.

A close inspection of Figure 4 reveals that the order of activity of Pt(*hkl*) for the ORR in H_2SO_4 increases in the sequence $Pt(111) < Pt(100) < Pt(110)$. The simplest and most logical explanation for the higher overpotential for O_2 reduction on the Pt(111) surface is that its activity is reduced by the relatively strong adsorption of tetrahedrally bonded (bi)sulfate anions on the (111) sites as observed in FTIR [37] and radiotracer experiments [38]. These differences of (bi)sulfate adsorption in terms of coverage and bonding between Pt(111) and the other two low-index faces are also apparent in the significant positive shift in the potential for the adsorption of oxygen-containing species. In addition, they are consistent with the significantly lower activity of Pt(111) for oxygen reduction compared to either Pt(100) or Pt(110) (Figure 4). The most active surface for oxygen reduction, however, is the Pt(110) plane, as was also found for O_2 reduction in 0.1-M $HClO_4$. The difference in activity between Pt(110) and Pt(100) in sulfuric acid electrolyte may derive from either subtle differences in (bi)sulfate adsorption or other structure-sensitive surface processes, such as O–O bond-breaking and/or O–H bond-making. The fact that the activity of all three low-index platinum planes is significantly higher in $HClO_4$ does, however, suggest that the major differences in H_2SO_4 electrolyte stem from the structure-sensitive adsorption of (bi)sulfate anions. It should be noted that although bisulfate adsorption onto Pt(*hkl*) surfaces inhibits the reduction of molecular O_2, probably by blocking the initial adsorption of O_2, it does *not* affect the pathway of the reaction, since no H_2O_2 is detected on the ring electrode for any of the surfaces in the kinetically controlled potential region.

A kinetic analysis is not complete without determination of the temperature effects and activation energies. Figure 6 summarizes some of the polarization curves for the ORR recorded at 333 K and 298 K; for details, see [41]. Clearly, results obtained at 333 K are qualitatively similar to the curves recorded at room temperature, and the order of activity remains the same as at room temperature, i.e., $Pt(111) < Pt(100) < Pt(110)$. As expected, currents for the ORR are higher at elevated temperatures in both the mixed diffusion-kinetic potential region and the hydrogen adsorption potential region. These higher currents reflect the temperature dependence of the chemical rate constant, which is approximately proportional to $e^{-\Delta H^{0\#}/RT}$, where $\Delta H^{0\#}$ is the apparent enthalpy of activation at the reversible potential (hereafter simply termed the activation energy), and T is the absolute temperature. All kinetic analysis, including extraction of the activation energies on Pt(*hkl*), is given below and summarized in Table 1. It is also important to note that the analysis of the polarization curves recorded for the ORR at elevated temperatures revealed that the Tafel slopes were directly proportional to the temperature, with the transfer coefficient, α, for the oxygen reaction being *independent of temperature* (see insert b in Figure 6).

The well-defined linear Tafel regions for the ORR on Pt(*hkl*) can be used to extrapolate the current to the theoretical equilibrium potential to obtain the exchange current densities for the ORR on "oxide"-free Pt. At the same temperature, the exchange current density increases in the order

$i_{0(111)} < i_{0(100)} < i_{0(110)}$, which gives the order of absolute kinetic activities of three surfaces. The activation energies for the ORR on each Pt(hkl) face were obtained from the exchange current density at different temperatures using the Arrhenius plots

$$\frac{d \log i_0}{d(1/T)} = -\frac{\Delta H^{0\#}}{2.3 \, R} \tag{1}$$

Insert c in Figure 6 shows that the same slope, $\approx 42 \, kJ/mol$, is obtained for all three single-crystal surfaces. Thus, while the catalytic activity varies strongly with the crystal face, essentially the same activation energies are obtained, implying that the structure sensitivity arises primarily from geometric factors in the pre-exponential term. As mentioned above, the extrapolation of the Tafel slopes in KOH was more uncertain than in H_2SO_4, i.e., depending on the fitting method the extrapolated i_0 may vary by about an order of magnitude. Consequently, the energy of activation for the ORR in alkaline solution falls in the range of $\approx 45\text{--}55 \, kJ/mol$ for all three surfaces; that is, it is possible to have a variation in activation energy with crystal face within this range of values.

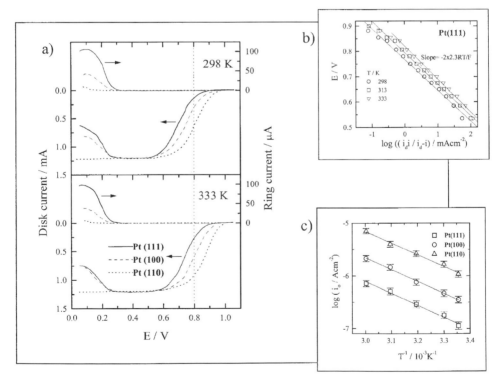

Figure 6 (a) Disk (I_D) and ring (I_R) currents during oxygen reduction on Pt(hkl) in 0.05-M H$_2$SO$_4$ at 298 K and 333 K. Rotation rate: 900 rpm. Sweep rate: 20 mV/s. (b) Tafel plots for the ORR on Pt(111) in 0.05-M H$_2$SO$_4$ at 298 K, and 313 K, and 333 K. (c) Arrhenius plots of the exchange current densities at the equilibrium potential for oxygen reduction on Pt(hkl) electrodes. E versus RHE.

These observations of structure sensitivity with single crystals do provide important new insight into particle-size effects reported for the ORR using supported Pt catalysts. Unfortunately, the observations are not as useful at present as they might be because the experimental conditions, e.g., temperature and electrolyte, used in most ORR kinetic measurements with supported catalysts are not those employed in the RRDE measurements with single crystals. Only one out of the many particle-size effect studies employed exactly the same conditions as those typically used for single crystals: the study by Peukart et al. [42], who report results in dilute sulfuric acid at ambient temperature. Fortunately, this was a very carefully done study using well-characterized supported Pt catalysts with (average) crystallite sizes ranging from 11.7 nm to 0.78 nm. As shown by Kinoshita (see Figure 4 in [33]), the crystallite size dependence in the data of Peukart et al. is essentially perfectly fit by a model that assumes (1) that the Pt particles are perfect cubo-octohedra and (2) that only the (100) face is active for the ORR. Below about 6 nm, the fraction of the surface having (100) facets drops rapidly, and actually goes to zero below 1.8 nm; particles below 2 nm are formed almost entirely from (111) facets and their intersections (edges and corners). Our ORR data from single-crystal Pt in sulfuric acid *fully* support this model, since we have observed that the (100) face is more active by two orders of magnitude versus the (111) face (in the model calculations, a factor of 100 between the two faces is sufficient to fit the data). By inference then, the crystallite size effect for the ORR on supported Pt in sulfuric acid at ambient temperature would be due to the structure-sensitive adsorption of (bi)sulfate anions, this adsorption effectively eliminating the contribution of the (111) facets to the overall rate.

Kinoshita has also shown that ORR data for supported catalysts in hot, concentrated H_3PO_4 (180 °C, 97–98% acid) reported in three different studies were also fit by this model. Since the physical basis for the crystallite size effect in sulfuric acid is anion adsorption, it would be a considerable reach to suggest that the same physical basis applies to this size effect, i.e., structure-sensitive anion adsorption. There are, nonetheless, indications that this is the case. Anion adsorption in dilute phosphoric [43] has a very similar structure sensitivity as sulfate adsorption, i.e., strongest adsorption on the (111) face, and on poly-Pt anion adsorption and/or neutral molecule adsorption in dilute phosphoric has a strongly inhibiting effect on the kinetics of the ORR [43]. Sattler and Ross [16] report a similar crystallite size dependence of the ORR on supported Pt in dilute phosphoric acid at ambient temperature as that found in hot, concentrated acid with the same catalysts. But it is unclear whether similar adsorption chemistry would exist in the extreme conditions of hot, concentrated phosphoric acid.

With differences in activity between crystal faces of only a factor of five or less for oxygen reduction in perchloric acid, the particle models of Kinoshita indicate that there would be little or no crystallite size effect for the ORR in a nonadsorbing acid, such as trifluoromethane sulfonic acid or closely related derivatives [44]. In KOH, because the (111) face is the most active, an *increase* in activity with the smallest particles (e.g., 1 nm)—ideally (uniform size, perfect geometry) would be predicted by the Pt(*hkl*) data—would be by about a factor of five. With real catalysts, where all particles are neither uniformly sized nor perfectly facetted, the effect might be much less, perhaps only a factor of two or so. Unfortunately, we do not know of any studies of crystallite size effects for supported Pt catalysts in either KOH or fluorosulfonic acids. Such studies would be of fundamental importance in furthering the connection between ideal studies with single crystals and the properties of real catalysts.

9.3.2 Correlation of the ORR Kinetics on Pt–Ni and Pt–Co Bulk Alloy Surfaces and Pt–Ni and Pt–Co Nanoparticle Catalysts

Recently in our laboratory we made a careful comparison of the kinetics of the ORR on Pt–Ni and Pt–Co polycrystalline bulk alloy surfaces and the activity relative to polycrystalline Pt. UHV surface preparation and analysis were used to control the surface composition. Argon ion sputtering followed by thermal annealing were used to vary the surface composition. Low-energy ion scattering (LEIS) was used to measure the surface composition. Both binary alloy systems exhibit strong surface enrichment of Pt [45]. Some selected results are summarized in Figure 10. The Pt_3Co bulk alloys were the most active, with a Pt-enriched surface produced by thermal annealing the most active. Curiously, the Pt-enriched Pt_3Ni surface was the least active, having even lower activity than pure Pt. It should be noted that the Pt-enriched surface produced by thermal annealing for both Pt_3Co and Pt_3Ni is essentially pure-Pt outermost layers with Co(Ni)-enriched second layers (>50% Co or Ni) [45]. The enhanced activity of the Pt–Co alloy surfaces could be attributed to an electronic effect of the intermetallic bonding between the Pt first-layer atoms and the Co second-layer atoms. Interestingly, the same intermetallic bonding in the Pt–Ni case is anticatalytic for the ORR.

In a companion study, we have compared the kinetics of the ORR on Pt, Pt–Ni, Pt–Co nanoparticle catalysts using the physical electrochemical characterization techniques described earlier in this chapter. Examples of the physical characterization by HRTEM are shown in Figures 7–9. All the catalysts have particle-size distributions centered about the 2.5–5-nm range. The alloy nanoparticles have the same cubo-octahedral shapes as the Pt, and elemental analysis of single particles by micro-XAS indicated uniform "bulk" composition. So we did not expect to find a significantly different trend in activity between the nanoparticle catalysts and the bulk alloy catalysts unless there is an unexpected surface composition of the nanoparticles versus the bulk alloys. The results of the ORR activity measurements are summarized in Figure 11. The trend in ORR activity was identical to that for the "sputtered" bulk alloys, i.e., surface composition equal to bulk. Interestingly, the activity of a 50% Co nanoparticle catalyst appeared to exhibit the properties of the Pt-enriched annealed Pt_3Co, which has 50% Co in the *second* atomic layer. The significance of this correlation is not clear at present. It is also interesting to note that the absolute activities of the nanoparticle catalysts and the bulk metals are quite close. For example, calculating the surface areas of the nanoparticles from the particle-size distribution, the metallic area in each of the catalysts is about $1\,cm^2$ ($+/-\,0.3$). The kinetic currents for the bulk metals in Figure 10 are already given normalized by the (nominal) metallic area, so to a first approximation we can compare the current scales in the two figures against one another to compare the absolute activities. The bulk metals have current densities at 0.85 V from 5–15 mA/ cm^2, while the nanoparticle catalysts at 0.85 V are 10–20 mA/cm^2, quite good agreements for measurements of this kind. This comparison also indicates that there is no crystallite size effect for the ORR on these Pt-based catalysts in a nonadsorbing

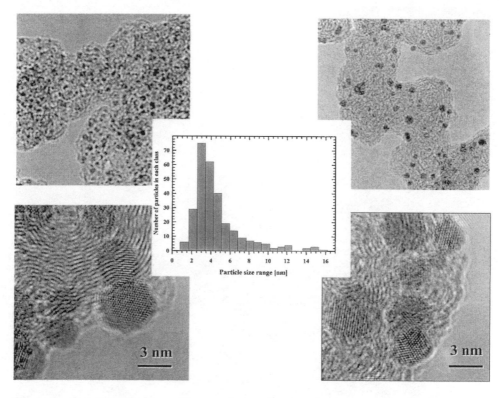

Figure 7 Top: TEM images at low resolution showing distribution of Pt nanoparticles on the carbon black support. Bottom: High-resolution TEM images showing generally facetted shapes of 3-nm characteristic dimension. Center: Particle-size distribution averaged over about 300 particles.

electrolyte (0.1-M $HClO_4$) for the particle sizes used here, e.g., >2 nm. This is consistent with the discussion presented in the previous section on the crystallite size effect. At least for these alloys, Pt–Ni and Pt–Co, for the ORR reaction Pt nanoparticle forms of the alloys in the <2-nm size range do *not* appear to exhibit any anomalous or unexpected properties relative to those of their bulk crystalline form.

9.4 ELECTROCATALYSIS OF THE OXIDATION OF H_2 AND H_2/CO MIXTURES ON PT AND PT-BIMETALLIC SURFACES

9.4.1 Structure Sensitivity of the HER and Implications for a Crystallite Size Effect

The hydrogen reaction on a platinum electrode is among the most widely studied electrochemical reactions [46]. Early kinetic studies of the HE/OR were carried out either on a polycrystalline platinum electrode [47] or on platinum single crystals that had poorly defined surface structures [48]. The first measurements of the kinetics of the HER on well-ordered, platinum single-crystal electrodes [49] reported that even

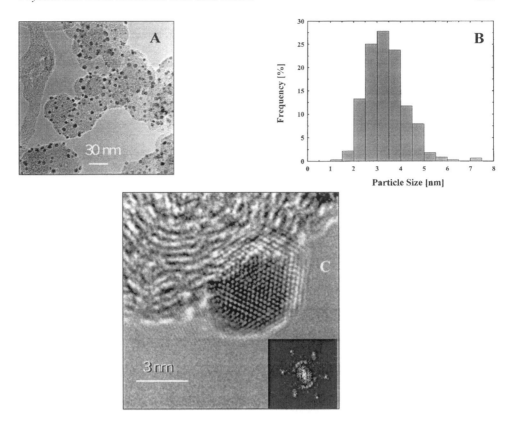

Figure 8 (A) TEM images at low resolution showing distribution of PtCo (3:1) bimetallic nanoparticles on the carbon black support. (B) Particle-size distribution averaged over about 300 particles. (C) HRTEM image of a single nanoparticle sitting on the surface of a carbon black primary particle and the corresponding microdiffraction pattern. Inner ring is diffraction from the carbon layer planes, spots are from (111) reflection from the single particle showing it is a single nanocrystal.

on well-ordered Pt(*hkl*) the kinetics of the HER at room temperature are insensitive to the crystallography of the surface. The significance of these results was, however, questionable since it is not clear that true kinetic rates were actually measured in any of these studies. As emphasized by Vetter [46a], and much earlier by Breiter [50], the hydrogen reaction on *active* Pt in acidic solutions is one of the fastest known electrochemical reactions (with an $i_0 > 10^{-3}\,A/cm^2$), and it is experimentally very difficult to measure anything but diffusion polarization. It becomes obvious, therefore, that in order to elucidate the true structural sensitivity of the HE/OR on Pt(*hkl*), the kinetics of this reaction should be reduced by more than one order of magnitude from that in acid solution at ambient temperature. Given that the rate of the HE/OR on a polycrystalline electrode is much slower in alkaline than in acid solution [51], and that the kinetics of any electrochemical reaction are significantly hindered at lower temperatures, the accurate measurement of kinetic rates of the

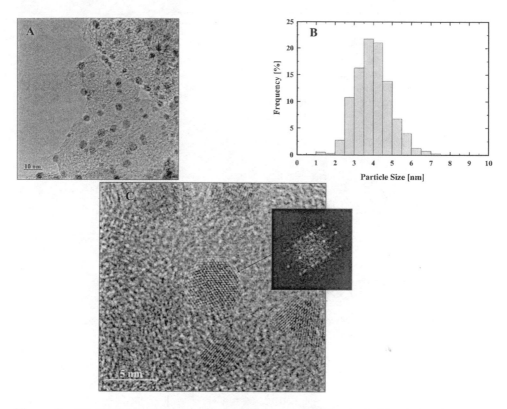

Figure 9 (A) TEM images at low resolution showing distribution of PtNi (3:1) bimetallic nanoparticles on the carbon black support. (B) Particle-size distribution averaged over about 300 particles. (C) HRTEM images of a nanoparticles and the microdiffraction pattern from the single nanoparticle in the center. Split spots indicate the particle is twinned.

HE/OR on Pt(*hkl*) appears more tractable in alkaline solution or at low temperature in acid solutions, as we illustrate below. Here we review the results and conclusions from work in our laboratory [52]. Kinetic results that in our view are not significantly different were reported by Barber and Conway [53a]. A review of the mechanism of the HE/OR on Pt(*hkl*) with a different perspective from that given here can be found in Conway [53b].

The HER in Alkaline Solution

Polarization curves over a wide range of overpotentials, both anodic and cathodic, for the hydrogen reaction at 1 atm on rotating Pt(*hkl*) disk electrodes at 298 K are shown in Figure 12. At low positive overpotentials, the order of activity for the HOR increased in the sequence (111)≈(100) <<(110). These differences in activity with crystal face can be attributed to the different state of adsorbed hydrogen and to different effects of these states on the mechanism of the hydrogen reaction [52,53]. The HOR on Pt(111) and (100) in alkaline solution is purely kinetically controlled

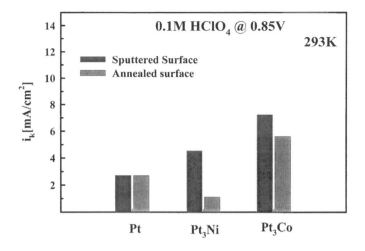

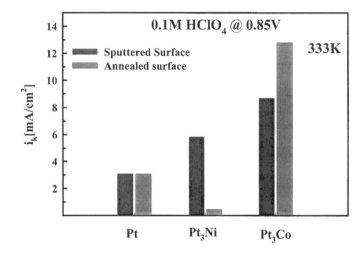

Figure 10 Bar graphs of the ORR activities of UHV-prepared surfaces of Pt, P₃Ni, and Pt₃Co bulk metal surfaces. Current densities are per unit geometric area.

over a relatively wide potential region, even at rotation rates as low as 400 rpm. It is under mixed control up to $\approx 0.4\,V$ of overpotential. On the (110) surface, the currents are under mixed control up to $\approx 0.2\,V$ of overpotential, providing a relatively wide window of overpotential where kinetic analysis can be performed; the kinetic parameters are summarized in Table 2. At high positive overpotentials, the activity of the HOR decreases in the sequence $(100) \ll (110) < (111)$. As discussed in previous sections for the ORR, the effects of surface crystallography on the HOR in this potential region can be attributed to the structural sensitivity of the adsorption

Table 2 Kinetic Parameters for HER and HOR on Pt(*hkl*) in 0.05 M H$_2$SO$_4$ at Different Temperatures*

Pt(*hkl*)	0.05 M H$_2$SO$_4$						0.1 M KOH			
	$\partial E/\partial \log i$ mV/dec	Exchange current density mA cm^{-2}			Activation energy^{-1} kJ mol^{-1}	Mechanism rds	$\partial E/\partial \log i$ mV/dec		i_0 mA cm^{-2}	Mechanism rds
		274 K	303 K	333 K			lcd	hcd		
Pt(110) 1 × 2	2.3RT/2F	0.65	0.98	1.35	9.5	Tafel-Volmer	55	140	0.70	Tafel-Volmer
Pt(100) 1 × 1	lcd: 2.3RT/3F hcd: 2×2.3RT/F	0.36	0.60	0.76	12	Heyrosky-Volmer	65	140	0.05	Tafel-Volmer
Pt(111) 1 × 1	≈2.3RT/F	0.21	0.45	0.93	18	Heyrosky-Tafel-Volmer	75	120	0.07	Tafel-Volmer

*Obtained from a linear polarization method.

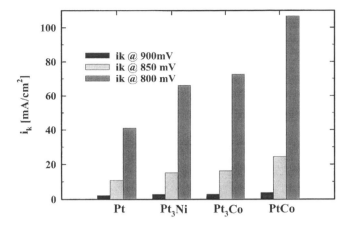

Figure 11 Bar graph of the ORR activities of Pt, PtNi (3:1), PtCo(3:1), and (1:1) nanoparticle catalysts. $0.14\,\mu g$ metal/cm^2.

of OH$_{ad}$ on Pt(hkl), with the irreversible form of OH$_{ad}$ having an inhibiting effect on the HOR.

The HER in Acidic Solution

In order to circumvent the problem of the fast kinetics of the HER in acid solutions at room temperature, the kinetics of these processes can be measured at the lowest temperatures possible with dilute acid solutions. Figure 13 provides typical examples by the series of polarization curves for the HER and the HOR at 1 atm H$_2$ in 0.05-M H$_2$SO$_4$ at 274 K. The exchange current densities (i_0) inferred from analysis of the "micropolarization" potential region (Figure 12b) plainly revealed that the HER in the low overpotential region is indeed a structure-sensitive process in acid solution, the activity increasing in the order $(111) < (100) < (110)$. While the structure sensitivity is very pronounced at low temperatures (274–293 K), in the temperature range $\approx$ 303–333 K the variation in the kinetics of the hydrogen reaction between crystal faces is less pronounced, especially between the (111) and (100) surfaces (for details, see [52]). This reconciles, at least in part, the absence of structure sensitivity in all previously published results for the HER on Pt(hkl) at room temperature [49]. For completeness, we should also mention that each crystal face has a unique, temperature-dependent Tafel slope for the HOR (Table 2), e.g., the transfer coefficient, α, for the hydrogen reaction being independent of temperature. Using the values of $i_{0(hkl)}$, assessed over the temperature range 274–333 K from either micropolarization region, the apparent activation energies for the hydrogen reaction on Pt(hkl) in 0.05-M H$_2$SO$_4$ can easily be determined from the slopes of the Arrhenius plots (Figure 12c). The apparent activation energy increases in order: 9.5 kJ/mol for Pt(110), 12 kJ/mol for Pt(100), and 18 kJ/mol for Pt(111). The fact that the most active surface also has the lowest activation energy is a "classical" result for electrocatalytic reactions and indicates that the energy of adsorption of the

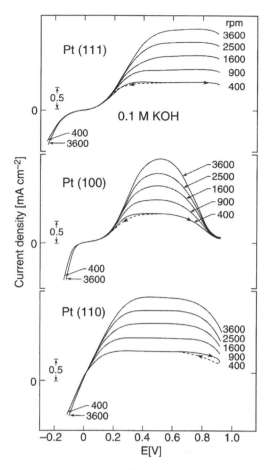

Figure 12 Polarization curves for the HER and the HOR on Pt(*hkl*) in 0.1-M KOH at sweep rate of 20 mV/sec. *E* versus RHE.

reactive intermediate plays the dominant role in the kinetics (versus the pre-exponential terms, which contain the coverage factor) [46c,d]. These differences in activation energy with crystal face are attributed to structure-sensitive heats of adsorption of the active intermediate whose physical state is unclear [52,53b].

In acid solution the HER is not strongly structure-sensitive, and thus no significant crystallite size effect is anticipated. We are not aware of any studies of the crystallite size effect for Pt nanoparticle catalysts. Given the high exchange-current density of the reaction even at 274 K, it would be difficult to obtain kinetic data free from mass-transfer effects with such an active catalyst. Even from a fundamental science perspective, there doers not appear to be a compelling motivation to undertake such a challenging experiment. In alkaline solution, however, the situation is quite different. The HER has a strong structure sensitivity, with (110) being about 10 times more active than either of the atomically "flatter" (100) or (111) faces. Following the classical geometric model for cubo-octahedral particles we discussed above for the ORR, in alkaline solution we would expect the HER to exhibit a

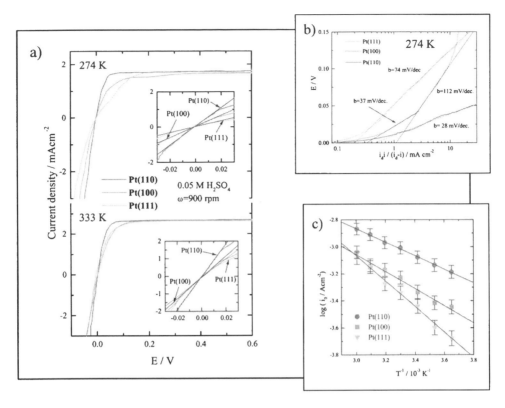

Figure 13   (a) Polarization curves for the HER and the HOR on Pt(*hkl*) in 0.05-M H$_2$SO$_4$ at 274 K and 333 K. Sweep rate: 20 mV/sec. Rotation rate: 900 rpm. (b) Tafel plots of mass-transfer corrected currents for the HOR on Pt(*hkl*) in 0.05-M H$_2$SO$_4$ at 274 K. (c) Arrhenius plots of the exchange current densities (i_0) for the HER on Pt(*hkl*) over the full range of temperatures (274–333 K). *E* versus RHE.

"positive" crystallite size effect, e.g., the activity per unit area Pt *increases* with decreasing particle size, if indeed it is the case that a (110) crystal face mimics the activity of edge and corner atoms in nanoparticles. Because the HER on Pt in alkaline solution is not an "infinitely fast" reaction, there is both a practical and scientific motivation to pursue measurements of the crystallite size effect in alkaline electrolyte.

The electro-oxidation of H$_2$/CO mixtures is a very complicated reaction with many variables, e.g., temperature, CO concentration, and pH. From a practical standpoint, the structure sensitivity is not so interesting, since at any realistic level of CO, e.g., >10 ppm CO, and temperature the pure-Pt surface is highly poisoned by adsorbed CO, and the electrode polarization is impractically large. It is, however, still of fundamental importance to know the structure sensitivity of the reaction on pure Pt in order to understand the properties of Pt-based alloy catalysts that are not so highly poisoned by the CO, i.e., so-called CO-tolerant catalysts. For our purposes here, we discuss only one characteristic measure of the structure sensitivity, shown in Figure 14. For a wider range of results, we refer the interested reader to [54] and references therein. Figure 14 shows the current for H$_2$ oxidation at 50-mV

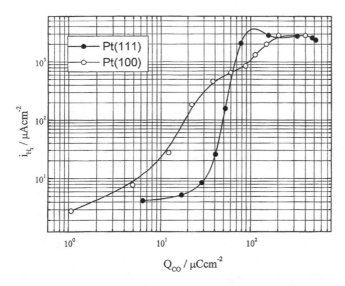

Figure 14 Correlation between the rate of the HOR (at 0.05 V versus RHE) and the stripping charge for the oxidative removal of CO_{ad} preadsorbed (at 0.05 V versus RHE) on Pt(100) and (111) surfaces in 0.5-M sulfuric acid.

polarization (298 K) versus the amount of CO removed oxidatively for Pt(111) and (100). In these experiments no CO is present in solution, the CO is preadsorbed from CO-saturated solution by holding the potential at 50 mV, the CO is purged from solution with argon, the coverage by CO is then adjusted by sweeping the potential positively, the oxidation charge is recorded, and the potential is returned to 50 mV. The solution is then saturated with H_2 and the kinetics of H_2 on the CO-poisoned surface recorded. For a given amount of CO removed, the H_2 current is substantially higher on (100) than on (111). For example, after removing $30 \, \mu C/cm^2$ of CO_{ad}, the rate is about 50 times higher on the (100) surface. Thus, the oxidation of H_2 on a CO-poisoned Pt surface is a strongly structure-sensitive reactive, and the oxidation of H_2/CO gas mixtures on Pt should be considered, in general, a structure-sensitive reaction. In our opinion, the reaction is too complicated to make any predictions of what the crystallite size effect might be, other than that one is expected.

9.4.2 Correlation Between the Kinetics on Pt–Ru and Pt–Mo Bulk Alloy Surfaces and Corresponding Nanoparticle Catalysts

The electro-oxidation of H_2 and H_2/CO mixtures has been studied extensively by our group using polycrystalline bulk alloys with surface compositions well-characterized using UHV surface analytical techniques [55]. These studies have produced the following general conclusions. The pure-polycrystalline Ru surface has a much lower activity for H_2 oxidation than Pt, but a much higher activity for CO oxidation than Pt. In both reactions, the differences in activity stem from the same property, a stronger interaction of Ru with water, leading to OH_{ad} formation at very low potential, i.e., close to the H_2 equilibrium potential. Because OH_{ad} is an intermediate in the rate-determining step for CO oxidation,

$$CO_{ad} + OH_{ad} = CO_2 + H^+ + 1e^-$$

this stronger interaction leads to enhanced catalysis for CO oxidation. On the other hand, OH_{ad} is not an intermediate in H_2 oxidation and is thus an inhibiting species for this reaction. With respect to the oxidation of pure CO, the addition of Pt to the Ru surface *does not produce any enhancement* in the reaction rate. However, for H_2/CO mixtures Pt–Ru alloy surface are much more active than either pure-Pt or pure-Ru surfaces, with the optimum surface composition being about 50% (for mixtures containing less than 0.1% CO). These conclusions are largely captured in the results of Figure 15, which is taken from [55b]. The mechanism for this enhancement is easily rationalized from the properties of the pure elements: Pt is needed to oxidize the H_2, Ru is needed to oxidize the CO at a rate that is higher than the rate of adsorption of CO, thus lowering the steady-state CO_{ad} coverage on the Pt sites. The optimum composition of 50% can also be rationalized by simple intuition: Pt–Ru pair sites, which are maximized at 50%, should have the maximum rate of CO_{ad} removal from the Pt sites.

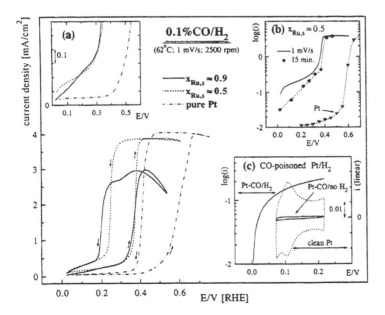

Figure 15 Potentiodynamic (1 mV/s) oxidation current densities for 0.1% CO/H_2 on sputter-cleaned Pt and Pt-Ru RDEs at 2500 rpm in 0.5-M sulfuric acid at 62 °C. Prior to electrochemical measurements, the electrode potential was held at 0.05 V at 2500 rpm for about 300 s. (a) Magnification of the low-current density region for the positive going sweeps. (b) Comparison of the potentiodynamic and potentiostatic (1000-s) oxidation current densities. (c) Potentiodynamic (20 mV/s) oxidation of pure H_2 on CO-poisoned Pt (Pt–CO/H_2), CO was adsorbed at ≈ 0.05 V, and then the electrode was cycled between ≈ 0.05 and 0.22 V in CO-free solution (Pt–CO/no H_2); the voltammetry of the unpoisoned Pt surface at the same conditions is added for reference.

Like the Pt–Co and Pt–Ni, the Pt–Ru binary system exhibits very strong surface segregation of Pt, but for fundamentally different reasons [45]. The thermodynamic driving forces for surface enrichment in Pt come from different components of the total energy. In all three systems, the heat of mixing is small, and actually even endothermic in the case of Pt–Ru. In Pt–Co(Ni), the heats of sublimation of the elements are not very different, so that the strain energies become determinant, i.e., the surface is enriched in the larger atom [56], in this case Pt. The surface-strain energy is minimized primarily by just exchange of atoms in the first two layers. In the Pt–Ru case, the heats of sublimation are very different, and the driving force is the classical one of enrichment in the element with the lower heat of sublimation (usually also the lower melting point). The endothermic heat of mixing actually drives the segregation of Pt even further, and the miscibility gap in the Ru-rich end of the bulk phase diagram is a further complication. The consequence is that it is very difficult, and possibly even impossible thermodynamically, to produce an *equilibrium* surface on a Pt–Ru bulk alloy that is 50% Ru [57]. And, in fact, we could only produce a 50% Ru surface by argon ion etching the thermally annealed surface of a 50% Ru alloy.

Fortunately for fuel-cell technology, Pt–Ru bimetallic nanoparticle catalysts can be prepared with nonequilibrium surfaces, or the thermodynamics of surface enrichment in Pt–Ru nanoparticles is fundamentally different. At the present time, it is not clear which of these is the case. It is nonetheless certain that Pt–Ru nanoparticle catalysts can be prepared whose catalytic properties for the oxidation of CO and H_2/CO mixtures mimic those of the sputtered surfaces Pt–Ru bulk alloys, i.e., nanoparticles having the same surface composition as the overall stoichiometry. This result is beautifully illustrated in Figure 16 from the study by Schmidt et al. [58]. Gasteiger et al. [55c] show that the the anodic stripping voltammetry of CO_{ad} on Pt–Ru alloy surfaces is unique to each surface composition. The stripping voltammetry for a Pt–Ru nanocluster with a 1:1 stoichiometry is compared to that of the 50% bulk alloy in Figure 16. The agreement is clearly excellent. The oxidation of a series of H_2/CO mixtures on the same 1:1 Pt–Ru nanoparticle catalyst is shown in Figure 17. Although the same H_2/CO mixture is not shown in Figure 15 for the bulk alloys, clearly the behavior of the 1:1 Pt–Ru nanoparticle catalyst is much closer to that of the 50% alloy than to a Pt-rich surface (>80%), which would be expected from the segregation thermodynamics. As with the ORR on Pt–Co and Pt–Ni alloys, the Pt–Ru nanoparticle catalysts, as prepared in the studies cited here, appear to exhibit the electrocatalytic properties of particles whose surface compositions are close to that of the overall stoichiometry of the metals present.

A different conclusion is reached, however, in the case of Pt–Mo catalyst. Bulk alloys of Pt–Mo were found to be even better electrocatalysts for the oxidation of H_2/CO mixtures than the 50% Pt–Ru alloy surface [59]. The optimum surface composition is on the Pt-rich side, close to 75% Mo. The Pt–Mo system exhibits a much smaller equilibrium segregation of Pt [59] to the surface than any of the other alloys discussed here. This alloy has a larger heat of mixing that counterbalances the lower heat of sublimation of Pt. The atomic sizes are also close, which reduces another of the segregation driving forces [56]. Interestingly, posttest analysis of immersed Pt–Mo bulk alloy surfaces by low-energy ion scattering (LEIS) did not indicate any oxidation-induced segregation of Mo to the surface [59]. More recently, we made a careful comparison [60] of the kinetics of oxidation of H_2/CO mixtures on

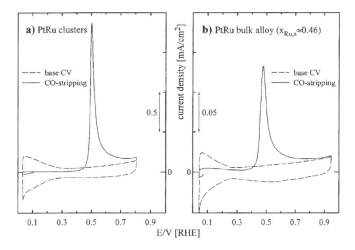

Figure 16 Comparison of the CO-stripping voltammetry on (a) glassy carbon supported PtRu clusters ($\approx 30\,\mu g/\,cm^2$) and (b) a UHV-prepared PtRu bulk alloy (polycrystalline) with a surface concentration of 46% (evaluated via low-energy ion scattering [55c]); solid curve CO-stripping, dashed-curve base voltammogram. CO adsorption from CO saturated solution at 0.1 V for 60 s; 20 mV/s, 25 °C, 0.5-M sulfuric acid. Note that the current density scale in (b) is magnified by a factor of 10 compared to (a). (From Ref. 58.)

the bulk alloys versus a series of Pt–Mo nanoparticle catalysts with stoichiometries of 1:1 (Pt:Mo) to 5:1 using the methods of characterization described earlier in this chapter. The preparation procedure used was essentially the same as used for the other bimetallics, although the final temperature of hydrogen reduction was higher. A key result is shown in Figure 18, which shows that the nanoparticles behave as if the surfaces were *enriched in Mo* relative to the overall stoichiometry.

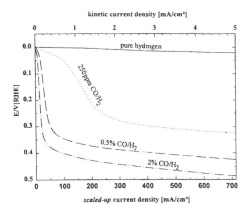

Figure 17 Potentiodynamic (0.1 V/s) oxidation current densities for several H_2/CO mixtures on a PtRu-colloid/Vulcan electrode (7 µg/cm²) at 60°C and 2500 rpm. The upper abscissa gives the kinetic current density while the lower uses a scale-up factor of 143 to simulate the performance of a fuel-cell electrode. The electrooxidation of pure H_2 at the same electrode is shown for reference. (From Ref. 58.)

We have been pursuing an explanation for this apparently anomalous result. The first possibility we sought to rule out was that the distribution of Mo in the nanoparticles was not uniform. In the most extreme case, if a significant fraction of particles had no Mo, then of course the particles that did have Mo would have concentrations greater than the overall stoichiometry. We were able to determine by use of analytical electron microscopy that this was, in fact, not the case. A sample result is shown in Figure 19 for the Pt:Mo (3:1) catalyst studied in [60], which shows both the secondary electron image and, from the same region, the energy-filtered Pt and Mo characteristic x-ray emission. It is clear that all particles in this region (approximately 30) contain both Pt and Mo. Point-to-point analysis revealed that all particles contained essentially the same amounts of Mo, so that the nonuniform distribution of Mo could not be the reason for the unexpected optimum concentration of Mo.

In work that is still ongoing and preliminary, there are indications that the explanation does lie in equilibrium segregation chemistry that is unlike that in bulk crystals and unique to nanoparticles. A hint of these phenomena can be seen, for example, in the Monte Carlo simulations pioneered by King and co-workers [61]. An example of the Monte Carlo simulation method applied to the Pt–Mo system is shown in Figure 20. We know that the Pt–Mo are highly facetted with the cubo-octahedral particle shapes seen for pure Pt and the other alloy systems discussed here. Figure 19 shows the expected atomic distribution of Pt and Mo in a particle containing approximately 2400 atoms with approximately 800 surface atoms. The bulk stoichiometry is Pt_3Mo. The equivalent sphere diameter is about 3 nm. The physical dimensions of the particle in the simulation are very typical of that in the real catalyst. The Monte Carlo simulation following the method of King [61] indicates that the total energy of the particle is minimized nearly entirely by the exchange of Mo atoms on the (100) facets with Pt atoms on the (111) facets rather

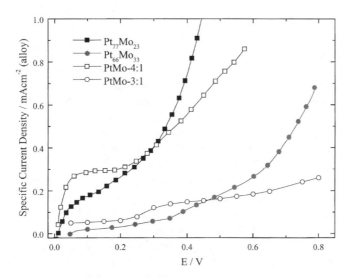

Figure 18 Polarization curves for the electrooxidation of H_2 containing 0.1% CO on the PtMo—4:1 and 3:1 catalyst RTLEs normalized by the alloy-specific surface area in the layer and compared to the curves for the bulk alloy RDEs. $T = 333$ K.

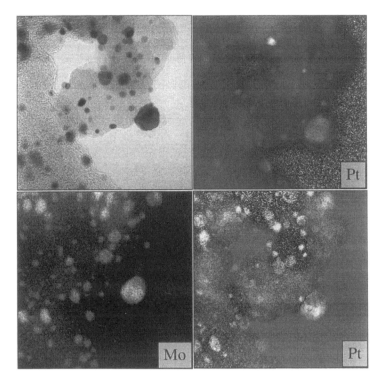

Figure 19 Top left: Secondary electron image from a region of a Pt–Mo nanoparticle catalyst. Bottom left: Energy-filtered image from Mo Kα x-ray emission from same region. Right: Energy-filtered image from Pt Lα x-ray emission from same region, top and bottom use reverse contrast for background. Note there is both Pt and Mo x-ray emission from all particles.

than between the first two atomic layers, as occurs in the bulk crystal. The result is that the (100) facets are pure Pt and the (111) facets are 33% Mo (!), but the total surface composition is about 80% Pt, the same (small) overall enrichment we observed for the bulk crystal [59]. Note that for the H_2/CO reaction, the (100) facets are totally poisoned by adsorbed and the reaction is catalyzed only by the (111) facet. Thus, judging from the reactivity alone, it appears that the nanoparticle surface is enriched in Mo, when in fact it is not. Minimization of the total energy by exchange of atoms between the (100) and (111) facets may be unique to the cubo-octahedral structure and would probably *not* occur in particles smaller than about 3 nm, as the fraction of surface atoms on (100) facets decreases rapidly below this size. In complete Monte Carlo simulations, one would want to vary the bimetallic particle geometry in this size regime to ensure that a true minimum energy structure is achieved for a given number of atoms of each metal. The simulation shown in Figure 20 appears to apply well to the size regime of interest here, the >3-nm particle, and provides at least a rationalization for the experimental results obtained with these nanoparticles.

a)   b)

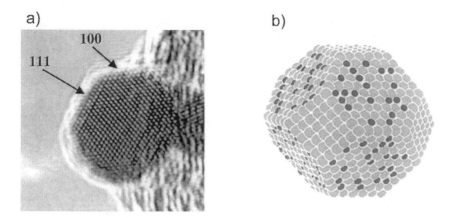

Figure 20   Left: High-resolution transmission electron micrograph image of a single PtMo (3:1) nanoparticle on the edge of a carbon black primary particle; (111) and (100) factes are clearly resolved. Right: Distribution Pt (light) and Mo (dark) atoms in an fcc cubo-octahedral particle containing 1806 Pt atoms and 600 Mo atoms from classical Monte Carlo simulation at 550 K.

9.5 FUTURE DIRECTIONS

Careful studies of Pt bimetallic nanoparticle catalysts using a toolkit of modern physical characterization techniques and well-characterized model systems have revealed only a few "nanoparticle anomalies," although the number of different ad-atoms studied in this manner is still relatively small. Generalizations to other bimetallic nanoparticle systems are undoubtedly risky, particularly to systems that do not have fcc structures. The anomalies are, however, of fundamental significance, and much is to be gained from understanding the underlying phenomena. We have discussed only two anomalous or unexpected results in this chapter: (1) Pt–Co, Ni, and Ru nanoparticles when prepared using certain (low-temperature) methods appear to have surface compositions very close to that of the overall stoichiometry of the metals, even though the bulk crystals show strong surface enrichment in Pt; (2) Pt–Mo nanoparticles when prepared using similar (low-temperature) methods appear to exhibit an anomalous surface enrichment in Mo versus surface enrichment in Pt observed with the bulk alloy. It may be that surface segregation thermodynamics can explain both anomalies in a consistent manner. It is clear that Monte Carlo simulations will play an important role in understanding the structure of bimetallic nanoparticles. It is also clear that we need new surface analytical tools to determine the surface composition of nanoparticles, not only the overall composition, but also the composition of individual facets. Here some type of scanning probe microscopy (SPM) appears attractive together with new synthetic methods to better control the structure of the particles on inert substrates so that an SPM can be used.

ACKNOWLEDGMENTS

The authors would like to acknowledge consistent support for our research on electrocatalysts and related surface science from the U.S. Department of Energy, both the Office of Science and the Office of Advanced Automotive Technologies under contract DE-AC03-76SF00098.

REFERENCES

1. J.H. Sinfelt, Bimetallic Catalysts: Discoveries, Concepts, and Applications, John Wiley and Sons, New York, 1983.
2. J.H. Sinfelt, J. Catal. 24 (1973) 250.
3. A.W. Dweydari, C.B.H. Mee, Phys. Status Solidi A 17 (1973) 247.
4. R.C. Baetzold, R. and J.F. Hamilton, Progr. Sol. State Chem. 15 (1983) 1; M.G. Mason, Phys. Rev. B 27 (1983) 748.
5. O.M. Poltorak, V.S. Boronin, Russ. J. Phys. Chem. 40 (1966) 1436.
6. M.A. Boudart, Adv. Catal. 20 (1969) 153.
7. H.R. Kunz, G.A. Gruver, J. Electrochem. Soc. 122 (1975) 1279.
8. F.J. Lucak, D.A. Landsman, Ordered Ternary Fuel Cell Catalysts Containing Platinum, Cobalt and Chromium, U.S. Patent 4,447,506, 1983.
9. J.H. Sinfelt, Y.L. Lam, J.A. Cusumano, A.E. Barnett, J. Catal. (1976) 227.
10. H.G. Petrow, R.J. Allen, U.S. Patent 3,992,331, 1974.
11. P. Stonehart, Ber. Bunsenges. Phys. Chem. 94 (1990) 913.
12. H. Boennemann, R. Brinkmann, P. Britz, U. Endruschat, R. Moertel, U. Paulus, G. Feldmayer, T. Schmidt, H. Gasteiger, R. Behm, J. New Mater. Electrochem. Sys. 3 (2000) 199.
13. A.K. Datye, D.J. Smith, Catal. Rev.-Sci. Eng. 34 (1992) 129.
14. M.J. Yacaman, M. Avalos-Borja, Catal. Rev.-Sci. Eng. 34 (1992) 55.
15. H. Poppa, Catal. Rev.-Sci. Eng. 35 (1993) 359.
16. M.L. Sattler, P.N. Ross, Ultramicroscopy 20 (1986) 21.
17. D.J. Smith, L.D. Marks, Ultramicroscopy 16 (1985) 101.
18. M.J. Yacaman, J.M. Dominguez, J. of Catalysis 64 (1980) 213.
19. J.C.H. Spence, Experimental High Resolution Electron Microscopy, Oxford University Press, Oxford (1980).
20. J. Cowley, Diffraction Physics, 2nd ed., North-Holland (1981).
21. R. Gronsky, In Treatise on Materials Science and Technology Series: Experimental Techniques, H. Herman (ed.), Academic Press, Vol. 19B (1983) 225.
22. M.A. O'Keefe, Resolution in high-resolution electron microscopy, Ultramicroscopy 47 (1992) 282–297.
23. V. Radmilovic, M.A. O'Keefe, Proc. 53rd MSA Meeting, SF Press, San Francisco, 1995, p. 590.
24. V. Radmilovic, M.A. O'Keefe, R. Kilaas, CD-ROM, EUREM'96, Dublin, 1996, file: M/M2/Radmilov.
25. J.W. Heckman, K.L. Klomparens, Scanning and Transmission Electron Microscopy: An Introduction, S.L. Flegler (ed.), W.H. Freeman and Co., New York, 1993.
26. W. Vogel, J. Lundquist, P.N. Ross, Jr., P. Stonehart, Electrochim. Acta. 20 (1974) 79.
27. P. Stonehart, P.N. Ross, Electrochim. Acta 21 (1976) 441.
28. S. Gojkovic, S. Zecevic, R. Savinell, J. Electrochem. Soc. 145 (1998) 3713.
29. U. Paulus, T. Schmidt, H. Gasteiger, R. Behm, J. Electroanal. Chem. 495 (2001) 134.
30. B. Grgur, N. Markovic, P.N. Ross, J. Electrochem. Soc. 146 (1999) 1613.
31. J.J. Burton, Catal. Rev. Sci. Eng. 9 (1974) 209; J.W. Cahn, Acta Met. 28 (1980) 1333.
32. G. Somorjai, Philos. Trans. R. Soc. London A318 (1986) 81.

33. K. Kinoshita, J. Electrochem. Soc. 137 (1990) 845.
34. F.T. Wagner, P.N. Ross, Surf. Sci. 160 (1985) 305.
35. N.M. Markovic, P.N. Ross, In Interfacial Electrochemistry—Theory, Experiments and Applications, A. Wieckowski (ed.), Marcel Dekker Inc., New York, 1999, pp. 821–841.
36. N.M. Markovic, H. Gasteiger, P.N. Ross, J. Phys. Chem. 99 (1995) 3411.
37. P. Faguy, N. Markovic, R. Adzic, E. Yeager, J. Electroanal. Chem. 289 (1990) 245; P. Faguy, N. Markovic, P.N. Ross, J. Electrochem. Soc. 140 (1993) 1638.
38. A. Kolics, A. Wieckowski, J. Phys. Chem. 105 (2001) 2588.
39. V. Stamenkovic, N. Markovic, P.N. Ross, J. Electroanal. Chem. 500 (2000) 44.
40. N.M. Markovic, H.A. Gasteiger, P.N. Ross, J. Electroanal. Chem. 467 (1999) 157.
41. B.N. Grgur, N.M. Markovic, P.N. Ross, Can. J. Chem. 75 (1997) 1465.
42. M. Peukart, T. Yoneda, M. Boudart, J. Electrochem. Soc. 13 (1986) 944.
43. F. El Kadiri, R. Faure, R. Durand, J. Electroanal. Chem. 301 (1991) 177.
44. H. Saffarian, P. Ross, F. Behr, G. Gard, J. Electrochem. Soc. 139 (1992) 2391. H. Saffarian, P. Ross, F. Behr, G. Gard, J. Electrochem. Soc. 137 (1990) 1345. K. Striebel, P. Andricacos, E. Cairns, P. Ross, F. McLarnon, J. Electrochem. Soc. 132 (1985) 2381. P. Ross, J. Electrochem. Soc. 130 (1983) 882.
45. P.N. Ross, In Electrocatalysis, J. Lipkowski, P.N. Ross (eds.), Wiley-VCH, New York, Chapter 2, and references therein.
46. (a) K.J. Vetter, Electrochemical Kinetics, S. Bruckenstein, B. Howard, Trans. Eds., Academic Press, New York, 1967, pp. 516–614. (b) B.E. Conway, J. O'M Bockris, J. Chem. Phys. 26 (1957) 532. (c) R. Parsons, Trans. Faraday Soc. 54 (1958) 1053. (d) H. Gerischer. Bull. Soc. Chim. Belg. 67 (1958) 506.
47. I.I. Physhnogreva, A.M. Skundin, By Vasiliev, V.S. Bagotski, Elektrokhimiya 6 (1970) 142, S. Schuldiner, M. Rosen, D. Flinn, J. Electrochem. Soc. 117 (1970) 1251.
48. F. Will, J. Electrochem. Soc. 112 (1965) 451.
49. K. Seto, A. Iannello, B. Love, J. Lipkowski, J. Electroanal. Chem. 226 (1987) 351, E. Protopopoff, P. Marcus, J. Chim. Phys. 88 (1991) 1423. H. Kita, S. Ye, Y. Gao, J. Electroanal. Chem. 334 (1992) 351.
50. M. Breiter, R. Clamroth, Z. Elektrochem. 58 (1954) 493–505.
51. V.S. Bagotzky, V. Osetrova, J. Electroanal. Chem. 43 (1973) 233–249.
52. (a) N.M. Markovic, B.N. Grgur, P.N. Ross, J. Phys. Chem. B 101 (1997) 5405. (b) N.M. Markovic, P.N. Ross, In Interfacial Electrochemistry: Theory, Experiment and Applications, A. Wieckowski (ed.), Marcel Dekker, New York and Basel, 1999, pp. 821–841.
53. (a) J.H. Barber, B.E. Conway, J. Electroanal. Chem. 461 (1999) 80. (b) B.E. Conway, In Interfacial Electrochemistry: Theory, Experiment and Applications, A. Wieckowski (ed.), Marcel Dekker, New York and Basel, 1999, pp. 131–150.
54. N.M. Markovic, C.A. Lucas, B.N. Grgur, P.N. Ross, J. Phys. Chem. B 103 (1999) 9616.
55. (a) H.A. Gasteiger, N.M. Markovic, P.N. Ross, J. Phys. Chem. 99 (1995) 8290. (b) J. Phys. Chem. 99 (1995) 16767. (c) H. Gasteiger, N. Markovic, P. Ross, E. Cairns, J. Phys. Chem. 98 (1994) 617.
56. G. Treglia, B. Legrand, Phys. Rev. B 35 (1987) 4338.
57. H.A. Gasteiger, P.N. Ross, E.J. Cairns, Surf. Sci. 293 (1993) 67.
58. T.J. Schmidt, M. Noeske, H.A. Gasteiger, R.J. Behm, P. Britz, W. Brijoux, H. Boennemann, Langmuir 13 (1997) 2591.
59. B.N. Grgur, N.M. Markovic, P.N. Ross, J. Phys. Chem. B 102 (1998) 2494.
60. B.N. Grgur, N.M. Markovic, P.N. Ross, J. Electrochem. Soc. 146 (1999) 1613.
61. T.S. King, In Surface Segregation Phenomena, P.A. Dowben, A. Miller (eds.), CRC Press, Boca Raton, FL, 1990, pp. 27–77.

10

Nanomaterials as Precursors for Electrocatalysts

HELMUT BÖNNEMANN and RYAN RICHARDS*

Max Planck Institut für Kohlenforschung, Mülheim an der Ruhr, Germany

CHAPTER CONTENTS

SUMMARY

This chapter elaborates on the methods and incentives for using nanomaterials as precursors to electrocatalysts. This "precursor" method facilitates tailoring of precursors with controlled structures and control of the interface between two metals. By use of this method homogeneous alloys, segregated alloys, layered bimetallics, and "decorated" paticles are all readily accessible. The incentive for the use of this concept is that we can prepreparc and thoroughly characterize the active components of electrocatalysts with the application of modern analytical techniques, including synchrotron radiation, electron microscopy, X-ray diffraction, and electrochemical examination of the surface.

---

* *Current affiliation:* International University of Bremen, Bremen, Germany

10.1 INTRODUCTION

Interest in the application of nanostructured catalysts stems from the unique electronic structure of the nanosized metal particles and their extremely large surface areas. Nanostructured metal colloids can be defined as isolable particles between 1 and 50 nm that are prevented from agglomerating by protecting shells. They can be prepared to be redispersed in both water ("hydrosols") and organic solvents ("organosols"). Here we hope to provide a synopsis of the wet chemical syntheses of these materials and their application as "precursors" of electrochemical catalysts.

A great deal of knowledge has been acquired about these materials through the efforts of several leading scientists [1,2]. Although highly dispersed mono- and bimetallic colloids have been used as precursors for a new type of catalyst that is applicable in both the homogeneous and heterogeneous phases, this contribution focuses only on those applications relevant to electrochemical catalysis and fuel cells [3,4].

An obvious advantage of the precursor concept over the conventional salt-impregnation method is that both the size and the composition of the colloidal metal precursors may be tailored independent of the support. The so-called precursor concept for manufacturing heterogeneous metal colloid catalysts was developed on this basis in the 1990s [2c,2d,2f,3c,5]. Further, the metal particle surface may be modified by lipophilic or hydrophilic protective shells and coated by intermediate layers, e.g., of oxide. The modification of the precursor by dopants such as SnO_2 and WO_3 is also possible. Preprepared nanometal colloids can easily be deposited on supports to give heterogeneous catalysts. As a result, so-called eggshell catalysts were obtained that contain the active metal particles as a thin layer (<250 nm) on the surface of the support. The catalyst is manufactured by dipping the supports into organic or aqueous media containing the dispersed precursor at ambient temperature to adsorb the preprepared particles. This has been demonstrated for supports such as charcoal, various oxidic support materials, and even low-surface materials, such as quartz, sapphire, and highly oriented pyrolitic graphite (HOPG). It is important to note that no subsequent calcination is required (see Figure 1).

A combination of X-ray and microscopy techniques including AFM, STM, and XPS [6] has been applied to reveal the interaction of platinum hydrosols with oxide (sapphire, quartz) and graphite single-crystal substrates (Figure 2). When dipped into aqueous Pt colloid solutions at 20 °C the metal core is immediately adsorbed onto the support surface. The carpetlike coat formed over the particles by the protecting shell and the support surface cannot be removed from the particle surface even by intense washing with solvent. Upon annealing at 280 °C and above in UHV, the organic protecting shell decomposes. By monitoring the thermal degradation by XPS and STM up to 800 °C it was shown that the Pt particles remain virtually unchanged up to about 800 °C, after which sintering processes are observed.

"Fuel-cell technology" allows the direct conversion of chemical energy into electricity [7]. The fuel cell is an electrochemical reactor where the catalyst systems are an important component. Among the wide-ranging applications of fuel cells are low-emission transport systems, stationary power stations, and combined heat and power sources. The classical studies were carried out in the early 1900s, and major innovations and improvements have been achieved over the last few years. The first "new electric cars" are expected to roll onto the market around the year 2005, but

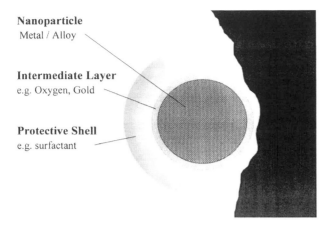

Nanoparticle
Metal / Alloy

Intermediate Layer
e.g. Oxygen, Gold

Protective Shell
e.g. surfactant

Figure 1   The precursor concept. (From Ref. 3c.)

further developments are still needed, notably in the catalyst sector. Hydrogen fuel-cell catalysts rely on pure Pt, whereas Pt–alloy electrocatalysts are applied for the conversion of reformer gas or methanol into electricity. The active component in the latter cases are small Pt-containing bi- or trimetallic particles of 1–3-nm size that scatter X-rays as nearly perfect "single crystals." These systems offer improved efficiency and tolerance against certain contaminations, especially CO in the anode feed [8]. It was clear from patents filed in the early 1970s that finely particulated colloidal platinum sols should be the ideal precursors for the manufacture of fuel-cell electrodes [8a]. The six main types of fuel cell are described in [9]. Since an average car fuel cell would need to generate about 70 KW, requiring about 19 g of Pt, the question has been raised of whether enough platinum reserves are in the ground to meet the future needs of fuel-cell vehicles [9b]. More than 95% of the platinum from a fuel-cell stack can be recovered, so that at most 5% of the platinum content of the average vehicle would require topping up in the recycling process. It has been estimated that the annual maintenance requirement for 1 billion automobiles would be less than 2 million troy ounces per year. This should be compared with the worldwide demand for catalytic converters in 1998, which required on the order of 1.4 million troy ounces [9c]. It is presently not completely clear what type of fuel will be the best to supply the cells in future vehicles. The three options are hydrogen, methanol, and synthetic fuel [9d]. In the long term (around the year 2030), engineers and clean-air experts in the petroleum industry favor hydrogen itself, pressurized in steel bottles at 700 bar. For the first fuel-cell cars that should enter the markets around 2005, automakers are banking on methanol because liquids are easier to distribute at conventional filling stations. The methanol can be either converted to hydrogen (and CO) in efficient on-board reformers or fed directly to methanol fuel cells. In present fuel-cell car prototypes the "tank to wheel efficiency" reaches about 30%, which is still economically insufficient. Before fuel-cell cars can begin to become a realistic alternative to the present combustion engines, their "tank to wheel efficiency" must be raised to 40–45%, the output of the membrane electrodes

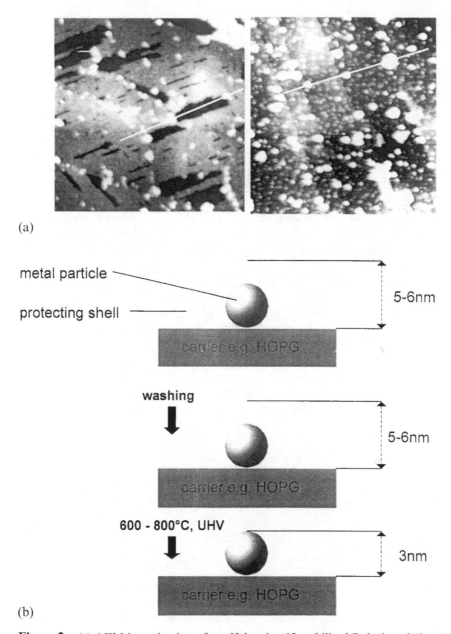

(a)

(b)

Figure 2 (a) AFM investigation of a sulfobetaine-12 stabilized Pt hydrosol (3 nm) absorbed on highly oriented pyrolytic graphite (HOPG) after dipping (left) and after additional washing (right). (b) Scheme of the Pt hydrosol adsorbation on HOPG derived from a combined STM and XPS study. (From Ref. 6b.)

should reach $700\,\mathrm{mV}$ at $1\,\mathrm{A/m^2}$, and the cost of the catalyst has to drop below $30\,\mathrm{USD/m^2}$.

10.1.1 Bimetallics

Bimetallic colloids are readily accessible by the controlled co-reduction of two different metal ions [2c,2d,3c,10m,11a–d,12,13]. The structural characterization and some catalytic aspects of bimetallic colloids have been recently reviewed [2g]. Nanostructured bimetallic colloids catalysts have opened the possibility for the study of the mutual influence that two different metals have on catalytic properties. Recent advances in the preparations of bimetallic colloids have provided catalysts of several compositions. By successive reduction of mixed-metal ions, bimetallic particles have been prepared that have composition gradients from the core to the shell [14b,15–18]. Truly layered particles consisting of, for example, a gold core plated by palladium or vice versa [10 m] have also been synthesized. The difference between the electronegativities of Pd and Au together with the combination of the partially filled d-band of Pd and the completely filled d-band of Au result in a novel electronic structure [19]. Bimetallic particles having a gradient metal distribution or a layered structure are most interesting for catalytic applications.

Bimetallic particles with layered structures have opened fascinating prospects for the design of new catalysts. Schmid et al. [10m] have applied the classical seed-growth method [20] to synthesize layered bimetallic Au/Pd and Pd/Au colloids in the size range of 20–56 nm. The sequential reduction of gold salts and palladium salts with sodium citrate allows the gold core to be coated with Pd. This layered bimetallic colloid is stabilized by trisulfonated triphenylphosphane and sodium sulfanilate. More than 90% metal can be isolated in the solid state and is redispersable in water in high concentrations.

Colloidal catalysts have been prepared in different particle sizes by the reduction of platinum tetrachloride with formic acid in the presence of different amounts of alkaloid.

The most active of these sol-gel entrapped Pd catalysts had a considerably higher activity than the commercially available Pd/Al_2O_3 samples. The preparation of a fully alloyed Pd–Au colloid of 3.0-nm particle size, by a modified sol-gel procedure using THF as the solvent, from the co-reduction of Pd and Au salts with tetraalkyl-ammonium-triethylhydroborate [2c], and its embedding in a silica have been described (21d). The integrity of the incorporated Pd–Au alloy particles remained virtually untouched. After the removal of the protecting surfactant, a mesoporous texture with a comparatively narrow pore distribution remained. According to the physical characterization by a combination of techniques, the SiO_2-embedded Pd–Au colloid preserves the size and the structural characteristics of the colloidal metal precursor. The material exhibits excellent catalytic properties in selective hydrogenation test reactions.

10.2 WET CHEMICAL PREPARATION METHODS

10.2.1 Wet Chemical Reduction

Nanostructured metal colloids have been obtained by both "top-down" and "bottom-up" methods. The so-called top-down method is generally accomplished by "tearing down" bulk samples and stabilizing the resulting smaller structures. An example of a top-down process would involve the mechanical grinding of bulk metals and subsequent stabilization of the resulting nanosized metal particles by the addition of colloidal protecting agents [22]. Another variation of this principle is the metal vapor technique [23], which has provided chemists with a very versatile route for the production of a wide range of nanostructured metal colloids on a preparative laboratory scale [24]. Metal vapor techniques have, however, been hindered by some difficulties. The installation of the metal vapor apparatus is demanding and it is difficult to adjust to obtain a narrow particle-size distribution. The bottom-up methods of wet chemical nanoparticle preparation basically rely on the chemical reduction of metal salts, electrochemical pathways, or the controlled decomposition of metastable organometallic compounds. A large variety of stabilizers, e.g., donor ligands, polymers, and surfactants, are used in order to control the growth of the primarily formed nanoclusters and to prevent them from agglomeration. The chemical reduction of transition metal salts in the presence of stabilizing agents to generate zerovalent metal colloids has become the most powerful synthetic method in this field [2g,h,25b]. This approach was first published in 1857 by Faraday [25a]. The first reproducible standard recipes for the preparation of metal colloids (e.g., for 20-nm gold by reduction of $[AuCl_4^-]$ with sodium citrate) were established by Turkevich [1]. In addition, he proposed a mechanism for the stepwise formation of nanoclusters based on nucleation, growth, and agglomeration [1a], which, in essence, is still valid. Data from modern analytical techniques and more recent thermo-dynamic and kinetic results [26,27] have been used to refine this model as illustrated in Figure 3 [28].

The metal salt is reduced to give zerovalent metal atoms in the embryonic stage of nucleation [26], which can collide in the solution with further metal ions, metal atoms, or clusters to form an irreversible "seed" of stable metal nuclei. The diameter of the "seed" nuclei can be well below 1 nm but is dependent on the strength of the metal–metal bonds, the difference of the redox potentials between the metal salt, and the reducing agent applied. It has been experimentally verified that stronger reducing agents produce smaller nuclei in the "seed" for silver [26].

These nuclei grow during the "ripening" process to give colloidal metal particles in the size range of 1–50 nm that have a narrow size distribution. It was assumed that the mechanism for the particle formation is an agglomeration of zerovalent nuclei in the "seed" or—alternatively—collisions of already formed nuclei with reduced metal atoms. Henglein et al. [27a] have followed the stepwise reductive formation of Ag_3^+ and Ag_4^+ clusters by spectroscopic methods. Their results strongly suggest that an autocatalytic pathway is involved in which metal ions are adsorbed and successively reduced at the zerovalent cluster surface.

More recently Henglein has shown that $Ag(CN)_2^-$ is reduced at a low rate by radiolytically generated hydroxy methyl radicals [27e]. It was shown that the reduction is much faster when colloidal silver seed particles are present in the solution, resulting in larger silver particles with a narrow size distribution. The reduction occurs on tiny nuclei formed by hydrolysis in solution. The mechanism proposed here involves CH_2OH radicals, which transfer electrons to the seed particles, and the stored electrons reduce the $Ag(CN)_2^-$ directly on the surface of the seeds. $Ag(CN)_2^-$ was also shown to be rapidly reduced by the organic radicals in the

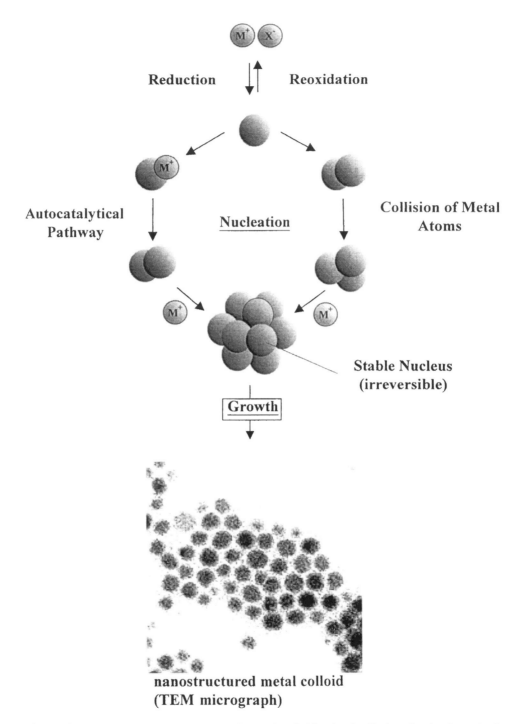

Figure 3 Formation of nanostructured metal colloids via the "salt-reduction" method. (From Ref. 42a.)

presence of colloidal platinum. In this way bimetallic particles of the $Pt_{core}Ag_{shell}$ type with a rather nonsymmetric shape of the shell are formed.

The formation of colloidal Cu protected by cationic surfactants (NR_4^+) was investigated by in-situ X-ray absorption spectroscopy showing the formation of an intermediate Cu^+ state prior to the nucleation of the particles [27d]. Although the processes during nucleation and particle growth cannot be analyzed separately, it is now generally accepted that the size of the resulting metal colloid is determined by the relative rates of nucleation and particle growth.

The main advantages of the method of salt reduction in the liquid phase are that it is easily reproduced and allows colloidal nanoparticles to be prepared on the multigram scale, which has a narrow size distribution. For example, the classical Faraday route via the reduction of $[AuCl_4]^-$ with sodium citrate is still used to prepare standard 20-nm gold sols for histological staining applications [1a,29]. In the last decades wet chemical reduction procedures have been applied to combine practically all transition metals with the different types of stabilizers and the whole range of chemical reducing agents successfully.

The "alcohol reduction process" by Hirai and Toshima [2g,11] is widely applicable for the preparation of colloidal precious metals stabilized by organic polymers such as poly(vinylpyrrolidone) (PVP), poly(vinyl alcohol) (PVA), and poly(methylvinyl ether). During the salt reduction, alcohols having α-hydrogen atoms are oxidized to the corresponding carbonyl compound (e.g., methanol to formaldehyde). The method for preparing bimetallic nanoparticles via the co-reduction of mixed ions has been evaluated in a recent review [2g]. Hydrogen is an efficient reducing agent for the preparation of electrostatically stabilized metal sols and of polymer-stabilized hydrosols of Pd, Pt, Rh, and Ir [30].

Using CO, formic acid, or sodium formate, formaldehyde, and benzaldehyde as reductants, allowed colloidal Pt in water [31,1b] to be obtained [32]. Silanes were found to be effective for the reductive preparation of Pt sols [33]. Duff, Johnson, and Baiker have successfully introduced tetrakis(hydroxymethyl)phosphoniumchloride (THPC) as a reducing agent, which allows the size- and morphology-selective synthesis of Ag, Cu, Pt, and Au nanoparticles from the corresponding metal salts [34]. Furthermore, hydrazine [35], hydroxylamine [36], and electrons trapped in, for example, $K^+ [(crown)_2K]^-$ [37] have been applied as reductants. In addition, BH_4^- is a powerful and valuable reagent for the salt-reduction method. However, a disadvantage is that transition metal borides are often found along with the nanometallic particles [38]. Tetraalkylammonium hydrotriorganoborates [2c,2d,2f,3c,5] offer a wide range of applications in the wet chemical reduction of transition metal salts. In this case the reductant $[BE_3H^-]$ is combined with the stabilizing agent (e.g., NR_4^+). The surface-active NR_4^+ salts are formed immediately at the reduction center in high local concentration and prevent particle aggregation. The advantage here is that trialkylboron is recovered unchanged from the reaction and no borides contaminate the products.

$$MX_v + NR_4(BEt_3H) \rightarrow Mcolloid + vNR_4X + vBEt_3 + v/2H_2\uparrow$$

M = metals of the groups 6–1; X = CI, Br; v = 1,2,3; and R = alkyl, C_6–C_{20}

(1)

As synthesized, the NR_4^+-stabilized metal "raw" colloids typically contain 6–12 wt.% of metal. "Purified" transition metal colloids containing about 70–85 wt.% of metal are obtained by workup with ethanol or ether and subsequent reprecipitation by a solvent of different polarity (see Table 9 in [2c]). The prepreparation of $[NR_4^+ BEt_3H^-]$ can be avoided when NR_4X is coupled to the metal salt prior to the reduction step. NR_4^+-stabilized transition metal nanoparticles can also be obtained from NR_4X-transition metal double salts. Because the local concentration of the protecting group is sufficiently high, a number of conventional reducing agents may be applied to give Eq. (2) [2d,3c].

$$(NR_4)_w MX_v Y_w +_v Red \rightarrow M_{colliod} +_v RedX +_w NR_4Y$$

$$M = metals; Red = H_2, HCOOH, K, Zn, LiH, LiBEt_3H, NaBEt_3H, KBEt_3H;$$

$$X, Y = Cl, Br;$$

$$v, w = 1\text{–}3 \text{ and } R = alkyl, C_6\text{–}C_{12} \tag{2}$$

The scope and limitations of this method have been evaluated in a recent review [2h]. Isolable metal colloids of the zerovalent early transition that are stabilized only with THF have been prepared via the $[BEt_3H^-]$ reduction of the preformed THF-adducts of $TiBr_4$ [Eq. 15], $ZrBr_4$, VBr_3, $NbCl_4$, and $MnBr_2$ [Eq. (3)].

$$x \cdot [TiBr4 \cdot 2THF + x \cdot 44K[BEt_3H] \xrightarrow{\text{THF, 2h, 20°C}}$$

$$Ti \cdot 0.5 THF]x + x \cdot 4 BEt_3 + x \cdot 4 KBr \downarrow + x \cdot 4H_2 \uparrow \tag{3}$$

Table 1 summarizes the results.

Detailed studies of [Ti $\cdot$ 0.5 THF] [39] show that it consists of Ti_{13} clusters in the zerovalent state, stabilized by six intact THF molecules (Figure 4).

$$M + R_4N^+R'CO_2^- \xrightarrow{50-90°C} M^0(R_4NR'CO_2)x + CO_2 + R'-R'$$

$$R = octyl, R' = Alkyl, Aryl, H \tag{4}$$

By analogy, [Mn $\cdot$ 0.3 THF] particles (1–2.5 nm) were prepared [40] and the physical properties have been studied [41]. The THF in Eq. (3) has been successfully

Table 1 THF-Stabilized Organosols of Early Transition Metals

Product	Starting material	Reducing agent	T (°C)	T (h)	Metal content (%)	Size (nm)
[Ti $\cdot$ 0.5THF]	$TiBr_4 \cdot 2THF$	$K[BEt_3H]$	rt	6	43.5	(<0.8)
[Zr $\cdot$ 0.4THF]	$ZrBr_4 \cdot 2THF$	$K[BEt_3H]$	rt	6	42	—
[V $\cdot$ 0.3THF]	$VBr_3 \cdot 3THF$	$K[BEt_3H]$	rt	2	51	—
[Nb $\cdot$ 0.3THF]	$NbCl_4 \cdot 2THF$	$K[BEt_3H]$	rt	4	48	—
[Mn $\cdot$ 0.3THF]	$MnBr_2 \cdot 2THF$	$K[BEt_3H]$	50	3	70	1–2.5

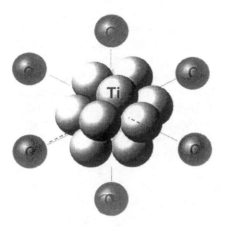

Figure 4 Ti$_{13}$ cluster stabilized by 6 THF–O atoms in an octahedral configuration.

replaced by tetrahydrothiophene (THT) for Mn, Pd, and Pt organosols; but attempts to stabilize Ti and V this way led to decomposition [3c].

A survey of the [BEt$_3$H$^-$] method is given in Figure 5.

The advantages of the method (Fig. 5) may be summarized as follows: the method is generally applicable to salts of metals in groups 4–11 in the periodic table; it yields extraordinarily stable metal colloids that are easy to isolate as dry powders;

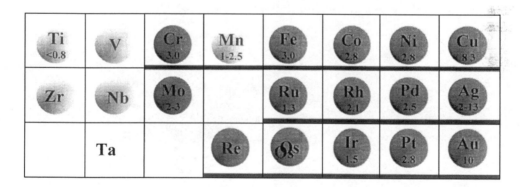

—— Nanometal powders

THF-stabilized nanometals

NR$_4^+$-stabilized nanometals

Figure 5 Nanopowders and nanostructured metal colloids accessible via the [BEt$_3$H$^-$]-reduction method [Eqs. (3), (5)] (including the mean particle sizes obtained). (From Ref. 2d.)

the particle-size distribution is nearly monodisperse; bimetallic colloids are easily accessible by co-reduction of different metal salts; and the synthesis is suitable for multigram preparations and easy to scale up. A drawback to this method is that the particle size of the resulting sols cannot be varied by altering the reaction conditions. Highly water-soluble hydrosols, particularly those of zerovalent precious metals, were made accessible using betaines instead of NR_4^+ salts as the protecting group in Eq. (1). In Eq. (2) a broad variety of hydrophilic surfactants may be used [5b,2d,3c].

A new method for the size- and morphology-selective preparation of metal colloids using tetraalkylammonium carboxylates of the type $NR_4^- R'CO_2^-$ (R = octyl, R' = alkyl, aryl, H) both as the reducing agent and as the stabilizer [Eq. (4)] was reported by Reetz and Maase [42].

The resulting particle sizes were found to correlate with the electronic nature of the R' group in the carboxylate with electron donors producing small nanoclusters while electron-withdrawing substituents R', in contrast, yield larger particles. For example, Pd particles of 2.2-nm size are found when $Pd(NO_3)_2$ is treated with an excess of tetra(*n*-octyl)ammonium-carboxylate-bearing R' = $(CH_3)_3CCO_2^-$ as the substituent. With R' = $Cl_2CHCO_2^-$ (an electron-withdrawing substituent) the particle size was found to be 5.4 nm.

The following bimetallic colloids were obtained with tetra(*n*-octyl) ammonium formate as the reductant: Pd/Pt (2.2 nm), Pd/Sn (4.4 nm), Pd/Au (3.3 nm), Pd/Rh (1.8 nm), Pt/Ru (1.7 nm), and Pd/Cu (2.2 nm). The shape of the particles was also found to depend on the reductant, e.g., with tetra(*n*-octyl) ammonium glycolate $Pd(NO_3)_2$, a significant amount of trigonal particles in the resulting Pd colloid was detected.

Organoaluminum compounds have been used for the "reductive stabilization" of mono- and bimetallic nanoparticles, presenting an interesting new method for the preparation of these colloids [see Eq. (5) and Table 2] [43].

$$MX_n + AlR_3 \xrightarrow{\text{Toluene}} [\text{nanoparticle}] + [R_2Alacac]$$

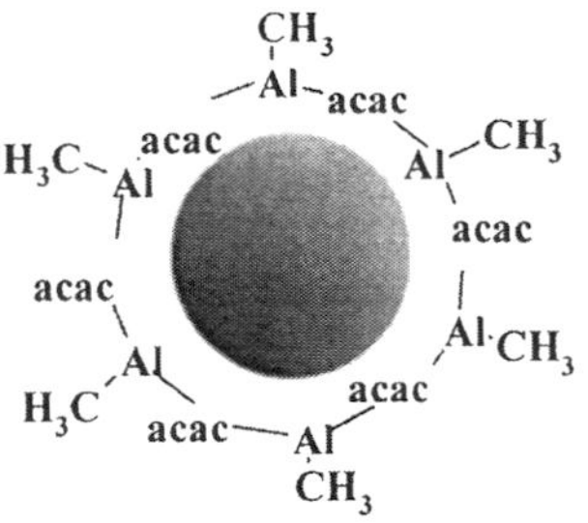

M = metals of groups 6–11 PSE

X = halogen, acetylacetonate, $n = 2$–4

R = C_1–C_8 alkyl

Particle sizes 1–12 nm

(5)

Colloids of zerovalent elements of groups 6–11 of the periodic table (and also of tin) may be prepared in the form of stable, isolable organosols as shown in Eq. (5). The analytical data available suggest that a layer of condensed organoaluminum

Table 2 Mono- and Bimetallic Nanocolloids Prepared via the Organo-Aluminum Route

Metal salt	g/mmol	Reducing agent	g/mmol	Solvent Toluene [ml]	T[°C]	t[h]	Product m[g]	Metal content wt.%	Particle size F[nm]
Ni(acac)$_2$	0.275/1	Al(i-but)$_3$	0.594/3	100	20	10	0.85	Ni: 13.8	2–4
Fe(acac)$_2$	2.54/10	Al(me)$_3$	2.1/30	100	20	3	2.4	n.d.	
RhCl$_3$	0.77/3./1	Al(oct)$_3$	4.1/11.1	150	40	18	4.5	Rh: 8.5 Al: 6.7	2–3
Ag- neodecanoate	9.3/21.5	Al(oc)t$_3$	8.0/21.8	1000	20	36	17.1	Ag: 11.8 Al: 2.7	8–12
Pt(acac)$_2$	1.15/3	Al(me)$_3$	0.86/7.6	150	20	24	1.45	Pt: 35.8 Al: 15.4	2.5
PtCl$_2$	0.27/1	Al(me)$_3$	0.34/3	125	40	16	0.47	Pt: 41.1 Al: 15.2	2.0
Pd(acac)$_2$ Pt(acac)$_2$	0.54/1.8 0.09/0.24	Al(et)$_3$	0.46/4	500	20	2	0.85	Pd: 22 Pt: 5.5 Al: 12.7	3.2
Pt(acac)$_2$ Ru(acac)$_3$	7.86/20 7.96/20	Al(me)$_3$	8.64/120	400	60	21	17.1	Pt: 20.6 Ru: 10.5 Al: 19.6	1.3
Pt(acac)$_2$ SnCl$_2$	1.15/2.9 0.19/1	Al(me)$_3$	0.86/12	100	60	2	1.1	Pt: 27.1 Sn: 5.2 Al: 14.4	n.d.

Source: Ref. 43.

species protects the transition metal core against aggregation as visualized in Eq. (5). However, the exact "backbone" of the colloidal organoaluminum protecting agent has not yet been completely established.

Quantitative protonolysis experiments have detected that unreacted organoaluminum groups (e.g., $Al-CH_3$, $Al-C_2H_5$) from the starting material are still present in the stabilizer. These active $Al-C$ bonds have been used for a controlled protonolysis by long-chain alcohols or organic acids ("modifiers") to give $Al-$alkoxide groups in the stabilizer [Eq. (6)].

$$MX_n \; + \; AlR_3 \quad \xrightarrow{\text{Toluene}} \quad + \; [R_2Alacac]$$

Modifiers: alcohola, carbonic acids, silanols, sugars, polyalcohols,

polyvinylpyrrolidone, surfactants, silica, alumina, etc. (6)

This "modification" [Eq. (6)] of the organoaluminum protecting shell can be used to tailor the dispersion characteristics of the original organosols. A vast spectrum of dissolubilities of the colloidal metals in hydrophobic and hydrophilic media (including water) has been achieved this way.

10.2.2 Electrochemical Synthesis

This very versatile preparation route for nanostructured mono- and bimetallic colloids has been further developed by Reetz and his group since 1994 [2e,12a,b]. The overall process of electrochemical synthesis [Eq. (7)] can be divided into six elemental steps (see Figure 6).

$$
\begin{aligned}
\text{Anode:} \quad &\text{bulk} \quad \rightarrow \quad M^{n+} + ne^- \\
\underline{\text{Cathode}: \; M^{n+} + ne^- &+ \text{stabilizer} \rightarrow M_{coll}/\text{stabilizer}} \\
\text{Sum } M_{bulk} \; &+ \; \text{stabilizer} \; \rightarrow \; M_{coll}/\text{stabilizer}
\end{aligned} \tag{7}
$$

1. Oxidative dissolution of the sacrificial Met_{bulk} anode
2. Migration of Met^{n+} ions to the cathode
3. Reductive formation of zerovalent metal atoms at the cathode
4. Formation of metal particles by nucleation and growth
5. Arrest of the growth process and stabilization of the particles by colloidal protecting agents, e.g., tetraalkylammonium ions
6. Precipitation of the nanostructured metal colloids

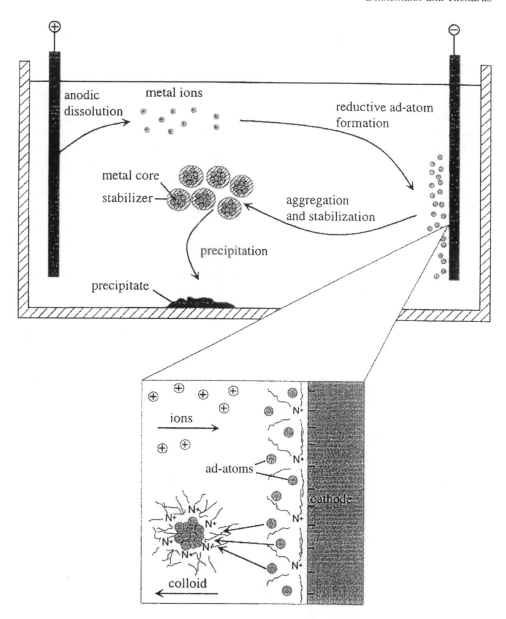

Figure 6   Electrochemical formation of $NR^{+4}\ Cl^-$ stabilized nanometal. (From Ref. 2e.)

The advantages of the electrochemical pathway are that contamination with byproducts resulting from chemical reduction agents are avoided and that the products are easily isolated from the precipitate. The electrochemical preparation also provides a size-selective particle formation. Reetz et al. have conducted several experiments using a commercially available Pd sheet as the sacrificial anode and the surfactant as the electrolyte and stabilizer. Analysis of the $(C_8H_{17})_4N^+$ Br-stabilized $Pd^{(0)}$ particles produced have indicated that the particle size depends on such parameters as solvent polarity, current density, charge flow, and distance between electrodes, and temperature can be used to control the size of Pd nanoparticles in the range of 1.2–5 nm [12a,j]. Similar experiments involving Ni particles are also reported.

The electrochemical synthesis yields almost monodisperse Pd(0) particles with sizes between 1 and 6 nm according to the parameters above. It has also been shown that the size of NR_4^+-stabilized Ni(0) particles [12c] can be adjusted at will. Using the electrochemical method, a number of monometallic organosols and hydrosols including Pd, Ni, Co, Fe, Ti, Ag, and Au have been successfully prepared on a scale of several hundred mg (yields >95%) [Eq. (19)] [12]. Using the electrochemical pathway, solvent-stabilized (propylene carbonate) palladium particles (8–10 nm) have also been obtained [12f]. When two sacrificial Met_{bulk} anodes are used in a single electrolysis cell, bimetallic nanocolloids (Pd/Ni–, Fe/Co–, Fe/Ni) are accessible [12g]. Anodically less readily soluble metals such as Pt, Rh, Ru, and Mo have been prepared using the corresponding metal salts, which were electrochemically reduced at the cathode (see lower part of Figure 6 and Table 3).

Tetraalkylammonium-acetate was used as both the supporting electrolyte and the stabilizer in a Kolbe electrolysis at the anode [see Eq. (8)] [12h].

$$\text{Cathode}: \quad Pt^{2+} + 2e^- \longrightarrow Pt^0$$
$$\text{Anode}: \quad 2CH_3CO_2^- \longrightarrow 2CH_3CO_2 + 2e^- \tag{8}$$

By combining the electrochemical methods in Eqs. (19) and (20), bimetallic nanocolloids can be prepared (see Table 4) [12h].

Layered bimetallic nanocolloids (e.g., Pt/Pd) have been synthesized by modifying the electrochemical method [12c,i]. Finally, the preparation of bimetallic colloids (Pt/Pt) electrochemically using a new strategy based on the use of a preformed colloid, e.g., $(n\text{-}C_8H_{17})_4N^+Br^-$ stabilized Pt particles and a sacrificial anode, e.g., Pd sheet, has recently been reported [12j] (Figure 7).

Table 3 Metal Salts

Metal salt	d (nm)	EA[1]
$PtCl_2$	2.5^2	51.21% Pt
$PtCl_2$	5.0^3	59.71% Pt
$RhCl_3 \cdot x\ H_2O$	2.5	26.35% Rh
$RuCl_3 \cdot x\ H_2O$	3.5	38.55% Ru
$OsCl_3$	2.0	37.88% Os
$Pd(OAc)_2$	2.5	54.40% Pd
$Mo_2(OAc)_4$	5.0	36.97% Mo
$PtCl_2 + RuCl_3 \cdot xH_2O$	2.5	41.79% Pt + 23.63% Rh[4]

[1] Based on stabilizer-containing material.
[2] Current density: 5.00 mAcm-2.
[3] Current density: 0.05 mAcm-2.
[4] Pt–Ru dimetallic cluster.

Table 4 Bimetallic Nanocolloids

Anode	Metal salt	d (nm)	Stoich. (EDX)
Sn	$PtCl_2$	3.0	$Pt_{50}Sn_{50}$
Cu	$Pd(OAc)_2$*	2.5	$Cu_{44}Pd_{56}$
Pd	$PtCl_2$	3.5	$Pd_{50}Pt_{50}$

* Electrolyte: 0.1-M $[(n\text{-octyl})_4N]OAc/THF$.

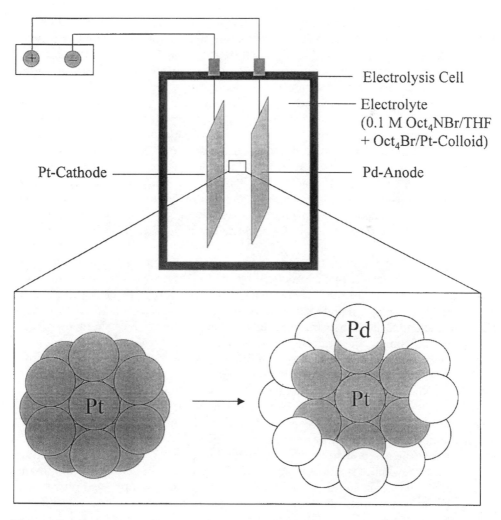

Figure 7 Modified electrolysis cell for the preparation of layered bimetallic Pt/Pd nanocolloids. (From Ref. 12c.)

The preformed Pt core may be regarded as a "living-metal polymer" on which the Pd atoms are deposited to give "onion-type" bimetallic nanoparticles (5 nm), the structure of which has been characterized by a combination of analytical methods.

10.2.3 Decomposition of Low-Valent Transition Metal Complexes

Short-lived nucleation particles of zerovalent metals can be generated in solution by the decomposition of low-valent organometallic complexes and several organic derivatives of the transition metals. This decomposition is usually induced by heat, light, or ultrasound, resulting in zerovalent metal particles, which may then be stabilized by colloidal protecting agents. For example, thermolysis [44] leads to the rapid decomposition of cobalt carbonyls to give colloidal cobalt in organic solutions [44a,b]. Colloidal Pd, Pt, and bimetallic Pd/Cu nanoparticles have been obtained by thermolysis of labile precious metal salts in the absence of stabilizers [44c]. The resulting particles, however, showed a broad size distribution. These results were much improved when the thermolysis was performed in the presence of stabilizing polymers, such as PVP [44d]. Microwave heating in a simple household oven was recently used to prepare nanosized metal particles and colloids [44e–h]. The electromagnetic waves can heat the substrate uniformly, leading to a more homogeneous nucleation and a shorter aggregation time.

Recent work by Lukehart et al. has demonstrated the applicability of this technique to fuel-cell catalyst preparation [44g,h]. Through the use of microwave heating of an organometallic precursor that contains both Pt and Ru, PtRu/Vulcan carbon nanocomposites have been prepared that consist of PtRu alloy nanoparticles highly dispersed on a powdered carbon support [44g]. Two types of these nanocomposites containing 16 and 50 wt.% metal with alloy nanoparticles of 3.4 and 5.4 nm, respectively, are formed with only 100 or 300 s of microwave heating time. The 50 wt.% supported nanocomposite has demonstrated direct methanol fuel-cell anode activity superior to that of a 60 wt.% commercial catalyst in preliminary measurements.

The energy induced in the sonochemical decomposition of metal salts and organometallic complexes [45] is produced by the formation, growth, and implosive collapse of bubbles in a liquid. Sonochemical preparations have been successfully developed by Suslick [45a] and Gedanken [45b–d]. Using this method Fe, Mo_2C, Ni, Pd, and Agnanoparticles in various stabilizing environments have been prepared. Within certain limits, the sonochemical decomposition also allows size control, which has been observed in the case of Pd nanoparticles immobilized on alumina [45e].

The growth of very clean colloidal metals has been achieved by the photolysis of organometallics and metal salts [46]. Henglein et al. [47a–i,k] and Belloni [47j] have studied the γ-radiolytic decomposition of metal ions to give nanostructured metal colloids. During this process short pulses of high-energy photons are applied to metal ions in solution, allowing the reduction process in the whole reaction medium to be "switched on" instantaneously. The process of nucleation and growth of the metal particles in solution can be monitored by spectroscopic methods. In the case of silver, the detailed mechanism of nucleation has been compared with the photographic process [47b]. Nanostructured Cd colloids [47] and even bimetallic systems [47e–g] have also been studied. In the case of gold, it has even been possible to control the particle size. Bimetallic Au_{core} Pt_{shell} and Pt_{core} Au_{shell} colloids have also been obtained [47k]. Using laser ablation the size-selective formation of silver colloids has recently been reported [48]. Radiolytic preparations of platinum nanoparticles have also been recently reported in which H_2PtCl_4 is reduced in the presence of protective water-soluble polymers [48c]. The limitation of photo-, γ-radiolysis, and laser irradiation methods lies in the restriction to low-metal concentrations in solution, rendering them unsuitable to prepare nanostructured metal colloids on a preparative scale.

Upon the addition of CO or H_2 in the presence of appropriate stabilizers, the controlled chemical decomposition of zerovalent transition metal complexes yields isolable products in multigram amounts [49]. The growth of metallic Ru particles from Ru(COT) (COD) (COT = cyclooctatetraene, COD = cycloocta-1,5-diene) with low-pressure dihydrogen was first reported by Ciardelli et al. [49a]. This material was, however, not well characterized, and the colloidal aspect of the ill-defined material seems to have been neglected in this work. Bradley and Chaudret [49b–l] have demonstrated the use of low-valent transition metal olefin complexes as a very clean source for the preparation of nanostructured mono- and bimetallic colloids.

Organometallic complexes of the type [M(dba)$_2$] (dba = dibenzylidene acetone, M = Pt, Pd) decompose under low-carbon monoxide pressure in the presence of polyvinylpyrrolidone (PVP) to give nanosized colloidal Pt or Pd particles having an fcc structure. Remarkably, the surface of the resulting particles undergoes no measurable interaction with the stabilizing polymer or with the solvent. Nanoparticles of Ru, Au, and Cu were obtained analogously from Ru(COT)(COD), (THT)AuCl (THT = tetrahydrothiophene), and (C$_5$H$_5$)Cu(Bu^{t}NC). Colloidal molybdenum, silver, and bimetallic Cu/Pd colloids were also reported, where a polymer matrix of PVP = polyvinylpyrrolidone, PPO = dimethylphenylene-oxide, or NC = nitrocellulose was used as the stabilizing agent [49b–f]. Chaudret and Bradley have described platinum and palladium nanoparticles stabilized by donor ligands such as carbonyl and phosphines in which the structure of the particles is influenced by the ligands [49g–i]. The cleavage of the complexed olefins from the coordination sphere of the metals and the simultaneous formation of nanosized colloidal particles have been shown to occur very smoothly. The hydrogenolysis of zerovalent organometallic olefin complexes has been established in the literature as a very clean and elegant method, e.g., for the manufacture of monometallic colloids via the decomposition of Fe(COT)$_2$, Ru(COT)(COD), Co(C$_8$H$_{13}$)(COD) (C$_8$H$_{13} = \eta^3$-cyclooctenyl), and Ni(COD)$_2$ under H_2. The hydrogenolysis method has been used to obtain a number of nanostructured Au, Pd, and Pt colloids and of bimetallic systems (Pt/Ru, Pt/Co, and Cu/Pd) in a polymer matrix [49j–l]. Ultrafine cobalt particles have been obtained by the hydrogenolytic decomposition of Co(η^3–C$_8$H$_{13}$)(η^4C$_8$H$_{12}$) in the presence of polyphenylphenylenoxide to give particles of 4.2-nm size; when the same procedure was carried out with polyvinylpyrrolidone, the size of the resulting particles was 1.4 nm. It is clear from these results that a close relationship exists between the synthesis conditions and the structure of the resulting particles [49m].

10.2.4 Preparation in Micelles, Reverse Micelles, and Encapsulation

Colloidal assemblies have widely been used as templates in order to "preform" the size, shape, and polydispersity of nanosized metal particles [50]. Surfactant micelles may enclose metal ions to form amphiphilic "microreactors" Figure 10a) or water-in-oil reverse micelles (Figure 10b), or larger vesicles may function in similar ways. On the addition of reducing agents nanosized metal particles are formed with the size and the shape of the products appearing to be "imprinted" by the constrained environments in which they are grown. Recently Chen et al. reported the preparation of bimetallic nanoparticles of Au/Pt (monodisperse 3–4.5 nm) by the co-reduction of chloroauric and chloroplatinic acid with hydrazine in the water in oil microemulsions of water/AOT/isooctane [50q]. Here the particle composition was found to be consistent with the feeding solutions, and XPS studies revealed that these nanoparticles had a structure of $Au_{core}Pt_{shell}$. Pileni [50a] was the first to grow cobalt rods with a length of 300–1500 nm and a diameter between 10 and 30 nm in colloidal micelles (Figure 8a). Colloidal metals have also been grown in reverse micelles (Figure 8b).

10.2.5 Colloidal Nanostructured Metal Oxides

This contribution focuses primarily on those metal oxides that are viewed as immediately relevant to electrocatalysis, but for additional information on the subject the reader is directed to a recent review article [2j]. It has been shown that after exposure to air, nanosized (e.g., 3–5 nm) colloidal $Fe^{(0)}$, $Co^{(0)}$, and $Ni^{(0)}$ particles cannot be redispersed because they are immediately oxidized both in solution and in powder form. When the addition of oxygen is precisely controlled, e.g., by using stoichiometric amounts of O_2 diluted in large excess of argon, an

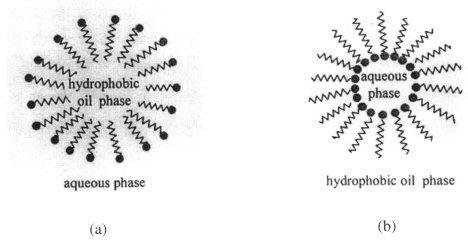

(a)   (b)

Figure 8 Intraparticle preparation of colloidal metals in (a) surfactant oil in water micelles or (b) water-in-oil reverse micelles as microreactors. (Reproduced after [2b], Figs. 6–10, p. 481.)

organic solution of a 3-nm iron$^{(0)}$-sol stabilized by $N(octyl)_4{}^+Br^-$ is transformed to a rusty brown solution of colloidal Fe^{3+}-oxide that is redispersible in THF [2d].

$N(octyl)_4{}^+Br^-$-stabilized $Co^{(0)}$ particles were oxidized in air to give colloidal CoO nanoparticles that have been characterized by HRTEM [51a]. This process has also been followed by UV-visible spectra and magnetic susceptibility measurements. The resulting CoO particles were supported on alumina. Small CoO and Co_3O_4 particles have been obtained in a polymer matrix dispersion by solid-state oxidation of 1.6-nm $Co^{(0)}$ particles. The structural changes that occur during the oxidation process were monitored using physical analytic methods, and it was shown that air oxidation at room temperature leads to surface passivation. The resulting particles have a metallic core surrounded by an oxide surface layer [51b]. The particle size of nanosized Pt dispersed on γ-Al_2O_3 is markedly influenced during the oxidation by O_2 [51c]. A new process was recently reported for the manufacture of a water-soluble PtO_2 colloid that is applicable as a water-soluble "Adams catalyst" [see Eq. (9)] [51d,e].

$$PtCl_4 \xrightarrow[\text{Stabilizer}]{H_2O/\text{base}} PtO_2\text{colloid}$$
$$\text{Stabilizer}: C_{12}H_{25}(CH_3)_2N^+(CH_2)_nCO_2^-$$
$$C_{12}H_{25}(CH_3)_2N^+(CH_2)_3SO_3^- \tag{9}$$

Using the simple hydrolysis/condensation of metal salts under basic aqueous conditions in the presence of carbo- or sulfobetaines, respectively, colloidal PtO_2 was obtained, the analytical data of which correspond to pure α-PtO_2 and to commercial samples of Adams catalyst. Bi- and trimetallic colloidal metal oxides as precursors for fuel-cell catalysts, e.g., colloidal Pt/RuO_x and $Pt/Ru/WO_x$, have also been prepared this way [51f].

10.3 PARTICLE-SIZE CONTROL

In general, metal colloid sols are referred to as "monodisperse" when the particle size deviates by less than 15% from the average value, and histograms with a standard deviation σ from the mean particle size of approximately 20% are described as showing a "narrow size distribution." The kinetics of the particle nucleation from atomic units and of the subsequent growth process cannot be observed directly by current physical methods. The tools of the preparative chemist to control the particle size in practice are size-selective separation [50c,52] and size-selective synthesis [42,12a,50,53].

The *size-selective precipitation* (SPP) was predominantly developed by Pileni [50c]. One example (SPP) is monodisperse silver particles (2.3 nm, $\sigma = 15\%$), which are precipitated from a polydisperse silver colloid solution in hexane by the addition of pyridine in three iterative steps. Recently, Schmid [52a] has reported the two-dimensional "crystallization" of truly monodisperse Au_{55} clusters. Chromatographic separation methods have thus far proven unsuccessful because the colloid decomposed after the colloidal protecting shell had been stripped off [42a]. The size-selective ultracentrifuge separation of Pt colloids has been developed by Cölfen [52b]. Although this elegant separation method gives truly monodisperse metal colloids, it still provides only milligram samples. Turkevich described the first size-selective colloid synthesis [1a,b]. Using the salt-reduction method, he was able to vary the particle size of colloidal Pd between 0.55 and 4.5 nm by altering the amount of the reducing agent applied and the pH value. According to the literature on the process of nucleation and particle growth the essential factors that control the particle size are the strength of the metal–metal bond [11d], the molar ratio of metal salt, colloidal stabilizer, and reduction agent [1a,32,34e,46f,53], the extent of conversion or the reaction time [32], and the temperature applied [1a,53b,m], and the pressure [53b]. The preparation of near-monodisperse nanostructured metal colloids using the salt-reduction pathway is well documented in the literature. In practice, however, the "control," e.g., the variation of particle sizes (and shapes), in wet chemical colloid synthesis is left to the intuition and imagination of the chemist.

The electrochemical synthesis developed by Reetz and co-workers offers at present the most rational method for control of particle size. Researchers have obtained at will almost monodisperse samples of colloidal Pd and Ni between 1 and 6 nm using variable-current densities and suitable adjustment of further essential parameters [12]. For thermal decomposition methods the resulting particle size has been found to depend on the heat source [44f]. Size control has also been reported for the sonochemical decomposition method [45e] and γ-radiolysis [48].

Preparation methods using constrained environments have enabled scientists to control the metal particle shape via the preformation of size and the morphology of the products in nano-reaction-chambers [50]. The controlled temperature-induced size and shape manipulation of 2–6-nm gold particles encapsulated in alkanethiolate monolayers [50n] has recently been reported. An enormous increase in the size of thiol-passivated gold particles up to about 200 nm has been induced by near-IR laser light [50n]. A new medium-energy ion scattering (MEIS) simulation program has successfully been applied to the composition and average particle-size analysis of Pt–Rh/α-Al$_2$O$_3$ [50o].

10.4 MODES OF STABILIZATION

Protective agents are required for the stabilization of nanostructured metals and to prevent their coagulation. Two basic modes of stabilization have been distinguished that meet these requirements by different means [25b] (Figure 9).

Electrostatic stabilization (see Figure 9a) employs the Coulombic repulsion between the particles caused by the electrical double layer formed by ions adsorbed at the particle surface (e.g., sodium citrate) and the corresponding counterions. An example is gold sols prepared by the reduction of [AuCl$_4^-$] with sodium citrate [1].

Steric stabilization (Figure 9b) is achieved by the coordination of sterically demanding organic molecules that act as protective shields on the metallic surface. The nanometallic cores are thus separated from each other, preventing coagulation. The main classes of protective groups selected from literature are polymers (11) and block copolymers (50); P, N, S donors (e.g., phosphines, amines, thioethers) [3c,10,49f–i,54], solvents such as THF [3c,39], THF/MeOH [55] or propylene carbonate [12f], long-chain alcohols [50,56], surfactants [2c–f,3c,3d,5,12,57]; organometallics (43,58,59). Lipophilic protective agents yield metal colloids that are soluble in organic media ("organosols") while hydrophilic agents yield water-soluble colloids ("hydrosols"). In Pd organosols stabilized by tetraalkylammonium

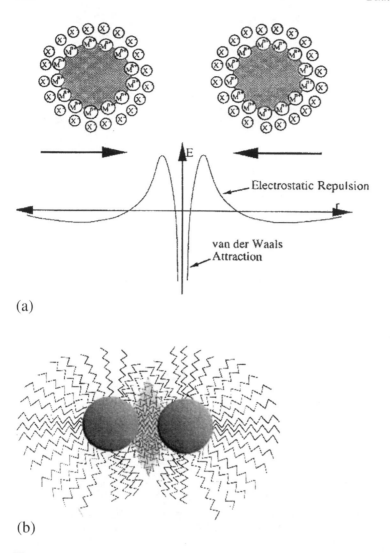

Figure 9   (a) Electrostatic stabilization of nanostructured metal colloids. (b) Steric stabilization of nanostructured metal colloids. (From Ref. 25b.)

halides, the metal core is protected by a monolayer of the surfactant coat (Figure 10) [60].

Metal hydrosols stabilized with zwitterionic surfactants able to self-aggregate are enclosed in organic double layers. After the application of uranylacetate as a contrasting agent, the TEM micrographs show that the colloidal Pt particles (average size = 2.8 nm) arem surrounded by a double-layer zone of the zwitterionic carboxybetaine (3–5 nm). The charged metal surface interacts with the hydrophilic head group of the betaine. The lipophilic tail is associated with the tail of a second surfactant molecule. As a result, a hydrophilic outer sphere is formed (see Figure 11) [61].

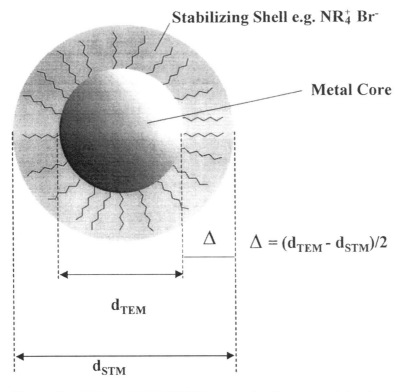

Figure 10 Differential TEM/STEM study of a Pd organosol showing that the metal core (size $= d_{TEM}$) is surrounded by a monolayer of the surfactant (thickness $\Delta = (d_{TEM} - d_{STM\ 2})$. (From Ref. 60.)

Pt or Pt/Au particles have also been hosted in the hydrophobic holes of nonionic surfactants, e.g., polyethylene monolaurate [46a,62]. The structure and diffusive dynamics of a colloidal palladium aggregate sol have been studied under dilute and semidilute conditions by high-resolution small angle X-ray scattering and X-ray photon correlation spectroscopy. When the sizes of the aggregates determined in the static structure and as derived from the diffusive dynamics at low concentration are consistent with each other.

The aggregates tend to overlap at high concentration. While the system remains in a liquidlike state, the apparent diffusion constant decreases. The structural features obtained by comparison of static and dynamic data are not accessible solely by one technique [63].

10.5 FUEL-CELL CATALYSTS

This survey focuses on recent catalyst developments in phosphoric acid fuel cells (PAFC), proton exchange membrane fuel cells (PEMFC), and the previously mentioned direct methanol fuel cell (DMFC). A PAFC operating at 160–220 °C uses orthophosphoric acid as the electrolyte; the anode catalyst is Pt and the cathode can

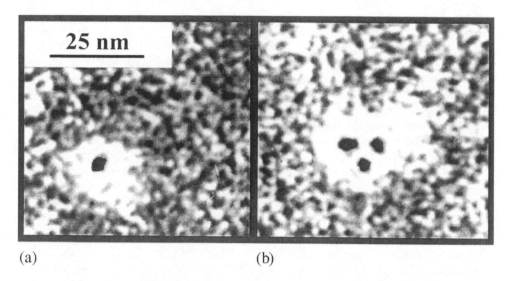

(a)   (b)

Figure 11 (a) TEM micrographs of colloidal Pt particles (single and aggregated, average core size $= 2.8$ nm) stabilized by Carboxybetaine 12 (3–5 nm, contrasted with uranylacetate against the carbon substrate). (b) Schematic model of the hydrosol stabilization by a double layer of the zwitterionic Carboxybetaine 12 ($=$ lipophilic alkyl chain; -o $=$ hydrophilic, zwitterionic head group.) (From Ref. 61.)

be, e.g., Pt/Cr/Co [9]. For this application a trimetallic colloidal precursor of the composition Pt_{50} Co_{30} Cr_{20} (size 3.8 nm) was prepared by the co-reduction of the corresponding metal salts [64]. According to XRD, the trimetallic particles are alloyed in an ordered fct structure. The electrocatalytic performance in a standard half-cell was compared with an industrial standard catalyst (trimetallic crystallites of 5.7-nm size) manufactured by co-precipitation and subsequent annealing to 900 °C. The advantage of the trimetallic colloid catalysts lies in its improved durability, which is essential for PAFC applications. After 22 h it was found that the potential had decayed by less than 10 mV [65].

PEM fuel cells use a solid proton-conducting polymer as the electrolyte at 50–125 °C. The cathode catalysts are based on Pt alone, but because of the required tolerance to CO a combination of Pt and Ru is preferred for the anode [8]. For low-temperature (80 °C) polymer membrane fuel cells (PEMFC) colloidal Pt/Ru catalysts are currently under broad investigation. These have also been proposed for use in the direct methanol fuel cells (DMFC) or in PEMFC, which are fed with CO-contaminated hydrogen produced in on-board methanol reformers. The ultimate dispersion state of the metals is essential for CO-tolerant PEMFC, and truly alloyed Pt/Ru colloid particles of less than 2-nm size seem to fulfill these requirements [4a,b,d,8a,c,66]. Alternatively, bimetallic Pt/Ru PEM catalysts have been developed for the same purpose, where nonalloyed Pt nanoparticles <2 nm and Ru particles <1 nm are dispersed on the carbon support [8c]. From the results it can be concluded that a Pt/Ru interface is essential for the CO tolerance of the catalyst regardless of whether the precious metals are alloyed. For the manufacture of DMFC catalysts, in contrast, Pt/Ru nano*powders* of 3–5-nm size or thin films are used as the precursors [8d–f]. For the electrocatalytic methanol oxidation a Pt metal or a Pt metal alloy catalyst has been developed where a Ru phthalocyanine complex is added as a dopant to reinforce the catalytic effect substantially [64c]. For comparison, the electrocatalytic activity was compared to that of a bimetallic Pt_{50}/Ru_{50}–$N(Oct_4)Cl$ colloid prepared by the salt–co-reduction method [2c,5] toward the oxidation of CO and a CO/H_2 gas mixture (simulated reformer gas) [4a]. According to high-resolution transmission electron microscopy (HRTEM) the mean particle diameter was 1.7 nm. The alloyed state of the particles was verified by point-resolved energy dispersive X-ray (EDX) analysis.

A preparation and characterization of new PtRu alloy colloids that are suitable as precursors for fuel-cell catalysts have been reported [43c]. This new method uses an organometallic compound both for reduction and as colloid stabilizer leading to a Pt/Ru colloid with lipophilic surfactant stabilizers that can easily be modified to demonstrate hydrophilic properties. The surfactant shell is removed prior to electrochemical measurements by reactive annealing in O_2 and H_2. This colloid was found to have nearly identical electrocatalytic activity to several other recently developed Pt/Ru colloids as well as commercially available Pt/Ru catalysts. This demonstrates the potential for the development of colloid precursors for bimetallic catalysts especially when considering the ease of manipulating the alloy composition when using these methods.

Recently Pt/Ru alloy catalysts were examined by X-ray absorption near edge structure spectroscopy (XANES) [125e]. An industrial (DMFC) alloy catalyst black and a carbon-supported Pt/Ru catalyst were both found to be predominantly in the form of Pt and Ru oxides. Both catalysts were reduced to the metallic form upon introduction into an electrochemical cell and brought to the potential region where methanol oxidation occurs. Glassy carbon-supported Pt_{50}/Ru_{50}–$N(Oct_4)Cl$ colloids were examined by CO-stripping voltammetry and the data were found to be essentially identical with those found in well-characterized bulk alloy electrodes. In a rotating disk electrode, the activity of the colloid toward the continuous oxidation of 2% CO in H_2 was determined at 25 °C in 0.5-M H_2SO_4, and the results led to the conclusion that these Pt/Ru colloids are very suitable precursors for high-surface-area fuel-cell catalysts [4d]. Structural information on the precursor was obtained by in-situ XRD via Debye function analysis. In addition, the XANES data support the bimetallic character of the particles. In-situ XRD has revealed the catalytic function of the alloyed Ru in the CO oxidation: surface oxide species are formed on the Ru surface at 280 °C which slowly coalesce to RuO_2 particles. After re-reduction the catalyst shows a pure-hcp ruthenium phase and larger platinum-enriched alloy particles [67]. Scanning probe microscopy (SPM) has been applied in order to characterize the real-space morphology of the electrode surfaces of supported nanostructured metal colloids on the nanometer scale [68]. Colloidal Pt_{50}/Ru_{50} precursors (<2 nm) raise the tolerance to CO, allowing higher CO concentrations in the H_2 feed of a PEMFC without a significant drop in performance [4b,d]. A selective Pt/Mo oxidation catalyst for the oxidation of H_2 in the presence of CO in fuel cells comprises Pt_xMo_y particles where x is 0.5–0.9 and y is 0.5–0.1 [69a]. The colloid method was found to be a highly suitable exploratory approach to finding improved formulations for binary and ternary anode electrocatalysts. The metals used include Pt, Ru, W, Mo, and Sn [4c]. Results of electrochemical measurements have demonstrated that the introduction of a transition metal oxide (WO_x, MoO_x, VO_x) leads to an improvement of catalytic activity toward methanol oxidation [4f]. The combinatorial screening method has successfully been applied to electrocatalysts [69b], and it is an obvious step to include colloids in these experiments. As an alternative to the reductive metal colloid synthesis, the so-called metal oxide concept was developed, which allows the fabrication of binary and ternary colloidal metal oxides as electrocatalyst precursors [Eq. (10)] [51d–f].

$$PtCl_4 + RuCl_3 \xrightarrow{\hspace{3cm}} PtRuO_x colloid \hspace{2cm} (10)$$

Colloidal Pt/RuO_x. ($1,5 \pm 0.4$ nm) stabilized by a surfactant was prepared by co-hydrolysis of $PtCl_4$ and $RuCl_3$ under basic conditions. The Pt:Ru ratio in the colloids can be between 1:4 and 4:1 by variation of the stoichiometry of the transition metal salts. The corresponding zerovalent metal colloids are obtained by the subsequent application of H_2 to the colloidal Pt/Ru oxides (optionally in the immobilized form). Additional metals have been included in the "metal oxide concept" [Eq. (10)] in order to prepare binary and ternary mixed metal oxides in the colloidal form. $Pt/Ru/WO_x$ is regarded as a good precatalyst especially for the application in DMFCs. Main-group elements such as Al have been included in multimetallic alloy systems in order to improve the durability of fuel-cell catalysts. $Pt_3AlC_{0.5}$ alloyed with Cr, Mo, or W particles of 4–7-nm size has been prepared by sequential precipitation on conductant carbon supports such as highly disperse Vulcan XC72® [70]. Alternatively, colloidal precursors composed of Pt/Ru/Al allow

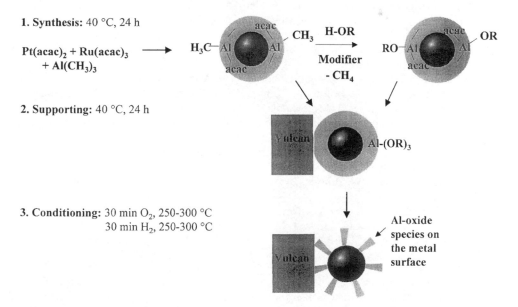

Figure 12   Scheme of the preparation of colloidal Pt/Ru/Al PEMFC anode catalysts (>20% metal on Vulcan XC72®) via the "precursor concept." (From Ref. 66.)

the manufacture of multimetallic fuel-cell catalysts (1–2 nm) having a metal loading of >20%. The three-step catalyst preparation is summarized in Figure 12.

The colloidal Pt/Ru/Al precursor is preprepared via the organoaluminum route. In the absence of stabilizers, the co-reduction of organic Pt and Ru salts using $Al(CH_3)_3$ gives halogen-free, multimetallic Pt colloids, e.g., $Pt_{50}Ru_{50}$/Al (size 1.2 ± 0.3 nm). When the stoichiometric ratio of the metal salts is changed, the ratio of Pt to the second metal in the colloid can be adjusted. The addition of alcohols or suitable surfactants allows the dispersivity of the colloidal precursor in organic media or water to be tailored without affecting the particle size. In the second step the Pt/Ru/Al colloid is adsorbed on high-surface-area carbon by treatment at 40 °C for 24 h. In the third step (conditioning) the dried Pt/Ru/Al Vulcan catalyst powders are exposed to O_2 and H_2 for 30 min each at 250–300 °C to remove the surfactants completely. The particle size of the Pt/Ru/Al colloid adsorbed on the support was found to be virtually untouched (1.3 ± 0.4 nm) and after the thermal treatment only a moderate growth was determined (1.5 ± 0.4 nm). The aluminum was found to be present on the Pt/Ru surface in an oxidized form. This accounts for the size stabilization observed in the Pt/Ru particles and for the improved durability of the resulting electrocatalysts. Recently, a fuel cell for generating electric power from a liquid organic fuel ("synfuel") was described. It comprises a solid electrolyte membrane directly supporting the anode and cathode layers, which contain 7–10% Pt and Ru, 70–80% of perfluorovinyl ether sulfonic acid and 15–20% polytetrafluoroethylene [66d]. In conclusion, nanostructured metal colloids are very promising precursors for manufacturing multimetallic fuel-cell catalysts that are truly nanosized (i.e. <2 nm) and have high metal loadings (20 wt.% of metal).

REFERENCES

1. (a) J. Turkevich, P.C. Stevenson, J. Hillier, Disc. Faraday Soc. 11, 1951, 55–75.
   (b) J. Turkevich, G. Kim, Science 169, 1970, 873.
   (c) J. Turkevich, Gold Bulletin 18, 1985, 86–91.
2. (a) G. Schmid, in Aspects of Homogeneous Catalysis, Vol. 7, R. Ugo, ed. Kluwer, Dordrecht, 1990, 1–33.
   (b) G. Schmid, ed., Clusters and Colloids, VCH, Weinheim, 1994.
   (c) H. Bönnemann, W. Brijoux, R. Brinkmann, R. Fretzen, T. Joussen, R. Köppler, P. Neiteler, J. Richter, J. Mol. Catal. 86, 1994, 129–177.
   (d) H. Bönnemann, G. Braun, W. Brijoux, R. Brinkmann, A. Schulze Tilling, K. Seevogel, K. Siepen, J. of Organometallic Chem. 520, 1996, 143–162.
   (e) M.T. Reetz, W. Helbig, S.A. Quaiser, In Active Metals, A. Fürstner, ed. VCH, Weinheim, 1996, 279–297.
   (f) H. Bönnemann, W. Brijoux, In Active Metals, A. Fürstner, ed. VCH, Weinheim, 1996, 339–379.
   (g) N. Toshima, T. Yonezawa, New J. Chem. 1998, 1179–1201.
   (h) J.D. Aiken III, R.G. Finke, J. Mol. Catal. A 145, 1999, 1–44.
   (i) B.F.G. Johnson, Coordination Chem. Rev. 190–192, 1999, 1269–1285.
   (j) M. Froba, A. Reller, Prog. Solid State Chem. 27, 1999, 1–27.
3. (a) G. Schmid, In Applied Homogeneous Catalysis with Organometallic Compounds, B. Cornils, W.A. Herrmann, eds. Wiley-VCH, Weinheim, Vol. 2, 1996, 636–644.
   (b) W.A. Herrmann, B. Cornils, In Applied Homogeneous Catalysis with Organometallic Compounds, Vol. 2, B. Cornils, W.A. Herrmann, eds. Wiley-VCH, Weinheim, 1996, 1171–1172.

(c)   H. Bönnemann, W. Brijoux, In Advanced Catalysts and Nanostructured Materials, Chapter 7, W. Moser, ed. Academic Press, San Diego, 1996, 165–196.

(d)   H. Bönnemann, W. Brijoux, In Metal Clusters in Chemistry, Vol. 2, P. Braunstein, L.A. Oro, P.R. Raithby, eds. Wiley-VCH, Weinheim, 1999, 913–931.

4.   (a)   T.J. Schmidt, M. Noeske, H.A. Gasteiger, R.J. Behm, P. Britz, W. Brijoux, H. Bönnemann, Langmuir 13, 1997, 2591–2595.

(b)   EP 09 24 784 Al (June 23, 1999), E. Auer, W. Behl, T. Lehmann, U. Stenke (to Degussa AG).

(c)   M. Götz, H. Wendt, Electrochimica Acta 43, 1998, 3637–3644.

(d)   T.J. Schmidt, M. Noeske, H.A. Gasteiger, R.J. Behm, P. Britz, H. Bönnemann, J. Electrochem. Soc. 145, 1998, 925–931.

(e)   W.E. O'Grady, P.L. Hagans, K.I. Pandya, D.L. Maricle, Langmuir 2001 ASAP article.

(f)   K. Lasch, L. Jorissen, J. Garche, J. Power Sources 84, 1999, 225–230.

5.   (a)   U.S. Pat. 5,580,492 (Aug. 26, 1993), H. Bönnemann, W. Brijoux, T. Joussen (to Studiengesellschaft Kohle mbH).

(b)   H. Bönnemann, W. Brijoux, R. Brinkmann, E. Dinjus, T. Joussen, B. Korall, Angew. Chem. Int. Ed. 30, 1991, 1344–1346.

(c)   U.S. Pat. 849,482 (Aug. 29, 1997), H. Bönnemann, W. Brijoux, R. Brinkmann, J. Richter (to Studiengesellschaft Kohle mbH).

6.   (a)   G. Witek, M. Noeske, G. Mestl, S. Shaikhutdinov, R.J. Behm, Catal. Lett. 37, 1996, 35–39.

(b)   S.K. Shaikhutdinov, F.A. Möller, G. Mestl, R.J. Behm, J. Catal. 163, 1996, 492–495.

7.   K. Kordesch, G. Simader, Fuel Cells and Their Applications, VCH, Weinheim, 1996.

8.   (a)   U.S. Pat. 4,044, 193 (Aug. 23, 1977), H.G. Petrow, R.J. Allen (to Prototech Comp.).

(b)   G.T. Burstein, C.J. Barnett, A.R. Kucernak, K.R. Williams, Catal. Today 38, 1997, 425–437.

(c)   EP 0 880 188 A2 (Nov. 25, 1998), E. Auer, A. Freund, T. Lehmann, K.-A. Starz, R. Schwarz, U. Stenke (to Degussa AG).

(d)   M.S. Wilson, S. Gottesfeld, J. Appl. Electrochem. 22, 1992, 1–7.

(e)   M.S. Wilson, X. Ren, S. Gottesfeld, J. Electrochem. Soc. 143, 1996, L 12.

(f)   S.C. Thomas, X. Ren, S. Gottesfeld, J. Electrochem. Soc. 146, 1999, 4354–4359.

9.   (a)   G.J.K Acres, J.C. Frost, G.A. Hards, R.J. Potter, T.R. Ralph, D. Thompsett, G.T. Burstein, G.J. Hutchings, Catal. Today 38, 1997, 393–400.

(b)   R.G. Cawthorn, South African J. of Science 95, 1999, 481–489.

(c)   Information available from "World Fuel Cell Council" E.V. Kroegerstrasse 5, D-60313 Frankfurt.

(d)   R.F. Service, Science 285, 1999, 682–685.

(e)   D.S. Cameron, Platinum Metals Rev. 43, 1999, 149–154.

10.   (a)   G. Schmid, R. Pfeil, R. Boese, F. Bandermann, S. Meyer, G.H.M. Calis, J.A.W. van der Velden, Chem. Ber. 114, 1981, 3634–3642.

(b)   G. Schmid, Polyhedron 7, 1988, 2321–2329.

(c)   L.J. de Jongh, J.A.O. de Aguiar, H.B. Brom, G. Longoni, J.M. van Ruitenbeek, G. Schmid, H.H.A. Smit, M.P.J. van Staveren, R.C. Thiel, Z. Phys. D: At., Mol. Clusters 12, 1989, 445–450.

(d)   G. Schmid, B. Morum, J. Malm, Angew. Chem., Int. Ed. Engl. 28, 1989, 778–780.

(e)   G. Schmid, N. Klein, L. Korste, Polyhedron 7, 1998, 605–608.

(f) T. Tominaga, S. Tenma, H. Watanabe, U. Giebel, G. Schmid, Chem. Lett. 1996, 1033.

(g) G. Schmid, Chem. Rev. 92, 1992, 1709–1727.

(h) H.A. Wicrenga, L. Soethout, I.W. Gerritsen, B.E.C. van do Leemput, H. van Kempen, G. Schmid, Adv. Mater. 2, 1990, 482.

(i) R. Houbertz, T. Feigenspan, F. Mielke, U. Memmert, U. Hartmann, U. Simon, G. Schön, G. Schmid, Europhys. Lett. 28, 1994, 641.

(j) G. Schmid, A. Lehnert, Angew. Chem. Int. Ed. Engl. 28, 1989, 780–781.

(k) G. Schmid, V. Maihack, F. Lantermann, S. Peschel, J. Chem. Soc., Dalton Trans. 1996, 589–595.

(l) G. Schmid, H. West, J.-O. Malm, J.-O. Bovin, C. Grenthe, Chem. Eur. J. 2, 1996, 1099.

(m) U. Simon, R. Flesch, H. Wiggers, G. Schön, G. Schmid, J. Mater. Chem. 8, 1998, 517–518.

(n) G. Schmid, S. Peschel, New J. Chem. 22, 1998, 669–675.

(o) G. Schmid, R. Pugin, J.-O. Malm, J.-O. Bovin, Eur. J. Inorg. Chem. 1998, 813–817.

(p) M. Giersig, L.M. Liz-Tarzan, T. Ung, D. Su, P. Mulvaney, Ber. Bunsenges. Phys. Chem. 101, 1997, 1617–1620.

11. (a) H. Hirai, Y. Nakao, N. Toshima, K. Adachi, Chem Lett. 1976, 905.

(b) H. Hirai, Y. Nakao, N. Toshima, Chem. Lett. 1978, 545.

(c) H. Hirai, Y. Nakao, N. Toshima, J. Macromol. Sci., Chem. A12, 1978, 1117.

(d) H. Hirai, Y. Nakao, N. Toshima, J. Macromol. Sci. Chem. A13, 1979, 727.

12. (a) M.T. Reetz, W. Helbig, J. Am. Chem. Soc. 116, 1994, 7401–7402.

(b) U.S. Pat. 5,620,564 (Apr. 15, 1997) and U.S. Pat. 5,925,463 (Jul. 20, 1999), M.T. Reetz, W. Helbig, S. Quaiser (to Studiengesellschaft Kohle).

(c) M.A. Winter, Ph.D. thesis, 1998, Verlag Mainz, Aachen, ISBN 3-89653-355.

(d) J.A. Becker, R. Schäfer, W. Festag, W. Ruland, J.H. Wendorf, Pebler, S.A. Quaiser, W. Helbig, M.T. Reetz, J. Chem. Phys. 103, 1995, 2520–2527.

(e) M.T. Reetz, S.A. Quaiser, C. Merk, Chem. Ber. 129, 1996, 741–743.

(f) M.T. Reetz, G. Lohmer, Chem. Commun. (Cambridge) 1996, 1921–1922.

(g) M.T. Reetz, W. Helbig, S.A. Quaiser, Chem. Mater. 7, 1995, 2227–2228.

(h) M.T. Reetz, S.A. Quaiser, Angew. Chem. 107, 1995, 2461–2463; Angew. Chem. Int. Ed. Engl. 34, 1995, 2240.

(i) U. Kolb, S.A. Quaiser, M. Winter, M.T. Reetz, Chem. Mater. 8, 1996, 1889–1894.

(j) M.T. Reetz, M. Winter, R. Breinbauer, T. Thurn-Albrecht, W. Vogel, Chemistry—a Eur. J. 7, 2001, 1084.

13. G. Schmid, A. Lehnert, J.-O. Malm, J.-O. Bovin, Angew. Int. Ed. Engl. 30, 1991, 874–876.

14. N. Toshima, T. Yonezawa, K. Kushihashi, J. Chem. Soc., Faraday Trans. 89, 1993, 2537.

15. H. Bönnemann, W. Brijoux, J. Richter, R. Becker, J. Hormes, J. Rothe, Z. Naturforsch. 50b, 1995, 333–338.

16. L.E. Aleandri, H. Bönnemann, D.J. Jones, J. Richter, J. Rozière, J. Mater. Chem. 5, 1995, 749–752.

17. N. Toshima. In Fine Particles Science and Technology, E. Pelizzetti, ed. Kluwer Academic Publishers, 1996, 371–383.

18. M. Harada, K. Asakura, N. Toshima, J. Phys. Chem. 98, 1994, 2653–2662.

19. (a) B.J. Joice, J.J. Rooney, P.B. Wells, G.R. Wilson, Discuss. Faraday Soc. 41, 1966, 223.

(b) J. Schwank, Gold Bull. 18, 1985, 2.

(c)   W. Juszczyk, Z. Karpinski, D. Lomot, J. Pielaszek, J.W. Sobczak, J. Catal. 151, 1995, 67.

(d)   A.O. Cinneide, J.K.A. Clarke, J. Catal. 26, 1972, 233.

(e)   S.H. Inami, W. Wise, J. Catal. 26, 1972, 92.

(f)   C. Visser, I.G.P. Zuidwijk, V. Ponec, J. Catal. 35, 1974, 407.

(g)   N. Toshima, J. Macromol. Sci. A 27, 1990, 1225.

(h)   A.O. Cinneide, F.G. Gault, J. Catal. 37, 1975, 311.

(i)   N. Toshima, H. Harada, Y. Yamazaki, K. Asakura, J. Phys. Chem. 96, 1992, 927.

(j)   H. Liu, C. Mao, S. Meng, J. Mol. Catal. 1992, 74, 275.

20.   J.B. Michel, J.T. Schwartz. In Catalyst Preparation Science IV, B. Delmon, P. Grange, P.A. Jacobs, G. Poncelet, eds. Elsevier, New York, 1987, 669–687.

21.   H. Bönnemann, U. Endruschat, B. Tesche, A. Rufinska, C.W. Lehmann, F.E. Wagner, G. Filoti, V. Pârvulescu, Eur. J. Chem. 2000, 819–822.

22.   (a)   E. Gaffet, M. Tachikart, O. El Kedim, R. Rahouadj, Mater. Charact. 1996, 36, 185–190.

(b)   A. Amulyavichus, A. Daugvila, R. Davidonis, C. Sipavichus, Fizika Metallov I Metallovedenie 85, 1998, 111–117.

23.   (a)   A. Schalnikoff, R. Roginsky, Kolloid Z., 43, 1927, 67–70.

(b)   J.R. Blackborrow, D. Young, Metal Vapor Synthesis, Springer-Verlag, New York, 1979.

(c)   K.J. Klabunde, Free Atoms and Particles, Academic Press, New York, 1980.

(d)   K.J. Klabunde, Y-X Li, B-J Tan, Chem. Mater. 1991, 3, 30–39.

24.   (a)   J.S. Bradley. In Clusters and Colloids, G. Schmid, ed. VCH, Weinheim, 1994, 477.

(b)   K.J. Klabunde, G.C. Cardenas-Trivino. In Active Metals, A Fürstner, ed. VCH, Weinheim, 1996, 237–278.

25.   (a)   M. Faraday, Philos. Trans. R. Soc. London 147, 1857, 145–153.

(b)   J.S. Bradley. In Clusters and Colloids, G Schmid, ed. VCH, Weinheim, 1994, 469–473.

26.   T. Leisner, C. Rosche, S. Wolf, F. Granzer, L. Wöste, Surf. Rev. Lett. 3, 1996, 1105–1108.

27.   (a)   R. Tausch-Treml, A. Henglein, J. Lilie, Ber. Bunsen-Ges. Phys. Chem. 82, 1978, 1335–1343.

(b)   M. Michaelis, A. Henglein, J. Phys. Chem. 1992, 96, 4719–4724.

(c)   M.A. Watzky, R.G. Finke, J. Am. Chem. Soc. 119, 1997, 10382–10400.

(d)   J. Rothe, J. Hormes, H. Bönnemann, W. Brijoux, K. Siepen, J. Am. Chem. Soc. 120, 1998, 6019–6023.

(e)   A. Henglein, Langmuir, ASAP article, 2001.

28.   A.I. Kirkland, P.P. Edwards, D.A. Jefferson, D.G. Duff, In Annual Reports on the Progress of Chemistry C, Vol. 87, R. Soc. Chem., Cambridge, 1990, 247–305.

29.   J.S. Bradley, In Clusters and Colloids, G. Schmid, ed. VCH, Weinheim 1994, 471.

30.   (a)   L.D. Rampino, F.F. Nord, J. Am. Chem. Soc. 63, 1941, 2745–2749.

(b)   L.D. Rampino, F.F. Nord, J. Am. Chem. Soc. 63, 1941, 3268.

(c)   L.D. Rampino, F.F. Nord, J. Am. Chem. Soc. 65, 1943, 2121–2125.

(d)   L. Hernandez, F.F. Nord, J. Colloid. Sci. 3, 1948, 363–375.

(e)   W.P. Dunsworth, F.F. Nord, J. Am. Chem. Soc. 72, 1950, 4197–4198.

31.   M.R. Mucalo, R.P. Cooney, J. Chem. Soc., Chem. Commun. 1989, 94–95.

32.   K. Meguro, M. Torizuka, K. Esumi, Bull. Chem. Soc. Jpn. 1988, 61, 341–345.

33.   (a)   L.N. Lewis, N. Lewis, J. Am. Chem. Soc. 108, 1986, 7228–7231.

(b)   L.N. Lewis, N. Lewis, Chem. Mater. 1, 1989, 106–114.

34.   (a)   A.C. Curtis, D.G. Duff, P.P. Edwards, D.A. Jefferson, B.F.G. Johnson, A.I. Kirkland, D.E. Logan, Angew. Chem. Int. Ed. Engl. 26, 1987, 676.

- (b) D.G. Duff, A.C. Curtis, P.P. Edwards, D.A. Jefferson, B.F.G. Johnson, D.E. Logan, J.C.S. Chem. Commun. 1987, 1264.
- (c) A.C. Curtis, D.G. Duff, P.P. Edwards, D.A. Jefferson, B.F.G. Johnson, A.I. Kirkland, A.S. Wallace, Angew. Chem. Int. Ed. Engl. 27, 1988, 1530.
- (d) D.G. Duff, P.P. Edwards, J. Evans, J.T. Gauntlett, D.A. Jefferson, B.F.G. Johnson, A.I. Kirkland, D.J. Smith, Angew. Chem. Int. Ed. Engl. 28, 1989, 590.
- (e) D.G. Duff, A. Baiker, P.P. Edwards, Langmuir 9, 1993, 2301–2309.
- (f) W. Vogel, D.G. Duff, A. Baiker, Langmuir 11, 1995, 401–404.
35. P.R. van Rheenen, M.J. McKelvey, W.S. Glaunsinger, J. Solid State Chem. 67, 1987, 151–169.
36. (a) D.G. Duff, A. Baiker, In Preparation of Catalysts VI, G. Poncelet, J. Martens, B. Delmon, P.A. Jacobs, P. Grange, eds. Elsevier Science, 1995, 505–512.
37. K.-L. Tsai, J.L. Dye, Chem. Mater. 5, 1993, 540–546.
38. (a) J. van Wonterghem, S. Mørup, C.J.W. Koch, S.W. Charles, S. Wells, Nature 322, 1986, 622.
- (b) G.N. Glavee, K.J. Klabunde, C.M. Sorensen, G.C. Hadjipanayis, Inorg. Chem. 32, 1993, 474–477.
39. R. Franke, J. Rothe, J. Pollmann, J. Hormes, H. Bönnemann, W. Brijoux, Hindenburg, J. Amer. Chem. Soc. 118, 1996, 12090–12097.
40. R. Franke, J. Rothe, R. Becker, J. Pollmann, J. Hormes, H. Bönnemann, W. Brijoux, R. Köppler, Adv. Mater. 10, 1998, 126–131.
41. J. Sinzig, L.J. de Jongh, H. Bönnemann, W. Brijoux, R. Köppler, Appl. Organomet. Chem. 12, 1998, 387–391.
42. (a) M. Maase, Ph.D. thesis, 1999, Verlag Mainz, Aachen, ISBN 3-89653-463-7.
- (b) M.T. Reetz, M. Maase, Adv. Mater. 11, 1999, 773–777.
- (c) J.S. Bradley, B. Tesche, W. Busser, M. Maase, M.T. Reetz, J. Am. Chem. Soc. 122, 2000, 4631–4636.
43. (a) H. Bönnemann, W. Brijoux, R. Brinkmann, U. Endruschat, W. Hofstadt, K. Angermund, Rev. Roum. Chim. 44, 1999, 1003–1010.
- (b) WO 99/59713 (Nov. 25, 1999), H. Bönnemann, W. Brijoux, R. Brinkman, (to Studiengesellschaft Kohle m.b.H.).
- (c) U.A. Paulus, U. Endruschat, G.J. Feldmeyer, T.J. Schmidt, H. Boennemann, R.J. Behm, J. Catal. 195, 2000, 383–393.
44. (a) P.H. Hess, P.H. Parker, J. Appl. Polymer. Sci. 10, 1966, 1915–1927.
- (b) J.R. Thomas, J. Appl. Phys. 37, 1966, 2914–2915.
- (c) K. Esumi, T. Tano, K. Torigue, K. Meguro, Chem. Mater. 2, 1990, 564.
- (d) J.S. Bradley, E.W. Hill, C. Klein, B. Chaudret, A. Duteil, Chem. Mater. 5, 1993, 254–256.
- (e) Y. Wada, H. Kuramoto, T. Sakata, H. Mori, T. Sumida, T. Kitamura, S. Yanagida, Chem. Lett. 1999, 607.
- (f) W. Yu, W. Tu, H. Liu, Langmuir 15, 1999, 6–9.
- (g) D.L. Boxall, G.A. Deluga, E.A. Kenik, W.D. King, C.M. Lukehart, Chem. Mater. 13, 2001, 891–900.
- (h) D.L. Boxall, C.M. Lukehart, Chem. Mater. 13, 2001, 806–810.
45. (a) K.S. Suslick, T. Hyeon, M. Fang, A. Cichowlas, In Advanced Catalysts and Nanostructured Materials, Chapter 8, W. Moser, ed. Academic Press, San Diego, 1996, 197–212.
- (b) A. Dhas, A. Gedanken, J. Mater. Chem. 8, 1998, 445–450.
- (c) Y. Koltypin, A. Fernandez, C. Rojas, J. Campora, P. Palma, R. Prozorov, A. Gedanken, Chem.—Mater. 11, 1999, 1331–1335.
- (d) R.A. Salkar, P. Jeevanandam, S.T. Aruna, Y. Koltypin, A. Gedanken, J. Mater. Chem. 9, 1999, 1333–1335.

(e)   K. Okitsu, S. Nagaoka, S. Tanabe, H. Matsumoto, Y. Mizukoshi, Y. Nagata, Chem. Lett. 1999, 271.

(f)   T.C. Rojas, M.J. Sayagués, A. Caballero, Y. Koltypin, A. Gedanken, L. Posonnet, B. Vacher, J.M. Martin, A. Fernández, J. Mater. Chem. 10, 2000, 715–721.

46.   (a)   N. Toshima, T. Takahashi, H. Hirai, Chem. Lett. 1985, 1245–1248.

(b)   Y. Yonezawa, T. Sato, M. Ohno, H. Hada, J. Chem. Soc., Faraday Trans. 83, 1987, 1559–1567.

(c)   Y. Yonezawa, T. Sato, S. Kuroda, K. Kuge, J. Chem. Soc., Faraday Trans. 87, 1991, 1905–1910.

(d)   T. Sato, S. Kuroda, A. Takami, Y. Yonezawa, H. Hada, Appl. Organomet. Chem. 5, 1991, 261–268.

(e)   K. Torigoe, T. Tano, K. Meguro, Chem. Mater. 2, 1990, 564–587.

(f)   K. Torigoe, K. Esumi, Langmuir 9, 1993, 1664–1667.

47.   (a)   A. Henglein, In Modern Trends in Colloid Science in Chemistry and Biology, H.F. Bicke, ed. Birkhauser Verlag, Stuttgart, 1985, 126.

(b)   B.G. Ershov, E. Janata, A. Henglein, A. Fojtik, J, Phys. Chem. 97, 1993, 4589–4594.

(c)   A. Henglein, J. Phys. Chem. 97, 1993, 5457–5471.

(d)   B.G. Ershov, E. Janata, A. Henglein, J. Phys. Chem. 97, 1993, 339–343.

(e)   A. Henglein, M. Giersig, J. Phys. Chem. 98, 1994, 6931–6935.

(f)   A. Henglein, M. Gutierrez, E. Janata, B. Ershov, J. Phys. Chem. 96, 1992, 4598–4602.

(g)   A. Henglein, P. Mulveney, A. Holzwarth, T.E. Sosebee, A. Fojtik, Ber. Bunsenges. Phys. Chem. 96, 1992, 2411.

(h)   P. Mulveney, M. Giersig, A. Henglein, J. Phys. Chem, 96, 1992, 10419–10424.

(i)   A. Henglein, D. Meisel, Langmuir 14, 1998, 7392–7396.

(j)   J. Belloni, M. Mostafavi, H. Remita, J.-L. Marignier, M.-O. Delcourt, New J. Chem. 22, 1998, 1239–1255.

(k)   A. Henglein, J. Phys. Chem. B. 104, 2000, 2201–2203.

48.   (a)   J.-S. Jeon, C.-S. Yeh, J. Chin. Chem. Soc. 45, 1998, 721–726.

(b)   F. Stietz, F. Träger, Physikalische Blätter 55, 1999, 57–60.

(c)   A.D. Belapurkar, S. Kapoor, S.K. Kuldhreshtha, J.P. Mittal, Mater. Res. Bull. 35, 2000, 143.

49.   (a)   F. Ciardelli, P. Pertici, Z. Naturforsch. 40b, 1985, 133–140.

(b)   J.S. Bradley, E.W. Hill, S. Behal, C. Klein, B. Chaudret, A. Duteil, Chem. Mater. 4, 1992, 1234–1239.

(c)   A. Duteil, R. Quéau, B. Chaudret, R. Mazel, C. Roucau, J.S. Bradley, Chem. Mater. 5, 1993, 341–347.

(d)   D. deCaro, V. Agelou, A. Duteil, B. Chaudret, R. Mazel, C. Roucau, J.S. Bradley, New. J. Chem. 19, 1995, 1265–1274.

(e)   F. Dassenoy, K. Philippot, T. Ould Ely, C. Amiens, P. Lecante, E. Snoeck, A. Mosset, M.-J. Casanove, B. Chaudret, New J. Chem. 19, 1998, 703–711.

(f)   C. Amiens, D. deCaro, B. Chaudret, J.S. Bradley, J. Am. Chem. Soc. 115, 1993, 11638–11939.

(g)   D. deCaro, H. Wally, C. Amiens, B. Chaudret, J. Chem. Soc., Chem. Comm. 1994, 1891–1892.

(h)   A. Rodriguez, C. Amiens, B. Chaudret, M.-J. Casanove, P. Lecante, J.S. Bradley, Chem. Mater. 8, 1996, 1978–1986.

(i)   M. Bardaji, O. Vidoni, A. Rodriguez, C. Amiens, B. Chaudret, M.-J. Casanove, P. Lecante, New. J. Chem. 21, 1997, 1243–1249.

(j)   J. Osuna, D. deCaro, C. Amiens, B. Chaudret, E. Snoeck, M. Respaud, J.-M. Broto, A. Fert, J. Phys. Chem. 100, 1996, 14571–14574.

(k) T. Ould-Ely, C. Amiens, B. Chaudret, E. Snoeck, M. Verelst, M. Respaud, J.M. Broto, Chem. Mater. 11, 1999, 526–529.

(l) J.S. Bradley, E.W. Hill, B. Chaudret, A. Duteil, Langmuir 11, 1995, 693–695.

(m) F. Dassenoy, M.-J. Casanove, P. Lecante, M. Verelst, E. Snoeck, A. Mosset, T. Ould Ely, C. Amiens, B. Chaudret, J. Chem. Phys. 112, 2000, 8137–8145.

50. (a) J. Tanori, M.P. Pileni, Langmuir 13, 1997, 639–646.

(b) M.P. Pileni, Langmuir 13, 1997, 3266–3276.

(c) M. Antonietti, C. Göltner, Angew. Chem. Int. Ed. Engl. 36, 1997, 910–928.

(d) M.P. Pileni, Supramol. Sci. 5, 1998, 321–329.

(e) M.P. Pileni, Adv. Mater. 10, 1998, 259–261.

(f) M. Antonietti, Chem.-Ing. Tech. 68, 1996, 518–523.

(g) S. Förster, Ber. Bunsen-Ges. 101, 1997, 1671–1678.

(h) J.J. Storhoff, R.C. Mucic, C.A. Mirkin, J. Cluster Sci. 8, 1997, 179–216.

(i) M. Möller, J.P. Spatz, Curr. Opin. Colloid Interface Sci. 2, 1997, 177–187.

(j) G.B. Sergeev, M.A. Petrukhina, Prog. Solid State Chem. 24, 1996, 183–211.

(k) J.P. Wilcoxon, P. Provencio, J. Phys. Chem. B 103, 1999, 9809–9812.

(l) T. Miyao, N. Toyoizumi, S. Okuda, Y. Imai, K. Tyjima, S. Naito, Chemistry Lett. 1999, 1125.

(m) S.T. Selvan, M. Nogami, A. Nakamura, Y. Hamanaka, J. Non-Crystalline Solids 255, 1999, 254–258.

(n) M.M. Maye, W. Theng, F.L. Leibowitz, N.K. Ly, C.J. Zhong, Langmuir 16, 2000, 490–497.

(o) Y. Niidome, A. Hori, T. Sato, S. Yamada, Chem. Lett. 2000, 310.

(p) I. Konomi, S. Hyodo, T. Motohiro, J. of Catal. 192, 2000, 11–17.

(q) M.L. Wu, D.H. Chen, T.C. Huang, ASAP article, Chem. Mater. 2001.

51. (a) M.T. Reetz, S. Quaiser, M. Winter, J.A. Becker, R. Schaefer, U. Stimming, A. Marmann, R. Vogel, T. Konno, Angew. Chem. Ind. Ed. 35, 1996, 2092–2094.

(b) M. Verelst, T. Ould Ely, C. Amiens, E. Snoeck, P. Lecante, A. Mosset, M. Respaud, J.-M. Broto, B. Chaudret, Chem. Mater. 11, 1999, 2702–2708.

(c) C.-B. Wang, C.-T. Yeh, J. Catal. 178, 1998, 450–456.

(d) PCT/EP 99/08594 (Nov. 9, 1999), M.T. Reetz, M. Koch (to Studiengesellschaft Kohle m.b.H.).

(e) M.T. Reetz, M. Koch, J. Am. Chem. Soc. 121, 1999, 7933–7934.

(f) M. Koch, Ph.D. thesis, 1999, Verlag Mainz, Aachen, ISBN 3-89653-514-5.

52. (a) G. Schmid, M. Bäumle, N. Beyer, Angew. Chem. Int. Ed. 39, 2000, 182–184.

(b) H. Cölfen, T. Pauck, Colloid Polym. Sci. 275, 1997, 175–180.

53. (a) T. Teranishi, M. Miyake, Chem. Mater. 10, 1998, 594–600.

(b) E. Papirer, P. Horny, H. Balard, R. Anthore, C. Pepitas, A. Martinet, J. Colloid Interface Sci. 94, 1983, 220–228.

(c) T. Teranishi, M. Hosoe, M. Miyake, Adv. Mater. 9, 1997, 65–67.

(d) T. Teranishi, I. Kiyokawa, M. Miyake, Adv. Mater. 10, 1998, 596–599.

(e) K. Esumi, H. Ishizuka, S. Masayoshi, K. Meguro, T. Tano, K. Torigoe, Langmuir 7, 1991, 457–459; RG DiScipio, Anal. Biochem. 236, 1996, 168–170.

(f) M.A. Watzky, R.G. Finke, Chem. Mater. 9, 1997, 3083–3095.

(g) T. Yonezawa, M. Sutoh, T. Kunitake, Chem. Lett. 7, 1997, 619–620.

(h) G. Frens, Nature 241, 1997, 20–22.

(i) X. Zhai, E. Efrima, Langmuir 13, 1997, 420–425.

(j) D.V. Leff, P.C. Ohara, J.R. Heath, W. Gelbart, J. Phys. Chem. 99, 1995, 7036–7041.

(k) C.H. Chew, J.F. Deng, H.H. Huang, F.C. Loh, G.L. Loy, X.P. Ni, K.L. Tan, G.Q. Xu, Langmuir 12, 1996, 909–912.

(l) G. Braun, H. Bönnemann, Angew. Chem. Int. Ed. Engl. 35, 1996, 1992–1994.

(m) M.B. Mohamed, Z.L. Wang, M.A. El-Sayed, J. Phys. Chem. A 103, 1999, 10255–10259.

(n) T. Teranishi, M. Miyake, Chem. Mater. 11, 1999, 3414–3416.

(o) G.W. Busser, J.G. van Ommen, J.A. Lercher, In Advanced Catalysts and Nanostructured Materials, Chapter 7, W Moser, ed. Academic Press, San Diego, 1996, 230–231.

(p) M. Antonietti, F. Gröhn, J. Hartmann, L. Bronstein, Angew. Chem. Int Ed. Engl. 36, 1997, 2080.

54. (a) M.N. Vargaftik, V.P. Zargorodnikov, I.P. Stolarov, I.I. Moiseev, D.I. Kochubey, V.A. Likholobov, A.L. Chuvilin, K.I. Zarnaraev, J. Mol. Catal. 53, 1989, 315–349.

(b) M.N. Vargaftik, V.P. Zargorodnikov, I.P. Stolarov, I.I. Moiseev, V.A. Likholobov, D.I. Kochubey, A.L. Chuvilin, V.I. Zaikosvsky, K.I. Zamaraev, G.I. Timofeeva, J. Chem. Soc., Chem. Commun. 1985, 937–939.

(c) V.V. Volkov, G. Van Tendeloo, G.A. Tsirkov, N.V. Cherkashina, M.N. Vargaftik, I.I. Moiseev, V.M. Novotortsev, A.V. Kvit, A.L. Chuvilin, J. Cryst. Growth 163, 1996, 377.

(d) I.I. Moiseev, M.N. Vargaftik, V.V. Volkov, G.A. Tsirkov, N.V. Cherkashina, V.M. Novotortsev, O.G. Ellett, I.A. Petrunenka, A.L. Chuvilin, A.V. Kvit, Mend. Commun. 1995, 87.

(e) V. Oleshko, V. Volkov, W. Jacob, M. Vargaftik, I.I. Moiseev, G. van Tendeloo, Z. Phys. D 34, 1995, 283.

(f) I.I. Moiseev, M.N. Vargaftik, T.V. Chernysheva, T.A. Stromnova, A.E. Gekhman, G.A. Tsirkov, A.M. Makhlina, J. Mol. Catal. A: Chem. 108, 1996, 77.

55. O. Vidoni, K. Philippot, C. Amiens, B. Chaudret, O. Balmes, J.-O. Malm, J.-O. Bovin, F. Senocq, M.-J. Casanove, Angew. Chem. Int. Ed. Engl. 38, 1999, 3736–3738.

56. D. Mandler, I. Willner, J. Phys. Chem. 91, 1987, 3600–3605.

57. J. Kiwi, M. Grätzel, J. Am. Chem. Soc. 101, 1979, 7214–7217.

58. WO 99/59713 (Nov. 25, 1999), H. Bönnemann, W. Brijoux, R. Brinkmann (to Studiengesellschaft Kohle).

59. J.S. Bradley, E.W. Hill, M.E. Leonowicz, H. Witzke, J. Mol. Catal. 41, 1987, 59–74.

60. M.T. Reetz, W. Helbig, S.A. Quaiser, U. Stimming, N. Breuer, R. Vogel, Science 267, 1995, 367–369.

61. A. Schulze Tilling, Ph.D. thesis, 1996, RWTH Aachen.

62. T. Yonezawa, N. Toshima, J. Mol. Catal. 83, 1993, 167–181.

63. T. Thurn-Albrecht, G. Meier, P. Müller-Buschbaum, A. Patkowski, W. Steffen, G. Grubel, D.L. Abernathy, O. Diat, M. Winter, M.G. Koch, M.T. Reetz, Phys. Rev. E: Stat. Phys., Plasmas, Fluids, Relat. Interdiscip. Top. 59, 1999, 642–649.

64. (a) U.S. Pat. 4,613,582 (Sept. 23, 1986), F.J. Luczak, D.A. Landsman (to United Technologies Corp.).

(b) W. Wittholt, Ph.D. thesis, R.W.T.H. Aachen, 1997.

(c) EP 0951 084 A2 (20.10.1999), H. Wendt, M. Götz (to Degussa-Hüls AG).

65. A. Freund, J. Lang, T. Lehmann, K.A. Starz, Catal. Today 27, 1996, 279–283.

66. (a) H. Bönnemann, In Extended Abstracts of the 3rd Int. Symposium on New Materials for Electrochemical Systems, O Savadogo, ed. Ecole Polytechnique de Montréal, Canada, ISBN 21-553-00739-6, 1999.

(b) H. Bönnemann, R. Brinkmann, P. Britz, U. Endruschat, R. Mörtel, U.A. Paulus, G.J. Feldmeyer, T.J. Schmidt, H.A. Gasteiger, R.J. Behm, J. New Mat. for Electrochem. Syst. 3, 2000, 199–206.

(c) EP 0 952 241 (27.10.1999), T. Itoh, J. Sato (to N.E. Chemcat Corporation).

(d) U.S. Pat. 5,945,231 (Aug. 31, 1999), S. Narayanan, S. Surampudi, G. Halpert (to California Institute of Technology).

67. W. Vogel, P. Britz, H. Bönnemann, J. Rothe, J. Hormes, J. Phys. Chem. B. 101, 1997, 11029–11036.

68. U. Stimming, R. Vogel. In Electrochemical Nanotechnology, W.J. Lorenz, W. Plieth, eds. Wiley-VCH, Weinheim, 1998, 73–86.

69. (a) WO 99/53557 (Apr. 14, 1999), J.R. Giallombardo, E.S. De Castro (to De Nora S.P.A.).

    (b) E. Reddington, A. Sapienza, B. Gurau, R. Viswanathan, S. Sarangapani, E.S. Smotkin, T.E. Mallouk, Science 280, 1998, 1735–1737.

70. EP 0 743 092 A1 (Nov. 20, 1996), A. Freund, T. Lehmann, K.-A. Starz, G. Heinz, R. Schwarz (to Degussa AG).

71. (a) P.M. Paulus, H. Bönnemann, A.M. van der Kraan, F. Luis, J. Sinzig, L.J. de Jongh, Eur. Phys. J.D. 9, 1999, 501–504.

    (b) WO 99/41758 (Aug. 19, 1999), H. Bönnemann, W. Brijoux, R. Brinkmann, M. Wagener (to Studiengesellschaft Kohle m.b.H.).

11

Preparation, Characterization, and Properties of Bimetallic Nanoparticles

TOSHIHARU TERANISHI

Japan Advanced Institute of Science and Technology, Ishikawa, Japan

NAOKI TOSHIMA

Tokyo University of Science, Yamaguchi, Japan

CHAPTER CONTENTS

SUMMARY

Controlling the primary structures of metal nanoparticles (i.e., size, shape, crystal structure, and composition) is one of the most important missions for colloid science, especially for nanoparticle science and technology because these structures determine the chemical and physical properties of metal nanoparticles. Here, chemical methods are dealt with to control the compositions and structures of various bimetallic nanoparticles by making use of the difference in the reduction (decomposition) rate or the reduction sequence of two kinds of metal species. When two kinds of metal ions are simultaneously reduced, the reduction rates of metal ions usually determine the final structure of bimetallic nanoparticles (i.e., a core/shell structure or an alloy structure). The successive reduction of two kinds of metal ions, however, generally gives bimetallic nanoparticles with a core/shell structure. As a representative of the chemical properties of such bimetallic nanoparticles, their catalytic properties for hydrogenation of olefins, hydration of acrylonitrile and photo-induced hydrogen generation are described, while the magnetic properties are presented as a novel physical property of bimetallic nanoparticles.

11.1 INTRODUCTION

11.1.1 Definition and Historical Aspects

Metal nanoparticles have attracted a great deal of interest in scientific research and industrial applications due to their unique properties based on large surface-to-volume ratio and quantum size effect [1–5]. Metal nanoparticles larger than a few ten nms in size show physical and chemical properties similar to the corresponding bulk metals. The upper limit of the size of metal nanoparticles that hold interesting properties different from those of bulk metals could be 20 nm at most. Therefore, the size of metal nanoparticles on which current scientists are focusing is restricted between 1 and 20 nm, consisting of 10 to 10^6 atoms. Speaking strictly from the viewpoint of the electronic structure, the metal nanoparticles smaller than 5 nm (often called *clusters* or *nanoclusters*) are the target materials, because it has been found from vigorous studies on metal nanoparticles that binding energies of metal nanoparticles consisting of less than 1000 atoms vary periodically due to the quantum size effect [6], meaning that metal nanoparticles are not just aggregates of metal atoms. In this size range metal will still hold secrets of the genesis. In this chapter, we deal with metal nanoparticles from 1 to 20 nm in size.

From not only the scientific but the technological point of view, bimetallic nanoparticles composed of two different metal elements are of greater interest and importance than monometallic nanoparticles [7,8]. Scientists have especially focused on bimetallic nanoparticles as catalysts because of their novel catalytic behaviors affected by the second metal element added. This effect of the second metal element can often be explained in terms of an ensemble and/or a ligand effect in catalyses. Such effects appear in bimetallic catalysts composed of both zerovalent metal atoms and another metal ions [9,10]. In this case, however, metal ions do not construct nanoparticles but are located close to them to exhibit an ensemble effect. This chapter covers the bimetallic nanoparticles composed of only zerovalent metals in homogeneous systems; the supported or heterogeneous systems of metal nanoparticles are not covered.

Before we come to the current investigation on the bimetallic nanoparticles, let us pause here to look briefly at the history of (bi)metallic nanoparticles. Since old times gold nanoparticles have been used for colored glass (stained glass) and red ceramics in Europe. The dawn of metallic nanoparticle science was brought by Michael Faraday in the mid-nineteenth century [11]. He prepared ruby red solutions of gold nanoparticles by the reduction of $[AuCl_4]^-$ solutions by using phosphorous as a reducing agent. In 1988 J.M. Thomas revealed by electron microscopy that these gold nanoparticles were in the 3- to 30-nm size range [12].

In the mid-twentieth century, an important paper was reported by Turkevich et al. [13]. Gold nanoparticles were prepared with various methods involving phosphorous and carbon monoxide reductions of $[AuCl_4]^-$ in solution and characterized by electron microscopy. It is noteworthy that gold nanoparticles with quite narrow size distributions and small mean diameters in the range of 10 to

20 nm could be produced by citrate reduction of $[AuCl_4]^-$ in solution; this method is adopted by many scientists even at present. Later, Pd/Au bimetallic nanoparticles with three different microstructures—alloyed Pd/Au particles (ca. 25 nm), Pd-coated Au particles (ca. 22 nm), and Au-coated Pd particles (ca. 30 nm)—were prepared by the simultaneous reduction of the corresponding metal salts and reductive deposition of a metal on a preformed core of another nanoparticle in the presence of citrate or hydroxylamine as a reducing agent, although precise structures were not characterized [14,15].

Miner et al. [16] describe the syntheses of Au/Pt and Pd/Pt alloys as monodispersed sols by simultaneous reduction of the corresponding salt mixtures at various molar ratios. They assert that Au/Pt alloys are formed at any atomic ratio, even if the two metals show a broad miscibility gap between 2 and 85 wt.% Au [17]. The homogeneous character of the various Au/Pt alloys has been proven by means of optical spectra, sedimentation measurements, and electron microscopy investigations.

Sinfelt has greatly contributed to the catalyses of bimetallic nanoparticles [18]. His group has thoroughly studied inorganic oxide-supported bimetallic nanoparticles for catalyses and analyzed their microstructures by an EXAFS technique [19–22]. Nuzzo and co-workers have also studied the structural characterization of carbon-supported Pt/Ru bimetallic nanoparticles by using physical techniques, such as EXAFS, XANES, STEM, and EDX [23–25]. These supported bimetallic nanoparticles have already been used as effective catalysts for the hydrogenation of olefins and carbon-skeleton rearrangement of hydrocarbons. The alloy structure can be carefully examined to understand their catalytic properties. Catalysis of supported nanoparticles has been studied for many years and is practically important but is not considered further here.

The precise control of the size, shape, composition, and crystal structure of bimetallic nanoparticles is one of the goals for the scientists of this field. In the last two decades quite a few preparative methods of bimetallic nanoparticles have been developed for this purpose. The development of the preparative methods enables us to elucidate the novel chemical and physical properties of bimetallic nanoparticles depending on their structures.

11.1.2 Current Focus

Intense research in the field of metal nanoparticles by chemists, physicists, and materials scientists is motivated by the search for new materials that hold novel physical (electronic, magnetic, optical) and chemical (catalytic) properties. Recently, the research on monometallic nanoparticles has been carried out in order to further miniaturize electronic devices [26–28] as well as to elucidate the fundamental question of how electronic properties of molecular aggregates evolve into novel properties with increasing size in this intermediate region between a molecule and a bulk [3,29–33]. Possible future applications include the areas of ultrafast communication and a large quantity of data storage [3,31,32,34].

On the other hand, bimetallic nanoparticles are very important as catalysts because of their high surface-to-volume ratios [29] and their novel catalytic behavior induced by the second metal element. The addition of a second metal can drastically change the catalytic properties of catalysts, even when the second metal is inactive for the reaction. Many industrial catalysts are bimetallic or multicomponent. Small amounts of additives (acting as promoters or inhibitors) are added to the catalysts in order to improve the catalytic performance. Details are presented in Section 11.3.1.

The second attractive feature of bimetallic nanoparticles is their magnetic property. In addition to the fact that the size control of bimetallic nanoparticles composed of noble and 3d-transition metals is easier than that of 3d-transition monometallic nanoparticles, recent technology makes it possible to control the compositions and crystal structures of such bimetallic nanoparticles. Some examples are provided in Section 11.3.2.

The last important feature of bimetallic nanoparticles is their optical property, including the change of surface plasmon resonance absorption [35–38]. Because nanoparticles of alkali metals, such as sodium, and coin metals (i.e., copper, silver, and gold) have broad adsorption bands in the visible region of the electromagnetic spectra, their optical properties have been thoroughly studied [39–47]. Solutions of these metal nanoparticles show very intense colors, which are absent for bulk materials as well as for atoms although alkali metal nanoparticles require nonaqueous solvents, of course. Their origin is attributed to the collective oscillation of the free conductive electrons induced by an interacting electromagnetic field. These resonances are also denoted as surface plasmons. When bimetallic nanoparticles composed of these metals are prepared, the solutions of alloy nanoparticles show a tunable single absorption peak between two wavelengths at which the solution of each monometallic nanoparticle shows an absorption peak, as shown in Figure 1 [36,37,40,48–51], and the solutions of core-shell bimetallic nanoparticles have two distinct plasmon absorption bands, whose relative intensities depend on the thickness of the shell [40,47,49,52,53]. These bimetallic nanoparticles will be applicable to the nonlinear optical devices, optical data storage devices [3,54], and surface-enhanced Raman scattering substrates [55].

11.2 PREPARATION AND CHARACTERIZATION OF BIMETALLIC NANOPARTICLES

11.2.1 Preparative Aspects

In principle, the preparations of metal nanoparticles could be classified into two categories: physical and chemical techniques. Evaporation [48,56–58] or laser ablation [59–62] from metal bulk samples is utilized to generate nanoparticles in the physical methods, while reduction of metal ions to neutral atoms, followed by particle growth, is the common strategy in chemical syntheses, including conventional chemical (one- or two-phase systems) [37,63–69], photochemical [70,71], sonochemical [72,73], electrochemical [74,75], and radiolytic reductions [76,77]. In general, the chemical methods have the significant advantage of being easily able to control the primary structures of nanoparticles, such as size, shape, and composition, as well as to achieve mass production. Preparative methods used for monometallic nanoparticles can be applied to the preparation of bimetallic nanoparticles. Because bimetallic nanoparticles are composed of two metal elements, preparations of bimetallic nanoparticles could be classified into two categories: simultaneous and successive reductions (or decompositions) of two metal salts. These approaches often give the different sizes and structures to the resulting bimetallic nanoparticles.

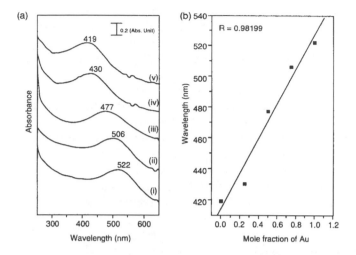

Figure 1 (a) UV/vis absorption spectra of dodecanethiol-stabilized Au-Ag bimetallic nanoparticles in *n*-hexane with nominal formulae of (i) Au, (ii) AuAg(3/1), (iii) AuAg(1/1), (iv) AuAg(1/3), and (v) Ag. (b) Position of the surface plasmon band is plotted with respect to the mole fraction of Au atoms in bimetallic nanoparticles. (Reprinted with permission from Ref. 36. Copyright 1998 Academic Press.)

11.2.2 Simultaneous Reduction or Decomposition

Bimetallic Nanoparticles Composed of Noble Metals

Because the noble metal ions can be easily reduced to the corresponding zerovalent metal atoms, research on the bimetallic nanoparticles consisting of two kinds of noble metals has been thoroughly investigated, with the goal of developing novel catalysts and optical materials. A simultaneous reduction of two noble metal ions with alcohol according to Eq. (1) is a simple and useful technique to prepare bimetallic nanoparticles:

$$x\, M_1^{m+} + y\, M_2^{n+} + (mx + ny)/2\ RCH_2OH \rightarrow$$
$$M_{1x}M_{2y} + (mx + ny)/2\ RCHO + (mx + ny)\ H^+ \tag{1}$$

Pd/Pt bimetallic nanoparticles can be prepared by refluxing the alcohol-water (1:1, v/v) mixed solution of palladium(II) chloride and hexachloroplatinic(IV) acid in the presence of poly(*N*-vinyl-2-pyrrolidone) (PVP) at $\sim 95°C$ for 1 h [78–80]. During the preparation, the color of the solution drastically changed from pale yellow to dark brown due to the reduction of precursors and the formation of nanoparticles. Thus, UV-vis spectroscopy is useful to confirm both the degree of consumption of precursors by monitoring their ligand-to-metal or metal-to-ligand charge-transfer transitions and the formation of band structures of nanoparticles by monitoring the broad tailing absorption in the range from UV to visible region derived from the inter- and intraband charge-transfer transitions. In the preparation of Pd/Pt bimetallic nanoparticles, an absorption peak at $\sim 260\,nm$ assigned to ligand-to-metal charge-transfer transitions of $[PtCl_6]^{2-}$ and $PdCl_2$ completely disappeared, and the broad tailing absorption was observed in the UV-vis region after 1 h of refluxing. The resulting brownish dispersions of Pd/Pt nanoparticles are quite stable against aggregation over several years. The size and shape of the resulting nanoparticles are easily determined by transmission electron microscopy (TEM). Preparation of a specimen for TEM is carried out by placing a drop of the solution of nanoparticles onto a carbon-coated copper grid, followed by evaporating the solvent. A particle size is controllable by the amount of protective polymer, PVP. When Pd/Pt nanoparticles were prepared in the presence of 40-fold molar amounts of PVP to metal ions, about 1.5 nm nanoparticles were produced [78–80], whereas when prepared in the presence of an equimolar amount of PVP, nanoparticles of about 2.5 nm in size were obtained [81]. The coordination of nitrogen and/or oxygen atoms of PVP to the surface Pd and/or Pt atoms of bimetallic nanoparticles was confirmed by FT-IR and XPS measurements. Therefore, the addition of a large amount of PVP prohibits the nanoparticles from growing, resulting in the production of small particles.

The composition of Pd/Pt nanoparticles can be controlled by the initially added amounts of two different metal ions. Thus, one can obtain Pd/Pt nanoparticles of arbitrary molar ratios of Pd/Pt from 1/0 to 0/1. Then our interests are focused on whether these Pd/Pt bimetallic nanoparticles have a core/shell structure or an alloy structure. A powerful spectroscopic method for structural analysis, which has been extensively applied to the problem of structure determination in nanoparticles, especially in bimetallic nanoparticles, is extended x-ray absorption fine structure analysis, or EXAFS analysis [82–85]. The technique is element-specific and structure-sensitive and gives information on the number and identity of neighboring atoms and their distances from the absorbing atom. The information most usually sought in an EXAFS experiment comprises the number of scattering atoms of each type and their distances from the absorbing atom. In other words, EXAFS experiments give information on the composition of the mean coordination sphere around the absorbing atom. If plural kinds of elements are involved, they can be analyzed both as an absorbing atom and as a scattering atom. Concentrations of up to 50% or more can be realized in dispersions of polymer-stabilized bimetallic nanoparticles, and much higher metal concentrations are accessible when ligand or surfactant-stabilized nanoparticles are used. Let us adopt EXAFS measurement to our samples, the PVP-protected Pd/Pt(4/1) and Pd/Pt(1/1) bimetallic nanoparticles prepared by a simultaneous reduction of $PdCl_2$ and H_2PtCl_6 with a mean diameter of ~ 1.5 nm, each nanoparticle composed of 55 metal atoms (magic number) [78]. In the case of Pd/Pt(4/1) bimetallic nanoparticles, the coordination number of Pt atoms around the Pt atom suggests that the Pt atom coordinates predominantly to the other Pt atoms. Moreover, the coordination numbers are quite different from those calculated for the random model [Figure 2(b)], where 42 Pd atoms and 13 Pt atoms are located completely at random. If 42 Pd atoms are located on the surface and the other 13 Pt atoms are at the center as a core of the fcc-structured nanoparticles [Figure 2(a)], then the Pd/Pt ratio is almost 4/1 and the coordination numbers calculated on the basis of the Pt-core model are quite consistent with the values observed from EXAFS. EXAFS analysis of Pd/Pt(1/1) also suggests not the random [Figure 3(b)] and separated models [Figure 3(c)] but the Pt-core/Pd-shell structure [Figure 3(a) and (d)]. Au/Pd [86,87], Au/Pt [88], Pd/Rh [89–91], and Pt/Rh [92] bimetallic nanoparticles can be prepared and characterized in a similar manner.

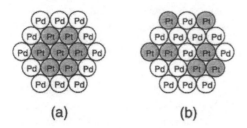

Figure 2 Cross section of the Pd/Pt(4/1) bimetallic nanoparticle models: (a) Pt core model and (b) random model. (Reprinted with permission from Ref. 78. Copyright 1991 American Chemical Society.)

Liu and co-workers also used PVP as a protective agent to prepare Pt/Ru and Pd/Ru bimetallic nanoparticles by $NaBH_4$ reduction of the corresponding mixed metal salts at room temperature [93]. These bimetallic nanoparticles were more stable than the PVP-protected Ru monometallic nanoparticles. Chaudret and co-workers report that the reaction of $Pt(dba)_2$ (dba = dibenzylidene acetone) with $Ru(COD)(COT)$ (COD = 1,5-cyclooctadiene, COT = 1,3,5-cyclooctatriene) in various proportions under H_2 in the presence of PVP led to the formation of Pt/Ru bimetallic nanoparticles of definite compositions resulting from the relative concentration of the two complexes in the initial solution [94]. The progressive incorporation of ruthenium into the platinum matrix led to a structural change from face-centered cubic packing (fcc) for high-platinum contents to hexagonal close-packing (hcp) for high-ruthenium contents. The most interesting observation has been made for the composition Pt/Ru = 1/3: a twinning is observed in the particles

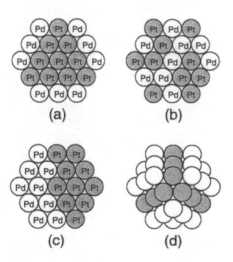

Figure 3 Cross section of the Pd/Pt(1/1) bimetallic nanoparticle models: (a) modified Pt core model, (b) random model, (c) separated model, and (d) the three-dimensional picture of the modified Pt core model. (Reprinted with permission from Ref. 78. Copyright 1991 American Chemical Society.)

(Figure 4) and is accompanied by a quasi-perfect monodispersity of the particle size (1.1 nm), suggesting a definite structure. As shown in Figure 4, high-resolution TEM (HRTEM) provides the atomic-resolution real-space images of nanoparticles using an accelerating voltage larger than 200 kV [95,96]. Crystal structures can be determined by x-ray, electron, and neutron diffraction. In addition, HRTEM is indispensable for the characterization of nanoparticles, particularly when the particle shape and composition are important. Today's TEM is a versatile tool that provides not only atomic-resolution lattice images but also chemical information at a spatial resolution of 1 nm or better, allowing direct identification of a single nanoparticle [97–100]. With a finely focused electron probe, the structural characteristics of a single nanoparticle can be fully clarified.

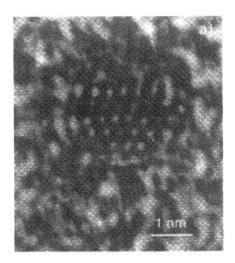

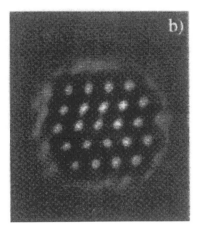

Figure 4 HRTEM micrograph (300 kV) of Pt/Ru(1/3) showing the twinning (a) and image simulation (b). (Reprinted with permission from Ref. 94. Copyright 1999 American Chemical Society.)

PVP was also used as a stabilizer by Esumi et al., who prepared Ag/Pd bimetallic nanoparticles by UV irradiation (253.7 nm) of silver perchlorate and palladium acetate in mixed solutions of acetone and 2-propanol in the presence of PVP [101]. An important feature of UV-vis measurement is to provide the information about the formation processes and the final structures of bimetallic nanoparticles. Au, Ag, Cu, and Hg nanoparticles have intense and rather sharp plasmon absorption bands in the visible range. The structure of bimetallic nanoparticles containing such metal(s), an alloy or a core-shell structure, can affect the appearance of the dispersions and the position of the plasmon absorption peak, as described in Section 11.1.2. The Ag/Pd nanoparticles, having an average size of about 1–4 nm, show the sharp optical absorption band, which shifts to longer wavelengths with increasing molar ratio of silver, indicating that these Ag/Pd bimetallic nanoparticles have alloy structures. Henglein and Brancewicz succeeded in preparing Ag/Hg bimetallic nanoparticles by the simultaneous reduction of corresponding ions with $NaBH_4$ in the presence of poly(ethyleneimine) as a stabilizer [38]. These nanoparticles were suggested to have an alloy structure, because their plasmon absorption band is blue-shifted with increasing mercury/silver ratio. The Ag/Hg alloy nanoparticles are stable up to a mercury-to-silver molar ratio of 2.

Other one-phase preparations of bimetallic nanoparticles include $NOct_4(BHEt_3)$ (Oct = octyl, Et = ethyl) reduction of platinum and ruthenium chlorides to give Pt/Ru(1/1) nanoparticles by Bönnemann et al. [102–104], sono-chemical reduction of gold and palladium ions to provide Au/Pd nanoparticles by Mizukoshi et al. [105,106], and $NaBH_4$ reduction of $PtCl_4^{2-}$ - and $PdCl_4^{2-}$ - dendrimer complexes to give dendrimer-stabilized Pd/Pt nanoparticles by Crooks et al. [107].

In 1994 a noteworthy preparative technique of alkanethiol-protected Au nanoparticles using a two-phase (toluene/water) reaction with a phase transfer catalyst was reported by Brust and co-workers [63]. The strategy of their method consists of growing the metal nanoparticles with the simultaneous attachment of self-assembled ligand monolayers on the growing nuclei. In order to allow the surface reaction to take place during metal nucleation and growth, nanoparticles are grown in a two-phase system. In this method, $AuCl_4^-$ ions are transferred from an aqueous solution to a toluene phase using tetraoctylammonium bromide as a phase transfer reagent and reduced with aqueous sodium borohydride in the presence of a long-chain thiol as a protective agent. Using this two-phase reaction, Hostetler et al. synthesized Au-based Ag, Pt, Pd, and Cu alloy bimetallic nanoparticles protected by dodecanethiol or didodecyl disulfide, whose thermal, optical, and electronic properties can be systematically altered [108]. Han et al. also prepared dodeca-nethiol-protected Ag/Au bimetallic nanoparticles with an average size of ∼4 nm via a two-phase synthetic route in water/toluene mixtures [36]. The position of the surface plasmon band was found to vary linearly between 419 and 522 nm as a function of mole fraction of Au (or Ag) content, which indicates that Ag/Au nanoparticles have an alloy structure. Decanethiol-protected Au/Pt alloy bimetallic nanoparticles of about 2.5 nm in size were prepared in a similar manner [109]. The preparations of Pd/Pt [110] and Au/Pd [111] bimetallic nanoparticles by using water-in-oil (w/o) microemulsions can be included in the two-phase reaction system, although a surfactant molecule itself, involved in the microemulsions, works as a stabilizer for nanoparticles in these cases as well.

Bimetallic Nanoparticles Composed of Noble and
3d-Transition Metals

Transition metals with 3d electrons, such as iron, cobalt, nickel, and copper, are of great importance for catalysis, magnetism, and optics. Although the reduction of 3d-metal ions to zerovalent metals is quite difficult because of their lower redox potentials than those of noble metal ions, a production of 3d-transition metal/noble metal bimetallic nanoparticles is not so difficult. In 1993 we successfully set up a new method for the preparation of PVP-protected Cu/Pd bimetallic nanoparticles according to Eq. (2) [112–114].

$$PVP + Cu^{2+} + Pd^{2+} \frac{(i) \ NaOH \ aq.}{PH = 9 - 11} >$$ PVP-protected Cu/Pd bimetallic hydroxide colloid

$$\frac{(ii) \ 198\,°C \ in \ glycol}{N_2 \ flow} >$$ PVP-protected Cu/Pd bimetallic nanoparticles   (2)

This method is characterized by the formation of bimetallic hydroxide colloid in the first step by adjusting the pH value with a sodium hydroxide solution before the reduction process, which is designed to overcome the problems caused by the difference in redox potentials. Thus, the bimetallic species are completely reduced by a polyol process [115] resulting in Cu/Pd bimetallic nanoparticles, as shown in Figure 5. One can adapt this method to the formation of Cu monometallic nanoparticles, but the size of Cu nanoparticles becomes larger than 100 nm, whereas that of Cu/Pd nanoparticles is smaller than 2 nm, indicating the formation of bimetallic nanoparticles. Auger electron spectroscopy (AES) and XPS proved Cu^0 species in Cu/Pd nanoparticles. The improved polyol method has an advantage of being possibly applied to the preparation of Cu-rich Cu/Pd nanoparticles. In the UV-vis spectrum of PVP-protected Cu/Pd(2/1) bimetallic nanoparticles, the absorption peak at about 560 nm, due to the plasmon resonance of Cu surface [116,117], was not detected, suggesting the formation of bimetallic nanoparticles. In the case of bimetallic nanoparticles, XRD is important to confirm their structures. Since the x-ray diffraction peaks of PVP-protected Cu/Pd(2/1) nanoparticles appeared between the corresponding diffraction lines for Cu and Pd nanoparticles, as shown in Figure 6, the bimetallic alloy phase was clearly found to be formed in Cu/Pd(2/1) bimetallic nanoparticles. As a result of an investigation on the Raman scattering behavior of several molecules adsorbed on the surface of Cu/Pd(4/1) bimetallic nanoparticles, the significant band shifts in the ring modes of *p*-aminobenzoic acid, thiophenol, and bis(3-carboxy-4-nitrophenyl) disulfide were observed, and the Raman signal intensities of the molecules were about $10–10^2$-fold more intense than normal Raman scattering in an aqueous solution [55]. PVP-protected Ni/Pd bimetallic nanoparticles could also be prepared in a similar manner by using nickel sulfate and palladium(II) acetate as precursors [118–120].

Uniform Cu/Pd bimetallic nanoparticles were prepared by thermal decomposition of mixtures of the corresponding acetates in high boiling solvents such as bromobenzene, xylenes, and methyl *iso*-butyl ketone in the absence of stabilizers [121]. The resulting agglomerated bimetallic nanoparticles often contained CuO in addition to the zerovalent metals. Smaller particle sizes and narrower size

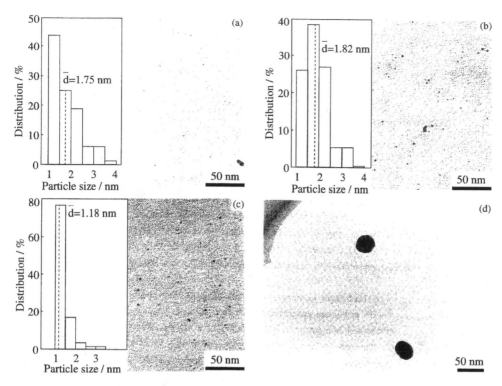

Figure 5 Transmission electron micrographs of PVP-protected (a) Cu/Pd bimetallic nanoparticles (Cu/Pd = 2); (b) Cu/Pd bimetallic nanoparticles (Cu/Pd = 1); (c) Pd nanoparticles; and (d) Cu nanoparticles. (Reprinted with permission from Ref. 112. Copyright 1993 The Chemical Society of Japan.)

distributions without oxide formation were reported by Bradley et al. as a result of an analogous reaction in 2-ethoxyethanol in the presence of PVP [122]. This method, however, could provide Cu/Pd bimetallic nanoparticles containing less than 50 mol% of copper.

The reaction of $Co(\eta^3-C_8H_{13})(\eta^4-C_8H_{12})$ and $Pt_2(dba)_3$ under H_2 in the presence of PVP led to Co/Pt bimetallic nanoparticles smaller than 2 nm in size [123]. It was found that platinum-rich particles adopted an fcc crystalline structure while cobalt-rich particles adopted a nonperiodic polytetrahedral arrangement.

11.2.3 Successive Reduction or Decomposition

Successive reduction (or decomposition) includes the reduction of the first metal ions followed by the reduction of the second metal ions. The second metal is usually deposited on the surface of the first metal due to the formation of the strong metallic bond, resulting in the preferable generation of core-shell structured bimetallic nanoparticles, as shown in Figure 7.

Our first attempt of a successive reduction method was adopted to PVP-protected Au/Pd bimetallic nanoparticles [87]. UV-vis spectroscopy and EXAFS measurement showed that an alcohol reduction of palladium ions in the presence of

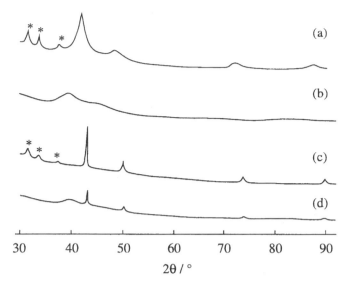

Figure 6 X-ray diffractograms of PVP-protected metal nanoparticles: (a) PVP-protected Cu/Pd (Cu/Pd = 2) bimetallic nanoparticles; (b) PVP-protected Pd nanoparticles; (c) PVP-protected Cu dispersion; (d) physical mixture of (b) and (c) (Cu/Pd = 2). (Reprinted with permission from Ref. 112. Copyright 1993 The Chemical Society of Japan.)

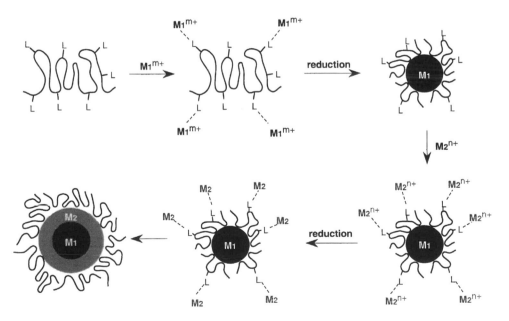

Figure 7 Schematic illustration of formation process of bimetallic nanoparticles prepared by a successive reduction method.

Au nanoparticles gave the mixtures of distinct Au and Pd monometallic nanoparticles, while an alcohol reduction of gold ions in the presence of Pd nanoparticles provided Au/Pd bimetallic nanoparticles. Unexpectedly, these bimetallic nanoparticles did not have a core-shell structure, which had been taken by the bimetallic nanoparticles prepared by a simultaneous reduction of the corresponding two metal ions. This difference in the structure may be derived from the redox potentials of palladium and gold ions. When gold ions are added in the solution of Pd nanoparticles, some palladium atoms on the Pd particles reduce the gold ions to gold atoms to form Pd ions. The oxidized palladium ions are then re-reduced by an alcohol to deposit on the particles. This process may form the particles with a cluster-in-cluster structure and does not produce Pd-core/Au-shell bimetallic nanoparticles. On the other hand, the formation of PVP-protected Pd-core/Ni-shell bimetallic nanoparticles proceeded by a successive alcohol reduction [124].

Schmid and co-workers succeeded in preparing ligand-stabilized Au-core/Pt- or Pd-shell bimetallic nanoparticles by a successive reduction [125–127]. Gold nanoparticles with a diameter of 18 nm [13,128] can be covered by platinum or palladium shells when an aqueous dispersion of Au nanoparticles is added to a solution of H_2PtCl_6 or H_2PdCl_4 and H_3NOHCl. The resulting nanoparticles were stabilized by water-soluble p-$H_2NC_6H_4SO_3Na$. The original red color of the gold nanoparticles then changes to brown-black due to the formation of a Pt- or Pd-shell on the surface of Au nanoparticles. In the case of Au/Pt nanoparticles, a gold core was surrounded by platinum crystals of about 5 nm, while in the case of Au/Pd nanoparticles, a gold core was surrounded by the shell of well-ordered palladium atoms. Energy-dispersive x-ray (EDX) analysis as well as electron energy-loss spectroscopy (EELS) analysis of nanoparticles are attractive methods for assessing the compositions and the valence states on constituent metal elements [125]. These core-shell structures were investigated by the EDX microanalysis, as shown in Figure 8. Analysis of the peripheral part of the particle gives a spectrum of almost pure palladium, while the central part of the particle consists of both palladium and gold. Applying this successive reduction method to large Pd nanoparticles produced the reverse Pd-core/Au-shell bimetallic nanoparticles as well, although addition of Au ions to fine Pd nanoparticles under reduction conditions did not produce the core/shell-structured bimetallic nanoparticles but cluster-in-cluster-structured ones.

Pd-core/Ag-shell bimetallic nanoparticles were prepared via the reduction of silver ions on the surface of Pd nanoparticles (mean radius: 4.6 nm) with formaldehyde by Michaelis and Henglein [129]. The particles became larger and the deviations from the spherical shape became more pronounced with the increasing ratio of the shell to the core. As shown in Figure 9, these Pd/Ag bimetallic nanoparticles possess a surface plasmon absorption band close to 380 nm when more than 10 monolayers of silver are deposited. When the shell thickness is less than 10 atomic layers the absorption band is located at shorter wavelengths. The band disappears when the thickness of the shell is below about three atomic layers.

For an improvement of catalytic and electronic properties of small bimetallic nanoparticles, a certain technique is required to control core-shell stuructures. However, some difficulties have to be overcome in order to obtain controllable core-shell structures. For example, the oxidation of the preformed metal core often takes place by the metal ions added for making the shell when the added metal ions have a high redox potential, and the production of large islands of shell metal proceeds on

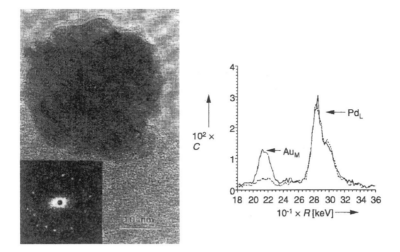

Figure 8 Left: HRTEM image of a "polycrystalline" Au/Pd particle recorded at 400 kV and with the corresponding optical diffractogram inserted. Right: A comparison of EDX spectra recorded from an Au/Pd particle. The counting time of the EDX spectrum recorded with the beam touching the edge of the particle is 10 times longer than the centered case. Dashed line: particle edge, solid line: particle center. C = count rate, R = measuring range, L and M signify electron transitions into the L and M shells, respectively, from higher states, not distinguishable if α or β. (Reprinted with permission from Ref. 125. Copyright 1991 Wiley-VCH.)

the surface of preformed metal core. Recently, we developed a so-called hydrogen-sacrificial protective strategy, which enables us to prepare the bimetallic nanoparticles in the size range of 1.5–5.5 nm with controllable core-shell structures [130]. As illustrated in Figure 10, noble metals like Pd, Pt, and many others have the ability to adsorb molecular hydrogen and split it to form metal hydrides on the metal surface. Hydrogen atoms adsorbed on noble metals have a very strong reducing ability, implying a very low redox potential. So, the second metal ions are easily reduced to metal atoms by the hydrogen atoms adsorbed on the first noble metal nanoparticles, and the obtained second metal atoms are deposited on the surface of the first metal nanoparticles to form bimetallic nanoparticles with a core-shell structure. With this strategy we obtained Pd-core/Pt-shell and Pt-core/Pd-shell bimetallic nanoparticles with various metal compositions. One of the methods for analyzing the surface composition of small bimetallic nanoparticles is IR measurement of carbon monoxide (CO) adsorbed on the metal surface (IR-CO probe measurement). CO is adsorbed on metals not only in on-top sites but also in twofold or threefold sites, depending on the variation of metal and surface structure. The wavenumber of adsorbed CO changes dramatically with the binding geometry [131–135]. We performed IR-CO probe measurement on core-shell Pd/Pt bimetallic nanoparticles and determined their structures [130]. Figure 11 represents the IR-CO probe spectra of Pd-core/Pt-shell bimetallic nanoparticles with different Pd/Pt ratios. In Figure 11a (Pd/Pt = 1/2) and 11b (Pd/Pt = 1/1), only the spectral feature of CO adsorbed on the Pt nanoparticles, i.e., a strong band at 2068 cm^{-1} and a very weak broad band at about 1880 cm^{-1}, can be observed, while that derived from CO

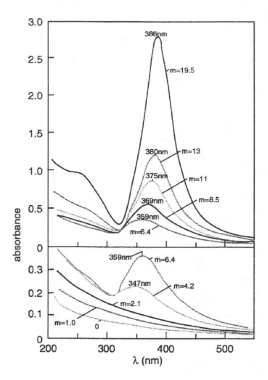

Figure 9 Absorption spectrum of the Pd nanoparticles before (0) and after deposition of various amounts of silver. m: number of monolayers of silver. (Reprinted with permission from Ref. 129. Copyright 1994 American Chemical Society.)

adsorbed on Pd at $1941 \ cm^{-1}$ is completely absent, proving that the Pd core has been completely covered by a Pt shell. The Pd-core/Pt-shell structure thus constructed is reverse in the order of elements in the core and shell to the Pt-core/Pd-shell structure, which is taken by Pd/Pt bimetallic nanoparticles prepared by simultaneous alcohol reduction of Pd and Pt ions in the presence of PVP [78]. Because the latter (Pt-core/Pd-shell) structure can be prepared by the simultaneous reduction, it is considered to be thermodynamically stable. In contrast, the former inverted core/shell structure can be constructed by a kinetically controlled technique, which means that it can be thermodynamically unstable. This is suggested by the observed change of catalytic properties of inverted core/shell-structured bimetallic nanoparticles by heat treatment.

11.3 PROPERTIES OF BIMETALLIC NANOPARTICLES

11.3.1 Catalytic Properties

Bimetallic nanoparticles are of great importance as catalysts because of their high surface-to-volume ratios and their novel catalytic behavior caused by the second metal element. The addition of a second metal can drastically change the catalytic property of nanoparticles, even when the second metal is inactive as a catalyst for the

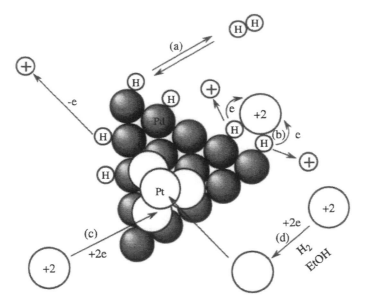

Figure 10 Hydrogen-sacrificial protective strategy for the preparation of bimetallic nanoparticles with a core/shell structure. (Reprinted with permission from Ref. 130. Copyright 1997 American Chemical Society.)

reaction. Many industrial catalysts are bimetallic or multicomponent in order to improve the catalytic properties. Industrial catalysts are generally used in heterogeneous systems, i.e., the nanoparticles are immobilized on certain supports, such as activated carbon, silica, alumina, and so on, prior to use for the sake of profits in processing and improvement in catalytic properties [136–138]. In this section, however, we focus on the catalysis of homogeneous dispersions of bimetallic nanoparticles in three examples. Other catalyses of homogeneous nanoparticles are reviewed elsewhere [8,139–141]. A homogeneous dispersion system can provide an advantage for cyclic usage of metal nanoparticles as catalysts. Usually metal nanoparticles are unstable during catalytic performance. Segregation and atomic rearrangement of metal nanoparticles can change the catalytic activity. In the homogeneous system, however, heat produced by the chemical reaction so easily diffuses into the solution that the structural changes of polymer-protected metal nanoparticles scarcely occur during the catalytic reaction. In other words, a homogeneous system provides such mild reaction conditions and the protection by polymers is so strong that the polymer-protected metal nanoparticle catalysts can usually be reused without loss of the activity.

Hydrogenation

Hydrogenation reactions have been the most extensively studies for measuring the catalytic activities of nanoparticles. Noble monometallic (Pd, Pt, Rh) nanoparticles protected by linear polymers like PVP or polyvinylalcohol have high catalytic activities for hydrogenation of olefins [142–144]. We applied PVP-protected Pd/Pt [79,80] and Au/Pd [86] bimetallic nanoparticles prepared by the simultaneous

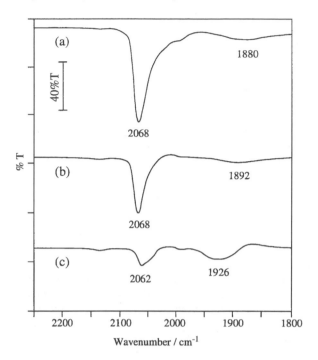

Figure 11 FTIR spectra of CO adsorbed on PVP-protected Pd-core/Pt-shell nanoparticles having a Pt shell: (a) Pt/Pd = 2; (b) Pt/Pd = 1; (c) Pt/Pd = 0.25. [Pd]: 0.1 mmol in 10 mL of CH_2Cl_2. (Reprinted with permission from Ref. 130. Copyright 1997 American Chemical Society.)

reduction of the corresponding metal salts with various compositions to the selective hydrogenation of 1,3-cyclooctadiene to cyclooctene. In both cases, without considering the particle size, the bimetallic nanoparticles with a Pd content of 80 mol% showed the highest activity, which is greater than that of Pd monometallic nanoparticles. Such bimetallic nanoparticles are found to have Pt- or Au-core/Pd-shell structures and Pt- or Au-cores are completely covered with a Pd monolayer. This improvement of catalytic activity can be interpreted only by a ligand effect of the core elements on the surface Pd.

To take the effect of particle size into consideration, the catalytic activity was normalized by the surface area, calculated from the mean diameters of Pd/Pt nanoparticles measured by TEM. The dependence of normalized activity on the metal composition shows a similar result. Because catalytic activity of palladium in hydrogenation of diene is high, while that of platinum is low, the selective existence of Pd atoms on the surface of bimetallic nanoparticles, confirmed by EXAFS and IR-CO, may account for the high catalytic activity. Figure 12 shows the relationship between the Pd composition and the catalytic activity of the surface Pd atoms. The activity of surface Pd atoms is almost constant for Pd/Pt bimetallic nanoparticles containing between 50–95 mol% of palladium. The normalized activity of surface Pd atoms in the nanoparticles is larger than that of monometallic Pd nanoparticles. As the work function of palladium (5.55 eV) is slightly smaller than that of platinum

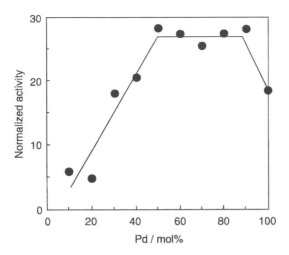

Figure 12 Normalized catalytic activity (in mmol-H_2 per mmol-surface Pd per s) as a function of metal composition of PVP-stabilized Pd/Pt bimetallic nanoparticles. The normalization was determined by the number of Pd atoms on the surface of the nanoparticle, assuming that Pd atoms exist selectively on the surface. (Reprinted with permission from Ref. 80. Copyright 1993 Royal Society of Chemistry.)

(5.64 eV), the electronic interaction between the Pt core and the surface shell Pd atoms results in the Pd atoms' being deficient in electron density. Since substrates with C=C bonds may favor the electron-deficient catalytic sites, the Pd atoms deficient in electron density have higher catalytic activity than normal surface Pd atoms. This is a practical example of the ligand effect of the Pt atom in the core upon the Pd atoms in the surface layer. The selectivity of diene to monoene produced by Pd/Pt and Au/Pd bimetallic nanoparticles was almost 100%. Pd/Rh bimetallic nanoparticles showed the highest catalytic activity for hydrogenation of *cis*-1,3-cyclopentadien to cyclopentene at a composition of Pd:Rh = 1:2 [90].

Bronstein and co-workers have studied catalytic properties of Pd/Pt(4/1), Au/Pd(1/4), and Pd/Zn(4/1) bimetallic nanoparticles, prepared by the simultaneous reduction in block copolymer micelles derived from polystyrene-*block*-poly-4-vinylpyridine (PS-*b*-P4VP), for dehydrolinalool (3,7-dimethylocten-6-yne-1-ol-3, DHL) hydrogenation [145]. It was found from IR-CO probe measurement and XPS that the Au/Pd bimetallic nanoparticles formed Au-core/Pd-shell structures, while Pd/Pt and Pd/Zn nanoparticles formed cluster-in-cluster structures, and that the second metal (Au, Pt, or Zn) acted as a modifier toward Pd, changing both its electronic structure and its surface geometry. This change provides higher catalytic activity of these bimetallic nanoparticles formed in PS-*b*-P4VP micelles than that of Pd nanoparticles, which can mainly be explained by the number of active sites on the particle surface. High selectivity of DHL hydrogenation (99.8% at 100% conversion) to linalool (3,7-dimethyloctadiene-1,6 ol-3) was achieved by all the bimetallic nanoparticle catalysts, the surface of which was chemically modified by pyridine units.

Ni/Pd alloy bimetallic nanoparticles, prepared by the improved polyol reduction method at high temperature, were applied to the catalysis for hydrogenation of nitrobenzene [119] and its derivatives [120]. They were proved to be excellent catalysts with a considerable air-resistance property for hydrogenation of nitrobenzene to aniline. Among the bimetallic nanoparticles with various Ni/Pd ratios, Ni/Pd(2/3) nanoparticles were found to be the most active catalyst for hydrogenation of nitrobenzene, as shown in Figure 13, whereas Ni/Pd(1/4) nanoparticles showed the highest activity for the hydrogenation of nitrobenzene derivatives. Two types of effects may be predicted for the hydrogenation of nitrobenzene and its derivatives on bimetallic Ni/Pd particles: (1) an increase of nickel on the surface may decrease the amount of palladium hydride (Pd-H) on the surface and consequently decrease the activity. (2) Since it is easier to form a charge-transfer complex of a benzene ring with nickel than with palladium, the increase of palladium on the surface may decrease the interaction between the substrate and the particle surface, and consequently decrease the activity. Thus, although the catalytic property of an alloy cannot be deduced from the sum of those of the two metals forming it, a combination of the above two effects may result in the present outcome.

Hydration of Acrylonitrile

From a catalytic point of view, bimetallic nanoparticles composed of 3d-transition metal and noble metal with specific structures will provide a great number of new candidates of catalysts for various chemical reactions, since the catalytic properties of bimetallic nanoparticles can be potentially tailored by both the ligand effect and

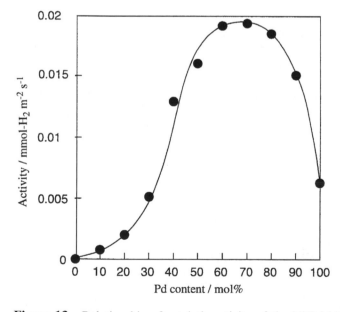

Figure 13 Relationship of catalytic activity of the Ni/Pd bimetallic nanoparticles for the hydrogenation of nitrobenzene, normalized by the surface area of the nanoparticles, versus Ni/Pd composition ratio. (Reprinted with permission from Ref. 119. Copyright 1999 American Chemical Society.)

the ensemble effect. Of particular interest are the excellent catalytic properties of Cu/ Pd alloy bimetallic nanoparticles for both the selective hydration of acrylonitrile [112–114], an important industrial process, and the hydrogenation of carbon–carbon double bonds under mild conditions. The rate of the catalytic hydration of acrylonitrile by the PVP-protected Cu/Pd(2/1) bimetallic nanoparticle catalyst is about seven times higher than that of the PVP-protected Cu monometallic nanoparticles. The activity increases with increasing Cu content within the range of Cu/Pd ratio from 1 to 3. The lack of cyanohydrine formation during the catalytic process suggests a nearly 100% selectivity for the amide [Eq. (3)].

$$CH_2=CH\text{-}CN + H_2O \rightarrow CH_2=CH\text{-}CONH_2 \tag{3}$$

A preliminary idea about the acceleration effect of Pd atoms on the surface on the hydration of acrylonitrile is shown schematically in Figure 14. The coordination of the C–C double bond of the acrylonitrile to the palladium atom in the bimetallic nanoparticles places the C–N triple bond close to the Cu species, thus facilitating the hydration catalyzed by the Cu species. This is a good example of the ensemble effect of bimetallic nanoparticles. The electronic effect of the neighboring atoms in the bimetallic nanoparticles upon the catalytic activity may also be important in this process.

Photo-induced Hydrogen Generation from Water

Conversion of solar energy is a very important theme for human beings. One of the potential methods for solar energy conversion is to construct an artificial photosynthetic system by mimicking natural photosynthesis. One of the most important and simplest concepts is visible light-induced decomposition of water to produce oxygen and hydrogen. Noble metal nanoparticles can work as a catalyst for visible light-induced hydrogen generation from water in the system like EDTA/tris- (bipyridine)ruthenium(II)/methyl viologen/nanoparticle catalyst as shown in Figure 15 [146–148], where metal nanoparticles work as electron mediators to accept

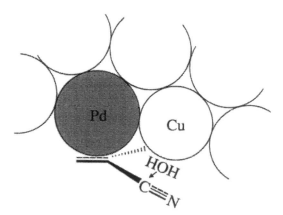

Figure 14   Schematic diagram of the acceleration effect of Pd on the hydration of acrylonitrile. (Reprinted with permission from Ref. 114. Copyright 1994 Wiley-VCH.)

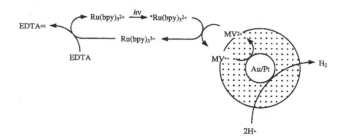

Figure 15 Schematic illustration of the mechanism of visible light-induced hydrogen generation catalyzed by Au/Pt bimetallic nanoparticles and the electron flow mechanism of the reaction. (Reprinted with permission from Ref. 8. Copyright 1998 Royal Society of Chemistry.)

electrons from methyl viologen cation radicals and donate them to protons to produce hydrogen molecules. Pt nanoparticles have been considered to be the best catalyst for this purpose. Recently five monometallic (Au, Pd, Pt, Ru, Rh) nanoparticles were investigated as electron mediators together with four core/shell bimetallic (Au/Pd, Au/Pt, Au/Rh, Pt/Ru) nanoparticles [149–152]. The linear relationship was observed between the electron transfer and the hydrogen generation rate constants, as shown in Figure 16. The bimetallic nanoparticles were generally more active than the corresponding monometallic nanoparticles. The highest catalytic activity was observed for Au/Rh and/or Pt/Ru bimetallic nanoparticles. In this reaction, it was found that the electron transfer from methyl viologen cation radicals to nanoparticles was a rate-determining step. Electron deficiency of surface metal atoms of core/shell bimetallic nanoparticles caused by the difference in the work functions of consisting metals might make it easier to transfer electrons from methyl viologen cation radicals to bimetallic nanoparticles than to monometallic ones.

11.3.2 Magnetic Properties

The bimetallic nanoparticles have long been investigated to optimize and maximize the activity and specificity of catalysts. One of the novel physical properties that the bimetallic nanoparticles show is the magnetic property. The bimetallic nanoparticle with a good magnetic property usually contains 3d-transition metal, such as iron, cobalt, and nickel, as one of the elements. Progress in ultrahigh-density magnetic recording is due in part to the development of metal thin-film media with smaller particles, narrower size distributions, and optimized compositions [153]. In principle, the size control of noble metal nanoparticles is relatively easy, while the formation of the uniform size of 3d-transition monometallic nanoparticles is quite difficult because of easy oxidation. So there are two approaches to prepare the small magnetic bimetallic nanoparticles with narrower size distributions. One is a successive reduction method including the formation of small noble metal nanoparticles, followed by the reduction of 3d-transition metal salts. The other includes the simultaneous reduction or decomposition of noble and 3d-transition

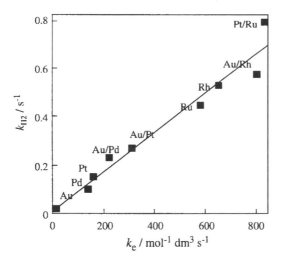

Figure 16   Relationship between electron transfer-rate constant k_e and hydrogen generation rate constant k_{H2}. (Reprinted with permission from Ref. 149. Copyright 2000 IUPAC.)

metal salts. Let us consider two examples of magnetic nanoparticles prepared by these procedures.

Ni/Pd Bimetallic Nanoparticles

Pd is an attractive element from the viewpoint of magnetism, because Pd does not polarize magnetically in bulk metal but has a giant magnetic moment in the presence of a small amount of a ferromagnetic 3d-transition metal. The appearance of the giant magnetic moment was first found by the measurements of paramagnetic susceptibility and ferromagnetic saturation magnetization [154–157]. For example, 1 mol% of Fe induces a strong polarization of Pd in a Pd/Fe alloy, resulting in a giant magnetic moment of about $10\,\mu_B$ per Fe atom [158] and the polarized region of about 1 nm around the Fe atom [159]. This magnetic enhancement can be investigated in detail by using bimetallic nanoparticles. Research on the Pd/3d-transition bimetallic nanoparticles is just at a starting point. The Pd/Fe [160], Pd/Ni [161,162], and Pd/Cu [112–114,134,163–166] nanoparticles have been prepared by the conventional gas evaporation method [160], the sol-gel method [161], simultaneous alcohol reduction of the corresponding metal salts in the presence of the linear polymer [162–165], and decomposition of metal salts [160]. For an accurate measurement of the giant magnetic moment induced on the Pd atoms by the 3d-metal impurities, it is important to prepare a series of the Pd/3d-metal nanoparticles of various compositions without changing their sizes, although this is quite difficult to achieve with previously reported techniques.

We have recently developed a novel synthetic route to produce a series of monodispersed Pd/Ni nanoparticles of similar size by using a successive reduction technique [124]. Nickel atoms were deposited on 2.5-nm Pd nanoparticles corresponding to the "magic number" of 561 atoms with a 5-shell structure [66] at various Ni/Pd molar ratios (1/561, 2/561, 10/561, 15/561, 38/561, 168/561, 281/561, and 561/561) by 1-propanol reduction of nickel acetate in the presence of PVP-protected Pd nanoparticles to produce Pd-core/Ni-shell bimetallic nanoparticles. The size distributions and mean diameters of Pd/Ni nanoparticles are very similar to those of Pd nanoparticles, indicating that the size of Pd nanoparticles was maintained by PVP in 1-propanol under reflux and that the Ni atoms were uniformly deposited on every Pd nanoparticle. In the magnetization measurements of a series of Pd/Ni nanoparticles, the remarkable threshold of the enhancement on the magnetic moment is observed at Ni/Pd = 38/561, and a large enhancement of the magnetic moment corresponding to a giant magnetic moment effect is found above this threshold, as shown in Figure 17 [124,167–169]. The XPS spectra near the Fermi level provide valuable information about the band structures of nanoparticles. Figure 18 shows the XPS spectra near the Fermi level of the PVP-protected Pd nanoparticles, Pd-core/Ni-shell (Ni/Pd = 15/561, 38/561) bimetallic nanoparticles, and bulk Ni powder. The spectra of the nanoparticles become close to the spectral profile of bulk Ni, as the amount of the deposited Ni increases. The change in the XPS spectrum near the Fermi level, i.e., the density of states, may be related to the variation of the band or molecular orbit structure. Therefore, the band structures of the Pd/Ni nanoparticles at Ni/Pd > 38/561 are close to that of the bulk Ni. The appearance of the giant magnetic moment induced on the Pd nanoparticles by the Ni additives is considered to be closely related to the change in the band structure of the Pd/Ni nanoparticles. Fabrication of two-dimensional superlattices of these Pd/Ni nanoparticles would provide a novel magnetic recording device.

Fe/Pt Bimetallic Nanoparticles

Fe/Pt alloys are an important class of materials in permanent magnetic applications because of their large uniaxial magnetocrystalline anisotropy $[K_u \approx 7 \times 10^6 \, \text{J/m}^3]$ [170] and good chemical stability. As the magnetic stability of individual particles scales with the anisotropy constant, K_u, and the particle volume, V, small Fe/Pt

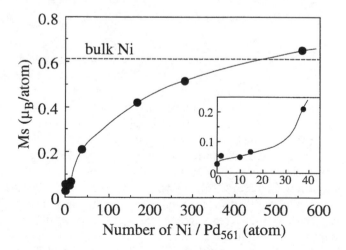

Figure 17 Saturation magnetization of a series of Pd/Ni nanoparticles. The dashed line shows a saturation magnetization of bulk Ni. (Reprinted with permission from Ref. 124. Copyright 1999 American Chemical Society.)

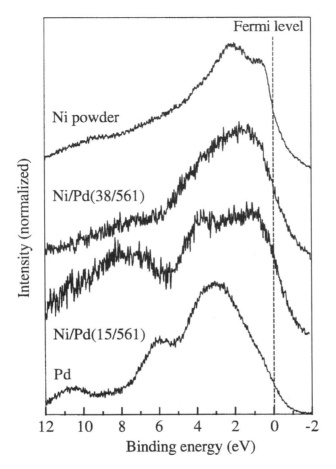

Figure 18 XPS spectra near the Fermi level of the PVP-Pd, PVP-Pd/Ni (Ni/Pd = 15/561, 38/ 561) nanoparticles, and bulk Ni powder. The broken line indicates Fermi level. (Reprinted with permission from Ref. 124. Copyright 1999 American Chemical Society.)

nanoparticles may be suitable for future ultrahigh-density magnetic recording media applications [171].

In 2000, Sun and co-workers succeeded in the synthesis of monodispersed iron-platinum (Fe/Pt) nanoparticles by the reduction of platinum acetylacetonate and decomposition of iron pentacarbonyl in the presence of oleic acid and oleyl amine stabilizers [34]. The Fe/Pt nanoparticle composition is readily controlled, and the size is tunable from 3 to 10 nm in diameter, with a standard deviation of less than 5%. These nanoparticles self-assemble into three-dimensional superlattices. Thermal annealing converts the internal particle structure from a chemically disordered face-centered cubic phase to the chemically ordered face-centered tetragonal phase and transforms the nanoparticle superlattices into ferromagnetic nanocrystal assemblies. The coercivity of these ferromagnetic assemblies is tunable by controlling annealing temperature and time, as well as the Fe:Pt ratio and particle size. The thin film of Fe/ Pt nanoparticles is one of the most promising candidates for future ultrahigh-density magnetic recording device at areal densities in the terabits-per-square-inch regime.

11.4 PROSPECTS

The 21st century is the age of nanotechnology. Because nanoparticles are applicable to a number of areas of technological importance, this research field will attract great scientific attention. It is certain that controlling the primary structures of metallic nanoparticles, i.e., size, shape, crystal structure, and composition, still provides one of the most important areas of research, because these structures determine the physical and chemical properties of metallic nanoparticles. Techniques are now readily available for detailed structural and spectroscopic analyses. Thus, the current interests of this field are focused on the fabrication of superlattices of metallic nanoparticles, which will give the promising materials especially for nanomagnetic, nano-optical, and nano-electronic devices. Bimetallic nanoparticles could be excellent materials to develop these physical properties as well as chemical properties. Development of the methodologies both to control the primary structure of bimetallic nanoparticles for further improvement of catalytic activity and selectivity and to assemble these bimetallic nanoparticles in desired patterns for fabricating the novel nanodevices will be the key to not only nanotechnology but also green chemistry in the 21st century.

ACKNOWLEDGMENT

This work is partly supported by a Grant-in-Aid for the Encouragement of Young Scientists (No. 09740519, TT) and a Grant-in-Aid for Scientific Research in Priority Areas (Nos. 11167281 and 13022274, NT) from the Ministry of Education, Culture, Sports, Science and Technology, Japan.

REFERENCES

1. R. Kubo, *J. Phys. Soc. Jpn.* **1962**, *17*, 975.
2. R.F. Marzke, *Catal. Rev.-Sci. Eng.* **1979**, *19*, 43.
3. G. Schmid, *Clusters & Colloids: From Theory to Application*, VCH: Weinheim, 1994.
4. A. Henglein, *Chem. Rev.* **1989**, *89*, 1861.
5. M.G. Bawendi, M.L. Steigerwald, L.E. Brus, *Annu. Rev. Phys. Chem.* **1990**, *41*, 477.
6. A. Kawabata, *J. Phys. Soc. Jpn.* **1970**, *29*, 902.
7. H. Hirai, N. Toshima, in *Tailored Metal Catalysts*, Y. Iwasawa ed. D. Reidel Pub: Dordrecht, 1986.
8. N. Toshima, T. Yonezawa, *New J. Chem.* **1998**, *22*, 1179.
9. T. Teranishi, K. Nakata, M. Iwamoto, M. Miyake, N. Toshima, *React. Funct. Polym.* **1998**, *37*, 111.
10. T. Teranishi, K. Nakata, M. Miyake, N. Toshima, *Chem. Lett.* **1996**, 277.
11. M. Faraday, *Phil. Trans. Roy. Soc.* **1857**, *147*, 145.
12. J.M. Thomas, *Pure and Appl. Chem.* **1988**, *60*, 1517.
13. J. Turkevich, P.C. Stevenson, J. Hillier, *Disc. Faraday Soc.* **1951**, *11*, 55.
14. J. Turkevich, G. Kim, *Science* **1970**, *169*, 873.
15. J.B. Michel, J.T. Schwartz, in *Catalyst Preparation Science, IV*, B. Delmon, P. Grange, P.A. Jacobs, G. Poncelet, eds. Elsevier: New York, 1987, p 669.
16. R.S. Miner, S. Namba, J. Turkevich, *Proc. 7th Intl. Cong. on Catalysis*, Kodansha: Tokyo, 1981.
17. M. Hansen, *Constitutions of Binary Alloys*, McGraw Hill: New York, 1958.
18. J.H. Sinfelt, *J. Catal.* **1973**, *29*, 308.

19. J.H. Sinfelt, *Acc. Chem. Res.* **1987**, *20*, 134.
20. G. Meitzner, G.H. Via, F.W. Lytle, J.H. Sinfelt, *J. Chem. Phys.* **1983**, *78*, 882.
21. G. Meitzner, G.H. Via, F.W. Lytle, J.H. Sinfelt, *J. Chem. Phys.* **1983**, *78*, 2533.
22. G. Meitzner, G.H. Via, F. Lytle, J.H. Sinfelt, *J. Chem. Phys.* **1985**, *83*, 4793.
23. M.S. Nashner, A.I. Frenkel, D.L. Adler, J.R. Shapley, R.G. Nuzzo, *J. Am. Chem. Soc.* **1997**, *119*, 7760.
24. M.S. Nashner, A.I. Frenkel, D. Somerville, C.W. Hills, J.R. Shapley, R.G. Nuzzo, *J. Am. Chem. Soc.* **1998**, *120*, 8093.
25. C.W. Hills, M.S. Nashner, A.I. Frenkel, J.R. Shapley, R.G. Nuzzo, *Langmuir* **1999**, *15*, 690.
26. T. Teranishi, M. Haga, Y. Shiozawa, M. Miyake, *J. Am. Chem. Soc.* **2000**, *122*, 4237.
27. G. Schmid, M. Bäumle, N. Beyer, *Angew. Chem. Int. Ed.* **2000**, *39*, 181.
28. C.A. Berven, L. Clarke, J.L. Mooster, M.N. Wybourne, J.E. Hutchison, *Adv. Mater.* **2001**, *13*, 109.
29. A. Henglein, *J. Phys. Chem.* **1993**, *97*, 8457.
30. A.P. Alivisatos, *J. Phys. Chem.* **1996**, *100*, 13226.
31. P.V. Kamat, D. Meisel, Studies in Surface Science and Catalysis, Vol. 103; *Semiconductor Nanoclusters—Physical, Chemical, and Catalytic Aspects*, Elsevier: Amsterdam, 1997.
32. A.S. Edelstein, R.C. Cammarata, *Nanoparticles: Synthesis, Properties and Applications*, Institute of Physics Publishing: Bristol, 1996.
33. M. Grätzel, in *Electrochemistry in Colloids and Dispersions*, R.A. Mackay, J. Texter, eds. VCH: Weinheim, 1992.
34. S. Sun, C.B. Murray, D. Weller, L. Folks, A. Moser, *Science* **2000**, *287*, 1989.
35. J. Bellone, M. Mostafavi, H. Remita, J.L. Marignier, M.O. Delcourt, *New J. Chem.* **1998**, 1293.
36. S.W. Han, Y. Kim, K. Kim, *J. Colloid Interface Sci.* **1998**, *208*, 272.
37. S. Link, Z.L. Wang, M.A. El-Sayed, *J. Phys. Chem. B* **1999**, *103*, 3529.
38. A. Henglein, C. Brancewicz, *Chem. Mater.* **1997**, *9*, 2164.
39. G. Mie, *Ann. Phys.* **1908**, *25*, 377.
40. U. Kreibig, M. Vollmer, *Optical Properties of Metal Clusters*, Springer: Berlin, 1995.
41. G.C. Papavassiliou, *Prog. Solid State Chem.* **1980**, *12*, 185.
42. J.A.A.J. Perenboom, P. Wyder, P. Meier, *Phys. Rep.* **1981**, *78*, 173.
43. A.E. Hughes, S.C. Jain, *Adv. Phys.* **1979**, *28*, 717.
44. M. Kerker, *The Scattering of Light and Other Electromagnetic Radiation*, Academic Press: New York, 1969.
45. C.F. Bohren, D.R. Huffman, *Absorption and Scattering of Light by Small Particles*, Wiley: New York, 1983.
46. J.A. Creighton, D.G. Eadon, *J. Chem. Soc., Faraday Trans.* **1991**, *87*, 3881.
47. P. Mulvaney, *Langmuir* **1996**, *12*, 788.
48. G.C. Papavassiliou, *J. Phys. F: Met. Phys.* **1976**, *6*, L103.
49. J. Sinzig, U. Radtke, M. Quinten, U. Kreibig, *Z. Phys. D* **1993**, *26*, 242.
50. L.M. Liz-Marzan, A.P. Philipse, *J. Phys. Chem.* **1995**, *99*, 15120.
51. B.K. Teo, K. Keating, Y.-H. Kao, *J. Am. Chem. Soc.* **1987**, *109*, 3494.
52. P. Mulvaney, M. Giersig, A. Henglein, *J. Phys. Chem.* **1993**, *97*, 7061.
53. T. Sato, S. Kuroda, A. Takami, Y. Yonezawa, H. Hada, *Appl. Organomet. Chem.* **1991**, *5*, 261.
54. A.S. Edelstein, R.C. Cammarata, *Nanoparticles: Synthesis, Properties and Applications*, Institute of Physics Publishing: Bristol, 1996.
55. P. Lu, J. Dong, N. Toshima, *Langmuir* **1999**, *15*, 7980.
56. K.J. Klabunde, Y.-X. Li, B.-J. Tan, *Chem. Mater.* **1991**, *3*, 30.
57. N. Sato, H. Hasegawa, K. Tsuji, K. Kimura, *J. Phys. Chem.* **1994**, *98*, 2143.

58. Y. Takeuchi, T. Ida, K. Kimura, *J. Phys. Chem. B* **1997**, *101*, 1322.
59. A. Fojtik, A. Henglein, *Ber. Bunsen-Ges. Phys. Chem.* **1993**, *97*, 252.
60. M.S. Sibbald, G. Chumanov, T.M. Cotton, *J. Phys. Chem.* **1996**, *100*, 4672.
61. A.M. Morales, C.M. Lieber, *Science* **1998**, *279*, 208.
62. M.S. Yeh, Y.S. Yang, Y.P. Lee, H.F. Lee, Y.H. Yeh, C.S. Yeh, *J. Phys. Chem. B* **1999**, *103*, 6851.
63. M. Brust, M. Walker, D. Bethell, D.J. Schiffrin, R. Whyman, *J. Chem. Soc., Chem. Commun.* **1994**, 801.
64. L.M. Liz-Marzán, A.P. Philipse, *J. Phys. Chem.* **1995**, *99*, 15120.
65. S.W. Han, Y. Kim, K. Kim, *J. Colloid Interface Sci.* **1998**, *208*, 272.
66. T. Teranishi, M. Miyake, *Chem. Mater.* **1998**, *10*, 594.
67. T. Teranishi, I. Kiyokawa, M. Miyake, *Adv. Mater.* **1998**, *10*, 596.
68. T. Teranishi, M. Hosoe, T. Tanaka, M. Miyake, *J. Phys. Chem. B* **1999**, *103*, 3818.
69. T. Teranishi, S. Hasegawa, T. Shimizu, M. Miyake, *Adv. Mater.* **2001**, *13*, 1699.
70. Y. Yonezawa, T. Sato, S. Kuroda, K.I. Kuge, *J. Chem. Soc., Faraday Trans.* **1991**, *87*, 1905.
71. H.H. Huang, X.P. Ni, G.L. Loy, C.H. Chew, K.L. Tan, F.C. Loh, J.F. Deng, G.Q. Xu, *Langmuir* **1996**, *12*, 909.
72. M.M. Mdleleni, T. Hyeon, K.S. Suslick, *J. Am. Chem. Soc.* **1998**, *120*, 6189.
73. B. Li, Y. Xie, J. Huang, Y. Liu, Y. Qian, *Chem. Mater.* **2000**, *12*, 2614.
74. M.T. Reetz, W.H. Helbig, *J. Am. Chem. Soc.* **1994**, *116*, 740.
75. Y.Y. Yu, S.S. Chang, C.L. Lee, C.R.C. Wang, *J. Phys. Chem. B* **1997**, *101*, 6661.
76. J.L. Marignier, J. Belloni, M.O. Delcourt, J.P. Chevalier, *Nature* **1985**, *317*, 344.
77. A. Henglein, *J. Phys. Chem. B* **2000**, *104*, 2201.
78. N. Toshima, M. Harada, T. Yonezawa, K. Kushihashi, K. Asakura, *J. Phys. Chem.* **1991**, *95*, 7448.
79. N. Toshima, K. Kushihashi, T. Yonezawa, H. Hirai, *Chem. Lett.* **1989**, 1769.
80. N. Toshima, T. Yonezawa, K. Kushihashi, *J. Chem. Soc., Faraday Trans.* **1993**, *89*, 2537.
81. N. Toshima, Y. Shiraishi, T. Teranishi, M. Miyake, T. Tominaga, H. Watanabe, W. Brijoux, H. Bönnemann, G. Schmid, *Appl. Organometal. Chem.* **2001**, *15*, 178.
82. J.H. Sinfelt, G.H. Via, F.W. Lytle, *Catal. Rev.-Sci. Eng.* **1984**, *26*, 81.
83. Y. Iwasawa, *X-ray Absorption Fine Structure for Catalysts and Surfaces (World Scientific Series on Synchrotron Radiation Techniques and Applications, Vol. 2)*, World Scientific: Singapore, 1996.
84. *X-ray Absorption: Techniques of EXAFS, SEXAFS and XANES*, D.C. Koningsberger, R. Prins, eds. J. Wiley & Sons: New York, 1988.
85. G.H. Via, K.F. Drake, Jr., G. Meitzner, F.W. Lytle, J.H. Sinfelt, *Catal. Lett.* **1990**, *5*, 25.
86. N. Toshima, M. Harada, Y. Yamazaki, K. Asakura, *J. Phys. Chem.* **1992**, *96*, 9927.
87. M. Harada, K. Asakura, N. Toshima, *J. Phys. Chem.* **1993**, *97*, 5103.
88. T. Yonezawa, N. Toshima, *J. Mol. Catal.* **1993**, *83*, 167.
89. M. Harada, K. Asakura, N. Toshima, *Jpn. J. Appl. Phys.* **1993**, *32*, 451.
90. B. Zhao, N. Toshima, *Chem. Express* **1990**, *5*, 721.
91. M. Harada, K. Asakura, Y. Ueki, N. Toshima, *J. Phys. Chem.* **1993**, *97*, 10742.
92. T. Hashimoto, K. Saijo, M. Harada, N. Toshima, *J. Chem. Phys.* **1998**, *109*, 5627.
93. M. Liu, W. Yu, H. Liu, J. Zheng, *J. Colloid Interface Sci.* **1999**, *214*, 231.
94. C. Pan, F. Dassenoy, M.-J. Casanove, K. Philippot, C. Amiens, P. Lecante, A. Mosset, B. Chaudret, *J. Phys. Chem. B* **1999**, *103*, 10098.
95. Z.L. Wang, *J. Phys. Chem. B* **2000**, *104*, 1153.
96. Z.L. Wang, *Adv. Mater.* **1998**, *10*, 13.
97. Z.L. Wang, *Characterization of Nanophase Materials*, Wiley-VCH: New York, 2000.

98. D.B. Williams, C.B. Carter, *Transmission Electron Microscopy*, Plenum Press: New York, 1996.

99. Z.L. Wang, Z.C. Kang, *Functional and Smart Materials—Structural Evolution and Structure Analysis*, Plenum Press: New York, 1988; Ch. 6.

100. R.F. Egerton, *Electron Energy-Loss Spectroscopy in the Electron Microscope, 2nd ed.*, Plenum Press: New York, 1996.

101. K. Esumi, M. Wakabayashi, K. Torigoe, *Colloids and Surfaces A: Physicochemical and Engineering Aspects* **1996**, *109*, 55.

102. T.J. Schmidt, M. Noeske, H.A. Gasteiger, R.J. Behm, P. Britz, W. Brijoux, H. Bönnemann, *Langmuir* **1997**, *13*, 2591.

103. T.J. Schmidt, M. Noeske, H.A. Gasteiger, R.J. Behm, P. Britz, H. Bönnemann, *J. Electrochem. Soc.* **1998**, *145*, 925.

104. W. Vogel, P. Britz, H. Bönnemann, J. Rothe, J. Hormes, *J. Phys. Chem. B* **1997**, *101*, 11029.

105. Y. Mizukoshi, T. Fujimoto, Y. Nagata, R. Oshima, Y. Maeda, *J. Phys. Chem. B* **2000**, *104*, 6028.

106. Y. Mizukoshi, K. Okitsu, Y. Maeda, T.A. Yamamoto, R. Oshima, Y. Nagata, *J. Phys. Chem. B* **1997**, *101*, 7033.

107. R.M. Crooks, M. Zhao, L. Sun, V. Chechik, L.L. Yeung, *Acc. Chem. Res.* **2001**, *34*, 181.

108. M.J. Hostetler, C.-J. Zhong, B.K.H. Yen, J. Anderegg, S.M. Gross, N.D. Evans, M. Porter, R.W. Murray, *J. Am. Chem. Soc.* **1998**, *120*, 9396.

109. Y. Lou, M.M. Maye, L. Han, J. Luo, C.-J. Zhong, *Chem. Commun.* **2001**, 473.

110. R. Touroude, P. Girard, G. Maire, J. Kizling, M. Boutonnet-Kizling, P. Stenius, *Colloids Surf.* **1992**, *67*, 9.

111. M.-L. Wu, D.-H. Chen, T.-C. Huang, *Chem. Mater.* **2001**, *13*, 599.

112. N. Toshima, Y. Wang, *Chem. Lett.* **1993**, 1611.

113. N. Toshima, Y. Wang, *Langmuir* **1994**, *10*, 4574.

114. N. Toshima, Y. Wang, *Adv. Mater.* **1994**, *6*, 245.

115. F. Fievet, J.P. Lagier, B. Blin, *Solid State Ionics* **1989**, *32/33*, 198.

116. H. Hosono, Y. Abe, N. Matsunami, *Appl. Phys. Lett.* **1992**, *60*, 2613.

117. D.M. Trotter, Jr., J.W.H. Schreurs, P.A. Tick, *J. Appl. Phys.* **1982**, *53*, 4657.

118. N. Toshima, P. Lu, *Chem. Lett.* **1996**, 729.

119. P. Lu, T. Teranishi, K. Asakura, M. Miyake, N. Toshima, *J. Phys. Chem. B* **1999**, *103*, 9673.

120. P. Lu, N. Toshima, *Bull. Chem. Soc. Jpn.* **2000**, *73*, 751.

121. K. Esumi, T. Tano, K. Torigoe, K. Meguro, *Chem. Mater.* **1990**, *2*, 564.

122. J.S. Bradley, E.W. Hill, C. Klein, B. Chaudret, A. Duteil, *Chem. Mater.* **1993**, *5*, 254.

123. T. Ould Ely, C. Pan, C. Amiens, B. Chaudret, F. Dassenoy, P. Lecante, M.-J. Casanove, A. Mosset, M. Respaud, J.-M. Broto, *J. Phys. Chem. B* **2000**, *104*, 695.

124. T. Teranishi, M. Miyake, *Chem. Mater.* **1999**, *11*, 3414.

125. G. Schmid, A. Lehnert, J.-O. Malm, J.-O. Bovin, *Angew. Chem. Int. Ed. Engl.* **1991**, *30*, 874.

126. G. Schmid, H. West, J.-O. Malm, J.-O. Bovin, C. Grenthe, *Chem. Eur. J.* **1996**, *2*, 1099.

127. A.F. Lee, C.J. Baddeley, C. Hardacre, R.M. Ormerod, R.M. Lambert, G. Schmid, H. West, *J. Phys. Chem.* **1995**, *99*, 6096.

128. G. Schmid, A. Lehnert, *Angew. Chem. Int. Ed. Engl.* **1989**, *28*, 780.

129. M. Michaelis, A. Henglein, *J. Phys. Chem.* **1994**, *98*, 6212.

130. Y. Wang, N. Toshima, *J. Phys. Chem. B* **1997**, *101*, 5301.

131. J.S. Bradley, J.M. Millar, E.W. Hill, S. Behal, B. Chaudret, A. Duteil, *Faraday Discuss. Chem. Soc.* **1991**, *30*, 1312.

132. J.S. Bradley, E.W. Hill, S. Behal, C. Klein, B. Chaudret, A. Duteil, *Chem. Mater.* **1992**, *4*, 1234.
133. D. de Caro, J.S. Bradley, *Langmuir* **1997**, *13*, 3067.
134. J.S. Bradley, E.W. Hill, B. Chaudret, A. Duteil, *Langmuir* **1995**, *11*, 693.
135. J.S. Bradley, G.H. Via, L. Bonneviot, E.W. Hill, *Chem. Mater.* **1996**, *8*, 1895.
136. B. Coq, P.S. Kumbhar, C. Moreau, P. Moreau, F. Figueras, *J. Phys. Chem.* **1994**, *98*, 10180.
137. G. Schmid, H. West, H. Methies, A. Lehnert, *Inorg. Chem.* **1997**, *36*, 891.
138. D.S. Shephard, *Stud. Surf. Sci. Catal.* **2000**, *129*, 789.
139. L.N. Lewis, *Chem. Rev.* **1993**, *93*, 2693.
140. J.S. Bradley, *Schr. Forschungszent. Juelich, Mater, Mater.* **1998**, *1*, D6.1.
141. H. Bönnemann, G. Braun, G.B. Brijoux, R. Brinkman, A.S. Tilling, K. Seevogel, K. Siepen, *J. Organomet. Chem.* **1996**, *520*, 143.
142. H. Hirai, Y. Nakao, N. Toshima, *Chem. Lett.* **1978**, 545.
143. H. Hirai, Y. Nakao, N. Toshima, *J. Macromol. Sci., Chem.* **1979**, *A12*, 1117.
144. H. Hirai, Y. Nakao, N. Toshima, *J. Macromol. Sci., Chem.* **1979**, *A13*, 727.
145. L.M. Brostein, D.M. Chernyshov, I.O. Volkov, M.G. Ezernitskaya, P.M. Valetsky, V.G. Matveeva, E.M. Sulman, *J. Catal.* **2000**, *196*, 302.
146. N. Toshima, T. Takahashi, H. Hirai, *Chem. Lett.* **1986**, 35.
147. N. Toshima, T. Takahashi, H. Hirai, *Chem. Lett.* **1987**, 1031.
148. N. Toshima, T. Takahashi, H. Hirai, *J. Macromol. Sci., Chem.* **1988**, *A25*, 669.
149. N. Toshima, *Pure Appl. Chem.* **2000**, *72*, 317.
150. N. Toshima, K. Hirakawa, *Polym. J.* **1999**, *31*, 1127.
151. N. Toshima, K. Hirakawa, *Appl. Surf. Sci.* **1997**, *121/122*, 534.
152. T. Yonezawa, N. Toshima, *J. Mol. Catal.* **1993**, *83*, 167.
153. J.S. Li, M. Mirzamaani, X.P. Bian, M. Doerner, S.L. Duan, K. Tang, M. Toney, T. Arnoldussen, M. Madison, *J. Appl. Phys.* **1999**, *85*, 4286.
154. H. Crangle, *Phil. Mag.* **1960**, *5*, 335.
155. J. Crangle, W.R. Scott, *J. Appl. Phys.* **1965**, *36*, 921.
156. R.M. Bozorth, P.A. Wolff, D.D. Davis, V.B. Compton, J.H. Wernick, *Phys. Rev.* **1961**, *122*, 1157.
157. A.M. Clogston, B.T. Matthias, M. Peter, H.J. Williams, E. Corenzwit, R.C. Sherwood, *Phys. Rev.* **1962**, *125*, 541.
158. G.J. Nieuwenhusy, *Adv. Phys.* **1975**, *24*, 515.
159. D. Shaltiel, J.H. Wernick, H.J. Williams, M. Peter, *Phys. Rev.* **1964**, *135*, A1346.
160. T. Taniyama, E. Ohta, T. Sato, M. Takeda, *Phys. Rev. B* **1997**, *55*, 977.
161. W. Mörke, R. Lamber, U. Schubert, B. Breitscheidel, *Chem. Mater.* **1994**, *6*, 1659.
162. N. Toshima, P. Lu, *Chem. Lett.* **1996**, 729.
163. Y. Wang. H.F. Liu, N. Toshima, *J. Phys. Chem.* **1996**, *100*, 19533.
164. L. Zhu, K.S. Liang, B. Zhang, J.S. Bradley, A.E. DePristo, *J. Catal.* **1997**, *167*, 412.
165. J.S. Bradley, E.W. Hill, C. Klein, B. Chaudret, A. Duteil, *Chem. Mater.* **1993**, *5*, 254.
166. S. Giorgio, C. Chapon, C.R. Henry, *Langmuir* **1997**, *13*, 2279.
167. N. Nunomura, T. Teranishi, M. Miyake, A. Oki, S. Yamada, N. Toshima, H. Hori, *J. Magn. Magn. Mater.* **1998**, *177*, 947.
168. N. Nunomura, H. Hori, T. Teranishi, M. Miyake, S. Yamada, *Phys. Lett. A* **1998**, *249*, 524.
169. H. Hori, T. Teranishi, T. Sasaki, M. Miyake, Y. Yamamoto, S. Yamada, H. Nojiri, M. Motokawa, *Physica B* **2001**, *294*, 292.
170. K. Inomata, T. Sawa, S. Hashimoto, *J. Appl. Phys.* **1998**, *64*, 2537.
171. D. Weller, A. Moser, *IEEE Trans. Magn.* **1999**, *35*, 4423.

12

Physicochemical Aspects of Preparation of Carbon-Supported Noble Metal Catalysts

P. A. SIMONOV and V. A. LIKHOLOBOV

Boreskov Institute of Catalysis at the Russian Academy of Sciences, Novosibirsk, Russia

CHAPTER CONTENTS

SUMMARY

Preparation of carbon-supported noble metal catalysts (Me/C) is usually based on supporting metal precursors on carbon followed by their transformation into the metal particles. Design of an appropriate catalyst implies an optimal selection of both the support and the method of synthesis of the active component, which requires understanding the following issues:

1.   Which physicochemical properties of the support determine its interaction with the metal particles?
2.   How do the initial metal compounds interact with the carbon surface?
3.   What are the state of the supported precursors and the manner of their fixation on the support?

4.   What are regularities of the transformations of these precursors to the metal particles during calcining and reduction steps?

Knowledge in the field of preparation of carbon-supported metal catalysts is expanded as experimental data are collected and physicochemical techniques for studying catalysts are developed, but it remains insufficient. Whereas development of catalytic compositions, even though based on some general empirical approaches (see *e.g.*, [1,2]), is still a matter of skill, the answers to the above-stated questions are often treated as secrets ("know-how") of the catalyst preparation. This is, evidently, the reason for the availability of only fragmentary and sometimes contradictory information regarding these aspects of synthesis of Me/C. In this connection the authors of this chapter see their goal not only as providing a survey of the published data on the problem of the catalyst preparation but also as creating model concepts to consolidate these data and to account for some existing contradictions.

12.1 PHYSICOCHEMICAL PROPERTIES OF CARBON SUPPORT AND INTERACTION OF METAL NANOPARTICLES WITH THE SUPPORT SURFACE

12.1.1 Factors Affecting Stabilization of Metal Particles on the Surface of Quasi-Graphitic Carbon Supports

The main role of support is to prevent recrystallization of metal particles, which usually proceeds via sintering, but sometimes by changing their shape or microstructure. Several approaches are currently in use to perceive how metal particles are attached to the surface of carbon support: (1) visualization of nanoparticles and/or their migration using high-resolution transmission electron microscopy (HRTEM) or scanning tunneling microscopy (STM); (2) quantum-chemical calculations of the interaction between the particles and the support; and (3) studies of the *variation* of metal dispersion upon high-temperature treatment.

It should be kept in mind, however, that the third approach is rather indirect and can give valuable information only provided that the following requirements are met. First, the migration of supported metal particles should be the limiting step of the sintering process. Second, it is of crucial importance that the initial state of the metal or its precursor (size distribution of the particles, distribution through the support grain and in the pores, as well as the surface concentration) in the series of samples to be compared is similar if not the same. (See ahead in this section for particle-size distribution as an independent factor determining sintering and Section 12.2 for the influence of the nature of supported precursors on the dispersion of Me/C.)

Several mechanisms of interaction between particles of solids are known [3]. *Mechanical adhesion* is achieved by flowing a metal into the support pores. The *molecular mechanism of adhesion* is based on the Van der Waals forces or hydrogen bonds, and the *chemical mechanism*—on the chemical interaction of the metal particles with the support. The *electric theory* relates adhesion to the formation of an electric double layer (EDL) at the adhesive–substrate interface. Finally, the *diffusion mechanism* implies interpenetration of the molecules and atoms of the interacting phases, which results in the interface blurring. These insights into the nature of adhesion can be revealed in the papers about the interaction of transition metal nanoparticles with the surface of quasi-graphitic carbon materials and are discussed below.

Influence of Textural Characteristics

The Role of Textural Properties of Carbons

Physical characteristics of a support, namely porosity and specific surface area, have long been understood to play a key role in stabilizing active components of the catalysts in dispersed state. Explicitly or implicitly, they reflect topological properties of the carbon surface, namely the nature and quantity of (1) "traps" (potential wells for atoms and metal particles), which behave as sites for nucleation and growth of metal crystallites and (2) hindrances (potential barriers) for migration of these atoms and particles [4,5]. An increase in the specific surface area and the micropore volume results, as a rule, in a decrease in the size of supported metal particles. Formal kinetic equations of sintering of supported catalysts always take into consideration these characteristics of a support [6].

The most effective mechanical trapping of a metal particle at a micropore mouth can be achieved only at a rather high temperature as a result of metal flowing into the pore. A similar type of particle trapping may also be observed for metals on nonmicroporous supports upon exposing them to a chemically reactive atmosphere (H_2, air, water vapor) at about $500°C$. Catalyzing carbon gasification, metal particles can "dig" small pits in the carbon support surface, where they are then trapped [7,8].

The Role of Surface Nanotexture

A carbon matrix is built up by quasi-graphitic crystallites; fragments of both basal and edge planes can be exposed to the surface. A surface built up solely by basal planes is rather uniform energetically and is considered *homogeneous*. Otherwise, it is *heterogeneous*. Both quantum chemical calculations [9,10] and experimental observations [8,11–17] indicate that heterogeneous surfaces can better stabilize metal in a highly dispersed state. Metal particles are arranged on carbon surfaces along the edges of graphene networks at intercrystalline boundaries or at steps. Each step comprises fragments of both basal and edge planes of carbon crystallites. Stabilization of metal nanoparticles at steps is also supported by the fact that, other conditions being equal, dispersion of Pd/C catalysts increases with an increase in the carbon surface heterogeneity [14–16].

The strength of metal particle bonding to the support surface can be described by the adhesion work (W_a), which in the first approximation is related to the surface tension at the support–gas ($\gamma_{s/g}$), metal–gas ($\gamma_{m/g}$), and support–metal ($\gamma_{s/m}$) interface and to the particle–support contact area (S) [18]:

$$W_a = \left(\gamma_{s/g} + \gamma_{m/g} - \gamma_{s/m}\right) S \tag{1}$$

Hence, the higher the surface tension of both the metal and the support and the lower the interface tension, the stronger the metal–support bonding is. Surface tension of quasi-graphite crystallites is anisotropic, being much higher at the basal plane than at the edge plane, since the former stems from the chemical interaction of carbon atoms within the graphite network, while the latter stems from the weaker Van der Waals interaction between these networks. Thus, from the first two terms of

Eq. (1) one might expect stronger bonding of metal particles to the basal than to the edge plane. However, the above-cited experimental data point to the opposite fact, suggesting that the gain in the particle–support interaction is provided by the last term of Eq. (1) namely by a considerable decrease in the surface tension at the metal–carbon interface and by an increase in the contact area due to the lateral interaction of the particle with the steps at the support surface.

Basically, the effect of the surface nanotexture on the strength of metal–carbon bonding may occur as a result of epitaxy or interdiffusion of atoms in the contact region of a metal crystallite and carbon support. However, information concerning these aspects of the metal–carbon interaction is scarce. Graphite-supported Pd and Pt crystallites are oriented their {202} for Pd [19] and {111} or {110} for Pt [20–22] planes parallel to the basal plane of graphite substrate, but this epitaxial interaction is relatively weak [19–21,23]. In contrast, Pd particles supported on amorphous carbons are in random orientation [19,25]. Hence, heterogeneous support surfaces comprise structurally different sites for metal-particle stabilization.

An influence of the interdiffusion of atoms in the contact region between metal and carbon crystallites has not been studied much. Based on HRTEM data [26], it has been proposed that Ru atoms in Ru/C catalysts percolate into the carbon matrix at the metal–carbon interface, so that metal particles become mechanically bonded by adjacent graphene layers. On the contrary, incorporation of carbon atoms into the metal crystallite lattice is characteristic, for example, of Pt and Pd, which may ultimately be transformed into carbidelike MeC_x ($x \leqslant 0.15$) structures detected by the XRD method [27–29]. Similar changes in the structure of Pt and Pd crystallites supported on carbon were observed in [30–33] as a result of heating the catalysts. A number of reasons may be responsible for the appearance of mobile carbon atoms on the carbon surface. They can be generated at temperatures as high as 900–1000 °C [34]. Indeed, PtC_x is observed under these particular circumstances [30]. However, PdC_x has been detected at far lower temperatures: from 30 °C [31] to 300 °C [32,33,35]. This is likely due to high reactivity of either amorphous carbon, formed upon decomposition of Pd organometallic precursor [32,33,35], or disordered fragments of carbon support [31,33,35]. Aliphatic species are known to be present on disordered carbon surfaces [34,36,37]. Under heating they may be decomposed catalytically on the metal surface to form carbon atoms [27–29]. Comparative data on the stability of Me/C and MeC_x/C samples to sintering are not available.

Influence of Chemical State of the Support Surface

Oxygen-Containing Groups as Anchors for Metal Nanoparticles

The ability of carbon supports to stabilize dispersed metals is often ascribed to the presence of the acidic oxygen-containing groups at the edges of graphene networks in analogy with anchoring metal particles (Lewis bases) to zeolite surfaces by means of acid–base interaction [38]. The results of [39,40] might indeed be taken as an argument in favor of this hypothesis. Thus, dispersion of Pd has been found to fall abruptly (by a factor of 5) [40] upon heating the catalysts supported on oxidized carbons (C_{ox}) in H_2 above the temperature of thermal decomposition of carboxyl groups (300–400 °C), as if these groups were indeed anchors of metal particles at temperatures below 300 °C. However, surprisingly the effect of any oxygen-containing groups on stabilization of carbon-supported Pd particles in the range between 150 and 300 °C is negligible (see Table 1). It is worth mentioning that carboxylic groups decompose catalytically in the presence of supported Pd (or Pt) particles and hydrogen, starting with 100–150 °C [40,42]. Hence, the opinion about their ability to prevent migration of the metal particles at higher temperatures appears to be controversial.

Taking into account the low thermal stability of carboxylic groups, Prado-Burguete et al. [43], to explain the experimental data on Pt/C catalysts sintering, attribute the dominant role in the interaction with Pt particles to phenol (weak acidic) and quinone (neutral) groups, which are more stable. Thus, Pt/C_{ox} prepared via adsorption of H_2PtCl_6 on oxidized active carbon and following reduction in H_2 at 350–500 °C demonstrated higher dispersion as well as higher stability to sintering than the catalysts supported on nonoxidized carbon. However, this study does not clarify how Pt particles may be fixed by these groups. First, according to the commonly accepted mechanism [38], an expected efficiency of these groups in preventing sintering of metal particles is low due to their low acidity. Second, the surface concentration of phenol and quinone groups attained in [43] was about $0.5 \, mmol/1000 \, m^2$ (i.e., less than one group per $2 \times 2 \, nm^2$). This means that a 1–2-nm Pt particle is coordinated by no more than one to four surface groups, which is obviously not sufficient to prevent its migration above 350 °C. Third, for Pt/C prepared using the same procedure, the effect of the concentration of the surface groups on platinum dispersion may be opposite [44] or volcanolike [46], or there may be none [45].

Table 1 Influence of the Chemical State of the Carbon Surface on Dispersion of Pd Particles Obtained by Reduction of Adsorbed $PdCl_2$ with H_2 at 250 °C/2 h (S1-series)* and 150 °C and 300 °C/1 h (A-series)[†]

Pd/C catalyst[1]	S_{BET} (m^2/g)	Carbon Support				Metal Dispersion (After the Reduction by H_2 at Different Temperatures)	
		Surface Oxygen Groups Content[2] (mmol/g)				HRTEM	CO Adsorption
		Acidic	Phenolic	Quinonic	Basic		
						250 °C	250 °C
1%Pd/S1	410	0.048	0.180	0.072	0.066	0.72	0.36
1.1%Pd/S1$_{ox}$−1 (H_2O_2)	380	0.060	0.258	0.132	0.030	0.76	0.34
1%Pd/S1$_{ox}$−2 ($KMnO_4$)	380	0.228	0.212	0.070	0.010	0.73	0.46
1%Pd/S1$_{ox}$−3 ($KMnO_4$)	360	0.304	0.084	0.235	0.00	0.71	0.45
1%Pd/S1$_{ox}$−4 ($KMnO_4$)	—	0.334	0.006	0.230	0.00	0.79	0.46
						150 °C	300 °C
2.1%Pd/A$_{HT}$	660	0.065	0.035	0.260	0.450	0.64	0.20
2.1%Pd/A	630	0.155	0.200	0.535	0.400	0.58	0.19
2.1%Pd/A$_{ox}$ (O_2, 380 °C)	650	0.560	0.380	0.490	0.300	0.73	0.20

[1] S1 = Sibunit (activated pyrolytic carbon), A = Anthralur (active carbon). Subscript ox means oxidized support (with the oxidant indicated in the brackets); HT the support calcined in inert atmosphere at 1000 °C.
[2] Determined from Na_2CO_3, NaOH, NaOEt, and HCl consumption in accordance with Boehm's method.
* *Source*: Ref. 16.
[†] *Source*: Refs. 41 and 42.

Discrepancies between the results of [43–46] are mainly accounted for by the influence of the oxygen containing groups on the ratio between the amounts of different metal precursors adsorbed on the carbon surface from H_2PtCl_6 solutions rather than on the mobility of Pt particles in the final Pt/C catalysts (see Section 12.2.1). Although the ability of the oxygen-containing groups to stabilize noble metal nanoparticles on a carbon surface has been under discussion for the last dozen years, we have not found any direct evidence of this point of view, such as a direct observation (for example, by XPS or FTIR spectroscopy) of the metal particles interacting with the surface groups. Many conclusions about the nature of the metal–support interaction are based on the indirect observations, such as a decrease in the metal dispersion under the reduction of Me/C catalysts in H_2. However, the *initial state of the metal precursors was not thoroughly specified*. The latter makes it difficult to correctly interpret the observed differences in sintering of metal supported on different carbons.

In this connection the data of Table 1 appear to be more appropriate to deduce a possible role of the surface groups in the stabilization of metal particles, since the state of the adsorbed $PdCl_2$ is well established [54–57]. Indeed, surface $PdCl_2$ complexes evenly distributed over the support surface are the only precursors for all catalysts of the Pd/S1 series, with no less than 90% of the precursors in the same state. Taking these data into account, we infer that the surface oxygen-containing groups seem to be incapable of stabilizing the highly dispersed state of the supported palladium via interaction with its particles. Furthermore, destruction of the oxygen-containing groups in Pd/C catalysts upon heating [16,39–42,47,48] may simply coincide with the metal sintering within the same temperature range or is a side effect of the sintering process (but not vice versa). It should be mentioned that anomalously high hydrogen uptake is observed for Pd/C catalysts heated in H_2 and attributed to hydrogen spillover [40,49–51]. For the catalysts on oxidized supports, it starts at $100\,^{\circ}C$ simultaneously with the decomposition of carboxylic groups [40] and at about $400\,^{\circ}C$ in the catalysts on nonoxidized carbons [49–52]. Heating of freshly prepared Pd/Sibunit catalysts in hydrogen at $250\,^{\circ}C$ and above results in an irreversible uptake of hydrogen, which is consumed, in particular, for hydrogenation of surface $>C=C<$ bonds of the carbon support matrix (Table 2). In this connection a question about the role of unsaturated C–C bonds of the support in anchoring metal nanoparticles arises.

π-Sites on Carbon Surface as Anchors for Metal Nanoparticles

Noble metals, both in zerovalent and in ionic states (in particular in the low-oxidation state), are prone to form stable π-complexes with multiple C–C bonds of organic compounds [53]. The stability of the π-complexes increases if the ligand contains electron acceptor groups (for example, oxygen-containing or aryl) in close proximity of the $>C=C<$ bond.

Fragments containing $>C=C<$ bonds compose the most numerous class of functional groups of carbons. However, methods for identification and quantification of these surface π-sites have not been adequately developed. Chemisorption of $PdCl_2$ on powdered carbon from aqueous H_2PdCl_4 [54–57] can, in principle, be adapted for this purpose. This method, based on data of mathematical simulation of equilibria of $PdCl_2$ adsorption on various carbon materials from H_2PdCl_4 solutions as well as XPS studies and eluent analysis of adsorbed Pd^{II} compounds, allows us to

Table 2 Concentration of π-Sites of A_2 Type on the Surface of Sibunit (S1)

#	Carbon Sample	Treatment in H_2	S_{BET} (m²/g)	H_2 Chemisorbed (mmol/g)	Acidic Groups[1] (mmol/g)	A_2 Sites (mmol/g)
1	S1	*No*	410	—	0.048	0.21
2	S1	350 °C, 5 h	400	0.075	0.020	0.18
3	2% Pd/S1	250 °C, 3 h	—	0.170	—	0.13[2]

[1] From Na_2CO_3 consumption.
[2] After Pd removal by washing of Pd/C with hot 4 M HCl in air.

distinguish π-sites of three types differing in the strength of their complexes with $PdCl_2$. These are sites of weak (A_1), strong (A_2), and very strong—or irreversible—(A_3) adsorption. Comparative studies [54–57,60,141] of a number of carbons (graphite, active carbon, carbon black, Sibunit, filamentous, onionlike carbon, and fullerene C_{60}) including also chemically modified (treated with oxidants, Cl_2 or H_2) samples revealed that an A_1 site is a hexagon in a graphene layer, which can be considered as a fragment of basal plane of carbon crystallite. A_2 sites belong to surface steps, being composed of >C=C< fragments of both basal and edge planes, and can be considered a chelating macroligand. A_3 sites are localized in micropores: their amount is an order of magnitude lower than that of A_1 and A_2.

It has been already mentioned that the metal nanoparticles contacting the edges of graphene networks are most strongly bound to the support. One can further assume that peripheral metal atoms of the metal particles form π-complexes with >C=C< ligands of A_2 sites. These coordinated ligands can be destroyed in the presence of H_2 (or O_2) at high temperatures through hydrogenation (or oxidation), which results in weakening of the metal–support bonds and favors migration of metal particles. So far, the arguments in favor of this hypothesis are available only for Pd/C catalysts [16,57–59] and are considered below. First, the electron structure of Pd clusters on the surface of polycrystalline graphite can be adequately described in terms of mixing Pd 4d and π^* states of graphite [58]. Direct similarity with the formation of Pd-olefin π-complexes can be noticed: a π-bond in such a complex is formed between an occupied 4d orbital of the metal atom and an unoccupied molecular π^*-orbital of the C=C fragment [53]. Second, sintering of supported palladium is accompanied by a considerable decrease in the concentration of A_2 sites due to hydrogenation of >C=C< bonds (Table 2).

Third, the average size of palladium particles and the width of their size distribution decrease as the content of A_2 sites on the support increases (Table 3).

Fourth, for a set of heterogeneous carbon supports that show nearly the same surface densities of A_2 sites, the average size of Pd particles and the width of the particle size distribution decrease with increasing the strength of A_2 sites measured as the value of the $PdCl_2$ adsorption equilibrium constant (Figure 1). This correlation between the strength of bonding catalyst precursor to the support surface and the mobility of Pd^0 particles is quite appropriate, since the stability of olefinic π-complexes of Pd^0 is known to vary along with that of Pd^{II} [53]. The strength of A_2 sites increases as the size of quasi-graphitic crystallites constituting the support matrix decreases [56]. Since finer carbon crystallites are stronger electron acceptors,

Table 3 Influence of Carbon Surface Nanotexture and Surface Concentration of A_2 Sites on the Average Size of Pd Particles[1]

Catalyst Sample	Carbon Support Characteristics				Mean Particle Size (HRTEM, nm)		
	S_{BET} (m^3/g)	V_{mi} (cm^3/g)	Surface Nanotexture[2]	Number of A_2 Sites (μmol/m^2)	d_n	d_s	d_v
Pd/Corax 3 graph	63	0.024	Basal planes	0.17	2.89	3.08	3.27
Pd/PM-105	110	—	Surface steps	0.92	1.62	1.73	1.83
Pd/CFC-2	85	0.028	Edge planes	1.57	1.09	1.12	1.15

[1] 1% Pd/C catalysts were prepared by reduction of PdCl$_2$/C in H$_2$ at 150 °C/1 h followed by sintering at 250 °C/3 h. (*Source*: Ref. 16.)
[2] Predominant orientation of fragments of quasi-graphitic crystallites planes at the support surface. (*Source*: Refs. 56 and 60.)

the strength of the metal–support interaction may be further assumed to increase with the efficiency of the charge redistribution between the metal particle and the coordinated >C=C< fragments of the support surface. (For example, the strength of π-complexes of Pt^0 with polycyclic aromatic molecules increases as these ligands

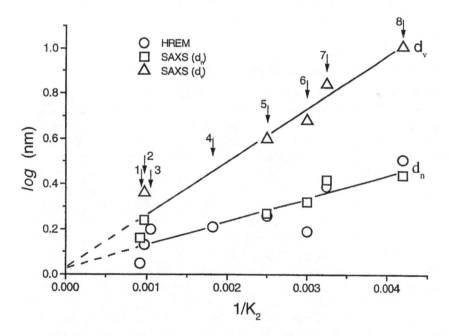

Figure 1 Relationship between the average size (d_n, d_v) of Pd particles in Pd/C catalysts and the equilibrium constant (K_2) of formation of the metal precursors (surface π-complexes of PdCl$_2$ with the A_2 sites). (From Ref. 16.) Pd/C catalysts are prepared by reduction of PdCl$_2$/C in flowing H$_2$ at 250 °C for 3 h; the metal loading is ~1 μmol/m^2 (S_{phenol}). *Supports*. Active carbons: Eponit 113H (1), PN (2), AR-D (3). Activated pyrocarbon: Sibunit (4,6,7). Carbon blacks: PME-800 (5), PM-105 (8).

become more electron-deficient due to a decrease in the number of hexagons in the molecules [61].)

Chemical Admixtures in Carbons as Anchors for Metal Nanoparticles

Introduction of multivalent metal cations, for example, Zr^{4+} [62], Pr^{3+} [63], Al^{3+} [64,65], or La^{3+} and Th^{4+} [65], into Pd/C catalysts hinders sintering of Pd. These cations may, in principle, play a double role. They can form islands of corresponding oxides, which provide a stronger stabilization of the noble metal particles than the carbon surface [66]. On the other hand, they are Lewis acids, which are able to interact with the metal nanoparticles (Lewis bases) to make them more electron-deficient [67]. In general, charging of supported metal particles results in an increase in their stability to sintering. Among Ru/C catalysts promoted with mono- (K^+) and divalent (Ba^{2+}) cations, the highest stability of Ru particles to heating is observed for Ba-Ru/C [68]. At variance, introduction of alkali or earth metals into Pd/C catalysts has a detrimental effect on the stability of Pd particles to sintering [65] for unknown reasons.

In this context it is worth mentioning that the influence of the natural ash content of the support on the dispersion of Me/C catalysts is rather ambiguous. It was proposed [69] that "*the inorganic matter of the carbon plays an important role in the stabilization of the average platinum particle size by making sintering more difficult.*" However, an inverse effect was observed for Pd/C [70]. An increase in the ash content in most cases leads to faster catalysts poisoning but only sometimes to the useful modification of their properties. Thus, inorganic impurities present in carbons are often washed off with acids (HCl, HF).

Role of Electrophysical Properties of the Carbon Matrix

Equalizing Fermi levels in two conductors brought in contact gives rise to the contact potential (V_k), which is proportional to the work function difference of these materials. Based on the work functions (Φ) of corresponding metals and quasi-graphitic carbons (Table 4), one would expect, for example, Fe, Au, and Ag particles to acquire a positive charge in contact with carbon, while Ir and Pt would get a negative. However, these are quite rough estimates. It should be mentioned that charging of metal particles is strongly dependent on the carbon nature. The value of Φ depends also on the type of crystalline plane, the presence of impurities adsorbed thereon, and the crystallite size as well. Because the values of Φ can vary for different surface fragments of the support, the sign and value of V_k will also be determined by the localization of the metal nanoparticles on the support.

Nevertheless, Mössbauer spectroscopic data demonstrate that Fe particles contacting carbon become electron-deficient [71]. There is numerous evidence that Pt particles may acquire either a negative [22,24,72,73] or positive [74,75] charge, with its value depending on the localization of the particles on the support surface [22]. XPS studies of Pd/C catalysts prepared by Pd vapor deposition on an amorphous carbon show a charge transfer from Pd clusters to the carbon support, the average charge of a metal atom in the cluster increasing with a decrease in the surface concentration of supported palladium [76]. (For Pd/C prepared by this method, a decrease in Pd loading leads to a decrease in the cluster size [19].) Furthermore, quantum-chemical calculations predict positive charging of silver clusters in contact

Table 4 Work Function (eV) for Some Metals,* Graphite,[†] and Carbon Materials[‡]

Metals	Polycrystals	Single Crystals {plane index}	
Fe	4.31		
Au	4.25–4.3	4.02 {100}	4.12 {111}
Ag	4.3	4.6–4.8 {100}	3.98 {111}
Ru	4.6–4.71	5.4 {0001}	4.52 {11$\bar{2}$4}
Os	4.7–4.8	5.59 {0001}	5.34 {10$\bar{1}$0}
Pd	4.8–4.99	—	—
PdC$_{0.2}$	4.5	—	—
Rh	4.75–4.8	—	—
Ir	4.7–5.4	5.2–5.5 {100}	5.7 {111}
Pt	5.32–5.39	5.79 {111}	5.1 {201}

Graphite and Quasi-Graphitic Carbons[§]	
Graphite (basal plane)	4.7–4.8
Graphitized carbon black (pH ~ 10)	~ 4.5
Polycrystalline carbon	4.2–5.2
Oxidized carbons (pH $= 6.8 \rightarrow 2.5$)	$4.3 \rightarrow 4.7$
Basic carbons (pH $= 7.2 \rightarrow 10.3$)	$4.35 \rightarrow 4.55$

* *Source*: Refs. 77–79.
[†] *Source*: Refs. 78 and 79.
[‡] *Source*: Refs. 78, 80, and 81.
[§] pH values of aqueous suspensions of carbons are given in parentheses.

with a model carbon support (naphthalene frame), but, at variance, negative charging of palladium clusters [9].

The contact potential between metal particles and the support is thought to influence considerably the processes of migration and coalescence of the particles. To account for the observed phenomena of coalescence and repulsion of Pt particles, as well as particle disintegration into smaller crystals, Chen et al. [82] suggest an equation describing the force (F) of the interaction between two metal particles of radii R_1 and R_2 on a carbon substrate at a distance r (edge to edge). This force is produced by a combination of repulsive forces due to the like charges and attractive forces due to the image charges:

$$F = 1.1 \times 10^{-4} \frac{V_k^2}{r_0^2} \left[\frac{R_1^2 R_2^1}{r^2} - \frac{R_1^4 R_2 r}{(r^2 - R_2^2)^2} - \frac{R_1 R_2^4 r}{(r^2 - R_1^2)^2} \right] \quad (\text{dyn}) \qquad (2)$$

where r_0 is the interatomic metal–carbon distance at the metal–carbon interface. According to this equation, two metal particles approaching each other will stick together if the energy barrier proportional to V_k^2 is transpassed. The closer the particles are in size ($R_1 \approx R_2$), the higher and wider the barrier is. This might be one of the reasons of the known phenomenon of higher stability of monodispersed catalysts to sintering in comparison to polydispersed. It is remarkable that the conclusions, which follow from [82], are surprisingly similar to those predicted by the theory of stability and coagulation of lyophobic dispersed systems by B.V. Derjaguin, L.D. Landau, E.J.W. Verwey, J.T.G. Overbeek (DLVO theory) [3].

Hence, there is some similarity in terms of the electrophysical approach between the stability of supported metal nanoparticles to coalescence and stability of sols of these metals to coagulation in solutions.

A phenomenological description of this kind of metal-active carbon interaction is given in [71] in the framework of the model of point diode. According to this model, "*the charge transfer* [from a metal to the support] *will be roughly equal in value to the number of surface states in the energy gap of the surface of the semiconductor* [carbon support]." In turn, "*the number of surface states at a perfect semiconductor surface is of the order of the number of surface atoms.*" It can be further deduced that the charge transfer from a metal particle to the support (and consequently the stability of metal particles to sintering) will increase with the number of surface carbon atoms with uncompensated bonds (for example, those at the edges of carbon networks) or, other conditions being equal, with an increase in the specific surface area of the support.

Electrophysical properties of quasi-graphitic carbons are directly related to their structure and chemical state of their surface [34,80,81,83–87]. Hence, the above conclusions [71,82] bring one to the understanding of the effect of *bulk* (i.e., *substructural*) properties of the carbon matrix on the state of the supported metal. Thus, the electrophysical approach to describing the stability of supported metal particles to coalescence is based on the *collective* effect of substructural properties of carbons, nanotexture, and the chemical state of their surface.

12.1.2 Blocking of Metal Nanoparticle Surface by Fragments of the Carbon Network

Blocking means making the surface of a metal particle inaccessible to adsorbates due to steric reasons and leads to a decrease in the catalytic activity. Its extent depends both on the specific orientation of the fragments of the support matrix relative to the metal particle surface and on the size of the adsorbate molecules. As a result, the *apparent* size of the particles (determined by chemisorption) is much higher than the *true* size (determined by physical methods, e.g., HRTEM). Chemical poisoning of metal particles with sulfur, iron, or other impurities, which are often present in carbon materials, leads to the same consequences. However, unlike blocking, the effects of poisoning can be diminished by chemical precleaning of the support surface.

Effect of Microporosity

If a metal particle resides in a pore of a similar size, a considerable part of its surface is in intimate contact with the pore walls and, therefore, inaccessible to adsorbates [6]. For Pd/C catalysts dispersed uniformly over the surface of a microporous support, the true (d_s) and apparent (d) sizes of Pd particles may indeed be very different (Table 5, nos. 1–6; Table 1). The influence of micropores is much lower when metal particles are stabilized in wider pores or in the form of rather coarse particles (Table 5, nos. 9, 10, 12, 14, 15). The same happens if the metal is nonuniformly distributed through the grain of microporous support (nos. 12–15), which is achieved, for example, by adsorption of colloidal catalyst precursors.

However, microporosity is not the only reason for blocking of the active component surface in Me/C catalysts. A relatively high blocking effect may be observed for some carbon supports almost free of micropores (nos. 7–9, 12). This

Table 5 True and Apparent Sizes of Pd Particles Depending on the Textural and Microstructural Properties of the Support and Catalyst Preparation Procedure*

| No. | Catalyst | Type | Carbon Support Characteristics | | | | | Particle Size of Pd | | |
| | | | Texture | | Microstructure[†] | | | d_s | d | d/d_s |
			S_{BET} (m²/g)	V_{mi} (cm³/g)	L_a (nm)	L_c (nm)	d_{002} (nm)	(HRTEM, nm)	(CO/Pd, nm)	
Catalysts prepared via adsorption of PdCl$_2$ on carbon from H$_2$PdCl$_4$ solution										
1	3.1%Pd/KGO-250	Channel black	450	0.06	2.0	1.5	0.368	1.21	3.9	3.2
2	1%Pd/S1	Sibunit	410	0.07	3.4	3.8	0.350	1.80	4.1	2.3
3	6%Pd/S1	Sibunit	410	0.07	3.4	3.8	0.350	1.93	3.6	1.9
4	9%Pd/Eponit 113H	Active carbon	850	0.15	2.1	1.1	0.367	1.35	3.2	2.4
5	1%Pd/S2	Sibunit	500	0.15	3.5	3.5	0.352	1.40	2.0	1.4
6	9%Pd/PN	Active carbon	950	0.38	2.0	1.1	0.375	1.42	3.1	2.2
7	2%Pd/Vulcan XC-72	Furnace black	150	0.01	1.2	1.6	0.366	1.13	3.6	3.2
8	1%Pd/PM-105	Furnace black	110	0.00	1.9	1.7	0.362	1.74	6.3	3.6
9	1%Pd/PM-105	Furnace black	110	0.00	1.9	1.7	0.362	3.60	4.5	1.3
10	3%Pd/S1	Sibunit	410	0.07	3.4	3.8	0.350	2.68	4.3	1.6
11	1%Pd/S3	Sibunit	415	0.00	3.6	3.4	0.349	1.61	1.6	1.0
Catalysts prepared via hydrolysis of H$_2$PdCl$_4$ in the presence of carbon										
12	1%Pd/Vulcan XC-72	Furnace black	150	0.01	1.2	1.6	0.366	2.51	3.7	1.5
13	1%Pd/S2	Sibunit	500	0.15	3.5	3.5	0.352	1.46	1.5	1.0
14	1%Pd/S2	Sibunit	500	0.15	3.5	3.5	0.352	4.69	3.4	0.7
15	4%Pd/S5	Sibunit	360	0.00	4.5	3.8	0.349	5.09	5.2	1.0

* Source: Ref. 16.

[†] Calculated from XRD data: L_a and L_c are the average sizes of the quasi-graphitic domains, and d_{002} is the average interlayer spacing.

phenomenon seems to result from specific features of their rather imperfect microstructures (cf. no. 11).

The Role of Amorphous Carbon in the Metal Surface Blocking

Amorphous carbon is characterized by a highly imperfect structure and high reactivity. This shows by a considerable amount of mobile carbon atoms at a surprisingly low temperature. Besides, a vast number of defects and small sizes of graphene sheets make the carbon matrix very labile. As a result, it may be deformed under the action of adsorbates. For example, granules of amorphous carbon swell [88,89] in water with concomitant changes in the carbon substructure and porosity [90,91]. These properties of the support weaken rapidly as its crystal structure becomes more perfect. The labile structure of amorphous carbon is responsible for at least two mechanisms of blocking of the surface of supported metal particles.

"High-Temperature-Condensation" Mechanism

Carbon atoms released by the support upon heating contaminate the surface of Pd particles to make them inaccessible to adsorbates [31–33,35]. At temperatures as high as 600–730 °C, carbon ad-atoms aggregate into extended graphitelike crystallites, which build up capsules around metal particles and prevent further sintering of the catalyst. Aggregation of carbon ad-atoms into graphite clusters at the Pt surface is observed at 630–930 °C in [73] and at 1100 °C in [30].

In order to remove these carbon contaminants, it was proposed to heat Pd/C catalysts at 300 °C and then to cool down to room temperature in $He^+4\%(vol)$ O_2 flow [33]. However, further heating of these samples in H_2 again causes contamination of the metal surface with carbon. For Pt/C catalysts, as well as for Pd/C, O_2 treatment (160 Torr, 350 °C) also leads to carbon removal from the metal surface [92]. Similar changes in the apparent metal dispersion upon similar treatments were reported in [42] for Pd/C and Pt/C. However, the authors attribute them to the known phenomenon of redispersion of supported metals in the oxygen-containing atmosphere [8,93], although HRTEM was not applied to validate the hypothesis.

"Shrinkage-Following-Swelling" Mechanism

Sometimes HRTEM reveals graphitelike structures around metal particles prepared below the temperature of carbon ad-atom aggregation at metal surfaces. Thus, encapsulation of some platinum particles (≈ 4 nm) into the support matrix has been observed in Pt/Vulcan XC-72 catalysts [34]. The carbon networks were oriented parallel to the metal surface in the vicinity of the metal particle but more randomly farther away from it.

Similar observations were reported for Pd/C [57]. A typical HRTEM image of encapsulated Pd particles is shown in Figure 2. Pd particles are arranged on the support in such a manner as if some force "pressed" them inside the carbon matrix. On closer inspection one may see a graphitelike capsule with a particular V-necked droplike shape. Formation of the graphitelike envelope seems surprising and improbable given the fact that Pd particles have been obtained as a result of spontaneous reduction of H_2PdCl_4 by the carbon surface at 20 °C in the liquid phase [94]. Such a phenomenon of "graphitization" and blocking of the surface of Pd particles, which may result in their encapsulation, is more pronounced the more

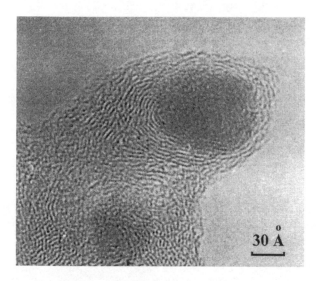

Figure 2   HREM image of a Pd^0 particle captured by the support matrix. The particle is formed via spontaneous reduction of aqueous H_2PdCl_4 by the surface of carbon black Vulcan XC-72 at 20 °C.

disordered the carbon structure is. According to RDF (radial distribution function method), XRD, and low-temperature nitrogen adsorption data [16,57], disordering of the support structure reveals a distortion of the structure and packing of the graphene sheets, a decrease in the carbon crystallite size, as well as an increase in the specific surface area and micropore content.

A likely reason for the encapsulation of the metal particles under mild preparation conditions is hydrostatic overpressure, called "disjoining pressure"* [3]. The latter arises in the pores when the support grains are impregnated with the metal salt solution and is higher the smaller the size of the pores is. This likely results in a considerable widening of the necks of the pores surrounded by the most disordered fragments of the carbon matrix [91], while the structure of the carbon matrix may become more perfect due to the induced matrix contraction [90,91]. Thus, carbon crystallites and separate carbon networks are thought to reorient in such a manner that extended graphitelike layers are formed along the pore walls [Figure 3(a)]. Then the solvent is evaporated and metal particles are formed inside the micropores. However, if the particles are larger in size than the diameter of the "dry" pore, full relaxation of the carbon structure is not possible. This leads to blocking particle surface and some ordering of the carbon structure in the vicinity of nanoparticle visualized in Figure 3(b).

---

* These forces arising in the pores and cracks result, in particular, from repulsion of double electric layers of their walls, as well as from disruption of the water structure. Besides, osmotic pressure may also contribute because the ion concentration inside the pores increases due to adsorption relative to that in the outside solution [3].

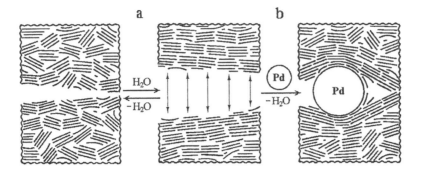

Figure 3 Probable mechanism of evolution of Pd particle blocking by the carbon support fragments: (a) enlargement of the pore space in the carbon matrix put into an electrolyte solution and (b) contraction of the pore size during drying to cause capsulation of the metal particle.

Note that the farther away the electric potential of the carbon surface is from the potential of zero charge point (φ_{PZC}), the higher the disjoining pressure is. In principle, this may result in a systematic variation of the support pore size in Me/C catalysts with potential (similar to the electrocapillary curve [96,97]) and consequently the efficiency of metal particle blocking by the pore walls. Such behavior of porous carbons obviously can influence the measurements of the electrochemically active surface area and might be one of the reasons for the observed correlation between the apparent dispersion of Pt/C catalysts, measured by cyclic voltammetry, and pH_{pzc} of the supports [95], whereas no noticeable difference in the particle size has been observed with HRTEM. Undoubtedly, this problem needs further investigation.

12.1.3 Conclusions

Fine metal particles are poorly stabilized by the homogeneous surface of graphitized carbons. Strong stabilization requires a considerable amount of structural defects in the carbon matrix. However, a very high concentration of defects results in a strong propensity of the carbon matrix to deformation and a high mobility of carbon atoms that favors encapsulation of the metal particles in the support body and contamination of their surface with carbon atoms under relatively mild conditions. As a result, a decrease in the apparent metal dispersion as determined by gas chemisorption is observed. Therefore, disordering of the carbon support structure must be optimal: that is, it must be appropriate for the formation and stabilization of metal particles of a required size but should not have a detrimental effect on the accessibility of the particle surface to reactants. An influence of chemisorbed oxygen (in the form of surface oxygen-containing species) on the mobility of supported metal particles (Pt, Pd) seems to be negligible in comparison to that of $>C{=}C<$ fragments at the edges of graphene layers.

A number of interrelated factors determining the interaction of metal nanoparticles with the support and, consequently, affecting their stability to sintering may operate. Evidently there is a hierarchy of these factors in the extent of their influence depending on the temperature and chemical environment of the Me/C catalysts. A simplified version of the hierarchy (without considering the role of electrophysical and microstructural properties of carbon supports) is given elsewhere [26]. It highlights a decrease in the influence of the chemical factors and an increase of mechanical factors regarding stabilization of metal particles on the surface of the carbon support along with an increase of temperature. The development of more consistent hierarchic models will undoubtedly stimulate further progress in mathematical modeling as applied to the theory of sintering of Me/C catalysts.

12.2   INTERACTION OF METAL COMPLEXES AS CATALYST PRECURSORS WITH CARBON SUPPORTS AND ITS INFLUENCE ON THE STATE OF SUPPORTED METAL NANOPARTICLES PRODUCED

The conventional method for synthesis of Me/C catalysts is supporting the active component precursors on the surface of a carbon material followed by transformation of the precursors into metal particles. Deposition is usually performed from the liquid phase, but sometimes (mainly with the flat supports) from the gas phase. The carbon surface may play either an active or passive role with respect to the metal compounds dissolved. In the former case the adsorption, resulting in removal of the metal ions from the solution, takes place. Adsorption maybe driven by physical (Van der Waals, electrostatic) forces, but in most cases it is of the chemical nature. The catalysts prepared this way are called *adsorbed* [99]. Solutions in contact with the "passive" supports do not undergo noticeable changes. Hence, physicochemical parameters of the whole system must be varied in such a way as to initiate deposition of the catalyst precursors on the carbon surface. The traditional procedure is evaporation of the solvent, leading to the so-called *impregnated* catalysts [99]. Otherwise, the deposition is achieved by varying temperature and the solvent composition, as well as by electrolysis or adding chemical compounds, which gives rise to the *deposited* catalysts.

Thus, Me/C catalysts can be classified* as *adsorbed, impregnated,* and *deposited* depending on the physicochemical nature of the processes used to support the precursors. This is a conventional classification, since in most cases support is not completely inert toward the solution.

Depending on the volume ratio between the solution (V_s) and support pores (V_p), the preparation methods can be divided into two groups:

---

* Notice that the term "impregnation" often relates exclusively to the technological approach to the catalyst preparation without any consideration of the nature and intensity of the processes of interaction of the solution constituents with the support. As a result, the same term is referred to catalysts, whose preparation is based on processes different in nature. G.K. Boreskov [99] imparts chemical sense to the term "impregnation", which makes it quite definite and usable along with the other terms, such as "ion exchange", "hydrolysis, electroless plating", etc., which imply the method for the catalyst preparation based on the chemical nature of the precursor and the mechanism of its interaction with the support. Unfortunately, comprehensive classification of the preparation has not been developed so far, though some attempts have been made [99–102].

1. *Excess solution* ($V_p < V_s$) is used for synthesis of powdered and granulated supported catalysts, if the adsorption is very intense, as well as for the preparation of powdered deposited catalysts.
2. *Incipient wetness impregnation* ($V_p \approx V_s$) is appropriate for the preparation of granulated catalysts but can also be used for preparation of powdered catalysts on the laboratory scale.

A significant difference between these methods is a different contribution of the gas phase to the surface chemical processes of formation of supported catalyst precursors, which has practically not been taken into consideration before. However, this may considerably affect the properties of the target Me/C catalysts (Section 12.2.1).

12.2.1 Me/C Catalysts Prepared by Adsorption of Metal Compounds

Discovery of redox, cation-, and anion-exchange properties of porous carbon materials dates back to the early 20th century. Two theories, *chemical* and *electrochemical*, were suggested shortly after in order to explain these properties. In terms of the *chemical theory*, carbon surface is considered as a set of fragments, whose structure and reactivity are considered similar to those of the functional groups of polycyclic organic compounds [37,98,103]. Cation-exchange properties of carbons are attributed to the presence of oxygen-containing acidic groups, while anion-exchange properties are attributed to chromene- or γ-pyronelike structures. Surface carbon atoms, as well as quinone and hydroquinone groups, are thought to bear responsibility for the redox properties of carbons. The studies in this field have traditionally focused on the chemical nature of surface oxygen-containing groups and on their influence on properties of the Me/C catalysts. Little attention has been paid to the graphene network fragments comprising unsaturated C–C bonds.

Carbons are typically ampholytes: they adsorb both anions and cations, depending on the pH of the electrolyte solution. An important characteristic of carbon is, therefore, the pH value of its isoelectric point (IEP). Anions are preferably adsorbed at pH < pH_{IEP} and cations at pH > pH_{IEP}, while the charge of the carbon surface changes from positive to negative. These concepts have been extensively applied in the last years to explain the observed regularities of formation of Me/C catalysts [44,102,104–107]. Unfortunately, the discussion is usually limited to the pH_{IEP} of the initial support and Coulomb interaction of the metal ions with it. However, numerous data indicating redox transformations of these ions and formation of their surface complexes are available. These processes can not only be independent of the ionic exchange regularities but also induce changes in the pH_{IEP} value of the support.

According to the *electrochemical theory*, when graphitelike carbons, as well as inert metals (Pt, Au), are immersed into an electrolyte solution, they reach thermodynamic equilibrium with the solution through establishing redox equilibrium of the system constituent compounds. In particular, carbon can behave as a gas (oxygen or hydrogen) electrode, the value and sign of its surface potential being dependent on the gas medium composition and pH of the solution [108]. As this takes place, the electrolyte ions are adsorbed in the electric double layer (EDL) at the carbon–electrolyte interface [37,108]. As a result, the solution can become strongly depleted with the ions, because porous carbon materials typically have high surface areas. This behavior of carbon can be schematically illustrated by Eqs. (3) (for acidic electrolytes) and (4) (for basic electrolytes),* which show an accumulation of ions at the interface to compensate the charge at the carbon surface ($\oplus$ or $\ominus$), induced by the redox reactions:

$$2C + 1/2O_2 + 2X^- + 2H^+ \rightarrow 2X^-/C^{\oplus} + H_2O \tag{3}$$

$$C + 1/2H_2 + Me^+ + OH^- \rightleftharpoons Me^+/C^{\ominus} + H_2O \tag{4}$$

Here, C stands for a site at the carbon surface.

In a chemically inert atmosphere, EDL is formed via establishing redox equilibriums of electrolyte ions [94,108–110], for example:

$$C + Fe^{3+} + X^- \rightleftharpoons X^-/C^{\oplus} + Fe^{2+} \qquad E^{\circ}_{Fe^{3+}/Fe^{2+}} = 0.771\,V \tag{5}$$

Reaction (5) and similar ones may occur at a relatively low redox potential of metal ions [109] (typically, the electrode potentials, E°, should not exceed 0.7 V). Thermodynamically favored oxidation of carbon to form surface oxides, CO or CO_2, does not occur as yet under these conditions, probably because of a considerable overvoltage of the carbon corrosion. Metal ions with higher oxidation potentials may oxidize a support surface to produce various oxygen-containing carbon compounds. The latter is always accompanied by a pH shift, while the pH is practically constant in the former case [109,111].

Assuming carbon to behave as an ideally polarizable electrode, the free charge (ΔQ) at the interface can be expressed as

$$\Delta Q = -FV \sum_i z_i \Delta C_i = S \int_{\varphi_{PZC}}^{\varphi} C_{EDL} d\varphi \tag{6}$$

where F is the Faraday, V is volume of the solution, C_i and z_i are the concentration of a compound involved in establishing redox equilibrium and the number of electrons consumed in its redox transformation, respectively, S is specific surface area of carbon, φ is the electric potential of the carbon surface, φ_{PZC} is the potential of zero (free) charge, and C_{EDL} is the differential capacitance of EDL. The value of C_{EDL} is a complex function of φ, temperature, carbon nanotexture, and porosity, nature and concentration of electrolyte ions, type of the solvent, as well as the potential of zero charge (φ_{PZC}), which, in turn, depends on the chemical state of the carbon surface [34,83,108,110].

EDL can be generated not only via redox interactions (path 1), exemplified by schemes (3)–(5), but also via physical adsorption of polar molecules (path 2), specific adsorption of surfactant ions (path 3), and chemisorption of heteroatoms or polar compounds (path 4). Thus, the electrochemical theory also takes into account chemical aspects of surface phenomena, even though they are not as detailed as in the chemical theory. When the adsorption processes occur according to paths 2–4,

---

*Note that Eqs. (3), (4) and those that follow in the next paragraphs do not make a distinction either between the type of adsorption (specific or nonspecific) or spatial charge distribution in the solid or liquid phase.

the value of φ_{PZC} is shifted and, correspondingly, C_{EDL} is changed. For this reason the level of the redox transformations of electrolyte ions [Eq. (6)] or their adsorption through the ion exchange in EDL appears dependent on the efficiency of all enumerated processes of the EDL formation.

Catalysts Synthesized from Ammonia Complexes of Metals

Adsorption of cations and cationic complexes of transition metals on carbons is usually accounted for by the presence of surface carboxyl or phenol groups [37]. The number of adsorbed $\mathrm{Ag^+}$ [112,113] or $[\mathrm{Pt(NH_3)_4}]^{2+}$ [12,114–116] ions indeed increases with the concentration of these acidic sites. The ion exchange on acidic sites ($\boxed{C}$–OH) occurs via substitution of protons by cation complexes:

$$z\boxed{C} - \mathrm{OH} + [\mathrm{Me(NH_3)}_n]\mathrm{X}_z \rightleftharpoons [\mathrm{Me(NH_3)}_n]^{z+}(\boxed{C} - \mathrm{O^-})_z + z\,\mathrm{HX} \qquad (7)$$

Obviously, the number of $[\mathrm{Me(NH_3)}_n]^{z+}$ complexes adsorbed depends both on the concentration of the acidic sites and on the nature of the $\mathrm{X^-}$ anion. Hence, solutions of ammonia complexes of metal hydroxides and oxidized carbon supports (C_{ox}) are used to prepare catalysts via adsorption [12,24,64,65,114,117–120]. Equilibrium (7) is significantly shifted to the left in the case of the adsorption of metal salts, $[\mathrm{Me(NH_3)}_n]\mathrm{X}_z$ ($\mathrm{X = Cl^-, NO_3^-}$), due to releasing strong acids HX. For this reason, the most frequently used method for supporting these salts is incipient wetness impregnation [41,42,121–127]. The completeness of the adsorption processes usually is not stated in this case, which makes it difficult to classify the prepared Me/C specimens as "impregnated" or "adsorbed" catalysts.

However, adsorption of the ammonia complexes seems more intricate than outlined above as evidenced by their ability to interact with the carbon surface deficient of acidic groups. Thus, the authors of [128] reached rather high Pd content (about 5 wt.%) in the catalysts prepared using both oxidized and initial active carbons through adsorption of $[\mathrm{Pd(NH_3)_4}]\mathrm{Cl_2}$ thereon. The amount of silver adsorbed from $[\mathrm{Ag(NH_3)_2}]^+$ solutions increased as the carbon surface became more heterogeneous [113], which allowed the authors to conclude that "*in this case ... the fixation of the cations is not a true ion-exchange but rather an adsorption followed by the decomposition of the adsorbed species.*" STM studies of Pt/graphite catalysts prepared by adsorption of $[\mathrm{Pt(NH_3)_4}](\mathrm{NO_3})_2$ [129] demonstrated very different states of platinum precursors when supported on the initial and oxidized graphite surface. Thus, it may be supposed that different adsorption mechanisms coexist at a shortage in surface acidic groups with respect to the number of supported metal ions. Hence, *the ratio of the number of the supported metal ions (Me^{n+}) to the number of surface acidic groups (AG) may be an important parameter in the synthesis of Me/C catalysts from ammonia complexes of metals.* The $\mathrm{Me^{n+}/AG}$ ratio is assumed to determine the relationship between different species of the supported precursors, thus affecting the state of the active component in the target catalyst.

Preparation of Me/C by Adsorption of Metal Hydroxide Ammoniates

This method requires an excess of ion-exchange groups compared to the number of the supported metal ions and thus relies on the utilization of oxidized carbon materials. Reduction of adsorbed precursors in $\mathrm{H_2}$ at 100–300 °C results in the generation of very fine metal particles, their size being 1.5–2.5 nm for

$[Pd(NH_3)_4](OH)_2$ precursor [64,118,119]; 0.8–1.8 nm for $[Pt(NH_3)_4](OH)_2$ [12,24,117]; subnanometer in size for $[Ru(NH_3)_6](OH)_2$; 1.5–2 nm for $[Ir(NH_3)_5Cl]$ $(OH)_2$; and 2–3 nm for $[Rh(NH_3)_5Cl](OH)_2$ [119]. Monomodal particle-size distribution is characteristic of these Me/C_{ox} catalysts [24,64,119,120]. Moreover, the average size of Pt particles formed on oxidized graphite (300 m^2/g) or on active carbon (1400 m^2/g) is independent of the support type and of the metal loading in the range between 1.1 and 11.4 wt. % [24]. This phenomenon has also been observed for oxidized graphitized carbon black Vulcan XC-72 containing 3–20 wt. % of platinum [114].

Preparation of Me/C by Supporting Salts of Ammonia Metal Complexes

As mentioned above, complex ammonia salts of noble metals are usually supported by incipient wetness impregnation. From literature data on Pd/C catalysts prepared by this method [41,42], complex changes in the metal dispersion are observed as the surface concentration of acidic oxygen-containing groups (AG) increases (Table 6). The dispersion first increases (nos. 1–6) to reach a maximum at the molar ratio $Pd/AG = 1.3$–1.5 (nos. 6, 7), which may result from an increase in the proportion of Pd^{II} adsorbed through ion exchange. At a considerable excess of the acidic sites over Pd ions, Pd dispersion appears somewhat lower than the maximal but changes very little with the Pd/AG ratio (nos. 8–14). The same range of dispersion variations is observed with ∼5% Pd/C catalysts prepared by *adsorption* of $[Pd(NH_3)_4]Cl_2$ on oxidized carbons, whereas an extremely low dispersion ($D_{CO/Pd} = 0.01$–0.03) is characteristic of the catalysts prepared by the same method on nonoxidized carbons [128]. The mechanism of interaction of $[Pd(NH_3)_4]^{2+}$ with the support responsible for formation of the coarse metal particles is not clearly understood in the latter case.

As to Pt/C catalysts prepared by treating supported salts $[Pt(NH_3)_4]X_2$ (X = Cl, NO$_3$) with H$_2$, the particles formed are very coarse whether oxidized ($D_{H/Pt} = 0.03$–0.07 [124,126]) or nonoxidized ($D_{H/Pt} = 0.02$–0.12 [121,124–127]) carbons are used. A higher dispersion of Pt is achieved through predecomposition of the precursors in He or vacuum at 350–400 °C ($D_{H/Pt} = 0.2$–0.7 [124–126]), as well as in air at 260 °C ($D = 0.3$–0.5 [122,123]). However, dispersion of Pt catalysts prepared by this method on oxidized carbon (Pt/C_{ox}) is usually two or three times lower than on nonoxidized [116,121,124,126,130]. This may be explained by observed [116,117,126,129] differences in states of platinum precursors on the surface of oxidized (C_{ox}) and initial (C) carbons.

Rodríguez-Reinoso et al. [121,124,125] claim that the necessity of decomposing supported precursor ammonia complexes stems from the fact that in the presence of H$_2$ they produce an intermediate labile hydride $[Pt(NH_3)_4]H_2$ [131], which leads to Pt agglomeration. However, the existence of that on a carbon surface is to a large extent speculative. In addition, this interpretation does not explain the fact that the calcination procedure is not needed to form fine Pt particles from adsorbed hydroxide $[Pt(NH_3)_4](OH)_2$ and does not cause any changes if protons of the surface acidic sites are substituted by Ca^{2+} ions prior to supporting $[Pt(NH_3)_4]Cl_2$ [126].

Catalysts Synthesized from Metal Chloride Complexes

Mechanism of Adsorption of Metal Chloride Complexes

Both simple Ru, Rh, Pd, Ir chlorides and complex chlorides of Au, Ru, Rh, Pd, Os, Ir, Pt are used for the catalyst preparation [100]. Adsorption of HAuCl$_4$

Table 6 Dispersion of Pd/C Catalysts Prepared by Supporting $Pd(NH_3)_4(NO_3)_2$ on Carbons with Different Surface Chemical Composition Followed by Reduction in H_2 at 150°C

| No. | Pd/C Catalyst* | Carbon Support | | | | Pd/C Catalyst |
| | | S_{BET} (m^2/g) | Surface Group's Content† | | Pd/AG ratio | Dispersion $(D_{CO/Pd})$ |
			Basic (μeq/g)	Acidic (AG) (μeq/g)		
1	2.1% Pd/Corax 3	84	35	26	7.69	0.15
2	1.05% Pd/Sutcliffe AR2–H	1370	350	19	5.26	0.18
3	2.1% Pd/Corax 3-ox	126	43	90	2.22	0.29
4	2.1% Pd/Norit-HT	1180	490	100	2.00	0.25
5	2.1% Pd/Anthralur-HT	660	450	101	1.98	0.38
6	2.1% Pd/Norit	1250	450	136	1.47	0.57
7	2.1% Pd/Anthralur-am	1000	740	159	1.26	0.52
8	1.05% Pd/Sutcl. AR2–am	1350	530	103	0.971	0.31
9	1.05% Pd/Anthralur-am	1000	740	159	0.629	0.19
10	2.1% Pd/Anthralur	630	400	355	0.563	0.30
11	2.1% Pd/Norit-ox	1210	390	391	0.512	0.42
12	1.05% Pd/Anthralur	630	400	355	0.282	0.27
13	2.1% Pd/Anthralur-ox	630	300	940	0.213	0.27
14	0.5% Pd/Norit-ox	1210	390	391	0.128	0.28
15	1.05% Pd/Anthralur-ox	630	300	940	0.106	0.15

* Indexes HT, H, am signify high-temperature-treated carbons (in an inert H_2 and NH_3 atmospheres, respectively), ox denotes carbons oxidized by air.
† From the method [98] based on consumption of HCl (by the basic groups) and NaOH (by the acidic groups).
Source: From Refs. 41 and 42.

[111,132–134], H$_2$PdCl$_4$ [16,54–57,59,94,111,135,136–144], H$_2$PtCl$_6$ [43–46,102,111,115,116,126,143–156], and H$_2$PtCl$_4$ [111,144,146,151] and their corresponding salts on carbon materials has been studied most comprehensively. Only a few publications are available on the adsorption of RuCl$_3$ [157], RuCl$_4$ [142], and RhCl$_3$ [142,158–160]. The extent of the metal adsorption in acidic solutions varies approximately as follows: Au^{III} > Pd^{II} ≈ Pt^{II} > Pt^{IV} ⩾ Ru^{IV} > Rh^{III}.

The diversity of opinions regarding the mechanism of the interaction of metal chloride complexes with the carbon surface is amazing. The suggested mechanisms may be classified as follows:

1. *Redox mechanism*, that is, oxidation of carbon by Pt^{IV} or Pt^{II} [44,46,111,115,116,126,143,144,146–148,150,151], Pd^{II} [16,59,94,111,136–144], Au^{III} [111,132,133], which results in the formation of gaseous and/or surface carbon oxides [44,54,116,132,133,136–140,143,144,146–148,150] or chlorination of the carbon surface [16,59,94,111], deposition of Pt^0 (Pd^0, Au^0), and adsorption of Pt^{II} and Pd^I chloride complexes

2. *Acid–base interaction* to form saltlike compounds of H$_2$PtCl$_6$ with the surface Lewis bases (Oxygen-containing groups [45] or C$_\pi$-sites on the basal planes of carbon crystallites [44,102,149] may behave as Lewis bases.)

3. *Coordination mechanism*, which implies substitution of oxygen-containing groups of the support for Cl$^-$ ions in the coordination spheres of Pt^{IV} and Pt^{II} chloride complexes [43] as well as formation of π-complexes with >C=C< ligands of the C$_\pi$-sites [161], for example, C$_\pi$–MeCl$_3^-$ and C$_\pi$–MeCl$_2$ for Pt^{II} [111,143,146,147] and Pd^{II} [16,54–57,59,135] or C$_\pi$–Me$_2$Cl$_2$ for Pd^I [139,143]

4. *Ion-exchange mechanism*, that is, substitution of Cl$^-$ ions adsorbed on intrinsic cationic sites of the support by complex chloride anions of Pt^{IV} [111], Pt^{II} [111,146], or Pd^{II} [16,59,111]

5. *Physical adsorption* of the metal compounds [134,150]

However, some of the above-mentioned mechanisms of H$_2$PdCl$_4$, H$_2$PtCl$_4$, and H$_2$PtCl$_6$ adsorption are hypothetical as they are based on a very restricted number of observations. To select a more adequate mechanism of the adsorption it is necessary to take into account the following well-established facts. First, the adsorption of these compounds proceeds mainly through transformations of metal complexes but does not depend on the pH of the solutions [54,136,145–147] or change its value [54,94,111,115,148] in the pH range between −1 and 2. This means that those mechanisms, which according to their stoichiometry should bring about strong variation in pH (oxidation of carbon to gaseous and surface carbon oxides, acid–base interaction, physical adsorption of molecules), must be excluded. Second, the adsorption of the metal ions decreases when oxidized [46,54,111,136,147] or graphitized [56,111] carbons are used instead of nonoxidized or heterogeneous materials, thus suggesting that the surface oxygen-containing groups and the basal planes of carbon crystallites play a minor role in this process. Third, XPS and chemical analysis data show that the reduction of these compounds ultimately leads to adsorption of chloride anions [16,59,94,111], while the adsorbed ionic metal species tends to form complexes with oxygen-free sites on the carbon surface [54,56,111,139,146]. Eventually, the *electrochemical mechanism* was suggested to describe adsorption of these compounds [16,111]. This mechanism combines (**1**) redox transformations of metal ions (8)–(10) and other compounds including ambient gases (3), (**2**) formation of complexes with the C_π-sites (11), and (**3**) ion exchange in the ionic part of EDL (12).

1. Reduction of the metal ions involves free electrons of the carbon matrix (C) and takes place immediately after carbon is brought in contact with the solution, giving rise to the ionic EDL. This can be illustrated by simplified equations (8)–(10) [see footnote following Eq. (5)]

$$2C + [PtCl_6]^{2-} \rightleftharpoons 2Cl^-/C^\oplus + [PtCl_4]^{2-}, \qquad E^\circ_{PtCl_6^{2-}/PtCl_4^{2-}} = 0.72\,V \tag{8}$$

$$3C + [MeCl_4]^{2-} \longrightarrow Me^0/C + 2Cl^-/C^\oplus + 2Cl^-,$$
$$E^\circ_{MeCl_4^{2-}/Me^0} = 0.73\,V\,(Pt); 0.64\,V\,(Pd) \tag{9}$$

$$4C + [AuCl_4]^- \longrightarrow Au^0/C + 3Cl^-/C^\oplus + Cl^-, \qquad E^\circ_{AuCl_4^-/Au^0} = 1.00\,V \tag{10}$$

where Me = Pt, Pd.

Reactions (8)–(10) are mainly localized at the periphery of the support grain (even if it is not larger than 0.2 mm) since the electron transfer is faster than the metal ion diffusion inside the grain. Gold is adsorbed only in the form of Au^0 [111,133]. Reaction (10) is the first, *fast* step of this process and occurs without essential changes in the solution pH. *Slow* adsorption of $HAuCl_4$ implies chemical corrosion of the carbon matrix to form carbon oxides and surface oxygen-containing groups and results in a strong acidification of the solution. No considerable change in the pH of the carbon suspension is observed during the adsorption of H_2PdCl_4, H_2PtCl_4, and H_2PtCl_6 [54,94,111,115,148].

2. Along with the reactions (8) and (9), formation of π-complexes of Pd^{II} and Pt^{II} with the surface fragments, which comprise unsaturated C–C bonds (called A- [54,146] or C_π- [161] sites), can take place:

$$C_\pi + [MeCl_4]^{2-} \rightleftharpoons C_\pi - MeCl_3^- + Cl^- \rightleftharpoons C_\pi - MeCl_2 + Cl^- \tag{11}$$

Complexes C_π–$PdCl_3^-$ predominate at a considerable Cl^- excess (>0.5 M) [54,55].

Three types of C_π-sites differing in the strength of their complexes with $PdCl_2$ can be distinguished [54–57,60,135]. These are sites of weak (A_1), strong (A_2), and very strong—or irreversible—(A_3) adsorption. For most unoxidized carbons used as catalyst supports, the concentration of A_1 and A_2 sites increases linearly with the surface area determined by phenol adsorption, with their surface densities being 1.27 and 1.06 $\mu mol/m^2$, respectively. Not obeying this rule is graphite, graphitized carbons, and fullerene or onionlike carbons, which show mainly A_1, and some types of filamentous carbons, which are usually rich in A_2 sites. The content of A_3 sites is determined by the micropore volume (about 0.23 $mmol/cm^3$). The concentration of all these sites decreases upon destruction of the surface >C=C< bonds (hydrogenation, chlorination, or oxidation), the sites of *weak* adsorption (A_1) being more resistant to chemical treatment than the sites of *strong* adsorption (A_2 and A_3). In most cases A_2 sites play a major role in the synthesis of 0.5–5% Me/C catalysts, since their concentration is higher by at least one order of magnitude than that of A_3 sites.

Processes (11) affect the whole carbon surface and are competitive to processes (8) and (9) if at certain conditions all supported metal ions can quickly access the carbon surface (for example, during incipient wetness impregnation of the granules with the metal compound solutions), formation of the metal phase is considerably suppressed. The reason is that chemisorbed metal chloride complexes are oriented in such a manner at the carbon–solution interface that give rise to the dipole potential jump of the same sign as that produced by process (8) or (9). In contrast, the yield of the metal phase increases when adsorption is achieved from the excess solvent. In this case the rate of process (11), which is governed by the internal diffusion, is slow, but its contribution to the whole adsorption process increases as soon as the rate of the metal ions' reduction drops down [due to an increase in the work function (Φ) when the carbon surface acquires a positive charge]. Moreover, the other paths of EDL formation, which impart a positive potential jump to the carbon surface, have a detrimental effect on the reduction of metal ions, but the common trend at increasing the total amount of the adsorbed species is that the content of the metal component among them rapidly reaches some constant value under given adsorption conditions [16]. Factors affecting the ratio of the metal and ionic adsorbed species have been most thoroughly analyzed for the system H_2PdCl_4—carbon support—atmosphere [16,94] and are listed in Table 7. Many of them have a similar effect on the reduction of gold and platinum chloride complexes.

In addition to data of Table 7, the Pd^{II}:Pd^0 ratio in the adsorbed compounds increases if the suspension is allowed to be stored long in air that leads to dissolving of Pd^0 particles under the action of O_2 and free HCl followed by re-adsorption of palladium in the form of $PdCl_2$.

3. Cl^- ions, accumulated in the EDL due to reactions (3), (8), or (9) can be subsequently substituted by complex anions such as $[MeCl_n]^{2-}$ or $[MeCl_{n-1}(H_2O)]^-$ (Me = Pd, $n = 4$; Me = Pt, $n = 4$, 6) via ion exchange, for example:

$$2Cl^-/C^\oplus + [MeCl_n]^{2-} \rightleftharpoons [MeCl_n]^{2-}/2C^\oplus + 2Cl^- \tag{12}$$

It should be emphasized, however, that a considerable number of primarily formed ionic pairs $Cl^-/C^\oplus$ can be rapidly transformed through the charge transfer into covalently bonded Cl—C [94,111]. For this reason not all chemisorbed chlorine can take part in the ion exchange with $[PdCl_4]^{2-}$ according to reaction (12) [16]. The quantity of $[PtCl_4]^{2-}$ and $[PtCl_6]^{2-}$ adsorbed through the ion exchange in the presence of air was found to increase, like the case of $[Au(CN)_2]^-$ [162], with the amount of the surface basic oxides.

Although the overall adsorption processes for these noble metal compounds are dissimilar at a first glance, they appear to have common grounds. This conclusion is supported by the similarity of the adsorption isotherms plotted in the canonical coordinates regardless of the type of carbon and the compound adsorbed Figure 4. This can be tentatively explained by the fact that different processes [e.g., (3), (5), and (11)] lead to formation of the double layers of similar structure: $C^\oplus \cdots Cl^-$ or C_π—$Me^{n+} \cdots Cl_n^-$. The 0.5–1.0 θ region of these isotherms is assigned to the so-called weak adsorption [54], which apparently comprises reactions (10)–(12) on the fragments of basal planes of quasi-graphitic crystallites. The 0–0.5 θ region corresponds to the *strong* adsorption on the edge planes of carbon crystallites in reactions (8)–(10) or (11) and results in the transfer of most of the metal from the solution to the support. The limiting value of the strong adsorption (A_{strong}) is roughly constant for a given support (for example, in the case of H_2PdCl_4 [16], $A_{strong} = A_2 + A_3 \approx A_2$), although the ratio of the reduced to oxidized species

Table 7 Factors Determining Pd^{II}:Pd^0 Ratio in the Compounds Adsorbed on Carbons from H_2PdCl_4 Solutions*

High $\longleftarrow$	Pd^{II} : Pd^0 ratio $\longrightarrow$ Low
1. Low temperature	1. High temperature
2. Fast penetration of Pd^{II} ions into support pores: powdered support; high concentration of H_2PdCl_4 (incipient wetness impregnation or adsorption from excess solution under vigorous stirring)	2. Diffusive retardation of the process of formation of the surface $PdCl_2$ complexes: granulated support; low concentration of H_2PdCl_4 (solvent in large excess)
3. High degree of the support surface coverage with the adsorbed palladium compounds	3. Low degree of the support surface coverage with the adsorbate
4. The presence of ligands diminishing the value of $E_{Pd^{2+}/Pd}$ (e.g., excess HCl)	4. The absence of the said ligands in the solution of H_2PdCl_4
5. The presence in the electrolyte solution of background with lyophilic anions (e.g., SO_4^{2-}, etc.), which are incapable of the specific adsorption onto carbon surface	5. The presence of extraneous electrolytes with surfactant anions, e.g., ClO_4^- (path 3)
6. The presence of extraneous oxidants, which may impart a positive charge to the surface of carbon being involved to side redox processes (paths 1): adsorption of acids on the support in the presence of O_2, H_2O_2, etc. before contacting the H_2PdCl_4 solution; introduction of oxidants directly to the H_2PdCl_4 solution	6. Absolute absence of the foreign oxidants in the system under consideration (degassed carbon, inert atmosphere)
7. Polar organic solvents as acetone, alcohol, etc. (path 2)	7. Water as the solvent
8. Low value of the electric capacitance of the EDL: low specific surface area of carbon; homogeneous surface (fragments of the basal planes of carbon crystallites are primarily exposed); solvents with a low dielectric constant (point 7); prechlorination or pre-oxidation of the surface of carbon (high content of the acidic groups) (path 4)	8. High electric capacitance of the EDL: high specific surface area of carbon; predominating contribution of the edge planes fragments to the support surface nanotexture; high dielectric constant of solvents (point 7); thermal destruction of surface acidic groups (high concentration of the basic groups)

* The paths of formation of EDL at the carbon surface are indicated.

adsorbed on the support may vary depending on the adsorption conditions (Table 7). The ratio of the total amount of metal to be supported (Me_{ads}) to A_{strong} at otherwise identical conditions will dictate the ratio of different supported precursors

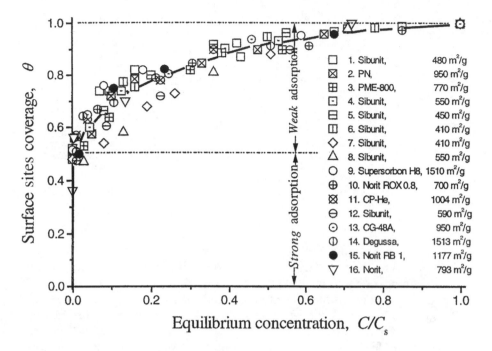

Figure 4 Canonic isotherms of adsorption of H_2PdCl_4 (1–6), H_2PtCl_4 (7), $HAuCl_4$ (8), and H_2PtCl_6 (9–16) on chemically unmodified carbon supports. Some of the isotherms are reproduced from literature data (9 [152], 10 [147], 11 [46], 14–16 [153]) and the others are obtained by the authors of the present review.

of the Me/C catalyst. Therefore, properties of the active component (dispersion, distribution through the support grain) may be a function of the Me_{ads}/A_{strong} ratio.

For the family of carbons similar in nature, the A_{strong} value increases with the specific surface area (S) but decreases to reach the value far lower than $0.5\,\theta$ when the support is subjected to oxidation [54–57,111,135]. For example, the A_{strong} value falls down linearly with increasing number of oxygen-containing groups [57]. The above speculations can be expressed in the simplest form as follows:

$$A_{strong} \approx (a^o_{strong} - \alpha O_{chem})S \qquad (13)$$

where α is a coefficient, O_{chem} is the surface concentration of oxygen-containing groups, a^o_{strong} is the specific value of the *strong* adsorption at $O_{chem} = 0$. Dependence of a_{strong} on textural parameters of nonoxidized carbons is demonstrated in Table 8.

Thus, the *electrochemical mechanism* of adsorption of noble metal chloride complexes combines a large set of different surface processes and establishes a relationship between them; the processes involving *components of the gas medium* are also included. Analysis of the mechanism allows the following conclusions, which are essentially different from the commonly accepted ideas:

First, the surface oxygen-containing groups affect the redox behavior of the support [16,94,111,136,147] and the ability of coordination of the surface >C=C< fragments by noble metal ions [54,55,57,59]. Hence, they influence

Table 8 Concentration of the Sites for the Strong Adsorption of Noble Metal Ions

| Carbon Sample | Textural Properties | | | *Strong* Adsorption, μmol/m^2 (S_{BET}), for Noble Metals Compounds | | | |
	S_{BET} (m^2/g)	Surface Nanotexture	V_{mi} (cm^3/g)	HAuCl$_4$*	H$_2$PdCl$_4$	H$_2$PtCl$_4$	H$_2$PtCl$_6$
Corax 3 (graphitizied)	72	Basal planes	0.00	1.41	0.17	0.40	0.13
CFC-2	85	Edge planes	0.028	5.97	1.57	1.58	1.01
Mogul L	107	Steps	0.00	2.30	0.78	0.75	0.46
PM-105	112	Steps	—	2.58	0.92	0.86	0.48
Eponit 113H	850	Steps	0.15	2.18	0.80	0.82	0.37
4GV-K (Japan)	1680	Steps	0.35	1.40	0.57	—	0.19

* The given values relate to the *fast* adsorption of HAuCl$_4$.
Adsorption conditions: inert atmosphere, 20 °C, 20 h (HAuCl$_4$, H$_2$PdCl$_4$), 120 h (H$_2$PtCl$_4$, H$_2$PtCl$_6$), powdered (<0.09 mm) carbons.

the interaction of gold, palladium, or platinum chloride complexes with the carbon surface *in an indirect way*, while their direct interaction with these metal compounds is negligible [54,56,111,139,146].

Second, the influence of the solvent nature and the surface oxide's content on distribution of the adsorbed species through the support grain may be accounted for by variations in the redox behavior of carbons and solutes, which determine the ratio of rates of different adsorption processes, but not by difference in the surface wettability (as commonly accepted [155,156]), since no differences in wettability of the initial and oxidized active carbons have been found [163].

Influence of the Nature of Adsorbed Precursors on Properties of Me/C

We shall try here to apply the electrochemical mechanism of adsorption for interpreting available literature data on metal dispersion and its distribution through the support grains in Me/C catalysts prepared by adsorption of metal chloride complexes. It is difficult to use earlier ideas on the mechanisms of formation of these catalysts [102,164] since newly obtained experimental data are inconsistent with these mechanisms: in particular, it is assumed that, in comparison with heterogeneous carbon surface, a more homogeneous one interacts more strongly with Pt particles or their precursors, which favors formation of more dispersed Pt/C catalysts [44,104], but the other concepts state otherwise [22,23,165]. This contradiction could be relieved to some extent if we suppose that the heterogeneous supports can comprise chemisorbed oxygen in different amounts, the influence of which was not taken into account before [23,165]. It was accepted later that the oxygen-containing groups behave as anchors for Pt particles or their precursors, and an increased dispersion of platinum supported on the oxidized carbons was explained this way [43,116,166]. However, some data indicate an antipodal effect [44,126], the absence of any effect [45], as well as a volcanolike dependence of the Pt dispersion on the concentration of chemisorbed oxygen [46]. One cannot relieve these contradictions even if one takes into account that the oxygen-containing groups favor a better wetting of the support surface with a solution of metal compound or make the value of pH$_{IEP}$ lower. In the former case, the precursor is supposed to distribute more uniformly over the carbon grains to provide a higher dispersion of the metal on oxidized supports [43,166];

however, it remains as yet unclear why the high dispersion is observed for Pt supported on pregraphitized carbons [44,104], which typically are more hydrophobic. In the latter case, electrostatic attraction of metal-containing anions to the carbon surface is suppressed, that is considered [102,126] to be the reason for the formation of coarse metal particles in the Pt/C_{ox} catalysts, but, if so, it becomes difficult to explain the above-mentioned results [43,116,166]. In our view, the above models are incomplete because they are based on assigning the key role in genesis of the supported metal to only *one* precursor while at least two kinds (metal and ionic) of the adsorbed platinum species have already been known [44,115].

The electrochemical mechanism of interaction of platinum (or palladium) chloride complexes with the carbon support implies *multiplicity* of states of adsorbed metal precursors, the relationship between them determined, on the one hand, by conditions of adsorption of the initial metal compounds and, on the other, by the Me_{ads}/A_{strong} ratio (see above). According to this mechanism, two origins of metal particles may occur in the Me/C catalyst [16]. Some particles, called the *former Me^0*, are generated at the early stage of the process of the catalyst preparation as a result of reactions (8) and (9). Rest particles (the *secondary Me^0*) are formed at the final stage under the action of a reducing agent on the surface π-complexes of Me^{II} (11) and on adsorbed complex anions (12).

Processes (9) and (10) produce very rough, micrograined particles of the former Pd^0 and Au^0 (from 10 nm to several micrometers in size) with nonuniform, egg-shell distribution through the catalyst grain [16,94,133,138,140]. Variations in the substructural properties of these particles obey regularities of electrolysis of metal salts including mechanisms of metal particle's nucleation and growth. For example, the average size of the former Pd^0 particles increases sharply with the temperature of H_2PdCl_4 adsorption, with an increase in the carbon grain size or O_2 partial pressure in the gas phase, and as a result of de-ashing of the support with mineral acids. Heating in H_2 at 250 °C leads to annealing of intercrystallite boundaries in these Pd particles but does not affect the average particle size [16]. On the contrary, the former Pt^0 particles much finer in size (2–4 nm and about 8 nm after their formation at 20 and 100 °C, respectively) are formed during the adsorption of H_2PtCl_6. They are usually grouped together (by 6 to 15 particles in one "nestle"), which results further in their fast sintering. When loaded in an amount no greater than 0.5 wt.%, the active component of Pd/C or Pt/C catalysts is often composed mainly of the former Me^0 (especially for the nonoxidized granulated supports with a high specific surface area). In the special case that preliminary grinded or oxidized carbons are used for the catalysts preparation, the contribution of the former Me^0 particles to the total number of supported metal particles decreases (Table 7), so that the dispersion of the obtained catalysts increases. An increase in the metal dispersion with increasing oxidation degree of the support surface (Pd/C [41], Pt/C [43,116,166]) may be associated with this fact.

Unlike the former Me^0 particles, π-complexes $MeCl_2 \cdot A$ (11) are distributed in a more uniform manner through the support surface [94]. At a certain combination of conditions of metal supporting (see, e.g., Table 7), π-complexes can be the only precursor of the Me/C catalysts. Their reduction in H_2 at 150–400 °C generates the secondary Me^0 particles, which are typically 1–5 nm in size for Pd [15,16,57,59,111,167] (see also Figure 5) and 1–3 nm for Pt. The average size is practically independent of the metal loading within the range of 1 to 10 wt.%.

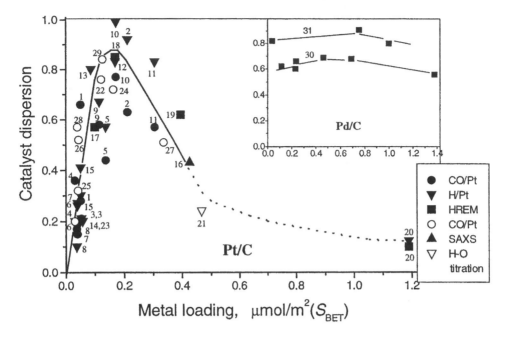

Figure 5 Dependence of metal dispersion on metal loading in Pt/C and Pd/C catalysts (compiled from the literature data and obtained by the authors of the present review). *Pt/C catalysts* were prepared by supporting platinum precursors from aqueous H_2PtCl_6 on the carbon surface followed by evaporation of water, drying and reduction in flowing H_2 at 250 °C/1 h (22–29), 350 °C/12 h (1–15, 24–26), 400 °C/2 h (16), 500 °C/4 h (20), 500 °C/10 h (21). *Pd/C catalysts* were prepared in the same way, the conditions of carbon impregnation with aqueous H_2PdCl_4 were chosen to prevent formation of the "former" Pd^0; supported $PdCl_2$ was then reduced in flowing H_2 at 250 °C/3 h (curves 30 and 31). *Powdered* carbon supports (original marks are used): (1) C (carbon black CC-40-220), (2) C (treated in He, 2273 K), (3) C (H_2, 1223 K), (4) T (carbon black T-10157), (5) T (He, 2273 K), (6) T (H_2, 1223 K), (7) W (active carbon), (8) W (H_2, 1223 K), (9) C (He, 1873 K), (10) C (He, 2073 K), (11) C (He, 2473 K), (12) $C2_{Ox}T$ (He, 773 K) [43,44,104,166]; (13) A (carbonized phenolformaldehyde polymer resin), (14) A4 (N_2, 800 K) [126]; (15) C3 (active carbon) [124]; (16) BAU (active carbon) [168], (17–19) Sibunit [111], (20) HI(A) (furnace black Vulcan XC-72) [45], (21) Vulcan 3G (graphitized carbon black, after activation to 48% burn-off) [165], (22,23) Sutcliffe AR-2, (24,25) Norit SGM (active carbons) [169]; (26) Sutcliffe AR-2, (27) Anthralur STA, (28) Norit SGM, (29) Sibunit [47]; (30) Sibunit, (31) Eponit 113H (active carbon) [unpublished data].

Dispersion of Pd/C catalysts synthesized from these precursors is also independent of the presence of oxygen-containing groups (Table 1) and of the former Pd^0 particles [16]. The dispersion increases as the surface concentration of A_2 sites (Table 3) or the constant of $PdCl_2 \cdot A_2$ complexes formation (Figure 1) increases. These characteristics of A_2 sites depend on the support surface nanotexture and on disordering of the carbon crystallites' structure [55–57]. Hence, these microstructural properties of carbons are key factors determining dispersion of palladium particles prepared by reduction of $PdCl_2$ adsorbed. The active component of these catalysts suffers from the blocking effect in the most cases [16,57,59] (see, e.g., Tables 1 and 5), in particular, due to localization of a part of its particles in fine pores.

The adsorbed ionic compounds (12) usually are impurities to the two main adsorbed species. They bear responsibility for generation of comparatively coarse *secondary* Me^0 particles. For example, 5–10 nm Pd particles are obtained under the action of H_2 on $[PdCl_4]^{2-}/2C^{\oplus}$ at 250 °C [16], which may be due to the nonuniform distribution of the precursors through the carbon surface because $C^{\oplus}$ sites are located at the outer rim of the carbon grain [94].

Coexistence of different paths (8)–(12) of formation of the metal precursors results in a wide particle-size distribution (up to a bimodal one) in final Me/C (*Me*=Pd, Pt) catalysts, as well as in the nonuniform distribution of the active component through the support grain [16,111]. The main reason for a lower dispersion of these catalysts is the appearance of the *former* Me^0 particles (8)–(9) and co-adsorbed metal-containing anions (2.10) among the catalyst precursors at a low and a high coverage of the support surface with the precursors, respectively. π-Complexes $MeCl_2 \cdot A_2$ are main precursors of the finest metal particles. Their contribution to the total amount of the adsorbed precursors is predominant when an *intermediate* value of the carbon surface coverage with the adsorbed precursors is attained. This phenomenon is responsible for the volcanolike dependence of the apparent dispersion of Pt/C catalysts on metal loading (Figure 5). However, such dependence becomes very weak or practically absent as soon as the adsorption conditions allow only the surface π-complexes to be formed (Figure 5, Pd/C).

The whole set of the above-discussed data [43–46,126,166] leads us to deduce that there is a volcanolike relationship between the apparent dispersion of the Pt/C_{ox} catalyst and the areal density of the oxygen-containing groups provided that the metal content is the same for all catalysts. Unfortunately, only the qualitative version of the dependence can be seen from the data of [46].

It is emphasized in Section 12.2.1 that metal dispersion for an Me/C catalyst may depend on the $Me_{ads.}/A_{strong}$ ratio, which governs the mass ratios between all supported catalyst precursors. With account of Eq. (13), one can easily see both curves (Pt dispersion versus metal loading and Pt dispersion versus number of surface oxides) to be special cases of this dependence. Hence, *the well-defined (optimal) value of the $Me_{ads.}/A_{strong}$ ratio corresponds to the position of the maximum of the catalyst dispersion.* In other words, a decrease in the carbon ability to reduce metal ions, for example, due to oxidation of the carbon surface, will make the maximum shifted toward a lower metal loading, and vice versa. Thus, numerous isolated and, at first glance, contradictory literature data on dependencies of dispersion of Me/C catalysts on textural and chemical properties of the support can be combined into a general dependence in the context of the above-discussed model for formation of supported precursors followed by their transformation into the active component of the catalyst.

12.2.3   Catalysts Me/C Prepared by Impregnation

This method for preparation of Me/C catalysts includes the following stages: (1) evacuation of gas from the support pores (often skipped); (2) supporting of a metal-containing solution (soaking, showering, spraying); (3) evaporation of the solvent; (4) final drying and calcination; (5) gas-phase reduction of the supported species. The support can be hydrophilic or hydrophobic depending on the surface nanotexture and on the content of chemisorbed oxygen, which usually influences the rate and depth of penetration of aqueous solutions into the support pores. This is the determining factor for choosing the solvent, especially in the case of homogeneous carbon surfaces. Supports with a heterogeneous surface seem to be more tolerant in respect to wetting with water. Thus, Knijff [163] studied the process of water elimination (50 °C, flowing N_2) from pores of Norit R3A peat-based, steam-activated extrudates and concluded that "*comparison of the drying behavior of the oxidized and the non-oxidized carbons indicated that the hydrophilic character of the investigated carbon was already appreciable without additional oxidation.*"

The support porous structure and the rate of solvent removal from the pores as well as the nature of solvent and metal compound dissolved can considerably influence both the distribution of the active component through the support grain and the catalyst dispersion [163,170–173]. As a rule, the resulting particles size of the active component will be smaller, the more liquid-phase ruptures caused by evaporation of the solvent from the support pores are attained before the solution saturation. Therefore, supports with an optimal porous structure are needed to prepare "impregnated" Me/C catalysts with the finest metal particles. As a result, carbon supports appropriate for synthesis of such catalysts are very limited in number. Besides, these catalysts will strongly suffer from the blocking effect (see Section 12.1.2) because some of the metal particles are localized in fine pores.

At otherwise identical conditions, the average size of metal particles in Me/C catalysts prepared with the use of $RuCl_3$/water [174,175], $RuCl_3$/acetone [176], $PdCl_2$/benzene+ethanol [8], $Pd_3(OAc)_6$/acetone [41], $Pt(NH_3)_2(NO_2)_2$/water [114], H_2PtCl_6/benzene+ethanol [165,177], H_2IrCl_6/water [178] solutions is very sensitive to the metal loading and increases rapidly as it increases. Certain of these dependencies built up in the logarithmic coordinates to produce linear anamorphoses are shown in Figure 6 (a)–(f). It is interesting to note that similar anamorphoses can also be observed with catalysts prepared by condensation of vaporized metal on the carbon surface [179,180]. Analytically, these anamorphoses can be easily obtained assuming that the whole mass (M) of the supported metal occurs as particles of diameter d that arise (without regard for sintering) from the crystallites of supported metal precursor, and their number (N) equals the number of special sites on the support surface where the precursors have been crystallized:

$$M = N\pi\rho d^3/6 \qquad \text{or} \qquad \log(M/\rho) = \log \pi/6 + \log N + 3\log d \qquad (14)$$

where ρ is the metal density. Correlation of this type usually indicates the presence of these sites in limited amounts. Any kind (structural or chemical) of surface defects can, in principle, behave as these sites. It is common knowledge that nucleation of a new phase is the easiest when its molecules exhibit chemical affinity to the support surface. Therefore, the most dispersed catalysts are synthesized when a quantity of the dissolved precursor is adsorbed on the support to behave as the sites for its following crystallization. This is probably an additional reason for an increase in the metal dispersion in 1% Pt/C catalysts [23,165] (H_2PtCl_6/benzene+ethanol, graphitized carbon black Vulcan 3G) with an increase in the carbon surface heterogeneity. Other reasons may be the evolution of a more optimal porous structure or strengthening of the metal-support interaction in the target catalyst as the carbon surface becomes more and more heterogeneous under the burn-off, as was supposed earlier [23,165]. Partial adsorption of metal salt also may be responsible for an

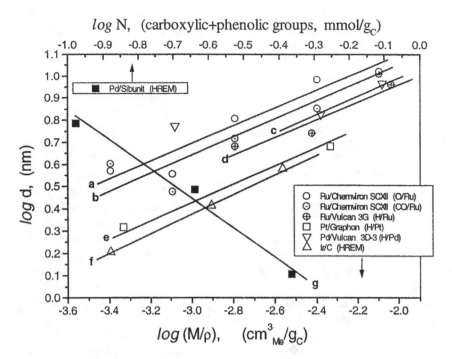

Figure 6 Dependence of the mean size (d) of the active component particles in Me/C catalysts on the total volume (M/ρ) of the supported metal (a–e) and the concentration of the surface acidic sites on the support (g). Literature data are used: (a,b) 0.5–10% Ru/C [175]; (d) 2.0–11.4% Ru/C [176]; (c) 1–10% Pd/C [8]; (e) 1–10% Pt/C [177]; (f) 0.9–6.1% Ir/C [178]; (g) 1% Pd/C [39].

increase in the dispersion of 1% Pd/Sibunit catalysts [39] [Pd$_3$(OAc)$_6$/benzene] when the concentration of acidic groups at the carbon surface increases [Figure 6(g)]. However, some data [41] argue the existence of the inverse dependence for the Pd/C catalysts prepared by supporting Pd$_3$(OAc)$_6$ from the *acetone* solutions onto chemically modified active carbons Anthralur and Sutcliffe. This effect may be caused either by the nature of the chosen solvent or by the textural properties of the support (unlike Sibunit, Anthralur and Sutcliffe are predominantly microporous carbons).

Sometimes the use of the impregnation procedure can even provide a high dispersion of the active components at unusually high values of the metal loading. This is especially true for the Pd/C catalysts synthesized from H$_2$PdCl$_4$. The reason is the formation of PdCl$_2$ clusters on the carbon surface in the course of drying [16,57,181]. The clusters are 1.8 ± 0.2 nm in size, which is practically independent of the nature of the carbon support and, over a very wide range, on the weight percentage of the supported palladium. The clusters are fixed at surface steps through both epitaxial and chemical interaction. The latter means formation of the surface π-complexes of PdCl$_2$ [16]. When reduced by H$_2$ at 250 °C, they originate metal particles 1–3 nm in size. This phenomenon was used to prepare high-dispersed catalysts based on platinum, a chemical analog of palladium. Electrocatalysts Pt/HOPG were prepared by supporting H$_2$PtCl$_4$ on the etched basal plane of HOPG followed by reduction of the supported $PtCl_2$ in H_2 at 300 °C. The metal loading in these catalysts reached about $75 \,\mu mol/m^2$ (i.e., approximately 100 times as high as the carbon adsorbability with respect to H_2PtCl_4, Table 8), the particle size (d_s) determined by HRTEM being 3.4 nm [182]. A strong influence of the carbon matrix structure on the electrocatalyst dispersion was discovered later: an increase in the concentration of intercrystallite boundaries and in disordering of the mutual orientation of the carbon crystallites led to the particle shrinkage to $d_s = 1.3$ nm. This was attributed to an increase in the concentration of the *strong* adsorption sites for H_2PtCl_4: binding $PtCl_2$, they behaved then as sites of $PtCl_2$ crystallization during evaporation of the solvent. On the contrary, the use of H_2PtCl_6 led to coarse Pt particles (10–50 nm plus spongelike aggregates) [182]. H_2PtCl_6 was found incapable of forming π-complexes with the ligands containing $>C{=}C<$ bonds [53], so as the low dispersion of Pt particles in the prepared samples could be only attributed to the low affinity of Pt^{IV} chloride complexes to the carbon surface. The relatively high metal dispersion independent of the metal loading (1.5–12%) was observed with Ru/C (H/Ru ∼ 0.7) [183] or Rh/C (H/Rh ∼0.12) catalysts [184] prepared by impregnation of the supports with $MeCl_3$ solutions, but the reason was not investigated.

Thus, dispersion of Me/C catalysts prepared by impregnation is in many respects dependent on the probability of side processes such as the metal compounds adsorption, even though their intensity is low that makes the proportion of the adsorbed species negligible among the supported precursors. In the general case, the detrimental effect on the metal dispersion caused by deficit of the sites for crystallization of the metal precursors can be compensated by means of repeated impregnation with low-concentrated salt solutions. An example is the preparation of Pt/C from $Pt(NH_3)_2(NO_2)_2$ [185].

12.2.4 Catalysts Me/C Prepared by Deposition

Deposition of Metal Hydroxides

Even though all noble metals can form insoluble hydroxides, this method is mostly employed for synthesis of Pd/C catalysts. Any water-soluble salts of Pd^{II} are appropriate [186–188], but the most usable are H_2PdCl_4 or Na_2PdCl_4. Hydroxides, carbonates, and bicarbonates of alkali metals and nitrogen-containing bases, most often Na_2CO_3, are used as alkali agents. The metal hydroxide deposited on the support is usually reduced in an alkali medium (H_2CO, NaOOCH, N_2H_4, KBH_4) and more rarely in a close-to-neutral medium (NaH_2PO_2, KBH_4). There are three basic versions of the deposition process:

1. An alkali solution is added to the support suspension in a Pd^{II} salt solution.
2. A Pd^{II} salt solution is added to the support suspension in an alkali solution.
3. Solutions of a Pd^{II} salt and an alkali agent are mixed in such a proportion to produce soluble Pd^{II} compounds of the colloidal nature, the so-called polynuclear hydroxocomplexes (PNHC), to make them contacting carbon.

These methods are practiced for preparation of powdered catalysts. Method 3 is also appropriate for supporting palladium on granulated carbons by incipient wetness impregnation. Although the idea of the method under consideration is to deposit metal hydroxide species formed in the solution on the carbon surface, in actual practice the process is never the only one. Adsorption of the initial Pd^{II} compounds to be hydrolyzed on the support surface takes place either before formation of PNHC in the solution (synthesis of Pd/C according to 1 and 2) or concurrently with deposition of PNHC (synthesis of Pd/C according to 3) [167,189,190]. Therefore, to understand the regularities of genesis of these catalysts it is necessary to distinguish *primary* processes (i.e., adsorption of initial Pd^{II} compounds followed by transformations of the adsorbed species upon variations in pH of the medium) and *secondary* ones (i.e., formation of PNHC in the solution and deposition of PNHC on the support) among all chemical processes taking place in the "Pd^{II} salt–alkali agent–carbon" system. Discussion of these processes is exemplified below.

Formation of Polynuclear Hydroxocomplexes of Pd(II).

Structural properties, as well as specific features of formation and aging of PNHC, are discussed in detail elsewhere [59,187,189,191–194]. Complexes [PdCl$_4$]$^{2-}$ and [PdCl$_3$(H$_2$O)]$^{-}$ are predominantly detected in the initial solutions of H$_2$PdCl$_4$ (pH 0.5–1.5). Addition of NaOH results first in the formation of Na$_2$PdCl$_4$ (pH 2.5–4.0) and in shifting the equilibrium of these Pd^{II} complexes toward predomination of [PdCl$_3$(H$_2$O)]$^{-}$ [191]. Further addition of the alkali leads to generation of PNHC particles [191,193]. Little changes in the pH of the solutions are observed with the ratio $\chi = $ NaOH/Na$_2$PdCl$_4$ because the added alkali is completely consumed for the formation of PNHC. The concentration of PNHC increases with χ, the proportion of Pd^{II} involved in PNHC particles being 0.5χ, but their size distribution remains practically unchanged [189,194]. The data obtained by pH measurements, UV, NMR (^{17}O, ^{23}Na, ^{35}Cl, ^{133}Cs), and EXAFS spectroscopic techniques [193,194], allowed the conclusion that the interaction of NaOH and Na$_2$PdCl$_4$ is intermediated by complex [PdCl$_2$(OH)$_2$]$^{2-}$, which readily enters the reaction of polycondensation to release Cl$^-$ ions and to form polymer threads formulated as [Pd(OH)$_2$]:

The thread can roll up into a ball due to the Coulomb attraction between some of its fragments and Na$^+$ cations, as well as through oxygen bridges. Thus formed, the PNHC particle is generally formulated as {Pd(OH)$_2$}$_n$$m$NaCl ($m \leqslant n$). It is free of Cl$^-$ ions directly bonded to palladium, while hydrated Cl$^-$ ions are arranged at the edge rims of the particle and compensate the positive charge induced by Na$^+$ ions occluded in the particles. Emergence of this kind of EDL on the surface of the PNHC particles makes these colloids resistant to coagulation. They are mainly detected as fine particles 0.8–1.2 nm ($n \sim 10 \div 20$) and 2.0–2.4 nm ($n \sim 100$) in size, with coarser particles formed by associated fine particles also observed [194]. PNHC particles comprising 10 to 1000 palladium atoms were also observed. [191]. RDF and

XRD studies of freshly prepared PNHC solutions [189] reveal a very disordered structure of these particles.

In long storage of PNHC solutions, slow chemical processes proceed inside the colloidal particles that lead to a decrease in their size [194]. Aging of a fine PNHC particle causes rearrangement of the threadlike structure of $\{Pd(OH)_2\}_n$ into a tighter three-dimensional structure of palladium oxide that is accompanied by the release of H^+ and, partly, Na^+ ions into the solution. In turn, aging of a PNHC *associate* leads to destruction of the solvate sheath of the involved particles and to closer contacts between them. In the first case this takes approximately 2.5 h at room temperature, whereas a longer period of time (a day or longer) is needed in the second one.

Analysis of data available from literature allows us to assume that both the pH region of resistance of PNHC colloids to coagulation and the rate of their aging depend on a number of not clearly understood factors, including the solution temperature, the nature of the alkali agent, the Pd^{II} concentration, and the presence of foreign electrolytes and surfactants. For example, formation of PNHC was observed at pH 3–3.5 [191], somewhat higher than pH 4 [189], or at pH 5.3–5.9 [187]. When the PNHC solutions are aged at pH 3, sedimentation is observed in 24 h [191], whereas pre-addition of NaCl shifts the sedimentation point toward pH 6 [192] or 6.8 [128]. It is recommended not to prepare PNHC solutions at a temperature higher than 30 °C [187]. Unfortunately, data on the formation of colloidal $Pd(OH)_2$ particles in the alkali medium (when the solution of a Pd^{II} compound is added to excess alkali) are practically unavailable.

Anyway, the PNHC solutions prepared by slow addition of aqueous NaOH or Na_2CO_3 to H_2PdCl_4 at room temperature are stable enough for them to be appropriate for synthesis of Pd/C catalysts in 2–4 h from the moment of their preparation [59,187].

Formation and Properties of Pd/C Catalysts

Method 1 includes three main stages. These are addition of the H_2PdCl_4 solution to the aqueous suspension of the support, hydrolysis of Pd^{II} compounds with an introduced alkali agent, followed by reduction. As pointed out at the beginning of Section 12.2.4, when Pd/C is synthesized by any of these methods, addition of the Pd^{II} solution to carbon slurry results in adsorption of a portion of Pd^{II} compounds on the support. Figure 7 (a) shows mononuclear Pd^{II} complexes to be better adsorbed from the acidic medium (method 1) than from the alkali one (method 2). Therefore, the above-mentioned primary processes will mostly affect the catalysts prepared by method 1. Hence, analysis of the literature data concerned with this method of the Pd/C catalysts preparation should be based on the data on studying the adsorption of H_2PdCl_4 on carbon materials.

In the case of carbons traditionally used as the supports, adsorption of H_2PdCl_4 is comparatively rapid [16], and the concentration of the *strong* adsorption sites $(A_2 + A_3)$ is higher than that of the supported metal. Hence, most of palladium transfers from the solution onto the support prior to the introduction of the alkali agent and forms there the former Pd^0 particles (9) and Pd^{II} chloride complexes (11). These surface compounds are chief precursors of the active component in the catalysts containing no more than 5% wt. Pd. A remarkable contribution of the deposition of PNHC particles from the solution is only observed for catalysts with a very high Pd loading or when supports with a very low specific surface area are used.

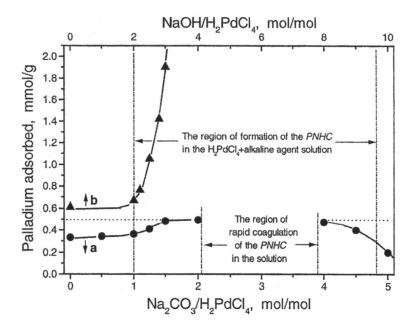

Figure 7 Adsorption of Pd^{II} compounds from aqueous solutions (obtained by mixing H$_2$PdCl$_4$ and an alkali agent) on carbon supports of the Sibunit family at 20 °C. (a) Distribution of Pd^{II} between the solution and the support surface ($S_{BET} = 270$ m^2/g) during preparation of 0.5 mmol Pd/C catalyst: Pd^{II} adsorbability from 0.025 M solution as a function of the Na$_2$CO$_3$/H$_2$PdCl$_4$ ratio. (According to Ref. 167.) (b) Dependence of adsorption capacitance of the support ($S_{BET} = 450$ m^2/g) in respect to Pd^{II} on the Na$_2$CO$_3$/H$_2$PdCl$_4$ ratio. (According to Ref. 189.)

As a cumulative result of all these processes, a bimodal Pd particle-size distribution is often observed in the target catalysts (Table 9): there are both coarse crystallites of the former Pd^0 generated as early as the stage of H$_2$PdCl$_4$ adsorption and round-shaped fine Pd^0 particles formed in the course of hydrolysis followed by reduction of the surface and solute palladium chloride complexes.

Emergence of x-ray-detectable Pd crystallites (larger than 10 nm in size) upon contact of the support with the solution of H$_2$PdCl$_4$ is a well-established feature of preparation of Pd/C [16,70,94,111,137–140,143,146,157,168,190] and, probably, the main reason for a decrease in the efficiency of these catalysts and the difference between the available data on the metal dispersion in the Pd/C catalysts prepared by method 1 (Table 9). The proportion of the coarse metal component among the supported precursors is determined by the carbon nature and the combination of conditions to achieve the contact between the support and the solution of H$_2$PdCl$_4$ (Table 7). In fact, the authors of numerous papers used some tricks enumerated in Table 7 [such as introduction of excess HCl [195] or NaCl [196,197] or addition of an oxidant (H$_2$O$_2$ [195]) to the H$_2$PdCl$_4$ solution, pre-oxidation [40,190,198] or treatment of the support surface with an acid solution [70,195], a longer contact of the solution with the support in air [190], the use of the support with certain textural properties [199,200], and grain size [199]] but often interpreted the obtained

Table 9 Comparative Analysis of Methods 1–3 for Preparation of Pd/C Catalysts. The Influence of the Nature of the Carbon Support and Preparation Conditions on the State of the Active Component of the Catalysts

Pd/C Preparation Mode	1	2	3
1. Particle-size distribution for palladium			
(a) On active carbon, carbon black or Sibunit	Bimodal[1]: fine + coarse components[2] [70,120,190,202]		Uniform[1] [14]
(b) On oxidized carbon	Uniform [40,190]	Uniform [190]	Uniform[1] [203]
2. Palladium distribution through the support grain			
(a) On active carbon, carbon black or Sibunit	At the grain circumference or on the surface of wide pores[3] [190], eggshell [70,167]	At the grain circumference [190]	Eggshell $\geqslant 20\mu$ [204], eggshell [167]
(b) On oxidized carbon	Wide, penetration of Pd into micropores [190]		
3. Average size of palladium particles, nm (by gas chemisorption)			
(a) On graphitized carbons	5–6 [70,190]	6 [190]	2–3 [203]
(b) On active carbon, carbon black, Sibunit etc.	1.4–1.7 [16,167], 2.5–3.5 [70,118], 4–6 [190], 1.5–6[1]	4–8 [190], 4–10 [186]	1.1–2.0 [128], 2.1–2.4 [167], 2.4–3.2 [14,16,203]
(c) On oxidized carbon	1.5 [190]	1.4–1.7 [190]	2–3 [203], 2.5–3.2 [128], 4–5[1,4]
4. Variations in the dispersion of Pd/C catalysts			
(a) At elevation of temperature of deposition of Pd[II] compounds on the support	Volcanolike, maximum at 40 °C for active carbons or at 20–30 °C for graphitized carbons [190], a decrease [188]	A very weak dependence in comparison with method[1] [190], a decrease [186]	
(b) With an increase in the specific surface area of carbon (S_{BET})	No effect [196] (730–2700 m^2/g)		Slight [128] (500–1500 m^2/g), no effect [201,204] (15–240 m^2/g, 70–700 m^2/g)
(c) With an increase in the metal weight concentration on the support[2] (the limits of Pd loading variations are indicated)	An increase[1] (1–5%)		Slight[1] (2–5%), slight volcanolike dependence (1–17%) with a gently sloping plateau[5] at 2.5–13% [128]
(d) With an increase in the degree of oxidation of the carbon surface	A strong increase[1] [190,198]	A strong increase [190]	A decrease [128], no effect [203]

[1] Observed (or confirmed) by the authors of the review but not published before.
[2] Prewashing of the support with aqueous HCl results in an increase in the dispersion of Pd due to a decrease in the amount of the *former* Pd^0. (From Refs. 16 and 70.)
[3] If the carbon was in long contact with H$_2$PdCl$_4$ before NaOH was added, the metal distribution becomes more uniform [167], and palladium penetrates into micropores [190].
[4] Formalin was used for reduction that may cause coarse palladium particles. (*Source*: Ref. 120.)
[5] The range of relative constancy of Pd dispersion depends on the type of carbon.

results in other ways that did not encompass the whole set of the available literature data on this method for preparation of Pd/C.

The properties of these catalysts also are strongly dependent on the nature of the alkali agent, its concentration in the solution, and the manner of adding to the carbon suspension [199,201]. A number of processes occur at this stage of Pd/C synthesis, including hydrolysis of Pd^{II} compounds (both adsorbed and dissolved), adsorption of produced PNHC from the solution, aging of palladium hydroxide particles on the support surface, as well as dissolution of these particles under the action of excess alkali and Pd^{II} transfer back to the solution. Evidently, the relationship among rates of these processes is affected by numerous factors. That is why a lot of different "optimal" procedures of the hydrolysis (especially in the patent literature) can be found, but most likely no universal one exists. Generally, it seems necessary to comply with the following:

1. The optimal rate of alkali feeding at the earliest stage of the catalyst synthesis, which allows maintenance of a higher rate of adsorption of generated PNHC in respect to the rate of their coagulation in the solution, thus suppressing formation of large spongelike Pd particles in the final catalyst.

2. The optimal period and temperature of aging of the supported palladium hydroxide particles at a pH close to the neutral point in order to suppress dissolution of these particles upon the further alkali addition before reduction, thus preventing enlargement of Pd particles. However, palladium hydroxide should not be aged too long, since in the course of this process its particles gradually lose the ability to be completely reduced in the liquid phase at mild conditions that, in turn, may result in desorption of palladium species when the final catalyst is washed with water to remove the salts produced during its synthesis.

Conditions of reduction of the supported palladium have a strong influence on the catalyst dispersion. The high dispersion is achieved by reduction with solutions of NaOOCH [196], NaBH$_4$ [118], and NaH$_2$PO$_2$ [128] in a moderately basic medium. The use of CH$_2$O needs a stronger basic medium and leads to comparatively coarser palladium particles [118,186,190], the bimodal particle-size distribution being possible due to the mass transfer and slow reduction of Pd^{II} [120]. Gas-phase reduction in H$_2$ produces a good result at 100–250 °C [14,16,118,167] but leads to remarkable metal sintering at 300–400 °C [40,196,203].

The briefly discussed specific features of the stages of hydrolysis and reduction of palladium compounds are in many respects coincided for all three versions of Pd/C synthesis discussed in the present section.

Method 2 is only scantily discussed in the literature, which makes it difficult to compare with methods 1 and 3 (see Table 9). There is only a slight dependence of the dispersion of these catalysts on the nature of the alkali agent (except hydro-carbonates) [186]. It goes through a maximum at elevation of the hydrolysis temperature [186,190]. However, the data on the maximum position are contradictory (ca. 30–40 °C [186] or 75 °C [190]). Although close dispersion may be observed with catalysts synthesized by methods 1 and 2, the catalytic activity of the latter is much higher. That is attributed to different distribution of the metal through the catalyst grain [190]: according to method 2, Pd^{II} is predominantly hydrolyzed in the solution, and PNHC particles are deposited on the external surface of the carbon grains.

Method 3 implies competitive adsorption of PNHC particles and mononuclear Pd^{II} complexes on the support [189], since PNHC solutions resistant to coagulation contain the initial monomer. Sorbability of PNHC is higher by approximately an order of magnitude than one of H$_2$PdCl$_4$ [189,204] [see also Figure 5 (b)]. Just adsorbed PNHC particles are practically of the same composition and size as the PNHC particles in the solution [189]. They do not tend to decorate the carbon surface and are uniformly distributed through the surface even in the case of graphitized carbon [189,201]. Properties of the Pd/C catalysts synthesized by this method reveal the lowest sensitivity to the nature of the carbon support (Table 9). A narrow eggshell distribution through the catalyst grain (even though the grains are as small as 0.1 mm) is always observed with these catalysts. In general, an increase in the proportion of PNHC among the adsorbed precursors of Pd/C results in narrowing of the active component distribution through the catalyst grain [167,204] as well as in weakening of the effect of blocking the metal surface by pore walls to make it inaccessible to the reactants [16,167] (see, e.g., Table 5).

12.3 CONCLUSIONS

The state of the active component in Me/C catalysts (dispersion, substructural properties of its particles, size distribution, and the localization on the support surface) is governed by the state of supported precursors, mechanisms of formation of these precursors, and their further evolution into the metal phase. In turn, the state of the catalyst precursors if often determined by the nature of the carbon support and initial metal compound dissolved, as well as by conditions of supporting the latter. For the Me/C catalysts prepared by adsorption, predomination of one type of adsorbed precursor species and the presence of a sufficiently large number of surface sites for their fixation favor generation of high-dispersed catalysts with their particle size slightly dependent on the metal loading and on the support surface nature. In the case of a deficiency of these sites, Me/C catalysts are prepared by impregnation or deposition. For the impregnated catalysts, the average size of the metal particles usually increases with the metal loading and depends on the physicochemical properties of the support affecting the concentration of these sites. However, these methods also allow the highly dispersed catalysts to be prepared over a very wide range of metal loading if the precursors formed from the initial metal compounds on the carbon surface are stable clusters of a certain size. In the case of co-existence of several routes to generation of the supported precursors, the catalyst dispersion is a complex function of some set of physicochemical properties of the support and conditions of the metal compound supporting, with a kind of volcanolike dependence on the metal loading usually observed.

The capacity of carbon support for behaving as a gas electrode and the ability of dissolved or supported metal ions, as well as the metal particles, to interact with the carbon surface through a charge transfer indicate the considerable influence of the atmosphere composition and of electrophysical properties of the carbon (the latter are determined both by the surface and *bulk* properties of the carbon matrix) on genesis of the supported metal catalysts. Surprisingly few papers deal with these aspects of scientific basis for preparation of Me/C catalysts. Undoubtedly, this gap should be bridged in the near future.

REFERENCES

1. J.F. LePage, In *Handbook of Heterogeneous Catalysis.*, Vol. 1. (G. Ertl, H. Knözinger, J. Weitkamp, eds.), Wiley-VCH, Germany, 1997, p. 49.

2. S.Z. Roginskii, *Geterogennyi Kataliz* (Heterogeneous Catalysis), Nauka, Moscow, 1979.
3. Y.G. Frolov, *Kurs Kolloidnoi Khimii. Poverkhnostnye Yavlenia i Dispersnye Sistemy*. (Comprehensive Colloid Chemistry: Surface Phenomena and Dispersed Systems), Khimiya, Moscow, 1988.
4. W.B. Phillips, E.A. Desloge, J.G. Skofronick, *J. Appl. Phys.* **39**, 3210 (1968).
5. J. Bett, K. Kinoshita, K. Routsis, P. Stonehart, *J. Catal.* **29**, 160 (1973).
6. P. Wynblatt, T.-M. Ahn, In *Sintering and Catalysis*, Vol. 10 (G.C. Kuczynski, ed.), Plenum, New York, 1975, p. 83.
7. R.T.K. Baker, J.A. France, L. Rouse, R.J. Waite, *J. Catal.* **41**, 22 (1976).
8. N. Martin, S. Fuentes, *Carbon* **26**, 795 (1988).
9. R.G. Baetzold, *J. Catal.* **29**, 129 (1973).
10. R.G. Baetzold, *Surf. Sci.* **36**, 123 (1973).
11. R.T. Yang, C. Wong, *J. Catal.* **82**, 245 (1983).
12. D. Richard, P. Gallezot, In *Preparation of Catalysts IV* (G. Poncelet et al., eds.), Elsevier, Amsterdam, 1987, p. 71.
13. K.L. Yeung, E.E. Wolf, *J. Catal.* **135**, 13 (1992).
14. V.A. Semikolenov, S.P. Lavrenko, V.I. Zaikovskii, *React. Kinet. Catal. Lett.* **51**, 507 (1993).
15. V.B. Fenelonov, L.B. Avdeeva, O.V. Goncharova, L.G. Okkel, P.A. Simonov, A.Y. Derevyankin, V.A. Likholobov, *6th Intl. Symp. Scientific Bases for the Preparation of Heterogen. Catalysts*, Louvain-la-Neuve, Belgium, 1994; Preprints, Vol. 3, p. 43.
16. P.A. Simonov, *Thesis*, Boreskov Institute of Catalysis, Novosibirsk, Russia, 2000.
17. Y.F. Komnik, *Fizika Metallicheskikh Plenok: Razmernye i Strukturnye Effekty* (Physics of Metallic Films: Size and Structure Effects), Atomizdat., Moscow, 1979, pp. 56–60.
18. A.W. Adamson, *Physical Chemistry of Surfaces*, Mack Printing Co., Easton, PA, 1963.
19. B. Tardy, C. Nuopa, C. Leclerco, J.C. Bertolini, A. Hoareau, M. Treilleux, J.P. Faure, G. Nikoul, *J. Catal.* **129**, 1 (1991).
20. N.L. Wu, J. Phillips, *Surface Sci.* **184**, 463 (1987).
21. M.J. Yacamán, J.M.E. Domínguez, *J. Catal.* **64**, 213 (1980).
22. P. Gallezot, D. Richard, G. Bergeret, In *Novel Materials in Heterogeneous Catalysis, ACS Symp. Series*, Vol. 437, (R.T.K. Baker and L.L. Murrell, eds.), Amer. Chem. Soc., Washington, 1990, pp. 150–159.
23. P. Ehrburger, P.L. Walker, Jr., *J. Catal.* **55**, 63 (1978).
24. D. Richard, P. Fouilloux, P. Gallezot, In *Proc. 9th Intl. Cong. on Catal.*, Vol. 3 (M.J. Phillips et al., eds.), The Chemical Institute of Canada, Ottawa, 1988, p. 1074.
25. W.F. Maier, S.J. Chettle, R.S. Rai, G. Tomas, *J. Amer. Chem. Soc.* **108**, 2608 (1986).
26. R. Schlögl, In *Handbook of Heterogeneous Catalysis*. Vol. 1. (G. Ertl, H. Knözinger, and J. Weitkamp, eds.), Wiley-VCH, Germany, 1997, p. 138.
27. A.I. Boronin, V.I. Bukhtiyarov, R. Kvon, V.V. Chesnokov, R.A. Buyanov, *Surf. Sci.* **258**, 289 (1991).
28. S.B. Ziemecki, G.A. Jones, *J. Catal.* **95**, 621 (1985).
29. S.B. Zicmccki, G.A. Jones, D.G. Swartzfager, R.L. Harlow, J. Faber, Jr., *J. Amer. Chem. Soc.* **107**, 4547 (1985).
30. R. Lamber, N. Jaeger, *Surf. Sci.* **289**, 247 (1993).
31. R. Lamber, N. Jaeger, G. Schulz-Ekloff, *Surf. Sci.* **227**, 15 (1990).
32. N. Krishnankutty, M.A. Vannice, *J. Catal.* **155**, 312 (1995).
33. N. Krishnankutty, J. Li, M.A. Vannice, *Appl. Catal.* A **173**, 137 (1998).
34. K. Kinoshita, *Carbon, Electrochemical and Physicochemical Properties*, Wiley-Interscience Publ., New York, 1988.
35. N. Krishnankutty, M.A. Vannice, *J. Catal.* **155**, 327 (1995).
36. P. Albers, K. Deller, B.M. Despeyroux, A. Schafer, K. Seibol, *J. Catal.* **133**, 467 (1992).
37. I.A. Tarkovskaya, *Okislennyi Ugol* (Oxidizied Carbon), Nauk. Dumka, Kiev, 1981.

38. W.M.H. Sachtler, A.Y. Stakheev. *Catal. Today* **12**, 283 (1992).
39. S.V. Gurevich, P.A. Simonov, A.S. Lisitsyn, V.A. Likholobov, E.M. Moroz, A.L. Chuvilin, V.N. Kolomiichuk, *React. Kinet. Catal. Lett.* **41**, 211 (1990).
40. D.J. Suh, T.-J. Park, S.-K. Ihm, *Carbon* **31**, 427 (1993).
41. M. Gurrath, T. Kuretzky, H.P. Boehm, L.B. Okhlopkova, A.S. Lisitsyn, V.A. Likholobov, *Carbon* **38**, 1241 (2000).
42. T. Kuretzky, *Ph.D. Thesis*, München University, Germany, 1993.
43. C. Prado-Burguete, A. Linares-Solano, F. Rodríguez-Reinoso, C. Salinas-Martinez de Lecea, *J. Catal.* **115**, 98 (1989).
44. F. Coloma, A. Sepúlveda-Escribano, J.L.G. Fierro, F. Rodríguez-Reinoso, *Langmuir* **10**, 750 (1994).
45. K.T. Kim, J.S. Chung, K.H. Lee, Y.G. Kim, J.Y. Sung, *Carbon* **30**, 467 (1992).
46. G.C. Torres, E.L. Jablonski, G.T. Baronetti, A.A. Castro, S.R. de Miguel, O.A. Scelza, M.D. Blanco, M.A. Peña Jiménez, J.I.G. Fierro, *Appl. Catal.* **A 161**, 213 (1997).
47. L.B. Okhlopkova, A.S. Lisitsyn, V.A. Likholobov, M. Gurrath, H.P. Boehm, *Appl. Catal.* **A 204**, 229 (2000).
48. Y.A. Ryndin, M.V. Stenin, A.I. Boronin, V.I. Bukhtiyarov, V.I. Zaikovskii, *Appl. Catal.* **54**, 277 (1989).
49. G. Chen, W.-T. Chou, C.-T. Yeh, *Appl. Catal.* **A 20**, 389 (1983).
50. C.E. Hunt, *J. Catal.* **23**, 93 (1971).
51. A. Guerrero-Ruiz, P. Badenes, I. Rodríguez-Ramos, *Appl. Catal.* **A 173**, 313 (1997).
52. A.J. Robell, E.V. Ballou, M. Boudart, *J. Phys. Chem.* **68**, 2748 (1964).
53. M. Herberhold, *Metal π-Complexes: Complexes with Mono-Olefinic Ligands*, Vol. 2 (1/2), Elsevier, Amsterdam, 1974.
54. P.A. Simonov, V.A. Semikolenov, V.A. Likholobov, A.I. Boronin, Y.I. Yermakov, *Izv. Akad. Nauk SSSR, Ser. Khim.* (Russ. ed.) **12**, 2719 (1988).
55. P.A. Simonov, A.L. Chuvilin, V.A. Likholobov, *Izv. Akad. Nauk SSSR, Ser. Khim.* (Russ. ed.) **9**, 1952 (1989).
56. P.A. Simonov, E.M. Moroz, V.A. Likholoboy, G.V. Plaksin, *Izv. Akad. Nauk SSSR, Ser. Khim.* (Russ. ed.) **7**, 1478 (1990).
57. P.A. Simonov, E.M. Moroz, A.L. Chuvilin, V.N. Kolomiichuk, A.I. Boronin, V.A. Likholobov, *6th Intl. Symp. Sci. Bases for the Preparation of Heterogen. Catalysts*, Sept. 5–8, 1994, Louvain-la-Neuve, Belgium. Preprints, Vol. 3, 1994, p. 201.
58. M. Cini, M. de Crescenzi, F. Patella, N. Motta, M. Sastry, F. Rochet, R. Pasquali, A. Balzarotti, C. Verdozzi, *Phys. Rev.* **B 41**, 5685 (1990).
59. P.A. Simonov, S.Y. Troitskii, V.A. Likholobov, *Kinet. Catal.* (Russ. ed.) **41**, 281 (2000).
60. P.A. Simonov, S.V. Filimonova, G.N. Kryukova, E.M. Moroz, V.A. Likholobov, T. Kuretzky, H.P. Boehm, *Carbon* **37**, 591 (1999).
61. J.L. Garnett, W.A. Sollich, *Austral. J. Chem.* **15**, 56 (1962).
62. Y.A. Ryndin, O.S. Alekseev, E.A. Paukshtis, V.I. Zaikovskii, A.V. Kalinkin, *J. Molec. Catal.* **68**, 355 (1991).
63. Y.A. Ryndin, Y.A. Nogin, E.A. Paukshtis, A.V. Kalinkin, A.L. Chuvilin, Y.A. Zverev, *J. Molec. Catal.* **62**, 45 (1990).
64. I. Furuoya, T. Yanagihara, T. Shirasaki, *Intl. Chem. Eng.* **10**, 333 (1970).
65. K. Morikawa, T. Shirasaki, M. Okada, In *Adv. Catal. and Relat. Subj.,* Vol. 20 (D.D. Eley, H. Pines, and P.B. Weisz, eds.), Academic Press, New York, 1969, P. 97.
66. R.T. Baker, E.B. Prestridge, R.L. Garten, *J. Catal.* **56**, 390 (1979).
67. P. Gallezot, *Surf. Sci.* **106**, 459 (1981).
68. Z. Kowalczyk, S. Jodzis, W. Raróg, J. Zieliński, J. Pielaszek, *Appl. Catal.* **A 173**, 153 (1998).
69. F. Rodríguez-Reinoso, C. Moreno-Castilla, A. Guerrero-Ruiz, I. Rodríguez-Ramos, J.D. López-González, *Appl. Catal.* **15**, 293 (1985).

70. P. Albers, R. Burmeister, K. Seibold, G. Prescher, S.F. Parker, *J. Catal.* **181**, 145 (1999).
71. E. Hegenberger, N.L. Wu, J. Phillips, *J. Phys. Chem.* **91**, 5067 (1987).
72. N.H. Sagert, R.M.L. Pouteau, *Platinum Met. Rev.* **19**, 16 (1975).
73. X.Q.D. Li, T. Radojicic, R. Vanselow, *Surf. Sci. Lett.* **225**, L29 (1990).
74. C.S. Nicolau, H.G. Thom, E. Pobitska, *Trans. Faraday Soc.* **55**, 1430 (1959).
75. P.N. Ross, K. Kinoshita, P. Stonehart, *J. Catal.* **32**, 163 (1974).
76. Z. Bastl, O. Přibyl, P. Mikušik, *Czech. J. Phys.* **B 34**, 981 (1984).
77. *Blagorodnye Metally, Spravochnik* (Handbook of Noble Metals) (E.M. Savitskii, ed.), Metallurgiya, Moscow, 1984, pp. 114–118.
78. V.S. Fomenko, Emissionnye Svoistva Materialov, *Spravochnik* (Handbook of Emission Properties of Materials), 4th ed., Nauk. Dumka, Kiev, 1981.
79. *CRC Handbook of Chemistry and Physics*, 73rd ed. (D.R. Lide, ed.), CRC Press, London, 1992–1993, pp. 12–108.
80. T.J. Fabish, D.E. Schleifer, *Carbon* **22**, 19 (1984).
81. R.O. Loutfy, *Carbon* **24**, 127 (1986).
82. E.G. Chen, M. Ohtsuki, A.V. Grewe, *Surf. Sci.* **144**, 465 (1984).
83. M.R. Tarasevich, *Elektrokhimiya Uglerodnykh Materialov* (Electrochemistry of Carbonaceous Materials), Nauka, Moscow, 1984.
84. I.L. Spain, In *Chemistry and Physics of Carbon*, Vol. 16. (P.L. Walker and P.A. Thrower, eds.), Marcel Dekker, New York, 1981, p. 119.
85. S. Mrozowski, *Phys. Rev.* **85**, 609 (1952).
86. E.E. Loebner, *Phys. Rev.* **102**, 46 (1956).
87. A. Marchard, *Carbon'86: Proc. 4th Intl. Carbon Conf.*, 1986, p. 9.
88. D.H. Bangham, N. Fakhoury, *Proc. Roy. Soc.* (London) **A 147**, 152 (1934).
89. J.R. Dacey, In *The Solid–Gas Interface*, Vol. 2 (E.A. Flood, ed.), Marcel Dekker, New York, 1967, p. 1019.
90. T.N. Ivanova, R.S. Vartapenyan, A.M. Volotschuk, In *Uglerodnye Adsorbenty i ikh Primenenie v Promyshlennosti* (Carbon Adsorbents and Their Industrial Application), 5th All-Union Conference, Abstracts, Perm, All-Union Research and Technological Institute of Carbon Adsorbents, 1991, p. 71.
91. K. Kaneko, C. Ishii, M. Ruike, H. Kuwabara, *Carbon* **30**, 1075 (1992).
92. M. Boudart, A.W. Aldag, M.A. Vannice, *J. Catal.* **18**, 46 (1970).
93. Y.F. Chu, E. Ruckenstein, *Surf. Sci.* **67**, 517 (1977).
94. P.A. Simonov, A.V. Romanenko, I.P. Prosvirin, E.M. Moroz, A.I. Boronin, A.L. Chuvilin, V.A. Likholobov, *Carbon* **35**, 73 (1997).
95. T. Torre, A.S. Arico, V. Alderucci, V. Antonucci, N. Giordano, *Appl. Catal.* **A 114**, 257 (1994).
96. D.H. Bangham, I.R. Razouk, *J. Chem. Soc.* **166**, 572 (1938).
97. A. Soffer, M. Folman, *J. Electroanal. Chem.* **38**, 25 (1972).
98. H.P. Boehm, In *Adv. Catal. and Relat. Subj.*, Vol. 16 (D.D. Eley, H. Pines, and P.B. Weisz, eds.), Academic Press, New York, 1966, p. 179.
99. G.K. Boreskov, *Geterogennyi Cataliz* (Heterogeneous Catalysis) (K.I. Zamaraev, ed.), Nauka, Moscow, 1988.
100. A.I. Bird, In *Catalyst Supports and Supported Catalysts* (A.B. Stiles, ed.), Butterworths, Boston, 1987, p. 107.
101. I.P. Mukhlenov, E.I. Dobkina, V.I. Deryuzhkina, V.E. Soroko, *Tekhnologiya Katalizatorov* (Technology of Catalysts), Khimia, Leningrad, 1989.
102. L.R. Radovic, F. Rodríguez-Reinoso, In *Chemistry and Physics of Carbon*, Vol. 25 (P.A. Thrower, ed.), Marcel Dekker, New York, 1997, p. 243.
103. J.-B. Donnet, *Carbon* **20**, 267 (1982).
104. F. Coloma, A. Sepúlveda-Escribano, F. Rodríguez-Reinoso, *J. Catal.* **154**, 299 (1995).
105. J.M. Solar, C.A. Leon y Leon, K. Osseo-Asare, L.R. Radovio, *Carbon* **28**, 369 (1990).

106. P.L. Antonucci, V. Alderucci, N. Giordano, D.L. Cocke, H. Kim, *J. Appl. Electrochem.* **24**, 58 (1994).

107. D.S. Cameron, S.J. Cooper, I.L. Dodgson, B. Harrison, J.W. Jenkins, *Catal. Today* **7**, 113 (1990).

108. A.N. Frumkin, *Potentsialy Nulevogo Zaryada* (Potentials of Zero Charge), 2nd ed., Nauka, Moscow, 1982.

109. E.S. Matskevich, D.N. Strazhesko, V.E. Goba, *Adsorptsiya i Adsorbenty* (Russ. ed.) **2**, 36 (1974).

110. V.S. Vilinskaya, A.A. Korobanov, R.K. Burshtein, A.V. Gerasimova, *Izv. Akad. Nauk SSSR, Ser. Khim.* (Russ. ed.) **5**, 1000 (1978).

111. P.A. Simonov, A.V. Romanenko, I.P. Prosvirin, G.N. Kryukova, A.L. Chuvilin, S.V. Bogdanov, E.M. Moroz, V.A. Likholobov, *Stud. Surf. Sci. Catal.*, Vol. 118 (G. Pocelet et al., eds.), Elsevier, Amsterdam, 1998, p. 15.

112. B.R. Puri, S. Singh, O.P. Mahajan, *Indian J. Chem.* **3**, 54 (1965).

113. P. Ehrburger, *Adv. in Colloid and Interface Sci.* **21**, 275 (1984).

114. K. Kinoshita, P. Stonehart, In *Modern Aspects of Electrochemistry*, no. 12 (J. O'M. Bockris and B.E. Conway, eds.), Plenum Press, New York, 1977, p. 183.

115. E. Von Czaran, J. Finster, K.H. Schnable, *Z. Anorg. Allg. Chem.* **443**, 175 (1978).

116. A. Sepúlveda-Escribano, F. Coloma, F. Rodríguez-Reinoso, *Appl. Catal.* **A 173**, 247 (1998).

117. P.A. Attwood, B.D. McNicol, R.T. Short, J.A. van Amstel, *J. Chem. Soc. Faraday Trans.* **176**, 2310 (1980).

118. A. Benedetti, G. Cocco, S. Enzo, F. Pinna, *React. Kinet. Catal. Lett.* **13**, 291 (1980).

119. P. Gallezot, R. de Mésanstourne, Y. Christidis, G. Mattioda, A. Schouteeten, *J. Catal.* **133**, 479 (1992).

120. A. Benedetti, G. Cocco, S. Enzo, G. Piccaluga, L. Schiffini, *J. Chim. Phys.* **78**, 961 (1981).

121. F. Coloma, A. Sepúlveda-Escribano, J.L. Fierro, F. Rodríguez-Reinoso, *Appl. Catal.* **A 150**, 165 (1997).

122. F.S. Kemp, M.A. Georg, U.S. Patent 3,857,737, Dec. 31, 1974.

123. J.A. Bett, K. Kinoshita, P. Stonehart, *J. Catal.* **35**, 307 (1974).

124. F. Rodríguez-Reinoso, I. Rodríguez-Ramos, C. Moreno-Castilla, A. Guerrero-Ruiz, J.D. López-González, *J. Catal.* **99**, 171 (1986).

125. C. Moreno-Castilla, A. Porcel-Jiménez, F. Carrasco-Marín, E. Utrera-Hidalgo, *J. Molec. Catal.* **66**, 329 (1991).

126. M.C. Román-Martínez, D. Gazorla-Amorós, A. Linares-Solano, C. Salinas-Martínez de Lecea, H. Yamashita, M. Anpo, *Carbon* **33**, 3 (1995).

127. M.B. Palmer, Jr., M.A. Vannice, *J. Chem. Tech. Biotech.* **30**, 205 (1980).

128. G.R. Heal, L.L. Mkayula, *Carbon* **26**, 815 (1988); 803 (1988).

129. K.L. Yeung, E.E. Wolf, *J. Vac. Sci. Technol.* **B 9**, 798 (1991).

130. F. Coloma, A. Sepúlveda-Escribano, F. Rodríguez-Reinoso, *Appl. Catal.* **A 123**, L1 (1995).

131. R.A. Dalla Betta, M. Boudart, *Proc. 5th Intl. Cong. on Catal.*, Vol. 1 (J. Hightower, ed.), North-Holland, Amsterdam, 1973, p. 1329.

132. J.B. Hiskey, X.H. Jiang, G. Ramadorai, *Proc. Gold'90 Symp.* (D.M. Hausen et al., eds.), Society for Mining, Metallurgy, and Exploration, Inc., Littleton, CO, 1990, p. 369.

133. S.A. Simanova, A.A. Lysenko, N.M. Burmistrova, A.V. Schyukarev, O.V. Astashkina, S.I. Timoshenko, *Zh. Prikl. Khim.* (Russ. ed.) **71**, 50 (1998).

134. A.J. Groszek, S. Partyka, D. Cot, *Carbon* **29**, 821 (1991).

135. Y.A. Ryndin, O.S. Alekseev, P.A. Simonov, V.A. Likholobov, *J. Molec. Catal.* **55**, 109 (1989).

136. Y.A. Tarasenko, A.A. Bagreev, V.V. Dudarenko, V.K. Mardanenko, Y.A. Solodovni-kov, *Ukr. Khim. Zh.* (Russ. ed.) **55**, 233 (1989).
137. Y.A. Tarasenko, G.V. Reznik, A.A. Bagreev, *Ukr. Khim. Zh.* (Russ. ed.) **55**, 249 (1989).
138. Y.A. Tarasenko, A.A. Bagreev, S.A. Gorban', G.V. Reznik, Y.A. Solodovnikov, *Ukr. Khim. Zh.* (Russ. ed.) **55**, 1269 (1989).
139. V.V. Strelko, Y.A. Tarasenko, A.A. Bagreev, E.D. Lavrinenko-Ometsinskaya, *Ukr. Khim. Zh.* (Russ. ed.) **57**, 920 (1991).
140. V.S. Kublanovskii, Y.A. Tarasenko, M.O. Danilov, S.P. Antonov, *Ukr. Khim. Zh.* (Russ. ed.) **51**, 948 (1985).
141. C.D. Keith, D.L. Bair, *U.S. Patent 3,138,560*, June 23, 1964.
142. I.A. Tarkovskaya, L.P. Tikhonova, A.N. Tomashevskaya, V.E. Goba, L.S. Antonova, I.P. Svarkovskaya, *Zh. Phiz. Khim.* (Russ. ed.) **70**, 1463 (1996).
143. Y.A. Tarasenko, A.A. Bagreev, V.V. Yatsenko, *Zh. Phiz. Khim.* (Russ. ed.) **67**, 2328 (1993).
144. S.A. Simanova, N.M. Burmistrova, A.A. Lysenko, A.V. Schyukarev, O.V. Astashkina, N.V. Khramova, *Zh. Prikl. Khim.* (Russ. ed.) **71**, 375 (1998).
145. Y.A. Tarasenko, V.K. Mardanenko, V.V. Dudarenko, A.A. Bagreev, V.A. Trikhleb, *Zh. Prikl. Khim.* (Russ. ed.) **62**, 305 (1989).
146. V.V. Strelko, Y.A. Tarasenko, A.A. Bagreev, V.V. Dudarenko, V.K. Mardanenko, *Zh. Prikl. Khim.* (Russ. ed.) **65**, 1742 (1992).
147. H.E. van Dam, H. van Bekkum, *J. Catal.* **131**, 335 (1991).
148. R. Fu, H. Zeng, Y. Lu, S.Y. Lai, W.H. Chan, C.F. Ng, *Carbon* **33**, 657 (1995).
149. A.J. Groszek, *Carbon* **35**, 1329 (1997).
150. R. Fu, Y. Lu, W. Xi, H. Zeng, *Carbon* **36**, 19 (1998).
151. S.A. Simanova, N.M. Burmistrova, A.A. Lysenko, A.V. Schyukarev, O.V. Knyaz'kov, T.V. Kuznetsova, *Zh. Prikl. Khim.* (Russ. ed.) **72**, 1630 (1999).
152. V. Machek, J. Hanika, K. Sporka, V. Ruzička, *Collect. Czech. Chem. Commun.* **46**, 1588 (1981).
153. V. Machek, M. Sourková, V. Ruzička, *Collect. Czech. Chem. Commun.* **46**, 2178 (1981).
154. V. Machek, J. Hanika, K. Sporka, V. Ruzička, J. Kunz, *Collect. Czech. Chem. Commun.* **46**, 3270 (1981).
155. J. Hanika, V. Machek, V. Němec, V. Ruzička, J. Kunz, *J. Catal.* **77**, 248 (1982).
156. V. Machek, J. Hanika, K. Sporka, V. Ruzička, J. Kunz, L. Janacek, In *Preparation of Catalysts III* (G. Poncelet et al., eds.), Elsevier, Amsterdam, 1983, p. 69.
157. Y.A. Tarasenko, V.K. Mardanenko, V.V. Dudarenko, A.A. Bagreev, *Zh. Prikl. Khim.* (Russ. ed.) **62**, 513 (1989).
158. Y.A. Tarasenko, S.P. Antonov, A.A. Bagreev, V.K. Mardanenko, G.V. Reznik, *Ukr. Khim. Zh.* (Russ. ed.) **57**, 385 (1991).
159. L. Daza, T. Gonzales, A. Yuso, S. Mendioroz, J.A. Pajares, *Appl. Catal.* **13**, 295 (1985).
160. L. Daza, S. Mendioroz, J.A. Pajares, *Carbon* **24**, 33 (1986).
161. C.A. Leon y Leon, J.M. Solar, V. Calemma, L.R. Radovic, *Carbon* **30**, 797 (1992).
162. E. Papirer, A. Polania-Leon, J.-B. Donnet, and P. Montangon, *Carbon* **33**, 1331 (1995).
163. L.M. Knijff, Ph.D. Thesis, University of Utrecht, The Netherlands, 1993.
164. F. Rodríguez-Reinoso, *Carbon* **36**, 159 (1998).
165. P. Ehrburger, O.P. Mahajan, P.L. Walker, Jr., *J. Catal.* **43**, 61 (1976).
166. C. Prado-Burguete, A. Linares-Solano, F. Rodríguez-Reinoso, C. Salinas-Martinez de Lecea, *J. Catal.* **128**, 397 (1991).
167. A.S. Lisitsyn, P.A. Simonov, A.A. Ketterling, V.A. Likholobov. In *Stud. Surf. Sci. Catal.*, Vol. 63 (G. Pocelet et al., eds.), Elsevier, Amsterdam, 1991, p. 449.
168. G.M. Plavnik, B. Parlits, M.M. Dubinin, *Dokl. AN SSSR* (Russ. ed.) **206**, 399 (1972).

169. M. Gurrath, T. Kuretzky, H.P. Boehm, L.B. Okhlopkova, A.S. Lisitsyn, V.A. Likholobov, *Carbon'95: 22nd Bienn. Conf. Carbon, Extended Abstracts and Program*, San Diego, 1995, p. 556.

170. V.B. Fenelonov, *Kinet. Catal.* **16**, 628 (1975).

171. A.V. Neimark, V.B. Fenelonov, L.I. Kheifets, *React. Kinet. Catal. Lett.* **5**, 67 (1976).

172. L.I. Kheifets, A.V. Neimark, V.B. Fenelonov, *Kinet. Catal.* **20**, 626 (1979).

173. J.W. Geus, J.A.R. van Veen, *Stud. Surf. Sci. Catal.*, Vol. 123 (R.A. van Santen et al., eds.), Elsevier, Amsterdam, 2000, p. 459.

174. S. Galvagno, C. Milone, G. Neri, A. Donato, R. Pietropaolo, *Stud. Surf. Sci. Catal.*, Vol. 78 (M. Guisnet et al., eds.), Elsevier, Amsterdam, 1993, p. 163.

175. S. Galvagno, G. Capannelli, G. Neri, A. Donato, R. Pietropaolo, *J. Molec. Catal.* **64**, 237 (1991).

176. A. Guerrero-Ruiz, J.D. López-González, I. Rodriguez-Ramos, *J. Chem. Soc. Chem. Commun.* 1681 (1984).

177. C.H. Barthomew, M. Boudart, *J. Catal.* **25**, 173 (1972).

178. I. Rodríguez-Ramos, A. Guerrero-Ruiz, *J. Catal.* **135**, 458 (1992).

179. J.F. Hamilton, D.R. Preuss, G.R. Apai, *Surf. Sci.* **106**, 146 (1981).

180. B. Tardy, C. Nuopa, C. Leclerco, J.C. Bertolini, A. Hoareau, M. Treilleux, J.P. Faure, G. Nikoul, *J. Catal.* **129**, 1 (1991).

181. E.M. Moroz, P.A. Simonov, S.V. Bogdanov, A.L. Chuvilin, *Materials Sci. Forum* **321–324**, 1074 (2000).

182. E.R. Savinova, N.P. Lebedeva, P.A. Simonov, G.N. Kryukova, *Russian J. Electrochem.* **36**, 952 (2000).

183. J.C. Ménézo, L.C. Hoang, C. Montassier, J. Barbier, *Appl. Catal.* **74**, L5 (1991).

184. J.S. Brinen, J.L. Schmitt, W.R. Doughman, P.J. Achorn, L.A. Siegel, W.N. Delgass, *J. Catal.* **40**, 295 (1975).

185. F.S. Kemp, M.A. George, *U.S. Patent 3,857,737*, Dec. 31, 1974.

186. K. Hattori, B. Miya, M. Matsuda, M. Ishii, H. Saito, H. Watanabe, H. Takizawa, U.S Patent 4,108,891, Aug. 22, 1978.

187. R.A. Kent, R.D. Evans, U.S. Patent 3,804,779, Ap. 16, 1974.

188. R.J. Card, J.L. Schmitt, J.M. Simpson, *J. Catal.* **79**, 13 (1983).

189. S.Y. Troitskii, A.L. Chuvilin, S.V. Bogdanov, E.M. Moroz, V.A. Likholobov, *Izv. Akad. Nauk, Ser. Khim.* (Russ. ed.) **6**, 1366 (1996).

190. D.J. Suh, T.-J. Park, S.K. Ihm, *Ind. Eng. Chem. Res.* **31**, 1849 (1992).

191. I.R. Wyatt, *Chem. Weekbland*, No. 62, 310 (1966).

192. N.K. Pshenitsyn, C.I. Ginzburg, *Izv. Sektora Platiny IONKh AN SSSR* (Russ. ed.), No. 28, 213 (1954).

193. S.Y. Troitskii, M.A. Fedotov, V.A. Likholobov, *Russ. Chem. Bull.* **42**, 634 (1993) (Engl. transl.).

194. S.Y. Troitskii, A.L. Chuvilin, D.I. Kochubei, B.N. Novgorodov, V.N. Kolomiichuk, V.A. Likholobov, *Russ. Chem. Bull.* **44**, 1822 (1995) (Engl. transl.).

195. M. Malentacchi, L. Cavalli, C. Rubini, EP 0879641, Nov. 25, 1998.

196. K.R. Chang, H.W. Chem, C.C. Wan, *J. Chem. Tech. Biotech.* A **34**, 237 (1984).

197. S.Y. Troitskii, V.A. Likholobov, P.A. Simonov, A.S. Lisitsyn, et al., Pat. USSR 1593009, May 15, 1990.

198. D. Pope, D.S. Walker, R.L. Moss, *J. Catal.* **28**, 46 (1973).

199. H. Takizawa, *Kagaku to Kogyo* **52**, 128 (1978)

200. P.A. Simonov, V.A. Semikolenov, V.A. Likholobov, S.Y. Troitskii, et al., Pat. USSR 1270939, July 15, 1986.

201. V.A. Semikolenov, S.P. Lavrenko, V.I. Zaikovskii, *Kinet. Catal.* (Russ. ed.) **35**, 622 (1994).

202. D. Pope, W.L. Smith, M.J. Eastlake, R.L. Moss, *J. Catal.* **22**, 72 (1971).

203. V.A. Semikolenov, S.P. Lavrenko, V.I. Zaikovskii, A.I. Boronin, *React. Kinet. Catal. Lett.* **51**, 517 (1993).

204. V.A. Semikolenov, *Zh. Prikl. Khim.* (Russ. ed.) **70**, 785 (1997).

13

NMR Investigations of Heterogeneous and Electrochemical Catalysts

Y.Y. TONG

Georgetown University, Washington, D.C., U.S.A.

J.J. VAN DER KLINK

Ecole Polytechnique Federale de Lausanne, Lausanne, Switzerland

CHAPTER CONTENTS

SUMMARY

Solid-state nuclear magnetic resonance (NMR), a canonical technique of chemistry and physics, possesses many versatile features such as, for example, elemental specificity and local structural, electronic, and motional sensitivity. In particular, NMR can characterize samples in most types of condensed matter, be it liquid or solid, single crystal or amorphous. Given adequate sensitivity it has, therefore, the unique ability of providing metal surface and adsorbate electronic and structural information on a molecular level and allows one to access motional information of adsorbate over a time range unattainable by any other single spectroscopic technique. In addition, solid-state NMR is nondestructive, technically versatile, and very applicable to oxide- as well as conducting carbon-supported high-surface-area transition metal catalysts of practical importance. It is also easily amendable under controlled environment close to real-world operating conditions. As such, metal NMR of supported metal catalysts has historically played an important role in illustrating many fundamental aspects relevant to gas-phase, heterogeneous catalysis at transition metal surfaces. In particular, the elemental specificity of NMR enables the catalytically relevant transition metal surfaces to be characterized either directly by NMR of transition metals themselves (if the surface signal is discernible) or, in more general cases, indirectly by NMR of adsorbates used as a molecular probe providing a unique flexibility in designing specific experiments. The most extensively investigated pair thus far has been the ^{195}Pt and ^{13}C NMR of platinum metal and adsorbed carbon monoxide. Fundamentally, the ability to access the (surface) Fermi level local density of states via metal NMR offers a promising approach to bridge the conceptual gap between widely used localized chemical descriptors (e.g., the active site or the surface bond) and the delocalized descriptors of condensed matter physics (e.g., the band structure of the metal surfaces). In this chapter, NMR investigations of heterogeneous and electrochemical catalysts will be discussed in the light of our current understanding of the matter.

13.1 INTRODUCTION

The scope of NMR applications to catalysis is probably as wide and variant as the field itself, as is illustrated, e.g., by the collection of articles under the title of *NMR Techniques in Catalysis* [1] or by the reflections on applications of NMR in surface chemistry and catalysis by Ken Packer [2], which are quite instructive to answer questions such as "why NMR" and "what kind of useful information is obtainable by NMR" in catalysis. A recent exciting development is the NMR of electrocatalysts in an electrochemically controlled environment [3,4].

Modern NMR spectrometers give access to (nearly) the whole Periodic Table, offering unmatched chemical specificity. The low-mass detection sensitivity of NMR is now less problematic, thanks to higher magnetic fields and improved electronics. In the catalytic context, NMR can work close to real-world conditions such as high pressure and high temperature, or active electrode potential control in an electrochemical environment. NMR can study both the catalytic metal itself and its adsorbates; the typical pair is platinum and carbon monoxide.

Slichter, Sinfelt, and co-workers have opened this line of research some two decades ago by discovering several unique features of the ^{195}Pt NMR of oxide-supported small platinum particles [5]. Specifically, they found that the overall ^{195}Pt NMR lineshape for supported catalysts is extremely broad, of the order of 40,000 ppm, extending downfield from the position of bulk platinum. They were able to correlate the intensity of the low-field feature in the ^{195}Pt spectrum with the dispersion (the fraction of atoms in the surface) of their particles. Later, these observations were confirmed by van der Klink and co-workers [6], who also determined that the signal from clean-surface Pt atoms was centered at about 34,000 ppm to low field from the bulk platinum signal. Most recently, Tong and co-workers have observed that nanoscale Pt particles on conducting carbon also show these very typical spectral characteristics, even in an electrochemical environment [7].

Interestingly, clean metal surface atoms have a Knight shift close to zero. This experimental observation has been confirmed in calculations by Weinert and Freeman [8]. They show that the surface shift results from a gradual drop in the d-like Fermi level local density of states (E_f-LDOS) on moving from the inside of the particle to the surface. This dominant E_f-LDOS effect discriminates the surface from bulk Pt atoms, making ^{195}Pt NMR unique in investigating the surface physics and chemistry of nanoscale platinum particles. While Pt is a favored catalytic material, it may also be the only transition metal showing such NMR surface specificity.

Chemisorption of CO has been widely used as a paradigm of metal surface reactivity, and ^{13}CO NMR has been extensively studied [9]. The 5σ forward- and $2\pi^*$ backward-donation scheme proposed by Blyholder [10] provides an intuitive, molecular orbital-based framework for the discussion of metal–adsorbate bonding. Some quite remarkable results of ^{13}CO NMR are the direct evidence for metallic characteristics of chemisorbed CO [11], and the correlation with the vibrational properties of chemisorbed CO and with the clean metal surface E_f-LDOS as determined from ^{195}Pt NMR [7,12]. These NMR observations are relevant to frontier-orbital theories [13] for chemisorption on metals [14,15]. The original theory for molecular reactivity [13] highlights the decisive role played by the highest occupied and the lowest unoccupied molecular orbitals (HOMO and LUMO, the frontier orbitals) in determining the reactive sites. On metal surfaces the closest equivalents to molecular HOMO and LUMO are the orbitals just below and just above the Fermi energy E_f. The density of these electrons on a given site is measured by the local density of states at E_f, or E_f-LDOS for short. This quantity can, in principle, be measured by metal NMR.

Several earlier review articles are relevant to our subject. Slichter reviews the work done in his laboratory [16], most of it concerned with atoms or molecules adsorbed on the metal clusters, and the experimental techniques used in such studies [17]. Duncan's review [9] pays special attention to the ^{13}C NMR of adsorbed CO. Most recently, one of us has given a rather detailed review of the field, in particular on metal NMR of supported metal catalysts [18]. While the topics and examples discussed in this chapter will inevitably have some overlap with these previous reviews, particular emphasis is directed toward highlighting the ability of metal NMR to access the E_f-LDOS at both metal surfaces and molecular adsorbates. The E_f-LDOS is an attractive concept, in that it contains information on both a spatial (local) and energy (electronic excitations) scale. It can bridge the conceptual gap between localized chemical descriptors (e.g., the active site or the surface bond) and the delocalized descriptors of condensed matter physics (e.g., the band structure of the metal surfaces).

This chapter is organized as follows: Section 13.2 gives a discussion of a multitude of concepts in this field between physics and chemistry: band structure, chemical bond, E_f-LDOS, ad-atoms, chemisorption, and metal surface reactivity. The next section presents a brief description of metal NMR theory, with which the values of E_f-LDOS can be deduced. In Section 13.4 selected examples from ^{195}Pt NMR are presented, and in Section 13.5 those for ^{13}CO. Section 13.6 offers brief conclusions.

13.2 CONCEPTS

The classic text on the N-electron system is *Theory of the Inhomogeneous Electron Gas* [19], and for our purposes specifically the chapter on atoms, molecules, and solids [20] and that on surfaces and adsorbates [21]. For general solid-state physics we refer to Ashcroft and Mermin [22], and for physical chemistry to Atkins [23].

By definition, the Hamiltonian of a system of identical particles is invariant under the interchange of all the coordinates of any two particles. The wavefunction describing the system must be either symmetric or antisymmetric under this interchange. If the particles have integer spin, the wavefunction is symmetric and the particles are called bosons; if they have half-integer spins, the wavefunction is antisymmetric and the particles are fermions. Our discussion will be restricted to electrons, which are fermions.

A single electron has three variable properties: position, momentum, and spin. A system of many electrons has as *only* observable ground-state properties the resulting charge density, momentum density, and magnetization density. Roughly, these three can be probed by elastic x-ray scattering, inelastic Compton scattering, and magnetic neutron scattering.

The density-functional theory formulated by Hohenberg and Kohn (1964) can be seen as a formal completion of the statistical theory of Thomas, Fermi, and Dirac for the inhomogeneous electron gas [19,24]. That theory attempts to find the electron number density $n(\mathbf{r})$ for a gas of N electrons moving in a common potential $V(\mathbf{r})$, without having recourse to the computation of wavefunctions. Hohenberg and Kohn have shown that the ground-state properties of the N-electron system are functionals only of the number density. This holds, e.g., for the energy, so that the electron distribution in the ground state of a molecule or a solid can be obtained from a variational type of calculation, where the number density is the varied quantity. The subsequent work by Kohn and Sham has shown that the exact ground-state charge density can be found from a set of single-particle Schrödinger equations. Formally, these "single particles" are nothing more than intermediaries in a calculation of the physically observable quantities, but many authors have pointed out that the Kohn–Sham orbitals are perfectly legitimate tools in the construction of orbital theories of chemistry [24–27].

Many experimentally important excited-state properties measure changes in energy, rather than in electron distribution. In certain special cases there is a one-to-one relation between these excitation energy spectra and the orbital eigenenergies of the (ground-state) Kohn–Sham theory. The theory for metals shows resemblances [20, p. 280] with Landau's theory of Fermi quasi-particles [22, p. 348]. In quasi-particle theory it is argued that, while neither the ground state nor the excited state can be described in terms of the independent uncorrelated electrons of the simplest Hartree theory, the *difference* between the two states can for many (but certainly not all) systems be so described [22, p. 345]. The most elementary requirement of the Pauli principle, that the occupation of any independent-electron state must be either 0 or 1 and that therefore its thermal average must be between these bounds, is expressed by applying Fermi–Dirac statistics to the orbital eigenenergies ε_i of the one-electron states.

13.2.1 Bands and Bonds

It is often said that the band description of one-electron states is in terms of itinerant electrons and is mainly useful for solids, while the bond description looks at localized electrons and is appropriate for molecules. Since our subject concerns interactions between molecules and solid surfaces, we need to establish our vocabulary clearly. We will consider an electron as localized if it cannot participate in (electrical) transport phenomena; otherwise it is itinerant. This is not the same as describing the one-electron orbitals by localized functions (such as the Wannier functions, introduced below) respectively by extended functions (such as the Bloch functions, see below). Nor is it simply a distinction between tight-binding orbitals constructed from (so-called localized) d-orbitals as opposed to those derived from (so-called extended) s-orbitals.

It is a basic consequence of the translational symmetry of a solid that its Kohn–Sham eigenfunctions can be uniquely labeled by four "quantum numbers," the band index n and a wavevector $\mathbf{k}$, as in $\psi_{n,\mathbf{k}}$. The diagram $\varepsilon(n, \mathbf{k})$ that represents the n, $\mathbf{k}$ dependence of the corresponding eigenenergies is called the band structure. The Bloch theorem asserts that the $\psi_{n,\mathbf{k}}(\mathbf{r})$ can be written in the form of a Fourier series

$$\psi_{n,\mathbf{k}}(\mathbf{r}) = N^{-1/2} \sum_{\alpha=1}^{N} \phi_n(\mathbf{r} - \mathbf{R}_\alpha) \exp(i\mathbf{k} \cdot \mathbf{R}_\alpha) \tag{1}$$

where the $\mathbf{R}_\alpha, \alpha = 1 \cdots N$, are the Bravais lattice vectors, which coincide with the nuclear positions in the case of a monatomic Bravais lattice. The $\mathbf{r} - \mathbf{R}_\alpha$ is a position vector in the unit cell at the origin. The $\psi_{n,\mathbf{k}}$ is called a Bloch function, and the ϕ_n a Wannier function.

The inverse series of Eq. (1) gives a Wannier function in terms of Bloch functions. This is perfectly general; but in the tight-binding (or LCAO, or extended Hückel) method we identify the ϕ_n with an (approximate) atomic wavefunction ($n \to s, p, d$) centered at the origin. It can be shown that the Wannier functions are normalized and mutually orthogonal. However, they are not quasi-particle states in the sense of Fermi–Dirac statistics, since in general they are not eigenfunctions of the Kohn–Sham problem. But if the index n refers to a completely filled or completely empty band (as in molecules or in insulating solids) the distinction becomes unimportant and the electron density $n(\mathbf{r})$ (do not confuse with the band index n) can be described at will in terms of localized or extended functions. This is no longer true when band n is partially occupied, because in that case the inverse Fourier transform of Eq. (1) has no physical meaning (the Wannier function is a superposition of Kohn–Sham eigenfunctions with different energies).

In the tight-binding method, the wavefunctions are constructed of localized atomic levels, but an electron in such a level will be found with equal probability in any cell of a monatomic crystal [22, p. 185]. Nevertheless, these freely moving electrons do not necessarily contribute to electric conduction, since in the semiclassical picture of a filled band half the electrons move "to the right" and half "to the left"; and this remains true even if an electric field is applied. For partially filled bands, the exact half/half-symmetry is broken by an electric field, and a current will flow. Note that half-filled bands can give an electric current but cannot be described with Fermi–Dirac occupation of localized Wannier functions; the opposite holds for filled bands.

In chemical usage [23, Section 14.11] an electron is said to be *delocalized* if its molecular orbital cannot be ascribed to a two-center bond; otherwise it is localized. It is, however, always possible, but perhaps rarely convenient, to describe the electron distribution in a molecule with delocalized orbitals only. The situation in a covalent insulator such as diamond is similar to the molecular case. There are four valence electrons per atom, and four neighbors. Therefore, it is possible to describe the structure with four two-center, two-electron bonds, and localized Wannier orbitals. But keep in mind that the only physical reality is the resulting charge distribution. This reality can also be described by freely moving Bloch electrons.

It will be useful to reserve the word "bond" to characterize the spatial arrangement of atomic nuclei and the electron distributions that keep them in place. A bond, therefore, is a spatial concept related to the topology of the electron distribution corresponding to a certain arrangement of nuclei. Bonding, on the other hand, will mean the comparison of the energy requirements for different arrangements. Thus, a heterogeneous catalytic process can be portrayed as a game of bonding in order to find the energetically most favorable bond.

In chemistry, solids are customarily classified according to the perceived nature of the cohesion or bonding, rather than by using features of the electronic charge distribution (but see [22, Ch. 19]), which is often a better descriptor in defining active sites in heterogeneous catalysis. Apart from the van der Waals interaction and hydrogen bonding, one usually distinguishes electrostatic-ionic, covalent, and metallic bonding. The standard physical classification of solids as insulators, semiconductors, or metals is based on their electrical conductivity, which is, however, a property that is not completely determined by the ground state [22, p. 396]. In this sense, a ground-state "metallic bond" may be impossible to define. Pauling [28, Section XI-55] considered the metallic and covalent bonds as similar in nature, and he suggested that all substances might possibly be classified on a linear scale, based on bond type, going from the extreme ionic first to the covalent and then to the metallic bond. The more general idea is the van Arkel–Ketelaar triangle [29], reproduced in Figure 1, that classifies compounds in terms of all three types of bonding. Recent discussions of this triangle again assert that there is no ground to consider bonding in metals as specifically different from covalent bonding [30,31].

Apart from the pictorial intuition traditionally expressed by stick-and-ball models in chemistry, how can we recognize a bond? The electronic charge distribution can (in principle) be measured by X-ray scattering [32]. Usually, the electron density in a bonded system is very similar to the superposition of free atomic densities. Their difference [33], called the *deformation density*, is widely used [32, p. 95] as a criterion to locate the bonds; where this density accumulates, there are the bonds. A more recent method, known as the *atom in a molecule (AIM) approach* [34], is based on an analysis of the Laplacian (the divergence of the gradient field) of the total electron density $n(\mathbf{r})$. A Laplacian is a second derivative, and therefore the method requires very good values for $n(\mathbf{r})$. Another technique, extending an idea of Lennard-Jones [35], analyzes an electron localization function, ELF, related to the two-particle density distribution $n_2(\mathbf{r}, \mathbf{r}')$ [36], which is the probability to find an electron in the volume $d\mathbf{r}$ around $\mathbf{r}$ and at the same time another in the volume $d\mathbf{r}'$ around $\mathbf{r}'$. A general discussion of ELF analysis of the bonds in bulk metals is given

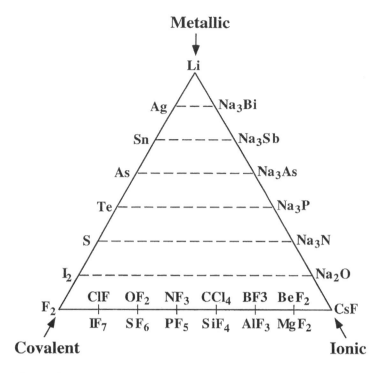

Figure 1 The van Arkel–Ketelaar triangle that classifies compounds in terms of three types of bonding: ionic, covalent, and metallic. (From Ref. 29.)

in [37]. To obtain the ELF, it is in practice necessary to have a one-electron orbital representation for the total charge density (e.g., from a Kohn–Sham calculation). Clearly then, the basic physical quantity is the local electron number density $n(\mathbf{r})$, while its Kohn–Sham representation allows us to construct both a theory of bands and a theory of bonds, unified by the theory of the inhomogeneous electron gas [19].

13.2.2 Three Views of Ad-Atom Bonds

An understanding of the bonding between an overlayer and a substrate is crucial in many fields besides catalysis. We are mainly interested in chemisorption and electrochemical adsorption of a monolayer of simple molecules on a metal substrate.

Since the pioneering work of Lang and Williams [38] the archetypical adsorbate atoms are Na (or sometimes Li), Cl, and Si. That paper considered single-atom adsorption on a jellium; later the same trio has been studied on a "real" Al substrate [39], and recently these ad-atom/jellium systems have been investigated using the electron localization function ELF [40].

Except in the ad-atom's vicinity, the metal screens out the effects of the ad-atom on the total charge density and on the Kohn–Sham potential, even though the disturbance in the individual Kohn–Sham single-particle wavefunctions is *not* short-ranged [38, p. 618]. That result is reproduced in Figure 2, together with the deformation density $\delta n(\mathbf{r}) = n^{ma}(\mathbf{r}) - n^m(\mathbf{r})$, where the index *ma* stands for the metal plus adsorbate system, and *m* for the bare metal. In a more formal way, this charge distribution can be characterized by its multipole moments: the point charge, the dipole moment, the quadrupole moment, etc. The variation of dipole moments with distance is shown in Figure 3. This plot suggests that in some way a negative charge is associated with adsorbed Cl, a positive charge with Na, and hardly any charge with Si. Going back to Figure 2, the cases of Cl and Si are easy to recognize, but the alkali (Li in this figure) is somewhat more complicated: it is not simply the mirror image of the halogen.

The physical description of the alkali charge transfer has been the subject of heated debate [39,41]. According to the Langmuir–Gurney model for low-coverage alkali adsorption, there should be partial transfer of charge from the atom to the substrate, followed by metallic screening of the ion so created. The other viewpoint [42,43] is that the bond contains important covalent contributions. It has been argued [38,39] that in the discussion additional information from the DOS curves should be used. With this additional information [which is not included in the $n(\mathbf{r})$ by itself] the charge transfer and screening picture is favored. These metallic-or-covalent

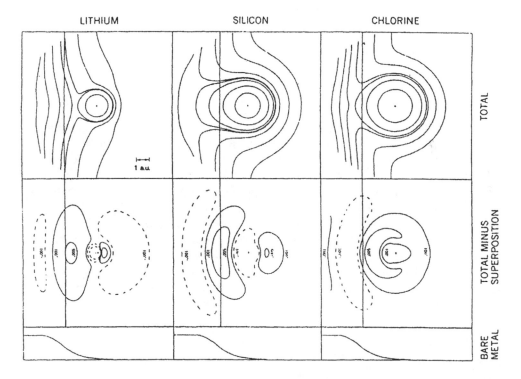

Figure 2 Electron-density contours for chemisorption. Upper row: contours of constant electron density in (any) plane normal to the metal surface containing the ad-atom nucleus (indicated by +). The metal is to the left of the solid vertical line. Center row: deformation charge density. The polarization of the core region, shown for Li, has been deleted for Si and Cl because of its complexity. Bottom row: The bare-metal electron-density profile, shown to establish the distance scale. (From Ref. 38.)

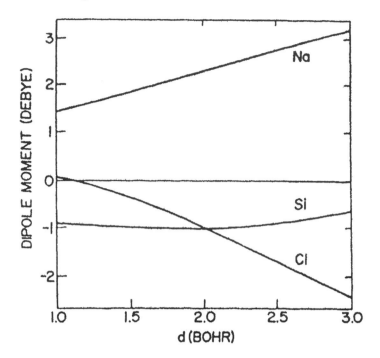

Figure 3 Dipole moment as a function of metal–ad-atom distance d for Na, Si, or Cl atom chemisorbed as in Figure 2. (From Ref. 38.)

disputes cannot be resolved by inspection of the charge density alone, and we refer back to Section 13.2.1, where Pauling's ideas on the subject are mentioned.

The third theoretical study of these adsorption systems avoids the use of deformation density maps, but discusses the ELF directly derived from $n_2^{ma}(\mathbf{r}, \mathbf{r}')$ [40]. For Si, a directional bond is found, with some sharing of charge. This fits with the idea from the earlier work of a mainly covalent bond. For Cl, the ELF graph points out that there are two charge regions in this system: the basin of electronic charge in the metal, and an almost closed region around the Cl, the latter suggesting a negatively charged ion with 0.38 excess electrons in its charge basin. The case of Li is again found to be somewhat more complicated. The electronic charge between the Li ad-atom and the surface is highly localized, whereas on the vacuum side of the ad-atom it is spread out, different from that in the isolated Li atom.

13.2.3 Chemisorption

Several energy scales are necessary when describing the formation of a surface bond between an incoming molecule and the metal surface. The typical picture of so-called strong chemisorption is the formation of a bonding–antibonding pair of surface orbitals, starting from an electron level on the atom and another in the surface. It is not immediately obvious, however, what the common zero of the energy scales should be when drawing such an interaction diagram. We will illustrate this on the example of a CO molecule and a platinum surface.

Let us begin by inspecting the ionization energies. In a Kohn–Sham type of calculation, these are given by the energy (w.r.t. e.g. the vacuum) of the highest occupied single-particle level; electron affinities are the ionization energy of the corresponding negatively charged entity. The Kohn–Sham energy of the HOMO 5σ level in CO is ≈ -13 eV, in good agreement with the experimental ionization energy, ≈ 14 eV. As expected, there is no relation between the Kohn–Sham $2\pi^*$ LUMO energy of -7.5 eV and the electron affinity: the experimental ionization energy of CO^- is -1.5 eV, indicating that the ion has a tendency to break up (experimentally, into $C + O^-$). For a metal surface, the ionization energy and the electron affinity are both equal to the workfunction, in the case of polycrystalline Pt approximately 5.65 eV. With these numerical values, no two-electron bonding interactions between the LUMO of a CO molecule and occupied metal orbitals are expected; but there may be a stable bonding interaction between the HOMO of CO^- and some empty metal orbitals, resulting in a CO with a fractional charge between 0 and $-e$. Such fractional charges must be interpreted in a statistical way, and therefore some chemical potential must be considered. The simplest image is that of two initially isolated reservoirs of electrons, the CO (or CO^-) and the sufface, that are connected by an imaginary "thin metal wire" so that they can exchange electrons.

When a contact is made between the neutral molecule and the surface, electrons will flow from the Pt Fermi level into the CO; and if it is between CO^- and the surface, the electrons will flow toward the Pt Fermi level. The energy of the Fermi level does not change, because the metal is a perfect source and sink of electrons. A simple linear interpolation says that we will end up with $CO^{-0.51}$. We have not yet considered a true chemical interaction: we have just supposed that we have two "reservoirs of electrons," that have been brought in diffusive contact and therefore exchange particles. For a given total (integer) particle count for the two reservoirs together, their common chemical potential will determine the average number of particles in each reservoir, a number that does not need to be an integer.

Usual orbital-mixing diagrams can now be drawn starting from this situation, where the adsorbate energy levels are said to be renormalized by the "weak chemisorption" interaction (the flow of electrons in our picture). For CO interacting with transition-metal surfaces, the renormalized $2\pi^*$ and 5σ levels are estimated to lie at $+2.5$ and -7 eV with respect to the common Fermi level [44]. The following step of "strong chemisorption" will cause additional changes of the excess charge on the CO, but we expect it to remain between zero and one electron. And since the (gaseous) CO^- is unstable with a tendency to yield negatively charged oxygen ions, we expect the C–O bond in the adsorbate to be weakened, and the CO dipole to have its negative end on the oxygen (while the free CO dipole has its negative end on the carbon).

After this simple energetic analysis, let us see if we can give some spatial reality to that imaginary "thin metal wire." The electron charge density at the metal surface does not fall abruptly to zero, but decreases exponentially with a characteristic length proportional to $W^{-1/2}$, with W the workfunction [45,46]. At some distance outside the metal surface, only the Fermi-level electrons contribute [47] to this so-called spill-out charge. Now, although the absolute value of $n(\mathbf{r})$ in the spill-out region is very small, all these electrons can be spatially rearranged at low energetic cost, which means that they are very reactive. The incoming molecule has a small number of discrete HOMO/LUMO energy levels, and electrons may hop back and forth between these levels and those in the far spill-out tail of the metal. As a result the molecular levels shift up in energy (the electrostatic part of the electrochemical potential increases) and broaden (due to lifetime effects of an electron temporarily residing on the CO). The rising of the energy levels can be seen as a barrier to be crossed before the formation of the "strong" chemisorption bond. In the barrier, the closed-shell molecule behaves as a noble gas atom [48]. Once the barrier has been crossed, the natural reference level of the energy scale is no longer the vacuum, but the common Fermi level for metal and molecule.

Upon closer approach of the metal surface, the renormalized levels start their bond-forming interaction, each yielding a bonding–antibonding pair of ad-atom–metal levels. Very schematically, the $2\pi^*$ now participates in a broadened ad-atom–metal bonding level that is at least partly below E_f, and the 5σ takes part in a broadened ad-atom–metal antibonding level that is at least partly above E_f. The result can be thought of as the molecular 5σ having lost a fraction of an electron, and the $2\pi^*$ having gained some (possibly different) fraction, which is Blyholder's model [10]. According to an early extended-Hückel calculation [49] the balance of this forward and backward donation is an increase of 0.36 electrons on the CO.

A typical more complete correlation diagram is shown in Figure 4 [27]. A recent review of theoretical treatments of molecule–metal surface interactions has been given in [50].

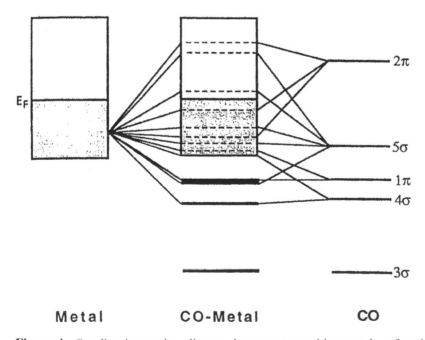

Figure 4 Bonding interaction diagram between a transition-metal surface band and CO molecular orbitals. (From Ref. 27.)

13.2.4 Local Density of States

From perspectives of metal NMR and its application in heterogeneous and electrochemical catalysis, it is important to distinguish the electron distribution in real space $n(\mathbf{r})$ and the distribution of electrons according to their energy. The special situation of bond formation on a metal, as compared to in a molecule, is of course that a very large number of metal orbitals is available in any energy interval, while only a spatially small part (near the surface) of the orbitals is directly involved. Therefore, the chemisorption bond should be analyzed not only in terms of single-orbital *energies*, but also considering the *spatial* characteristics of these orbitals.

The result of a Kohn–Sham calculation is a collection of one-electron orbitals $\psi_i(\mathbf{r})$ and one-electron energies ε_i. Most physically interesting properties of an ordered, crystalline solid are determined by the periodicity of the crystal lattice, and then it is useful to arrange the results such that the index i expresses the Bloch character of the wavefunctions: $i \rightarrow (n, \mathbf{k})$. An integral representation of the bandstructure diagram $\varepsilon(n, \mathbf{k})$ is the density of states curve $D(\varepsilon)$, that says how many $(n, \mathbf{k})$ combinations have their energies in an interval $d\varepsilon$ around ε. Note that, following a well-established convention, we call this a "density," but it is not something "per unit volume," but "per unit energy interval." One step further, the local density of states $D(\varepsilon, \mathbf{r})$ actually is something per unit volume *and* per unit energy interval: it gives the electron density in point $\mathbf{r}$, $n(\mathbf{r})$, broken up along the energy axis. In the rest of this paper we will frequently need to refer to the local density of states at the Fermi energy E_f on a given site, and write E_f-LDOS for short. While the band structure $\varepsilon(n, \mathbf{k})$ of the metal describes the one-electron energy distribution in reciprocal space, the local density of states $D(\varepsilon, \mathbf{r})$ gives that distribution in real space. The integral of $D(\varepsilon, \mathbf{r})$ over all $\mathbf{r}$ in the sample brings back the density of states $D(\varepsilon)$, while the integral over all ε up to the Fermi energy E_f again yields the number density $n(\mathbf{r})$.

A very clear distinction between the total DOS $D(\varepsilon)$ and the LDOS $D(\varepsilon, \mathbf{r})$ is shown by calculations on binary alloys [51]. Look, e.g., at the (partly hypothetical) series of isoelectronic 1:1 alloys TcTc, MoRu, NbRh, ZrPd, YAg, where the alloying partners have nominal valence differences between 0 (for pure Tc) and 8 (for YAg). As shown in Figure 5, the overall density of states curves look more or less alike for all five alloys, but the partial densities on the sites of the individual partners are very different. Such curves also show why the so-called collective electron model does not work for catalytic activity [52, p. 458], or even for alloy properties in general.

In some specific situations, it may turn out that all one-electron wavefunctions of the system at energy E_f have intensity zero in a certain site $\mathbf{R}_\alpha$. From the definition, the E_f-LDOS on that site then is zero, and NMR people say that the site is not metallic; but note that this implies nothing about the *total* charge density at this site. Furthermore, from this LDOS argument alone, it is impossible to subdivide the "nonmetallic" sites into ionic, covalent, etc. In our language, therefore, an adsorbate can be simultaneously ionic (as seen from the total charge distribution, including the electrons of all energies) and metallic (considering its E_f-LDOS, i.e., only the electrons in the highest occupied levels). Such situations are also found on the Ketelaar triangle of Figure 1.

The E_f-LDOS is not simply a mathematical aid in the theory of metal surfaces or of magnetic resonance. It has a physical sense wherever quasi-particle theory is

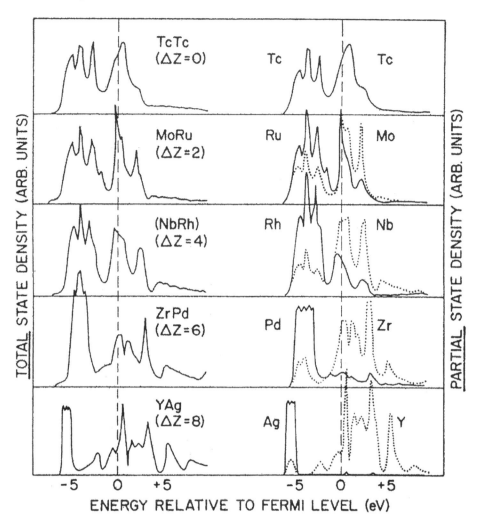

Figure 5 Total and partial DOS curves for the (partly hypothetical) series of isoelectronic alloys TcTc, MoRu, NbRh, ZrPd, YAg, where the alloying partners have nominal valence differences between 0 (for pure Tc) and 8 (for YAg). The overall DOS curves look more or less alike for all five alloys, but the partial densities on the sites of the individual partners are very different. (From Ref. 51.)

valid (and that is the case for nearly all metals used in catalysis), and it can, in a certain approximation, even be pictured. In the Tersof–Hamann description, the tunneling conductance in STM experiments is directly proportional to the E_f-LDOS of the metallic substrate at the center position of the STM tip. For a review of this model, and STM on metals in general, see [53].

13.2.5 Reactivity Theory

It is intuitively clear that the "reactivity" of a molecule, however defined, is unlikely to be a ground-state property in the Hohenberg–Kohn sense: it probably cannot be deduced from the ground-state charge density alone. So for all these things that Atkins [23] has under the heading "change" (including the change from atoms to molecules), knowledge of the ground state alone is not sufficient. Indeed, in the pioneering paper on the reactivity of aromatic hydrocarbons [13], it is pointed out that site-specific reactivities cannot be understood from a consideration of the total charge density of π symmetry, since this is the same on all sites. The new idea introduced in that paper was to consider only the highest occupied π-orbital. Fukui et al. [13] actually compare not only the relative intensity of the frontier orbital on different sites of the same molecule, but also the absolute intensities in (slightly) different molecules. They observe, e.g., that triphenylene is more stable to oxidation than naphtacene, and phenantrene more than anthracene; they conclude that, *at a constant number* of carbons and π-electrons, the more reactive molecule has the higher absolute value of frontier-orbital intensity.

We have proposed an analogy for metal surface reactivity: among metal particles *with similar surfaces* (made of the same metal, with comparable sizes, etc.) those that have the higher local density of states at the Fermi energy on their surface sites will be the more reactive [54,55]. This is a weaker statement than the sometimes-heard proposal that the E_f-LDOS is a useful yardstick to compare more widely different systems, e.g., a series of transition metals.

As described in Section 13.2.3, the E_f-LDOS (together with the workfunction) determines the intensity of the far tail of the metal's surface charge, which plays an important role in the "renormalization" of the HOMO/LUMO energies of the reactant molecule and is also involved in the barrier formation between some physisorbed and some weakly chemisorbed state. However, the subsequent formation of pairs of molecule-surface bonding–antibonding levels ("strong chemisorption") involves a range of metal surface electron energies of $10\,eV$ or so [27]. The variation of the electronic structure in such an energy range may be poorly parameterized by changes in the E_f-LDOS alone. Our main interest in this review is the *restricted* E_f-LDOS effect, when relatively small modifications are made to a given metal surface, since the NMR technique allows us to measure it for a favorable metal, platinum.

13.3 NUCLEAR MAGNETIC RESONANCE

In its simplest form, NMR can be thought of as a technique to measure the magnetic field that is present at the nucleus. In "nonmagnetic" substances in zero applied field, the magnetic field at the nucleus is zero as well. But in general, the field at the nucleus is not *exactly* equal to the applied field, and even if the applied field could be made perfectly homogeneous over the sample volume, the resulting field would remain inhomogeneous at the molecular or atomic level. It is useful to treat separately the orbital and the spin magnetism of the sample. All liquids and all solids show orbital magnetism to some extent, but many have zero spin magnetism. Spin magnetism is important in metals, but it can also occur in liquids, a simple example being liquid oxygen.

The origin of the chemical shift is orbital magnetism. It is best understood through the additional magnetic field created by a current distribution, induced in the electronic charge distribution of the sample by the vector potential of the externally applied field. The usual expression for this additional field consists of a sum of two terms, one negative (*diamagnetic*) and one positive (*paramagnetic*). Although it is often convenient to discuss the two terms separately, the separation is not unique and only their sum is a physical observable. For metals, it is usual to take as reference state the chemical shift originating from the filled electron bands (the ionic cores). The chemical shift resulting from the partially filled bands (the conduction electrons) is often called the orbital Knight shift.

The Knight shift (in the narrow sense) is the response to electron spin magnetism. The original theory of spin magnetism in "nonmagnetic" metals is the Pauli theory, which provided the first explanation for the temperature independence of the susceptibility. At a closer look, however, the numerical value of the susceptibility according to the original theory does not fit very well to experiment, even for the alkali metals. The required refinement is to take the "molecular field" into account: inside the metal the total field sensed by a given electron spin is the sum of the externally applied field plus the magnetic field created by all other electron spins. (The same molecular field is used to explain ferromagnetism in Stoner theory.) A formal theory in terms of Kohn–Sham orbitals within the local density approximation of density functional theory has been proposed in [56,57]. The corresponding theory for metal NMR, including the orbital Knight shift, has been given in [58]; we will use results from that paper in the form appropriate for transition metals.

It is useful to separate the total density of states at the Fermi level of transition metals into contributions $D_{lm}(E_f)$ of different orbital angular momentum l, m inside a Wigner–Seitz cell (approximated by a sphere). We restrict ourselves to cubic metals. Under certain approximations [58] the equations for the spin susceptibility, the Knight shift, and the relaxation rate can be written as sums over (l, m)-like parts, with $l = 0, 1, 2 \ldots$ corresponding to $s, p, d \ldots$ in a tight-binding picture. The derivation of the equations itself, however, does not use the tight-binding approximation such as [58].

The expression for the uniform spin susceptibility takes the form

$$\chi = \mu_0 \mu_B^2 \Omega^{-1} \sum_l \sum_{m=-l}^{l} \frac{D_{lm}(E_f)}{1 - \alpha_l} = \sum_{lm} \chi_{lm} \tag{2}$$

where the last equality defines the partial susceptibilities χ_{lm} for later use. Ω is the atomic volume. The enhancement of the susceptibility with respect to the Pauli value is given by the l-like partial Stoner factor $\alpha_l = I_l D_l(E_f)$, where the I_l is the l-like exchange integral. For a paramagnetic system we should have $0 \leqslant \alpha_l \leqslant 1$.

The Knight shift K can be written as

$$K = \sum_{lm} \chi_{lm} \frac{\Omega B_{hf,l}}{\mu_0 \mu_B} = \sum_{lm} K_{lm} \tag{3}$$

where the effective l-like hyperfine field $B_{hf,l}$ can be nonzero also for non-s electrons, $l \neq 0$, and its value may be negative. In the formalism [58] these effective hyperfine fields can be given a precise meaning in terms of a generalized nonlocal susceptibility. That formalism is not suitable for computations, however, and in practice the values of hyperfine fields are obtained by fitting to experimental data. When comparing shifts of the same nucleus in different environments, it is usually assumed that the hyperfine fields are (more or less) atomic quantities, independent of environment, whereas the susceptibilities change, through changes in the $D_{lm}(E_f)$. The exchange integrals I_l are also treated in practice as atomic quantities, which do not change when the atom is placed in different environments.

The equation for the relaxation rate T_1^{-1} can be brought to the form

$$S(T_1 T)^{-1} = \sum_{lm} k_l K_{lm}^2 \tag{4}$$

where the partial contribution to the Knight shift K_{lm} have been defined in Eq. (3), and where the k_l are called "disenhancement factors." The Korringa constant S is given by

$$S = (2\mu_B)^2/(4\pi\hbar k\gamma^2) \tag{5}$$

where γ is the gyromagnetic ratio of the nucleus under consideration. If χ is independent of temperature, then so is the product $T_1 T$. This latter result was derived by Heitler and Teller in 1936 [59], well before the discovery of NMR. In the Pauli approximation, where all $\alpha_l = 0$ and all $k_l = 1$, for simple metals where $l = 0$ is the most important contribution, Eq. (4) leads to the Korringa [60] relation

$$S(T_1 T)^{-1} = K^2 \tag{6}$$

Note that Eq. (6) can be applied only when there is one (dominant) term in Eq. (3), and susceptibility enhancement effects in Eq. (2) can be neglected. Therefore, it cannot be used for transition metals.

For the transition metals, contributions with $l = 0, 1, 2$ are considered, and usually the s and p are combined together. In cubic metals, the five values of m corresponding to $l = 2$ come together in a triplet t_{2g} and a doublet e_g. So instead of

$$D_d(E_f) = \sum_{m=-2}^{2} D_{2m}(E_f) \tag{7}$$

we should write

$$D_d(E_f) = D_{t2g}(F_f) + D_{e_g}(E_f) \tag{8}$$

and similarly for the sum of squares that occurs in the expression for the relaxation rate

$$\sum_{m=-2}^{2} D_{2m}^2(E_f) = 3(D_{t2g}(E_f)/3)^2 + 2(D_{e_g}(E_f)/2)^2 = R_d D_d^2(E_f) \tag{9}$$

where the last equality defines the "orbital reduction factor" R_d. There is a similar, but slightly differently defined, reduction factor R_{orb} that appears in the expression for the orbital relaxation rate [the last term in Eq. (14)], which is associated with fluctuations in the orbital Knight shift. We will not go into the details of its derivation [61], but we mention for reference

$$R_d = \frac{2D_{t_{2g}}^2 + 3D_{e_g}^2}{6D_d^2} \tag{10}$$

$$R_{\text{orb}} = \frac{2D_{t_{2g}}(D_{t_{2g}} + 6D_{e_g})}{9D_d^2} \tag{11}$$

In the above paragraphs, we have already introduced several approximations in the description of the shift and relaxation rates in transition metals, the most severe being the introduction of the three densities of states $D_{sp}(E_f)$, $D_{t_{2g}}(E_f)$, and $D_{e_g}(E_f)$. The advantage is that these values can be supplied by band structure calculations and that the d-like hyperfine field can sometimes be found from experiment. We have no reliable means to calculate the effective Stoner factors α_l that appear in Eq. (2), and the disenhancement factors k_l in the expression for the relaxation rate, Eq. (4), are also unknown. It is often assumed that k_l can be calculated from some l-independent function of the Stoner parameter $k(\alpha)$, thus $k_l = k(\alpha_l)$. A few models exist to derive the relation $k(\alpha)$, all of them for simple metals [62–65]. For want of something better they have sometimes been applied to transition metals as well [66–69]. We have used the Shaw–Warren result [64], which can be fitted to a simple polynomial in α. There is little fundamental justification for doing so, but it leads to a satisfactory description of, e.g., the data for bulk Pt and Pd.

The equations used to analyze the ^{195}Pt NMR in catalytic particles are

$$\chi = \mu_0 \mu_B^2 \Omega^{-1} \left(\frac{D_{sp}(E_f)}{1 - \alpha_{sp}} + \frac{D_d(E_f)}{1 - \alpha_d} \right) + \chi_{\text{orb}} = \chi_{sp} + \chi_d + \chi_{\text{orb}} \tag{12}$$

$$K = (\chi_{sp} B_{\text{hf},sp} + \chi_d B_{\text{hf},d} + \chi_{\text{orb}} B_{\text{hf},orb}) \Omega / (\mu_0 \mu_B) = K_{sp} + K_d + K_{\text{orb}} \tag{13}$$

$$S(T_1 T)^{-1} = k(\alpha_{sp}) K_{sp}^2 + k(\alpha_d) K_d^2 R_d + (\mu_B D_d B_{\text{hf},orb})^2 R_{\text{orb}} \tag{14}$$

$$\alpha_i = I_i D_i(E_f) \tag{15}$$

$$k(\alpha_i) = (1 - \alpha_i)(1 + \alpha_i/4) \tag{16}$$

where only $D_{sp}(E_f)$ and $D_d(Ef)$ are supposed to be site-dependent. It is usual practice to simplify the index sp to s. The terms with index orb come from the orbital susceptibility.

A set of parameters fitted to experimental data for bulk platinum has been given in Table 7 of [58]. There are two typographical errors in that table. The value of $R_{\text{orb}} = 0.39$ and $\chi_{\text{total}} = \chi_{\text{exp}} = 325 \times 10^{-6}$. The experimental susceptibility has been corrected for an estimated diamagnetic contribution of 41×10^{-6}; for the Pd data in Table 8 a similar correction of 57×10^{-6} has been applied.

Metal NMR is interesting in catalysis because of the relation of the (spin) Knight shift with the Pauli susceptibility, which in turn is related to the local density of states at the Fermi energy at the site of the nucleus. In practice, a detailed analysis of spin-shift-related effects requires that the orbital shift be relatively weak. This is not the general case for transition metals, but fortunately it applies to the catalytically important metal platinum. This is the subject of the next section.

13.4 NMR OF THE CATALYTIC METAL

The catalytic metal most widely studied by NMR is platinum; the first observation of oxide-supported [195]Pt was made by Slichter and co-workers 20 years ago. Nearly all of the metal-NMR results in this review are concerned with this nucleus. Some data for [103]Rh will be discussed also. Recently, ^{99}Tc NMR spectra have been published of small Tc particles (average diameter 2.3 nm, but a rather wide size distribution) on alumina [70]. The spectra were taken between 120 and 400 K. While bulk technetium has the hcp structure, these small particles are cubic, and their ^{99}Tc shift (around 7400 ppm) is about 600 ppm larger than the isotropic part of the bulk shift. The linewidth varies with support material and method of preparation, but remains amazingly small (15–75 ppm). This linewidth/shift ratio of about 0.5% is much less than that found for small particles of rhodium or platinum and is comparable to that found for silver [71]. It is unlikely, however, that the linebroadening mechanisms in small particles of silver and of technetium are the same.

13.4.1 Lineshape

For reasons that have to do with the penetration of radiofrequency fields, most NMR experiments on "bulk" metals actually use powders with grain sizes of a few tens of microns. At that kind of size the overwhelming majority of atoms still is in the same electronic environment, and they have the same NMR shift and relaxation rate. This is no longer the case when the grain size comes down in the nanometer range. Imagine that we build up a face-centered cubic (fcc) metal by starting with a single atom, and adding successive layers to it, to form particles of a cubo-octahedral shape. The smallest such particle contains 13 atoms; the next-larger ones 55, 147, and so on. To characterize the size of such particles, it is usual to give the diameter of the sphere that has a volume of $13, 55, 147 \ldots$ times the atomic volume of the bulk. For platinum, these diameters are $0.72, 1.165, 1.617 \ldots$ nm. Therefore, a particle of 1.6-nm diameter has about 60% of its atoms in the surface. Slichter and co-workers were the first to show experimentally a size dependence in the NMR of such small metal particles, using alumina-supported platinum [5]. They found that with decreasing particle size, the NMR intensity decreases at the field/frequency ratio (the inverse of the "effective" gyromagnetic ratio) 1.138 G/kHz typical of bulk metal platinum, while a new feature at 1.10 G/kHz appears and increases in intensity. Later it was shown that these size-dependent spectral shapes were independent of the kind of oxidic support (alumina, silica, titania) used [72]. The same spectral shapes have been found for platinum particles on conducting carbon black [4], as shown in Figure 6.

A simple parameterization of these spectral shapes is provided by the "exponential healing" [73] or "NMR layer" model [74]. The size of the metal particles (or, more precisely, their size distribution) can be measured by electron microscopy. Using the cubo-octahedral model, we can then calculate what fraction of atoms is in the surface layer, in the subsurface layer, and so on. The local density of states in all sites of the surface layer is not quite the same, but nevertheless clearly different from those in the subsurface layer, and so on. Therefore, the spectrum is decomposed as a superposition of (for convenience) Gaussians, each representing the collection of sites in a layer. The integrals must be proportional to the fraction of atoms in the layer. The maximum of the Gaussian corresponding to the nth layer

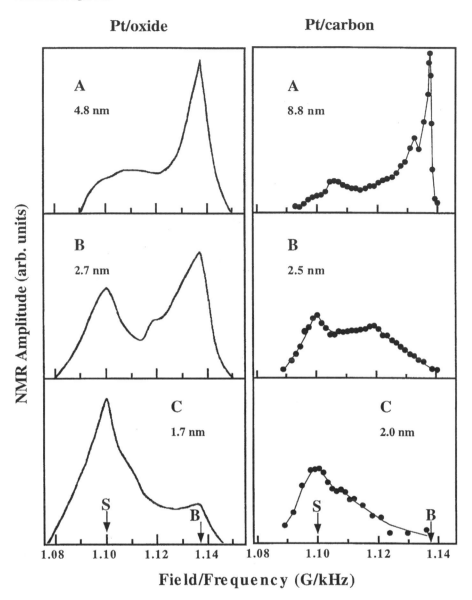

Figure 6 ^{195}Pt NMR spectra as a function of the size of Pt particles supported on insulating oxides, left-side panels, and on conducting carbon black, right-side panels. The arrows indicate the surface and bulk resonance positions.

(where layer 0 is the surface, and n increases when going inwards) is assumed to occur at a Knight shift K_n given by

$$K_n - K_\infty = (K_0 - K_\infty) \exp(-n/m) \tag{17}$$

where the parameter m gives the healing length in units of a layer thickness, K_0 is the Knight shift at the surface, and K_∞ that of the bulk.

Two examples [74] of such a deconvolution are shown in Figure 7, where $m = 1.35$ and $K_0 = 0$. The same model can be applied to the ^{195}Pt spectra for electrochemically cleaned Pt/carbon immersed in an electrolyte [7] in Figure 8, with $m = 2.0$ and $K_0 = 0$. The change in m says that the change from oxidic to conducting support is not only sensed by the Pt atoms in the surface of the metal particle, but also deeper inside. In a similar way, the important change of spectral shape that occurs in the ^{195}Pt/oxide NMR spectrum upon hydrogen chemisorption can be parameterized by $m = 2.6$ and $K_0 = 2400$ ppm [74]. Again, the presence of hydrogen is sensed deeper inside the particle than that of the clean surface; we may also think of this as saying that the hydrogen forms bonds not only with the platinum surface atoms, but also with those in the deeper layers. The chemisorption effect is not limited to hydrogen: as an example we show in Figure 9 a Pt/carbon sample, covered with CO by methanol electrodecomposition and immersed in electrolyte [7].

The signals occurring at $K = 0$ raise an interesting question. If the Knight shift is zero, can we speak of a metallic NMR signal? The answer is given by the spin-lattice relaxation: if the Heitler–Teller product $T_1 T$ is independent of temperature [59], then the NMR is metallic. The Knight shift of platinum has a positive contribution from the s-like LDOS, and a negative from the d-LDOS. They may accidentally sum up to zero, but the relaxation rate is proportional to a sum of squares and is therefore nonzero. Even after chemisorption of hydrogen or carbon monoxide, this $T_1 T$ criterion says that the surface platinum atoms retain their

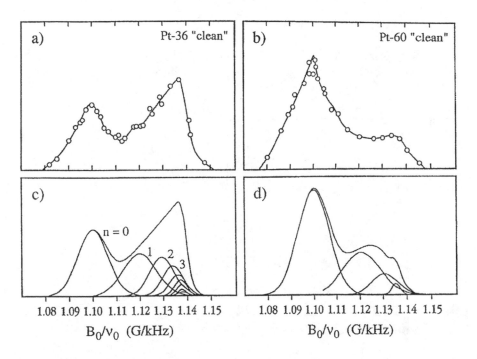

Figure 7 Point by point ^{195}Pt NMR spectra, under clean-surface conditions, for oxide-supported samples with average diameter of (a) 2.7 and (b) 1.7 nm, and the corresponding NMR layer model deconvolutions.

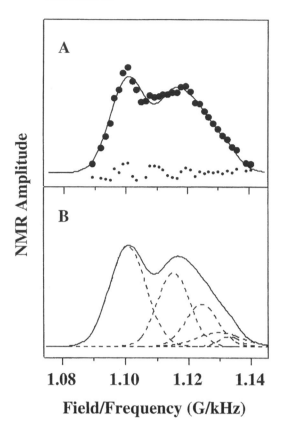

Figure 8 Point-by-point ^{195}Pt NMR spectrum, (A), and its layer model deconvolution, (B), for electrochemically cleaned Pt particles supported on conducting carbon and immersed in electrolyte. The average particle size is 2.5 nm.

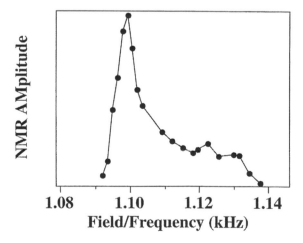

Figure 9 Change in the spectrum shown in Figure 8 when CO is adsorbed from electrochemical decomposition of methanol. The high-field intensity diminishes, and a rather sharp feature appears at low field.

metallic character. We stress that this conclusion is purely experimental, very different in character from assertions based on *calculated* electron densities, Section 13.2.1, and close to the intuitive idea that a metal must conduct electricity. Usual metallic conduction is related to a nonzero density of states at the Fermi energy, which also causes the relaxation rate T_1^{-1} to be proportional to T. Calculations for a five-layer platinum slab in a magnetic field [8] also find a value close to zero for the Knight shift in a clean platinum surface. There are three distinct sites in such a slab, and the shift changes in a monotonic (rather than oscillatory) way when going into the slab. This supports the use of an exponential function in the layer model, although just three points cannot prove it, of course.

As we have seen in Figure 5 the local densities of state in ordered 1:1 alloys AB are different on the two different sites (one site for constituent A, the other for B). It is easy to understand that in a disordered alloy of the same overall composition, there will be many different sites for each of the constituents, depending on, e.g., the composition of the surrounding layer of atoms. Indeed, the NMR lines in alloys are always broader than those in pure metals. A further complication is added when we make small particles of disordered alloys. The catalytic activity of $Pt_{1-x}Pd_x$ protected by films of poly(N-vinyl-2-pyrrolidone) (PVP) varies strongly with composition [75]. One would like to study both metals by NMR, but unfortunately ^{105}Pd NMR is very difficult and has not been attempted on small particles. Some information on small palladium particles is available from susceptibility measurements on catalysts [76] and on cluster molecules [77]. In both cases it was concluded that the susceptibility of the surface atoms is less than that of the bulk, just as is found for platinum particles from their NMR data. It is then reasonable to assume that in these systems the Pd susceptibility "heals back" when going from the surface to the interior, and that also for $Pt_{1-x}Pd_x$ particles an expression like Eq. (17) is valid.

In the pure Pt/PVP, the ^{195}Pt NMR spectrum and relaxation times are practically the same as those in Pt/oxide samples of similar dispersion. The spectrum predicted by using the layer distribution from Figure 10 and the Gaussians from Figure 7 shows qualitative agreement with the observed spectrum in Figure 11(a). This implies that the interaction of the polymer with the surface platinums is weak and/or restricted to a small number of sites. The layer distributions for the three samples in Figure 10 are very similar, and therefore the differences between the ^{195}Pt spectra in Figure 11 must be due to the alloying. The rightmost arrows in that figure give the average resonance position in the corresponding bulk materials (the disorder broadens the bulk resonance lines considerably). The high-field edges of the small-particle spectra follow these positions very well, as would be expected from Eq. (17) for layers with $n > 2m$, and the spin-lattice relaxation rates tend to the bulk values as well. This shows that on the scale of one or two healing lengths the composition of the interior of the particles corresponds to the bulk formula.

The ^{103}Rh NMR of small rhodium particles evolves in a different way when the particle size is changed. The spectra in Figure 12 are approximately centered at the bulk resonance position, and they broaden rather symmetrically when the particle size decreases. Such results have been found for several systems (Rh/titania, Rh/ alumina, Rh/PVP) [78]. There is no well-defined surface region in the spectrum, and chemisorption of hydrogen has a measurable, but small effect (see Figure 13). There

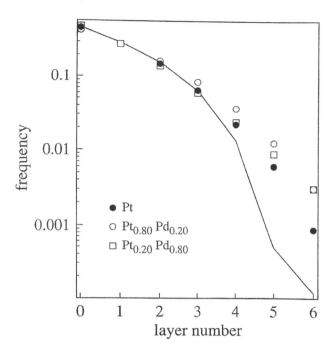

Figure 10 Layer statistics according to the NMR layer model for PVP protected $Pt_{1-x}Pd_x$ samples with $x = 0$, 0.2, and 0.8. The solid curve represents the layer statistics of a hypothetic sample with a mono-dispersed size of 2.35 nm.

are two reasons for this different behavior: one in the NMR of Rh metal, the other in the electronic structure of the Rh surface. There is no generally agreed set of parameters to describe bulk ^{103}Rh NMR. The approximate analysis given in [79] contains numerical errors and is inconsistent. A consistent set of parameters has been proposed in [78]. According to this set, the Knight shift and the relaxation rate are dominated by the orbital parts, with additional contributions from the spins of the d-like electrons and nearly negligible s-like parts. This is very different from Pt or Pd, where the d-like spin part dominates, and the orbital parts are nearly negligible [58]. In small particles of Pt (and presumably also of Pd) the magnetic behavior can be reasonably well described by considering only the site-to-site variation of the d-like spin part, and as it happens this part of the susceptibility is measurably smaller on surface sites than in the bulk. For that reason, a healing model as in Eq. (17) is successful for Pt (and probably would also be in Pd, if the experiment could be done). From calculations for a rhodium slab [80] it can be concluded that the site-to-site variation of the (spin) Knight shift is rather small: therefore, the spectral shapes in Figure 12 are probably dominated by competition between (positive) orbital shifts and (negative) spin shifts. On some surface sites the net result is negative, on others positive; and no clear surface signal can be found in the spectra.

When going to metals further to the left of Rh in the periodic system, the importance of the orbital parts in shift and relaxation rate grows further. It is therefore very likely that ^{195}Pt NMR will remain the most interesting nucleus for catalytic studies.

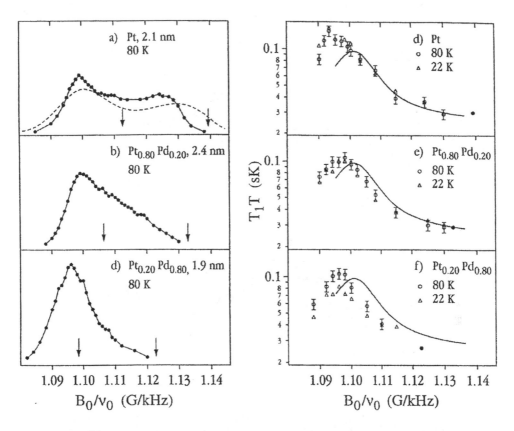

Figure 11 ^{195}Pt NMR spectra and spin-lattice relaxation times across the spectra for the three Pt_{1-x} Pd_x samples whose layer statistics are shown in Figure 10. Since the size distributions are nearly identical, all differences must be due to alloy formation. (a–c) Spectra. The rightmost arrow is the resonance position in the bulk alloy. The left arrow is the center of the spectra (divides the integral in halves). The surface peak does not seem to shift with composition. (d–f) The product T_1T across the spectrum at two temperatures (open symbols). The ^{195}Pt are in a metallic environment. The full curve sketches the values found for Pt/TiO$_2$ catalysts. The open symbols extrapolate well to the filled circle in each panel, which is the corresponding bulk value.

13.4.2   Densities of State at E_f on Different Atomic Sites

The extraction of numerical values for the local densities of state at the Fermi energy from NMR resonance position and relaxation rate requires of course a number of hypotheses. Some of them (such as knowledge of the resonance position corresponding to zero total shift; the breakup of the density of states into parts of different symmetry, etc.) already come into play when we try to parameterize data for the bulk metal [58]. Here we mention only the additional ones used to go to the local version of the equations. It is assumed that the hyperfine fields and exchange integrals are a kind of atomic properties that do not vary when the atom is put in one environment or another, whether it is deep inside the particle or on its surface. The approximation is probably reasonable when the atomic volume stays approximately

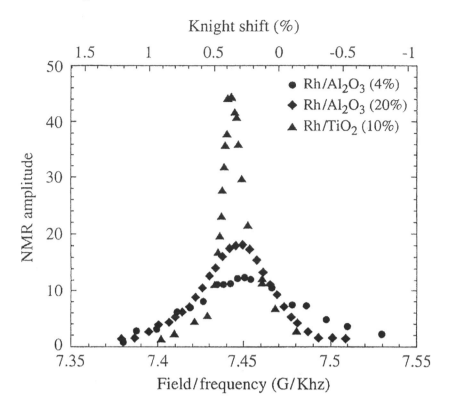

Figure 12 Point-by-point ^{103}Rh NMR spectra for oxide-supported, clean-surface Rh samples, taken at 80 K. The remarkable result is that the spectrum broadens in both directions and not only to low field, as for platinum. The broadening is more important for the sample with smaller particles. (From Ref. 78.)

the same: but, e.g., under heavy hydrostatic compression, the hyperfine field in a bulk metal can change considerably [58]. We neglect the shift due to dipole–dipole coupling between the electronic and nuclear spins. This coupling vanishes by symmetry in a cubic environment, but not, e.g., on a surface. In the case of Pt (but not for Rh), the site-to-site variation of the orbital susceptibility is neglected. The remaining local variables are the densities of state of different *symmetry* $D_{lm}(E_f)$. The site-to-site variation in the ratios of D_{2m}, as expressed by the R-factors in Eq. (10) and Eq. (11), are neglected. This leaves only D_{sp} and D_d as local variables. As an illustration of how to solve the equations, Figure 14 shows curves of constant resonance position and of constant relaxation rate in a (D_{sp}, D_d)-plane. Measure the relaxation rate at a certain resonance position, look up the intersection of the two corresponding lines in Figure 14, and you have the local densities of state for the nuclei you have measured. A further combination with the layer model in Figure 7 says how deep in the interior these nuclei are.

While the different contributions to the shift may not always be easy to estimate, one expects the relaxation rate in any transition metal to be dominated by D_d. From relaxation-rate measurements across the spectrum, as, e.g., in Figure 11 for ^{195}Pt in platinum/palladium alloys, we can immediately say that D_d is higher in

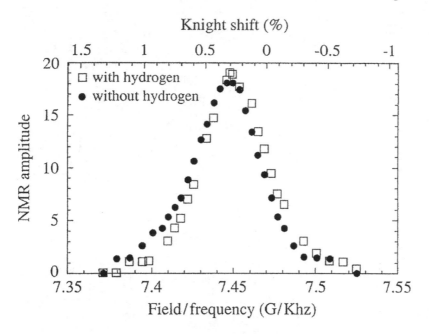

Figure 13 Point-by-point ^{103}Rh NMR spectra for an oxide-supported Rh sample with (open squares) and without (solid circles) hydrogen. Chemisorption of a monolayer of hydrogen has a measurable, but much smaller effect than on Pt surface. It mainly shifts the whole line slightly upfield. (From Ref. 78.)

the bulk (rightmost points) than in the surface (leftmost points). Such a simple argument is especially useful when the relaxation curves are nonexponential: it is often easy to see that one nonexponential decay curve is "faster" than another. As is immediately clear from Figure 14, a range of different relaxation rates, corresponding to different atomic sites, may be found at one and the same resonance position. If this happens indeed, the experimental relaxation curves are nonexponential. From measurements at different temperatures and by using time/temperature scaling, it can nevertheless be checked that the individual components still have a temperature-independent $T_1 T$, confirming the metallic character of all contributing sites. A lack of such scaling, as has been found for a part of the ^{195}Pt relaxation curves for Pt/carbon immersed in electrolyte [7], shows that some of the signal comes from "nonmetallic" platinum sites. This fraction of the signal is not modified by adsorption of CO (from electrochemical decomposition of methanol), which suggests that it is related to the Pt–carbon interface. The rest of the Pt signal has the metallic signature, and its T_1^{-1} drops by a factor of two going from the clean surface to the CO-covered one. This drop shows that after the adsorption the D_d on the platinum surface sites has lowered. If we want to give a numerical value for the change, we need the additional hypotheses listed above, which may or may not lead to accurate values; but the direction of the change is qualitatively clear from the relaxation-rate behavior. Therefore, we believe that, while in general the numerical values for local densities of state derived with this procedure must be treated with some caution, the trends will usually be correct.

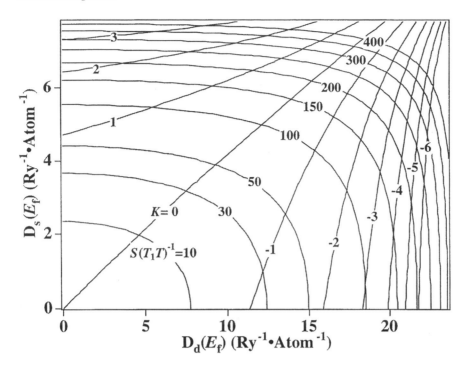

Figure 14 Contours of constant (spin) Knight shift (dashed) and of constant relaxation rate (full lines) for [195]Pt, as a function of the *s*-like and *d*-like E_f-LDOS. The relaxation curves are marked in units of 10^{-6}, the shift in units of 10^{-2} (percent). The relaxation contour marked "10" corresponds to $T_1 T \approx 0.57$ sK.

The drop in D_d on the platinum surface sites upon adsorption has also been found when CO is chemisorbed on "dry" Pt/alumina [17] and for Pt/titania after hydrogen chemisorption [69]. There are no calculations of the effect of chemisorption on the densities of state in a platinum surface, but there are results for hydrogen on palladium [81,82] and rhodium [82]. For Pd, the clean-surface density of states at the Fermi energy is lower than that of the bulk, and it drops further upon hydrogen chemisorption [81]. Exactly this behavior is found experimentally from NMR for Pt. The calculated effect of chemisorption of hydrogen on Rh is much smaller than that on Pd [82] (see Figure 15), and experimentally the effect on the [103]Rh spectrum in Figure 13 is negligible as compared to what is found in the [195]Pt spectrum (the effect of hydrogen on Pt looks very much like that of CO in Figure 9). Remarkable is the inverse result from ^{1}H NMR: the shift of the proton NMR line on Rh is much larger than it is on Pt or Pd [18]. This may be related to the higher calculated density of states at the Fermi energy in the H/Rh layer, as compared to the H/Pd layer, in Figure 15.

13.4.3 Frontier-Orbital Reactivity Theory

Using the methods described in the previous section, we can deduce values for the E_f-LDOS on the platinum sites of clean or adsorbate-modified surfaces from NMR

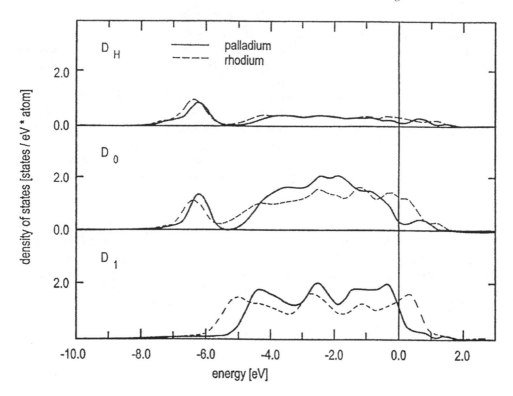

Figure 15 Layer-resolved density of states (DOS) for hydrogen-covered metal slabs of palladium and rhodium. Top: DOS on the hydrogen layer; middle: on the metal surface layer; bottom: on the metal subsurface layer. (From Ref. 82.)

data. As explained in Section 13.2.5, the metal frontier-orbital model predicts that, among a series of similarly prepared, but slightly different surfaces, the most "reactive" will be the one that has (before the reaction) the highest E_f-LDOS. A rather generally used yard stick of reactivity is carbon monoxide chemisorption. If the chemisorption is strong, the intramolecular carbon–oxygen bonding is weak; and if that bonding is weak, the frequency of the infrared stretching frequency is low. Ideally, we should have a series of clean Pt surfaces, measure the [195]Pt NMR, next chemisorb a single CO molecule on each of them, and measure the stretching frequencies. If the clean-surface E_f-LDOS was low, the stretching frequency will be high.

The restriction to a single molecule in this idealized case guarantees that there is no macroscopic change in the surface LDOS upon chemisorption and that there is no dipole–dipole interaction in the infrared result. Experimentally, this is approximated by measuring the stretching frequency at a number of coverages θ and extrapolating to $\theta = 0$.

Such [195]Pt NMR and the CO infrared experiments were performed on five Pt/zeolite catalysts with varying zeolite acidity. It was checked that varying the dispersion of the platinum particles in the same zeolite did not change the NMR data. Qualitatively, it was found that the spin-lattice relaxation rate in the surface region of the NMR spectrum is slower when the zeolite acidity increases [83]. It is known (although not necessarily understood) that when the acidity increases, the stretch frequency of the CO chemisorbed on the platinum particles increases, as was also found for these particular samples [12]. Therefore a lower clean-surface E_f-LDOS is correlated to a higher CO frequency. The numerical values obtained are shown in Figure 16. The result can be related immediately to the frontier-orbital interpretation of the Blyholder model [10] in which a higher E_f-LDOS means that more metal electrons are available to dump into the antibonding LUMO of the CO molecule, and also that more metal holes can pull electrons out of the bonding HOMO. The result is an increased weakening of the molecule. Indeed, the elongation of the C–O bond after chemisorption has been determined by $^{13}C^{17}O$ double resonance [84].

This difference in reactivity is often described not in terms of Fermi-level quantities, but in terms of total charge. If the metal particle is "electron-deficient," it has fewer electrons available for the Blyholder mechanism. Initially, the expression referred to the experimental fact that catalysts of platinum or palladium in zeolites, when compared to the same metals on oxides, often seemed to behave as their left-hand neighbors in the periodic system, iridium and rhodium. In a rigid-band picture

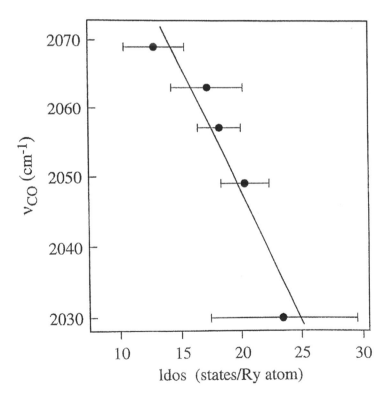

Figure 16 E_f-LDOS on the clean metal surface and (extrapolated) infrared stretch frequency of CO adsorbed on those same surfaces. The samples are Pt/zeolite with different zeolite acidity. This plot says that the C–O intramolecular bond after chemisorption will be weaker if the initial clean-surface E_f-LDOS is higher.

the principal difference is that these have one electron less available to fill the band, and it was thought that Pt or Pd somehow lost an electron through particle–support interactions. The idea of sizable charge transfers between metal particles of the dimensions considered here and their support has been severely criticized in a recent book [52, Ch. 5], but also the collective-electron model of catalytic activity was "sent to the trash can of science," because it does not account, e.g., for the fact that metal atoms in an alloy retain distinct properties [52, p. 451]. That model does not distinguish between DOS (density of states on the energy scale, a global property of a particle) and E_f-LDOS, a local property, as made clear by Figure 5. Similarly, the charge-deficiency models failed to recognize that not only is charge itself important, but also the energy cost at which it can be made available. Again, the E_f-LDOS is a quantity that can express this requirement. Actually, the E_f-DOS for the bulk metals goes down as well when a step is made to the left [85]; from 32.2 per Ry and per atom in Pd to 18.7 in Rh, and from 29.9 in Pt to 12.7 in Ir. A recent calculation for three-layer slabs of Pd and Rh shows differences for the surface E_f-LDOS at the Fermi energy similar to those in the bulk DOS [86].

13.4.4 Promotion and Poisoning

The promoting effect of the addition of alkali on the catalytic performance of many transition-metal-based catalysts is experimentally well known, but there is no general agreement on its theoretical explanation. The same holds for the opposite effect: the poisoning of catalysts by, e.g., the adsorption of sulphur.

In relation to the ^{195}Pt NMR experiments to be discussed below, a calculation of the surface electronic structure of a two-layer Rh(001) film, modified by adsorbing quarter monolayers of Li (or other atoms) [46], is very interesting. It was found that perturbations of the charge density are small beyond nearest neighbors of the adsorbate, but that the E_f-LDOS is increased over a wide region of the surface, while that quantity decreases with Cl or S adsorbates. It was suggested that there is a relation between the adsorbate-induced LDOS variations and the promoting, respectively poisoning effect of these atoms on the CO adsorption and dissociation. Similar LDOS influences have been invoked to explain the large changes in CO adsorption when thin overlayers of a late-transition metal such as Pd are deposed on early-transition metal substrates such as Ta or Nb [87]. Related conclusions have been reached in a calculation of the effect of S on the H_2 dissociation over Pd {100} [88]. Close to the sulphur atom the poisoning effect is related to direct adsorbate–adsorbate interactions, but at larger distances one of the S-induced changes in the surface-Pd electronic structure is a very strong reduction in the (average) E_f-LDOS on Pd surface atoms (see Figures 3b, 3c in [88]). However, other than LDOS effects have been invoked by other authors; see, e.g., [89] for the coadsorbate system CO+S on Rh, or [90] for transition-metal overlayers.

For the NMR experiments, material from a batch of Pt/titania that has been often used for ^{195}Pt NMR was impregnated with lithium or potassium salts and re-reduced afterwards [101]. It is not known how much of the alkali ends up on the metal particles. The NMR spectra are not visibly modified by the impregnation, but the spin-lattice relaxation becomes faster on the surface and is unchanged in the bulk: see Figure 17. The relaxation curves could be described by simple exponentials and Figure 17(b) shows that the ^{195}Pt atoms in the impregnated surface still have

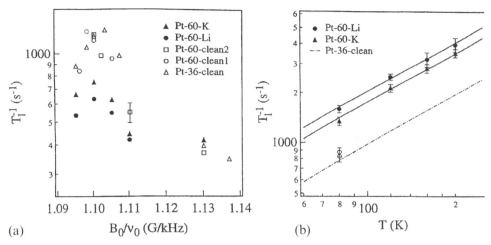

Figure 17 Effect of alkali impregnation on ^{195}Pt spin-lattice relaxation for Pt/TiO2. (a) Spin-lattice relaxation rates across the NMR spectrum for several clean-surface (open symbols) and alkali-impregnated (filled symbols) samples. The changes are important near 1.10 G/kHz (the surface signal) and undetectable at 1.13 G/kHz. (b) Korringa relationship for the spin-lattice relaxation at the surface peak of the spectrum. The alkali impregnation does not change the metallic character, but increases the E_f-LDOS on the metal surface. The dash-dotted straight line is an extrapolation of earlier clean-surface data obtained at lower temperatures for comparison.

metallic character. The impregnation increases the relaxation rate (at a fixed position in the spectrum). This is direct evidence that the E_f-LDOS in the surface has increased and supports the LDOS-based mechanism for alkali promotion.

An interesting, somewhat related experiment is atomic-beam NMR of lithium on ruthenium [91]. The electronic structure of such alkali-atom/transition-metal surface complexes has been debated for quite some time. These experiments have found that between 200 K and 1250 K the lithium nuclear spin relaxation has a contribution with T_1T independent of temperature. This immediately shows that in this system the alkali is in a metallic environment. It is somewhat surprising that up to a coverage of $\theta = 0.15$ the value of T_1T does not change. Apparently any given Li atom does not see its E_f-LDOS perturbed by the presence of the other atoms. This situation seems similar to that of ^{1}H NMR, where the coverage dependence of the shift is explained by assuming the existence of sites with intrinsic shifts that are coverage-independent [18].

In a somewhat speculative extension of the connection between surface E_f-LDOS and promotion effects, the difference in chemisorptive behavior of hydrogen and oxygen has been related to ^{195}Pt surface-NMR as function of coverage [55]. When the amount of hydrogen increases, the surface E_f-LDOS, averaged over both occupied and unoccupied sites (the hydrogens probably move around rapidly, and the NMR only sees an average), decreases monotonically. Values derived from data in [92] are shown by the filled circles in Figure 18. In the case of partial oxygen coverage, two kinds of sites seem to be created: metallic and nonmetallic ones. The number of nonmetallic Pt surface sites per oxygen drops from about 2.7 at a

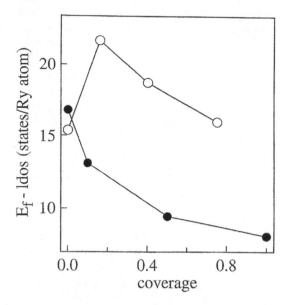

Figure 18   Variation of the metal surface E_f-LDOS for Pt/TiO$_2$ with hydrogen (filled circles) and oxygen (open circles) coverage. The determination of the hydrogen points is rather straightforward, but the reasoning for the oxygen case is somewhat involved.

coverage $\theta = 0.16$ to 1.5 at $\theta = 0.40$, and to zero at saturation, $\theta = 0.75$. On the other hand, it is known that the surface E_f-LDOS on the remaining sites is enhanced over the clean-surface value; these values are shown by the open circles in Figure 18 [93]. In most platinum catalysts, the oxygen chemisorption isotherm becomes flat already below 1 Torr, while the hydrogen isotherm joins the asymptote above 100 Torr. The speculation is that the continuous decrease of the surface E_f-LDOS with increasing hydrogen coverage makes the remaining surface less and less reactive, while the enhanced E_f-LDOS on the oxygen-free sites leads to an autopromotion effect for oxygen chemisorption.

The catalytic activity of Pt$_{1-x}$ Pd$_x$ polymer-protected bimetallics has been found to vary strongly with composition [75]. As shown in Figure 11, such catalysts have been studied by ^{195}Pt NMR. However, because the palladium NMR has not been observed, the average surface electronic properties can be determined only indirectly and tentatively. The ^{195}Pt NMR in Figure 11 has shown that the interior of the alloy particles is bulklike. In the bulk alloys the E_f-LDOS on both Pt and Pd sites varies rapidly with composition around $x = 0.8$ [54]. It is supposed, but not proven, that on the surfaces of the alloy particles the E_f-LDOS changes strongly with composition as well and that this explains the variation in catalytic activity.

13.5   NMR OF ADSORBED ^{13}CO

It is mentioned in Section 13.4.2 that chemisorption of carbon monoxide or hydrogen from the gas phase, or adsorption of CO by electrochemical methods, does not change the metallic character of the surface of the platinum particles, since there remains a sizable density of Fermi-energy electrons on the surface sites. Perhaps one could say that the surface sites are a bit "less metallic" than the bulk, because their E_f-LDOS is less than that of the bulk (although a more general interpretation of that statement is not obvious). A similar general result is that the plot of the ^{13}CO relaxation rates T_1^{-1} as a function of temperature T is a straight line. An example is shown in Figure 19 for CO adsorbed by electrochemical decomposition of methanol on carbon black-supported Pt particles with an average diameter of 8.8 nm. As Charles Slichter has repeatedly put it, such straight lines show that "the CO is really a piece of the metal." (For hydrogen, the experimental situation is less clear, but a T-independent $T_1 T$ has been found for hydrogen on copper [94].)

We argue in Section 13.2.1 that one should be careful when discussing bonds and bonding in extended systems; but likewise we should avoid too local a picture, where one adsorbate molecule forms a bond with one metal surface atom. If this picture were correct, the NMR shift of a ^{13}CO on the surface of a disordered AB alloy should take one of two values, depending on whether it "forms a bond" with A or with B (diffusional motion will then usually average over the two sites; for an exception see Section 13.5.5 below). For the particular case of $Pt_{1-x}Rh_x$, the experimental data are detailed enough [95] to show that such a viewpoint is not correct [96]. As another example, we mention in Section 13.4.1 that the "healing length" parameter of ^{195}Pt spectra changes when hydrogen or CO is on the surface. If we believe that this parameter represents some physical reality, this says that the "chemisorption bond" extends more than one metal atom layer deep.

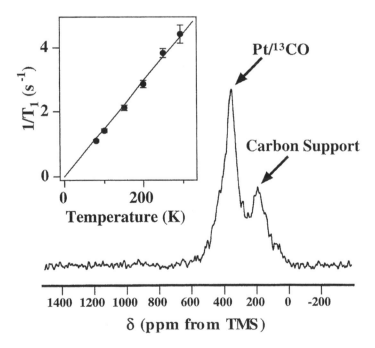

Figure 19 Typical ^{13}C NMR spectrum and temperature-dependent T_1 data for an 8.8-nm Pt/C sample. The straight line through the origin is characteristic of the metallic state.

The discussion here will be restricted to ^{13}CO on platinum (or sometimes palladium) and gives special attention to electrochemically obtained samples. For a review of results on other metals, see [18].

13.5.1 Lineshape and Position

The ^{13}C chemical shift tensor in the CO molecule is anisotropic with zero asymmetry, reflecting the axial symmetry of the molecule. The isotropic shift with respect to TMS is 185 ppm, and the anisotropy in the solid is 408 ppm. There is no known carbonyl based on zerovalent Pt or Pd, but for such carbonyls of other metals the shift tensor elements change by a few tens of ppm with respect to the free molecule. In most discussions of chemisorbed CO, the orbital Knight shift is taken comparable to the isotropic shift in the corresponding carbonyls. For CO on Pt, it has been proposed to take the calculated value for a Pt$_7$CO cluster, 160 ± 20 ppm [7].

The for a large part unpublished ^{13}CO/Pt spectra obtained by the Slichter group at 77 K can all reasonably well be fitted to Gaussians with an average position of 325 ± 25 ppm and an average full width at half-maximum (FWHM) of 360 ± 40. That width is not related to the anisotropy of the chemical shift of the isolated molecule, since with increasing temperature these lines narrow to values around 50 ppm, because of diffusion [97]. It remains, however, rather unclear why the anisotropy of the chemical shift of the isolated molecule disappears from the experimental observations; an explanation based on cancellation of the anisotropies in the chemical shift and in the dipolar Knight shift has been proposed [9]. With increasing temperature, the ^{13}CO/Pt line also shifts to lower field, probably reflecting a change in the type of bonding [97].

The ^{13}CO adsorbed on subnanometer particles in methylcyclohexane solutions of isobutylaluminoxane [98] shows at room temperature a line at 195 ppm with an FWHM of 50 ppm. The line position indicates that these Pt particles are too small to be metallic. Rapid tumbling of the Pt particles in the solution is supposed to average out the chemical shift anisotropy, and the remaining linewidth represents a distribution of isotropic shifts.

Yahnke et al. [99] have obtained spectra for ^{13}CO adsorbed from solution onto a commercial Pt/carbon fuel-cell electrode. They could vary the coverage in situ by connection to a potentiostat, but the actual NMR measurements were performed under open-cell conditions. They found an excellent correlation between the quantities of CO determined from the NMR spectra and by coulometry.

An interesting relation between the E_f-LDOS on clean Pt surfaces and the ^{13}CO shift found after chemisorption (or adsorption by electrochemical means) of CO on these surfaces has been mentioned in [17]; see Figure 20. For oxide-supported platinum, the ^{195}Pt relaxation rate is independent of dispersion and support, and the total surface E_f-LDOS is found as 14.8 ± 1.5 states per atom and per Rydberg [69]. The corresponding ^{13}CO line position at low temperatures is the 325 ± 25 ppm mentioned above. For electrochemically charged Pt/carbon immersed in electrolyte, both the clean-surface ^{195}Pt and the ^{13}CO NMR results (at constant coverage) vary with the platinum particle size. The NMR experiments have been performed for particles with average diameters of 2.5 and 8.8 nm. These results, together with the proposed reference ^{13}CO at 160 ppm when E_f-LDOS is zero, indicate that when the E_f-LDOS increases, the ^{13}CO Knight shift in Figure 20 increases as well. Loosely

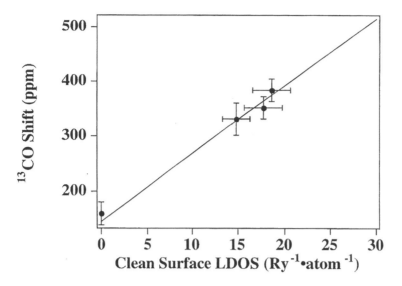

Figure 20 Correlation between the total E_f-LDOS found on clean platinum catalyst surfaces and the Knight shift of ^{13}CO chemisorbed afterward.

speaking, this figure says that when the clean Pt surface is "more metallic," then the adsorbed CO will be "more metallic" as well: in the spirit of the quotation from Charles Slichter, Section 13.4, the metallic overlayer of CO follows the metallic behavior of the underlying Pt. This correlation might be useful to find the trends in E_f-LDOS for different clean surfaces when it is experimentally difficult to detect the metal NMR itself: in such cases, adsorb ^{13}CO on the clean surface, and study the trends in the ^{13}C shift.

13.5.2 Local Densities of State

For these ^{13}CO that are "a piece of the metal" we should be able to write equations for the (local) Knight shift and relaxation rate, just as for a layer in a metal particle. Writing down the equations is rather easy: the local density of states near E_f can be projected on one kind of σ-like and two kinds of π-like orbitals. It is slightly problematic to write an expression for the orbital relaxation, but it is likely to be dominated by the π-like E_f-LDOS. Assuming that both π-like orbitals have the same E_f-LDOS [the equivalent of setting $R_d = 0.2$ in Eq. (10) for equal occupation of all five d-like orbitals] and absorbing the orbital Knight shift in the definition of the zero of the shift scale:

$$^{13}K = \mu_B \left(\frac{D_\sigma(E_f)}{1 - \alpha_\sigma} \, ^{13}B_{hf,\sigma} + \frac{D_\pi(E_f)}{1 - \alpha_\pi} \, ^{13}B_{hf,\pi} \right) = K_\sigma + K_\pi \tag{18}$$

$$^{13}S(^{13}T_1 T) = k(\alpha_\sigma)K_\sigma^2 + \frac{1}{2}k(\alpha_\pi)K_\pi^2 + CD_\pi^2(E_f) \tag{19}$$

where the constant C takes care of details in the expression for the orbital and dispolar relaxation rate, and the meaning of the other symbols can be found from Eq. (12) to Eq. (16) by analogy. Reasonable estimates for the magnitude of the hyperfine fields are available, but the sign of $^{13}B_{hf,\pi}$ is not known, and the values of the exchange integrals appearing in α_σ and α_π are difficult to guess; the constant C even more so [18,100]. With the particular choice made in [100] for the parameters, the equivalent diagram of Figure 14 looks like Figure 21. In the present review, however, we will not argue on the basis of the numerical values given by this diagram, but rely instead on qualitative arguments based on increasing/decreasing relaxation rates. For a wide range of parameter values, Eq. (19) says that an increase in relaxation rate corresponds to an increase in E_f-LDOS, and vice versa.

13.5.3 Potential Tuning of the ^{13}CO NMR Shift

Certainly the most interesting electrochemical NMR experiments are those under active potentiostatic control. They require careful attention to many experimental details [3], and only a few reports have appeared in the literature. There are data for ^{13}CO [4,101] and ^{13}CN [3] on Pt black electrodes (see Figure 22) and for ^{13}CO on a nanostructured palladium electrode [102].

For a good understanding of the interpretation of such data, it is useful to start with remembering classical experiments on NMR of organic molecules aligned in

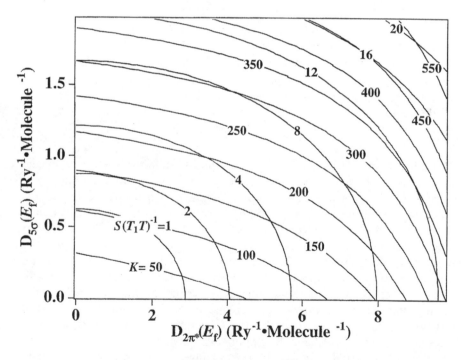

Figure 21 Contours of constant (spin) Knight shift (dashed) and of constant relaxation rate (full lines) for ^{13}CO, as a function of the σ-like and π-like E_f-LDOS. The relaxation curves are marked in units of 10^{-8}, the shift in units of 10^{-6} (ppm). The relaxation contour marked "1" corresponds to $T_1 T \approx 420$ sK. Note that the parameters used to calculate these curves 100 are not as well known as those used, e.g., in Figure 14.

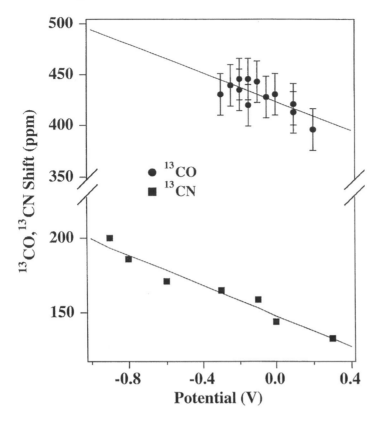

Figure 22 Electrode potential dependence of ^{13}C shifts for ^{13}CO (circles) and ^{13}CN (squares) chemisorbed on a 10-nm Pt electrocatalyst in an electrochemical environment. (From Refs. 3 and 4.)

externally applied electric fields [103]. The neat liquid sample is placed between the plates of a capacitor. There is no supporting electrolyte, and the liquid need not be in direct contact with the electrodes. For samples such as nitrobenzene, field strengths of several times 10^7 V/m could be successfully applied. But even at such field strengths, the main observable effect is an alignment, i.e., a loss of isotropy, so that dipolar and quadrupolar couplings become (faintly) visible. To really change the isotropic chemical shift, still-higher fields would be needed. Calculations indicate that it is experimentally impossible to create *external* fields of the required strength. For a single oriented ^{13}CO molecule this NMR shift tuning, as well as the accompanying change in the infrared stretch frequency (the vibrational Stark effect), has indeed been calculated. The ratio of the ^{13}CO shift to infrared stretch tuning effects is found between -0.14 and -0.23 ppm/cm^{-1} [104,105].

It is often argued [106] that the observed change in stretch frequency of CO adsorbed on an electrochemical electrode with change of potential ϕ is a form of vibrational Stark effect, caused by the very strong electric field in the double layer. A typical tuning rate on polycrystalline Pt is $d\nu/d\phi = 30$ cm^{-1}/V. Therefore, if the ^{13}CO shift tuning is caused by this same electric field, the expected tuning rate would be of the order of $d\delta/d\phi \approx -10\,\text{ppm/V}$. This is significantly smaller than the best-fit value, $-71 \pm 20\,\text{ppm/V}$, to the data in Figure 22. However, from the temperature-independent $T_1 T$ found from (ex-situ) ^{13}CO NMR of similar samples, Figure 19, it follows that the adsorbates are really a part of the electrode, and not a part of the ionic double layer. The importance of this observation has not always been appreciated, and we will try to emphasize it by using elementary reasoning based on electrostatics.

If we want to increase the electrostatic potential of a conducting sphere in vacuo, we must increase its positive charge, i.e., pull out electrons. According to continuum electrostatics, this positive charge appears at the surface of the sphere. For a resulting electric field $\mathbf{E}$ just outside the metal, the surface-charge density will be $\varepsilon_0 |\mathbf{E}|$ Coulomb per square meter. In a jellium model, the surface charge appears partly in the spill-out region, and partly in the region of Bardeen–Friedel oscillations inside the metal [107]. Outside this surface-charge region, the electrostatic potential behaves as predicted by macroscopic electrostatics. This does not mean, however, that in this model an adsorbed molecule remains unaffected by the applied field. An adsorbed CO has dimensions comparable to the thickness of the spill-out region, so it will participate in the charge rearrangements at the surface.

The NMR data show that such charge rearrangements have an important effect on the local density of state curves. In the simplest approach, we would use a rigid-band argument and say that the Fermi level moves up and down when electrons are added or taken away by changes in the potential. Assuming that the low-energy tail of the π-like density of states curve decreases monotonously toward lower energy (and that the σ-like DOS can be neglected), an increasing ϕ corresponds to a decreasing π-like E_f-LDOS, and thus (if the hyperfine field is positive) to a decrease in Knight shift, as is indeed observed. However, the resulting variations in E_f-LDOS (and therefore in Knight shift) should be unmeasurably small, since the amount of charge involved is so tiny. Even for the fields supposed to exist in the double layer, perhaps of the order of 10 GV/m, the surface-charge density is less than a thousandth of an electron per surface atom. Assuming an E_f-LDOS of 1 state per eV and per atom, the filled levels should shift up or down by at most 1 meV. We would not expect to see any structure in the density of states curves on this energy scale, and therefore the rigid-band model cannot explain the observed $d\delta/d\phi$ either.

The conclusion is therefore that the charge rearrangements invalidate the rigid-band model. Indeed, this is not unreasonable when we think of the delicate balance between positive nuclear charges and the screening effect of electron distributions that exists for the molecule on an "electrically neutral" surface. The experimental data suggest that the applied potential ϕ changes the shape of the CO-type local density of state curves $D_\pi(E, \phi)$. With increasing ϕ, the density of states in the tail diminishes, while at the same time the bottom of the tail sinks deeper below the Fermi energy, ensuring that the total charge does not drop by more than a fraction of an electron.

The potential-dependent $^{13}\text{CO}/\text{Pd}$ relaxation data [102] have rather large error bars but nevertheless indicate that T_1^{-1} increases when the potential becomes more negative; the same trend is observed for the $^{13}\text{CO}/\text{Pd}$ shift. If the relaxation is dominated by $D_\pi(E_f)$, then the sign of $B_{\text{hf},\pi}$ is probably positive, as assumed in [100]. The π-like density of states at the Fermi energy on the adsorbed CO decreases when electrons are taken out of the surface. At the same time, the intramolecular C–O bond grows stronger, as seen from an increasing IR frequency: this shows that (most) electrons are pulled out of π-like levels.

13.5.4 CO Infrared Spectroscopy and ^{13}CO Relaxation

A somewhat similar relation between a decreasing ^{13}CO relaxation rate and an increasing CO stretch frequency, but now based on differences in electrode properties, has been found for ^{13}CO/Pt/carbon. Potential-dependent in-situ infrared reflection spectroscopy [108] and ex-situ ^{13}CO NMR measurements [100] have been performed using different samples of fuel-cell-grade platinum electrocatalysts, with CO adsorbed from electrochemical decomposition of methanol. The main difference between the five samples was the average Pt particle size: 2.0, 2.5, 3.2, 3.9, and 8.8 nm. The CO coverage of the NMR samples was determined as (in the same order) 0.64 ± 0.04, 0.56 ± 0.03, 0.67 ± 0.05, 0.69 ± 0.05, and 0.65 ± 0.04. During the IR experiments, the coverage for all samples was 0.61 ± 0.05. As can be seen in Figure 23(a) the infrared frequency decreases when the NMR relaxation rate increases. For this comparison, the IR frequencies were extrapolated to zero potential with respect to an Ag/AgCl (3M) reference electrode. Note that these results are for (nearly) maximum CO coverage, whereas those shown in Figure 16 for "dry" catalysts were extrapolated to zero coverage. Nevertheless, the trends are the same: a higher ^{13}CO relaxation rate corresponds to a higher E_f-LDOS of the underlying clean surface (by our loose interpretation of Figure 20), which corresponds (according to Figure 16)

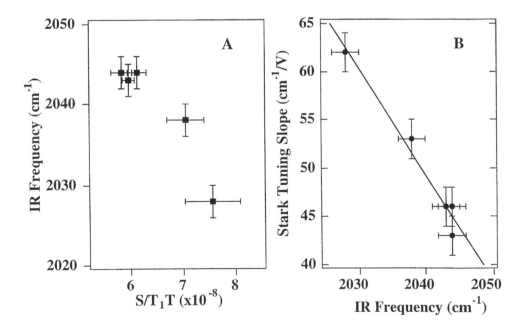

Figure 23 Ex-situ ^{13}CO NMR and in-situ CO infrared data for CO adsorbed on Pt/carbon with different Pt particle sizes. (A) Correlation between the extrapolated IR frequency (see text) and the ^{13}CO relaxation rate. (B) The Stark tuning rate and the extrapolated IR frequency.

to a lower infrared frequency; the overall result is the NMR/IR correlation in Figure 23.

The Stark tuning rate $dv/d\phi$ was determined also, and the relation with the extrapolated infrared frequency is shown in Figure 23(b). When the intramolecular bond strength is smaller (lower IR frequency), it is more easily changed by the potential (higher tuning rate). While the range of infrared frequencies is not unusual, the Stark tuning rate on the smallest Pt particles is roughly double the typical value on polycrystalline Pt. This may be related to the higher clean-surface E_f-LDOS for these carbon-supported particles shown in Figure 20 [108].

13.5.5 ^{13}CO on Ru-Promoted Pt-Black Electrodes

Actual fuel-cell electrodes, like actual heterogeneous catalysts, are rarely composed of a single metal. An interesting promoting effect in direct methanol oxidation is provided by the decoration of platinum electrocatalysts by ruthenium. The decoration consists essentially of monolayered islands of Ru on the Pt particles. It is generally assumed that under working cell conditions, the Ru islands are mainly populated by OH groups, and the exposed Pt surface by CO [109,110]. Oxidation occurs mainly in the Pt/Ru transition regions, and the process is controlled by diffusion of CO on the exposed Pt surface. A series of four samples with overall surface compositions $Pt_{1-y}Ru_y$, $y = 0, 0.14, 0.35$, and 0.52 has been studied by cyclic voltammetry, potentiostatic current generation, and ex-situ ^{13}CO NMR [111]. The NMR samples were prepared by prolonged CO adsorption from methanol decomposition; their total CO coverage was determined by voltammetry as ≈ 0.9, indicating that under these conditions the Ru islands are covered with CO as well.

This conclusion from voltammetry was confirmed by NMR. As the Ru concentration y increases, a high-field shoulder appears on the ^{13}CO NMR line [111]. The signal can be deconvoluted into two Gaussians with distinct spin-lattice relaxation behavior. The intensity of the two signals correlates well with the fractions of CO on Ru and on pure-Pt sites, as obtained from electrochemical stripping, Figure 24. It is rather amazing that on the NMR timescale (microseconds to milliseconds) there is no exchange of CO between the two types of metal surface.

The resonance position of ^{13}CO/Ru and its relaxation rate vary clearly with Ru content y, whereas the changes for the ^{13}CO/Pt signal are smaller. The ^{13}CO/Ru relaxation rates show around room temperature a contribution from diffusional motion, which is not seen for ^{13}CO/Pt. It seems that, because of the overall high coverage and the "slow" diffusion on the Pt surface, the "fast" diffusing CO on Ru cannot spill over onto the Pt, thereby maintaining two distinct populations. This relatively low mobility of CO/Pt might explain how under working cell conditions hardly any CO is left on the Ru islands, while the current is limited by the diffusion on Pt.

13.6   CONCLUSION

We have argued in this chapter that the E_f-LDOS (on which the nuclear spin-lattice relaxation rate of metals depends) is a useful concept to discuss variations in surface reactivity, bonding, and electrode potential effects among a series of related catalysts, in heterogeneous as well as in electrochemical catalysis. The E_f-LDOS is a

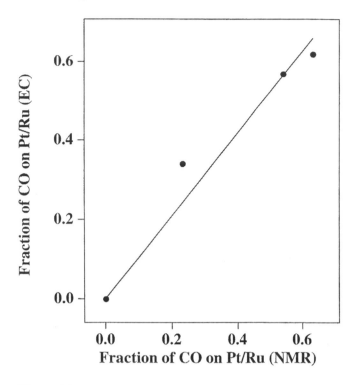

Figure 24 Correlation between the fractions of CO on the Ru/Pt sites as obtained from electrochemical CO stripping and from deconvolution of the NMR signal.

physical quantity that can be measured by metal NMR on the one hand and calculated by modern density functional methodologies on the other hand, therefore offering an ideal playground for closely correlating the experimental and theoretical efforts.

Phenomenologically, we have seen that Blyholder's ideas for CO chemisorption can also be discussed in terms of E_f-LDOS on the CO. When, in a series of closely related catalysts, changes in IR stretch frequency indicate an increase (decrease) in the occupation of the π-like levels, the ^{13}CO NMR sees an increase (decrease) in the π-like E_f-LDOS. We have indicated that this correlation cannot be explained by a simple rigid-band argument, and it would be very interesting if a generally valid theoretical argument could be found.

At a larger scale of diversity among catalysts, the metal-to-metal variation of reactivity can be theoretically understood from the position of the Fermi energy in the d-band [15, p. 681]. Just as the surface metal-to-metal bonds are different (bonding, nonbonding, antibonding) in different energy regions across the band, so the interactions between the molecule and the surface vary as well [27]. What will actually happen is therefore strongly dependent on which d-levels are empty and which are filled (for a pictorial representation see, e.g., Section XII of [14]). It has been shown by Nørskov and co-workers that a very convenient single-parameter summary of these effects is given by the energy difference between the center of the d-band and the Fermi energy [112].

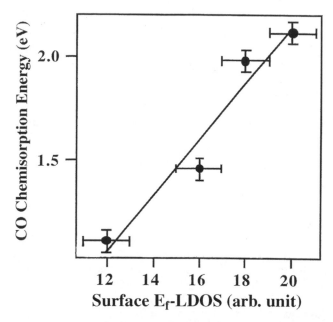

Figure 25 Correlation between the Pt surface E_f-LDOS before CO adsorption and the CO chemisorption energy for surfaces with different openness or orientation.

We should not expect, of course, that the variation of any single attribute of a clean surface, be it the d-band center, the E_f-LDOS, the workfunction, etc., will in all circumstances correctly predict the variation in its chemisorption properties, but we may find that some of these numbers are more useful for the purpose than others. It is instructive to point out that for transition-metal surface whose Fermi energy cuts the rising tail of the d-band the E_f-LDOS is expected to correlate with the d-band center; the lower the d-band center, the smaller the corresponding E_f-LDOS. Indeed, based on the results of calculations in [113], a correlation can be found between the surface E_f-LDOS (before CO adsorption) of Pt surfaces with different openness (i.e., different orientation) and the CO adsorption energy on these surfaces, of which the latter is conrelated to the d-band center; see Figure 25. It would be very beneficial to the field of catalysis if a more general relationship between these two quantities could be established.

On the experimental side, we expect NMR to remain an important tool in the analysis of the nanoscale metallic materials widely used in heterogeneous and electrocatalysis. Its unique molecular/electronic information will complement that from other techniques such as IR, x-ray, or STM, etc. This is particularly true in the field of interfacial electrochemistry, where many electron-based spectroscopies are technically inapplicable.

ACKNOWLEDGMENTS

Support by the Georgetown University Startup fund (to YYT) and by the Swiss National Science Foundation (grant no. 20–53637.98 to vdK) is gratefully acknowledged.

REFERENCES

1. A.T. Bell, A. Pines, eds. NMR Techniques in Catalysis, Marcel Dekker Inc., New York, (1994).
2. K.J. Packer, Topics in Catal. **3**, 249 (1996).
3. J.J. Wu, J.B. Day, K. Franaszczuk, B. Montez, E. Oldfield, A. Wieckowski, P.A. Vuissoz, J.P. Ansermet, J. Chem. Soc., Faraday Trans. **93**, 1017 (1997).
4. Y.Y. Tong, E. Oldfield, A. Wieckowski, Anal. Chem. **70**, 518A (1998).
5. H.E. Rhodes, P.-K. Wang, H.T. Stokes, C.P. Slichter, J.H. Sinfelt, Phys. Rev. B **26**, 3559 (1982).
6. J.J. van der Klink, J. Buttet, M. Graetzel, Phys. Rev. B **29**, 6352 (1984).
7. Y.Y. Tong, C. Rice, N. Godbout, A. Wieckowski, E. Oldfield, J. Am. Chem. Soc. **121**, 2996 (1999).
8. M. Weinert, A.J. Freeman, Phys. Rev. B **28**, 6262 (1983).
9. T.M. Duncan, Colloids Surf. **45**, 11 (1990).
10. G. Blyholder, J. Phys. Chem. **68**, 2772 (1964).
11. S.L. Rudaz, J.P. Ansermet, P.-K. Wang, C.P. Slichter, J.H. Sinfelt, Phys. Rev. Lett. **54**, 71 (1985).
12. Y.Y. Tong, P. Meriaudeau, A.J. Renouprez, J.J. van der Klink, J. Phys. Chem. B **101**, 10155 (1997).
13. K. Fukui, T. Yonezawa, H. Shingu, J. Chem. Phys. **20**, 722 (1952).
14. R. Hoffmann, Rev. Mod. Phys. **60**, 601 (1988).
15. R.A. van Santen, Theoretical Heterogeneous Catalysis, World Scientific, Singapore (1991).
16. C.P. Slichter, Ann. Rev. Phys. Chem. **37**, 25 (1986).
17. J.P. Ansermet, C.P. Slichter, J.H. Sinfelt, Progr. NMR Spectroscopy **22**, 401 (1990).
18. J.J. van der Klink, Adv. Cat. **44**, 1 (2000).
19. S. Lundqvist, N.H. March, Theory of the Inhomogeneous Electron Gas, Plenum, New York (1983).
20. A.R. Williams, U. von Barth, In ref. [19].
21. N.D. Lang, In ref. [19].
22. N.W. Ashcroft, N.D. Mermin, Solid State Physics, Holt Rinehart and Winston, New York (1976).
23. P.W. Atkins, Physical Chemistry, Oxford, Univ. Press, New York (1990).
24. W. Koch, M.C. Holthausen, A Chemist's Guide to Density Functional Theory, Wiley-VCH, New York (2000).
25. F.M. Bickelhaupt, E.J. Baerends, Rev. Comput. Chem. **15**, 1 (2000).
26. R. Stowasser, R. Hoffmann, J. Am. Chem. Soc. **121**, 3414 (1999).
27. P. Hu, D.A. King, M.-H. Lee, M.C. Payne, Chem. Phys. Lett. **246**, 73 (1995).
28. L. Pauling, Nature of the Chemical Bond, Cornell Univ. Press, Ithaca, NY (1945).
29. J.A.A. Ketelaar, Liaisons et Propriétés Chimiques, Dunod, Paris (1960); Chemical Constitution, Elsevier, Amsterdam (1958).
30. W.P. Anderson, J.K. Burnett, P.T. Czech. J. Am. Chem. Soc. **116**, 8808 (1994).
31. L.C. Allen, J.F. Capitani, J. Am. Chem. Soc. **116**, 8810 (1994).
32. P. Coppens, X-Ray Charge Densities and Chemical Bonding, Oxford Univ. Press, New York (1997).
33. M. Roux, R. Daudel, Compt. Rend. Acad. Sci. **240**, 90 (1955).
34. R.F.W. Bader, Atoms in Molecules, Oxford Univ. Press, New York (1990).
35. J.E. Lennard-Jones, J. Chem. Phys. **20**, 1024 (1952).
36. A.D. Becke, K.E.J. Edgecombe, J. Chem. Phys. **92**, 5397 (1990).
37. B. Silvi, C. Gatti, J. Phys. Chem. A **104**, 947 (2000).
38. N.D. Lang, A.R. Williams, Phys. Rev. B **18**, 616 (1978).

39. J. Bormet, J. Neugebauer, M. Scheffler, Phys. Rev. B **49**, 17242 (1994).

40. N. Bonini, M.I. Trioni, G.P. Brivio, J. Chem. Phys. **113**, 5624 (2000).

41. R.D. Diehl, R. McGrath, J. Phys.: Condens. Matter **9**, 951 (1997).

42. H. Ishida, Phys. Rev. B **38**, 8006 (1988).

43. D.M. Riffe, G.K. Wertheim, P.H. Citrin, Phys. Rev. Lett. **64**, 571 (1990).

44. B. Hammer, Y. Morikawa, J.K. Nørskov, Phys. Rev. Lett. **76**, 2141 (1996).

45. A.K. Gupta, K.S. Singwi, Phys. Rev. B **15**, 1801 (1977).

46. P.J. Feibelman, D.R. Hamann, Surf. Sci. **149**, 48 (1985).

47. C. Werner, F.K. Schulte, H. Bross, J. Phys. C **8**, 3817 (1975).

48. P.K. Johansson, Surf. Sci. **104**, 510 (1981).

49. S. Sung, R. Hoffmann, J. Am. Chem. Soc. **107**, 578 (1985).

50. G.P. Brivio, M.I. Trioni, Rev. Mod. Phys. **71**, 231 (1999).

51. V.L. Moruzzi, C.B. Sommers, Calculated Electronic Properties of Ordered Alloys, World Scientific, Singapore (1995).

52. V. Ponec, G.C. Bond, Catalysis by Metals and Alloys, Elsevier, Amsterdam (1995).

53. F. Besenbacher, Rep. Prog. Phys. **59**, 1737 (1996).

54. Y.Y. Tong, T. Yonezawa, N. Toshima, J.J. van der Klink, J. Phys. Chem. **100**, 730 (1996).

55. Y.Y. Tong, A.J. Renouprez, G.A. Martin, J.J. van der Klink, Studies in Surf. Sci. and Catal. **101**, 901 (1996).

56. S.H. Vosko, J.P. Perdew, Can. J. Phys. **53**, 1385 (1975).

57. O. Gunnarson, J. Phys. F: Met. Phys. **6**, 587 (1976).

58. J.J. van der Klink, H.B. Brom, Prog. NMR Spectrosc. **36**, 89 (2000).

59. W. Heitler, E. Teller, Proc. Roy. Soc. A **155**, 629 (1936).

60. J. Korringa, Physica **16**, 601 (1950).

61. Y. Obata, J. Phys. Soc. Jpn. **18**, 1020 (1963).

62. T. Moriya, J. Phys. Soc. Jpn. **18**, 516 (1963).

63. A. Narath, H.T. Weaver, Phys. Rev. **175**, 373 (1968).

64. R.W. Shaw, Jr., W.W. Warren, Jr., Phys. Rev. B **3**, 1562 (1971).

65. P. Bhattacharyya, K.N. Pathak, K.S. Singwi, Phys. Rev. B **3**, 1568 (1971).

66. F.Y. Fradin, D.D. Koelling, A.J. Freeman, T.J. Watson-Yang, Phys. Rev. B **12**, 5570 (1975).

67. M. Shaham, U. El-Hanany, D. Zamir, Phys. Rev. B **17**, 3513 (1978).

68. L.M. Noskova, E.V. Rozenfeld, Y.P. Irkhin, Sov. Phys. Solid State **26**, 1686 (1984).

69. J.P. Bucher, J.J. van der Klink, Phys. Rev. B **38**, 11038 (1988).

70. V.P. Tarasov, Y.B. Muravlev, K.E. German, N.N. Popova, Doklady Physical Chem. **377**, 71 (2001).

71. J.J. Bercier, M. Jirousek, M. Graetzel, J.J. van der Klink, J. Phys.: Condens. Matter **5**, L571 (1993).

72. J.P. Bucher, J. Butter, J.J. van der Klink, M. Graetzel, E. Newton, T.B. Truong, J. Mol. Catal. **43**, 213 (1987).

73. C.D. Makowka, C.P. Slichter, J.H. Sinfelt, Phys. Rev. B **31**, 5663 (1985).

74. J.P. Bucher, J. Buttet, J.J. van der Klink, M. Graetzel, Surf. Sci. **214**, 347 (1989).

75. N. Toshima, T. Yonezawa, K. Kushihashi, J. Chem. Soc., Faraday Trans. **89**, 2537 (1993).

76. S. Ladas, R.A. Dalla Betta, M. Boudart, J. Catal. **53**, 356 (1978).

77. D.A. van Leeuwen, J.M. van Ruitenbeek, G. Schmid, L.J. de Jongh, Phys. Lett. A **170**, 325 (1992).

78. S. Burnet, Ph.D. thesis no. 2333, EPFL Switzerland (2001).

79. P.A. Vuissoz, T. Yonezawa, D. Yang, J. Kiwi, J.J. van der Klink, Chem. Phys. Lett. **264**, 366 (1997).

80. A. Eichler, J. Hafner, J. Furthmüller, G. Kresse, Surf. Sci **346**, 300 (1996).

81. S.G. Louie, Phys. Rev. Lett. **42**, 476 (1979).
82. R. Löber, D. Hennig, Phys. Rev. B **55**, 4761 (1997).
83. Y.Y. Tong, J.J. Billy, A.J. Renouprez, J.J. van der Klink, J. Am. Chem. Soc. **119**, 3929 (1997).
84. S.E. Shore, J.P. Ansermet, C.P. Slichter, J.H. Sinfelt, Phys. Rev. Lett. **58**, 953 (1987).
85. D.A. Papaconstantopoulos, Handbook of the Bandstructure of Elemental Solids, Plenum Press, New York, 1986.
86. W. Mannstadt, A.J. Freeman, Phys. Rev. B **55**, 13298 (1997).
87. M.W. Ruckman, M. Strongin, Acc. Chem. Res. **27**, 250 (1994).
88. S. Wilke, M. Scheffler, Phys. Rev. Lett. **76**, 3380 (1996).
89. C.J. Zhang, P. Hu, M.-H. Lee, Surf. Sci. **432**, 305 (1999).
90. J.A. Rodriguez, D.W. Goodman, Acc. Chem. Res. **28**, 477 (1995).
91. H.D. Ebinger, H.J. Jansch, C. Polenz, B. Polivka, W. Preiss, V. Saier, R. Veith, D. Fick, Phys. Rev. Lett. **76**, 656 (1996).
92. Y.Y. Tong, J.J. van der Klink, J. Phys. Chem. **98**, 11011 (1994).
93. Y.Y. Tong, J.J. van der Klink, J. Phys.: Condens. Matter **7**, 2447 (1995).
94. M.A. Chesters, K.J. Packer, H.E. Viner, M.A.P. Wright, J. Phys. Chem. B **101**, 9995 (1997).
95. Z. Wang, J.P. Ansermet, C.P. Slichter, J.H. Sinfelt, J. Chem. Soc. Faraday Trans I **84**, 3785 (1988).
96. J.J. van der Klink, J.P. Ansermet, J. Catal. **170**, 211 (1997).
97. L.R. Becerra, C.P. Slichter, J.H. Sinfelt, J. Phys. Chem. **97**, 10 (1993).
98. J.S. Bradley, J.M. Millar, E.W. Hill, S. Behal, J. Catal. **129**, 530 (1991).
99. M.S. Yahnke, B.M. Rush, J.A. Reymer, E.J. Cairns, J. Am. Chem. Soc. **118**, 12250 (1996).
100. Y.Y. Tong, C. Rice, A. Wieckowski, E. Oldfield, J. Am. Chem. Soc. **122**, 1123 (2000).
101. J.B. Day, P.A. Vuissoz, E. Oldfield, A. Wieckowski, J.P. Ansermet, J. Am. Chem. Soc. **118**, 13046 (1996).
102. P.A. Vuissoz, J.P. Ansermet, A. Wieckowski, Phys. Rev. Lett. **83**, 2457 (1999).
103. C.W. Hilbers, C. MacLean, NMR Princ. Progr. **7**, 1 (1972).
104. J.D. Augspurger, C.E. Dykstra, E. Oldfield, J. Am. Chem. Soc. **113**, 2447 (1991).
105. A.C. de Dios, E.M. Earle, J. Phys. Chem. A **101**, 8132 (1997).
106. D.K. Lambert, Electrochim. Acta **41**, 623 (1996).
107. N.D. Lang, W. Kohn, Phys. Rev. B **7**, 3541 (1973).
108. C. Rice, Y.Y. Tong, E. Oldfield, A. Wieckowski, F. Hahn, F. Gloaguen, J.M. Léger, C. Lamy, J. Phys. Chem. B **104**, 5803 (2000).
109. A. Hamnett, In Interfacial Electrochemistry, A. Wieckowski ed., Marcel Dekker, New York (1999).
110. M.T.M. Koper, J.J. Lukkien, A.P.J. Jansen, R.A. van Santen, J. Phys. Chem B **103** (1999).
111. Y.Y. Tong, H.S. Kim, P.K. Babu, P. Waszczuk, A. Wieckowski, E. Oldfield, J. Am. Chem. Soc. 124, 468 (2002).
112. B. Hammer, J.K. Nørskov, Adv. Catal. **45**, 71 (2000).
113. B. Hammer, O.H. Nielsen, J.K. Nørskov, Catal. Lett. **46**, 31 (1997).

14

In-Situ X-Ray Absorption Spectroscopy of Carbon-Supported Pt and Pt-Alloy Electrocatalysts: Correlation of Electrocatalytic Activity with Particle Size and Alloying

SANJEEV MUKERJEE

Northeastern University, Boston, Massachusetts, U.S.A.

CHAPTER CONTENTS

SUMMARY

Carbon supported Pt and Pt-alloy electrocatalysts form the cornerstone of the current state-of-the-art electrocatalysts for medium and low temperature fuel cells such as phosphoric and proton exchange membrane fuel cells (PEMFCs). Electrocatalysis on these nanophase clusters are very different from bulk materials due to unique short-range atomic order and the electronic environment of these cluster interfaces. Studies of these fundamental properties, especially in the context of alloy formation and particle size are, therefore, of great interest. This chapter provides an overview of the structure and electronic nature of these supported nanoclusters from the perspective of surface as well as bulk characteristics as studied using synchrotron based X-ray absorption spectroscopy (XAS). XAS with its two complimentary spectral regions provides chemical information and the corresponding changes in the short range atomic order. With its in-situ capability and element specificity, this technique has provided some very important insights into the role of cluster composition and size of electrocatalysis. These are discussed in the context of the cathodic oxygen reduction reaction (ORR). Variations in activity for ORR with particle size are treated in the context of variation of electronic and short-range atomic order with electrode potential. An alternative viewpoint, ascribing these activity variations to interparticle diffusive interference, is also discussed. Models are used to correlate the surface morphology changes with particle size. Changes in the particle size from $10\,\text{Å}$ to $80\,\text{Å}$ results in tremendous variation in the population of sites with small coordination numbers (6–9) and, thus, in the corresponding electronic characteristics of the surface. These have been shown to correlate well with observed variation in activity with particle size. Attempts to enhance ORR activity by modification of the surface electronic and short-range atomic order was also attempted as a consequence of alloying Pt with base transition elements. Recent studies have shown that alloying results in a shift in the onset of potential of water activation and, thus, formation of surface Pt-OH, which acts as a poison for adsorption of molecular oxygen. This has been shown to enable a lowering of ORR overpotential losses by approximately $50\,\text{mVs}$ with some Pt alloys. In this endeavor, however, the surface nature of the cluster is very important and must be predominantly Pt with the inner core as the alloy. Thus, the methodologies of preparation of these metal clusters is of prime importance.

14.1 INTRODUCTION

14.1.1 Highly Dispersed Carbon-Supported Pt Electrocatalysts: Need and Evolution

The development of supported Pt electrocatalysts came as a result of intensive research on fundamental and applied aspects of electrocatalysis [especially for kinetically difficult oxygen reduction reaction (ORR)] fueled by attempts at commercialization of medium-temperature phosphoric acid fuel cells (PAFCs) in the late 1960s and early 1970s. The prime motivator was the lowering of the metal loading achieved by increased surface area as a consequence of dispersion of the metal clusters. Dispersion of metal crystallites in a conductive carbon support resulted, however, in significant improvements in all three polarization zones (activation, ohmic, and mass transport). The lowering of ohmic and mass transport overpotential losses was a natural function of the conductive carbon substrate and the extension of the electrode–electrolyte interface further into the electrode structure, thereby increasing the catalyst, and reactant utilization. In the case of PAFCs, this was achieved by varying the hydrophobicity of the catalyst-containing reaction layer on the electrode. In the case of PEMFCs, this is a more complicated function of ink formulations for creating the reaction layer, composed of the catalyst, binder, and solubilized ionomers.

It is the improvements in the activation overpotentials that, however, have been most difficult to comprehend and quantify. The underlying nature of the electrocatalysis has been the focus of electrochemists and spectroscopists over the last three decades. The focus of this chapter therefore resides primarily in critically examining the current state of our understanding of the physics and chemistry of the electrocatalysis at these supported catalyst interfaces. Most of the discussion will use the oxygen reduction reaction (ORR) as a case scenario for examining the electrocatalysis from the perspective of the two principal fundamental parameters, the electronic state and short-range atomic order (bond distance and closed-shell coordination numbers). The critical issue is to understand these two parameters as a function of cluster size, environment (such as support synergy), and alloying Pt with other transition elements.

14.1.2 Synchrotron-based X-ray and IR Techniques as an In-Situ Probe of the Electrochemical Interface

The prospect of probing an electrochemical interface in situ, under actual operating conditions, and the ability to map both the substrate (in terms of its electronic and short-range atomic order) as well as the substrate–adsorbate interactions have tremendous technological implications. Synchrotron-based far infrared and x-ray techniques (x-ray absorption, and scattering) offer such an opportunity as a consequence of the unique characteristics of the synchrotron source. These include higher intensity (10^4 higher), collimation, polarization, and pulse time structure enabling true in-situ interfacial measurements. While far infrared spectroscopy is an emerging technique, x-ray absorption and scattering have recently evolved [1] as a true in-situ probe for electrochemical interface with both model and commercially relevant nano-dispersed materials. Among these techniques, x-ray absorption spectroscopy is the most mature and relevant from the perspective of supported nanocluster electrocatalysts. While x-ray scattering has been used to study electrochemical interfaces, it use is limited to investigations of adsorbed species on well-defined surfaces and hence outside the scope of this chapter. Far infrared spectroscopy is yet to emerge, and developments in optics and detectors are being pursued for enabling this as a mature technique.

The primary advantages of in-situ XAS is the ability to study changes to both the electronic (such as the Pt *d-orbital* vacancy), using near-edge region, XANES and short-range atomic order (first shell bond distance and coordination numbers) using extended fine structure region (EXAFS). Further, electrochemical interfaces can be probed in situ, with element specificity.

14.2 NANO-PHASE METAL REACTION CENTERS SUPPORTED ON CARBON: A UNIQUE ELECTROCATALYTIC ENVIRONMENT

14.2.1 Need for Higher Surface Areas in the Context of ORR Electrocatalysis

The major deficiency of the oxygen electrode reaction is its low exchange current density (about 10^{-9} A/cm^2 on a smooth surface) in acid electrolytes on even the best-known electrocatalyst (a platinum-chromium alloy). This value is about six orders of magnitude lower than that for the hydrogen electrode reaction in the same electrolyte. The reaction is about three orders of magnitude faster on smooth platinum or nickel oxide surfaces in an alkaline medium as compared to acid. The considerably lower rate in an acidic medium is because of anion adsorption, which blocks sites for the adsorption of oxygen on the electrocatalyst surface. The oxygen reduction is considerably slower in phosphoric acid than in a perfluorosulfonic acid medium because of the considerably lower anion adsorption and higher oxygen solubility in the latter medium. With such low values of the exchange current density, there is a vital need for use of high-surface-area electrocatalysts (high roughness factors) to enable the attainment of higher current densities, based on the geometric area of the electrode.

As stated in the preceding subsection, ultrahigh roughness factors of the electrodes is necessary to minimize activation overpotentials for the fuel-cell reactions in acid or alkali electrolytes. This is considerably more so for the oxygen reduction than the hydrogen oxidation reaction. In order to achieve this goal, high-surface-area nanocrystallites of the electrocatalysts are essential. For this purpose, in the early 1960s, unsupported platinum or platinum alloy electrocatalysts were used in the active layer. The diameter of these particles was of the order of 100 to 200 Å. There were two disadvantages of this approach—one was that in order to reduce the overpotentials to low enough values, high loading of the noble metal (at least $4\,mg/cm^2$ or higher) was necessary. The second was that due to particle–particle interaction, agglomeration of the particles occurred, reducing the effective surface area and causing a loss of electrocatalytic activity. The revolutionary step, which was taken in PAFC, R&D to overcome these problems, was to develop supported electrocatalysts, as was previously done in gas-phase heterogeneous catalysis [2]. Work conducted in the 1970s showed that it was possible to increase the surface area of the electrocatalysts from $20\,m^2/gm$ to $100\,m^2/gm$ by supporting the Pt nanocrystallites on a conducting matrix such as graphite or turbostratic carbon blacks (furnace blacks). For this purpose, a serious effort was made to identify carbon blacks with high surface areas and corrosion resistance. Most of the early work in this endeavor was conducted under the TARGET program in the United States at places such as UTC, Energy Research Corporation, Engelhard, etc. (see [3] for a review of these early efforts in the 1970s and 1980s). Using this approach, the particle sizes were reduced to about 20 to 50 Å and the platinum loading reduced to $0.4\,mg/cm^2$, in PAFC electrodes. Further, because the electrocatalyst particles were supported on electronically conducting high-surface-area carbon supports (e.g., Vulcan-XC72R.a furnace black), the agglomeration of the particles was also greatly hindered but not eliminated.

14.2.2 Issue of Particle Size

Assuming that the nanocrystals are spherical and have uniform size, the surface area (or roughness factor) may be expressed by

$$S = 6/\rho d \tag{1}$$

where ρ is the density of the particles and d is the diameter of the particles. It is clear from this equation that the smaller the particle diameter, the greater the value of S. In principle, platinum crystallites with a diameter of 1.4 to 1.5 nm can be prepared on carbon support, resulting in a BET surface area of about $180–200\,m^2/g$. However, in general, the diameter of the platinum crystallites on high-surface-area carbon support is about 3–5 nm, corresponding to a BET surface area of $60–100\,m^2/g$. The effect of particle size on the electrocatalytic activity, as defined by its turnover number (activity per surface metal atom per unit time), has been of interest to both the heterogeneous and electrocatalytic communities. In electrocatalysis this is especially important when dealing with kinetically difficult multielectron processes such as the oxygen reduction and direct methanol oxidation. In terms of geometric considerations, conventional carbon-supported electrocatalysts with crystallites in the range of 10–70 Å show a large variation in terms of the number of surface to bulk atoms. These crystallites can therefore have surface sites with coordination numbers well below the bulk coordination of 12 (based on an fcc lattice, such as those for Pt bulk). A detailed description of geometric characteristics as a function of particle size has been reviewed by Kinoshita [4,5] and Watanabe [6]. For these analyses, spherical cluster models, such as the cubo-octahedron, were used. Analyses by these investigators was based on earlier work by Romanowski [7], who showed that there is an exponential decrease in the low coordination sites (6 to 9, based on edge and kink sites) as a function of increase in particle size in the range 10 to 90 Å. In addition, the bulk coordination sites such as $\langle 111 \rangle$ achieve steady-state maxima with an increase of particle size within the same range.

Correlation of the effect of particle size on the oxygen reduction reaction has been reviewed extensively by Kinoshita [4,5], Stonehart [8], and Mukerjee [9]. The general consensus, based on a large number of steady-state polarization measurements in several electrolytes, is that ORR exhibits a strong particle size effect in the range of 10 to 50 Å cluster sizes. The maximum in ORR activity as a function of particle size, based on previous reviews by Kinoshita [4,5], is in the range of 30–40 Å. Figure 1 (taken from a review of this effect on ORR by Kinoshita [5]) shows the specific activity of oxygen reduction as a function of particle size. Also superimposed is the surface-averaged distribution of $\langle 100 \rangle$ sites. The remarkable correlation clearly demonstrates the influence of surface sites with different coordination numbers in determining ORR activity. The attempts to understand the true nature of these effects are reviewed later in this chapter (see Section 14.4).

14.2.3 Nanophase Electrocatalysts: Surface Structure of Small Particles

Well-defined geometric models simulating the relationship between particle size and population of crystallographic surface sites associated with atoms at edges, kinks, and vertices have been used to understand the particle-size effect. However, an accurate representation of the structure of small catalyst particles using a well-defined geometric model is debatable. Using concepts of localized metal bonds, Romanowski [7] suggests a cubo-octahedron structure as a possible model where surface energy is minimized. The basic cubo-octahedron consists of eight octahedral $\langle 111 \rangle$ and six $\langle 100 \rangle$ crystal faces, bound by edge and corner atoms. Romanowski [7] has analyzed the relationship between the particle size and the statistics of various crystallographic sites such as the $\langle 111 \rangle$, $\langle 100 \rangle$, and the low coordination sites associated with edge, kink, and corner sites. In these models the particle diameter has been defined as

$$ d = \frac{6(d_{\mathrm{at}}\sqrt{2})^3}{\pi 4} N(t)^{1/3} = 1.11 d_{\mathrm{at}} \sqrt[3]{N(t)} \tag{2} $$

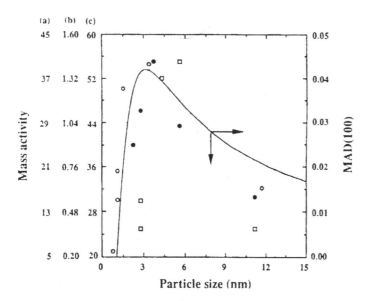

Figure 1   Correlation of specific activity for oxygen reduction reaction with particle size measured as steady-state polarization values with different electrolytes. (a) 98% H_3PO_4, 180 °C, (b) 0.5 M H_2SO_4, 25 °C, and (c) 97% H_3PO_4, 177 °C. Also superimposed is the surface-averaged distribution (SAD) of $\langle 100 \rangle$ sites (solid line). (From Ref. 4.)

where d_{at} is the diameter of a Pt atom (0.276 nm) and $N(t)$ is the total number of Pt atoms in the particle. Kinoshita [4,5] has used these models to map the mass-averaged and surface-averaged distributions (MAD and SAD, respectively) of various crystallite planes such as $\langle 111 \rangle$, $\langle 100 \rangle$; for edge and corner sites, these are shown in Figure 2. Recently Benfield [10] has suggested algorithms that are based on the use of concentric shells of atoms to formulate models corresponding to cubo-octahedron or the closely related icosahedron clusters. This model allows calculation of the distribution of various sites with different coordination numbers as well as crystallographic planes for both cubo-octahedron and icosahedron models. In addition, parameters such as the average coordination number and surface-to-bulk atomic ratio can be calculated. Figure 3 shows the distribution of average coordination number, ratio of surface to bulk sites, and the variation of the number of interstitial sites with a coordination number of 12.

In light of the different preparative techniques, however, it is unlikely that the structure of the highly dispersed Pt would correspond exactly to such geometric models. At this juncture it is essential to appreciate that the measurement techniques for estimating the particle size are many and varied, each with their associated pros and cons. Among the most widely used techniques are chemisorption with CO and H_2, either via adsorbing and then measuring the desorbed gas or via progressive surface titration of pre-adsorbed oxygen on Pt with hydrogen at room temperature. Cautious handling is essential when applying these methods in order to avoid overestimation of surface area due to spillover effects and variations of stoichiometry with particle size. Figure 4 shows the ratio of surface atoms obtained

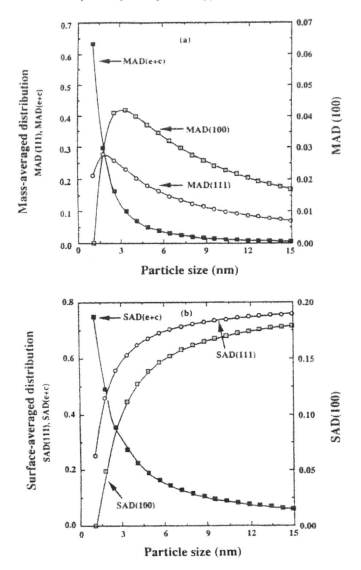

Figure 2 The (a) mass- and (b) surface-averaged distribution of atoms on the ⟨111⟩ and ⟨100⟩ crystal faces and on the edges and corner sites of a cubo-octahedral cluster model. (From Ref. 4.) Mass-averaged and surface-averaged distributions are based on calculations using cubo-octahedron cluster model and represent number of different crystallographic planes divided by the (a) mass or (b) the surface area of the cluster (at the corresponding particle size). Hence (e+c) in (a) represents edge and kink positions and (100) & (111) the normal cubic crystal planes.

from chemisorption of hydrogen (assuming one hydrogen atoms H_s chemisorbed per surface metal atom) to the total number of atoms on the surface (N_t). It is evident that for the ratio H_s/N_t to be 0.5, the Pt surface area must be greater than $140 \, m^2/g$. This corresponds to a particle size of 2.5 nm. Extrapolating further, a particle size of

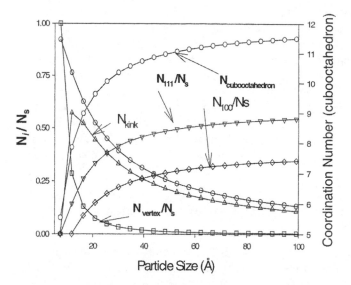

Figure 3 A plot of variation of ratio of the number of sites with different coordination number with total surface atoms, and overall coordination number ($N_{cubo-octahedron}$) as a function of particle size. The calculations are based on algorithms developed previously by Benfield (Ref. 35) based on cubo-octahedron cluster model. Therefore, N_{vertex}/N_s represents the ratio of the number of sites at the vertex position in the cluster to the total number of surface sites. The other ratios based on different cluster positions [such as Kink sites, (100) and (111)] are similarly marked.

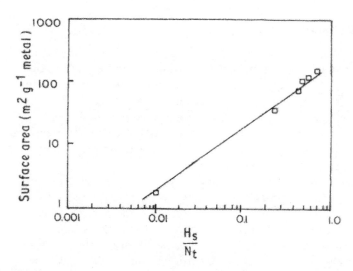

Figure 4 Relationship between surface area and the fraction of the surface atoms on the particles of noble metals determined by hydrogen chemisorption for Pt metals. H_s/N_t represents the ratio of surface area obtained from hydrogen chemisorption to the total number of atoms in the cluster of a particular cluster size. (From Ref. 9.)

less than 1 nm has a surface-to-bulk ratio of 1 ($H_s/N_t = 1$). This particle size, however, approaches a quantum dot, where the metallic character of the crystallite is very different from the bulk. Electrochemical methods, on the other hand, have problems associated with surface rearrangements and roughening due to potential cycling. In addition, the electrode must be designed to maximize electrocatalyst utilization (preferably in a flooded state). X-ray linebroadening analysis, which provides good data on particle size, has limited applicability to very small particles because in such cases limited long-range order exists. This causes the line broadening to be large and with significant contributions from the background. TEM analysis provides a good measure of particle size; however, care should be taken to obtain a good statistical average.

14.2.4 Microstructure of Supported Catalysts and Interaction with Support Material

Electronic interaction and synergistic effects between catalysts and the support material have been investigated in the context of fuel-cell electrocatalysts. Electron spin resonance (ESR) has been used to demonstrate the electron donation by Pt to carbon [11] support. This has been further supported by XPS studies [12], which show that the metal acts as an electron donor to the support, their interaction depending on their respective Fermi levels. Bogotsky and Snudkin [13] have shown that the characteristics of the electrical double layer formed between the microdeposit (Pt) and the support depends to a certain extent on the difference in the work function of Pt (5.4 eV) and carbon support (pyrolytic support 4.7 eV), thereby resulting in an increase of the electron density of Pt. However, the rise in the electron density can be significant only when the particle size of the microdeposit is comparable to the thickness of the double layer.

14.2.5 Alloy Electrocatalysts

As mentioned in the previous subsections, the principal cause of the poor kinetics of the four-electron oxygen reduction reaction can be attributed mainly to the low-exchange current density of the oxygen reduction reaction. The high cathodic overpotential loss of ~ 220 mV, even at open circuit potentials with the current state-of-the-art low Pt loading electrocatalysts, is attributed to a mixed potential that is set up at the oxygen electrode. Since the activation overpotential stretches over three orders of magnitude in current density, the problem of significant improvement in O_2-reduction activity is primarily electrocatalytic.

This problem was tackled in the late 1970s and early 1980s by alloying Pt with other transition elements, earliest reports being from United Technologies (now IFC) where refractory metal combinations with Pt were considered for ORR activity enhancement. The first choice reported was with vanadium [14]; subsequently a multitude of other elements were considered (see a review of this early effort by Stonehart [15]). The prime chronology of development in the early 1980s was based on the attempts to enhance the activity of ORR in an PAFC fuel-cell environment. With concerns of stability of supported PtV alloy electrocatalysts (67 w/o of the vanadium was reported to be dissolved at 0.9 V, 194 °C in 90% H_3PO_4), it was immediately followed by PtCr [16], which showed significantly higher stability. Later

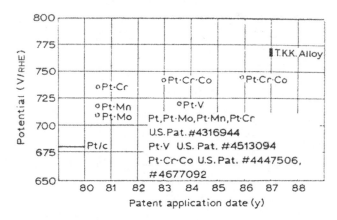

Figure 5 Comparison of oxygen reduction activity at $200\,mA/cm^2$ from patent literature, showing chronology of initial developments in Pt alloy electrocatalysts for PAFC's technology. Electrodes contain $0.5\,mg/cm^2$ Pt at $190\,°C$ in H_3PO_4. KKK is Tanaka Kikinzoku Kogyo Company (K.K.K., Japan). (From Ref. 15.)

addition of Co to form ternary alloys was patented [17], which allowed further stabilization of the alloy and boosted the ORR activity. Figure 5 shows the chronology of these early developments in the PAFC technology development as cited by Stonehart [15]. A review of patent literature after these early developments has shown a wide variety of combinations of elements; some of these alloy combinations include PtGa, PtFe, PtCu, PtFeCo, PtCoGa, etc. (see review by Stonehart [8,15] for details of patents). These developments led some commercial manufacturers to choose ternary alloys as the cathode electrocatalyst such as IFC (PtCrCo/C ternary).

Accompanying the initial developments in the alloy electrocatalysts were investigations on the nature of particle-size sensitivity on the alloys as compared to highly dispersed Pt. Buchanan et al. [18] studied the effect of alloying 3d-transition elements with Pt and reported an average 1.5 times enhancement over Pt alone. Of greater importance was their finding of greater stability toward degradation via sintering and dissolution of the more oxidizable alloying component. In addition, they also reported a similar particle-size effect as seen previously with Pt based on a correlation of electrochemically active surface areas (cm^2/g Pt) versus the mass activity for ORR (Figure 6).

Subsequent to these early developments of alloy electrocatalysts in the PAFC technology have been attempts to use the same in pefluorinated sulfonic acid fuel cells (solid-state membranes such as Nafion® from Dupont, Dow, Asahi, and others). Yeager et al. [19] have reviewed the effect of different electrolytes on the ORR electrocatalysis. The summary of this work is that the solid-state perfluorinated acid environment offers significant advantages over phosphoric acid. These are

1. Higher oxygen solubility and diffusion coefficient.
2. Significantly lower anion adsorption from the electrolyte.

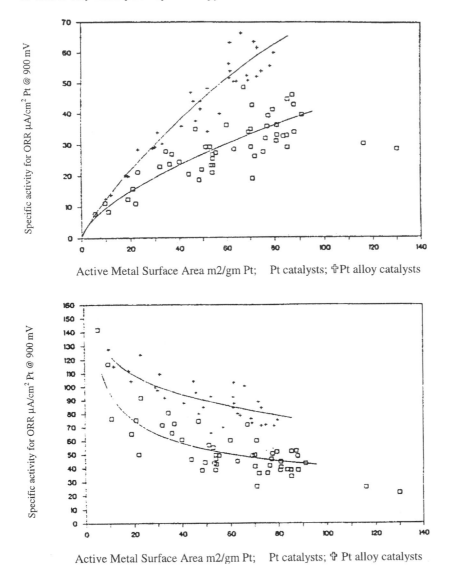

Active Metal Surface Area m2/gm Pt;   Pt catalysts; ⷬ Pt alloy catalysts

Figure 6 Comparison of the mass activity for ORR at 0.9 V as a function of electrochemically active surface area (m²/gm Pt) for supported Pt and Pt alloy electrocatalysts. (From Ref. 18.)

3. Hydrated environment in contrast to lower levels for higher-temperature H_3PO_4 (PA)-based systems (dependent on the concentration and temperature of operation of the PA-based systems). This means a higher degree of water activation and formation of Pt-OH species above 0.8 versus RHE on the Pt surface.

4. Lower temperature of operation (>100 °C, preferably at 85 °C) as compared to PA systems (<150 °C), resulting in different ORR kinetics.

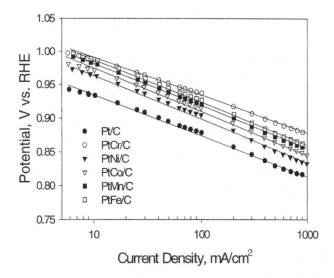

Figure 7 Steady-state *i*R corrected Tafel plots of cathodic ORR performance of several binary Pt alloy electrocatalysts at 90 °C and 5-atm pressure. Performance for a Pt/C electrocatalyst is shown for comparison. The electrodes had 0.3 mg/cm² metal loading and the loading of the metal on carbon support was 20%. The humidifaction temperature for the anode and cathode gas streams were kept at 10 and 5 °C above the cell temperature.

5. Higher proton conduction as compared to PA-based systems due to the hydrated nature of the electrolyte membrane.

Taking these differences into consideration, an automatic correlation of enhancement could not be made for the lower-temperature PEMFC-based fuel cells. The first reported systematic study of the ORR in these systems was conducted by Mukerjee et al. [20,21]. Five binary alloys of Pt with first-row transition elements ranging from Cr to Ni were investigated. All the alloy electrocatalysts were found to enhance ORR activity albeit to different extents. The highest active binary alloy was found to be PtCr, where a threefold enhancement was reported. The activation energy for ORR was reported to be approximately half that of Pt; however, the reaction order with respect to molecular oxygen was reported to be one, the same as Pt/C. Hence, the conclusion was that the surface nature of these alloys was more kinetically facile toward higher ORR kinetics, with the initial adsorption geometry remaining unchanged. Figure 7 shows the comparison of ORR activity as *i*R-corrected Tafel plots for oxygen reduction taken at 95 °C (5 atm pressure) in a PEMFC environment. It is interesting to note that the particle size for all the electrocatalysts are similar (within limits of error in their determination); however, the level of performance is significantly improved. In addition, results of a long-term test were also reported [20], where stable performance for PtCr/C was shown up to a period of 1200 h (steady-state operation at 50 °C and 200 mA/cm²). Publications in the 1990s have further confirmed these earlier reports in both the PA and PEM environments. In PEMFCs, ternary alloys have shown enhanced ORR activity with

PtCr and PtCrCu electrocatalysts (twofold enhancement) [22]. However, lifetime tests on these electrocatalysts showed an excessive loss of Cu over a 300-h test operating at $200 \, mA/cm^2$. Hydrodynamic voltammograms with rotating disks were used by Watanabe et al. [23] to study thin layers of PtNi sputter deposits in $HClO_4$, wherein a tenfold activity enhancement was reported with 30% Ni content. A subsequent report by Watanabe et al. examined the effect of Pt alloys with Fe, Ni, and Co, in a similar sputter deposition mode using $HClO_4$, where maximum activity was reported with 30, 40, and 50 atom % Ni, Co, and Fe, respectively. Enhancement of the order of 10, 15, and 20 times higher kinetic current was reported for these alloys, respectively. Besides these results there were several new reports in the 1990s on electrocatalysis in PAFCs. These are reviewed in more detail in Section 14.4, where the true nature of ORR activity on Pt alloys is discussed.

14.2.6 Inhibiting Mass Transport Limitations and Ohmic Limitations in Porous Gas Diffusion Electrodes

Besides the activation overpotential, mass transport losses is an important contributor to the overall overpotential loss, especially at high current density. By use of such high-surface-area electrocatalysts, activation overpotential is minimized. But since a three-dimensional reaction zone is essential for the consumption of the fuel-cell gaseous reactants, it is necessary to incorporate the supported electrocatalysts in the porous gas diffusion electrodes, with optimized structures, for aqueous electrolyte fuel-cell applications. The supported electrocatalysts and the structure and composition of the active layer play a significant role in minimizing the mass transport and ohmic limitations, particularly in respect to the former when air is the cathodic reactant. In general, mass transport limitations are predominant in the active layer of the electrode, while ohmic limitations are mainly due to resistance to ionic transport in the electrolyte. For the purposes of this chapter, the focus will be on the role of the supported electrocatalysts in inhibiting both mass transport and ohmic limitations within the porous gas diffusion electrodes, in acid electrolyte fuel cells. These may be summarized as follows:

1. *Electrocatalysts*: One of the positive features of the supported electrocatalyst is that stable particle sizes in PAFCs and PEMFCs of the order of 2–3 nm can be achieved. These particles are in contact with the electrolyte, and since mass transport of the reactants occurs by spherical diffusion of low concentrations of the fuel-cell reactants (hydrogen and oxygen) through the electrolyte to the ultrafine electrocatalyst particles, the problems connected with diffusional limiting currents are minimized. There has to be good contact between the electrocatalyst particles and the carbon support to minimize ohmic losses and between the supported electrocatalysts and the electrolyte for the proton transport to the electrocatalyst particles and for the subsequent oxygen reduction reaction. This electrolyte network, in contact with the supported electrocatalyst in the active layer of the electrodes, has to be continuous up to the interface of the active layer with the electrolyte layer to minimize ohmic losses.

2. *Electrocatalyst support*: The main functions of the carbon support are to (a) disperse the ultrafine electrocatalyst particles, (b) bind strongly with the

electrocatalyst particles and with the neighboring carbon support, and (c) facilitate mass transport of the reactant gases to the active sites.

3. *Thin active layers*: In order to attain high current densities in acid electrolyte fuel cells, thin active layers are necessary. Further, there should be enough electrocatalyst sites to maximize the electrochemically active surface area. From about the time that PAFCs were developed, 10 wt.% Pt/C-supported electrocatalysts were used [8]. With a noble metal loading of $0.4\,mg/cm^2$ and Vulcan-XC72R as the support, the thickness of the active layer is about $100\,\mu m$. Since the state-of-the-art PAFCs operate at a relatively low current density of $300\,mA/cm^2$ (because of significant activation overpotential for the oxygen reduction reaction and ohmic resistance of the electrolyte), the mass transport and ohmic overpotential losses in the active layer of the electrodes are negligible. However, in the case of PEMFCs, the goals are to operate at high current densities in the range of 500 to $1000\,mA/cm^2$. The novel approach to reach this goal was to use a higher percent of the noble metal or alloy electrocatalyst on the carbon support for the same loading of the electrocatalyst [24]. In this way by, for example, using 20% Pt/C and the same noble metal loading $0.4\,mg/cm^2$ as in PAFCs, the thickness of the active layer was reduced by 50%. It was thus possible to minimize the mass transport and ohmic overpotentials in the active layer, as well as to have a high utilization of the surface area of the electrocatalysts. The followup research activities, to minimize the mass transport and ohmic overpotentials at higher current densities in PEMFCs, included the preparation of the active layer constituents (Nafion with the proton in the sulfonic acid group of solubilized Nafion replaced by the sodium tetra butyl ammonium in an emulsion and directly depositing it on a Nafion membrane) [25]. The reason for the conversion to the sodium or tetra butyl ammonium form was to heat-treat the sample of the active layer to a desired temperature for good bonding between the two layers. After this heat treatment, the ion exchange was carried out to convert the membrane and the electrode assembly to the protonic form. An alternative was to deposit the active layer on a decal and then bond it on to the Nafion membrane by hot pressing. This approach was highly successful in the preparation of very thin active layers (10 to $20\,\mu m$), with ultralow platinum loading and high utilization of the platinum electrocatalyst in PEMFCs.

4. *Electrocatalyst/support electrolyte network in active layer*: Having an optimized structure and composition of the active layer is vital for minimizing activation and more so for mass transport and ohmic overpotentials in the active layers. This is particularly essential for attaining the desired high-power densities in PEMFCs. Thus, within the past 5 to 6 years, attention has been focused in this direction to (1) use electrocatalyst particles with the ideal minimum size (about 2–4 nm), (2) use high-surface-area carbon with the desired internal porosity, (3) have excellent contact between the electrocatalyst and the support, (4) have intimate contacts between the supported electrocatalysts and also between the supported electrocatalysts and uncatalyzed carbon particles (often used to improve the electronic conductivity in active layer), (5) have a network of continuous pathways of electrolyte in the active layer leading to the membrane electrolyte layer, (6) have some gas channels (in the active layer if it is not thin enough) to have ready access of gas to the active electrocatalytic site, via the electrolyte films, and (7) include some Teflon in the active layer (again if not thin enough), to have some hydrophobicity for gas access and product water removal [26].

14.3 SYNCHROTRON METHODS AS A PROBE OF ELECTROCHEMICAL INTERFACE

14.3.1 In-Situ X-Ray Absorption Spectroscopy: The Current State of the Art

The use of synchrotron-based in-situ x-ray absorption spectroscopy (XAS) for the study of catalysis, both heterogeneous and electrocatalysis, has matured over the last decade with simultaneous efforts in the United Sates, European Union, and Japan. An excellent review of these efforts has recently been published [27]. Application of in-situ XAS for investigating battery and fuel-cell materials has recently been reviewed [28,29], and an extensive database of literature on its application to catalysis can be accessed electronically [30]. Detailed aspects on the application of the technique and methodology used for data analysis have recently been published [31].

Being a core-level spectroscopy, XAS by its very nature is bulk averaging; even measurements of total electron yield (Auger electrons) have penetration depths of several hundred angstroms [32] for moderate energies such as those corresponding to Ni K and Pt L edges. Application for studying highly dispersed materials such as electrocatalysts are therefore approached with two different motivations. The first is the study of nano-clusters from a short-range atomic order perspective, which provides important details such as bond distances, coordination numbers, and Debye Waller factors. This thus provides important structural information in clusters, which lack long-range order and hence is an important probe for studying metal alloy clusters such as those proposed in this research effort. These measurements are typically carried out under in-situ electrochemical control under conditions akin to the potential of zero charge. The potential control is necessitated by the need to ensure clean cluster surfaces without the effect of adsorbates. Since the information in this case is not limited by being bulk-averaged, the particle-size limitation depends only on self-absorption effects, which occur typically at three absorption lengths (1–5 µm), a size range one order of magnitude larger than the nano-clusters proposed in this research effort. Application of EXAFS for investigation of short-range atomic order in nano-clusters have evolved with better algorithms and new powerful theoretical fitting software available such as FEFFFIT (6) from the University of Washington [33]. The second motivation is to investigate the effect of adsorbents under in-situ electrochemical conditions; being surface-sensitive, this requires small particle sizes with relatively large surface-to-volume ratios. Its efficacy in investigating clusters with greater than 80 Å diameter is therefore limited and has never been the focus of our research. Effect of Pt particle size and the application of XAS technique to investigate its influence on electrocatalytic performance have been the focus of prior research by the proposer [34]. The choice of experimental conditions and methodology for analysis are very important for the successful application of XAS measurements to study highly dispersed clusters under in-situ electrochemical conditions. In all these investigations it is important to determine with great accuracy the particle size. This is usually conducted using a variety of methods such as x-ray linebroadening analysis, transmission electron microscopy, and compared with particle size based on coordination numbers determined from EXAFS data using well-known algorithms

[35]. The XAS data can thus be normalized to the average number of surface atoms for these nano-clusters.

The advantage is that this technique offers the ability to study in situ the changes to both electronic and short-range atomic order in the nano-clusters with element specificity under conditions of actual fuel-cell operation. Hence, despite being a bulk-averaging technique, it has been extensively used for studying highly dispersed catalysts both for classical heterogeneous gas-phase applications as well as for electrocatalysis. Two important considerations are (1) particle size must be below 80 Å, and (2) the particle-size distribution should be narrow.

14.3.2 Overview of the Underlying Principle and Data Analysis

X-ray absorption spectroscopy (XAS) constitutes the determination of the absorption coefficient (either as a transmission or fluorescence measurement) of a core-level transition. Typical measurements involve transitions from p or d orbitals (such as $2p^{3/2}$ or $5d^{5/2}$), which constitute the K or L absorption edge, respectively, the selection rule for the transition being ($2J \pm 1$). The typical XAS spectra shown in Figure 8 (Pt L_3 and L_2 edge) comprise two complementary parts. The near-edge part (± 50 eV), referred to as x-ray absorption near-edge structure, provides information on the electronic states of the absorbing atom, while the extended part of the spectra (50 to 1200 eV), referred to as the extended x-ray absorption fine structure, provides information on the short-range atomic order (bond distances, coordination numbers, etc.). Measurements at the specific core-level absorption edge depend on the energy range of the transition; for low Z elements this is typically at the K edge, while for higher Z elements such as Pt etc., the L edge is more convenient. Most synchrotrons have an energy range between 4 to 40 KeV, which enables the study of a wide swath of the Periodic Table.

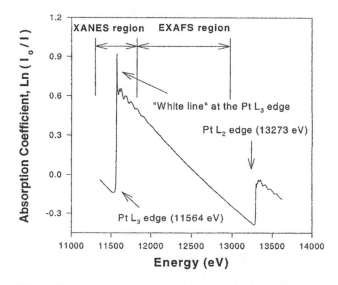

Figure 8 Typical raw XAS spectra for Pt/C at 0.54 V versus RHE in 1 M $HClO_4$ at the Pt L_3 and L_2 edge. The figure shows the various regions of the XAS spectra, namely the XANES and the EXAFS region.

X-Ray Absorption Near-Edge Structure

The near-edge part of the XAS spectrum constitutes the XANES (x-ray absorption near-edge structure), as shown in Figure 9, which represents the Pt L_3 edge data for carbon-supported Pt/C and PtCo/C. This fine structure is due to core-level transition of the ejected photoelectron, which due to its low kinetic energy probes the unoccupied states in the vicinity of the Fermi level. These photoelectrons have long mean free paths and result in multiple elastic scattering around the excited absorbing atom. Due to these long-range interactions, the shape of the XANES spectra can on occasion provide important insights into the changes in long-range order, such as a octahedral to tetrahedral transition. In cases such as Pt where a wealth of theoretical data exists, it is possible to determine the d-band occupancy, which is very useful in determining parameters such as the extent of perturbation of electronic states as a result of alloying. This methodology, first developed by Mansour and co-workers [36,37], has been used extensively to determine the extent of perturbation of electronic states as a result of alloying in supported binary alloy electrocatalysts such as those published later by McBreen and Mukerjee et al. [21]. Due to their ability to probe the electronic states, XANES spectra have been used extensively to determine the oxidation states of individual components of a complex compounds such as a multicomponent alloy or in metalloproteins, etc. These have been demonstrated in the XAS literature on battery materials such as those published earlier [38], for metal hydride alloys, and for lithium ion batteries [39]. Besides this XANES is a very good alternative method for quantitative determination of the extent of corrosion. This serves as a very good complement to the electrochemical methods, for studying

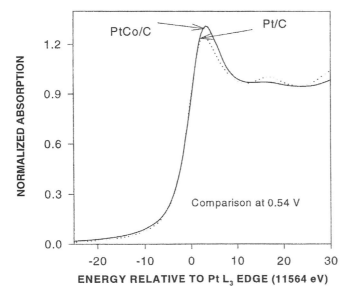

Figure 9   XANES spectra at the Pt L_3 edge for Pt/C and PtCo/C at 0.54 V versus RHE in 1 M HClO$_4$, showing the differences in the white line intensity. The observed differences are due to changes in the density of states as a result of alloying Pt with Co.

corrosion of multicomponent systems such as binary and ternary supported alloy clusters and multicomponent compounds and alloys [40].

The physical mechanism of XANES can be understood from the Fermi Golden rule, where the probability of a transition (w) from an initial state $|I>$ to a final state $<f|$ by the action of a perturbation H' is given by

$$w = \frac{2\pi}{\hbar} |<f|H'|i>|^2 \rho(E_i) \tag{3}$$

where $\rho(E_I)$ is the density of final states at the energy of the initial state E_I and $H' = Ae^{ikr}$, the x-ray photon. In the dipole approximation the transition is restricted by the selection rule $\Delta L = \pm 1$ and $\Delta J = 0, \pm 1$, where L and J are the orbital and total angular quantum numbers $(J = L \pm 1/2)$. This implies that at the K edge, the transition is mostly from the s to p levels such as those involving $1s$ to $4p$ states for the first row transition elements. However, weaker quadrupole transitions such as those involving $(\Delta L = \pm 2)$ are often observed such as those for first-row transition elements, appearing as pre-edge peaks, as a consequence of inversion of symmetry. In these cases this involves a $1s$ to $3d$ transition. From the perspective of electrocatalysts, for low- and medium-temperature acid-based fuel cells such as PEMFC, Pt XAS is important. The Pt L_3 edge corresponds to transitions to $s_{1/2}$ and d states $(d_{3/2}, d_{5/2})$. The transition to s-symmetric final state is normally very small and is spread over a wide range of energies, whereas the d-symmetric portion of the density of final states is large and confined to a narrow range of energies. Assuming no significant hybridization, the L_2 edge probes empty final states characterized by $J = 3/2$, while the L_3 edge probes the states with $J = 3/2$ and $5/2$. Spin orbit coupling causes a splitting of the final d states. This causes the $d_{5/2}$ states to be shifted to higher energies with a corresponding shift for $d_{3/2}$ state to lower energies. As a result the probability of empty states in the vicinity of the Fermi level is predominantly with $J = 5/2$. This accounts for the L_3 edge of Pt, for example, to be higher in intensity as compared to the L_2 edge. The magnitude of the spectrum at any edge is usually referred to as the white line, reminiscent of the days when photographic detectors were used.

Extended X-Ray Absorption Fine Structure (EXAFS)

EXAFS spectrum is a result of photoelectrons that have enough energy to be ejected out of the absorbing atom to have inelastic interaction with the surrounding atoms, which are called scatterers. The spectrum, therefore, is a result of interference due to a backscattered photoelectron wave, where the incoming wave represents a free induction decay pattern. Fortunately, this phenomenon can be explained on the basis of single scattering theory and has been worked out in detail [41]. The EXAFS is, therefore, a final-state interference effect whereby, depending on energy E, the backscattered wave interferes constructively or destructively with the outgoing wave. The EXAFS can be expressed in energy space in terms of $\chi(k)$ with the expression

$$\chi(k) = \frac{\mu(E) - \mu_0(E)}{\mu_0(E)} \tag{4}$$

where μ and μ_0 are the x-ray absorption coefficients of the absorber in the sample and that in the free state, respectively. The difference therefore reflects on the local structure and represents the EXAFS. The ratio with respect to absorption cross section in the free state therefore normalizes the spectrum on a per-atom basis. The spectrum is extracted by background removal procedures using separate routines for pre-edge and extended part of the spectrum; these methods are described in detail in [41]. The extracted EXAFS is usually converted from the energy space to the wavevector space k using the relationship

$$k = \sqrt{\frac{2m}{\hbar^2}(E - E_0)} \tag{5}$$

where $\hbar = h/2\pi$, m is the mass of the electron, and E_0 is the threshold energy of the core shell of the excited atom. The EXAFS spectrum is the superimposition of the contribution from different coordination shells to the backscattering process. The theoretical expression, which relates the measured EXAFS to the structural parameters, is given by

$$\chi(k) = \sum_j A_j(k) \sin(2kR_j + \varphi_j(k)) \tag{6}$$

where j refers to the jth coordination shell, R_j is the average distance between the absorbing atoms and the neighboring scatterers in the jth shell. $\varphi(k)$ is the total phase shift suffered by the photoelectron in the scattering process. $A_j(k)$ is the amplitude function, which is expressed as

$$A_j(k) = \frac{N_j}{kR_j^2} S_0^2(k) F_j(k) e^{-2(R_j - \Delta)/\lambda_e - \sigma_j^2 k^2} \tag{7}$$

where N_j is the average coordination number and $F_j(k)$ is the backscattered amplitude of the atoms in the jth shell. σ_j^2 is the Debye Waller term, which accounts for the static and the thermal disorder present in the materials. $S_0^2(k)$ is an amplitude reduction term that accounts for the relaxation of the absorbing atom and multielectron excitations (shake up and off) at the absorbing atoms. λ is the mean free path of the photoelectron and is a correction to Δ since S_0^2 and $F_j(k)$ already account for the photoelectron losses in the first shell. Data analysis of the EXAFS involves the determination of R_j, N_j, and σ^2. The other parameters φ_j and E_j are obtained from standards that could be experimental or theoretical.

The contribution of individual shells to the EXAFS can be isolated by Fourier transformation of the signal in wavevector space to frequency domain (as a radial distribution function) using a typical Fourier integral represented as

$$\Theta = \frac{1}{\sqrt{2\pi}} \sum_{k_{min}}^{k_{max}} k^n \chi(k) e_{ikr}^2 dk \tag{8}$$

The result is peaks in r space that correspond to interatomic distances between the central absorber atom and the surrounding scatterers in the individual coordination shells. As is typical in all Fourier transformations, the resolution of this transform depends on the range EXAFS signal used. In other words, it is important to have a good EXAFS signal without noise up to a χ range of 14 or 15. The r value is shifted lower than actual coordination distances because of the phase shift term $\varphi(k)$. Since the spectrum in r space is the sum of the χ's of the individual shells, it is thus possible to extract information of an individual shell by back-transformation from r to χ space. For this, windows in r space are set carefully. Structural parameters can then be determined for each individual shell using either empirical or theoretical phase and amplitude parameters. The final fitting between the measured EXAFS phase and amplitude functions and those from empirical or theoretical origins is done using nonlinear least-square fitting routines to yield short-range atomic parameters, N (coordination number), R (bond distance), $\Delta\sigma^2$ (Debye Waller factor with respect to some standard) and ΔE_0 (a sliding parameter). Both N and $\Delta\sigma^2$ are strongly correlated but can be distinguished; the same is true for R and ΔE_0. These effects are usually decoupled by conducting the k^1 and k^3 weighted fits in both k and r space [42]. The effects on N and R are the same in terms of fits in both weightings; however, the effects on R and ΔE_0 are such that the magnitude and the imaginary parts of the transform are reversed. These provide for several checks, which help at arriving at a unique solution for fits between the standard and experimental EXAFS data.

In the context of supported electrocatalysts typically used in the current state-of-the-art PEMFCs, the in-situ XAS spectroscopy has three important functions.

1. Determination of short-range atomic order of the cluster: this is independent of particle size, as it is bulk-averaged information. The important aspect in this measurement is to ensure that the measurements are done in situ at potentials close to the double-layer region ($\sim 0.54\,\mathrm{V}$ versus RHE) to avoid interference from surface adsorbed species.

2. Determination of the changes in the short-range atomic order and Pt d-band vacancies as a result of changes in fuel-cell operating conditions. Since effect of surface-adsorbed species is an important component of this measurement, normalization of the data with population of surface sites is important. For these purposes, cluster synthesis with a high degree of monodispersity, an important component of this proposed research, in the particle size range 25–40 Å will be an asset for the application of this technique.

3. Determination of the extent of corrosion of an individual component in a muticomponent alloy cluster, both ex situ and in situ as a function of various operating conditions.

14.4 STRUCTURAL AND ELECTRONIC EFFECTS ON ELECTROCATALYSIS

14.4.1 Structural Effects on Electrocatalysis: Effect of Particle Size

One of the first investigations toward elucidating particle size and structural dependence was by Zeliger [43] and Bett et al. [44]. These studies concluded that platinum atom at the vertices, edges, and kink sites or dislocations are not more active than atoms on the platinum crystal faces, and hence specific activity for oxygen reduction is independent of particle size. This conclusion was despite the fact that Zeliger's work involved platinum deposited on asbestos, while Bett's study was on commercial platinum blacks and platinum supported on graphitized carbon (average particle size in the range of 3 to 4 nm, corresponding to surface area between 9–6 m^2/g). Further support to this conclusion was provided by Gruver [45] and Vogel et al. [46], who showed a similar independence of particle size on ORR activity. Furthermore, Kunz and Gruver [45] concluded that the interaction between carbon and platinum was not significant since the activity of smooth Pt and Pt supported on carbon (76 m^2/g) were found to be approximately the same in both 96% H_3PO_4 at 160 °C and 20% H_2SO_4 at 70 °C.

On the other hand, Blurton et al. [47] and Bregoli [48] found a decrease in the specific activity for oxygen reduction with diminishing particle size from 12 nm to 3 nm and 2 nm, respectively. In their studies on highly dispersed platinum on carbon in 20% H_2SO_4 at 70 °C, Blurton et al. [47] attributed the results to either a particle-size effect or an effect of the support interaction, or a combination of both these factors. Bregoli's study [48] involved highly dispersed catalysts in 99% H_3PO_4 at 177 °C. He found the specific activity for oxygen reduction to vary by a factor of 2 in the range of particle size studied (12 nm to 2 nm). Later, Peuckert et al. [49], in their investigations with highly dispersed Pt/C (particle size in the range, 12 to 1 nm) in 0.5 M H_2SO_4, showed a constant site-time yield for particle sizes above 4 nm. However, as the particle size decreased to ∼1 nm, the site-time yield dropped 20-fold. It was inferred that as particle size decreased, a larger fraction of platinum atoms participates in the surface reactions. However, for particle size below 4 nm, the analysis in terms of activity per surface atom showed an exponential drop, which was in contrast to expectations based on changes to surface area (or dispersion). This decrease in the site-time yields for ORR was consistent, however, with prior published reports. Bett et al. [44] did not see an effect of particle size, since crystallite size range was larger than 3 nm. Investigation by Bregoli [48], where the range was between 25 to 3 nm, showed a twofold decrease in activity for ORR as one approached the lower limit of particle size. Blurton [47] found a 20-fold decrease in activity with a particle-size range of 1.7 nm to 10 nm. The last two reports, however, had a greater scatter of data. An alternative viewpoint to the observed particle-size effect was presented by Stonehart [8], where the particle-size effect was instead attributed to an interparticle diffusive interference between the platinum crystallites. This report pointed out that the specific activities of platinum particles with varying sizes were the same on different carbon supports when the crystallite separations on the carbon supports were the same. Mutual interaction between Pt particles were considered as a factor toward setting up diffusive interference. This viewpoint has been supported by Watanabe [6] based on a study of various supported electrocatalysts with different supports, platinum loading, carbon surface area, and platinum crystallite size. When the crystallite separations were greater than 17 nm, the specific activity for ORR was reported to be constant; however, they reported a decrease in activity when the interparticle separation was less than 17 nm. The rationale was based on the creation of a mutual interference of platinum particles for diffusion of oxygen to the individual crystallite surfaces. The fundamental issue, therefore, was whether the observed difference in ORR activity with different specific surface areas of platinum was due to an inherent change in the surface property of the crystallites or due to diffusional limitations brought about as a function of electrocatalyst dispersion. Investigations by Giordano et al. [50,51] have attempted to differentiate between these issues, using variables such as (1) Pt concentration on carbon support, (2) time and temperature of electrocatalyst activation, (3) PTFE concentration, (4) platinum loading, and (5) sintering temperature of the electrodes. These investigations revealed that the adsorptive morphology of the crystallites was paramount in the observed variations of ORR activity with specific surface areas of platinum. These results supported the earlier contention by Kinoshita [4], where a correlation was made with the statistics of the presence of sites with different coordination numbers as a function of particle size. A later publication by the same group provided further support, with an expanded range of particle size and choice of carbon support material.

The original explanation of the specific activity for ORR based on the statistical variations of different coordinate sites with particle size (using cluster models such as cubo-octahedron) (see Figures 2 and 3) was remarkably effective. Peuckert et al. [49] offered a similar explanation based on the increase in the concentration of low coordination sites with decreased particle size. This interpretation was partly based on a publication by Ross et al. [52], who found no difference in activity on Pt$\langle 111 \rangle$ and $\langle 100 \rangle$ surfaces. A recent investigation on the effect of particle size on the Pt electronic (d-band vacancy/atom) and short-range atomic order (Pt–Pt bond distance and coordination numbers) was conducted using in-situ XAS methods by Mukerjee et al. [53]. In this investigation the Pt crystallite size was varied in the range of 30 to 90 Å. This study showed direct spectroscopic evidence of the effect of electronic and short-range atomic order as a function of particle size. Figure 10 shows the effect of change in the Pt d-band vacancy/atom (Δ Pt d-band vacancy/atom normalized for number of surface atoms) as a function of particle size when the electrode potential is increased from 0.0 V to 0.54 V (1 M HClO$_4$ at room-temperature sealed electrochemical cell). The plot in Figure 10 therefore probes the sensitivity of the surface electronic states toward H-adsorption as a function of particle size (i.e., switching from a potential of 0.0 V in the hydrogen region to 0.54 V, which is close to the potential of zero charge). As is evident from the plot, a crystallite size less than 50 Å has a progressively larger change in the electronic states. Comparison of the change in the bulk coordination numbers of Pt with particle size exhibited similar behavior. In this case, there was a larger sensitivity of the small crystallites to morphological change as compared to those where particle sizes are larger than 50 Å. This aspect has been discussed in more detail elsewhere (see [54] and publications cross-referenced within). A similar comparison was made for change of potential from 0.84 V (a potential where activation of water results in well-known coverage of Pt–OH) to 0.54 V. Once again there is a great sensitivity toward change of Pt d-band vacancy/atom with particle size (shown in Figure 11, as a Δ Pt d-band vacancy/atom normalized with respect to the number of surface atoms). From this analysis it is evident that smaller particle sizes (<50 Å) with their higher proportion of low coordination sites (and hence greater number of surface sites with higher Pt d-band vacancies) adsorb species such as H, OH, and C$_1$ moieties more strongly. These observations are in support of indirect evidence of the same alluded to originally by Kinoshita [4,5] and later by Giordano [50] and others on Pt cluster interactions on zeolites and alumina support (see cross-references in previous report [53]).

14.4.2 Electronic and Structural Effects on Electrocatalytic Properties of Platinum Alloys

Attempts to understand the nature of ORR activity enhancement using Pt alloys started after the publication of the first patent in this area by Jalan [14]. In his

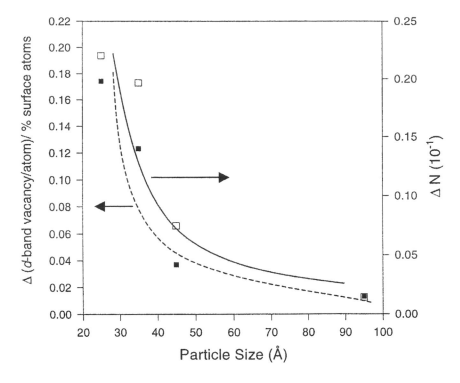

Figure 10 The effect of particle size on the change in the normalized Pt d-band vacancies (d-band vacancies/% surface atoms) ($\dots$) in going from 0.54 to 0.0 V. The change in the Pt–Pt coordination number as determined from the Pt L_3 edge EXAFS analysis is also shown (—). The ordinate axis refers to the change in the d-band vacancy/atom determined from XANES spectra normalized with respect to total number of surface atoms present (based on cluster calculations on a cubo-octahedron model).

analysis [55], a correlation is drawn between specific activity for ORR and bulk interatomic distance in the alloy, which is related to the strength of the HO_2 bond. The function used was the bulk interatomic distance in the alloy, which was related to bond tightness and hence the strength of the $HO_{2(ads)}$ bond. In this case $HO_{2(ads)}$ was considered as the most probable product of the rate-determining step. Jalan's correlation therefore was a fine-tuning of the basic volcano plot. Using this correlation with interatomic distances (Pt–Pt bond distances and nearest-neighbor interactions), Jalan proposed that contraction of lattice parameters was beneficial for the initial adsorption of molecular oxygen. A more favorable dual-site adsorption model for the alloyed Pt surfaces was considered to offer a more facile environment as compared to the unalloyed surface. Glass et al. [56], in their investigation on bulk alloys of PtCr (the binary alloy on top of the volcano plot) of different compositions, however, found no enhancement of ORR activity in phosphoric acid. This study therefore suggests the possibility of different ORR activity between bulk and supported nanocrystallites. A subsequent study on PtCo electrocatalyst [52] revealed the possibility that particle termination, primarily at $\langle 100 \rangle$ vicinal planes in the supported alloy electrocatalyst, is the reason for enhanced ORR activity. This was

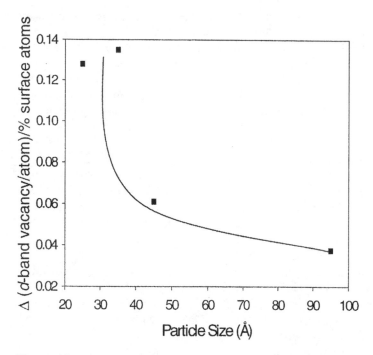

Figure 11 Variation in the normalized Pt *d*-band vacancies (*d*-band vacancies/% surface atoms) (...) in going from 0.54 and 0.84 V in 1 M HClO$_4$ as a function of particle size. The ordinate axis refers to the change in the *d*-band vacancy/atom determined from XANES spectra normalized with respect to total number of surface atoms present (based on cluster calculations on a cubo-octahedron model).

based on their finding of a higher ORR activity on a ⟨100⟩ surface as compared to a ⟨111⟩ surface, which demonstrates that surface morphological changes, resulting from particle-size changes, could account for different ORR activities. Pafett et al. [57] attribute the higher ORR activity to surface roughening and increased Pt surface area due to dissolution of the more oxidizable alloying element. This study shows that potential excursion, specially above 1.25 V versus RHE, results in selective depletion of chromium [present as Cr(III) oxide or hydroxide on the surface] as Cr(IV) species in solution. In contrast to these results on bulk alloys, supported alloy electrocatalysts have been reported to retain their nonnoble alloying element in the electrode during long periods (6000–9000 h) of operation in PAFCs [44] and PEMFCs [20].

Recent studies on the issue of surface roughening in the context of PAFC operating conditions have been investigated by Watanabe et al. [58] and Kim et al. [59]. Watanabe and co-workers have shown that under PAFC conditions, supported PtCo electrocatalysts do undergo an initial dissolution of the more oxidizable alloying element (in this case CO). However, depending on the nature of the nanocluster (ordered alloy such as Pt$_3$Co or disordered), there are different extents of this dissolution. The resulting cluster contains a Pt skin with a core alloy in the bulk. The depth of the Pt skin is dependent on the ordered–disordered nature of the nanocluster. In this study, a higher specific activity for ORR was exhibited by the disordered alloy. A later study by Kim et al. [59] showed a similar effect of surface roughening and Pt skin effect with ordered–disordered PtCr alloy nano-clusters, where a disordered PtCr alloy electrocatalyst was found to have the highest ORR activity (mass activity) enhancement. The first attempt to relate the fundamental aspects of electronic (such as Pt d-band vacancy/atom) and short-range atomic order (such as the Pt–Pt, Pt–M, M–M, and M–Pt bond distances and coordination numbers) in nano-clusters of Pt and Pt alloys (with first-row transition elements ranging from Cr to Ni) was conducted by Mukerjee et al. [21]. These studies show the following:

1. Alloying of Pt with transition elements increases the Pt d-band vacancy/atom and decreases the Pt–Pt bond distance. The extent of the change is shown to be dependent on the electronegativity of the transition element. Correlation of the ORR activity in terms of current density at 0.9 V with the d-band vacancy/atom of Pt and the Pt–Pt bond distance shows a familiar volcano-type behavior with respect to the various binary alloys. This clearly indicates the importance of both these parameters in the rate-determining step of ORR.

2. X-ray absorption near-edge data (XANES) also shows that the onset of OH adsorption, which is known to typically occur above 0.8 V on Pt/C, is shifted to higher potentials in the alloy (extent of shift dependent on the alloying element) leading to a lowering of overpotential losses by about 60 mV. Spectroscopic evidence for the inhibition of the formation of Pt–OH species, which are known to poison the Pt surface near OCV, creating a mixed potential region and a consequent loss of $\sim$220 mV, is an important discovery [21,22], as shown in Figure 12. In a comparison of Fourier transforms of the EXAFS at 0.84 V versus 0.54 V for Pt/C and PtCr/C, Pt/C at 0.84 V shows the peaks for Pt–OH along with the Pt–Pt interactions, while the corresponding spectra for PtCr/C show no evidence of Pt–OH interactions at this potential. This explains the lowering of overpotential loss by $\sim$50 mV for this alloy electrocatalyst relative to Pt. This is a pivotal discovery and lends credence to the assertion that further enhancements can be achieved by a discriminate selection of the alloying element.

3. XANES analysis also confirmed that the outer surfaces of the ORR active alloy crystallites are almost entirely Pt; the alloying element was relegated to the subsurface layers. This study likewise verifies *insignificant* dissolution of the alloying element, even at potentials as high as 0.9 V, after the initial formation of the Pt skin as reported by others under PAFC's operating conditions [58,59].

4. The influence of alloying conditions (e.g., alloying temperature, choice of reducing agent) was also investigated. The results indicate a close correlation of preparative route with the extent of perturbation of the Pt electronic and short-range atomic order as a result of alloying.

14.5 CONCLUSIONS

Development of supported Pt electrocatalysts came as a result of intensive research on fundamental and applied aspects of electrocatalysis [especially for kinetically difficult oxygen reduction reaction (ORR)] fueled by attempts at commercialization of medium-temperature phosphoric acid fuel cells (PAFCs) in the late 1960s and early 1970s. Dispersion of metal crystallites in a conductive carbon support resulted in significant improvements in all three polarization zones (activation, ohmic, and

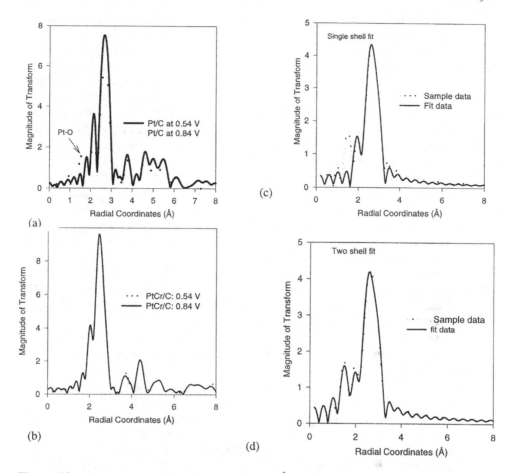

Figure 12 Comparison of Fourier transforms (k^3 weighted) at 0.54 (—) and 0.84 (…) V versus RHE for (a) Pt/C and (b) PtCr/C alloy (JMRC) electrocatalyst. Fits to the data (…) for a Pt/C catalyst at 0.84 V in *r* space: (c) a single Pt–Pt shell fit (—) and (d) a two-shell, Pt–Pt and Pt–O fit (—) (k^3 weighted).

mass transport). The lowering of ohmic and mass transport overpotential losses were a natural function of the conductive carbon substrate and the extension of the electrode–electrolyte interface further into the electrode structure, thereby increasing the catalyst, and reactant utilization. In the case of PAFCs this was achieved by varying the hydrophobicity of the catalyst containing the reaction layer on the electrode. In the case of PEMFCs this was a more complicated function of ink formulations for creating the reaction layer, composed of the catalyst, binder, and solubilized ionomers.

The improvements in the activation polarization defined as either mass-specific activity or site-specific activity (activity/number of specific crystal planes on the surface) were reported, especially for the kinetically difficult ORR. Wealth of prior data on both ORR as well as direct methanol oxidation (both multielectron reduction and oxidation processes) showed clear particle-size effects. Bulk of these prior results on ORR activity variations with Pt cluster size showed a sharp drop in activity when cluster size dropped below 20 Å. Two schools of thought have emerged to explain the effect of cluster size and morphology (for binary and other muticomponent metal reaction centers). One of these ascribed the activity variations to interparticle diffusive interference. This viewpoint therefore correlates the observed activity variations to mass transport effects when small clusters are packed differently on the carbon support. An alternative viewpoint relates the activity variations with actual changes in the cluster electronic and short-range atomic order (coordination number, bond distances, etc.). These correlations have used well-known cluster models, mostly cubo-octahedron and icosahedron, to correlate the surface morphology changes with particle size. These models have therefore been used to ascertain the variations in the number of sites with different coordination numbers, crystal planes, etc. with change in the particle size. Changes in the particle size from 10 Å to 80 Å result in tremendous variation in the population of sites with small coordination numbers (6–9) and hence in the population of different crystal planes. The population of crystal planes such as $\langle 111 \rangle$ and $\langle 100 \rangle$ for the face-centered cubic Pt changes from negligible at ~ 7 Å to a steady state at 35 Å. These crystal planes possess a very different work function and hence catalytic activity. Small particle size >20 Å have surface sites that are predominantly edge, kink, or steps, corresponding to sites with a low coordination number.

Attempts to enhance ORR activity by modification of the surface electronic and short-range atomic order were also attempted as a consequence of alloying Pt with base transition elements. The first patents were issued in pursuance with attempts to improve ORR activity for PAFCs in the early 1980s. As reviewed in this chapter a large number of binary and ternary alloys were patented and continue to be patented for both anodic CO tolerance and methanol oxidation as well as cathodic oxygen reduction. In the early 1990s, alloys were also shown to enhance ORR activity in the lower-temperature PEMFCs. Even though both PAFCs and PEMFCs are acid-based systems, they differ significantly. The prime difference is the nature and function of water activation, which occurs on Pt at approximately 0.7–0.8 V versus RHE (depending on the cluster size). The presence of surface Pt–OH formed as a consequence of water activation acts as a poison for adsorption of molecular oxygen. In PAFCs, depending on the temperature of operation, the levels of surface coverage by Pt–OH may be lower than PEMFCs, but the phosphate anion from the electrolyte acts as a severe poison, eliminating any potential advantage of lower water content. The other differences are the nature of proton transport, operating temperature, and associated kinetics of ORR.

In this endeavor synchrotron spectroscopy has played an important role in understanding the effect of fundamental parameters such as electronic density of states and short-range atomic order. The primary advantages of using the synchrotron are (1) the ability to probe these parameters in situ while the interface is under electrochemical control and (2) the fact that these can be measured with element specificity. The latter is particularly useful when investigating multi-component alloy clusters. In addition, this technique lends itself to systems with limited long-range order, which is typical for these nanoclusters used in fuel-cell electrode interface. This chapter describes some recent results with in-situ X-ray absorption spectroscopy, which has provided a direct probe into the variations of the Pt *d*-band vacancy (normalized with respect to number of surface atoms) between different potential regions. These include regions where anion adsorption occurs (such as near the hydrogen evolution potential, hence formation of $Pt-H_{ads}$) and those beyond which water activation occurs (formation of surface $Pt-OH$). The results of these studies showed that for a particle size smaller than 20 Å, anion adsorption is very strong and hence detrimental to electrocatalysis; this is because of the relatively large fraction of the surface atoms possessing low coordination and hence higher susceptibility to changes in their d-band vacancies with potential.

Recently in-situ XAS spectra have shown that alloying of Pt with base transition elements such as Co and Cr enable changes in the electronic properties of Pt that in turn shift the onset of $Pt-OH$ formation to higher potentials. This has been shown to enable a lowering of ORR overpotential losses by approximately 50 mVs. In this endeavor, however, the surface nature of the cluster is very important and has to be predominantly Pt with the inner core as the alloy. Hence, the methodologies of preparation of these metal clusters are of prime importance.

REFERENCES

1. J. McBreen, S. Mukerjee, X.O. Yang, *Synchrotron Radiation News* **11**, 18 (1998).
2. H.G. Petrov, R.J. Allen, U.S. Patent 3,992, 512, Nov. 16, 1976, Catalytic Platinum Metal Particles on a Substrate and Method of Preparing the Catalyst.
3. A.J. Appleby, In *Assessment of Research Needs for Advanced Fuel Cells* (Department of Energy, USA, Advanced Working Group Report (SS Penner, ed.), U.S. Department of Energy, Office of Energy Research (Contract # DE-AC01-84ER30060) (1985).
4. K. Kinoshita, *Modern Aspects of Electrochemistry* **14**, 557 (1982).
5. K. Kinoshita, *J. Electrochem. Soc.* **137**, 845 (1990).
6. M. Watanabe, H. Sei, P. Stonehart, *J. Electroanal. Chem.* **261**, 375 (1989).
7. W. Romanowski, *Surf-Sci.* **18**, 373 (1969).
8. P. Stonehart, *ber.Bunsengos, Phys.Chem.* **94**, 913 (1990).
9. S. Mukerjee, *J. Appl. Electrochem.* **20**, 537 (1990).
10. R.E. Benfield, *J. Chem Soc., Faraday Trans.* **88**, 1107 (1992).
11. L.L. Hillenbrand, J.W. Lacksonen, *J. Electrochem. Soc.* **112**, 249 (1965).
12. J. Escard, C. Leclerc, J.P. Conour, *J. Catal.* **29**, 31 (1973).
13. V.S. Bogotsky, A.M. Snudkin, *Electrochemica Acta* **29**, 757 (1984).
14. V.M. Jalan, *U.S. Patent* 4,202,934 (1980).
15. P. Stonehart, *J. Appl. Electrochem.* **22**, 995 (1992).
16. D.A. Landsman, F.J. Luczak, *U.S. Patent* 4,316,944 (1982).
17. D.A. Landsman, F.I. Luczak, *U.S. Patent* 4,447,506 (1984), and *U.S. Patent*, 4,677,092 (1987).
18. J.S. Buchanan, L. Keck, J. Lee, G.A. Hards, N. Scholey, *Proc. First Intl. Fuel Cell Workshop* (Tokyo, 1989) (M. Watanabe and P. Stonehart, eds.).
19. E. Yeager, M. Razak, D. Gervasio, A. Razak, D. Tryk, In *Structural Effects in Electrocatalysis and Oxygen Electrochemistry*, Proc. Vol. 92–11. (D.D. Scherson, D. Tryk, M. Daroux, X. Xing, eds.) The Electrochemical Society, Pennington, NJ (1992), p. 440.
20. S. Mukerjee, S. Srinivasan, *J. Electroanal. Chem.* **357**, 210 (1993).
21. S. Mukerjee, S. Srinivasan, M.P. Soriaga, J. McBreen, *J. Electrochem. Soc.* **142**, 1409 (1995); S. Mukerjee, S. Srinivasan, M.P. Soriaga, J. McBreen, *J. Phys. Chem.* **99**, 4577 (1995).
22. G. Tamizhmani, G.A. Capuano, *J. Electrochem. Soc.* **141**, 968 (1994); *J. Electrochem. Soc.* **141**, L132-L134 (1994).

23. T. Toda, H. Igarashi, M. Watanabe, *J. Electrochem. Soc.* **145**, 4185 (1998); T. Toda, H. Igarashi, H. Uchida, M. Watanabe, *J. Electrochem. Soc.* **146**, 3751 (1999).

24. E.A. Ticianelli, C.R. Deronin, S. Srinivasan, *J. Electroanal. Chem.* **251**, 275 (1988).

25. M.S. Wilson, S. Gottesfeld, *J. Appl. Electrochem.* **22**, 1 (1992).

26. M. Uchida, Y. Aoyama, N. Eda, A. Ohta, *J. Electrochem. Soc.* **142**, 4143 (1995).

27. P.A. Montano, H. Oyanagi, *MRS Bulletin* **24**, 13 (1999).

28. J. McBreen, S. Mukerjee, X.Q. Yang, *Synchrotron Radiation News* **11**, 18 (1998).

29. J. McBreen, S. Mukerjee, In *Interfacial Electrochemistry* (A. Wiekowski, ed.), Marcel Decker, New York (1998).

30. Searchable data base on EXAFS literature at www.exafs.chem.msu.su/∼papers.

31. *X-Ray Absorption: Principles, Applications, Techniques of EXAFS, Sexafs and XANES Absorption Principles X-ray*, R. Prins (ed.), John Wiley & Sons (1998).

32. W.T. Elam, J.P. Kirkland, R.A. Neiser, P.D. Wolf, *Phys. Rev. B* **38**, 26 (1998).

33. A.I. Frenkel, *J. Synch. Rad.* **6**, 293 (1999); C.W. Hills, M.S. Nasher, A.I. Frenkel, J.R. Shapely, R.G. Nuzzo, ᵛ*Langmiur* **15**, 690 (1999); M.S. Nasher, A.I. Frenkel, D. Somerville, C.W. Hills, J.R. Shapely, R.G. Nuzzo, *J. Am. Chem. Soc.* **120**, 8093 (1998).

34. S. Mukerjee, J. McBreen, *J. Electroanal. Chem.* **448**, 163 (1998).

35. R.E. Benfield, *J. Chem. Soc. Faraday Trans.* **88**, 1107 (1992).

36. A.N. Mansour, *X-ray Absorption Studies of Silica Supported Pt Catalysts*, Ph.D Thesis, Department of Physics, University of North Carolina, Raleigh, NC, (1983).

37. A.N. Mansour, J.W. Cook, Jr., D.E. Sayers, *J. Phys. Chem.* **88**, 2330 (1984); A.N. Mansour, J.W. Cook, Jr., D.E. Sayers, R.J. Emrich, J.R. Katzer, *J. Catal.* **89**, 464 (1984).

38. S. Mukerjee, J. McBreen, J.J. Reilly, J.R. Johnson, G.D. Adzic, K. Petrov, M.P.S. Kumar, W. Zhang, S. Srinivasan, *J. Electrochem. Soc.* **142**, 2278 (1995).

39. Y. Ein-Eli, S.H. Lu, M.A. Reznik, S. Mukerjee, X.O. Yang, J. McBreen, $LiCu_xMn_{2-x}O_4$ spinels $(0.1 \leqslant x, \leqslant 0.5)$: A new class of cathode material for Li batteries. II In situ measurements, *J. Electrochem. Soc.* **145**, 3383 (1998).

40. S. Mukerjee, J. McBreen, G.D. Adzic, J.R. Johnson, J.J. Reilly, M.R. Marrero, M.P. Soriaga, M.S. Alexander, A. Visinti, S. Srinivasan, *J. Electrochem. Soc.* **144**, L258 (1997).

41. B.K. Teo, *EXAFS: Basic Principles and Data Analysis*, Springer Verlag, Berlin (1986); DC Koningsberger, R Prins, *X-ray Absorption: Principles, Applications, Techniques of EXAFS, SEXAFS and XANES*, John Wiley and Sons, New York (1988); P.A. Lee, P.H. Citrin, P. Eisenberger, B.M. Kincaid, *Rev. Mod. Phys.* **53**, 769–806 (1981); T.M. Hayes, J.B. Boyce, *Sol St. Phys.* **37**, 173–351 (1982); E.A. Stern, S.M. Heald, In *Handbook of Synchrotron Radiation* **Vol. 1**, Elsevier Science Publishers, Amsterdam, pp 955–1014, (1983); H. Winick, S. Doniach, *Synchrotron Radiation Research*, Plenum Press, New York (1980).

42. D.C. Kinongsberger, In *Synchrotron Techniques in Interfacial Electrochemistry* (C.A. Melendres, C.A. Tadjeddine, eds.), Kluwer Academic Press, Dordrecht, Netherlands, pp 181–198 (1994).

43. H. Zeliger, *J. Electrochem. Soc.* **114**, 144 (1967).

44. J. Bett, L. Lundquist, E. Washington, P. Stonehart, *Electrochim. Acta.* **18**, 343 (1973).

45. H.R. Kunz, G.A. Gruver, *J. Electrochem. Soc.* **112**, 1279 (1975).

46. W.M. Vogel, J.M. Baris, *Electrochim. Acta.* **22**, 1259 (1977).

47. K.F. Blurton, P. Greenburg, G.H. Oswin, D.R. Rutt, *J. Electrochem. Soc.* **119**, 559 (1972).

48. L.J. Bregoli, *Electrochim. Acta.* **23**, 489 (1978).

49. M. Peukert, T. Yoneda, R.A. Dalla Betta, M. Boudart, *J. Electrochem. Soc.* **113**, 944 (1996).

50. N. Giordani, E. Passalacqua, V. Recupero, M. Vivaldi, E.J. Taylor, G. Wilemsky, *Electrochim. Acta.* **35**, 1411 (1990).

51. N. Giordano, E. Passalacqua, L. Pino, A.S. Arico, V. Antonucci, M. Vivaldi, K. Kinoshita, *Electrochim. Acta.* **36**, 1979 (1991).

52. B.C. Beard, P.N. Ross, *J. Electrochem. Soc.* **130**, 3368 (1990).

53. S. Mukerjee, J. McBreen, *J. Electroanal. Chem.* **448**, 163 (1998).

54. S. Mukerjee, J. McBreen, *J. Electrochem. Soc.* **143**, 2285 (1996).

55. V.M. Jalan, E.J. Taylor, *J. Electrochem. Soc.* **130**, 2299 (1983); V.M. Jalan, E.J. Taylor, In *Proc. Symposium on the Chemistry and Physics of Electrocatalysis* **84 (12)**, The Electrochemical Society, pp 547–557 (1984).

56. J.T. Glass, G.L. Cahen, G.E. Stoner, *J. Electrochem. Soc.* **134**, 58 (1987).

57. M.T. Paffett, J.G. Beery, S. Gottesfeld, *J. Electrochem. Soc.* **135**, 1431 (1988).

58. M. Watanabe, K. Tsurumi, T. Mizukami, T. Nakamura, P. Stonehart, *J. Electrochem. Soc.* **141**, 2659 (1994).

59. K.T. Kim, Y.G. Kim, J.S. Chung, *J. Electrochem. Soc.* **142**, 1531 (1995).

15

STM and Infrared Spectroscopy in Studies of Fuel Cell Model Catalysts: Particle Structure and Reactivity

J.A. COLLINS and U. STIMMING

Technical University of Munich, Munich, Germany

CHAPTER CONTENTS

SUMMARY

Model electrodes with a defined mesoscopic structure can be generated by a variety of means, e.g., electrodeposition, adsorption from colloidal solutions, and vapor deposition and on a variety of substrates. Such electrodes have relatively well-defined physico-chemical properties that differ significantly from those of the bulk phase. The present work analyzes the application of in-situ STM (scanning tunneling microscopy) and FTIR (Fourier Transformed infrared) spectroscopy in determining the mesoscopic structural properties of these electrodes and the potential effect of these properties on the reactivity of the fuel cell model catalysts. Special attention is paid to the structure and catalytic behavior of supported metal clusters, which are seen as model systems for technical electrocatalysts.

15.1 INTRODUCTION

The development of electrochemical power sources for the cleaner generation of electricity in fuel cell power stations and for automobile applications is a major challenge both for the present and future, with a huge increase in demand for the generation of power at high efficiency or to store electric power in batteries. Presently, some types of fuel cells approach commercial feasibility. Fuel cell technology is now becoming applicable for a large variety of technical areas. The present work does not focus on direct developments in fuel cell technology—there are a number of reviews in the literature on the state-of-the-art in the fields of methodology, catalysis, catalyst characterization, polymer electrolytes as well as assessment and interpretation of cell and electrode performance [1–4].

Paramount to the development of this technology is the development of effective and efficient electrocatalysts—as Bowker [5] puts it, the "Holy Grail" of catalysis has been to identify what Taylor [6] describes as the active site, that is, that ensemble of atoms responsible for the surface reactions involved in catalytic turnover. Many advances in the knowledge of the effect of electrode structure on electrochemical processes have been achieved due to the utilization of structurally defined electrodes, usually low-index single-crystal surfaces, and to the application of surface-sensitive in-situ monitoring techniques for electrode surface characterization. The influence of atomic order on processes occurring at the solid–liquid interface have become accessible through the utilization of well-defined single-crystal surfaces. Although an extremely powerful approach, structural effects on catalytic activity in general may not only be related to the local atomic-scale order, but are also to a large extent dependent on mesoscopic structural properties.

Furthermore, electrocatalysts for real applications often contain supported metal clusters in the nm-size range. However, the correlation between catalytic properties and the structure and morphology of technical electrodes is difficult due to their structural complexity, and in the past, this has led to contradictive results. By controlling such mesoscopic properties, it should be possible to evaluate their impact on electrochemical reactivity. For example, recent investigations at model electrodes with reduced complexity [7] established a correlation of particle size and reactivity, e.g., for the oxygen reduction reaction, and highlighted the necessity to control the structural properties of model electrodes on the mesoscopic scale. The possibility of characterizing the real-space morphology of electrode surfaces on the nm scale using scanning tunneling microscopy (STM) gives access to knowledge of the influence of mesoscopic structures on electrochemical processes. Defined mesoscopic structures can be regarded as model electrodes for electrocatalysts used in technical applications, including various types of fuel cells. They are investigated by STM, in-situ Fourier transform infrared (FTIR) spectroscopy (both of which are discussed in the present work), as well as cyclic voltammetry, electro-oxidation of adsorbed CO, and TEM (transmission electron microscopy), topics that are not dealt with in much detail here.

Atomic surface order can be adequately described in terms of a simple unit cell, and techniques for the preparation of surfaces with defined atomic order are well established. Describing mesoscopic structures is not as easy. But single crystals do not always consist of only well-defined terrace structure—they also contain steps, and possibly other irregular surface features such as kinks and defects. The mobility of surface metal atoms is much greater than that of similar atoms in the bulk phase, the vibrations of surface atoms being highly asymmetric as they lack neighbors on one side. It is worth nothing that there is now explicit evidence from STM work [8] that, from a catalytic viewpoint, singularities, e.g., low coordination surface metal atoms (ad-atoms), which should not exist at ideally ordered low-index single-crystal surfaces, are of major importance; such atoms are evidently the prime components in active sites at surfaces. Thus, a single-crystal surface can have mesoscopic properties such as terrace widths and step densities, and with dispersed electrodes, these properties are influenced by the size and distribution of particles. To characterize such mesoscopic properties, it is necessary to ascertain real-space information under in-situ electrochemical conditions, by applying scanning probe techniques, which give high-resolution real-space images of electrode surfaces.

The objective of this chapter is to show that particles in the mesoscopic regime have very different properties to the bulk phase and, specifically, to demonstrate how in-situ STM and FTIR spectroscopy have been successfully employed to determine information on the structure of model catalysts based on modification of substrate electrodes with metal particles of mesoscopic dimensions, and the effect of this structure on reactivity. It will be shown that studying these model electrodes helps provide a link between single-crystal electrodes, which have provided a wealth of useful information, and electrodes for real application. FTIR has long been invaluable as a probe for localized particle reaction on surfaces in electrochemical processes, and the present work will show how it can complement STM in providing excellent characterization of mesoscopic properties.

15.2 TECHNIQUES EXPLORED IN THE PRESENT WORK

15.2.1 STM: An Overview

Scanning probe microscopies (SPMs), including STM and atomic force microscopy (AFM), are powerful tools for probing the microscopic properties of surfaces. Using these techniques, it has become increasingly possible to elucidate atomic-scale structural and electronic properties characteristic of the bulk of a material as well as the surface. By combining such measurements on particle structure with chemical synthesis and various other means of characterization and analysis, it is possible to bring attention to relationships between composition, structure, and physical properties including catalytic activity, thus leading to an improved understanding of the chemical basis of material properties. Since the invention of the scanning tunnel microscope by Binnig and Rohrer in the early 1980s [9], the theory and operating principles of STM are now well known and widely reviewed [9–13], since STM is now a key tool in many areas of research, including the study of fuel cell model catalysts.

The nonperiodic structure of surface defects such as steps makes them very difficult, if not impossible, to investigate by commonly employed diffraction techniques, and real-space imaging becomes mandatory. In this respect, STM, with its capability of imaging electrode surfaces in situ with atomic resolution, provides a unique possibility of studying processes for which surface imperfections play a key role, such as metal deposition and dissolution [14–20], oxide formation [21–24], and corrosion [25–29]. The additional capability of STM to control material properties

on the atomic and nanometer levels has led to the exciting field of generation of
nanometer-sized structures on surfaces [30–37].

15.2.2 Brief IR Status Overview

Fourier transform infrared (FTIR) and in-situ FTIR spectroscopy are among many
modern instrumental tools of analytical chemistry well established in fuel-cell-related
electrochemistry [1]. In general, FTIR spectroscopy is a valuable tool in the
characterization of fuel cell technical electrodes, where the nature of surface groups
can be identified, since such electrodes are rather difficult solid surfaces on which to
work. FTIR is among the methods less commonly used for the characterization of
dispersed catalysts and supports, but as a technique is able to give an idea about the
nature of the surface groups on carbon supports and on the structure of adsorbed
species on noble metal clusters.

From both an environmental and efficiency viewpoint, the development of an
ambient temperature fuel cell for mobile power applications is highly desirable. One
of the main barriers to progress [especially in the case of cells designed to operate
with liquid fuels, e.g., the direct methanol/air fuel cell (DMFC)] is the lack of highly
effective, economically viable electrocatalysts. The DMFC performance has been
limited for long periods of time by the performance of both electrodes, with the
anode reaction in particular proving difficult to optimize. The literature contains
many articles on DMFC research, which can be consulted for more general
information on the principles involved in DMFC operation, and an overview of
recent developments in technical catalysts and fuel cell technology [1,2,4,38–54].
There have been significant advances both in the understanding of the electro-
oxidation reaction itself at a molecular level, although the mechanism is not fully
understood, and in the development of more active electrocatalysts, to such a point
that now a small increase in performance of the anode could make DMFCs
attractive commercially.

Since platinum in its pure state, and either alloyed or in mixtures with other
metals/metal oxides, (which act as promoters), is among the most active materials for
methanol oxidation, much attention has been devoted to the nature of, and
mechanism involved in, the methanol oxidation reaction on platinum. As such,
platinum has served as a useful model system illustrating the general features of
metal electro-oxidation in an aqueous environment. There are many postulated
mechanisms for the oxidation of methanol, and detailed descriptions of the same can
be found in the literature [55–59] and will not be discussed in the present work,
except from the point of view of contributions of IR and STM toward the
understanding of the overall picture of electrocatalysis at model electrodes.

Rigorous control of the electrode surface before the experiment and especially
during the transfer procedure has led to significant progress in the understanding of
the reaction mechanism of electro-oxidation of methanol. For fundamental studies,
single-crystal electrodes are ideal model systems, and single-crystal research has laid
the foundations for some basic understanding of methanol electrocatalysis and has
created a wealth of data from which technical catalysts should benefit. In-situ FTIR
has been used to great effect to study adsorbed species formed at smooth electrodes
allowing the detection of both volatile and nonvolatile products, with separation of
adsorbed and nonadsorbed species using IR-light of different polarization. Despite

the undisputed fact that PtRu is a better catalyst for methanol oxidation than Pt alone, IR investigations on methanol adsorption kinetics at different surfaces are scarce, and the issue is seldom studied dynamically for materials other than Pt [60].

Generally, the majority of investigators agree that catalysis by small particles is different from that at smooth electrodes [61–63]. Christensen et al. [64] believe that it is very dangerous to extrapolate from bulk to particle electrodes. Thus, it seems that some rules for the application of single-crystal electrochemistry to technical electrodes have also to be established before meaningful conclusions can be drawn. The study of model catalysts may help in finding an answer as to how to extract the best-possible performance out of a given amount of noble metals incorporated into a particular catalyst.

Such information is vital, to avoid a purely empirical approach in the search for better catalysts, and consequently for the improvement of fuel cell performance. It is clear from the literature [65,66] that methanol anode electrocatalysis depends greatly upon understanding the methanol adsorption process in detail, e.g., knowing the steric, kinetic, and energetic details of all reaction, dissociation, and diffusion steps involved. It is also known that methanol adsorption is sensitive to surface geometry. Therefore, the use of FTIR in studying the role of modified surfaces as fuel cell model catalysts in this process will be addressed in the present work. The study of the structure of model electrode systems affords the opportunity to successfully relate quantities such as size effects and surface coverage to catalytic reactivity. Generally, it is reported that the onset of methanol oxidation on PtRu is between 0.2 and 0.25 V (RHE) [62,67]. It seems as if the limit of the optimization of PtRu catalysts has been reached as far as the diminution of the methanol oxidation overpotential is concerned [56]. However, there is still a vast amount of work to do with maximization of the active surface area, in which the characterization of fuel cell model catalyst structure plays a vital role.

15.3 SMALL METAL PARTICLES AND ACTIVE SITES

15.3.1 The Nature and Novel Properties of Small Particles

Small metal particles on the nanometer dimension are of particular interest regarding their solid-state properties as well as their important application as catalysts. Such particles offer a useful model allowing the study of structural effects with relevance for a variety of applications in chemistry and physics. The key point of interest lies in the fact that such particles often possess very different and sometimes novel properties compared to those of bulk materials, i.e., their physical properties (spectroscopic, electronic, magnetic) differ from those of the bulk phase and are particle size-dependent [68]. One such property concerns the variation in the electrochemical redox potential between metals in a dispersed and bulk state, as was shown by Plieth [69], when he demonstrated that the redox potential depended on the radius (r) of a metal particle on nanometer dimension (the radius was assumed to vary from 1 to 10 nm) according to the following equation:

$$\Delta\varphi_d = -2\gamma V_m/zFr \tag{1}$$

where $\Delta\varphi_d$ is the potential shift associated with the reduction of a metal in a dispersed state and is seen to be inversely proportional to the particle radius, r (or the smaller the particle size (down to 100 atoms) the more negative the equilibrium potential); γ is the surface tension, V_m is the molar volume, z is the number of electrons involved in the reduction, and F is the Faraday constant. A cathodic shift of redox potential of approximately 20 to 60 mV for a 10-nm cluster depending on surface energy, γ, was calculated [69]. For a smaller cluster of 1-nm size, a larger shift in potential from 0.2 to 0.6 V, again depending on γ, was determined. The generality of the approach does not extend to clusters of less than 100 atoms, since for growing clusters, the condition of equilibrium will rarely be satisfied and in general nonequilibrium values of γ and r are to be expected. A similar approach was employed by Fleischmann et al. to model deposition (nucleation) and electrocatalytic reactivity at single-atom catalyst sites on metal surfaces [70]. It is now generally accepted that the metal atoms in ultrafine metal particles are unusually electropositive; the effect may be attributed to quantum confinement effects, or the reduction in lattice stabilization energy, or a lower work function, resulting in a lower electrochemical potential. In these minute metal particles, most of the atoms exist at the surface, where their lattice stabilization energy and lattice coordination number are low. For silver [71] it is been shown by Henglein that on going from the bulk metal (Ag_∞, $U^0 = +0.8$ V) to a single atom (Ag_1, $U^0 = -1.8$ V) the electropositive character increases dramatically (especially for values of $n < 15$).

It has also been well established that metals in the very finely divide state are unusually reactive. For example, Parmigiani and co-workers [72] have demonstrated that at low oxygen pressure, supported platinum clusters oxidize at room temperature, whereas the bulk metal reacts at the same pressure only around 800 K. Clearly, the active sites, based on small particle sizes and high defect densities, have implications aside from electrochemistry.

15.3.2 Determination of the Active Site

The notion of active sites is fundamental to heterogeneous catalysis. Modern approaches to solid surface disorder provide a more rigorous way of defining roughness at solid surfaces. An understanding of the operation of a catalyst depends partly on developing a fundamental knowledge of the surface properties for adsorption and reaction. The topography of these surfaces can be determined by STM at different scales and described by means of the dynamic scaling theory applied to STM imaging. As far back as 1987, Arvia et al. used STM in an ex-situ determination on the nanometer scale of the topography of electrochemically highly activated platinum electrodes [73]. The term "roughness" usually implies the existence of both macropores (macroroughness), which to a great extent are responsible for additional diffusional relaxation, and micropores (microroughness), which concern the effective catalytic area. Although there is a close relationship between microroughness and catalytic activity, many real systems involve complex macro- and micropore structures that make direct determination of microroughness a difficult task. The increase in the number of reacting sites can be achieved in three different ways, namely, by increasing the surface roughness, by promoting close-packed crystallographic structures, and by creating a volume structure that is equally accessible to reactants. The STM data observed by Arvia et al. [73] suggest that neither of the first two possibilities by themselves explains the large increase in activity of the treated electrodes. The behavior of activated platinum electrodes can be explained by assuming that each pebblelike electroreduced crystallite can be represented by sphere of 10-nm radius. The sphere ensemble is made by piling the spheres in a volume defined by the geometric area times the average thickness of the electroreduced platinum layer. This thickness is obtained from the corresponding oxide reduction charge; its value for a 100 times activation can be estimated as $5 \cdot 10^2$ nm. The number of spheres of 10-nm radius per cm^2 that can be accommodated within the volume of the activated electrode is about 10^{12}. Therefore, provided that the entire surface of each sphere is catalytically active, the large activation factor, of the order of 10^2, deduced from the voltammetric charge can be immediately understood. The proposed model explains the important fact that the electrocatalytic activation remains practically the same for electrochemical processes occurring in different potential ranges with reactants and intermediates of different sizes. Thus, the STM patterns furnish direct information about the microtopography of active platinum electrodes measured on the nanometer scale. They also indicate a correlation between the activation of the electrode and the change in surface topography.

15.3.3 Electrochemical Processing to Produce Model Electrodes

Because surface imperfections can act efficiently as nucleation centers, the defect structure of the surface onto which the metal is plated plays an important role in determining the morphology of the deposit. On a microscopic level, the amount of metal deposited onto a given area depends largely on the density of nuclei, which in return is greatly influenced by both the number of defects and their local arrangement across the surface. Thus, preferential deposition of metal at atomic steps is often seen.

The role of atomic steps in the nucleation process for bulk copper deposition has been demonstrated [17,18]. Kolb et al. [17] have used in-situ STM to study the nucleation-and-growth processes of copper deposition onto Au(111) and Au(100) electrodes, comprising the formation of a copper monolayer at underpotentials as well as nucleation and growth of small Cu clusters at overpotentials. It was seen from STM studies that Cu is preferentially deposited at monoatomically high steps on Au(111), while deposition on the atomically flat terraces commences at a much later stage. After a potential step into the bulk deposition region, Cu clusters are seen to grow preferentially at a step that starts at a screw dislocation with the high density of kink sites, whereas at an upper step with low density of kink sites, hardly any copper nuclei are formed, which is also true for the terraced regions. In particular, cluster growth is prominent at the screw dislocation.

It has also been demonstrated that the rims of atom high islands (average diameter 3–10 nm) brought about by surface diffusion, during reconstruction of the Au(111) surface, are very effective nucleation centers for Cu on Au(111), containing as they do a large number of kink sites [32]. Nucleation at deliberately induced surface defects, i.e., by modifying the surface structure by the tunneling tip (e.g. by applying potential jumps to the working electrode), lead to the generation of small protrusions (from the tip material) on the surface. Figure 1 shows a flat Au(111) terrace, (a) before and (b) after a train of potential pulses of − 1 V amplitude with 10-ms duration and manual repetition had been applied to the Au(111) electrode, while the *y*-scan of the tip had been stopped [74]. During the pulse treatment, a potential of

$-20\,mV$ versus Cu/Cu^{2+} was constantly applied to the Au(111) electrode for bulk deposition. Afterward, normal STM operation was resumed to image the surface. The result is a linear array of copper clusters, about 2 nm high, exactly aligned in the x-direction, which is the direction of movement of the tip during pulse treatment, and distributed over 100 nm, the width of the x-scan. Figure 1 demonstrates the capabilities of STM for surface modifications on the nanometer scale, which are discussed in detail in Section 15.5.

15.4   THE MESOSCOPIC REGIME

15.4.1   Introduction

So the question arises as to what extent are the mechanisms of various reactions determined by the atomic arrangement on the terrace, and how much do the presence and density of steps on the surface influence the reactions taking place? Many scientists have approached this problem, studying various reactions using different single-crystal faces, i.e., the terrace arrangement is varied as the crystal quality is kept constant. This involves studying reactivity on the surface at a macroscopic or bulk level, and although it is quite successful in extrapolating data for many important reaction processes, it has not ultimately led to the detection of what exactly occurs at the active site during catalysis. Another approach is to vary the density of steps present on a given crystal terrace; this approach of nanostructuring at electrode surfaces has become very popular in recent years. Recently, however, people have begun focusing on what is described as the "mesoscopic regime," or the nanometer dimension. Studying processes taking place on nanometric surfaces, roughly 50 to 100 nm in diameter, and sorting out many of the complexities involved in reaction at this dimension may provide some knowledge that will act as a bridge between the molecular level on one hand and the bulk regime on the other. By looking at the fundamental properties of colloidal systems, and the effect on them of varying well-defined physical parameters, proper structural characterization and definition of reactivity may follow more easily. At this stage, both the fundamental characterization of the nature of catalytic sites and the study of their reactivity (e.g., by electrochemical means) are poorly understood and require much research.

Taking the breakdown of the physical system one step further—and looking at single nanoparticles—allows one to eliminate complexities and associated problems often observed with groups of clusters, such as agglomeration, side-on-side interactions, etc., and facilitates the study of reactions at single nanoparticles. This is discussed in detail in Section 15.5.2, studying the hydrogen evolution reaction as a model process.

15.4.2   Average Mesoscopic Properties

Small metal particles as model systems for electrocatalysts are crucial to unravel the influence of electronic or geometrical structure on the catalytic activity. The effect of metal particle size on electrochemical reactivity has been proposed to exist for the electro-oxidation of alcohols as well as for the reduction of oxygen [7,75,76], both vital processes that require much deeper understanding for the development and

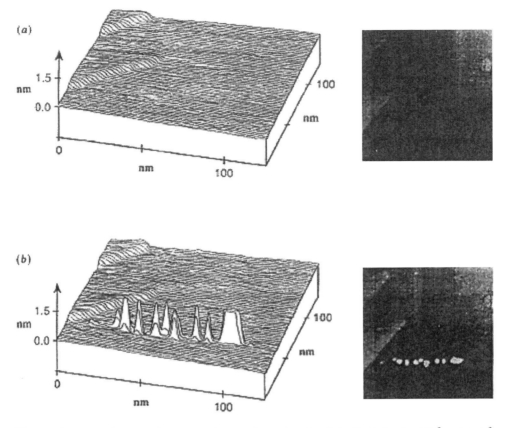

Figure 1 STM images (3D scan plots and topviews) of Au(111) in $5 \times 10^{-3}\,\mathrm{mol\,dm^{-3}}$ $H_2SO_4 + 5 \times 10^{-5}\,\mathrm{mol\,dm^{-3}}$ $CuSO_4$, demonstrating a tip-induced nucleation of Cu clusters on a flat terrace. (From Ref. 74.)

widespread application of fuel cell technology. A "size effect," if present, may be an important criterion for the design of practical electrodes for technical applications. Even more interestingly, some observations indicate significant substrate–particle interactions that may alter catalytic properties.

Dispersed Particle Electrodes

An important class of technical electrodes are those based on dispersed catalyst particles. For these, the relation between structure and reactivity is very important, but unraveling the complexities involved is slowed by the problem that reactivity is usually referred to a macroscopic sample while the structural characterization gives local information. Thus, the modification of conductive surfaces by nanometer-sized particles, i.e., to control the size and distribution of the catalyst particles on the substrate, allows one to control the average mesoscopic structure of these electrodes. Recently a technique for the preparation of catalyst particles with a narrow size distribution was developed by Reetz et al. [77], yielding colloidal palladium clusters stabilized by a shell of tetra-alkylammonium surfactants, which when adsorbed onto suitable substrates result in model electrodes for dispersed electrocatalysts [78]. The variety of particle sizes results in significantly different mesoscopic structures, which of course has implications on their reactivity as catalysts. Such data also raise the question of particle–particle interactions, and how such interactions vary with particle size, and density on model electrode surfaces. The difference between the mean diameter determined by STM and that measured by TEM allows the determination of the protective surfactant layer.

Similar model electrodes were prepared using highly oriented pyrolytic graphite (HOPG) as the substrate [77]. In these examples, imaging of clusters was only possible when they were attached to defects, which are rather scarce on a freshly cleaved HOPG electrode. Clusters were also present on the defect-free terraces, but these were easily swept away by the tunneling tip [79,80]. Thus, it was concluded that the defect density was vital for these model electrodes. Electrochemical oxidation of the surface at 1.9 V (RHE) was seen to increase the defect density, and it is possible to follow the process with STM in real time, since the process is controlled by the electrochemical potential, thus allowing direct interactive control on the surface morphology of the substrate.

Adsorbed and Electrodeposited Platinum

During work in the authors' group, model platinum electrodes with defined mesoscopic structure exhibiting a small size distribution were prepared by a variety of methods: by adsorption of tetra-alkylammonium surfactant-stabilized Pt clusters from colloidal solution on well-characterized Au(111) supports or HOPG [81]; by adsorption from colloidal solutions prepared using the well-known citrate method [82,83] (citrate acts as both the reducing agent as well as the stabilizing surfactant) [84]; and by electrochemical deposition of platinum on gold substrates [81]. These model electrodes have been characterized by STM, electro-oxidation of adsorbed CO, and FTIR spectroscopy. It is worth noting that both substrates used have the advantage that they are inert with respect to the adsorption of CO. Thus, the oxidation of a CO monolayer on the particles at saturation coverage could be investigated.

The morphology of the deposits was characterized with STM using a single-crystal substrate Au(111). STM images of electrochemically deposited Pt on Au(111) are shown in Figure 2: (a) low loading, and (b) high loading. The two electrodes show vastly different substrate coverages of Pt clusters. At the lower coverage (a), isolated Pt particles are observed, which mainly populate step edges of the substrate as well as the rims of small Au islands. Only a small density of clusters can be seen on the terraces. This seems to indicate the importance of step edges as sites for reactive centers on catalyst surfaces. At the higher coverage (b), similarly sized particles (~ 6 nm) are aggregated to form a network on the substrate surface. The characterization with STM showed that the clusters remain intact upon adsorption. No significant structural changes were observed on the time scale of hours. However, due to agglomeration effects on the surface, the size distribution of the particles varies and nonuniform particle distributions are observed. The adhesion of particles is improved on such defective surfaces, when compared to freshly cleaved graphite, but force interactions between tip and sample are still evident in the measurements.

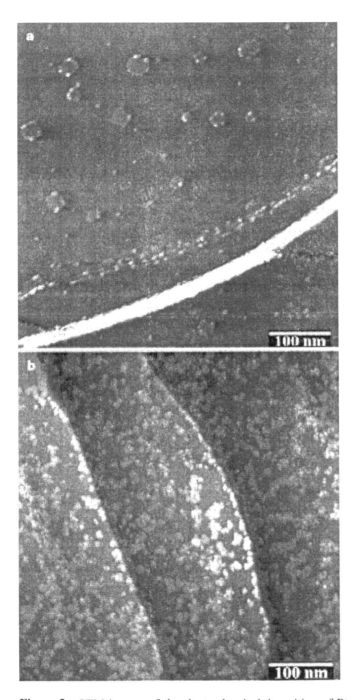

Figure 2 STM images of the electrochemical deposition of Pt on Au(111) electrode; (a) at low loading: $\sim 0.05\,\mu g\,Pt/cm^2$; (b) at high loading: $\sim 0.2\,\mu g\,Pt/cm^2$. (From Ref. 81.)

From cyclic voltammetry (CV) data on the adsorbed platinum particles [81], the peak potential for Pt oxide reduction is shifted to more negative potentials (0.7 V) as compared to a polycrystalline Pt bulk electrode (0.76 V), both against reference hydrogen electrode (RHE). This is a general result that several authors found for small supported metal particles [61,72]. Compared to polycrystalline Pt (oxidation at 0.73 V), the CO electro-oxidation at the Pt clusters is shifted to more positive potentials and shows two oxidation peaks (0.78 V and 1.03 V).

Figure 3 shows CVs of two different platinum electrodeposits on polycrystalline gold (solid line), the CVs for the oxidation of a monolayer of adsorbed CO (dashed line), and the corresponding IR spectra of adsorbed CO (inserts).

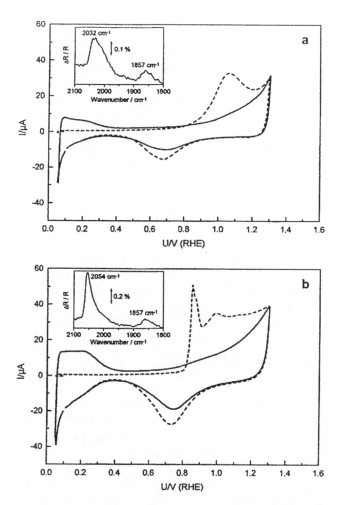

Figure 3 Cyclic voltammograms of electrochemical deposition of Pt on polycrystalline gold in 0.1 mol dm^{-3} HClO$_4$; scan rate 0.05 V/s; potential range 0.05–1.3 V; (—) without CO; (---) with CO; adsorption potential 0.1 V. In-situ FTIR spectra (inserts) taken at 0.1 V; 500 scans; resolution 4 cm^{-1}; (a) at low loading: $\sim 0.05\,\mu$g Pt/cm^2; (b) at high loading: $\sim 0.2\,\mu$g Pt/cm^2. (From Ref. 81.)

Figure 3(a) corresponds to a low loading $U_{dep} = 0.68$ V_{RHE}, $t_{dep} = 5$ s(~ 0.05 μg Pt/cm^2), (b) to a high loading $U_{dep} = 0.68$ V_{RHE}, $t_{dep} = 80$ s(~ 0.2 μg Pt/cm^2). Both voltammograms show similar current features as the Pt colloid model electrode described above. The different loading is qualitatively reflected in different charges for the hydrogen adsorption/desorption region and the oxide formation/reduction on Pt corresponding to varying amounts of Pt sites. Once again, for both loadings, the Pt oxide peak is shifted to more negative potentials than for a polycrystalline Pt bulk electrode, with values in (a) and (b) of 0.67 V and 0.74 V, respectively. A strong dependence on the catalyst loading is observed for the CO electro-oxidation. For the low loading the CO oxidation occurs at much more positive potentials (1.05 V); for the higher loading, two oxidation peaks are observed at less positive potentials (0.85 V and 0.98 V). The FTIR spectra also reveal differences depending on the different deposition parameters. A band assigned to CO_{ads} in a linear bonded (on-top) coordination appears at 2032 cm^{-1} for the low loading and at 2054 cm^{-1} for the high loading, both shifted to lower wavenumbers as compared to bulk Pt (2074 cm^{-1}) [85]. In both spectra, a second band around 1857 cm^{-1} can be observed and is assigned to bridge-bonded CO_{ads} on Pt. This band position does not depend on the loading.

Thus, it is obvious that the catalytic properties of supported nm-scale Pt clusters on Au differ significantly from the properties of bulk Pt electrodes. Clusters electro-oxidize CO_{ads} at much more positive potentials, and the vibrational frequency of linearly bonded (on-top) CO_{ads} is considerably lower as compared to bulk Pt. A pronounced dependence on the catalyst loading is seen, indicating that at high loading the properties of the clusters are more similar to those of polycrystalline Pt. Such trends are observed for several adsorbed surfactant-stabilized Pt clusters as well as for electrodeposited Pt on Au, thus indicating that the surfactants do not significantly affect the electrocatalytic properties. It is assumed that the macroscopic electrode characteristics reflect the properties of individual catalyst particles at very low catalyst loading. At higher coverages, cooperative phenomena emerge, which may result from agglomeration. The dipole-coupling of adsorbed species as well as their diffusion characteristics may be different on agglomerates and on isolated clusters, thus affecting the FTIR spectra and electro-oxidation kinetics.

These results are also confirmed by recent work by Friedrich et al. [84], illustrating the effect of Pt particle coverage, prepared from colloidal solution, and supported on polycrystalline gold substrates, on the oxidation of a monolayer of CO and on the characteristic vibrational bands of CO_{ads}. In order to reveal the properties of isolated particles, it is necessary to prepare model electrodes with small particle coverage, which, however, inherently involves weak signals. Thus, the use of the Au substrate has the advantage that the high IR reflectivity of gold means that distinct features in the IR spectra corresponding to the stretching vibration of CO are indicative of the structural properties of the colloid particles.

The model electrodes are prepared from aqueous colloidal Pt solutions prepared according to the citrate method mentioned above, with particle size determined from TEM (transmission electron microscopy) measurements [84]. After several months in solution, the colloids exhibit an aging effect evident in a change of ultra-violet/visible (UV/VIS) spectra and in the TEM images. Figure 4 shows TEM micrographs of the Pt colloids on Au (a) freshly prepared colloid and (b) aged Pt colloid prepared with gelatin. While the freshly prepared colloids consist of particles

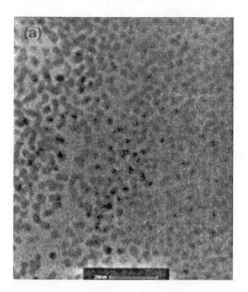

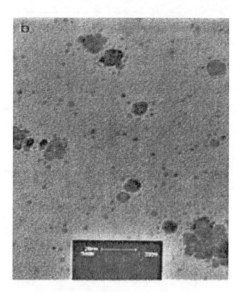

Figure 4 (a) TEM micrograph of freshly prepared Pt colloid particles on gold in the range of 3 nm; magnification 540,000; (b) TEM micrograph of an aged Pt colloid prepared with gelatin. The dark features correspond to Pt. Three different sizes can be observed: very small (primary) particles ~ 2 nm; larger, isolated particles of ~ 8 nm; and aggregates of these larger particles; magnification 540,000. Line in box represents 20 nm. (From Ref. 91.)

with a narrow size distribution (mean size: 2.8 nm, stan. dev.: 0.47 nm), these aged colloids clearly show three distinguishable sizes: isolated particles showing two main sizes with smaller (2.2 nm), and larger (8.5 nm) particles. Furthermore, aggregates that seem to consist mainly of the 8-nm particles are also observed. These aggregates show a very broad size distribution and can reach sizes up to ~ 50 nm. The isolated 8-nm particles are clearly a minority species.

In Figure 5, the IR spectra for CO-covered particle electrodes with increasing coverage from $\Gamma_{Pt} = 0.06$ to $\Gamma_{Pt} = 0.38$ at 80 mV are shown. An intense IR band due to the stretching vibration of adsorbed CO in the range 2010–2060 cm^{-1} is detected for all surfaces, increasing in intensity with increasing Pt coverage.

The observed bands are assigned to CO adsorbed on Pt clusters. The existence of CO adsorbed on gold can be confirmed by IR spectroscopy and has been reported in the literature [86,87]. Bands that can be assigned to CO_{ads}/Au are only found for quite different experimental conditions than were used for the measurements in Figure 5. Vibrational bands attributed to CO_{ads}/Au at 2080 cm^{-1} and 1990 cm^{-1} are found only in CO-saturated electrolytes—the latter band only for exposure times longer than 1 h. For CO-saturated electrolytes, the intensities of the CO_{ads}/Au bands are smaller than 0.1% absorption. The frequencies of the band for CO_{ads}/Pt are significantly lower than the frequency found for polycrystalline Pt (2068 cm^{-1}) at this potential. The vibrational frequencies increase linearly with increasing Pt coverage, as can be seen from the inset to Figure 5, where the peak frequencies of the vibrational bands are plotted versus Γ_{Pt}. For an electrode prepared by drying up a drop of colloidal solution, the vibrational frequency of polycrystalline Pt is obtained,

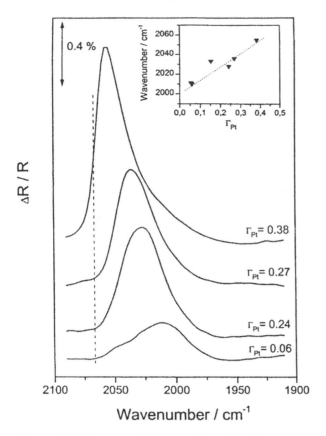

Figure 5 IR spectra of the stretching vibration of CO adsorbed on Pt particles at different particle coverages. The dashed line indicates peak frequency of on-top CO on polycrystalline platinum. All spectra are recorded at $80\,mV_{RHE}$. Inset: Dependence of the vibrational peak frequency ω on the Pt particle coverage. (From Ref. 84.)

although the bandwidth is much broader and asymmetrical for the dispersed system. An ad-hoc explanation for the changes in vibrational frequency in terms of the usual 5σ and $2\pi^*$ bonding interactions is to assume an increase of backbonding of d electrons of the metal to the empty antibonding $2\pi^*$ orbital of CO, leading to a stronger binding of the molecule to the surface. However, additional interactions, e.g., the interaction of the vibrating molecules with each other through their oscillating dipole field interactions, may exert a significant influence on the vibrational frequencies.

The comparison of the IR spectra of adsorbed CO on particles with rather homogeneous size distributions with those of electrodes prepared from an aged colloidal solution with a very inhomogeneous size distribution is very instructive regarding the influence of particle coverage and structural information provided. Summarizing the information from TEM images, the aged colloid contains particles of about 2 nm which should exhibit bulk-deviating properties as in the case of the original colloid and much larger particles that should manifest properties similar to that of macroscopic polycrystalline Pt. A count of the distinct particles from TEM images of the aged colloids gives a total ratio between smaller and larger particles of $N_{2.2}/N_{8.5} = 21.2$ (including the aggregates), and from this value an estimate of the ratio of the areas, assuming a spherical shape. This estimate yields a value of $A_{2.2}/A_{8.5} \approx 1.4$. However, the Pt contact areas of the aggregates are not taken into account, and thus this estimate yields a lower limit for the ratio and this value may well be higher by a factor of 2.

IR spectra of electrodes with different particle coverages and prepared from aged colloid solution and gold substrates are shown in Figure 6. A strong dependence of the spectra on Pt particle coverage is again evident in the spectra. At low coverage ($\Gamma_{Pt} = 0.06$ and $\Gamma_{Pt} = 0.11$) the inhomogeneous size distribution is visible in the IR spectra since two vibrational bands assignable to linearly bonded CO are seen to be present in the spectra.

The lower-frequency band at about $2015\,\mathrm{cm}^{-1}$ coincides with the frequency observed for the electrodes prepared from the original colloid at low coverages. This band is therefore assigned to CO adsorbed on isolated 2.2-nm particles. Consequently, the higher-frequency band at $2046\,\mathrm{cm}^{-1}$ is assigned to CO on larger particles and agglomerates of particles. For the spectrum of the lowest Pt coverage, the integrated band intensities of the two bands are determined by fitting the band. The ratio between the intensity of the lower-frequency band to the intensity of the higher-frequency band is 2.3, in acceptable agreement with the particle ratios from the TEM images. At intermediate particle coverage, only one band is clearly seen in the spectra, but the shoulder at lower frequencies may still relate to the second band. At high coverages ($\Gamma_{Pt} = 0.29$ and $\Gamma_{Pt} = 0.42$) the band is broad and asymmetrical with a peak frequency about $15\,\mathrm{cm}^{-1}$ lower than on polycrystalline Pt. Such a CO stretching vibrational band is typical of very dispersed Pt electrodes. From Figure 6, it is evident that the lower-frequency band shifts noticeably with increasing Pt coverages, whereas the higher-frequency band shows only a very small change. In this respect, the band at $2015\,\mathrm{cm}^{-1}$ again mirrors the Pt coverage dependence of the band in Figure 5.

In order to discuss the IR spectra in terms of the structure and size of the particles, it is necessary to assess the importance of the adsorbate–surface and intermolecular interactions for the internal vibrational modes. The interactions discussed in the literature to cause frequency shifts are dipole–dipole field interactions between the vibrating molecules, the chemical interaction mediated directly by the electronic structure of the metal (including the interaction with the image dipole of the molecule), and the indirect molecule–molecule interaction mediated via the substrate lattice [88]. The singleton frequency of a molecule-metal system corresponding to the vibration of a single adsorbed molecule in the absence of any coupling to other CO molecules or of any influence of other adsorbed molecules is very important in this respect. However, some studies include the chemical contribution, therefore giving rise to a singleton frequency that depends on CO coverage [89]. In the study by Friedrich et al. [84], singleton wavenumbers are defined to be the values obtained in the low-CO coverage limit. Although numerous other studies exist that have determined the vibrational frequency of CO on single-crystal electrodes [90], Friedrich's study did not find a value corresponding to polycrystalline platinum under the experimental conditions mentioned. Hence, the vibrational frequency of linearly adsorbed CO on polycrystalline Pt as a function of

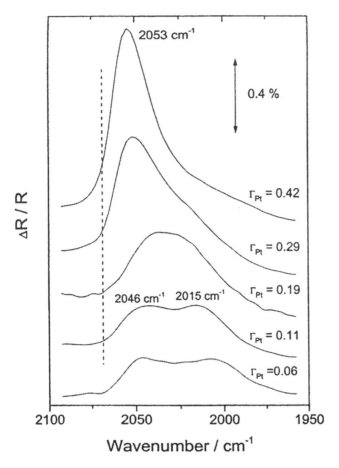

Figure 6 IR spectra of the stretching vibration of CO adsorbed on Pt particles supported on gold and prepared from an aged colloid with an inhomogeneous size distribution at different particle coverages. All spectra are recorded at $100\,\text{mV}_{\text{RHE}}$. The dashed line indicates the peak frequency of the stretching vibration of CO bonded linearly on polycrystalline Pt. (From Ref. 84.)

CO coverage was measured. Figure 7 shows the IR spectra of the stretching vibration of linearly adsorbed CO (at 80 mV) versus decreasing CO coverage, which is evident in the accompanying decrease in the band intensity.

From Figure 7 it is obvious that the vibrational frequency is lowered considerably from 2066 to $2001\,\text{cm}^{-1}$, with decreasing CO coverage. The integrated intensity is determined by fitting the bands with a Voigt profile. The peak wavenumber of the vibrational bands is plotted versus the IR intensity (as a measure of CO coverage) in the inset in Figure 7. Although the relationship between frequency and IR intensity is nonlinear and complex at high coverages, it becomes linear when approaching the low coverage limit, and thus a value of $1995\,\text{cm}^{-1}$ for the singleton frequency can be accurately determined. This number is also used for an estimate of the dipole field contribution to the vibrational frequency, presuming that this singleton frequency is similar for the supported particles on gold.

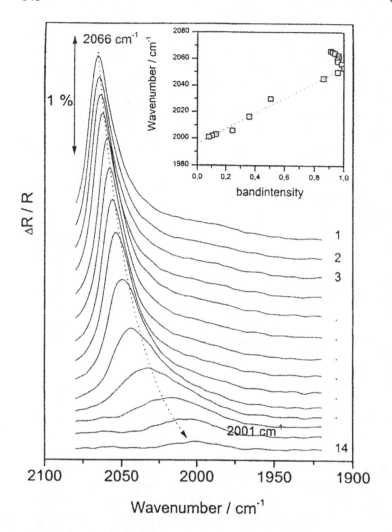

Figure 7 IR spectra of the stretching vibration of a partial CO coverage on a polycrystalline Pt electrode. The partially filled CO layer is created in a potential sweep of 100 mV/s with varying positive potential limits. All spectra are recorded at 80 mV$_{RHE}$. Spectrum 1 corresponds to a measurement without potential changes. Spectra 2–14 correspond to measurements after a potentiodynamic cycle with changing positive limits from 255 mV to 675 mV. Inset: Dependence of the vibrational peak frequency ω on the CO coverage. Determination of the singleton frequency ω_s by extrapolating to zero CO coverage. (From Ref. 84.)

Because of this strong CO coverage dependence, it is important for the interpretation of the IR spectra to estimate the CO saturation coverage of the Pt particles. This can be done by measuring the CO oxidation charge (normalized versus the Pt oxide reduction charge) determined from the CVs of the electrodes as a function of Γ_{Pt}. Values for the CO oxidation charges at high particle coverages are comparable to polycrystalline Pt, but much higher values are seen for low particle coverages. This latter observation is presently not understood and needs further investigation. However, it can be concluded that the CO coverage on the particles corresponds at least to the saturation value of polycrystalline Pt, and therefore the IR spectra of the CO saturated particles (e.g., vibrational frequencies) should be correlated with the spectra of CO on polycrystalline Pt at saturation coverage.

Thus, the main divergences between the CO-covered Pt particles compared to polycrystalline Pt are observed in the CO oxidation at higher potentials and in lower vibrational frequencies ($\Delta\omega = -50\,\mathrm{cm}^{-1}$). In addition, an influence of particle size is evident in the spectra of the electrodes prepared from a colloid with an inhomogeneous size distribution. These deviations, however, are only evident at low coverage. A detailed theoretical background to the significantly lower vibrational frequencies found on the particles at low particle density, the resultant size effect of the IR spectra, and a reasoning for the changed catalytic properties of the electrode with respect to the oxidation of CO_{ads} are presented in the literature [84,91] and are not discussed in detail here. In summary, it is obvious that the bulk-deviating properties of the particles are exposed at low coverages. Such observations can be explained either by simple dipole–dipole interactions of the adsorbate layer or by more refined electronic interactions, e.g., the degree of coordination of the Pt bonding site or substrate–particle interactions leading to charge transfer. An estimate of the dipole coupling of the adsorbates shows that this interaction can explain neither the low frequency nor the size effect. Similarly, the dependence of vibrational frequency on the coordination number of the adsorbate site is too weak to account for the measured wavenumbers. Therefore, an increased back donation to the antibonding CO $2\pi^*$ orbitals is assumed, and the prominent coverage dependence of the frequency favors a substrate contribution. The CO oxidation on the supported particles proceeds at a higher potential compared to polycrystalline Pt in a potential sweep experiment. This changed catalytic property is mainly attributed to a higher CO coverage on the small particles at low particle coverage compared to polycrystalline Pt, indicating a higher adsorption energy of CO, but this needs further work and investigation.

From a follow up study involving mainly TEM measurements and electro-chemical measurements on citrate-stabilized Pt particles on gold [91], it is seen that small particles (2–5 nm in size) are single crystalline, but the TEM images reveal that the larger particles (8–20 nm) consisting of aggregates of smaller particles resemble polycrystalline Pt. The CV of the CO monolayer oxidation on 3-nm Pt particles exhibits three oxidation peaks located at more positive potentials compared to a polycrystalline Pt electrode. The observed difference is especially clear for low particle coverages. On the contrary, the larger aggregate particles show CO monolayer oxidation potentials, which resemble the behavior of polycrystalline Pt. Particles that are clearly polycrystalline and about 10 nm in size still show significant variations in properties compared with polycrystalline Pt. These varying properties of the small particles of different sizes are evident in the CO oxidation transients. The time-dependent oxidation on the 3-nm particles differs from the classical Langmuir–Hinshelwood mechanism and can be better described by an Eley–Rideal mechanism [91,92]. This can be explained either with the absence of oxygen adsorption or by the absence of transportation processes on the electrode surface. The larger particles exhibit bulk behavior in this respect. However, it is still an open question whether the novel properties of the small particles are due to a pure size effect or are dominated by the structural properties of the particle (single-crystal versus polycrystal). Recent preparation procedures for obtaining larger single-crystal colloidal Pt particles [93,94] should mean that the investigation of electrodes prepared with such particles with regard to CO electro-oxidation will enable one to distinguish surface structure from particle-size effects.

So far we have demonstrated how IR can be used to investigate adsorbates on well-defined model electrode surfaces. Since the understanding of the interaction of CO with Pt is theoretically well advanced, it is possible to interpret the spectra of structurally complex catalysts in terms of the simpler model systems discussed here, by comparison of the IR spectra of adsorbed CO as a probe molecule on technical catalysts and on monodisperse Pt particles. Figure 8 shows the IR spectra of the stretching vibration of CO adsorbed on 10 wt.% carbon-supported Pt ETEK catalyst at different recording potentials.

There are many similarities to the results observed for the model catalysts. The main features to be noted are that it is mostly on-top bonded CO that is also observed for the technical catalyst, with two bands assigned to linearly bonded CO

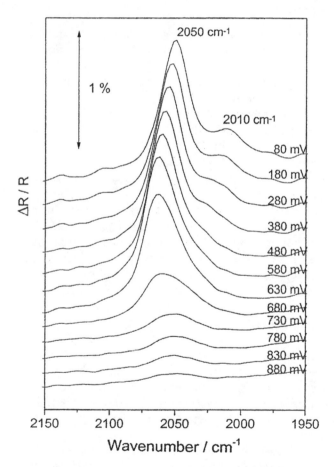

Figure 8 IR spectra of the stretching vibration of CO adsorbed on carbon supported 10 wt.% Pt ETEK catalysts, as a function of measurement potential versus RHE.

found at 2050 cm^{-1} (main band) and 2010 cm^{-1} (satellite band). The low-frequency band at 2010 cm^{-1} is associated with isolated particles (2–4 nm, from TEM measurements), while the higher vibrational frequencies of 2050 cm^{-1} are associated with agglomerated particles. These vibrational frequencies are very close to those observed for the model Pt particles on the Au system (Figure 6). Inhomogeneous band broadening is also observed, due to interaction of the CO molecules on different adsorption sites (terraces, kinks, steps, etc.). Higher frequencies are observed with increasing potential, which is believed to be due to the back-donation effect described above for model catalysts. Once again, as for the model catalysts, a strong dependence of the vibrational frequency on particle coverage is observed for the technical system, which can be interpreted in terms of dipole–dipole interactions. This dependence is a consequence of increasing agglomeration with particle coverage. Therefore, it is obvious that the spectra of the complex technical electrocatalysts can be interpreted in terms of the model systems, and such experiments show the versatility of IR spectroscopy, not only for studies of the fundamental science of model electrodes, but also for application in technical catalyst systems.

The PtRu Anode Electrocatalyst

Since on pure platinum, methanol oxidation is strongly inhibited by poison formation, bimetallic catalysts such as PtRu or PtSn, which partially overcome this problem, have received renewed attention as interesting electrocatalysts for low-temperature fuel cell applications, and consequently much research into the structure, composition, and mechanism of their catalytic activity is now being undertaken at both a fundamental and applied level [62,77]. Presently, binary PtRu catalysts for methanol oxidation are researched in diverse forms: PtRu alloys [55,63,95], Ru electrodeposits on Pt [96,97], PtRu codeposits [62,98], and Ru adsorbed on Pt [99]. The emphasis has recently been placed on producing high-activity surfaces made of platinum/ruthenium composites as a catalyst for methanol oxidation [100].

It is worth noting that despite the diversity of methods for catalyst preparation, all these materials present an enhanced activity toward methanol oxidation. Although the enhancement effect of Ru on methanol oxidation has been well known for decades and has long been considered in the development of fuel cells, many details concerning the enhancement in catalytic activity, especially concerning the mechanisms on the atomic scale, are not yet well understood. Apart from the electronic effect of Ru on the bond strengths of the adsorbates [101], a bifunctional mechanism is considered to be responsible for the enhancement effect [102]. The latter effect involves the adsorption of some oxygen-containing species on ruthenium atoms at, compared to platinum, lower potentials. This species, in turn, is necessary for the oxidation of intermediates such as CO or COH. By this means, the onset of methanol oxidation is shifted from around 450 mV for pure platinum catalysts down to 250 mV for PtRu electrodes. This can easily be verified by electrochemical online mass spectroscopy (DEMS), which allows the detection of the onset of CO_2 formation at the respective potentials [103].

The newest trend has been to deposit controlled amounts of ruthenium on the well-defined platinum single-crystal substrates of different crystallographic orientations. This approach allows one to investigate surface-structure effects in PtRu methanol oxidation electrocatalysis. In all the studies addressed in the present work, only the Pt(111) surface [out of other Pt(*hkl*) surfaces] is investigated because, when covered with ruthenium, it is the most active catalyst for methanol oxidation known to date. Much of the data presented here are not only significant for increasing the understanding of methanol oxidation mechanisms on mixed-metal, catalytic electrodes, but also further the understanding of noble metal on noble-metal deposition processes.

Ru Evaporation onto Platinum in UHV Conditions [Pt(111)/Ru]

Several methods of preparation and analysis of PtRu model electrodes [63,97,102], some of which involve transfer between UHV and electrochemical environments, have been studied. In order to relate the surface structure and catalytic behavior of PtRu electrocatalysts investigated so far, distinct crystal faces and preparation methods were often used, the latter leading to, respectively, different surface and ad-atom distributions. Many of the preparative techniques, however, allow only limited control over the resulting surface structure, and generally the Ru concentration rather than its geometric distribution may be controlled. The situation is different for UHV studies of Pt, Ru, or mixed model surfaces. Most of the UHV systems offering STM analysis do not offer the possibility of a clean transfer into an electrochemical environment. Therefore, the respective surface characterization of the catalytic behavior is often restricted to gas-phase experiments. But especially in the case of methanol oxidation, the reaction mechanisms in the gas phase differ strongly from those in an electrochemical environment. To fill this gap, Vielstich et al. [104] designed a UHV/electrochemical cell apparatus. Thus, both the preparation of Pt(111)/Ru surfaces in UHV and their structural analysis by in-situ STM, as well as study of the electrocatalytic activity, become possible.

Evaporation of ruthenium onto Pt(111) surfaces leads to the formation of 3D islands having a width of 5–10 nm, as shown in Figure 9. The number of Ru edge sites can be increased until island coalescence occurs. The second layer of the islands is filled to a higher extent than kinetically necessary. Ru atoms that fall on one of the pre-existing islands are not able to overcome the barrier at the island edge and will thus stay on the island; however, in the second layer one encounters more material than is expected from the total coverage. Ru tends to nucleate at the ascending steps of each terrace, however, in contrast to most other known systems presenting epitaxial growth, and this decoration also provides nucleation centers for islands at the upper terrace, which can be seen in the higher island density near the descending step, i.e., the steps are decorated from both sides. In Figure 9(b), the number of PtRu pair sites is higher than in Figure 9(a), and this nearly leads to island coalescence. Similar behavior has been reported for Ru spontaneously adsorbed on Pt(111) [105] and is discussed ahead.

By ion sputtering at room temperature, an artificial increase in the number of steps prior to Ru evaporation is achieved. Removing about 1 ML—i.e., digging grooves into the terraces with a depth of up to 3 crystal layers—keeps the surface STM analyzable. The resulting morphology is shown in Figure 10(a).

At room temperature, only the movement of single Pt atoms and small clusters on the terraces and islands is activated, but no interlayer mass transport or even diffusion occurs along step edges. Thus, most of the large defects created by the ion bombardment are stable (apart from atom diffusion and formation of small islands),

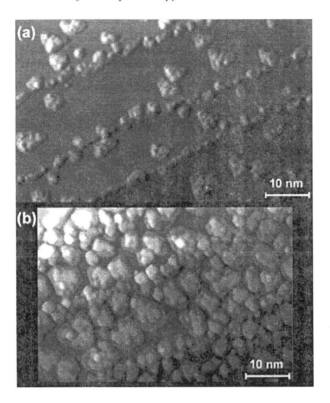

Figure 9 Ru island growth on Pt(111), preparation by Ru vapor deposition in UHV, STM images taken in the same chamber. Ru coverage: in (a) about 25%, in (b) about 60%, estimated by height statistics. (From Ref. 104.)

and no preferential island geometry is visible. After evaporation of Ru onto the surface, very few islands are formed (necessitating larger terraces) and all material is decorating the Pt steps and islands. Figure 10(b), which involves use of the "statistical differencing" method [106], shows that it is possible to visualize the increased roughness of the step edges due to the Ru decoration. It can be seen that the overall number of PtRu neighbor sites is higher than that for the other shown surfaces. The ruthenium distribution on the surface where ion bombardment was carried out after ruthenium deposition could not be visualized, as the entire surface including the ad-atom islands was disordered, with Pt and Ru randomly intermixed.

Iwasita et al. [107] also studied the electro-oxidation of methanol on Ru-evaporated Pt(111) modified electrodes with different Ru coverage. The surface compositions were characterized by cyclic voltammetry and Auger spectroscopy. The topography of the UHV-prepared deposits was observed by STM. Figure 11 shows STM data for a Pt(111) electrode without (a) and with (b–f) Ru layers formed by vapor deposition.

Several pertinent remarks concerning the surface topography can be made. Ru is deposited in the form of islands having a height of 0.21–0.22 nm. Given a Ru–Ru distance of 0.214 nm in the (001) direction, it is obvious that under these conditions,

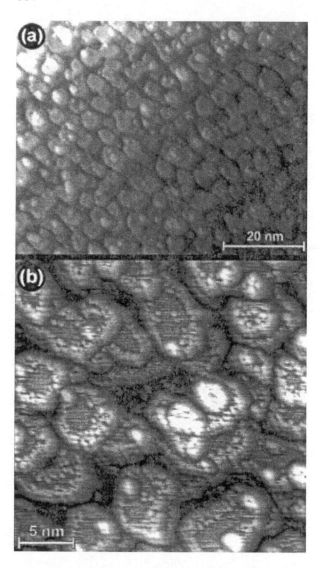

Figure 10 Preparation of a Pt(111)/Ru surface with increased number of defects and Pt–Ru pair sites. (a) After ion bombardment with 500 eV Ar$^+$ at 300 K; (b) after deposition of Ru on the surface in (a). Island edges are decorated by Ru, indicated by the slightly increased roughness. Contrasts have been increased by image processing techniques for better visualization in printed form. (From Ref. 104.)

the islands are monoatomic. With increasing coverage the diameter increases and a second layer is formed on top [e.g., Figure 11(d), for $\Gamma_{Ru} = 0.25$, the Ru amount is 0.4 ML, or a fraction of a monolayer]. Their average diameter is 1–6 nm, depending on the coverage. The island sizes have been determined using line profiles, assuming a broadening of about 1 nm caused by the tip. For the catalytic properties under investigation, only the order of magnitude of the island is of interest. In the whole range of Ru coverage, the density of the islands remains nearly constant, a result obtained from a plot of island density versus the amount of deposited Ru. These model electrodes have been tested as catalysts for methanol oxidation [107].

In conclusion, it seems that for the catalytic activity of the Pt(111)/Ru surface toward methanol oxidation, the total number of PtRu neighbor sites is an essential quantity that may be specifically increased, creating additional surface defects by ion sputtering before or after Ru deposition. In contrast, smooth PtRu surface alloys prepared by implanting Ru into Pt(111) terraces show surprisingly low activity, which exhibits a fast exponential decay toward zero [104]. This strongly emphasizes the importance of low-coordinated Ru sites at edge positions for the performance of the bifunctional mechanism at Pt(111)/Ru surfaces. These and other differences found with rough PtRu electrodes freshly prepared by electrochemical codeposition [105] show that behavioral differences between smooth and porous surfaces are most likely due to structural and/or electronic effects: other local atomic configurations may hamper a reaction pathway, leading to deactivation, and the electronic properties of very rough surfaces are also likely to be different.

Ru Electrodeposited on Pt(111)

Several studies have used Pt electrodes modified by Ru electrodeposition as a model Pt/Ru electrocatalyst, which show an enhanced electrochemical activity for the oxidation of H_2/CO gas mixtures as compared to pure platinum [108,109]. The current efficiency of Ru deposition from $RuCl_3$ solutions is low; consequently, this allows easy control of the amount of deposition in the submonolayer regime onto Pt(111) or polycrystalline Pt. The details of the deposition procedure are described elsewhere [108]. The electrodeposition of Ru from solutions of $RuCl_3$ in H_2SO_4 was investigated over a wide range of deposition potentials [110]. For potentials positive of 0.9 V (RHE), the voltammetry of the Pt(111) surface remains unaffected, indicating that no deposition occurs in this potential range. Between 0.8 V and 0.3 V, a gradual disappearance of the typical features of the voltammogram of the Pt(111) electrode is evident, and around 0.2 V, a transition to the interfacial characteristics of pure polycrystalline Ru occurs, i.e., a massive growth of Ru metal occurs. Once again, in the submonolayer regime, the surface coverage is seen to increase linearly with decreasing potential, from 0.2 ML at 0.8 V, to 0.51 ML at 0.25 V.

Initial attempts to monitor the deposition reaction in real time using STM under in-situ electrochemical conditions gave inconsistent results. These were rationalized by assuming that the tunneling tip locally enhances the metal deposition reaction. This assumption is supported by the observation that after a long experiment, a brown spot of about 0.5-mm diameter was visible on the crystal surface at the location of the tip. Much work will be necessary before such an approach can become common practice. Thus, the deposition was performed in a conventional electrochemical cell, and the modified surface was subsequently imaged, after transfer to the STM cell.

STM results performed with a Pt(111) substrate are shown in Figure 12. The electrode surfaces for these experiments were prepared in a standard electrochemical cell and then, protected with a drop of water, transferred to the STM cell.

Figure 12(a) corresponds to a blank experiment, showing the Pt(111) surface without Ru deposit. Terraces larger than 50 nm are clearly identified and are separated by monoatomic steps. A moderate density of islands on the terraces can be attributed to Pt islands. Figure 12(b) and (c) show two examples of Ru-modified

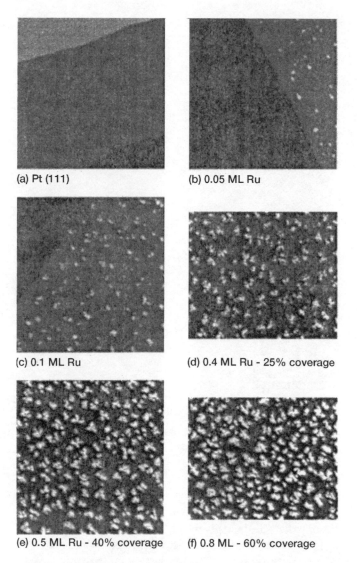

Figure 11 STM images (93 nm × 93 nm) showing (a) a clean Pt(111) surface and (b–f) island formation after Ru evaporation on Pt(111); ML amount of Ru expressed as fractions of a monolayer, percentage of covered surface as indicated. Note the 3D formation for 0.4 ML: all images have been obtained with +0.4 V sample bias and 1 nA tunneling current. (From Ref. 107.)

Pt(111) electrodes. Quite a high density of islands can be seen in Figure 12(b): the islands are between 2 and 5 nm in diameter, and their height is close to the height of the monoatomic step of the substrate. The islands are statistically distributed on the substrate surface. The fraction of the surface covered with Ru islands, Γ_{Ru}, has been estimated for several images and yields values between 0.2 and 0.4, with a mean value of 0.25. This is reasonably consistent with $\Gamma_{Ru} = 0.3$, which was determined for the polycrystalline electrode subjected to the same deposition conditions. These

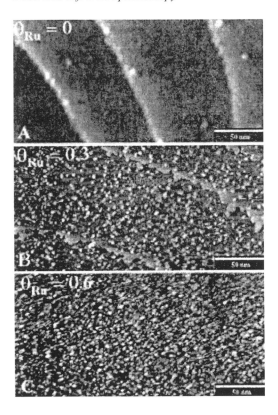

Figure 12 In-situ STM images of Pt(111) electrodes in $0.1 \, mol \, dm^{-3}$ $HClO_4$ at an applied potential $E = 0.5 \, V$ versus RHE, $E_t = 0.8 \, V$, $I_t = 0.4 \, nA$. (A) Without Ru deposit; (B) after Ru deposition at 0.6 V, and $\Gamma_{Ru} = 0.3$; and (C) after Ru deposition at 0.3 V, and $\Gamma_{Ru} = 0.6$. (From Ref. 108.)

results seem to confirm Watanabe and Motoo's approach [111] and the conclusion that the observed islands represent the Ru deposit, which has also been confirmed with x-ray surface diffraction data (XRD) [112], and x-ray photoelectron spectroscopy (XPS) experiments [108,113]. Because the structural data obtained from STM are not always unambiguous (e.g., difficulties in distinguishing Pt and Ru on the surface), independent XRD measurements provide a three-dimensional structure model of the interface and is a sensitive technique for measuring heights of samples to fractions of angstroms. It was seen from a series of XRD experiments on Ru deposition onto Pt(100) surfaces that the Ru atoms form monoatomic islands on the Pt surface [112]. A monoatomic height of the Ru deposit was found. This represents important information for the interpretation of the STM images, since it will be shown later that different structural results were obtained for Ru spontaneously adsorbed onto platinum [105]. It is also clear that the rate of deposition of Ru is similar for both polycrystalline Pt and Pt(111). No influence of the deposition time or deposition potential was detected. This indicates a limitation of the island growth due to either thermodynamic or kinetic reasons. If the island growth is thermodynamically limited, then one would expect a potential dependent deposition rate. In view of the small lattice mismatch between Pt and Ru of around 2.5% [110], thermodynamic reasons should not be dominant. Thus, a kinetic limitation, where Cl^- adsorption may play an important role, is considered more likely. To further check the results, a Ru-modified Pt(111) electrode, with $\Gamma_{Ru} \approx 0.7$, was imaged and is shown in Figure 12(c). From a comparison of the images (b) and (c), it looks likely that the Ru coverage increases by about a factor of two.

The possibility of determining the Ru surface concentration indirectly by IR spectroscopy has also been investigated [108]. This is accomplished by adsorbing a monolayer of the CO on the composite electrodes. In-situ IR spectroscopy of the CO stretching vibration shows distinct bands for CO adsorbed on Pt and on Ru and thus can be used to quantitatively determine the coverage of Ru deposits on the composite electrode surfaces.

Figure 13(a) shows IR reflection absorption spectra from CO-covered Pt(111) surface with different Ru coverages at an applied potential of $400\,mV_{RHE}$.

On the clean Pt(111) surface in Figure 13(a), the bands at about 2070 and $1790\,cm^{-1}$ are the well-known bands assigned to CO_{ads} in a linear and threefold hollow coordination, respectively [114]. In Figure 13(a) part (2), referring to Pt(111) with $\Gamma_{Ru} \approx 0.25$, a new vibration band around $2010\,cm^{-1}$ is observed. Since the intensity of this band increases with increasing Ru coverage and its integrated intensity is higher than that of the linearly bonded CO_{ads}/Pt, for the surface with the highest Ru coverage $\Gamma_{Ru} \approx 0.6$, Figure 13(a) part (4), this band is assigned to CO linearly bonded on Ru. Quantitative coverage determinations from the integrated intensities of surface vibrational bands are often intricate due to the strong lateral dipole interaction leading to intensity transfers from low-frequency bands to high-frequency bands [115]. Nevertheless, an independent estimate of the extent of Ru coverage on the surface is attempted by integrating the CO_{ads}/Ru band intensities and relating them to the CO_2 band intensity (the product present after electro-oxidation of CO), in order to account for experimental differences in the measurement. Because the composite model electrode consists of separate metallic regions of mesoscopic sizes (see Figure 12), it is possible to assume that the long-range dipole interaction, between adsorbates on Pt and Ru, may not be important. Figure 13(b) shows the ratio of the integrated band intensity of the CO_2 band versus deposition potential, revealing a linear dependence. The insert in Figure 13(b) shows the typical normalized spectra with the vibrational bands used in this procedure. The linear relationship between the intensity ratio and deposition potential (equivalent to that found from XPS analysis) shows that direct correlation with Ru coverage is possible. Therefore, it is possible to state that apparently the dipole–dipole interactions of adsorbates on Pt and Ru are not important, and because the deposition rate of Ru on Pt is independent of potential, the CO saturation coverage on Ru is seen to be independent of the Ru coverage on Pt(111).

An investigation, using electrochemical techniques and IR spectroscopy, of the electrocatalytic properties of submonolayer electrodeposits of Ru on Pt substrates led to an observation of enhanced catalytic properties of Pt(111) by the Ru islands. This is seen as evidence that surface diffusion of the adsorbed CO is essential for understanding the electro-oxidation kinetics, and it provides a basis for understanding the bifunctional mechanism [111].

From chemical and structural analysis [110], one may expect—as a first approximation—that the catalytic properties of the Ru-modified Pt(111) surface are

(a)

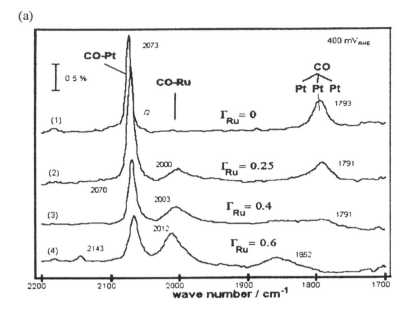

(b)

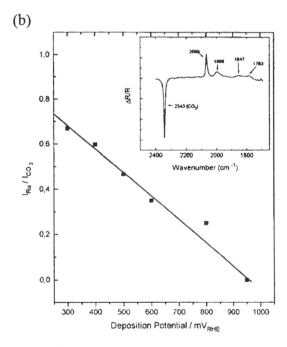

Figure 13 (a) In-situ IR spectra of Ru-modified Pt(111) with varying Ru coverages, measured in $0.1 \, mol \, dm^{-3}$ $HClO_4$ after CO adsorption at $90 \, mV$; 400 scans signal averaging at a resolution of $4 \, cm^{-1}$; (b) integrated band density corresponding to CO_{ads}/Ru related to the band intensity of CO_2 plotted versus Ru deposition potential. The insert illustrates a typical normalized spectrum with positive vibrational bands for the CO adsorbates present at the applied potential and a negative band at $2343 \, cm^{-1}$ corresponding to CO_2 (reaction product) present at the reference potential of $800 \, mV$. (From (a) Ref. 110, (b) Ref. 108.)

a superposition of the properties of both metals. In fact, vibrational spectroscopy of adsorbed carbon monoxide shows the typical band for CO adsorbed on Ru around $2000\,cm^{-1}$, superimposed with the well-known spectral features of CO on the Pt(111) surface, as shown in Figure 13(a).

For the electro-oxidation of an adsorbed monolayer of CO, however, synergistic effects between Pt and Ru are obvious from the time dependence of the IR spectra, i.e., there is a coupling of the reaction on the composite surface. Figure 14(a) shows the temporal variation of integrated band densities after a potential step from 90 mV to 450 mV. At 450 mV no CO is oxidized on Pt(111), and one would expect an uneven concentration of CO on Ru or Pt, reflected in the IR–CO stretching band.

However, the intensities for on-top bonded CO on Ru and on Pt sites decrease simultaneously on the time scale of minutes. This is supported by a simultaneous increase of a band at $2345\,cm^{-1}$, attributed to CO_2 in the solution phase. If only superimposition of the catalytic properties occurred, then a significantly faster oxidation of CO on the Ru sites than on the Pt sites would be expected. A similar result is seen from potential dependence data of the IR spectra, with the intensities for on-top bonded CO on Ru and Pt decreasing at the same potential, approximately 450 mV; Figure 14(b). The intensity for bridge-bonded CO on Pt also diminishes at this potential. The adsorbate structure of CO on the Pt areas is disturbed, since the threefold hollow site disappears and the bridge site appears. An unchanged CO adlayer structure is assumed compared to Pt(111), and thus patches of Pt(111)/CO_{ads} and Ru islands are on the surface [116].

In blank experiments with the unmodified Pt(111) surface, no changes in the IR spectra occurred on the same time scale, even when the step potential was increased to 0.55 V. This is clear evidence that Pt(111) is not active for the electro-oxidation of CO_{ads} in this potential range. Hence, the Pt areas in between the Ru islands behave quite differently with respect to CO oxidation than the unmodified substrate.

Dissociative water adsorption, which is a prerequisite for CO electro-oxidation, starts at more negative potentials on the composite Ru/Pt(111) electrode than on the Pt(111) surface. A plausible explanation is that an activation of the Pt surface due to long-range electronic interactions with the Ru islands occurs, facilitating OH adsorption at lower potentials. However, because IR spectroscopy of CO_{ads} is very sensitive to electronic influences, but gives no indications for such an effect, this explanation is rather unlikely.

As a second possibility, one may assume that at low potentials OH adsorption occurs exclusively on Ru sites, in agreement with the bifunctional mechanism for CO oxidation on PtRu alloy surfaces. The observed depletion of CO on the Pt sites can then only be explained by rapid surface diffusion of the adsorbed CO molecules to the reactive Ru islands. Because the separation of the active Ru islands is on the nm scale, even a low mobility of the adsorbed CO can account for the observed catalytic enhancement.

In order to evaluate details about CO surface diffusion, the transient oxidation of absorbed CO monolayers under potentiostatic conditions was investigated and compared with the behavior of the Pt(111) surface. Several current transients following a potential step from 0.1 V, where a CO monolayer is adsorbed, to varying oxidation potentials $0.45\,V \leqslant E_{ox} \leqslant 0.6\,V$ in CO-free solution are shown in Figure 15.

(a)

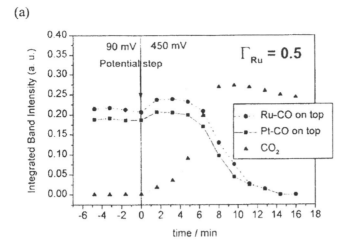

(b)

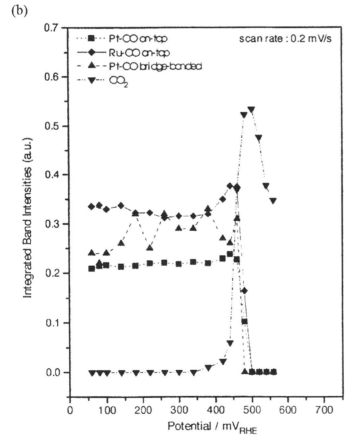

Figure 14 (a) Time dependence of integrated IR band densities of CO on Pt [linear (on-top)], CO on Ru [linear (on-top)] and solution phase CO_2 ($\Gamma_{Ru} = 0.5$), following a potential step from 90 mV to 450 mV; (b) potential dependence of integrated band intensities of the same stretching vibrations on Ru/Pt(111) ($\Gamma_{Ru} = 0.6$), scan rate 0.2 mV/s; both (a) and (b) recorded in 0.1 M $HClO_4$. (From (a) Ref. 110, (b) Ref. 116.)

It should be noted that the time axis is displayed on a logarithmic scale and the currents are normalized to the peak currents. Essentially the same shape with only different displacements on the log t-axis is observed for all transients investigated. From Figure 15 it is obvious that at the same oxidation potential of 0.6 V, the CO_{ads} oxidation is faster by 3 orders of magnitude on the Ru-modified surface than on the unmodified Pt(111) substrate. Even at a potential 0.15 V lower, the reaction is still a factor of 5 faster on the Ru-modified electrode.

A lower limit for the surface diffusion coefficient of CO on Pt(111) can be estimated from Figure 15. At a potential of 0.6 V, essentially all CO molecules are oxidized on the Pt(111)/Ru electrode within 1 s, whereas in the same time interval only a negligible fraction of CO is oxidized at the Pt(111) surface. Thus, one can approximate that, within 1 s, CO molecules are able to travel half the distance between neighboring Ru islands. For the surface considered with $\Gamma_{Ru} = 0.25$, the mean island separation is about 4 nm, and hence a diffusion coefficient $D_{CO} \geqslant 4 \cdot 10^{-14}$ cm^2 s^{-1} results as the lowest limit. If diffusion of CO adsorbate is rate-limiting, then the CO diffusion coefficient must be higher than $1 \cdot 10^{-12}$ cm^2 s^{-1}. The transients in Figure 15 can be compared with the measurements of Gasteiger et al. [117], who performed similar measurements with a smooth polycrystalline $Pt_{54}Ru_{46}$ alloy electrode. For this surface, a statistical distribution of Pt and Ru was assumed; thus, the distance for surface diffusion of CO_{ads} to reactive sites should be small. The current transients measured on the alloy electrode are very similar to the transients of Figure 15, except that the maxima are shifted to shorter times by a factor of 2. The decay of the transient current after the maximum is essentially the same on both surfaces. Assuming that the diffusion distance on the alloy surface is negligible compared to the characteristic 2 nm for the Ru-modified Pt(111), this similarity indicates that the mobility of CO is significantly higher than the calculated lower-limit value for the diffusion coefficient. Values for D_{CO} in the range 10^{-13}–

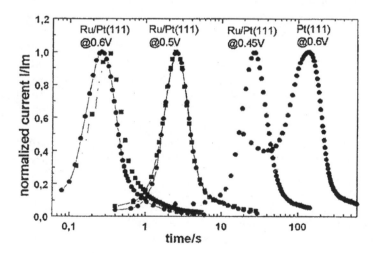

Figure 15 Current transients of potentiostatic CO monolayer oxidation in 0.1 mol dm^{-3} HClO$_4$ for Ru/Pt(111) ($\Gamma_{Ru} = 0.25$) and Pt(111) at oxidation potentials of $0.45\,\text{V} \leqslant E_{ox} \leqslant 0.6\,\text{V}$. For the two faster transients, data from different electrodes with the same preparation are superimposed in order to indicate the reproducibility. (From Ref. 110.)

$10^{-12}\,\text{cm}^2\,\text{s}^{-1}$ seem reasonable. It is worth mentioning that in a recent study on the electro-oxidation of CO at Pt/Ru electrodes, Petukhov et al. [118] estimate that with diffusion coefficients in this range, a single CO molecule makes at least 10^5 jumps between nearest adsorption sites during the time $\sim 100\,\text{s}$ needed for the reaction. The reaction is seen to accelerate with increasing defect density.

To conclude, the Pt(111) surfaces are seen to be modified by Ru islands of $\approx 3\,\text{nm}$ in size and monoatomic height. An ex-situ XPS analysis of the Ru-modified surfaces yields a linear dependence of the Ru coverage with the deposition potential over a wide range of 0.5 V [108,110], allowing a precise tuning of the nm-scale structural properties of the Ru/Pt(111) model electrode. On the Ru-modified Pt(111) surfaces, a monolayer of adsorbed CO has been investigated by IR spectroscopy and distinct bands can be assigned to CO adsorbed on Pt and on Ru. An equivalent linear dependence on deposition potential is found for the integrated band intensity of the CO/Ru band, indicating that it is possible to determine the coverage of deposits from adsorbate IR spectroscopy. The linearity between deposition potential and Ru surface coverage over a wide range of 0.5 V allows a precise tuning of the nm-scale structural properties of the Ru/Pt(111) model electrode. Therefore, the CO adsorbate may be convenient as a probe molecule for area determination of metal deposits in electrochemistry, assuming the investigated deposits form areas that are in the nm range or larger. In spite of the complexity of the interfacial processes, the electrodeposition process is a versatile and easily controlled tool for the preparation of mesoscopically defined model electrocatalytically active surfaces. IR spectroscopy of CO_{ads}, in agreement with the structural characterization, shows that distinct areas of Ru metal exist on the otherwise undisturbed well-ordered Pt(111) surface. Nevertheless, CO oxidation is enhanced on the Pt sites due to the nearby presence of Ru islands. Time- and potential-dependent IR spectroscopy shows that the oxidation of CO on Pt sites and Ru sites is coupled. Vibration bands assigned to CO/Pt and CO/Ru disappear simultaneously upon oxidation. There is evidence that CO_{ads} molecules are mobile on the electrode surface and diffuse from unreactive Pt(111) domains to the reactive Ru islands. A characteristic displacement of 2 nm/s $(D_{CO} \geqslant 4 \cdot 10^{-14}\ \text{cm}^2\,\text{s}^{-1})$ is determined as the lower-limit value from transient CO monolayer oxidation on the Ru/Pt(111) electrode.

These results indicate that surface diffusion of adsorbed molecules is important for understanding interfacial electrochemical reactions, with, in this case, initiation of CO oxidation on or around the Ru islands in combination with a high CO_{ads} mobility. Thus, as a result of this mobility, the structural properties of the electrode surface on the nm scale, i.e., increased number of defects or deposited Ru ad-atoms, is seen to be crucial for electrocatalytic performance. The observed results are consistent with the bifunctional mechanism [111], where oxygen-containing species adsorb preferentially on Ru surface atoms and a sufficient surface mobility of $CO_{ads}/$ Pt is seen in order to explain the coupling between the distinct regions on the composite electrodes. This structure sensitivity of anodic CO oxidation was confirmed in a recent second harmonic generation (SHG) study of Pt electrode surfaces, by Akemann et al. [119], with step defects seen to significantly enhance the reaction rate, with the CO oxidation reaction enhanced at lower potentials on Pt(997) compared to Pt(111). The terraces present on the Pt(997) surface are about 15 nm wide, and thus the CO has a shorter distance over which to diffuse to oxygen-carrying species (step defects in this case) than on Pt(111) [118]. In contrast to results seen for Pt(111), the reaction rate on Pt(997) is not directly related to the absence or presence of a stable CO adlayer, indicating an intricate interrelationship between step and terrace sites with regard to the catalytic activity of model electrodes. The overall result from these studies is that different oxygen-carrying defects (either Ru deposits or steps) are seen to enhance CO oxidation, i.e., oxidation occurs in an accelerated manner at lower potentials.

Ru Island Formation on Pt by Spontaneous Adsorption

Interestingly, Wieckowski et al. [105] found that spontaneous adsorption from a RuCl$_3$ solution onto a Pt(111) electrode surface does not develop a surface structure accessible to STM. However, when such species are subjected to a brief voltammetric treatment, the surface displays an array of Ru islands that are of nanometer size and largely monoatomic. An adlayer of adsorbed iodine (I) protects the electrode from contamination during the STM treatment, but more importantly since the electrode structure in solution is represented by iodine surrounding the ruthenium islands, the STM images obtained in air are a replica of the in-situ distribution of the islands, with molecularly resolved iodine structure aiding the resolution of Pt(111) terraces perturbed by the presence of Ru features. An STM image of the Pt(111) electrode obtained after ruthenium deposition for 90 s is shown in Figure 16. The presence of Ru features is seen by the clear white spots in the image, since, otherwise, there would be no perturbation of the crystallographically perfect, I-covered Pt(111) surface at this STM resolution. This is clear evidence that Ru is deposited as surface islands. The islands appear on the STM image even without iodine, thus showing their inherent stability.

As adsorption time is varied from 10 to 90 s, there is an increase in the uptake of Ru, as shown in Table 1. For the low coverage, 0.08 ML, obtained after 20 s of adsorption, the island distribution is not uniform, with some areas of the electrode displaying higher coverage. There is no significant enhancement in the island density at the steps, indicating that spontaneous deposition is not nucleated by the crystallographic defects at the surface. At low coverages, the islands appear to be rounded without any particular orientation with respect to the surface, and there seems to be no exclusion zone for island growth, with the formation of a new island occurring at any distance from a preexisting island. Some islands are seen to collapse while growing, forming a bigger island, which no longer has a rounded shape. The major outcome of Wieckowski's work is that he observes that at and above 0.14 ML

Table 1 Ruthenium Coverage and the Ru Island Heights as Determined by STM as a Function of Deposition Time

Deposition time (s)	Ru coverage (ML)	Relative height in island center (nm)
10	0.01	0.22
20	0.08	0.22
40	0.14	0.22 and 0.45
90	0.19	0.22 and 0.45
120	0.19	0.22 and 0.45

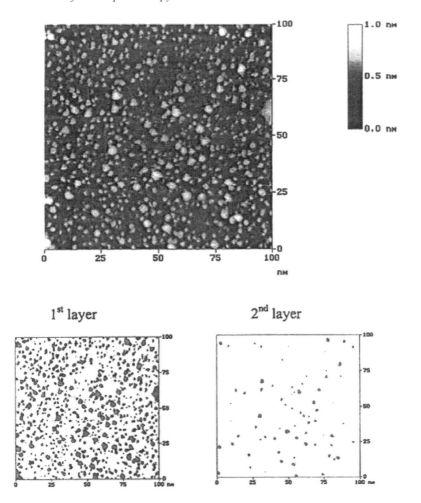

Figure 16 STM images of Pt(111) electrode (at 100-mV bias) obtained after ruthenium deposition for 90 s (top). Shown also are results of grain-size analysis of ruthenium island distribution for all islands (bottom left) and those from the analysis of bilayer ruthenium islands (bottom right). Ruthenium coverage is 0.19 ML. (From Ref. 105.)

(Figure 16), in addition to the monolayer islands in the image, there is also a second layer deposit on top of the inner layer, although the inner monoatomic layer has not yet been completed (and is far from approaching a monolayer coverage). This formation of bilayers at such low total surface coverage is surprising and in contrast to the data of the present authors' group obtained by electrodeposition on Pt(111) [108], by independent XRD data on the same system [112], and by Iwasita et al. using vapor deposition of Ru on Pt(111) in UHV conditions [107]. As such, it is a novel result that may provide a more detailed description of the surface properties of Ru deposits on Pt(111). For example, if Ru_2O_3 is formed as one of the components, the Ru_2O_3 surface molecule may assume the orientation requiring the Ru–Ru-axis to be parallel to the surface normal.

Thus, only monoatomic islands are found at the Ru coverage lower than 0.14 ML. At coverages of 0.14 and 0.19 ML, the bilayer character of the islands is quite distinct, especially at 0.19 ML. The height-resolved data at 0.19 ML indicate that the area of the surface covered by the bilayer islands is about 10% of the overall island population. Wieckowski believes that since this is a small percentage of an already low total ruthenium coverage, the bilayer islands may actually exist but escape detection at Ru coverages lower than 0.14 ML. It should be noted that from STM grain-size analysis, it appears that the development of the second Ru layer is restricted to islands that have a surface area of at least 2 nm^2. Extending the exposure beyond 90 s does not result in a higher ruthenium uptake (Table 1). This confirms previous data, that increasing Ru coverage above 0.20 ML by spontaneous deposition is practically impossible [97,99].

Thus, in summarizing this STM data, unlike the electrolytic deposition discussed in the previous section, where up to 0.7 monolayer (ML) coverage of ruthenium is deposited as mainly monoatomic islands with a tendency to create three-dimensional deposits as the coverage increases, when spontaneous deposition is used, about 10% of the islands are no longer monoatomic. Instead, such islands have a bilayer character, displaying a second monolayer deposit over the first monolayer. The result that such bilayer islands are formed at low coverage is related to the composition and morphology of the ruthenium deposits formed under a variety of electrochemical conditions.

A molecular insight into the catalytic oxidation of methanol was given by an in-situ FTIR spectroscopic study of these systems. Since methanol electro-oxidation is a surface-sensitive process, measuring infrared spectroscopy using single-crystal surfaces covered with submonolayers of ruthenium is an obvious approach toward the possibility of modeling the electrocatalytic process. Figure 17(a) shows the in-situ FTIR spectra obtained during oxidation of methanol on Pt(111), Pt(111)/Ru prepared by spontaneous adsorption, and a PtRu alloy (85:15), which is not discussed here but is included for the sake of comparison. For Pt(111), the main features are due to solution species, at 2341 cm^{-1} (CO$_2$) and 1710 cm^{-1} (formic acid/ methyl formate [65]). The bands at 2059 and 1826 cm^{-1} are due, respectively, to linearly and bridge-adsorbed CO [120].

In the presence of Ru, no band for bridge-bonded CO is observed. From all three materials, the alloy shows the largest production of CO$_2$. This electrode also presents larger intensities for linear-bonded CO at low potentials, due to the ease of dissociative adsorption of this material. It is noteworthy that despite a strong production of CO$_2$ at 0.5 V, the band intensity for CO does not diminish. This can only be possible if either (1) CO is not oxidized at all, the production of CO$_2$ taking place over a parallel reaction pathway, or (2) CO is being oxidized to CO$_2$ but the rate of readsorption of methanol (to form CO) is high enough to keep its concentration at a given level. Option (1) can be discarded since it is well established that CO can be oxidized at 0.5 V on PtRu alloys [117]. Also, a parallel pathway, although discussed by Baltruschat et al. [121,122], can be neglected under the experimental conditions employed by Iwasita et al. [107]. Option (2) is more likely to occur, and the constant value of adsorbed CO is in agreement with its role as a reaction intermediate.

The influence of the procedure for Ru adsorption is demonstrated by the spectra shown in Figure 17(b). A comparison is made between the electrode prepared

(a)

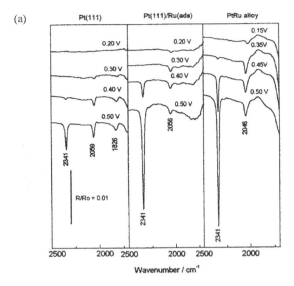

(b)

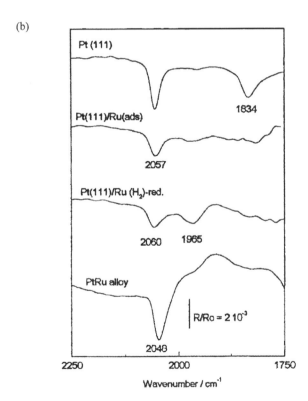

Figure 17 (a) In-situ FTIR spectra for Pt(111), Pt(111)/Ru (39%), and PtRu alloy (85:15) in $0.5\,mol\,dm^{-3}$ $CH_3OH + 0.1\,mol\,dm^{-3}$ $HClO_4$. Potentials as indicated in each spectrum; reference spectrum taken at 0.05 V; (b) comparison of the CO features in FTIR spectra for methanol obtained as those in (a), for Pt(111), Pt(111) with spontaneously adsorbed Ru (39%), Pt(111) with Ru reduced by hydrogen, and PtRu alloy (85:15). (From Ref. 107.)

by spontaneous adsorption and another prepared by bubbling H_2, the Ru composition being 20% for both electrodes. For the latter electrode a band at $1965\,cm^{-1}$ is clearly observed and is characteristic of CO adsorbed on Ru. This band appears only as a weak feature on the electrode prepared by spontaneous adsorption alone and is absent in the alloy. Recently Weaver et al. [123] showed that for PtRu materials, Ru segregation into islands with a minimum size of about 8 atoms is necessary for the CO feature at Ru sites to appear. The results in Figure 17(b) can thus be interpreted in terms of the formation of islands of different size, depending on the experimental procedure. Larger islands seem to be formed via the H_2-reduction procedure. This justifies the low catalytic activity of the latter material. Iwasita concludes from examining the electrochemical data for these systems [107] that a catalyst composition of Ru:Pt between 10% and 40% provides enough Pt and Ru sites for the reaction of methanol. The kinetic limitation is caused by the reaction between adsorbed CO and RuOH, and, in this respect, a homogeneous distribution of Pt and Ru atoms must have a strong influence on the reaction rate. Therefore, the optimum distribution seems to be that of the alloys. For other materials forming Ru islands, the method of preparation must be chosen to reduce the diameter of the islands as much as possible. These findings are in complete agreement with the IR data. The spectra in Figure 17(b) suggest that Ru islands formed via spontaneous adsorption are much smaller than those obtained by using H_2 as a reducing agent. Thus, it can be concluded that the reaction rate of methanol oxidation on Ru-modified Pt(111) electrodes depends on the size and structure of the Ru islands, as prepared by the different procedures. Using Pt(111) as a substrate, spontaneously adsorbed Ru presents higher catalytic activity than Ru reduced by H_2 or deposited via UHV evaporation. This result can be rationalized in terms of the formation of smaller islands in the case of spontaneously adsorbed Ru.

15.5 ELECTROCHEMICAL NANOTECHNOLOGY

15.5.1 Surface Modification by STM

The first example of an atomic-scale surface modification was reported by Becker et al. for the creation of structures such as the Ge(111) surface [33]. The surface was modified by holding the tip over a surface site, raising the tip-sample bias to $-4.0\,V$ (20 pA) current, and rapidly withdrawing the tip by $\sim 0.1\,nm$. Following this procedure, images of the surface displayed isolated protrusions about 0.8 nm wide and $\sim 0.1\,nm$ high. Since this process occurred with a higher success rate when the tip had previously contacted the surface, it was proposed that the Ge atoms, acquired from previous tip–surface contact, were being deposited from the tip onto the surface.

SPMs, especially in-situ STM, are useful tools for manipulating matter on the atomic to nanometer scales, leading to constructing individual nanostructures [10,124], i.e., for the deliberate modification of electrode surfaces on the mesoscopic scale [37]. By applying appropriate signals to an STM tip, local metal deposition reactions can be induced from metal-ion-containing solutions. Such work was initiated by Eigler et al., where, for example, they constructed quantum corrals from small metal clusters of nanometer radii, which were arranged in circles with the STM [125]. Penner et al., for example, have used an STM to create defects in highly oriented pyrolytic graphite (HOPG), which then act as nucleation centers in a metal deposition reaction [126].

Kolb et al. have demonstrated that Au and Ag surfaces can be modified on a nanometer scale by tip-induced deposition of Cu clusters [127,128]. For this purpose, Cu was deposited onto the STM tip by choosing a tip potential slightly negative of the respective bulk deposition potential. Then Cu was transferred from the tip to the surface by an appropriate tip approach. This approach was achieved by applying a potential step to the sample during which the tunneling voltage U_T changed its sign, that is, U_T passed through zero. The momentary null value of U_T causes the tip to approach the surface.

When an externally generated voltage pulse is applied directly to the z-piezo of the STM, this leads to a defined movement of the STM tip in the z-direction. The suggested mechanism for the tip-induced metal deposition is shown in Figure 18(a). The details of this mechanism can be found in [127,129]. This new technique of nanostructuring electrode surfaces provides the possibility of a microprocessor-controlled metal deposition at preselected positions [129]. By additionally modifying the x- and y-piezo signals, it is possible to move the tip across the surface by external control. At every preselected position, a voltage pulse is applied to the z-piezo and a Cu cluster is formed at that position. For example, to illustrate the microprocessor-controlled nanomodification, an array of about 100 Cu clusters was deposited onto a Au(111) surface [Figure 18(b)]. The average cluster height is 0.9 nm and has a full width at half-maximum of 3.5 nm, although some clusters were as high as 1.7 nm.

These clusters are remarkably stable, even at potentials slightly positive of the Nernst potential of Cu/Cu^{2+}. Aside from the precise positioning of individual Cu clusters, it is also possible to vary within certain limits the size of the deposited clusters in an extremely controlled fashion. In general, the cluster height increases with increasing the tip approach; see Figure 18(c).

The influence of the underlying substrate on the shape and size of the cluster has also been studied, and this has given some initial clues about the mechanism of material transfer between tip and substrate. The mechanism is not restricted to Cu on Au(111), but has also been shown to work for Pb on Au(111) and Ag on Au(111). In all three systems, however, metal deposition starts by forming a monolayer at underpotentials (UPD) [129] before the bulk phase nucleates on top of it. Accordingly, the tip-induced Cu clusters are placed onto a full Cu monolayer on gold rather than onto bare gold. Comparing this to Cu deposition onto Ag(111), where a layer-by-layer growth is reported [130] and the result is a two-dimensional island with a second, smaller two-dimensional island on top of it, there is a striking contrast to the nucleation-and-growth behavior of bulk Cu on the (UPD-covered) Au(111), where three-dimensional cluster growth was found to occur from the very beginning [129]. In addition, the amount of Cu transferred during tip approach is significantly larger for Cu on Ag(111) than for Cu on Au(111). Cu on Ag(111) seems to spread over the surface much more readily than on the UPD-covered Au(111), suggesting a strong interaction between Cu and Ag. The absence of any UPD for this system, despite this strong interaction, has not yet been explained.

More quantitative information can be drawn from the time dependence of the Cu clusters on Ag(111), because of their two-dimensional shape. The number of copper atoms in each such cluster is precisely determined, since the convolution between tip and cluster in the imaging process is negligible for a monoatomic high

(a)

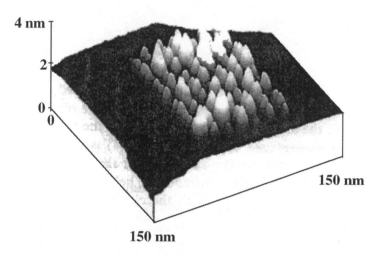

(b)

Figure 18 (a) Sequence of sketches, illustrating the mechanism of tip-induced metal deposition; (b) STM of an array of tip-induced Cu clusters on Au(111) in 0.05 mol dm^{-3} H$_2$SO$_4$ + 1 mmol dm^{-3} CuSO$_4$, consisting of 7 cluster rows 10 nm apart. They were generated by 7 potential steps from E_{sample} = +9 to − 31 mV at E_{tip} = −30 mV and I_T = 20 nA. The mean cluster height is 0.9 nm, with some as high as 1.7 nm. (c) Mean cluster height, as derived from 15 experiments, as a function of tip displacement Z_{dp} used to generate the Cu clusters. E_{sample} = +10 mV : E_{tip} = −50 mV versus Cu/Cu^{2+} : I_{tunnel} 0.2 nA. In general, the cluster height increases with increasing tip displacement, but above a value of about Z_{dp} = 1.2 nm, the cluster height does not increase any further, and holes appear in the surface instead. This suggests that the tip displacement becomes too large and the tip hits the substrate. (From Ref. 127.)

island with a large diameter. Ostwald ripening behavior, where smaller clusters are dissolved faster than bigger ones and large clusters grow at the expense of the smaller ones, is also observed with this system [131]. As a result, small Cu clusters on Ag(111) are unstable even at −10 mV in the vicinity of larger clusters. Calculating local current densities from the observed mass changes with time leads to a method

(c)

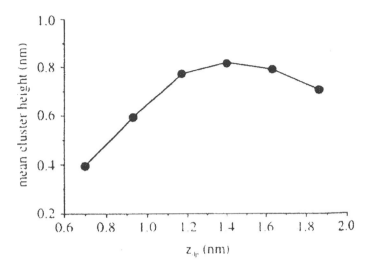

Figure 18 Continued.

of estimating the cluster size. Selective dissolution of single clusters from the array, by working with positive tip potentials at close proximity to the cluster in question, was also achievable [129]. Since the clusters next to the one that was reduced do not show any significant change in size, it can be concluded that the process is very localized and that the material removed was anodically dissolved.

As a second example of nanofabrication on electrode surfaces, Kolb et al. [132] recently fabricated small palladium clusters by tip-induced metal deposition and patterned them on an Au(111) electrode with the help of an STM. An array of 400 clusters, which had an average height of 0.4 nm, was generated at a rate of 80 clusters/s and was evenly distributed over the atomically flat terrace of the electrode. By changing the tip approach, the cluster height was varied between 0.2 and 1.0 nm. The fact that the technique also works with palladium opens up new directions for the study of size effects in catalysis. An electrode potential is chosen that is positive enough to prevent UPD of Pd on gold, and at the same time negative enough to prevent rapid dissolution of the Pd clusters. Hence, unlike the deposition of the copper on Au(111), the Pd clusters are deposited on the bare, catalytically inactive gold surface. Thus, this approach allows for the use of tip-induced nanodecoration in electrocatalytic investigations, since a Pd monolayer could mask any current contribution from the nanoclusters, which represent only a tiny fraction of the total surface. Figure 19 shows STM images (top view and 3D plot) of an array of 20×20 Pd clusters, on Au(111), the clusters generated via the tip and arranged in a square fashion at a distance of about 9 nm. Since the apparent cluster width of about 6 nm reflects the tip resolution rather than the actual cluster size, an upper limit of about 100 Pd atoms per cluster can only be estimated. Although the size distribution is not yet as narrow as the one achieved for Cu clusters, it appears to be sufficient for future applications in electrocatalytic studies.

The nanofabrication method works fast and reproducibly and, with the help of the microprocessor, the positioning of the clusters is fully automated. Since many more Pd clusters on a gold surface are needed to detect an electrochemical reaction occurring at the clusters, Kolb has fabricated a micrometer-size array from 10^4 Cu clusters on Au(111), the limit determined by the availability of flat areas on gold of macroscopic dimensions. Thus, the production of similarly large numbers of Pd

a)

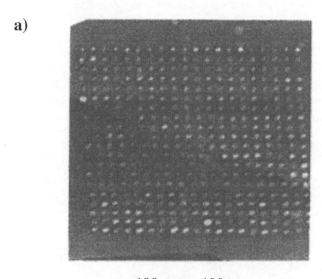

190 nm x 190 nm

b)

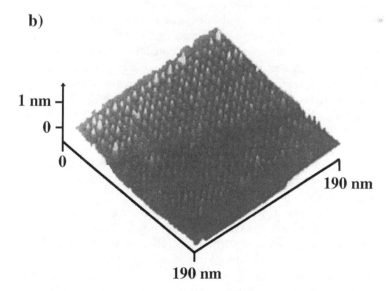

Figure 19 STM images [(a) top view and (b) three-dimensional plot] of an array of 400 tip-generated Pd clusters on Au(111) in in $0.05\,mol\,dm^{-3}$ $H_2SO_4 + 1\,mmol\,dm^{-3}$ H_2PdCl_4. $E_{sample} = +0.66\,V$ versus SCE. (From Ref. 132.)

clusters to perform studies on the size dependence of their electrocatalytic activity is possible and is discussed in the next section. Such images represent a highly defined model electrocatalyst, since there is only one type of catalyst particle with known morphology on the electrode. By preparation of arrays of catalyst particles, the surface area of the catalyst can be increased in a very controlled fashion, and scanning electrochemical microscopy (SECM) utilized for the investigation of the reactivity, reducing the substrate area contribution to a great extent. Such experiments are still in their infancy, but the field is developing rapidly.

15.5.2 Electrocatalysis at Nanostructured Single Particles

When the catalytic properties of supported clusters are measured by standard electrochemical methods such as cyclic voltammetry or oxidation transient measurement, only the average properties of the entire distribution of active particles on the electrode surface can be measured. A range of properties of supported nanoparticles, e.g., their geometric structure, their electronic and magnetic properties, as well as their catalytic activity, depends on the size of the particles. Geometric as well as electronic effects have been used to explain particle-size effects in electrocatalysis.

Colloidal nanoparticles have been successfully investigated, but the particles have the tendency to agglomerate during the adsorption process, and therefore the supported particles exhibit broader size distribution compared to the colloidal particles in solution [91]. An accurate control of the particles size is vital in investigations into physical and chemical properties of supported particles. Typically, as has been demonstrated in Section 15.5.1, a large number of particles is investigated in order to obtain sufficiently high signals. In order to investigate the local reactivity of clusters and to obtain a correlation with its structure, the STM tip electrode can be used as a local sensor using hydrogen evolution and oxidation as a model reaction. To avoid interference from different particles and to be sure to characterize only the hydrogen produced by a single particle, it is essential to use electrodes with a very low particle density on an inert surface. An all-in-situ method, consisting of generation of a single supported nanoparticle on a nonreactive substrate by an STM, and using the same instrument for characterization of the structure of the particle and subsequently for the measurement of the particle reactivity, has now been developed [38]. This is the ultimate example of nanofabrication at electrode surfaces and allows the study of structural and size effects of particles in electrocatalysis, without interference from neighboring particles.

Normally the current generated at a single particle is too low to be extracted from the background signal of the surface. Due to their local nature, detection by the STM tip is extremely selective. The principle of the method is outlined in Figure 20. The modified procedure is summarized as follows: the clusters are deposited and subsequently scanned by the STM in constant current mode. The tip is then positioned above one cluster. The potentials of the substrate and the STM tip are kept at a value where no reactions take place. After switching off the feedback regulation and retracting the tip out of the tunneling distance, the potential of the particle is pulsed to a value where hydrogen evolution occurs. The hydrogen produced at the particle can subsequently be oxidized at the tip, and the current for this reaction is measured.

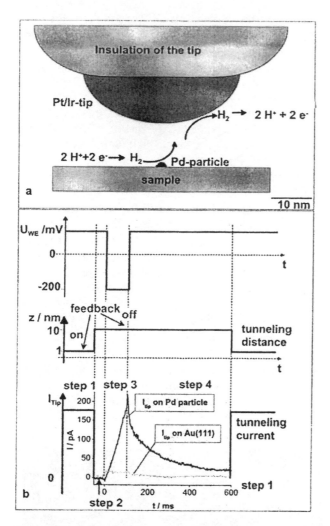

Figure 20 (a) Illustration of the proton reactivity measurement; (b) principle of the method: (1) the structure of the palladium particle is characterized; (2) the tip is positioned over the cluster and retracted while the feedback control is switched off; (3) a potential pulse of 100-ms duration is applied to the sample while the tip potential is kept constant and the hydrogen oxidation current on the STM tip is recorded. Finally, the feedback control is switched on and the STM tip is returned to tunneling mode. (From Ref. 38.)

Preliminary experiments involving the in-situ deposition of a single platinum nanoparticle on HOPG using the STM tip demonstrate the feasibility of this approach [133]. At the position of the platinum particle, a tip current due to the oxidation of hydrogen of almost 1 nA is measured, while 200-nm away, over the clean graphite surface, no tip current could be detected, indicating that the method is selective and can be used to relate the activity of single clusters to their atomic properties. A strong dependence of the detected current on substrate potential is observed. This indicates that it is indeed possible to measure the reactivity of single particles and deduce kinetic information regarding the reactions.

Recent results have demonstrated the application of the method for the quantitative analysis of reactivity at a single particle. Samples with only one palladium particle on Au(111) samples were prepared in the authors' group, by modification of Kolb's technique [38]. The deposition of the palladium is performed in a separate electrochemical cell containing palladium sulfate solution, in order to prevent palladium contamination of the Au(111) surface, with palladium-free 0.1 M H_2SO_4 used in the in-situ electrochemical STM cell. Figure 21 shows images of tip-induced particles of different sizes, on which reactivity measurements were performed. Figure 21(a) shows an in-situ STM of a single palladium particle on Au(111) in 0.1 M H_2SO_4, while three small Pd particles were generated in Figure 21(b). The height of the particles is approximately 0.5 nm. The in-situ STM image of a larger palladium particle (2.5 nm in height) is seen in Figure 21(c).

From the kinetic measurements of the hydrogen evolution reaction at the particles [38], the current density for the large particle of 200-nm diameter is 17 times higher than that reported for the bulk electrode. It is important to point out that for the bulk electrode, the actual surface structure is uncertain, while in the measurements involving the single particles, the actual surface area determined by STM results in much less uncertainty. All other particles deposited exhibited higher current densities than the response for a reference particle of 200-nm diameter. Several explanations for the higher reactivity of the individual nanoparticles are possible: (1) a high density of low-coordinated atoms at the surface of the nanoparticles; (2) the electronic properties of the substrate could result in the particle being charged due to a difference in work functions; (3) a modified double-layer structure interface of the nanoparticles could lead to higher reactivity; and (4) the effect of structure-induced changes in electronic properties of the surface may

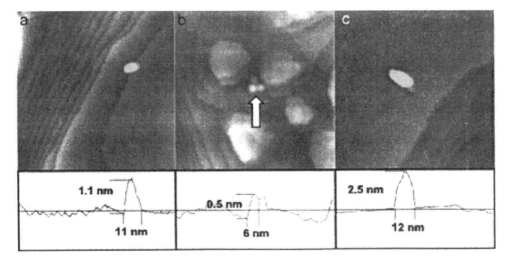

Figure 21 (a) In-situ STM image of one tip-induced palladium particle of 1.1 nm height; (b) in-situ STM image of three tip-induced palladium particles generated by one tip approach; (c) in-situ STM image of one larger tip-induced palladium particle of 2.5 nm height. All images are $100 \times 100 \, nm^2$. All particles are on Au(111) in 0.1 M H_2SO_4, $I_{tunnel} = 1 \, nA$, $U_{WE} = 400 \, mV$, $U_{tip} = 500 \, mV$. (From Ref. 38.)

enhance the reactivity. So far, the results point to the last explanation as being most likely, according to the model proposed by Norskøv et al. [134,135], to describe reactivity changes in heterogeneous catalysis, where compressive or tensile strain in the surface is seen to lead to a downshift or upshift of the d-band center. Upshifting the d-band means that antibonding metal d-adsorbate states are pushed above the Fermi level and the metal–d-adsorbate interaction becomes net attractive. The calculated d-band center of an expanded Pd monolayer on Au is shifted 350 meV toward the Fermi level, increasing the binding energy of adsorbates. If the Pd islands grow pseudo-morphically on the gold surface, the Pd lattice is expanded by around 4.8%. In such a case it is reasonable to expect a stronger adsorbate interaction with the surface. This expectation would also be consistent with a faster proton discharge.

15.6 CONCLUSION

Well-defined mesoscopic structures can be regarded as model electrodes for electrocatalysts used in many technical applications, including various types of fuel cells. Structurally characterizing electrode surfaces at the mesoscopic scale is an important factor in elucidating mesoscopic effects on electrochemical reactivity since it is known that particles of the mesoscopic regime have interesting and sometimes novel physicochemical properties compared to bulk electrodes. To characterize such mesoscopic properties, it is necessary to ascertain real-space information under in-situ electrochemical conditions, as in the present work—for example, by applying STM—which give high-resolution real-space images of electrode surfaces and in-situ FTIR spectroscopy.

Many interesting insights from STM studies into structural characterization and its effect on electrocatalytic activity have been demonstrated. Such measurements give a visualization of the reactivity of the sample area under investigation, under very controlled conditions. The interpretation of the observed enhancement in catalytic activity with particles on the mesoscopic phase has been complicated by the unknown structure and surface coverage of the surface deposits. Thus, the use of in-situ STM to structurally characterize Ru electrodeposits on a Pt(111) single crystals, for example, is a major step forward. Different observations were made with various STM studies. The present authors' group found for electrodeposited Ru on Pt(111) that up to 0.7 ML coverage of ruthenium is deposited as mainly monoatomic islands of 3–5-nm diameter, with a tendency to create three-dimensional deposits as the coverage increases [108]. These observations were backed up by independent XRD data [112]. Wieckowski et al. noted for Ru particles adsorbed on the same substrate that about 10% of the islands are not monoatomic, but such islands have a bilayer character, displaying a second monolayer deposit over the first monolayer [105]. These results make for interesting further work. It is certain that low-coordinated Ru sites at edge positions are the best sites for the performance of a bifunctional mechanism at Pt(111)/Ru surfaces.

FTIR spectroscopy has been shown to be a useful tool in the characterization of fuel cell model catalysts. It has helped elucidate much information on the electronic and geometrical structure of surfaces, which may help in the explanation of unusual size effects on electrocatalysis. Surface diffusion of the adsorbed molecules has been seen from time- and potential-dependent IR spectroscopy showing that the oxidation of CO on Pt sites and Ru sites are coupled. There is evidence that CO_{ads} molecules are mobile on the electrode surface and diffuse from unreactive Pt(111) domains to the reactive Ru islands. Vibration bands assigned to CO/Pt and CO/Ru disappear simultaneously upon oxidation. These results indicate that surface diffusion of adsorbed molecules is important for understanding interfacial electrochemical reactions, with, in this case, initiation of CO oxidation on/or around the Ru islands in combination with a high CO adsorbate mobility. Thus, the structural properties of the electrode surface on the nanometer scale are crucial for their electrocatalytic performance. Another factor of interest to electrocatalysis discussed here is the pronounced dependence of performance on catalyst loading, with clusters on highly loaded catalysts having properties approaching polycrystalline metal behavior. The dependence of the vibrational frequency for CO_{ads} oxidation on particle size is discussed in terms of substrate–particle interaction, with a high CO coverage on small particles at low particle coverage compared to the bulk phase, indicating a higher adsorption energy of CO. The structure sensitivity of CO oxidation has also been confirmed. It is also clear that IR is a versatile tool for the interpretation of the spectra of complex technical electrocatalysts in terms of the model systems discussed, helping to bridge the gap between the fundamental science of these model electrodes and technical catalyst systems.

The final section of the chapter discusses a uniquely novel method that has been developed, using STM, to study the reactivity of supported single nanoparticles. This is, in effect, an adaptation of the application of STM to undertake nanometer surface modification. This method allows investigation of reaction kinetics, in a potential regime and at high current densities, where traditional methods fail due to sensitivity problems and mass transfer limitations. This indicates that it is indeed possible to measure the reactivity of single particles and deduce kinetic information regarding the reactions. Thus, truly, STM has shown its absolute applicability to determining the structure of model electrodes and the effect of structure and size on chemical reactivity.

In conclusion, the combined use of the two techniques in studying mesoscopic surface properties has been shown to be mutually complementary, with knowledge of the surface properties from the STM studies facilitating the interpretation of results from IR spectroscopy and conventional electrochemical techniques, since these are affected by the surface structure but do not contain any information about the morphology. Such studies should prove very useful in helping to tailor the properties of small metal clusters and could prove to be universally applicable for a variety of metals, and thus may be quite useful in the design of nanostructured materials for catalytic applications. These model catalysts can be used to study the unusual and as yet not fully explained catalytic properties of nanosize particles, helping to bridge the structure gap between physicochemical properties on the mesoscale, and bulk dimensions.

ACKNOWLEDGMENTS

Dr. J.A. Collins gratefully acknowledges support from a Marie Curie Fellowship of the European Community program, "Energy, Environment and Sustainable Development," under contract number ERK5-CT-1999-50002. The authors thank the DFG (German Science Foundation) for funding under the Key Target Program

"New Layer Structures for Fuel Cells," Sti 74/8-1. The authors would also like to thank Dr. Andreas Friedrich, for helpful advice and discussion during the preparation of this chapter.

REFERENCES

1. S. Wasmus, A. Kuver, J. Electroanal. Chem. 461:14 (1999).
2. L. Carrette, K.A. Friedrich, U. Stimming, Chem. Phys. Chem. 1:162 (2000).
3. G.J.K. Acres, J.C. Frost, G.A. Hards, R.J. Potter, T.R. Ralph, D. Thompsett, G.T. Burstein, G.J. Hutchings, Catal. Today 38:393 (1997).
4. L. Carrette, K.A. Friedrich, U. Stimming, Fuel Cells 1:5 (2001).
5. M. Bowker, Stud. Surf. Sci. Catal. Vol. 101, 1996, p. 287.
6. H.S. Taylor, Proc. R. Soc. London A, 108:105 (1925).
7. Y. Takasu, N. Ohashi, X.-G. Zhang, Y. Murakami, H. Mingagawa, S. Sato, K. Yahikozawa, Electrochim. Electrochim. Acta. 41:2595 (1996).
8. T. Zambelli, J. Wintterlin, J. Trost, G. Ertl, Science 273:1688 (1996).
9. G. Binnig, C. Rohrer, E. Gerber, E. Weibel, Phys. Rev. Lett. 49:57 (1982).
10. D. Sarid, Scanning Force Microscopy, Oxford University Press, New York, 1994.
11. C.J. Chen, Introduction to Scanning Tunneling Microscopy, Oxford University Press, New York, 1993.
12. C.M. Lieber, J. Liu, P.E. Sheehan, Angew. Chem. Intl. Ed. Engl. 35:687 (1996).
13. P.A. Christensen, Chem Soc Rev. 21: 197 (1992).
14. R.C. Jaklevic, L. Elie, Phys. Rev. Lett. 60:120 (1988).
15. R.J. Behm, W. Hoesler, E. Ritter, G. Binnig, Phys. Rev. Lett. 56:228 (1986).
16. R. Sonnenfeld, B.C. Schardt, Appl. Phys. Lett. 49: 1172 (1986).
17. B. Batina, T. Will, D.M. Kolb, Faraday Discuss. 94:93 (1992).
18. R.J. Nichols, D.M. Kolb, R.J.J. Behm, Electroanal. Chem. 313:109 (1991).
19. U. Schmidt, W. Obretenov, W.J. Lorenz, G. Staikov, E. Budevski, D. Carnal, U. Müller, H. Siegenthaler, E. Schmidt, Faraday Discuss. 94:107 (1992).
20. N. Kimizuka, K. Itaya, Faraday Discuss. 94:117 (1992).
21. F. Jensen, F. Besebbacher, E. Loegsgarrd, L. Stensgaard, Phys. Rev. B42:9206 (1990).
22. M. Valden, X. Lai, D.W. Goodman, Science 281:1647 (1998).
23. Y. Iizuka, Catal. Today 36:115 (1997).
24. S.K. Shaikhutdinov, E.R. Savinova, A. Scheybal, K. Doblhofer, R. Schoegl, J. Electroanal. Chem. 500:208 (2001).
25. I.C. Oppenheim, D.J. Trevor, C.E.D. Chidsey, P.L. Trevor, K. Sieradski, Science 254:687 (1991).
26. N. Breuer, U. Stimming, R. Vogel, Electrochim. Acta 40:1401 (1995).
27. W. Suggs, A.J. Bard, J. Am. Chem. Soc. 116:10725 (1994).
28. M.P. Ryan, R.C. Newman, G.E. Thompson, J. Electrochem. Soc 141:L164 (1994).
29. K. Sashikata, Y. Mabui, K. Itaya, M.P. Soriaga, J. Phys. Chem. 100:20027 (1996).
30. O.V. Cherstiouk, S.N. Pron'kin, A.L. Chuvilin, A.N. Salanov, E.R. Savinova, G.A. Tsirlina, O.A. Petrii, Russ. J. Electrochem. 36:741 (2000).
31. T. Hachiya, K. Itaya, Ultramicroscopy 42–44:445 (1992).
32. R. Michaelis, Ph.D. Thesis, Free University, Berlin, 1991.
33. R.S. Becker, G.S. Higashi, Y.J. Chabal, A.J. Becker, Phys. Rev. Lett. 65:1917 (1990).
34. H.J. Mamin, P.H. Guenther, D. Rugar, Phys. Rev. Lett. 65:2418 (1990).
35. I.-W. Lyo, P. Avouris, Science 253:173 (1991).
36. N.J. DiNardo, In Nanoscale Characterization of Surfaces and Interfaces (P.D. Gregory, D. Hollis, U. Anton, eds.), VCH, Weinheim, 1994.
37. D.M. Eigler, E.K. Schweizer, Nature 344:524 (1990).
38. J. Meier, K.A. Friedrich, U. Stimming, Faraday Discuss. 121:365 (2002).

39. L.P.L. Carrette, J.A. Collins, A.J. Dickinson, U. Stimming, Bunsenmagazin der Deutsche Bunsen-Geschellschaft für Physikalische Chemie 2:27 (2000).

40. S.G. Chalk, J.F. Miller, F.W. Wagner, J. Power Sources 86:40 (2000).

41. M. Watanabe, H. Sei, P. Stonehart, J. Electroanal. Chem. 261:375 (1989).

42. K. Kinoshita, J. Electrochem. Soc. 137:845 (1990).

43. Y. Morimoto, E.B. Yeager, J. Electroanal. Chem. 444:95 (1998).

44. A. Hamnett, B.J. Kennedy, Electrochim. Acta 33:1613 (1988).

45. A.S. Arico, Z. Poltarzewski, H. Kim, A. Morana, N. Giordano, V. Antonucci, J. Power Sources 55:159 (1995).

46. M. Goetz, H. Wendt, Electrochim. Acta 43:3637 (1998).

47. A.S. Arico, A.K. Shukla, K.M. El-Khatib, P. Creti, V. Antonucci, J. Appl. Electrochem. 29:671 (1999).

48. L. Liu, R. Pu, R. Viswanathan, Q. Fan, R. Liu, E.S. Smotkin, Electrochim. Acta. 43:3657 (1998).

49. R. Ramkumar, S. Dheenadayalan, R. Pattabiraman, J. Power Sources 69:75 (1997).

50. W. Vielstich, A. Kuver, M. Krausa, A.C. Ferreira, K. Petrov, S. Srinivasan, In Symp. Batteries and Fuel Cells for Stationary and Electric Vehicle Applications (A.R. Landgrebe, Z.-I. Takehara, eds.), Honolulu, HI, 1993, p. 269.

51. X. Ren, T.E. Springer, S. Gottesfeld, J. Electrochem. Soc. 147:92 (2000).

52. A.K. Shukla, P.A. Christensen, A.J. Dickinson, A. Hamnett, J. Power Sources 76:54 (1998).

53. N. Alonso-Vante, H. Tributsch, O. Solorza-Feria, Electrochim. Acta 40:567 (1995).

54. O. Solorza-Feria, K. Ellmer, M. Giersig, N. Alonso-Vante, Electrochim. Acta 39:1647 (1994).

55. T. Iwasita, F.C. Nart, W. Vielstich, Ber. Bunsenges. Phys. Chem. 94:1030 (1990).

56. A. Hamnett, Catal. Today 38:445 (1997).

57. J. Munk, P.A. Christensen, A. Hamnett, E. Skou, J. Electroanal. Chem. 401:215 (1996).

58. R. Parsons, T.J. Van der Noot, J. Electroanal. Chem. 257:9 (1988).

59. B. Beden, F. Hahn, C. Lamy, J.M. Leger, N.R. De Taconi, R.O. Lezna, A.J. Arvi, J. Electroanal. Chem. 307:129 (1991).

60. A. Papoutis, J.-M. Leger, C. Lamy, J. Electroanal. Chem. 234:315 (1987).

61. T. Frelink, W. Visscher, J.A.R. van Veen, J. Electroanal. Chem. 382:65 (1995).

62. J.P. Iudice de Souza, T. Iwasita, F.C. Nart, W. Vielstich, J. Appl. Electrochem. 30:43 (2000).

63. H.A. Gasteiger, P.N. Ross, E.J. Cairns, Surf. Sci. 293:67 (1993).

64. P.A. Christensen, A. Hamnett, J. Munk, G.L. Troughton, J. Electroanal. Chem. 370:65 (1994).

65. X.H. Xia, T. Iwasita, F. Ge, W. Vielstich, Electrochim Acta 41:711 (1996).

66. E. Herrero, K. Franaschuk, A. Wieckowski, J. Phys. Chem. 98:5074 (1994).

67. S. Wasmus, W. Vielstich, J. Appl. Electrochem. 23:120 (1993).

68. A.N. Goldstein, ed., Handbook of Nanophase Materials, Marcel Dekker, New York, 1997.

69. W.J. Plieth, J. Phys. Chem. 86:3166 (1982).

70. M. Fleischmann, S. Pons, J. Souza, J. Ghoroghchian, J. Electroanal. Chem. 366:171 (1994).

71. A. Henglein, Ber. Bunsenges. Phys. Chem. 99:903 (1995).

72. F. Parmigiani, E. Kay, P.S. Bagus, J. Electron. Relat. Phenom. 50:39 (1990).

73. L. Vasquez, J. Gomez, A.M. Baro, N. Garcia, M.L. Marcos, J. Gonzalez Velasco, J.M. Vara, A.J. Arvia, J. Presa, A. Garcia, M. Aguilar, J. Am. Chem. Soc. 109:1730 (1987).

74. J. Wintterlin, J. Wiechers, H. Brune, T. Gritsch, H. Hofer, R.J. Behm, Phys. Rev. Lett. 62:59 (1989).

75. K. Yahikozawa, Y. Fujii, K. Matsuda, K. Nishimura, Y. Takasu, Electrochim. Acta. 26:973 (1991).
76. N.P. Lebedeva, G.N. Kryukova, S.V. Tsybulya, A.N. Salanov, E.R. Savinova, Electrochim. Acta 44:1431 (1998).
77. M.T. Reetz, W. Helbig, J. Am. Chem. Soc. 116:7401 (1994).
78. M.T. Reetz, W. Helbig, S.A. Quaiser, U. Stimming, N. Breuer, R. Vogel, Science 267:367 (1995).
79. N. Breuer, U. Stimming, R. Vogel, Surf. Coat. Technol. 67:145 (1994).
80. U. Stimming, R. Vogel, In Electrochemical Nanotechology, In-Situ Probe Techniques at Electrochemical Interfaces (W.J. Lorenz, W. Plieth, eds.), Wiley-VCH, Weinheim, 1998, p. 73.
81. K.A. Friedrich, A. Marmann, U. Stimming, W. Unkauf, R. Vogel, J. Fresenius, Anal. Chem. 358:163 (1997).
82. D.N. Furlong, A. Launikonis, W.H.F. Sasse, J. Chem. Soc. Faraday Trans. 80:571 (1984).
83. A. Harriman, G.R. Millward, P. Neta, M.C. Richoux, J.M. Thomas, J. Phys. Chem. 92:1286 (1988).
84. K.A. Friedrich, F. Henglein, U. Stimming, W. Unkauf, Colloids and Surf. A: Physicochemical and Engineering Aspects 134:193 (1998).
85. K. Kunimatsu, H. Seki, W.G. Golden, J.G. Gordon II, M.R. Philpott, Surf. Sci. 158:596 (1985).
86. D.S. Corrigan, P. Gao, L.W.-H. Leung, M.J. Weaver, Langmuir 2:744 (1986).
87. A. Kunimatsu, A. Aramata, H. Nakajima, H. Kita, J. Electroanal. Chem. 207:293 (1986).
88. R. Rydberg, In Advances in Chemical Physics (K.P. Lawley, ed.), Wiley, New York, 1989.
89. C.W. Olsen, R.I. Masel, Surf. Sci. 201:444 (1988).
90. S.-C. Chang, M.J. Weaver, Surf. Sci. 238:142 (1990).
91. K.A. Friedrich, F. Henglein, U. Stimming, W. Unkauf, Electrochim. Acta 45:3283 (2000).
92. B. Love, J. Lipowski, In Electrochemical Surface Science: Molecular Phenomena at Electrode Surfaces (M.P. Soriaga, ed.), ACS Symposium Series, Washington, DC, 1988, p. 484.
93. Z.L. Wang, T.S. Ahmadi, M.A. El-Sayed, Surf. Sci. 380:302 (1997).
94. T.S. Ahmadi, Z.L. Wang, T.C. Green, A. Henglein, M.A. El-Sayed, Chem. Mater. 8:1161 (1997).
95. H.A. Gasteiger, N. Markovic, P.N. Ross, E.J. Cairns, Electrochim. Acta 39:1825 (1994).
96. K.A. Friedrich, K.-G. Geyzers, U. Linke, U. Stimming, J. Stumper, J. Electroanal. Chem. 402:123 (1996).
97. W. Chrzanowski, A. Wieckowski, Langmuir 14:1967 (1998).
98. M.P. Hogarth, J. Munk, A.K. Shukla, A. Hamnett, J. Appl. Electrochem. 24:85 (1994).
99. W. Chrzanowski, H. Kim, A. Wieckowski, Catal. Lett. 50:69 (1998).
100. P.N. Ross, In Frontiers of Electrochemistry (Electrocatalysis) (J. Lipowski, P.N. Ross, eds.), Vol. 4, Wiley-VCH, New York, 1998.
101. B. Hammer, J.K. Norskøv, In Chemisorption and Reactivity on Supported Clusters and Thin Films (R.M. Lambert, G. Pacchioni, eds.), Kluwer Academic, Dordrecht, 1997, p. 289.
102. N.M. Markovic, H.A. Gasteiger, P.N. Ross, X.D. Jiang, I. Villegas, M.J. Weaver, Electrochim. Acta 40:91 (1995).
103. N. Fujiwara, K.A. Friedrich, U. Stimming, J. Electroanal. Chem. 472:120 (1999).

104. H. Hoster, T. Iwasita, H. Baumgaertner, W. Vielstich, Phys. Chem. Chem. Phys. 3:337 (2001).

105. E. Herrero, J.M. Feliu, A. Wieckowski, Langmuir 15:4944 (1999).

106. W.K. Pratt, Digital Image Processing, Wiley, New York, 1978.

107. T. Iwasita, H. Hoster, A. John-Anacker, W.F. Lin, W. Vielstich, Langmuir 16:522 (2000).

108. S. Cramm, K.A. Friedrich, K.-P. Geyzers, U. Stimming, R. Vogel, Fresenius J. Anal. Chem. 358:189 (1997).

109. V.M. Schmidt, P. Broeckerhoff, B. Hoehlein, R. Menzer, U. Stimming, J. Power Sources 49:299 (1994).

110. K.A. Friedrich, K.P. Geyzers, A. Marmann, U. Stimming, R. Vogel, Zeitschrift für Physik. Chemie 208:137 (1999).

111. M. Watanabe, S. Motoo, J. Electroanal. Chem. 60:275 (1975).

112. M. Cappadonia, J. Schmidberger, W. Schwegle, U. Stimming, Proc. 189th Meeting of the Electrochemical Society, Los Angeles, 1996.

113. R. Koetz, H.J. Lewerenz, S. Stuki, J. Electrochem. Soc. 130:825 (1983).

114. I. Villegas, M.J. Weaver, J. Chem. Phys. 101:1648 (1994).

115. B.N.J. Persson, R. Rydberg, Phys. Rev. B24:24 (1981).

116. K.-P. Geyzers, Investigation of CO Electrooxidation on Ruthenium Modified Platinum Electrodes, Ph.D. Thesis, Friedrich Wilhelm University, Bonn, 1997.

117. H.A. Gasteiger, N. Markovic, P.N. Ross, E.J. Cairns, J. Phys. Chem. 98:617 (1994).

118. A.V. Petukhov, W. Akemann, K.A. Friedrich, U. Stimming, Surf. Sci. 402–404:182 (1998).

119. W. Akemann, K.A. Friedrich, U. Linke, U. Stimming, Surf. Sci. 402–404:571 (1998).

120. B. Beden, C. Lamy, A. Bewick, K. Kunimatsu, J. Electroanal. Chem. 121:343 (1981).

121. H. Wang, C. Wingender, H. Baltruschat, M. Lopez, M.T. Reetz, J. Electroanal. Chem. 509:163 (2001).

122. E. Jusys, H. Massong, H. Baltruschat, J. Electrochem. Soc. 146:1093 (1999).

123. S. Zou, I. Villegas, C. Stuhlmann, M.J. Weaver, Electrochim. Acta 43:2811 (1998).

124. P. Avouris, ed., Atomic and Nanometer-Scale Modification of Materials: Fundamentals and Applications, Kluwer, Boston, 1993.

125. M.F. Crommie, C.P. Lutz, D.M. Eigler, Science 262:218 (1992).

126. W. Li, J.A. Virtanen, R.M. Penner, J. Phys. Chem. 96:6529 (1992).

127. R. Ullmann, T. Will, D.M. Kolb, Ber. Bunsenges. Phys. Chem. 99:1414 (1995).

128. R. Ullmann, T. Will, D.M. Kolb, Chem. Phys. Lett. 201:238 (1993).

129. D.M. Kolb, R. Ullmann, T. Will, Science 275:1097 (1997).

130. T. Will, M. Dietterle, D.M. Kolb, In Nanoscale Probes of the Solid-Liquid Interface, Nato-ASI (A.A. Gewirth, H. Siegenthaler, eds.), Kluwer, Dordrecht, 1995.

131. O.M. Magnussen, J. Hotlos, R.J. Behm, N. Batina, D.M. Kolb, Surf. Sci. 296:310 (1993).

132. G.E. Bagelmann, J.C. Ziegler, D.M. Kolb, J. Electrochem. Soc. 145:L33 (1998).

133. U. Stimming, R. Hiesgen, J. Meier, Proc. of the 196th Meeting of the Electrochemical Society, Hawaii, 1999.

134. B. Hammer, J.K. Norshøv, Adv. Catal. 45:71 (2000).

135. A. Ruban, B. Hammer, P. Stoltze, H.L. Skriver, J.K. Norskøv, J. Mol. Catal. A 115:421 (1997).

16

Electrochemical and Chemical Promotion by Alkalis with Metal Films and Nanoparticles

RICHARD M. LAMBERT

Cambridge University, Cambridge, England

CHAPTER CONTENTS

SUMMARY

Alkali metal compounds are sometimes employed as promoters of metal catalysts, a fact that has stimulated much research on alkali adsorption. Much of this work has made use of single-crystal model systems that are amenable to investigation by the methods of surface science. Most of these single-crystal studies have been carried out under conditions of ultrahigh vacuum. Although in favorable cases such data can be directly related to the behavior of practical catalysts working at atmospheric pressure, in general the linkage between the two is more tenuous. Here we describe and illustrate the application of an interesting electrochemically induced promoter effect: *electrochemical promotion* (EP). The phenomenon of EP is described, elucidated, and exploited in the investigation of catalytic reaction mechanisms.

Time-resolved and spatially resolved photoelectron spectroscopic data along with reactor measurements demonstrate that EP of thin-film metal catalysts deposited on solid electrolyte supports is the result of spillover phenomena at the three-phase boundary between the electrolyte, the catalyst, and the gas phase. Ions from the electrolyte are discharged at the catalyst–electrolyte interface and migrate to cover the catalyst surface, whose properties are thereby strongly altered. This is illustrated by reference to a variety of metal-catalyzed reactions. Reaction mechanisms and the mode of promoter action are deduced, and it is shown how this understanding may be exploited to develop improved nano-particulate supported metal catalysts.

16.1 INTRODUCTION: ALKALI METAL PROMOTERS IN CATALYSIS

Alkali metals, or more precisely alkali metal compounds, are sometimes used as promoters of heterogeneous catalysts whose performance they may enhance with respect to activity, selectvity, or both. Practical supported metal catalysts are relatively complex materials whose structure and composition are optimized by empirical means. Although these procedures are effective, they provide little insight into the underlying phenomena that determine performance: in particular, they rarely shed any light on the mechanism of promoter action. This technologically important use of alkali compounds, which has a huge impact on the economics of the world's petrochemicals and chemical industries, has stimulated much research on idealized systems involving alkali adsorption on single-crystal surfaces, investigated by the methods of surface science [1,2,3]. Most of these studies have been carried out under ultrahigh vacuum conditions. Although in favorable cases such data [4] can be directly related to the behavior of practical catalysts working at atmospheric pressure [5], in general the linkage between the two is often more tenuous.

Let us pause to examine more closely the often-stated view that "alkali metals are important in catalysis, therefore fundamental studies of adsorbed alkalis on metal surfaces are of major significance with respect to catalysis, and so forth. . . . " This argument has been used as the motivation for innumerable research proposals and actual investigations. One is sometimes left with the impression that the certainty with which it is propounded is in inverse proportion to knowledge of the facts. There *are* good reasons for fundamental theoretical and experimental studies of alkali overlayers on metal surfaces. Their relative simplicity makes for rigorous calculations and precise measurements; in turn, such results can provide real insight into the nature of more complex problems, some of which are genuinely relevant to issues of promotion in heterogeneous catalysis.

But how ubiquitous actually are alkalis in the promotion of reactions catalyzed at metal surfaces? An examination of recent authoritative sources [6,7] shows that the majority of medium-to large-scale processes do not employ alkali promoters, even when one includes nonmetallic (i.e., metal oxide) catalysts. In a number of cases (e.g., steam reforming of naphtha) it seems clear that the role of alkali is simply to reduce the acidity of the oxide support. There are famous cases, of course, where the presence of alkali species on the catalytically active *metal* surface is critically important to the chemistry. Notable are ethene epoxidation (Ag-Cs), ammonia synthesis (Fe-K), acetoxylation of ethene to vinyl acetate (Pd, Pd/Au-K), and Fischer–Tropsch synthesis (Fe, Co, Ru-K). The first three are major industrial processes carried out using mature technologies. In the cases of ethene epoxidation [8–11] and ammonia synthesis [12,13], fundamental work, including EP studies [14,15], has generated a good understanding of reaction mechanisms and the modes of promoter action: recent reviews of both topics are available [16,17]. Fischer–Tropsch synthesis is not yet commercially viable in most locations (exceptions include the Shell plant at Bintulu and Sasol plants in Africa and the Middle East). The reaction mechanism remains controversial [18], though recent EP results have shed some light on the way in which alkali promotion affects selectivity [19]. Very little work of any kind has been carried out on the mechanism of vinyl acetate synthesis, which is therefore unclear. Some indication has been obtained about the nature of possible reaction intermediates [20], partially supported by single-crystal vibrational spectroscopy data [21]. However, the role of the potassium promoter remains obscure.

In summary, alkali promotion of supported metal catalysts is an interesting subject that does have important technological implications in those cases where the presence of alkali has a pivotal influence on the surface chemistry of the metal phase. Fundamental studies of such systems are certainly justified. However, we should maintain a sense of proportion. Alkalis find relatively limited use as promoters in practical catalysis—indeed in some cases they act as powerful poisons. And we should not lose sight of the fact that what is actually present at the surface of the working catalyst is not an alkali metal, but some kind of alkali surface compound. This chapter deals with the application of alkali promoters to catalysis by metals, as opposed to catalysis by oxides, and, in particular, the technique of electrochemical promotion (EP), which enables us to address some pertinent issues.

EP is an elegant way of controlling alkali concentration at the surface of a working metal catalyst by varying its electric potential. The technique is implemented by depositing porous thin-film metal catalysts on solid electrolyte supports. Noteworthy features of EP systems are as follows:

1. They permit *in-situ control* of catalyst performance under realistic conditions of steady-state turnover.
2. The polycrystalline metal film/oxidic solid electrolyte system provides a reasonable approximation to nanocrystalline metal/metal oxide catalysts.
3. Such systems may be operated at atmospheric pressure and above.
4. EP samples are amenable to study by electron spectroscopy and related methods.

As a consequence of these features, one may use electrochemical promotion to

Elucidate reaction mechanisms [22,23] by studying the response of activity and selectivity to controlled changes in alkali promoter levels.

Develop new nanostructured catalytic systems [24–26] based on the insight provided by the corresponding EP analogs.

It is in this sense that EP provides a valuable way of modeling and elucidating important aspects of the behavior of practical heterogeneous catalysts. There are additional cogent reasons for fundamental research on EP. First, one would like to understand the phenomenon in as much detail as possible so that a quantitative theoretical framework can be established. Second, it is desirable to extend the range of systems that can be addressed by EP so as to diversify the catalytic chemistry [27];

among other things, this calls for the application of new solid electrolytes. Third, the development of robust EP systems capable of technological implementation is an attractive prospect: recently, there have been encouraging developments in this direction [28]. Here, however, we shall focus on certain fundamental aspects of EP illustrated by reference to several different catalytic reactions subject to promotion by alkalis. As we shall see, a combination of reactor measurements and electron spectroscopic data can provide important insight into the mechanism of alkali promotion, the chemical state of the promoter species under reaction conditions, and the identity of the active surface sites. In turn, this understanding may be exploited to develop conventional dispersed catalysts with improved performance [24–26,29–31].

EP was discovered more than a decade ago [32] by Vayenas and co-workers, who subsequently developed the methodology and laid down the basis for interpreting the effect. Comprehensive reviews are available, for example, [33,34]; and for an authoritative view of the subject the reader is referred to Vayenas's chapter in this volume. In addition, Pacchioni et al. have provided a quantum chemical treatment of the topic [35]. As noted above, the technique entails electrochemical pumping of ions from a solid electrolyte to the surface of a porous, catalytically active metal film with which it is in contact. It is helpful to consider a specific example, e.g., the use of Na β'' alumina (a Na^+ conductor) as the solid electrolyte. Under forward bias (catalyst working electrode negative relative to counterelectrode) Na^+ ions are transported to the working electrode, where they are discharged at the metal electrode–solid electrolyte–gas three-phase boundary. The resulting species (Na) spills over onto the surface of the metal catalyst, strongly altering its electronic properties. The precise chemical state of the spilt-over alkali promoter under reaction conditions is an issue we address later. It is noteworthy that STM has been used to demonstrate the electrochemically controlled migration of Na between the solid electrolyte and the catalyst [36]; as we shall see later, in-situ electron spectroscopy and photoelectron microscopy also provide compelling evidence concerning the reversible spatial and temporal dependence of electro-pumped alkali overlayers on metal catalyst surfaces. The resulting alkali-induced changes in surface electronic properties are manifested by concomitant changes in (1) catalyst work function, (2) adsorption enthalpies of adsorbed species, and (3) activation energies of reactions involving these species. These phenomena give rise to electrochemical promotion.

In practice, the state of the catalyst surface is conveniently monitored by following the changes in catalyst potential measured with respect to a reference electrode (ΔU_{WR}). Experiment shows that ΔU_{WR} is usually close to $\Delta\phi$, although the two quantities are not always identical. The question has recently been examined in detail by Tsiplakides and Vayenas [37]: the origin of any measured differences between ΔU_{WR} and $\Delta\phi$ is the subject of current research and not of importance for present purposes.

The effects of EP on activity and selectivity are generally reversible, reflecting the reversibility of the electrode reaction involving the promoter species. It is important to appreciate from the outset that one is not dealing with conventional electrocatalysis, which is a strictly Faradaic process: one mole of charge transported through the electrolyte results in one mole of chemical reaction at the working electrode. Under EP conditions, the electrochemically induced catalytic rate changes are typically 10^3–10^5 times greater than the rate of supply of promoter species. That is, one mole of charge transported results in 10^3–10^5 turnovers. Thus, EP is strongly non-Faradaic, which is why its discoverers called the effect *non-Faradaic electro-chemical modification of catalytic activity*—NEMCA [32].

The EP phenomenon has been reported for over 50 catalytic reactions taking place on Pt, Rh, Pd, Ag, Ni, RuO_2, and IrO_2 catalyst films electro-pumped by a variety of solid electrolytes, including oxygen [38–40], sodium [41–44], fluorine [45], and hydrogen ion conductors [46]. An exciting recent development in this field is the demonstration by Marwood and Vayenas [47] that EP may also be induced in a so-called wireless configuration, i.e., without any direct electrical connection being made to the working electrode. This points the way to practical applications of EP with dispersed catalysts systems. Recently Comninellis et al. demonstrated the viability of the wireless approach by electrochemically promoting ethene oxidation in a multichannel thin-film tubular reactor [28]. More recently, Yiokari et al. applied the same principle to the electrochemical promotion of a commercial, *dispersed* ammonia synthesis catalyst operated at elevated pressure [15].

16.2   ALKALI ADSORPTION AND CO-ADSORPTION WITH ALKALIS ON METAL SURFACES

Understandably, alkali adsorption at metal surfaces was one of the first problems in chemisorption to be investigated by simple theoretical methods [48]. It was the prototypical problem examined by means of the so-called ionic model of chemisorption, which is plausible, intuitively appealing, and easy to understand. The simple procedure for estimating the adsorption enthalpy at zero coverage (ΔH_0) is as follows. One imagines (1) ionizing the alkali ad-atom in field-free space, at infinite separation from the metal surface (energy cost = first ionization energy, I), (2) transferring the electron to the Fermi level of the metal (energy gain = $e\phi$), (3) moving the alkali ion to its equilibrium separation r from the metal surface. The energy gain due to (3) is approximated by means of an image charge potential between ion and metal surface, resulting in an estimated adsorption enthalpy given by

$$\Delta H_0 = \frac{e^2}{4}r + \phi - I \tag{1}$$

Semiquantitative agreement with experiment is found, subject to uncertainty in the value to be used for r; this arises from the impossibility of satisfactorily defining a mathematical plane that corresponds to truncation of the metal. It is also the case that the image charge potential law is not valid at the very small separations involved. Despite these shortcomings, the model has considerable didactic value. It teaches us that in practice the actual degree of charge transfer between alkali and metal surface must correspond to much less than one complete electronic charge. It correctly predicts that ΔH_0 for close-packed metal surfaces will be higher than for more open planes, so that alkali adsorption on the former will always be preferred. And it provides a basis for rationalizing quantitatively [49] the coverage dependence of $\Delta\phi$ with increasing coverage due to mutual repulsion between—and concomitant depolarization of—the ad-atom dipoles.

More advanced treatments employing the methods of condensed matter physics provide deeper insight [50], but the essential picture remains unchanged: a strongly polar chemisorption bond is formed that exerts a pronounced effect on the work function of the metal. With increasing alkali coverage depolarization ("metallization") of the alkali overlayer occurs: this may manifest itself by changes in XP binding energy [51] and/or the appearance of characteristic surface plasmon losses [51]. Alkali overlayers also exhibit very interesting structural properties [52], including order–disorder phase transitions [53], which have attracted the attention of both experimentalists and theoreticians, and for which a rich literature exists. Comprehensive reviews are available [1,54] and although these issues are important in their own right, they are relatively peripheral to our purposes. On the other hand, co-adsorption of alkalis with other species is directly relevant to catalytic promotion: this we now consider.

The earliest interpretations of the effect of adsorbed alkalis on co-adsorbed diatomic molecules, typically CO, are based on the classical model proposed by Blyholder for the mode of bonding of such molecules to transition-metal surfaces. According to Blyholder, the bonding of CO to a platinum surface may be considered entirely analogous to that of a CO ligand in a Pt cluster compound, as described by the well-known Dewar–Chatt model. Two components in the electronic interaction are invoked: donation of charge from the CO 2σ orbital to the metal; and so-called back donation of charge from metal d-orbitals to the CO π^* orbital. This localized description of the bonding provides a satisfactory way of rationalizing many observations and is supported by modern theoretical investigations [55,56]. Within this framework, the effect of adsorbed alkali on (say) CO is interpreted in terms of lowered metal work function and a concomitant increase in the $d \rightarrow \pi^*$ back bonding. This results in an increase in the metal–C bond strength and a decrease in the C–O bond strength. As a result, one would expect an increase in the rates of those reactions for which C–O dissociation is rate-limiting.

In 1985 Lang et al. provided a detailed theoretical analysis using the methods of condensed matter physics [57]. They considered the effect on an adsorbed diatomic molecule (X_2) of an adjacent alkali cation. The field due to the alkali cation lowers the energy of the X_2 antibonding orbital with respect to the Fermi level, thus increasing charge transfer from the metal catalyst to the $X_2\pi^*$ orbital, resulting in increased metal–X bond strength and decreased X–X bond strength. This explanation is equivalent to Blyholder's more traditional "chemical" explanation. However, the electrostatic view (essentially an example of the Stark effect) is more satisfactory in that it also provides a transparent explanation for the relative unimportance of anion effects in catalyst promotion. In principle, the co-adsorbed anions (which, with the alkali ions constitute the alkali promoter compound) should exert the opposite influence to cations. This effect should, however, be much weaker for the following reasons. The anions are generally larger than the cations, thus preventing close approach of co-adsorbates to the center of net charge. Therefore, the electrostatic field at the co-adsorbate due to the anion is much smaller. As a result, promotion by cations should overwhelm poisoning by anions and, to a first approximation at least, the identity of the anion should not matter. Direct experimental evidence for substantial lowering of co-adsorbate electrostatic potential by alkalis is provided by the work of Jannssens et al., who use photoemission from Xe, Kr, and Ar to probe local surface potential [58,59]. Direct evidence for alkali-induced NO dissociation on metal surfaces comes from single-crystal studies on a variety of metals including Pt [60,61], Rh [62–64], and Ru [65]. The most elegant demonstration of cause and effect is provided by the work of Lee et al., who use He metastable quenching spectroscopy to detect directly the enhanced electron population of the CO $2\pi^*$ induced by co-adsorbed potassium [66].

16.3 ELECTROCHEMICAL PROMOTION BY ALKALIS

16.3.1 Methodology and Implementation

As noted above, EP involves controlling the coverage of an electro-active species, supplied from a solid electrolyte, on the surface of a metal thin-film catalyst electrode with which it is in contact. In our case, the electro-active species is alkali ions and the electrolyte is Na (or K) β'' alumina. A schematic of the experimental arrangement is shown in Figure 1. The thin-film metal catalyst must be both continuous and porous.

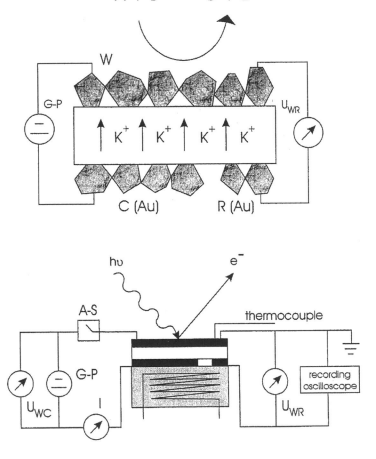

Figure 1 Schematic of EP sample and experimental setup. G-P, galvanostat-potentiostat; A-S, analog switch; W,C,R, working, counter-, and reference electrodes, respectively.

The first criterion is absolutely necessary and set by the requirement that all parts of the catalyst should be at the same electric potential. The second is highly desirable in order to generate the longest possible three-phase boundary electrolyte–electrode–gas: this minimizes the response time of the EP sample to changes in catalyst potential. The system may be operated in two modes—potentiostatic or galvanostatic. The former involves maintaining the catalyst potential, measured with respect to a reference electrode, at a predetermined value: this is the standard three-electrode arrangement well known in liquid-state electrochemistry [67]. In the galvanostatic mode, a constant current is imposed between the counter- and working electrodes; the electro-active species continuously accumulates at (or is depleted from) the catalyst electrode whose potential relative to the reference electrode consequently changes with time. Both modes are of value in the investigation of catalytic reactions. Upon arrival at the catalyst electrode, the alkali ions are discharged ($Na^+ + e^- \rightarrow$ Na), where, upon interaction with the ambient gas, they are converted to an alkali surface compound that acts as the promoter phase.

In practice, the metal thin-film catalyst electrode may be deposited on the solid electrolyte by thermal decomposition of a suitable precursor compound or by a sputtering technique. With some metals (e.g., Pd) sputtering is to be preferred because of superior adherence of the resulting thin film to the solid electrolyte. Both techniques yield morphologically rough films whose actual surface areas are usually about two orders of magnitude greater than their geometric surface areas. This is advantageous from the viewpoint of experimental sensitivity. It also means that in terms of metal particle size, EP samples are intermediate between macroscopic single-crystal specimens and nanoscopic dispersed metal catalysts.

Actual measurement of the active metal area of an EP sample is nontrivial and important because knowledge of this parameter is necessary for the quantitative determination of turnover frequencies. It may be determined by two different approaches:

1. A suitable titration technique; e.g., in the case of Pt, Rh, Pd ... , the CO methanation technique developed by Komai et al. [68] works well. Here, a sensitive FID detector is used to monitor the conversion of CO chemisorbed on metal sites to methane.
2. The electrochemical technique developed by Vayenas et al. [69], which involves measuring galvanostatic transients. A fixed negative current was applied between the working and counterelectrodes (i.e., sodium supply to the catalyst) while measuring the resulting changes in catalyst potential as a function of time. From Faraday's law and the Helmholz equation, one may show that

$$A = \frac{P_0 I}{\varepsilon_0 \frac{d\phi}{dt}} \qquad (2)$$

where A is the catalyst surface area (m^2), P_0 is the initial dipole moment of Na on the metal surface, I is the constant current applied between the working and counterelectrodes, and $d\phi/dt$ is the initial slope of the work function versus time curve. $d\phi/dt$ cannot be directly determined experimentally; instead one measures dU_{WR}/dt and uses the theoretical [37] or experimentally determined relationship between dU_{WR} and $d\phi$.

Whatever the details, typical EP samples are well suited to investigation by electron spectroscopic methods under conditions of ultrahigh vacuum, as we shall see. The form of the solid electrolyte may be that of a closed tube that also acts as the reaction vessel [70] or, more commonly, a wafer, as illustrated in Figure 1. In the latter case the wafer sample is suspended in a well-mixed reactor with all electrodes exposed to the reactant gas mixture. As a result, the Au reference electrode is more correctly described as a pseudo-reference electrode since it is usually subjected to a variable- (as opposed to fixed-) gas composition. However, the errors so introduced are negligible. A galvanostat-potentiostat is used in order to maintain a given potential difference (U_{WR}) between the working and reference electrodes (potentiostatic mode) or to supply a constant current to the catalyst electrode (galvanostatic mode).

16.3.2 Phenomenology of Electrochemical Promotion Spectroscopy and Microscopy

The interpretation of EP reactor data is carried out subject to the following assumptions:

1. Altering the catalyst potential causes the electro-active species (the promoter) to be pumped reversibly to or from the *surface* of the catalyst working electrode.
2. For a given gas environment, there is a fixed relationship between the measured potential of the working electrode (U_{WR}) and the extent to which it is covered by the promoter species.

Related issues that arise are

What is the relationship between changes in catalyst potential and the associated changes in its work function induced by spillover of the electro-active promoter species to or from the solid electrolyte?

How uniform is the spatial distribution of the electro-pumped promoter, and what is the time scale for promoter migration to or from the catalyst's surface?

What is the chemical state of the alkali promoter under reaction conditions?

These questions can be addressed directly by means of appropriate spectroscopic and microscopic measurements, as we now show.

Surface Coverage, Catalyst Potential, and Work Function Changes: XPS and UPS

Figure 2 shows Na 1s XP spectra obtained from the surface of a $Rh/Na - \beta''$ alumina sample at 580 K as a function of catalyst potential (U_{WR}) under ultrahigh vacuum conditions [71]. The +1000-mV spectrum corresponds to the electrochemically cleaned sample; increasingly negative values of U_{WR} correspond to increasing amounts of electro-pumped Na on the catalyst surface. These observations demonstrate directly that EP works by supplying alkali promoter species to the

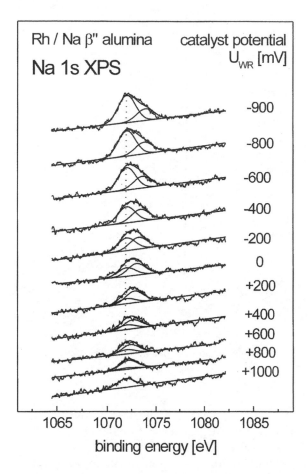

Figure 2   Na 1s XP spectra obtained from the surface of a Rh/Na-β'' alumina sample at 580 K as a function of catalyst potential (U_{WR}) under ultrahigh vacuum conditions.

(rhodium) catalyst surface. The spectral behavior was reversible and reproducible as a function of U_{WR}: as we shall see, this is consistent with the corresponding reversible and reproducible catalytic response observed under reaction conditions at atmospheric pressure. Note the residual Na 1s emission from the electrochemically cleaned sample. In order to understand this, we recall that the rhodium film is both thin and porous. The Na 1s emission therefore comprises two components: one due to alkali present on the surface of the Rh film, the other due to Na present in the underlying β'' alumina electrolyte, visible to XPS through the cracks and pores in the metal film. The component due to the Na on the Rh catalyst exhibits invariant binding energy (BE) and increases in intensity with decreasing U_{WR}. In contrast with this, the other component exhibits constant intensity and a systematic shift in apparent BE (ΔE_B). This shift is numerically equal to the change in U_{WR}, as shown in Figure 3, where the apparent binding energy shift of the Na component (ΔE_B) ascribed to the Na β'' alumina has been plotted against ΔU_{WR}. These results confirm that this component of the Na 1s emission arises from the solid electrolyte whose

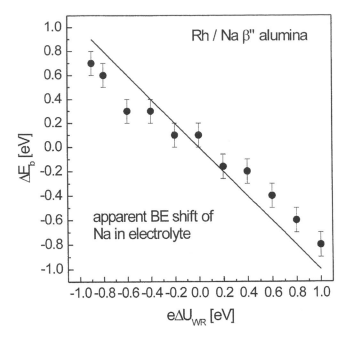

Figure 3 Rh/Na-β'' alumina: *apparent* binding energy shift of the Na in the solid electrolyte as a function of ΔU_{WR}.

electrostatic potential changes from that before potential application by the amount ΔU_{WR}. The filled circles in Figure 4 show the integrated Na 1s emission intensity versus U_{WR}. For $U_{WR} < \sim 1000\,\mathrm{mV}$, decreasing the catalyst potential caused a linear increase of the sodium coverage with potential. Note that over this same regime, the work function also varied linearly with U_{WR} (Figure 5). From these results, a value for the sodium coverage (ϑ_{Na}) may be estimated, given the measured true surface area of the metal film (see above), by making use of the relationship derived by Carley and Roberts [72]:

$$\vartheta_{Na} = \frac{Y_m \sigma_s N \rho \lambda \cos\psi}{Y_s \sigma_m M_s} \tag{3}$$

where Y_m is the integrated Na 1s photoelectron intensity, Y_s is the integrated Rh 3d intensity, σ_m [73] and σ_s [74] are the corresponding photoionization cross sections, M_s is the relative atomic mass of the substrate, ρ is the density of the substrate, λ is the escape depth of the Rh 3d photoelectrons, ψ is the detection angle relative to the sample normal, and N is Avogadro's number.

In the present case this procedure yields a calculated sodium coverage of 0.013 monolayers (ML) at $U_{WR} = -900\,\mathrm{mV}$. This value is in good accord with the observed linear dependence of work function on ϑ_{Na} (apparent from inspection of Figures 4 and 5), as discussed below.

Work function data (measured here by the UPS low-energy cutoff method) provide complementary information about the surface processes occurring during

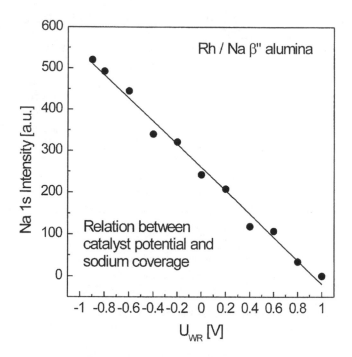

Figure 4 Rh/Na-β'' alumina: the integrated Na 1s emission intensity due to sodium on the Rh surface versus U_{WR}.

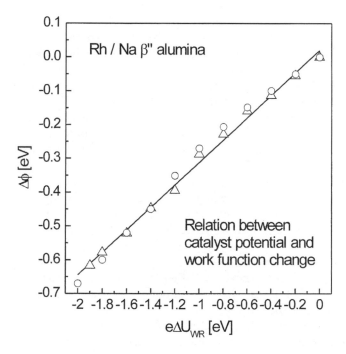

Figure 5 Rh/Na-β'' alumina: linear variation of catalyst work function with ΔU_{WR}.

the spillover of electro-pumped Na. In Figure 5 the open circles show the work function change ($\Delta\phi$) when the catalyst potential was progressively decreased from $+1000\,mV$ to $-1000\,mV$ at $580\,K$; the open triangles show corresponding data obtained when the process was reversed. Thus, the value of $\Delta\phi$ corresponding to a particular value of U_{WR} is independent of sample history. This indicates that (1) the spillover process is indeed reversible and (2) a given U_{WR} corresponds to a given sodium coverage. It is apparent from Figure 5 that $\Delta\phi$ varies linearly with both catalyst potential and ϑ_{Na}. This is a clear indication that we are in the low-alkali coverage regime, where such behavior is to be expected [75]. Specifically, ϑ_{Na} must be less than $\sim 0.1\,ML$. In this case the maximum sodium coverage achieved caused a work function decrease of $-0.65\,eV$. Comparison with single-crystal data obtained for the Rh(111)/Na system indicates that in the present case $U_{WR} = -900\,mV$ corresponds to $\vartheta_{Na} \sim 0.025\,ML$, i.e., about twice the value based on XPS intensity analysis given earlier. This is a very satisfactory level of agreement, given the approximations in our treatment. The key point is that in EP experiments with alkalis, the coverage lies in the submonolayer range—as required for effective promoter action.

Spatial Distribution, Transport, and Chemical State of the Alkali Promoters Species

Scanning tunneling microscopy has been successfully used to follow the reversible migration of electro-pumped Na across macroscopic distances from Na β'' alumina to the surface of a Pt{111} single crystal where a well-ordered (12 × 12) alkali oxide structure was formed [76]. This was the first microscopic imaging of the transport of alkali between electrolyte and catalyst—and is still the only case in which atomic-level resolution of an EP sample surface has been achieved. On a larger distance scale, photoelectron microscopy has been used to image the spatial distribution and time dependence of electro-pumped Na at the polycrystalline Pt surface of a practical EP sample at temperatures relevant to catalysis. Figure 6(a) shows the up-welling and subsequent spreading out of Na at a grain boundary that acts as a source of electro-pumped alkali when negative potential is applied to the metal film [77]. The effect occurs on a time scale of seconds, which is commensurate with the observed response time of an EP catalyst to changes in its potential. When the potential is reversed, the grain boundary source becomes a sink and the alkali migrates back to the electrolyte below, again on a time scale of seconds [Figure 6(b)].

The photoelectron microscopy results described above were necessarily acquired in vacuum. Under such conditions there is no doubt that the Na is present on the Pt surface as sodium metal. However, under reaction conditions, this cannot be the case: it is to be expected that the alkali would be present as a submonolayer quantity of surface compound, and indeed this is just what is observed. Furthermore, also in accordance with expectation, the nature of the alkali promoter compound is dependent on the composition of the gas atmosphere. Figure 7 shows postreaction XPS and XANES spectra acquired from Pt/Na β'' alumina EP samples after exposure to reaction conditions and *without exposure to laboratory atmosphere* for the Pt-catalyzed reactions NO + propene [Figure 7(a)] and O_2 + propene [Figure 7(b)], respectively. In the first case the promoter phase consists of a mixture of $NaNO_2$ and $NaNO_3$, in the second case it consists of Na_2CO_3. This is important

(a)

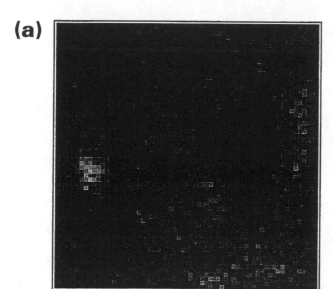

Na 2p / +1 V

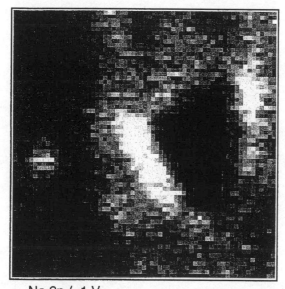

Na 2p / -1 V

Figure 6 Photoelectron microscopy of Na electro-pumped to a Pt catalyst film: (a) showing up-welling of Na from a surface source upon switching from positive to negative catalyst potential and (b) showing time dependence of Na spillover.

information, unavailable in studies with conventional catalysts, where the nature of the alkali promoter phase is generally unknown.

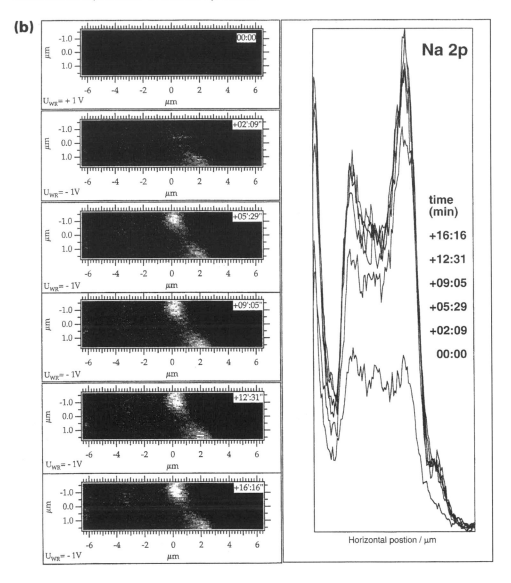

Figure 6 Continued.

The range of catalytic systems that can be addressed and the mechanistic information that may be obtained will now be illustrated by reference to specific examples.

16.4   EXAMPLES OF ELECTROCHEMICAL PROMOTION BY ALKALIS

16.4.1   Selective Hydrogenation of Acetylene

This is the first example of the use of EP in catalytic selective hydrogenation [78]. The system exhibits strong promotion by spillover Na or K supplied from a β''-alumina

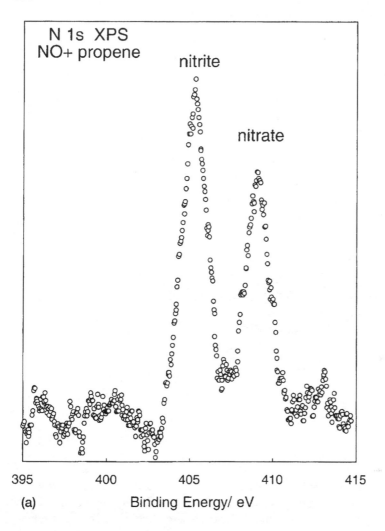

Figure 7 (a) Postreaction XPS (NO + propene) and (b) postreaction XANES (oxygen + propene) spectra acquired from Pt/Na β'' alumina EP samples.

solid electrolyte. In the hydrogen-rich regime, reversible, potential-dependent increases in ethene selectivity of an order of magnitude are achievable under the influence of ∼0.02 monolayers Na, while maintaining high acetylene conversion (Figure 8). Thus, EP by Na or K induces Pt, normally a poor selective hydrogenation catalyst, to behave like Pd. Butenes, butadiene, and benzene are observed as minor products whose formation is also promoted by Na. Again, these are reactions characteristic of Pd rather than (unpromoted) Pt. A simple reaction model based on the destabilizing effect of Na on co-adsorbed adsorbed ethene (an electropositive adsorbate) accounts for the principal findings. That is, co-adsorbed alkali weakens the Pt–olefin bond, reducing the surface lifetime of ethene and favoring its desorption at the expense of further hydrogenation to ethane. Consistent with this view [78], galvanostatic measurements indicated that adsorbed Na interacted

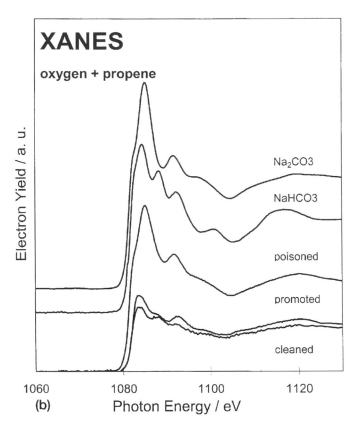

Figure 7 Continued.

relatively weakly with the species formed by acetylene adsorption, and very strongly with the adsorbed species formed by acetylene adsorption in the presence of hydrogen. Interestingly, these measurements also showed [78] that Na interacted strongly with H_a, suggesting that in this case the alkali promoter was present as a hydridelike compound. Related to these findings, Cavalco and Haller observed that sodium inhibited benzene hydrogenation over Pt under EP by Na [79], an effect understandable in similar terms. In this case Na destabilizes adsorbed benzene (also an electropositive adsorbate), reducing its surface lifetime and hence retarding the hydrogenation rate.

16.4.2 Ethene and Propene Epoxidation

Heterogeneously catalyzed alkene epoxidation is of great interest from both academic and technological points of view. Ethene epoxidation, for which Ag is an apparently unique catalyst, is a strategically important large-scale industrial process. As a result, elucidation of the mechanism of Ag-catalyzed epoxidation has attracted a great deal of academic research, and many key features of the reaction may be regarded as well understood. Gaseous and solid-state promoters play a vital role in ethene epoxidation catalysis. In practice, adsorbed chlorine, usually supplied

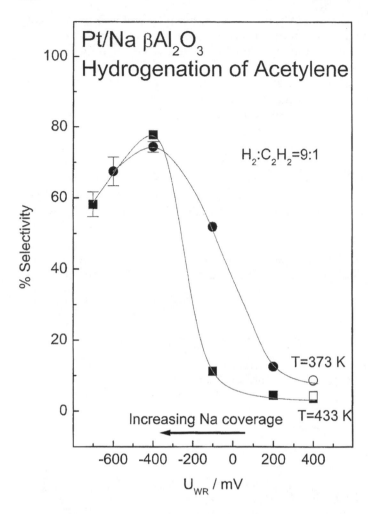

Figure 8 Selective hydrogenation of acetylene with Pt/Na. Selectivity as a function of catalyst potential.

by addition of ppm levels of gaseous dichloroethane (DCE), is always used as a promoter. Additionally, alkalis are almost always added as co-promoters in commercial catalysts, where they influence both the metal chemistry and the support chemistry. Here, we are of course concerned with effects on the metal chemistry. The mode of action of adsorbed chlorine and alkali [80] promoters has been elucidated, and a recent review of progress in the field is provided by [16]. Karavasilis et al. investigate EP by Na of ethene epoxidation over Ag, all data being acquired in the presence of gaseous dichloroethane [81], although no results are reported for the effect of Na in the absence of surface chloriding. Their findings are in good accord with the behavior of conventional Ag/α-alumina catalysts operated under "normal" conditions.

In the 1980s a series of patents appeared [82–85] in which it is claimed that addition of ppm levels of NO_x or NO_x precursors to the feedstream leads to selectivities even higher than those delivered by conventional doubly promoted Ag (Cl, alkali). However, the mechanism of NO_x promotion was not known. We used EP with Ag/K-β'' alumina to shed light on this ultraselective epoxidation catalysis induced in the presence of gaseous NO [86]. In particular, we were able to answer this question: What are the minimum necessary conditions for the phenomenon to appear? It was shown that in the absence of other promoters, K alone served only to poison both the activity and selectivity of Ag in the epoxidation of ethene. However, in the presence of ppm levels of gaseous NO, the situation was transformed: supply of alkali to the catalyst strongly promoted selectivity (Figure 9), and it was demonstrated that the minimum necessary and sufficient conditions for the appearance of NO_x promotion are ppm levels of gaseous NO and sub-monolayer quantities of alkali on the silver surface. It seems highly likely that surface alkali nitrate is present under reactions conditions [22] and that this species participates in direct oxygen transfer to the alkene [87], this new reaction pathway accounting for the ultraselective behavior.

On the other hand, propene epoxidation over Ag/α-alumina is notably inefficient, yielding epoxide selectivities in the order of only a few percent. As a

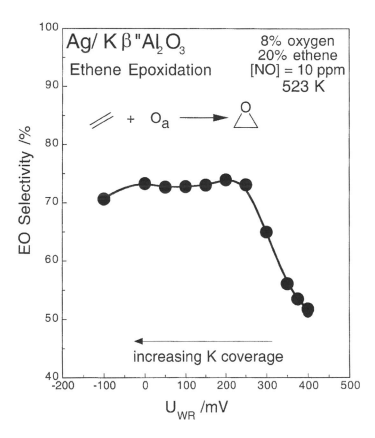

Figure 9 Ethene epoxidation in the presence of NO with Ag/K. Selectivity as a function of catalyst potential.

result, efficient heterogeneous selective oxidation of propene is a key technological goal that remains to be realized. Current technology employs either (1) the old chlorohydrin process, environmentally unfriendly because it involves the use of chlorine or (2) a newer homogenous route that involves co-production of propene epoxide and styrene. It has been proposed, and seems to be generally agreed, that the difficulty with propene epoxidation resides in the ease with which an allylic hydrogen atom may be stripped from the molecule, a process that presumably shuts off the epoxidation channel and results in combustion.

Quite recently we used EP by potassium in a comparative study of the epoxidation of propene and ethene [88]. We found that the two systems respond in very different ways to the addition of K, NO, Cl, and combinations of these modifiers, pointing to important differences in reaction mechanisms. Interestingly, the results suggest that in the case of propene epoxidation adsorbed alkali superoxides or peroxides may act as epoxidizing agents, in addition to or instead of O_a. Further conversion of propene epoxide appeared not to be an issue: the problem of low selectivity resides in the primary chemistry and probably hinges on the presence of allylic hydrogen atoms in the alkene. Furthermore, postreaction XPS data suggested that the state of oxygenation of the Ag subsurface region is likely to be significant in determining the very different selectivities observed in ethene and propene epoxidation.

16.4.3 Fischer–Tropsch Synthesis

Recently [19] we carried out an EP investigation of Fischer–Tropsch synthesis (FTS) with Ru/Na-β'' alumina, motivated by three principal objectives:

> First, to broaden the applicability of EP in the study of heterogeneous catalysis: EP had previously not been applied to C–C bond-forming reactions.
>
> Second, to investigate the effects of Na promotion on FTS with a view to understanding of the reaction mechanism, which remains controversial in regard to the role played by alkali promoters. Ensemble effects [89], electronic effects [90,91], and a decrease in the surface coverage and in the mobility of H ad-atoms [92,93] have been variously invoked, whereas Lahtinen and Somorjai [94] propose that the principal effect of K is to inhibit graphite formation.
>
> Third, to establish whether such EP data are of relevance to the behavior and understanding of classical dispersed FTS catalysts.

It was shown that Na strongly suppressed the methanation rate, thus resulting in a marked increase in selectivity toward higher hydrocarbons, accompanied by an increase in the alkene:alkane ratio. The results obtained with this EP model system, including the "C_2 anomaly," are in close agreement with those found for classically promoted conventional dispersed catalysts. It was found that alkali substantially increased the probability of chain growth (Figure 10) and that CO dissociation is not rate-controlling: effects in addition to purely geometrical ones must be taken into account. These may include electronic effects due to alkali enhancement of CO chemisorption at the expense of hydrogen chemisorption. Very recently, this work has been extended to higher pressures (25 bar), which permits us to examine the

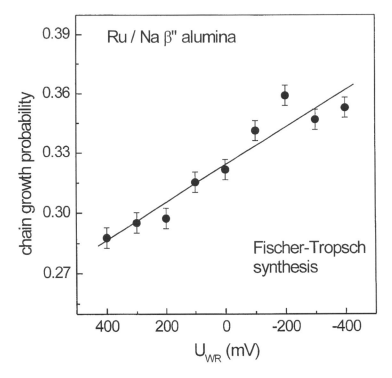

Figure 10 Fischer–Tropsch synthesis with Ru/Na. Effect of catalyst potential (U_{WR}) on the probability of hydrocarbon chain growth.

formation of hydrocarbons $>C_{30}$ under conditions of electrochemical promotion [95].

16.4.4 Another Important Class of Cationic Conductors: EP by Proton Conductors in Ammonia Synthesis and Alkene Isomerization

Proton-conducting solid electrolytes are closely related to alkali ion conductors and are likely to receive increasing attention in EP applications with respect to both fundamental issues and technical applications. Both aspects are therefore illustrated in this section.

Formally, ammonia synthesis is closely related to Fischer–Tropsch synthesis. Industrial operation involves the use of an iron catalyst promoted with calcium and potassium oxides. However, the reason we consider this process here is not directly in connection with alkali promotion of the catalyst. We are concerned with a remarkable achievement reported by Yiokari et al. [15], who use a *proton*-conducting electrolyte to achieve electrochemical promotion of a fully promoted ammonia synthesis catalyst operated at elevated pressure. Specifically, they make use of a fully promoted industrial catalyst that was interfaced with the proton conductor $CaIn_{0.1}Zr_{0.9}O_{3-\alpha}$ operated at $\sim 700\,K$ and 50 bar in a multipellet configuration. It was shown that under EP the catalytic rate could be increased by a factor of 13 when protons were supplied to the catalyst. This work represents a milestone in some important respects. It is the first time that an industrial dispersed metal catalyst has been electrochemically promoted, let alone with such spectacular results. And it corresponds to the first successful scale-up of an EP process. It is particularly interesting that the industrial catalyst, already fully promoted for conventional use, responds so strongly to EP. The authors argue persuasively [15] that protons supplied to the metal catalyst act in an analogous way to alkalis: the work function is lowered so that the kinetics and energetics of N_2 dissociative chemisorption are both favorably affected. The implication is that EP by protons acts synergistically with the promotional effects induced by the conventionally supplied promoters.

We cannot leave the subject of electrochemical promotion by proton conductors without mentioning the intriguing results produced by Smotkin and his co-workers [96,97], who exploited Nafion/Pd assemblies in a fuel-cell configuration. They demonstrate that under appropriate conditions protons supplied from Nafion to the Pd electrode can induce very strong electrochemical promotion of the isomerization of 1-butene to cis- and trans-butene. Huge rate-enhancement factors (~ 40) relative to the open circuit rate were achieved at low temperatures ($\sim 340\,K$) and a very plausible reaction mechanism is proposed. In effect, the Nafion acts as a superacid, supplying protons that undergo Markovnikov addition to the C1 carbon, followed by proton elimination at C3 to yield the isomerized product. The authors elegantly confirm the essential correctness of their model by means of deuterium isotope tracing and FTIR spectroscopy [97]. Further important developments in this area seem highly likely, not least because it involves promotion of relatively subtle organic reactions carried out on delicate molecules at low temperatures.

16.4.5 NO$_x$ Reduction by CO and by Propene

Reactions involving the catalytic reduction of NO are of major environmental importance [98]. Accordingly, we have used EP by Na to study NO reduction by CO and by propene in the absence and presence of oxygen over Pt, Rh, and Cu catalysts. In every case, the results indicate that NO dissociation is the crucial reaction-initiating step and that Na strongly promotes the reaction by triggering dissociation of NO. We now illustrate this important class of reactions in more detail by considering a specific case: the EP by Na of the catalytic reduction of NO by propene over platinum.

NO reduction by propene exhibits strong electrochemical promotion under Na pumping to the catalyst (Figure 11). As discussed above, the promotional effect is due to *enhanced NO chemisorption and dissociation on the Pt surface* induced by electrochemically pumped Na. Recall that the polycrystalline Pt film consists mainly of large crystallites ($\sim 1\,\mu m$) whose external surfaces are dominated by low-index planes: such low-index planes of Pt are known to be relatively ineffective for NO dissociation [99]. Na-induced NO dissociation produces the O_a species that are then responsible for initiating the ensuing oxidation reactions of adsorbed propene and propene fragments. In the region of strong electrochemical promotion (0 to $-350\,mV$), the activity toward formation of all products shows an exponential increase with (decreasing) catalyst potential, as predicted by Vayenas's model [33,34]. Eventually, at around $-350\,mV$, there is a precipitous fall in rates as the amount of Na species increases beyond a critical value; i.e., the regime of

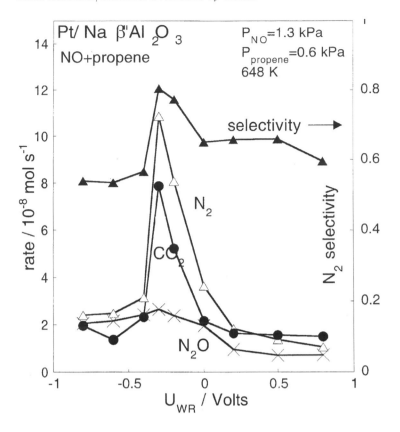

Figure 11 NO reduction by propene with Pt/Na showing promotion of rate and N_2 selectivity within a certain window of catalyst potential.

electrochemical promotion is followed by a regime of strong poisoning. This poisoning behavior is discussed below.

The dependence of activity and selectivity on catalyst potential may be rationalized by considering the effects of the Na promoter species on the surface chemistry of propene and NO. At high positive U_{WR} the Pt surface is free of sodium and covered mainly with propene. We postulate that the low clean-surface residual rate is due to dissociation of NO at defect sites, $NO_a \rightarrow N_a + O_a$ being the reaction-initiating step.

As the catalyst potential is decreased (pumping Na to the Pt) the electronic effect of the sodium promoter, discussed above, strengthens the Pt–N bond (increasing NO coverage) and weakens the N–O bond (facilitating NO dissociation). Thus, the effect of the alkali promoter is to (1) activate previously inactive crystal planes on the catalyst surface toward NO dissociation, (2) decrease self-poisoning due to overadsorption of propene. The former effect increases progressively with the amount of Na pumped, and the exponential dependence of rate on U_{WR} in the promoted regime suggests that changes in catalyst work function are indeed strongly correlated with the greatly increased activity. In addition, Na also enhances the selectivity toward N_2 formation from 60% to 80%. This quantity is determined by the competition between the following reactions that occur on the Pt surface:

$$N_a + N_a \rightarrow N_2 \tag{4}$$
$$N_a + NO_a \rightarrow N_2O \tag{5}$$

The observed increase in selectivity is a consequence of the increased dissociative chemisorption of NO, which is strongly promoted by Na. The alkali strengthens the binding of NO to the metal surface but also decreases the activation energy for its dissociation. As a result, for a given total uptake of NO, the fraction dissociated is increased [61]. Under reaction conditions, this corresponds to an alkali-induced increase in coverage by NO and its dissociation products, the latter favored at the expense of the former. This decreases and increases, respectively, the amounts of molecularly adsorbed NO and atomic N on the surface, thus favoring the first reaction over the second. The systematic effect of U_{WR} on the Langmuir–Hinshelwood (L–H) reaction kinetics is also understandable in terms of the relative effects of the Na promoter on the adsorption of NO and propene. For example, it is found that as the sodium coverage increases at constant P_{NO}, the propene partial pressure at the L–H rate maximum shifts systematically to higher values, reflecting an increase in chemisorption bond strength of NO relative to propene with increasing Na coverage [23]. Such behavior is exactly what one would expect in the case of an electropositive promoter: the chemisorption strength of electron donors (propene) should be decreased, whereas the chemisorption of electron acceptors (NO and its dissociation products) should be enhanced.

The strong poisoning behavior observed at high Na loading is understandable in terms of the postreaction Na 1s XPS and KLL Auger spectra, which indicate the presence of large amounts of Na compounds at very negative catalyst potentials. In other words, coverage of active Pt sites by overloading with promoter is a major cause of poisoning. A more detailed analysis of these spectra that takes account of the very different sampling depths in the two cases [electron kinetic energies $\sim 180\,eV$ (1s) and $\sim 990\,eV$ (KLL), respectively] leads to the conclusion that on the poisoned surface the Na compounds are principally present as 3D crystallites. XPS also reveals the chemical identity of the surface chemical compounds formed by Na electro-pumping in reaction gas. Both $NaNO_2$ and $NaNO_3$ are produced in the present case: i.e., nitrite and nitrate are the principal counterions, which, with Na, constitute the promoter phase. These spectroscopic data also show that the surface nitrate is efficiently reduced by propene, raising the interesting possibility that NO_{3-}/NO_{2-} may itself constitute a system that is involved in the catalytic turnover, thus

hydrocarbon + nitrate → oxidation product + nitrite

nitrite + O_a (from NO dissociation) → nitrate

If this were the case, it would mean that we are dealing with a promoter system in which the cation triggers the primary chemistry while the anion facilitates subsequent oxidative reactions. Further work is in progress to investigate this possibility.

16.5 DEVELOPMENT OF ALKALI-PROMOTED NANOPARTICLE CATALYSTS BASED ON UNDERSTANDING DERIVED FROM ELECTROCHEMICAL PROMOTION

In the introduction to this chapter, we state that one of the reasons for studying EP is that it offers the possibility of developing new nanostructured catalytic systems based on the insight provided by the corresponding EP analogs. In the case of EP by alkalis, this is best illustrated by the subject of the previous section: the catalytic reduction of NO_x for purposes of environmental protection. Early investigations of the effect of alkali promoters on dispersed Pt metal catalysts used for NO_x reduction led to the inclusion that alkalis were ineffective at best, poisons at worst [100]. On the other hand, our EP studies provide strong reasons for supposing that alkalis should in principle be capable of very substantially improving the performance of such practical catalysts, in terms of both overall activity and selectivity for N_2 (as opposed to N_2O) formation.

Recent work in our laboratory has confirmed that alkali promotion of practical catalysts consisting of metal nanoparticles (Pt, Pd, Rh) supported on γ-alumina is indeed a highly effective strategy for improving performance in regard to both activity and selectivity [101–103]. Relevant results for sodium promotion of the reduction of NO_x by propene over Rh/γ-alumina dispersed catalysts are shown in Figure 12. The key contribution made by the EP work to the development of these catalysts was the demonstration that the alkali loading is critically important. Too little, and the effect is negligible. Too much, and the result is poisoning rather than promotion.

16.6 CONCLUSIONS AND FUTURE PROSPECTS

The phenomena that underlie electrochemical promotion may now be regarded as rather well understood. Equally, we have a sound theoretical basis for interpreting EP data. This enables us to exploit the technique as a means for elucidating the mechanisms of catalytic reactions, a particular merit of the approach being that EP samples are amenable to reactor studies at elevated pressures and to investigation by advanced spectroscopic and microscopic techniques. New, nanodispersed practical catalysts for NO_x reduction which outperform those in current use have been developed as a direct result of EP investigations. So where may future developments be expected? We have already referred to the work of Marwood and Vayenas, who demonstrate "wireless" EP [47], and of Wodiunig et al. [28], who show that such bipolar configurations could be exploited to improve thin-film catalyst performance in tubular reactors. The very recent work of Yiokari et al. [15] represents a major step forward along this road—they show that the activity of a commercial, fully promoted, ammonia synthesis catalyst could be greatly enhanced by electrochemical promotion from a proton conductor. We may anticipate more important developments of this kind. The obvious and crucial feature of an EP catalyst is that its performance is controlled electrically. This opens the door to using such systems in applications where the characteristics of the working catalyst can be varied, perhaps automatically, by a sensor-generated feedback signal, to respond to changes in operating conditions.

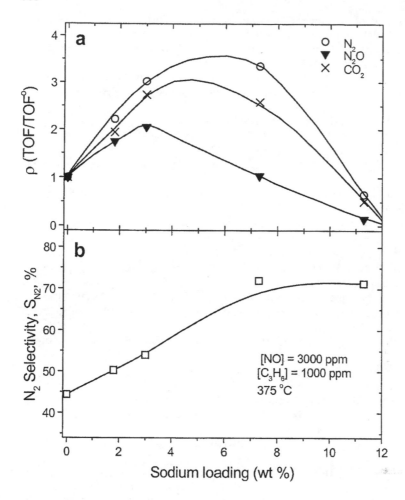

Figure 12 Na promotion of a nano-particulate Rh/γ alumina dispersed catalyst. The improvements in rate and selectivity mirror those observe when Rh thin-film catalysts are electrochemically promoted by Na.

ACKNOWLEDGMENTS

I am indebted to my students and co-workers whose many contributions made this chapter possible. And it is a pleasure to acknowledge helpful comments from Geoffrey Bond, Noel Cant, Renee Diehl, Hans Niemantsverdriet, Gianfranco Pacchioni, and Neville Richardson.

REFERENCES

1. M.P. Kiskinova, In *Poisoning and Promotion in Catalysis Based on Surface Science Concepts and Experiments*, Elsevier, Amsterdam (1992).
2. J.W. Niemantsverdriet, *Appl. Phys. A.* **61** (1995) 503.
3. S.J. Thomson, *J. Chem. Soc., Faraday Trans.* **83** (1987) 2001.

4. R.M. Ormerod, R.M. Lambert. In *Surface Reactions*, Springer Series in Surface Science, Vol. 34, R.J. Madix, ed. Springer-Verlag, Berlin (1994) pp 89–131.

5. A.F. Lee, C.I. Baddeley, C. Hardacre, R.M. Ormerod, R.M. Lambert, G. Schmid, H. West, *J. Phys. Chem.* **99** (1995) 6096.

6. R.A. Meyers, *Handbook of Petroleum Refining Processes*, McGraw-Hill, Boston (1996).

7. H.F. Rase, *Handbook of Commercial Heterogeneous Catalysts*, CRC Press, Boca Raton, FL (2000).

8. R.B. Grant, R.M. Lambert, *J. Catal.* **92** (1985) 364.

9. S.A. Tan, R.B. Grant, R.M. Lambert, *J. Catal.* **100** (1986) 383.

10. R.B. Grant, R.M. Lambert, *J. Catal.* **93** (1985) 92.

11. R.B. Grant, C.A. Harbach, S.A. Tan, R.M. Lambert, *J. Chem. Soc. Faraday Trans. I* **83** (1987) 2035.

12. P. Stolze, In *Ammonia Synthesis: Catalysis and Manufacture*, A.E. Nielsen, ed. Springer-Verlag, Berlin (1995) p 21.

13. D.R. Strongin, G.A. Somorjai, In *Catalytic Ammonia Synthesis, Fundamentals and Practice*, J.R. Jennings, ed. Plenum, New York (1991) p 133.

14. A. Palermo, A. Husain, R.M. Lambert, *Catal. Lett.* **69** (2000) 175.

15. C.G. Yiokari, G.E. Pitselis, D.G. Polydoros, A.D. Katsaoounis, C.G. Vayenas, *J. Phys. Chem. A* **104** (2000) 10600.

16. J.G. Serafin, A.C. Liu, S.R. Seyedmonir, *J. Mol. Catal. A: Chemical* **131** (1998) 157.

17. G. Ertl, In *Catalytic Ammonia Synthesis, Fundamentals and Practice*, J.R. Jennings, ed. Plenum, New York (1991) p 109.

18. P.M. Maitlis, R. Quyoum, H.C. Long, M.L. Turner, *Appl. Catal. A: General* **186** (1999) 363.

19. F.J. Williams, R.M. Lambert, *Catal. Lett.* **70** (2000) 9.

20. S.M. Augustine, J.P. Blitz, *J. Catal.* **142** (1993) 312.

21. R.D. Haley, M.S. Tikhov, R.M. Lambert, *Catal. Lett.* **76** (2001) 125.

22. I.V. Yentekakis, A. Palermo, N.C. Filkin, M.S. Tikhov, R.M. Lambert, *J. Phys. Chem. B* **101** (1997) 3759.

23. N.C. Filkin, M.S. Tikhov, A. Palermo, R.M. Lambert, *J. Phys. Chem. A* **103** (1999) 2680.

24. I.V. Yentekakis, R.M. Lambert, M.S. Tikhov, M. Konsolakis, V. Kiousis, *J. Catal.* **176** (1998) 82.

25. M. Konsolakis, A. Palermo, M. Tikhov, R.M. Lambert, I.V. Yentekakis, *Ionics* **4** (1998) 148.

26. M. Konsolakis, L. Nalbantian, N. McLeod, I.V. Yentekakis, R.M. Lambert, *Appl. Catal. B: Environ.* **22** (1999) 123.

27. S. Tracey, A. Palermo, J.P. Holgado Vazquez, R.M. Lambert, *J. Catal.* **179** (1998) 231.

28. G. Foti, S. Wodiunig, C. Comninellis, *Curr. Top. Electrochem.* **7** (2000) 1.

29. N. Macleod, J. Isaac, R.M. Lambert, *J. Catal.* **198** (1) (2001) 128.

30. M. Konsolakis, I.V. Yentekakis, A. Palermo, R.M. Lambert, *Appl. Catal. B: Environ.* **33** (2001) 293.

31. N. Macleod, J. Isaac, R.M. Lambert, *Appl. Catal. B: Environ.* **33** (2001) 335.

32. S. Bebelis, C.G. Vayenas, *J. Catal.* **118** (1989) 125.

33. C.G. Vayenas, M. Jaksic, S. Bebelis, S.G. Neophytides, In *Modern Aspects of Electrochemistry*, O'M Bockris, B.F. Conway, E. White, eds. Plenum, New York (1995) pp 57–202.

34. C.G. Vayenas, S.G. Neophytides, In *Catalysis*, Vol. 12, The Royal Soc. Chem., Cambridge (1996) pp 195–253.

35. G. Pacchioni, F. Illas, S. Neophytides, G.C. Vayenas, *J. Phys. Chem.* **100** (1996) 16653.

36. M. Makri, C.G. Vayenas, S. Bebelis, K.H. Besocke, C. Cavalca, *Surf. Sci.* **369** (1996) 351.

37. D. Tsiplakides, C.G. Vayenas, *J. Electrochem. Soc.* **148** (2001) 189.

38. C. Pliangos, I.V. Yentekakis, X. Verykios, C.G. Vayenas, *J. Catal.* **154** (1995) 124.

39. I.V. Yentekakis, S. Bebelis, *J. Catal.* **137** (1992) 278.

40. C. Karavasilis, S. Bebelis, C.G. Vayenas, *J. Catal.* **160** (1994) 190.

41. I.V. Yentekakis, G.D. Moggridge, C.G. Vayenas, R.M. Lambert, *J. Catal.* **146** (1994) 292.

42. O.A. Marina, I.V. Yentekakis, C.G. Vayenas, A. Palermo, R.M. Lambert, *J. Catal.* **166** (2) (1997) 218.

43. R.M. Lambert, M.S. Tikhov, A. Palermo, I.V. Yentekakis, C.G. Vayenas, *Ionics* **5** (1995) 366.

44. A. Palermo, R.M. Lambert, I.R. Harkness, I.V. Yentekakis, O. Marina, C.G. Vayenas, *J. Catal.* **161** (1996) 471.

45. I.V. Yentekakis, C.G. Vayenas, *J. Catal.* **149** (1994) 238.

46. T.I. Politova, V.A. Sobyanin, V.D. Belyaev, *React. Kinet. Catal. Lett.* **41** (1990) 321.

47. M. Marwood, C.G. Vayenas, *J. Catal.* **178** (2) (1998) 429.

48. A. Clark, In *The Theory of Adsorption and Catalysis*, Chap. 3, Academic Press, New York (1970).

49. M.W. Roberts, C.S. McKee, In *Chemistry of the Metal Gas Interface*, Oxford Univ. Press, New York (1978).

50. M. Scheffler, C. Stampfl, In *Handbook of Surface Science, Vol. 2: Electronic Structure*, K. Horn, M. Scheffler, eds. Elsevier, Amsterdam (2000), pp 286–356.

51. R.D. Diehl, R. McGrath, In *Alkali Metals on Metals*, in Landolt-Boernstein Vol. III/42A, Physics of alkali covered solid surfaces, H.P. Bonzel, ed. Springer, Berlin (2001), pp 131–177.

52. R.D. Diehl, R. McGrath, *Surf. Sci. Rep.* **23** (1996) 43.

53. S. Chandavarkar, R.D. Diehl, *Phys. Rev. B* **40** (1989) 4651.

54. H.P. Bonzel, *Surf. Sci. Rep.* **8** (1987) 43.

55. G. Pacchioni, P.S. Bagus, *Phys. Rev. B* **40** (1989) 6003.

56. P.S. Bagus, G. Pacchioni, *Surf. Sci.* **236** (1990) 233.

57. N.D. Lang, S. Holloway, J.K. Norskov, *Surf. Sci.* **150** (1985) 24.

58. T.W.V. Jannssens, J.W. Niemantsverdriet, R.A. van Santen, *J. Chem. Phys.* **101** (1994) 2995.

59. T.V.W. Jannssens, G.R. Castro, K. Wandelt, J.W. Niemantsverdriet, *Phys. Rev. B* **49** (1994) 14599.

60. E.L. Garfunkel, J.J. Maj, M.H. Farias, G.A. Somorjai, *J. Phys. Chem.* **87** (1983) 3629.

61. I.R. Harkness, R.M. Lambert, *J. Chem. Soc. Faraday Trans.* **93** (1997) 1425.

62. L. Bugyi, F. Solymosi, *Surf. Sci.* **188** (1987) 475.

63. L. Bugyi, J. Kiss, K. Revesz, F. Solymosi, *Surf. Sci.* **233** (1990) 1.

64. H. Hochst, E. Colavita, *J. Vac. Sci. Technol. A* **4** (1986) 1442.

65. H.J. Zhang, Z.J. Yan, H.Y. Li, P.M. He, S.N. Bao, J. Wang, C.Y. Xu, Y.B. Xu, *Acta Phys. Sinica* **49** (2000) 577.

66. J. Lee, C.P. Hanrahan, J. Arias, R.M. Martin, H. Metiu, *Phys. Rev. Lett.* **51** (1983) 1803.

67. R.D. Armstrong, M. Todd, In *Solid State Electrochemistry*, P.G. Bruce, ed. Cambridge Univ. Press, Cambridge (1995) p 277.

68. S. Komai, T. Hattori, Y. Murakami, *J. Catal.* **120** (1989) 370.

69. S. Ladas, S. Bebelis, C.G. Vayenas, *Surf. Sci.* **251/252** (1991) 1062.

70. I.V. Yentekakis, S. Bebelis, *J. Catal.* **137** (1992) 278.

71. F.J. Williams, A. Palermo, M.S. Tikhov, R.M. Lambert, *J. Phys. Chem. B* **104** (50) (2000) 11883.

72. A.F. Carley, M.W. Roberts, *Proc. R. Soc. Lond. A* **363** (1978) 403.

73. J.J. Yeh, I. Lindau, *Atomic Data and Nuclear Data Tables* **32** (1985) 1.

74. D.R. Penn, *J. Electron Spect. and Rel. Phen.* **9** (1976) 29.

75. R.W. Verhoef, M. Asscher, *Surf. Sci.* **391** (1997) 11.

76. M. Makri, C.G. Vayenas, S. Bebelis, K.H. Besocke, C. Cavalca, *Surf. Sci.* **369** (1996) 351.

77. M.S. Tikhov, S. Günter, M.P. Kiskinova, R.M. Lambert, In preparation.

78. S. Tracey, A. Palermo, J.P. Holgado Vazquez, R.M. Lambert, *J. Catal.* **179** (1) (1998) 231.

79. C.A. Cavalca, G.L. Haller, *J. Catal.* **177** (2) (1998) 389.

80. R.B. Grant, R.M. Lambert, *Langmuir* **1** (1985) 29.

81. C. Karavasilis, S. Bebelis, C.G. Vayenas, *J. Catal.* **160** (1996) 205.

82. P. Hayden, R.C. Clayton, J.R. Bamforth, A.F.G. Cope, Eur. Patent 003642A2 (1980) to ICI.

83. D.W. Johnson, Eur. Patent 0119710 (1984) to ICI.

84. P. Hayden, Eur. Patent 0176253 (1986) to ICI.

85. M. Nakajima, H. Kuboyama, T. Komiyama, and K. Yoshida, British Patent 2161480 A (1986) to Mitsui Toatsu Chemicals.

86. A. Palermo, A. Husain, R.M. Lambert, *Catal. Lett.* **69** (2000) 175.

87. D. Bird, A. Palermo, R.M. Lambert, In preparation.

88. A. Palermo, A. Husain, M.S. Tikhoy, R.M. Lambert, Submitted to *J. Catal.* (2001).

89. M.M. McClory, R.D. Gonzalez, *J. Catal.* **89** (1984) 392.

90. T. Okuhara, H. Tamura, M. Misono, *J. Catal.* **95** (1985) 41.

91. T.E. Hoost, J.G. Goodwin, Jr, *J. Catal.* **137** (1992) 22.

92. T. Komaya, A.T. Bell, Z. Weng-Sieh, R. Gronsky, F. Engelke, T.S. King, M. Pruski, *J. Catal.* **152** (1995) 350.

93. D.O. Uner, M. Pruski, B.C. Gerstein, T.S. King, *J. Catal.* **146** (1994) 530.

94. J. Lahtinen, G.A. Somorjai, *J. Mol. Catal. A: Chem.* **130** (1998) 255.

95. F.J. Williams, A. Urquhart, R.M. Lambert, In preparation.

96. L. Ploense, M. Salazar, B. Gurau, E.S. Smotkin, *J. Amer. Chem. Soc.* **119** (1998) 11550.

97. L. Ploense, M. Salazar, B. Gurau, E.S. Smotkin, *Sol. State Ionics* **136** (2000) 713.

98. K.C. Taylor, *Catal. Rev.-Sci. Eng.* **35** (1993) 457.

99. R.I. Masel, *Catal. Rev.-Sci. Eng.* **28** (2,3) (1986) 335.

100. R. Burch, T.C. Watling, *Appl. Catal B: Environ.* **11** (1997) 207.

101. M. Konsolakis, N. Macleod, J. Isaac, I.V. Yentekakis, R.M. Lambert, *J. Catal.* **193** (2000) 330.

102. I.V. Yentekakis, R.M. Lambert, M.S. Tikhov, M. Konsolakis, V. Kiousis, *J. Catal.* **176** (1998) 82.

103. N. Macleod, J. Isaac, R.M. Lambert, *J. Catal.* **193** (2000) 115.

17

Metal-Support Interaction in Low-Temperature Fuel Cell Electrocatalysts

A.S. ARICO and V. ANTONUCCI

Institute for Transformation and Storage of Energy, Messina, Italy

P.L. ANTONUCCI

University of Reggio Calabria, Reggio Calabria, Italy

CHAPTER CONTENTS

SUMMARY

Metal-support interactions in fuel cell electrocatalysts, with particular emphasis for methanol oxidation and oxygen reduction, are discussed. The interaction of the active metal with the carbonaceous support strongly influences the physico-chemical and electrochemical properties of the catalysts and, as a consequence, its performance under fuel cell operation. The analogies of such interactions with the Bronsted–Lewis acid-base theories have been interpreted and rationalized in terms of the pHzpc of the surface, as one of the main parameters able to predict and determine the level of interaction. The latter often results in a wide range of metal particle size, depending upon the acid–base nature of carbon functional groups that act as anchoring centers for the noble metal particles. Dispersion, however, also influences the Pt surface fraction of crystallographic plans with low Miller index, which is thought to be connected with the specific activity of the catalysts. Moreover, carbon microstructure plays an important role by affecting the access of the reactant to the catalytic sites and the removal of reaction products. In conclusion, the complex interrelationships occurring between the active metal phase and its support need to be interpreted in order to formulate and design new, better performing catalysts for low-temperature fuel cells.

17.1   INTRODUCTION

Low-temperature fuel cells have reached a mature level of technology. These systems appear now ready for electro-traction, portable power sources, remote power generation, and stationary applications [1]. The high thermodynamic efficiency and the near-zero emission levels make them an attractive alternative to internal combustion engines, batteries, and thermal combustion power plants. However, one of the main drawbacks that still limits the commercial exploitation of these devices is due to the sluggishness of the electrochemical reactions in proximity of their reversible potentials. The electrochemical processes involved in low-temperature fuel cells need noble metal-based electro-catalysts (Pt or Pt-alloys) to occur at significant rates; this is one of the issues that make fuel cells not actually competitive with respect to other current power sources [1]. Different routes can address such drawbacks: (1) increase of operation temperature, still maintaining a solid polymer electrolyte configuration; (2) search for low-cost nonnoble metal-based electro-catalysts; (3) decrease of noble metal loading. The latter point appears, at present, one of the most successful ways to be pursued.

In the last decades, the Pt loading in low-temperature fuel cells has been reduced by almost one order of magnitude. This has been made possible by the development of high-surface-area catalysts and by increasing the catalyst–electrolyte interface [1]. Preparation of high-surface-area catalysts implies the dispersion of catalytically active nanoparticles on electrically conductive supports [2]. Suitable dispersions are generally achieved by using high-surface-area carbon black supports, which, at the same time, ensure good electrical conductivity through their π-graphitic basal planes [2]. Because of the peculiar nature of Pt or Pt-alloy nanoparticles and the synthesis procedures used for catalyst preparation, the chemical interaction between active phase and support determines a modification of the electrocatalytic properties with respect to the bulk metal. The metal–support interaction involves a chemical bond between the metal atoms and the carbon functional groups. This charge transfer mechanism can be described in terms of electron accepting/donating characteristics, and the effects of this interaction can be reconducted to the Bronsted–Lewis acid–base theory [2]. Various techniques allow one to get information on this interaction. An evaluation of the carbon-supported catalyst acid–base properties, including determination of concentration and strength of functional groups on the support, as well as the zero point of charge (pHzpc), allows one in some cases to predict the level of interaction and the influence on the electrocatalytic properties. An important technique used for the determination of nature and properties of the species present on the surface, in terms of oxidation states, is x-ray photoelectron spectroscopy (XPS). The nature of carbon surface species (functional groups) is also disclosed by Fourier transform infrared spectroscopy (FTIR); moreover, additional information may be gained by the temperature programmed desorption (TPD) in conjunction with online mass spectrometry (MS) detection. All the information derived from these techniques provides a basis to understand the nature of the active phase and to draw relationships between some of these characteristics and the observed electrocatalytic behavior.

Beside the surface properties, bulk and morphological aspects play a significant role [1]; as an example, both oxygen reduction and methanol oxidation reactions involve a stage of electrosorption and, as generally occurs in these cases, they are structure-sensitive reactions. Electrocatalysis is a surface phenomenon, but it is quite influenced by the bulk properties of the materials, since in most cases the surface reflects the nature of bulk arrangements (e.g., occurrence of particular crystal-lographic planes, compositional enrichment for alloys). In addition, morphology is often dictated by the various steps of catalyst preparation where reactions between the active species and support occur. Accordingly, techniques like x-ray diffraction, electron diffraction, and transmission electron microscopy are of significant importance in the interpretation of bulk properties and their relationships with the surface characteristics. The electrochemical behavior of supported nanoparticles is influenced by material properties. Two main aspects must be taken into account. One is the effect of the support on the electrochemically active surface area, and the other is connected with its influence on the electrocatalytic activity; these are often correlated. Cyclic voltammetry (CV) is a powerful tool to get insight on these aspects; steady-state polarizations experiments are carried out to assess the electrocatalytic activity of the materials under investigation.

In the present work, metal–support interactions in fuel cell electrocatalysts are discussed and investigated using various physicochemical techniques. Some insights on the effects of such interactions on the electrocatalytic activity toward oxygen reduction and methanol oxidation are presented.

17.2 ACID–BASE PROPERTIES OF CARBON-SUPPORTED NOBLE METAL CATALYSTS

The performance of a catalyst is determined by its intrinsic activity and dispersion. This latter is defined as the ratio between the number of surface and bulk atoms. Metal nanoparticles supported on a high-surface-area carbon are thermodynamically unstable and give rise to sintering phenomena upon thermal treatment. In this regard, the influence of the nature of the support on the properties of dispersed catalysts has been continuously investigated [2]. For fuel cell catalysts, carbon support provides a framework that allows electron conduction and enhances the dispersion of the active phase.

Although the acid–base properties of carbon supports and fuel cell catalysts play an important role in determining the dispersion, activity, and sintering behavior of the metal phase, there is only a limited number of reports in the literature on such aspects. In the following, a brief description of the principle of measurement and the experimental procedure generally adopted are reported [2].

17.2.1 Acid–Base Titrations and ZPC Measurements

In all potentiometric titrations, both slope and height of the pH variation determine the detectable limit of the amount and strength of functional groups. In differential curves, this is expressed by the height and sharpness of the peaks in proximity of the equivalence points. In particular, in the titration of weak acids or bases, such as the surface functional groups of carbon, at halfway to the inflection point the concentration is almost equal to the functional groups to be titrated. For example, for a base $-C-OH$:

$$-C - OH \leftrightarrow C^+ + OH^- \tag{1}$$

$$Kb = \frac{[-C^+] \cdot [OH^-]}{[-C - OH]} \tag{2}$$

$$[OH^-] = \frac{[-C - OH] \cdot Kb}{[-C^+]} \tag{3}$$

$$2H_2O \leftrightarrow H_3^+O + OH^- \tag{4}$$

$$Kw = [H_3^+O] \cdot [OH^-] \tag{5}$$

at half-titration $[-C-OH] = [-C^+]$, and therefore $[OH^-] = Kb$, and $pH = pKw - pKb$.

For an acid titrated halfway to its equivalent point, $pH = pKa$. For mixtures of acids and bases, and hence for carbons having functional groups of different acid or basic strength, this holds true as well. For weak acid and base groups, the effect of water dissociation is significant around $pH = 7$. Therefore, a simple potentiometric titration can give information about the dissociation constants and neutralization equivalence of the carbon. In several cases these indications can be sufficient to determine the nature of the functional groups and provide a comprehensive description of the behavior of carbon in terms of acidity and basicity. A differential plot of the titration curve can be considered in the same way as a conventional absorption spectrum of the sample. Acidity or basicity constants are then calculated at half-titration, as $pH = pKw - pKb$ for a base and $pH = pKa$ for an acid.

The differential plots may be combined to the extra adsorption curves [2] to provide a more complete picture. These latter curves describe, as a function of pH, the excess adsorption of H^+ or OH^- ions on the surface; thus, they determine the pHzpc at which the electroneutrality is achieved on the surface.

17.2.2 Experimental Details

The acid–base properties of carbon support and catalysts can be evaluated by potentiometric titrations in aqueous suspension according to Parks et al. [3,4]. Some experimental details are reported in the following. The apparatus consists of a Pyrex glass cell with an external jacket for circulation of the thermostatic liquid, a glass electrode for pH measurement, and a probe for temperature control, the latter two connected to a pH/mV meter. The cell is covered with a Teflon cap provided with an inlet for CO_2 removal by N_2 flux. Titrations are carried out in KNO_3 electrolytic solution (500 ml) at different concentrations (0.1, 0.01, and 0.001 M) containing the catalyst powder (5 g), under continuous stirring. A 0.1 N HNO_3 solution is introduced (in increments of 5 ml up to 20 ml) to the suspension in the electrolytic solution; the titrant, 0.1 N KOH, is added with an automatic burette. After immersion, complete equilibration of the samples generally requires from 5 to 150 min; pH values are recorded after 3 min of stabilization [i.e., after which the pH value displayed by the instrument is stable (± 0.02 pH units for almost 3 min)].

The extra adsorption function (i.e., $\Gamma H^+ - \Gamma OH^-$, or the adsorption density in micromoles/g) is calculated numerically with a Pc [2]. The concentration (Ca, Cb) and strength (pKa, pKb) of surface sites from titration curves are calculated from the difference at the equivalent points, in the presence and absence of the sample powder, by considering that at halfway to each equivalent point, pH = pKa or pH = pKw − pKb for acidic and basic sites, respectively.

17.3 SURFACE GROUPS ON CARBONS AND CATALYST PREPARATION PROCEDURES

Adsorption and dispersion of Pt metal precursors on carbon supports are largely influenced by the acid–base properties of the substrate and by the pH of the catalyst precursor solution [5,6]. A suitable surface-charge density on the support, in combination with the appropriate charge of the ionic precursor, favors the electrostatic interaction between the two phases, thus affecting the metal dispersion. Clearly, the surface-charge density of a carbon at a specific solution pH is determined by the acid–base behavior of the adsorption sites present on the carbon surface (Figure 1). The metal dispersion is further enhanced by the presence of

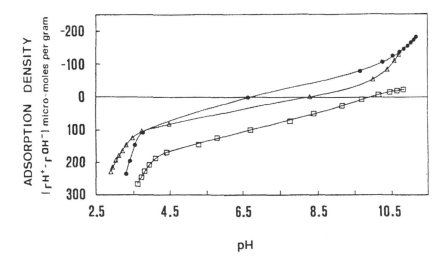

Figure 1 Adsorption density vs. pH curves of carbon supports in 0.1 M KNO_3 solution. □, Ketjenblack EC; △, Sibunit; ●, CC1.

reactive sites at the edges of the basal plane structures, Cπ [7–9], which give rise to the formation of strongly bound oxygen ligand complexes, thus hindering diffusion of Pt during the metal particles growth [6]. The concentration of surface adsorption sites generally increases with the increase of the surface area of carbons; as mentioned above, these adsorption sites govern the uptake of the precursor, and subsequently the Pt dispersion on the carbon surface.

The propensity to acquire a positively charged surface at low pH, which in turn implies a strong tendency to establish electrostatic interaction with the metal anionic precursor (e.g., $PtCl_6^{2-}$), appears to be greater for carbons with larger surface areas (see Figure 1 and Table 1). Hence, large-area supports seem to be more suitable for preparation of high Pt loading catalysts. The large and homogeneous dispersion in a carbon-supported Pt catalyst is determined by several different conditions. As an example, a high-surface-area carbon black with a well-defined basic character such as Ketjenblack (KB) produces a strong interaction with the H_2PtCl_6 precursor [2]. At low pH values, KB adsorbs a significant amount of H^+ ions from the solution (Figure 1). The anionic precursor $PtCl_6^{2-}$ is electrostatically forced to move toward the carbon surface to balance this electrical charge; hence, the deposition kinetics appear to be controlled, in this catalyzation procedure, by the carbon surface, and consequently the dispersion is favored by large-surface-area carbon blacks. Although the $PtCl_6^{2-}$ complex is quite stable in aqueous solution, the carbon black may behave as a reducing agent toward this compound. Depending on its reducing strength, the carbon black support can produce a first reaction step for Pt reduction [6,10]. The reduction process is generally completed by adding suitable liquid reducing agents or by gas-phase H_2 reduction. Generally, if any formation of surface Pt^{2+} species occurs, this induces an electrostatic repulsion between adjacent particles, thus hindering the particle growth; a complete reduction process is required to achieve good intrinsic activity. From the analysis of metal surface area (MSA) data of the catalysts (Table 2), it appears that such values are not directly related to the pHzpc values of the final catalysts if different procedures are used for catalyst preparation (see Figure 2 and Table 2); likely, MSA values appear to depend mostly

Table 1   Physico-Chemical Characteristics of Some Carbon Supports

Carbon	La[a] (nm)	Lc[a] (nm)	D(002)[b] (nm)	X[c] (%)	S_{BET} (N_2) ($m^2 \cdot g^{-1}$)	pHzpc	Basic Functional Group Dissociation Constants (pKb)	Basic Functional Group Amounts (meq · g^{-1})
Ketjenblack EC	1.0	<1.0	0.366	7	950	9.85	3.75 6.22	Cb1 = 0.05 Cb2 = 0.05
Sibunit	3.5	3.6	0.349	60	640	8.28	3.42	Cb1 = 0.09
CC1	6.0	6.0	0.343	60	115	6.64	3.58	Cb1 = 0.08
Graphite	27	15	0.339	100	8	—	—	—

[a] Lattice parameters of graphitelike carbon crystals along a e c directions.
[b] Interlayer spacing.
[c] Relative mass of ordered carbon.

Table 2 Physico-Chemical Characteristics of Various Pt/C Catalysts

SAMPLE	Pt Loading	Preparation Procedure	Activation Temperature	Activation Time	pHzpc	pKa	meq$_a$	pKb	meq$_b$	MSA$_{CV}$	MSA$_{TEM}$
1	—	—	—	—	9.85					—	—
2	19.6	A	200	4 h	2.74	3.68 / 6.1	0.57 / 0.18	3.75 / 6.22	0.05 / 0.05	—	—
3	18.3	B	200	4 h	3.0	3.92 / 5.84	0.385 / 0.169	—	—	—	—
4	20.1	C	200	4 h	2.8	3.75 / 6.3	0.4 / 0.15	—	—	—	—
5	20.7	B	300	4 h	3.9	5.16	0.19	—	—	—	—
6	19	B	400	4 h	5.0	6.1	0.073	—	—	—	—
7	20.5	B	500	4 h	4.16	5.67	0.1	—	—	—	—
8	20.3	B	600	4 h	4.36	5.52	0.085	—	—	—	—
9	17.3	B	600	4 h	5.41	5.24	0.05	—	—	57	116
10	16.7	B	600	4 h	5.56	6	0.089	—	—	69	140
11	16.9	B	600	4 h	5.9	6.36	0.046	—	—	58	121
12	16.8	B	600	4 h	4.7	5.36	0.09	—	—	75	127
13	16.6	B	700	4 h	6.5			—	—	84	117
14	20	A	700	4 h	3.9	5.2	0.15	—	—	90	83
15	19.8	A	700	4 h	4.3	5.1	0.16	—	—	98	89
16	16.1	B	850	2 h	8.6			3.7	0.122	58	60
17	17.2	C	900	45 min	8			3.63	0.095	75	74
18	18.1	C	900	1 h	9.3			3.71	0.149	67	—
19	16.1	C	900	45 min	8.75			3.31	0.125	60	70
20	19.1	B	900	1 h	9.68			3.2	0.2	55	—
21	13.8	B	900	1 h	10.6			3.48	0.225	52	62

Sample 1 is the uncatalyzed carbon black.
(A) Adsorption of a Pt colloid onto carbon surface followed by liquid-phase reduction.
(B) Adsorption of a Pt precursor onto carbon surface followed by in-situ colloidal particle formation and subsequent liquid-phase reduction.
(C) Impregnation of a Pt precursor onto carbon surface followed by liquid-phase reduction.

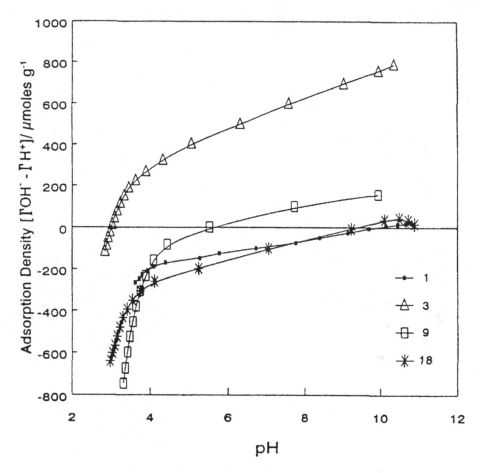

Figure 2   Adsorption density versus pH curves for exemplary Pt/C catalysts in Table 2.

on the surface properties of the support, such as adsorption properties and BET (N_2) area of carbon, using the same preparation method.

An exemplary case is represented by the preparation of a high-surface-area fuel cell catalyst supported on a graphitic carbon such as Ketjenblack (KB) (Table 1). This carbon is characterized by a significant basicity (0.1 meq/g), a d_0 value of 3.57 Å, and a high surface graphiticity index (SGI = $43.1 \cdot 10^{-3} J_b m^{-2}$) [11]; such basic sites undergo a strong interaction with H_2PtCl_6.

The preparation procedure generally involves chemical reactions and thermal activation treatments that strongly modify the original surface properties of the carbon black. As can be seen from Figure 3, interaction with H_3O^+ ions coming from the dissociation of the Pt precursor (H_2PtCl_6) determines, on a sample activated at 110 °C, a significant modification (increased acidity) of the original properties of the carbon. Evidently, extra acidic sites are generated by the interaction of HCl originated from the decomposition of H_2PtCl_6, as the Pt loading increases, the increase in acidity is even more pronounced, in agreement with results reported for other classes of catalysts [12]. However, as the severity of the activation (thermal)

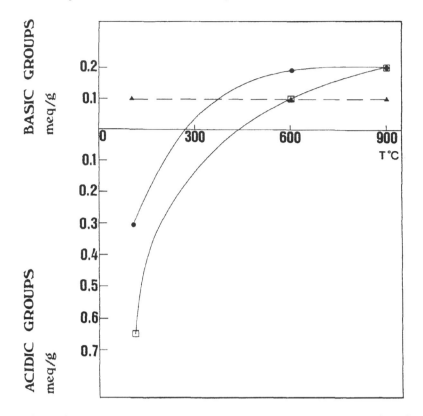

Figure 3   Effect of activation temperature on the relative concentration of acid–base surface groups of Pt/KB catalysts. (▲) Uncatalyzed carbon, (●) 13%, and (□) 20% Pt loadings on carbon.

treatment in nitrogen atmosphere is increased, the catalyst regains and even surpasses its initial basic characteristics. It is evident from Figure 3 that at a certain *T*, depending upon the Pt loading, the basicity of the catalyst exceeds that of the equivalent heat-treated support (KB) until a limiting value is reached at 900 °C; under said conditions, the total basicity is two fold that of the uncatalyzed KB.

Figure 4 shows the variation of surface acid–base properties of various Pt/C catalysts, reflected by the pHzpc, as a function of the activation temperature. The observed behavior is qualitatively similar to that reported in Figure 3. Figure 5 shows that an increase of acidic or basic surface groups (Ca, Cb) produces a corresponding shift of the pHzpc toward acidic or basic regions, respectively. Thus, the pHzpc clearly reflects the concentration of surface functional groups.

Two different processes determine the adsorption properties of the activated Pt/C samples. According to the premises, a first process arises from the reaction of the carbon with H_2PtCl_6, causing an increase in acidic sites that are likely to be enriched with oxygen-containing surface groups. The number of these new acidic sites is almost directly proportional to the Pt concentration, at least in the range of temperature <300–400 °C. As similarly described for different classes of catalysts [13], it is likely that at these relatively low activation temperatures the chlorine atoms, directly bonded to carbon black edges (carbon bonded to an oxygen atom is

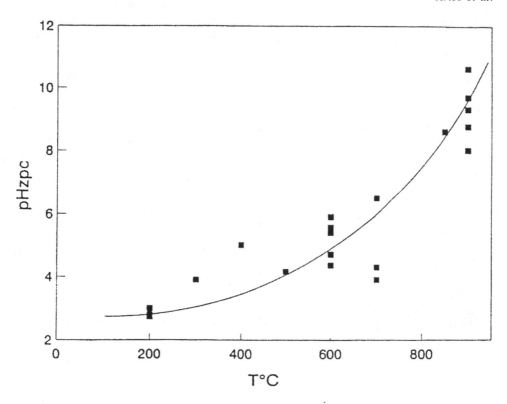

Figure 4 Variation of the pHzpc of Pt/C catalysts versus the activation temperature.

considered a good nucleophilic center for a chloride ion), influence the polarization of hydrogen–oxygen bonds of the surface groups, thus increasing their dissociation constants by means of an electron-withdrawing inductive effect $(-I)$. Afterward, the treatment at higher temperatures causes the onset of a second process that progressively clears the surface from the oxygen-containing groups, leaving behind a higher degree of surface graphitization and therefore a greater number of basic sites. On the other hand, according to Boehm et al. [14], a thermal treatment in inert gas is likely to cause the pyrolysis of acidic groups and thereupon the formation of pyrone groups upon exposure to air. Differently from other oxygen-containing surface groups, pyrone groups act as basic sites. It is thus suggested that, depending on the catalyst reactivity or the time of exposure to open air, a rearrangement of the surface groups toward a pyrone structure could occur. As reported by several investigators [15–17], the uptake of water is greater on acidic carbon blacks by virtue of their larger surface oxygen content, held in the form of carboxylic, lactone, and phenolic surface groups [18]. A graphiticlike surface structure, such as that of Ketjenblack (KB), can be partially converted into an acidic one by using acidic solutions [19] in wet preparation procedures. The subsequent stepwise temperature treatment in inert gas flux re-establishes the basic behavior (Figure 3); however, after exposure to air and humidity from the environment, some oxygen functionality can easily adsorb on the surface. For such an activated catalyst, only a small influence of water vapor is

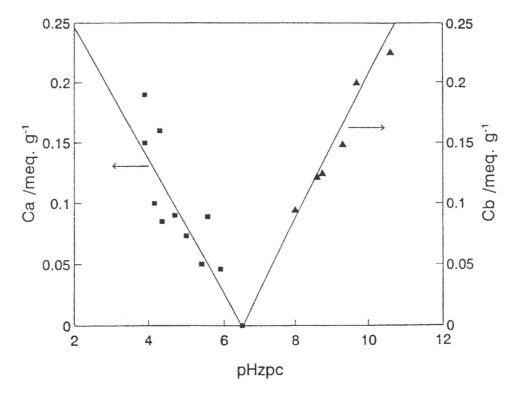

Figure 5 Variation of acidic (■) and basic (▲) functional groups concentration on Pt/C catalysts versus pHzpc.

expected because, as said above, basic carbons contain many fewer oxygen surface groups than acidic carbons. Nevertheless, the humidity content in the environment could cause considerable variations in the surface behavior. In this respect, since the observed dissociation constants of the detectable functional groups for differently exposed samples are very similar, it is suggested that the variations connected with water or oxygen uptakes do not involve the creation of new basic groups with different strength but, more likely, an increase or a decrease of those same groups of sites having an electron-releasing character. This event might occur via the above-cited rearrangement of functional groups toward a pyrone structure. The formation of a pyrone structure on carbons derives from the reaction with oxygen of the carbonyl groups remaining on the surface [19], because of their rectifying behavior toward thermal decomposition at 900 °C. As carbon atoms in the carbonyl groups are positively polarized, they can interact with the electron pairs of atmospheric oxygen, thereby giving rise to a surface complex that has proved to have a less basic character. Furthermore, a time-dependent cyclization toward a more stable pyrone structure can occur due to the stability gain caused by the resonance energy that arises from conjugation of the oxygen electrons in the pyrone ring [20].

The mechanism describing the basic behavior of the edge pyrone groups in the graphitic basal planes of the carbon black is depicted below; it derives from the

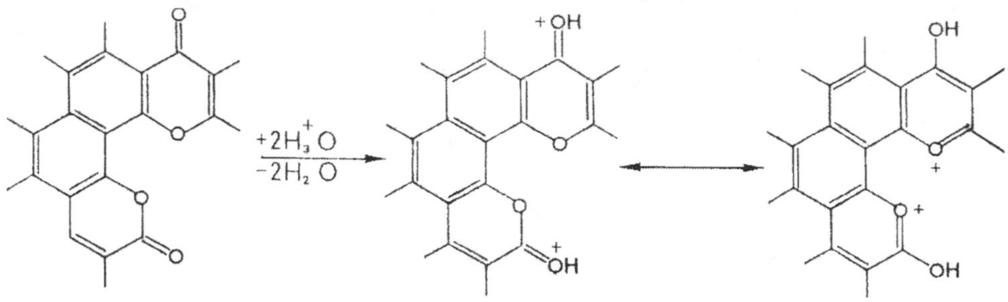

Figure 6 Acid–base reaction involving edge pyrone groups in the graphitic basal plane of a carbon black.

nucleophilic action promoted by the carbonylic oxygen and the resonance delocalization of charge (Figure 6). However, in the absence of oxygen uptake, the basic properties of carbon blacks are mostly related to the Lewis base behavior of the graphitic basal planes [21]; only if basic oxygen-containing groups (e.g., pyrone [14] or chromene [22]) are present in significant amount on the carbon surface, they may play an important role in neutralizing the aqueous protons. Furthermore, alcohol-type groups provide polar adsorption sites through which the water molecules are strongly held on the carbon surface via hydrogen bonds; thus, they become able to chemisorb water than other electron-releasing centers, especially the $C\pi$ sites of the graphitic basal planes. Such a mechanism of water chemisorption on pyrone or alcohol sites, for instance, reduces the electron-releasing character of these groups, causing a decrease in the catalyst basic properties.

The above considerations suggest that the pHzpc is an important index of the propensity of a solid to acquire a positively or negatively charged surface, and they provide a description of the surface ion-exchange behavior of carbon-based catalysts. In fact, the electrokinetic phenomena occurring in fuel cell gas electrodes are strongly influenced by the adsorption effects associated with the electrical double-layer formation [23]. Thus, in developing electrodes with zero charge or with specific adsorption properties, one cannot leave out of consideration the knowledge of aspects related to the precursor catalyst.

17.4 SPECTROSCOPIC INVESTIGATION OF FUNCTIONAL GROUPS ON THE SURFACE OF A CARBON-SUPPORTED CATALYST: FTIR AND TPD

Some important insights into the nature of surface functionalities are given by Fourier transform infrared spectroscopy analysis and temperature programmed desorption in combination with mass-spectrometry detection. FTIR spectra (Figure 7) and TPD measurements [Figure 8(a) and (b)] were collected on Pt-loaded carbon samples activated at different temperatures. Figure 7 shows a comparison of the FTIR spectra for the Ketjenblack carbon and Pt/KB catalysts subjected to increasing activation temperatures. The most interesting features are observed in the region from 370 to $1900\,cm^{-1}$. The shape of the absorption band in the 1480–

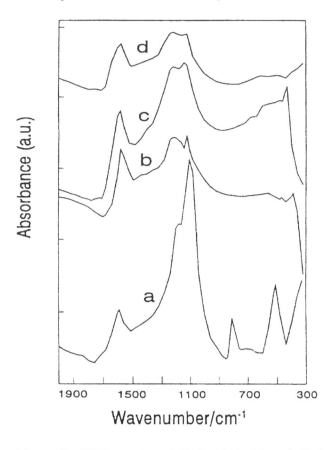

Figure 7 FTIR spectra of Ketjenblack (a) and Pt/C catalysts activated at various temperatures (b) 200 °C, (c) 600 °C, (d) 900 °C. The catalysts correspond to samples 3, 8, 20 in Table 2, respectively.

1700-cm^{-1} range appears to be similar in all samples (Figure 7), with maxima centered in a very narrow interval (1577–1586 cm^{-1}). This band could be ascribed to the stretching vibration of a highly conjugated C=O group in a quinone configuration and to skeletal modes of the aromatic structure due to the highly graphitic character of Ketjenblack [24]. The broad absorption band observed in the 840–1500-cm^{-1} region (Figure 7, curve a) is attributed to the C–O stretching mode of ethers and esters that give rise to strong absorption bands in the region 1000–1300 cm^{-1} [25,26]. It is observed in Figure 7 (curve b) that a variation of the fine structure of the band at 840–1500 cm^{-1} occurs on the Pt-loaded carbon treated at low temperature. The band appears broadened and composed of two distinct peaks occurring at about 1125 and 1235 cm^{-1}, with the IR intensity increasing at higher frequencies and accounting for the formation of carboxylic surface groups (C–O stretching and OH bending vibrations). This view is strengthened by the evidence of strong acidic surface groups in the potentiometric titrations and from the CO$_2$ desorption at low temperature (200 °C) in the TPD spectrum of the platinized carbon

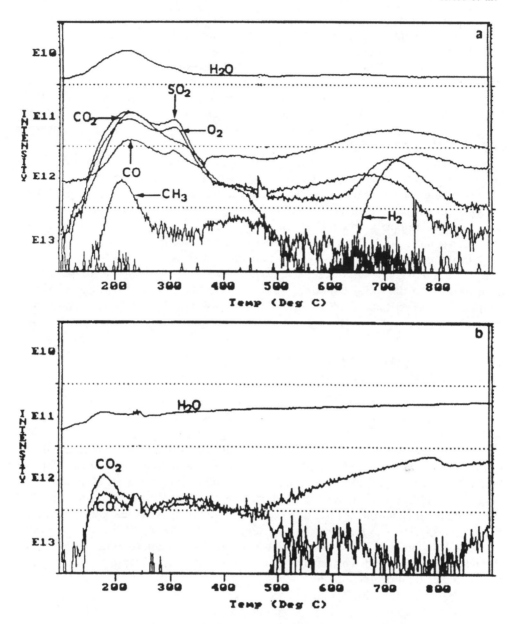

Figure 8 TPD spectra of Pt/C catalysts activated at different temperatures (a) 200 °C, (b) 900 °C. The catalysts correspond to samples 3 and 20 in Table 2, respectively.

treated at 200 °C [Figure 8(a)]. The IR spectrum of Pt-loaded carbon activated at 600 °C shows sharper bands in the 900–1500-cm^{-1} region, probably accounting for the decomposition of carboxylic groups (Figure 7, curve c). The peak at a lower frequency (1125 cm^{-1}) is enhanced in intensity in the 600 °C-treated sample, whereas the higher-frequency peak (1235 cm^{-1}) is larger in the 900 °C sample (Figure 7, curve d). A rearrangement toward a pyrone structure has been hypothesized to occur at this latter temperature as proven by a large increase in basicity; this seems to be confirmed by the CO_2 detection in the TPD spectrum of the 900 °C-activated Pt/C catalyst [Figure 8(b)]. The TPD spectrum of bare Ketjenblack showed no CO or CO_2 signals, confirming the low surface-oxygen content. A considerable evolution of CO and CO_2 is observed in the sample activated at 200 °C [Figure 8(a)], accounting for the formation of carboxylic and lactone or ketone groups upon chemical interaction of the Pt precursor with the carbon surface and subsequent air exposure. The peaks of CO_2, CO, and O_2 observed at both low and high temperatures [Figure 8(a)] account for the formation of various oxygen-containing surface groups with different strength. It was pointed out that oxygen may interact with carbonyl surface groups, giving rise to a rearrangement toward a more stable pyrone structure [2]. The presence of the pyronelike and ketonelike structures in the high-temperature activated Pt/C sample seems to be confirmed by the CO_2 and CO evolved in the TPD spectrum [Figure 8(b)]; these groups could have been formed during air exposure after heat treatment. The SO_2 signal appearing in the low-activation temperature sample (200 °C) is due to the reducing agent, $Na_2S_2O_4$, used in the preparation procedure or to the sulfur contained in the carbon black. The SO_2 is desorbed at temperatures lower than 500 °C, and it is not observed in the TPD spectrum of the sample activated at 900 °C [Figure 8(b)].

17.5 X-RAY PHOTOELECTRON SPECTROSCOPY ANALYSIS OF PLATINIZED CARBONS

X-ray photoelectron spectroscopy is an important technique for investigating the metal–support interaction effects and obtaining further information on the nature of surface groups [27]. The XP Pt 4f, C 1s, and O 1s spectra of typical Pt/C catalysts are shown in Figure 9. The binding energy (B.E.) shift (the difference between the B.E. of Pt $4f_{7/2}$ in the sample and that of the Pt metal used as an internal standard [28]) is taken into account to evaluate the magnitude of interaction between platinum and the functional groups of the carbon support. Depending on the oxidation state of Pt, the Pt 4f signal is derived from 2 or 3 pairs of platinum peaks (Figure 9). The most intense peaks (71–74 eV) are attributed to metallic Pt. The second set of doublets, 1.1–3.5 eV higher than Pt^0, can be assigned to the Pt^{2+} chemical state in PtO and $Pt(OH)_2$-like species [29]. The third doublet is observed at a B.E. approximately 3.6–5.4 eV higher than the metallic Pt; these B.E. shifts are attributed to the presence of Pt^{4+} in PtO_2–xH_2O or $Pt(OH)_4$-like species [30].

The C 1s spectra appear to be composed of graphitic carbon (284.6 eV) and –C=O (285.4–287.6 eV) species [Figure 10(a)]. A peak is also observed at about 289 eV [Figure 10(a)]; this seems to be due to either –COO or carbonatelike species [28]. The O 1s signal observed at around 531.7 eV in the XP spectra can be attributed to Pt–O-like (530 eV) and Pt–OH species (531.5 eV) [31]; the peak observed at 532.5 eV is ascribed to –C=O [Figure 10(b)].

The B.E. shift of the Pt $4f_{7/2}$ signal increases in the catalyst samples with respect to the metal Pt reference with an increase of acidity or basicity; accordingly, the fraction of oxidized Pt increases with the corresponding B.E. shift toward higher values. The oxidation of Pt in strongly basic catalysts could be ascribed to the coordination of Pt clusters by oxygenated basic functional groups, such as

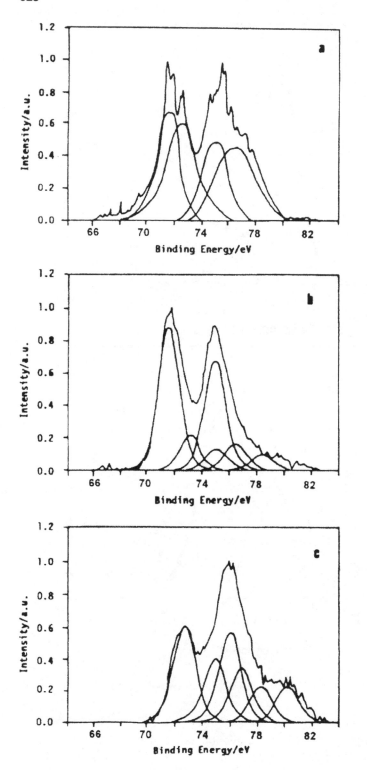

Figure 9 Pt-4f XP spectra of platinized carbons 8 (a), 16 (b), 20 (c) in Table 2.

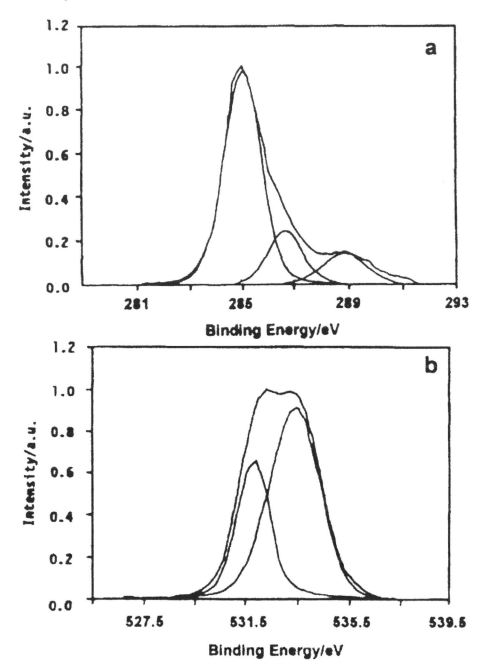

Figure 10 C 1s (a) and O 1s (b) XP spectra of a platinized carbon (sample 20 in Table 2).

pyronetype. This is strengthened by the presence of a high-B.E. species (289 eV) in the XP C 1s spectrum [Figure 9(a)] that could be assigned to the effect of such oxygenated species; graphitic basal plane sites, Cπ, having electron-releasing behavior generally produce lower Pt B.E. shifts.

17.6 EFFECTS OF THE METAL–SUPPORT INTERACTION ON THE METAL SURFACE AREA

Figure 11 shows the variation in metal surface area (MSA), determined by cyclic voltammetry (CV), as a function of the pHZPC. The trend of data suggests that MSA decreases with the increase of the pHZPC for each series of acidic or basic Pt/C catalysts (Figure 11). Moreover, it is observed that catalysts with a medium content of basic groups show higher metal surface area than the corresponding acidic samples with a medium content of surface groups.

The formation of a large content of oxygenated functional groups acting as anchoring centers for Pt clusters, for samples activated at low temperatures, probably limits the metal particle growth. The progressive decomposition of these acidic surface groups with increasing temperature allows a fast growth of the Pt particles; simultaneously, the number of Pt sites blocked by coordination of surface groups decreases. The formation of Cπ sites with electron-releasing behavior on the surface of weakly basic catalysts probably accounts for the increased availability of electrochemically active Pt sites in the basic range. The XPS analysis shows that the lowest fraction of oxidized Pt sites is found for weakly basic catalysts, whereas Pt oxidation suddenly increases for weakly acidic and strongly basic catalysts; the electrochemically active surface area correspondingly increases in the weakly basic

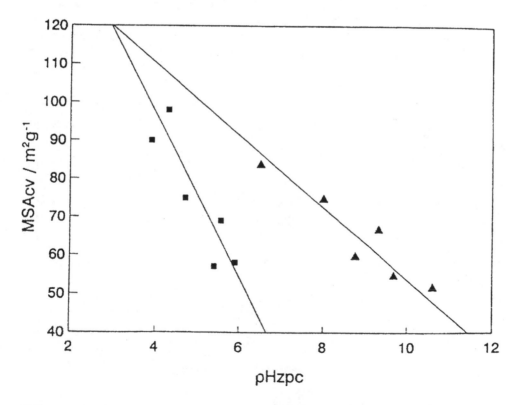

Figure 11 Variation of MSAcv values for acidic (■) and basic (▲) Pt/C catalyst series as a function of pHzpc.

range. The large increase in Pt particle size for samples activated at high temperature and the formation of oxygen-containing pyrone groups, which are probably coordinated to the Pt clusters, reduce the number of electrochemically active Pt sites in samples with strong basic behavior.

It can be concluded that both the sintering process and the metal–support interaction determine the electrochemically active surface area of Pt on a carbon black support. The role exerted by surface functional groups, as evidenced by the above discussion, is quite complex. These groups act both as anchoring sites for Pt clusters and as surface-blocking coordination ligands, slowing down the sintering process, in the first case, and producing an increase of metal–support interaction in the second case.

17.7 INFLUENCE OF PARTICLE SIZE AND METAL–SUPPORT INTERACTION ON THE OXYGEN ELECTRO-REDUCTION REACTION

A procedure of general use to obtain Pt/C catalysts with different particle sizes is based on the variation of the time and temperature of activation. Figures 12 and 13 show the mass activity (mA g^{-1} Pt) and specific activity (μA cm^{-2} Pt), respectively, for O_2 reduction at 900 mV, in 98% phosphoric acid at 170 °C. The mass activity for oxygen reduction appears to reach a maximum at a Pt particle size of 3 nm, corresponding closely to the particle size at which the maximum in the fraction of (111) and (100) surface atoms on Pt particles of cubo-octahedral geometry occurs [32]. The specific activity increases gradually with an increase in Pt particle size and closely follows the trend observed between the surface fraction of (111) and (100) Pt

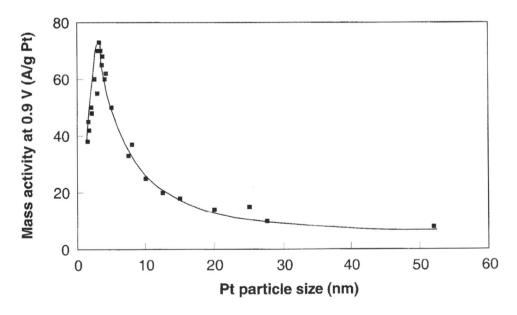

Figure 12 Variation of mass activity (mA g^{-1} Pt) for O_2 reduction at 900 mV, in 98% phosphoric acid and 170 °C as a function of the average Pt particle size.

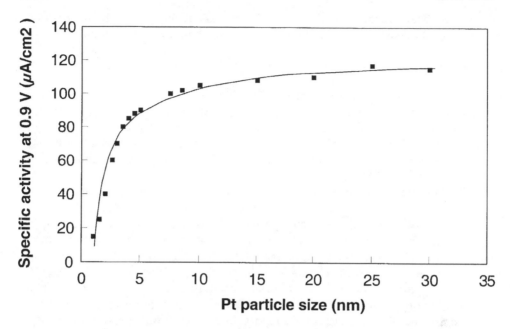

Figure 13 Variation of specific activity ($\mu A\,cm^{-2}$ Pt) for O_2 reduction at 900 mV, in 98% phosphoric acid and 170 °C as a function of the average Pt particle size.

atoms and the particle size. This indicates that the (111) and (100) surface atoms are more electrocatalytically active than Pt atoms located on high Miller index planes. As pointed out by Kinoshita [32], platinum atoms at edge and corner sites are considered to be less active than Pt atoms on the crystal faces. These phenomena were also observed in other liquid protonic electrolytes and in solid polymer electrolyte fuel cells [1].

The following reaction mechanism for O_2 reduction (although not proven) has been proposed to account for the role of specific active sites [33]:

$$O_2 + Pt \rightarrow Pt\text{–}O_2 \tag{6}$$
$$Pt\text{–}O_2 + H^+ + e^- \rightarrow Pt\text{–}HO_2 \tag{7}$$
$$Pt\text{–}HO_2 + Pt \rightarrow Pt\text{–}OH + Pt\text{–}O \tag{8}$$
$$Pt\text{–}OH + Pt\text{–}O + 3H^+ + 3e^- \rightarrow 2\,Pt + 2\,H_2O \tag{9}$$

Equation (8) indicates a pathway with a dual-site reaction, and if it is the rate-determining step, this is more affected by catalyst particle size than the reaction step described by Eq. (7). When the particle size becomes very small, only the inactive edge and corner atoms will be present, and dual sites of proper orientation would not be available; thus, the activity of the particle should be lower.

According to a different theoretical approach, the kinetics of oxygen on Pt catalysts were interpreted on the basis of the separation distance between metal nanoparticles. Pt crystallites, separated by less than about 20 nm, were observed to exhibit lower electrocatalytic activity because of either diffusion or a mutual interaction. The interparticle distance, X, was calculated for carbon-supported catalysts by Watanabe et al. [34,35] on the basis of the following equation:

$$X = (\pi \sigma d^3 Sc(100 - L)/[3^{3/2}L])^{1/2} \tag{10}$$

where X is the Pt interparticle distance on the carbon support, σ is the density of Pt, d is the mean particle diameter, Sc is the specific surface area of carbon support, and L is the Pt loading. Accordingly, Stonehart et al. [36] suggest a limit of $80\,\mu A\,cm^{-2}$ Pt in the specific activity for O_2 reduction at $900\,mV$ in 100% H_3PO_4 at $180\,°C$, which is comparable to the limit of about $90\,\mu A\,cm^{-2}$ Pt that was also observed and reported by Buchanan et al. [37].

The interpretations given by Kinoshita et al. [32] and Watanabe et al. [34,35] to the peculiar trend of the variation of specific activity as a function of particle size for oxygen reduction reflect two different approaches and two different ways to explain the same behavior. In their analysis, the relationships derived are based on the fitting of curves that actually show significant scattering. Since the experimental procedures to determine the active surface areas and the specific activities are quite standard, this would indicate the presence of additional phenomena.

One factor that may be important, but not systematically investigated, is the influence of the Pt electrocatalyst–support interactions on the electrocatalytic activity for O_2 reduction. In Figure 14, an attempt to incorporate the pHzpc as a qualitative measure of the importance of carbon surface chemistry and metal–support interaction on the electrocatalytic activity of Pt is reported. The trend of the data in Figure 14 suggests that the specific activity for oxygen reduction increases as the pHzpc of the surface becomes more basic; this effect may be related to the parallel increase of the particle size with the pHzpc of the catalyst. At this stage, one

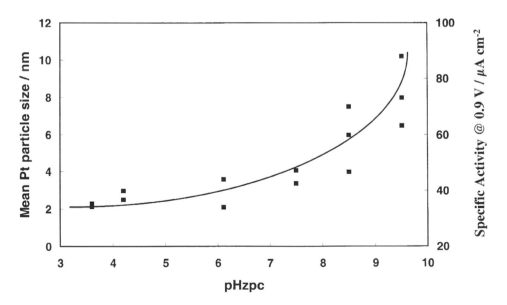

Figure 14 Variation of the average Pt particle size and specific activity ($\mu A\,cm^{-2}$ Pt) for O_2 reduction as a function of pHzpc.

may infer that the effect of metal–support interaction particularly influences the mean particle size, and thus the electrocatalytic activity that is strictly related to this latter; but it is conceivable that an electrocatalyst–support interaction may exist which has a stronger influence on the electrocatalytic activity of smaller Pt particles. In this case, the electrocatalytic activity of small Pt particles may possibly be lower than that of large particles. It should be noted that Jalan [38] reported that high-resolution electron microscopy showed Pt particles on carbon with a two-dimensional shape (i.e., flat, raftlike, disclike), which indicated the presence of a strong interaction between the Pt atoms and the carbon substrate; a transmission electron micrograph of Pt nanoparticles dispersed on a carbon black support is shown in Figure 15. Furthermore, McBreen et al. [39] investigated the influence of the carbon support on Pt dispersion, as well as the sintering rate and performance of supported Pt electrocatalysts. Of the five carbon blacks used in their study, Vulcan XC-72 and Regal 6008 yielded a higher Pt dispersion than that obtained with Monarch 1300, CSX98, and Mogul L. These results were attributed to the high internal porosity of Vulcan XC-72, resulting in a high-Pt dispersion. On the other hand, the high-Pt dispersion on Regal 6008 is attributed to the surface properties of the carbon, which results in a strong Pt–carbon interaction. The origin of this effect may involve an interaction with the oxygenated surface groups of carbon. From all these evidences it is inferred that metal–support interaction and dispersion of Pt crystallites are strictly related; the specific activity clearly decreases with the decrease of Pt particle size. Accordingly, in very small particles (10–30 Å) a significant number of surface Pt atoms, with respect to the total number of surface atoms, is involved in chemical bonds with the carbon functional groups. The level of this interaction is especially significant in the presence of oxygenated groups with acidic properties; on the contrary, basic groups, such as $C\pi$ sites, producing an electron-realeasing behavior, does not significantly affect the oxygen chemisorption properties of Pt sites.

17.8 INFLUENCE OF METAL–SUPPORT INTERACTION ON METHANOL ELECTRO-OXIDATION AT PT-BASED CARBON SUPPORTED CATALYSTS

The most active catalysts for methanol oxidation are presently based on bifunctional systems such as Pt–Ru. Yet the evaluation of a metal–support interaction would require the analysis of a simpler catalytic system (i.e., a monometallic catalyst); this would avoid the interference of all those aspects such as degree of alloying, changes in crystallographic parameters, chemical state of the promoting element, which significantly affect the activity, and thus a comprehensive interpretation of the data actually available.

There is no agreement in the literature on the role of the catalyst particle size for methanol oxidation. Two main evidences have been reported: one is related to the presence of a maximum in the mass activity as a function of particle size located, for some authors [40], at 3 nm and at 2 nm for others [41]; moreover, for some authors there is no influence of the interparticle distance and particle size on specific activity. If these parameters are not controlling, an increase in mass activity should be observed as the metal surface area is increased [42]. In this context, the metal–

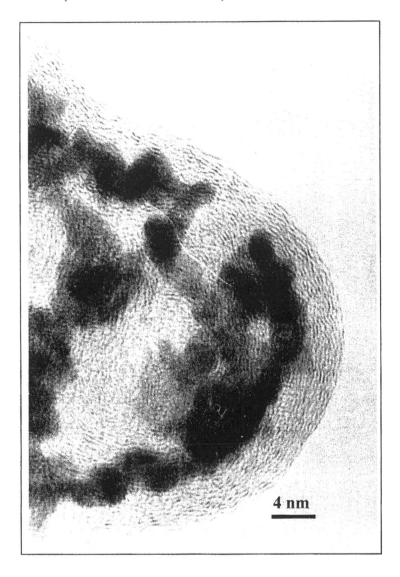

Figure 15 TEM micrograph of a platinized carbon showing Pt particles of irregular shape and an irregular carbon black lattice originated by graphitic basal planes.

support interaction plays a significant role by influencing the variation of specific activity with particle size.

The influence of the metal–support interaction on the electrocatalytic activity for methanol oxidation has been evaluated by analyzing the voltammetric behavior of various electrocatalysts differing in terms of surface properties, as expressed by their pHzpc but having similar average particle size (around 3 nm). These data were combined with the information gained by cyclic voltammetry (CV) experiments in sulphuric acid electrolyte. In particular, the potential at which the onset of a Pt–OH layer formation on the surface occurs in the anodic sweep and the potential at which reduction of the Pt–oxide layer occurs in the cathodic sweep were correlated with the activity for methanol electro-oxidation.

The onset potential for Pt–OH layer formation was found to increase very slightly with the pHzpc of the catalyst. The reduction peak of Pt oxide, occurring at about 0.8 V (NHE) on smooth Pt electrodes [43], is cathodically shifted to various extents depending on the nature of the electrocatalyst [Figure 16(a)] Similar evidence, previously observed by Lowde et al. [44], was attributed to metal–support interaction effects.

The potential values for Pt oxide reduction as a function of pHzpc show a volcano-shaped curve with the maximum in potential, close to that of pure Pt, at about pHzpc = 6 [Figure 16(b)]. According to Lowde et al. [44], this latter curve suggests that a strong interaction between Pt particles and the carbon support occurs for the most acidic and basic catalysts, whereas such an interaction appears to be lower in the neutral region.

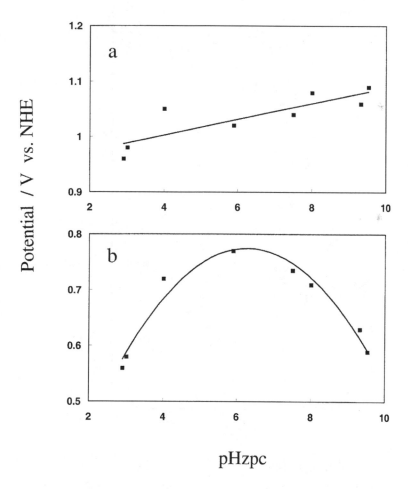

Figure 16 Electrochemical behavior of Pt/C electrocatalysts in sulphuric acid (2.5 M H_2SO_4) as a function of the pHzpc. (a) potential for the Pt–OH layer formation in the anodic sweep; (b) peak potential for Pt–oxide reduction in the cathodic sweep.

An exemplary CV of a Pt/C catalyst in presence of methanol is shown in Figure 17. There are two voltammetric peaks for methanol oxidation, in the anodic as well as in the cathodic sweep. As shown in Figure 18, both peak potentials vary with pHzpc; in particular, the peak in the anodic sweep shows a minimum at pHzpc close to neutrality [Figure 18(a)]. This behavior is very similar to that observed in pure sulphuric acid, confirming the influence of the metal–support interaction.

The peak related to methanol oxidation in the cathodic sweep is due to the presence of labile Pt-bonded oxygen [Figure 18(b)]. After partial reduction of the Pt oxide layer, a significant presence of labile Pt–OH groups occurs on the surface; this peak is cathodically shifted for low pHzpc catalysts. This phenomenon has been interpreted on the basis of an electronic interaction between the surface groups and

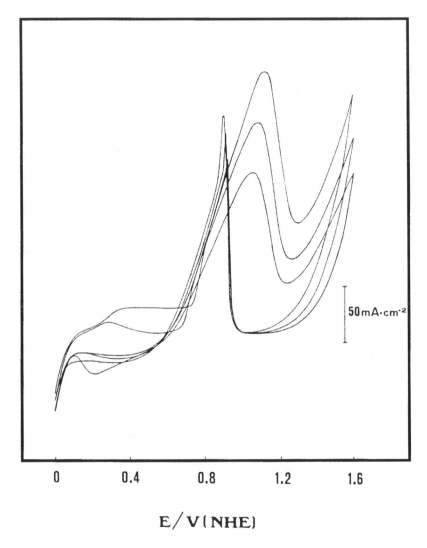

Figure 17 Cyclic voltammograms of a Pt/C catalyst in 1 M CH_3OH, 2.5 M H_2SO_4 solution at increasing sweep rates (10, 20, and $50\,m\,Vs^{-1}$).

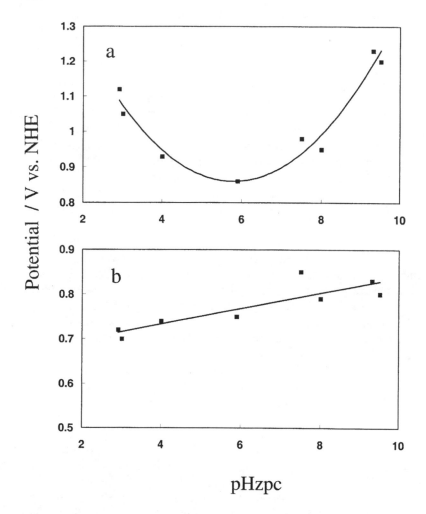

Figure 18 Electrochemical behavior of the electrocatalysts for the methanol oxidation in sulphuric acid (1 M CH_3OH, 2.5 M H_2SO_4) as a function of the pHzpc; (a) oxidation peak potentials of methanol in the anodic sweep; (b) oxidation peak potentials of methanol in the cathodic sweep.

Pt atoms. In fact, oxygen-containing acidic groups probably share the Pt electron density in a way to favor the adsorption of electron-donating OH groups at lower potentials on the Pt sites which are depleted in electronic charge density. In other words, the acidic groups alter the Pt work function, thus influencing the double-layer capacitance and causing a shift of the electrode zero charge potential toward more negative values; water dipoles hence interact with the electrode surface at lower potentials.

As in the case of catalysts for oxygen reduction, the surface properties of Pt catalysts for methanol oxidation are governed by both the preparation method and the nature of carbon support. As discussed above, the presence of surface groups such as carboxylic, carbonyl, phenolic, lactone, and pyrone functionalities determine the acid–base characteristics of the catalysts, and hence their adsorption properties. These groups are easily formed at the edge planes of carbon during preparation, thermal treatments, and exposure to air of the catalysts [2,8]. Both oxygen-containing groups and basal plane sites ($C\pi$) of the graphitic structure [9] behave as acid–base sites; thus they are suitably monitored by potentiometric titrations. It was discussed that oxygen-containing functional groups and the edge sites with electron localization may establish coordination bonds with the Pt particles on the surface, thus determining the strength of the metal–support interaction [45]. In other words, Pt particles can be anchored to the carbon surface by functional groups [46], especially in catalysts characterized by low and high pHzpc values. The metal–support interaction level increases as the amount and strength of surface groups increase, either for acidic or basic samples. In this way, the metal–support interaction could influence the CH_3OH oxidation, as revealed by the variation of the peak potential values as a function of the pHzpc of catalysts.

Besides the effects of the typical carbon functional groups, the role of nitrogen and sulfur functionalities, introduced on carbons by chemical and thermal treatments, on the electrochemical performance of Pt catalysts for oxygen reduction in direct methanol fuel cells was investigated [47]. Once again, the metal–support interaction influences the size and chemical state of platinum particles and, as a consequence, the electrocatalytic activity. The introduction of nitrogen and sulphur functionalities was reported to improve the catalytic activity, but this result was mainly ascribed to the Pt particle size.

In conclusion, even if the CH_3OH oxidation occurs for supported electrocatalysts in a similar way as on smooth Pt electrodes, the surface characteristics of the catalysts, besides Pt dispersion, may play a significant role in the reduction of the overpotential for this process. The acid–base functional groups influence the oxidation mechanism as they establish the level of metal–support interaction and the surface adsorption behavior. Thus, the optimization of such parameters can significantly improve the activity of Pt electrocatalysts for CH_3OH oxidation.

In the present evaluation, monometallic Pt/C catalysts were considered only because the effects of metal–support interaction have to be distinguished from the influence of any promoting element and its interaction with Pt. At present, the considerations made on Pt/C catalysts could be extended to the Pt–Ru system, provided that degree of alloying, dispersion, and chemical state of Ru are the same for catalysts with a different level of metal–support interaction. Samples with these characteristics cannot be easily prepared. Recently, it was shown that samples containing both Pt and Ru atoms in the zerovalent state perform better in DMFC devices with respect to similar samples having these elements in a partially oxidized state [48]. Accordingly, if one considers that the principal effect of a strong metal–support interaction is to increase the concentration of oxidized sites, the same conclusions derived for the Pt/C system may be extended to bifunctional catalysts.

17.9 SOLID POLYMER ELECTROLYTE FUEL CELL SYSTEMS

Some important aspects that should also be taken into consideration in the choice of the proper carbon support should also rely on the catalyst–electrolyte interface characteristics; these are mainly connected with the carbon microstructure, as this influences the access of the reactants to the catalytic sites and the removal of the

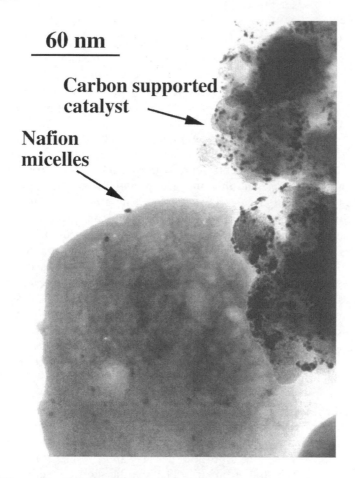

Figure 19   TEM micrograph of catalyst–Nafion interface showing metal particles supported on carbon agglomerate and Nafion ionomer micelles.

reaction products [4,9]. In the most recently developed electrode design for solid polymer electrolyte fuel cells, the electrolyte ionomer is mixed with the catalyst and penetrates inside the catalyst pores to create an extended interface (Figure 19). Due to the dimension of ionomer micelles ($\approx 50\,\text{nm}$), high-surface-area carbon supports with a large amount of micropores may give rise to diffusional problems. Location of very fine metal particles occurs much more easily on the sites where a strong metal–support interaction exists; yet, because these particles are not easily accessible to reactants, they do not actively participate to the reaction. Generally, carbon blacks with a low amount of micropores and a small surface area possess a small concentration of functional groups; they do not produce strong metal–support interactions but give rise to a low degree of dispersion.

These aspects need to be jointly evaluated in order to select the best support able to give rise to an optimal compromise in terms of dispersion, metal–support interaction, and microstructure.

17.10 CONCLUSIONS

The role of metal–support interaction on the catalytic activity of carbon-supported Pt nanoparticles toward oxygen reduction and methanol oxidation was analyzed. It was observed that both dispersion and specific activity are influenced by the interaction of the active phase with the support, determining well-defined relationships that may be used for interpreting the electrochemical behavior of new, more advanced catalytic systems.

Generally, surface acidic oxygen containing functional groups may act as anchoring centers for the metal particles, limiting their growth and enhancing the dispersion. This effect produces a strong metal–support interaction that affects the electronic nature of platinum sites. Carbon functional groups with basic behavior are mainly Cπ and pyrone-type sites. Pyrone groups behave similarly to the acidic groups, whereas Cπ groups possess an electron-releasing behavior, producing a lower level of metal–support interaction. These findings have been strongly supported by XPS analysis, which shows significant amounts of surface-oxidized Pt in the presence of strong metal–support interaction. On the other hand, a large content of surface functional groups does not necessarily indicate an increase in Pt dispersion. In many cases, it has been shown that an increase in the content of basic functional groups or degree of graphitization of the carbon support is accompanied by an increase in metal particle size. Such effects have been easily observed by subjecting a Pt/C catalyst to various thermal treatments. In the case of oxygen reduction reaction, the specific activity increases almost exponentially with the increase of the pHzpc. This behavior cannot be univocally attributed to the metal–support interaction, since the increased pHzpc is often accompanied by an increase in average particle size. In fact, because oxygen reduction is a structure-sensitive reaction, it is strongly influenced by the occurrence of particular crystallographic planes on the surface. An increase in the average particle size corresponds to a change in the relative fraction of Pt surface atoms on the (111) and (100) faces of the Pt particle, assuming a cubo-octahedral geometry. The specific activity increases gradually with an increase in the Pt particle size and closely follows the increase in surface fraction of crystallographic planes with low Miller indexes. A maximum in mass activity was observed at an average particle size of 3 nm; this would represent the best compromise between active surface area and specific activity. Beside this, there is an alternative interpretation based on the role of interparticle distance which considers the effects of either diffusion or mutual interactions that limit the electrocatalytic activity in the particles separated by less than 20 nm.

In the case of methanol oxidation, it has been reported that the catalytic activity of carbon-supported Pt catalysts with similar particle size shows a maximum at about pHzPC = 6. This behavior has been interpreted by considering that, besides enhancing dispersion, the metal–support interaction reduces the number of active sites for methanol oxidation. In fact, as the pHzpc deviates from neutrality, the number of acid or basic functional groups on the catalyst progressively increases. This was confirmed by the shift of the Pt-oxide reduction peak potential in the catalysts with high metal–support interaction, as determined by pHzpc measurements.

REFERENCES

1. S. Srinivasan, R. Mosdale, P. Stevens, C. Yang, *Ann. Rev. Energy Environ.* **24**, 281 (1999).
2. A.S. Aricò, V. Antonucci, L. Pino, P.L. Antonucci, N. Giordano, *Carbon* **28**, 599 (1990).
3. G.A. Parks, *Chem. Rey.* **65**, 177 (1965).
4. G.A. Parks, P.L. de Bruyn, *J. Phys. Chem.* **66**, 967 (1962).
5. J.M. Solar, C.A. Leon y Leon, K. Osseo-Asarc, L. Radovic, *Carbon* **28**, 369 (1990).
6. H.P. Boehm, *Adv. Catal. Relat. Subj.* **16** 179 (1966).
7. E. Papirer, S. Li, J.B. Donnet, *Carbon* **22**, 243 (1987).
8. T.J. Fabish, D.E. Shleifer, *Carbon* **22**, 19 (1984).
9. F. Rositani, P.L. Antonucci, M. Minutoli, N. Giordano, *Carbon* **25**, 325 (1987).
10. D. Pletcher, V. Solis, *Electrochim. Acta* **27**, 775 (1982).
11. A.J. Groszek, *Carbon* **25**, 717 (1987).
12. P. Pichat, J.M. Herrmann, B. Jenny, Y. Disdier, H. Courbon, N. Yaffrezic, In *Advances in Catalysis Science and Technology* (T.S.R. Prasada Rao, ed.), Wiley Eastern Limited, New Delhi, 741 (1985).
13. M. Tanaka, S. Ogasawara, *J. Catal.* **16**, 157 (1970).
14. H.P. Boehm, M. Voll, *Carbon* **9**, 473 (1971); *Carbon* **9**, 481 (1971).
15. L.B. Adams, C.R. Hall, R.L. Holmes, R.A. Newton, *Carbon* **26**, 451 (1988).
16. V.M. Jalan, C.L. Bushnell, U.S. Patent 4, 136,059, Jan. 23, 1979.
17. M.M. Dubinin, *Carbon* **18**, 355 (1980).
18. H.P. Boehm, *Adv. Catal. Relat. Subj.* **16**, 179 (1966).
19. E. Papirer, S. Li, J.B. Donnet, *Carbon* **25**, 243 (1987).
20. E.H. Rodd, *Rodd's Chemistry of Carbon Compounds* (S. Coffey, ed.), Elsevier, Amsterdam, Volume III parts (A, H) (1979).
21. T.J. Fabish, D.E. Shleifer, *Carbon* **22**, 19 (1984).
22. V.A. Garten, D.E. Weiss, *Australian J. Chem.* **10**, 309 (1957).
23. A.N. Frumkin, O.A. Petrii, B.B. Damaskin, In *Comprehensive Treatise of Electrochemistry* (J.O.M. Bockris, B.E. Conway, E. Yeager, eds.), Plenum Press, New York, Vol. 1, p. 221 (1980).
24. J.P.R. Vissers, S.M.A. Bouwens, V.H.J. de Beer, R. Prins, *Carbon* **25**, 485 (1987).
25. C. Prado-Burguete, A. Linares-Solano, F. Rodriguez-Reinoso, C. Salinas Martinez De Lecea, *J. Catal.* **115**, 98 (1989).
26. B.J. Meldrum, C.H. Rochester, *Chem. Soc., Faraday Trans.* **86**, 861 (1990).
27. P. Ehrburger, *Adv. Colloid Interface Sci.* **21**, 275 (1984).
28. P. Albers, K. Deller, B.M. Desseyroux, A. Schafer, K. Seibold, *J. Catal.* **133**, 467 (1992).
29. K.S. Kim, N. Winograd, R.E. Davis, *J. Am. Chem. Soc.* **93**, 6296 (1974).
30. P.J.M. Dykgraaf, H.A.M. Duisters, B.F.M. Kuster, K. van der Wiele, *J. Catal.* **112**, 337 (1988).
31. M. Peuckert, *Electrochim. Acta* **29**, 1315 (1984).
32. K. Kinoshita, *J. Electrochem. Soc.* **137**, 845 (1990).
33. H.R. Kunz, *Proc. Symposium on Electrode Materials and Processes for Energy Conversion and Storage*, p. 607 (J.D.E. McIntyre, S. Srinivasan, F.G. Will, eds.), The Electrochemical Society, Pennington, NJ (1977).
34. M. Watanabe, H. Sei, P. Stonehart, *J. Electroanal. Chem.* **261**, 375 (1989).
35. M. Watanabe, S. Saegusa, P. Stonehart, *Chem. Lett.* **9**, 1487 (1988).
36. P. Stonehart, J. Baris, J. Hockmuth, *Preparation and Evaluation of Advanced Catalysts for Phosphoric Acid Fuel Cells*, Final Technical Report, DOE/NASA/0176-10 (NASA CR 168223), Morgantown, WV (July 1984).
37. J.S. Buchanan, L. Keck, J. Lee, G.A. Hards, H. Scholey, *Proc. 1st Intl. Fuel Cell Workshop on Fuel Cell Technology Research and Development*, p. 29 (M. Watanabe,

P. Stonehart, K. Ota, eds.), Organizing Committee of the 1st Intl. Fuel Cell Workshop (Sept. 1989).

38. V. Jalan, *Extended Abstracts*, Electrochemical Society, Los Angeles, CA, Fall Meeting, Abstract 192 (Oct. 1979).

39. J. McBreen, H. Olender, S. Srinivasan, K. Kordesch, *J. Appl. Electrochem.* **11**, 787 (1981).

40. B.D. McNicol, D.A.J. Rand, K.R. Williams, *J. Power Sources* **83**, 15 (1999).

41. J.B. Goodenough, A. Hamnett, B.J. Kennedy, S.A. Weeks, *Electrochim. Acta* **32**, 1233 (1987).

42. X. Ren, P. Zelenay, S. Thomas, J. Davey, S. Gottesfeld, *J. Power Sources* **86**, 111 (2000).

43. M.R. Tarasevich, A. Sadkowski, E. Yeager. In *Comprehensive Treatise of Electro-chemistry*, Vol. 7, (B.E. Conway, J.O'M. Bockris, E. Yeager, S.U.M. Khan, R.E. White, eds.), Plenum Press, New York, p. 315 (1983).

44. D.R. Lowde, J.O. Williams, B.D. McNicol, *Appl. Surface Science* **1**, 215 (1978).

45. J.B. Goodenough, A. Hamnett, B.J. Kennedy, R. Manoharan, S.A. Weeks, *J. Electroanal. Chem.* **240**, 133 (1988).

46. P. Ehrburger, *Adv. in Colloid Interface Sci.* **21**, 275 (1984).

47. S.C. Roy, P.A. Christensen, A. Hamnett, K.M. Thomas, V. Trapp, *J. Electrochem. Soc.* **143**, 3073 (1996).

48. A.S. Aricò, G. Monforte, E. Modica, P.L. Antonucci, V. Antonucci, *Electrochem. Commun.* **2**, 466 (2000).

49. M. Uchida, Y. Fukuoka, Y. Sugawara, N. Eda, A. Ohta, *J. Electrochem. Soc.* **143**, 2246 (1996).

18

Effects of Size and Contact Structure of Supported Noble Metal Catalysts in Low-Temperature CO Oxidation

MASATAKE HARUTA

National Institute of Advanced Industrial Science and Technology, Tsukuba, Japan

SUSUMU TSUBOTA

National Institute of Advanced Industrial Science and Technology, Ikeda, Japan

CHAPTER CONTENTS

SUMMARY

Low-temperature CO oxidation has many applications such as indoor air quality control, automobile exhaust gas treatment, and CO removal from hydrogen for polymer electrolyte fuel cells. This chapter focuses on CO oxidation over supported noble metal catalysts at temperatures below 150 °C. A general mechanism for CO oxidation over metal surfaces is reviewed and the catalytic activity order in terms of the critical CO/O_2 molecular ratios is presented: $Ru > Rh > Ir > Pd > Pt$. Over these 8th group noble metals, CO oxidation is a structure insensitive reaction giving almost identical turnover frequency with specific rate of reaction per surface exposed atom. Size and support effect is very dramatic in supported gold catalysts, which exhibit surprisingly high activity even at a temperature as low as -76 °C. In this case, it is proposed that the perimeter interfaces act as reaction sites for CO adsorbed on the surfaces of Au particles and O_2 is adsorbed on the support surfaces. In order to prepare highly active noble metal catalysts, coprecipitation, deposition-precipitation, and grafting by use of organometallic compounds, are effective methods because they can give strongly interacted metal particles with the support.

18.1 INTRODUCTION

The catalytic oxidation of CO with O_2 has been investigated most extensively and intensively in heterogeneous catalysis [1,2]. Because of its simplicity that the product is only CO_2, which is gaseous and hardly sticks to metal surfaces, CO oxidation can now be understood at the atomic scale through a variety of surface-science approaches using single-crystal metal surfaces [3–6].

Due to growing concerns of environmental protection, CO oxidation is a very important reaction, finding many applications. Pollutants in automobile exhaust gases, NO_x, CO, and hydrocarbons are simultaneously removed by catalytic converters in the absence of excess O_2. However, during idling in winter, CO emission still remains a problem. Indoor air-quality control requires the removal of CO, HCHO, and NH_3, etc. at ambient temperature. Similar needs of polluted-air treatment are also growing in smoking areas and tunnels of motorways. Recently, polymer electrolyte fuel cells (PEFC) were planned to be placed on automobiles. Because PEFCs are operated at temperatures below 120 °C, CO present at trace concentrations in hydrogen fuel deactivates Pt electrodes. It is one of the solutions for reducing CO concentration to a level below 100 ppm [7] to selectively oxidize CO in H_2 stream by adding a minimum amount of O_2. In sealed CO_2 laser, CO_2 tends to decompose into CO and O_2, which depress laser output power. To prevent this, the catalytic regeneration of CO_2 has widely been studied at the U.S. National Aeronautics and Space Administration (NASA) [8].

In view of the above applications, CO oxidation at low temperatures, from room temperature to 150 °C, is of immense importance. At such low temperatures, transition-metal oxides are more active than supported noble metals. However, a serious constraint is that base metal oxide catalysts are heavily deactivated by moisture [9,10]. Over noble metals supported on thermally stable and insulating metal oxides such as Al_2O_3 and SiO_2, CO oxidation takes place only at temperatures above 150 °C, because CO is so strongly adsorbed on the metal surfaces inhibiting the adsorption of O_2. In contrast to base metal oxides, moisture enhances the catalytic activity of noble metals through promoting the desorption of CO and/or activating O_2. In addition, noble metals can be tuned to be active at lower temperatures by being deposited on reducible metal oxides, maintaining the beneficial promotion by moisture [11]. From the viewpoint of practical applicability, this article deals with supported noble metal catalysts that are active for CO oxidation at low temperatures, focusing on the effect of the support metal oxides, and the shape, contact structure, and particle size of noble metal crystallites.

18.2 MECHANISM FOR CO OXIDATION OVER METAL SURFACES

The oxidation of CO over noble metal surfaces takes place at a temperature range of 150 °C to 300 °C, in most cases by the Langmuir–Hinshelwood mechanism.

$$CO(g) + S_{ad} \rightleftarrows S_{ad}-CO$$
$$O_2(g) + 2S_{ad} \rightleftarrows S_{ad}-O_2-S_{ad} \rightarrow 2[S_{ad}-O]$$
$$S_{ad}-CO + S_{ad}-O \rightarrow CO_2(g) + 2S_{ad}$$

where g is the gas phase and S_{ad} is the adsorption site on metal surfaces. The adsorption of CO and O_2 takes place competitively on the same sites over the metal surfaces. The heat of dissociative adsorption of O_2 is larger than the heat of molecular adsorption of CO. However, CO adsorption takes place preferentially, because the area required for the dissociative adsorption of O_2 is eight times larger than the area required for the linear adsorption of CO. A general mechanism is proposed, based on density function theory calculations for CO oxidation, which involves the activation of oxygen atoms from hollow sites to bridge sites and the approach of CO molecules to the activated oxygen atoms from the correct direction at an appropriate time [12,13].

The reaction mechanism of CO oxidation over metal surfaces changes depending on reaction conditions: partial pressure of CO, CO/O_2 molar ratio, and catalyst temperature. The reaction takes place in two distinct temperature regimes at a certain CO/O_2 molar ratio or two distinct CO/O_2 molar ratio regimes at a certain temperature. The regimes are separated by an ignition temperature or a critical CO/O_2 molar ratio above or below which, respectively, CO oxidation is self-sustained without the need of external heating. Figure 1 shows critical partial-pressure ratios of CO/O_2 below which the dramatic increase in the rate of CO oxidation occurs as a function of reciprocal absolute temperature [14]. The site preference on a stepped Pt(113) surface was found to switch from the (111) site to (001) sharply in the boundary between the active and inhibited region above the critical pressure ratio of CO/O_2 [15].

Simultaneous in-situ SFG (sum frequency generation) and gas chromatography studies by Somorjai revealed that over Pt(111) the ignition temperature for 40 Torr CO/100 Torr O_2 was about 320 °C [4]. As shown in Figure 2, below this temperature on-top bonded CO is the major species that inhibits the reaction by preventing oxygen adsorption. Therefore, the rate-limiting step is the desorption of the on-top CO. The activation energy for CO oxidation is very large, 42 kcal/mol, which corresponds to the desorption energy of CO. At temperatures above 320 °C, where turnover rates are much larger, the on-top CO disappears and is replaced by in-commensurate CO. The rate-limiting step is the reaction between incommensurate CO and oxygen on the Pt surface and presents smaller activation energy of 14 kcal/mol. The activation energy is dependent on the coverage of CO and oxygen and is increased to 25 kcal/mol at low coverage.

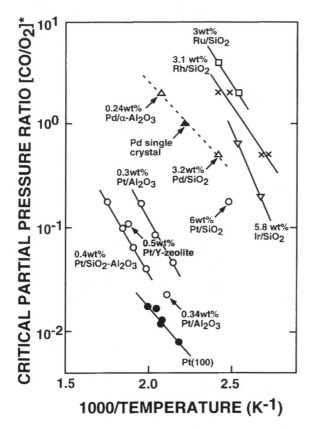

Figure 1 Critical partial-pressure ratios [CO/O$_2$] as a function of reciprocal absolute temperature. (From Ref. 14.)

Figure 3 shows STM images for the progress of CO oxidation over Pt(111) surface observed in nm scale by Ertl [3]. The reaction proceeds not at random, but predominantly at the boundaries between O$_{ad}$ and CO$_{ad}$ islands, and hence, the rate is expected to be proportional to the total length of the domain boundaries. This conclusion was obtained by the careful analysis of the reaction rates as a function of the coverage of CO and oxygen and of the boundary length. The reaction boundary dynamically changes with time on the Pt surfaces, which was clearly observed in the μm scale by ellipsomicroscopy imaging of reactant and product local coverages [16].

18.3 CATALYTIC ACTIVITY ORDER AND STRUCTURE SENSITIVITY OF NOBLE METALS

In view of numerous practical applications of CO oxidation, it is desirable to obtain a rough idea of which metal is the most active and which is the least active. Since the ignition temperature is closely related to the replacement of on-top CO and to the adsorption of oxygen, it should change with the affinity of metals toward oxygen. As shown in Figure 1, metals with stronger affinity to oxygen (larger metal–oxygen

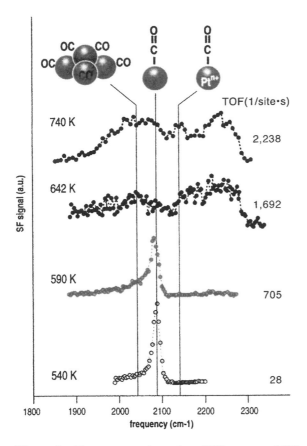

Figure 2 Temperature-dependent SFG spectra of CO oxidation on Pt(111) at 40 Torr CO–100 Torr O_2 together with turnover frequencies (TOF) at each temperature. (From Ref. 4.)

bonding) show higher CO/O_2 ratios at the same ignition temperature, or lower ignition temperatures for the same CO/O_2 ratio [14]. The critical CO/O_2 molar ratios of noble metals change only by a factor of 10 over a very wide pressure range of 10^{-5} to 10^2 torr [17] and decrease in the following order:

$$Ru > Rh > Ir > Pd > Pt$$

In the stoichiometric region and above the critical CO/O_2 molar ratio or under a low total pressure of 0.002–0.004 torr [18], the catalytic activity data reported so far are mostly in agreement with this order. Under an excess of O_2, metal oxide layers that are less reactive are formed on the metal surfaces, so that the catalytic activity order changes from that under CO-rich condition. In particular, Ru is deactivated appreciably with time-on-stream. While Rh is more stable and resistive during CO oxidation, it is deactivated with the progress of oxidation from Rh^0, Rh^+, to Rh^{3+}. Even Pt, which is the most resistant to oxidation, is subjected to gradual deactivation by surface oxidation when supported on SiO_2. The surfaces of nanoparticles of noble metals, highly dispersed over metal oxide supports, are more readily subjected to oxidation, especially at steps, edges, and corners, causing a decrease in catalytic

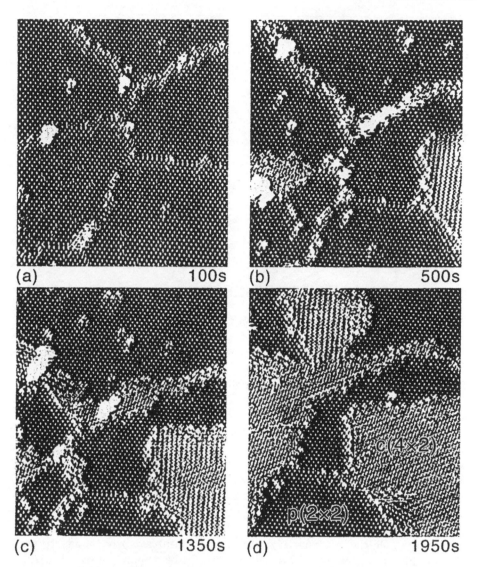

Figure 3 STM images for the progress of CO oxidation over Pt(111) at 244 K. (From Ref. 3.) The surface precovered with O_{ad} was exposed to 5×10^{-7} mbar CO, which resulted in the substitution of the $p(2 \times 2)$–$O_{ad}(+ CO_{ad})$ areas by the $c(4 \times 2)$–CO_{ad} areas.

activity through site blocking. The catalytic activity order is mainly defined in this case by the activity of the oxidized surface layer, and this may account for the discrepancies in the catalytic activity orders reported so far [6,19].

Gold, which is one of the Ib metals, having a fully occupied d-band should be distinguished from the Pt group metals of VIII group. Carbon monoxide and O_2 are adsorbed only on steps, edges, and corners of Au particles, and only weakly [20]. Therefore, CO oxidation is a demanding or structure-sensitive reaction over Au surfaces. The apparent activation energy for CO oxidation is nearly zero and the rate is zero order, both with respect to CO and O_2 [21]. Due to these features, the position

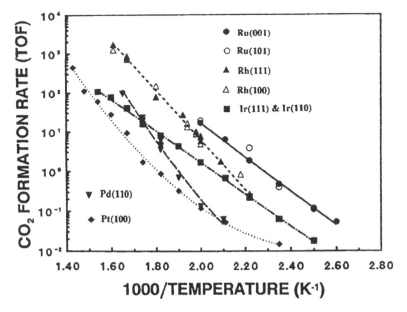

Figure 4 Effect of crystallographical planes on CO oxidation over noble metals. (From Ref. 5.)

of the catalytic activity of Au depends on the size of Au particles and reaction temperature. For example, gold with smooth surfaces is the least active, while small Au particles are the most active at low temperatures, below room temperature. Above 150 °C, even small Au particles are less active than Pt group metals.

Another valuable piece of information concerning catalysis by metals is whether specific activity changes depending on the crystal planes or not. As typically shown in Figure 4 [5], CO oxidation is a facile, structure-insensitive reaction showing identical reaction rates for different planes of single crystals of Pt group metals [5,6,19]. Metal particles of Ru, Rh, Pd, Ir, and Pt supported on Al_3O_3 or SiO_2 also give almost identical TOF (turnover frequency), specific rate of reaction per surface exposed atom [5]. On the other hand, at higher temperatures, where O_2 adsorption prevails (roughly above 160 °C). CO oxidation turns to be demanding over Rh [22]. Over Pt supported on Al_2O_3, TOF increases with a decrease in the diameter of Pt particles at 245 °C while it remains unchanged at 160 °C [23], indicating that the structure sensitivity of CO oxidation depends on reaction temperature. It is also suggested by density functional theory calculations that CO oxidation on a closed-packed surface, Pd(111), is strongly coverage-sensitive, while less sensitive on a more open surface, Pd(100) [24].

18.4 SIZE EFFECT OF NOBLE METAL PARTICLES

The third concern of those who are searching for better catalysts might be the metal crystallite size effect. As shown in Figure 5, with a decrease in the diameter of metal particles the fractions of edges, corners, and surface exposed atoms increase, markedly in the range below 2 nm. The electronic state of metal particles may also

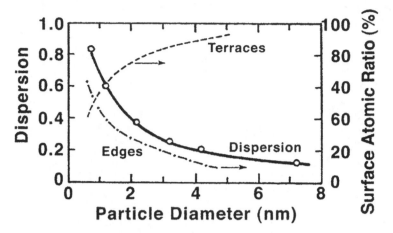

Figure 5 Surface atomic ratios of terrace and edge together with metal dispersion as a function of particle diameter.

tend to change toward a nonmetallic one with a decrease in diameter. It should be noted that this occurs at a diameter around 2 nm for noble metals, where the fraction of surface-exposed metal atoms exceeds 50% [48–52].

The turnover frequency (TOF) of CO oxidation may change with the diameter of noble metal particles following four patterns: independent, increase, decrease, and parabolic, having a maximum with a decrease in diameter. Parabolic with a minimum has not yet been observed. A maximum is usually observed at 2–5 nm in metal particle diameter. Table 1 summarizes the relevant data [25]. Except for Au, most noble metals show independent or decreasing TOF with decreasing particle diameter. Therefore, from the practical point of view, it is reasonable to disperse Pt-group metals as nanoparticles of intermediate (i.e., 5 nm) diameter over the support because minimization of particle size does not result in improvement of the catalytic activity per unit mass but reduces thermal stability.

Figure 6 shows the rates of CO oxidation per exposed surface area of Au, which are equivalent to TOFs, measured for unsupported Au powder and Au supported on TiO_2 [26,27]. In both cases, the rate increases by an order of two thirds with decreasing the diameter of Au particles, indicating that minimization of the size of Au particles is markedly rewarding, because both TOF and exposed surface area (atoms) increase. The difference in the reaction rate by an order of magnitude at 0 °C corresponds to the support effect of TiO_2. Because the apparent activation energy differs, being 0 kcal/mol for unsupported Au powder and 8 kcal/mol for Au/TiO_2, the rate difference becomes larger at higher temperatures.

Goodman fabricated model Au/TiO_2 catalysts by vacuum vapor deposition of Au clusters on a single crystal of TiO_2 and used STM (scanning tunnel microscopy) and STS (scanning tunnel spectroscopy) to measure TOF and electron band gap as a function of the size of Au clusters [28,29,40]. As shown in Figure 7, TOF reaches a maximum at a diameter of Au islands of 3.5 nm (2 or 3 atoms thick), where Au partially loses its metallic nature. It is suggested that this transition to nonmetallic state might be correlated to the high catalytic activity. This explanation is based on a mechanism assuming that CO oxidation takes place at the surfaces of Au particles.

Table 1 Dependence of TOFs for CO Oxidation on the Particle Diameter of Noble Metal Catalysts for CO Oxidation

Catalyst	Loading (wt. %)	System[a]	Total pressure (Pa)	Ratio of O_2/CO	Temperature (°K)	Particle size (nm)	Disperson (%)	TOF (s^{-1})	Type	Ref.
Rh/Al_2O_3	0.03–3	C	?	1	423, 428	1–67.6	?	0.0331–0.0654	1	30
Rh/Al_2O_3	?	V	2×10^{-8}	?	300–480	2.5, 5	?	?	3	31
Pd/Al_2O_3	?	V	2.5×10^{-4}	1.1	455	1.5–8	?	0.01	1	24
Pd/Al_2O_3	?	V	2.5×10^{-4}	1.1	518	1.5–8	?	0.1–0.5	3	24
Pd/Al_2O_3	?	V	2.6×10^{-4}	1.2	550–650	1.5–10	?	0.2–0.8	3	32
Pd/MgO	?	V	4.8×10^{-5}	0.37	400–600	2.8–13	?	0.01–0.14	3	33
Ir/SiO_2	0.3–5	F	10^5	0.5	450	?	4–100	0.004–0.09	2	34
Pt/Al_2O_3	0.035	F	10^5	5000	481–553?	2.8–100	?	0.1–10?	2	35
Pt/Al_2O_3	0.035	F	10^5	50–100	481–553?	2.8–100	?	40	1	35
Pt/Al_2O_3	?	V	2660	1	600–700	1.7, 1.4	?	1–300	2	36
Pt/SiO_2	0.48–1.89	F	10^5	0.4	450	?	6.0–81	0.008–0.017	1	37
Pt/SiO_2	0.825–1.97	F	10^5	~1	437	?	6.0–81	0.0073–0.0163	2	38
Au/Co_3O_4	~1.2?	F	10^5	20	273	~3	?	0.1–0.2	3	39
Au/Fe_2O_3	~1?	F	10^5	20	273	3–7	?	~0.15	3	39
Au/TiO_2	3.0–7.4	F	10^5	20	273	3.5–18.2	?	~0.12	3	39
Au/TiO_2	?	V	5320	5	350–450	2.5–6.0	?	1–4	4	40
Au/Al_2O_3	0.94–5.3	F	10^5	21	273	2.4, 3.5	?	0.006–0.01	3	41
Au/SiO_2	2.9–6.6	F	10^5	21	273	4.0–6.6	?	0.02–0.05	1	42

[a] C, closed circulatory reactor; V, *in vacuo*; F, flow reactor.

Another, more probable, explanation is based on a mechanism assuming that the reaction takes place at the periphery of the Au particles. Since the sample used for catalytic activity measurements was composed of Au islands with a certain size distribution, whereas the band gap was measured for a specific Au island by STS, a more reasonable interpretation is that the ratio of the number of surface atoms of Au

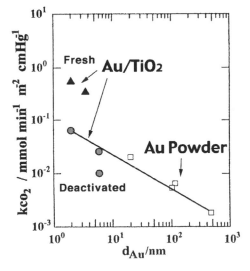

Figure 6 Rate of CO oxidation as a function of mean diameter of Au particles for unsupported Au powder and Au/TiO_2.

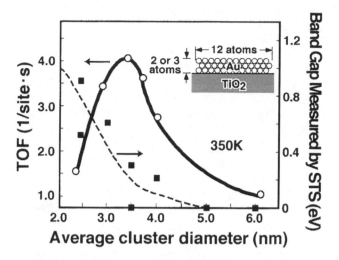

Figure 7   TOF and band gap measured by STS as a function of the diameter of Au islands deposited on TiO₂(110). (From Ref. 28.)

flat islands having metallic nature to the number of Au atoms at the periphery in contact with the TiO₂ support reaches an optimum at a mean diameter of 3.5 nm. Because the adsorption of CO on oxidic Au sites is inhibited by moisture, which is adsorbed more strongly, catalytic activity enhancement of Au/TiO₂ by moisture [43] strongly suggests that metallic Au particles are indispensable for the adsorption of CO under practical conditions.

Among supported noble metal catalysts, Au supported on Mg(OH)₂ is the most active for CO oxidation at − 77 °C giving 100% conversion at an hourly space velocity of 20,000 h⁻¹ ml/g-cat. However, it totally deactivates after 3 to 4 months, losing activity even at 200 °C. Our DFA (Debye functional analysis) study of the x-ray scattering for the fresh and aged catalysts suggests that the active Au species are 13 atom Au clusters and, when they grow to 55 atom clusters with truncated decahedral structure, the catalytic activity is completely lost [44]. Among the two structures of 13 atom clusters, the icosahedron structure is suggested to be active, whereas the cubo-octahedron structure is inactive. It is hard to explain why such a drastic difference is observed. The coordination number of Au atoms is 5 for icosahedron and 4 for cubo-octahedron. The active Au/Mg(OH)₂ catalyst, which is mainly composed of icosahedral Au clusters of 13 atoms, showed negative apparent activation energy in the temperature range of −77 °C to 0 °C [45]. This can be explained by the enhanced transformation of icosahedron into cubo-octahedron with a rise in reaction temperature. It has also been reported that 8 and 11 atom clusters are the smallest and the second-smallest cluster to exhibit catalytic activity of Au over the MgO support [46,47].

The reactivity of free Au clusters (cations, neutrals, and anions) was reported by Cox and his co-workers showing that specific small Au clusters exhibited a pronounced variation in their reactivity toward H₂, CH₄, and O₂ [27]. Ozin et al. reported that one single Au atom was reactive with equimolar mixtures of CO and O₂ even at −263 °C and liberated CO₂ at −243 °C and −233 °C successively [27].

18.5 PREPARATION OF NOBLE METALS STRONGLY INTERACTING WITH METAL OXIDE SUPPORTS

Supported noble metal catalysts have usually been prepared by the impregnation method. The interaction between noble metal particles and the supports is relatively weak, because noble metal salts, such as chlorides, are simply loaded on the support surfaces and are decomposed in air. It is proven that the chloride ion accelerates the coagulation of metal particles and poisons CO oxidation. After the discovery of surprisingly high catalytic activity of gold catalysts prepared by coprecipitation [53], it has been found that noble metal particles can also strongly interact with the support by using coprecipitation and deposition-precipitation of hydroxides [54] and grafting of organo complexes in gas or in liquid phase [41,42].

The first characteristic feature in the above three methods is that the precursors of noble metals can be strongly interacted with the metal oxide supports. The second feature is that these methods do not cause the deactivation of active metal oxides such as TiO_2, Co_3O_4, Fe_2O_3 during the course of catalyst preparation. The capability of these metal oxides for the activation of oxygen is destroyed by chloride ions remaining on the surfaces when the metal oxide supports are impregnated with the aqueous solution of noble metal chlorides.

A typical example is shown in Figures 8 and 9, for Au deposited on TiO_2 (Degussa, p. 25) [55,56]. When Au/TiO_2 is prepared by deposition-precipitation, Au

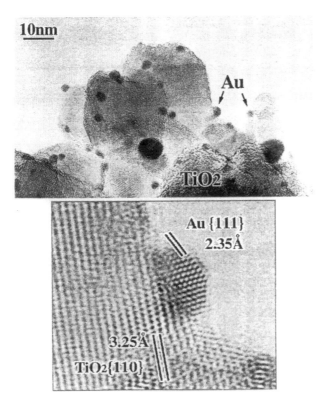

Figure 8   TEM micrographs of Au/TiO_2 prepared by the deposition-precipitation method.

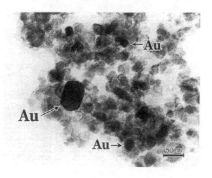

Figure 9   TEM micrograph of Au/TiO$_2$ prepared by the impregnation method.

particles are very small with a diameter around 3 nm and are dispersed uniformly on each TiO$_2$ particle. On the other hand, when the impregnation method is used, Au particles are larger than TiO$_2$ particles and simply loaded on the aggregates of TiO$_2$ particles.

Coprecipitation is advantageous to prepare catalysts with high metal loadings, for example, above 10 wt.%, while deposition precipitation is suitable to prepare catalysts with metal loadings from 5 wt.% to 0.01 wt.%. A limitation to co-precipitation is that it is not applicable to Al$_2$O$_3$ and SiO$_2$ supports, because these metal oxide or hydroxide precursors cannot be coprecipitated with noble metal hydroxides. For SiO$_2$ support the sol-gel method can replace coprecipitation.

Deposition precipitation is applicable to metal oxides of basic nature with point of zero charge (PZC) above pH 5. Noble metal hydroxides cannot be precipitated on the surfaces of acidic metal oxide supports with PZC below pH 5. Figure 10 shows PZC for a variety of metal oxides.

A typical procedure for Au deposition is shown in scheme 1. By adjusting pH, temperature, and concentration of noble metal salts, hydroxide precipitates can be deposited exclusively on the surfaces of the supports without forming precipitates in solution. Then the precursor is washed several times to remove Na$^+$ and Cl$^-$ ions and is finally calcined in air at temperatures above 250 °C after drying overnight.

18.6   SUPPORT EFFECT OF METAL OXIDES

As summarized in Table 2 base metal oxides can be classified into three groups: *p*-type semiconductors, *n*-type semiconductors, and insulators, the catalytic activity of which for CO oxidation decreases in this order. As support for Pt group metals, except for Au, semiconductor metal oxides are more effective than insulating metal oxides because the former are partly reducible by CO and H$_2$ at temperatures below 200 °C. However, the degree of reducibility is not directly reflected on the support effect. Although *n*-type semiconductor metal oxides themselves are less active than *p*-type semiconductor ones, their support effect is usually larger. The reason is not clear. However, it seems to be curious that noble metals supported on *p*-type semiconductors have not been extensively studied so far, except for a few cases.

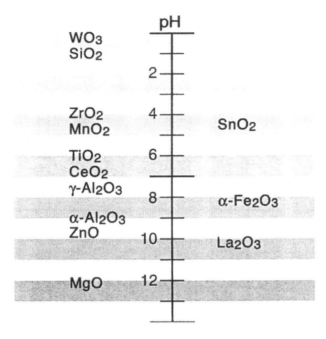

Figure 10 Isoelectric points (points of zero charge, PZC) of metal oxides. (Data are presented in Parks, J. Phys. Chem., 66 (1962) 967.)

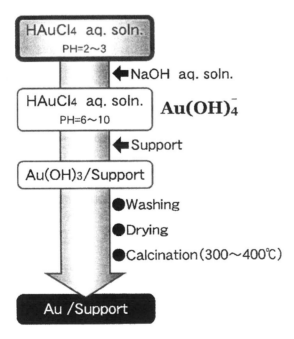

Scheme 1 Procedures in the deposition precipitation method for the preparation of supported Au catalysts.

Table 2 Classification of Base Metal Oxides for CO Oxidation

Type	I	II	III
Electrical property	*p*-Type semiconductor	*n*-Type semiconductor	Insulator
Metal Oxides	Ag_2O^a, CO_3O_4, MnO_2,	Fe_2O_3, SnO_2, CdO,	V_2O_5, M_O3O_4, WO_3, Bi_2O_3,
(in the order activity)	CuO, NiO, $Cr_2O_3^b$	ZnO, TiO_2, CeO_2	MgO, Al_2O_3, SiO_2
Working Temperature	Low (RT–473 °K)	Middle (473–773 °K)	High (>773 °K)
M–O Bond Energy			
(kcal/g · O atom)	40–65		>65
Active oxygen species	O_2^-, O^-	O^-	O^{2-}
(ML)	10^{-2}–1	10^{-3}–10^{-2}	(lattice oxygen)
Mechanism	Concerted (RT) to stepwise (473 °K)	Stepwise	Reduction–Oxidation
Rate expression under			
excess O_2	$r = kP_{CO}^0 P_{O2}^0$ to $r = kP_{CO}$	$r = kP_{CO}^0/P_{O2}^{0.5}$ (α-Fe_2O_3)	$r = kP_{CO}^{0.63}/P_{O2}^{0.22}$ (MgO)
Effect of moisture	Appreciable deactivation	Little	None

Exceptions of the M–O bond energy: [a] 8 kcal/g · O atom; [b] 91 kcal/g · O atom.

The most effective supports for Pt-group metals are SnO_2, TiO_2, and CeO_2 [57–64]. These metal oxides make Pt-group metals active for CO oxidation even at room temperature. SnO_2 especially has been most extensively studied since the 1970s, and Pd and Pt supported on SnO_2 are commercially used for sealed CO_2 lasers and semiconductor-type gas leak detectors [65–70]. Optimization of practical catalyst, which is the most active among Pt-group metal catalysts for CO oxidation in CO_2 laser, is as follows [70]:

1. Composition: Pt (17 wt. %) + Pd (5 wt. %) supported on SnO_2, carried on SiO_2 with high specific surface area of about 200 m²/g. Fe_2O_3 (below 1 wt. %) and P_2O_5 are added as promoters.
2. Method of preparation: inherently clean method to eliminate Cl^- (deposition-precipitation?).
3. Pretreatment: in a stream of 5 vol % CO in He at 125 °C for 1 h. This is assumed to form Pt/Sn alloy particles having surface hydroxyls.

For Au a variety of metal oxides work as effective supports. Even insulating Al_2O_3 and SiO_2 works as good supports. The only exceptions are acidic metal oxides like Al_2O_3-SiO_2, and WO_3. It is an essential question why CO oxidation does not take place at temperatures below 150 °C under acidic circumstances, as depicted in Figure 11. In electrochemical oxidation of CO, the Au electrode is more active than the Pt electrode by about 1000 times, but this happens only in alkaline solution [84,85]. These features are scientific obstacles for making active anodes for direct methanol fuel cells. At room temperature, TOFs of supported Au catalysts dispersed on Al_2O_3, SiO_2, and TiO_2 do not differ appreciably for the same size of Au particles. However, more active metal oxides lead to activity at lower temperatures, down to −70 °C. Moisture enhances the catalytic activity of supported Au catalysts in comparison with that under dry condition (100 ppb H_2O) [71]. A major difference among different metal oxide supports is seen in the effect of moisture in the range from 0.1 to 1000 ppm.

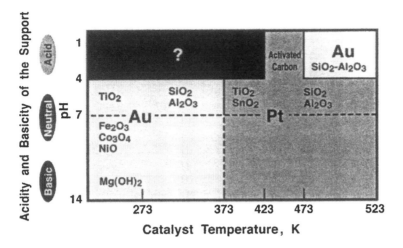

Figure 11   Correlation of typical reaction temperatures of CO oxidation over supported Au and Pt with the acidity (basicity) of the support oxides.

18.7   EFFECT OF CONTACT STRUCTURE OF NOBLE METAL PARTICLES WITH THE SUPPORTS

Even though the component and size of metals and metal oxide support are defined, the catalytic activity for CO oxidation often markedly changes depending on the contact structure of noble metal particles with the supports. In particular, Pd, Ir, and Au exhibit high catalytic activity when they are deposited on reducible metal oxides by coprecipitation, deposition-precipitation, and grafting. Goulanski has classified supported metal catalysts for low-temperature oxidation into three groups [72]. There are three possible active sites: metal surfaces with metal oxide as a simple support; metal oxide thin layer underneath of which metal particles are buried; and the perimeter interfaces around noble metal particles.

In the first case where metal surfaces provide active oxygen species to the support contact structure is not critical. The second case is often observed when supported metal catalysts are prepared by coprecipitation or sol-gel methods. Noble metals whose oxides are more stable than Pt oxides such as Pd and Ir are more readily buried in the bulk of metal oxide supports, and the metal oxide overlayers of a thickness of about a few monolayers are modified in their electronic and redox properties by underlying noble metal nanoparticles to become active at lower temperatures.

The third case is typically evidenced by supported gold catalysts. As shown in Figure 12, TOF for Au/TiO_2 prepared by DP is larger by 4 orders of magnitude than that prepared by impregnation, followed by washing and drying to eliminate chloride ions. An explanation is that when spherical particles of Au are simply loaded on the support (when a catalyst is prepared by impregnation method; Figure 8), the length of the perimeter interface is very short, while it is much longer when hemispherical particles are strongly contacted with the supports at their flat planes (when prepared by deposition-precipitation method; Figure 9). Since it is very likely that reaction takes place between CO adsorbed on the Au surfaces and oxygen

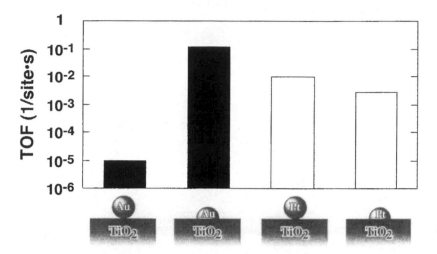

Figure 12 Effect of metal particle shape on the TOF at 300 K for CO oxidation over Au/TiO$_2$ and Pt/TiO$_2$. Spherical particles corresponds to the sample prepared by impregnation and hemisperical particles by deposition-precipitation method. (Data are presented in Bamwenda, Catal. Lett. 44 (1997) 83.)

adsorbed on the support surface or the perimeter interface, the latter contact structure gives much higher catalytic activity. In the case of Pt, CO oxidation takes place mainly on the metal surfaces, or the contact structure of Pt particles with the support might not differ as much as that of Au particles due to better wetting through Pt oxide formation. TOFs for the two different shapes of Pt particles are almost similar to each other.

A more delicate item may be the effect of alloy formation between noble metals and the metal components of the supports. In the case of Pt/SnO$_2$, low-temperature (at 120 °C) reduction is required, which leads to both the formation of Pt–Sn alloys and the formation of surface hydroxyls at the perimeter [68]. On the other hand, in the case of Au/TiO$_2$, vacuum evacuation or reduction dramatically suppresses the initial catalytic activity, which can be recovered gradually during CO oxidation in excess O$_2$. The removal of oxygen species at the perimeter interface is deleterious to supported Au catalysts.

Over Pt-group metal particles supported on reducible metal oxides or Au metal particles on most of metal oxides including Al$_2$O$_3$ and SiO$_2$, reaction pathways appear to be similar at temperatures between −70 °C and 150 °C. The only difference is that at temperatures below −70 °C, CO oxidation takes place only on the surfaces of Au particles with nearly 0 apparent activation energy; at temperatures above 150 °C, the reaction takes place only on the surfaces of Pt-group metal particles with apparent activation energy of about 20–40 kcal/mol. At the intermediate temperature, CO oxidation takes place at the perimeter interface; CO is adsorbed on the metal surfaces and oxygen is adsorbed, most likely as negatively charged species, O$_{2-}$, at the perimeter surfaces of the support. These pathways are summarized in scheme 2 for Au/TiO$_2$ as a typical catalyst [83]. The perimeter pathways involving bifunctional sites are also presented for bimetallic systems such as Au/Pt(335), Pt/Rh, and Pt group metals supported on reducible metal oxides [65,73–75].

1) Reaction at the Au Surfaces
Below 200 K

$$Au(s)\text{-}CO + Au^{\delta+}(s)\text{-}O_2{}^{\delta-} \longrightarrow CO_2(g) + Au(s) + Au^{\delta+}(s)\text{-}O^{\delta-}$$

$$Au(s)\text{-}CO + Au^{\delta+}(s)\text{-}O^{\delta-} \longrightarrow CO_2(g) + 2Au(s)$$

$$Au(p)\text{-}CO + Au^{\delta+}(p)\text{-}O_2{}^{\delta-} \longrightarrow CO_2(g) + Au(p) + Au^{\delta+}(p)\text{-}O^{\delta-}$$

$$Au(p)\text{-}CO + Au^{\delta+}(p)\text{-}O^{\delta-} \longrightarrow CO_2(g) + 2Au(p)$$

2) Reaction at the Perimeter Interfaces
Above 300 K

$$Au(p)\text{-}CO + TiO_2(p)\text{-}O_2{}^- \longrightarrow CO_2(g) + TiO_2(p) + e^- + Au^{\delta+}(p)\text{-}O^{\delta-}$$

$$Au(p)\text{-}CO + Au^{\delta+}(p)\text{-}O^{\delta-} \longrightarrow CO_2(g) + 2Au(p)$$

Below 300 K: accompanied by

$$Au(p)\text{-}CO + TiO_2(p)\text{-}O_2{}^- + e^- \longrightarrow Au(p) + TiO_2(p)\text{-}CO_3{}^{2-}$$

(s): step, (p): perimeter, (g): gas phase

Scheme 2   Probable pathways for CO oxidation over supported Au catalyst.

18.8   SELECTIVE CO OXIDATION IN THE PRESENCE OF HYDROGEN

When hydrogen is produced by steam reforming of methanol or hydrocarbons, CO is the second-largest byproduct component (about 1 vol %) after CO_2. In application to PEFC, to remove CO to a concentration of 100 ppm, CO should be reduced to methane or oxidized to CO_2. Oxidation is advantageous because it potentially consumes less hydrogen.

Series of supported Ru, Pt, and Au catalysts have been tested for the selective oxidation of CO. Table 3 shows a comparison of three typical catalysts reported recently. Since Pt group metals are essentially more active for H_2 oxidation than for CO oxidation, support materials [77,78] and bimetallic systems [79] were exploited to suppress H_2 oxidation. Better selectivities to CO oxidation (around 50%) are

Table 3   Comparison of Noble Metal Catalysts for CO Removal from H_2

Catalyst	Metal loading (wt. %)	CO (vol %)	O_2 (vol %)	H_2 (vol %)	SV (h^{-1})	Temperature (°K)	CO conversion (°K)	Ref.
Au/MnO_x powder	5	1.0	1.0	98	1×10^4 (ml/g)	323–353	97–99	76
Ru/Al_2O_3, Rh/Al_2O_3 pellet	0.5	0.09	0.09	0.85	2×10^4	403–453	100	77
Pt/A-zeolite powder	6.4	1.0	1.0	98	3×10^4 (ml/g)	473–523	91–92	78

obtained over Pt/zeolite and Pt$_3$Sn/Vulcan XC carbon than over Pt/γ-Al$_2$O$_3$, at temperatures from 150 °C to 200 °C.

Gold is potentially more active for CO oxidation; however, stability and reactivity of the perimeter interface between Au particles and the support may not be maintained in H$_2$ stream. Behm tested a variety of supported Au catalysts and chose Au/Fe$_2$O$_3$ as the best candidate [80,81]. Gold on MnOx prepared by coprecipitation with LiCO$_3$ followed by calcination in air at 400 °C gives sufficiently high conversion of CO at temperatures from 50 °C to 80 °C [76]. This catalyst is very stable in H$_2$ stream containing CO$_2$ and H$_2$O and might be superior to Au/Fe$_2$O$_3$, in catalytic activity.

It should be noted that supported Au catalysts are the most active for CO concentrations above 0.1 vol % and supported Pt-group metal catalysts are the most active for CO concentrations lower than 0.1 vol %. A combination of the two types of catalysts might present satisfactory performance in the practical process to remove CO from H$_2$-containing streams.

18.9 CONCLUSIONS

Since supported gold catalysts prepared by coprecipitation were found to be active for CO oxidation even at temperatures far below room temperature, attempts are increasing to prepare other noble metal catalysts by coprecipitation, deposition-precipitation, and grafting methods, which were used for the preparation of active supported gold catalysts. Although the affinity to CO is markedly different between Pt-group metals and Au supported on selected metal oxides, the contribution of metal–support interactions to the enhancement of low-temperature catalytic activity for CO oxidation appears to be similar, namely, the enhancement of oxygen activation at the perimeter interface. This line of approach may be valid to seek for a new type of catalysts active at lower temperatures for reactions other than CO oxidation [82,83].

The major conclusions of this chapter are as follows:

1. Carbon monoxide is adsorbed too strongly on the surfaces of Pt-group metals, while it is adsorbed too weakly on the surfaces of Au.
2. Due to the strong adsorption of CO on the surfaces of Pt-group metals, the rate of CO oxidation abruptly increases at a certain temperature, which ranges from 150 °C to 300 °C for conventional Pt-group metals catalysts supported on Al$_2$O$_3$ or SiO$_2$.
3. Due to the weak adsorption of CO and O$_2$ on the surfaces of Au, the rate of CO oxidation changes only little with temperature, giving rise to nearly 0 activation energy.
4. The catalytic activity order of metals for CO oxidation in terms of temperature for 50% conversion in a fixed bed flow reactor is as follows:

    $$Ru > Rh > Ir > Pd > Pt > Au$$

5. As for the size effect of noble metal particles, only gold prefers smaller particles but down to 2 nm in diameter. Pt-group metals give constant or decreasing TOF with a decrease in the particle diameter.

6. To prepare supported metal catalysts for low-temperature CO oxidation, coprecipitation, deposition-precipitation, and grafting methods are effective, because they can give strongly interacting metal particles with the support.

7. Some semiconductive metal oxides are effective in lowering the temperature for CO oxidation. For Pt-group metals, SnO_2, TiO_2, and CeO_2 and for Au, Fe_2O_3 and TiO_2 are especially effective.

8. Perimeter interface around metallic particles appears to be the most important part in the metal catalysts active at low temperatures. For Pt/SnO_2, Pt-Sn alloy formation (thus, reducing pretreatment) is crucial for the enhancement in catalytic activity, while for Au/TiO_2, oxidic Au formation (thus, oxidizing pretreatment) is assumed to be crucial.

9. The practical application of CO oxidation is growing, especially in relation to the development of polymer electrolyte fuel cells. An ongoing attempt is focused on the selective CO oxidation in H_2 stream. The key challenging question related to the development of direct methanol fuel cells is whether CO oxidation can proceed at low temperatures, even under a strongly acidic environment.

ACKNOWLEDGMENTS

The authors wish to thank Dr. Masakazu Daté for his kind help in making figures and for his critical reading of this article.

REFERENCES

1. A. Bielanski, J. Haber, In Oxygen in Catalysis, Marcel Dekker, New York, 1991, pp. 211–276.
2. T. Engel, G. Ertl, In The Chemical Physics of Solid Surfaces and Heterogeneous Catalysis, vol. 4, D.A. King, D.P. Woodruff, eds., Elsevier, 1982, pp. 73–93.
3. G. Ertl, Chem. Record (2000) 33–45.
4. G.A. Somorjai, CATTECH 3 (1999) 84–97.
5. C.H.F. Peden, In Surface Science of Catalysis—In situ Probes and Reaction Kinetics, ACS Symp. Ser. 482, D.J. Dwyer, F.M. Hoffmann, eds., Am. Chem. Soc., Washington D.C., 1992, pp. 143–159.
6. D. Böcker, R.D. Gonzalez, In Closed-Cycle, Frequency-Stable CO_2 Laser Technology, NASA Conf. Publication 2456, C.E. Batten, I.M. Miller, G.M. Wood, Jr., D.V. Willetts, eds., 1987, pp. 85–102.
7. S. Kawatsu, J. Power Sources 71 (1998) 150.
8. NASA Conf. Pub. 3076. Low-Temperature CO-Oxidation Catalysts for Long-Life CO_2 Lasers, D.R. Schryer, G.B. Hoflund, eds., NASA, 1990.
9. V.D.J. Berg, V.A.J. Dillen, V.D.J. Meijden, J.W. Geus, In Surface Properties and Catalysis by Non-Metals, J.P. Bonnelle, B. Delmon, E. Derouane, eds., D. Reidel Pub. Co., Dordrecht, 1983, pp. 493–532.
10. D.A.H. Cunningham, T. Kobayashi, N. Kamijo, M. Haruta, Catal. Lett. 25 (1994) 257–264.
11. G. Croft, M.J. Fuller, Nature 269 (1977) 585–586.
12. C.J. Zhang, P. Hu, J. Am. Chem. Soc. 122 (2000) 2134–2135.
13. C. Zhang, P. Hu, A. Alavi, J. Am. Chem. Soc. 121 (1999) 7931–7932.
14. M. Haruta, Shokubai (Catalyst & Catalysis) 36 (1994) 310–318.

15. G. Cao, Y. Seimiya, Y. Ohno, T. Matsushima, Chem. Phys. Lett. 294 (1998) 419–424.
16. J. Wolff, A.G. Papathanasiou, I.G. Kevrekidis, H.H. Rotermund, G. Ertl, Science 294 (2001) 134–137.
17. J.P. Dath, J.E. Dauchot, J. Catal. 97 (1986) 100–107.
18. G.W. Coulston, G.L. Haller, In Surface Science of Catalysis—In Situ Probes and Reaction Kinetics, ACS Symp. Ser. 482, D.J. Dwyer, E.M. Hoffmann, eds., Am. Chem. Soc., Washington D.C., 1992, pp. 58–70.
19. P.J. Berlowitz, C.H.F. Peden, D.W. Goodman, J. Phys. Chem. 92 (1988) 5213–5221.
20. F. Boccuzzi, A. Chiorino, M. Manzoli, P. Lu, T. Akita, S. Ichikawa, M. Haruta, J. Catal. 202 (2001) 256–267.
21. Y. Iizuka, T. Tode, T. Takao, K. Yatsu, T. Takeuchi, S. Tsubota, M. Haruta, J. Catal. 187 (1999) 50–58.
22. M. Bowker, Q. Guo, Y. Li, R.W. Joyner, Catal. Lett. 18 (1993) 119–123.
23. S. Ladas, H. Poppa, M. Boudart, Surf. Sci. 102 (1981) 151–171.
24. C.J. Zhang, P. Hu, J. Am. Chem. Soc. 123 (2001) 1166–1172.
25. FY1998 Survey Report for Speedy Formation of Intellectual Platform, Hi99-8(1), Chapter 6, New Materials Center of Osaka Science & Technology Center, March 2000.
26. Osaka National Research Institute Report No. 393, Abilities and Potential of Gold as a Catalyst, 1999.
27. M. Haruta, In The Physics and Chemistry of Clusters, E.E.B. Campbell, M. Larsson, eds., World Scientific, Singapore, 2001, pp. 99–109.
28. M. Valden, X. Lai, D.W. Goodman, Science 281 (1998) 1497–1650.
29. C. Xu, W.S. Oh, G. Liu, D.Y. Kim, D.W. Goodman, J. Vac. Sci. Technol. A. 15 (1997) 1261–1268.
30. S.H. Oh, C.C. Eickel, J. Catal. 128 (1991) 526–536.
31. V. Nehasil, I. Stará, V. Matolin, Surf. Sci. 352–354 (1996) 305–309.
32. F. Rumpf, H. Poppa, M. Boudart, Langmuir 4 (1988) 722–728.
33. C. Becker, C.R. Henry, Surf. Sci. 352–354 (1996) 452–462.
34. N.W. Cant, P.C. Hicks, B.S. Lennon, J. Catal. 54 (1978) 372–383.
35. E. McCarthy, J. Zahradnik, G.C. Kuczynski, J.J. Carberry, J. Catal. 39 (1975) 29–35.
36. G.S. Zafiris, R.J. Gorte, J. Catal. 140 (1993) 418–423.
37. N.W. Cant, J. Catal. 62 (1980) 173–175.
38. M. Herskowitz, R. Holliday, M.B. Cutlip, C.N. Kenney, J. Catal. 74 (1982) 408–410.
39. M. Haruta, S. Tsubota, T. Kobayashi, H. Kageyama, M.J. Genet, B. Delmon, J. Catal. 144 (1993) 175–192.
40. M. Valden, S. Pak, X. Lai, D.W. Goodman, Catal. Lett. 56 (1998) 7–10.
41. M. Okumura, S. Nakamura, S. Tsubota, T. Nakamura, M. Azuma, M. Haruta, Catal. Lett. 51 (1998) 53–58.
42. M. Okumura, S. Tsubota, M. Iwamoto, M. Haruta, Chem. Lett. 4 (1997) 315–316.
43. M. Haruta, M. Daté, Y. Iizuka, F. Boccuzzi, Shokubai (Catalyst and Catalysis) 43 (2001) 125–127.
44. D.A.H. Cunningham, W. Vogel, H. Kageyama, S. Tsubota, M. Haruta, J. Catal. 177 (1998) 1–10.
45. D.A.H. Cunningham, W. Vogel, M. Haruta, Catal. Lett. 63 (1999) 43–47.
46. W.T. Wallace, R.L. Whetten, J. Phys. Chem. B104 (2000) 10964–10968.
47. U. Heiz, W.-D. Schneider, J. Phys. D:Appl. Phys. 33 (2000) R85–R102.
48. S. Ichikawa, H. Poppa, M. Boudart, In Catalytic Materials: Relationship Between Structure and Reactivity, ACS Sym. Ser. 248, pp. 440–451, T.E. Whyte, Jr., R.A. Della Betta, E.G. Derouane, R.T.K. Baker, eds., Am. Chem. Soc., Washington, D.C., 1984.
49. I. Meusel, J. Hoffmann, J. Hartmann, J. Libuda, H.-J. Freund, J. Phys. Chem. B 105 (2001) 3567–3576.
50. S. Abbet, U. Heiz, H. Häkkinen, U. Landman, Phys. Rev. Lett. 86 (2001) 5950–5953.

51. L. Piccolo, C. Becker, C.R. Henry, Eur. Phys. J. D9 (1999) 415–419.
52. U. Heiz, A. Sanchez, S. Abbet, W.-D. Schneider, J. Am. Chem. Soc. 121 (1999) 3214–3217.
53. M. Haruta, N. Yamada, T. Kobayashi, S. Iijima, J. Catal. 115 (1989) 301.
54. S. Tsubota, D.A.H. Cunningham, Y. Bando, M. Haruta, Preparation of Catal, G. Poncelet, J. Martens, B. Delmon, P.A. Jacobs, P. Grange eds., Elsevier, Amsterdam, 1995, pp. 227–235.
55. T. Akita, K. Tanaka, S. Tsubota, M. Haruta, J. Electron Microscopy 49 (2000) 657–662.
56. T. Akita, P. Lu, S. Ichikawa, K. Tanaka, M. Haruta, Surf. Interface Anal. 31 (2001) 73–78.
57. T. Bunluesin, H. Cordatos, P.J. Gorte, J. Catal. 157 (1995) 222–226.
58. A.D. Logan, M.T. Paffett, J. Catal. 133 (1992) 179–190.
59. A.V. Kalinkin, V.I. Savchenko, A.V. Pashis, Catal. Lett. 59 (1999) 115–119.
60. L.S. Sun, S.Y. Li, B.L. Li, React. Kinet. Catal. Lett. 62 (1997) 151–156.
61. P. Thormählen, M. Skoglundh, E. Fridell, B. Andersson, J. Catal. 188 (1999) 300–310.
62. G. Dong, J. Wang, Y. Gao, S. Chen, Catal. Lett. 58 (1999) 37–41.
63. Y.J. Mergler, A. van Aalst, J. van Delft, B.E. Nieuwenhuys, Appl. Catal. B: Environmental 10 (1996) 245–261.
64. A.E. Aksaylu, M. Madalena, A. Freitas, J.L. Figueiredo, Catal. Today, 62 (2000) 337–346.
65. R.K. Herz, A. Badlani, D.R. Schryer, B.T. Upchurch, J. Catal. 141 (1993) 219–238.
66. D.R. Schryer, B.T. Upchurch, B.D. Sidney, K.G. Brown, G.B. Hoflund, R.K. Herz, J. Catal. 130 (1991) 314–317.
67. S.D. Gardner, G.B. Hoflund, D.R. Schryer, B.T. Upchurch, J. Phys. Chem. 95(2) (1991) 835–838.
68. D.R. Schryer, B.T. Upchurch, J.D.V. Norman, K.G. Brown, J. Schryer, J. Catal. 122 (1990) 193–197.
69. S.D. Gardner, G.B. Hoflund, M.R. Davidson, D.R. Schryer, J. Catal. 115 (1989) 132–137.
70. 3rd Intl. DRA/NASA Confl. Long-Life CO_2 Laser Techn. Nov. 10–12, 1992, Malvern, Worcs, U.K.
71. M. Daté, M. Haruta, J. Catal. 201 (2001) 221–224.
72. S. Goulanski, R. Rajaram, N. Hodge, G.S. Hutchings, C.J. Kiely, Proc. Cat. Gold. Intl. Conf., Cape Town, April 2001, p. 11.
73. D.C. Skelton, R.G. Tobin, D.K. Lambert, C.L. Di Maggio, G.B. Fisher, J. Phys. Chem. 103 (1999) 964–971.
74. R.H. Nibbelke, M.A.J. Campman, J.H.B.J. Hoebink, G.B. Marin, J. Catal. 171 (1997) 358–373.
75. M. Sheintuch, J. Schmidt, Y. Lecthman, G. Yahav, Appl. Catal. 49 (1989) 55–65.
76. R.M. Torres Sanchez, A. Ueda, K. Tanaka, M. Haruta, J. Catal. 168 (1997) 125–127.
77. S.H. Oh, R.M. Sinkevitch, J. Catal. 142 (1993) 254–262.
78. M. Watanabe, et al. Chem. Lett. (1995) 21–22.
79. M.M. Schubert, M.J. Kahlich, G. Feldmeyer, M. Hüttner, S. Hackenberg, H.A. Gasteiger, R.J. Behm, Phys. Chem. Chem. Phys. 3 (2001) 1123–1131.
80. K. Kahlich, H.A. Gasteiger, R.J. Behm, J. Catal. 182 (1999) 430–440.
81. M.M. Schubert, S. Hackenberg, A.C. van Veen, M. Muhler, V. Plzak, R.J. Behm, J. Catal. 197 (2001) 113–122.
82. W.J. Shen, M. Okumura, Y. Matsumura, M. Haruta, Appl. Catal. A: General 213 (2001) 225–232.
83. M. Haruta, M. Daté, Appl. Catal. A: General, 222 (2001) 427–437.
84. H. Kita, H. Nakajima, K. Hayashi, J. Electroanal. Chem. 190 (1985) 141–156.
85. L.D. Burke, D.T. Buckley, J.A. Morrissey, Analyst 119 (1994) 841–845.

19

Promotion, Electrochemical Promotion, and Metal–Support Interactions: The Unifying Role of Spillover

CONSTANTINOS G. VAYENAS, C. PLIANGOS, S. BROSDA, and D. TSIPLAKIDES

University of Patras, Patras, Greece

CHAPTER CONTENTS

SUMMARY

Promotion of catalyst nanoparticles, electrochemical promotion (NEMCA) of porous and of single-crystal catalyst films, and metal nanoparticle–support interactions are three, at a first glance, independent phenomena that can all dramatically affect catalytic activity and selectivity on metal and metal oxide catalyst surfaces.

Recent experimental and theoretical work has shown that the three phenomena are intimately related via the action of spillover to the point that one may conclude that the three phenomena are functionally identical and only operationally different.

In this chapter we review some of the key phenomenological aspects of promotion, electrochemical promotion, and metal–support interactions, underline their kinetic similarities and common fundamental origin on the basis of surface spectroscopic and theoretical investigations, including the new concept of the absolute potential of supports, and discuss some key experiments that prove their identical nature, i.e., catalysis in presence of a controllable double layer at the catalyst–gas interface.

We then discuss the recently established rules of promotion and electrochemical promotion and an extension of Langmuir–Hinshelwood kinetics, based on an effective medium double-layer isotherm model, which is in good qualitative agreement with experiment and allows one to make predictions about the effect of promoters, but also of catalyst supports, on the kinetics of different catalytic reactions.

19.1 INTRODUCTION

19.1.1 Scope

Promotion and metal–support interactions play a key role in the design of successful commercial finely dispersed nanoparticle catalysts [1–5]. The detailed molecular mechanism of promotion [6] and particularly of metal–support interactions [7,8] is still a subject of intensive study and dispute.

During the last 15 years the closely related phenomenon of electrochemical promotion [9–12], or non-Faradaic electrochemical modification of catalytic activity, NEMCA effect [9–12], has been discovered and studied for more than 60 catalytic reactions [13,14].

The goal of this chapter is to summarize and systematize the phenomenology of the three phenomena, i.e., classical promotion, electrochemical promotion, and metal–support interactions, present their striking similarities and some common rules that govern them, and demonstrate their intimate link and common molecular mechanism.

19.1.2 Basic Principles and Terminology

Promotion

The term "promotion," or classical promotion, is used to denote the action of a substance, called *promoter*, which when added to a catalyst, usually on its surface, enhances its catalytic performance, i.e., it increases the rate of a catalytic reaction or the selectivity to a desired product, or the useful lifetime of the catalyst. For example, K or K_2O is a promoter of Fe for the synthesis of ammonia or for the production of higher hydrocarbons in the Fischer–Tropsch synthesis.

Broadly speaking, promoters can be divided into *structural promoters* and *electronic promoters*. In the former case, they enhance and stabilize the dispersion of the nanoparticle-dispersed active phase on the catalyst support. In the latter case, they enhance the catalytic properties of the active phase itself. This stems from their ability to modify the chemisorptive properties of the catalyst surface and to significantly affect the chemisorptive bond strength of reactants and intermediates. At the molecular level this is the result of direct ("through the vacuum") and indirect ("through the metal") interactions. The term "through the vacuum" denotes direct electrostatic, Stark type, attractive or repulsive interactions between the adsorbed reactants and the local electric field created by the coadsorbed promoter. The term "through the metal interaction" refers to changes in the binding state of adsorbed reactants due to promoter-induced redistribution of electrons near the Fermi level of the metal [6,14].

Denoting by θ_p the coverage of a promoter on a catalyst surface and by p_j the partial pressures of the reactants, j, of the catalytic reaction being promoted, we can formulate mathematically the above definition of a promoter as

$$\left(\frac{\partial r}{\partial \theta_p}\right)_{P_j} > 0 \Leftrightarrow \text{P is a promoter} \qquad \left(\frac{\partial r}{\partial \theta_p}\right)_{P_j} < 0 \Leftrightarrow \text{P is a poison} \qquad (1)$$

The promotional propensity of a promoter, P, can be quantified by defining [13,14] a promotional index, PI_p, from

$$PI_P = (\Delta r/r_0)/\Delta \theta_P \qquad (2)$$

where r_0 is the unpromoted catalytic rate.

Thus, the promotional index PI_P is positive for promoters and negative for poisons. In the latter case the definition of PI_P coincides with that of the "toxicity" defined by Barbier several years ago [15]. In the case of pure site blocking, it is $PI = -1$. Values of $PI_{O^{2-}}$ up to 150 and $PI_{Na^{\delta+}}$ up to 6000 have been measured [13,14].

Another useful parameter for quantifying the promotional action is the promotional rate enhancement ratio, ρ_P, defined from

$$\rho_P = r/r_0 \qquad (3)$$

Promotional rate enhancement ratios, ρ_P, values of the order 10–100 are rather common [6,14], as we shall also see in this chapter.

A promoter is not, in general, consumed during a catalytic reaction. Many millions of catalytic turnovers usually take place on a promoted site of a classically promoted catalyst before the promoter is deactivated. The ratio of the promoter average lifetime on the catalyst surface, τ_{PR}, over that of the catalytic reactants, τ_R, is usually in excess of 10^7.

$$\tau_{PR}/\tau_R > 10^7 \Leftrightarrow \text{classical promotion} \qquad (4)$$

When the promoter is consumed at a faster rate, which is still, however, smaller than that of the catalytic reactant, then the promoter is termed *sacrificial promoter* [13,14]. This is the case, as we will see, in electrochemical promotion utilizing O^{2-} conducting solid electrolytes. The promoting O^{2-} species is introduced via a Faradaic process on the catalyst surface at a rate of $I/2F$, where I is the applied current and F is Faraday's constant. At steady state, $I/2F$ also equals the rate of consumption of the sacrificial promoter O^{2-} species on the catalyst surface.

Using the standard definition [9–14] of the Faradaic efficiency, Λ, of an electrochemically promoted process:

$$\Lambda = \Delta r/(I/2F) \qquad (5)$$

one can easily show that

$$\tau_{PR}/\tau_R = \Lambda \rho_P/(\rho_P - 1) \qquad (6)$$

When $\rho_P \gg 1$, as is very frequently the case, Eq. (6) reduces to

$$\tau_{PR}/\tau_R = \Lambda \tag{7}$$

which shows that the Faradaic efficiency, Λ, of electrochemically promoted reactions expresses the ratio of the average lifetimes of promoting species (O^{2-}, $Na^{\delta+}$) and of the key reactants on the catalyst surface [13,14].

When a promoter is added continuously to the reactive gas mixture, as, e.g., in the case of a few ppm $C_2H_2Cl_2$ addition to C_2H_4 and O_2 during C_2H_4 epoxidation on Ag catalysts, this promoter, which is a sacrificial promoter, is also sometimes referred to as a moderator.

Promoters are usually added to a catalyst during its preparation (classical or chemical promotion). Thus, if they somehow get lost (evaporation) or deactivated during prolonged catalyst operation, this leads to significant catalyst deterioration. Their concentration cannot be controlled in situ, i.e., during catalyst operation. One of the most important advantages of electrochemical promotion is that it permits direct in-situ control of the amount of the promoter on the catalyst surface.

Electrochemical Promotion

Electrochemical promotion of catalysis (EPOC) or non-Faradaic electrochemical modification of catalytic activity (NEMCA) is the phenomenon whereby application of small current density (1–$10^4\,\mu A/cm^2$) or potential ($\pm 2\,V$) between a conductive catalyst, deposited on a solid electrolyte, and a second (catalytically inert) electrode, also deposited on the solid electrolyte, enhances the catalytic performance of the catalyst [9–14,16–25] (Figure 1).

In electrochemical terms the catalyst also acts as the *working* electrode of the solid electrolyte cell that is formed, the second (catalytically inert) electrode is the

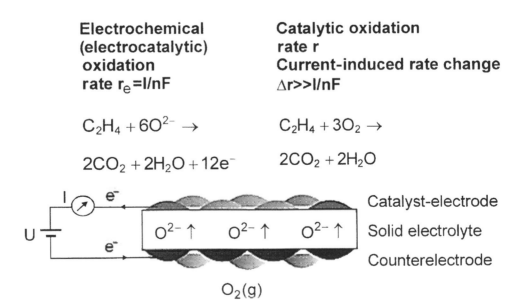

Electrochemical (electrocatalytic) oxidation rate $r_e = I/nF$

$$C_2H_4 + 6O^{2-} \rightarrow$$

$$2CO_2 + 2H_2O + 12e^-$$

Catalytic oxidation rate r Current-induced rate change $\Delta r \gg I/nF$

$$C_2H_4 + 3O_2 \rightarrow$$

$$2CO_2 + 2H_2O$$

Catalyst-electrode

Solid electrolyte

Counterelectrode

$O_2(g)$

Figure 1 Experimental setup used in NEMCA experiments.

counterelectrode, while a third metal film, acting as a *reference* electrode, is also useful to deposit on the solid electrolyte for fundamental EPOC studies. In this way one can study the dependence of the catalytic rate, *r*, on the catalyst-working electrode as a function of the potential difference, U_{WR}, between the catalyst-working (*W*) electrode and the reference (*R*) electrode [13,26–30].

The counter- and the reference electrodes can be in a separate gaseous compartment (fuel-cell-type design) or can be in the same gaseous compartment with the catalyst-electrode (single-chamber-type design) [13,27–30].

The magnitude of electrochemical promotion can be quantified by the following three parameters:

1. The Faradaic efficiency, Λ, already defined in Eq. (5):

$$\Lambda = \Delta r/(I/2F) \tag{5}$$

where *r* is expressed in mol O, or more generally:

$$\Lambda = \Delta r/(I/F) \tag{8}$$

where *r* is expressed in g-equivalent.

A reaction is electrochemically promoted, or exhibits NEMCA, when $|\Lambda| > 1$. When $\Lambda > 1$, i.e., when the reaction rate is enhanced with positive current and increasing catalyst potential U_{WR}, the reaction is termed *electrophobic*. When $\Lambda < -1$, i.e., when the rate is enhanced with negative current and decreasing catalyst potential, the reaction is termed *electrophilic*.

Clearly, the limits of electrocatalysis are $|\Lambda| \leqslant 1$ [13]. Faradaic efficiency, Λ, values up to $3 \cdot 10^5$ and down to $-3 \cdot 10^4$ have been measured for different catalytic systems (Table 1) [9,10,13,17–21,23,26,28–86].

2. The rate enhancement ratio, ρ, which, similar to the case of classical promotion [Eq. (3)], is defined from

$$\rho = r/r_0 \tag{9}$$

As shown in Table 1, ρ values as high as 150 and as low as zero (complete catalyst poisoning) have been measured for different catalytic reactions.

3. The promotional index, PIp [Eq. (2)]. After the establishment, via the use of surface spectroscopy (XPS [87,88], UPS [89], TPD [90], PEEM [91], STM [92], work function measurements [93]) but also electrochemistry (cyclic voltammetry [90], potential programmed reduction [94], AC impedance spectroscopy [43,95]), that electrochemical promotion is due to the potential-controlled migration (reverse spillover or backspillover) [13] of promoting ionic species (O^{2-}, Na^+, H^+, F^-) from the solid electrolyte to the gas-exposed catalyst surface, it became clear that electrochemical promotion is functionally very similar to classical promotion and that the promotional index PI_p, already defined in Eq. (2), can be used interchangeably, both in classical and in electrochemical promotion.

As already mentioned, one important operational difference between classical and electrochemical promotion is that in the latter case one can control in situ, via the applied potential, the coverage of the promoting species on the catalyst surface. Apart from the potential practical significance of such in-situ tuning of the catalytic activity on a working catalyst, there is a second advantage for the fundamental study of promotion: one can now examine in situ, i.e., under constant gaseous

Table 1 Electrochemical Promotion Studies Classification based on the Type of Solid Electrolyte

1. EP studies utilizing YSZ (promoting ion: O^{2-})

Reactants								
Electron Donor (D)	Electron Acceptor (A)	Products	Catalyst	$T\ (°C)$	$\Lambda_{max}(>0)$ or $\Lambda_{min}(<0)$	$\rho_{max}(>1)$ or $\rho_{min}(<1)$	$PI_{O^{2-}}$	Ref.
C_2H_4	O_2	CO_2	Pt	260–450	3×10^5	55	55	10,26,31
C_2H_6	O_2	CO_2	Pt	270–500	300	20	20	32
					-100	7	—	
CH_4	O_2	CO_2	Pt	600–750	5	70	70	13,33
CO	O_2	CO_2	Pt	300–550	2×10^3	3	2	34
					-500	6	—	
CO	O_2	CO_2	Pt	468–558	1000	5	5	13,35,36
CH_3OH	O_2	H_2CO,CO_2	Pt	300–500	1×10^4	4,15*	3	13,37
C_3H_6	O_2	CO_2	Pt	350–480	-3×10^3	6	—	38
CH_3OH		H_2CO,CO,CH_4	Pt	400–500	-10	3*	—	29,37
C_2H_4	NO	CO,CO_2,N_2,N_2O	Pt	380–500	-50	7	—	39
C_2H_4	O_2	CO_2	Rh	250–400	5×10^4	90	90	28,30,40
H_2	CO_2	CH_4,CO	Rh	300–450	200	3*	2	13
C_3H_6	NO,O_2	N_2,N_2O,CO_2	Rh	250–450	1×10^3	150*	15	41
CO	NO,O_2	N_2,N_2O,CO_2	Rh	250–450	20	20*	20	42
C_2H_4	O_2	CO_2	Pd	290–360	10^4	2	—	43,44
CO	O_2	CO_2	Pd	400–550	1×10^3	2	1	13,29
H_2	CO	$C_xH_y,C_xH_yO_z$	Pd	300–370	10	3*	2	13
H_2S		S_x,H_2	Pt	600–750	—	11	10	13,45
CH_4	O_2	CO_2	Pd	380–440	2×10^3	90	90	43,46
H_2	CO_2	CO	Pd	500–590	-50	10	—	13,29
CO	NO	CO_2,N_2,N_2O	Pd	320–480	-700	3	—	47,48
CO	N_2O	CO_2,N_2	Pd	440	-20	2	—	47
C_2H_4	O_2	C_2H_4O,CO_2	Ag	320–470	300	30*	30	9,49–51
C_3H_6	O_2	C_3H_6O,CO_2	Ag	320–420	300	2*	1	13,52
CH_4	O_2	CO_2,C_2H_4,C_2H_6	Ag	650–850	5	30*	30	13,53
CO	O_2	CO_2	Ag	350–450	20	15	15	13,54
CH_3OH		H_2CO,CO,CH_4	Ag	550–750	-25	6*	—	13,55
CH_3OH	O_2	H_2CO,CO_2	Ag	500	-95	2	—	23
CH_4	O_2	C_2H_4,C_2H_6,C	Ag	700–750	-1.2	8*	—	56,57
CO	O_2	CO_2	Ag-Pd	450–500	30	5	4	58
CH_4	H_2O	CO,CO_2	Ni	600–900	12	2*	—	13,59
CO	O_2	CO_2	Au	450–600	-60	3	—	56,57
CH_4	O_2	CO_2	Au	700–750	-3	3*	—	56,57,60
C_2H_4	O_2	CO_2	IrO_2	350–400	200	6	5	18,61
C_2H_4	O_2	CO_2	RuO_2	240–500	4×10^3	115	115	62

2. EP studies utilizing F^- conductors

Reactants									
Electron Donor (D)	Electron Acceptor (A)	Products	Catalyst	Solid Electrolyte	$T\ (°C)$	$\Lambda_{max}(>0)$ or $\Lambda_{min}(<0)$	$\rho_{max}(>1)$ or $\rho_{min}(<1)$	PI_{F^-}	Ref.
CO	O_2	CO_2	Pt	CaF_2	500–700	200	2.5	1.5	13,63

Table 1 Continued

3. EP studies utilizing mixed conductors

Electron Donor (D)	Electron Acceptor (A)	Products	Catalyst	Solid Electrolyte	T (°C)	$\Lambda_{max}(>0)$ or $\Lambda_{min}(<0)$	$\rho_{max}(>1)$ or $\rho_{min}(<1)$	PI_{F^-}	Ref.
C_2H_4	O_2	CO_2	Pt	TiO_2 (TiO_x^+, O^{2-})	450–600	5×10^3	20	20	64
C_2H_4	O_2	CO_2	Pt	CeO_2 (CeO_x^+, O^{2-})	500	-10^5	3	—	65
C_2H_4	O_2	CO_2	Pt	YZTi10#	400–475	-250	2	—	66
C_3H_6	O_2	CO_2	Pt	YZTi10#	400–500	1000 −1000	2.4	—	66

4. EP studies utilizing Na$^+$ conductors

Electron Donor (D)	Electron Acceptor (A)	Products	Catalyst	Solid Electrolyte	T (°C)	$\Lambda_{max}(>0)$ or $\Lambda_{min}(<0)$	$\rho_{max}(>1)$ or $\rho_{min}(<1)$	PI_{F^-}	Ref.
C_2H_4	O_2	CO_2	Pt	β''-Al_2O_3	180–300	5×10^4	0.25	−30	13,67
CO	O_2	CO_2	Pt	β''-Al_2O_3	300–450	1×10^5 -1×10^5	0.3 8	−30 250	13,68
H_2	C_6H_6	C_6H_{12}	Pt	β''-Al_2O_3	100–150	—	~0	−10	20,69
H_2	C_2H_2	C_2H_4,C_2H_6	Pt	β''-Al_2O_3	100–300	—	—	—	70
C_2H_4	NO	CO_2,N_2,N_2O	Pt	β''-Al_2O_3	280–400	—	∞	500	19
CO	NO	CO_2,N_2,N_2O	Pt	β''-Al_2O_3	320–400	—	13*	200	71–73
C_3H_6	NO	CO_2,N_2,N_2O	Pt	β''-Al_2O_3	375	—	10	—	72–74
H_2	NO	N_2,N_2O	Pt	β''-Al_2O_3	360–400	—	30	6000	75
H_2	C_2H_2,C_2H_4	C_2H_4,C_2H_6	Pd	β''-Al_2O_3	70–100	—	0.13	—	76
C_2H_4	O_2	C_2H_4O,CO_2	Ag	β''-Al_2O_3	240–280	—	—	40	77
CO	O_2	CO_2	Ag	β''-Al_2O_3	360–420	—	2	—	13
C_2H_4	O_2	CO_2	Pt	$Na_3Zr_2Si_2PO_{12}$	430	—	10	300	78

5. EP studies utilizing K$^+$ conductors

Electron Donor (D)	Electron Acceptor (A)	Products	Catalyst	Solid Electrolyte	T (°C)	$\Lambda_{max}(>0)$ or $\Lambda_{min}(<0)$	$\rho_{max}(>1)$ or $\rho_{min}(<1)$	PI_{F^-}	Ref.
NH_3		N_2,H_2	Fc	$K_2YZr(PO_4)_3$	500–700	—	4.5	—	79

6. EP studies utilizing H$^+$ conductors

Electron Donor (D)	Electron Acceptor (A)	Products	Catalyst	Solid Electrolyte	T (°C)	$\Lambda_{max}(>0)$ or $\Lambda_{min}(<0)$	$\rho_{max}(>1)$ or $\rho_{min}(<1)$	PI_{F^-}	Ref.
C_2H_4	O_2	CO_2	Pt	$CaZr_{0.9}In_{0.1}O_{3-a}$	385–470	-3×10^4	5	—	80
H_2	N_2	NH_3	Fc	$CaZr_{0.9}In_{0.1}O_{3-a}$	440	6	∞	6	81
NH_3		N_2,H_2	Fe	$CaZr_{0.9}In_{0.1}O_{3-a}$	530–600	150	3.6	—	79
CH_4		C_2H_6,C_2H_4	Ag	$SrCe_{0.95}Yb_{0.05}O_3$	750	—	8*	10	13,82
H_2	C_2H_4	C_2H_6	Ni	$CsHSO_4$	150–170	300	2	12	13,17
H_2	O_2	H_2O	Pt	Nafion	25	20	6	5	83
1-C_4H_8		C_4H_{10}, 2-C_4H_8 (cis,trans)	Pd	Nafion	70	−28	40*	—	21

Table 1 Continued

7. EP studies utilizing aqueous alkaline solutions

	Reactants								
Electron Donor (D)	Electron Acceptor (A)	Products	Catalyst	Solid Electrolyte	T (°C)	$\Lambda_{max}(>0)$ or $\Lambda_{min}(<0)_x$	$\rho_{max}(>1)$ or $\rho_{min}(<1)$	PI_{OH}	Ref.
H_2	O_2	H_2O	Pt	H_2O – 0.1N KOH	25–50	20	6	20	13,84,85

8. EP studies utilizing molten salts

	Reactants							
Electron Donor (D)	Electron Acceptor (A)	Products	Catalyst	Solid Electrolyte	T (°C)	$\Lambda_{max}(>0)$ or $\Lambda_{min}(<0)_x$	$\rho_{max}(>1)$ or $\rho_{min}(<1)$	Ref.
SO_2	O_2	SO_3	Pt	V_2O_5-K_2SO_4	350–450	–100	6	86

*Change in product selectivity observed.
#4.5 mol% Y_2O_3—10 mol% TiO_2—85.5 mol% ZrO_2.

composition, the effect of the coverage of the promoting species on the catalytic rate in an efficient and systematic manner.

Another important operational advantage of electrochemical promotion is that if the promoting species lifetime, τ_{PR}, on the catalyst surface is too short for any realistic classical promotion application (i.e., if τ_{PR} is of the order of h or min or even s), one can still use it by continuously replenishing it on the catalyst surface at a rate I/nF, where n is the ion charge. In this case, the current, I, must be chosen to satisfy the equation

$$I/nF = N_G/\tau_{PR}$$

where N_G, in mol, is the catalyst surface area. In this way some extremely effective but short-lived promoting species, such as O^{2-} originating from Y_2O_3-stabilized-ZrO_2 (YSZ) or TiO_2 [13,14], and which were totally unknown from classical promotion studies [6], can now be utilized in a continuous and efficient manner.

Another important parameter in electrochemical promotion studies is the characteristic rate relaxation time, τ, needed for the catalytic rate to reach steady state upon imposition of a constant current (galvanostatic transient). As one would expect and as experiment has clearly shown [13,14], τ is always of the order of $2FN_G/I$ (Figure 2):

$$\tau \approx 2FN_G/I \qquad (10)$$

The parameter $2FN_G/I$ expresses the time required to form a monolayer of O^{2-} on a catalyst surface with N_G sites when O^{2-} is supplied at a rate $I/2F$, as is the case in electrochemical promotion experiments. Equation (10) was the first evidence to suggest that electrochemical promotion is due to the electrochemically controlled migration (backspillover) of promoting O^{2-} species on the metal catalyst surface

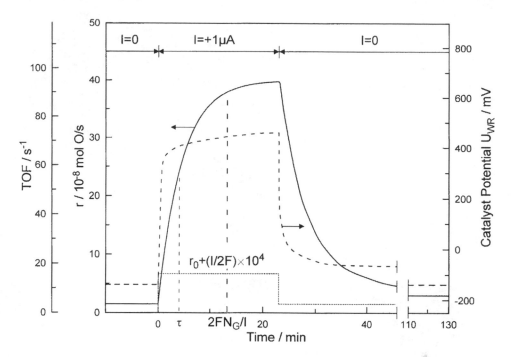

Figure 2 Rate and catalyst potential response to step changes in applied current during C_2H_4 oxidation on Pt deposited on YSZ, an O^{2-} conductor. $T = 370\,°C$, $p_{O_2} = 4.6\,kPa$, $p_{C_2H_4} = 0.36\,kPa$. The catalytic rate increase, Δr, is 25 times larger than the rate before current application, r_0, and 74,000 times larger than the rate $I/2F$ [26] of O^{2-} supply to the catalyst. N_G is the Pt catalyst surface area, in mol Pt, and TOF is the catalytic turnover frequency (mol O reacting per surface Pt mol per s). (Reprinted with permission from Academic Press.)

[26], a view that has been confirmed by numerous surface spectroscopic techniques [13,14,43,87–95].

Metal–Support Interactions

In commercial catalysts the catalytically active phase is usually dispersed on a highly porous and of high ($>100\,m^2/g$) surface-area support. This high-surface-area support, also frequently termed "carrier," has pores as small as $10\,\text{Å}$ and allows for the use of the active phase in a highly dispersed form. The pores are termed "macropores" when their diameter, d, is larger than $200\,\text{Å}$. When d is smaller than $20\,\text{Å}$, the pore is termed "micropore." IUPAC recommends the term "mesopore" when d is between 20 and $200\,\text{Å}$. In most supported commercial catalysts the size of the supported crystallites of the active phase is of the order $10\text{–}50\,\text{Å}$ (Figure 3). This implies that each crystallite consists of few, typically 10 to 1000, atoms. It also implies that a significant portion of the active-phase atoms are on the gas-exposed surface of the crystallites and are thus catalytically active. This portion (percentage) is termed catalyst dispersion, D_c, and is defined from

$$D_c = \left(\frac{\text{number of surface catalyst atoms}}{\text{total number of catalyst atoms}} \right) 100 \qquad (11)$$

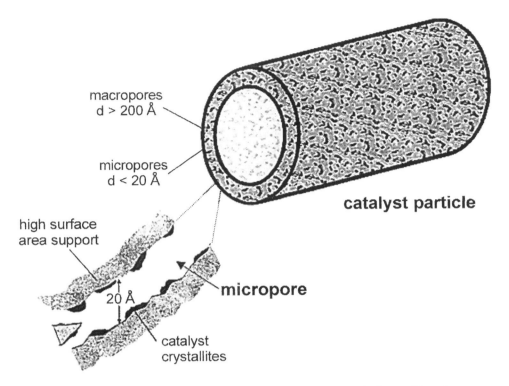

Figure 3 Schematic of a commercial supported catalyst pellet and of one of its micropores.

Since the catalytically active phase is frequently quite expensive (e.g., noble metals), it is clear that it is in principle advantageous to prepare catalysts with high, approaching 100%, catalyst dispersion, D_c. This can usually be accomplished without much difficulty by impregnating the porous carrier with an aqueous solution of a soluble compound (acid or salt) of the active metal followed by drying, calcination, and reduction [1].

Successful and reproducible preparation of highly dispersed catalysts crucially depends on the state of the carrier surface and on the concentration and pH of the impregnating solution. It is an art and a science for which several good books and reviews exist [1–5].

In the early days of catalysis the porous high-surface-area support was usually thought to be inert. It soon became obvious, however, that the catalytic activity, or turnover frequency, of a catalytic reaction on a given active phase is quite often seriously affected by both the crystallite size and the nature of the support.

The former phenomenon is usual referred to as *particle-size effect* and is pronounced for structure-sensitive reactions [1,2], i.e., catalytic reactions where the rate and/or selectivity is significantly different from one crystallographic plane to another. Structure-sensitive reactions (e.g., isomerizations) frequently occur on catalytic sites consisting of an "ensemble" of surface atoms with specific geometry. It is thus reasonable to expect that as the active-phase crystallite size decreases, there will be a different distribution of crystallographic planes on the catalyst surface, with the possible disappearance of "ensemble" sites, so that both the catalyst activity and selectivity will be significantly affected. On the other hand, *structure-insensitive*, also termed *"facile"* [1,2], reactions (e.g., most hydrogenations, some oxidations) are hardly affected by particle-size effects.

The second phenomenon, i.e., the change in catalytic activity or selectivity of the active phase with varying catalyst support, is usually termed *metal–support interaction*. It manifests itself even when the active phase has the same dispersion or average crystallite size on different supports. Metal–support interactions can influence in a very pronounced way the catalytic and chemisorptive properties of metal and metal oxide catalysts. Typical and spectacular examples are

1. The phenomenon of strong metal–support interactions (SMSI) discovered by Tauster et al. [96] which attracted worldwide attention for many years [7].
2. The effect of dopant-induced-metal–support interactions (DIMSI) studied for years by Verykios and co-workers [97].
3. The relatively recent discovery of the highly active Au/SnO$_2$ oxidation catalysts by Haruta and co-workers [98–100].

In all these cases the support has a dramatic effect on the activity and selectivity of the active phase. In classical terminology [1] all these are Schwab effects *of the second kind*, where an oxide affects the properties of a metal. Schwab effects *of the first kind*, where a metal affects the catalytic properties of a catalytic oxide, are less common although in the case of the Au/SnO$_2$ oxidation catalysts [98,99] it appears that most of the catalytic action takes place near the metal-oxide-gas three-phase boundaries.

The nature of metal–support interactions has been the focal point of extensive research and dispute, particularly after the discovery by Tauster et al. [96] of the phenomenon of strong metal–support interactions (SMSI). Although particle-size effects and electronic interactions between the metal particles and the support have been known for years to play a role, the SMSI effect was finally shown to be due to migration of ionic species from the support onto the catalyst particle surface ("decoration") [7]. There have been some interesting recent experimental and theoretical advances [101–108], but a thorough understanding of metal–support interactions is one of the greatest challenges in heterogeneous catalysis.

Although SiO$_2$ and γ-Al$_2$O$_3$ are the most common high-surface-area industrial catalyst supports (considered in general to give rise to weak metal–support interactions), in recent years there has been an increasing tendency to replace these supports for numerous catalytic applications, mostly oxidations, but also NO reduction, with TiO$_2$ or CeO$_2$ or ZrO$_2$-based porous supports [109]. Little information exists in the open literature as to why this gradual substitution is taking place [106,110], but it is a common understanding that these supports, generally believed to lead to stronger metal–support interactions, result in increased activity [106,110], selectivity, and useful lifetime of the metal particles deposited on them.

The extent of metal–support interactions for a given catalytic system can again be quantified by defining a metal–support interactions rate enhancement ratio, ρ_{MSI}, from

$$\rho_{MSI} = r/r_0 \tag{12}$$

where r_0 is the catalytic rate per unit surface area of the active catalyst in its unsupported form and r is the same catalytic rate per unit surface area of the supported catalyst.

Clearly, ρ_{MSI} is expected to vary from one support to another even for facile reactions, while for structure-sensitive reactions ρ_{MSI} can be reasonably expected to be a function not only of the support but also of the particle size of the active phase.

Figure 4 shows measured ρ_{MSI} values for the case of C_2H_4 oxidation on finely dispersed (0.5 wt. %) Rh on five different supports: ρ_{MSI} values exceeding 100 are obtained for this interesting system. As we shall see, the sequence of increasing activity of the five supports coincides with the sequence of their absolute potentials or work functions.

Spillover-Backspillover Phenomena

Main Features

The effect of spillover plays an important role in heterogeneous catalysis and was extensively studied during recent years. It was first noticed in the 1950s by Kuriacose [111]. Work in this area has been reviewed by Teichner [112] and by Conner et al. [113].

The spillover effect can be described as the mobility of sorbed species from one phase on which they easily adsorb (donor) to another phase where they do not directly adsorb (acceptor). In this way a seemingly inert material can acquire catalytic activity. In some cases, the acceptor can remain active even after separation from the donor. Also, quite often, as shown by Delmon and co-workers [114–116], simple mechanical mixing of the donor and acceptor phases is sufficient for spillover to occur and influence catalytic kinetics leading to a "remote control mechanism," a term first introduced by Delmon [114]. Spillover may lead not only to an improvement of catalytic activity and selectivity, but also to an increase in lifetime and regenerability of catalysts.

The effect of spillover was observed for different species such as H [117], O [118], N [119], NO [113], or CO [118]. Most of the research has been carried out with hydrogen spillover.

Bond [118] reported one of the first examples of spillover of nonhydrogen species. They observed a spillover of O and CO from Pd onto SnO_2 during the oxidation of CO. On pure SnO_2 the rate-limiting step is the oxidation of CO by SnO_2. Adding Pd to the SnO_2 leads to a change in the rate-limiting step toward reoxidation of the SnO_2 due to spillover of O and CO. The rate of CO oxidation was greatly increased by adding Pd to SnO_2.

The simplest example of oxygen spillover is found in the adsorption of oxygen on carbon. The spillover oxygen migrates from the basal carbon (donor) to carbon atoms exposed at steps between layers of the graphite surface, where it reacts with the edge carbons (acceptor) [120]. In this case the donor and acceptor phase consist of the same material with different surface properties.

Examples of reverse spillover (or backspillover) are the dehydrogenation of isopentane and cyclohexane on active carbon. Deposition of a transition metal on the active carbon accelerates the recombination of H to H_2 due to a reverse spillover or backspillover effect [121].

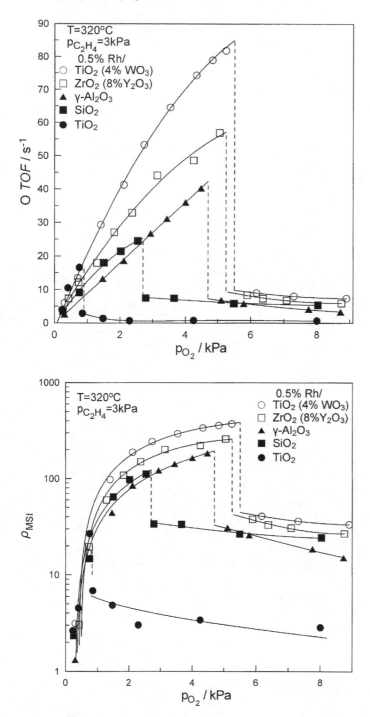

Figure 4   (top) Effect of p_{O_2} on the rate (TOF) of C_2H_4 oxidation on Rh supported on five supports of increasing Φ. Catalyst loading 0.5 wt. % [137,152]. (bottom) Effect of p_{O_2} on the metal–support interactions rate enhancement ratio, ρ_{MSI}. (Reprinted with permission from Elsevier Science.)

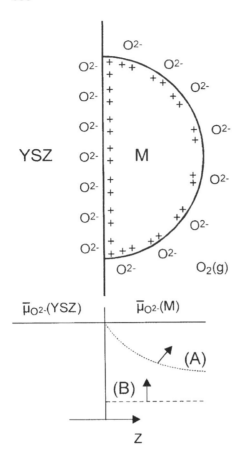

Figure 5 Spatial variation of the electrochemical potential, $\bar{\mu}_{O^{2-}}$, of O^{2-} in YSZ and on a metal electrode surface under conditions of spillover (broken lines A and B) and when equilibrium has been established. In case (A) surface diffusion on the metal surface is rate-limiting while in case (B) the backspillover process is controlled by the rate, I/nF, of generation of the backspillover species at the three-phase boundaries. This is the case most frequently encountered in electrochemical promotion (NEMCA) experiments. (From Refs. 13 and 14.)

Spillover-Backspillover in Metal–Solid Electrolyte Systems: Thermodynamics

We start from Eq. (13), which, under open-circuit conditions, is at equilibrium at the three-phase boundaries (tpb) metal (M)–gas–YSZ (Figure 5):

$$1/2O_2(g) + 2e^-(M) \leftrightarrows O^{2-}(YSZ) \tag{13}$$

The equilibrium condition is

$$1/2\mu_{O_2}(g) + 2\bar{\mu}(M) = \bar{\mu}_{O^{2-}}(YSZ) \tag{14}$$

where $\mu_{O_2}(g)$ is the chemical potential of O_2 in the gas phase (assumed constant), $\bar{\mu}(M)$ is the electrochemical potential (or Fermi level) [13,14] of electrons in the metal (which is also constant over the entire metal electrode), and $\bar{\mu}_{O^{2-}}(YSZ)$ is the electrochemical potential of O^{2-} in the YSZ.

We then note that the equilibrium condition for reaction (13) now taking place not only at the tpb (three-phase boundaries), but over the entire gas-exposed metal electrode surface, is very similar to Eq. (14), i.e.,

$$1/2\mu_{O_2}(g) + 2\bar{\mu}(M) = \bar{\mu}_{O^{2-}}(M) \tag{15}$$

where now $\bar{\mu}_{O^{2-}}(M)$ is the electrochemical potential of O^{2-} on the gas-exposed metal electrode surface. Thus at equilibrium, i.e., when spillover-backspillover is not kinetically frozen, one has

$$\bar{\mu}_{O^{2-}}(M) = \bar{\mu}_{O^{2-}}(YSZ) \tag{16}$$

Thus the driving force for O^{2-} backspillover from YSZ to the gas-exposed, i.e., catalytically active, electrode surface exists and equals $\bar{\mu}_{O^{2-}}(YSZ) - \bar{\mu}_{O^{2-}}(M)$. It vanishes only when O^{2-} backspillover has taken place and established the "effective" double layer over the entire metal–gas interface (Figure 5).

Kinetics of Backspillover

The kinetics of ion backspillover in metal–solid electrolyte systems depends on two factors: on the rate, I/nF, of overall neutral backspillover species formation at the tpb and on their surface diffusivity, D_s, on the metal surface. Experiment has shown that the rate of electrochemically controlled ion backspillover is normally limited by I/nF, i.e., the slow step is their creation at the tpb [13]. Surface diffusion is usually fast. For the case of Pt electrodes, reliable surface O diffusivity data exist for the Pt(111) and Pt(110) surfaces, obtained by Gomer and Lewis several years ago [122], and thus a conservative estimate for the surface diffusivity of O^{2-} can be made:

$$D_s = \delta^2 v \exp(\Delta S/R) \exp(-E/RT) \tag{17}$$

with $\delta = 3\,\text{Å}$, $\Delta S = 17\,\text{cal/mol K}$, $v = 10^{12}\,\text{s}^{-1}$, $E = 34.1\,\text{kcal/mol}$. From this expression one computes that D_s is at least $4 \times 10^{-11}\,\text{cm}^2/\text{s}$ at $400\,°\text{C}$ and thus an O^{2-} ion can move at least $1\,\mu\text{m}$ per s on a Pt(111) or Pt(110) surface. Therefore, ion backspillover from solid electrolytes onto an electrode surface is not only thermodynamically feasible, but can also be quite fast on the electrode surface.

19.2 CLASSICAL PROMOTION

19.2.1 Some Examples

Promoters play a key role in heterogeneous catalysis. Their use is of paramount importance in the design of successful commercial catalysts. As already mentioned, they can be divided into structural promoters and electronic promoters. Here we deal with the latter case, i.e., with promoters that affect the chemisorptive and catalytic properties of the active phase itself.

A classical example of promotion is the use of alkalis (K) on Fe for the ammonia synthesis reaction. Coadsorbed potassium (in the form of K_2O) significantly enhances the dissociative adsorption of N_2 on the Fe surface, which is the crucial and rate-limiting step in the ammonia synthesis (Figure 6).

There is a very rich literature, a comprehensive book [6], and a recent chapter [14] on the role of promoters in heterogeneous catalysis. The vast majority of studies

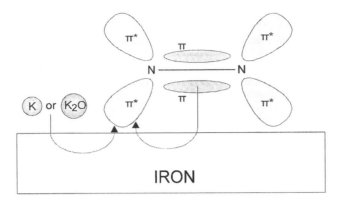

Figure 6   Spatial distribution of the main orbitals of N_2 involved in molecular chemisorption on iron promoted by potassium (K or K_2O). Arrows indicate the direction of transfer of electron density. (From Ref. 155.)

refers to the adsorption of promoters and to the effect of promoters on the chemisorptive state of coadsorbed species, usually on well-characterized single-crystal surfaces. A nice example is shown in Figure 7 for the case [123] of coadsorption of CO and K on Pt(100). Increasing K dosing on the Pt(100) surface causes CO desorption at high temperatures, i.e., the Pt–carbon monoxide chemisorptive bond is strengthened. At higher K coverages a new chemisorption state appears corresponding to dissociatively chemisorbed CO. The Pt–carbon bond has been strengthened to the point that the CO bond has been broken [123].

19.2.2   Work Function of Surfaces, Electropositive and Electronegative Promoters

The work function, Φ, of solid surface, in eV/atom, is the minimum energy an electron must have in order to escape from the Fermi level, E_F, of that solid through that surface, when the surface is electrically neutral, to a distance of a few μm outside the surface so that image charge forces are negligible:

$$\Phi = -E_F(= -\bar{\mu}) \tag{18}$$

where $\bar{\mu}$ is the electrochemical potential of electrons in the solid. The latter always equals ($\bar{\mu} = E_F$) the Fermi level of the solid [13,14,124–127].

When the solid surface under consideration carries a *net* charge, then the work function Φ and electrochemical potential $\bar{\mu}$ (or Fermi level E_F) are related via (Figure 8)

$$-\bar{\mu} = \Phi + e\Psi \tag{19}$$

where Ψ is the outer (Volta) potential of the surface. The energy $e\Psi$ is known to surface physicists as the "vacuum level." Equation (19) presents the surface-science approach of counting the energy difference between the zero energy state of electrons [always taken in this chapter as the energy of an electron at its ground state and

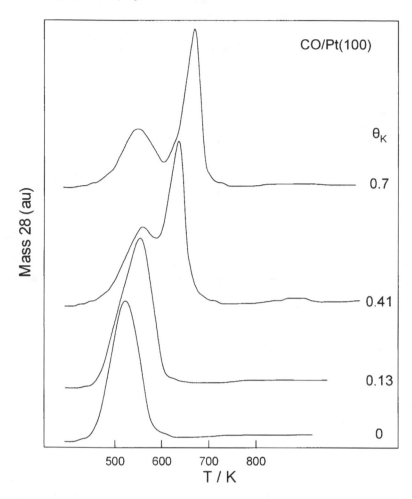

Figure 7 Thermal desorption spectra of carbon monoxide, measured mass spectrometrically at mass 28 (atomic units, a.u.), on a platinum (100) surface upon which potassium has been pre-adsorbed to a surface coverage of θ_K (From Ref. 123. Reprinted with permission from Elsevier Science.)

"infinite" distance from any solid (Figure 8)]. There is a second, electrochemical, way of counting this energy difference:

$$\bar{\mu} = \mu + (-\text{e})\varphi \tag{20}$$

where μ is the chemical potential of electrons in the metal and the inner or Galvani potential, φ, is the electrostatic potential of electrons in the solid.

The surface-science approach [Eq. (19)] has the important advantage that both Φ and Ψ are measurable quantities. This is not the case for the electrochemical approach [Eq. (20)] since neither μ nor φ are absolutely measurable quantities. Only changes in φ are measurable [13].

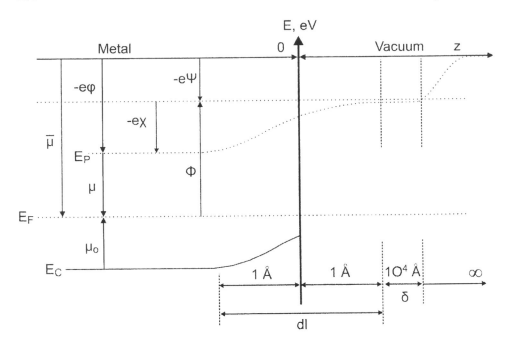

Figure 8 Schematic representation of the definitions of work function Φ, chemical potential of electrons μ, electrochemical potential of electrons or Fermi level $\bar{\mu} = E_F$, surface potential χ, Galvani (or inner) potential φ, Volta (or outer) potential Ψ, Fermi energy μ_o, and of the variation in the mean effective potential energy E_P of electrons in the vicinity of a metal–vacuum interface according to the jellium model. E_C is the energy at the bottom of the conduction band and dl denotes the double layer at the metal–vacuum interface.

The quantities $\bar{\mu}, \mu$, and φ are bulk properties of the metal [4,124–127]. The quantities Φ, and of course Ψ, are surface properties that can vary on metal surfaces from one crystallographic plane to the other. Such variations are typically on the order of 0.1 eV but can be as high as 0.5 eV. The measured work function, Φ, of a polycrystalline metal is an average of the Φ values for different crystallographic planes [13,14].

The work function of clean metal surfaces, which we denote by Φ_0, varies between 2 eV for alkalis up to 5.5 eV for transition metals such as Pt. In general, it increases as one moves to the right on the Periodic Table [128], but deviations exist (Figure 9).

When atoms or molecules (e.g., promoters or reactant adsorbates) adsorb on a metal surface, they change its work function. Electronegative (electron acceptor) adsorbates such as O or Cl can increase the Φ of a metal surface up to 1 eV. Electropositive (electron donor) adsorbates such as H or, particularly, alkalis can decrease the Φ of a metal surface up to 3 eV.

Thus, depending on the change a promoter induces on the work function, Φ, of a catalyst surface, a major distinction can be made between *electropositive* (*electron donor*) and *electronegative* (*electron acceptor*) promoters.

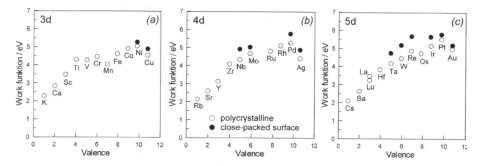

Figure 9 Experimental work function values, Φ_0, for the 3d, 4d, and 5d series including the alkali, alkaline-earth, and noble metals for polycrystalline surfaces (open circles) and for single-crystal surfaces (filled circles). (From Ref. 128. Reprinted with permission from the American Physical Society.)

In the same way catalytic reactant adsorbates, and thus catalytic reactants, can be divided into *electron donor* and *electron acceptor* adsorbates or reactants. Hydrocarbons, and in particular unsaturated ones, always behave as electron donors (D), while O, Cl, and in most cases CO and NO, behave as electron acceptors (A). Adsorbates, such as H, CO, and NO, which, depending on the catalyst surface and the nature of the other coadsorbates, can change their behavior between electron donor and electron acceptor are called *amphoteric* adsorbates [13,14,129,130].

The variation in Φ of a metal with the coverage j of an adsorbate, promoter, or adsorbed reactant is described by the Helmholtz equation:

$$\Delta\Phi = \frac{eN_M}{\varepsilon_0}\Delta P_j\theta_j \tag{21}$$

where e is the electron charge ($1.6 \ 10^{-19}$C), N_M is the surface atom density (atom/m^2), ε_0 is the electric permeability of vacuum ($\varepsilon_0 = 8.85 \ 10^{-12} \, C^2/Jm$), and P_j(Cm) is the dipole moment of the adsorbate j in the adsorbed state. Typically P_j is of the order of 10^{-29} Cm or 3D (Debye). The Debye unit, D, equals $3.36 \ 10^{-30}$ Cm. The dipole moments of adsorbates are by convention taken positive in this chapter when the adsorbate dipole vector, $\tilde{P}_j$, is pointing to the vacuum (electronegative adsorbates, e.g., O$^{\delta-}$, Cl$^{\delta-}$) and negative when $\tilde{P}_j$ is pointing to the surface (electropositive adsorbates, e.g., Na$^{\delta+}$).

19.2.3 Basic Promoter Selection Criteria

In selecting a promoter for classical (chemical) promotion of a catalytic reaction two necessary, but not sufficient, criteria must be satisfied [129,130]:

1. The promoter must not be consumed in the catalytic reaction and must not be deactivated by the reactants and products.
2. The promoter must have a large absolute dipole moment value, $|P_j|$, so that large variations, $\Delta\Phi$, in the catalyst work function, Φ, can be induced by relatively small coverages, θ_j, of the promoter [Eq. (21)]. The latter is

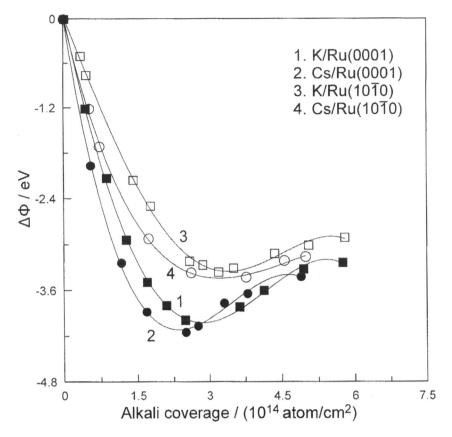

Figure 10   Work function changes, $\Delta\Phi$, as a function of K and Cs coverages for Ru(0001) (1 and 2) and for Ru($10\bar{1}0$) (3 and 4). (From Ref. 131. Reprinted with permission from Springer-Verlag GmbH & Co.)

necessary in order to minimize site-blocking effects caused by high promoter coverage.

Alkali metals satisfy both of these criteria [131,132] as can be seen in Figures 10 and 11. Due to their absolutely large dipole moments, $|P_{alk}| \approx 5$–$10\,D$, alkali coverages, θ_{alk}, of the order of 0.1 suffice to decrease the catalyst work function Φ by more than 2.5 eV.

Which types of catalytic reactions can be promoted (accelerated) by such a pronounced decrease in catalyst work function? They are called *electrophilic*, and we discuss them in the next section, together with their counterpart, *electrophobic* reactions, which are promoted by increasing catalyst work function.

At this point it is important to note that even when alkali promoters react with the catalytic reactants and products on a catalyst surface forming oxides, hydroxides, carbonates, or nitrates, they do not lose their promoting ability [13,88]. This is because the large dipole moment of the alkali cation is maintained when such surface compounds are formed, and thus the alkali cation can still exert its promoting action.

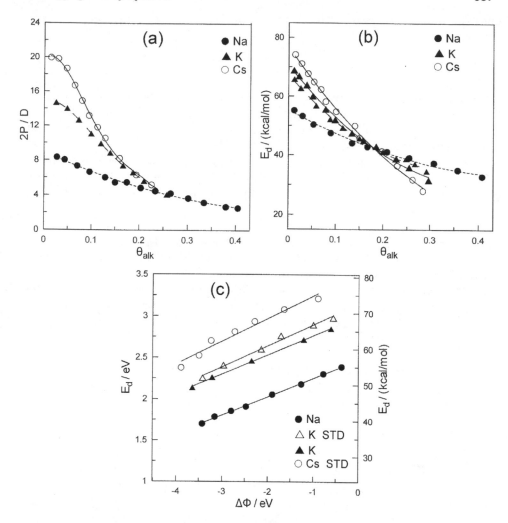

Figure 11 Effect of alkali coverage on (a) the alkali ad-atom dipole moment and alkali desorption energy (b) for Na, K and Cs adsorbed on Ru (0001) and corresponding effect of work function change $\Delta\Phi$ on the alkali desorption energy (c). (From Ref. 132. Reprinted with permission from Elsevier Science.)

19.2.4 Effect of Promoters on Chemisorption

What is the mechanism of the promoting action? To start discussing this, we can first focus on Figure 11 and observe that as the alkali coverage, θ_{alk}, is increased, the alkali dipole moment decreases and so does the alkali chemisorptive bond strength, E_d. This weakening in the alkali chemisorptive bond strength is, to a large extent, due to *repulsive lateral interactions* between the adsorbed, partly positively charged, alkali atoms.

In fact, as can be seen in Figure 11, upon crossplotting E_d versus the work function change, $\Delta\Phi$, induced by the alkali ad-atoms, a straight line with a positive slope is obtained. For nonactivated adsorption, as is the case here, this implies a linear decrease in the heat of chemisorption of the alkali atoms $|\Delta H_{ad}|(= E_d)$ with decreasing Φ:

$$\Delta|\Delta H_{ad}| = \alpha_H \Delta\Phi \tag{22}$$

where the parameter α_H has in this case a value near 0.25. This equation is of rather broad significance as it approximates, in most cases with reasonable accuracy [14,130], the observed variation in heats of adsorption of adsorbates (promoters but also catalytic reactants and products) with varying Φ.

Another example of Eq. (22) is shown in Figure 12, which depicts the effect of $\Delta\Phi$, induced by varying the coverage of Na [133,134], on the heat of chemisorption of CO, which, in the presence of alkalis, always behaves as an electron acceptor. Here α_H is negative and equals -0.2.

Equation (22) is in good agreement with recent rigorous quantum mechanical calculations using metal clusters [135,136] and can also be derived [129], as we will see in the next section, by simply taking into account the electrostatic ("through the vacuum") lateral interactions in the adsorbed layer.

The parameter α_H is positive for electropositive (electron donor) adsorbates and negative for electronegative (electron acceptor) adsorbates. Even when deviations from linearity exist, the main features of Eq. (22) remain valid and

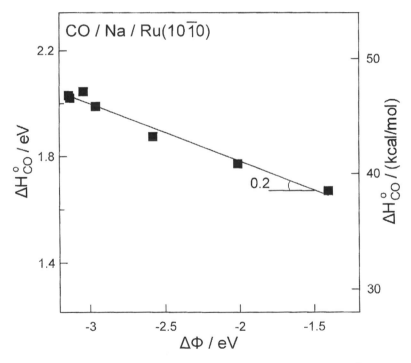

Figure 12 Dependence of the initial heats of CO adsorption. ΔH^0_{CO}, on the alkali coverage, as estimated from the CO TPD spectra at very low CO coverages assuming invariable frequency factor (from Refs. 133 and 134) (a) and on the corresponding work function change $\Delta\Phi$ (from Refs. 133 and 134) (b). (Reprinted with permission from Elsevier Science.)

form the *basis* for understanding the *main kinetic features of classical and electrochemical promotion* [13,14,130]:

> Increasing catalyst work function causes an increase in the heat of adsorption (thus chemisorptive bond strength) of electropositive (electron donor) adsorbates and a decrease in the heat of adsorption (thus chemisorptive bond strength) of electronegative (electron acceptor) adsorbates.

As a corollary of the above general chemisorption promotional rule, the following two rules are immediately derived [14,130]:

> Rule C1: Electropositive adsorbates strengthen the chemisorptive bond of electron acceptor (electronegative) adsorbates and weaken the chemisorptive bond of electron donor (electropositive) adsorbates.
> Rule C2: Electronegative adsorbates weaken the chemisorptive bond of electron acceptor (electronegative) adsorbates and strengthen the chemisorptive bond of electron donor (electropositive) adsorbates.

These rules must be supplemented by the following two "amphotheric" rules [14,130] which supplement the definition of electron acceptor and electron donor adsorbates.

> Rule C3: In the presence of a strong electron donor (electropositive) adsorbate (e.g., K, Na), a weaker electron donor [e.g., NO on Pt(111)] behaves as an electron acceptor.
> Rule C4: In the presence of a strong electron acceptor (electronegative) adsorbate (e.g., O) a weaker electron acceptor [e.g., CO on Ni(111)] behaves as an electron donor.

Two molecular mechanisms lead to the above rules:

> Direct electrostatic ("through the vacuum") dipole attraction or repulsion, which, in the case of attraction, may lead even to surface compound formation.
> Indirect ("through the metal") interaction due to the redistribution of electrons in the metal. In this case an electropositive promoter decreases the work function of the surface and this in turn weakens the chemisorptive bond of electropositive (electron donor) adsorbates and strengthens the chemisorptive bond of electronegative (electron acceptor) adsorbates.

The extent of the contribution of each of these two mechanisms varies from one system to the other as recent quantum mechanical calculations have shown [135,136]. In either case, however, linear variations are often obtained in the change in heat of adsorption versus the change in the work function, with slopes on the order of ± 1, in good agreement with experiment [14,129].

19.2.5 Promotional Kinetics and Rules

Upon varying the coverage of a promoter on a catalyst surface, one is varying the catalyst work function Φ (in the negative direction when the promoter is electropositive and in the positive direction when the promoter is electronegative). As one might expect, such a variation in promoter coverage leads to four main types of catalytic rate, r, versus Φ behavior (Figure 13) [14,129,130] where p_A and p_D denote

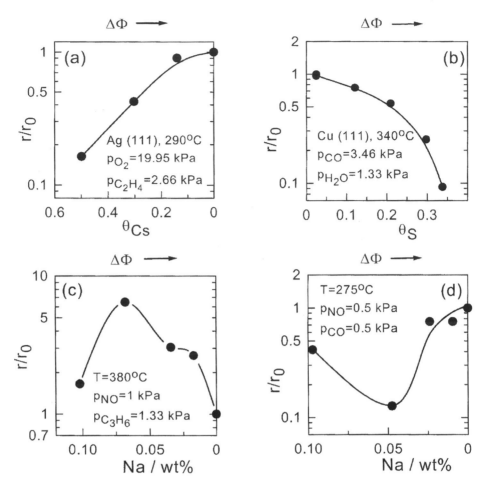

Figure 13 Examples for the four types of global classical promotion behavior. Work function increases with the *x*-axis. (a) Steady-state (low conversion) rates of ethylene oxide (EtO) and CO_2 production from a mixture of 20 torr of ethylene and 150 torr of O_2 for various Cs predosed coverages on Ag(111) at 563 K (from Ref. 156). (b) Rate of water-gas shift reaction over Cu(111) as a function of sulphur coverage at 612 K, 26 Torr CO, and 10 Torr H_2O (from Ref. 157). (c) Effect of sodium loading on NO reduction to N_2 by C_3H_6 on Pd supported on YSZ at $T = 380\,°C$ (from Ref. 158). (d) Effect of sodium loading on the rate of NO reduction by CO on Na-promoted 0.5 wt. % Rh supported on TiO_2 (4% WO_3). (From Ref. 159.)

the partial pressures of the electron acceptor (A) and electron donor (D) reactants:

1. Purely electrophobic, $(\partial r/\partial \Phi)_{p_A, p_D} > 0$
2. Purely electrophilic, $(\partial r/\partial \Phi)_{p_A, p_D} < 0$
3. Volcano type (*r* exhibits a maximum at a value Φ_M)
4. Inverted volcano-type (*r* exhibits a minimum at a value Φ_M)

Can we predict a priori which one of the above four types of global *r* versus Φ behavior a given catalytic reaction will exhibit? The answer, as shown recently

[14,129,130], is yes, provided we know the kinetics (i.e., the r versus p_A and p_D dependence) on the unpromoted surface!

The corresponding rules are valid both for classical (chemical) and electrochemical promotion and are intimately related to the previously discussed chemisorption promotional rules. Their extraction became possible due to easy continuous and systematic experimental Φ variation, which electrochemical promotion offers, but we present them in summary here since, as already noted, they are valid both for classical and for electrochemical promotion [14,129,130].

A summary of the rules is the following:

G1. Purely electrophobic behavior is obtained when A is very strongly adsorbed on the catalyst surface.

G2. Purely electrophilic behavior is obtained when D is very strongly adsorbed on the catalyst surface.

G3. Volcano-type behavior is obtained when both A and D are strongly adsorbed on the catalyst surface.

G4. Inverted volcano-type behavior is obtained when both A and D are weakly adsorbed on the catalyst surface.

More than 60 examples of the above rules can be found in the next section. These rules enable one, in a straightforward manner, to formulate the following three practical rules for promoter selection with respect to rate maximization [14,129].

Rule P1: If a catalyst surface is predominantly covered by an electron acceptor adsorbate, then an electron acceptor (electronegative) promoter is to be recommended.

Rule P2: If a catalyst surface is predominantly covered by an electron donor adsorbate, then an electron donor (electropositive) promoter is to be recommended.

Rule P3: If a catalyst surface has very low coverages of both electron acceptor and electron donor adsorbates, then both an electron acceptor and electron donor promoter will enhance the rate.

The above practical promotional rules are applicable for modest (e.g., < 0.2) coverages of the promoting species so that site-blocking by the promoter does not become the dominant factor limiting the catalytic rate [14,129].

19.3 ELECTROCHEMICAL PROMOTION

19.3.1 Similarities with Classical Promotion and Metal–Support Interactions

Electrochemical promotion bears fundamental similarities with classical (chemical) promotion. Once the promoting species ($O^{\delta-}$ in the case of O^{2-}-conducting solid electrolytes, $Na^{\delta+}$ in the case of Na^+-conducting solid electrolytes) have been introduced electrochemically at the catalyst–gas interface (via current or potential application and concomitant ion backspillover from the solid electrolyte to the metal–gas interface) then they act as classical promoters. In fact, as shown in another chapter of this book, where Lambert and co-workers review electrochemical promotion with cationic conductors, elegant use of XPS has shown that sodium introduced on Pt surfaces from the gas phase (classical promotion) is indistinguishable from sodium introduced on the same Pt surface via electrical current application between the Pt catalyst and a counterelectrode deposited on the same $\beta'' \text{-} Al_2O_3$ component (electrochemical promotion) [88]. Thus it is to be well expected that *classical and electrochemical promotion obey the same rules* regarding the catalytic rate dependence on catalyst work function [14,129]. Classical and electrochemical promotion are *functionally identical* [14,129] (Table 2).

There exist, however, important *operational differences* between classical and electrochemical promotion, as already discussed in the introduction:

1. The ability to *control in situ, i.e., during catalyst operation, the coverage of the promoting species* on the catalyst surface.
2. The ability to *continuously replenish the promoting species* on the catalyst surface, allowing thus for the use of short-lived (10–10^3 s) but extremely effective promoters such as $O^{\delta-}$ (most likely O^{2-}), which is 10–10^5 times less reactive than normally chemisorbed atomic oxygen and which was totally unknown from classical promotion studies [14,129]. This is because

Table 2 Relationship Between Classical (Chemical) and Electrochemical Promotion

	Chemical Promotion	Electrochemical Promotion
Control variables	(Initial) promoter coverage θ_P	Catalyst potential U_{WR} or Fermi level E_F ($e\Delta U_{WR} = -\Delta E_F$) Rate, I/nF, of promoter supply to catalyst surface
Effect	Change in catalytic rate r Change in work function Φ Change in adsorption enthalpies $|\Delta H_j|$	
Action	Addition of electronegative promoter	Increase in U_{WR} Decrease in E_F
Result	$\Delta r > 0$: Electrophobic reactions $\Delta r < 0$: Electrophilic reactions $\Delta\Phi > 0$ $\Delta|\Delta H_j| > 0$: Electron donor adsorbates (D) $\Delta|\Delta H_j| < 0$: Electron acceptor adsorbates (A)	
Action	Addition of electropositive promoter	Decrease in U_{WR} Increase in E_F
Result	$\Delta r < 0$: Electrophobic reactions $\Delta r > 0$: Electrophilic reactions $\Delta\Phi < 0$ $\Delta|\Delta H_j| < 0$: Electron donor adsorbates (D) $\Delta|\Delta H_j| > 0$: Electron acceptor adsorbates (A)	

of its short lifetime, which makes it unsuitable for any classical promotion application, but, even more importantly, because it can be only introduced on metal catalyst surfaces interfaced with O^{2-}-conducting solid electrolytes (e.g., YSZ) or mixed O^{2-}-electronic conductors (e.g., TiO_2, CeO_2).

Yet, as recently discovered [137,138] and as analyzed in the next section, this very efficient anionic promoter, O^{2-}, has for many years been promoting via backspillover the surface of commercial nanoparticle noble metal catalysts deposited on porous ZrO_2-, TiO_2-, and CeO_2-containing catalyst supports for many catalytic reactions, as it is the main species mediating the phenomenon of metal–support interactions (MSI) with these supports. Its extremely important role, however, was very difficult to discover with classical surface spectroscopic techniques since its XPS signal, for example, is practically identical with that of the O^{2-} of the support.

It was only the use of larger metal particles deposited on these supports, up to 1 µm in size, comparable to the spatial resolution of XPS, that enabled researchers to understand the origin of electrochemical promotion and, at the same time, to discover the O^{2-} backspillover mechanism of metal–support interactions [137,138].

We will examine this in more detail in the next sections. Here it is important to discuss thoroughly the origin of electrochemical promotion with O^{2-} conductors. The similar case of cationic conductors is treated in another chapter of this book.

19.3.2 Molecular Origin of Electrochemical Promotion

In order to understand the molecular origin of electrochemical promotion, a whole arsenal of surface-science, catalytic, and electrochemical techniques have been employed during the last few years (Figure 14). These include analysis of time constants during galvanostatic NEMCA transients, work function measurements via the Kelvin probe technique and via UPS (ultra-violet photoelectron spectroscopy), temperature programmed desorption (TPD), cyclic voltammetric investigations, AC impedance spectroscopic investigations, XPS (x-ray photoelectron spectroscopy) investigations, UPS (ultra-violet photoelectron spectroscopy) investigations, AES (Auger electron spectroscopic) investigations, SERS (surface enhanced Raman spectroscopy) investigations, PEEM (photoelectron emission microscopy) investigations, scanning tunneling microscopy (STM) investigations, and ab initio quantum mechanical calculations [14].

All these techniques have provided a unanimous answer to the molecular origin of NEMCA [13,14]. A combination of the results of any two or three of them would have sufficed to put together the puzzle. But each one of them has something new to offer, some new facet of the surface chemistry to reveal. So first we will reexamine the galvanostatic transient of Figure 2 in light of parallel TPD and cyclic voltammetry investigations of Pt deposited on YSZ (Figure 15) [14]. These two techniques are very well understood among surface scientists and electrochemists, respectively, and thus Figure 15 can help every reader to immediately grasp the underlying molecular phenomena.

A Galvanostatic NEMCA Transient Revisited

We can now concentrate on Figure 15, which shows a galvanostatic catalytic rate transient during ethylene oxidation on Pt/YSZ at 370 °C [Figure 15(a)] [26] together

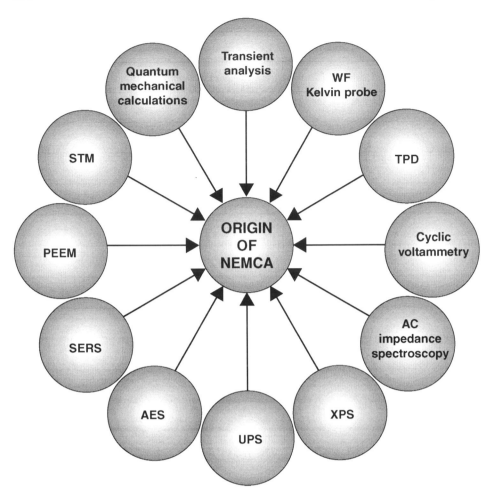

Figure 14 Surface-science, catalytic, and electrochemical techniques employed during the last few years to determine the origin of electrochemical promotion.

with (a) oxygen TPD spectra obtained on Pt/YSZ [90] upon exposure at 400 °C to $p_{O_2} = 4 \times 10^{-6}$ Torr for 1800 s (7.2 kilolangmuirs) followed by electrochemical O^{2-} supply ($I = 15 \, \mu A$) for various time periods t_I, rapid cooling to 300 °C followed by a linear increase in T at a heating rate β (K/s) under open circuit to obtain the TPD spectra of Figure 15(b). (b) Cyclic voltammograms obtained at 400 °C and various holding times at a positive potential ($U_{WR} = 0.8$ V) under UHV conditions [90] [Figure 15(c)].

In comparing Figure 15(a), (b), and (c), it is worth noting that (b) and (c) have been obtained with a Pt film having a true surface area $N_G = 2 \times 10^{-7}$ mol Pt [90] while Figure 15(a) has been obtained on a Pt film with true surface area $N_G = 4.2 \times 10^{-9}$ mol Pt [26]. Thus, for the three different experiments of Figure 15(a), (b), and (c), the NEMCA time constants $\tau (= 2FN_G/I)$ are 800 s, 2500 s, and 1200 s, respectively, i.e., they are of the same order of magnitude.

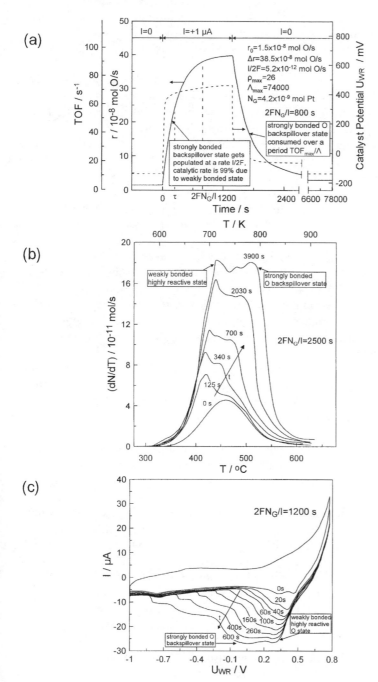

Figure 15 NEMCA and its origin on Pt/YSZ catalyst electrodes. Transient effect of the application of a constant current (a, b) or constant potential U_{WR} (c) on (a) the rate, r, of C_2H_4 oxidation on Pt/YSZ (also showing the corresponding U_{WR} transient) (from Ref. 26). (b) The O_2 TPD spectrum on Pt/YSZ (from Refs. 90 and 139) after current ($I = 15\,\mu A$) application for various times t. (c) The cyclic voltammogram of Pt/YSZ (from Refs. 90 and 139) after holding the potential at $U_{WR} = 0.8\,V$ for various times t.

Both the TPD spectra [Figure 15(b)] and the cyclic voltammograms [Figure 15(c)] clearly show the creation of *two* distinct oxygen adsorption states on the Pt surface [versus only one state formed upon gas-phase O_2 adsorption, Figure 15(b), $t = 0$].

The weakly bonded O adsorption state is populated almost immediately [Figure 15(b) and (c)]. The strongly bonded O adsorption state is populated over a time period of the order $2FN_G/I$. This is exactly the time period the catalytic rate needs to reach its electrochemically promoted value [Figure 15(a)].

One can then ask: what is the rate at which the strongly bonded state is populated during the TPD and cyclic voltammetric experiments of Figure 15(b) and (c)? The answer is clear: it is the rate of O^{2-} supply to the catalyst, i.e., $I/2F$.

The next question is: what is the rate at which the strongly bonded oxygen state is supplied and populated during the catalytic transient of Figure 15(a)? Clearly, it is again of the order $I/2F$, since in Figure 15(a)–(c), the time constants are all similar, i.e., $2FN_G/I$.

Finally, one may ask: what is the rate at which the strongly bonded oxygen state is removed from the catalyst surface at steady state, i.e., when the rate has reached its maximum value? It clearly also has to equal $I/2F$. What removes the strongly bonded state from the catalyst surface? Clearly, it is reaction with C_2H_4 (although desorption to the gas phase also plays a role that can become dominant in absence of C_2H_4). Consequently, at steady state the strongly bonded oxygen state reacts with C_2H_4 at a rate $I/2F$.

At this point (steady state) the weakly bonded oxygen is reacting with ethylene at a rate $r - I/2F \approx r (= 3.85 \cdot 10^{-7}\ \text{mol O/s})$, which is 74,000 $(=\Lambda)$ times larger than $I/2F$. So, what is the physical meaning of the Faradaic efficiency Λ? It is simply the ratio of the reactivity (with C_2H_4) of the weakly bonded and strongly bonded oxygen state [13].

The latter acts as a *sacrificial promoter*. It is a promoter, as it forces oxygen to populate the weakly bonded (and highly reactive) oxygen adsorption state. It is also "sacrificed" as it is consumed by C_2H_4 at a rate $I/2F$, equal to its rate of supply.

In view of the above physical meaning of Λ, it is clear why Λ can approach "infinite" values when Na^+ is used as the sacrificial promoter (e.g., when using β''-Al_2O_3 as the solid electrolyte) to promote reactions such as CO oxidation or NO reduction by H_2 (Table 1). In this case Na on the catalyst surface is not consumed by a catalytic reaction, and the only way it can be lost from the surface is via evaporation. Evaporation is very slow below 400 °C so Λ can approach "infinite" values.

Returning to the Figure 15(a), we can further discuss the physical meaning of Λ to gain some more physical insight and to prove the validity of the sacrificial promoter concept.

Since Λ expresses the ratio of the rates of consumption of the two oxygen states by C_2H_4, one has

$$\Lambda = \frac{r_1}{r_2} = \left(= \frac{\Delta r}{(I/2F)} \right) = \frac{\text{TOF}_1}{\text{TOF}_2} \tag{23}$$

where TOF_1 and TOF_2 are the turnover frequencies of the two reactions (s^{-1}), both based on the total Pt catalyst surface area or maximum oxygen uptake

($N_G = 4.2 \times 10^{-9}$ mol Pt). The values of the two TOF at steady state in Figure 15(a) are

$$\text{TOF}_1 = 95.2\,\text{s}^{-1} (= r_1/N_G) \tag{24}$$

$$\text{TOF}_2 = 1.2 \cdot 10^{-3}\,\text{s}^{-1} (= r_2/N_G) \tag{25}$$

The inverse of these numbers express roughly the average lifetimes of oxygen at the two adsorption states at steady state, i.e.,

$$\tau_1 \approx 0.01\,\text{s} \qquad \tau_2 \approx 1000\,\text{s} \tag{26}$$

Can these two oxygen adsorption states, clearly manifested by TPD and cyclic voltammetry in Figure 15(b) and (c), have indeed such different reactivities with C_2H_4? A look at Figure 16 can convince us about this: the figure shows how the population of the two states evolves in time, in absence of C_2H_4, if we just let oxygen desorb by stopping the applied current, i.e., Figure 16 shows the results of isothermal desorption [90,139]. It is clear that the weakly bonded state desorbs much faster, at least 50 times faster, than the strongly bonded one. It is thus also reasonable to expect that it will react with C_2H_4 much faster than the strongly bonded one.

Can we identify the strongly bonded oxygen state as the "backspillover" oxygen originating from the solid electrolyte? Since the strongly bonded state is

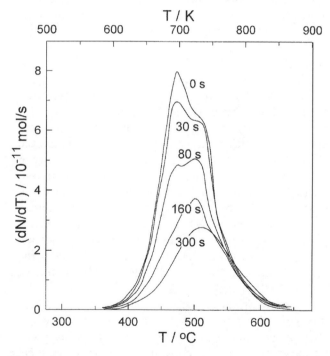

Figure 16 Oxygen thermal desorption spectra after electrochemical O^{2-} supply to Pt/YSZ at 673 K ($I = +12\,\mu\text{A}$ for 1800 s) followed by isothermal desorption at the same temperature at various times as indicated on each curve. (Reprinted from Ref. 139 with permission from Academic Press.)

occupied over a time period $2FN_G/I$ (Figure 15) the answer is, for all practical purposes, yes. But we should also keep in mind that oxygen atoms can exchange between the two states: at 400 °C and in presence of gaseous $^{18}O_2$, ^{18}O tracer experiments have shown isotopic scrambling between the two states to be almost complete after 30 s.

A straightforward, qualitative, but strong confirmation of the electrochemical promotion mechanism described above is obtained when we try to predict what will happen to the promoted catalytic rate after the current is interrupted [Figure 15(a)]: if the average lifetime of the promoting strongly bonded oxygen species is indeed 1000 s $[= TOF_2^{-1} = (r_2/N_G)^{-1}]$, then one would expect that upon current interruption the promoted catalytic rate will decay to its unpromoted (open-circuit) values within a time τ_D of roughly 1000 s. This is exactly what Figure 15(a) shows upon current interruption. Thus by simply knowing the TOF of the electrochemically promoted catalytic reaction $(= TOF_1)$ and estimating the TOF of the reaction of the promoting species $(= TOF_2)$ from the time decay of the rate upon current interruption, one can estimate the Faradaic efficiency Λ from

$$\Lambda = TOF_1/TOF_2 = \tau_D/(r_1/N_G) \tag{27}$$

without knowing the value of the applied current I [14]!

Thus the picture that emerges is quite clear (Figure 17): at steady state, before potential (or current) application, the Pt catalyst surface is covered, to a significant extent, by chemisorbed O and C_2H_4. Then upon current (and thus also potential) application, O^{2-} ions arriving from the solid electrolyte at the tpb at a rate $I/2F$ react at the tpb to form a backspillover ionically strongly bonded species:

$$O^{2-}(YSZ) \rightarrow [O^{\delta-}-\delta^+](Pt) + 2e^- \tag{28}$$

The exact value of δ is not yet known, but useful information can be extracted from the surface spectroscopic techniques described in the continuation of this chapter. Both XPS [87] and dipole moment measurements [139] suggest $\delta \approx 2$, so that $O^{\delta-}$ is O^{2-}, at least for Pt [138]. Nevertheless it is still safer to maintain the symbolism $O^{\delta-}$. The symbolism $[O^{\delta-}-\delta^+]$ emphasizes that the backspillover oxygen species is overall neutral, as it is accompanied by its compensating image charge in the metal (δ^+). At the same time the applied potential and concomitant high-oxygen chemical potential creates on the Pt surface two new oxygen adsorption states: a strongly bonded one and weakly bonded one. The backspillover $[O^{\delta-}-\delta^+]$ species migrate over the entire Pt catalyst surface. Due to the repulsive interaction of the $[O^{\delta-}-\delta^+]$ dipoles, diffusion on the Pt surface is fast and the rate of spreading of the $[O^{\delta-}-\delta^+]$ species on the Pt surface is controlled by the rate, $I/2F$, of their creation at the tpb and not by their surface diffusivity [13,14].

The backspillover oxygen species primarily occupies the strongly bonded oxygen chemisorption state. Oxygen adsorbing from the gas phase is forced to populate primarily the weakly bonded (and highly reactive) state (Figure 17). Consequently, the catalytic rate starts increasing dramatically up to the point where a steady-state coverage is established for the strongly bonded oxygen. At this point the (electrochemically promoted) catalytic rate is at its new steady-state value (Figures 15 and 17).

Catalysis

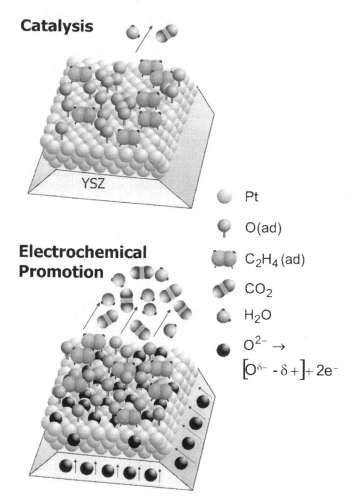

Electrochemical Promotion

- Pt
- O(ad)
- C_2H_4 (ad)
- CO_2
- H_2O
- $O^{2-} \rightarrow$
  $\left[O^{\delta-} - \delta+\right] + 2e^-$

Figure 17 Atomic visualization of NEMCA during ethylene C_2H_4 on Pt/YSZ. The backspillover $[O^{\delta-}-\delta+]$ species forces O(ad) to a more weakly bonded and more reactive state.

Which surface sites are occupied by the strongly bonded oxygen state? This is not known yet. It is likely that they correspond to threefold hollow sites while the weakly bonded state corresponds to bridge-bonded or even on-top sites. This is only a proposition at this time, but STM should soon be able to provide useful information. The point is that, as shown by the TPD spectra of Figure 15(b), electrochemical promotion forces large amounts of oxygen (near monolayer coverages) to remain adsorbed on the catalyst surface under conditions where gaseous adsorption (many kilolangmuirs) leads to coverages of the order of 0.05 [Figure 15(b)].

The same figure proves unambiguously that electrochemically controlled backspillover of oxygen from the solid electrolyte onto the catalyst surface takes place and is the cause of NEMCA.

Figure 15(b), as well as 15(c), also demonstrates the enormous power of electrochemistry to create new adsorption states on a catalyst surface.

Other Techniques

Although the use of TPD and cyclic voltammetry described above suffice to clarify the origin of electrochemical promotion with O^{2-} conductors, we survey here some of the key results obtained with work function measurements, AC impedance spectroscopy, XPS, PEEM, and STM, together with some additional results obtained from TPD in conjunction with rigorous quantum mechanical calculations. For experimental details, results with other techniques of Figure 14 and a more detailed analysis of the results surveyed here, the reader is referred to a recent book [14] and to the original papers.

Work Function Measurements

One of the first key steps in understanding the origin of electrochemical promotion was the realization that solid electrolyte cells with metal electrodes are both work function probes and work function controllers for the gas-exposed surfaces of their electrodes (Figure 18):

$$eU_{WR} = \Phi_W - \Phi_R \tag{29}$$

$$e\Delta U_{WR} = \Delta\Phi_W \tag{30}$$

These important equations have been established using both the Kelvin probe (vibrating capacitor) technique [11,140] and UPS [89] (electron cutoff energy). At the molecular level the variation of Φ with catalyst-electrode potential U_{WR} is due to the spillover-backspillover of O^{2-} (or Na^+) from the solid electrolyte onto the catalyst–gas interface.

In view of the fact that in any electrochemical cell it is [13,14,140]

$$eU_{WR} = \bar{\mu}_R - \bar{\mu}_W = \Phi_W - \Phi_R + e\Psi_W - e\Psi_R \tag{31}$$

$$e\Delta U_{WR} = \Delta\Phi_W + e\Delta\Psi_W \tag{32}$$

the experimental Eqs. (29) and (30) imply that in solid electrolyte cells with ion spillover-backspillover one has

$$\Psi_W = \Psi_R = C \tag{33}$$

where the constant C is zero for overall neutral cells [140], which is usually the case. Equation (33) and its mathematically equivalent Eqs. (29) and (30) simply reflect the presence of an overall neutral electrochemical double layer at the metal–gas interface [140–142]. This double layer is usually termed the *effective double layer* [140–142] and is overall neutral, as is every other double layer in electrochemistry (Figure 19).

The effective double-layer concept is an important one, as it shows that electrochemical promotion, but also promotion and, as we shall see, metal–support-

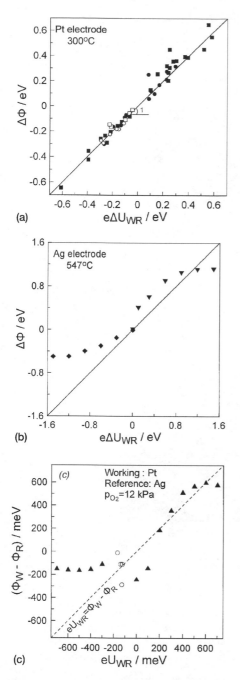

Figure 18 Effect of catalyst-electrode potential U_{WR} on the work function Φ of the gas-exposed catalyst-electrode surface. (a) Pt/YSZ, $T = 300\,°C$ (squares), Pt/β''-Al$_2$O$_3$, $T = 240\,°C$ (circles), filled symbols: closed-circuit operation, open symbols: open-circuit operation, O$_2$, C$_2$H$_4$/O$_2$, and NH$_3$/O$_2$ mixtures. (From Refs. 11 and 93.) (b) Ag/YSZ, $T = 547\,°C$. (From Ref. 89.) (c) Dependence of $\Phi_{W(Pt)}$-$\Phi_{R(Ag)}$ on potential U_{WR} for the system Pt(W)-Ag(R). (From Ref. 140.) Open symbols: Open-circuit operation. Filled symbols: Closed circuit operation $T = 673\,K$.

Gas-exposed catalytically active surface
surface area A_G

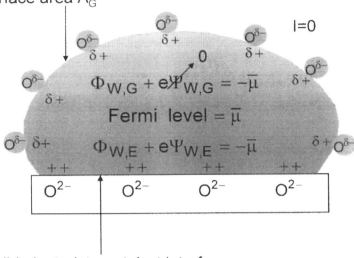

Solid electrolyte-catalyst interface
surface area A_E

Figure 19 Schematic representation of a metal crystallite deposited on YSZ and of the changes induced in its electronic properties upon polarizing the catalyst–solid electrolyte interface and changing the Fermi level (or electrochemical potential of electrons) from an initial value $\bar{\mu}$ to a new value $\bar{\mu}$-eη. (From Ref. 142. Reprinted with permission from Elsevier Science.)

interactions-promoted catalysis, is *catalysis in presence of a double layer* [14]. In the case of electrochemical promotion the double layer is in situ controllable.

Equations (29), (30), and (33) are valid both at open-circuit and under closed-circuit, i.e., NEMCA, conditions, in the presence or absence of catalytic reactions as long as the effective double layer is present at the metal–gas interface [140].

Deviations from Eqs. (29), (30), and (33) occur when the effective double layer at the metal–gas interfaces is destroyed [140]. This is the case for (1) very low temperatures ($<250\,°C$ for YSZ, $<100\,°C$ for β''-Al_2O_3) where ion spillover-backspillover is kinetically frozen, or (2) very high temperatures ($>500\,°C$ for YSZ, $>400\,°C$ for β''-Al_2O_3) where the effective double layer desorbs; (3) fast diffusion-controlled catalytic reactions, which again destroy the double layer; (4) formation of insulating carbonaceous or oxidic deposits at the metal–gas interface which allow for the storage of electric charge [140].

Nevertheless, the experimental range of stability of the effective double layer and thus validity of Eqs. (29), (30), and (33) is very broad and practically coincides with the range of electrochemical promotion itself.

Absolute Potential of Supports

Another important consequence of Eq. (29) is that it automatically defines by itself an absolute potential scale in solid-state electrochemistry. *The absolute potential is simply the extraction potential, Φ/e, of an electrode (any electrode) in contact with the solid electrolyte. The reference state of the electrons is not that in another (reference) electrode, but rather that of an electron at its ground state outside the electrolyte surface* (or at its ground state at "infinity" as long as the cell is overall neutral; thus $\Psi = 0$) [140].

An amazing, yet straightforward, fact is that *the absolute potential does not depend on the electrode material, but is only a property of the electrolyte and of the gas phase* [140]. This is why Frumkin [143] called the absolute potential (in aqueous electrochemistry) energy of solvation of an electron into the electrolyte.

This can be understood easily as follows:

Consider the equilibrium

$$O_2 + 4e^- \rightleftharpoons 2O^{2-} \text{ (YSZ)} \tag{34}$$

which takes place not only at the three-phase boundaries (tpb) metal–solid electrolyte–gas but also over the entire metal–gas interface as long as O^{2-} backspillover is in equilibrium ($\bar{\mu}_{O^{2-} \text{(YSZ)}} = \bar{\mu}_{O^{2-} \text{(M)}}$). The resulting equilibrium conditions are

$$\bar{\mu} = \frac{1}{2}\bar{\mu}_{O^{2-} \text{(YSZ)}} - \frac{1}{4}\mu_{O_2\text{(gas)}} \tag{35}$$

where $\bar{\mu}$ is the electrochemical potential (Fermi level) of electrons in the metal.

As long as the double layer is present at the metal–gas interface, it is $\Psi = 0$, thus $-\bar{\mu} = \Phi$, and thus

$$U_{\text{abs}} = \Phi/e = \frac{1}{4e}\mu_{O_2\text{(g)}} - \frac{1}{2e}\bar{\mu}_{O^{2-} \text{(YSZ)}} \tag{36}$$

and one therefore sees that indeed the absolute potential, U_{abs}, is only a property of the electrolyte and of the gaseous oxygen chemical potential [140].

Experiment has shown [140] that for standard conditions in solid-state electrochemistry ($p_{O_2} = 1$ atm, $T = 673$ K) the corresponding standard absolute potential of YSZ is

$$U^{\text{o}}_{\text{abs}} = \Phi^{\text{o}}/e = -\frac{1}{2e}\bar{\mu}_{\text{O}^2\ (\text{YSZ})} = 5.14 \pm 0.05 \text{ V} \tag{37}$$

This, in view of Eq. (36), means that the standard electrochemical potential of O^{2-} in YSZ is

$$\bar{\mu}_{\text{O}^2\ (\text{YSZ})} = -10.28 \pm 0.1 \text{ eV/atom} \tag{38}$$

It is thus now clear that one can assign to each O^{2-}-conducting support, or to any ionic or mixed ionic-electronic conducting support two new properties, i.e., its *absolute potential* and its *standard absolute potential*. For O^{2-}-conducting or mixed O^{2-}-electronic conductors the defining equations are Eq. (36) and Eq. (37), respectively, where the numerical value corresponds to 8 mol% Y_2O_3-ZrO_2 [140].

In view of Eq. (37) one may conclude that the more stable O^{2-} is in the support (carrier), i.e., the lower $\bar{\mu}_{\text{O}^2}$ is, the higher are the work function and absolute potential of the support.

It must be emphasized that due to Fermi-level pinning ($\bar{\mu} = E_F$) between the metal and the solid electrolyte, $\bar{\mu}$ in Eq. (35) is also the Fermi level of the support [140]. Thus *the absolute potential of a support equals the negative of its Fermi level.*

The importance of these considerations for metal–support interactions with ZrO_2-, TiO_2-, and CeO_2-containing supports is obvious and will be discussed in the last section.

A last practical question: how can one easily measure the absolute potential of a support? The answer is simple: by simply depositing a metal (any metal) electrode on it and measuring the work function of the backspillover-modified electrode [140].

AC Impedance Spectroscopy

The technique of AC impendance spectroscopy, one of the most common techniques in aqueous and solid-state electrochemistry, was used recently to confirm the formation of the effective double layer on metal surfaces interfaced with YSZ [43,95]. An example for the case of Pd/YSZ is shown in Figure 20. The semicircle labeled C_1 is associated with the charge-transfer reaction

$$O^{2-} \rightleftharpoons O(a) + 2e^- \tag{39}$$

taking place at the tpb metal–YSZ–gas. The semicircle labeled C_2, which forms in this case only under anodic (positive) polarization, i.e., supply of O^{2-} to the catalyst, corresponds to reaction (39) taking place over the entire metal–gas interface. The computed capacitance of the double layer at the metal–gas interface is of the order of 200–500 μF/cm^2 of solid electrolyte area or [43] 10–25 μF/cm^2 of metal/gas interface area, i.e., comparable to that of a dense metal/aqueous electrolyte double layer.

XPS

The key XPS results confirming O^{2-} backspillover as the origin of electrochemical promotion both with YSZ [87] and with TiO_2 [64] are shown in Figure 21. The figure shows the Ols spectrum of porous polycrystalline Pt films deposited on YSZ and on TiO_2, first under open-circuit conditions [Figure 21(a)A, (b)A] and then under positive current and potential application [Figure 21(a)B, (b)B]. Figure 21(a)C and

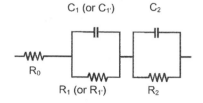

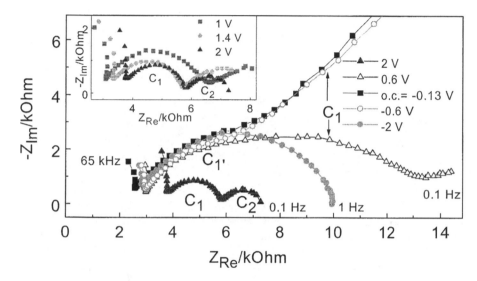

Figure 20 Complex impedance spectra (Nyquist plots) of the CH_4, O_2, Pd|YSZ system at different Pd catalyst potentials. Open-circuit potential $U^o_{WR} = 0.13$ V.

(b)C show the difference spectra. In both cases XPS clearly shows the presence of the O^{2-} double layer, even under open-circuit conditions, and also clearly confirms the electrochemically controlled backspillover of O^{2-} from the YSZ or TiO_2 support onto the catalyst surface. Note that the binding energy of the backspillover O species is in both cases near 529 eV, which confirms its strongly anionic (probably O^{2-}) state.

Another point worth discussing in Figure 21 is the following: part of the O1s spectrum under open-circuit conditions [Figure 21(a)A, (b)A] corresponds to O^{2-} in YSZ at the vicinity of the Pt electrode visible through microcracks. Upon potential application ΔU_{WR} between the grounded Pt electrode and a reference electrode, the O1s signal of the O^{2-} of YSZ species shifts to lower binding energies by ΔU_{WR}. This is the well-known *electrochemical shift* in XPS. Note, however, that the electrochemically shifted O^{2-} of YSZ has practically the same binding energy with the backspillover O^{2-} on the Pt electrode. This means that O^{2-} in YSZ at the vicinity of the Pt electrode has the same electrochemical potential with O^{2-} on the Pt electrode, i.e., the two species are energetically indistinguishable and the Pt electrode is "wetted" by a sea of O^{2-} ions both from the electrolyte side and from the metal–gas interface side [14].

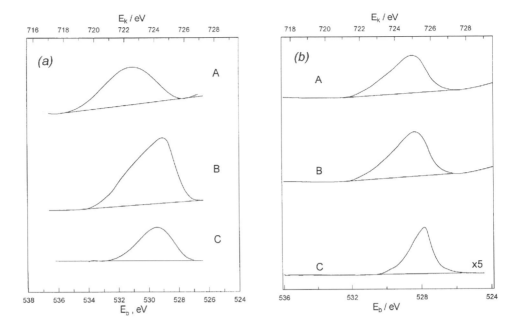

Figure 21 XPS confirmation of $O^{\delta-}$ backspillover as the mechanism of electrochemical promotion on Pt films deposited on YSZ (a) and on TiO_2 (b). (Adapted from Refs. 64,87.) In both cases A is the open-circuit O1s spectrum, B is the O1s spectrum under anodic ($I > 0$, $\Delta U_{WR} > 0$) polarization and C is the difference spectrum. (Reprinted with permission from the American Chemical Society [(a), ref. 87] and from Academic Press [(b), 64].)

Photoemission Electron Microscopy

The photoelectron emission microscopy (PEEM) investigations of Imbihl and co-workers [91] (Figure 22) have nicely confirmed not only the potential-controlled variation in the work function of model Pt electrodes deposited on YSZ but also the Fermi-level pinning between Pt and YSZ.

Figure 22 shows a Pt electrode (light) deposited on YSZ (dark). There are three circular areas of bare YSZ connected by very narrow bare YSZ channels. The rest of the surface is Pt. Note that the Fermi levels of the Pt film and of the YSZ solid electrolyte in the vicinity of the Pt film are equal. The YSZ, however, appears in the PEEM images much darker than the Pt film since YSZ has a negligible density of states at its Fermi level in comparison to a metal such as Pt.

Figure 22 clearly shows how the application of potential changes the brightness and thus the work function, Φ, of the grounded Pt catalyst-electrode and of the YSZ surface, in accordance to the above discussed alignment (pinning) of the two Fermi levels.

Increasing U_{WR} increases Φ, according to Eq. (30), thus causes local darkening (fewer electrons are emitted), as expected. Decreasing U_{WR} decreases Φ, again according to Eq. (30), thus the brightness increases as expected (Figure 22).

The significant point is that PEEM, as clearly presented in Figure 22, has shown conclusively that Φ reversibly follows the applied potential and has provided the basis for space- and time-resolved ion spillover studies of electrochemical

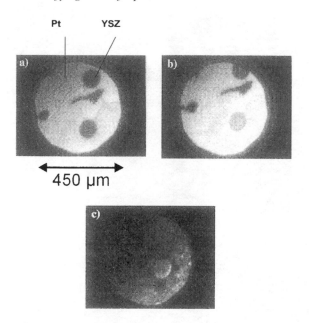

Figure 22 PEEM image of a Pt/YSZ microstructure showing three circular YSZ domains connected via channels surrounded by a Pt film. (a) $U_{WR} = 0$ at 695 K; (b) $U_{WR} = -2$ V for 2 min at 695 K; (c) $U_{WR} = 2$ V for 2 min at 737 K. (From Ref. 91. Reprinted with permission from Wiley-VCH.)

promotion. It has also shown that the Fermi level and work function of the solid electrolyte in the vicinity of the metal electrode follow the Fermi level and work function of the metal electrode [91,140].

Scanning Tunneling Microscopy

Scanning tunneling microscopy (STM) has also nicely confirmed ion backspillover as the origin of NEMCA [92,144]. Figure 23 shows an unfiltered STM image of a Pt single-crystal surface consisting mainly of Pt(111) terraces interfaced with β''-Al_2O_3, a Na^+ conductor, following negative current application between the Pt single crystal and a counterelectrode also deposited on β''-Al_2O_3. Electrochemically controlled Na backspillover leads to the formation of an ordered Pt(111)-(12 × 12)-Na adlattice. Sodium can be reversibly pumped back and forth on the Pt catalyst surface via current or potential application [92,144].

Temperature Programmed Desorption

In addition to investigating the dual nature of surface oxygen under electrochemical promotion conditions on Pt/YSZ [90], temperature programmed desorption, TPD, was also used to investigate the effect of catalyst potential and work function on the binding energy of the weakly bonded atomic oxygen species on Pt [90,139], Ag [145], and Au [146]. The last case is noteworthy since oxygen was found to chemisorb on Au only electrochemically and not when added from the gas phase [146].

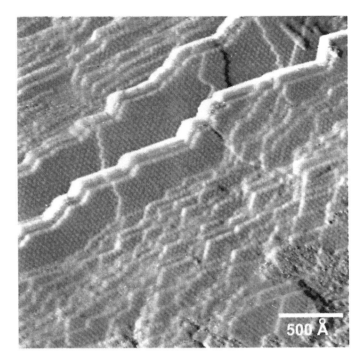

Figure 23 STM image (unfiltered) of a Pt single-crystal surface consisting mainly of Pt(111) terraces and covered by a Pt(111)-(12 × 12)-Na adlattice formed via electrochemical Na^+ supply (from a β''-Al_2O_3 Na^+ conductor interfaced with the Pt single crystal) on a Pt(111)-(2 × 2)-O adlattice. Each sphere on the image corresponds to a Na atom. (From Refs. 14,140,144.)

The key results are shown in Figure 24. The chemisorptive binding energy of O decreases linearly with Φ in all three cases, in excellent agreement with Eq. (22). The slopes are -1 for Pt and Ag and -4 for the case of Au.

Quantum Mechanical Calculations

Pacchioni and Illas performed rigorous quantum mechanical calculations using Cu_{34} and Pt_{25} clusters to model O chemisorption on the Cu(100) and Pt(111) surfaces [135,136]. The adsorbed oxygen atom was surrounded by ions or point charge on the cluster to simulate electrochemical promotion conditions.

The key results for Pt are shown in Figure 25. The binding energy of O on the Pt(111) surface decreases linearly with the cluster HOMO, thus with the work function of the Pt(111) surface. The slope is -0.5. This is in excellent qualitative agreement with the experimental value -1 (Figure 24).

Equally important is the observation that, as Figure 25 shows, the chemisorptive binding energy–work function behavior can be described almost quantitatively by taking into account only the electrostatic (Stark) interactions, i.e., only the "through the vacuum" electrostatic interactions in the double layer and neglecting the "through the metal" interactions, i.e., the redistribution of electron states near the Fermi level of the cluster. This is an important result as it provides

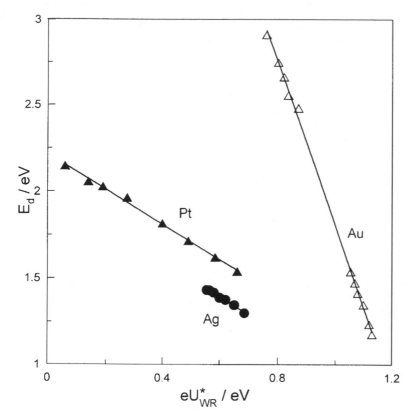

Figure 24 Effect of catalyst potential on the oxygen desorption activation energy, E_d, calculated from the modified Redhead analysis for Pt, Ag, and Au electrodes deposited on YSZ. (Reprinted from Ref. 146 with permission from the Institute for Ionics.)

strong support for the effective double-layer kinetic model of promotion and catalysis presented in the next section.

19.3.3 The Effective Double Layer

The picture that emerges from all the catalytic, electrocatalytic, and surface-science techniques reviewed in this section is clear [14]. At sufficiently high temperatures for ionic motion in the solid electrolyte, the entire gas-exposed surface of the metal is covered, to a significant extent, by ionic species that migrate there from the solid electrolyte. These ionic species are accompanied by their compensating (screening) charge in the metal; thus an overall neutral dipole layer is formed (Figure 26). This double layer has practically all the properties of the classical metal–solid electrolyte double layer. Like in every double layer, the field strength, $\tilde{E}$, within the effective double layer is typically of the order of 1 V/Å. This parameter, which equals $d\varphi/dz$ within the double layer, plays an important role in the quantitative description of electrochemical and classical promotion.

The unique characteristic of the "effective double layer," in comparison to all classical metal/electrolyte double layers, is that it is directly accessible to gaseous reactants. Thus electrochemical promotion is catalysis *in the presence of a*

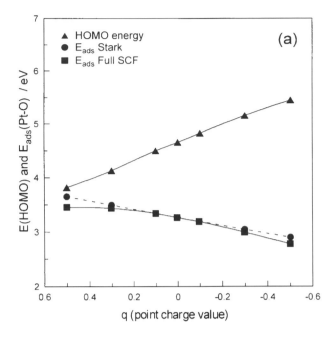

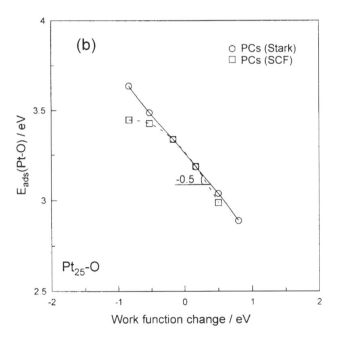

Figure 25 (a) Dependence of the position of the HOMO in Pt_{25} and of the Pt_{25}/O adsorption energy, E_{ads}, at the Stark and full self-consistent field (SCF) levels, as a function of the presence of point charge q above and below the cluster first layer. (b) Oxygen adsorption energy, E_{ads} versus work function change, as measured by the cluster HOMO, for Pt_{25}/O. The curves refer to the cluster with point charges (PC). Both Stark and full SCF curves are shown. (From Ref. 135. Reprinted with permission from the American Chemical Society.)

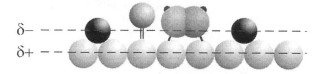

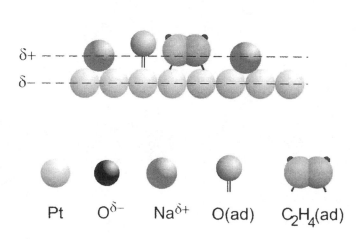

Figure 26 Schematic of the effective double layer during C_2H_4 oxidation on Pt/YSZ (top) and Pt/β''-Al_2O_3.

controllable (via current and potential) electrochemical double layer. The theoretical implications and practical opportunities are obvious and numerous [14].

19.3.4 Dependence of Catalytic Rates on the Catalyst Potential and Work Function

Similar to the case of classical promotion, four types of r versus U_{WR}, thus in view of Eq. (29) r versus Φ, behavior are observed experimentally (Figure 27) [4,129,130]:

1. Purely electrophobic
2. Purely electrophilic
3. Volcano-type
4. Inverted-volcano type

In the first two cases the r versus U_{WR} (or r versus Φ) dependence frequently conforms to the equation

$$\ln(r/r_0) = \alpha F \Delta U_{WR}/RT = \alpha \Delta \Phi / k_b T \qquad (40)$$

where the "NEMCA coefficient" α is positive (and of the order of 0.2 to 1) for electrophobic reactions and negative (and of the order of -0.2 to -1) for electrophilic reactions.

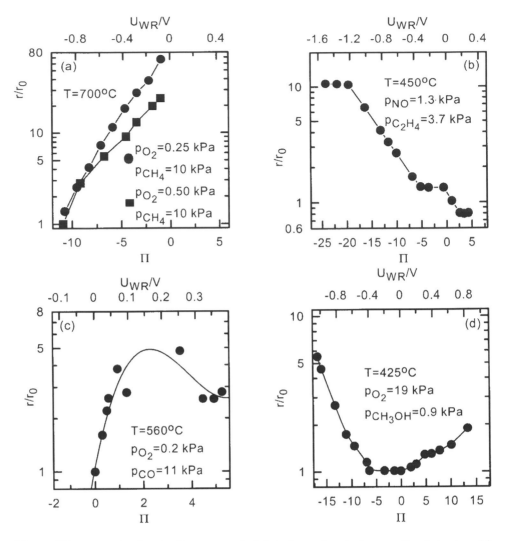

Figure 27 Examples for the four types of global electrochemical promotion behavior: (a) electrophobic, (b) electrophilic, (c) volcano-type, (d) inverted volcano-type. (a) Effect of catalyst potential and work function change (versus $I = 0$) for high (20:1) and (40:1) CH_4 to O_2 feed ratios, Pt/YSZ (from Ref. 33). (b) Effect of catalyst potential on the rate enhancement ratio for the rate of NO reduction by C_2H_4 consumption on Pt/YSZ (from Ref. 39). (c) NEMCA generated volcano plots during CO oxidation on Pt/YSZ (from Ref. 34). (d) Effect of dimensionless catalyst potential on the rate constant of H_2CO formation, Pt/YSZ (from Ref. 37). $\Pi = FU_{WR}/RT(= \Delta\Phi/k_bT)$.

Rules of Promotion

Table 3 lists all published electrochemical promotion studies on the basis of their global r versus Φ behavior. The term "global" denotes over the entire accessible eU_{WR} and $\Delta\Phi$ range (typically -1.5 to 1.5 eV) as opposed to "local," which denotes the r versus Φ behavior at a specific Φ value. The table also provides, wherever

Table 3　Classification of Electrochemical Promotion Studies on the Basis of Global r Versus Φ Behavior

A. Purely electrophobic reactions

Reactants (D)	(A)	Catalyst	Solid Electrolyte	p_A/p_D	T (°C)	Kinetics in D $(\partial r/\partial p_D)_\Phi$	Kinetics in A $(\partial r/\partial p_A)_\Phi$	Rule	Ref.
C_2H_4	O_2	Pt	$ZrO_2(Y_2O_3)$	12–16	260–450	+	0	G1	10,26,31
C_2H_4	O_2	Pt	$\beta''\text{-}Al_2O_3$	238	180–300	+	0	G1	13,67
C_2H_4	O_2	Pt	TiO_2	3.5–12	450–600	+	0	G1	64
C_2H_4	O_2	Rh	$ZrO_2(Y_2O_3)$	0.05–2.6	250–400	+	0	G1	28,30,40
C_2H_4	O_2	Pd	$ZrO_2(Y_2O_3)$	0.2–10	290–360	+	$\leqslant 0$	G1	43
C_2H_4	O_2	Ag	$ZrO_2(Y_2O_3)$	0.2–1.1	320–470	+	0	G1	9,49–51
C_2H_4	O_2	IrO_2	$ZrO_2(Y_2O_3)$	300	350–400	+	0	G1	18,61
C_2H_4	O_2	RuO_2	$ZrO_2(Y_2O_3)$	155	240–500	+	$\leqslant 0$	G1	62
CO	O_2	Pt	CaF_2	11–17	500–700	+	0	G1	13,63
CO	O_2	Pd	$ZrO_2(Y_2O_3)$	500	400–550	?	?	?	13,29
CH_4	O_2	Pd	$ZrO_2(Y_2O_3)$	0.2–4.8	380–440	+	0	G1	43,46
C_3H_6	O_2	Ag	$ZrO_2(Y_2O_3)$	20–120	320–420	+	$\leqslant 0$	G1	13,52
CH_4	O_2	Ag	$ZrO_2(Y_2O_3)$	0.02–2	650–850	+	0	G1	13,53
C_6H_6	H_2	Pt	$\beta''\text{-}Al_2O_3$	0.02–0.12	100–150	$\geqslant 0$	~ 0	G1	20,69
C_2H_2	H_2	Pt	$\beta''\text{-}Al_2O_3$	1.7–9	100–300	?	?	?	70
H_2	CO_2	Rh	$ZrO_2(Y_2O_3)$	0.03–0.7	300–450	+	0	G1	13
H_2	$C_2H_2,$ C_2H_4	Pd	$\beta''\text{-}Al_2O_3$	0.1–5.9#	70–100	$\geqslant 0$	0	G1	76
H_2S	—	Pt	$ZrO_2(Y_2O_3)$	—	600–750	?		?	13,45
CH_4	—	Ag	$SrCe_{0.95}Yb_{0.05}O_3$	—	750	?		?	13,82
NH_3	—	Fe	$CaZr_{0.9}In_{0.1}O_{3-\alpha}$	4–12 kPa	530–600	+		G1	79
NH_3	—	Fe	$K_2YZr(PO_4)_3$	4–12 kPa	500–700	+		G1	79
CH_4	H_2O	Ni	$ZrO_2(Y_2O_3)$	0.05–3.5	600–900	+	$\leqslant 0$	G1	13,59

$^{\#}$ $p_D = p_{C_2H_2} + p_{C_2H_4}$

B. Purely electrophilic reactions

Reactants (D)	(A)	Catalyst	Solid Electrolyte	p_A/p_D	T (°C)	Kinetics in D $(\partial r/\partial p_D)_\Phi$	Kinetics in A $(\partial r/\partial p_A)_\Phi$	Rule	Ref.
C_2H_4	O_2	Pt	$CaZr_{0.9}In_{0.1}O_{3-\alpha}$	4.8	385–470	—	+	G2	80
C_2H_4	O_2	Pt	CeO_2	1.6–3.7	500	—	+	G2	65
C_2H_4	O_2	Pt	YZTi10	3	400–475	?	?	?	66
C_2H_4	O_2	Ag	$\beta''\text{-}Al_2O_3$	0.3–0.4	240–280	—	+	G2	77
CO	O_2	Ag	$\beta''\text{-}Al_2O_3$	0.1–10	360–420	0	+	G2	13
C_3H_6	O_2	Pt	$ZrO_2(Y_2O_3)$	0.9–55	350–480	$\leqslant 0$	+	G2	13,38
CH_3OH	O_2	Ag	$ZrO_2(Y_2O_3)$	0–2	500	?	+	G2	23
CH_4	O_2	Au	$ZrO_2(Y_2O_3)$	0.1–0.7	700–750	0	+	G2	56,57,60
H_2	N_2	Fe	$CaZr_{0.9}In_{0.1}O_{3-\alpha}$	0–3	440	?	?	?	81
H_2	C_2H_4	Ni	$CsHSO_4$	1	150–170	?	?	?	13,17
	CH_3OH	Pt	$ZrO_2(Y_2O_3)$	—	400–500	—	?	?	13,37
	CH_3OH	Ag	$ZrO_2(Y_2O_3)$	0–6 kPa	550–750	—	+	G2	13,55
C_2H_4	NO	Pt	$ZrO_2(Y_2O_3)$	0.2–10	380–500	0	+	G2	39
C_2H_4	NO	Pt	$\beta''\text{-}Al_2O_3$	0.1–1.1	280–400	?	?	?	19
CO	NO	Pt	$\beta''\text{-}Al_2O_3$	0.3–5	320–400	$\leqslant 0$	+	G2	71–73
CO	NO	Pd	$ZrO_2(Y_2O_3)$	0.5–6.5	320–480	~ 0	+	G2	47,48
CO	N_2O	Pd	$ZrO_2(Y_2O_3)$	2–50	440	—	+	G2	47
	$1\text{-}C_4H_8$	Pd	Nafion	—	70	—	?	G2	21

Table 3   Continued

C. Volcano-type reactions

Reactants						Kinetics in D	Kinetics in A		
(D)	(A)	Catalyst	Solid Electrolyte	p_A/p_D	T (°C)	$(\partial r/\partial p_D)_\Phi$	$(\partial r/\partial p_A)_\Phi$	Rule	Ref.
C_2H_4	O_2	Pt	$Na_3Zr_2Si_2PO_{12}$	1.3–3.8	430	−	+	G3	78
CO	O_2	Pt	$ZrO_2(Y_2O_3)$	0.2–55	468–558	+	−	G3	13,35,36
CO	O_2	Pt	β''-Al_2O_3	0.5–20	300–450	−	+	G3	13,68
H_2	O_2	Pt	H_2O-0.1N KOH	0.3–3	25–50	+	−	G3	13,84,85
H_2	O_2	Pt	Nafion	0.2–5	25	+	−	G3	83
SO_2	O_2	Pt	V_2O_5-$K_2S_2O_7$	1.8	350–450	?	?	?	86
C_3H_6	NO	Pt	β''-Al_2O_3	2–70	375	−	+	G3	72–74
H_2	NO	Pt	β''-Al_2O_3	0.3–6	360–400	−	+	G3	75

D. Inverted volcano reactions

Reactants						Kinetics in D	Kinetics in A		
(D)	(A)	Catalyst	Solid Electrolyte	p_A/p_D	T (°C)	$(\partial r/\partial p_D)_\Phi$	$(\partial r/\partial p_A)_\Phi$	Rule	Ref.
C_2H_4	O_2	Pt	TiO_2	0.2–0.3[#]	450–600	+	+	G4	64
C_3H_6	O_2	Pt	YZTi10	5	400–500	?	?	?	66
CO	O_2	Ag	$ZrO_2(Y_2O_3)$	0.6–14	350–450	+	+	G4	13,54
CO	O_2	Ag-Pd alloy	$ZrO_2(Y_2O_3)$	3.5–12.5	450–500	+	+	G4	58
CO	O_2	Au	$ZrO_2(Y_2O_3)$	3–53	450–600	+	≥ 0	G4	56,57
C_2H_6	O_2	Pt	$ZrO_2(Y_2O_3)$	0.06–7	270–500	+	+	G4	32
CH_4	O_2	Pt	$ZrO_2(Y_2O_3)$	0.02–7	600–750	+	+	G4	13,33
CH_3OH	O_2	Pt	$ZrO_2(Y_2O_3)$	3–45	300–500	+	?	?	13,37
H_2	CO_2	Pd	$ZrO_2(Y_2O_3)$	0.2–1.1	500–590	+	+	G4	13,29
C_3H_6	NO,O_2	Rh	$ZrO_2(Y_2O_3)$	0.08–8[$]	250–450	+	NO: + O_2: 0	G4	41
CO	NO,O_2	Rh	$ZrO_2(Y_2O_3)$	0.33[$]	250–450	+	NO: + O_2: 0	G4	42

[$] p_A/p_D is the ratio $p_{NO}/p_{C_3H_6}$ and p_{NO}/p_{CO}. The p_{O_2} range is between 0–6 kPa.
[#] low p_A, p_D region, (?): No data available.

available, the observed open-circuit kinetics, i.e., the dependence of r on the partial pressures, p_D and p_A, of the electron donor and electron acceptor reactant, respectively.

   The rules of electrochemical promotion follow directly from Table 3 [4,129]: for example, as shown in Table 3, all purely electrophobic reactions are positive order in D and zero or negative order in A. All purely electrophilic reactions are positive order in A and zero or negative order in D. Volcano-type reactions are always positive order in one reactant and purely negative order in the other. Inverted volcano-type reactions are positive order in both reactants. Thus the following promotional rules can be formulated.

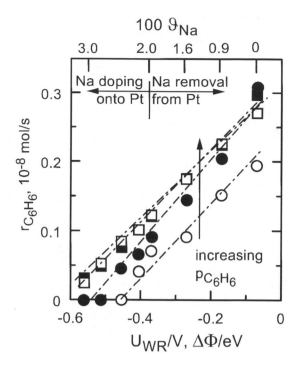

Figure 28 Example of rule G1 (electrophobic behavior): Effect of Na coverage and concomitant work function change on the rate of C_6H_6 hydrogenation on Pt deposited on β''-Al$_2$O$_3$ at 130 °C. Note that the rate is positive order in C_6H_6 (D). It is also near zero order in H_2. (From Refs. 20 and 69.)

Electrophobic Reactions

Inspection of Table 3 shows the following rule for electrophobic reactions:

> Rule G1: A reaction exhibits purely electrophobic behavior $[(\partial r/\partial\Phi)_{p_A,p_D} > 0]$ when the kinetics are positive order in the electron donor (D) reactant and negative or zero order in the electron acceptor (A) reactant.

Table 3 provides 22 such examples. Sufficient information (kinetics) exist in 18 cases. Rule G1 is valid in all 18 cases. There are no exceptions. A typical example from the literature [20] is shown in Figure 28.

An equivalent formulation of rule G1 is the following:

> Rule G1′: A reaction exhibits purely electrophobic behavior $[(\partial r/\partial\Phi)_{p_A,p_D} > 0]$ when the electron acceptor reactant (A) is strongly adsorbed and much more strongly adsorbed on the catalyst surface than the electron donor reactant (D).

In terms of adsorption equilibrium constants the latter can be expressed as

$$k_A p_A \gg 1 \quad \text{and} \quad k_A p_A \gg k_D p_D \Rightarrow (\partial r/\partial\Phi)_{p_A,\, p_D} > 0 \tag{41}$$

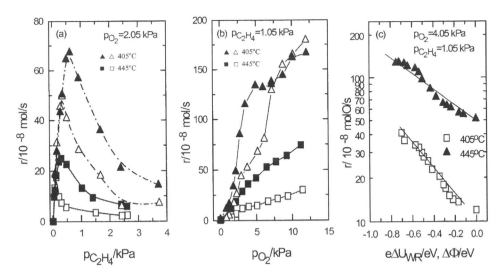

Figure 29 Example of rule G2 (electrophilic behavior): Effect of $p_{C_2H_4}(= p_D)$ (a), $p_{O_2}(= p_A)$ (b), and $\Delta\Phi$ (c) on the rate of C_2H_4 oxidation on Pt films interfaced with $CaZr_{0.9}In_{0.1}O_{3-\alpha}$, a H^+-conductor (from Ref. 80). Note that Figure 29(c) is obtained under gaseous composition where the rate is positive order in O_2 and negative order in C_2H_4 [(a) and (b)]. (Reprinted with permission from the Institute for Ionics.)

where k_A, k_D are the adsorption equilibrium constants of A and D on the catalyst surface.

Electrophilic Reactions

Inspection of Table 3 shows the following rule for electrophilic reactions:

> Rule G2: A reaction exhibits purely electrophilic behavior $[(\partial r/\partial\Phi)_{p_A,p_D} < 0]$ when the kinetics are positive order in the electron acceptor (A) reactant and negative or zero order in the electron donor (D) reactant.

Table 2 provides 18 such examples. Sufficient information (kinetics) exists in 13 cases. Rule G2 is satisfied in all cases. A typical example from the electrochemical promotion literature [80] is shown in Figure 29.

An equivalent formulation of rule G2 is the following:

> Rule G2': A reaction exhibits purely electrophilic behavior $[(\partial r/\partial\Phi)_{p_A,p_D} < 0]$ when the electron donor reactant (D) is strongly adsorbed and much more strongly adsorbed on the catalyst surface than the electron acceptor reactant (A).

In terms of adsorption equilibrium constants the latter can be expressed as

$$1 \ll k_D p_D \quad \text{and} \quad k_A p_A \ll k_D p_D \Rightarrow (\partial r/\partial\Phi)_{p_A, p_D} < 0 \qquad (42)$$

Volcano-type Reactions

Inspection of Table 3 shows that reactions exhibiting volcano-type (maximum type) behavior with respect to Φ are those where the kinetics also exhibit a maximum with

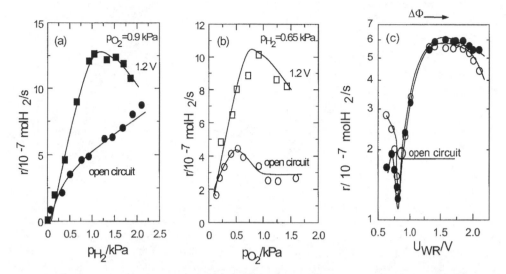

Figure 30 Example of rule G3 (volcano-type behavior): Effect of $p_{H_2}(= p_D)$ (a), $p_{O_2}(= p_A)$ (b), and potential U_{WR} and $\Delta\Phi$ (c) on the rate of H_2 oxidation on Pt/graphite (a and b) and Pt/black (c) in aqueous 0.1 M KOH solutions (from Refs. 84 and 85). Note that under the p_{H_2}, p_{O_2} conditions of (c) the open-circuit rate is positive order in H_2 (a) and negative order in O_2 (b) and that the orders are reversed with the applied positive potential ($U_{WR} = 1.2$ V). At this potential the rate passes through its maximum (volcano) value (c). [Reprinted with permission from McMillan Magazines Ltd (Ref. 84) and from the American Chemical Society (Ref. 85).]

respect to A and D so that the rate is always positive order in A or D and at the same time negative (not zero) order in D or A, respectively. Thus the following rule is derived:

> Rule G3: A reaction exhibits volcano-type behavior when both the electron donor D and electron acceptor A are strongly adsorbed on the catalyst surface.

Table 3 provides eight such examples. Sufficient information exists in seven cases. There are no exceptions. Some typical examples from the literature [84,85] are shown in Figure 30. In terms of adsorption equilibrium constants, rule G3 can be expressed as

$$k_A p_A, k_D p_D \gg 1 \Rightarrow \begin{cases} (\partial r/\partial \Phi)_{p_A, p_D} > 0; \ \Phi < \Phi_M \\ (\partial r/\partial \Phi)_{p_A, p_D} = 0; \ \Phi = \Phi_M \\ (\partial r/\partial \Phi)_{p_A, p_D} < 0; \ \Phi > \Phi_M \end{cases} \tag{43}$$

where Φ_M is the work function value at the rate maximum (volcano-maximum).

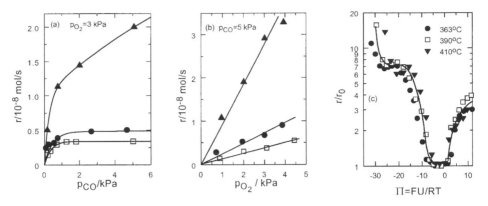

Figure 31 Example of rule G4 (inverted volcano-type behavior): Effect of p_{CO} ($= p_D$) (a), $p_{O_2}(= p_A)$ (b), and $\Delta\Phi$ and dimensionless work function Π ($= \Delta\Phi/k_bT$) (c) on the rate of CO oxidation on Ag films deposited on YSZ (from Ref. [54]); $T = 415\,°C$. (a) $p_{O_2} = 3\,kPa$, $\square$: open-circuit, $\bullet$: $U_{WR} = 475\,mV$, $\blacktriangle$: $U_{WR} = -1300\,mV$. (b) $p_{CO} = 5\,kPa$, $\square$: open-circuit, $\bullet$: $U_{WR} = 475\,mV$, $\blacktriangle$: $U_{WR} = -1300\,mV$. (c) $p_{O_2} = 3\,kPa$, $p_{CO} = 5\,kPa$, $\bullet$: $T = 363\,°C$, $r_0 = 2.7\,nmol$ O/s, $\square$: $T = 390\,°C$, $r_0 = 3.4\,nmol$ O/s, $\blacktriangledown$: $T = 410\,°C$, $r_0 = 5.5\,nmol$ O/s. (Reprinted with permission from Trans. Tech. Publications.)

Inverted Volcano (Minimum) Type Reactions

Inspection of Table 3 shows the following rule for inverted volcano-type reactions:

> Rule G4: A reaction exhibits inverted volcano (minimum rate) type behavior when the kinetics are positive order in both the electron acceptor (A) and electron donor (D) reactant.

Table 3 provides 11 such examples. Sufficient information exists in nine cases. There are no exceptions. Some typical examples from the literature [54] are shown in Figure 31.

In terms of adsorption equilibrium constants, rule G4 can be expressed as

$$k_A p_A, k_D p_D \ll 1 \Rightarrow \begin{cases} (\partial r/\partial\Phi)_{p_A,p_D} < 0; & \Phi < \Phi_m \\ (\partial r/\partial\Phi)_{p_A,p_D} = 0; & \Phi = \Phi_m \\ (\partial r/\partial\Phi)_{p_A,p_D} > 0; & \Phi > \Phi_m \end{cases} \qquad (44)$$

where Φ_m is the work function value at the rate minimum.

Complementary Rules

Three additional rules, also derived from experiment [14,129], can be formulated:

> Rule G5: The above rules G1–G4 also apply when D and A are both electron acceptors or electron donors. In this case D is always the stronger electron donor or weaker electron acceptor and A is always the weaker electron donor or stronger electron acceptor.
>
> Rule G6: A monomolecular reaction is electrophobic for an electron donor adsorbate and electrophilic for an electron acceptor adsorbate.
>
> Rule G7: The maximum rate modification (ρ_{max}/ρ_{min}) obtained under electrochemical promotion conditions increases for every fixed overpotential

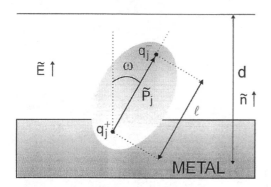

Figure 32 Schematic of an adsorbate, modeled as a dipole, in the presence of the metal–gas effective double layer.

with increasing difference in the electron acceptor-electron donor character of the two reactants.

On the basis of the published literature on NEMCA or electrochemical promotion as of this writing (2001) *there appear to be no exceptions to the above rules.*

19.3.5 Mathematical Modeling of Promotional Kinetics

Recently a rigorous quantitative model was developed in order to describe promotional and, more generally, catalytic kinetics [130,147]. The model can be viewed as an extension of classical Langmuir–Hinshelwood–Hougen–Watson (LHHW) kinetics.

The basis of the model is shown in Figure 32. Each adsorbate is modeled as a dipole characterized by the distance, λ, between the positive and negative charge q and by the partial charge-transfer parameter λ_j defined from the equilibrium:

$$S_j(g) \rightleftharpoons [S_j^{\lambda_j^+} - \lambda_j e^-] \tag{45}$$

where S_j denotes a reactant, j, and the rhs of Eq. (45) denotes the adsorbate dipole. The dipole interacts with the field strength, $\tilde{E}$, of the double layer, so that its electrochemical potential, $\bar{\mu}_j(\text{ad})$, can be written as

$$\bar{\mu}_j(\text{ad}) = \mu_j(\text{ad}) + N_{AV}\tilde{P}_j \cdot \tilde{E} \tag{46}$$

where $\mu_j(\text{ad})$ is the chemical potential of the adsorbed species and N_{AV} is Avogadro's number.

Assuming uniform adsorption sites and using the fact [14] that

$$\tilde{E} = (\Delta\Phi/\text{ed})\tilde{n} \tag{47}$$

where $\Delta\Phi$ is the deviation of Φ from its value at the point of zero charge (pzc) [14], one obtains the isotherm

$$k_j p_j = (\theta_j/(1 - \theta_j)) \exp(-\lambda_j \Pi) \tag{48}$$

with

$$\Pi = \Delta\Phi\left(\frac{\lambda}{2d} \cos \omega\right)/k_b T \tag{49}$$

$$k_j = \exp((\mu_j^o(g) - \mu_j^o(ad))/RT) \tag{50}$$

The first success of the effective double-layer isotherm [Eq. (48)] is that it predicts the experimentally observed linear variation of isosteric heat of adsorption or chemisorptive binding energy E_j with work function Φ (Figures 11, 12, 24, 25):

$$E_{b,j} = E_{b,j}^o + (\lambda_j \lambda/2d)\Delta\Phi \tag{51}$$

Thus for an electron acceptor adsorbate ($\lambda_j < 0$), Eq. (51) predicts a linear decrease in $E_{b,j}$ with Φ while for an electron donor adsorbate ($\lambda_j > 0$), Eq. (51) predicts a linear increase in $E_{b,j}$ with Φ, both in excellent agreement with experiment (Figures 11, 12, 24) and with rigorous cluster quantum mechanical calculations (Figure 25).

The second, and most important, success of the effective double-layer isotherm is that when considering the coadsorption and rate-controlling surface reaction of two reactants, A and D, one obtains the following *analytical* expression for the catalytic rate:

$$r = k_R \theta_D \theta_A = \frac{k_R k_A k_D p_D p_A \exp[(\lambda_D + \lambda_A)\Pi]}{(1 + k_D p_D \exp(\lambda_D \Pi) + k_A p_A \exp(\lambda_A \Pi))^2} \tag{52}$$

which provides an excellent semiquantitative fit to the experimentally observed promotional kinetics as shown in Figure 33. This simple expression predicts indeed purely electrophobic behavior for strong adsorption of A ($k_D = 10^{-2}$, $k_A = 10^2$), purely electrophilic behavior for strong adsorption of D ($k_D = 10^2$, $k_A = 10^{-2}$), and volcano-type behavior for strong adsorption of A and D ($k_D = k_A = 10^2$). In the case of weak adsorption of A and D, since repulsive lateral interactions can be neglected [14,129], Eq. (52) is replaced by

$$r = \frac{k_R k_A k_D p_A p_D \exp\{[\max(0, \lambda_D \Pi)] + [\max(0, \lambda_A \Pi)]\}}{(1 + k_D p_D \exp[\max(0, \lambda_D \Pi)] + k_A p_A \exp[\max(0, \lambda_A \Pi)])^2} \tag{53}$$

which also nicely predicts inverted volcano behavior for $k_A = k_D = 10^{-2}$ [Figure 33(d)].

The success of the model can be appreciated from Figure 34, which compares model predictions [top, Figure 34(a) and (b)] with some truly interesting and complex experimental results [bottom, Figure 34(c) and (d)] obtained during C_2H_4 oxidation on Pt/TiO_2 [64,130]. As shown in Figure 34(c) and (d) (bottom) the rate dependence on U_{WR} and Π shifts from inverted volcano [Figure 34(c)] to purely electrophobic [Figure 34(d)] as $p_{C_2H_4}(=p_D)$ is decreased by a factor of 10 at fixed p_{O_2}. As shown in Figure 34(a) and (b) (top) the model predicts the shift in global behavior

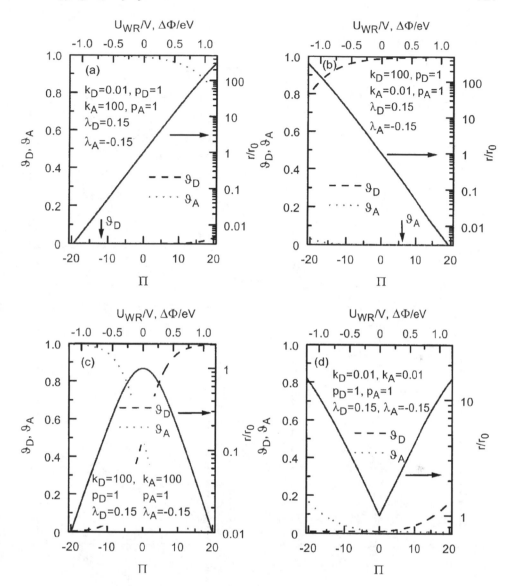

Figure 33 Model predicted electrochemical promotion behavior: (a) electrophobic, (b) electrophilic, (c) volcano-type, and (d) inverted volcano-type.

in a truly impressive semiquantitative manner and in fact with very reasonable λ_D and λ_A values ($\lambda_D > 0$, $\lambda_A < 0$).

Finally, the success of the model can be judged from Figure 35(a) and (b), which show the experimental and model-predicted rate dependence on p_{CO} and work function during CO oxidation on Pt/β''-Al$_2$O$_3$ [68]. Note the transition from a classical Langmuir–Hinshelwood to a positive order rate dependence on p_{CO} with decreasing work function. Also notice that on every point of the experimental or

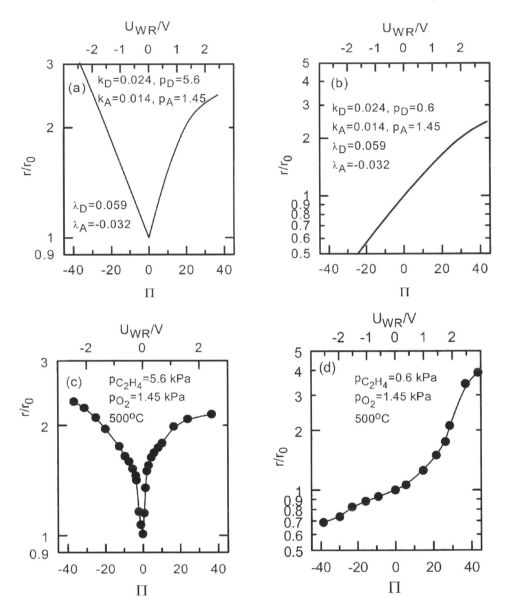

Figure 34 Experimentally observed (bottom) and model predicted (top) transition from inverted volcano to electrophobic behavior upon increasing the O_2 to ethylene (i.e., A/D) ratio by a factor of 10, C_2H_4 oxidation on Pt/TiO_2. (From Ref. 64. Reprinted with permission from Academic Press.)

model-predicted rate dependence, the basic promotional rule

$$\left(\frac{\partial r}{\partial \Phi}\right)_{p_A, p_D} \left(\frac{\partial r}{\partial p_D}\right)_{\Phi, p_A} > 0 \tag{54}$$

is strictly obeyed. The optimal λ_D and λ_A values are again quite reasonable

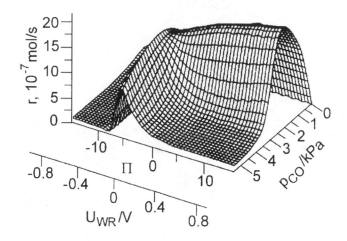

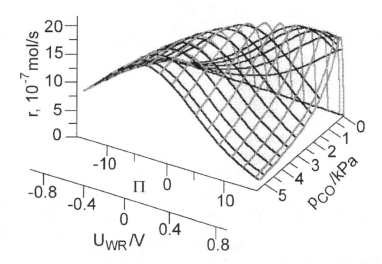

Figure 35 Experimental (from Ref. 68) (top) and model-simulated (from Ref. 147) (bottom) dependence of the rate of CO oxidation on Pt deposited on β''-Al_2O_3 as a function of p_{CO}, catalyst potential U_{WR}, and dimensionless catalyst work function $\Pi \, (= \Delta\Phi/k_b T)$ at $p_{O_2} = 6\,kPa$ (from Ref. 68). Parameters used in Eqs. (52) and (53): $k_A = 9.133$, $k_D = 8.715$, $\lambda_A = -0.08$, $\lambda_D = 0.09$, $\lambda_R = 0$, $k_R = 6.19 \cdot 10^{-6}$. (Reprinted with permission from Academic Press.)

$(\lambda_D, \lambda_A < 0)$. The large optimal k_A and k_D values (~ 9) are also quite reasonable as they indicate strong adsorption of both CO (=D) and oxygen (=A), which is the necessary and sufficient condition (rule G3) for the appearance of volcano-type behavior.

In general, Figures 33 to 35 show, beyond any reasonable doubt, that the effective double-layer model of promotion, expressed mathematically by Eq. (52), grasps the essence of promotional kinetics [14].

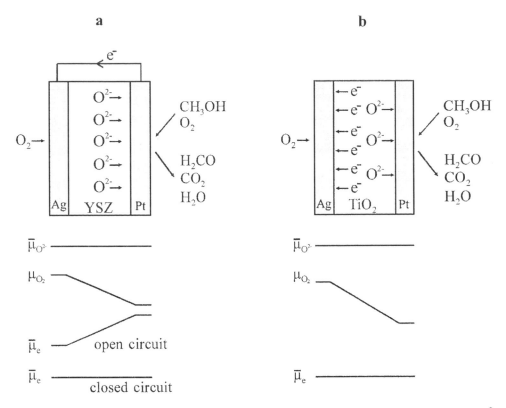

Figure 36 Self-driven electrochemical promotion of CH_3OH oxidation on Pt using O^2 conductor (YSZ) with external catalyst-counterelectrode short-circuiting (a) and a mixed ionic (O^{2-})-electronic conductor (TiO_2) with internal short-circuiting.

19.4 THE MECHANISM OF METAL–SUPPORT INTERACTIONS

19.4.1 Self-Driven NEMCA: The Bridge Between Electrochemical Promotion and Metal–Support Interactions

Before discussing in the next section the series of experimental results that proved the functional identity of electrochemical promotion and metal–support interactions with O^{2-} conductions (YSZ) and mixed O^{2-}-electronic conducting (TiO_2, CeO_2) supports, it is useful to first review an experiment reported in 1993 by Cavalca, Larsen, Vayenas, and Haller [141] in the course of their investigation of electrochemical promotion of methanol oxidation on Pt deposited on YSZ in a single-chamber reactor (Figure 36). Cavalca et al. [141] noticed (as one easily observes in any electrochemical promotion study using the single-chamber design) that a *negative* potential appears ($U_{WR} \approx 0.3\,V$) between the Pt catalyst and the Ag counter- and reference electrodes *under open-circuit conditions*.

This negative potential is easy to understand and is actually used today in the *single-chamber fuel cell* [14]: oxygen is consumed by the catalytic reaction of CH_3OH oxidation much faster on the Pt catalyst-electrode than on the Ag counterelectrode (Ag is also a catalyst for CH_3OH oxidation and partial oxidation, but much less active than Pt). Consequently the observed open-circuit potential is a Nernst potential:

$$U^o_{WC} = (1/4e)(\mu_{O_2(Pt)} - \mu_{O_2(Ag)}) = \frac{RT}{4F} \ln(\alpha^2_{O(Pt)}/\alpha^2_{O(Ag)}) \tag{55}$$

where the oxygen chemical potential and activity on Pt are significantly smaller than on Ag (due to the faster catalytic reaction on Pt, which forces the oxygen adsorption step to be not in equilibrium), thus U^o_{WC} is negative.

Cavalca et al. [141] utilized this self-created potential and the fact that methanol oxidation on Pt exhibits *electrophobic behavior*, i.e., is promoted by O^{2-} supply *to* the catalyst, to induce NEMCA (electrochemical promotion) on Pt *without an external power source (galvanostat or potentiostat) by just short-circuiting the catalyst and the counterelectrode* [Figure 36(a)].

What is the mechanism of electrochemical promotion under these short-circuit conditions? It is very simple, as shown in Figure 36. Upon short-circuiting, the Fermi levels in the Pt catalyst and Ag counterelectrode become equal and O^{2-} are being continuously supplied to the Pt catalyst, inducing NEMCA and attempting to establish there an oxygen chemical potential $\mu_{O_2(Pt)}$ equal to that on Ag. Due to the fast catalytic reaction on Pt, this is not possible and thus the O^{2-} flux to the Pt catalyst continuous ad infinitum. The spent O^{2-} are continuously replenished by gaseous O_2 at the Ag counterelectrode. This is a *self-driven NEMCA system*, suitable for promoting *any electrophobic reaction*.

To compensate the O^{2-} flux and close the circuit there is an electric current flowing through the short-circuiting wire [Figure 36(a)]. Now let us imagine that YSZ is replaced by TiO_2, which has both ionic (O^{2-}) and electronic conductivity [Figure 36(b)]. In this case the need for the short-circuiting wire is eliminated as the electronic current can flow in the TiO_2 pellet itself. Under these conditions there is no net current flowing in the TiO_2 pellet but the Pt catalyst is electrochemically promoted. This is again a self-driven NEMCA system suitable for any electrophobic reaction and having no *net* current flowing anywhere in it (the O^{2-} current in the TiO_2 is equal and opposite to the electronic current).

The similarity with TiO_2-supported dispersed Pt catalyst is obvious. For if O^{2-} backspillover can take place over mm distances on the porous Pt films used for electrochemical promotion studies with YSZ and TiO_2 (as proven conclusively by XPS and several other techniques [148]) it can certainly take place over nm distances on the supported Pt nanoparticles of commercial highly dispersed YSZ and TiO_2-supported catalysts.

Do we really need an Ag counterelectrode in this self-driven NEMCA (or metal–support interaction promoted) system? The answer is no, for two reasons:

1. With the usual size distribution of nanoparticles in a commercial ZrO_2- or TiO_2- or CeO_2-supported noble metal catalyst, each two dissimilar size crystallites create a small local galvanic cell able to electrochemically promote one of the two nanoparticles.
2. In mixed conducting supports (TiO_2, CeO_2) but also on doped ZrO_2 *direct* replenishment of O^{2-} by gaseous O_2 is rather efficient:

$$O_2 + 4e^- \rightarrow 2O^{2-} \text{ (YSZ)} \tag{56}$$

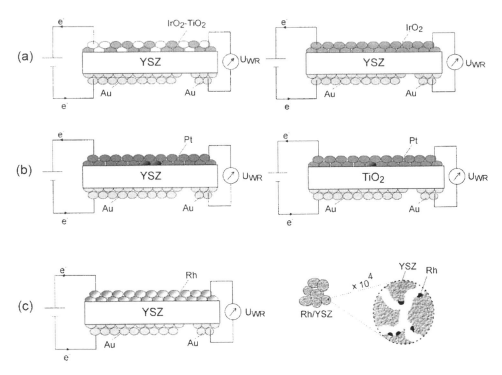

Figure 37 Schematic of the experimental setup used (a) to induce electrochemical promotion (via YSZ) on IrO_2 and IrO_2-TiO_2 porous catalyst films (b) to compare the electrochemical promotion induced on Pt via YSZ and via TiO_2 and (c) to compare the electrochemical promotion behavior induced by varying U_{WR} on a Rh porous catalyst film (left) and on a fully dispersed Rh catalyst supported on porous ($80\,m^2/g$) YSZ support. (From Ref. [137].)

Thus on the basis of the experiments of Cavalca et al. [141], in conjunction with the XPS studies surveyed previously, it is reasonable to anticipate that the mechanism of metal–support interaction promotion of catalytic oxidations is the same with the mechanism of self-driven electrochemical promotion systems: *self-induced migration of promoting O^{2-} species on the nanoparticle catalysts surface* [137].

19.4.2 Confirmation of the Self-driven NEMCA Metal–Support Interaction Mechanism

Three independent systems were used by Nicole, Tsiplakides, Pliangos, Verykios, Comninellis, and Vayenas [137] to show the mechanistic equivalence of NEMCA and metal–support interactions (Figure 37).

Here we discuss the results obtained for the model reaction of C_2H_4 oxidation on IrO_2, Pt, and Rh, but similar conclusions are reached when using other model reactions such as CO oxidation or NO reduction by CO [137]. The three systems shown in Figure 37 were used to compare

1. The open-circuit and NEMCA-induced catalytic activity of IrO_2 (which is a metal-type conducting metal oxide [149]) and of mixed IrO_2-TiO_2 catalysts consisting of micro- and nanoparticles of IrO_2 (active phase) and TiO_2 (inert support) in intimate contact [Figure 37(a)] [150].

2. The open-circuit and NEMCA-induced catalytic activity of Pt films deposited on YSZ [26] and on TiO_2 [137]. In this case XPS was also used in vacuum [87,137] to quantify the coverage of the backspillover O^{2-} species on the Pt surface [Figure 37(b)].

3. The catalytic rate enhancement induced on porous Rh films via electrochemical promotion with YSZ [Figure 37(c), left [40]] and that induced on dispersed Rh nanoparticles upon varying the porous, high-surface-area ($\sim 100\,m^2/g$) catalyst support (TiO_2, SiO_2, γ-Al_2O_3, YSZ, and TiO_2 doped with 4 mol% WO_3) [151]. In all five cases the Rh metal loading was 0.5 wt. % [151].

In view of the title of this section the reader can anticipate and predict the results of these key experiments [137]:

1. There is similar, roughly 12-fold, maximum rate enhancement induced on the IrO_2 catalyst via NEMCA ($\rho \approx 12$, pure IrO_2, Figure 38) and via metal–support interactions of IrO_2 with TiO_2 ($\rho_{MSI} \approx 13$, $X_{IrO_2} \approx 0.5$, Figure 38). The parameter ρ_{MSI} is defined from

$$\rho_{MSI} = r/r_u \tag{12}$$

    where r_u is the (unpromoted) catalytic rate per unit mass of the active catalyst and r is the same (promoted) catalytic rate, enhanced due to the metal–support interaction.

    Moreover, as also shown in Figure 38, there is practically *no electrochemical promotion* ($\rho < 1.5$) of the mixed IrO_2-TiO_2 catalyst. It is thus clear that IrO_2 in the IrO_2-TiO_2 catalyst is already in an electrochemically promoted state. It thus becomes apparent that TiO_2 is constantly supplying O^{2-} to the IrO_2 surface. This ingenious experiment is due to Nicole and Comninellis [150]. Note that pure TiO_2 ($X_{IrO_2} = 0$) is always inactive (Figure 38).

2. There is similar transient and steady-state electrochemical promotion behavior of Pt on YSZ and Pt on TiO_2 [Figure 39(a) and (b)] and similar O^{2-} backspillover mechanism of Pt on YSZ and of Pt on TiO_2 as manifested by XPS (Figure 21).

    In particular:

    In both cases imposition of a positive current I (with a concomitant rate, $I/2F$, of supply of O^{2-} to the catalyst for the case of Pt/YSZ and also an concomitant increase in catalyst potential U_{WR}) causes a pronounced, 25-fold in Figure 39(a), 22-fold in Figure 39(b), increase in catalytic rate ($\rho = 26$ and $\rho = 23$, respectively).

    The Faradaic efficiency Λ is $74 \cdot 10^3$ in Figure 39(a) (YSZ) and $1.88 \cdot 10^3$ in Figure 39(b) (TiO_2), suggesting that only a fraction $f (\approx 2.5\%)$ of the current I in TiO_2 is anionic (O^{2-}), the rest being electronic, in good agreement with the literature [137]. This is nicely confirmed by comparing

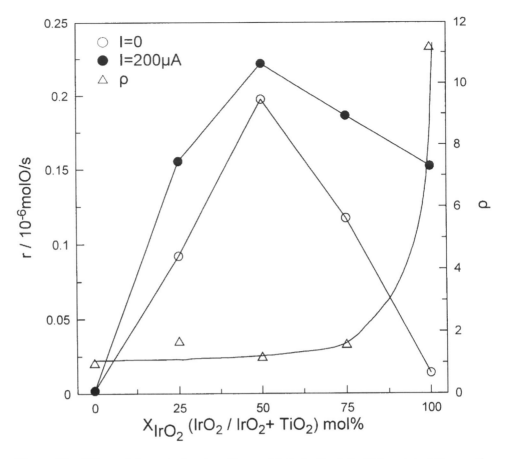

Figure 38   Effect of the mole fraction, X_{IrO_2}, of IrO_2 in the IrO_2-TiO_2 catalyst film on the rate of C_2H_4 oxidation under open-circuit conditions (open circles) and under electrochemical promotion conditions (filled circles) via application of $I = 200\,\mu A$, $T = 380\,°C$, $p_{C_2H_4} = 0.15\,kPa$, $p_{O_2} = 20\,kPa$. Triangles indicate the corresponding electrochemical promotion rate enhancement ratio ρ values. (From Refs. 137 and 150.)

the time, τ, required for the rate increase to reach 63% of its steady-state value with the parameter $2FN_G/I$ (Figure 39). In the case of YSZ, τ is shorter than $2FN_G/I$, while in the case of TiO_2 τ is longer than $2FN_G/I$, again suggesting that only a fraction of the current I in TiO_2 is ionic. By comparing the ratio $\tau/(2FN_G/I)$ in both cases, one may conclude that $f \approx 0.05$, in qualitative agreement with the value estimated from the Faradaic efficiency, Λ, values.

3.   There is similar electrochemical promotion behavior of Rh films on YSZ and similar metal–support interaction-induced behavior of *dispersed* Rh on different supports for the model reaction of C_2H_4 oxidation on Pt (Figure 40). In particular, there are very similar ρ values ($\rho \approx \rho_{MSI} \approx 120$) upon increasing the potential and work function of the Rh film or upon increasing the work function (or absolute potential) of the support of the dispersed Rh catalyst (Figure 40).

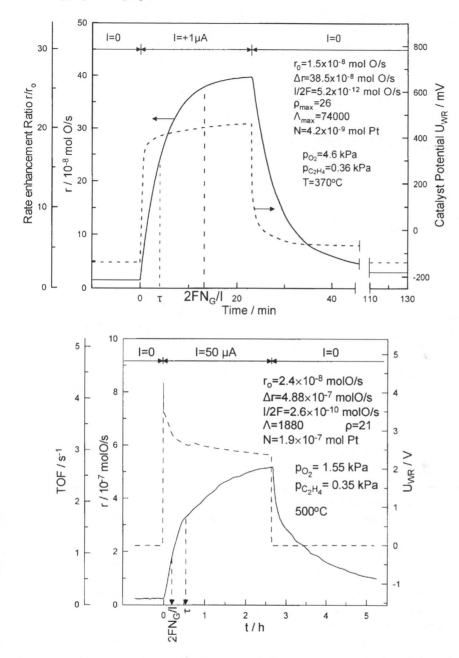

Figure 39 Galvanostatic catalytic rate transients showing the equivalence of electrochemical promotion when using YSZ (from Ref. 26) (a) or TiO_2 (from Ref. 64) (b) as the Pt metal film support. See text for discussion. (From Ref. 137. Reprinted with permission from Academic Press.)

In more detail: Figure 40 shows the rate dependence on p_{O_2} for the dispersed Rh catalysts deposited on TiO_2, SiO_2, YSZ (8 mol% Y_2O_3 in ZrO_2), and TiO_2 doped with 4 mol% WO_3. In all five cases the Rh metal loading is 0.05 wt. % [137,152].

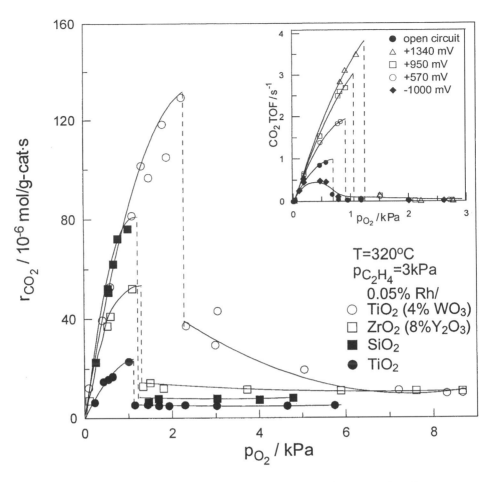

Figure 40   Effect of p_{O_2} on the rate of C_2H_4 oxidation on Rh supported on four supports of increasing Φ. Catalyst loading 0.05 wt. %. (From Refs. 137 and 152.) Inset: Electrochemical promotion of a Rh catalyst film deposited on YSZ: effect of potentiostatically imposed catalyst potential U_{WR} on the rate and TOF dependence on p_{O_2} at fixed $p_{C_2H_4}$. [Reprinted with permission from Elsevier Science (Ref. 137) and Academic Press (Ref. 152).]

The inset of Figure 40 shows the rate dependence on p_{O_2} (at the same $p_{C_2H_4}$ and T) for the Rh *film* deposited on YSZ at various imposed potentials U_{WR}. The similarity between Figure 40 and the inset of Figure 40 is striking and underlines the equivalence of metal–support interactions and electrochemical promotion: for low p_{O_2} values the rate is first order in p_{O_2} followed by a sharp decrease at a characteristic p_{O_2} value denoted by $p_{O_2}^*$ (U_{WR}^*), which depends on the support (Figure 40) or on the potential (inset of Figure 40).

Thereafter the rate becomes very low and negative order in p_{O_2}. It has been shown that this sharp rate transition is due to the formation of a catalytically inactive surface Rh oxide [137]. As shown in Figure 40 (inset) increasing U_{WR} and thus Φ causes a pronounced increase in $p_{O_2}^*$ and thus a dramatic rate increase at intermediate p_{O_2} values with ρ values up to 100. Exactly the same behavior is

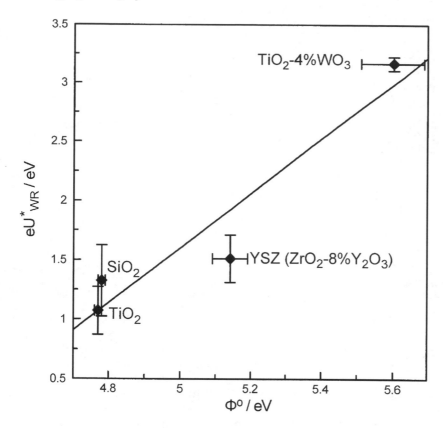

Figure 41 Correlation between the equivalent potentials of the supports defined by Figure 40 and of the work function or absolute potential (from Ref. 140) of the supports measured via the Kelvin probe technique in $p_{O_2} = 1$ atm at 400 °C. (From Ref. 137.)

obtained (Figure 40) upon varying the dispersed catalyst support in the sequence TiO_2, SiO_2, γ-Al_2O_3, YSZ, TiO_2 (4%WO_3). For intermediate p_{O_2} values the metal–support interaction rate enhancement ratio, ρ_{MSI}, is up to 120 versus $\rho \approx 100$ for the electrochemically promoted system.

This destabilization of surface Rh oxide formation with increasing catalyst potential or work function has been shown to be due to strong lateral repulsive interactions of the backspillover O^{2-} species and normally chemisorbed oxygen, which causes a pronounced, up to 1 eV, decrease in the chemisorptive bond strength of normally chemisorbed O [13,14].

By comparing Figure 40 and the characteristic $p_{O_2}^*$ (U_{WR}^*) rate breaks of the inset of Figure 40 one can assign to each support an equivalent potential U_{WR}^* value (Figure 41). These values are plotted in Figure 41 versus the actual work function $\Phi°$ measured via the Kelvin probe technique for the supports at $p_{O_2} = 1$ atm and $T = 400$ °C, i.e., versus the absolute potential of the supports.

The good qualitative agreement between eU_{WR}^* variation and $\Phi°$ variation shown in Figure 41 for the various supports used underlines again the common promotional mechanism of electrochemically promoted and metal–support interaction promoted metal catalysts.

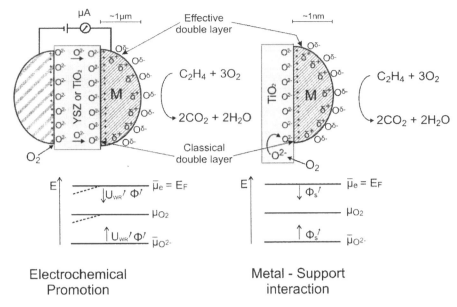

Figure 42 Schematic of a metal grain ($\sim \mu m$) in a metal catalyst film deposited on YSZ or TiO$_2$ under electrochemical promotion conditions (left) and of a metal nanoparticle ($\sim$ nm) deposited on a porous TiO$_2$ support (right) showing the locations of the classical double layers formed at the metal–support interface and of the effective double layers formed at the metal–gas interface. The energy diagrams (bottom) indicate schematically the spatial constancy of the Fermi level E_F (or electrochemical potential $\bar{\mu}_e$) of electrons, of the chemical potential of oxygen, and of the electrochemical potential of O^{2-}. Note that under electrical bias application (left) $\bar{\mu}_{O^{2-}}$ remains spatially constant but $\bar{\mu}_e$ and μ_{O_2} both bend in the solid electrolyte support (dashed lines). The Fermi level $\bar{\mu}_e$ of the metal can be affected by varying U_{WR} (left) or by varying via doping the Fermi level of the support (right). (From Refs. 137 and 138.)

The common underlying principle is shown in Figure 42. The electrochemical potential of electrons $\bar{\mu}_e$ ($= E_F$, the Fermi level) in the metal catalyst is fixed at that of the Fermi level of the support [140]. This is valid both for electrochemically promoted model catalysts (left) and for semiconducting or ion-conducting-supported metal nanoparticles (right).

In electrochemical promotion experiments (Figure 42, left) one can vary $\bar{\mu}_{e(M)} = \bar{\mu}_{e(S)}$ by varying U_{WR} and thus also [Eq. (29)] Φ. In this way via Eq. (35) one can also vary the electrochemical potential and thus coverage of backspillover O^{2-} on the catalyst-electrode surface.

In dispersed metal–support systems (Figure 42, right), one can vary $\bar{\mu}_{e(M)} = \bar{\mu}_{e(S)}$ by varying the support or by doping the support with aliovalent cations. This is known in the literature as dopant-induced metal–support interactions (DIMSI) [153,154]. Thus one can again vary the electrochemical potential and thus the coverage of backspillover O^{2-} on the supported catalyst surface.

This simple model (Figure 42) can account for the observed equivalence between electrochemical promotion and metal–support interaction-induced promotional phenomena. In both cases O^{2-} backspillover to the catalyst surface is the dominant promotional mechanism. It should be noted that according to the above equivalence only electrophobic reactions can be promoted by metal–support interactions on YSZ and TiO$_2$, i.e., only reactions in which the rate increases with potential or work function. This is nicely confirmed by experiment, since C$_2$H$_4$ oxidation both on IrO$_2$ and on Rh is an electrophobic reaction ($\partial r/\partial U_{WR} > 0$). Detailed investigation of C$_2$H$_4$ oxidation on different metals and ion-conducting supports (Tables 1 and 3) has shown that it always exhibits electrophobic behavior under oxygen-rich conditions [31].

In conclusion, electrochemical promotion is an electrically controlled metal–support interaction. The corollary is that metal–support interactions on YSZ, TiO$_2$, and doped TiO$_2$ supports are similar to the "wireless" self-driven NEMCA configurations tested already on YSZ pellets (for the electrophobic reaction of CH$_3$OH oxidation on Pt) by short-circuiting the Pt catalyst and Ag counterelectrodes. The carrier continuously supplies promoting O$^{\delta-}$ species to the catalyst surface while spent O^{2-} in the support is continuously replenished by gaseous O$_2$.

Mathematical Modeling and Range of Operability of Electrochemical Promotion and of Metal–Support Interactions

In view of the functional similarity of electrochemical promotion and metal–support interactions, a mathematical promoter reaction-promoter diffusion model has been developed recently [138] in order to identify the dimensionless groups, which govern both phenomena and quantify their limits of applicability.

The model geometries are shown in Figure 43, and the basic dimensionless promoter (O^{2-}) reaction-diffusion equation, governing both phenomena, is

$$\frac{d^2\theta_i}{d\xi^2} - \Phi_P^2 \theta_i = 0 \tag{57}$$

with boundary conditions

$$\xi = 0 \qquad \frac{d\theta_i}{d\xi} = -J\Phi_P^2(1 - \theta_i) \tag{58}$$

$$\xi = 1 \qquad d\theta_i/d\xi = 0 \tag{59}$$

where ξ is the dimensionless distance across the electrochemically promoted catalyst film or nanoparticle, θ, is the coverage of the backspillover O^{2-} species, and Φ_P and J are the promotional Thiele modulus and dimensionless current, respectively, defined from

$$\Phi_P = L\sqrt{k/D_s} \tag{60}$$

$$J = I/(2FkC_{i,max}A_c) \qquad \text{for electrochemical promotion} \tag{61}$$

$$J = (r/A_c)/(\Lambda k C_{i,max}) \qquad \text{for MSI} \tag{62}$$

Here L is the catalyst film thickness or nanoparticle size, k is the rate constant for depletion (reaction or desorption) of the promoting O^{2-} species, $C_{i,max}$ is its maximum possible surface concentration on the catalyst or nanoparticle surface, A_c is the metal–gas interface area of the film or nanoparticle, r is the promoted catalytic rate, and Λ is the Faradaic efficiency of the catalytic reaction being promoted.

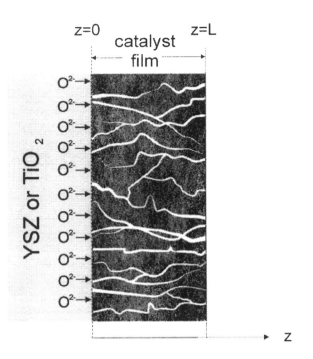

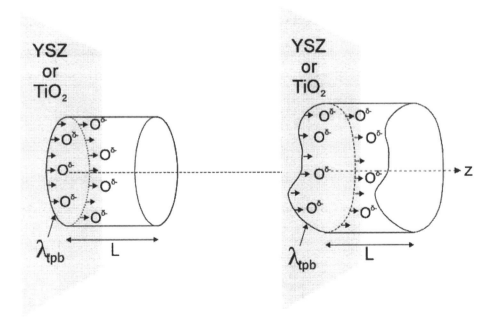

Figure 43 Schematic of an electrochemically promoted metal catalyst film supported on an O^{2-} conductor (top) and schematic of cylindrical or, more generally, fixed cross-section nanoparticles deposited on an O^{2-} conducting support (bottom). (From Ref. 138.)

Solution of Eq. (57) with boundary conditions (58) and (59) leads to the following expressions for the rate enhancement ratio ρ in terms of the dimensionless dipole moment, Π, of the promoting species:

$$\rho = \exp(\Pi\eta_P) \tag{63}$$

$$\Pi = \frac{\alpha e N_M P_i}{\varepsilon_0 k_b T} \tag{64}$$

where η_P is the promotional effectiveness factor defined from

$$\eta_P = \int_0^1 \theta_i(\xi)d\xi \tag{65}$$

and computed from

$$1/\eta_P = 1/J + \Phi_P/\tanh\Phi_P \tag{66}$$

In order to have significant electrochemical promotion of metal–support interaction promotion of a catalytic reaction, η_P must be at least 0.2. Equation (66) and the corresponding Figure 44 show the range of Φ_P and J values that allow for this to happen: the Thiele modulus Φ_P must be smaller than 5. This means small film thickness or catalyst particle size, small kinetic constant k for promoter destruction, and finite surface diffusivity, D_s, of the promoter. Also the dimensionless current J must be larger than 2, and this again dictates a small k value for promoter destruction, a finite current for electrochemical promotion, and a fast catalytic rate, r, for metal–support interactions.

Numerical Examples

Electrochemically Promoted Films
In order to estimate η_P in actual electrochemical promotion experiments we use here typical values [138] of the operating parameters (Table 4) to calculate J and Φ. The

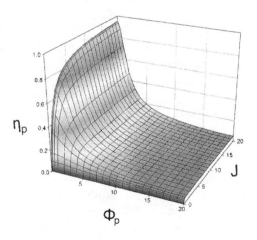

Figure 44 Dependence of promotional effectiveness factor, η_p, on Thiele modulus Φ_p and dimensionless current J. (From Ref. 138.)

Table 4 Typical Operating Parameters in Electrochemical Promotion Studies (Electrolyte surface area $A_E = 1 \text{ cm}^2$, $T = 400°C$)

$I = 100 \, \mu A$
$k = 10^{-2} \, s^{-1}$
$A_c C_{i,max} = 10^{-7} \, \text{mol}$
$L = 3 \, \mu m$
$D_s = 4 \cdot 10^{-11} \, \text{cm}^2/\text{s}$

Computed Parameters
$J = 0.5$
$\Phi_p = 4.8$
$\eta_P = 0.15$

Condition for $\Phi < 1$
$\Phi_p < 1 \Rightarrow L < 0.6 \, \mu m$

Table 5 Typical Operating Parameters in a Supported Catalyst

$r/A_c = 10^{-6} \, \text{mol/s} \cdot \text{cm}^2$
$\Lambda = 10^2$
$k = 10^{-2} \, s^{-1}$
$C_{i,max} = 10^{-7} \, \text{mol/cm}^2$
$L = 3 \, \text{nm}$

Computed Parameters
$J = 10$
$\Phi_p = 4.8 \cdot 10^{-3}$
$\eta_P = 0.91$

Condition for $\Phi < 1$
$\Phi_p < 1 \Rightarrow L < 0.6 \, \mu m$

value of k is estimated on the basis of typical NEMCA galvanostatic transients which show that the lifetime of the promoting $O^{\delta-}$ species on the catalyst surface is typically 10^2 s at temperatures 350–400 °C.

The surface diffusivity D_s is computed (conservatively) from the diffusivity measurements of Lewis and Gomer [122] for O on Pt(111) and Pt(110) near 400 °C [Eq. (17)].

The computed J, Φ_P, and η_P results (Table 4) show the significance of this analysis [138]. It is very likely that many of the published electrochemical promotion data have been obtained under promoter diffusional-controlled conditions, i.e., the actual measured (quite large) ρ values may not correspond to full utilization of the promoting species ($\eta_P = 1$), so that one could obtain even larger ρ values [$= \exp \Pi$, Eq. (63)] if thinner catalyst films are used. This is corroborated by the fact that the highest ρ value for C_2H_4 oxidation on Pt supported on YSZ ($\rho \approx 60$) has been reported so far for a Pt film with surface area corresponding to 4.2×10^{-9} mol Pt [26] ($L < 0.1\,\mu\text{m}$), whereas significantly smaller ρ values (~ 10–20) have been reported [31] for the same reaction on Pt films with surface areas corresponding to 10^{-7} mol Pt ($L \approx 3\,\mu\text{m}$) [31].

Dispersed Supported Catalysts

In order to estimate η_P in actual fully dispersed Pt, Rh, and Pd catalyst deposited on highly porous Y_2O_3-doped-ZrO_2, WO_3-doped-TiO_2, TiO_2, and γ-Al_2O_3 supports used for CO and light hydrocarbon oxidation we use typical operating parameter values similar to those used in Table 4 and assume $\Lambda = 100$, which is a rather conservative value [13], to compute J via Eq. (62) and Φ via Eq. (60). The results show $\eta_P \approx 0.91$, which implies that the O^{2-} backspillover mechanism is fully operative under oxidation reaction conditions on nanoparticle metal crystallites supported on ionic or mixed ionic-electronic supports, such as YSZ, TiO_2, and CeO_2. This is quite reasonable in view of the fact that, as already mentioned, an adsorbed O atom can migrate $1\,\mu\text{m}$ per s on Pt at $400\,^{\circ}\text{C}$. So unless the oxidation reaction turnover frequency is higher than $10^3\,\text{s}^{-1}$, which is practically never the case, the $O^{\delta-}$ backspillover double layer is present on the supported nanocrystalline catalyst particles [137].

19.5 INTERRELATION OF PROMOTION, ELECTROCHEMICAL PROMOTION, AND METAL–SUPPORT INTERACTIONS: THE DOUBLE-LAYER MODEL OF CATALYSIS

Promotion, electrochemical promotion, and metal–support interactions are, at a first glance, three independent phenomena that can affect catalyst activity and selectivity in a dramatic manner. However, as recently discovered, promotion, electrochemical promotion, and metal–support interactions on ion-conducting and mixed-conducting supports are three different facets of the same phenomenon. All three are linked via the phenomenon of spillover-backspillover. And all three are due to the same underlying cause: the interaction of adsorbed reactants and intermediates with an effective double layer formed by promoting species at the metal–gas interface (Figure 45).

For time scales shorter than that of a catalytic turnover (typically 10^{-2} to 10^2 s) the three phenomena are indistinguishable. Looking at the Na-promoted Pt surface in Figure 23 and imagining that CO oxidation is taking place on that surface, there is no way to distinguish if this is a classically promoted surface where Na has been added from the gas phase, or an electrochemically promoted one where Na originated from β''-Al_2O_3 interfaced with the Pt crystal, or finally if it is the surface of a larger crystallite deposited on a porous β''-Al_2O_3 carrier where Na has spontaneously migrated on the Pt surface (metal–support interaction). The oxidation of CO will be equally promoted in all three cases.

Similar would be the situation on a Pt surface decorated with O^{2-}, the only difference being the experimental difficulty of introducing O^{2-} with classical promotion and its short lifetime on the catalyst surface, only Λ times longer than the catalytic turnover.

Consequently, the proven functional identity of classical promotion, electrochemical promotion, and metal–support interactions should not lead the reader to

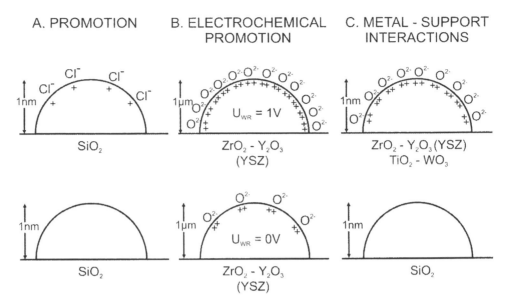

Figure 45 Schematic of the promotional mechanism for electrophobic reactions: A: via electronegative promoter addition of long lifetime (classical promotion), B: via potential- or current controlled O^{2-} backspillover (electrochemical promotion), C: via self-driven O^{2-} backspillover (metal–support interactions).

pessimistic conclusions regarding the practical usefulness of electrochemical promotion. Operational differences exist between the three phenomena, and it is very difficult to imagine how one can use metal–support interactions with conventional supports to promote an electrophilic reaction or how one can use classical promotion to generate the strongest electro-negative promoter, O^{2-}, on a catalyst surface. Furthermore, there is no reason to expect that a metal–support-interaction-promoted catalyst is at its "best" electrochemically promoted state. Thus the experimental problem of inducing electrochemical promotion on fully dispersed catalysts remains important [14].

Having discussed the functional equivalence of classical promotion, electrochemical promotion, and metal–support interactions on O^{2-}-conducting and mixed electronic-ionic conducting supports, it is useful to also address and systematize their operational differences. This is attempted in Figure 46: the main operational difference is the promoter lifetime, τ_{PR}, on the catalyst surface (Figure 46).

For any practical classical promotion application in a fixed-bed catalytic reactor, τ_{PR} must be longer than a year ($\sim 3 \times 10^7$ s). But even for lab-scale classical promotion experiments τ_{PR} values in excess of 10^6 s are required (Figure 46). On the other hand, electrochemical promotion is not subject to any such restrictions regarding τ_{PR} (Figure 46). Thus when using O^{2-} conductors or H^+ conductors, τ_{PR} is 10^2–10^4 s, but when using Na^+ conductors τ_{PR} can be well in excess of 10^7 s at low T, but also in the range 10^4–10^6 s for higher T.

This is an important operational advantage of electrochemical promotion: it permits the use of a wide variety of sacrificial promoters (e.g., O^{2-}, H^+) that have too short lifetimes for classical promotion applications.

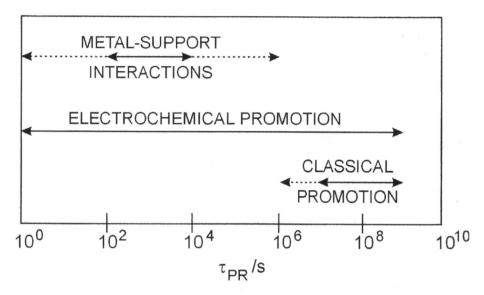

Figure 46 Operational range of classical promotion, electrochemical promotion, and metal–support interactions in terms of the promoter lifetime on the catalyst surface.

ACKNOWLEDGMENT

We thank BASF and the TMR Programme of the EU for financial support and Drs. H. Pütter and F. Kalhammer for numerous helpful discussions.

REFERENCES

1. M. Boudart, G. Djega-Mariadassou, *Kinetics of Heterogeneous Catalytic Reactions*, Princeton Univ. Press, Princeton, NJ (1984).
2. L.L. Hegedus, R. Aris, A.T. Bell, M. Boudart, N.Y. Chen, B.C. Gates, W.O. Haag, G.A. Somorjai, J. Wei, *Catalyst Design: Progress and Perspectives*, John Wiley & Sons, New York (1987).
3. C.N. Satterfield, *Heterogeneous Catalysis in Industrial Practice*, McGraw-Hill, Inc. (1991).
4. R.J. Farrauto, C.H. Bartholomew, *Fundamentals of Industrial Catalytic Processes*, Chapman & Hall, London (1997).
5. G. Ertl, H. Knötzinger, J. Weitcamp, *Handbook of Catalysis*, VCH Publishers, Weinheim (1997).
6. M. Kiskinova, Poisoning and promotion in catalysis based on surface science concepts and experiments, in *Stud. Surf. Sci. Catal.* **70**, Elsevier, Amsterdam (1992).
7. G.L. Haller, D.E. Resasco, *Adv. in Catal.* **36**, 173–235 (1989).
8. See, e.g., chapter by X. Verykios in this book.
9. M. Stoukides, C.G. Vayenas, *J. Catal.* **70**, 137–146 (1981).
10. C.G. Vayenas, S. Bebelis, S. Neophytides, *J. Phys. Chem.* **92**, 5083–5085 (1988).
11. C.G. Vayenas, S. Bebelis, S. Ladas, *Nature* **343**, 625–627 (1990).
12. J. Pritchard, *Nature* **343**, 592 (1990).
13. C.G. Vayenas, M.M. Jaksic, S. Bebelis, S.G. Neophytides, The electrochemical activation of catalysis, in *Modern Aspects of Electrochemistry*, J.O.M. Bockris, B.E.

Conway, R.E. White, eds., Kluwer Academic/Plenum Publishers, New York (1996), pp. 57–202.

14. C.G. Vayenas, S. Bebelis, C. Pliangos, S. Brosda, D. Tsiplakides, *Electrochemical Activation of Catalysis: Promotion, Electrochemical Promotion and Metal–Support Interactions*, Kluwer/Plenum Press, New York (2002).

15. E. Lamy-Pitara, L. Bencharif, J. Barbien, *Appl. Catal.* **18**, 117–131 (1985).

16. H. Baltruschat, N.A. Anastasijevic, M. Beltowska-Brzezinska, G. Hambitzer, J. Heitbaum, *Ber. Buns. Phys. Chem.* **94**, 996–1000 (1990).

17. T.I. Politova, V.A. Sobyanin, V.D. Belyaev, *Reaction Kinetics and Catalysis Letters* **41**(2), 321–326 (1990).

18. E. Varkaraki, J. Nicole, E. Plattner, C. Comninellis, C.G. Vayenas, *J. Appl. Electrochem.* **25**, 978–981 (1995).

19. I. Harkness, R.M. Lambert, *J. Catal.* **152**, 211–214 (1995).

20. C.A. Cavalca, G.L. Haller, *J. Catal.* **177**, 389–395 (1998).

21. L. Ploense, M. Salazar, B. Gurau, E.S. Smotkin, *JACS* **119**, 11550–11551 (1997).

22. J. Poppe, S. Voelkening, A. Schaak, E. Schuetz, J. Janek, R. Imbihl, *Phys. Chem. Chem. Phys.* **1**, 5241–5249 (1999).

23. J.K. Hong, I.-H. Oh, S.-A. Hong, W.Y. Lee, *J. Catal.* **163**, 95–105 (1996).

24. D.A. Emery, P.H. Middleton, I.S. Metcalfe, *Surf. Sci.* **405**, 308–315 (1998).

25. J.O.M. Bockris, Z.S. Minevski, *Electrochim. Acta* **39**(11/12), 1471–1479 (1994).

26. S. Bebelis, C.G. Vayenas, *J. Catal.* **118**, 125–146 (1989).

27. C.G. Vayenas, S. Bebelis, I.V. Yentekakis, H.-G. Lintz, *Catal. Today* **11**(3), 303–442 (1992).

28. C.G. Vayenas, S. Bebelis, *Catal. Today* **51**, 581–594 (1999).

29. C.G. Vayenas, S. Neophytides, Electrochemical activation of catalysis: In situ controlled promotion of catalyst surfaces, in *Catalysis-Special Periodical Report*, Royal Society of Chemistry, Cambridge (1996), pp. 199–253.

30. C.G. Vayenas, I.V. Yentekakis, Electrochemical modification of catalytic activity, in *Handbook of Catalysis*, G. Ertl, H. Knötzinger, J. Weitcamp, eds., VCH Publishers, Weinheim (1997), pp. 1310–1338.

31. S. Bebelis, M. Makri, A. Buekenhoudt, J. Luyten, S. Brosda, P. Petrolekas, C. Pliangos, C.G. Vayenas, *Solid State Ionics* **129**, 33–46 (2000).

32. A. Kaloyannis, C.G. Vayenas, *J. Catal.* **171**, 148–159 (1997).

33. P. Tsiakaras, C.G. Vayenas, *J. Catal.* **140**, 53–70 (1993).

34. C.G. Vayenas, S. Bebelis, I.V. Yentekakis, P. Tsiakaras, H. Karasali, *Platinum Metals Review* **34**(3), 122–130 (1990).

35. I.V. Yentekakis, C.G. Vayenas, *J. Catal.* **111**, 170–188 (1988).

36. H. Karasali, C.G. Vayenas, *Materials Science Forum* **76**, 171–174 (1991).

37. C.G. Vayenas, S. Neophytides, *J. Catal.* **127**, 645–664 (1991).

38. A. Kaloyannis, C.G. Vayenas, *J. Catal.* **182**, 37–47 (1998).

39. M. Marwood, A. Kaloyannis, C.G. Vayenas, *Ionics* **2**, 302–311 (1996).

40. C. Pliangos, I.V. Yentekakis, X.E. Verykios, C.G. Vayenas, *J. Catal.* **154**, 124–136 (1995).

41. C. Pliangos, C. Raptis, T. Badas, C.G. Vayenas, *Solid State Ionics* **136**/**137**, 767–773 (2000).

42. C. Pliangos, C. Raptis, T. Badas, C.G. Vayenas, *Ionics* **6**, 119–126 (2000).

43. A.D. Frantzis, S. Bebelis, C.G. Vayenas, *Solid State Ionics* **136–137**, 863–872 (2000).

44. K. Yiokari, S. Bebelis, *J. Appl. Electrochem.* **30**, 1277–1283 (2000).

45. H. Alqahtany, P.H. Chiang, P. Eng., M. Stoukides, A.R. Robbat, *Catal. Lett.* **13**, 289 (1992).

46. A. Giannikos, A.D. Frantzis, C. Pliangos, S. Bebelis, C.G. Vayenas, *Ionics* **4**, 53–60 (1998).

47. M. Marwood, C.G. Vayenas, *J. Catal.* **170**, 275–284 (1997).
48. G.L. Haller, S. Kim, *ACS Petroleum Division Preprints, Symposium in Catalytic Combustion* in *213th National ACS Meeting*, 155–158 (1997, April 13–17) San Francisco, CA.
49. S. Bebelis, C.G. Vayenas, *J. Catal.* **138**, 570–587 (1992).
50. S. Bebelis, C.G. Vayenas, *J. Catal.* **138**, 588–610 (1992).
51. C. Karavasilis, S. Bebelis, C.G. Vayenas, *J. Catal.* **160**, 190–204 (1996).
52. M. Stoukides, C.G. Vayenas, *J. Electrochem. Soc.* **131**(4), 839–845 (1984).
53. P. Tsiakaras, C.G. Vayenas, *J. Catal.* **144**, 333–347 (1993).
54. C. Karavasilis, S. Bebelis, C.G. Vayenas, *Materials Science Forum* **76**, 175–178 (1991).
55. S. Neophytides, C.G. Vayenas, *J. Catal.* **118**, 147–163 (1989).
56. O.A. Mar'ina, V.A. Sobyanin, *Catal. Lett.* **13**, 61 (1992).
57. O.A. Mar'ina, V.A. Sobyanin, V.D. Belyaev, V.N. Parmon, *Catalysis Today* **13**, 567–570 (1992).
58. T.I. Politova, G.G. Gal'vita, V.D. Belyaev, V.A. Sobyanin, *Catal. Lett.* **44**, 75–81 (1997).
59. I.V. Yentekakis, Y. Jiang, S. Neophytides, S. Bebelis, C.G. Vayenas, *Ionics* **1**, 491–498 (1995).
60. O.A. Mar'ina, V.A. Sobyanin, V.D. Belyaev, V.N. Parmon, The effect of electrochemical pumping of oxygen on catalytic behaviour of metal electrodes in methane oxidation, in *New Aspects of Spillover Effect in Catalysis for Development of Highly Active Catalysts*, Elsevier Science Publishers B.V. (1993).
61. D. Tsiplakides, J. Nicole, C.G. Vayenas, C. Comninellis, *J. Electrochem. Soc.* **145**(3), 905–908 (1998).
62. S. Wodiunig, C. Comninellis, *J. Eur. Ceramic Soc.* **19**, 931–934 (1999).
63. I.V. Yentekakis, C.G. Vayenas, *J. Catal.* **149**, 238–242 (1994).
64. C. Pliangos, I.V. Yentekakis, S. Ladas, C.G. Vayenas, *J. Catal.* **159**, 189–203 (1996).
65. P.D. Petrolekas, S. Balomenou, C.G. Vayenas, *J. Electrochem. Soc.* **145**(4), 1202–1206 (1998).
66. P. Beatrice, C. Pliangos, W.L. Worrell, C.G. Vayenas, *Solid State Ionics* **136–137**, 833–837 (2000).
67. C.G. Vayenas, S. Bebelis, S. Despotopoulou, *J. Catal.* **128**, 415–435 (1991).
68. I.V. Yentekakis, G. Moggridge, C.G. Vayenas, R.M. Lambert, *J. Catal.* **146**, 292–305 (1994).
69. C.A. Cavalca, *Ph.D. Thesis*, Yale University (1995).
70. S. Tracey, A. Palermo, J.P.H. Vazquez, R.M. Lambert, *J. Catal.* **179**, 231–240 (1998).
71. A. Palermo, R.M. Lambert, I.R. Harkness, I.V. Yentekakis, O. Mar'ina, C.G. Vayenas, *J. Catal.* **161**, 471–479 (1996).
72. A. Palermo, M.S. Tikhov, N.C. Filkin, R.M. Lambert, I.V. Yentekakis, C.G. Vayenas, *Ionics* **1**, 366–372 (1995).
73. A. Palermo, M.S. Tikhov, N.C. Filkin, R.M. Lambert, I.V. Yentekakis, C.G. Vayenas, *Stud. Surf. Sci. Catal.* **101**, 513–522 (1996).
74. R.M. Lambert, M. Tikhov, A. Palermo, I.V. Yentekakis, C.G. Vayenas, *Ionics* **1**, 366–376 (1995).
75. O.A. Mar'ina, I.V. Yentekakis, C.G. Vayenas, A. Palermo, R.M. Lambert, *J. Catal.* **166**, 218–228 (1997).
76. A. Giannikos, P. Petrolekas, C. Pliangos, A. Frenzel, C.G. Vayenas, H. Putter, *Ionics* **4**, 161–169 (1998).
77. C. Karavasilis, S. Bebelis, C.G. Vayenas, *J. Catal.* **160**, 205–213 (1996).
78. P.D. Petrolekas, S. Brosda, C.G. Vayenas, *J. Electrochem. Soc.* **145**(5), 1469–1477 (1998).
79. G. Pitselis, P. Petrolekas, C.G. Vayenas, *Ionics* **3**, 110–117 (1997).

80. M. Makri, A. Buekenhoudt, J. Luyten, C.G. Vayenas, *Ionics* **2**, 282–288 (1996).
81. C.G. Yiokari, G.E. Pitselis, D.G. Polydoros, A.D. Katsaounis, C.G. Vayenas, *J. Phys. Chem.* **104**, 10600–10602 (2000).
82. P.H. Chiang, D. Eng, M. Stoukides, *J. Catal.* **139**, 683–687 (1993).
83. D. Tsiplakides, S. Neophytides, O. Enea, M.M. Jaksic, C.G. Vayenas, *J. Electrochem. Soc.* **144**(6), 2072–2088 (1997).
84. S. Neophytides, D. Tsiplakides, P. Stonehart, M. Jaksic, C.G. Vayenas, *Nature* **370**, 292–294 (1994).
85. S. Neophytides, D. Tsiplakides, P. Stonehart, M.M. Jaksic, C.G. Vayenas, *J. Phys. Chem.* **100**, 14803–14814 (1996).
86. I.M. Petrushina, V.A. Bandur, F. Cappeln, N.J. Bjerrum, *J. Electrochem. Soc.* **147**(8), 3010–3013 (2000).
87. S. Ladas, S. Kennou, S. Bebelis, C.G. Vayenas, *J. Phys. Chem.* **97**, 8845–8847 (1993).
88. R.M. Lambert, F. Williams, A. Palermo, M.S. Tikhov, *Topics in Catalysis* **13**, 91–98 (2000).
89. W. Zipprich, H.-D. Wiemhöfer, U. Vöhrer, W. Göpel, *Ber. Buns. Phys. Chem.* **99**, 1406–1413 (1995).
90. S.G. Neophytides, C.G. Vayenas, *J. Phys. Chem.* **99**, 17063–17067 (1995).
91. J. Poppe, A. Schaak, J. Janek, R. Imbihl, *Ber. Buns. Phys. Chem.* **102**, 1019–1022 (1998).
92. M. Makri, C.G. Vayenas, S. Bebelis, K.H. Besocke, C. Cavalca, *Surf. Sci.* **369**, 351–359 (1996).
93. S. Ladas, S. Bebelis, C.G. Vayenas, *Surf. Sci.* **251/252**, 1062–1068 (1991).
94. Y. Jiang, I.V. Yentekakis, C.G. Vayenas, *J. Catal.* **148**, 240–251 (1994).
95. D. Kek, M. Mogensen, S. Pejovnik, *J. Electrochem. Soc.* **148**(8), A878–A886 (2001).
96. S.J. Tauster, S.C. Fung, R.L. Garten, *JACS* **100**, 170–175 (1978).
97. E.C. Akubuiro, X.E. Verykios, *J. Catal.* **113**, 106–119 (1988).
98. M. Haruta, A. Ueda, S. Tsubota, R.M.T. Sanchez, *Catal. Today* **29**, 443–447 (1996).
99. Y. Iizuka, H. Fujiki, N. Yamauchi, T. Chijiiwa, S. Arai, S. Tsubota, M. Haruta, *Catal. Today* **36**, 115–123 (1997).
100. S. Tsubota, D.A.H. Cunningham, Y. Bando, M. Haruta, Preparation of nanometer gold strongly interacted with TiO_2 and the structure sensitivity in low-temperature oxidation of CO, in *Preparation of catalysts VI*, G. Ponchelet, ed. (1995), pp. 227–235.
101. Z. Hong, K.B. Fogash, J.A. Dumesic, *Catalysis Today* **51**, 269–288 (1999).
102. Y.D. Kim, A.P. Seitsonen, H. Over, *Surf. Sci.* **465**, 1–8 (2000).
103. D.G. Barton, M. Shtein, R.D. Wilson, S.L. Soled, E. Iglesia, *J. Phys. Chem.* **103**(4), 630–640 (1999).
104. G. Meitzner, E. Iglesia, *Catalysis Today* **53**, 433–441 (1999).
105. B.I. Mojet, J.T. Miller, D.E. Ramaker, D.C. Koningsberger, *J. Catal.* **186**, 373–386 (1999).
106. S. Tagliaferri, R.A. Koeppel, A. Baiker, *Appl. Catal. B* **15**, 159–177 (1998).
107. A.Y. Stakheev, L.M. Kustov, *Appl. Catal. A* **188**, 3–35 (1999).
108. A. Cimino, D. Gazzoli, M. Valigi, *J. Electron Spectroscopy and Related Phenomena* **104**, 1–29 (1999).
109. S. Rossignol, C. Micheaud-Especel, D. Duprez, *Stud. Surf. Sci. Catal.* **130**, 3327–3332 (2000).
110. R.M. Ferrizz, T. Egami, J.M. Vohs, *Surf. Sci.* **465**, 127–137 (2000).
111. J. Kuriacose, *Industrial J. Chem.* **5**, 646 (1957).
112. S.J. Teichner, *New Aspects of Spillover Effect in Catalysis for Development of Highly Active Catalysts* in *Third International Conference on Spillover*, 27 (1993), Amsterdam: Elsevier.
113. W.C. Conner, G.M. Pajonk, S.J. Teichner, *Adv. Catal.* **34**, 1–79 (1986).

114. B. Delmon, H. Matralis, The remote control mechanism, general phenomena, possible consequences concerning unsteady state processes, unsteady state processes in catalysis, Y.S. Matros, ed., USF, Utrecht, The Netherlands (1991), p. 25.

115. T. Rebitzki, B. Delmon, J.H. Block, *AlChE Journal* **41**, 1543 (1995).

116. B. Delmon, G.F. Froment, *Catal. Rev.-Sci. Eng* **38**(1), 69–100 (1996).

117. T. Inui, T. Takeguchi, *Catal. Today* **10**, 95–106 (1991).

118. G.C. Bond, In *Proc. 6th Int. Congr. Catalysis*, p. 356 (1977).

119. Z. Knor, J. Sotola, *Coll. Czech. Chemical Comm.* **53**, 2399–2411 (1988).

120. R.T. Yang, C. Wong, *J. Chem. Phys.* **75**, 4471 (1981).

121. K. Fujimoto, S. Toyoshi, *Proc. 7th Int. Congr. Catalysis*, p. 235 (1981).

122. R. Lewis, R. Gomer, *Surf. Sci.* **12**, 157–176 (1968).

123. J.C. Bertolini, P. Delichere, J. Massardier, *Surf. Sci.* **160**, 531–541 (1985).

124. H. Reiss, *J. Phys. Chem.* **89**, 3783–3791 (1985).

125. P.M. Gundry, F.C. Tompkins, in *Experimental Methods in Catalyst Research*, R.B. Anderson, ed., Academic Press, New York (1968), pp. 100–168.

126. J. Hölzl, F.K. Schulte, Work function of metals, in *Solid Surface Physics*, Springer-Verlag, Berlin (1979), pp. 1–150.

127. S. Trasatti, The Work Function in Electrochemistry, in *Advances in Electrochemistry and Electrochemical Engineering*, H. Gerisher, C.W. Tobias, eds., Journal Wiley and Sons (1977).

128. H.L. Skriver, N.M. Rosengaard, *Physical Review B* **45**(16), 9410–9412 (1992).

129. C.G. Vayenas, S. Brosda, C. Pliangos, *J. Catal.* **203**, 329–350 (2001).

130. C.G. Vayenas, S. Brosda, *Stud. Surf. Sci. Catal.* **138**, 197–204 (2001).

131. S. Surnev, M. Kiskinova, *Appl. Phys.* **46**, 323–329 (1988).

132. G. Rangelov, L. Surnev, *Surf. Sci.* **185**, 457–468 (1987).

133. M. Kiskinova, *Surf. Sci.* **182**, 150–160 (1987).

134. M. Kiskinova, M. Tikhov, *Surf. Sci.* **194**, 379–396 (1988).

135. G. Pacchioni, F. Illas, S. Neophytides, C.G. Vayenas, *J. Phys. Chem.* **100**, 16653–16661 (1996).

136. G. Pacchioni, J.R. Lomas, F. Illas, *Molecular Catalysis A: Chemical* **119**, 263–273 (1997).

137. J. Nicole, D. Tsiplakides, C. Pliangos, X.E. Verykios, C. Comninellis, C.G. Vayenas, *J. Catal.* **203**, in press (2001).

138. C.G. Vayenas, G. Pitselis, *I&EC Research* **40**(20), 4209–4215 (2001).

139. S. Neophytides, D. Tsiplakides, C.G. Vayenas, *J. Catal.* **178**, 414–428 (1998).

140. D. Tsiplakides, C.G. Vayenas, *J. Electrochem. Soc.* **148**(5), E189–E202 (2001).

141. C. Cavalca, G. Larsen, C.G. Vayenas, G. Haller, *J. Phys. Chem.* **97**, 6115–6119 (1993).

142. C.G. Vayenas, *J. Electroanal. Chem.* **486**, 85–90 (2000).

143. A. Frumkin, B. Damaskin, *J. Electroanal. Chem.* **79**, 259–266 (1977).

144. A. Frantzis, *Ph.D. Thesis*, University of Patras (2002).

145. D. Tsiplakides, C.G. Vayenas, *J. Catal.* **185**, 237–251 (1999).

146. D. Tsiplakides, S. Neophytides, C.G. Vayenas, *Ionics* **3**, 201–208 (1997).

147. S. Brosda, C.G. Vayenas, *J. Catal.* **208**, 38–53 (2002).

148. C.G. Vayenas, R.M. Lambert, S. Ladas, S. Bebelis, S. Neophytides, M.S. Tikhov, N.C. Filkin, M. Makri, D. Tsiplakides, C. Cavalca, K. Besocke, *Stud. Surf. Sci. Catal.* **112**, 39–47 (1997).

149. J. Nicole, C. Comninellis, *J. Appl. Electrochem.* **28**, 223–226 (1998).

150. J. Nicole, *Ph.D. Thesis*, EPFL (1999).

151. C. Pliangos, *Ph.D. Thesis*, Department of Chemical Engineering, University of Patras (1997).

152. C. Pliangos, I.V. Yentekakis, V.G. Papadakis, C.G. Vayenas, X.E. Verykios, *Appl. Catal. B* **14**, 161–173 (1997).

153. K.E. Karakitsou, X.E. Verykios, *J. Phys. Chem.* **97**, 1184–1189 (1993).

154. T. Ioannides, X.E. Verykios, *J. Catal.* **161**, 560–569 (1996).

155. C.T. Campbell, *Catalysis at Surfaces*, Chapman and Hall (1988).

156. C.T. Campbell, *J. Phys. Chem.* **89**(26), 5789–5795 (1985).

157. C.T. Campbell, B.E. Koel, *Surf. Sci.* **183**, 100–112 (1987).

158. I.V. Yentekakis, R.M. Lambert, M.S. Tikhov, M. Konsolakis, V. Kiousis, *J. Catal.* **176**, 8292 (1998).

159. F.A. Alexandrou, V.G. Papadakis, X.E. Verykios, C.G. Vayenas, *The promotional effect of Na on the NO reduction by CO on supported Pt, Pd and Rh catalysts in Proc. 4th Intl. Congress on Catalysis and Automotive Pollution Control* **2**, 1–16 (1997).

160. D. Tsiplakides, S. Neophytides, C.G. Vayenas, *Solid State Ionics* **136–137**, 839–847 (2000).

161. D. Tsiplakides, S, Neophytides, C.G. Vayenas, *Ionics* **7**, 203–209 (2001).

20

Support Effects on Catalytic Performance of Nanoparticles

XENOPHON E. VERYKIOS

University of Patras, Patras, Greece

CHAPTER CONTENTS

SUMMARY

Interactions between semiconductive support materials and highly-dispersed metal particles, affecting chemisorptive and kinetic parameters, are induced by aliorvalent cation doping of the carrier. Doping is achieved by high-temperature diffusion of the foreign cation and it is confirmed by measurements of electrical conductivity and activation energy of electron conduction. It was observed that upon doping TiO_2 with higher valence cations, the adsorption capacity of Group VIII metal nanoparticles and the strength of adsorption bonds was significantly altered.

Turnover frequencies and kinetic parameters of hydrogenation, oxidation, and reduction reactions were also influenced to a significant extent. These phenomena are interpreted on the basis of the metal–semiconductor boundary layer theory, which proposes that, at thermodynamic equilibrium, the Fermi energy level of the two solids in contact is at equal height. This implies that charge is transported across the interface, until the electrochemical potential is uniform throughout the system. As a result, long-range, or short-range interactions, affecting the electronic structure of the gas–metal–support interface atoms are induced. An alternative explanation is based on the theory of electrochemical promotion. According to this, the charge carriers that are transported from the support to the metal nanoparticles are oxygen ions which diffuse to the surface, altering the work function and, subsequently, the chemisorptive and catalytic parameters.

20.1 INTRODUCTION

A large fraction of practical catalysts consists of transition-metal or metal oxide nanoparticles dispersed onto the surface of insulator or semiconductor oxides that function as support materials. For industrial applications, the supports employed are selected on the basis of their surface area (high surface area is usually, but not always, desirable), high thermal and hydrothermal stability, chemical stability, and mechanical strength.

The criteria listed above ignore the fact that in many cases the support may play a direct or indirect role, significantly influencing the chemisorptive and catalytic properties of the metal-carrier ensemble ([1] and references therein). The direct role refers to a bifunctional type of catalytic mechanism in which a particular function is provided by the dispersed metal particles and a distinctive function is provided by the carrier. Alternatively, the catalytic function may occur at sites created at the interface or at the point of contact between the metal particles and the carrier material [2]. Obviously, the nature and strength of these catalytic sites are defined, to a large extent, by the support material. The indirect catalytic role refers to the fact that the support may influence the size (dispersion) and morphology of the metal crystallites, their surface and bulk electronic properties, as well as their resistance to sintering and poisoning. These parameters define catalytic performance, in terms of both fundamental scientific considerations as well as industrial and economic aspects.

Many investigations have been inspired by the possibility that an interaction between metal crystallites and support materials can alter the chemisorptive and catalytic properties of the former. Strong evidence for this concept has been provided by Sinfelt and his co-workers [3] and later by many other investigators who showed that kinetic parameters of certain catalytic systems may be influenced by the particular support employed. In more recent years, the concept of strong metal–support interactions (SMSI) was introduced by Tauster et al. [4], who observed that Group VIII metal particles supported on titania and other reducible oxides lose their ability to chemisorb H_2 and CO when they are reduced at temperatures higher than 773 K. Significant alterations in kinetic parameters in hydrogenation, dehydrogenation, and hydrogenolysis reactions have also been observed over SMSI catalysts [5–7].

A number of models have been proposed to explain the mechanism of this phenomenon, which include formation of intermetallic bonds with localized charge transfer from the reduced titania support to the metal crystallites [8–10], and diffusion of a suboxide species to the surface of the metal crystallites, rendering them partially inaccessible to chemisorption and catalytic action [1,11,12]. The geometric decoration model has been shown by many investigators to be mostly responsible for the SMSI state [13–18]. Direct experimental evidence for the occurrence of metal decoration by TiO_x species has been provided by HREM [16–19]. In some cases, electronic effects may also be playing an important role in the phenomenon [20,21]. Certain similarities but also significant differences with the M/TiO_2 catalytic system are observed with the M/CeO_2 system [22–27], in which support effects are mainly attributed to electronic type of interactions [27].

The general subject of metal–support interactions is discussed in detail in another chapter of this volume. In the present article we concentrate on interactions of the electronic type and their influence on chemisorptive and kinetic parameters. More specifically, we concentrate on electronic metal–support interactions induced by doping of semiconductive support materials with cations of valence lower or higher than that of the host cation.

20.2 METAL–SUPPORT INTERACTIONS OF THE ELECTRONIC TYPE

The origin of electronic type of interactions between dispersed metal particles and the support material has been viewed in two distinctly different ways: (1) as the result of collective interactions between the metal particle and the support, defined primarily by the bulk electronic structure of the two phases, and resulting in short-range or long-range charge transport to or from the metal nanoparticles; (2) as the result of the interaction of the metal clusters with local sites of the support, possessing specific properties, such as acidity or basicity.

The concept of electronic interactions originating from the bulk electronic properties of the two phases in contact (metal crystallites and support materials) was first proposed by Schwab ([28] and references therein) and Solymosi ([29] and references therein), followed by other investigators at the time [30]. This concept is based on the metal–semiconductor boundary-layer theory, according to which thermodynamic equilibrium develops at a metal–semiconductor contact (metal crystallites-support, in the present case). Equilibrium requires that the Fermi energy level of electrons of the two solids is at equal height. In order for this to be achieved, charge must be transferred from one material to the other. In cases where the work function of electrons of the metal is higher than the work function of the semiconductor, charge will flow from the semiconductor to the metal, until the Fermi level at the interface is equilibrated. The opposite is also true. This charge-transfer process causes significant alterations in catalytic properties of the dispersed metal particles and also of the support, if it has a distinctive catalytic function. In fact, the phenomenon of the metal particles' modifying the electron structure of the semiconducting support has been termed the "Schwab effect of the first kind," while the modification of the electron density of the dispersed metal particles by charge transport from the carrier has been termed the "Schwab effect of the second kind" [31]. The methodology of these investigators [28,29] was based on the alteration of the electronic structure of semiconducting carriers by doping with aliovalent cations, thus raising or lowering the Fermi energy level of the semiconducting carrier. They observed that the activation energy of certain reactions, such as formic acid decomposition, was significantly affected by doping of the support.

The concept of electronic interactions of metal nanoparticles with local sites of the support, primarily Broensted or Lewis acid sites or oxide entities, has been mostly applied to metal/zeolite systems and has recently been reviewed [32]. In the same field, the idea of metal-proton adduct has been proposed [33–36] according to which protons of the support are delocalized on the metal particles forming a proton adduct that withdraws electron density from metal atoms. Metal particles forming metal-proton adducts with electron-deficient metal have been detected in Rh/Yzeolite [37–39], Ru/Yzeolite [40], and Au/Yzeolite [41]. The effect of Broensted acidity on the electronic structures of Pt/Yzeolite has also been investigated [42]. The creation of electron-deficient metal particles was also detected in the case of Pt/mordenite [43,44] as well as Pt dispersed on L zeolite [45–47]; Pd dispersed on the same support [48], and Ir dispersed on zeolite ZSM-5 [49]. Similar interactions, leading to electron-deficient metal clusters, have been observed in Pt, Pd, Rh, and Ni dispersed on sulfated zirconia, which is a superacidic support [50,51]. The opposite effect, i.e., negative charging of metal clusters when these are dispersed on basic supports, has been attributed to enhancement of the electron density of the oxygen atom of the support with increasing alkalinity [33,52,53]. Electron transfer between the oxygen atoms of the support and metal particles leads to enhanced electron density of the metal particles. Such effects have been observed in various catalytic systems [54–57].

A new model describing electronic metal–support interactions in noble metal catalysts supported on acidic or alkaline supports has recently been proposed by Mojet et al. [58], based on EXAFS structural analysis, XPS, FTIR of adsorbed CO, XANES shape resonance, and AXAFS. This model is based on the change in the energy position of the metal valence orbitals as a result of the metal–support interaction [58–60]. A decrease of the metal ionization potential with increasing alkalinity of the carrier occurs, i.e., a shift to lower binding energy due to the electrostatic Coulomb interaction between the two solids in contact. The changes in the energy of the metal valence orbitals alter the chemisorption energy, a factor that influences catalytic activity [58].

Electronic interactions at the metal–support interface have been used to interpret alterations in catalytic parameters in other systems as well. As mentioned earlier, although the SMSI phenomenon has been found to be, in its greatest part, due to migration of reduced support species onto the metal surface, electronic interactions have also been proposed to account for some of the experimental observations. Evidence for electron transfer from the support to the metal has been provided by use of electron spectroscopics and other techniques [61–68]. Similarly, electronic-type metal–support interactions have also been detected in the Rh/, Pd/, and Pt/CeO$_2$ catalytic systems when they are reduced at moderate temperatures [27]. The decrease of the adsorption strength of CO on Pt/ and Ni/TiO$_2$ [69,70] was also attributed to electronic interactions. Electron transfer has also been proposed in the case of Ru supported on alkaline earth [71] and different kinds of oxides [72]. The adsorptive and catalytic behavior of the Ru crystallites was correlated with the electronegativity of the support, which is a measure of its electron-donating properties. Similarly, enhanced activity of Cu supported on high work function semiconductive oxides (73) under CO hydrogenation conditions was attributed to electron transfer from Cu to the support.

A quantitative understanding of metal–support interactions, especially those involving charge transport at the metal–support interface, has been attempted through theoretical calculations [74–77]. These calculations provide information concerning the nature of chemical bonds at the surface, the direction and magnitude of charge transport at the metal–support interface, and the influence of charge transport on the strength of adsorption bonds.

It is apparent from the discussion above that metal–support interactions of the electronic type have been proposed to explain a large variety of catalytic phenomena. The exact nature and mechanism of this interaction depend on the particular catalytic system, i.e., the support material, the nature of the active phase, the size of the metal cluster or particle, the gaseous atmosphere under which the catalyst operates, and possibly other parameters.

In subsequent sections of this chapter we concentrate on electronic-type metal–support interactions that are attributed to the collective electronic structure of the support and the metallic phases. The approach is based on the early work by Schwab [28] and Solymosi [29], and the interpretation of the observed chemisorptive and catalytic phenomena is, to a large extent, based on this early work. In order to obtain a more comprehensive understanding of the proposed interpretations, a brief review of important concepts of the electronic structure of metals and semiconductors, as well as of their contacts at equilibrium, is presented.

20.3 THE ELECTRON STRUCTURE OF METALS AND SEMICONDUCTORS

The electron potentials related to the electron structure of a metal are shown schematically in Figure 1(a). Free electrons inside the metal possess kinetic energy, and the energy of the highest occupied level is defined as the Fermi energy, E_F. Electrons are bound inside the bulk by the action of an attractive potential, named V_b. The net stabilizing energy corresponds to the chemical potential of the electrons, μ_e [78,79].

$$\mu_e = E_F + V_b \tag{1}$$

A metal surface is characterized by the presence of a dipole layer with its negative end pointing away from the surface. The potential due to this dipole is called surface potential, χ. The real potential, α_e, of the electron is defined as [78]

$$\alpha_e = \mu_e - e\chi \tag{2}$$

while the minimum energy required for transferring an electron out of the metal, namely the work function, Φ, is [78–80]

$$\Phi = -\alpha_e = -\mu_e + e\chi \tag{3}$$

The work function incorporates the contribution of a term related to the bulk, μ_e, and a term related to the surface, $e\chi$. If the metal surface contains a net macroscopic charge, there is an electrostatic (or outer) potential, Ψ, corresponding to this charge. The energy binding the electrons inside the metal in this general case is called the

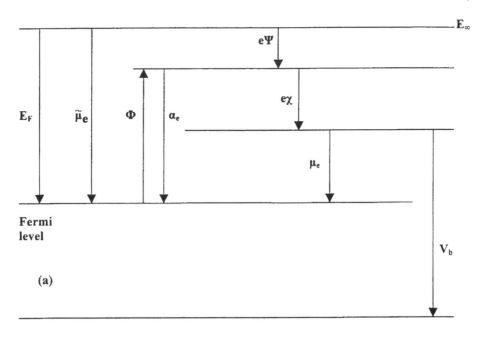

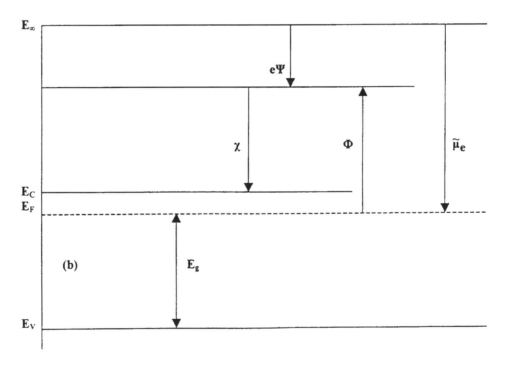

Figure 1 Electron potentials related to (a) the metal and (b) the semiconductor electron structure.

electrochemical potential of the electrons, $\bar{\mu}_e$ [78]:

$$\bar{\mu}_e = \alpha_e - e\Psi = \mu_e - e\chi - e\Psi \tag{4}$$

The condition of thermodynamic equilibrium requires that the electrochemical potential should be the same throughout the metal. In a polycrystalline sample with many crystal planes exposed (which have, in general, different work functions), the condition of equilibrium requires the development of net charge in each plane.

The electron potentials relating to the electronic structure of a semiconductor are shown schematically in Figure 1(b). The upper edge of the valence band is of energy E_v, while the lower edge of the conduction band is of energy E_C. Thus, semiconductors are characterized by the existence of an energy gap, E_g, between the highest occupied energy level (valence band) and the lowest unoccupied empty level (conduction band). The magnitude of this energy gap is used to differentiate metal oxides as semiconductors and insulators. If the magnitude of E_g is such that thermal excitation is capable of transferring an appreciable number of electrons into the conduction band, the solid is considered to be a semiconductor, otherwise an insulator. Generally, solids whose band gap is less than $4\,eV$ are classified as semiconductors.

Electrons in a solid are distributed in available energy states following the Fermi–Dirac statistics. The Fermi energy, E_F, is defined as the energy at which the probability of occupancy is 0.5. For intrinsic or ionic conducting solids, the Fermi level is located at the center of the energy gap and shifts toward the empty zone for an *n*-type semiconductor or the filled band for a *p*-type semiconductor. As in the case of metals, the Fermi energy is equal to the electrochemical potential ($E_F = \bar{\mu}_e$) as shown in Figure 1 for both metals and semiconductors.

The work function, Φ, is defined as the average work needed to move an electron from the solid and place it at a distance x from the surface where the classical image-force potential, $-e^2/4x$, is just negligible. A distance of $10^{-4}\,cm$ is sufficient to meet this requirement. Its value is equal to the difference between the Fermi level and the electrostatic potential, Ψ, just outside the semiconductor. (The vacuum level is the energy level of an electron just outside the semiconductor with zero kinetic energy.)

$$\Phi = -\bar{\mu}_e - e\Psi \tag{5}$$

Another important parameter is the electron affinity, χ_S, which is defined as the difference between the lower end of the conduction band and the electrostatic potential just outside the semiconductor:

$$\chi_S = -E_c \quad e\Psi \tag{6}$$

The electron affinity is a critical surface parameter that is not dependent on the nature and concentration of dopants in a semiconductor. In contrast, the work function depends on doping as well as on temperature.

For an intrinsic semiconductor the electron concentration in the conduction band, n, is given by the simplified expression (when $E_F - E_C \gg kT$)

$$n = N_C \exp\left(-\frac{E_C - E_F}{kT}\right) \tag{7}$$

Similarly, the hole concentration in the valence band, p, is

$$p = N_v \exp\left(-\frac{E_F - E_v}{kT}\right) \tag{8}$$

where, N_C and N_V are the densities of states in the conduction and valence bands, respectively [81].

The Fermi level in an intrinsic semiconductor is located near or at the middle of the band gap. The position of the Fermi level can be manipulated by doping, which creates donor or acceptor levels in the band gap. Focusing on the semiconductor that is of interest in this study, i.e., TiO_2, doping with cations of valence higher than $+4$ (W^{+6}, for example) leads to the creation of a donor level near the conduction band. As a result, the Fermi level moves upward close to the conduction band and the work function decreases. This has been verified experimentally by electrical conductivity [82] and work function measurements.

20.4   METAL–SEMICONDUCTOR CONTACT THEORY

The condition of thermodynamic equilibrium in a metal–semiconductor contact states that the electrochemical potential should be uniform throughout the system. If, before contact, the metal and the semiconductor have different electrochemical potential, then upon contact, charge will flow to the material with the smaller potential, until the potentials are equalized. When the two materials have no net charge, $\bar{\mu}_e^M = \Phi_M$ and $\bar{\mu}_e^S = \Phi_S$, where M and S denote the metal and the semiconductor, respectively. If $\Phi_M > \Phi_S$, the electron flux will be toward the metal. At equilibrium, the common electrochemical potential will be

$$\bar{\mu}_e^{eq} = -\Phi_M - e\Psi_M = -\Phi_S - e\Psi_S \tag{9}$$

and

$$\Delta\Phi = -e\Delta\Psi \tag{10}$$

that is, a contact potential difference equal to the work function difference has developed. Ψ_M and Ψ_S refer to the outer potential of the metal and the semiconductor, respectively. The contact of a metal with an n-type semiconductor, in the situation in which $\Phi_M > \Phi_S$, is shown schematically in Figure 2. In this case, charge is transferred from the semiconductor to the metal. Electrons transferred to the metal are contained at the interface atoms, while the region in the semiconductor depleted of electrons is characterized by bending of the valence and conduction bands [83–86]. The Schottky barrier, Φ_{SB}, at the interface is

$$\Phi_{SB} = (\Phi_M - \Phi_S)/e \tag{11}$$

Covalent semiconductors, such as Si or GaAs, have a large density of surface states in the band gap, causing pinning of the barrier [83]. On the other hand, ionic semiconductors, such as ZnO, have low density of surface states in the band gap, and the barrier height does indeed depend on the work function difference between the metal and the semiconductor. It has been shown that for strongly ionic semiconductors the barrier height will indeed be equal to or proportional to the

(a) before contact

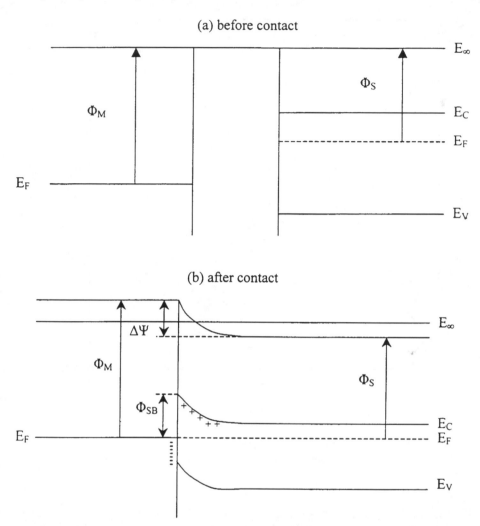

(b) after contact

Figure 2 Contact of a metal with work function Φ_M, with an *n*-type semiconductor with work function, Φ_S; $\Phi_M > \Phi_S$: (a) before and (b) after contact.

work function difference [83,84], as Eq. (11) indicates. Because TiO_2 is an ionic semiconductor, it is logical to assume that the ideal Schottky theory can be used to describe a metal–TiO_2 contact; experimental evidence also supports this hypothesis. The barrier height in contacts of TiO_2 with metals, such as Pd, Pt, or Au under air, was experimentally found to be equal to the work function difference [87], as Eq. (11) predicts.

A mathematical model of metal–semiconductor contacts has been employed to estimate the quantity of charge transferred through the interface, based on parameter values that pertain to the M/TiO_2 system [88]. The direction of electron flux in a metal–semiconductor contact depends on the relative values of the work function of the two materials. The work function of the semiconductor is a function of the kind (valence) and concentration of the dopant and of temperature. Doping of

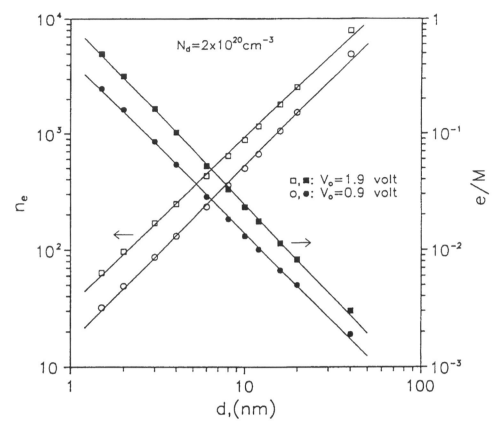

Figure 3 Total number of electrons transferred, n_e, and electrons transferred per metal atom, e/M, as a function of the metal crystallite diameter, d. (From Ref. [88].)

the TiO_2 carrier with cations of higher valence (e.g., W^{6+}) significantly enhances its electrical conductivity [82], which can be attributed to the creation of donor levels near the lower edge of the conduction band. As a consequence, the Fermi level of TiO_2 is shifted upward, closer to the conduction band, and the work function decreases. Therefore, the Schottky barrier will be higher in the case of TiO_2 doped with higher-valence cations.

The number of electrons transferred, n_e, and the number of electrons transferred per metal atom, e/M, were estimated as a function of the mean diameter of the metal crystallites, d, employing two models pertaining to infinite and finite interfaces between the metal and the support [88]. The parameters employed correspond to metal with work functions of 5.0 or 6.0 eV (the corresponding contact potential differences, V_0, being 0.9 and 1.9 V, respectively), in contact with TiO_2 doped with a donor impurity (W^{6+}, for example) with donor concentration of 2×10^{20} cm^{-3}. The results obtained are shown in Figure 3, in which the number of transferred electrons, n_e, and the number of electrons transferred per metal atom, e/M, are plotted as a function of d [88]. The number of transferred electrons ranges from about 8000 for a 40-nm metal particle to approximately 60 electrons for a

Table 1 Prediction of Magnitude of Electrostatic Field Developing at the Metal–Support Interface

N_d (cm^{-3})	V_0 (V)	D (nm)	(V/cm)
2×10^{20}	1.9	10	7.9×10^6
		2	2.5×10^7
2×10^{20}	0.9	10	4.5×10^6
		2	1.3×10^7
2×10^{19}	0.9	10	3.3×10^6
		2	1.1×10^7

Source: Ref. [88].

1.5-nm particle. It can be seen that the ratio e/M is less than 0.01 for crystallites larger than 10 nm, but can be as high as 0.5 for 1.5-nm crystallites. The quantities n_e and e/M are almost directly proportional to $d^{-1.5}$.

The presence of a fairly strong dipole at the metal–support interface leads to the creation of strong electrostatic fields in this region. Predictions of the magnitude of the electrostatic field are shown in Table 1 [88] for different metal-particle diameters, for metals of different work function, corresponding to different contact potential (V_0) and for different donor concentrations (N_d). Typical estimated values are in the range of 10^6–10^7 V/cm or 0.1–1 V/nm. The electric field is dependent on metal-particle diameter, being more intense for small particles, due to the curvature of the metal surface. It is also dependent on the contact potential difference and, less intensely, on donor concentration. The fraction of metal atoms at the three-phase boundary (metal–support–gas) affected by the electric field and that participate in the catalytic action is proportional to metal dispersion, D. Because, however, the electric field increases with increasing dispersion, the total effect will be proportional to D^a, where $1 < a < 2$.

It should also be pointed out that any process or mechanism that alters the Fermi level of the support (or of the metal), and thus the contact potential difference, will also influence the magnitude of the electric field. For example, alterations in the acidity of the surface of the support influence the Fermi level, or adsorption or spillover of adsorbed species from the metal to the support. Obviously then, the magnitude of the electric field at the metal–support interface may depend strongly on the atmosphere (i.e., reaction conditions, temperature) in which the catalyst exists.

20.5 PROBING ALTERATIONS OF SUPPORTS AND METAL PARTICLES UPON DOPING

As mentioned earlier, electrical conductivity measurements have been employed to gain an insight into the modification of the electronic structure of TiO_2 carriers upon doping with aliovalent cations [28,29,82]. Results of electrical conductivity measurements conducted in vacuum over 0.5% Pt/TiO_2 are summarized in Figure 4. The presence of Pt on the samples is not expected to influence electrical conductivity measurements because of the high dispersion of the metal. It is apparent that both electrical conductivity and activation energy of electron conduction are affected by

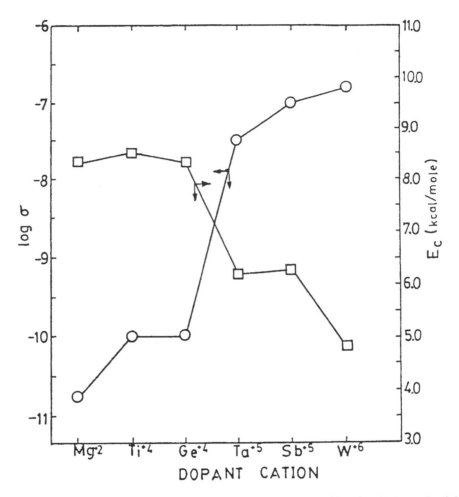

Figure 4 Influence of aliovalent cation doping on specific electrical conductivity and activation energy of electron conduction of 0.5% $Pt/TiO_2(D)$ at 333 K, in vacuum. (From Ref. 82.)

the doping process. The electrical conductivity of TiO_2 doped with cations of higher valence (Ta^{5+}, Sb^{5+}, W^{6+}) is two to three orders of magnitude higher than that of undoped TiO_2, while the electrical conductivity of lower-valence (Mg^{2+})-doped TiO_2 is one order of magnitude lower. The conductivity of Ge^{4+}-doped TiO_2 is approximately the same as that of the undoped sample.

Since the absolute values of specific electrical conductivities may be influenced by grain boundary effects, the true effect of modification of the electronic structure of TiO_2 by altervalent cation doping is given by the activation energy of electron conduction, E_C, which is not influenced by grain boundary effects. It is observed in Figure 4 that E_C is decreased by 30 to 45% when TiO_2 is doped with cations of higher valence, whereas no significant changes are observed for lower-(Mg^{2+}) or equal (Ge^{4+}) valence doping. Similar results were obtained under other environments, such as H_2 and CO [82] or when measurements were conducted with TiO_2 [89].

The measurements of electrical conductivity and activation energy of electron conduction [82,89] confirm that dopant cations are indeed incorporated into the crystal matrix of TiO_2 and those of higher valence create a donor level below the conduction band of TiO_2, as expected for an *n*-type semiconductor. These results further suggest that doping with higher-valence cations shifts the Fermi energy level closer to the conduction band. The fact that the doping cations are incorporated into the crystal matrix of the host material has also been shown by high-resolution ESCA, which, in most cases, failed to detect surface segregation of the dopants [90].

The influence of doping of the support with altervalent cations on the chemical and electronic structure of the dispersed metal nanoparticles has been investigated employing various techniques, such as XPS, FTIR, and XANES [91–95]. The effects of modification of the support (TiO_2) on the oxidation state of Ru under conditions of partial oxidation of methane to synthesis gas have been investigated employing XPS and FTIR techniques [94,95]. XPS spectra obtained in the Ru(3d) region from the unmodified Ru/TiO_2 catalyst are shown in Figure 5(a). It is observed that the oxidation state of ruthenium depends strongly on reaction temperature and, under reaction conditions, the fraction of metallic Ru increases with increasing temperature. Aliovalent cation doping of TiO_2 results in significant changes in the oxidation state of supported Ru [Figure 5(b)–(c)]. In particular, doping with W^{6+} cations results in the stabilization of a portion of Ru in its oxide forms under reaction conditions, while doping of the carrier with Ca^{2+} cations results in the stabilization of Ru in its metallic form.

These results may be explained using the theory of metal–semiconductor contacts as follows: metal crystallites may be viewed as electron donors while the adsorbed oxygen molecules behave as surface electron acceptors. It is then expected that in the case of the W^{6+}-doped catalyst, where charge transfer from the carrier to the supported Ru is increased, the oxygen chemisorption process will be favored. In addition, doping TiO_2 with W^{6+} cations will also favor the formation of active oxygen species, such as O_2^-, O_2^{2-}, and O^-, which are generated by electron acceptance on the metal surface. Therefore, the stabilization of Ru in its oxidized forms observed over the W^{6+}-doped catalyst under reaction conditions (Figure 5) may be attributed to the stronger interaction of oxygen with the Ru surface, compared to the other two catalysts, which originates from the W^{6+}-induced increase of charge transfer from the carrier to the Ru crystallites.

The opposite behavior is expected from the Ca^{2+}-doped catalysts under similar experimental conditions, i.e., Ru should remain in the metallic state under reaction conditions and should be more easily reduced compared to the undoped catalyst. This is because doping TiO_2 with Ca^{2+} cations results in reduced charge transfer to the Ru crystallites. Thus, it is expected that Ru on Ca^{2+}-doped TiO_2 will exhibit lower density of states at the Fermi level.

In-situ FTIR spectroscopy [95] over the Ru/TiO_2 family of catalysts showed that an adsorbed CO species, located at 1985 (or 1990) cm^{-1}, exists on the surface of the unmodified and Ca^{2+}-doped catalysts under reaction conditions, even at temperatures as high as 1073 K. This species, which was not observable over the W^{6+}-doped catalyst, was attributed to CO species linearly bonded on reduced Ru sites [95]. It was concluded that the presence and population of this adsorbed CO species depend on the oxidation state and the electronic properties of the catalyst, which, in turn, depend on the type of the support employed to disperse the Ru

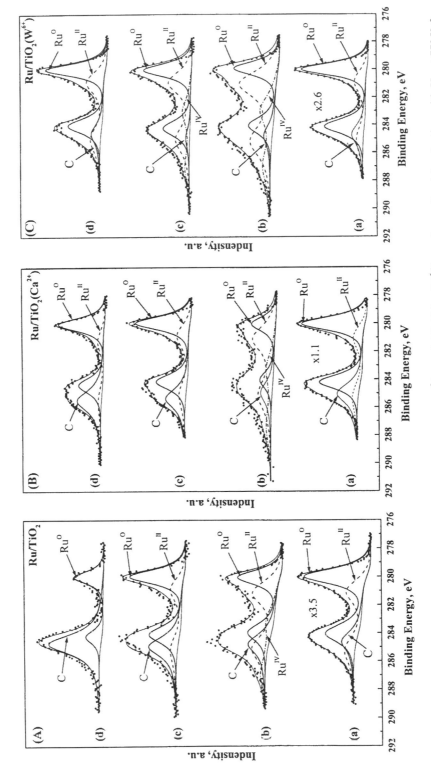

Figure 5 XPS spectra obtained from (A) Ru/TiO₂, (B) Ru/TiO₂ (Ca²⁺), and (C) TiO₂ (W⁶⁺) catalysts after (a) reduction with H₂ at 773 K for 20 min and exposure to a 10% CH₄–5% O₂ in Ar mixture at (b) 673 K, (c) 823 K, and (d) 973 K. (From Ref. 95.)

crystallites, in particular the type of doping cation incorporated into the crystal matrix of TiO_2.

20.6 INFLUENCE OF CARRIER DOPING ON ADSORPTION PARAMETERS

20.6.1 Equilibrium Chemisorption Measurements

Equilibrium chemisorption of H_2 and CO at room temperature is a standard technique used for the estimation of the metal dispersion of supported metal catalysts. The values of H/M and CO/M (number of hydrogen atoms or CO molecules adsorbed at equilibrium over the total number of metal atoms in the catalyst) determined from the monolayer uptakes of H_2 and CO are presented for the M/TiO_2 (*D*), M: Pt, Rh catalysts in Table 2. In the case of Pt catalysts, doping with higher-valence cations leads to a decrease, by about 70–80%, of hydrogen and CO uptake, while doping with lower-valence cations has no significant effect. For Rh catalysts supported on higher-valence doped TiO_2, on the other hand, exactly the opposite behavior is observed. The hydrogen uptake increases monotonically with W^{6+} dopant content in the TiO_2 support by up to 1000%, while CO uptake exhibits a maximum at a 0.22-at.% W^{6+} dopant content. Transmission electron microscopy (TEM) experiments showed that this behavior is not due to a different particle size of Pt or Rh on doped catalysts [90,96]. The particle size of Pt or Rh crystallites on the doped catalysts was found not to differ appreciably from that observed over the undoped catalysts. These alterations are suggested to be manifestations of an electronic interaction between the doped TiO_2 carrier and the supported metal. The observed increase in the H_2 uptake of Rh catalysts when they are dispersed on W^{6+} doped carriers is too large to be solely attributed to enhanced adsorption on the Rh surface. There are experimental indications, obtained from IR spectroscopy of adsorbed CO, supporting the idea that the enhanced hydrogen uptake is due to increased hydrogen spillover on the doped TiO_2 surface [97]. The enhanced CO uptake has been found by IR to be due to the enhancement of the disruption of Rh crystallites induced by CO. As a result of the electronic interaction at the support–

Table 2 Equilibrium H_2 and CO Uptakes over the $M/TiO_2(D)$ Catalysts Expressed as H/M and CO/M Ratios

Catalyst	H/M	CO/M
0.5% Pt/TiO_2	0.73	0.92
TiO_2 (0.66% Mg^{2+})	0.68	0.83
TiO_2 (0.12% Ta^{5+})	0.12	0.19
TiO_2 (0.16% Sb^{5+})	0.14	0.28
0.5% Rh/TiO_2	0.10	0.09
TiO_2 (0.11% W^{6+})	0.13	0.26
TiO_2 (0.22% W^{6+})	0.62	0.37
TiO_2 (0.45% W^{6+})	0.81	0.17
TiO_2 (0.67% W^{6+})	0.99	0.13

Source: Ref. 96.

metal interface, the Rh–Rh bond strength at the periphery of the metal particles is weakened and the formation of gem-dicarbonyl [Rh$^+$(CO)$_2$] species on isolated Rh sites is facilitated [97].

20.6.2 Infrared Spectroscopy of Adsorbed CO

IR spectroscopy was used to probe the adsorption mode of CO, at room temperature, on M/TiO$_2$ (*D*) catalysts and possible effects of the doping process on adsorption characteristics. In the case of Pt catalysts, adsorbed CO was found to be mainly in the linear form. In Rh catalysts, linear, bridged, and gem-dicarbonyl CO IR bands were observed. A blue shift toward higher wavenumbers of the linear CO band was found in both Pt and Rh catalysts supported on higher-valence doped TiO$_2$. The magnitude of this blue shift, at full coverage, was found to be 15–22 and 9 cm^{-1} for Pt and Rh catalysts, respectively [90,97].

For the linearly adsorbed CO, the shift of the C–O vibrational frequency upon carrier doping may be related to three factors: (1) effects of the surface electronic states of the Rh crystallites; (2) Rh particle size (due to the Rh–Rh bond disruption); and (3) changes of CO coverage. In order to eliminate any shift due to factors other than electronic parameters, the position of the linear CO bands on the doped and the undoped catalysts obtained at small and equal CO coverage are compared. Low θ_{CO} was obtained by careful control of CO exposure to the fresh sample. Figure 6 shows the spectral features of CO ($\theta_{CO} = 0.08$–0.09) adsorbed on the doped (*x* = 0.67 at.%) and the undoped catalysts, following CO exposure at 300 K. The linear CO band is observed at 2028 cm^{-1} on the undoped catalyst and at about 2037 cm^{-1} on the doped catalyst, showing a blue shift of about 9 cm^{-1} upon carrier doping. Under these experimental conditions, the observed blue shift should be exclusively attributed to alterations of the surface electronic states of the Rh crystallites via electronic interactions between the Rh particles and the carrier at the interface.

According to the Blyholder–Dewar–Chatt model [98–100], the bonding of CO to a metal surface occurs through formation of a σ bond by overlapping of an unoccupied d orbital and a lone electron pair on the carbon atom of CO and formation of a π bond between a filled metal d orbital and an empty antibonding π orbital ($2\pi^*$) of CO. The σ and π bondings between the metal and CO are coordinative to each other. The π bonding cannot be efficiently formed without a proper σ bonding [98]. Earlier studies of CO adsorption on metal surface (e.g., Pt, Cu, Pd, Ni) [101,102] suggested that the π back donation was more important in forming a metal–CO bond than the bonding between the CO 5σ molecular orbitals and the unoccupied metal electronic states. In this case, the transfer of electrons from the carrier to the supported metal particles would favor the π back donation to the antibonding π orbital of CO, which would result in strengthening the metal–CO bond. However, evidence indicating the significance of the σ bond between a metal and CO has been presented by several authors. In a study of CO adsorption over Ni/TiO$_2$, Raupp and Dumesic [103] observed a correlation between weakening of the Ni–CO bond and charge transfer from the reduced TiO$_2$ to metallic Ni. They explain that the transfer of charge to Ni would result in a saturated electronic configuration of Ni, onto which the bonding of CO is significantly weakened. In the study of CO adsorption over M/TiO$_2$ (W^{6+}) catalysts [97], the strengthening of the C–O bond, which is revealed by the blue shift over the doped catalysts, was interpreted to imply

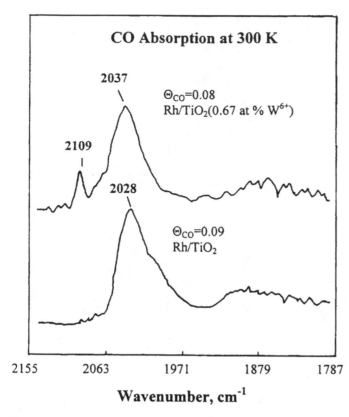

Figure 6 Infrared spectra of Rh/TiO$_2$(D) catalysts following exposure to CO at 300 K: (a) θ_{CO} = 0.09, Rh/TiO$_2$; (b) θ_{CO} = 0.08 Rh/TiO$_2$ (0.67 at.% W^{6+}). (From Ref. [97].)

that the platinum or rhodium particles dispersed on the doped carrier have an electronic configuration that does not favor the formation of M–CO bonds, as compared to the undoped catalysts. This result can be explained by the reasoning discussed above, i.e., the intensified transfer of charge from the doped carrier to the metal crystallites, which then possess a nearly saturated electronic configuration, which resembles that of silver or gold. Consequently, the bonding between the CO molecule and the Pt or Rh sites is weakened due to the unavailability of unoccupied states of the metal to form the σ bond. Without a proper σ bonding, back donation of electrons from the d orbitals of the metal to the $2\pi^*$ orbital of CO (i.e., the π bonding) cannot be efficiently conducted even if the metal is enriched with electrons in the d orbitals [98,104]. Alternatively stated, the σ bonding becomes important for the metal with a nearly saturated electron configuration.

20.6.3 TPD of Adsorbed CO

Temperature programmed desorption (TPD) is a very useful technique for studying the interactions between metal surfaces and various adsorbates. TPD profiles of desorbed CO, CO$_2$, and H$_2$ were obtained following room-temperature adsorption of CO on Rh/TiO$_2$ (W^{6+}) catalysts. The CO and CO$_2$ profiles are shown in Figure 7

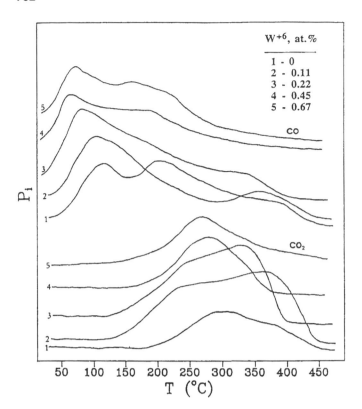

Figure 7 TPD profiles of CO and CO_2 following CO adsorption at room temperature on Rh/TiO_2 ($x\%$ W^{6+}) catalysts. (From Refs. [89] and [96].)

[89,96]. CO_2 desorption, accompanied by H_2 desorption, was attributed to reaction of desorbing CO with hydroxyl groups on the carrier surface, although a small amount of desorbed CO_2 might be due to CO dissociation.

Peak temperatures observed during CO TPD over the entire family of Rh/TiO_2 (W^{6+}) catalysts are summarized in Table 3. It is apparent that the CO peaks are shifted toward lower temperatures as the W^{6+} content of the supports increases. The

Table 3 Peak Temperatures of TPD Profiles from the 0.5% Rh/TiO_2 ($x\%$ W^{6+}) Family of Catalysts Following CO Adsorption at 25 °C

x at.%	T_M (°C)	
	CO	CO_2
0	119, 202, 380	300, 380
0.11	104, —, 360	240, 360
0.22	85, —, 330	250, 330
0.45	70, 180, —	278
0.67	77, 160, —	270

Source: Refs. 89 and 96.

main CO peak is shifted from 119 °C for the undoped catalyst to 70–77 °C for the catalyst containing 0.45–0.67 at.% W^{6+}. The peak at 202 °C in the undoped catalyst is not apparent in the 0.11 and 0.22 at.% W^{6+} catalysts, but it appears again at lower temperatures (160–180 °C) when the dopant concentration is further increased. The small peak at 380 °C is also shifted toward lower temperatures and disappears at higher levels of doping. The broad CO_2 peak seems to consist of two overlapping peaks at 300 and 380 °C in the undoped catalyst which shift to 240, 360 °C at low levels of doping, while at higher levels the two peaks seem to merge into a single one at 270–278 °C.

These observations indicate that the adsorption of CO on Rh supported on higher-valence-doped titania is weakened and that the degree of alteration of the strength of the Rh–CO adsorption bond is a function of the concentration of the dopant in the TiO_2 matrix. Similar weakening was not observed in the case of H_2 adsorption. It was also found that, because of weakening of the Rh–CO bond in doped catalysts, CO does not displace preadsorbed hydrogen from the Rh surface as easily as in the case of undoped Rh/TiO_2 [96]. This implies that hydrogen competes more favorably with CO for active sites.

20.7 INFLUENCE OF CARRIER DOPING ON KINETIC PARAMETERS

The probe reactions of CO and CO_2 hydrogenation, CO and C_2H_4 oxidation, partial oxidation of CH_4, aromatic hydrocarbons hydrogenation, NO reduction in the absence and in the presence of oxygen, and photocatalytic decomposition of water have been employed to study the effects of doping of the TiO_2 carrier on the catalytic performance of Pt, Rh, and Ru catalysts. Because of the anomalous hydrogen and CO adsorption behavior of doped catalysts, specific activities (turnover frequencies) cannot be estimated with certainty. There are, however, experimental indications (TEM) showing that doped catalysts have approximately the same dispersion as that of the corresponding undoped catalysts. Therefore, specific activities were calculated on the basis of the measured dispersion of the undoped catalysts, assuming that the doped catalysts exhibit the same dispersion as the corresponding undoped ones.

20.7.1 CO and CO_2 Hydrogenation

Doping of the TiO_2 carrier with higher-valence cations was found to have a negative influence on the activity of Pt in the CO/H_2 reaction. As shown in Table 4, [104] Pt catalysts supported on higher-valence doped TiO_2 are about two orders of

Table 4 Summary of Kinetic Results of the CO/H_2 Reaction over $Pt/TiO_2(D)$ Catalysts

Catalyst	TOF_{CH_4}	TOF_{CO_2}	E_{CH_4}	E_{CO_2}
	$(10^3 \, s^{-1})$		$(kJ \, mol^{-1})$	
0.5% Pt/TiO_2	5.78	3.95	70	76
$Pt/TiO_2 \, (Sb^{5+})$	0.05	0.04	79	82
$Pt/TiO_2 \, (Ta^{5+})$	0.04	0.03	78	83

$T = 548 \, K$, $H_2/CO = 3$. *Source*: Ref. 104.

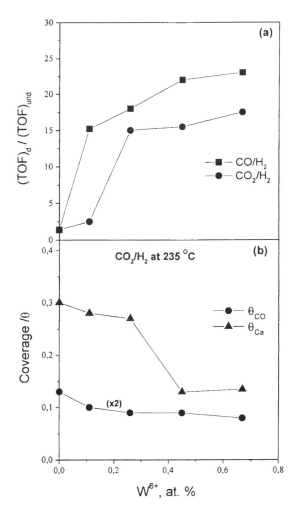

Figure 8 (a) Activity enhancement of Rh catalysts for methane formation upon carrier doping, expressed as the ratio $(TOF)_d/(TOF)_{und}$, for CO/H_2 and CO_2/H_2 mixtures and (b) dependence of coverage of adsorbed CO, θ_{CO}, and surface carbon species, θ_{Ca}, on W^{6+} dopant content of Rh/TiO_2 (W^{6+}) catalysts. (From Refs. [107] and [109].)

magnitude less active than the undoped Pt/TiO_2 catalyst [104,105]. The dramatic reduction in the activity of Pt catalysts supported on higher-valence doped TiO_2 is accompanied by a small increase in the apparent activation energy of the reaction. In all cases the main products were CH_4 and CO_2, although small quantities of ethane and propane were observed over the doped catalysts.

In contrast, significant increases in the specific activity of Rh in the CO/H_2 and CO_2/H_2 reactions were observed when the metal was dispersed on W^{6+}-doped TiO_2 carriers [106–108]. The kinetic results are summarized in Figure 8(a), which shows the ratio of the turnover frequency (TOF) of the doped catalysts over that of the undoped catalysts as a function of dopant content in the TiO_2 carrier. Doped catalysts were found to be up to 20 times more active in CO_2 methanation. The activity enhancement, in both cases, was found to be a similar function of the dopant content in the carrier and is significant even at low levels of dopant content in the TiO_2 matrix. In all catalysts, CH_4 was the main product, with a selectivity of about 83%. The selectivity, however, toward higher hydrocarbons decreased on doped catalysts (the probability of carbon chain growth decreased), while the selectivity toward C_2 and C_3 saturated hydrocarbons increased at the expense of the corresponding unsaturated ones. In the CO_2/H_2 reaction, the selectivity toward CH_4 formation was observed to increase on doped catalysts at the expense of CO formation. At 508 K and a ratio of H_2/CO_2 of 4, the CH_4 selectivity was approximately 55% on the undoped catalyst, while on doped catalysts it was as high as 90%.

The apparent activation energy of both reactions was found to decrease when they were carried out over the doped Rh catalysts. The decrease was found to be about $20 \, kJ \, mol^{-1}$ for the CO/H_2 reaction and $25–30 \, kJ \, mol^{-1}$ for the CO_2/H_2 reaction, almost independent of the W^{6+} dopant content in the TiO_2 support. Doped Rh catalysts also exhibited superior stability with time on stream in both reactions. While the activity of the undoped Rh catalyst declined by about 30–50% in the first 3 h on stream, the activity of the doped catalysts remained practically constant even after 4–5 h on stream. This implies that carbon deposition is retarted on doped catalysts, a phenomenon also investigated by temperature programmed surface reaction techniques [107].

This dramatically different behavior of Pt and Rh catalysts when they are supported on higher-valence doped TiO_2 is, at a first glance, a contradiction. The decrease in Pt activity can be explained, considering the lowering of its adsorption capacity for H_2 and CO and the weakening of the Pt–CO bond, which implies that under steady-state reaction conditions the surface coverage of CO and hydrogen is very low.

Concerning the behavior of Rh when it is supported on TiO_2 (W^{6+}), temperature programmed surface reaction (TPSR) and oxidation (TPO) techniques were employed to further investigate the kinetic results and their cause [107]. TPSR of preadsorbed CO with H_2 revealed that adsorbed CO is less active toward hydrogenation on doped catalysts. The amount of surface carbon formed during exposure to CO or CO/H_2 mixtures, as determined by TPO, was also found to be less on doped catalysts, which implies that CO has a smaller tendency toward dissociation on doped catalysts. The activity enhancement under steady-state reaction conditions must then be sought in the increased surface coverage of hydrogen under these conditions, which is a consequence of the weakened Rh–CO bond. All experimental observations concerning the CO and CO_2 hydrogenation over Rh/TiO_2 (W^{6+}) catalysts are in harmony with this hypothesis. For example, the observed changes in the selectivity in the CO/H_2 and CO_2/H_2 reactions over the doped catalysts are consistent with the hypothesis of reduced surface coverage of CO and carbon and increased coverage of hydrogen. This has indeed been observed experimentally, as shown in Figure 8(b) [109], which shows the influence of W^{6+} dopant content on concentration of adsorbed CO and carbon on Rh/TiO_2 (W^{6+}) catalysts. Reduced formation of surface carbon during reaction is expected to lead to smaller deactivation (less active sites blocked by carbon), as was indeed observed in the case of doped catalysts. The reduced activation energy of both reactions on doped catalysts might reflect the reduced energy barrier for product formation in a surface where hydrogen competes favorably for active sites.

20.7.2 Hydrogenation of Aromatic Compounds

Pt/TiO_2 (W^{6+}) and Rh/TiO_2 (W^{6+}) catalysts were also tested under conditions of benzene and naphthalene hydrogenation in a batch, slurry-phase reactor. Similar trends to those reported earlier for the CO and CO_2 hydrogenation reactions over the same catalysts were observed [110,111]. Thus, Pt was found to lose more than 90% of its hydrogenation activity when dispersed on higher-valence doped TiO_2 carriers, while the specific hydrogenation activity of Rh was found to be significantly enhanced. Rh supported on W^{6+}-doped TiO_2 was found to be approximately 10 times more active for benzene hydrogenation and more then 30 times more active for naphthalene hydrogenation. The rate enhancement was observed to be a function of the dopant content in the TiO_2 carrier. Doped Rh catalysts were also found to be more resistant to sulfur poisoning. In the presence of 50 ppm of thiophene in the reaction mixture, the Rh/TiO_2 (0.45% W^{6+}) catalyst retained about 60% of its hydrogenation activity compared to 20% for the undoped Rh/TiO_2 catalyst. In the presence of thiophene, the doped catalyst was more than two orders of magnitude more active in naphthalene hydrogenation than the undoped catalyst.

20.7.3 CO and C_2H_4 Oxidation

Generally, the trends observed in hydrogenation reactions were also observed in the case of CO oxidation. Pt catalysts supported on higher-valence doped TiO_2 were about 3–4 times less active than the undoped Pt/TiO_2 catalyst [105]. The decrease of specific activity upon carrier doping was not as dramatic as in the case of CO hydrogenation. The apparent activation energy of the reaction was found to be about $20\,kJ\,mol^{-1}$ higher when Pt was dispersed on higher-valence doped carriers. W^{6+} doping of the TiO_2 carrier was found to lead to an increase of the activity of supported Rh catalysts in the CO/O_2 reaction [107]. The kinetic results are presented in Figure 9, where the ratio $(TOF)_d/(TOF)_{und}$ is plotted as a function of W^{6+} dopant content in the TiO_2 carrier for two different gas-phase compositions. The activity enhancement appears to be rather small (50–90%) in the region of the gas-phase composition where CO inhibits the reaction ($p_{CO} = 0.02\,bar$, $p_{O2} = 0.06\,bar$, $T = 393\,K$). These variations in activity could even be attributed to slightly different dispersion of doped catalysts. Greater enhancement of the specific activity, by up to 8 times, was observed in the composition region where the maximum rate is observed ($p_{CO} = 0.001\,bar$, $p_{O2} = 0.06\,bar$, $T = 393\,K$). The activity enhancement exhibits a maximum of about 0.2 at.% W^{6+} dopant content. This trend is qualitatively similar to the one observed for the CO equilibrium uptake over the same catalysts. The apparent activation energy of CO oxidation was found not to be affected by doping in the gas-phase composition where CO inhibits the reaction. At low CO pressures ($p_{CO} = 0.001\,bar$), however, an increase in the activation energy by about $25\,kJ\,mol^{-1}$ was observed on doped catalysts.

The important issue of surface oxidation of the metal particles under reaction conditions was investigated in the cases of C_2H_4 oxidation over $Rh/TiO_2(D)$ catalysts [112] and CH_4 partial oxidation over $Ru/TiO_2(D)$ catalysts [93–95]. The dependence of turnover frequency on oxygen and ethylene partial pressures is shown in Figure 10(a) and (b), respectively, for the Rh/TiO_2 and Rh/TiO_2 (0.45% W^{6+}) catalysts. At low oxygen pressures the reaction is approximately first order in oxygen

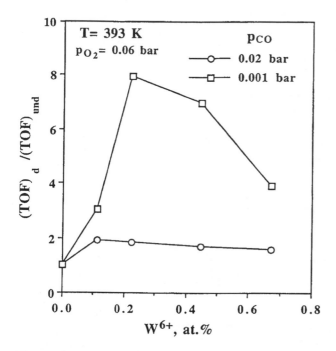

Figure 9 Effect of gas-phase composition on the activity enhancement ratio $(TOF)_d/(TOF)_{und}$ in the CO/O_2 reaction over Rh catalysts as a function of dopant content. (From Ref. 107.)

pressure. There exists, however, a very narrow oxygen pressure region in which the catalysts abruptly lose a large fraction of their activity. This can also be seen in Figure 10(b), where the effect of ethylene pressure on activity is shown. In this case, at low ethylene pressures, the catalysts have low activity, and they "light off" abruptly at a very narrow region of ethylene pressure. At higher ethylene pressures, zero or slightly negative order with respect to ethylene pressure is observed. The doped catalyst was found to be significantly more active in ethylene oxidation, especially in the region of intermediate oxygen and ethylene pressures, where the activity enhancement is more than 100-fold. The sudden drop of catalytic activity as the oxygen pressure is increased, or, alternatively, the sudden "light-off" as the ethylene pressure is increased, is attributed to oxidation or, respectively, reduction of the Rh surface. This transition takes place at higher O_2/C_2H_4 ratios in the case of the doped catalysts, which means that oxidation of Rh on doped catalysts takes place at higher oxygen pressures. This is in accordance with what was observed in the case of CO oxidation and suggests that Rh catalysts supported on W^{6+}-doped TiO_2 are more resistant to deactivation resulting from oxidation of the Rh surface.

20.7.4 Partial Oxidation of Methane

At relatively low temperatures ($< 800\,°C$), over most catalysts investigated, the catalytic partial oxidation of methane to synthesis gas is thought to follow the so-called indirect scheme, according to which initially methane is combusted until all

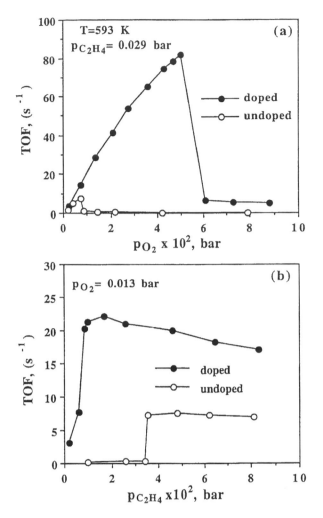

Figure 10 Dependence of the rate of C_2H_4 oxidation on the oxygen (a) and ethylene (b) partial pressure for W^{6+}-doped and undoped Rh/TiO_2 catalysts. (From Ref. [112].)

available oxygen is consumed, followed by reforming of the unconverted methane with CO_2 and H_2O produced primarily. It has been shown [91–96,113] that TiO_2-supported Ru exhibits a unique catalytic behavior by promoting the direct-conversion scheme, i.e., direct formation of synthesis gas without the intermediate production of CO_2 and H_2O, even at relatively low temperatures. This behavior has been correlated with the ability of TiO_2 to stabilize dispersed Ru in its metallic state under reaction conditions [93]. It was also found that the selectivity of Ru in catalyzing direct formation of CO and H_2 in the presence of oxygen is improved upon doping of TiO_2 with small amounts of Ca^{2+} cations, while the opposite is true upon doping with W^{6+} cations. Typical results, obtained over Ru/TiO_2, Ru/TiO_2 (Ca^{2+}), and Ru/TiO_2 (W^{6+}) catalysts at 1073 K, are presented in Figure 11, in which selectivities toward CO (S_{CO}) and H_2 (S_{H_2}) are shown as functions of CH_4

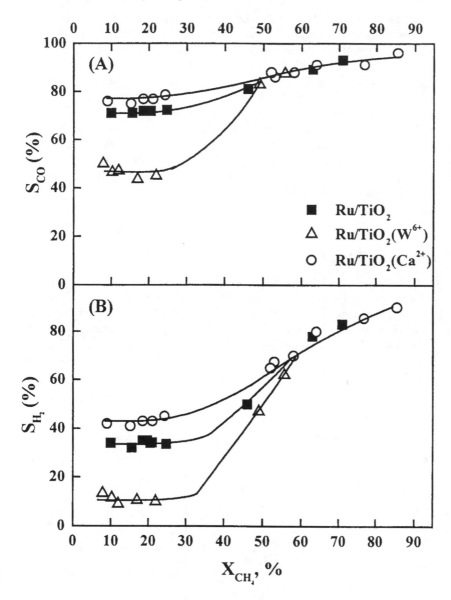

Figure 11 Selectivities toward CO and H_2 formation, obtained over $Ru/TiO_2(D)$ catalysts, as a function of methane conversion at $1073°$ K under conditions of partial oxidation of methane. (From Ref. 95.)

conversion (X_{CH_4}). In all cases, two distinct regions can be observed: one for $X_{CH_4} < 35\%$, where S_{CO} and S_{H_2} remain constant, and one for $X_{CH_4} > 35\%$, where selectivities toward both products increase with increasing methane conversion. It is important to note that in the first region ($X_{CH_4} < 35\%$), analysis of the gas at the effluent of the reactor shows that oxygen is not fully consumed ($X_{O_2} < 100\%$) while conversion varies between 10 and 90%. In this region, selectivities toward both CO and H_2 depend strongly on the nature of the support: doping TiO_2 with Ca^{2+} cations results in increased selectivities, while the opposite is true upon doping TiO_2 with W^{6+} cations. The activity of the catalysts was also found to depend on the carrier and to follow a pattern similar to that of selectivity, i.e., the Ca^{2+}-doped catalyst exhibits highest activity while the W^{6+}-doped catalyst exhibits the lowest (Table 5). A qualitatively similar behavior was observed over a wide temperature range (973–1073 K). These results correlate well with the oxidation state of Ru under reaction conditions [93–95]. As illustrated in Figure 5, doping of the TiO_2 carrier with W^{6+} cations results in the stabilization of a portion of Ru in its oxide forms and, concomitantly, in lower methane conversions and lower selectivities toward CO and H_2. In contrast, doping of TiO_2 with Ca^{2+} cations results in stabilization of a larger fraction of Ru in its metallic form and, thus, enhanced activity and selectivity for CO and H_2 formation.

20.7.5 Reduction of NO by CO

Detailed investigation of the NO + CO reaction over Rh catalysts supported on TiO_2 and TiO_2 (W^{6+}) carriers [114–115] has shown that W^{6+}-doping of TiO_2 results in expansion of the temperature window of operation toward lower temperatures by more than 50 °C (Figure 12). The mechanism of the reaction has been examined employing FTIR and transient-MS techniques. It was found that the formation of N_2 is related to the dissociation of negatively charged [Rh-NO$^-$] species while the formation of N_2O is related to the presence of a dinitrosyl [Rh(NO)$_2$] complex on the

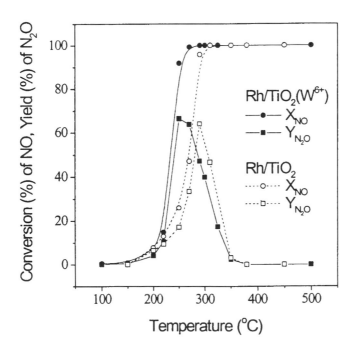

Figure 12 Conversion of NO and yield of N_2O obtained over Rh/TiO$_2$(D) catalysts as a function of reaction temperature. (From Refs. [114] and [115].)

Table 5 Results of Activity and Selectivities Toward CO (S_{CO}) and H_2 (S_{H_2}) Formation, at Low Oxygen Conversion, Under Conditions of Partial Oxidation of Methane

Reaction Temp. (K)	Ru/TiO$_2$			Ru/TiO$_2$ (Ca^{2+})			Ru/TiO$_2$ (W^{6+})		
	*R_{CH_4} (mol. g^{-1} s^{-1})	S_{CO} (%)	S_{H_2} (%)	*R_{CH_4} (mol. g^{-1} s^{-1})	S_{CO} (%)	S_{H_2} (%)	*R_{CH_4} (mol. g^{-1} s^{-1})	S_{CO} (%)	S_{H_2} (%)
973	8.0×10^{-5}	61	27	8.9×10^{-5}	70	35	6.8×10^{-6}	30	6
1023	10.1×10^{-5}	64	31	11.6×10^{-5}	75	37	8.1×10^{-6}	37	9
1073	11.9×10^{-5}	71	34	14.9×10^{-5}	81	42	1.4×10^{-5}	45	11

*The rate of CH$_4$ consumption is referred to the region where $X_{CH_4} < 10\%$. *Source*: Refs. 95 and 113.

catalytic surface. The presence of reduced Rh^0 sites was found to be necessary for the production of both reduction products [114–115]. The positively charged species [Rh-NO$^+$], which is present over partially oxidized surfaces, is inactive. Rh-NCO species do not interact directly with CO and O$_2$. On the contrary, isocyanates react with NO to yield N$_2$O, thus providing an alternative route for the production of nitrous oxide. The role of CO is to remove atomic oxygen formed by dissociation of NO, thus regenerating the catalytically active sites.

The beneficial effect of doping TiO$_2$ with W^{6+} cations on the catalytic performance has been attributed to alterations of the chemisorptive behavior of Rh toward NO and CO induced by doping. In particular, it was found that doping results in blue shifts in the stretching frequencies of N–O and C–O bonds contained in Rh-NO$^+$, Rh(NO)$_2$, Rh-CO, and Rh(CO)$_2$ species, indicating a weaker bonding of the adsorbed molecules with the surface (Table 6). In contrast, the N–O bond of the Rh-NO$^-$ species is weakened by doping, resulting in higher rates of dissociation and, therefore, in higher transient yields of N$_2$ production in the gas phase, compared to the undoped catalyst.

The strengthening of the C–O bond, which is revealed by the blue shift over the doped catalysts, implies that the rhodium particles have an electronic configuration that does not favor the formation of Rh–CO bonds, as compared to the undoped

Table 6 Vibrational Frequencies of CO and NO Species Adsorbed over Rh Catalysts Supported on Undoped and W^{6+}-Doped TiO$_2$, at 250 °C

Adsorbed Species	Band Frequency (cm^{-1})		Blue Shift (Δv, cm^{-1})
	Rh/TiO$_2$	Rh/TiO$_2$ (W^{6+})	
Rh-NO$^+$	1910–1912	1920–1922	10
Rh(NO)$_2$	1725 (asym)	1750	25
	1830–1840 (sym)	1870–1880	40
Rh-NO$^-$	1770		
	1660	Not clearly distinguishable	
Rh-CO	2045	2050	5
Rh(CO)$_2$	2094 (asym)	2106	12

Source: Refs. 114 and 115.

catalysts. This result has been attributed to the intensified transfer of charge from the doped carrier to the Rh crystallites so that the supported Rh crystallites have a nearly saturated electronic configuration, which resembles that of silver or gold, as described earlier.

As described in detail earlier, the binding energy of CO to a metal surface occurs through formation of a σ bond by overlap of an empty d orbital and a lone pair of the carbon atom of CO, and formation of a π bond between a filled metal d orbital and an empty antibonding π orbital (π^*) of CO. Then, if transfer of electrons from the doped carrier to the supported metal particles is assumed to partially fill the empty d orbitals of the surface atoms, the overlapping of empty Rh d orbitals and a lone pair of the C atom will be less effective in forming a Rh–CO bond. On the other hand, the ability of filled d orbitals to backdonate electrons to the $2\pi^*$ orbitals of adsorbed CO will be enhanced due to the greater availability of electrons. Consequently, the bonding between the CO molecule and the Rh site is weakened due to the unavailability of unoccupied states of Rh metal to form a σ bond. Without a proper σ bonding, back donation of electrons from the d orbitals of the metal to the $2\pi^*$ orbital of CO (i.e., the π bonding) cannot be efficiently conducted even if the metal is enriched with electrons in the d orbitals.

NO has a very similar electronic structure as CO and the corresponding molecular orbitals (2π and 5σ) involved in the formation of the metal–NO bond are the same. The only difference is that NO possesses an unpaired electron in the antibonding 2π orbital, which is the reason for a higher probability for dissociation of NO, compared to CO. The cationic species (Rh-NO$^+$) is produced by donation of an electron antibonding (π^*_{2pz}) of NO to d orbital of the metal to strengthen the N–O bond, while the anionic one (Rh-NO$^-$) is produced by transfer of a d electron of the metal to an antibonding π^*_{2pz} orbital of NO to weaken the N–O bond. Due to these differences in the characteristics of bonding of the two species, W^{6+}-doping of TiO$_2$ is expected to affect in a different manner the positively and negatively charged species. The blue shift of Rh-NO$^+$ and Rh(NO)$_2$ bands may be explained in a way similar to that used for the Rh-CO and Rh(CO)$_2$ species discussed above.

Concerning the catalytically active Rh-NO$^-$ species, doping results in further weakening of the N–O bond. Assuming that the rate-determining step is that of NO dissociation, the higher initial yield of N$_2$ formation observed over Rh/TiO$_2$ (W^{6+}) as compared to the Rh/TiO$_2$ catalyst can then be attributed to the larger value of the rate constant associated with this step.

It was also found that isocyanate species adsorbed on Rh are significantly stabilized over the W^{6+}-doped Rh/TiO$_2$ catalyst and are present on the catalyst surface under all experimental conditions that lead toward formation of N$_2$O in the gas phase. As a result, reduction of NO by CO over doped catalysts occurs at lower temperatures ($\approx 50\,^\circ$C) compared to the undoped catalyst mainly due to expansion of the temperature window of N$_2$O formation.

20.7.6 Other Reactions

The influence of aliovalent cation doping of the support (TiO$_2$) on the catalytic properties of supported Pt and Rh crystallites was also investigated under other reaction conditions, among which the photocatalytic cleavage of water [116] and the reduction of NO by propylene [117] in the presence or absence of oxygen. In the case of photocatalytic water cleavage, it was found that doping of TiO_2 with cations of higher valence has a positive influence in the rate of hydrogen production, while doping with cations of lower valence has a weak negative influence. In the case of $NO/C_3H_6/O_2$ reaction, it was observed that the activity of supported Rh is weakly affected when the TiO_2 carrier is doped with aliovalent cations.

20.8 INTERPRETATION OF CHEMISORPTIVE AND KINETIC PHENOMENA ON THE BASIS OF THE METAL–SEMICONDUCTOR BOUNDARY-LAYER THEORY

According to the metal–semiconductor boundary-layer theory, when a Group VIII metal is dispersed on the surface of TiO_2, charge is being transferred from the support to the metal particles. The driving force for charge transfer is the difference in the work function of the two materials or, alternatively, the difference in the electrochemical potentials. The driving force is enhanced upon doping TiO_2 with higher-valence cations. The maximum number of electrons transferred has been estimated [88] to be in the range of 0.1–$0.4\,e/atom$, for metal crystallites in the range of 1–$2\,nm$ (Figure 3). This significant amount of charge transferred is partly due to the very high static dielectric constant of TiO_2, compared to other semiconductors.

The charge transferred is expected to lie at the metal–semiconductor interface, on the side of the metal (Figure 13). This can be explained by the following argument: a metal surface is an equipotential surface. Because the dielectric constant of TiO_2 is 130, the charge located at the interface will be 130 times larger than the charge at the free metal surface, even without taking into account the attractive contribution of the positively charged depletion region in the semiconductor. Therefore, the charge at the free metal surface is negligible. The negative charge on the metal side of the metal–semiconductor interface is balanced by a positive charge on the side of the semiconductor, which is created by the transport of electrons away from this region. The distribution of positive charge within the semiconductor follows the Poisson distribution, and the volume of the semiconductor depleted from electrons is very large. In fact, the depletion length W (Figure 13) can be as high as a few hundred nanometers [88].

It is still unclear in what way the electrons transferred to the metal–semiconductor interface can perturb the electronic structure of the metal crystallites, leading to modifications of their chemisorptive and catalytic properties. Intuitively, one expects that whatever the interaction is, it will be more intense for small metal crystallites, for which the ratio e/M was found to be significant (Figure 3) and the fraction of interface metal atoms is large. Two different modes of electronic interaction can be considered, namely:

1. Long-range interactions, affecting the whole metal crystallite
2. Short-range interactions, affecting the atoms at the gas–metal–support three-phase boundary

The former case implies modifications of the electronic structure of the metal crystallite, such as the following:

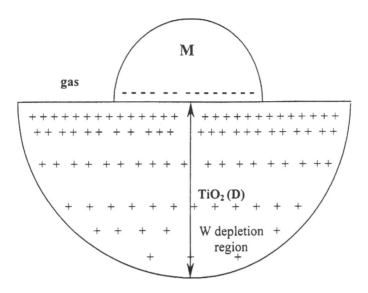

Figure 13 Physical model of a metal crystallite in intimate contact with an *n*-type semiconductor whose work function is lower than that of the metal. Charge is transported from the semiconductor to the metal and is contained at the interface metal atoms, while a large volume of the semiconductor is depleted of electrons.

1. Changes in the d-band population. Chen et al. [118] have proposed that electronic modifications might be due to rehybridization of electron orbitals in the presence of charge at the interface.
2. Shifting of the Fermi level and, as a consequence, of the work function. Theoretical calculations by Ward et al. [119] have shown that the Fermi level of Rh, Pd, and Pt is shifted when they are present as mono- or three-layer slabs on alumina surfaces. This shift is a consequence of charge transfer between the metal and the support. The charge transferred was found to be of the order of 0.1–0.2 e/atom.
3. Changes in the density of states at the Fermi level.

It should be emphasized at this point that charge transported across the interface cannot propagate through the metal as a result of very effective screening by metal electrons [90,120]. Furthermore, metals have a high concentration of free electrons and, as a result, only minimal perturbations of the electron state of the metal atoms via electron transfer from the support would be expected. In fact, if electrical conductivity is used as a measure of free electron concentration, the difference between Pt metal and TiO_2 is of the order of 10^7 [90]. However, in addition to the relative concentration of free electrons, the "affected volume" of each solid must also be considered. The electrons that cross the metal–semiconductor interface into the metal form a thin sheet of negative charge contained within a distance of approximately $0.5\,\text{Å}$ (Thomas–Fermi screening distance). On the other hand, the "depletion region" in the semiconductor is of the order of many hundreds of Å [90], giving a difference in the affected volumes of the two solids in the order of 10^6. This difference is similar in magnitude to the difference in free electron concentration between the metal and the semiconductor.

The short-range interaction can be considered as a consequence of the strong electric fields, which are present at the interface. The existence of strong electric fields (> 0.1 V/nm) can have a significant impact on the adsorptive and catalytic properties of the metal atoms. The effect of an electric field on the adsorption of NO on Pt(111) and Rh(111) has been studied theoretically and experimentally [121–123]. It was shown that for a field pointing away from the surface the binding energy of NO on Rh(111) is reduced by 15% (or about 4 kcal/mol) at a field of 3 V/nm. In stronger fields, even the dissociation of NO on the surface can be induced. Theoretical calculations show that an electrostatic interaction can explain many aspects of the promoting or poisoning effect of several preadsorbed atoms on the metal surface [124]. Vanselow and Mundschau [125] have observed the formation of high work function islands on a Pt surface on which TiO_2 has been deposited and has been heated above 800 K. They estimate that the induced electric field by the interaction of these islands with the rest of the surface is in the range of 0.06–6 V/nm. It should be noted that our own estimates (Table 1) are in the same range.

The transfer of electrons from the support to the metal particles, or vice versa, also influences the electron structure of the support at the periphery of the metal particles. This part of the support can be either depleted or enriched in electrons. The affected region of the support, as deduced from the magnitude of the depletion length [88], can be quite large and might respond differently to various adsorbates. For example, in the depletion region of an *n*-type semiconductor, the adsorption of electron donor adsorbates might be enhanced, while the adsorption of electron acceptor adsorbates might be hindered. This phenomenon can affect the catalytic behavior of the metal–semiconductor system via one of the following mechanisms:

1. Creation of new catalytic sites on the support surface at the periphery of the metal particle
2. Enhancement of the spillover of intermediate species of the reaction from the support to the metal, or vice versa

The work function is an important parameter that influences the catalytic behavior of the metal surface. The work function effect has been clearly demonstrated in the case of the NEMCA (non-Faradaic electrochemical modification of catalytic activity) effect, where it has been shown experimentally that catalytic reaction rates depend exponentially on the work function change of the metal surface ([126–130] and references therein). Work function changes in the case of NEMCA are caused by backspillover of O^{2-} anions on the metal surface, and they reflect changes in the surface potential.

Another quantity that has to be taken into consideration is the density of states at the Fermi level, $g(E_F)$, and any alterations caused to it by the charge transfer process. The importance of $g(E_F)$ on the adsorptive and catalytic properties of a metal surface has been stated by some investigators [131–133]. More specifically, the density of states defines the ability of the surface to respond to the presence of an adsorbate [132]. Theoretical calculations for the density of states function, $g(E)$, have been reported in the literature for certain metals [134]. The $g(E)$ function for the d metals Ru, Rh, and Pd is characterized by the participation of the d electrons. All three metals have a high density of states at the Fermi level (1.13 for Ru, 1.35 for Rh, and 2.31 for Pd). The position of the Fermi level with respect to the d band gives an indication of the d unoccupied density. Thus, for Ru there are many unoccupied d states in the region up to 2 eV above the Fermi level, while for Pd the Fermi level is situated very close to the end of the d band [134]. It can then be speculated that significant charge transfer to a Pd crystallite can lead to filling of all the unoccupied d levels, so that the Fermi level will now be located at the region where the s electrons predominate, which is characterized by a low density of states, as in the case of silver. Essentially, what is proposed is that a similar phenomenon to the one found in alloys of Pd or Ni with electropositive elements is at work [135]. In these alloys, the Ni and Pd d bands become filled and the density of states at the Fermi level is greatly decreased. This effect does not require significant charge transfer, as it is mainly a hybridization effect. Based on this approach, the observed suppression in the adsorptive and catalytic activity of Pt, when it is supported on higher-valence doped TiO_2 [90,104,105], can be related to the decrease in the density of states of Pt at the Fermi level because of the electron-transfer process. On the other hand, the opposite effect has been observed in the case of Rh [105–107]. Rh is different from Pt, in the sense that it has a large fraction of unoccupied d states, so that the charge transfer should be much more significant to produce a similar effect. An estimation from [134] gives a value of about 1.1 d unoccupied states per Rh atom, whereas for Pt the d unoccupied states are about 0.2–0.3. The maximum charge transfer predicted by our model, for comparison, is also about 0.2–0.4 e/atom (Figure 3). It should also be stressed that changes induced at the Fermi-level local density of states by an electronic perturbation of a metal surface can have a long-range character, despite the fact that the electronic perturbation is localized [132,133].

20.9 AN ALTERNATIVE INTERPRETATION OF METAL–SUPPORT INTERACTIONS INDUCED BY CARRIER DOPING

Recently [136], ethylene oxidation over IrO_2 and Rh was used as a probe reaction to establish comparative kinetic modifications, induced either by electrochemical promotion (NEMCA) or by metal–support interactions, employing pure and doped TiO_2 and other carriers whose work function was determined. Electrochemical promotion refers to the electrochemical supply of O^{2-} ions from the Yttria–stabilized-Zirconia (YSZ) solid electrolyte to the surface of the metal film in contact with the solid electrolyte, and the concomitant increase of the metal work function [127–130]. In the case of IrO_2, it was observed that a maximum enhancement of catalytic rate by a factor of 12 could be achieved by electrochemical promotion, while exactly the same rate enhancement was achieved by addition of TiO_2 to the IrO_2 catalyst. It was also found that IrO_2-TiO_2 mixtures could be only marginally promoted electrochemically.

The same reaction was also investigated over Rh nanoparticles dispersed over pure and doped TiO_2 and other porous supports of known work functions [136]. It was established that the influence of electrochemical promotion on kinetic parameters of the model reaction is identical to the influence of metal–support interactions, under conditions at which the change of the work function of the catalyst is the same, regardless of the means by which the alteration in the work function is achieved.

These results indicate that, at least under certain conditions, the mechanism of metal–support interactions may be identical to that of electrochemical promotion, i.e., the charge exchanged between the metal particles and the support material, caused by the difference in the electrochemical potential of the two solids, may be oxygen ions, as in the case of electrochemical promotion. This, of course, could be complementary, or in addition, to electrons or positive holes or other charge carriers, depending on the support material and gas-phase conditions.

In order for this process (backspillover of oxygen ions from the support material to the surface of the metal crystallites) to be effective, the support material must possess sufficient ionic conductivity. It has been shown that TiO_2 is a mixed conductor, i.e., it possesses n-type semiconductivity at low-oxygen partial pressures and p-type semiconductivity at high-oxygen partial pressures, while, at intermediate oxygen pressures, it exhibits ionic conductivity, which is due to migration of O^{2-} ions or vacancies [137,138] and/or protons [139]. The ionic conductivity of TiO_2 has been confirmed by Pliangos et al. [140], who show that it can be used as a solid electrolyte to induce electrochemical promotion to Pt catalysts by supply of O^{2-} ions to its surface, by application of a current or potential. The alteration of oxygen ion conductivity of TiO_2 upon doping with a lower-valence cation was studied by Efstathiou et al. [141], who measured the mobility of lattice oxygen by performing $^{18}O_2$ transient isotopic experiments. The activation energy of bulk lattice oxygen diffusion was found to increase upon doping with 1 wt.% Li_2O and to decrease at high levels of Li doping. The influence of various dopants on oxygen ions conductivity of metal oxides is discussed by Zhang et al. [142].

The concepts discussed earlier are still valid: the thermodynamic driving force for charge transport is due to equilibration of the Fermi level of the two solids in contact. This is true for both the electrochemically promoted catalysts and the dispersed metal particles on an ion-conducting support. This interpretation of metal–support interactions, which involves oxygen ion backspillover from the support to the metal crystallites, is shown schematically in Figure 14. In the same figure, the concept of electrochemical promotion, which is induced by application of a current or potential (NEMCA), is also shown. The difference between electrochemical promotion and metal–support interactions is the driving force for the transport of oxygen ions from the carrier to the metal particles.

At the three-phase boundary (metal–support–gas) the following charge-transfer equilibrium is established:

$$O_2\ (g) + 4e^-\ (M) \leftrightharpoons 2O^{2-}\ (S) \tag{12}$$

where M denotes the metal particles and S the support. Then, the electrochemical potentials of electrons, $\bar{\mu}_e$, oxygen ions, $\bar{\mu}_{O^{2-}}$, and gaseous oxygen, μ_{O_2}, are related as follows [136]:

$$\bar{\mu}_{O^{2-}}(S) = 2\bar{\mu}_e(M) + (1/2)\mu_{O_2}(g) \tag{13}$$

If we assume that the oxygen ions are mobile on the metal surface and do not desorb rapidly or do not participate in fast catalytic reactions, then the equilibrium described by Eq. (12) is also established at the catalyst surface, in which case,

$$\bar{\mu}_{O^{2-}}(M) = 2\bar{\mu}_e(M) + (1/2)\mu_{O_2}(g) \tag{14}$$

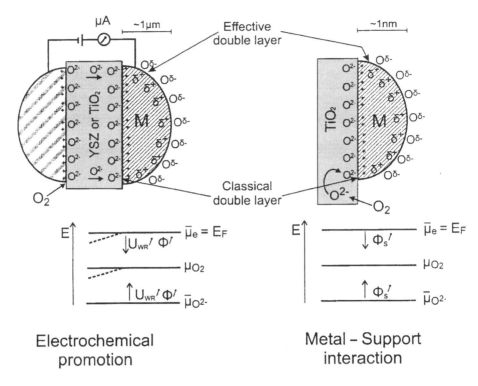

Figure 14 Alternative approach to metal–support interaction involving backspillover of oxygen ions from the doped TiO₂ carrier to the metal nanoparticles. The locations of the classical double layers formed at the metal–support interface and of the effective double layers formed at the metal–gas interface are shown. The bottom diagram shows the spatial constancy of the Fermi level, E_F (or electrochemical potential, $\bar{\mu}_e$) of electrons, of the chemical potential of oxygen, and of the electrochemical potential of O^{2-}. (From Ref. [136].)

Thus, in the metal–support system, the thermodynamic driving force for O^{2-} backspillover from the support to the metal surface is the difference in the electrochemical potentials of oxygen ions, $\bar{\mu}_{O^{2-}}(S) - \bar{\mu}_{O^{2-}}(M)$ [130,143,144]. Alternatively stated, the Fermi level of the two solids in contact is equalized by transport of charged carriers, such as oxygen ions. This could occur simultaneously with the transport of electrons or positive holes, depending on the extent of electronic and ionic conductivities of the semiconducting support. In any event, in dispersed metal-semiconducting support catalytic systems the extent of this phenomenon can be altered by doping the support with aliovalent cations, thus altering its electrochemical potential (or Fermi level) and subsequently the electrochemical potential of the metal particles since, under all conditions $\bar{\mu}_e(M) = \bar{\mu}_e(S)$.

20.10 EPILOGUE

Significant alterations in the chemisorptive and catalytic parameters of supported metal nanoparticles are observed upon doping semiconductive carriers with cations of valence different than that of the host cations. These alterations may be interpreted on the basis of the theory of metal–semiconductor contact, according to which the Fermi energy level of the two solids in contact must be at equal height. This requires transport of charge across the interface, until the electrochemical potential is uniform throughout the system. Thus, electrons may be transferred from the semiconductor to the metal, they are contained at the interface metal atoms, and a strong dipole develops at the interface. This may result in long-range interactions, affecting the entire metal crystallite, or in short-range interactions, affecting the atoms at the gas–metal–support three-phase boundary.

An alternative interpretation of the phenomenon of metal–support interactions induced by doping of semiconductive carriers with aliovalent cations is based on the theory of electrochemical promotion or the NEMCA effect. According to this interpretation, the charge carriers transported from the carrier to the metal particles are oxygen ions, which diffuse to the surface of the metal particles, thus altering the surface work function and, subsequently, chemisorptive and catalytic parameters. Work is currently in progress to elucidate the mechanism of induction of metal–support interactions by carrier doping.

REFERENCES

1. G.L. Haller, D.E. Resasco, Adv. Catal. 36 (1989) 173.
2. K. Hayek, R. Kramer, Z. Paal, Appl. Catal. A: General, 162 (1997) 1.
3. J.H. Sinfelt, J. Phys. Chem. 68 (1964) 344; W.F. Taylor, D.L.C. Yates, J.H. Sinfelt, J. Phys. Chem. 68 (1964) 2962; D.J.C. Yates, J.H. Sinfelt, W.F. Taylor, Trans. Faraday Soc. 61 (1965) 2044; W.F. Taylor, J.H. Sinfelt, D.J.C. Yates, J. Phys. Chem. 69 (1965) 3857.
4. S.J. Tauster, S.C. Fung, J. Catal. 55 (1978) 29; S.J. Tauster, S.C. Fung, R.L. Garten, J. Amer. Chem. Soc. 100(1) (1978) 170; S.J. Tauster, S.C. Fung, R.T.K. Baker, J.A. Horsley, Science 217 (1981) 1121.
5. M.A. Vannice, R.L. Garten, J. Catal. 56 (1979) 236.
6. B. Imelik, C. Naccache, G. Coudurier, H. Praliaud, P. Meriaudeau, P. Gallezot, G.A. Martin, J.C. Vedrine (eds.), Studies in Surface Science and Catalysis, Vol. 11, Metal-Support and Metal-Additive Effects in Catalysis, Elsevier, Amsterdam, 1982.
7. D.E. Resasco, G.L. Haller, J. Catal. 82 (1983) 279.
8. P. Meriaudeau, J.F. Dutel, M. Dufaux, G. Naccache, In Metal-Support and Metal-Additive Effects in Catalysis (B. Imelik et al., eds.), Elsevier, Amsterdam, 1982, p. 95.
9. A. Dauscher, W. Müller, G. Maire, Catal. Lett. 2 (1989) 139.
10. A. Dauscher, G. Maire, J. Mol. Catal. 60 (1991) 259.
11. R. Burch, In Hydrogen Effects in Catalysis (Z. Paal, P.G. Menon, eds.), Marcel Dekker, New York, 1988, p. 347.
12. G.C. Bond, R.R. Rajaram, R. Birch, J. Phys. Chem. 90 (1986) 4877.
13. H. Sadeghi, V.E. Henrich, J. Catal. 87 (1984) 279.
14. A.J. Simeons, R.T.K. Baker, D.J. Dwyer, R.J. Madon, J. Catal. 86 (1984) 359.
15. C.S. Ko, R.J. Gorte, J. Catal. 90 (1984) 59.
16. A.D. Logan, E.J. Braunschweig, A.K. Datye, D.J. Smith, Langmuir 4 (1988) 827.
17. S. Bernal, J.J. Calvino, M.A. Cauqui, G.A. Cifredo, A. Jobacho, J.M. Rodriguez-Izquierdo, Appl. Catal. 99 (1993) 1.
18. S. Bernal, F.J. Botana, J.J. Calvino, C. Lopez, J.A. Perez Omil, J.M. Rodriguez-Izquierdo, J. Chem. Soc. Faraday Trans. 92 (1996) 2799.
19. A.K. Datye, D. Kalakkad, M.H. Yao, D.J. Smith, J. Catal. 155 (1995) 148.
20. J.P. Belzunegui, J.M. Rojo, J. Sanz, J. Phys. Chem. 95 (1991) 3464.

21. J. Sanz, J.P. Belzunegui, J.M. Rojo, J. Am. Chem. Soc. 114 (1992) 6749.

22. P. Meriaudeau, J.F. Dutel, M. Dufaux, C. Naccache, Stud. Surf. Sci. Catal. 11 (1982) 95.

23. S. Bernal, F.J. Botana, J.J. Calvino, G.A. Cifredo, R. Garcia, J.M. Rodriguez-Inquierdo, Catal. Today 2 (1988) 653.

24. Y. Zhou, M. Nakashima, J.M. Whiter, J. Phys. Chem. 92 (1988) 812.

25. D.W. Daniel, J. Phys. Chem. 92 (1988) 3891.

26. C. Sudhakar, M.A. Vannice, Appl. Catal. 14 (1985) 47.

27. S. Bernal, J.J. Calvino, M.A. Cauqui, J.M. Gatica, C. Larese, J.A. Perez Omil, J.M. Pintado, Catal. Today 50 (1999) 175.

28. G.-M. Schwab, In Advances in Catalysis; Academic Press: Orlando, 1978; Vol. 27, p. 1.

29. F. Solymosi, Catal. Rev. 1 (1967) 233.

30. R.F. Baddour, M.C. Deibert, J. Phys. Chem. 70 (1966) 2173.

31. M. Boudart, G. Djega-Mariadassou, Kinetics of Heterogeneous Catalytic Reactions, Princeton University Press, Princeton, NJ, 1984.

32. A.Y. Stakheev, L.M. Kustov, Appl. Catal. A: General, 188 (1999) 3.

33. G. Larsen, G.L. Haller, Catal. Lett. 3 (1989) 103.

34. W.M.H. Sachtler, Z.C. Zhang, Adv. Catal. 39 (1993) 129.

35. W.M.H. Sachtler, A.Y. Stakheev, Catal. Today 12 (1992) 283.

36. W.M.H. Sachtler, Ber. Bunsenges, Phys. Chem. 99 (1995) 1295.

37. D.C. Tomczak, V.L. Zholobenko, H. Trevino, G.D. Lei, W.M.H. Sachtler, Stud. Surf. Sci. Catal. 84 (1994) 893.

38. P. Reyes, I. Concha, M.E. Konig, J.L.G. Fierro, Appl. Catal. A: 103 (1993) 5.

39. T.J. McCarthy, G.D. Lei, W.M.H. Sachtler, J. Catal. 159 (1996) 90.

40. T.J. McCarthy, C.M.P. Marques, H. Trevino, W.M.H. Sachtler, Catal. Lett. 43 (1997) 11.

41. D. Guillemot, V. Yu, Borovkov, V.B. Kazansky, M. Polisset-Thfoin, J. Fraissard, J. Chem. Soc. Faraday Trans. 93 (1997) 3587.

42. B.I. Boyanov, T.I. Morrison, J. Phys. Chem. 100 (1996) 16310.

43. V.L. Zholobenko, G.D. Lei, B.T. Carvill, B.A. Lerner, W.M.H. Sachtler, J. Chem. Soc. Faraday Trans. 90 (1994) 233.

44. S.N. Reifsnyder, M.M. Otten, H.H. Lamb. Catal. Today 39 (1998) 317.

45. M. Vaarkamp, J.T. Miller, F.S. Modica, G.S. Lane, D.C. Koningsberger, T. Uematsu, G.L. Haller, V. Haensel, K. Klier, A. Renouprez, R.W. Joyner, R. Prins, Stud. Surf. Sci. Catal. 75 (1993) 809.

46. M. Vaarkamp, B.L. Mojet, M.J. Kappers, J.T. Miller, D.C. Koningsberger, J. Phys. Chem. 99 (1995) 16067.

47. C. Dossi, R. Psaro, L. Sordelli, M. Bellatreccia, R. Zanoni, J. Catal. 159 (1996) 435.

48. B.L. Mojet, M.J. Kappers, J.C. Muijsers, J.M. Niemantsverdriet, J.T. Miller, F.S. Modica, D.C. Koningsberger, Stud. Surf. Sci. Catal. 84 (1994) 909.

49. T.V. Voskoboinikov, E.S. Shpiro, H. Landmesser, N.I. Jaeger, G. Schulz-Ekloff, J. Mol. Catal. A: Chem. 104 (1996) 299.

50. K. Ebitani, T. Tanaka, H. Hattori, Appl. Catal. A: 102 (1993) 79.

51. L.M. Kustov, T.V. Vasina, A.V. Ivanov, O.V. Masloboishchikova, E.V. Khelkovskaya-Sergeeva, P. Zeuthen. Stud. Surf. Sci. Catal. 101 (1996) 821.

52. A. de Mallmann, D. Burthaneuf, J. Chim. Phys. 87 (1990) 535.

53. M. Sugimoto, H. Katsuma, T. Hayasaka, N. Ishikawa, K. Hirasawa, Appl. Catal. A: General, 102 (1993) 167.

54. T. Tatsumi, L.X. Dai, H. Sakashita, Catal. Lett. 27 (1994) 289.

55. N.D. Triantafillou, J.T. Miller, B.C. Gates. J. Catal. 155 (1995) 131.

56. R.M. Watwe, B.E. Spiewak, R.D. Cortight, J.A. Dumesic, Catal. Lett. 51 (1998) 139.

57. A.Y. Stakheev, E.S. Shpiro, N.I. Jaeger, G. Schulz-Ekloff. Catal. Lett. 34 (1995) 293.

58. B.L. Mojet, J.T. Miller, D.E. Ramaker, D.C. Koningsberger, J. Catal. 186 (1999) 373.
59. D.C. Koningsberger, J. de Graaf, B.L. Mojet, D.E. Ramaker, J.T. Miller, Appl. Catal. A: General, 191 (2000) 205.
60. J.T. Miller, B.L. Mojet, D.E. Ramaker, D.C. Koningsberger, Catal. Today 62 (2000) 101.
61. S.C. Fung, J. Catal. 76 (1982) 225.
62. B.A. Sexton, A.E. Hughes, J. Foger, J. Catal. 77 (1982) 85.
63. C.-C. Kao, S.-C. Tsai, Y.-W. Chung, J. Catal. 73 (1982) 136.
64. H.R. Sadeghi, V.E. Henrich, J. Catal. 109 (1988) 1.
65. D.E. Resasco, R.S. Weber, S. Sakellson, M. McMillan, G.L. Haller, J. Phys. Chem. 92 (1988) 189.
66. J.M. Herrmann, J. Catal. 89 (1984) 404.
67. F. Solymosi, I. Tombacz, M. Kocsis, J. Catal. 75 (1982) 78.
68. P. Meriaudeau, O.H. Ellestad, M. Dufaux, C. Naccache, J. Catal. 75 (1982) 243.
69. G.B. Raupp, J.A. Dumesic, J. Catal. 97 (1986) 85.
70. D.N. Belton, Y.-M. Sun, J.M. White, J. Phys. Chem. 88 (1984) 5172.
71. K. Aika, A. Ohya, A. Ozaki, Y. Inoue, I. Yasumori, J. Catal. 92 (1985) 305.
72. T. Ishihara, K. Harada, K. Eguchi, H. Arai, J. Catal. 136 (1992) 161.
73. H.-W. Chen, I.M. White, J.G. Ekerdt, J. Catal. 99 (1986) 293.
74. H. Haberlandt, In Theoretical Aspects of Heterogeneous Catalysis (J.B. Moffat, ed.), Van Nostrand Reinhold, New York, 1990.
75. G. Pacchioni, N. Rosch, Surf. Sci. 306 (1994) 169.
76. F.S. Ohuchi, R.H. French, R.V. Kasowski, J. Appl. Phys. 62 (1987) 2286.
77. G. Blyholder, J. Mol. Catal. A: Chemical 119 (1997) 11.
78. S. Trasatti, In Comprehensive Treatise of Electrochemistry, (J.O.'M. Bockris, B.E. Conway, E. Yeager, eds.), Plenum Press: New York, 1980, Vol. 1, p. 45.
79. N.W. Ashcroft, N.D. Mermin, Solid State Physics, Holt, Rinehart and Winston, New York, 1976.
80. W.F. Egelhoff, J. Surf. Sci. Reports 6 (1986) 253.
81. S.M. Sze, Physics of Semiconductor Devices, 2nd ed., John Wiley & Sons, New York, 1981.
82. E.C. Akubuiro, X.E. Verykios, J. Phys. Chem. Solids, 50 (1989) 17.
83. M.S. Tyagi, In Metal–Semiconductor Schottky Barrier Junctions and Their Applications (B.L. Sharma, ed.), Plenum, New York, 1984.
84. L.J. Brillson, Surf. Sci. Rep. 2 (1982) 123.
85. H.K. Henisch, In Electrode Processes in Solid State Ionics (M. Kleitz, J. Dupay, eds.), Reidel, Dordrecht, Holland, 1976.
86. M.J. Sparnaay, Surf. Sci. Rep. 4 (1985) 101.
87. N. Yamamoto, S. Tonomura, H. Tsubomura, J. Electrochem. Soc. 129 (1982) 444.
88. T. Ioannides, X.E. Verykios, J. Catal. 161 (1996) 560.
89. T. Ioannides, X.E. Verykios, Chem. Eng. Technol. 18 (1995) 25.
90. E.C. Akubuiro, X.E. Verykios, J. Catal. 103 (1987) 320; 113 (1988) 106.
91. Y. Boucouvalas, Z.L. Zhang, A.M. Efstathiou, X.E. Verykios, Stud. Surf. Sci. Catal. 101 (1996) 443.
92. Y. Boucouvalas, Z.L. Zhang, A.M. Efstathiou, X.E. Verykios, Catal. Lett. 40 (1996) 189.
93. C. Elmasides, D.I. Kondarides, W. Grunert, X.E. Verykios, J. Phys. Chem. 13, 103 (1999) 5227.
94. C. Elmasides, D.I. Kondarides, S.G. Neophytides, X.E. Verykios, Stud. Surf. Sci. Catal. 130 (2000) 3083.
95. C. Elmasides, D.I. Kondarides, S.G. Neophytides, X.E. Verykios, J. Catal. 198 (2001) 195.

96. T. Ioannides, X.E. Verykios, J. Catal. 145 (1994) 479.

97. Z.L. Zhang, A. Kladi, X.E. Verykios, J. Phys. Chem. 98 (1994) 6804.

98. G. Blyholder, J. Phys. Chem. 68 (1964) 27722; 79 (1975) 756.

99. F. Stoop, F.J.C.M. Toolenaar, V. Ponec, J. Catal. 73 (1982) 50.

100. E.L. Garfunkel, M.H. Farias, G.A. Somorjai, J. Am. Chem. Soc. 107 (1985) 349.

101. G.W. Graham, J. Vac. Technol. A: 4 (1986) 760.

102. R.A. Campbell, J.A. Rodriguez, D.W. Goodman, New Frontiers in Catalysis. In Proc. 10th Intl. Congress on Catalysis, Part A, Budapest, Hungary, Elsevier Science Publishers B.V., Amsterdam, 1992, p. 333.

103. G.B. Raupp, J.A. Dumesic, J. Catal. 97 (1986) 85.

104. E.C. Akubuiro, T. Ioannides, X.E. Verykios, J. Catal. 116 (1989) 590.

105. E.C. Akubuiro, X.E. Verykios, T. Ioannides, Appl. Catal. 46 (1989) 297.

106. F. Solymosi, I. Tombacz, J. Koszta, J. Catal. 95 (1985) 578.

107. T. Ioannides, X.E. Verykios, M. Tsapatsis, C. Economou, J. Catal. 145 (1994) 491.

108. Z.L. Zhang, A. Kladi, X.E. Verykios, J. Catal. 148 (1994) 737.

109. Z.L. Zhang, A. Kladi, X.E. Verykios, J. Catal. 156 (1995) 37.

110. M. Koussathana, N. Vamvouka, M. Tsapatsis, X.E. Verykios, Appl. Catal. 80 (1992) 99.

111. M. Koussathana, N. Vamvouka, X.E. Verykios, Appl. Catal. 95 (1993) 211.

112. C. Pliangos, I.V. Yentekakis, X.E. Verykios, C.G. Vayenas, J. Catal. 154 (1995) 124.

113. C. Elmasides, T. Ioannides, X.E. Verykios, AIChE J. 46 (2000) 1260.

114. T. Chafik, D.I. Kondarides, X.E. Verykios, J. Catal. 190 (2000) 446.

115. D.I. Kondarids, T. Chafik, X.E. Verykios, J. Catal. 191 (2000) 147; 193 (2000) 303.

116. K.E. Karakitsou, X.E. Verykios, J. Phys. Chem. 97 (1993) 1184.

117. T. Chalkides, D.I. Kondarides, X.E. Verykios, Catal. Today 73 (2002) 213.

118. H.-W. Chen, J.M. White, J.G. Ekerdt, J. Catal. 99 (1986) 293.

119. T.R. Ward, P. Alemany, R. Hoffmann, In Environmental Catalysis (J.N. Armor, ed.), ACS Symposium Series 552, Am. Chem. Soc. Washington, DC, 1994, p. 140.

120. V. Ponec, St. Surf. Sci. Catal. 11 (1982) 63; 64 (1991) 117.

121. H.J. Kreuzer, L.C. Wang, J. Chem. Phys. 93 (1990) 6065.

122. J.H. Block, H.J. Kreuzer, L.C. Wang, Surf. Sci. 246 (1991) 125.

123. R.P. Madenach, G. Abend, M.S. Mousa, H.J. Kreuzer, J.H. Block, Surf. Sci. 266 (1992) 56.

124. N.D. Lang, S. Holloway, J.K. Nørskov, Surf. Sci. 150 (1985) 24.

125. R. Vanselow, M. Mundschau, J. Catal. 103 (1987) 426.

126. C.G. Vayenas, S. Bebelis, I.V. Yentekakis, H.-G. Lintz, Catal. Today 11 (1992) 303.

127. C.G. Vayenas, S. Bebelis, Catal. Today 51 (1999) 581.

128. C.G. Vayenas, S. Neophytides, In Electrochemical Activation of Catalysis: In Situ Controlled Promotion of Catalyst Surface, no. 12, pp. 199–253, The Royal Society of Chemistry, Cambridge, 1996.

129. C.G. Vayenas, I.V. Yentekakis, In Electrochemical Modification of Catalytic Activity (G. Ertl, A. Knozinger, J. Weitcamp, eds.), VCH Publishers, Weinheim, 1997, pp. 1310–1338.

130. C.G. Vayenas, S. Bebelis, S. Pliangos, C. Brosda, D. Tsiplakides, In The Electrochemical Activation of Catalysis, in press, Plenum Press, 2001.

131. J.J. Prinsloo, P.C. Gravelle, J. Chem. Soc. Faraday Trans. I 76 (1980) 512.

132. P.J. Feibelman, D.R. Hamann, Phys. Rev. Letters 52 (1984) 61.

133. P.J. Feibelman, D.R. Hamann, Surf. Sci. 149 (1985) 48.

134. V.L. Moruzzi, J.F. Janak, A.R. Williams, Calculated Electronic Properties of Metals, Pergamon Press, New York, 1978.

135. J.C. Fuggle, F.U. Hillebrecht, R. Zeller, Z. Zolnierek, P.A. Bennett, C. Freiburg, Phys. Rev. B: 27 (1982) 2145.

136. J. Nicole, D. Tsiplakides, C. Pliangos, X.E. Verykios, C. Comninellis, C.G. Vayenas, J. Catal. 204 (2001) 23.
137. D.L. Douglas, In Selected Topics in High Temperature Chemistry (O. Johannessen, A.G. Andersen, eds.), Elsevier, New York, 1989, pp. 192–194.
138. P.J. Gellings, H.J.M. Bouwmeester, Catal. Today 12 (1992) 1.
139. T. Norby, Solid State Ionics, 40/41 (1990) 857.
140. C. Pliangos, I.V. Yentekakis, S. Ladas, C.G. Vayenas, J. Catal. 159 (1996) 189.
141. A.M. Efstathiou, D. Papageorgiou, X.E. Verykios, J. Catal. 141 (1993) 612.
142. Z. Zhang, X.E. Verykios, M. Baerns, Catal. Rev.-Sci. Eng. 36 (1994) 507.
143. C.G. Vayenas, J. Electroanal. Chem. 486 (2000) 85.
144. C.G. Vayenas, D. Tsiplakides, Surf. Sci. 467 (2000) 23.

21

Abnormal Infrared Effects of Nanometer-Scale Thin Film Material of Platinum Group Metals and Alloys at Electrode–Electrolyte Interfaces

SHI-GANG SUN

Xiamen University, Xiamen, China

SUMMARY

In this chapter, the investigations of abnormal infrared effects (AIREs) of
nanometer-scale thin film material of platinum group metals and alloys were
reviewed. The AIREs describe the ensemble of abnormal IR features that have been
discovered recently for CO and other molecules (CN^{-1}, SCN^{-1}, PoPD, etc.)
adsorbed on nanostructured films. It was observed on nanometer-scale thin film of

platinum group metals and alloys that, in comparison with IR features acquired on corresponding massive metals, the direction of IR bands of adsorbates was inverted, the IR absorption of adsorbates was significantly enhanced, and the full width at half maximum (FWHM) of IR bands of adsorbates was considerably increased. It was revealed that the AIREs relate mainly to the structure and properties of nanometer scale thin film material, and reflect the collective interaction of adsorbates with nanoparticles that come into contact with each other in the thin film. Studies of scanning tunneling microscopy (STM) illustrated that the thin film prepared by electrodeposition under cyclic voltammetric conditions consisted of uniformly distributed layer crystallites, and may be considered as a two-dimensional nanomaterial. Such a layer structure of the film is different from the island structure that is essential to initiate the so-called surface-enhanced infrared absorption (SEIRA), in which only enhancement of IR absorption and increase in FWHM of IR bands of adsorbates are presented, and the inversion of the direction of IR bands of adsorbates is absent. A comparison of AIREs with the SEIRA was also done. The significance of the AIREs are straightforward not only in studying the properties of two-dimensional nanostructured film material, but also in applications of catalysis, electrocatalysis, and surface analysis.

21.1 INTRODUCTION

The platinum group metals and alloys are excellent catalytic materials that are widely employed in the field of electrocatalysis [1,2]. It is well known that electrocatalytic reactions occur at electrode–electrolyte interfaces and depend strongly on the surface structure (chemical composition, atomic arrangement, electronic structure, etc.) of the electrocatalyst. Different types of material from the platinum group metals including polycrystalline [3], single-crystal planes [3–7], and bulk alloys [8–10] have been extensively studied in both fundamental aspects and applications of electrocatalysis. The studies, especially those employing single-crystal planes, helped to reveal the intrinsic relationship between the electrocatalytic activity of electrocatalysts and their surface structure. The knowledge gained from fundamental studies is of importance in designing practical electrocatalysts employed in diverse applications. Electrocatalysis is the key subject in high-technology fields such as energy conversion (fuel cells) and green synthesis (electrosynthesis), and electrocatalysts employed in these fields are often prepared by dispersing catalytic material on conductive substrates [11,12]. Since electro-catalytic effects depend mainly on the interaction of the reagent molecule with electrode surface, the efforts to reduce the cost and to enhance the performance of electrocatalysts were conducted mainly in two aspects: (1) to disperse, as large as possible, the catalytic material on the substrate in order to minimize the loading; and (2) to design and control the surface structure for optimizing the electrocatalytic properties. In recent years the investigation of nanomaterial has become a rapidly expanding area. Novel effects such as the nanosize effect [13], the surface effects [14], and the macro quantum tunneling effect [15], etc. make nanomaterials more promising as the next revolution in science and technology and have attracted the attentions of almost all disciplines [16]. In accordance with this trend the design and fabrication of different nanostructured catalytic materials have become important subjects of study [17,18].

In 1996 it was revealed for the first time that the nanometer-scale thin films of platinum supported on glassy carbon or other substrates exhibited abnormal IR features in CO adsorption [19], i.e., the inversion of IR band direction, the significant enhancement of IR absorption, and the increase of the full width at half-maximum (FWHM) of IR band for adsorbed species in comparison with the same species adsorbed on a massive Pt electrode. These abnormal IR properties of nanometer-scale thin film material were later called *abnormal infrared effects* (*AIREs*) [20]. AIREs were also observed on platinized platinum electrodes [21] and thin film electrodes of other platinum group metals. The thin film deposited on a conductive substrate is, in fact, a sort of two-dimensional nanomaterial, and the AIREs are associated effectively to nanosize effects. It is evident that systematic studies of the AIREs will contribute not only to reveal the origin of nanosize effects of two-dimensional nanomaterial, but also to explore applications of the nanometer-scale thin film material in electrocatalysis. The present chapter reviews the main progresses in the investigation of AIREs in platinum group metals and alloys at electrode–electrolyte interfaces as well as the particular electrocatalytic properties of these film materials.

21.2 EXPERIMENTAL SECTION

21.2.1 In Situ FTIR Spectroscopy

The in situ FTIR spectroscopic experiments were carried out either on a Nicolet 730 FTIR spectrometer or on a Nexus 870 apparatus (Nicolet), both equipped with liquid-nitrogen-cooled MCT detectors and a Gopher IR source or an Ever-Glo™ IR source. A thin-layer configuration IR cell was constructed by pushing the working electrode against a CaF_2 disk IR window, so as to form a thin layer of electrolyte between the electrode and the IR window for in-situ FTIR measurements. Two procedures were employed in the investigations: (1) the subtractively normalized interfacial FTIRS (SNIFTIRS) [22,23], in which the interferograms were collected alternatively at the sample potential, E_S, and at the reference potential, E_R. The data acquisition process may be repeated a few times, and interferograms collected at E_S and E_R were respectively co-added and Fourier-transformed into single-beam spectra $R(E_S)$ and $R(E_R)$; (2) the multistep FTIRS (MSFTIRS) [24,25], in which a series of single-beam spectra, $R(E_{S,i})$, were acquired first at different sample potentials, then the single-beam spectrum of reference, $R(E_R)$, was collected at the reference potential. In both procedures 400 interferograms in total were collected at both E_S and E_R, and the resulting spectrum is defined as the relative change in reflectivity of the electrode [26,27] and is calculated as

$$\frac{\Delta R}{R} = \frac{R(E_S) - R(E_R)}{R(E_R)} \tag{1}$$

In order to study only adsorbates on the electrode surface, the adsorption of CO was conducted in such a way that during potential cycling between -0.20 and $0.05\,V$ (SCE) in sulphuric acid solution CO gas of high purity was introduced into the IR cell, and after the adsorption of CO on the electrode surface has reached saturation, the solution CO species were removed by bubbling the solution with nitrogen gas of high purity. In the study using the SNIFTIRS procedure the E_S and E_R were set at potentials where the adsorbed CO (CO_{ad}) are stable on the electrode surface. According to Eq. (1), bipolar bands will appear in IR spectra with their negative-going peak corresponding to IR absorption of CO_{ad} at E_S and their positive-going peak for IR absorption of CO_{ad} at E_R. In the investigation employing the MSFTIRS procedure, the $E_{S,i}$ and E_R were regulated in such a way that CO_{ad} was stable at $E_{S,i}$ and oxidized at E_R. As a consequence, negative-going bands relating to IR absorption of CO_{ad} at $E_{S,i}$ and positive-going CO_2 bands will both appear in IR spectra.

21.2.2   Preparation of Electrodes of Nanometer-Scale Thin Films of Platinum Group Metals and Alloys

The glassy carbon (GC) substrate was polished using sand paper, alumina powder of 5, 1, 0.3, and 0.05 μm with a polishing cloth. It was cleaned in an ultrasonic bath and then subjected to electrochemical cleaning through potential cycling between -0.25 and 0.40 V (SCE) in 0.5 M H_2SO_4 solution. The nanometer-scale thin film of platinum group metals was deposited on the prepared GC substrate in 0.5 M H_2SO_4 solution containing 2 mM metal ions under cyclic voltammetric conditions. The electrochemical co-deposition method was employed in the preparation of nanometer-scale thin film of alloys on GC substrate, in which the 0.5 M H_2SO_4 solution contains a mixture of metal ions. The thickness of the film was controlled by altering the number of cycles of potential cycling applied to deposition or co-deposition. The electrodes of nanometer-scale thin films of platinum group metals and alloys prepared in this way were denoted as nm-M/GC and nm-M_1M_2/GC, respectively.

21.2.3   Other Methods

A model P4-18-SPM scanning tunneling microscope (NT-MDT, Russia) was employed to investigate the structure, in atmosphere, of nanometer-scale thin film materials and also to measure the thickness of the film. A setup [28] combining electrochemical studies and X-ray photoelectron spectroscopic (XPS) analysis served to characterize the surface composition of alloy films.

Electrode potentials reported throughout this chapter were referenced to the saturated calomel electrode (SCE) scale.

21.3   ABNORMAL INFRARED EFFECTS IN CO ADSORPTION ON ELECTRODES OF NANOMETER-SCALE THIN FILM OF PLATINUM

21.3.1   In-Situ FTIRS Features of CO Adsorbed on a Massive Pt Electrode

In order to define the abnormal infrared effects, it may be convenient to first examine in-situ FTIR spectra of CO adsorbed on massive Pt electrode. Figure 1 shows a series of MSFTIR spectra [29] of CO adsorbed on a massive Pt electrode recorded at different E_S that varied from -0.20 to 0.30 V. It is known that the adsorbed CO (CO_{ad}) species are stable at these potentials of E_S [30]. The E_R was set at 0.70 V, where CO_{ad} was oxidized immediately into CO_2 species. The negative-going band

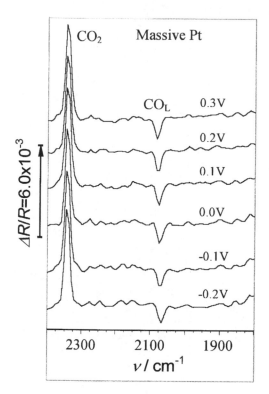

Figure 1 MSFTIR spectra of CO adsorbed on a massive Pt electrode in 0.1 M H_2SO_4 solution, $E_R = 0.70$ V, E_S is indicated for each spectrum. (From Ref. 29. Reprinted by permission of the publisher.)

near $2075 \, \text{cm}^{-1}$ is observed in all spectra in Figure 1. This band is assigned to IR absorption of linearly bonded CO (CO_L) [29–31] at E_S, and its negative-going direction fits in with the prediction of Eq. (1). It can be observed also as a positive-going band near $2345 \, \text{cm}^{-1}$, which is ascribed to IR absorption of CO_2 species at E_R. The center of the CO_L band (ν_{COL}) is shifted positively following the increase of E_S, manifesting a linear variation (Figure 2). The Stark effect that represents the effect of electric field in the double layer on the adsorbed molecule [32], i.e., $d\nu/dE$, can be determined from the slope of the linear relationship between ν and E_S. A value of $31 \, \text{cm}^{-1} \, \text{V}^{-1}$ is measured from Figure 2, which is in good agreement with those reported in the literature [30,31,33]. The full width at half-maximum (FWHM) of the CO_L band is measured to be around $20 \, \text{cm}^{-1}$ from the spectra in Figure 1.

As indicated in the experimental section, only adsorbed CO on the electrode surface is subjected to MSFTIRS investigation, since the solution is free of CO. The CO_2 species were thus derived uniquely from the oxidation of CO_{ad} at E_R when the electrode potential was stepped from the last E_S to E_R. It is known that within a relatively short time window the CO_2 species can all be retained in the thin layer between electrode and IR window, because diffusion from the thin layer to bulk solution is very slow. As a consequence, the integrated intensity of the CO_2 band

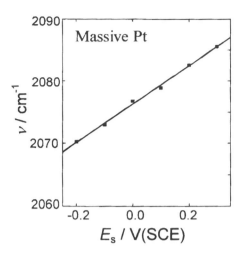

Figure 2 Variation of ν_{CO_L} versus E_S, data measured from Figure 1.

(A^{CO_2}) can be taken as a measure of the quantity of CO adsorbed on the electrode surface, and the normalized intensity of CO_{ad} bands $(A_{N,COad})$ may be defined as

$$A_{N,COad} = \frac{\sum_i A^{CO_i}}{A^{CO_2}} \tag{2}$$

The term $\sum_i A^{CO_i}$ is the sum of the integrated intensity of IR bands of different adsorbed CO species. In Figure 1 only the CO_L band appeared, the $\sum_i A^{CO_i}$ is equal to A^{CO_L}, and the $A_{N,COad}$ has been measured to be 0.21 on a massive Pt electrode.

21.3.2   In-Situ FTIRS Studies of CO Adsorption on Nm-Pt/GC Electrodes—Abnormal Infrared Effects (AIREs)

Characterization of Nm-Pt/GC Electrode by Cyclic Voltammetry

The cyclic voltammograms [20] recorded on massive Pt and nm-Pt/GC electrodes are compared in Figure 3. The voltammogram recorded on nm-Pt/GC displays two distinct pairs of current peaks at −0.02 and −0.16 V, which are characteristic of hydrogen adsorption-desorption on a polycrystalline Pt surface. Nevertheless, the current in the potential region between 0.1 and 0.5 V due to double-layer charging is much larger than that recorded on a Pt surface, signifying the influence of the GC substrate. The charge of hydrogen adsorption per geometric area of GC substrate ($712\,\mu C\,cm^{-2}$) integrated from the voltammogram recorded on the nm-Pt/GC electrode is close to the value obtained on the Pt surface ($663\,\mu C\,cm^{-2}$), indicating that the two electrodes have a comparable number of surface sites for hydrogen adsorption. The roughness of the nm-Pt/GC and the mechanically polished Pt surfaces can be estimated from the comparison of above electric charge densities with the value of $210\,\mu C\,cm^2$ for hydrogen adsorption on a perfect, smooth

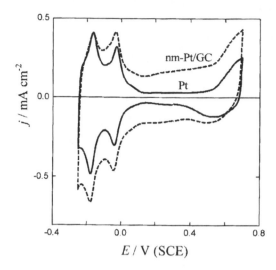

Figure 3 Comparison of cyclic voltammograms recorded on a massive Pt and an nm-Pt/GC electrode in 0.5 M H_2SO_4 solution, sweep rate 50 mV s^{-1}. (From Ref. 20. Reprinted by permission of the publisher.)

polycrystalline Pt electrode; the values of 3.39 and 3.15 have been obtained respectively for the roughness of nm-Pt/GC and mechanically polished Pt surfaces.

MSFTIRS Studies of CO Adsorbed on Nm-Pt/GC Electrode

Figure 4 displays a series MSFTIRS spectra [29] of CO adsorbed on an nm-Pt/GC electrode at different E_S. Three positive-going bands appear in all spectra. As previously discussed the band near 2345 cm^{-1} corresponds to IR absorption of CO_2 species at E_R that were derived from oxidation of CO_{ad} and were in solution. The band around 2075 cm^{-1} is obviously assigned to IR absorption of linearly bonded CO species at E_S. A positive-going broadband can be observed in all the spectra in Figure 4 near 1850 cm^{-1}, which is ascribed to IR absorption of bridge-bonded CO (CO_B) species. In comparison with CO adsorption on a massive Pt electrode (Figure 1), three abnormal IR features of CO_{ad} bands are manifested: (1) the inversion of the direction of CO_{ad} bands. Both CO_L and CO_B bands are in the positive-going direction in Figure 4, i.e., they appear in the same direction as the CO_2 band; (2) the increase of the FWHM of CO_{ad} bands. The FWHM of the CO_L band is measured at about 32 cm^{-1}, i.e., 12 cm^{-1} larger than the value measured from spectra of CO adsorption on a massive Pt electrode; (3) the significant enhancement of IR absorption of CO_{ad} species. The appearance of the CO_B band in the spectra is definitely due to enhancement effects, since the population of this species is too small to yield enough IR absorption to be determined on a massive Pt electrode. The normalized intensity of CO_{ad} bands, $A_{N,CO_{ad}}$, has been measured from Figure 4 to be 4.30 for CO adsorbed on an nm-Pt/GC electrode. The enhancement factor (Δ_{IR}) of IR absorption of CO adsorbed on an nm-Pt/GC electrode is defined as the comparison of the intensity of IR absorption of the same quantity of CO species adsorbed on an nm-Pt/GC with that on a massive Pt, which is equal to the ratio of

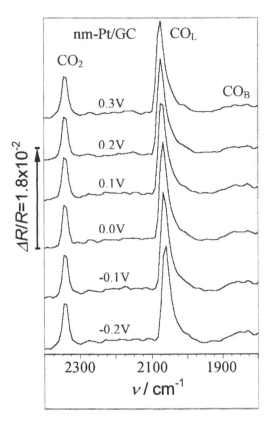

Figure 4 MSFTIR spectra of CO adsorbed on nm-Pt/GC electrode in 0.1 M H_2SO_4 solution, $E_R = 0.70$ V, E_S is indicated for each spectrum. (From Ref. 29. Reprinted by permission of the publisher.)

$A_{N,COad}$ measured on an nm-Pt/GC versus that acquired on a massive Pt, i.e.,

$$\Delta_{IR} = \frac{(A_{N,CO_{ad}})nm - Pt/GC}{(A_{N,CO_{ad}})_{Pt}} \tag{3}$$

In Figure 4 the Δ_{IR} is measured to be 20.5 for CO adsorption on an nm-Pt/GC electrode. We have named the above abnormal IR features observed on electrodes of nanometer-scale thin films of platinum in CO adsorption as the abnormal infrared effects (AIREs) [20,29,34,35]. It may be worthwhile to point out that the Stark effect of CO_L on nm-Pt/GC measured from spectra of Figure 4 is 32 cm^{-1} V^{-1}, which is close to the value obtained on a massive Pt electrode.

It is interesting to compare the single-beam spectra [35] (Figure 5) recorded on an nm-Pt/GC electrode at 0.0 and 0.70 V, respectively. In normal cases, such as CO adsorption on a massive Pt electrode, IR bands of CO adsorbates often cannot be observed in single-beam spectrum since the IR absorption of CO_{ad} is usually too small and is buried in the strong background, only the CO_2 band that is derived from CO_{ad} oxidation and is in the thin layer may appear due to the strong IR absorption

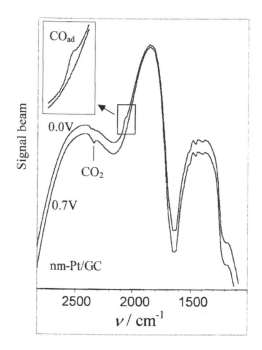

Figure 5 Comparison of single-beam spectra of CO adsorbed on nm-Pt/GC electrodes recorded at 0.0 V and 0.7 V, respectively. (From Ref. 35.)

of this species. We can observe from the single-beam spectrum at 0.70 V a negative-going band near 2345 cm^{-1} due to IR absorption of CO_2, and a positive-going band centered around 2070 cm^{-1} from the single-beam spectrum at 0.0 V. It is evident that the band centered near 2070 cm^{-1} can be assigned to the enhanced IR absorption of CO adsorbed on the nm-Pt/GC electrode and that the direction of the band has been inverted because in normal cases all IR bands in a single-beam spectrum are in the negative-going direction, as previously stated. The results obtained from the single-beam spectra provided further evidence of the AIREs.

SNIFTIRS Studies of CO Adsorbed on Nm-Pt/GC Electrode: Effects of Substrate Material on the AIREs

In order to examine IR features of CO_{ad} at both E_S and E_R, the SNIFTIRS procedure was used to record the spectra. Figure 6 illustrates a series of SNIFTIR spectra [35] of CO adsorbed on nanometer-scale thin film of Pt supported on different substrates, which are graphite, gold, platinum, conducting polymer of pyrrole on GC (ppy/GC) and GC. A spectrum recorded on a massive Pt electrode is also displayed in the figure for comparison. Because the CO_{ad} species are stable on electrode surface at both E_R (-0.20 V) and E_S (0.10 V), the IR absorption at the two potentials yields a bipolar band in the resulting spectrum. The bipolar band in the spectrum recorded on the massive Pt electrode consists of a positive-going peak near 2067 cm^{-1} and a negative-going peak around 2084 cm^{-1}. According to Eq. (1) the negative-going peak corresponds to IR absorption of CO_L at E_S, and the positive-going peak is associated with IR absorption of CO_L at E_R. However, the peak of the

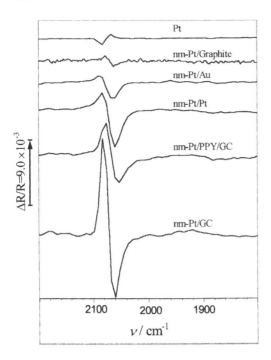

Figure 6 SNIFTIR spectra of CO adsorbed on electrodes of nanometer-scale thin film of platinum supported on different substrate materials, 0.1 M H_2SO_4, $E_R = -0.20$ V, $E_S = 0.10$ V. (From Ref. 35.)

bipolar band for IR absorption of CO_L at E_R (at low wavenumbers, according to the Stark effect) becomes negative-going and that at E_S (at high wavenumbers) becomes positive-going in the spectra recorded on all Pt thin film electrodes. The peak-to-peak intensity of the CO_L bipolar band on nm-Pt/GC is 11 times larger than that on the massive Pt. It is evident that the inversion of the IR band direction and the enhancement of IR absorption of CO_{ad} are sustained in the SNIFTIRS spectrum recorded on the nm-Pt/GC electrode. It can be observed also that the peak-to-peak intensity of the bipolar band varies with different substrate materials, descending in the order of nm-Pt/GC > nm-Pt/ppy/GC > nm-Pt/Pt > nm-Pt/Au > nm-Pt/graphite. It is evident that the inversion of the band direction in the AIREs is maintained on all Pt film electrodes and is independent of substrate material. However, the enhancement of IR absorption of CO_{ad} in the AIREs depends strongly on the substrate material; among the substrate material studied, GC manifests the largest enhancement factor of IR absorption of CO_{ad}.

Recently Gong et al. [36] discovered that a nanostructured film grown on a microplatinum electrode also exhibits AIREs in CO adsorption. They applied a fast potential cycling treatment (between 0.0 V ∼ 1.58 V at a scan rate of 30V·s⁻¹) to a micro-Pt electrode ($r = 100 \mu$m) and found that a nanostructured film ranging from a few nm to a few tens of nm in thickness was formed at the surface of Pt upon the treatment. The CO on such nanostructured film surface of micro-Pt electrode was detected by employing an in-situ FTIR microscope [37]. Similar abnormal IR features to those previously reported on an nm-Pt/GC electrode were also observed on the electrode of nanostructured film grown on a Pt surface, implying that the AIREs are related mainly to the properties of film materials.

21.4   ABNORMAL INFRARED EFFECTS IN CO ADSORPTION ON ELECTRODES OF NANOMETER-SCALE THIN FILM OF PALLADIUM

21.4.1   Cyclic Voltammetric Characterization of Nm-Pd/GC Electrodes

The cyclic voltammogram of an nm-Pd/GC electrode and the CV recorded on a massive palladium electrode are compared in Figure 7 [38,39]. Using STM the thickness of Pd film was measured to be 7.4 nm. The CV features of the nm-Pd/GC electrode are much different from those of the massive Pd electrode. We can observe two pairs of current peaks near -0.04 and -0.12 V from the voltammogram of nm-Pd/GC, which are ascribed to hydrogen adsorption-desorption on the surface of the Pd film. Nevertheless, the current appearing in the voltammogram of the Pd electrode is mainly the reduction current corresponding to hydrogen absorption by Pd matrices. Although the pair of current peaks near -0.04 V can be perceived in voltammogram c of the Pd electrode when the E_L is increased to -0.07 V, the current of hydrogen absorption is always large. This result demonstrates that the main process occurring on an nm-Pd/GC is the hydrogen adsorption rather than the hydrogen absorption, which is the principal reaction on a massive Pd electrode. Clavilier et al. [40] and Kolb et al. [41] have both reported that the hydrogen adsorption process is the main process on ultrathin film Pd electrodes, which were fabricated by epitaxial growth of one or a few atomic monolayers of Pd on single-crystal surfaces. It is evident that the Pd film formed by electrochemical deposition on GC is much thicker than the ultrathin film of Pd grown by epitaxial growth on single-crystal surfaces, i.e., the size (thickness) of the two-dimensional Pd film of the nm-Pd/GC electrode is much larger. Cai et al. [38] have demonstrated that hydrogen adsorption is in fact the main process on the electrode surface of a thin film of Pd. As shown in Figure 8, the two pairs of hydrogen adsorption-desorption current appear in all voltammograms of the nm-Pt/GC electrode where the Pd film thickness varies from 7.4 to 27.2 nm.

21.4.2   In-Situ FTIRS Studies of CO Adsorption on Nm-Pd/GC Electrodes in Acid Solutions and Effects of Thickness of Pd Film on the AIREs

The nanometer-scale thin film may be regarded as a two-dimensional grown film on the GC substrate, so the thickness of the film is an important parameter characterizing the dimension of the film. The effects of film thickness on the AIREs were investigated quantitatively on nm-Pd/GC electrodes. Figure 9 shows a series of MSFTIR spectra recorded on nm-Pd/GC electrodes of different film thickness (d_m) measured by using an STM [38,39]. We can observe that following the increase of d_m the intensity of the CO_2 band is increased continuously, while the intensity of the

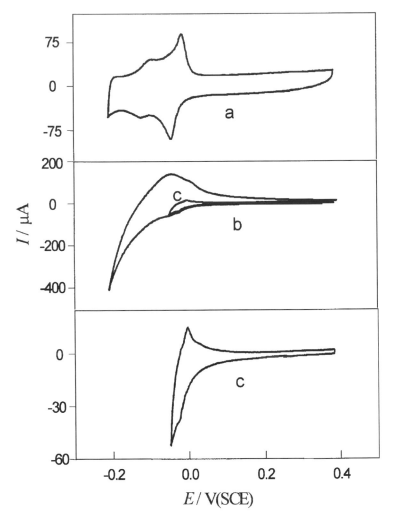

Figure 7 Comparison of cyclic voltammogram of an nm-Pd/GC electrode (a, d = 7.4 nm) with those of a massive Pd electrode (b, c) in 0.5 M H_2SO_4, sweep rate 50 mV s^{-1}. (From Ref. 38. Reprinted with permission of the publisher.)

CO_{ad} bands is augmented at first to a maximum and then declines progressively. The quantitative results are listed in Table 1. The A^{CO_2} and A^{CO_B} are the integrated intensity of CO_2 and CO_B IR bands, respectively. The Δ_{IR} is calculated with Eq. (3) by taking into account the appearance of both CO_L and CO_B bands in the spectra of nm-Pd/GC electrodes, so that the normalized intensity of CO_{ad} bands $(A_{N,COad})_{nm-Pd/GC}$ is calculated as $A_{N,COad} = (A^{CO_L} + A^{CO_B})/A^{CO_2}$. The R_{IR} is the surface roughness of the nm-Pd/GC electrode that is defined as

$$R_{IR} = \frac{(A^{CO_2}/S)_{nm-Pd/GC}}{(A^{CO_2}/S)_{Pd/GC}} \cdot \left(\frac{A^{CO_B}}{A^{CO_B} + A^{CO_L}}\right)_{nm-Pd/GC} \tag{4}$$

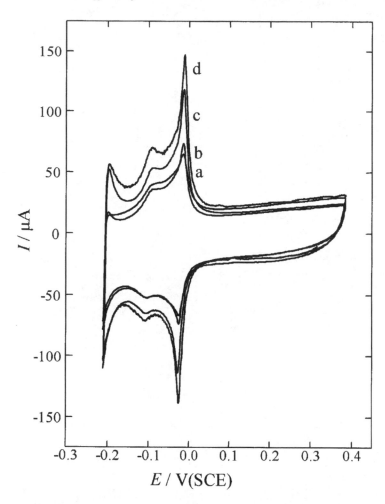

Figure 8 Cyclic voltammograms of nm-Pd/GC electrode of different Pd film thickness: (a) 7.4 nm, (b) 12.2 nm, (c) 20.3 nm, and (d) 27.2 nm, 0.5 M H_2SO_4, sweep rate 50 mV s^{-1}. (From Ref. 38. Reprinted with permission of the publisher.)

where S is the geometric area of massive Pd or nm-Pd/GC electrode. Since only bridge-bonded CO is determined on a massive Pd electrode, the term $\left(A^{CO_B}/A^{CO_B} + A^{CO_L}\right)_{nm-Pd/GC}$ gives the proportion of CO_B to the total CO_{ad} species on nm-Pd/GC and is introduced to correct the deviation of using A^{CO_2} that is in direction proportional to the total quantity of CO_{ad} species. The data listed in Table 1 confirm quantitatively the variation of the A^{CO_2} and A^{CO_B} with d_m and illustrate that at $d_m = 12.2$ nm the Δ_{IR} reaches its maximum value of 42.6. It is interesting to see that R_{IR} is increasing continuously with the increase of d_m and that R_{IR} is always small. R_{IR} is only 2.7 even as the Pd film thickness reaches 32.5 nm, signifying that the Pd film electrodeposited under cyclic voltammetric conditions on GC substrate has a low roughness. Similar results concerning the surface roughness have been also

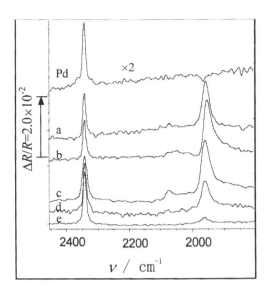

Figure 9 In-situ MSFTIR spectra of CO adsorbed on Pd and nm-Pd/GC electrodes. The d_m of nm-Pd/GC is (a) 4.7, (b) 6.2, (c) 8.5, (d) 10.7, and (e) 14.5 nm; $E_S = 0.2$ V, $E_R = 0.7$ V, 0.1 M H_2SO_4. (From Ref. 38. Reprinted with permission of the publisher.)

Table 1 List of A^{CO_2}, A^{CO_B}, R_{IR}, and Δ_{IR} for Nm-Pd/GC Electrodes with Different d_m

d_m (nm)	A^{CO_2} (a.u.)	A^{CO_B} (a.u.)	R_{IR}	Δ_{IR}
4.7	12.7	26.9	1.0	19.8
6.2	17.7	74.9	1.3	42.6
8.5	20.8	54.3	1.5	26.3
10.7	23.3	24.4	1.8	9.8
14.5	34.4	5.9	2.7	1.6

Source: Ref. 38. (Reprinted by permission of the publisher.)

obtained by comparing the reduction charge of surface oxide formed on the nm-Pd/ GC in a positively going potential scan until 1.1 V with results from cyclic voltammetry studies on massive Pd electrode [38,43].

21.4.3 In-Situ FTIRS Studies of CO Adsorption on Nm-Pd/GC Electrodes in Alkaline Solutions

The reaction steps of CO adsorption and oxidation in alkaline solutions may be a bit complex due to the retention of CO_2 by the solution. The adsorption and electro-oxidation of CO on a Pd surface may be written as

$$CO + 2Pd \rightleftharpoons Pd_2CO_B \tag{5}$$

$$CO + Pd \rightleftharpoons PdCO_L \tag{6}$$

$$Pd_2CO_B + H_2O \longrightarrow CO_2 + OH^- + 2Pd + 2e^- \tag{7}$$

$$PdCO_L + H_2O \longrightarrow CO_2 + OH^- + Pd + 2e^- \tag{8}$$

$$CO_2 + 2OH^- \longrightarrow CO_3^{2-} + H_2O \tag{9}$$

The MSFTIR spectral features for CO adsorption and oxidation in alkaline solutions on a nm-Pd/GC electrode are shown in Figure 10 [44]. The main bands appearing in the spectra within the given wavenumber region (2100–1200 cm^{-1}) are two positive-going bands situated in the region from 1905 to 1930 cm^{-1} and near 1390 cm^{-1}, respectively. The first band between 1905 and 1930 cm^{-1} is obviously assigned to IR absorption of CO_B species at E_S. It has been negatively shifted by about 35 cm^{-1} in comparison with the CO_B band center measured in 0.1 M H_2SO_4 solution (Figure 9). A linear relationship between the center of this band and E_S is illustrated by the insert to Figure 10, which yielded a Stark effect at 59 cm^{-1} V^{-1} that is much larger than the Stark effect (38 cm^{-1} V^{-1}) measured in 0.1 M H_2SO_4 solution [39]. The FWHM of this band is 28 cm^{-1}. The second band near 1390 cm^{-1} can be ascribed to IR absorption of carbonate species [45,46] that are generated according to Eq. (9) during CO_{ad} oxidation at E_S. The center of this band remains unchanged regardless of the variation of E_S, signifying that the carbonate species are in solution of the thin layer between the nm-Pd/GC electrode and CaF_2 IR window. Figure 10 clearly illustrates that one of the main features of the AIREs in CO adsorption on nm-Pd/GC in alkaline solutions is that the CO_{ad} bands appear in the same direction of the IR band of solution carbonate species.

21.5 ABNORMAL INFRARED EFFECTS IN CO ADSORPTION ON ELECTRODES OF NANOMETER-SCALE THIN FILM OF RUTHENIUM

21.5.1 Acid Solutions

A series of MSFTIR spectra [47] of CO adsorbed on the nm-Ru/GC electrode of $d_m = 86$ nm in 0.1 M H_2SO_4 solution are shown in Figure 11(a). Three positive-going bands appear in the spectra, i.e., the CO_2 band near 2345 cm^{-1}, a strong broadband around 2006 cm^{-1}, and a weak broadband close to 1810 cm^{-1}. The latter two bands can be assigned to IR absorption of CO_L and CO_B species, respectively. The FWHM of CO_L and CO_B bands are 53 and 80 cm^{-1}, respectively. It is evident that the nm-Ru/GC electrode exhibits AIREs in CO adsorption. If we take a massive Pt electrode as reference (Figure 1), the direction of CO_{ad} bands is inverted; the enhancement factor of IR absorption for CO adsorbed on the nm-Ru/GC electrode has been calculated at 25.5 according to Eq. (3). Lin et al. [49] have studied the adsorption and oxidation of CO on a massive Ru electrode by using in-situ FTIRS. They observed only the CO_L band around 2000–2020 cm^{-1}, the CO_B band being absent in their spectra. Moreover, the CO_L band appeared in the opposite direction of the CO_2 band, signifying that the IR features of CO_{ad} on a massive Ru electrode are normal.

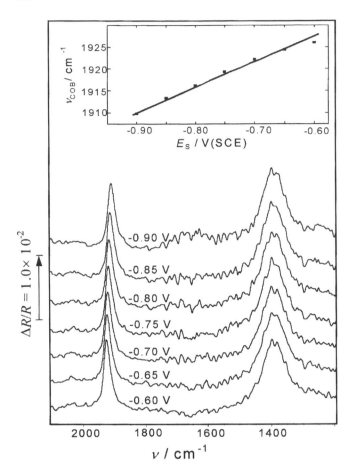

Figure 10 MSFTIR spectra of CO adsorbed on nm-Pd/GC electrode ($d_m = 20.3\,nm$), $E_R = 0.0\,V$, E_S is indicated for each spectrum, 0.1 M NaOH solution. (From Ref. 44. Reprinted with permission of the publisher.)

The fact that both the CO_{ad} bands and the CO_2 band appeared in the same direction on an nm-Ru/GC electrode illustrates one of the main characteristics of the AIREs. It may be worth noting that the CO_L band center measured in Figure 11(a) is close to that reported by Lin et al. but is about $15\,cm^{-1}$ lower than the value for CO adsorbed on a massive Ru electrode in CO saturated 0.5 M $HClO_4$ solution studied by using electromodulated infrared spectroscopy (EMIRS) [50]. It is evident that the appearance of the CO_B band in the spectrum of the nm-Ru/GC electrode is due to significant enhancement of IR absorption.

The positive-going CO_2 band appears in all spectra, and its intensity maintains a constant value along with the variation of E_S. It can be seen that following the increase of E_S the intensities of the CO_L and the CO_B bands also keep constant, but their center is blue-shifted. The variation of the center of the CO_L band (ν_{CO_L}) versus E_S is plotted in Figure 11(b). We may draw two straight lines on the variation of ν_{CO_L} with E_S. One is for $E_S < -0.1\,V$, which yields a slope of $34\,cm^{-1}\,V^{-1}$ and

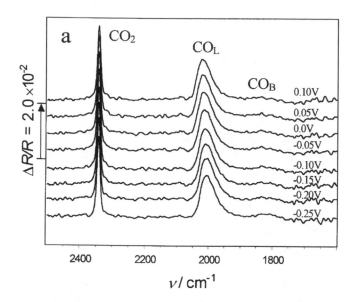

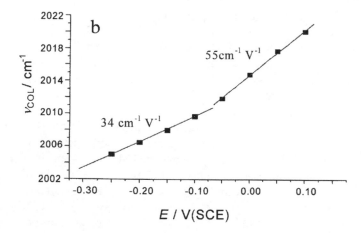

Figure 11 (a) MSFTIR spectra of CO adsorbed on nm-Ru/GC ($d_m = 86$ nm) electrode for E_S varying from -0.25 to 0.10 V and $E_R = 0.5$ V in 0.1 M H_2SO_4 solution; (b) potential dependence of IR band center of CO_L (ν_{CO_L}) on the nm-Ru/GC electrode. (From Ref. 47. Reprinted with permission of the publisher.)

corresponds to a Stark shift rate of 34 cm^{-1} V^{-1}. The second linear part of the ν_{CO_L} versus E_S plot is manifested in the potential range between -0.05 and 0.1 V, from which a Stark shift rate of 55 cm^{-1} V^{-1} has been measured. In comparison with the values of Stark effect of 52 cm^{-1} V^{-1} for lower E_S and 72 cm^{-1} V^{-1} for higher E_S measured on a massive Ru electrode [49], the smaller values of Stark shift rate measured in Figure 11(b) may be attributed to the particular structure of Ru thin film. On an nm-Ru/GC electrode of $d_m = 20$ nm the Stark shift rates were measured to be 21.1 cm^{-1} V^{-1} for lower E_S and 82.6 cm^{-1} V^{-1} for higher E_S [47], indicating

Table 2 Quantitative Results of IR Features for CO Adsorbed on
Nm-Ru/GC Electrodes

d_m (nm)	$A^{CO_{ad}}$ (a.u.)	A^{CO_2} (a.u.)	Δ_{IR}
5	6.1	5.1	9.2
10	11.5	9.0	9.8
20	43.1	18.7	17.7
35	63.6	24.0	20.4
43	72.7	26.9	20.8
74	123.3	38.2	24.8
86	141.1	42.5	25.5
187	211.1	76.2	21.3
397	56.5	154.3	2.8

Source: Ref. 47. (Reprinted by permission of the publisher.)

furthermore that the IR features of CO_{ad} depend strongly on the thickness of the Ru film of the nm-Ru/GC electrode.

Figure 12(a) shows a series of MSFTIR [47] spectra of CO adsorbed on nm-Ru/GC electrodes of different thickness for $E_S = 0.0\,V$ and $E_R = 0.5\,V$. The IR features of AIREs can be observed from all spectra displayed in this figure. The quantitative results measured from spectra in Figure 12(a) are listed in Table 2. It can be seen that, following the increase of Ru film thickness, A^{CO_2} continuously increases. This increase of A^{CO_2} implies an increase in the quantity of CO adsorbed on the nm-Ru/GC electrode surface, corresponding to an increase of surface roughness of the nm-Ru/GC electrode since the geometric area of GC substrate is a fixed value of $0.28\,cm^2$, as stated previously. However, the enhancement factor Δ_{IR} increases first as the thickness of d_m is increased up to 86 nm, then decreases for nm-Ru/GC electrodes with a higher film thickness. The maximum value of Δ_{IR} was 25.5, found for the nm-Ru/GC electrode of $d_m = 86$ nm. In Figure 12(a) a large negative-going band appearing near $1645\,cm^{-1}$ in the spectra h and i corresponding to CO adsorbed on thick Ru film electrodes ($d_m = 180, 397$ nm, respectively) can be observed. As the quantity of CO adsorbed on these nm-Ru/GC electrodes is large (this can be noticed from the large values of A^{CO_2} listed in Table 2), the oxidation of CO_{ad} will consume a corresponding large quantity of H_2O, i.e.,

$$CO_{ad} + H_2O \longrightarrow CO_2 + 2H^+ + 2e^- \tag{10}$$

As a consequence, the large negative-going band near $1645\,cm^{-1}$ due to H_2O bonding appears.

The variation of the enhancement factor Δ_{IR} versus the thickness d_m of the Ru film is plotted in Figure 12(b) [47], which shows an asymmetrical volcano curve. As stated previously, similar results of variation of Δ_{IR} versus d_m were observed on nm-Pt/GC and nm-Pd/GC electrodes. The maximum values of Δ_{IR} measured on the three kind of electrodes can be sorted in a descending order of nm-Pd/GC > nm-Ru/GC > nm-Pt/GC.

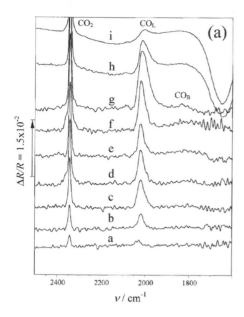

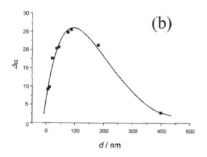

Figure 12   (a) MSFTIR spectra for CO adsorbed on nm-Ru/GC electrodes of different thickness (a: 5 nm, b: 10 nm, c: 20 nm, d: 35 nm, e: 43 nm, f: 74 nm, h: 180 nm, g: 86 nm, i: 397 nm) in 0.1 M H_2SO_4 solution, $E_S = -0.0\,V$, $E_R = 0.5\,V$. The spectra h and i have been multiplied respectively by factors of 0.5 and 0.2 for displaying in the figure; (b) variation of enhancement factor (Δ_{IR}) of IR absorption of CO adsorbed on nm-Ru/GC electrode versus the thickness d_m of Ru film. (From Ref. 47. Reprinted with permission of the publisher.)

21.5.2 Alkaline Solutions

Figure 13 displays MSFTIR spectra [51] of CO adsorbed on an nm-Ru/GC electrode in 0.1 M NaOH solution. Three positive-going bands can be observed. The strong band near 1400 cm^{-1} is assigned to IR absorption of solution carbonate species, which are formed according to Eqs. (7)–(9) due to oxidation of CO_{ad} at E_R. The other two bands are located near 1960 and 1780 cm^{-1} and may be ascribed to IR absorption of CO_L and CO_B species. The centers of CO_L and CO_B bands have been red-shifted about 50 and 60 cm^{-1}, respectively, in comparison with CO adsorption on the nm-Ru/GC electrode in 0.1 M H_2SO_4 solution (Figure 11). From results of

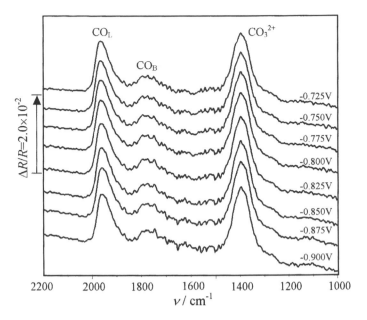

Figure 13 MSFTIR spectra of CO adsorbed on an nm-Ru/GC in 0.1 M NaOH. The E_S is indicated on each spectrum, $E_R = -0.2$ V. (From Ref. 51. Reprinted with permission of the publisher.)

CO adsorption on an nm-Pd/GC electrode in alkaline solution (Section 21.4.3) we know that the nm-Ru/GC exhibited also AIREs in alkaline solution. The enhancement factor Δ_{IR} of IR absorption of CO_{ad} is 33, which is larger than the Δ_{IR} measured in acid solutions. We can see that both the CO_L and CO_B bands are shifted linearly to higher wavenumbers when the E_S is increased positively. The Stark effect of the CO_L band is 34 cm^{-1} V^{-1}. It may be noteworthy that in an early study [50] of CO adsorption on a massive Ru electrode in alkaline solutions, only one CO_{ad} band near 1970 cm^{-1} was observed and assigned incorrectly to IR absorption of bridge-bonded CO species.

Besides nanometer-scale thin film materials of Pt, Pd, and Ru studied above, some preliminary studies on nanometer-scale thin film materials of three other platinum group metals, Rh [20,52], Os [53], and Ir [54], have illustrated the AIREs of all these metals in CO adsorption. Based on studies so far, we can conclude that the AIREs are general phenomena of platinum group metals in chemisorptions of CO and other molecules [20].

21.6 ABNORMAL INFRARED EFFECTS IN CO ADSORPTION ON ELECTRODES OF NANOMETER-SCALE THIN FILM OF ALLOYS

Bimetallic alloy materials often present good electrocatalytic properties and have been abundantly investigated [55]. The surface composition of an alloy can be conveniently monitored, leading one to vary the synergy of the surface. As a consequence the electrocatalytic activity may be easily adjusted. The most commonly employed bimetallic electrocatalysts are platinum alloys. Both bulk alloys prepared by metallurgical methods [56–62] and surface alloys made by chemical or electrochemical codeposition [63–67] were studied extensively. Since the activity of an electrocatalyst depends strongly on its surface composition and structure, the thin film alloy materials are most frequently used in diverse applications such as fuel cells, electrosynthesis, etc. This section describes studies of nanometer-scale thin film materials of PtRu and PtPd alloys, which are of particular importance in fuel-cell applications.

21.6.1 Nanometer-Scale Thin Film Materials of PtRu Alloy

Electrocatalytic Properties

The material of PtRu alloy exhibits good properties for CO tolerance in polymer electrolyte membrane fuel cells (PEMFC) [68] and has been studied extensively in recent years [69]. Particular interest has been focused on the application of the PtRu alloy materials as anodes in methanol fuel cells (MFC) for electric vehicles [70]. The most convenient way to alter the surface composition of a PtRu alloy is to employ the electrochemical co-deposition method in the preparation of the alloy. Richcharz and co-workers have studied the surface composition of a series of PtRu alloys using X-ray photoelectron spectroscopy (XPS) and low-energy ion spectroscopy (LEIS) [71]. They found that the composition of electrodeposited PtRu alloys showed a linear dependence on the concentrations of the deposition solution, but an enrichment of the Pt-component both in the bulk and at the surface. Zheng et al. [72] have tested that the composition of PtRu alloy supported on glassy carbon depends also on the ratio of Pt^{4+} and Ru^{3+} concentration in the electrodeposition solution under fixed deposition conditions (i.e., 5 cycles of potential cycling between -0.5 V and 0.4 V at sweep rate $50\,mV\,s^{-1}$). As it is not convenient to measure accurately the content of Pt and Ru in a PtRu alloy film, the composition of nm-PtRu/GC electrodes is represented thereafter by specifying the concentration of Pt^{4+} and Ru^{3+} ions in the electrodeposition solution. Thus an nm-PtRu/GC(4:1) indicates that the electrode was prepared from electrodeposition solution containing $4\,mM$ H_2PtCl_6 and $1\,mM$ $RuCl_3$ in $0.1\,M$ H_2SO_4.

The oxidation of adsorbed carbon monoxide may be employed as a probe reaction to assess the activity of nm-PtRu/GC electrodes. Figure 14 displays cyclic voltammograms [72] of CO oxidation on nm-Pt/GC, nm-Ru/GC, and nm-PtRu/GC of different alloy composition. Significant catalytic characteristics are observed on nm-PtRu/GC and nm-Ru/GC electrodes. The most evident catalytic effect is the negative shift of peak potential (E_p) of the oxidation current peak. E_p measured from the voltammogram recorded on an nm-Pt/GC electrode is $0.575\,V$. However, for nm-Ru/GC and nm-PtRu/GC electrodes it is shifted negatively to the potential region between 0.3 and $0.4\,V$. The main parameters that characterize the catalytic properties of nanometer-scale thin film electrodes toward CO_{ad} oxidation are the potential (E_p), the current density (j_p), and the full width at half-maximum (FWHM) of the current peak. The saturation coverage of CO adsorption has been achieved on each electrode under experimental conditions. As a consequence, the quantity of electric charge of CO_{ad} oxidation (Q_{CO}^{OX}) can be taken as a measure of the

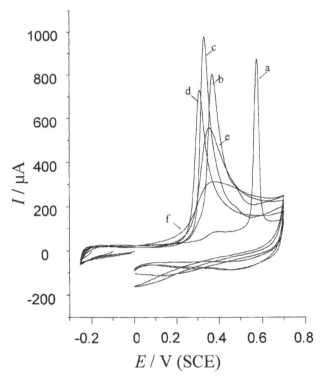

Figure 14 j_E curves for CO_{ad} oxidation on different nanometer thin film electrodes. (a) nm-Pt/GC, (b) nm-PtRu/GC(4:1), (c) nm-PtRu/GC(4:2), (d) nm-PtRu/GC(4:6), (e) nm-PtRu/GC(4:10), (f) nm-Ru/GC. 0.1 M $HClO_4$ solutions, sweep rate 50 mV s^{-1}. (From Ref. 72. Reprinted with permission of the publisher.)

surface roughness of the electrode, since the geometric area of the GC substrate has a defined value of 0.28 cm^{-2}. The relative roughness of Ru and PtRu alloy film electrodes with regard to the nm-Pt/GC electrode may be defined as $(Q_{CO}^{OX})/(Q_{CO}^{OX})_{nm-Pt/GC}$, and j_p/Q_{CO}^{OX} represents the normalized rate of CO_{ad} oxidation. Table 3 lists parameters measured on different nanometer-scale thin film electrodes. It is interesting to observe that, following the increase of Ru in PtRu alloy film, the FWHM increases progressively from 27 mV on nm-Pt/GC to 212 mV on nm-Ru/GC, and j_p/Q_{CO}^{OX} decreases from 1.35 s^{-1} on nm-Pt/GC to 0.24 s^{-1} on nm-Ru/GC. It is important to note that the variation of E_p showed a minimum at 308 mV on an nm-PtRu/GC(4:6) electrode. Gasteiger et al. [60] demonstrated that the minimum E_p of CO_{ad} oxidation was measured on a 50:50 PtRu alloy electrode. The present results confirmed that the composition of nm-PtRu/GC(4:6) electrode might be close to a 50:50 PtRu alloy, since it was prepared in a solution containing 4 mM Pt^{4+} and 6 mM Ru^{3+}, which is near the solution composition of electrodeposition to produce 50:50 PtRu alloy [66,67]. The relative roughness of nm-Ru/GC and nm-PtRu/GC electrodes is always larger than 1 but smaller than 2.6, signifying that these electrodes have a low surface roughness.

Table 3 List of Parameters of Cyclic Voltammograms Recorded on Different Thin Film Electrodes

Electrode	E_p (mV)	FWHM (mV)	j_p ($\mu A\,cm^{-2}$)	Q_{CO}^{OX} ($\mu C, cm^{-2}$)	j_p/Q_{CO}^{OX} (s^{-1})	$Q_{CO}^{OX}/(Q_{CO}^{OX})_{nm\text{-}Pt/GC}$
nm-Pt/GC	575	27	874.9	646.7	1.35	1.0
nm-PtRu/GC(4:1)	370	67	804.1	1666	0.48	2.57
nm-PtRu/GC(4:2)	331	56	978.4	1555	0.63	2.40
nm-PtRu/GC(4:4)	335	53	796.5	1223	0.65	1.89
nm-PtRu/GC(4:6)	308	63	732.1	1364	0.54	2.11
nm-PtRu/GC(4:8)	356	88	562.0	1184	0.48	1.83
nm-PtRu/GC(4:10)	355	128	557.1	1626	0.34	2.51
nm-PtRu/GC(4:12)	357	160	340.1	1163	0.29	1.80
nm-Ru/GC	378	212	289.2	1190	0.24	1.84

Source: From Ref. 72. (Reprinted with permission of the publisher.)

In-Situ FTIRS Studies of CO Adsorption on Nm-PtRu/GC Electrodes

Figure 15 displays the spectra [48] recorded on nm-PtRu/GC electrodes of different film composition at the same $E_S(0.0\,V)$ and $E_R(0.5\,V)$. Three positive-going bands appeared in all spectra, i.e., the CO_2 band, the CO_L and the broad CO_B bands. It is interesting to see that the center of the CO_L band is shifted to a lower wavenumber direction as the Ru component is increased in the PtRu alloy film. The ν_{COL} is measured respectively at 2067, 2064, 2051, 2044, and 2043 cm^{-1} on PtRu/GC(4:0.5), PtRu/GC(4:2), PtRu/GC(4:4), PtRu/GC(4:6), and PtRu/GC(4:8) electrodes. These values of ν_{COL} are between the ν_{COL} measured on nm-Pt/GC (2076 cm^{-1}) and that on nm-Ru/GC (2009 cm^{-1}) under the same conditions. The FWHM of the CO_L band also increases with the increase of the Ru component in the PtRu alloy film. As described in previous sections, the FWHM of the CO_L band measured on an nm-Pt/GC is 20 cm^{-1} and that on an nm-Ru/GC electrode is 55 cm^{-1}. The values of the FWHM of the CO_L band measured on the five nm-PtRu/GC electrodes are 24, 30, 35, 43, and 53 cm^{-1}, respectively. The IR features of CO_{ad} in Figure 15 clearly illustrate that the nanometer-scale thin film of PtRu alloy exhibited abnormal infrared effects in CO adsorption. The IR absorption of CO_{ad} has been significantly enhanced, and the enhancement factor Δ_{IR} has been measured to be 10.5 on nm-PtRu/GC(4:2), 12.8 on nm-PtRu/GC(4:6), and 13.1 on nm-PtRu/GC(4:8).

The variations of ν_{COL} versus E_S on the five nm-PtRu/GC electrodes are shown in Figure 16 [48]. The Stark shift for CO_L on all nm-PtRu/GC alloy film electrodes was measured to be around 34 $cm^{-1}\,V^{-1}$. In contrast to the nm-Ru/GC electrode, where two Stark shift rates (a small value in the low potential region and a large value in the high potential region) were obtained, only one straight line can be drawn through the experimental data points. The results clearly demonstrate that the properties of an nm-PtRu/GC electrode are not a simple combination of the properties of nm-Pt/GC and those of nm-Ru/GC. The fact that the band center, the FWHM, and the Stark effect of the CO_L band all lie in between the values of nm-Pt/GC and those of nm-Ru/GC confirmed that the alloy of PtRu thin film was formed by electrochemical co-deposition under cyclic voltammetric conditions.

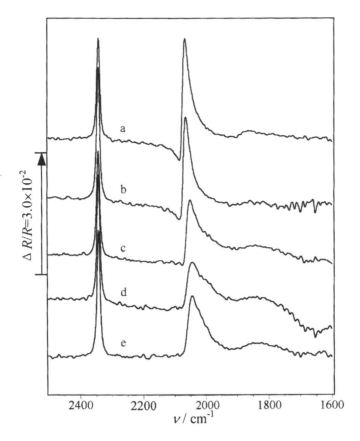

Figure 15 In-situ MSFTIR spectra of CO adsorbed on electrodes of nanometer-scale thin film of PtRu alloys of different compositions: (a) PtRu/GC(4:0.5), (b) PtRu/GC(4:2), (c) PtRu/GC(4:4), (d) PtRu/GC(4:6), (e) PtRu/GC(4:8) electrodes; $E_S = 0.0$ V, $E_R = 0.5$ V, 0.1 M H_2SO_4. (From Ref. 48.)

In-Situ FTIRS Studies of CO Adsorption on Pt-Modified Nm-Ru/GC and Ru-Modified Nm-Pt/GC Electrodes

In order to study the influence of the surface composition of the thin film electrode on IR features of CO adsorption, nm-Pt/GC and nm-Ru/GC electrodes were modified with Ru and Pt, respectively. Modifications to the Pt on the nm-Ru/GC surface and the Ru on the nm-Pt/GC surface were carried out by employing an electrodeposition method [48]. The nm-Pt/GC or nm-Ru/GC electrode was introduced into 0.1 M H_2SO_4 solution containing 1 mM Ru^{3+} or Pt^{4+} ions, and the potential was cycled between -0.25 and 0.40 V at a scan rate of 50 mV s^{-1}. The quantity of Pt deposited on the nm-Ru/GC surface or Ru deposited on the nm-Pt/GC surface was controlled by varying the number of potential cycles that initiated the deposition.

Figure 17(a) shows MSFTIR spectra [48] of CO adsorbed on an nm-Ru/GC electrode modified with different quantities of Pt on the surface. It is observed that

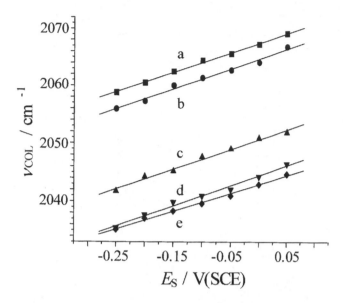

Figure 16 Potential dependence of IR band center of CO_L (ν_{COL}) on (a) PtRu/GC(4:0.5), (b) PtRu/GC(4:2), (c) PtRu/GC(4:4), (d) PtRu/GC(4:6), and (e) PtRu/GC(4:8) electrodes in 0.1 M H_2SO_4 solution. (From Ref. 48.)

on all electrodes the CO_{ad} bands (one or two CO_L bands and a broad CO_B band) and the CO_2 band are in the same positive-going direction, and the intensity of CO_{ad} bands is significantly enhanced, indicating that the AIREs are maintained on all Pt-modified nm-Ru/GC electrodes. The CO_L band in the spectrum of the nm-Ru/GC electrode is located near 2025 cm^{-1} with an FWHM of 42 cm^{-1}. The enhancement factor of IR absorption Δ_{IR} is measured to be 9.7. Following the increase in the quantity of Pt on the nm-Ru/GC surface, the most significant changes in IR features are the increase of the intensity of the CO_2 band and the variation of the CO_L band. The increase of the CO_2 band intensity is obviously due to the increase of electrode surface roughness. It is interesting to observe, in spectrum b, two CO_L bands, one near 2019 cm^{-1} and another around 2069 cm^{-1}. It is evident that the CO_L band near 2019 cm^{-1} can be assigned to IR absorption of CO adsorbed on surface Ru sites, and the CO_L band around 2069 cm^{-1} to IR absorption of CO adsorbed on surface Pt sites. When the quantity of Pt is progressively increased, the band of CO_L on Ru sites (CO_L-Ru) becomes a shoulder peak and finally disappears in spectrum e. However, the band of CO_L on surface Pt sites (CO_L-Pt) gradually develops into the main band in the spectra and is measured near 2063 cm^{-1} in spectrum e. Two points may be remarked from the above results: (1) two kind of surface sites, i.e., the Pt and Ru sites, can be distinguished in CO adsorption when the nm-Ru/GC surface is not completely covered by Pt. This is very different from the case of nm-PtRu/GC alloy electrodes, where only one CO_L band was observed, indicating that the surface Pt or Ru sites could not be discerned on PtRu alloy surface [49,62,73]; (2) the intensity of the CO_{ad} bands decreases with the increase of the quantity of Pt deposited on nm-Ru/GC surface. The Δ_{IR} measured from spectrum e is only 2.2. Goodman et al. [74,75] have pointed out, in FT-IRAS studies of CO adsorption on Cu or Ag

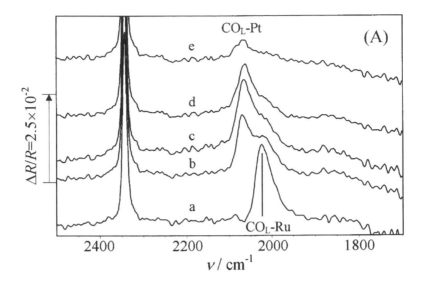

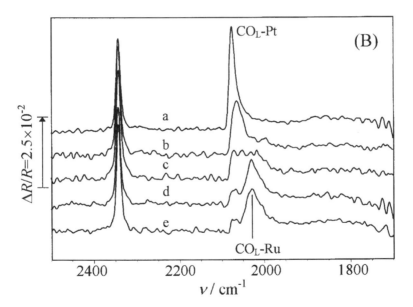

Figure 17 (A) In-situ MSFTIR spectra for CO adsorbed on Pt modified nm-Ru/GC electrode. The number of potential cycling used in Pt deposition: 5(e), 3(d), 2(c), 1(b), and 0(a), which represents the quantity of Pt deposited on nm-Ru/GC surface (see text in detail); $E_S = 0.0\,V$, $E_R = 0.5\,V$; (B) in-situ MSFTIR spectra for CO adsorbed on Ru modified nm-Pt/ GC electrode. The number of potential cycling used in Ru deposition: 10(e), 7(d), 4(c), 1(b), and 0(a); $E_S = 0.0\,V$, $E_R = 0.5\,V$ (0.75 V for nm Pt/GC electrode). (From Ref. 48.)

ultrathin film modified Pt(111) surface, that the Cu-bonded CO molecules or Ag-bonded molecules screen CO molecules bonded to Pt, and they found that this "screening" effect depends on the morphology and polarizability of the Cu or the Ag in the overlayer. The decrease in the enhancement factor of IR absorption in the AIREs upon the deposition of Pt may thus be attributed to the "screening" effect together with the increase in film thickness.

A series of MSFTIR spectra of CO adsorbed on nm-Pt/GC and Ru-modified nm-Pt/GC electrodes are illustrated in Figure 17(b) [48]. The AIREs are manifested in all spectra. We observe two CO_L bands from spectra c, d, and e; one is the CO_L-Pt band near $2065\,cm^{-1}$ and another is the CO_L-Ru band close to $2025\,cm^{-1}$. The CO_L-Ru band appeared as a shoulder peak in spectrum b. It can be seen that the intensity of the CO_L-Pt band progressively decreases and the intensity of the CO_L-Ru band increases with the increase of the quantity of Ru deposited on the nm-Pt/GC surface. Nevertheless, the CO_L-Ru band remains discernible in spectrum e for 10 Ru deposition potential cycles, which may indicate that the nm-Pt/GC surface is still partially covered by Ru. The results imply that the deposition of Ru on an nm-Pt/GC surface is less efficient than the inverse process, i.e., the deposition of Pt on an nm-Ru/GC surface. Similar results have been reported concerning in-situ FTIRS studies of CO adsorption on Ru ad-atom or Ru nanoparticle modified Pt(111) single-crystal electrodes [76–78], in which a CO_L-Pt band near $2070\,cm^{-1}$ and a CO_L-Ru band around $2010\,cm^{-1}$ were observed in the spectra.

21.6.2 Nanometer-Scale Thin Film Materials of PtPd Alloy

According to the phase diagram [79], Pt and Pd can dissolve each other in any proportion to form a binary alloy. Thus it may be interesting to take PtPd alloy as a kind of model alloy electrocatalyst, for which the ratio of surface Pt sites versus surface Pd sites may be varied continuously [80]. PtPd alloy is a good catalyst material [81] and exhibits high tolerance to poisoning by sulfur [82]. It has been revealed that the nanometer-scale thin film material of PtPd alloy also presented AIREs in CO adsorption [52,83–85]. A series of spectra [85] for CO adsorbed on an nm-PtPd/GC electrode at different E_S is illustrated in Figure 18. The positive-going CO_2 band near $2345\,cm^{-1}$ appears in all spectra. Two positive-going CO_{ad} bands are observed, one near $2047\,cm^{-1}$ and another close to $1933\,cm^{-1}$, corresponding to the IR absorption of CO_L and CO_B, respectively. It is obvious that the CO_L and CO_B species are adsorbed on surface Pt and Pd sites, respectively, since, as previously demonstrated, the main band appearing in the spectra of CO adsorbed on Pt or the nm-Pt/GC electrode is the CO_L band and that on Pd or the nm-Pd/GC surface is the CO_B band. The FWHM of the CO_L band is measured to be $54\,cm^{-1}$ and that of the CO_B band is $48\,cm^{-1}$ from the spectra in Figure 18. These FWHM values are three times and twice as large as the FWHM of CO_L ($14\,cm^{-1}$) on massive Pt and CO_B ($21\,cm^{-1}$) on Pd electrode, respectively, implying an increase in the discreteness of vibration energy states of CO_{ad} on the surface of the nm-PtPd/GC electrode. It is apparent from the figure that the C–O stretching frequency of both CO_L and CO_B species is shifted to higher wavenumbers when E_S is increased. From the linear variation of the band center versus E_S the Stark shift rate is measured to be $28\,cm^{-1}\,V^{-1}$ for the CO_L band and at $43\,cm^{-1}\,V^{-1}$ for the CO_B band. These values

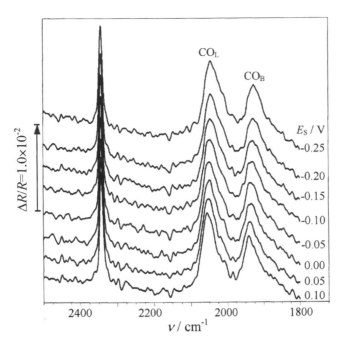

Figure 18 In-situ MSFTIR spectra of CO adsorbed on nm-PtPd/GC ($n = 2$) electrode in 0.5 M H_2SO_4 + CO (saturated) solution, $E_R = 0.75$ V, E_S are listed in the figure. (From Ref. 85. Reprinted with permission of the publisher.)

are close to the values of the Stark effect measured for CO_L adsorbed on a massive Pt electrode and for CO_B on a massive Pd electrode.

The enhancement factor for the IR absorption of CO on nm-PtPd/GC is calculated using the following equation:

$$\Delta_{IR} = \frac{\left(\frac{A^{CO_L} + A^{CO_B}}{A^{CO_2}}\right)_{nm-PtPd/GC}}{\left(\frac{A^{CO}}{A^{CO_2}}\right)_{Pt-Pd}} \tag{12}$$

where A^{CO_L}, A^{CO_B}, and A^{CO_2} refer to the integrated intensity of CO_L, CO_B, and CO_2 bands, respectively. $\left(A^{CO}/A^{CO_2}\right)_{Pt-Pd}$ is the mean value of A^{CO_L}/A^{CO_2} and A^{CO_B}/A^{CO_2} measured respectively from the spectrum of a massive Pt electrode and the spectrum of a massive Pd electrode. $\left(A^{CO}/A^{CO_2}\right)_{Pt-Pd}$ was measured to be 0.152 [85]. A Δ_{IR} value of 38.4 has been measured from Figure 18 for CO adsorbed on an nm-PtPd/GC electrode that was prepared by applying two cycles of potential cycling at a scan rate of 50 mV s^{-1} between 0.20 and 0.40 V in a solution containing 2×10^{-3} M K_2PtCl_6 + 2×10^{-3} M $PdCl_2$ + 0.5 M H_2SO_4.

It has been demonstrated in previous sections that the enhancement factor for IR absorption in the AIREs strongly depends on the thickness of the film material of the platinum group metals. This dependence is also valid in nanometer-scale thin film materials of alloys. Preliminary results concerning this point are shown in

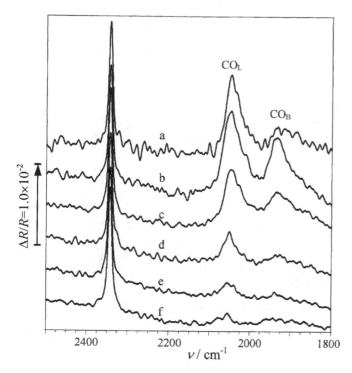

Figure 19 In-situ MSFTIR spectra of CO adsorbed on nm-PtPd/GC electrodes in 0.5 M $H_2SO_4 + CO$ (saturated) solution. The numbers of potential cycling in codeposition are (a) 1, (b) 2, (c) 3, (d) 5, (e) 8, (f) 12; $E_S = 0.0$ V, $E_R = 0.75$ V. (From Ref. 85. Reprinted with permission of the publisher.)

Figure 19 [85], in which in-situ MSFTIR spectra of CO adsorbed on nm-PtPd/GC of deferent film thickness (indicated by the number of potential cycles of co-deposition) are displayed. The IR data measured from the spectra of Figure 19 are listed in Table 4. The co-deposition charge Q_{codep} and the estimated film thickness d_e calculated from Q_{codep} by using a model of cubic-closest packing are also listed in the table. It may be shown from STM images of the PtPd alloy film (see the following section) that the co-deposited PtPd alloy could not be packed in a cubic-closest way, so the

Table 4 IR Data of CO Adsorbed on nm-PtPd/GC of Different PtPd Film Thickness

n	Q_{codep} (mC)	d_e (nm)	A^{CO_2} (a.u.)	A^{CO_L} (a.u.)	A^{CO_B} (a.u.)	A^{CO_B}/A^{CO_L}	Δ_{IR}
1	4.15	4.5	11.9	30.7	29.8	0.97	33.4
2	6.50	7.1	18.7	53.1	55.8	1.05	38.3
3	8.46	9.2	22.6	38.0	44.7	1.18	24.1
5	11.85	13.0	22.3	21.7	27.3	1.26	14.5
8	16.31	17.8	33.2	12.8	22.8	1.78	7.1
12	21.70	23.7	32.4	12.2	17.6	1.44	6.1

Source: Ref. 83. (Reprinted with permission of the publisher.)

real thickness should be larger than d_e, but it will still be within a few tens of nanometers [38]. The data listed in Table 4 illustrate that an increase of d_e leads to an increase of A^{CO_2}. As stated previously, the A^{CO_2} can therefore be taken as a measure of roughness of the surface of the nm-PtPd/GC; the results indicate that the roughness of the nm-PtPd/GC surface becomes larger with the increase of the film thickness. As a consequence, the quantity of adsorbed CO is continuously increased. However, after reaching their maximum values at $n = 2$, the A^{CO_L} and the A^{CO_B} decrease gradually, yielding a volcano-type variation of Δ_{IR} with the increasing n. When $n = 2$, Δ_{IR} reaches its maximum value of 38.3. It may be interesting to see from Table 4 that the ratio A_{CO_B}/A_{CO_L} is increased continuously with the increase of n, indicating that the bridge-bonded CO trends to increase its proportion on the nm-PtPd/GC surface.

21.7 PRELIMINARY INVESTIGATIONS ON THE ORIGIN OF AIREs

21.7.1 In-Situ MSFTIRS Studies of Adsorption of Molecules Other Than CO on Electrodes of Nanometer-Scale Thin Film Material of Platinum Group Metals

Adsorption of CN⁻ Ions on Nm-Pd/GC Electrode

CN^- can adsorb irreversibly on metal surfaces, and the study of the adsorption of CN^- is of importance in revealing the properties of the electrolyte–electrode interface [86]. The adsorption of CN^- on the platinum electrode has been investigated extensively by using in-situ IR spectroscopy [87–89], while only a few papers concern in-situ IR spectroscopic studies of CN^- adsorption on the palladium electrode [88,90] since IR absorption of CN^- on Pd is very weak. Figure 20 shows a series of in-situ MSFTIR spectra of CN^- adsorbed on an nm-Pd/GC electrode in 0.1 M NaOH solution [39,44]. A broad positive-going band with FWHM equal to $67\,cm^{-1}$ appears around $1970\,cm^{-1}$ in all spectra. This band can be assigned to IR absorption of bridge-bonded CN^- on the nm-Pd/GC surface. The band center is shifted to high wavenumbers as E_S is increased (see the insert to Figure 20), yielding a Stark effect of $24\,cm^{-1}\,V^{-1}$. It is evident that AIREs are maintained for CN^- adsorption on the nm-Pd/GC electrode.

It may be interesting to study the coadsorption of CO with CN^-. The CN^- compound holds the same number of covalent electrons as CO but possesses one unit of negative charge. In-situ FTIR studies of the coadsorption may reveal the interaction between CO and CN^- on the electrode surface. In the coadsorption process the nm-Pd/GC electrode was first put in a 1.0 mM KCN solution to adsorb CN^- and then transferred to a CO-saturated 0.1 M KOH solution to adsorb CO. The in-situ MSFTIR spectra are shown in Figure 21 [44]. The IR features of the spectra in Figure 21 are similar to those of CO adsorbed on nm-Pd/GC (Figure 10). We can observe two positive-going bands in the spectra, one near $1910\,cm^{-1}$ assigned to IR absorption of CO_B and another around $1390\,cm^{-1}$ ascribed to IR absorption of CO_3^{2-} species. Although the CN^- band could not be discerned in the spectra, the effects of the interaction of CN^- with CO on the nm-Pd/GC surface may be determined in the following four aspects in comparison with IR features of CO adsorption alone: (1) the CO_B band center has been red-shifted about $6\,cm^{-1}$; (2) the FWHM of the CO_B band is measured to be $35\,cm^{-1}$, i.e., an increase of about 25%;

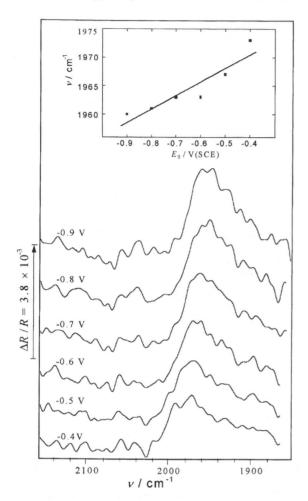

Figure 20 In-situ MSFTIR spectra of CN⁻ adsorbed on nm-Pd/GC electrode. $d_m = 20.3$ nm, $E_R = 0.5$ V, E_S is indicated for each spectrum, 0.1 M NaOH solution. (From Ref. 44. Reprinted with permission of the publisher.)

(3) the Stark effect of the CO_B band is measured from the insert to Figure 21 to be 67 cm^{-1} V^{-1}, which is 8 cm^{-1} V^{-1} larger than that determined from Figure 10; (4) a higher enhancement factor for IR absorption has also been determined; it is augmented for about 60% of that measured in Figure 10. The last point may be also confirmed by perceiving a weak band near 2025 cm^{-1} that is attributed to IR absorption of CO_L species. Such a CO_L band could not be observed in the spectra of CO adsorption alone on the nm-Pd/GC electrode in Figure 10.

Adsorption of SCN⁻ Ions on Nm-Pt/GC Electrode

It is generally agreed that SCN⁻ ions can adsorb on Pt via either the nitrogen atom or the sulfur atom. The N-bound species are dominant at negative potentials, while

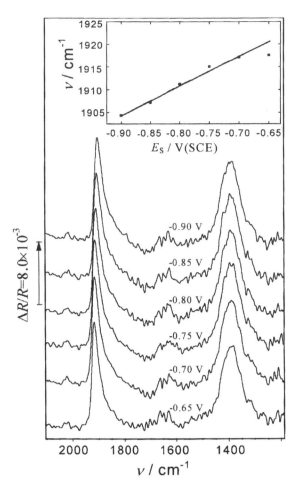

Figure 21 In-situ MSFTIR spectra of CN⁻ and CO adsorbed on nm-Pd/GC electrode. $d_m = 20.3$ nm, $E_R = 0.5$ V, E_S is indicated for each spectrum, 0.1 M NaOH solution. (From Ref. 44. Reprinted with permission of the publisher.)

the S-bound species are the main adsorbates at more positive potentials [91–93]. Figure 22 displays a series of spectra [20] for SCN⁻ adsorbed on an nm-Pt/GC electrode. We observed a positive-going band near 2085 cm⁻¹ for E_S at −0.8 V. This band is assigned to the stretch of C≡N in SCN⁻ that is coordinated with the N-end to the surface of the nm-Pt/GC. The center of this band is shifted linearly to higher wavenumbers with the increase of electrode potential, yielding a Stark effect of 40 cm⁻¹ V⁻¹. A small negative-going band around 2063 cm⁻¹ can be observed in the spectra. It has been attributed to solution SCN⁻ species in the thin layer, since the center of this small band is independent of the electrode potential. The assignment of this negative-going band has been confirmed by adding additional NaSCN to the solution. As the concentration of SCN⁻ increases, the negative-going band near 2063 cm⁻¹ becomes the dominant band in the spectra and does not shift with the variation of electrode potential. The appearance of a positive-going band due to the

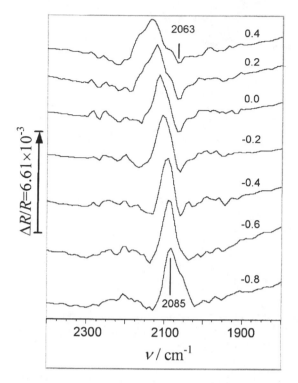

Figure 22　In-situ MSFTIR spectra of SCN$^-$ adsorbed on nm-Pt/GC electrode. $E_R = 0.5$ V, E_S is indicated for each spectrum, in 0.1 M NaOH solution. (From Ref. 20. Reprinted with permission of the publisher.)

IR absorption of adsorbed SCN$^-$ species confirmed the abnormal IR effects of nm-Pt/GC in SCN$^-$ adsorption [20]. It is interesting to observe the broadening of the positive-going band from the spectra recorded at potentials above 0.2 V in Figure 22. A shoulder peak is growing to the high wavenumber side of the positive-going band. These observations suggest that some of the N-bound SCN$^-$ species have changed their orientation to form S-bound SCN$^-$ species when the electrode potential is increased above 0.2 V.

It may be worthy to mention some studies concerning the AIREs of the nm-Pt/GC electrode in the adsorption of other molecules: (i) a polymer of poly-o-phenylenediamine (PoPD) covered on an nm-Pt/GC surface [20]. All IR bands of PoPD in the fingerprint region, i.e., the bands near 1550, 1488, and 1306 cm^{-1} corresponding to the stretching of the C=N bond, the skeletal breathing of aromatic rings, and the C–N stretching of the C–N–H group of the PoPD, respectively, appear in the opposite direction of the bands of PoPD covered on a massive Pt electrode; (ii) the adsorbed species derived from the dissociative adsorption of glycine [94]. As illustrated by in-situ SNIFTIR spectra in Figure 23, the bipolar band with its two peaks around 2076 and 2104 in the spectra recorded on nm-Pt/GC appear in the opposite direction of the bipolar band in spectra acquired on a massive Pt electrode, and the intensity of the bipolar band has been enhanced about 14 times on an nm-Pt/GC electrode. The bipolar band in Figure 23 was assigned to IR absorption of CN$^-$

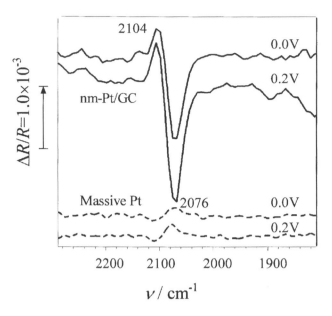

Figure 23 In-situ FTIR spectra of irreversibly adsorbed glycine on massive Pt and nm-Pt/GC electrodes. $E_R = -0.6\,V$, E_S is indicated for each spectrum, 0.1 M Na_2SO_4 solution. (From Ref. 20. Reprinted with permission of the publisher.)

species, since glycine is dissociated in alkaline solutions on Pt surface according to following equation:

$$CH_2(NH_2)COO^- + Pt + 6OH^- \longrightarrow Pt\text{-}CN^- + CO_3^{2-} + 5H_2O + 5e^- \qquad (12)$$

21.7.2 Structure of Nanometer-Scale Thin Film of Platinum Group Metals and Alloys Investigated Using STM

The structure of thin film prepared by electrodeposition was studied using scanning tunneling microscopy. STM images of an nm-Ru/GC electrode are shown in Figure 24 [47]. Figure 24(a) illustrates a typical STM image of the Ru thin film deposited on a GC substrate. It shows that the electrodeposited Ru film is composed of layered Ru crystallites appearing in a hexagonal form with a uniform size of about 250 nm in dimension. The thickness of each Ru crystallite can be measured from the profile [Figure 24(b)] of a cross section S–S in Figure 24(a). An average thickness of about 30 nm has been measured. It is interesting to see that the electrodeposited Ru thin film manifests a particular structure, which consists of Ru crystallites of hexagonal shape in layered stacks. It is obvious that the hexagonal shape of Ru crystallites may be attributed to a hexagonal system of crystallography.

An STM image of nm-Pt/GC and an STM image of nm-Pd/GC are shown in Figure 25 [72,38]. The nm-Pt film is composed of crystallites in a cubic form [Figure 25(a)], clearly showing a face-centered cubic (fcc) structure. The width and the length of each crystallite are around 200 and 250 nm; nevertheless, the thickness (or the height) ranges only between 20 and 25 nm. The nm-Pd film is also respectively composed of layered crystallites [Figure 25(b)]. The average size of the Pd crystallite

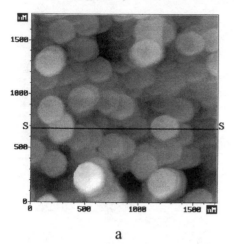
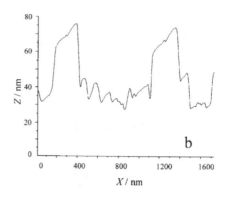

a

Figure 24 (a) STM image of Ru thin film supported on GC, $V_b = 0.160$ V, $I_t = 0.122$ nA. (b) The profile of A–A section in (a). (From Ref. 47. Reprinted with permission of the publisher.)

is measured to be 120 nm in length, 60 nm in width, and a few nm in height. Figures 24 and 25 demonstrate that the films of Ru, Pt, and Pd electrodeposited on GC substrate under cyclic voltammetric conditions were grown in a specific way that led to form nanostructured material; not only is the thickness of the film in nanometer range, but also the film consists of crystallites in layered stacks.

The STM images of the nm-PtRu alloy films of different composition are illustrated in Figure 26 [72]. The crystallites of the PtRu(4:1) [Figure 26(a)] alloy appear to be hexagonal in form. The dimension of each crystallite is around 300 nm, with a thickness of about 10 nm only. In comparison with the STM image of nm-Ru film in Figure 24, some fine structures over the PtRu alloy crystallites can be

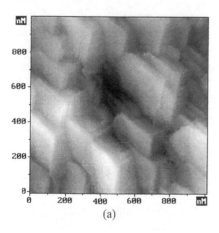

(a)

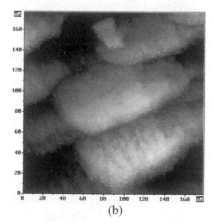

(b)

Figure 25 STM images of (a) nm-Pt/GC, $I_t = 0.274$ nA, $V_b = 0.130$ V. (From Ref. 72. Reprinted with permission of the publisher.) (b) nm-Pd/GC, $I_t = 0.049$ nA, $V_b = 0.020$ V. (From Ref. 38. Reprinted with permission of the publisher.)

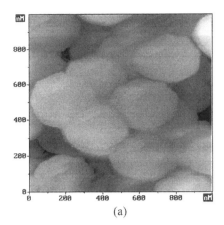 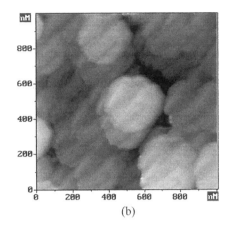

(a)   (b)

Figure 26 STM images of (a) nm-PtRu/GC(4:1), $I_t = 0.504$ nA, $V_b = 0.060$ V, and (b) nm-PtRu/GC(4:8), $I_t = 0.093$ nA, $V_b = 0.170$ V. (From Ref. 72. Reprinted with permission of the publisher.)

observed. When the component of Ru is increased in the nm-PtRu thin film, i.e., for an electrode of nm-PtRu/GC(4:8) [Figure 26(b)], the crystallites are still in a hexagonal-like form of dimension around 300 nm and thickness about 15 nm. It can be observed that one or two scratches appear on the top of each crystallite of nm-PtRu(4:8) alloy. This may imply that some sort of twin crystallite of PtRu alloy is formed. Richard et al. [71] report that the surface composition of PtRu alloy prepared by electrochemical co-deposition depends on the percentage of Pt^{4+} ions (Pt%) in the solution, and a linear relationship was obtained when the Pt% in solution was above 20%. They found by X-ray photoelectron spectroscopy (XPS) and low-energy ion spectroscopy (LEIS) analysis that the molar fraction of Pt in the surface is always higher than that of the deposition solution. The above STM results indicate that the crystallites of electrodeposited PtRu alloy tend to grow in the form of a hexagonal layered structure, even with a composition of 4:1 (Pt:Ru) at which the fcc structure was determined for bulk PtRu alloy with x-ray diffraction [95]. Similar layered structure of nm-PtPd alloy film was also observed with STM [85].

The STM investigations reveal that the nanometer-scale thin film material of platinum metals and alloys prepared by electrochemical co-deposition under cyclic voltammetric conditions presents a common feature, i.e., the film is composed of layered crystallites. This kind of particular nanostructure may be one of the origins to produce the abnormal infrared effects.

21.7.3 Comparison of AIREs with the Phenomenon of Surface-Enhanced IR Absorption (SEIRA)

Since the enhancement of IR absorption is one of the important characteristics in abnormal IR effects, it may be interesting to compare the AIREs with the phenomenon of surface-enhanced infrared absorption (SEIRA) reported in the literature. The SEIRA was discovered by Hartstein et al. [96] in the early 1980s and describes the phenomenon of the enhancement of IR absorption for some specific molecules adsorbed on film materials of island structures. In order to generate surface enhancement of IR absorption, vacuum-evaporated films of Au and Ag on IR-transparent substrates were employed in attenuated total reflection (ATR) mode and transmission mode studies, and p-nitrobenzoic acid (PNBA) and molecules of similar structure were typical compounds used in the investigations [96–107]. An enhancement factor of 1 to 3 orders of magnitude for IR absorption of some particular vibration modes have been reported in the SEIRA. The absorbate of organic molecules over the island metal surface was prepared by depositing an aliquot of solution containing the organic molecules and allowing the solvent to be evaporated. The origin of the surface enhancement has not been completely understood yet and has been discussed by similar enhancement mechanisms as those proposed for surface-enhanced Raman spectroscopy (SERS), i.e., the electromagnetic (EM) effect and the chemical effects [100,101]. As for the EM mechanism, excitation of surface plasmon polarization (SPP) caused by metal islands or rough metal surfaces is introduced for the interpretation of the SEIRA phenomenon. Although SEIRA with thin island films [101] has been reported, almost all observations were conducted on island films of coinage-group metals (Cu, Ag, Au) as those employed in the SERS studies, where the significant SPP effect is involved. Obviously, for the investigation of molecules adsorbed on coinage-group metals, the SEIRA is much less popular and also much less significant than SERS, especially in the in-situ electrochemical studies where infrared spectroscopy is facing additional difficulties. In fact, there are only very few in-situ electrochemical SEIRA investigations on electrode–electrolyte interfaces. Osawa and co-workers have reported SEIRA for the water molecule on gold [104] and the heptylviologen on silver [105] thin films. More recently, Sun et al. reported the first SEIRA study of CO adsorption on gold films [108].

From the above discussions, it is obvious that the AIREs differ from SEIRA in the following aspects:

1. The AIREs exhibit other distinct characters not including the enhanced IR absorption. The most significant character is the inversion of the direction IR bands that was not present in the SEIRA.
2. In the AIREs, the nanometer thin film supported on glassy carbon or other conductive substrates was prepared by convenient electrodeposition method of cyclic voltammetery that led to form nanometer-scale thin film of layered structure. The film of island structure is nevertheless fabricated by evaporation method and employed in the SEIRA.
3. AIREs were observed with platinum group metals and alloys, which are of particular interest in wide applications, especially in electrocatalysis.

It may be worthwhile to point out that the term "abnormal infrared effects" refers to the phenomenon of different spectral properties of adsorbates at thin film electrodes in comparison with those at massive metal electrodes. The origin of the AIREs relates certainly to the particular properties of thin film material and the interaction of adsorbates with the nanometer-scale thin film. A change in optical properties of the adsorbate–thin film system may be consequently expected. Bjerke et al. [21] have simulated the variation of IR features of CO adsorption on platinized platinum electrodes, which may have thrown a light on interpreting the phenomenon. However, thoroughly understanding the origin of the AIREs is yet a subject of an ongoing scientific discussion, which may involve efforts of investigation from different related disciplines.

21.8 CONCLUSIONS

The present chapter reviews the discovery of abnormal infrared effects of nanometer-scale thin film material of platinum group metals and alloys at electrode–electrolyte interfaces. AIREs describe the abnormal IR features for CO or other molecules adsorbed on electrode surfaces of thin film nanostructured materials. In comparison with the same molecule adsorbed on an electrode of the corresponding massive metal, the direction of IR bands of chemisorbed species is inverted, the IR adsorption is significantly enhanced, and the FWHM of the IR bands is increased to a large extent on a nanometer-scale thin film electrode. The inversion and the increase of the FWHM of IR bands of chemisorbed species were observed on all electrodes of nanometer-scale thin film material studied so far and manifested the typical unambiguous characteristics of AIREs. The enhancement of IR absorption in the AIREs nevertheless depends strongly on the thickness (d) of the thin film material, and a volcano relationship has been revealed between the enhancement factor (Δ_{IR}) and the d. The maximum values of Δ_{IR} (Δ_{IR}^{max}) were measured at about 10 to 40, associating in general with electrodes of thin film material of thickness at a few tens of nanometers. The different platinum group metals investigated in the current chapter can be sorted as nm-Pd/GC > nm-Ru/GC > nm-Pt/GC in descending order of Δ_{IR}^{max}.

The studies employing an STM revealed that the thin film prepared by electrodeposition under cyclic voltammetric conditions is a sort of nanostructured material. Either the thin films of platinum group metals or the thin films of alloys consist of uniformly distributed layered crystallites and can be considered as a kind of two-dimensional nanomaterial. The AIREs are, in fact, common effects of nanometer-scale thin film materials of platinum group metals and alloys and are related directly to nanosize effects of the two-dimensional nanomaterial. Unlike the phenomenon of so-called surface-enhanced IR absorption (SEIRA), which is exhibited solely on coinage metals (Au, Ag, and Cu), the AIREs are more significant in diverse applications, especially in electrocatalysis. As stated previously, the platinum group metals and alloys are excellent catalytic materials; thin films of these materials supported on carbon or other conductive substrates can be conveniently prepared by electrochemical deposition or co-deposition methods under cyclic voltammetric conditions and can serve directly as electrocatalysts employed in fuel cells or electrosynthesis. The AIREs also provided the ability to easily determine the surface-adsorbed species according to the inversion of IR band direction and considerably improved the determining sensitivity of in-situ infrared spectroscopy in surface analysis through the enhancement of IR absorption.

It is evident that the study of AIREs will contribute directly to finding the origin of nanosize effects of material, and play an important role in the development of the fundamentals of nanometer science and technology. The discovery of the AIREs has initiated a new research field in many disciplines, including electrochemistry, IR spectroscopy, surface science, material science, etc. The progresses reviewed in the current chapter are just preliminary studies at the beginning stage of this new field. Further investigation on the AIREs may be carried out in the following directions: (1) systematic study of the platinum group metals and alloys, especially with attention on alloys of different compositions that are very important for developing applications for nanometer thin film materials; (2) extend the scope to transition metals other than the platinum group, in order to gain knowledge concerning the influence of the electronic structure of the metal on the AIREs; (3) explore the use of nanometer-scale thin film materials in electrocatalytic systems, especially for direct fuel cells, electrosynthesis, etc.; (4) reveal the origin of the AIREs, such as the inversion of the IR band direction, the enhancement IR absorption, which are related directly to fundamentals of optics, surface physics, and material science. The enhancement of IR absorption is undoubtedly related to the general optical enhancement by nanomaterials such as surface enhancement of Raman scattering (SERS) [109], surface-enhanced second harmonic generation (SESHG) [110], and surface-enhanced sum frequency generation (SESFG) [111].

ACKNOWLEDGMENTS

The studies reviewed in the present chapter were supported financially by National Natural Science Foundation of China (NNSFC) and Education Ministry of China through grants. Scientific contributions from my former graduates in different periods, especially Dr. Guo-Qiang Lu, Dr. Li-Rong Cai, and Mr. Ming-Sen Zheng, are appreciatively indebted. The correction in English expressions done kindly by Dr. Ian Burgass is gratefully acknowledged.

REFERENCES

1. J. Lipkowski, P.N. Ross, (eds), *Electrocatalysis*, Vol. 4 of Frontiers in Electrochemistry, Wiley-VCH, Inc., New York, 1998.
2. A. Wieckowski, (ed), *Interfacial Electrochemistry, Theory, Experiment, and Application*, Marcel Dekker, Inc., New York, 1999, pp 769–954.
3. B. Beden, J.M. Leger, C. Lamy, *Electrocatalytic oxidation of oxygenated aliphatic organic compounds at noble metal electrodes*, Chapter 2 in Modern Aspects of Electrochemistry, Vol. 22, Bockris J.O.'M. et al. (eds), Plenum Press, New York, 1992.
4. S.G. Sun, J. Clavilier, *J Electroanal Chem* 1987, **236**: 95.
5. S. Motoo, N. Furuya, *Ber Bunsenges Phys Chem* 1987, **91**: 457.
6. J. Clavilier, R. Albalat, R. Gomez, J.M. Orts, J.M. Feliu, *J Electroanal Chem* 1993, **360**: 325.
7. D. Kolb, *Angew Chem Int Ed* 2001, **40**: 1162.
8. A. Capon, R. Parsons, *J Electroanal Chem* 1975, **65**: 285.
9. F. Kadirgan, B. Beden, J.M. Lager, C. Lamy, *J Electroanal Chem* 1981, **125**: 89.
10. B.N. Grgur, N.M. Markovic, P.N. Ross, *Electrochim Acta* 1998, **43**: 3631.
11. H. Wendt, C. Lamy, (eds), *Abstracts volume of workshop for electrocatalysis in indirect and direct methanol fuel cells*, Portorose, Solvenia, Sep. 12–14 (1999).
12. A. Wieckowski, K. Itaya, (eds), *Electrode Processes VI*, The Electrochemical Society, Inc., New Jersey, PV 96–8, pp 322–480.
13. W.P. Halperin, *Rev of Moden Phys* 1986, **52**: 532.
14. P. Bell, L. Garwin, *Nature* 1992, **355**: 761.
15. L.D. Zhang, J.M. Mu, *Physics* 1992, **21**: 167.
16. H.G. Craighead, *Science* 2000, **290**: 1532.
17. N.M. Markovic, P.N. Ross, *Electrochim Acta* 2000, **45**: 4101.
18. J. Zhu, G.A. Somorjai, *Nano Lett* 2001, **1**: 8.

19. G.Q. Lu, S.G. Sun, S.P. Chen, N.H. Li, Y.Y. Yang, Z.W. Tian, *Chapter 36 of Electrode Processes VI*, A. Wieckowski, K. Itaya, (eds), PV 96–8, The Electrochemical Society, Inc., New Jersey, 1996, pp 436–445.

20. G.Q. Lu, S.G. Sun, L.R. Cai, S.P. Chen, Z.W. Tian, K.K. Shiu, *Langmuir* 2000, **16**: 778.

21. A.E. Bjerke, P.R. Griffiths, W. Theiss, *Anal. Chem.* 1999, **71**: 1967.

22. S. Pons, *J Electroanal Chem* 1983, **150**: 495.

23. S. Pons, T. Davison, A. Bewick, *J Electroanal Chem* 1984, **160**: 63.

24. W.F. Lin, S.G. Sun, *Electrochim Acta* 1996, **41**: 803.

25. W.F. Lin, S.G. Sun, *Electrochemistry* 1996, **2**(1): 20.

26. S.G. Sun, D.F. Yang, Z.W. Tian, *J Electroanal Chem* 1990, **289**: 177.

27. S.G. Sun, Chapter 6 in *Electrocatalysis*, J. Lipkowski, P.N. Ross, (eds), Wiley-VCH, Inc., New York, 1998, pp 243–290.

28. S.G. Sun, S.P. Chen, N.H. Li, G.Q. Lu, B.Z. Chen, F.C. Xu, *Colloids and Surface A: Physicochemical and Engineering Aspects*, 1998, **134**: 207.

29. G.Q. Lu, S.G. Sun, S.P. Chen, L.R. Cai, N.H. Li, Z.W. Tian, *Chemical Journal of Chinese Universities*, 1997, **18**(9): 1491.

30. K. Kunimatsu, H. Seki, W.G. Golden, J.G. Gorden II, M.R. Philpott, *Surface Science* 1985, **158**: 596.

31. B. Beden, A. Bewick, K. Kunimatsu, C. Lamy, *J Electroanal Chem* 1982, **142**: 345.

32. D.K. Lambert, *Electrochim Acta* 1996, **41**(5): 623.

33. J.W. Rusell, J. Overend, K. Scanlon, M. Severson, A. Bewick, *J Phys Chem* 1983, **87**: 293.

34. G.Q. Lu, S.G. Sun, S.P. Chen, L.R. Cai, *J Electroanal Chem* 1997, **421**: 19.

35. G.Q. Lu, *Ph.D. Thesis*, Xiamen University, 1997.

36. H. Gong, S.P. Chen, Z.Y. Zhou, S.G. Sun, *Chinese Science Bulletin* 2001, **46**: 996.

37. S.G. Sun, S.J. Hong, S.P. Chen, G.Q. Lu, H.P. Dai, X.Y. Xiao, *Science in China B* 1999, **42**(3): 261.

38. L.R. Cai, S.G. Sun, S.Q. Xia, F. Chen, M.S. Zheng, S.P. Chen, G.Q. Lu, *Acta Physico-Chimica Sinica* 1999, **15**(11): 1023.

39. L.R. Cai, M.Sc. *Thesis*, Xiamen University, 1999.

40. J. Clavilier, M.J. Llorca, J.M. Feliu, A. Aldaz, *J Electroanal Chem* 1991, **310**: 429.

41. M. Baldauf, D.M. Kolb, *Electrochim Acta* 1993, **38**: 2145.

42. R.J. Lipert, B.D. Lamp, M.D. Porter, Chapter 3 in *Modern Techniques in Applied Modern Spectroscopy*, Mirabella F.M., (ed), John Wiley & Sons, New York, 1999, pp 83–126.

43. R. Wood, *Electroanalytical Chemistry: A Series of Advances*, Vol. 9, Bard A.J., (ed), Marcel Dekker, New York, 1976, pp 109–117.

44. S.G. Sun, L.R. Cai, Z.Y. Zhou, S.P. Chen, *Spectroscopy and Spectral Analysis* 2000, **20**(5): 601.

45. S.G. Sun, A.C. Chen, *J Electroanal Chem* 1992, **323**: 319.

46. D. Lin-Vien, N.B. Colthup, W.G. Fateley, J.G. Grasselli, *The Handbook of Infrared and Raman Characteristic Frequencies of Organic Molecules*, Academic Press, Inc., Boston, 1991.

47. M.S. Zheng, S.G. Sun, *J Electroanal Chem* 2001, **500**(1–2): 223.

48. M.S. Zheng, M.Sc. *Thesis*, Xiamen University, 2000.

49. W.F. Lin, T. Iwasita, W. Vielstich, *J Phys Chem B* 1999, **103**: 3250.

50. C. Gutiérrez, J.A. Caram, B. Beden, *J Electroanal Chem* 1991, **305**: 289.

51. M.S. Zheng, S.P. Chen, S.G. Sun, *Spectroscopy and Spectral Analysis* 2000, **20**(6): 755.

52. S.G. Sun, M.S. Zheng, G.Q. Lu, L.R. Cai, S.P. Chen, X.Y. Xiao, *Electrochemistry* 2000, **6**(1): 25.

53. G. Drozco, C. Gutierrez, *J Electroanal Chem* 2000, **487**: 64.

54. R. Ortiz, A. Cuesta, O.P. Marquez et al., *J Electroanal Chem* 1999, **465**(2): 234.

55. P.N. Ross, Chapter 2 in *Electrocatalysis*, J. Lipkowski, P.N. Ross, (eds), Wiley-VCH, Inc., New York, 1998, pp 43–74.
56. C.T. Campbell, *Annu Rev Phys Chem* 1990, **41**: 775.
57. A. Capon, R. Parsons, *J Electroanal Chem* 1975, **65**: 285.
58. B.E. Conway, H. Angerstein-Kozlowska, G. Czartoryska, *Z Phys Chem NF, Bd* 1978, **112**: 195.
59. N.R. De Tacconi, J.M. Leger, B. Beden, C. Lamy, *J Electroanal Chem* 1982, **134**: 117.
60. H.A. Gasteiger, N. Markovic, P.N. Ross, E.J. Cains, *J Phys Chem* 1994, **98**: 617.
61. A. Kabbabi, R.R. Faure Durand, B. Beden, F. Hahn, J.M. Leger, C. Lamy, *J Electroanal Chem* 1998, **444**: 41.
62. T. Iwasita, F.C. Nart, W. Vielstich, *Ber Bunsenges Phys Chem* 1990, **94**: 1030.
63. C.B. Roy, S.C. Das, H.R. Kundu, *Ind J Chem* 1976, **14**: 315.
64. D.F.A. Koch, D.A.J. Rand, R. Woods, *J Electroanal Chem* 1976, **70**: 73.
65. M. Watanabe, M. Uchida, S. Motoo, *J Electroanal Chem* 1987, **229**: 395.
66. F. Richarz, B. Wohlmann, U. Vogel, H. Hoffschulz, K. Wamdelt, *Surf Sci* 1995, **335**: 361.
67. E. Jusys, H. Massong, H. Baltruschat, *J Electrochem Soc*, 1999, **146**: 1093.
68. S. Gttesfeld, "Polymer Electrolyte Fuel Cells," chapter in *Advances in Electrochemical Science and Engineering*, Vol. 5, R.C. Alkire, H. Gerischer, D.M. Kolb, Tobias C.W., (eds), Wiley-VCH, New York, 1997.
69. H. Wendt, C. Lamy, (eds), Abstracts volume of *Workshop for Electrocatalysis in Indirect and Direct Methanol Fuel Cells*, Portorose, Solvenia, Sep. 1999.
70. C. Lamy, J.M. Léger, S. Srinivason, Direct methanol fuel cells from a 20th century electrochemists dream to a 21st century engineering technology, in *Modern Aspects of Electrochemistry*, J. O.'M. Bockris, (ed), Vol. 34, Chap. 3, Plenum, New York, 1999.
71. F. Richarz, B. Wolmann, U. Vogel, H. Hoffschulz, K. Wandely, *Surf Sci* 1995, **335**: 361.
72. M.S. Zheng, S.G. Sun, S.P. Chen, *J Appl Electrochem* 2001, **31**: 749.
73. A. Kabbabi, R. Faure, R. Duramed, B. Beden, F. Hahn, J.M. Leger, C. Lamy, *J Electroanal Chem* 1998, **444**: 41.
74. J.A. Rodriguez, C.M. Truong, D.W. Goodman, *J Chem Phys* 1992, **96**: 7814.
75. J.A. Rodriguez, C.M. Truong, D.W. Goodman, *Surf Sci* 1992, **271**: L331.
76. K.A. Friedrich, K.P. Geyzers, U. Linke, U. Stimming, J. Stumper, *J Electroanal Chem* 1996, **402**: 123.
77. S. Cramm, K.A. Friedrich, K.P. Geyzers, U. Stimming, R. Vogel, *Fresenius J Anal Chem* 1997, **358**: 189.
78. W.F. Lin, M.S. Zei, M. Eiswirth, G. Ertl, T. Iwasita, W. Vielstich, *J Phys Chem B* 1999, **103**: 6968.
79. J.Q. Yu, W.Y. Yi, B.D. Chen, H.J. Chen, (eds), *Atlas of Phase Diagram of Binary Alloys*, Shanghai Science and Technology Press, Shanghai, 1987.
80. F. Kadirgam, B. Beden, C. Lamt, et al., *J Electroanal Chem* 1981, **125**: 89.
81. K. Nomura, K. Noro, Y. Nakamura, et al., *Catalysis Letters* 1998, **53(3–4)**: 167.
82. H. Yasuda, T. Kameoka, T. Sato, et al., *Applied Catalysis A, General* 1999, **185**: 199.
83. G.Q. Lu, L.R. Cai, S.G. Sun, J.X. He, *Spectroscopy and Spectral Analysis* 1998, **18(4)**:19.
84. G.Q. Lu, L.R. Cai, S.G. Sun, J.X. He, *Chinese Science Bulletin* 1999, **44(16)**: 1470.
85. Z. Chen, S.G. Sun, N. Ding, Z.Y. Zhou, *Chinese Science Bulletin* 2001, **46**: 1439.
86. T. Iwasita, F.C. Nart, Chapter in *Advances in Electrochemical Science and Engineering*, Gerischer H., Tobias C.W., (eds), Vol. 4, VCH, New York, 1995.
87. H. Kawashima, Y. Ikezawa, T. Takamura, *J Electroanal Chem* 1991, **317**: 257.
88. A. Ashley, M. Lazaga, M.G. Samant, H. Seki, M.R. Philpott, *Surf Sci* 1989, **219**: L590.
89. F. Kitamura, M. Takahashi, M. Ito, *Chem Phys Lett* 1986, **130**: 181.

90. K. Ashley, F. Weinert, M.G. Samant, H. Seki, M.R. Philpott, *J Phys Chem* 1991, **95**: 7409.

91. K. Ashley, M.G. Samant, H. Seki, M.R. Philpott, *J Electroanal Chem* 1989, **270**: 349.

92. A. Tadjeddine, P. Guyot-Sionnest, *Electrochim Acta* 1991, **36**: 1849.

93. Z.Q. Tian, B. Ren, B.W. Mao, *J Phys Chem B* 1997, **101**: 1338.

94. X.Y. Xiao, S.G. Sun, Q.H. Wu, Z.Y. Zhou, S.P. Chen, *Chemical Journal of Chinese Universties* 2000, **21(8)**: 1288.

95. H.A. Gasteiger, P.N. Ross, E.J. Cairns, *Surf Sci* 1993, **293**: 67.

96. A. Hartstein, J.R. Kirtly, C. Tsang, *Phys Rev Lett* 1980, **45**: 201.

97. T. Kamata, A. Kato, J. Umemura, T. Takenaka, *Langmuir* 1987, **3**:1150.

98. S. Badilescu, P.V. Ashrit, V. Truong, II. Badilescu, *Appl Spectrosc* 1989, **43**: 549.

99. Y. Nishikawa, K. Fujiwara, T. Shima, *Appl Spectrosc* 1990, **44**: 691.

100. M. Osawa, M. Ikeda, *J Phys Chem* 1991, **95**: 9914.

101. Y. Nishikawa, T. Nagasawa, K. Fujiwara, M. Osawa, *Vib Spectrosc* 1993, **6**: 43.

102. Y. Nishikawa, K. Fujiwara, K. Ataka, M. Osawa, *Anal Chem* 1993, **65**: 556.

103. M. Osawa, K. Yoshii, K. Ataka, T. Yotsuyanag, *Langmuir* 1994, **10**: 640.

104. K. Ataka, T. Yotsuyanagi, M. Osawa, *J Phys Chem* 1996, **100**: 10664.

105. M. Osawa, K. Yoshii, *Appl Spectrosc* 1997, **51**: 512.

106. E. Johnson, R. Aroca, *J Phys Chem* 1995, **99**: 9325.

107. G.T. Merklin, P.R. Griffiths, *Langmuir* 1997, **13**: 6159.

108. S.G. Sun, W.B. Cai, L.J. Wan, M. Osawa, *J Phys Chem B* 1999, **103(13)**: 2460.

109. B. Pettinger, G. Picardi, R. Schuster, G. Ertl, *Electrochemistry* 2000, **68(12)**: 942.

110. T.A. Leskova, M. Leyva-Lucero, E.R. Mendez, A.A. Maradudin, I.V. Novikov, *Optics Communications* 2000, **183(5–6)**: 529.

111. S. Baldelli, A.S. Eppler, E. Anderson, Y.R. Shen, G.A. Somorjai, *J Chem Phys* 2000, **113(13)**: 5432.

22

Design of Electrocatalysts for Fuel Cells

MASAHIRO WATANABE

University of Yamanashi, Kofu, Japan

CHAPTER CONTENTS

SUMMARY

Designing alloy electrocatalysts by the so-called "ad-atom method," and by alloy sputtering for oxidation of CH_3OH and CO, and for CO tolerance in H_2 oxidation, respectively, as well as for O_2 reduction are discussed. Many years of experience are summarized and collaborations with other groups are highlighted. The particle size effect in electrocatalysis by small particle electrodes, and the effect of corrosion of carbon-black supported nanoparticles on the electrocatalytic activity are also discussed. All these factors, as well as catalyst lifetimes, are very important in fuel cell performance and in the final cost estimates for the practical fuel cell applications.

22.1 FACTORS CONTROLLING ELECTROCATALYTIC PROPERTIES

Electrocatalytic reactions occur at active surface sites of electrodes. When an electrode surface is not active enough to promote a catalytic reaction, it is usually activated by coating or interdispersing the proper catalyst or additive. For any

827

electrode reaction, current density drawn under an overpotential, I (amperes/ projected electrode surface area), is expressed by the following equation;

$$I = jsw \tag{1}$$

where j, s, and w stand for the specific current density (amperes/real catalyst surface area), the specific surface area of catalyst (real catalyst surface area/g), and the amounts of loaded catalyst (g/projected electrode surface area), respectively. It is essential to increase j and/or s without increasing w, especially when precious metal catalysts are used. The j value can be enhanced by cooperation of different metals, for example, by means of ad-atom addition or alloying. The s value can be increased by dispersing the catalyst as nanoparticles (2–5 nm in diameter) on high-surface-area support surfaces (10–50 nm in diameter), e.g., Pt alloy catalysts supported on carbon black (CB) in gas-diffusion electrodes for fuel cells. In fuel cells, enhancement of the diffusion rates of reactants or products onto/from the nanoparticle catalysts is also important in order to minimize the performance loss due to concentration polarization. In this article, however, we focus mainly on increases of the j and s described above, i.e., catalytically controlled performances, and on related problems with the anodic and cathodic reactions in fuel cells.

22.2 ENHANCEMENT OF INTRINSIC PROPERTIES OF PRECIOUS METAL CATALYSTS FOR FUEL-CELL REACTIONS

22.2.1 Ad-Atom Method for the Design of Anode Electrocatalysts

Watanabe and Motoo extensively studied electrocatalysis by a series of precious metal alloy blacks [1,2] and precious metals with various ad-atoms [3–11] for the anodic oxidation of CO and small carbonaceous compounds such as CH_3OH and its oxidative derivatives. They first proposed a "bifunctional mechanism" for the electrocatalysis at binary or ternary alloys [1–4], consisting of two types of A and B metals with different adsorbing properties. The former, such as Pt, Ir, Pd, or Rh, have an affinity for adsorption of the above reactants and show relatively high catalytic activity for their oxidation. On the other hand, Ru, Os, or Re (both in acid and alkaline electrolytes), or Au and Ag (in alkaline electrolyte), are inert to such reactions but have an affinity for adsorption of oxygen-containing species (weakly bonded O or OH) in a relatively less positive potential region. In the mechanism, the rate-determining step (r.d.s.) of the oxidation is either an introduction of oxygen-containing species on the B sites or a successive surface reaction between the oxygen-containing species and the organic species adsorbed on the A sites. The rate of the r.d.s. considered above can be enhanced by a synergetic action between species on the neighboring A and B sites. Relationships between catalytic activities for methanol oxidation and compositions of Pt-Ru alloy blacks (≈ 8 nm in diameter) [1], or monolayer alloys of Pt with Ru ad-atoms [3], or Pt-Ru nanoparticles (≈ 3 nm in diameter) supported on carbon black [5] are shown in Figure 1. The superior consistency in the catalytic behavior among these catalysts having completely different morphologies demonstrates the usefulness of the ad-atom method for the catalyst design and/or the study of mechanism in either wet or dry systems in various areas of chemistry or physics. This is because the alloyed surfaces of well-defined composition and easy to be surface-analyzed can be easily obtained as compared

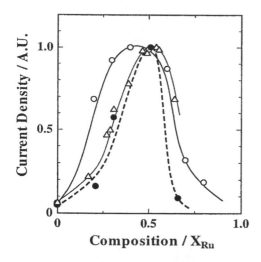

Figure 1 Relationships between the catalytic activities for methanol oxidation, normalized at the maximum of each type alloy, and the composition of Pt-Ru alloy blacks (8 nm in particle mean diameter, $\bigcirc$), momolayer alloys of Pt with Ru adatoms (electrodeposited from 10^{-5} M $RuCl_3$ solution, $\triangle$), or CB supporting Pt-Ru nonoparticles (3 nm in mean diameter, $\bullet$). Steady activities were measured in 1 M CH_3OH/1M H_2SO_4 solution at 60 °C ($\bigcirc$), 40 °C($\triangle$), and 60 °C ($\bullet$).

with those prepared via other means. For instance, a maximum activity about 20 times higher than that of pure Pt was achieved at around PtRu (50:50). The enhancement was ascribed to the contribution of the Ru component to the introduction of oxygen species at less positive potential; and the decrease of activity at a high Ru content was ascribed to the lowered adsorption rate of methanol (r.d.s.) due to the decrease of Pt sites. This was based on in-situ measurements of the coverage by organic and oxygen species by applying fast single potential pulses [3]. The work using the ad-atom method was extended further to design new electrocatalysts consisting of precious metals and nonprecious metals for various oxidation reaction [3–9]. Oxygen-adsorbing metal ad-atoms belonging to the IVth and Vth groups of the Periodic Table were found to enhance the oxidation of CO and small organic molecules [3–7], particularly Ge and Sn on the same level as the Pt-Ru system. These electrodes with nonprecious components need to be held at a less positive potential to avoid a formation of oxide formation or oxidative dissolution. Other electrocatalyses were also found. A typical example is suppression in formation of self-poisoning species [8] on Pt by no O-adsorbing ad-atoms via formation of small reactive domains, which consist of a few Pt sites unfavorable to poison formation, such as CO at formic acid oxidation. This was originally proposed as a "third body effect" for Hg/Pt [9]. Further research also found enhancement of Pt by the VIth group elements (S, Se, and Te), explained by the effect of their electronegativity on the ad-atom surfaces [10]. Adzic and his co-workers have also contributed extensively to this field [11].

Can information obtained by the ad-atom method be applied to practical nano-sized alloy catalysts? Watanabe et al. examined the enhanced CO oxidation on model surfaces and compared the reactivity of the ad-atom catalysts to that of

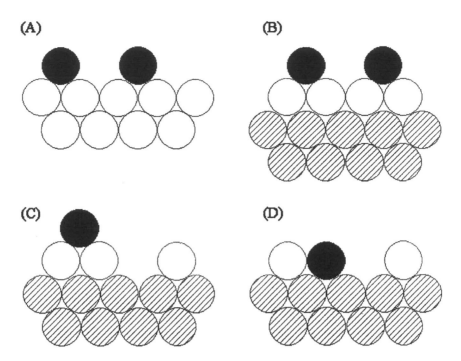

Figure 2 Schematic explanation of types of Pt-M catalysts on Pt or Au substrates, Pt($\bigcirc$), M ($\bullet$), Au($\oslash$). (A) M/Pt(subs); (B) M-Pt/Au(subs), $\theta_{Pt} = 1.0$; and (C) M-Pt(cluster)/Au(subs), $1 > \theta_{Pt} > 2/3$, (D) M/Pt(cluster)/Au(subs), $\theta_{Pt} < 1/3$.

practical clusters by using an Au substrate as an ideal inert catalyst support, as shown schematically in Figure 2 [12,13]. At least for the reactions and catalysts shown in these papers, almost the same relationship between the activity and the alloy composition was obtained. This observation was made for a monolayer of Pt alloy on bulky Pt, or a monolayer of Pt on the Au support, or a single Pt alloy cluster on the Au support. Recently, many mechanistic studies have been reported on the enhancement of CO or CH_3OH oxidation on surfaces containing either Pt or Ru ad-atoms. These are frequently deposited chemically or electrochemically on Ru or Pt single-crystal electrodes and are characterized by modern experimental methods such as STM, FTIR, or LEED-HEED [14–18]. In these studies, an islandlike deposition of Ru ad-atoms has been observed by STM, which corresponds to the observation of separated CO bands, assigned to CO adsorbed on Pt and Ru sites. However, a single band was observed on the Pt-Ru alloy [19]. In spite of such a difference of neighboring states of Pt and Ru sites on their surfaces, the onset potential of CO oxidation did not show much difference, probably due to enhanced mobility of adsorbed CO on the surface and high catalytic activity at the boundary of Pt-Ru or the surface of Ru island. As we showed previously [4,20], the activity increases with an increasing number of Ru on Pt surface similar to the Pt-As system, where the adsorption of O-containing species required for the CO oxidation is the r.d.s., regardless of the dispersion of ad-atoms. On the other hand, Pt-Sn showed a maximum at Sn coverage about 0.5, which was explained by the adsorption rate of

O-containing species being high enough and the activity being proportional to the number of Pt-Sn pairs [7,12]. Thus, we think, each r.d.s. may be changed, depending on the reactant and catalyst combinations, and a difference of surface states might be a possible origin of the inconsistency in terms of the optimum composition of Pt-Ru catalysts for methanol oxidation between different groups. (Other factors would, of course, be different experimental methods and conditions such as temperature or reactant concentration, etc.) Likewise, a clear statement as to what constitutes an active site needs to be investigated further experimentally and theoretically. From this viewpoint, the recent work by Wieckowski et al. on Pt nanoparticles, supported on CB, with spontaneously deposited Ru [16] or vice versa [21] is very interesting, also as an extension of previous research on Pt black with Sn ad-atoms [22].

22.2.2 Design of Anode Electrocatalysts for the CO-Tolerant H_2 Oxidation and CH_3OH Oxidation

Polymer electrolyte fuel cells (PEFC) are a potential power source for electric vehicles (FCEV), residential cogeneration systems, or portable electric devices. PEFCs require the use of platinum or Pt alloy catalysts at the anode and cathode in order to achieve acceptable performance at relatively low temperatures. The high costs of platinum serve as the motivation for research in the further design and mechanistic understanding of fuel-cell catalysts. A major drawback of conventional PEFCs using pure Pt anode catalysts operating on reformates is the drastic decrease in performance due to poisoning of Pt sites [23–25]. There is no need for a reforming system in the direct methanol fuel cells (DMFC), allowing their use in portable applications. However, they require 10 times more Pt than reforming-gas-fed PEFCs. It is essential, therefore, to develop high-performance, low-cost electro-catalysts for H_2 oxidation with high CO tolerance, for direct methanol oxidation (DMO) as a competitor to the relatively expensive Pt-Ru alloy, which does show superior performance in both reactions [26].

In order to screen promising new combinations for Pt alloy catalysts, Watanabe et al. used a rotating disk electrode (RDE) to test for CO tolerance for H_2 oxidation. By sputtering several metals simultaneously, they prepared a series of Pt alloys on a glass disk with electric lead wires to be used as the RDE. The experiment was performed in 0.1 M $HClO_4$ saturated with 100 ppm CO/H_2 at room temperature [27–30]. Prior to the performance measurement, the samples were pretreated by several potential sweeps ($10\,Vs^{-1}$, 0.05–1.2 V) until their cyclic voltammograms were stable. The cystallographic structure was determined by grazed incidence ($\theta = 1°$) XRD, while bulk or surface composition was determined by EDX or XPS. These measurements were done both before and after electrochemical experimentation. In addition, cyclic voltammetry was used to determine surface areas and CO coverage (θ_{CO}).

Figure 3 shows typical polarization curves for H_2 oxidation on pure Pt and $Pt_{85}Fe_{15}$ electrodes at 1500 rpm in the presence and the absence of 100 ppm CO. After 2 hours of CO adsorption, the oxidation current of H_2 was hardly observable on pure Pt. Conversely, no CO poisoning for H_2 oxidation was observed on $Pt_{85}Fe_{15}$. Based on the data at various rotating velocities, kinetic currents were evaluated on each electrode at 20 mV. Some examples are shown in Figure 4 as a function of the CO poisoning time. The Pt-Fe, Pt-Ni, Pt-Co, and Pt-Mo alloys

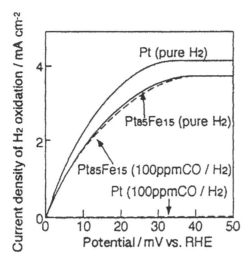

Figure 3   Steady polarization curves for H_2 oxidation on pure Pt and Pt-Fe alloy electrodes at 2000 rpm in 0.1 M $HClO_4$ saturated with H_2 or 100 ppm CO/H_2 balance.

exhibit no current degradation for a prolonged time period, although the current on pure Pt declined completely within 30 min. CO tolerance of these alloys was found independent of the catalyst composition. As shown in Figure 9, alloying Pt with only 5 atomic % Fe resulted in excellent stability. The results for all combinations examined are summarized in Table 1. The same CO tolerance is expected on Pt and Os or Re alloy, not being tested, similarly to those mentioned above since both elements behave as Ru described in Section 22.2.1, and Rh, Pd, or Ir are also very promising as they behave like Pt.

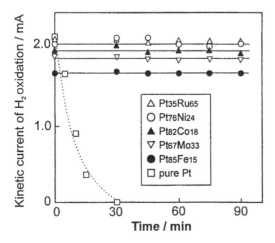

Figure 4   The change of H_2 oxidation current on alloy RDE at 20 mV in 0.1 M $HClO_4$ saturated with 100 ppm CO/H_2 balance at room temperature.

Table 1   Rank of the Evaluation of Alloying with Pt Toward CO-Tolerant H_2 Oxidation

4a	5a	6a	7a	8			1b	2b	3b	4b	5b
Ti *P*	V	Cr *P*	Mn *G*	Fe *E*	Co *E*	Ni *E*	Cu *P*	Zn *G*	Ga	Ge *P*	As
Zr	Nb *P*	Mo *E*	Tc	Ru *E*	Rh	Pd *P*	Ag *G*	Cd	In *P*	Sn *G*	Sb *P*
Hf	Ta	W *P*	Re	Os	Ir	Pt	Au *P*	Hg	Tl	Pb *P*	Bi *P*

E: No degradation, *G*: degradation with time, *P*: immediate degradation.

Figure 5 shows examples of where the CO coverage on Pt and the alloy surfaces was kept constant during H_2 oxidation in the presence of 100 ppm CO. The CO coverage on the CO-tolerant alloys is suppressed to values less than 0.6, while on the other alloys and on pure Pt the surfaces are almost completely covered with CO [27–30]. One may therefore ask why the number of the CO-free sites on the CO-tolerant surfaces is higher, and how do these sites behave for H_2 adsorption and CO oxidation.

Cyclic voltammograms (CV) of all of the alloy electrodes prior to the experiments mentioned above resembled that of pure Pt, strongly indicating that the nonprecious metals were leached out from the surface and that a Pt skin-layer was formed. This was confirmed to be correct using XPS before and after the electrochemical measurement (Figure 6). Despite the appearance of a superior CO tolerance discussed above, it is clear from the XPS spectra that nonprecious metals such as Fe disappear almost completely from the alloy surface, in contrast to the precious metal Ru alloying the Pt surface. All of the nonprecious metal alloys exhibit similar phenomena and are leached out to the solution. However, the XPS spectra for Pt in the alloys revealed only a negligible change before and after the

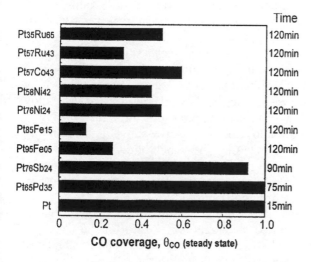

Figure 5   CO coverage on various surfaces of alloy electrodes under a steady H_2 oxidation condition at 20 mV vs. RHE in 0.1 M $HClO_4$ saturated with 100 ppm CO/H_2 balance at room temperature.

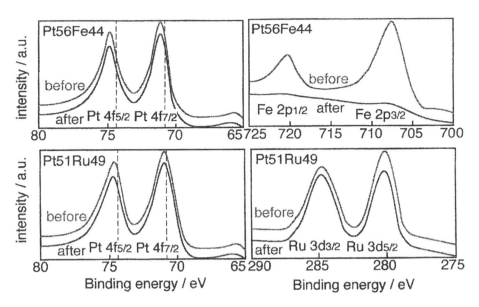

Figure 6 Examples of XPS of Pt in virgin Pt alloys and those used for electrochemical measurements after stabilizing. Dotted lines show peak potentials of the corresponding energies of pure Pt.

electrochemical pretreatment by potential sweeps, as shown in Figure 6. This gives us clear evidence for the formation of a Pt skin layer on Pt alloys with nonprecious metals. The thickness of the Pt layers was estimated to be about 1 to 2 nm, based on the residual XPS intensity for nonprecious metal permeated through the Pt layer. The result was confirmed by EQCM measurements [31]. A formation of the (111)-oriented Pt skin layer has also been confirmed by STM after the CV pretreatment of the Pt-Fe alloy [32]. In terms of the reactivity, this is contrary to the "bifunctional theory," at least for the present nonprecious alloy systems, because of the absence of second metal atoms on the surface layer.

The differences between the pure-Pt surface and the Pt skin layer were also examined. One of the differences appeared in the binding energy of Pt4f in $Pt_{56}Fe_{44}$ as well as $Pt_{51}Ru_{49}$, exhibiting a positive shift compared to that of pure Pt shown by dotted lines in Figure 6. The shifts of the binding energy for typical alloy combinations displaying high and low CO tolerance are shown in Figure 7 [27–30]. The data reveal that all CO-tolerant alloys show positive shifts in the binding energies for both Pt4d and Pt4f electrons. On the other hand, non-CO-tolerant ones showed negative shifts [27–30]. The higher binding energies for core levels of the Pt skin layers are clear evidence of the modification of the electronic structure by the underlying alloys and are distinctively different from that of pure Pt.

Figure 8(a) demonstrates the dependence of the normalized CO intensities at pure Pt ($\theta_{CO} = 0.48$) and Pt-Fe (saturated coverage, $\theta_{CO} = 0.55$) on the electrode potential. The difference of the electronic structure between pure Pt and Pt skin layer on the CO-tolerant alloys, mentioned above, is reflected in the infrared spectra of chemisorbed CO, i.e., the multibonded CO, which might require the contribution of the 5d electron, was hardly found on the skin layer [27–30]. This is in contrast to the

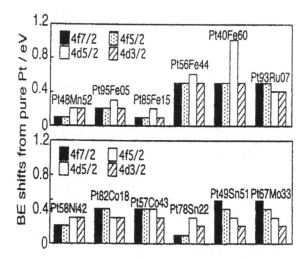

Figure 7 Binding energy shifts of Pt skin layers on various alloys.

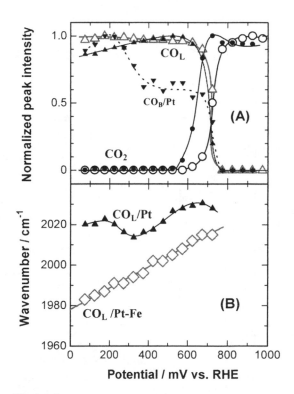

Figure 8 (A) Normalized IR intensity for adsorbed species on $Pt_{27}Fe_{73}$alloy ($\theta_{CO} = 0.5$) and pure $Pt(\theta_{CO} = 0.48)$ as a function of potential, $\triangle CO_L/Pt$-Fe, $\bigcirc CO_2/Pt$-Fe, $\blacktriangle CO_L/Pt$, $\blacktriangledown CO_B/Pt$, $\bullet CO_2/Pt$; and (B) change of wavenumber for CO_L on $Pt_{27}Fe_{73}$alloy ($\diamond$) and pure Pt ($\blacktriangle$) as a function of potential.

case of pure Pt, on which multibonded CO presents all the time together with atop-CO, except the existence of the latter alone at $\theta_{CO} = 1$ [33]. The bridged CO on the pure Pt decreases as the electrode potential increases. This is probably due to the lowering of the Fermi level of the 5d valence band. However, the bridged CO on pure Pt does not disappear until the potential increases up to nearly 0.6 V, although no bridged CO exists on the Pt-Fe in any potential region. This seems to indicate lowered back donation of the d-electrons to CO molecules due to higher 5d-vacancy of Pt in the skin layer.

The wavenumber of the CO_L band at the saturated coverage on the Pt-Fe alloy electrode is plotted (□) as a function of the electrode potential [Figure 8(b)] in comparison with that on pure Pt. It was found that the wavenumbers exhibit extremely small values, e.g., ≈ 2000 cm^{-1} at 0.4 V, which is 20 cm^{-1} and 60 cm^{-1} lower than for pure Pt at $\theta_{CO} = 0.48$ and 1.0 [28], respectively. The redshift for CO vibrations on the skin layer from that for a pure-Pt reference indicates that the simple model with the d-electron transfer cannot explain all the results observed, and further experimental analyses, e.g., using UPS, are necessary.

On the Pt skin of the alloy electrode and at potentials higher than about 0.6 V, the surface occupancy by the linear CO (△) decreases steeply with the progress in oxidation, which is accompanied by an increase in CO_2 intensities (○) [Figure 8(a)]. The oxidation potential is almost the same as that of the CO_B (▼) and CO_L (▲) on pure Pt, and, moreover, the formation of CO_2 was observed at rather positive potentials on the Pt skin layer. Therefore, it is clear that the excellent CO tolerance on the alloys specified above cannot be explained by the enhancement of the direct CO oxidation via the bifunctional mechanism originally proposed by Watanabe and Motoo [1,3,4].

Figure 9 shows the steady kinetic currents (J_k) on various CO-tolerant alloys as a function of the CO coverage in comparison with those obtained previously on pure Pt. The J_k values on the pure Pt decrease with increasing CO coverage, i.e., $J_k = k(1 - \theta_{CO})$ at $\theta < \approx 2/3$, and $J_k = k(1 - \theta_{CO}^2)$ at $\theta > \approx 2/3$, as shown by the dotted line [34]. The relations indicate that the mobility of adsorbed CO is relatively small

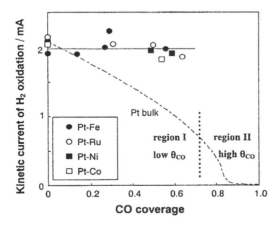

Figure 9 Dependency of steady H_2 oxidation current at 20 mV vs. RHE on CO coverage at various alloy surfaces with Pt skin layer in 0.1 M $HClO_4$ saturated with 100 ppm CO/H_2 balance at room temperature. Dotted line: pure Pt.

on pure Pt so that dissociative H_2 adsorption followed by atomic H ionization becomes the rate-determining step for the H_2 oxidation. On the other hand, it is clear that the J_k values barely depend on the CO coverage at the CO-tolerant alloys, as mentioned above. The CO coverage is suppressed on the alloys, and the surface can support the same large currents as that of pure Pt without poisoning by CO, regardless of the alloy combination or composition. The independence on this CO coverage, distinctive from that of pure Pt, cannot be explained unless high mobility of adsorbed CO and weakening of Pt–C bonding are taken into account. The weakening of Pt–C bond strength and the increase in the mobility can be explained by the theory proposed recently by Nørskov et al. [35], where it is claimed that H_2 dissociation and successive oxidation may be unaffected by remaining CO molecules on the surface.

Recently, there has been extensive research on the CO tolerance for Pt-Ru, Pt-Mo, and Pt-Sn systems [36–39]. It was claimed that the CO tolerance results from the increased CO-free sites due to the enhanced CO oxidation via bifunctional mechanism. On Pt-Ru alloys and Pt surfaces covered by Ru ad-atoms, the CO oxidation has been found at lower positive potential, i.e., around 0.3–0.4 V, in many works. However, the H_2 oxidation occurs near the RHE potential. At this potential, O or OH species required for the oxidation of CO are not present due to negligibly small equilibrium coverage and due to O or OH reaction with H_2 present in large excess. Assuming that a small fraction of the CO-free sites was present, then they may only account for very small J_k indeed, as in pure Pt, unless the CO mobility is taken into account. Consequently, the major factor for the CO tolerance in H_2 oxidation of the Pt-Ru system must be ascribed to weakening of the Pt–CO bond in addition to lowering CO coverage, caused by the modification of the electronic structure when alloyed with a second metal.

22.2.3   Design of Pt Alloy Catalysts for the O_2 Electroreduction

PEFCs and DMFCs cathodes share a large part of the energy loss in the fuel cells. It is essential, therefore, to develop high-performance, low-cost electrocatalysts for the O_2 reduction reaction (ORR). Watanabe et al. prepared Pt-Ni, Pt-Co, and Pt-Fe alloy films with various compositions in a solid solution phase of the fcc structure on the glassy substrates by a sputtering method. They examined the catalytic activity for ORR using a rotating disk electrode in 0.1 M $HClO_4$ saturated with O_2, and at ambient temperature [40–42]. (As in the CO-tolerant studies, see section 22.2.2.) There were no noticeable changes in the roughness factors of the electrode surfaces, which lie in the range of 4.0 to 7.7 independent of the alloy composition. The effect of the roughness factor among the electrodes (in the evaluation of an enhancement effect) was eliminated by using a normalized kinetic current density, J_k. The dependency of J_k on the alloy composition of these electrodes is shown in Figure 10. For the first time, we report that Pt-Ni, Pt-Co, and Pt-Fe exhibit a maximum activity by alloying around 30, 40, and 50 at.%, and the enhancement factors are around 10, 15, and 20, respectively. Previously, an enhancement on fine Pt alloy particles in the gas-diffusion electrodes was reported under PEFC operation condition, but the factors were all small. For example, noticeable values among them were 1.4 and 2.6 for PtFe and PtCr alloys, respectively [43].

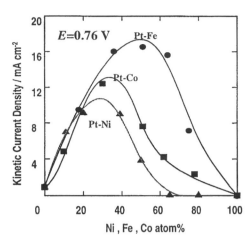

Figure 10 Dependency of kinetic current density on the alloy composition at Pt-Ni, Pt-Co and Pt-Fe electrodes in 0.1 M HClO$_4$ solution saturated with O$_2$ at room temperature.

It was found that the Pt skin layers were formed on the alloy surfaces and exhibited positive chemical shifts of the binding energies in 4f and 4d orbitals depending on the composition, which indicates a modification of the electronic structure by alloying, similarly to those for CO-tolerant anode catalysts. The thickness of the skin layer was estimated to be 1–3 nm, based on the change in the XPS intensities of Ni, Co, and Fe before and after formation of the layer. The effect of the thickness of the layer on ORR was examined at Pt$_{50}$Fe$_{50}$ with various thickness of Pt skin formed by sputtering/deposition of pure Pt. The result is shown in Figure 11. Below 1 nm in the thickness, where no change of the positive shift (+0.4 eV) in the Pt4d5/2 binding energy was observed, the activity for the ORR shows a maximum value. But it decreases with the further increase in the thickness and coincides with that of pure Pt beyond 4 nm, where the chemical shift was completely lost. Therefore, it is clear that the enhancement of the ORR by alloying with the nonprecious d-metals results from the modification of the electronic structure, i.e., increased 5d-vacancy in the Pt skin layer.

Mechanisms for ORR on pure Pt have been proposed but still remain elusive [43–48]. As one of the possible mechanisms, Yeager et al. considered that the rate-determining step (r.d.s.) in acid electrolytes is an adsorption step of O$_2$ molecule as side-on or bridged types accompanied by electron transfer [49]. Let us consider this mechanism on the alloy catalysts (see Figure 12 [42]), where the r.d.s. is

$$O_2(\text{solution}) \mid e^- \rightarrow O_2^-(\text{adsorbed}) \quad \text{r.d.s.} \tag{2}$$

These types involve a lateral interaction of the π-orbitals of the O$_2$ with empty d$_{z^2}^2$ orbitals of a surface Pt atom or with empty d_{xz} and d_{yz} orbitals of dual Pt atoms, respectively, with back bonding from the partially filled orbitals of the Pt to the π^* orbitals of the O$_2$. Increased d-vacancy of Pt on the electrode surface, brought about by alloying, must lead to the strong metal–oxygen interaction. Such a strong interaction brings an increase in O$_2^-$ weakening of the O–O bond and an increase in the bond length, which results in fast bond scission and/or a new bond formation

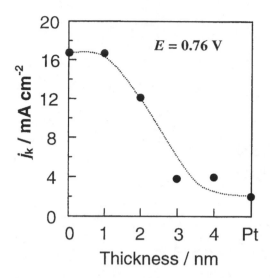

Figure 11 The effect of Pt skin thickness on the ORR at 0.76 V. The Pt layer was formed on a virgin $Pt_{50}Fe_{50}$ alloy surface by sputtering Pt target.

between the O atom and H^+ in electrolyte. Thus, the enhanced catalysis by alloying, found in the present work, can be well explained in this mechanism. The r.d.s. for the ORR will not be changed by alloying because Pt and the Pt alloys exhibit the same Tafel slope, i.e., 120 mV/decade, even in the high current density region. Beyond the Fe content of the maximum enhancement, a back donation of electrons from Pt atoms to adsorbed O_2 molecules might not be enough due to too much d-vacancy, resulting in the slowdown of the ORR. A review of similar issues by other groups is given in [42].

Supported binary or ternary alloy catalysts are still used in practical PEFCs as the cathode catalysts in spite of the presence of dealloying. It is believed that the dispersion of such catalyst particles is preserved in comparison to pure Pt for a long time period, although no clear evidence has been reported in that regards. The stabilization of fine catalyst particles may also be achieved if the change in the electronic structure and resulting increase in d-vacancies of the bulk alloy affect the Pt skin layer toward reducing surface energy.

22.3 NANOPARTICLE CATALYSTS SUPPORTED ON CARBON BLACK

22.3.1 Particle Size Effect

From the viewpoint of cost reduction or conservation of catalyst resources, it is essential to support the catalysts as nanoparticles with high specific surface area, s, on high-surface-area CB, even if the catalyst has a high specific activity, J_k. A method for producing such a high dispersion was first developed by Allen et al. [50]. It is now not difficult to achieve a mean particle diameter $d < 2$ nm, $s > 140 \, m^2/gPt$, or the dispersion of particles (number of atoms exposed to the surface/number of total atoms) $D > 0.5$ [51]. Unfortunately, however, such highly dispersed Pt catalysts

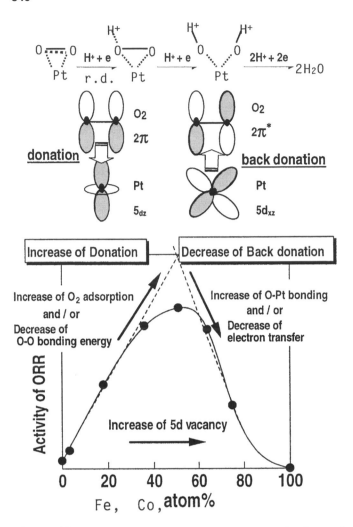

Figure 12 Schematic explanation of oxygen reduction enhancement at Pt skin layer underlined by Pt alloy.

have not shown the same mass activity (A/g Pt) as bulky or larger-size catalysts ($>$ca. 5 nm) [52,53]. In this small range of particle sizes, there are significant variations, not only in the dispersion but also in the nanoparticle morphology. Typical is a presence of atoms with different coordination numbers; e.g., on (111), (100) crystallite planes, there are sites at the edge and kink positions of the cluster [54]. A change in the electronic structure of Pt with decreasing particle sizes has also been indicated; e.g., a steep increase in the 5d vacancy sites was reported [55].

Bregoli indicated that as the surface area of Pt crystallites supported on CB increases, the J_k decreases in practical PEFCs operated around 190 °C [52]. This effect was observed for Pt surface areas ranging from 20 to 80 m^2g^{-1} on CB. Ross extended this work to higher-surface-area Pt and observed a similar but more pronounced effect in the region larger than 80 m^2g^{-1}, or $d < 3$ nm. This is called the "particle-size effect" [53]. Takasu et al. observed the effect of nanoclusters on a glassy carbon surface for the ORR or for the oxidation of small organic molecules at ambient temperature in H_2SO_4 solution. They attributed the particle-size effect to too strong binding of reactants on the catalyst surfaces [56]. Mukerjee arrived at the same conclusion based on the in-situ XAS analysis [55]. Kinoshita proposed the "active surface theory" to explain the effect from the similarity of the relationship between the Pt mass activity I and d, and that between I and the number of (100) sites calculated on cubo-octahedron nanoparticles [57]. Durand et al. used this theory to explain their experimental results on the ORR at ambient temperature [58]. Watanabe et al. proposed a new explanation and formulated the "territory theory" [51]. When Pt particles are close together within some critical region, there must be a mutual influence on the diffusion, or some other parameters, such that not all of the Pt surface sites are usable. As a working concept, if we can separate the Pt particles by using high-surface-area CB supports or loading small amount of Pt, so that they do not influence each other, then we should increase the mass activity I. Some experiments were performed using CB supports with various surface areas (60 to $1250\,m^2g^{-1}$) with different amounts of Pt (5–40 wt. %) loaded in the gas-diffusion electrodes for the measurement of the ORR in hot phosporic acid at 190 °C and in H_2SO_4 solution at 60 °C. Care was taken to ensure that the potential was never higher than 0.8 V to avoid particle coagulation. It became obvious that (1) when the intercrystallite distance was greater than an average of 20 nm, the I was fully achieved for all catalyst dispersions and (2) the "crystallite size effect" was not truly dependent on the crystallite sizes, but on the interparticle distances. As predicted by this theory, even the intercrystallite distance effect was not observed for the direct methanol oxidation, because of the sufficiently small territory of nanoparticle catalysts for the reactant in 1 M CH_3OH/1 M H_2SO_4 solution. The results were obtained at elevated temperature operations leading to higher reactivity or lowered poisoning, indicating that they should be discussed separately from the catalyst stability.

Recently, Takasu et al. successfully prepared Pt-Ru nanoclusters of various particle sizes supported on CB [59]. The activity for methanol oxidation at 60 °C was independent of the particle sizes—so that the higher the dispersion, the higher the I— and exhibited a maximum at the composition of 1:1, which is consistent with the author's previous work [3,5].

The bond strength measured or estimated at a constant temperature or at a specified facet of single-crystal planes are, of cause, important factors in the discussion of electrocatalysis. But, if they are obtained under different conditions from the practical ones, the conclusion might not be correct. The author believes that the general understanding on the size effect still needs further experimental and theoretical studies. Particularly careful attention should be paid to the operation temperature, the coagulation of nanoparticles during the experiment, or the "territory." Of course, evaluation on the stability of such small particles is essential in the practical application.

22.3.2 Mechanism of the Corrosion and the Degradation of Activity at Nanoparticle Electrocatalysts

Alloying with some proper metals can enhance the J_k value. Such alloy nanoparticles are in single-crystal phases similar to pure metal particles. They expose crystallite planes with a small surface energy such as (111) and (100) planes at Pt alloys with fcc lattice structure, as indicated by icosahedron or cubo-octahedron structures, consisting of 20 (111) facets or 6 (100) and 8 (111), respectively [54]. The nanoparticles highly dispersed on supports coagulate easily into larger ones, and dealloy within a short time, if a nonprecious metal is included as a component, particularly at small-size particles. However, the mechanism of corrosion, as well as the way of the suppression, has not yet been elucidated. Using a mother Pt-Ni-Co (ca. 2:1:1) alloy supported on CB, Watanabe et al. prepared a well-ordered alloy (in a phase of intermetallic compound, f.c. tetragonal) and a disordered one (in a phase of solid solution, f.c.c. α), respectively, without a change in their particle sizes ($\approx 6\,nm$) [60]. They compared the behaviors of the catalysts against the corrosion and the degradation of activity for the ORR in hot phosphoric acid at 190 °C before and after occurrences of accelerated corrosion. It became obvious that the disordered alloy shows better corrosion resistance than the ordered alloys. The result is a suppressed degradation in the catalytic activity at the disordered alloy. Based on the results by a chemical analysis, XRD, SEM with EPMA, the authors proposed a

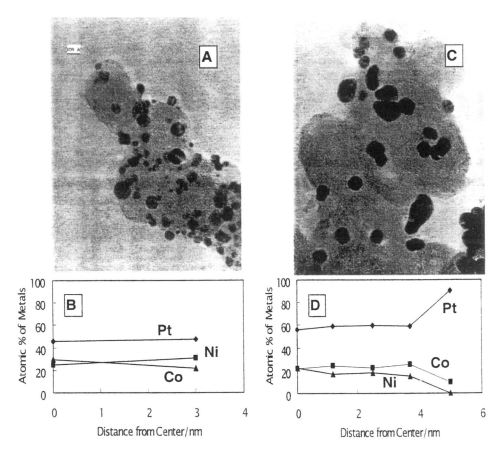

Figure 13   TEM photographs of Pt-Ni-Co supported on carbon black before (A) and after (C) corrosion (in 100% H_3PO_4 at 240 °C and 0.9 V vs. RHE for 48 hrs), and distributions of alloy components from the center to the periphery on their representative particles, shown in (B) and (D).

corrosion mechanism; once after the corrosion proceeds by the simultaneous dissolution of every alloy components from the surfaces at small particles, for example, less than 5 nm, layer by layer, only Pt redeposit on larger alloy particles existing as their neighbors, resulting in the formation of Pt skin layer on the surfaces [60].

Figure 13 shows photographs of Pt-Ni-Co alloys supported on CB and the component distribution in the representative particles before and after the occurrence of accelerated corrosion (at 240 °C and 0.9 V versus RHE in 100% for 48 h), which were taken by Field-Emitting-type TEM with EPMA at an electron-beam probe size around 1 nm [61]. After corrosion occurred, it is clear that all small particles disappear and some larger particles with a similar size distribution appear. A representative small particle before corrosion shows a uniform distribution of the alloyed components throughout the particle. On the other hand, large particles subjected to corrosion have a Pt skin layer on the core particle and the uniform chemical distribution inside. As described in sections 22.2.2 and 22.2.3, the effect of the modification of the electronic structure by underlying alloy composition declines when the thickness of the skin layer increases more than 2 or 3 nm, resulting in the lowered activity for the CO tolerance or the ORR, respectively. In general, therefore, the specific activity J_k of such larger particles must be smaller than that of the fine alloy particles; in other words, the degradation of the catalytic activity of these alloys results from the disappearance of small particles with high specific surface area s and high specific activity J_k, and the increased abundance of larger particles with lower J_k. It has been demonstrated that nanoparticles in the solid solution phase, or in atomically disordered alloy particles, are more stable against corrosion. These result in higher resistance against activity degradation, in comparison with the intermetallic compound phase or atomically ordered alloy particles [60].

The major factor for the accelerated corrosion of CB supports and Pt catalysts is found when the temperature is higher than 220 °C and the potential is higher than 0.9 V versus RHE [61], respectively. It is essential to evaluate the stability as well as the activity under such accelerated corrosion evaluation conditions.

REFERENCES

1. M. Watanabe et al., *Denki Kagaku* (presently *Electrochemistry*), **38**, 927–932 (1970).
2. M. Watanabe et al., *Denki Kagaku* (presently *Electrochemistry*), **39**, 394–399 (1971); **40**, 205–210 (1972); **40**, 210–215 (1972); **41**, 190–196 (1973); **43**, 147–151 (1975); **43**, 153–158 (1975).
3. M. Watanabe, S. Motoo, *J. Electroanal. Chem.* **60**, 267–273 (1975).
4. M. Watanabe, S. Motoo, *J. Electroanal. Chem.* **60**, 275–283 (1975).
5. M. Watanabe, M. Uchida, S. Motoo, *J. Electroanal. Chem.* **229**, 395–406 (1987).
6. S. Motoo, M. Watanabe, *J. Electroanal. Chem.* **69**, 429–431 (1976); **111**, 261–268 (1980); S. Motoo, M. Shibata, M. Watanabe, **110**, 103–109 (1980); **191**, 367–375 (1985); **202**, 125–135 (1986); M. Shibata et al., **267**, 163–170 (1989); **344**, 389–393 (1993).
7. M. Watanabe et al., *J. Electroanal. Chem.* **187**, 161–174 (1985).
8. S. Motoo, M. Watanabe, *J. Electroanal. Chem.* **98**, 203–211 (1979); **111**, 261–268 (1980); M. Watanabe et al., **250**, 117–125 (1988); M. Shibata et al., **249**, 253–264 (1985); **263**, 97–108 (1989).
9. H. Angesstein-Kozlowska, B. MacDougall, B.E. Conway, *J. Electrochem. Soc.* **120**, 756 (1973).
10. M. Watanabe, S. Motoo, *J. Electroanal. Chem.* **194**, 275–285 (1985).

11. R.R. Adzic, H. Gerischer, C.W. Tobias (eds.), *Advances in Electrochemistry and Electrochemical Engineering* Vol. 13, John Wiley, New York, 1985, p. 159.

12. S. Motoo, M. Watanabe, *J. Electroanal. Chem.* **98**, 203–211 (1979).

13. S. Motoo, M. Shibata, M. Watanabe, *J. Electroanal. Chem.* **110**, 103–109 (1980).

14. K.A. Friedrich et al., *J. Electroanal. Chem.* **402**, 123–128 (1996).

15. W. Chrzanowski, A. Wieckowski, *Langmuir* **13**, 5974–5978 (1997); **14**, 1967–1970 (1998).

16. P. Waszczuk, et al., *J. Catalysis* **203**(1), 1–6 (2001).

17. W.F. Lin et al., *J. Phys. Chem. B* **103**, 3250–3257 (1999); **103**, 6968–6977 (1999).

18. E. Herrero, J.M. Feliu, A. Wieckowski, *Langmuir* **15**, 4944–4948 (1999).

19. T. Iwashita et al., *Electrochim. Acta* **39**, 1817–1823 (1994).

20. S. Motoo, M. Watanabe, *J. Electroanal. Chem.* **111**, 261–268 (1980).

21. R.R. Adzic et al., *Electrochem. & Solid-State Lett.* **4**(12), A 217–220 (2001).

22. M. Watanabe, Y. Furuuchi, S. Motoo, *J. Electroanal. Chem.* **191**, 367–375 (1985).

23. S. Srinivasan et al., *J. Power Sources* **29**, 367 (1990).

24. A.J. Appleby, F.R. Foulkes, *Fuel Cell Handbook* Van Nostrand Reinhold, New York (1989).

25. R.A. Lemons, *J. Power Sources* **29**, 251 (1990).

26. M. Watanabe, S. Motoo, *Denki Kagaku* (presently *Electrochemistry*) **44**, 602–607 (1976).

27. M. Watanabe, H. Igarashi, T. Fujino, *Electrochemistry* **67**, 1194–1196 (1999).

28. M. Watanabe, Y. Zhu, H. Uchida, *J. Phys. Chem. B* **104**, 1762–1768 (2000).

29. M. Watanabe, Y. Zhu, H. Igarashi, H. Uchida, *Electrochemistry* **68**, 244–251 (2000).

30. H. Igarashi, T. Fujino, Y. Zhu, H. Uchida, M. Watanabe, *Phys. Chem. & Chem. Phys.* **3**, 306–314 (2001).

31. H. Uchida, H. Ozuka, M. Watanabe, *J. Electrochim. Acta 47*, 3629–3636 (2002).

32. L.J. Wan, T. Moriyama, M. Ito, H. Uchida, M. Watanabe, *Chem. Comm.* 58–59 (2002).

33. Y. Zhu, H. Uchida, M. Watanabe, *Langmuir* **15**, 8757–8764 (1999).

34. H. Igarashi, T. Fujino, M. Watanabe, *J. Electroanal. Chem.* **391**, 119–123 (1995).

35. E. Christoffersen, P. Liu, A. Ruban, H.L. Skriver, J.K. Nørskov, *J. Catal.* **199**, 123 (2001).

36. B.N. Grgur, G. Zhuang, N.M. Markovic, P.N. Ross, *J. Phys. Chem.* **101**, 3910 (1997).

37. H.A. Gasteiger, N. Markovic, P.N. Ross, Jr., E.J. Cairns, *J. Phys. Chem.* **98**, 617 (1994).

38. H. Gasteiger, N. Markovic, P. Ross, *J. Phys. Chem.* **99**, 8945 (1995).

39. T.A. Zawodzinski, T.E. Springer, S. Gottesfeld, *The 1997 Joint International Meeting of ECS and ISE, Meeting Abstract* **97–2**, 1228 (1997).

40. T. Toda, H. Igarashi, M. Watanabe, *J. Electrochem. Soc.* **145**, 4185–4188 (1998).

41. T. Toda, H. Igarashi, M. Watanabe. *J. Electroanal. Chem.* **460**, 258–262 (1999).

42. T. Toda, H. Igarashi, H. Uchida, M. Watanabe, *J. Electrochem. Soc.* **146**, 3750–3756 (1999).

43. S. Mukerjee, S. Srinivasan, M.P. Soriaga, J. McBreen, *J. Electrochem. Soc.* **142**, 1409 (1995).

44. V.M. Jalan, E.J. Taylor, *J. Electrochem. Soc.* **130**, 2299 (1983).

45. J.T. Glass, G.L. Cahen, G.E. Stoner, *J. Electrochem. Soc.* **134**, 58 (1987).

46. M.T. Paffet, J.G. Beery, S. Gottesfeld, *J. Electrochem. Soc.* **135**, 1431 (1988).

47. B.C. Beard, P.N. Ross, Jr. *J. Electrochem. Soc.* **130**, 221 (1990).

48. J. Appleby, *Energy* **11**, 13 (1986).

49. E. Yeager et al. In D. Scherson et al. (eds.), *Structural Effects in Electrocatalysis and Oxygen Electrochemistry* PV 92–11, The Electrochem. Soc., Pennington, NJ, p. 440 (1992).

50. H.G. Petrow, A.J. Allen, U.S. Patent 4,044,193, (Aug. 23, 1977).

51. M. Watanabe, H. Sei, P. Stonehart, *J. Electroanal. Chem.* **261**, 375–387 (1989).

52. L.J. Bregoli, *Electrochim. Acta* **23**, 489 (1978).

53. M.L. Sattler, P.N. Ross, *Ultramicroscopy* **20**, 21 (1986).
54. R. van Hardeld et al., *Surf. Sci.* **15**, 189 (1969); **4**, 396 (1969).
55. S. Mukerjee, J. McBreen, *J. Electronal. Chem.* 163–171 (1998).
56. Y. Takasu et al., *Electrochim. Acta* **41**, 2595–2600 (1996).
57. K. Kinoshita, *J. Electrochem. Soc.* **137**, 845 (1990).
58. A. Kabbabi et al., *J. Electroanal. Chem.* **373**, 251–254 (1994).
59. Y. Takasu et al., *J. Electrochem. Soc.* **147**, 4421–4427 (2000).
60. M. Watanabe et al., *J. Electrochem. Soc.* **141**, 2659–2668 (1994).
61. N. Nakajima, H. Miyoshi, T. Kitai, H. Uchida, M. Watanabe, *Extended Abs. of Fuel Cell Seminar* pp. 230–233, Nov. 17–20, Orlando, FL (1996).

23

Effects of Particle Size and Support on Some Catalytic Properties of Metallic and Bimetallic Catalysts

BERNARD COQ

ENSCM-CNRS, Montpellier, France

FRANÇOIS FIGUERAS

Institut de Recherche sur la Catalyse, Villeurbanne, France

CHAPTER CONTENTS

SUMMARY

Some aspects of the particle size, "alloying" effect, and metal–support interaction in nano-sized supported metal particles are presented for the oxidation of ethylene, the hydrogenolysis of alkanes, and the hydrogenations of unsaturated hydrocarbons and α,β-unsaturated aldehydes. The influence of these phenomena is highlighted on the

847

kinetics, the mechanism, the nature of adsorbed species and intermediates, and the consequences on turnover frequency and selectivity. It is shown, in particular, how the strength of the chemisorption bond, the topology of the active site, and the coordination of the metal atoms can determine the nature of the surface intermediates and, *in fine*, the orientation of the reaction. Some elements are also provided about the preparation of supported mono- and "bimetallic" nanoparticles.

23.1 INTRODUCTION

From the pioneering Sabatier's works on hydrogenation of several organic and inorganic substrates on powdered metals, the tremendous developments of metal catalysts during the 20th century can be accounted for by the appearance of supported pure and "alloyed" metal nanoparticles. At the beginning the first aim when supporting metal was the stabilization of "small" particles (10–20 nm) against aggregation. Actually, it soon appeared that the catalytic properties, activity, selectivity, and even stability could be modified when modulating the size of metal particles below 10 nm, thanks to the occurrence in various proportions of sites with different reactivity, depending on the reaction. The first to recognize this concept was H.S. Taylor in 1925 [1], who said: "The amount of surface which is catalytically active is determined by the reaction catalyzed. There will be all extremes between the case in which all the atoms in the surface are active and that in which relatively few are so active" The most distinctive feature of a small particle is anisotropy. At the surface this is manifested by surface atoms with different values of their coordination number C_n. Chemical intuition dictates that the rate of a catalytic reaction should depend on anisotropy of structure [2]. The classical calculations from Boronin and Poltorak [3] and van Hardeveld and Hartog [4] showed the most drastic changes of proportion between facets, edges, corners, and microdefects at the surface occur between 1.0–1.5 and 5 nm. These changes of proportion of sites with different topologies tend to disappear for particles larger than 5 nm, making coarser crystals of little value for studying structural effects in catalysis. In parallel to these changes of the geometrical features of the surface, there are also strong modifications of the electronic properties of the metal atoms. Metal particles having less than 200 atoms (1.2–1.4 nm) exhibit electronic properties differing from those observed for the bulk metal and move from a band structure to the appearance of discrete levels. Moreover, the surface sites exhibit specific local density of states (LDOS) depending on their coordination.

A second breakthrough in catalysis by metals was the discovery of "alloyed" catalysts. Formally, alloyed particles should only be considered when a solid solution or a definite compound is formed. However, we wish to extend the terminology of "alloy", taking the proposition of Ponce and Bond [5], we will adopt this term to describe "... any metallic system containing two or more components, irrespective of their intimacy of mixing or the precise manner in which their atoms are disposed.... " The benefit brought by "alloying" two active metals, or an inactive with an active one, was recognized early, and systematic studies were carried out between the First and Second World Wars. From the first finding about stabilization of Pt gauzes in the NH_3 oxidation by adding Pd, some of the best achievements in catalysis by alloying metals can be found in the refining industry with the very classical reforming Pt-based catalysts promoted by Ir, Re, Ge, Sn, and selective hydrogenations in petrochemistry with promotion of Pd with Sn, Au, Cu, Pb, and so on.

At the beginning, the behavior of alloy catalysts was interpreted by both the electronic theory of catalysis and the rigid band theory (RBT) of alloys. The former stated that molecules are activated during the adsorption process by either releasing or accepting one electron, whereas the latter postulated that the number of d-band holes in one metal could be decreased by alloying this metal with another one, containing more valence electrons, due to the electron transfer. Actually, the interpretation of the alloying effect was found to be much more complex later, though this initial concept has remained stimulating ever since. There was thus an evolution from pure electronic considerations to geometrical ones. The milestone was an understanding that some reactions needed an ensemble of several neighbor surface atoms in a dedicated arrangement to proceed, and that dilution of the surface by any inactive component reduced the rate by disruption of this specific surface arrangement. For instance, Dautzenberg et al. [6] reported that the rate of C–C hydrogenolysis in alkanes on Pt is equally affected by addition of C, S, Sn, or Re, which should have different effects on the LDOS of surface Pt.

It appeared finally that the support can no longer be considered as a pure inert stabilizing partner. Actually, the support acts as a supramolecular ligand and has been claimed to promote specific electronic properties and/or geometrical features of the nano-sized supported metal particles. Any metal–support interaction (MSI) does occur in any case when small particles are deposited on a carrier. However, the extent of their interaction depends on the nature of the metal, but much more on the size of the particles and the nature of the support.

With insulator carriers like Al_2O_3 or SiO_2, the interaction is generally weak with the remarkable exception of metal in zeolite molecular sieves. The similar order of magnitude between the sizes of zeolite channels and cavities and the metal particles makes the latter particularly sensitive to induction of specific morphology and to the electrostatic field induced by the cations [7]. The MSI increases with semiconductors with the possibility of electronic modifications [8]. Moreover, some reducible metal oxides like TiO_2 and Nb_2O_5, initiate after activation at high temperature in H_2 the so-called strong metal–support interaction (SMSI) [9], which is discussed later. Finally, the interaction between metals and conducting carriers, like graphite, can lead to significant electronic modifications of the metal particles. We also have to consider that the support can be a partner to the metal particles in the chemical transformation of the reactants. Once occurring, the catalyst then operates as a bifunctional material.

It is moreover obvious that the size of metal particles should have a great influence on the extent of MSI. In general, properties of particles larger than 4–5 nm are influenced little by MSI, except in the remarkable case of SMSI. In contrast, particles on the nanometer size will be much more sensitive to this effect.

These three aspects in catalysis by metals enter in the general frame of "structure-activity" relationships. They have been the subject of reviews dealing with the (1) particle-size and plane-structure sensitivities [10], (2) ensemble-size sensitivity [11], and (3) metal–support interaction [9]. Depending on whether or not the turnover frequency (TOF), or rate per unit surface area or per accessible metal atom, is affected by the structure of the particle surface, the reactions have been called structure-sensitive or structure-insensitive [12]. The structure-activity relationships have been and still are of paramount importance for a better understanding of the physicochemical phenomena underlying the catalytic transformation. Ultimately, the goal of structure-activity relationships is to act as a tool for the design of novel or improved formulations for very selective catalysts.

In the following we illustrate the particle-size, alloying, and MSI effects with some selected examples of catalytic transformations. We shall first present very briefly some fundamentals about the so-called electronic and geometric models in catalysis by metal and how they help to understand the structure-activity relationships. These fundamental aspects are described in detail in other chapters of this book.

23.2 THE ELECTRONIC AND GEOMETRIC MODELS OF CATALYSIS BY METALS

As mentioned above, the structure-activity relationships are often explained on the basis of electronic and/or geometric effects. Indeed, with metals and alloys especially, it has also been proposed to talk more about ligand effect than electronic factors to designate the influence on a given site of nearest-neighbor sites [2,13]. Geometric and electronic influences cannot often be separated as independent parameters. For instance, decreasing the size of metallic particles results in an electron bandwidth decrease, the appearance of discrete levels, and a shift-up of valence band centroid. But the nature of the exposed planes and the topology of the surface sites change as well.

Moreover, we have to take care of a simple view of a geometric basis when explaining structure sensitivity. There is the possibility, under high-temperature conditions, or very reactive atmospheres, that surface mobility levels down the influence of geometric and structural features on small particles. In many cases the catalyst surface will be very fluid under reaction conditions, and adsorbate-induced surface reconstruction, fragmentation or agglomeration, and surface segregation may also create special active centers. In an interesting study by HRTEM of alumina-supported Au clusters (>1 nm), Ijima [14] has convincingly demonstrated that surface Au atoms exchange their locations at the minute level when heated, by the flux of electrons, at 500–600 K. Surface-induced reconstruction has been implicated in the oscillatory behavior of CO oxidation on Pt [15]. This concept of flexible surface [15,16], and the correlation between reactivity and restructuring ability, appear extremely puzzling, though fruitful; this is particularly true with reactant-induced segregation in catalysis by alloys.

Finally, the kinetics approximations cannot be forgotten. Usually, in order to put in evidence structure-activity relationships, a simple parameter, the TOF, is used. The TOF, which reflects the rate per accessible site, contains the combination of all the adsorption and surface reaction elementary steps. Each of these steps is dependent on adsorption and/or rate constants. For that reason, the significance of TOF dependence as a function of structural parameters, e.g., the particle size, is not obvious since the rate equation can be particle-size-dependent [17]. Moreover, the adsorption and surface reaction steps may exhibit very different sensitivities to electronic and geometrical features.

23.2.1 The Electronic Model

To be succinct, the key point in this model lies in the interaction between the d-band orbitals of the surface sites with the molecular orbitals of reactants and products. Historically, the premises of this concept were suggested by Sabatier [18], who said that the intermediate compound formed by one reactant at the surface must be stable enough to be formed but not too stable since it must decompose to yield the products. This can be correlated with the heat of adsorption of reactants and products, mainly governed by the electronic factors, which should be neither too strong nor too weak to give the optimum coverage for species competing at the surface, or for the products to desorb. The ammonia synthesis, competition between N_2, H_2, and NH_3, and the selective hydrogenation of unsaturated hydrocarbons, competition between H_2 and hydrocarbons, are good examples for that. This is illustrated for the hydrogenation of ethylene on metal wires (Figure 1); the rate is maximum for Rh, for which the heat of adsorption of ethylene provides the optimum coverage in C_2H_4 and H_2.

23.2.2 The Geometric Model

The first element of this model lies in the pioneering works from Kobozev [20] and Boronin and Poltorak [3]. They showed that some reactions need more than one surface atom to proceed. Moreover, a specific arrangement between these atoms fitting the molecular feature of the reactant will even be required to generate the active site [4]. This geometric model was more recently revisited by several authors [21–23] and named the "ensemble-size" model. The basic idea is that the rate is the function of the probability to find an ensemble of n free and neighbor atoms in a specific arrangement on which the adsorption of the reactant(s), and the further transformations, can occur. In the illustration of Figure 2, the hydrogenolysis of ethane needs several neighbor metal atoms to accommodate the multibonded HCC* reactive intermediate and the 5 H* abstracted, 12 atoms on Ni catalysts according to

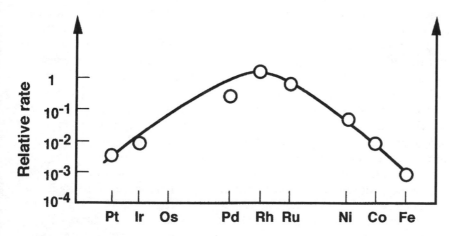

Figure 1   Relative rate of C_2H_4 hydrogenation on some metal wires. The heat of ethylene adsorption increases from the left to the right. (From Ref. 19.)

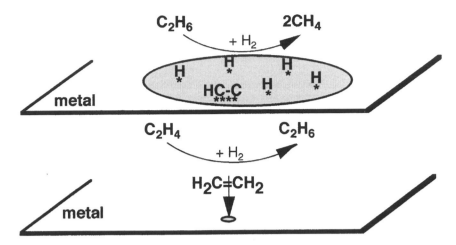

Figure 2 Schematic representation of the ensemble size required for the hydrogenolysis of C_2H_6 and the hydrogenation of C_2H_4.

Dalmon et al. [23]. In contrast, the reactive adsorption of ethylene could proceed on a single atom.

23.2.3 Particle-Size Effect

This topic was reviewed in detail by Che and Bennett [10]. In most of the supported metal catalysts, the size of the metal particles varies in the critical range between 1 to 10 nm. The first point addresses the question as to below which particle size the metallic properties are lost; "how small is a metal?" [24]. The answer is not so simple and depends on the properties we are looking for. Bulk properties of metal, like the melting temperature, are not reached below 10 nm [24].

Regarding the electronic properties, the critical size above which band structure appears is about 2 nm (several hundred atoms). At a lower size, the d-band becomes narrower with the appearance of discrete levels. XPS probes the binding energy of core electrons, and it has been well proved that small metal particles bind their core electron more tightly (Figure 3). This has been accepted by many as evidence of electron deficiency of small particles. However, the phenomenon of higher binding energies may be due to various effects other than electron deficiency. On the other hand, the occurrence of a possible charge transfer between the metal and the support may be accounted for by a change in the binding energies observed in XPS spectra. Finally, the electronic properties of sub-nanometer-sized clusters change nonmonotonously with the number of atoms [25].

Regarding the geometrical features, the topology of the surface suffers great changes until 5 nm [4], with evolution in proportion of the various facets, edges, corners, and defects. Moreover, the lattice parameter generally decreases (by a few percent) with cluster size due to surface stress [26]. However, the lattice may also be expanded if the cluster grows pseudo-morphically on the substrate lattice. These variations of the interatomic distances induce changes in the adsorption energies and the dissociation barrier. The shape of the cluster is also partially controlled by

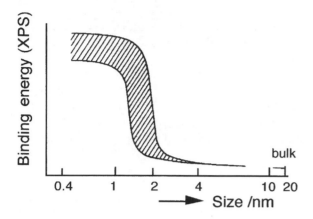

Figure 3   Binding energy of core levels of metal particles as a function of size (schematically).

cluster–support interfacial energy, which can change depending on the atmosphere composition.

A second very important point when discussing effects of particle size is the distribution in size of the clusters. The latter is generally wide in real catalysts due to the methods of preparation (*vide infra*). The observed rate is thus an average of the behavior of each entity of the population. Methods for the preparation of model supported catalysts, with very narrow particle-size distribution, are now developed and proceed through cluster vapor deposition [27]. Even in this case, the behavior of an individual particle is not always simple to interpret because the rate is not simply the average between the intrinsic rates on the different facets of the cluster. In fact, the facets are connected through edges that can accelerate the rate, or the reverse.

Taking into account all these points, depending on whether the reaction is more or less affected by electronic or geometric parameters, the zone where particle-size effect would occur will be different.

23.2.4   The "Alloying" Effect

For both technical and academic reasons the way by which the second component modifies the properties of pure metals is a subject of interest and is still not well elucidated. Depending on the reaction and the operating conditions considered, geometrical and/or electronic effects were claimed (see [11,28] for a review). Electronic effects were first claimed to operate due to the popularity of the RBT of alloys in which the d-band holes of one metal are filled by the valence electrons of a second metal. For instance, the valence shell electronic configuration of Pd in atomic form is of the $4d^{10}5s^0$ type; in metallic Pd, ≈ 0.36 electrons of the d-shell are in the s-band. This leaves an equivalent number of holes in the d-band. The filling of the 4d shell on alloying can therefore arise either by a flow of charge from s-band of Pd or by a charge transfer from a second component. Actually, it was recognized later that the extent of electron transfer from one component to the other upon alloying is very small, if any. From quantum chemical calculations on Pd_4Cu_6 clusters, Fernandez-Garcia et al. [29] proposed a redistribution of charge from 4d to 5s Pd levels without any transfer of charge from Cu.

Since the simple and attractive hypothesis of RBT in alloys has failed to stand the test of time, the geometric or "ensemble-size" model became very helpful to interpret the behavior of alloys. When applied to alloys between an active component A and an inactive one B, the ensemble-size model in its simplest form reflects the dilution in ensembles of smaller size of the active surface A by B; these smaller ensembles of A are less prone to activate the reactant(s). The immediate consequence of this phenomenon is a sharp decrease of the TOF. Obviously, the fall of TOF should be all the more so if preferential segregation to the surface of the inactive component B occurs. In the classical treatment of diluting an infinite surface A by B, the probability to find an ensemble of neighboring n A atoms is given by [5,30,31]

$$P = (1 - \theta_B)^n \qquad \theta_B : \text{surface coverage by B} \tag{1}$$

This ideal relationship is true when A and B are randomly distributed on an infinite surface. The situation becomes more complicated when the distribution of A and B is not random. If ΔH_f (enthalpy of alloy formation) is positive, components A and B tend to form separated phases; at variance when ΔH_f is negative, ordering of B into A prevails. Random distribution of A and B is favored when ΔH_f is slightly negative. Another limitation to the ideality represented by Eq. (1) is the finite size of metal particles in real catalysts [31]. Moreover, the surface of small metal particles is composed of sites with different topologies: facets, kinks, edges, corners, etc. The distribution of the different components of bimetallic particles can be ordered, and one component can preferentially occupy sites of a given topology [32,33]. This was demonstrated by Monte Carlo calculations for Pt alloyed with Cu, Ag, or Au [34] and by density functional theory calculations for Rh alloyed with Ge, Sn, or Pb [32].

Nevertheless, even if there is clear evidence that the extent of electron transfer is generally low upon alloying, the consequence of diluting the active metal into smaller ensembles by a second component is the appearance of discrete levels in the valence band and in many cases the rehybridization of the orbitals.

In any case, understanding the alloying effect would require the best possible knowledge for the following points: (1) what is the chemical state of the two components? (2) are the two elements intimately interacting in the same aggregates or not? (3) does surface segregation of one component occur? (4) are the two elements randomly distributed in the surface layer or not? The first and the second points mainly address the method used for preparing the catalysts and will be shortly presented. In order to avoid superimposition of particle-size and alloying effects, the preparation method through selective deposition of the second component on a parent monometallic catalyst should be preferred (Section 23.3).

23.2.5 The Metal–Support Interaction Effect

An electronic interaction does occur at the interface between a metal particle and a support, and it may physically be detected, e.g., by x-ray absorption or x-ray photoelectron spectroscopies [35–38]. However, in many cases it does not significantly alter the chemical and catalytic properties of the metal particles except for those metal atoms at the interface or directly contiguous to the support. If the Fermi level (E_F) of the metal and the support are different, an electron transfer does occur when the metal–support interface is created. The electrons will diffuse until the

E_F is equalized at the interface. But this electron transfer will be quite small. From solid-state physics, it has been estimated from calculations that no more than 0.003 electrons per metal atom will be transferred at the interface [39]. For cubic shaped particles of 216 atoms having 36 atoms in contact with the support, one can expect a transfer of 0.1 electrons per particle. It is obvious that topology of the particle should, however, affect the extent of electron transfer, and "raftlike" 2D particles, in which all the metal atoms are in contact with the support, would suffer higher electron transfer. DFT calculations of Pd atoms supported on Mg_nO_n model clusters ($n = 10, 13, 16$) have shown that no electron transfer between Pd and MgO occurs [40]. At variance, there is a strong interaction between Pd and O^{2-} with polarization of Pd orbitals and promotion of Pd 4d electrons to 5s and 5p levels. Calculations on Ir_4 clusters also led to the conclusion that metal particles of this size can be polarized by the electric field of Mg^{2+} [41]. It appears that there is no definite direct proof from physical techniques that a significant electron transfer occurs at the metal–support interface. However, even if a clear electron transfer does not occur, a strong modification of the metal d-band with rehybridization of the orbital may occur.

There are two aspects of geometrical features induced by MSI. One is concerned with the changes in the morphology of metal particles, the second with the ensemble-size effect on the active metal surface and arising upon initiation of the aforementioned SMSI. The MSI can induce some specific shapes of the metal particles, with changes in the proportion of exposed crystallographic planes and of the coordination number of surface metal atoms, and pseudo-morphous growth of cluster on the substrate lattice [27]. Raftlike structures may expose a higher proportion of highly coordinated atoms than a small polyhedron with the same number of surface atoms. In addition, more atoms are close to the support, leading to possible alterations of the electronic states.

A very interesting phenomenon is the SMSI [42] reviewed by Haller and Resasco [9], which may occur when metal particles are supported on some reducible supports, e.g., TiO_2. The symptom is a strong decrease of H_2 chemisorption upon reduction of the sample without any significant sintering of the metal particles. This is exemplified in Figure 4 for a series of Group VIII metals supported on TiO_2. However, even if some electronic modifications of the metal in the SMSI state cannot be neglected, the main feature of SMSI is the decoration of the metal particles by migrating TiO_x species (Figure 5).

Finally, one cannot forget that with supported metal catalysts, besides any other modifications of the cluster properties by the support, this latter may also play a role in the catalytic transformation through its own active sites. In this situation, we could speak about metal–support cooperation, since both active sites of the support and the metal cluster operate in a synergistic fashion. Some examples will be described by the end of this chapter in bifunctional catalysis and spillover of adsorbed species.

23.3 PREPARATION OF SUPPORTED METAL CATALYSTS

There are two main steps in the preparation of real supported metal catalysts through chemical processes. The first consists of deposing the active component precursors in a divided form on the support, whereas the second consists of transforming these precursors into the required active phase. It is worth noting that

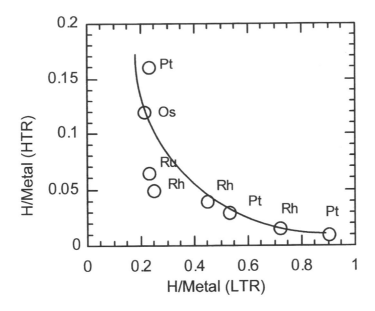

Figure 4 Hydrogen uptake (H/metal) after reduction at 773 K (HTR) over a series of metals supported on TiO$_2$ as a function of H/metal after reduction at 473 K (LTR). (From Ref. 9.)

industrially a large majority of deposition methods involve aqueous solutions and liquid–solid interface. The activation of the deposited precursor(s) will obviously have a clear impact on the chemical nature and the mutual interaction between the metal and the support, and between the two components in alloy catalysts. One can anticipate that treatments in a mild reducing medium at a low temperature will maintain the ionic states to some extent, whereas reduction at a high temperature in pure hydrogen will promote stronger MSI and/or pure alloy formation. The metal precursors generally used are inorganic or organic transition-metal complexes (TMC), e.g., Pt(NH$_3$)$_4$Cl$_2$, PtCl$_6$H$_2$, or Pt acetylacetonate [Pt(acac)$_2$]. The method of preparation will be based on the different kinds of interaction between TMC and the support, through the concept of the interfacial coordination chemistry (ICC) [43]. It is generally accepted that the stronger the interaction in supported system, the smaller the metal particles size generated after the thermal reduction of the

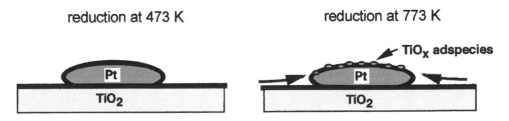

Figure 5 Schematic representation of Pt/TiO$_2$ reduced at 473 K (left) and at 773 K (right, SMSI state).

precursor. This sentence should, however, be mitigated since a strong interaction between TMC and support needs a higher reduction temperature, which results in a faster surface diffusion of the atoms obtained. This should eventually lead to larger particles. Moreover, high-temperature reduction of TMC on reducible supports (TiO_2, Nb_2O_5) can lead to the SMSI state. The deposition of TMC on support can be simply categorized within two methods: (1) impregnation and (2) ion or ligand exchange [44,45].

In the course of impregnation, the support is contacted with a solution of TMC, and the excess solution is then evaporated. The weakness of the interaction in this deposition procedure usually leads after activation to metal particles with sizes larger than 4–5 nm.

The ion or ligand exchange method results from an electrostatic or chemical interaction between TMC and the support. The support surface plays the role of supramolecular counter-ion or ligand. The resulting interaction is generally strong and leads to nano-sized (1–2-nm) particles after activation.

For the preparation of supported M1M2 alloy catalysts, the deposition method of the two precursors of components M1 and M2 can be tentatively categorized within three general methods:

1. The nonselective deposition (NSD) in which there is no interaction between the two precursors during the deposition process
2. The selective deposition (SD) in which the precursor of "co-metal" M2 is selectively deposited onto the metallic particles of the other component M1
3. The deposition of heterobinuclear organometallic complexes, or inorganic salts, of M1 and M2

The NSD of precursors mainly applies when M1 and M2 are transition metals. The basic principle of NSD of precursors, together or successively, is such that the interaction of both precursors with the solid surface was stronger than between the two precursors. On that account the two deposited precursors are separated on the support, and surface diffusion will be necessary to yield the "bimetallic" aggregates during the activation process. NSD will use the same deposition methods as for monometallic catalysts (*vide supra*).

The SD of M2 precursor covers methods of wide applicability ranging from the deposition of clear nonmetal, e.g., Si or Se, to early transition metals, e.g. Cr, on supported M1 particles. In contrast to NSD, the SD aims at creating an interaction between M1 and precursor of M2 much stronger than that of M2 precursor with the support. This is generally achieved through a highly selective interaction between the M2 precursor, inorganic salt, or organometallic complex, with a prereduced supported M1 catalyst. The SD methods can be categorized within two types of approaches. The first group of methods, reviewed by Barbier [46], involves inorganic salts in aqueous solution and are based on the principle of "redox" reactions. They are mainly applied to transition metals. Methods of the second type, occurring in gas phase or in a polar organic solvent, involve organometallic complexes and are based on surface chemical reactions [47]. The great interest of these methods when studying the alloying effect in model catalysts is to start from a monometallic parent catalyst of a definite metal particle size of M1 and to retain a very similar size of M1M2 alloys particles after doping with various components M2.

Elements for the deposition of heterobinuclear complexes for preparing bimetallic catalysts have been reported by Ichikawa [48]. The use of heterobinuclear complexes as precursors for bimetallic catalysts was aimed at the fine-tuned control of a homogeneous composition in the nuclearity of bimetallic particles. As compared to catalysts prepared by NSD methods, mixed-metal cluster-derived catalysts should retain more bimetallic ensembles. However, the use of heterobinuclear complexes is limited by the availability of high-nuclearity clusters and by the narrow choice of metallic couples. The different types of interaction of these complexes with the support are the same as previously described for the NSD methods.

23.4   A PARTICLE-SIZE EFFECT: THE EPOXIDATION OF ETHYLENE

The epoxidation of ethylene on silver is an important industrial process in which the practical yield is lower than 20%, with a selectivity of about 80%. It is remarkable that only Ag selectively catalyzes this reaction. The mechanism of epoxidation originally proposed by Twigg [49] for this reaction can be written as

$$2Ag + O_2 \rightarrow Ag_2O_2$$
$$Ag_2O_2 + C_2H_4 \rightarrow C_2H_4O + Ag_2O$$
$$4Ag_2O + C_2H_4 \rightarrow 2CO + 2H_2O + Ag$$

This scheme seems consistent with the experimental results since it predicts a maximum selectivity of 85.7% and has found theoretical support. Recent ab initio calculations using a cluster approach predict that ethylene can react with the superoxide ion or the atomically adsorbed oxygen [50,51]. However, while the total oxidation involving the superoxide is forbidden by the large energy barrier, the attack of adsorbed oxygen is nonselective and yields CO_2 [52]. It was suggested later that the reaction may occur by a different mechanism [53]. Grant and Lambert [54] showed that ethylene oxide could be formed as a result of temperature programmed reaction of ethylene with a silver surface preliminarily treated with the reaction mixture. According to them, evacuation (UHV conditions) at $T > 300\,K$ after the pretreatments excluded the existence of a weakly bonded molecular oxygen before subsequent adsorption of ethylene. The description of oxygen adsorption, and the natures of adsorbed species, is therefore a central problem.

The adsorption of oxygen on Ag, Au, Pt, Pd, and Rh usually follows the scheme

$$O_2 + e^- \rightarrow O_2^- \qquad O_2^- + e^- \rightarrow 2O^- \qquad O^- + e^- \rightarrow O^{2-}$$

with metal remaining in the zerovalent state or being oxidized only at the surface layer [55]. The most stable form at intermediate temperatures is dissociated O^{2-}, as shown by the isotopic exchange reaction: $^{16}O_2 + {}^{18}O_2 \rightarrow 2\,^{16}O^{18}O$, which is catalyzed by most metals [56]. This reaction has been found structure-sensitive with large differences of activity between the different group VIII metals [57].

In a study on ethylene epoxidation on Ag catalysts, Kilty and Sachtler [58] discussed the various forms of adsorbed oxygen in order to establish the presence of diatomic oxygen at the surface. Two sets of results supported this proposal:

1. SIMS analysis of the ions emitted by the surface has shown the presence of Ag_2O^-, Ag_2O^2, O_2^-, and O_2^+ ions [59,60].
2. IR spectroscopy of the adsorption of ethylene onto preadsorbed oxygen on Ag provided evidence of the formation of a peroxo-complex [61]. This finding was later confirmed by isotopic labeling, giving unequivocal evidence for the formation of a peroxidic compound $Ag-O-O-C_2H_4$ [62]. A careful study of the kinetics of oxygen chemisorption further suggested that the initial fast adsorption at low coverage was dissociative, whereas the further adsorption was nondissociative. The dissociative adsorption of oxygen was represented as the reaction

$$O_2 + 4Ag \rightarrow 4Ag^+ + 2O^{2-}(ads.)$$

requiring an ensemble of four adjacent Ag atoms, while the nondissociative adsorption was written as

$$O_2 + Ag \rightarrow Ag^+ + O_2^-(ads.)$$

with a single silver atom. Therefore, this model might suggest a dependence of the selectivity of epoxidation depending on the number of quartets of Ag atoms at the surface (ensemble-size effect). The addition of chlorine, a usual promotor of epoxidation, decreases the number of these ensembles and increases the selectivity for epoxidation, supposedly, by an ensemble-size effect.

An investigation about the nature of Ag–O bonding for different oxygen states on Ag(111) and polycrystalline foils by XPS and UPS has been recently reported [63]. Besides the adsorbed atomic oxygen with ionic Ag–O bonding, a quasi-molecular adsorbed oxygen was proposed with "covalent" Ag–O bonding, the latter being more thermostable. With the help of quantum chemical calculations on Ag_{12} and $Ag_{12}O_3$ clusters, a more detailed picture of this quasi-molecular species was proposed [64]. It resembles the C_{2v} structure $-Ag-O-O_{ep}-O-Ag-$ ("metal ozonide" species, O_{ep} for epoxidizing oxygen) with O_{ep} interacting with a surface defect (cation vacancy).

An alternative description of the adsorbed oxygen states based on different ionicity of the Ag–O bond has also been proposed: one "ionic" or nucleophilic oxygen [$E_b(O1s) = 528.4\,eV$] associated to Ag^+ ions as sites for ethylene adsorption, and a "covalent" or electrophilic species [$E_b(O1s) = 530.5\,eV$] [65]; both species are of atomic nature [66]. The covalent species exists only after activation by the reaction mixture [65], which suggests a role of subsurface oxygen. It was further established that the relative populations of the nucleophilic and electrophilic oxygen species were strongly dependent on silver particle size, the nucleophilic oxygen being only observed for clusters larger than 30 nm [67].

This structure sensitivity of the respective population between the various states of adsorbed oxygen could explain why it has long been found that ethylene oxidation is structure-sensitive, with higher selectivity, up to 60–70%, for nonpromoted catalysts with sizes of about 40–50 nm [68,69]. A similar result has been recently reported for model Ag/α-Al_2O_3 catalysts of very narrow size distribution (Table 1) [70]. The particle-size sensitivity has been attributed to a change of the equilibrium Ag crystal shape with cluster size (geometric effect) [71]. This leads to a change in the relative concentration of facets and edges at the metal

Table 1 Intrinsic Rate of Ethylene Epoxidation and Selectivity Toward Ethylene Oxide as a Function of Mean Silver Particle Size for Unpromoted Ag/Al$_2$O$_3$ Supported Catalysts

Particle Size (nm)	Rate (molecules m^2 s^{-1}) ÷ 10^{-17}	Selectivity to Ethylene Oxide (mol%)
16	0.18	7
22	0.21	8
40	2.0	22
56	6.3	47
100	4.8	41

Source: Ref. 70.

surface, which affects adsorption of C$_2$H$_4$ and O$_2$, and thus the reaction rate; the active sites are those responsible for the ethylene π-adsorption.

Cs has a larger positive effect on small particles and was proposed to stabilize the active sites at the Ag surface [71]. An alternative explanation was put forward for the promoting effect of Cs by Epling et al. [72]. They proposed that Cs coats the support material during the preparation and acts as a binder between the Ag and the support, resulting in a larger Ag coverage of the α-alumina. The total oxidations of ethylene and ethylene oxide to CO$_2$ occur primarily on the alumina surface. The enhancement in selectivity toward ethylene epoxidation may then result from the fact that Cs addition favors the formation of a thin Ag film covering the alumina [73].

In conclusion, it should be mentioned that notwithstanding the vast amount of publications on oxygen chemisorption and ethylene epoxidation on Ag, the mechanisms of these important processes are still not fully understood.

23.5 A PARTICLE-SIZE EFFECT: THE HYDROGENATION OF UNSATURATED HYDROCARBONS ON Pd-BASED CATALYSTS

The hydrogenation of unsaturated hydrocarbons continues to attract attention by reason both of the practical importance and of the theoretical interest of this system. This is particularly true for the selective hydrogenation of alkadienes and alkynes in alkenes rich cuts [74]. Fundamental studies [75] have shown that Pd is the most active and selective metal for these reactions. Some reasons for the better performances exhibited by Pd-based catalysts can be learned from the mechanism and the surface complexes involved in these transformations.

Taking the hydrogenation of ethyne as a case study, it is well established now that the interaction of ethyne and/or ethene with a metal surface leads to a wide variety of adsorbed species, the proportion of which depends on the hydrocarbon pressure, the temperature, and the nature of the metal [76]. Earlier identifications of such species—in particular, the ethylidyne $\equiv$C–CH$_3$—have been reported first on Pt(111) single crystals [77,78], then on Pd/Al$_2$O$_3$ [79]. With all available spectroscopic and kinetic data at various conditions, a nearly complete reaction scheme for the hydrogenation of ethyne on Pd(111) has recently been proposed [80] and illustrated in Figure 6. The choice between the two routes through ethylidyne or π-bonded adsorbed species obviously depends on the nature of the metal, but also on some

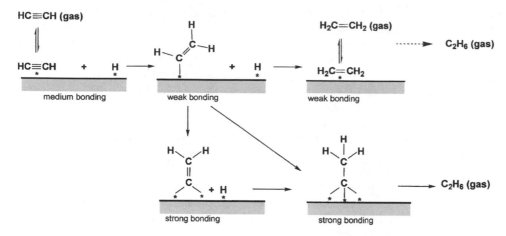

Figure 6 A proposed reaction scheme for the hydrogenation of ethyne on Pd and Pt surfaces.

specific geometric requirements between π-bonded and multibonded species to be accommodated by the surface. The choice between these two pathways is called the *mechanistically* or *kinetically determined selectivity* [5].

With Pd, and with Pt to a lesser extent, the pathway through the ethylidyne species, which leads to ethane directly, makes a negligible contribution. As a consequence, the selectivity to ethene remains high as long as ethyne coverage stays large enough to reject ethene in the gas phase. Ethyne displaces the weakly bonded ethene and keeps the selectivity to ethene high; this is called *thermodynamic determined selectivity* [5]. The thermodynamic selectivity is mainly dictated by the ratio of adsorption constants, which are dominated by electronic factors.

The dependence on Pd particle size of the hydrogenations of alkenes, alkadienes, and alkynes has been studied in the gas and liquid phases and for pure reactants or simulated real feedstocks. This is illustrated by some typical examples later.

The liquid-phase hydrogenation of 1-butene, 1,3-butadiene, and 1-butyne under 2 MPa H_2 pressure has been reported on Pd supported on SiO_2, α- and γ-Al_2O_3 catalysts of widely varying dispersion [81]. Regardless of the support, the TOF for 1-butene hydrogenation has been found constant regardless of the particle size. In contrast, the TOF for 1,3-butadiene and 1-butyne hydrogenations remained constant for particles larger than 3 nm but decreased noticeably at higher Pd dispersion. Moreover, the selectivity to the unwanted deep hydrogenation to ethane increased on large Pd particles.

The TOF also decreased in the gas-phase hydrogenation of but-1-yne-3-ene (vinylacetylene) at high Pd dispersion regardless of the nature of the support, Al_2O_3, SiO_2, or carbon [82,83]. In the gas phase and large excess of ethene and H_2 pressure, the hydrogenation of ethyne was also found particle-size-dependent with a decrease of TOF upon increasing Pd dispersion [84]. Deep hydrogenation to ethane is lower on smaller particles. Otherwise, kinetics has shown that the reaction order with respect to ethyne is zero on large Pd particles but -0.5 on the smallest. Very similar behaviors were reported for the gas-phase hydrogenation of 1-butyne [85] and of but-1-ene-2-methyl-3-yne or 2-methyl-1,3-butadiene [86]. The TOF for the hydrogenation of 1,3-cyclooctadiene to cyclooctene decreased also as the size of Pd in pumice-supported catalysts decreased [87].

On model-supported Pd catalysts prepared from in-situ chemical vapor deposition (CVD) on carbon and graphite [88] or on SiO_2 [89], it was also put into evidence a decrease of TOF in the hydrogenation of 1,3-butadiene hydrogenation when the Pd particle sizes decreased.

From all the studies reported above, two main facts come out: (1) the low particle-size sensitivity of the alkenes hydrogenation, and (2) in contrast, the very clear structure sensitivity of alkadienes and alkynes hydrogenations. The latter finds expression in a decrease of TOF with a concurrent higher selectivity for the half-hydrogenation when the size of Pd particles decreases. Several proposals were put forward to explain this structure sensitivity. Regarding activity, the ability of Pd to form a hydride phase could explain to some extent catalytic properties [75]. The presence of β-PdH phase in the catalyst considerably increases the acetylene hydrogenation rate to ethane [75,90], and electronic effects mainly accounted for the interpretation of this behavior. The β-PdH phase tends to disappear as the size of Pd particles decreases to the nanometer scale [91]. The electronic properties of Pd are of utmost importance in these reactions, as shown by the correlation between the increase of E_b Pd 3d levels, the decrease of particle size, and the TOF for hydrogenation [82]. This decrease of E_b has been generally interpreted as an "electron-deficient" character of the metal particles. However, the real significance of E_b with respect to particle size has to be taken with care due to the interference of support effects [87,92]. There is a general agreement, with the exception of Pd on Na,K-promoted pumice [87], that the decrease of TOF is correlated to a strong chemisorption of alkynes or alkadienes. This is reflected by the order of -0.5 on small particles and of zero on large particles in the hydrogenation of ethyne [84]. This effect results in self-inhibition of the catalyst. The too-strong "complexation" of Pd by vinylacetylene can even result in Pd dissolution from the catalyst support [74]. From calorimetric measurements the differential heat of adsorption of propyne was found to be more than 50% higher for the highly dispersed Pd/Al_2O_3 catalyst than for the catalyst of low dispersion, at coverages ranging from 0.2 to 0.7 [93].

Regarding selectivity, the large Pd particles and the presence of β-PdH phase were claimed to promote the deep hydrogenation of ethyne to ethane [90,92]. Moses et al. [94] proposed that β-PdH favors the direct hydrogenation of the adsorbed ethylidyne species to ethane (Figure 6). From a geometrical point of view, one can indeed anticipate that the formation of these multibonded adsorbed species are easier on the facets of large particles. This loss of selectivity will be kinetically determined; in contrast, the hydrogenation on small particles will follow quasi-exclusively the π-adsorbed species (Figure 6) and the selectivity will be thermodynamically determined (*vide supra*).

23.6   AN ALLOYING EFFECT: THE HYDROGENOLYSIS OF C–C BONDS IN ALKANES OVER Ru BIMETALLIC CATALYSTS

Multimetallic formulations are currently in use to improve the properties of metal catalysts, in hydrocarbons processing, Fischer–Tropsch synthesis, and so on [28]. The hydrogenolysis of alkanes is an unwanted reaction in naphta reforming, and has

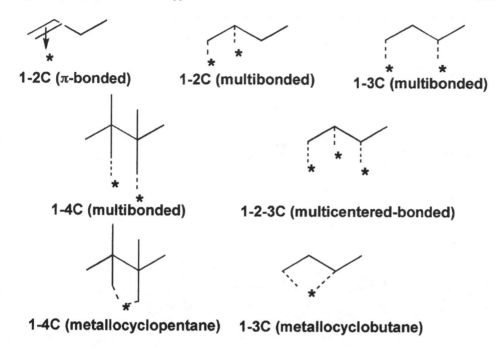

1-2C (π-bonded)   **1-2C (multibonded)**   **1-3C (multibonded)**

1-4C (multibonded)   **1-2-3C (multicentered-bonded)**

1-4C (metallocyclopentane)   **1-3C (metallocyclobutane)**

Figure 7 Some possible adsorbed species in the case of substituted butanes.

thus been from the last decades one of the most documented reactions for both practical and fundamental reasons. C–C bonds breaking has long been considered as the archetype of a structure-sensitive reaction in both alkanes and cycloalkanes series. Depending on the alkane the C–C bond hydrogenolysis can occur via a great variety of surface complexes [95–97], which accommodate different degrees of bonding with the surface. Various adsorbed complexes have thus been proposed for C–C hydrogenolysis, 1–2C, 1–3C, 1–4C and 1–5C; these species are illustrated in Figure 7 for different alkanes. The probability for each species to occur depends on (1) the nature of the alkane, (2) the nature of the metal and its dispersion, and (3) the number of free and neighbor surface metal atoms. However, some general tendencies can be drawn. The 1–2C π-bonded and the 1–3C metallocyclobutane species can occur on a small active center (a single atom?). In contrast, the 1–2C, 1–3C, and 1–4C multibonded species are favored on large ensembles of atoms. It has been pointed out that metals with high hydrogenolytic activity also show a high propensity to form metal–carbon multiple bonds [98]. This was later fully confirmed by works from Ponec's group, which also established that small particles are more reluctant to form multiple bonds than the large ones [99]. The propensity of metals to form multiple bonds with carbon atoms follows the order

Pt, Pd ≪ Rh, Ir < Ni, Co < Ru

One can speculate that (1) a certain electronic structure of metal favors the multiple bonds formation, and (2) the active center can actually be considered as an

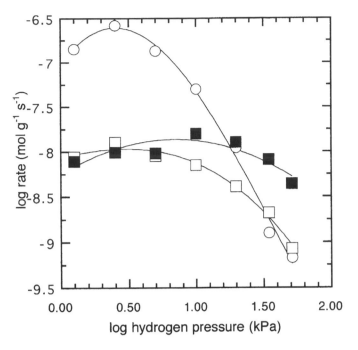

Figure 8   Hydrogen pressure dependence for the rate of 1–3C (full symbols) and 1–4C (open symbols) routes in TeMB hydrogenolysis at 463 K on Ru/Al$_2$O$_3$ of large particles (○, particle size = 4 nm) and small particles (□,■,particle size = 1 nm).

ensemble of atoms. To demonstrate the ensemble-size effect in the hydrogenolysis of alkanes, bimetallic Ru-based model catalysts have been therefore chosen many times since Sinfelt's pioneering work on the hydrogenolysis of ethane on RuCu/SiO$_2$ [100].

The kinetics of alkanes hydrogenolysis on metal obeys a Langmuir–Hinshelwood mechanism with adsorption equilibria for H$_2$ dissociation and alkane dehydrogenation leading to the C$_n$H$_{2n-x}$ active surface species; the latter then suffering C–C bond rupture by addition of one H atom, this is usually considered the rate-determining step [101]. The kinetic study of 2,2,3,3-tetramethylbutane (TeMB) hydrogenolysis has put in evidence structure sensitivity, even subtle, in mono- and bimetallic alumina-supported Ru catalysts [102]. Figure 8 shows the rate dependence of the routes via the 1–3C and 1–4C complexes as a function of H$_2$ pressure on large and small Ru particles. The kinetic analysis of these plots allowed us to derive a rate expression that showed that (1) on small particles ($\approx$ 1 nm) three H atoms were abstracted from TeMB to form the 1–4C complex before C–C bond-breaking occurred, but only two H atoms to form the 1–3C complex; (2) on large Ru particles (4–5 nm) five H atoms were removed from TeMB to form the 1–4C complex. The surface complexes shown in Figure 9 have thus been proposed for TeMB hydrogenolysis on Ru/Al$_2$O$_3$ catalysts [102]. This is a typical illustration of an ensemble-size effect where the accomodation of the surface complex requires an

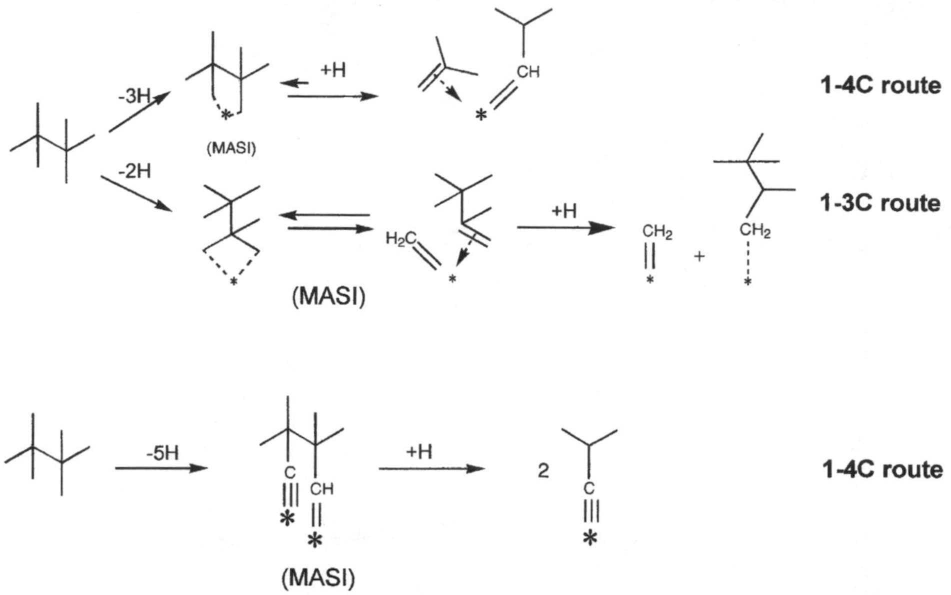

Figure 9 1–3C and 1–4C routes in the hydrogenolysis of TeMB on large and small Ru particles.

increased number of free neighboring Ru atoms when the surface complex becomes more and more dehydrogenated. The kinetics of TeMB hydrogenolysis on a series of Ru/Al_2O_3 modified by Sn or Ge, by selective deposition (*vide supra*), has been studied [102]. The bimetallic catalysts were prepared from the same parent Ru/Al_2O_3 (particle size ≈ 1 nm) to give $RuSn/Al_2O_3$ and $RuGe/Al_2O_3$ (particle size ≈ 1 nm). Figure 10 shows the rate dependence as a function of H_2 pressure for the 1–3C and 1–4C routes in TeMB hydrogenolysis on these catalysts. It has to be noticed that while selectivity changes significantly with alloying, TOF was only marginally affected since the addition of Sn did not change the specific activity, and the addition of Ge decreased it by a factor of 2–3 only. The effect was stronger on the 1–4C route when Ge was added, but the reverse was true for Sn. It was concluded that hydrogenolysis of TeMB through 1–3C or 1–4C routes occurs on specific Ru sites and that a different location of Ge and Sn atoms exists at the surface. The 1–4C route takes place on the sites of highest coordination of a given particle (dense plane), whereas the 1–3C route takes place on low-coordination sites (e.g., kinks, edges, corners). As a result, the preferential location of Sn at defect sites, corners, and edges in RuSn nanoparticles was postulated (Figure 11), as it was earlier proposed for RhSn nanoparticles [32]. The same approach, developed for RuGe nanoparticles, showed that Ge does not have such a definite site preference and is distributed in a more random fashion in the surface layer.

A similar interpretation was proposed for the hydrogenolysis of *n*-butane on the same bimetallic Ru catalysts [103]: a loss of activity and changes of selectivity

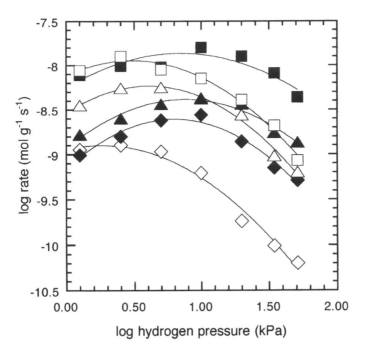

Figure 10 Hydrogen pressure dependence for the rate of 1–3C (full symbols) and 1–4C (open symbols) routes in TeMB hydrogeolysis at 463 K on small metallic particles of Ru/Al$_2$O$_3$ ($\square$,$\blacksquare$), RuSn/Al$_2$O$_3$ ($\blacktriangle$,$\triangle$), and RuGe/Al$_2$O$_3$($\blacklozenge$,$\lozenge$).

(more ethane) due to dilution by Ge species of Ru ensembles and Sn located at low-coordination sites. This behavior is consistent with the EXAFS experiments [104] and quantum chemical calculations on bimetallic RuSn and RuGe model clusters [105,106].

Alterations of catalytic properties in *n*-butane, ethane, and dimethylpropane hydrogenolysis have also been interpreted by changes in the topology of the Ru surface in Ru alloyed with Cu, Au, Ag, and Pb [107–112]. From these studies, clear achievements appear: (1) with the exception of Ge and Si, the modifier (Cu, Ag, Pb) is located at defect sites of small particles, and (2) the decrease of activity upon alloying becomes larger when short alkanes are processed due to an increase of the ensemble size. In particular, from kinetics of ethane hydrogenolysis on RuCu/SiO$_2$ it was concluded that the active center was composed of 12 neighbor Ru atoms [112]. Moreover, it was also proposed that blocking of Ru defect sites by Cu alters the H$_2$ chemisorption, because these sites could be the "portal sites" for the rapid dissociative adsorption of highly mobile surface hydrogen [113]. The incidence of alloying Ru upon hydrogen adsorption, strength, mobility, location, and so on has long been considered with respect to geometric [114,115] and electronic [116] features and its consequenes on alkanes hydrogenolysis [103].

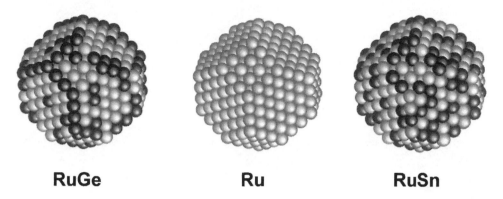

RuGe   **Ru**   **RuSn**

Figure 11 Representation of the surface distribution of Ru, Sn and Ge on Ru, RuSn, and RuGe model clusters (Sn, Ge/Ru = 0.2, at/at) with a perfect cubooctahedral shape (particle size ≈ 3.0 nm).

23.7 A METAL–SUPPORT INTERACTION EFFECT: THE HYDROGENATION OF α,β-UNSATURATED CARBONYL COMPOUNDS ON SUPPORTED Pt CATALYSTS

The switch from homogeneous to heterogeneous processes has long been achieved in the manufacture of commodities, but still remains at stake for fine chemicals production. In this field, the hydrogenation of C=O bonds constitutes a very important class of reactions, especially the hydrogenation of those compounds containing α,β-unsaturated carbonyls; the target compound is usually the unsaturated alcohol. This is illustrated in the hydrogenation of cinnamaldehyde (CAL) (Figure 12). In this formal reaction scheme, the selectivity to cinnamyl alcohol (COL) is determined by the initial attack of CAL at the C=C or C=O

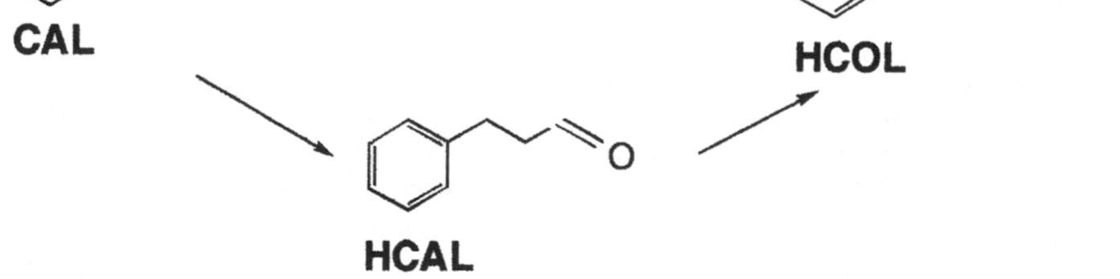

Figure 12 Formal reaction scheme for the hydrogenation of cinnamaldehyde.

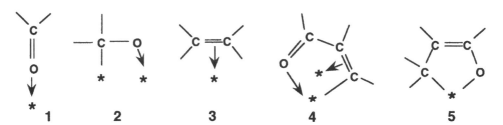

Figure 13   Some surface complexes formed in the interaction of α,β-unsaturated carbonyls with metals.

double bonds, and the desorption of COL or its further hydrogenation to hydrocinnamyl alcohol (HCOL).

Figure 13 shows some complexes formed in the interaction of unsaturated C=C–C=O systems with metal surfaces [117]. To reach a high selectivity in the selective hydrogenation of C=O bonds, it is necessary to promote the η_2(C–O) and/or the η_4-species (Figure 13, species 2 and 5). The latter was claimed to be a common intermediate to both the unsaturated alcohol and saturated aldehyde [118,119]. The selective hydrogenation to unsaturated alcohol through the η_4-species was proved in deuteriation of acrolein on Pt/TiO$_2$ with formation of HDC=CHCH$_2$OD [120]. Unfortunately, the normal behavior of the metals is to interact preferentially with the C=C bonds (Figure 13, species 3) to yield the saturated carbonyl. To reach the challenging goal of C=O hydrogenation, the MSI is of great concern, and some aspects have been recently reviewed by Gallezot et al. [121] and Ponec [117]. Some examples of MSI effects on the hydrogenation of α,β-unsaturated aldehydes are presented in Table 2. For a sake of clarity in the comparison, the selectivity to the target compounds, the unsaturated alcohols, is only given.

From Table 2, it clearly appears that, e.g., Pt/TiO$_2$,Nb$_2$O$_5$,CeO$_2$ catalysts in an SMSI-like state are much more prone to hydrogenate the C=O bonds than most of the others. The promoting action is seen as a strong interaction between the carbonyl group and a positively charged center of TiO$_x$, Nb$_2$O$_x$, Ce^{n+} species (Figure 14). The interaction of unsaturated carbonyls through intermediate η_2(C–O), shown in this scheme, or η_4-species will be impeded in the case of unsaturated ketones by the higher steric hindrance around the carbonyl group; this would explain the large difference in promoter effect for hydrogenating aldehydes or ketones [117,127]. The specific activation of the C=O bond by TiO$_x$ adspecies on metals has been put in evidence by Williams et al. [128] in the hydrogenation of CO to CH$_4$ on Rh foil coated with increasing coverage of TiO$_2$. The rate went to a maximum value for a fractional coverage of 0.5.

This promotion of C=O activation in these systems cannot be accounted neither for a geometric effect nor for an electronic effect in an SMSI-like state. Actually, this is better explained by the creation of new catalytic sites, mixed sites, very active for C=O hydrogenation (Figure 14). However, electronic effects were also considered as operating at a second level, in Pt/ZnO catalysts, for instance [122]. Upon high-temperature reduction, Pt$^{\delta-}$-Zn$^{\delta+}$ entities form; an electron enrichment of Pt would occur, which repels the C=C bond farther from the Pt surface.

Table 2 Selectivity to Unsaturated Alcohols in the Hydrogenation of α,β-Unsaturated Aldehydes on Metal Catalysts

Reactant	Selectivity to Unsaturated Alcohol	Ref.
Acrolein[a]	Pt/SiO$_2$: 0%; Pt/TiO$_2$ reduced at 773 K (SMSI): 40%	120
Crotonaldehyde[a]	Pt/SiO$_2$: 0%; Pt/TiO$_2$ reduced at 773 K (SMSI): 37%	119
Crotonaldehyde[a]	Pt/ZnO reduced at 473 K: 21%; but reduced at 673 K: 49%	122
Crotonaldehyde[a]	Pt/CeO$_2$ reduced at 473 K: 7.5%, but reduced at 973 K: 83%	123
Cinnamaldehyde[b]	Pt/carbon: 39%; Pt/TiO$_2$ reduced at 773 K (SMSI): 87%	124
Cinnamaldehyde[b]	Pt/SiO$_2$: 0%; Pt/SiO$_2$ + K: 70%	125
Cinnamaldehyde[b]	Pt/charcoal: 0%; Pt/graphite: 72%	121
Cinnamaldehyde[b]	Pt/SiO$_2$-C$_{60}$: 89%	126

[a] Gas phase.
[b] Liquid phase.

The higher selectivity to COL found in CAL hydrogenation on graphite-supported Pt, with respect to those supported on charcoal, cannot be interpreted in the same terms, i.e., the occurrence of mixed sites. It was proposed that graphite acts as an "electron-donating" macroligand for the nano-sized metal particles [129]. The adsorption of CAL through the C=C bond will be thus weakened and the selectivity to COL will increase. Physical proofs for a strong electronic modification of small Pt clusters (≈ 1.3 nm) on graphite were provided by the radial electron distribution (RED) obtained from X-ray diffraction spectroscopy [129]. The lattice expansion of Pt in Pt/graphite, indicated by the RED, has been attributed to an "electron transfer" from graphite to metal, which would increase the population of the antibonding levels and thus decrease the cohesive energy of the lattice. In this case the intervention of electronic effect can be invoked to explain the better performance of Pt/graphite due to the MSI effect.

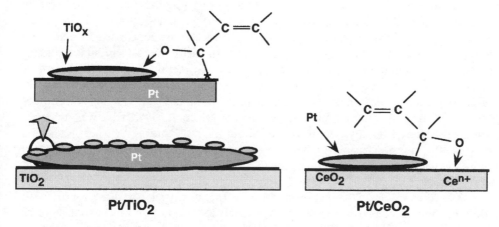

Figure 14 Schematic representation of the activation of the C=O bond in α,β-unsaturated carbonyls on Pt/TiO$_2$ (SMSI state) and Pt/CeO$_2$ (reduced at high temperature).

A silica chemically grafted with [60] fullerene, SiO_2–C_{60}, was fabricated and Pt particles were supported on the two-dimentional layer of grafted [60] fullerene [126]. This material, tested in the hydrogenation of CAL, exhibited fairly high selectivity to COL, 89% at 80% CAL conversion. This behavior was assigned to a joint adsorption of CAL on both [60] fullerene and Pt particles at their borderline. In this picture, the phenyl ring will be attracted by [60] fullerene favoring thus an orientation of the α,β-unsaturated system for an adsorption by the C=O on the Pt particles.

23.8 METAL–SUPPORT COOPERATION

Besides the interaction between metal particles and the support, and without excluding it, a cooperation between them for the catalytic transformation can exist. This concerns (1) the occurrence of pre-existing active sites on both metal and support, which operate in synergy (bifunctional catalysis), and (2) the creation of active sites on the support by spillover of adsorbed species activated at the metal sites.

When talking about bifunctional catalysis, one thinks immediately of catalysts possessing metal and acid functions. It is well known that traces of olefins accelerate the acid-catalyzed conversion of hydrocarbons and that such a catalysis usually results in rapid deactivation. More stable catalytic activity for the isomerization of paraffins is achieved by bifunctional catalysis, i.e., the association of a hydrogenation function of a metal with an acidic function of a support. In this case, the amount of olefins is controlled by the hydrogenation–dehydrogenation equilibrium. This topic has received considerable attention and has been earlier reviewed by Weisz [130]. However, bifunctional catalysis cannot be restricted to catalysts composed of metal and support with acid sites, but also with supports possessing acid–base pairs, basic or redox sites [131]. This is illustrated by some upcoming short examples.

The condensation of ketones and aldehydes is an important industrial process for producing higher homologues. This is a multistep transformation, which needs acid, basic, and hydrogenating functions. Catalysts prepared from layered double hydroxydes (LDH) as precursors comprise these functionalities in one material. A very typical example is the "one-pot" synthesis of 4-methyl-2-pentanone (methyl isobutyl ketone, MIBK) from acetone on Pd(Ni) particles supported on mixed oxides Mg(Al)O derived from LDH [132–134]. Acetone is condensed to diacetone alcohol on basic sites; diacetone alcohol is, in turn, dehydrated to mesityl oxide on acid sites, which is finally hydrogenated to MIBK on metal particles. The extent of the acid function is not limiting in these materials, but the selectivity to MIBK mainly depends on the balance between the basic and metal sites. At 448 K, acetone/$H_2 = 0.17$ and atmospheric pressure, the highest selectivity to MIBK was achieved with a Pd/Mg(Al)O catalyst composed of about 4×10^{-6} mol g^{-1} surface Pd atoms and 6×10^{-4} mol g^{-1} basic sites. Moreover, a size of Pd particles lower than 3–5 nm favors the selective hydrogenation of the C=C bond at the cost of the C=O bond in mesityl oxide [134].

A nice example of cooperation between metal particles and a support with redox properties concerns the three-way catalysts for automotive exhaust gases treatment composed of a noble metal (Pt, Pd, Rh) on a support with ceria additive. The cooperation was put in evidence in the $CO/NO/O_2$ reactions [135–137] and would operate in both stoichiometric and lean-burn conditions. Ceria additive confers some promoting effect to the catalysts by generating additional bifunctional sites at the metal–support interface. Under an oxygen-deficient atmosphere, CO can migrate at the metal–support interface, where it is oxidized by a ceria lattice oxygen with concurrent oxygen vacancy formation. The so-created oxygen vacancies can act as NO dissociation sites to yield N_2 and N_2O. Cataluna et al. [138] have proposed the formation of N_2O over ceria as follows:

$$2[Ce^{3+} - (_v)] + 2NO \leftrightarrow Ce^{4+} - (ONNO)^{2-} - Ce^{4+}$$
$$\leftrightarrow (Ce^{4+} - O^{2-}) + [Ce^{4+} - (_v)] + N_2O$$

Finally, the last example of metal–support cooperation is the so-called spillover phenomenon, which is concerned with the diffusion of adsorbed species from the catalyst surface onto the surface of the support. Spillover includes a wide variety of phenomenon, the most widespread being hydrogen spillover. Hydrogen spillover, first reviewed by Sermon and Bond [139], is concerned when H atoms formed by dissociation of H_2 on metal particles may easily migrate to another phase of the catalyst that has hydrogen acceptor sites. The process has chiefly been observed using Pd or Pt as the metal, which does not bind H atoms so strongly. The spilled-over hydrogen may (1) reside on the surface of the support, (2) diffuse into it, and (3) lead to a partial reduction of the support. In each of these states, spilled-over hydrogen has important implications and potential utilization in catalysis. It has, for example, been claimed that it can increase the catalytic activity of the support. This can be illustrated by very recent works concerning the conversion of *n*-hexadecane over Pt-zeolite catalysts [140]. The occurrence of a maximum in the rate as a function of the H_2 pressure has been revealed and cannot be explained by the classical bifunctional mechanism aforementioned. It was proposed that molecular dihydrogen is homolytically dissociated on metal sites. The hydrogen atoms then migrate to paired Lewis/framework-oxygen sites (AB) to form hydride species and Brønsted acid sites, thus leading to active site generation. This phenomenom of Brønsted site generation by hydrogen spillover has already been put in evidence on a wide variety of catalysts ([131,141] and references therein).

23.9 CONCLUDING REMARKS

There is no single interpretation to explain the effects of particle size, "alloying," and metal–support interaction on the chemisorption and catalytic properties of supported metal particles. Depending on the particle size, the nature of co-metal and support, and the nature of the reaction, the change of chemisorption and catalytic properties can be interpreted in terms of geometric features, electronic modifications, and/or mixed sites. This is due to the formation of various adsorbed species and intermediates. Moreover, in many cases, the promotion of catalytic properties will be directly related to the method of catalyst preparation, which affects the architecture of the active site, with respect to chemical and electronic states of components and topology.

REFERENCES

1. H.S. Taylor, *Proc. Roy. Soc. London A108* **105** (1925).
2. M. Boudart, Concepts in heterogeneous catalysis, in *Interaction Metal Surfaces* (R. Gomer, ed.), Springer-Verlag, Berlin, 1975, p. 275.
3. O.M. Boronin, V.S. Poltorak, *Int. Chem. Eng. 7* **452** (1967).
4. R. van Hardeveld, F. Hartog, *Surf. Sci. 15* **189** (1969).
5. V. Ponec, G.C. Bond, *Catalysis by Metals and Alloys*, Elsevier, Amsterdam, 1995.
6. F.M. Dautzenberg, J.N. Helle, P. Biloen, W.M.H. Sachtler, *J. Catal. 63* **119** (1980).
7. P. Gallezot, In *Metal Clusters*, John Wiley & Sons, 1986, p. 219.
8. G.M. Schwab, In *Advances in Catalysis*, Academic Press, San Diego, 1978, p. 1.
9. G.L. Haller, D.E. Resasco, In *Advances in Catalysis*, Academic Press, San Diego, 1989, p. 173.
10. M. Che, C.O. Bennett, In *Advances in Catalysis*, Academic Press, San Diego, 1989, p. 36.
11. V. Ponec, In *Advances in Catalysis*, Academic Press, San Diego, 1983, p. 32.
12. M. Boudart, In *Advances in Catalysis*, Academic Press, San Diego, 1969, p. 20.
13. Y. Soma-Noto, W.M.H. Sachtler, *J. Catal. 32* **215** (1974).
14. S. Iijima, *J. Electron. Microsc. 34* **249** (1985).
15. G. Ertl, In *Advances in Catalysis*, Academic Press, San Diego, 2000, p. 1.
16. G.A. Somorjai, *Langmuir 7* **3176** (1991).
17. G.C. Bond, *Chem. Soc. Rev. 20* **441** (1991).
18. P. Sabatier, *La Catalyse en Chimie Organique*, Librairie Polytechnique, Ch Béranger, Paris, 1913.
19. J. Cosyns, In *Catalyse par les Métaux* (B. Imelik, G.-A. Martin, A.-J. Renouprez, eds.), Editions du CNRS, Paris, 1984, p. 971.
20. N.I. Kobozev, *Acta Physicochim. URSS 9* **1** (1938).
21. W.M.H. Sachtler, R.A. van Santen, In *Advances in Catalysis*, Academic Press, San Diego, 1977, p. 69.
22. R. Coekelbergs, A. Frennet, G. Lienard, P. Resibois, *J. Phys. Chem. 39* **604** (1963).
23. J.A. Dalmon, G.A. Martin, *J. Catal. 66* **214** (1980).
24. T. Stace, *Nature 331* **116** (1988).
25. K.J. Taylor, C.L. Pettiete-Hall, O. Chesnowsky, R.E.K. Smalley, *Phys. Rev. B 96* **3319** (1992).
26. C.R. Henry, *Cryst. Res. Technol. 33* **1119** (1998).
27. C.R. Henry, *Surf. Sci. Rep. 31* **235** (1998).
28. J.K.A. Clarke, A.C.M. Creaner, *Ind. Chem. Prod. Res. Dev. 20* **574** (1981).
29. M. Fernández-García, J.C. Conesa, A. Clotet, J.M. Ricart, F. López, F. Illas, *J. Phys. Chem. 102* **141** (1998).
30. E. Miyazaki, I. Yasumori, *J. Math. Phys. 18* **215** (1977).
31. G.A. Martin, J.A. Dalmon, C. Mirodatos, *Proc. 8th International Congress Catalysis*, Berlin, 1984, Verlag Chemie, Weinheim, 1984, Vol. IV, p. 371.
32. B. Coq, A. Goursot, T. Tazi, F. Figueras, D. Salahub, *J. Am. Chem. Soc. 113* **1485** (1991).
33. C.A. Balseiro, J.L. Moran-Lopez, *Surf. Sci. 156* **404** (1985).
34. M.W. Smale, T.S. King, *J. Catal. 119* **441** (1989).
35. S.-H. Chien, B.N. Shelimov, D.E. Resasco, E.H. Lee, G.L. Haller, *J. Catal. 77* **301** (1982).
36. B.A. Sexton, A.E. Hughes, K. Foger, *J. Catal. 77* **85** (1982).
37. D.E. Resasco, R. Weber, S. Sakellson, M. McMillan, G.L. Haller, *J. Phys. Chem. 92* **189** (1988).
38. D. Bazin, C. Mottet, G. Tréglia, J. Lynch, *Appl. Surf. Sci. 164* **140** (2000).
39. V. Ponec, *Stud. Surf. Sci. Catal. 64* **118** (1991).

40. B. Coq, A. Goursot, L. Pedocchi, L.C. de Ménorval, D. Tichit, Poster presentation at *11th International Congress on Catalysis*, Baltimore, July 1996.

41. E. Sanchez-Marcos, A.P.J. Jansen, R.A. van Santen, *Chem. Phys. Lett.* 162, **399** (1990).

42. S.J. Tauster, S.C. Fung, R.L. Garten, *J. Am. Chem. Soc. 100* **170** (1978).

43. M. Che, In *Proceedings 10th International Congress on Catalysis*, Elsevier, Amsterdam, 1992, p. 31.

44. M. Che, O. Clause, C. Marcilly, In *Handbook of Heterogeneous Catalysis* (G. Ertl, H. Knözinger, J. Weitkamp, eds.), Wiley-VCH, Weinheim, 1997, p. 191.

45. C. Louis, M. Che, In *Handbook of Heterogeneous Catalysis* (G. Ertl, H. Knözinger, J. Weitkamp, eds.), Wiley-VCH, Weinheim, 1997, p. 207.

46. J. Barbier, In *Handbook of Heterogeneous Catalysis* (G. Ertl, H. Knözinger, J. Weitkamp, eds.), Wiley-VCH, Weinheim, 1997, p. 257.

47. Y.A. Ryndin, Y.I. Yermakov, In *Surface Organometallic Chemistry: Molecular Approaches to Catalysis* (J.-M. Basset, B.C. Gates, J.-P. Candy, A. Choplin, M. Lecomte, F. Quignard, C. Santini, eds.), Kluwer, Dordrecht, 1988, p 127.

48. M. Ichikawa, In *Tailored Metal Catalysts* (Y. Ywasawa, ed.), D. Reidel, Dordrecht, 1985, p. 183.

49. G.H. Twigg, *Proc. Roy. Soc. London A188* **92** (1946).

50. A.I. Boronin, V.I. Avdeev, S.V. Koshcheev, R.T. Murzakhmetov, S.F. Ruzankin, G.M. Zhidomirov, *Kinet. Catal. 40* **653** (1999).

51. V.I. Avdeev, A.I. Boronin, S.V. Koscheev, G.M. Zhidomirov, *J. Mol. Catal. A: Chemical 154* **257** (2000).

52. H. Nakatsuji, H. Nakai, K. Ikeda, Y. Yamamoto, *Surf. Sci. 384* **315** (1997).

53. R.A. van Santen, H.P.C.E. Kuipers, In *Advances in Catalysis*, Academic Press, San Diego, 1987, p. 265.

54. R.B. Grant, R.M. Lambert, *J. Catal. 92* **364** (1985).

55. H.H. Voge, C.R. Adams, In *Advances in Catalysis*, Academic Press, San Diego, 1967, p. 151.

56. G.K. Boreskov, *Disc. Faraday Soc. 41* **236** (1966).

57. C. Descorme, D. Duprez, *Appl. Catal. A: General 202* **231** (2000).

58. P.A. Kilty, W.M.H. Sachtler, *Catal. Rev. Sci. Eng. 10* **1** (1974).

59. Y.M. Fogel, B.T. Nadykto, V.I. Shvachko, V.F. Rybalko, *Zh. Fiz. Khim. 38* **2397** (1964).

60. R.E. Honig, *J. Appl. Phys. 29* **549** (1958).

61. S.V. Gerej, K.M. Kholyavenko, M.Y. Rubanik, *Ukr. Kkim. Zh. 31* **449** (1965).

62. P.A. Kilty, N.C. Rol, W.M.H. Sachtler, In *Proc. 5th International Congress Catalysis*, North-Holland Amsterdam, 1973, Vol. 2, p. 929.

63. A.I. Boronin, S.V. Koscheev, G.M. Zhidomirov, *J. Electr. Spec. Rel. Phenom. 96* **43** (1998).

64. V.I. Avdeev, A.I. Boronin, S.V. Koscheev, G.M. Zhidomirov, *J. Mol. Catal. A: Chemical 154* **257** (2000).

65. V.I. Bukhtiyarov, A.I. Boronin, L.P. Prosvirin, V.I. Savchenko, *J. Catal. 150* **262** (1994).

66. V.I. Bukhtiyarov, M. Havecker, V.V. Kaichev, A. Knop-Gericke, R.W. Mayer, R. Schlogl, *Catal. Lett. 74* **121** (2001).

67. V.I. Bukhtiyarov, A.F. Carley, L.A. Dollard, M.W. Roberts, *Surf. Sci. 381* **L605** (1997).

68. J.C. Wu, P. Harriott, *J. Catal. 39* **395** (1975).

69. M. Jarjoui, P.C. Gravelle, S.J. Teichner, *J. Chim. Phys. Phys.-Chim. Biol. 75* **1069** (1978).

70. V.I. Bukhtiyarov, I.P. Prosvirin, R.I. Kvon, S.N. Goncharova, B.S. Bal'zhinimaev, *J. Chem. Soc., Faraday Trans. 93* **2323** (1997).

71. B.S. Bal'zhinimaev, S.N. Goncharova, E.A. Paukshtis, *Appl. Catal. A: General 126* **67** (1995).

72. W.S. Epling, G.B. Hoflund, D.M. Minahan, *J. Catal. 171* **490** (1997).

73. D.M. Minahan, G.B. Hoflund, W.S. Epling, D.W. Schoenfeld, *J. Catal. 168* **393** (1997).

74. J.-P. Boitiaux, J. Cosyns, M. Derrien, G. Léger, *Hydrocarbon Processing 64* **51** (1985).

75. G.C. Bond, P.B. Wells, In *Advances in Catalysis*, Academic Press, San Diego, 1964, Vol. 15, p. 92.

76. G.C. Bond, *Appl. Catal. A: General 149* **3** (1997).

77. H. Ibach, S. Lehwald, *J. Vac. Sci. Technol. 15* **407** (1978).

78. L.L. Kesmodel, L.H. Dubois, G.A. Somorjai, *J. Chem. Phys. 70* **2180** (1979).

79. T.P. Beebe, Jr., M.R. Albert, J.T. Yates, Jr., *J. Catal. 96* **1** (1985).

80. S. Azad, M. Kaltchev, D. Stacchiola, G. Wu, W.T. Tysoe, *J. Phys. Chem. B 104* **3107** (2000).

81. J.-P. Boitiaux, J. Cosyns, S. Vasudevan, *Appl. Catal. 6* **41** (1983).

82. Y.A. Ryndin, L.V. Nosova, A.I. Boronin, A.L. Chuvilin, *Appl. Catal. 42* **131** (1988).

83. Y.A. Ryndin, M.V. Stenin, A.I. Boronin, V.I. Bukhtiyarov, V.I. Zaikovskii, *Appl. Catal. 54* **277** (1989).

84. H.R. Adùriz, P. Bodnariuk, M. Dennehy, C.E. Gigola, *Appl. Catal. 58* **227** (1990).

85. S. Hub, L. Hilaire, R. Touroude, *Appl. Catal. 36* **307** (1988).

86. H.R. Adùriz, P. Bodnariuk, B. Coq, F. Figuéras, *J. Catal. 129* **47** (1991).

87. G. Deganello, D. Duca, A. Martorana, G. Fagherazzi, A. Benedetti, *J. Catal. 150* **127** (1994).

88. B. Tardy, C. Noupa, C. Leclercq, J.C. Bertolini, A. Hoareau, M. Treilleux, J.P. Faure, A. Nihoul, *J. Catal. 129* **1** (1991).

89. V. De Gouveia, B. Bellamy, Y. Hadj Romdhane, A. Masson, M. Che, *Z. Phys. D— Atoms, Molecules and Clusters 12* **587** (1989).

90. A. Borodzinski, R. Dus, R. Frak, A. Janko, W. Palczewska, In *Proc. 6th International Congress on Catalysis*, The Chemical Society, London, 1977, p. 151.

91. M. Boudart, H.S. Wang, *J. Catal. 39* **44** (1975).

92. J. Goetz, M.A. Volpe, R. Touroude, *J. Catal. 164* **369** (1996).

93. A. Guerrero, M. Reading, Y. Grillet, J. Rouquerol, J.-P. Boitiaux, J. Cosyns, *Z. Phys. D—Atoms, Molecules & Clusters 12* **583** (1989).

94. J.M. Moses, A.H. Weiss, K. Matusek, L. Guczi, *J. Catal. 86* **417** (1984).

95. J.R. Anderson, In *Advances in Catalysis*, Academic Press, San Diego, 1973, p. 1.

96. F.G. Gault, In *Advances in Catalysis*, Academic Press, San Diego, 1980, p. 1.

97. G. Leclercq, R. Maurel, *Bull. Soc. Chim. Belg. 88* **599** (1979).

98. C. Kemball, In *Proc. 4th International Congress on Catalysis*, Moscow, 1968, Akademiai Kiado, Budapest, 1972, Vol. 2, p. 53.

99. V. Ponec, *Appl. Catal. A: General 133* **221** (1998).

100. J.H. Sinfelt, *J. Catal. 29* **308** (1973).

101. S.B. Shang, C.N. Kenney, *J. Catal 134* **134** (1992).

102. B. Coq, E. Crabb, F. Figuéras, *J. Mol. Catal. A: Chemicals 96* **35** (1995).

103. G.C. Bond, J.C. Slaa, *J. Mol. Catal. A: Chemicals 106* **135** (1996).

104. M.C. Sanchez Sierra, J. Garcia Ruiz, M.G. Proitti, J. Blasco, *J. Mol. Catal. A: Chemicals 96* **65** (1995).

105. A. Goursot, L. Pedocchi, B. Coq, *J. Phys. Chem. 98* **8747** (1994).

106. A. Goursot, L. Pedocchi, B. Coq, *J. Phys. Chem. 99* **12718** (1995).

107. B. Coq, A. Bittar, R. Dutartre, F. Figueras, *J. Catal. 128* **275** (1991).

108. S. Galvagno, J. Schwank, G. Parravano, F. Garbassi, A. Marzi, G.R. Tauszik, *J. Catal. 69* **283** (1981).

109. A.J. Rouco, G.L. Haller, J.A. Olivier, C. Kemball, *J. Catal. 84* **297** (1983)

110. M.W. Smale, T.S. King, *J. Catal. 120* **335** (1990).

111. M. Sprock, X. Wu, T.S. King, *J. Catal.* *138* **617** (1992).
112. B. Chen, J.G. Goodwin, Jr. *J. Catal.* *158* **228** (1996).
113. D.P. VanderWiel, M. Pruski, T.S. King, *J. Catal.* *188* **186** (1996).
114. T. Okuhara, T. Jin, Y. Zhou, J.M. White, *J. Phys. Chem.* *92* **4141** (1988).
115. X. Wu, B.C. Gerstein, T.S. King, *J. Catal.* *121* **271** (1990).
116. R. Liu, B. Tesche, H. Knözinger, *J. Catal.* *129* **402** (1991).
117. V. Ponec, *Appl. Catal. A: General* *149* **27** (1997).
118. J. Simonik, P. Beranek, *Coll. Czech. Chem. Comm.* *37* **353** (1972).
119. M.A. Vannice, B. Sen, *J. Catal.* *115* **65** (1989).
120. H. Yoshitake, Y. Iwazawa, *J. Phys. Chem.* *96* **1329** (1992).
121. P. Gallezot, A. Giroir-Fendler, D. Richard, In *Catalysis of Organic Reactions*, Marcel Dekker, New York, 1991, p. 1.
122. M. Consonni, D. Jokic, D.Y. Murzin, R. Touroude, *J. Catal.* *188* **165** (1999).
123. M. Abid, G. Erhet, R. Touroude, *Appl. Catal. A: General* *217* **219** (2001).
124. A.B. da Silva, E. Jordao, M.J. Mendes, P. Fouilloux, *Appl. Catal. A: General* **148**, 253 (1997).
125. A. Wagray, J. Wang, R. Oukaci, D. Blackmond, *J. Phys. Chem.* **96**, 5964 (1992).
126. B. Coq, V. Brotons, J.M. Planeix, L.C. de Menorval, R. Dutartre, *J. Catal.* **176**, 358 (1998).
127. J. Kaspar, M. Graziani, G.P. Escobar, A. Trovarelli, *J. Mol. Catal.* *72* **243** (1992).
128. K.J. Williams, A.B. Boffa, M.E. Levin, M. Salmeron, A.T. Bell, G.A. Somorjai, *Catal. Lett.* **5**, 385 (1990).
129. D. Richard, P. Gallezot, D. Neibecker, I. Tkatchenko, *Catal. Today* **6**, 171 (1989).
130. P.B. Weisz, In *Advances in Catalysis*, Academic Press, San Diego, 1962, p. 137.
131. E. Iglesia, D.G. Barton, J.A. Biscardi, M.J.L. Gines, S.L. Soled, *Catal. Today* *38* **339** (1997).
132. Y.Z. Chen, C.M. Huang, C.W. Liaw, *Appl. Catal. A: General* *169* **207** (1998).
133. R. Unnikrishnan, S. Narayanan, *J. Mol. Catal. A: Chemicals* *144* **173** (1999).
134. N. Das, D. Tichit, R. Durand, P. Graffin, B. Coq, *Catal. Lett.* *71* **181** (2001).
135. B. Harrison, A.F. Diwell, C. Hallet, *Plat. Metals Rev.* *32* **73** (1988).
136. G. Leclercq, C. Dathy, G. Mabilon, L. Leclercq, In *Automotive Pollution Control II* (A. Crucq, ed.), Elsevier, Amsterdam, 1991, p. 181.
137. G. Djéga-Mariadassou, F. Fajardie, J.-F. Tempère, J.-M. Manoli, O. Touret, G. Blanchard, *J. Mol. Catal. A: Chemicals* *161* **179** (2000).
138. R. Cataluna, A. Arcoya, X.L. Scoane, A. Martinez-Arias, J.M. Coronado, J.C. Conesa, J. Soria, L.A. Petrov, In *Catalysis and Automotive Pollution Control III* (F. Frennet, J.M. Bastin, eds.), Elsevier, Amsterdam, 1995, p. 215.
139. P.A. Sermon, G.C. Bond, *Catal. Rev. 8* **211** (1973).
140. L. Perrotin, A. Finiels, F. Fajula, T. Cholley, In *Proc. 13th International Zeolite Conference*, Elsevier, Amsterdam, 2001.
141. K. Fujimoto, I. Nakamura, In *Spillover and Migration of Surface Species on Catalysts*, Elsevier, Amsterdam, 1997, p. 29.

24

Conductive Metal-Oxide Nanoparticles on Synthetic Boron-Doped Diamond Surfaces

I. DUO and C. COMNINELLIS

Swiss Federal Institute of Technology EPFL, Lausanne, Switzerland

SERGIO FERRO and ACHILLE DE BATTISTI

Università di Ferrara, Ferrara, Italy

CHAPTER CONTENTS

SUMMARY

The electrochemical properties of IrO$_2$ and RuO$_2$ nanoparticles, deposited on synthetic boron-doped diamond (BDD) surfaces, are discussed. After a description of the preparation procedure and the morphological characterization of BDD/IrO$_2$ and BDD/RuO$_2$ samples, the dispersion efficiency of these oxides on BDD was estimated for different loading, using cyclic voltammetry.

The electrochemical behavior of 1,4-benzoquinone/hydroquinone redox couple, oxygen evolution, and organics oxidation reactions have been investigated at BDD/IrO$_2$ samples; in turn, the chlorine evolution reaction has been studied at BDD/RuO$_2$ electrodes. "Noncatalytic" and "catalytic" models are proposed to explain the modification of the electrochemical behavior of BDD by low- and high-oxide loading, respectively.

24.1 INTRODUCTION

The determination of the electrochemical properties of supported oxide particles of nanometer size is of great interest for both fundamental and applied research. The fundamental interest mainly ensues from their different and sometimes novel properties compared with that of bulk materials, while the applied interest arises because oxides are commonly adopted as active catalysts in technical electrodes.

In particular, the catalytic active components for O_2 and Cl_2 evolution electrodes are IrO_2 and RuO_2, respectively [1,2]. These oxide catalysts are usually deposited on titanium base metal by the thermal decomposition technique, using an appropriate precursor [3]. The synthesis involves several steps: dissolution of the precursor (H_2IrCl_6, $RuCl_3$) in an appropriate solvent (e.g., isopropanol), deposition of the precursor solution on the pretreated titanium substrate (a common pretreatment is represented by an etching in hot and concentrated HCl), evaporation of the solvent, and, finally, thermal decomposition at 350–450 °C, leading to the formation of a few microns' layer of the oxide on the titanium substrate (DSA® electrodes). This procedure is very sensible to preparation conditions and, as a consequence of the thermal treatment, the titanium base metal itself is partially oxidized, forming a thin layer of TiO_X between the Ti base metal and the overlying oxide catalyst [4]. The Ti base metal can also strongly influence the electrochemical behavior of the oxide catalyst (segregation of TiO_X on the surface of the oxide catalyst) and the service life of the electrode (formation of TiO_X at the Ti–electrocatalyst interface during anodic polarization). In order to avoid these problems of Ti–electrocatalyst interaction, an inert substrate must be used.

Glassy carbon (GC) and gold have usually been adopted as supports in the investigation of nano-sized electrocatalyst particles [5,6]. However, these supports pose problems due to a low anodic stability (GC) and the possible formation of oxides (Au) at high anodic potentials.

The recent achievements in the preparation of highly boron-doped diamond on silicon substrate (BDD) seem to supply an interesting solution to the above limitations. BDD does not form a macroscopic oxide layer on its surface, when brought into contact with an aqueous solution. Other advantages of the BDD electrode are a high chemical and electrochemical stability, a low background current, and a wide electrochemical window of polarizability [7–11]. Moreover, the BDD electrode can be heated in air up to 500 °C without undergoing a significant surface oxidation. These features make BDD an interesting candidate to use as an inert substrate for electrocatalysts investigation.

Moving from the above considerations, research has recently been carried out on the electrochemical properties of nanoparticles of IrO_2 and RuO_2, synthesized on highly conductive BDD thin films. Amounts of oxide electrocatalyst on BDD were between 6.4×10^{13} and 1.3×10^{16} molecules cm^{-2} (geometric surface area), in the case of iridium dioxide [12], and between 1.2×10^{13} and 2.9×10^{16} molecules cm^{-2} in the case of ruthenium dioxide [13]. Such amounts correspond to a nominal surface coverage (Γ) ranging approximately between 0.01 and 10 (assuming that 10^{15} oxide molecules cm^{-2} correspond to a monolayer surface coverage, namely $\Gamma = 1$).

24.2 EXPERIMENTAL

Boron-doped diamond (BDD) thin films were synthesized at CSEM (Neuchâtel, Switzerland) by the hot filament chemical vapor deposition technique (HF CVD) on p-type, low-resistivity (1–3 mΩ cm), single-crystal, silicon wafers (Siltronix). The temperature of the filament was between 2440 and 2560 °C and that of the substrate was monitored at 830 °C. The reactive gas was a mixture of 1% methane in hydrogen, containing trimethylboron as a boron source (1–3 ppm, with respect to H_2). The reaction chamber was supplied with the gas mixture at a flow rate of 51 min^{-1}, giving a growth rate of 0.24 μm h^{-1} for the diamond layer. The obtained diamond film has a thickness of about 1 μm ($\pm 10\%$) and a resistivity of 15 mΩ cm ($\pm 30\%$). This HF CVD process produces columnar, random textured, polycrystalline films [9].

An "as-grown" BDD film is typically hydrogen-terminated and its surface is hydrophobic; anodic polarization at 10 mA cm^{-2} during 30 min results in the transformation of the BDD surface from hydrophobic to hydrophilic. Alternatively, a quite hydrophilic surface can be obtained by thermally treating the BDD support at 400 °C for 30 min under oxygen.

IrO_2 and RuO_2 nanoparticles were deposited on hydrophilic BDD surfaces (1 cm^2) by the above-described traditional sol-gel, thermal decomposition technique [1,3], applying an isopropanolic solution of a precursor salt ($H_2IrCl_6 \cdot 6H_2O$ or $RuCl_3 \cdot xH_2O$, with $x \approx 3$). Five microliters of the precursor solution were applied on the BDD surface, then the solvent was evaporated at 60 °C; finally, the sample was treated in air during 1 h at the target temperature (350–450 °C). Blank experiments with Si/BDD electrodes showed that the above thermal treatment does not modify the CV response of the diamond film.

Electrochemical measurements were made in a conventional three-electrodes cell, using a computer-controlled EcoChemie potentiostat, model Autolab PGSTAT 30.

The properties of the thermally treated BDD support and of the modified surfaces were investigated in situ by cyclic voltammetry (CV) in H_2SO_4 and $HClO_4$ (IrO_2), and in H_2SO_4 and $HClO_4/NaClO_4/NaCl$ (RuO_2). The ex-situ characterization was carried out by x-ray photoelectron spectroscopy (XPS), scanning electron microscopy (SEM), transmission electron microscopy (TEM), and atomic force microscopy (AFM).

24.3 BDD/IrO₂ ELECTRODES

24.3.1 Morphological Characterization

Figure 1(a) shows an SEM image of a polycrystalline BDD film, deposited on a p-Si substrate. The boron-doped diamond layer appears to be continuous and consisting of randomly arranged microcrystals, having a size variable from 0.2 to 0.6 μm and facets mainly {111} oriented, with some {100} orientation. Twining can also be observed on some crystals. Raman spectrum analyses have shown that the ratio between nondiamond and diamond is less than 1%, indicating films of good quality and with a significant diamond character (i.e., sp^3 bonding).

Figure 2 shows XPS measurements that have been carried out for the characterization of the iridium-based samples; the binding energy values of Ir4f$_{5/2}$ (65 eV) and Ir4f$_{7/2}$ (62 eV) are evidence of the IrO_2 formation.

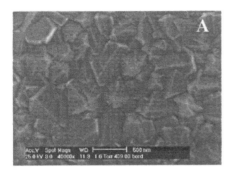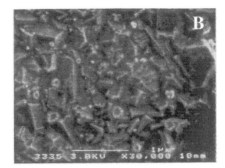

Figure 1   SEM images of (A) BDD, (B) BDD/IrO$_2$ ($\Gamma = 13$, sample prepared at 450 °C).

As shown by SEM investigations carried out at IrO$_2$-modified BDD samples, the noble metal oxide particles are mainly localized at the grain boundaries of the BDD microcrystals. At low IrO$_2$ loading ($\Gamma < 1$), very small particles are formed and they are not detectable by SEM; in contrast, at higher IrO$_2$ loading ($\Gamma = 13$), the particles have dimensions varying between 20 and 200 nm [Figure 1(b)].

Finally, TEM investigations of BDD/IrO$_2$ samples, with two different oxide loadings, are reported in Figure 3(a) and (b). Nanoparticles having dimensions between 2 and 5 nm are present on the sample with the lower IrO$_2$ loading [$\Gamma = 0.6$; Figure 3(a)], while IrO$_2$ clusters with dimensions of 10–30 nm are present on the sample with the higher IrO$_2$ loading [$\Gamma = 6.4$; Figure 3(b)].

24.3.2 Voltammetric Characterization

Cyclic voltammetry in the potential range of water and supporting electrolyte stability provides a sort of "electrochemical spectrum," giving evidence of surface

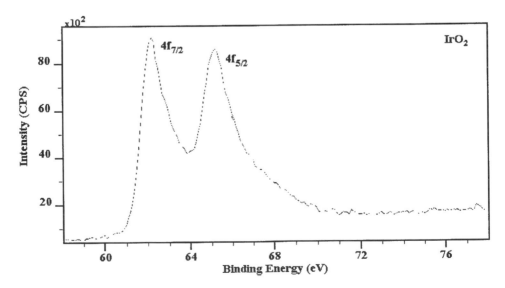

Figure 2   XP spectrum of a BDD/IrO$_2$ sample ($\Gamma = 6.4$, sample prepared at 450 °C).

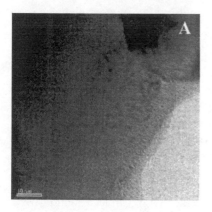
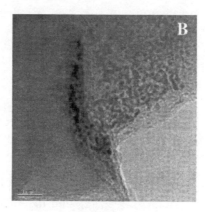

Figure 3 TEM images of BDD/IrO$_2$ samples with different IrO$_2$ loading: (A) $\Gamma = 0.6$; (B) $\Gamma = 6.4$. Samples prepared at 450 °C.

redox transitions. Furthermore, the anodic voltammetric charge (q^*), obtained by integration of the anodic part of voltammetric curves (Figure 4), turns out to be proportional to the number of surface active sites, thus providing a relative measure of the electrochemically active surface area. Figure 4 shows voltammetric curves for BDD and BDD/IrO$_2$ (1.3×10^{15} IrO$_2$ molecules cm^{-2}, corresponding to $\Gamma = 1.3$) obtained in 0.5 M H$_2$SO$_4$, at a scan rate of 3 V s^{-1}. It can be appreciated how the deposition of small amounts of IrO$_2$ on BDD strongly increases the anodic voltammetric charge (q^*), indicating an enhanced electrochemical activity for the modified electrode.

As shown in the figure, two well-resolved peak couples are observed, probably related to the solid-state redox couple Ir(III)/Ir(II), at 0.4 V, and Ir(IV)/Ir(III) at

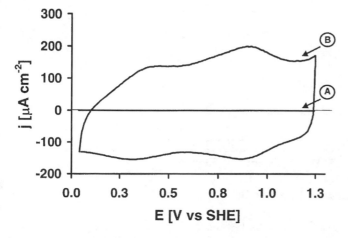

Figure 4 Cyclic voltammetric curves for (A) BDD and (B) BDD/IrO$_2$ ($\Gamma = 1.3$, sample prepared at 450 °C) in 0.5 M H$_2$SO$_4$. Scan rate: 3 V s^{-1}; $T = 25$ °C.

0.95 V (both values are referred to SHE). Such distinct oxidation state changes are not observed on oxide coatings deposited on Ti substrate. In that case, broad, flat maxima are generally obtained. Figure 5(a) and (b) show the dependence of anodic voltammetric charge (q^*) on the potential scan rate (v), for different IrO_2 loading, prepared at 350 °C and 450 °C, respectively. These figures show that q^* decreases rapidly with the scan rate then, above about 0.1 V s^{-1}, q^* remains almost constant. The influence of the scan rate on q^* has to be related to the existence of less accessible (inner) surface regions in the IrO_2 deposits (pores, cracks, and grain boundaries). At high scan rates (> 0.1 V s^{-1}), the voltammetric charge q^* is related only to the more accessible (outer) surface area (A_o), while, at very low scan rates (≈ 0.01 V s^{-1}), the voltammetric charge is proportional to the whole electroactive surface (A_t). Thus, the "inner" surface (A_i) can be calculated as the difference between the "total" surface (A_t) and the "outer" surface (A_o). Table 1 summarizes these definitions.

The influence of IrO_2 surface coverage (Γ) on the ratio between the "inner" surface (A_i) and the "total" surface (A_t), for IrO_2 electrodes prepared at 350 °C, is

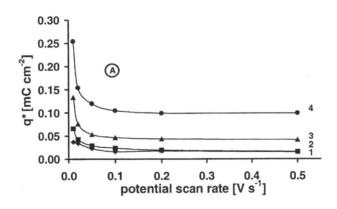

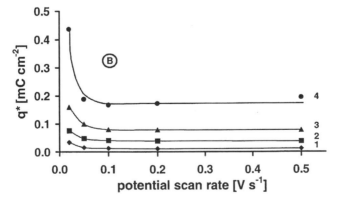

Figure 5 Voltammetric charge (q^*) of BDD/IrO_2 samples in 0.5 M H_2SO_4 as a function of scan rate; $T = 25$ °C. (1) $\Gamma = 0.6$; (2) $\Gamma = 1.3$; (3) $\Gamma = 2.5$; (4) $\Gamma = 6.4$. Samples prepared at (A) 350 °C; (B) 450 °C.

Table 1 Estimation of the Total (A_t), Outer (A_o), and Inner (A_i) Surface Area of the IrO$_2$ Coating from the Anodic Charge (q^*) Obtained at Different Scan Rates (v)

Total surface area	$A_t = q^*(v = 0.01 \text{ V s}^{-1})$
Outer surface area	$A_o = q^*(v > 0.1 \text{ V s}^{-1})$
Inner surface area	$A_i = [q^*(v = 0.01 \text{ V s}^{-1}) - q^*(v > 0.1 \text{ V s}^{-1})]$

reported in Figure 6. According to the definitions, the A_i/A_t value represents the fraction of surface area that is of difficult access, and the figure shows that it decreases with increasing the IrO$_2$ loading. This implies that the morphology of the deposit changes considerably with the IrO$_2$ loading because of the three-dimensional growth of IrO$_2$ crystallites and clusters formation, as has been shown in TEM measurements (Figure 3). Similar results have been obtained for IrO$_2$ deposited at 450 °C.

The dispersion efficiency (γ) of IrO$_2$ on BDD can be calculated as the ratio between the measured voltammetric charge $q^*(v > 0.1 \text{ V s}^{-1})$ (related to the outer surface area A_o only) and the theoretical charge q_{th}^*, obtained assuming that all oxide sites are involved in the charging process [Eq. (1)].

$$\gamma = \frac{q^*(v > 0.1 \text{ V s}^{-1})}{q_{th}^*} \tag{1}$$

Assuming that only the redox steps from Ir(II) to Ir(IV) (i.e., two electron exchanges) are involved in the charging process, the theoretical voltammetric charge can be estimated from Faraday's law (3.2×10^{-16} mC per IrO$_2$ molecule).

The dependency of the IrO$_2$ dispersion efficiency (γ) on the IrO$_2$ surface coverage (Γ) for two deposition temperatures (350 °C and 450 °C) is shown in

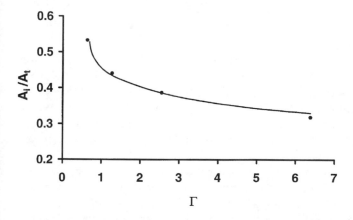

Figure 6 Influence of IrO$_2$ nominal surface coverage (Γ) on the fraction of surface area of difficult access for BDD/IrO$_2$ samples prepared at 350 °C. A_i = "inner" surface; A_t = "total" surface. Electrolyte: 0.5 M H$_2$SO$_4$. $T = 25$ °C.

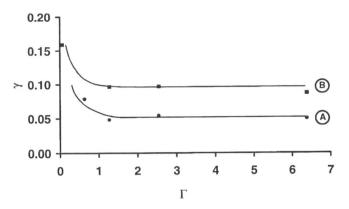

Figure 7 Influence of IrO_2 nominal surface coverage (Γ) on the IrO_2 dispersion efficiency (γ) for BDD/IrO_2 samples. Electrolyte: 0.5 M H_2SO_4; $T = 25\,°C$. Samples prepared at (A) 350 °C and (B) 450 °C.

Figure 7 (curves A and B). This figure shows that, for the lowest preparation temperature, the dispersion efficiency (γ) decreases with the IrO_2 loading. Above about 10^{15} molecules cm^{-2} (corresponding approximately to one IrO_2 monolayer, i.e., $\Gamma = 1$), the dispersion efficiency remains almost constant. The increase in the deposition temperature also leads to an increase in the dispersion efficiency: the highest dispersion obtained in this work ($\gamma = 0.16$) is for the coating prepared at 450 °C, with a nominal coverage $\Gamma = 0.06$.

24.3.3 Investigation of Redox Couples at BDD and BDD/IrO_2 Electrodes

The 1,4-benzoquinone/hydroquinone (Q/H_2Q) redox couple, which is known to be quasi-reversible at noble metals and oxide electrodes [14], has been investigated at BDD and BDD/IrO_2 electrodes.

For this couple, both electron and proton transfer reactions are involved [Eq. (2)].

$$Q + 2H^+ + 2e^- \rightleftharpoons H_2Q \tag{2}$$

Figure 8(a) shows a cyclic voltammetric curve obtained at BDD electrode in 0.5 M H_2SO_4. The fact that the separation between the cathodic and the anodic peaks (ΔE_p) is very high (about 0.9 V) indicates that the Q/H_2Q system is irreversible at the boron-doped diamond electrode. Furthermore, the apparent equilibrium redox potential of the couple Q/H_2Q($E_0 = 0.65\,V$) is much closer to the anodic peak potential than to the cathodic one.

Figure 8(b)–(d) show voltammetric curves for the Q/H_2Q couple, obtained at BDD/IrO_2 electrodes with different IrO_2 loading. The reversibility of the couple increases with the IrO_2 loading. At high IrO_2 loading, the behavior of the system approaches that of a continuous film deposited on a Ti substrate [Ti/IrO_2; Figure 8(e)], while maintaining a background current much lower than the Ti/IrO_2 electrode.

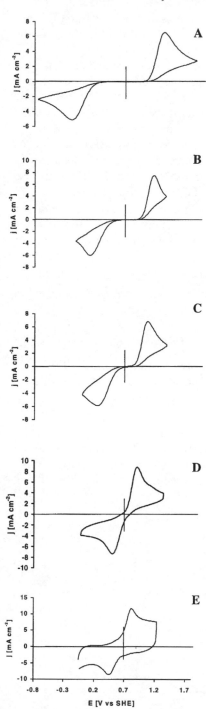

Figure 8 Voltammetric curves of Q/H_2Q in $0.5\,M$ H_2SO_4 obtained at electrodes with different IrO_2 loading. (A) BDD; (B) BDD/IrO_2($\Gamma = 0.6$); (C) BDD/IrO_2($\Gamma = 6.4$); (D) BDD/IrO_2($\Gamma = 30$); and (E) IrO_2 continuous film ($1\,\mu m$) on Ti substrate. Scan rate: $100\,mV\,s^{-1}$; $T = 25\,°C$.

The exact role of IrO_2 is not clear; as a proposal, both "catalytic" and "noncatalytic" models can be discussed to explain the modifications observed in the electrochemical behavior of BDD after deposition of IrO_2 particles.

According to the catalytic model, only the IrO_2 particles participate in the electrochemical reaction, acting as mediators in the redox process. In the noncatalytic model, we speculate that the IrO_2 particles promote the BDD electrode surface. This can be achieved either by introducing surface states that facilitate the electron transfer reaction (this model is similar to that proposed by Swain [8], according to which carbon sp^2 impurities act as surface states at the BDD surface) or by the spillover of electrogenerated active intermediates from the IrO_2 particles to the BDD surface. These intermediates can be further discharged at the BDD surface.

The fact that at high IrO_2 loading the BDD/IrO_2 electrode behaves very similarly to the Ti/IrO_2 electrode [Figure 8(e)] can be explained by the catalytic model, in which IrO_2 acts as mediator in the electron transfer reaction. However, the catalytic model cannot explain the increase in the reversibility of the Q/H_2Q couple at very low IrO_2 loading (6.0×10^{14} IrO_2 molecules cm^{-2}, corresponding to a nominal surface coverage $\Gamma = 0.6$); in that case, the noncatalytic model should be preferred.

24.3.4 Oxygen Evolution at BDD and BDD/IrO_2 Electrodes

The oxygen evolution reaction (OER) has been investigated at both BDD and BDD/IrO_2 electrodes, using perchloric and sulfuric acid aqueous test solutions.

Oxygen Evolution in 1 M $HClO_4$

Figure 9 shows cyclic voltammograms for BDD and BDD/IrO_2 electrodes ($\Gamma = 6.4$), obtained in 1 M $HClO_4$. The presence of IrO_2 particles on the BDD surface results in a considerable overpotential decrease for the OER (about 1 V at

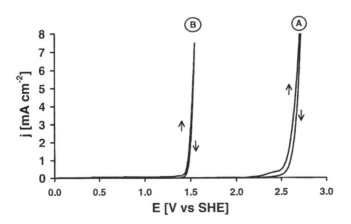

Figure 9 Voltammetric curves for (A) BDD and (B) BDD/IrO_2($\Gamma = 6$) in 1 M $HClO_4$. Scan rate: $100\,mV\,s^{-1}$; $T = 25\,°C$.

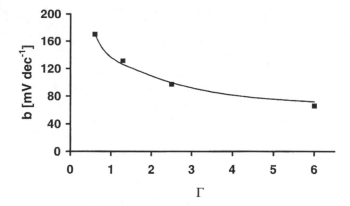

Figure 10 Influence of IrO_2 nominal surface coverage (Γ) on the value of Tafel slope for BDD/IrO_2 electrodes with different IrO_2 loading.

1 mA cm^{-2}). As shown in Figure 10, the Tafel slope for the reaction is also strongly influenced. The value obtained at the BDD electrode (186 mV dec^{-1}) decreases strongly with the amount of IrO_2: the value (66 mV dec^{-1}) obtained at a BDD/IrO_2 electrode with a high IrO_2 loading ($\Gamma = 6.4$) approaches the value reported for continuous IrO_2 films (1–2 μm), deposited on a Ti substrate [14].

This behavior of BDD and BDD/IrO_2 electrodes can be explained on the basis of the generalized phenomenological model, previously proposed for oxygen evolution in acidic medium [10,14,15]. According to this model (Figure 11), the initial step is represented by the oxidation of water molecules to form adsorbed hydroxyl radicals (reaction *a*). Next, two different mechanisms for the oxygen

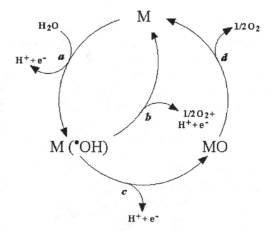

Figure 11 Schematic representation of the reaction pattern for the OER: (a) formation of adsorbed hydroxyl radicals (˙OH) by water discharge; (b) oxygen evolution by electrochemical oxidation of (˙OH) species; (c) formation of a metal oxide; (d) oxygen evolution by chemical decomposition of the oxide. (From Ref. 10. Copyright 1999, The Electrochemical Society, Inc. Reprinted with permission.)

evolution reaction are proposed. In the first case, the electrode is referred as "nonactive" and the oxygen evolution occurs via the electrochemical oxidation of hydroxyl radicals (reaction *b*). In the second case, the electrode is referred as "active" and the anodic reaction follows a redox catalysis mechanism, in which the electrode participates via the higher oxide formation. The BDD electrode, because of its inert surface, behaves like a nonactive electrode, leading to a high overvoltage for oxygen evolution. Concerning the BDD/IrO_2 electrode, Figure 9 shows that, at this electrode, oxygen evolution starts at about 1.4 V (SHE), which is a potential value close to that of the IrO_3/IrO_2 redox couple. This indicates an active electrode behavior, in which the oxygen evolution takes place through a redox catalysis mechanism.

According to this mechanism, water molecules adsorbed on the IrO_2 particles are oxidized, forming, in a first step, hydroxyl radicals [Eq. (3)] that chemically interact with IrO_2, leading to the higher oxide IrO_3 [Eq. (4)]. This step is followed by the chemical decomposition of IrO_3, to oxygen and IrO_2 [Eq. (5)].

$$IrO_2 + H_2O \longrightarrow IrO_2(OH) + H^+ + e^- \tag{3}$$

$$IrO_2(OH) \longrightarrow IrO_3 + H^+ + e^- \tag{4}$$

$$IrO_3 \longrightarrow IrO_2 + 1/2 O_2 \tag{5}$$

In order to normalize the polarization measurements with respect to the IrO_2 loading, we have defined the turnover frequency (TOF), as given by relation (6):

$$TOF = \frac{j}{F \times m} \tag{6}$$

where

> j: current density (geometric) measured at a given potential (A cm^{-2})
> m: IrO_2 loading (mol cm^{-2})
> F: Faraday's constant

Figure 12 shows that the slopes of potential-log (TOF) plots have very similar values for different IrO_2 loadings; this indicates that oxygen is evolved by the same mechanism, independently on the IrO_2 loading. This result is in favor with the catalytic model, in which IrO_2 acts as a mediator, via redox catalysis, in the oxygen evolution reaction. However, the striking feature of BDD/IrO_2 electrodes is that already at low IrO_2 loading (about 10^{15} IrO_2 molecules cm^{-2}, which is equivalent to a monolayer) the overpotential for oxygen evolution is decreased by about 1.0 V (at 1 mA cm^{-2}), compared to BDD. This can be explained by the noncatalytic model, in which the IrO_2 particles promote the BDD electrode surface, either by the introduction of new surface states on the BDD surface [8] or by the spillover of active intermediates (hydroxyl radicals) from the IrO_2 particles to the BDD surface. These hydroxyl radicals can be further discharged at the BDD surface.

Oxygen Evolution in 0.5 M H$_2$SO$_4$ as Electrolyte

The cyclic voltammograms recorded at BDD and BDD/IrO_2 electrodes ($\Gamma = 6.4$) in 0.5 M H_2SO_4 (Figure 13) are very similar to those obtained in 1 M $HClO_4$ [12]. However, different reaction products are obtained during preparative electrolyses at

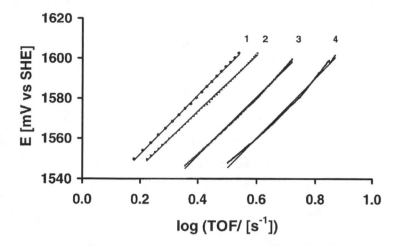

Figure 12 Potential-log (TOF) plot in 1 M $HClO_4$ at BDD/IrO_2 electrodes with different IrO_2 loading. (1) $\Gamma = 0.6$; (2) $\Gamma = 1.3$; (3) $\Gamma = 2.5$; (4) $\Gamma = 6.4$.

BDD and BDD/IrO_2 electrodes. In fact, during galvanostatic electrolysis using a BDD anode, peroxodisulfate [Eq. (8)] is formed, beside the oxygen evolution reaction [Eq. (7)].

$$2H_2O \longrightarrow O_2 + 4H^+ + 4e^- \qquad E° = 1.23\,V \qquad (7)$$

$$2SO_4^{--} \longrightarrow S_2O_8^{--} + 2e^- \qquad E° = 2.01\,V \qquad (8)$$

The prevalence of one reaction over the other depends on the applied current density. In fact, it has been found that high current efficiencies of peroxodisulfate formation at BDD electrodes can be achieved at relatively low current densities [16].

In the case of BDD/IrO_2 electrodes, no traces of peroxodisulfate formation are detected and the oxygen evolution reaction [Eq. (7)] is the unique anodic reaction. The inhibition of the oxidation of sulfate to peroxodisulfate, induced by the presence of IrO_2 particles on the BDD surface, can be explained by the catalytic action of IrO_2 toward the oxygen evolution reaction. As previously discussed, IrO_2 particles act as a redox catalyst, strongly decreasing the oxygen evolution overpotential. In this case, the potential for the oxidation of sulfate to peroxodisulfate cannot be achieved, even at high current densities.

24.3.5 Oxidation of Organic Compounds at BDD and BDD/IrO₂ Electrodes

Different organic compounds have been investigated at BDD and BDD/IrO_2 electrodes by cyclic voltammetry and preparative electrolysis. As model organic compounds, simple alcohols (methanol, ethanol, n-propanol, isopropanol, and ter-butanol) and simple carboxylic acids (formic acid, oxalic acid, and maleic acid) have been investigated. Two mechanisms can be distinguished for the organics oxidation:

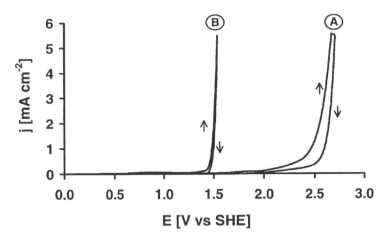

Figure 13 Voltammetric curves for (A) BDD and (B) BDD/IrO$_2$(Γ = 6.4) electrodes in 0.5 M H$_2$SO$_4$. Scan rate: 50 mV s^{-1}; T = 25 °C. (From Ref. 12. Copyright 2000, The Electrochemical Society, Inc. Reprinted with permission.)

a direct oxidation and, alternatively, an indirect oxidation via surface intermediates at the anode surface.

The cyclic voltammetry measurements have shown that, in the potential region of water stability, the investigated organic compounds are electrochemically inactive at both BDD and BDD/IrO$_2$ electrodes. Furthermore, the overpotential for water discharge decreases in the presence of organic compounds [12]. This is in favor of the indirect oxidation mechanism; however, the direct oxidation of organics cannot be completely excluded. A typical example is the isopropanol oxidation in 0.5 M H$_2$SO$_4$ at BDD and BDD/IrO$_2$ electrodes (Figure 14). For both cases, the presence of isopropanol in the electrolyte decreases the overpotential for water discharge. This may indicate that the intermediates of the oxygen evolution reaction are involved in the organics oxidation. The fact that at a BDD electrode the relative decrease in water discharge overpotential, in the presence of organics, is more important than at a BDD/IrO$_2$ electrode (Figure 15) may indicate that more reactive intermediates are formed at BDD than at BDD/IrO$_2$ electrodes.

This behavior of BDD and BDD/IrO$_2$ electrodes can be explained on the basis of the generalized phenomenological model previously proposed for organics oxidation in an acidic medium [14,15]. According to the model (Figure 16), the initial step is represented by the oxidation of water molecules to form adsorbed hydroxyl radicals (reaction *a*). In the case of BDD, a nonactive anode, the hydroxyl radicals can be involved in the combustion of organics (reaction *e*) or they can recombine at the anode producing O$_2$ (reaction *b*). In the case of BDD/IrO$_2$ electrodes (active anodes), the adsorbed hydroxyl radicals may interact with the IrO$_2$ particles forming the higher oxide IrO$_3$ (reaction *c*), which is likely to be the active intermediate in the partial oxidation of organics (reaction *f*).

This again shows that the IrO$_2$ particles on the BDD surface act as a redox catalyst in the oxidation of organics. Such reaction is in competition with the side oxygen evolution reaction (pathway *d*).

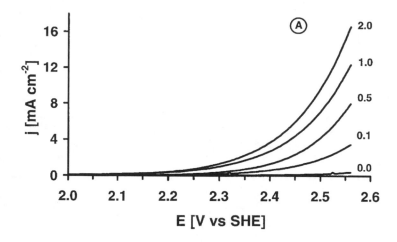

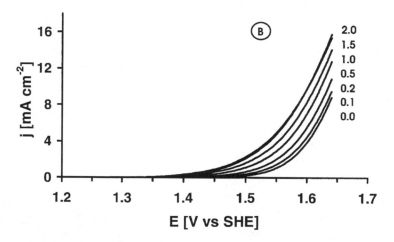

Figure 14 Voltammetric curves for (A) BDD and (B) BDD-IrO$_2$($\Gamma = 6.4$) electrodes in 0.5 M H$_2$SO$_4$ and different isopropanol concentrations (0 to 2.0 M). Scan rate: 20 mV s^{-1}; $T = 25\,^{\circ}$C. (Picture from Ref. 12. Copyright 2000, The Electrochemical Society, Inc. Reprinted with permission.)

24.4 BDD/RuO$_2$ ELECTRODES

24.4.1 Morphological Characterization

A characterization of BDD surfaces, modified with different oxide loadings, has been carried out by XPS. The Ru3d$_{5/2}$ signal, partially superimposed with the C1s, and the O1s signal are shown for the oxide loadings 2.2×10^{13} ($\Gamma \approx 0.02$, curve 1), 3.7×10^{15} ($\Gamma \approx 3.7$, curve 2), and 2.9×10^{16} ($\Gamma \approx 29$, curve 3) in Figure 17(a)–(c). The binding energy of the signal Ru3d$_{5/2}$(280.6 eV) and the value of the ratio between the intensity of the O1s and Ru3p$_{3/2}$ signals (2.0 ± 0.2) are evidence for the formation of an RuO$_2$ phase in all the investigated samples.

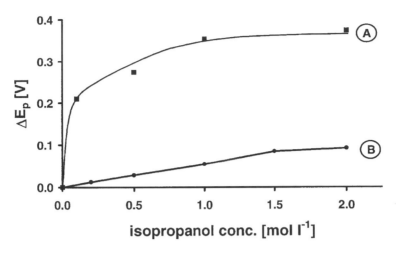

Figure 15 Influence of isopropanol concentration on the decrease in water discharge overpotential at $1\,mA,cm^{-2}$. (A) BDD and (B) BDD/IrO_2($\Gamma = 6.4$) electrodes.

On a more qualitative basis, from the data in Figure 17(a), it can be observed that the ruthenium-related signal does not increase linearly with the oxide loading. This may be taken as an indication that, with increasing the amount of oxide deposited, the dispersion of the phase decreases, which may be justified by accumulation of excess oxide in the "valleys" of the diamond surface texture and increase in the particle size.

A surface texture of oxide deposits has been also carried out in [13], making use of AFM, and images are shown in Figure 18(a) and (b) (oxide loading: 2.9×10^{16}

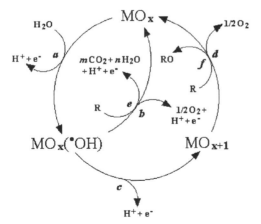

Figure 16 Schematic representation of the reaction pattern for the oxidation of organic compounds with simultaneous oxygen evolution at metal-oxide anodes: reactions (*a*), (*b*), (*c*), (*d*) as in Figure 11; (*e*) combustion of the organic compound R via electrochemical oxidation mediated by physisorbed hydroxyl radicals; (*f*) selective chemical oxidation of the organic compound at the higher metal oxide surface sites. (From Ref. 15. Copyright 1994, Pergamon Press Ltd. Reprinted with permission.)

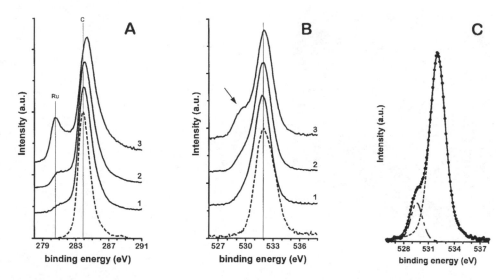

Figure 17   XP spectra for an unmodified BDD (dashed line) and for BDD/RuO$_2$ samples with different metal-oxide loading: (1) 2.2×10^{13}, (2) 3.67×10^{15}, and (3) 2.87×10^{16} molecules cm^{-2}. The curves in (A) report C1s and Ru3d$_{5/2}$ signals; data for O1s are reported in (B) and (C) (magnification of the curve for sample 3).

molecules cm^{-2}, $\Gamma \approx 29$) and Figure 19(a) and (b) (oxide loading: 3.7×10^{15} molecules cm^{-2}, $\Gamma \approx 3.7$). When the amount of deposited oxide is large enough (Figure 18), there is evidence for the formation of clusters of spherical particles, whose diameter can reach the size of 50–60 nm. This occurrence is certainly much less evident in the case of the lower loading (Figure 19) and becomes quite disregardable below this limit. These data qualitatively confirm the conclusion drawn on the basis of XP spectra, of a decreased degree of dispersion of the oxide phase with increasing amounts of deposit. In fact, the combination of the two pieces of experimental evidence suggests that particle cluster formation possibly takes place prior to the accomplishment of an oxide monolayer and conditions any further growth.

24.4.2   Voltammetric Characterization

Cyclic voltammograms obtained at as-prepared BDD/RuO$_2$ electrodes reached a stable shape after a limited number of cycles. Typical CV data are shown in Figure 20(a) and (b) for samples with different oxide loadings. The main features of the voltammogram are already present for the lowest oxide deposit (Figure 20(a), dashed line) and they remain unchanged at electrodes with higher oxide loading (Figure 20(a), solid line), whose only effect is the increase of voltammetric current. Two solid-state redox couples, between 0.00 V and 1.25 V (versus SCE), are clearly identified and related to the oxidation-state transitions Ru(III)/Ru(II) and Ru(IV)/Ru(III). At RuO$_2$ thick-film electrodes, these pairs partially convolute into broad, flat maxima [17,18], the voltammograms affording the typical "brick" shape, desired for supercapacitor applications. As shown in Figure 20(b), evidence for the above

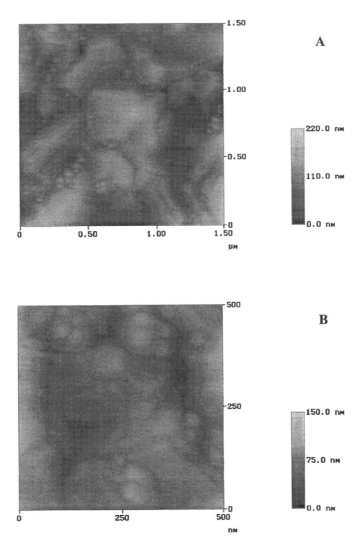

Figure 18 AFM images of a BDD/RuO$_2$ sample, with a nominal deposit of 2.65×10^{16} molecules cm^{-2}. (Photograph B reprinted with permission from A. De Battisti, S. Ferro, M. Dal Colle. *J. Phys. Chem. B*, **105**, 1679 Copyright 2001, The American Chemical Society.)

signal convolution can be found at oxide loadings as low as 2.65×10^{16}, where the agglomeration of isolated oxide nanoparticles, accompanying the film growth, evidently leads to heterogeneity of redox sites. This poor peak resolution has possibly justified an overestimation of the purely capacitive contribution to the charge–storage capacity of oxide electrodes. Injection–ejection of protons, involved in solid-state redox equilibria [13,17,19], can better account for the very high capacities measured, the proton exchange being much faster than double-layer rearrangements caused by changes of electrode potential.

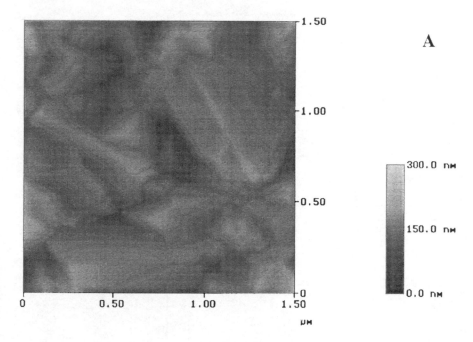

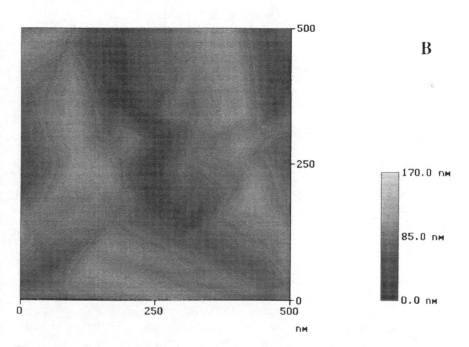

Figure 19 AFM images of a BDD/RuO$_2$ sample, with a nominal deposit of 3.67×10^{15} molecules cm^{-2}.

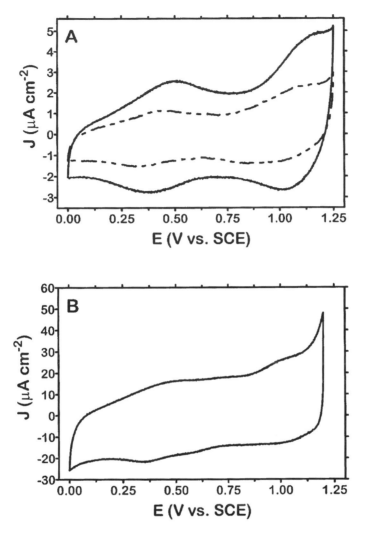

Figure 20   Cyclic voltammetric curves for three RuO$_2$-modified BDD electrodes, 1.2×10^{13} and 6.0×10^{14} molecules, cm^{-2} (A), and 2.65×10^{16} molecules, cm^{-2} (B), recorded in 1 M H$_2$SO$_4$. Scan rate: 300 mV s^{-1}. Electrode (geometric) area $= 0.785$ cm^2. (Photograph (A) reprinted with permission from A. De Battisti, S. Ferro, M. Dal Colle. *J. Phys. Chem. B*, **105**, 1679. Copyright 2001, The American Chemical Society.)

In this connection, it is interesting to observe that the RuO$_2$ rutile crystal structure consists of a three-dimensional array of distorted RuO$_6$ octaedra, with void channels normal to the 001 plane (Figure 21). Because the inner walls of these channels are oxygen-based, the above electrochemical ion (proton) exchange equilibria may extend several Å below the oxide surface through, e.g., a Grotthus-type mechanism [20]. The intergrain–interparticle boundaries, often considered the reason for the large capacities of the RuO$_2$ films, would rather magnify an intrinsic property of the structure of this oxide.

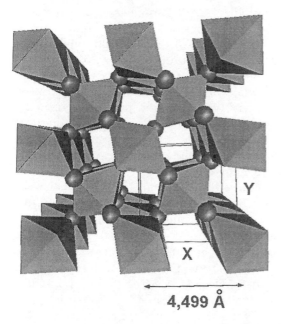

Figure 21 Three-dimensional array of distorted octahedral in the rutile structure. View along the 001 direction. (Reprinted with permission from A. De Battisti, S. Ferro, M. Dal Colle. *J. Phys. Chem. B*, **105**, 1679. Copyright 2001, The American Chemical Society.)

From the voltammetric charge and the amount of RuO_2 deposit, it has been possible to evaluate the number of electrons exchanged per ruthenium ion. Following this procedure, a value of 2.28 electrons per oxide molecule could be evaluated for the lowest oxide loading (1.2×10^{13} molecules cm^{-2}, $\Gamma \approx 0.01$). Assuming an electronicity of 2 for the global charging process, this result indicates that partial oxide coverage allows a complete participation of ruthenium sites to the charging process. An increase of the oxide loading to 6.0×10^{14}, which approximately represents one oxide monolayer ($\Gamma \approx 0.6$) and then to 2.65×10^{16} molecules cm^{-2} is accompanied by an increase of voltammetric charge to only $6.77\,\mu C$ (almost twice the previous value, $3.44\,\mu C$) and $50.62\,\mu C$, respectively. This evident decrease in charge–storage efficiency certainly reflects the building-up of a three-dimensional oxide phase, in agreement with the observations made on the basis of XPS and AFM data. Referring again to Figure 18, a simple calculation, based on the average number of the approximately spherical particles and their size, leads to an estimated charge of $24.90\,\mu C$, in qualitative agreement with the experimental stored charge.

24.4.3 Chlorine Evolution at BDD and BDD/RuO_2 Electrodes

Ruthenium dioxide-based materials represent the catalyst for the industrial anodic production of chlorine. There is no doubt that, in the case of dimensionally stable anodes (DSA®), the technological innovation was adopted by industry [21] a few decades before the beginning of academic studies [22,23]. Despite the fact that RuO_2 constitutes the catalytically active material, the composition of the electrode is much more complex, involving other oxides (e.g., TiO_2, SnO_2) which should stabilize and disperse the active material, allowing less loading of a noble metal. As a consequence, the mechanism of the chlorine evolution reaction (chl.e.r.) is not well clarified yet, being dependent on material preparation and composition. As introduced above, a given composition can be the result of a specific experimental design, or the consequence of interactions with the electrode support, during the oxide-film preparation. Such interactions cannot be avoided with "standard" supporting materials, but they should be extremely limited by the great chemical and electrochemical properties of conductive diamond.

It is worth mentioning that neither the electrode material under discussion now (BDD/RuO_2) nor the chl.e.r. has been deeply characterized so far, and the investigation of the latter could represent a way for going inside both problems.

In Figure 22 a set of cyclic voltammetric curves, recorded in 0.01 M perchloric acid $+ x$M sodium chloride and $(4 - x)$M sodium perchlorate (in order to maintain a constant ionic strength of four), is reported, concerning a sample of BDD modified with a low loading of RuO_2 (*SMD, slightly modified diamond*; nominally, 1.2×10^{13} molecules cm^{-2}, $\Gamma \approx 0.01$). Although the metal-oxide loading is well below the amount required for a monolayer, the catalytic effect of RuO_2 can be appreciated by comparing the right y-axis of the figure with those of Figure 23, reporting analogous cyclic voltammetries for an "unmodified" diamond thin film (3.2×10^{15} "C sites" cm^{-2}, assuming a unitary roughness coefficient). Clearly, a comparison

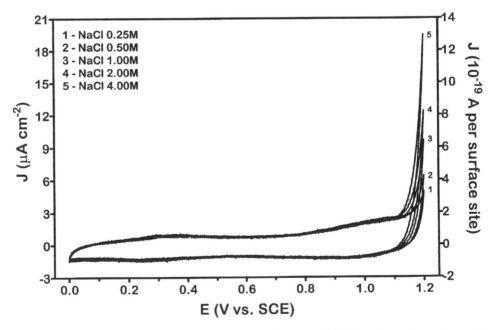

Figure 22 Cyclic voltammetric curves for a BDD/RuO_2 electrode with 1.2×10^{13} RuO_2 molecules, cm^{-2}, in 0.01 M $HClO_4/x$ M $NaCl/(4 - x)$ M $NaClO_4$. Scan rate: $100 \, mV \, s^{-1}$. Electrode (geometric) area $= 0.785 \, cm^2$.

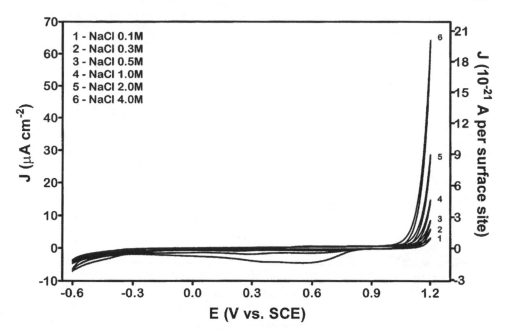

Figure 23 Cyclic voltammetric curves for a BDD electrode in 0.01 M HClO$_4$ + xM sodium chloride and $(4 - x)$M sodium perchlorate. Scan rate: $100\,\mathrm{mV\,s^{-1}}$. Electrode (geometric) area $= 0.785\ \mathrm{cm^2}$.

between different materials requires normalization of current values to the number of surface atoms at which the reaction under discussion takes place.

Extending the comparison to the parameters of the chl.e.r. kinetics, the catalytic activity of the modified electrode becomes more appreciable. Data, obtained in 1 M NaCl/3 M NaClO$_4$/0.01 M HClO$_4$, are shown in Figures 24 and 25, in terms of polarization curves and Tafel plots, respectively. The slope of the Tafel plot for the BDD electrode is about 0.150 V, giving evidence for a rate-determining discharge of chloride ions [11,24]; as a further support to this mechanism, the reaction was found to be of the first order with respect to the chloride concentration [11]. At the sample slightly modified with RuO$_2$, the Tafel slope has a value of 0.060 V and the reaction order with respect to Cl$^-$ is about 0.74. It is clear that the mechanism of the chl.e.r. is not the same. Another interesting difference concerns the role of pH: at BDD [11], the chlorine evolution is strangely stimulated by a higher acidity, the reaction order with respect to H$^+$ being ≈ 0.5 (on the contrary, at the oxide anodes generally used for chlorine evolution, the acidity has a depressing effect on the reaction rate), while a just-measurable amount of RuO$_2$ changes the reaction order with respect to H$^+$ to about -0.6, thus approaching the behavior reported in the literature for this kind of material [25]. The fact that such a behavior cannot be easily accounted for by considering simple mechanistic paths suggests that also in the case under discussion there would be complications arising from the diamond surface itself. This problem has been considered by the present authors, and a possible explanatory mechanism has been

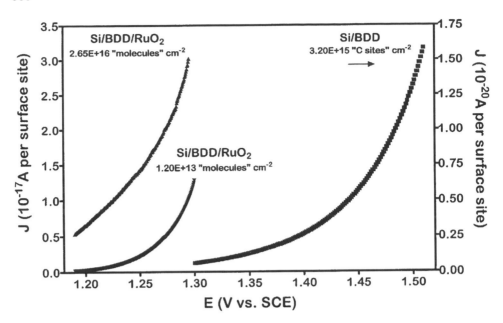

Figure 24 Current-potential curves for chlorine evolution in 1 M NaCl/3M NaClO₄/0.01 M HClO₄. Current values are normalized to the number of surface sites. Scan rate: $0.5\,\mathrm{mV\,s^{-1}}$. Electrode (geometric) area $= 0.785\,\mathrm{cm^2}$. (Reprinted with permission from A. De Battisti, S. Ferro, M. Dal Colle, *J. Phys. Chem. B*, **105**, 1679. Copyright 2001, The American Chemical Society.)

given [13,26], involving the spillover of chlorine radicals from the donor phase (RuO₂) to the acceptor one (diamond). Accordingly, an electrochemical desorption* (Volmer–Heyrovsky mechanism [24]) could easily take place at the latter surface, due to the lack of stabilizing interaction with the C(diamond) surface. Alternatively, a chemical desorption (Volmer–Tafel mechanism [24]) can occur at the boundaries between the two materials.

The necessity to consider the above-described mechanism of spillover can be better justified in light of results obtained at higher RuO₂-modified BDD surfaces (*HMD, highly modified diamond*). Figure 26 shows cyclic voltammetric curves recorded in 0.01 M perchloric acid $+\, x$M sodium chloride and $(4-x)$M sodium perchlorate at the sample with 2.65×10^{16} RuO₂ molecules cm^{-2} ($\Gamma \approx 26.5$). As expected, the CV currents are noticeably higher than those previously discussed (Figures 22 and 23), evidence for a higher amount of catalyst, present at the BDD surface. The onset of chlorine evolution shifts to less positive potentials as the

---

*The mechanistic path, with a rate-determining second step, can be illustrated as follows, where S and S′ represent two different sites at the electrode surface (and $* = \triangle$ radical specie):

$$S + Cl^- \rightleftharpoons S - Cl^{\bullet} + e \tag{I}$$
$$S - Cl^{\bullet} + S' \longrightarrow S + S' - Cl^{\bullet} \tag{II}$$
$$S' - Cl^{\bullet} + Cl^- \longrightarrow S' + Cl_2 + e \tag{III}$$

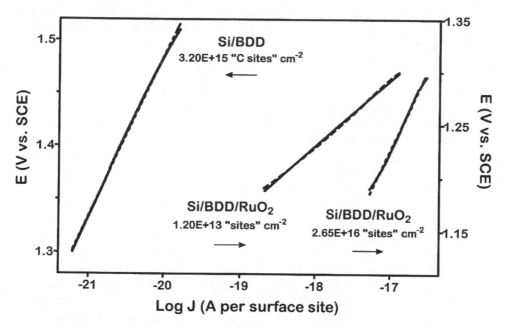

Figure 25 Tafel plots for chlorine evolution in 1 M NaCl/3 M NaClO$_4$/0.01 M HClO$_4$. Current values are normalized to the number of surface sites. Electrode (geometric) area = 0.785 cm^2. (Reprinted with permission from A. De Battisti, S. Ferro, M. Dal Colle, *J. Phys. Chem. B*, **105**, 1679. Copyright 2001, The American Chemical Society.)

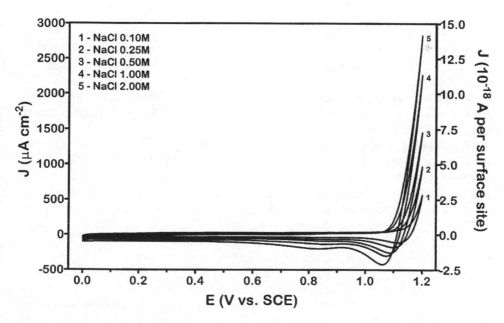

Figure 26 Cyclic voltammetric curves for a BDD/RuO$_2$ electrode with 2.65×10^{16} RuO$_2$ molecules, cm^{-2}, in 0.01 M HClO$_4$/xM NaCl/(4 $-$ x)M NaClO$_4$. Scan rate: 100 mV s^{-1}. Electrode (geometric) area = 0.785 cm^2.

chloride concentration increases; in the cathodic sweep, the catalytic activity of RuO_2 toward the chl.e.r. is manifested by the good reversibility of the reaction, the reduction of chlorine (produced during the anodic sweep) taking place at a potential close to that at which the chloride ion is oxidized. In Figure 22 such cathodic signals could not be observed, as a consequence of too small a chlorine production; on the contrary, analogous signals are present in Figure 23, for the case of BDD. The presence of two cathodic peaks in cyclic voltammetry has been observed also in the case of DSA-type (titanium-supported) RuO_2 film electrodes [27], and related to different species of dissolved and electroactive chlorine, i.e., the chlorine itself and hypochlorous acid formed by hydrolysis of anodically formed Cl_2 or by direct electro-oxidation of Cl^- (the potential threshold for this reaction decreases with increasing the solution pH and/or the chloride concentration).

As far as the chl.e.r. mechanism is concerned, the same, previously described, investigation has been performed and Figures 24 and 25 respectively report the polarization curve and the Tafel plot (currents normalized to the number of active sites at the electrode surface), for the case of a 1 M NaCl/3 M $NaClO_4$/0.01 M $HClO_4$ test solution. The measured Tafel slope has a value of 0.149 V, and the reaction order with respect to Cl^- is about 0.7; the values of b and R both agree well with a Volmer–Heyrovsky mechanism [24], with a rate-determining electrochemical desorption, provided a value of about 0.7 is assumed for the coverage by the intermediate chlorine radicals [28]:

$$S + Cl^- \underset{k_{-1}}{\overset{k_1}{\rightleftharpoons}} S-Cl^{\bullet} + e \tag{I}$$

$$S - Cl^{\bullet} + Cl^- \overset{k_2}{\longrightarrow} S + Cl_2 + e \tag{II}$$

It seems that the modified BDD surface has now a number of RuO_2 sites sufficiently high to sustain the required passage of current (which is a function of the applied potential), and accordingly the mechanism does not need any involvement of the diamond surface. This conclusion is further supported by the dependence on pH: the Volmer–Heyrovsky sequence of steps does not involve protons and, in fact, a zero reaction order with respect to H^+ has been found for the electrode material under discussion (HMD) [13,26].

The pH independence of the reaction taking place at HMD has to be compared with analogous results obtained at single-crystal faces of RuO_2 [25]; as discussed in [26], the explanation of such a behavior could be related to the homogeneity of sites that characterizes the two materials. On the contrary, a pH dependence of the chl.e.r. has been found at unmodified BDD electrodes [11], at polycrystalline RuO_2 (relatively thick deposits on titanium substrates [25]), and at SMD, as above discussed. Focusing on the two RuO_2-containing materials, the similarity can be justified with the presence of active sites having a different energy content. This evidence, together with the nonlinear decrease of activity of the modified BDD surface, with decreasing the oxide loading, can be understood only assuming a strong diamond–ruthenium oxide interaction, more important for thinner deposits. Apparently, the catalytic activity at the oxide surface, requiring a rearrangement of oxidation state of the metal ions in the oxide lattice, is enhanced by the existence of underlying oxide layers, where such rearrangements can take place as well. On the

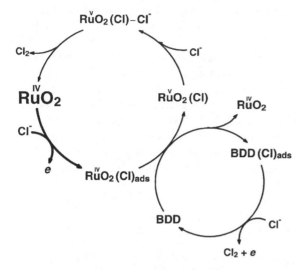

Figure 27 Schematic representation of the reaction pattern for the chlorine evolution reaction at BDD/RuO$_2$ electrodes. At low RuO$_2$ coverage, radical spillover can take place. (Reprinted with permission from A De Battisti, S Ferro, M Dal Colle. *J. Phys. Chem. B*, **105**, 1679. Copyright 2001, The American Chemical Society.)

basis of this interpretation, the chl.e.r. at the electrode with the lowest oxide loading can be tentatively described by Figure 27, which takes into consideration the two possible routes of the chlorine radical consumption. The assumption of adsorbed radicals interaction with higher oxidation states of the metal ion at the surface would necessarily imply the participation of the whole near-surface region of the oxide film. On the contrary, the involvement of the diamond surface can represent an escape route for the case where the ruthenium site is unable to change its oxidation state. According to this mechanistic path, both the catalytic and the noncatalytic models, previously proposed (Section 24.3.3), seem to be also involved in the BDD/RuO$_2$ electrodes. The discharge of chloride ions takes place at the RuO$_2$ particles (catalytic model) and the electrogenerated chlorine radicals spill over from the RuO$_2$ particles to the BDD surface (noncatalytic model). The noncatalytic model seems to be dominant in the presence of the lowest RuO$_2$ loading.

24.5 CONCLUSIONS

The aim of this work was to investigate the electrochemical behavior of IrO$_2$ and RuO$_2$ nanoparticles deposited on synthetic boron-doped diamond (BDD) surfaces by the thermal decomposition technique.

BDD/IrO$_2$ Electrodes

Iridium dioxide nanoparticles have been deposited on BDD surface by the thermal decomposition technique, using H$_2$IrCl$_6$ · 6H$_2$O as precursor. The XPS characterization of the oxide-modified BDD samples confirmed the formation of IrO$_2$ on the BDD surface (BDD/IrO$_2$ electrode). SEM investigations have shown that the IrO$_2$ particles are mainly localized at the grain boundaries of the BDD microcrystals. At low IrO_2 loading ($\Gamma < 1$), very small particles are formed and they are not detectable by SEM; in contrast, at higher IrO_2 loading ($\Gamma = 13$), the particles have dimensions varying between 20 and 200 nm.

The TEM study has confirmed that IrO_2 nanoparticles, having dimensions between 2 and 5 nm, are present at low IrO_2 loading ($\Gamma = 0.6$), while IrO_2 clusters with dimensions of 10–30 nm are present at relatively high IrO_2 loading ($\Gamma = 6.4$).

Cyclic voltammetry experiments showed the presence of two well-resolved peak couples, related to the solid-state redox transitions Ir(III)/Ir(II) and Ir(IV)/Ir(III).

The voltammetric charge measurements have shown that the dispersion efficiency decreases with the oxide loading; the highest dispersion obtained in this work ($\gamma = 0.16$) is for the coating prepared at 450 °C, with a nominal coverage $\Gamma = 0.06$.

Electrode reactions such the 1,4 benzoquinone/hydroquinone redox couple (Q/H_2Q), oxygen evolution (in $HClO_4$ and H_2SO_4), and organics oxidation have been investigated at BDD/IrO_2 samples. The following results could be obtained:

> The reversibility of the Q/H_2Q couple increases with the IrO_2 loading.
>
> The presence of IrO_2 nanoparticles on the BDD surface causes a considerable decrease in the overpotential for oxygen evolution, in the inhibition of sulfate to peroxodisulfate oxidation, and in the modification of the mechanism of organic oxidation.

Finally, a "noncatalytic" and a "catalytic" model have been proposed to explain the modification of the electrochemical behavior of BDD by low- and high-oxide loading, respectively.

BDD/RuO_2 Electrodes

The preparation of deposits of ruthenium dioxide at BDD surfaces has been carried out as described above for the IrO_2 deposits. The XPS characterization of the RuO_2-modified BDD surfaces confirmed that the oxygen-to-metal ratio in the deposits is with good approximation equal to 2. The intensity of the Ru-related signals do not increase linearly with the amount of oxide deposited, already in the sub-monolayer range, which is possibly evidence for a clustering of the oxide phase around a given number of nucleation sites. This view is confirmed by data obtained for the higher amounts of deposited oxide (around 10^{16} molecules per cm^2).

The AFM study has shown that, for loading around 10^{16} RuO_2 molecules per cm^2 (formally equivalent to about 10 monolayers), the oxide phase is essentially organized in cluster of particles of approximately spherical shape. The size of the clusters may reach values of 50 to 60 nm. For lower loadings no particles can be detected, possibly because of their quite small size.

The cyclic voltammetry experiments have confirmed the presence of species at the BDD surface across all the loading ranges, from 10^{13} to 10^{16}. The two peak pairs, related to the Ru(III)/Ru(II) and Ru(IV)/Ru(III) solid-state couples, are well separated only for smaller oxide loadings. When the amount of oxide deposited reaches the value of 10^{16} molecules per cm^2, the shape of the cyclic voltammogram is very close to that of RuO_2 thick films, with the well-known peak convolutions. The charge–storage capacity of oxide deposits increases with increasing the oxide loading, but the higher the loading, the lower the increase rate.

The study of the chlorine evolution reaction at the RuO_2-modified BDD surfaces conforms to the schemes proposed in the literature for thicker oxide films based on the same catalyst. However, for lower oxide loading, the nanoparticle size and distribution on the support surface cause a somewhat different reaction path, possibly related to the occurrence of chlorine radical spillover. Voltammetric tests on the electrodes after prolonged chlorine evolution experiments showed that the oxide modifications at BDD were quite stable.

The combined information obtained by the different characterization methods applied allows the conclusion that deposits of ruthenium oxide at BDD, ranging from approximately one hundredth of a monolayer, maintain the physicochemical properties of RuO_2, which proves the very limited degree of chemical interaction with the support. The deposits are most probably organized in nanoparticles growing around nucleation sites. When particles and clusters of particles reach a size of 50–60 nm, their charge–storage and catalytic behavior closely resembles that of thick oxide films.

ACKNOWLEDGMENTS

The authors thank the CSEM, Centre Suisse d'Electronique et de Microtechnique SA (Neuchâtel, Switzerland) for preparing the boron-doped diamond electrodes and for financial support.

I.D. and Ch. C. are very grateful to Mr. B. Senior, Mrs. D. Laube, and Prof. P. Buffat, for SEM and TEM investigations and helpful discussion, and to N. Xanthopoulos for his contribution to the XPS measurements.

S.F. and A.D. thank Dr. M. Dal Colle for XPS measurements and helpful discussion.

REFERENCES

1. S. Trasatti, G. Lodi, In *Electrodes of conductive metallic oxides, Part B*; S. Trasatti, ed.; Elsevier Sci. Publ. Co.: Amsterdam, 1981; pp. 521–626.
2. G.N. Martelli, R. Ornelas, G. Faita, *Electrochim. Acta* **39**, 1551–1558 (1994).
3. Ch. Comninellis, G.P. Vercesi, *J. Appl. Electrochem.* **21**, 335–345 (1991).
4. G.P. Vercesi, J. Rolewicz, J. Hinden, Ch. Comninellis, *Thermochim. Acta* **176**, 31–47 (1991).
5. V.S. Bagotzky, A.M. Skundin, *Electrochim. Acta* **29**, 951–956 (1984).
6. K.A. Friedrich, F. Henglein, U. Stimming, W. Unkauf, *Electrochim. Acta* **45**, 3283–3293 (2000).
7. Q. Chen, M.C. Granger, T.E. Lister, G.M. Swain, *J. Electrochem. Soc.* **144**(11) 3806–3812 (1997).
8. G.M. Swain, A.B. Anderson, J.C. Angus, *MRS Bull.* **23**, 56–60 (1998).
9. D. Gandini, P.-A. Michaud, I. Duo, E. Mahé, W. Haenni, A. Perret, Ch. Comninellis, *New Diamond Front. Carbon Technol.* **9**(5), 303–316 (1999).
10. G. Foti, D. Gandini, Ch. Comninellis, A. Perret, W. Haenni, *Electrochem. and Solid State Letters* **2**(5), 228–230 (1999).
11. S. Ferro, A. De Battisti, I. Duo, Ch. Comninellis, W. Haenni, A. Perret, *J. Electrochem. Soc.* **147**, 2614 (2000).
12. I. Duo, P.-A. Michaud, W. Haenni, A. Perret, Ch. Comninellis, *Electrochem. and Solid-State Lett.* **3**(7), 325–326 (2000).

13. S. Ferro, A. De Battisti, *J. Phys. Chem. B* **105**, 1679 (2001).
14. G. Foti, D. Gandini, C. Comninellis, *Current Topics in Electrochem.* **5**, 71–91 (1997).
15. C. Comninellis, *Electrochim. Acta* **39**, 1857–1862 (1994).
16. P-A. Michaud, E. Mahé, W. Haenni, A. Perret, C. Comninellis, *Electrochem. and Solid-State Lett.* **3**(2), 77–79 (2000).
17. S. Trasatti, G. Lodi, In *Electrodes of Conductive Metallic Oxides, Part A*; S. Trasatti, ed.; Elsevier Sci. Publ. Co.: Amsterdam, 1980; pp. 301–358.
18. A. Benedetti, P. Riello, G. Battaglin, A. De Battisti, A. Barbieri, *J. Electroanal. Chem.* **376**, 195 (1994).
19. C. Bock, V.I. Birss, *J. Electroanal. Chem.* **475**, 20 (1999) and literature therein.
20. J.O'M. Bockris, A.K.N. Reddy, In *Modern Electrochemistry*, Plenum/Rosetta: New York, 1973; pp. 474–476.
21. H.B. Beer, Brit. Patent 1,147,442 (1965).
22. G.F. Tantardini, Thesis, University of Milan (1968).
23. S. Trasatti, G. Buzzanca, *J. Electroanal. Chem.* **29**, App. 1 (1971).
24. S. Trasatti, *Electrochim. Acta* **32**, 369 (1987).
25. V. Consonni, S. Trasatti, F. Pollak, W.E. O'Grady, *J. Electroanal. Chem.* **228**, 393 (1987).
26. S. Ferro, A. De Battisti, *J. Phys. Chem. B*, **106**, 2249 (2002).
27. L. Tomcsányi, A. De Battisti, G. Hirschberg, K. Varga, J. Liszi, *Electrochim. Acta* **44**, 2463 (1999).
28. B.E. Conway, B.V. Tilak, *Electrochim. Acta* **37**, 51 (1992).

25

Electrocatalysis with Electron-Conducting Polymers Modified by Noble Metal Nanoparticles

C. LAMY and J.-M. LÉGER

Université de Poitiers-CNRS, Poitiers, France

CHAPTER CONTENTS

SUMMARY

Different electron-conducting polymers (polyaniline, polypyrrole, polythiophene) are considered as convenient substrates for the electrodeposition of highly dispersed metal electrocatalysts. The preparation and the characterization of electron-conducting polymers modified by noble metal nanoparticles are first discussed. Then, their catalytic activities are presented for many important electrochemical reactions related to fuel cells: oxygen reduction, hydrogen oxidation, oxidation of C1 molecules (formic acid, formaldehyde, methanol, carbon monoxide), and electro-oxidation of alcohols and polyols.

25.1 INTRODUCTION

Electron-conducting polymers (ECP) are conjugated π systems obtained by (electro)-polymerization of monomers containing some double bonds, such as those found in an aromatic ring or in a furanoic ring. This leads to quasi-metallic materials, the conductivity of which can be as high as a few thousand $S\,cm^{-1}$ in a given potential window [e.g., from 0.1 V to 0.9 V versus RHE for a polyaniline- (PAni) conducting polymer]. Some of these polymers, e.g., PAni, Polypyrrole (PPy), Polymethylthiophene (PMeT), are very stable under ambient conditions, and also in contact with strong acid electrolytes, including Nafion® film. Therefore, electron-conducting polymers were proposed many years ago as good electronic-conducting substrates to support dispersed electrocatalysts, particularly noble metals (Pt, Pd, Rh) [1–8]. Due to their controlled structure and to their high porosity, electron-conducting polymers allow electrocatalytic materials to be dispersed at the molecular level in a three-dimensional structure. Therefore, each catalytic center will be reached by the reacting species and will be involved in the electrocatalytic reaction. In this way the reaction zone is displaced from the electrode surface (classical heterogeneous electrocatalysis) to the bulk material, i.e., the electrochemical reaction will become relevant to homogeneous catalysis, and the catalytic process will be more efficient [9]. Provided that the dispersion of the metallic catalyst is high and the metal loading low, the degree of utilization of the catalysts will approach 100%.

25.2 PREPARATION OF ELECTRON-CONDUCTING POLYMER MODIFIED BY NOBLE METAL NANOPARTICLES

Different ways can be used to prepare ECP modified by noble metals [9 and references therein]:

1. Dispersion of the electrocatalysts after the polymerization process. A polymer film can be synthesized on an inert conducting substrate, e.g., glassy carbon (GC), by electropolymerization of the corresponding monomer. This can be carried out at a fixed electrode potential, or by potential cycling. Potential cycling between 0 and 1.0 V versus RHE at a sweep rate of $50\,mV\,s^{-1}$ in a $0.5\,M$ H_2SO_4 solution containing 0.1 M aniline is a very convenient way to prepare PAni film of controlled porosity and thickness (0.1 to a few µm).

Then the electrodeposition of noble metals (e.g., Pt, Ru) or transition metals (e.g., Co, Cu, Fe, Mo, Ni, Sn, W) is realized at a fixed potential (e.g., 0.1 V versus RHE for Pt). With a low concentration of the metallic salt (10^{-4} to $10^{-5}\,M$) and a controlled time of deposition, very small quantities of noble metals are deposited, ranging from a few $\mu g\,cm^{-2}$ to a few hundred $\mu g\,cm^{-2}$.

This procedure usually leads to very active films for many electrocatalytic reactions: reduction of protons, or of dioxygen, oxidation of dihydrogen, and oxygenated molecules.

2. Inclusion of the electroactive species as counter-ion during the electropolymerization of the monomer. Another way to prepare noble metal modified electron-conducting polymers is to incorporate into the film a noble metal complex (e.g., $[PtCl_6]^{2-}$) as a counter-ion during the electropolymerization process. Then reduction or electroreduction of the metallic salt gives the corresponding metal particles. This procedure usually gives not very active Pt particles, maybe because the

particles are embedded into some polymer films, preventing them from activating the adsorption of the electroreactive species.

25.3 CHARACTERIZATION OF THE MODIFIED ELECTRON-CONDUCTING POLYMERS

25.3.1 Electrochemical Techniques

Cyclic voltammetry is a very useful technique to monitor the growth of an electropolymerized film and to evaluate its thickness. During the continuous growth of the polymer layer at an inert substrate (gold or glassy carbon electrodes) the current density at the redox peaks of the electroactive polymer increases continuously [Figure 1(a)]. Then the film thickness can be evaluated from the maximum current density (anodic or cathodic peaks) or from the quantity of electricity involved in the redox process [Figure 1(b)]. Empirical relations were proposed by Stilwell et al. [10] giving the film thickness of a PAni film as

$$d(\mu m) = (0.059 \pm 0.007)j_{\text{peak A}}(\text{mA/cm}^{-2}) + (0.04 \pm 0.09)$$

$$d(\mu m) = (7.2 \pm 1.5)Q_{\text{anodic}}(\text{C/cm}^{-2}) - (0.08 \pm 0.16)$$

Furthermore, the deposition of platinum-based metals can be made inside the polymer film by the electroreduction of the corresponding metallic salts at a constant electrode potential or during a voltammetric sweep. By using diluted solution of metallic salts (e.g., 10^{-4} M K_2PtCl_6, or 10^{-4} M K_2RuCl_5) and by controlling the deposition time (from a few seconds to a few minutes) it is easy to prepare electron-conducting polymer electrodes modified by a very small amount of catalytic metals (from a few $\mu g\,cm^{-2}$ to a few tenths of $mg\,cm^{-2}$). The metal loading can be estimated from the quantity of electricity involved in the electrodeposition process, as obtained by integration of the current versus time curves (chronoamperometry). For very small Pt loadings the voltammogramms are only slightly modified, so that the true surface area of the Pt-modified polymer is impossible to evaluate from the coulometry of the hydrogen adsorption–desorption region. Therefore, the CO stripping technique is preferably used. It consists of adsorbing a monolayer of CO from gaseous CO dissolved in the supporting electrolyte, followed by pure nitrogen bubbling to eliminate the excess CO in the electrochemical cell. Then an anodic sweep is applied to the electrode surface allowing us to estimate the true surface area from the quantity of electricity involved in the electro-oxidation peak of CO, assuming $420\,\mu C\,cm^{-2}$ for the oxidation of a full monolayer of adsorbed CO (Figure 2).

25.3.2 In-Situ Infrared Reflectance Spectroscopies

An important point to monitor the catalytic properties of modified ECP is to understand the mechanism of an electrocatalytic reaction, i.e., to identify the adsorbed intermediate species. During an electrocatalytic reaction the key steps always involve species coming from the dissociative chemisorption of reactants (organic molecules and solvent). These species can react at the electrode surface to give the reaction products (generally also adsorbed), but some of them can remain

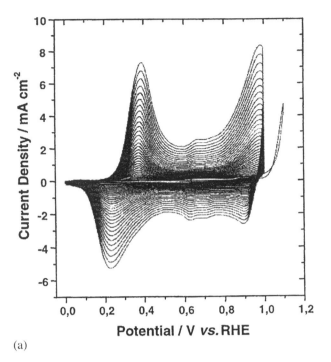

(a)

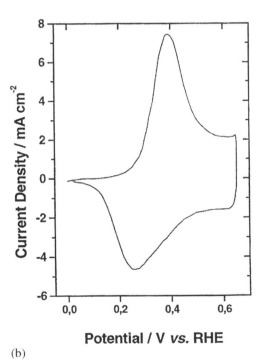

(b)

Figure 1 Cyclic voltammograms recorded at 50 mV/s during the electropolymerization on a gold surface of 0.1 M aniline in 0.5 M sulphuric acid (a), and of the polyaniline film in the supporting electrolyte (0.5 M sulphuric acid) alone after polymerization (b).

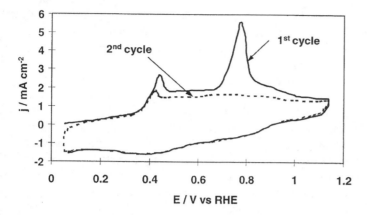

Figure 2 Cyclic voltammograms of a platinum-modified polyaniline–glassy carbon stationary electrode in the supporting electrolyte under nitrogen atmosphere after adsorption of a CO monolayer (0.5 M H_2SO_4; 300 µg Pt cm^{-2}; 20 mV s^{-1}; 25 °C).

strongly adsorbed (poisoning species). The determination of the nature and structure of the different species and above all of their coverage versus potential is essential to understand the overall process. This can be obtained by using in-situ infrared reflectance techniques designed to get information on the adsorbed layer during the electrochemical reaction. Such techniques are now used routinely, even if such experiments are still difficult to do because of the weakness of the reflectivity change due to the adsorbed layer, particularly at rough surfaces, and because of the strong absorption due to the water layer for experiments performed in aqueous medium.

Typically, in-situ infrared reflectance spectroscopic measurements are carried out with a Fourier transform infrared spectrometer, using an electrochemical cell specially designed to allow us to carry out electrochemical experiments in a thin layer of electrolyte. The working electrode surface is pushed against the transparent window (e.g., CaF_2) of the spectro-electrochemical cell. Two acquisition modes are generally used during the spectroscopic measurements: (1) single potential alteration infrared reflectance spectroscopy (SPAIRS); (2) subtractively normalized interfacial Fourier transform infrared spectroscopy (SNIFTIRS). The first mode consists of the acquisition of interferograms at several electrode potentials, during the first voltammetric sweep, which is conducted at a very low sweep rate (e.g., 1 mV s^{-1}). The resulting spectra are calculated as the relative reflectivity change $\Delta R/R = (R_{(E)} - R_{Eref.})/R_{Eref.}$, where $R_{Eref.}$ is the reflectivity taken at a reference potential, corresponding generally to the lowest potential limit of the forward scan. The second mode consists of the acquisition of interferograms at two different potentials, E_1 and E_2. Spectra are also calculated as the relative reflectivity change $\Delta R/R = [R(E_2) - R(E_1)]/R(E_1)$. The potential range between 0 and 900 mV versus RHE is usually investigated, keeping the difference $E_2 - E_1$ constant (e.g., $E_2 - E_1 = 0.2$ V). According to these calculations, reaction products appear as negative bands, while reactants appear as positive bands.

25.3.3 Other Characterizations

Among the specific characterizations needed for such modified conducting polymers, the determination of the metallic loading and of the atomic composition of the metallic particles inserted in the organic matrix are essential.

The observation of the structure and determination of the composition of particles can be carried out by transmission electron microscopy (TEM) in connection with energy dispersive x-ray (EDX) analysis.

The direct analysis of a modified conducting polymer is rather difficult to carry out except when the metallic loading is high. For other cases, it is generally necessary to prepare the sample in order to have a better observation. It consists of dissolving the conducting polymer (in ethanol in the case of polyaniline) with an ultrasonic stirring bath. Part of the suspension was then deposited on a copper grid and the ethanol was evaporated before any measurement.

25.4 KINETICS OF ELECTROCATALYTIC REACTIONS ON ECP MODIFIED BY PT-BASED CATALYSTS

Most of the electrocatalytic reactions involved in low-temperature fuel cells (i.e., the dioxygen reduction, the dihydrogen oxidation, or the electro-oxidation of small organic molecules, such as methanol and C1 molecules) occur on platinum and platinum-based catalysts dispersed on a convenient substrate. Graphite, carbon black, and carbon powders are the most favorite substrates used in fuel cells, because of their large specific surface area (up to $2000\,m^2\,g^{-1}$), their good electronic conductivity, and their low cost. In addition, electron-conducting polymers are good substrates to highly disperse the catalytic material. The catalytic electrodes are usually realized by reduction process or Doctor blade process, which are not very well controlled in terms of properties and reproducibility. In the case of electron-conducting polymers used as substrates, both the polymer film and the metal deposit can be realized by electrochemical techniques, such as cyclic voltammetry and/or chronoamperometry allowing us to control the film thickness and structure, and the metal dispersion and loading.

25.4.1 Oxygen Reduction Reaction

Metallic particles, particularity Pt, have been dispersed in an electron-conducting polymer, such as PAni and PPy, in order to improve their catalytic behavior toward the oxygen reduction reaction (ORR).

Apparently, Holdcroft and Funt were the first to disperse platinum microparticles into a conducting film of PPy in order to activate the oxygen reduction reaction [11]. Different methods of preparation of the PPy/Pt electrodes were developed leading to a film thickness ranging from 0.08 to 0.8 μm with a Pt loading ranging from $0.8\,\mu g/cm^2$ to $30\,\mu g/cm^2$. The oxygen electroreduction was investigated at different pHs (from 2 to 12) with a rotating ring-disk electrode using a glassy carbon disc on which the modified Pt/PPy electrodes were deposited and a Pt ring. Levich plots of the oxygen reduction current density deviates from linear at low

Pt loadings and a large film thickness. This results mainly from a low O_2 permeation through the film, particularly for PPy films containing homogeneously dispersed Pt. Later on, Vork and Barendrecht [12] investigated again the ORR in 0.5 M H_2SO_4 at a rotating ring-disk electrode, with a glassy carbon disk first covered by a PPy film of 0.28-μm thickness, then modified by platinum electrodeposition at a constant cathodic current density (Pt loading from 25 to 200 μg cm^{-2}). The results depended on the magnitude of the current density used to deposit Pt. At a high current density most of the Pt particles are concentrated near the polypyrrole–electrolyte interface, leading to a four-electron reduction to water, as obtained with bulk platinum. At low current densities the Pt particles are more evenly distributed in the whole film, leading to a low conversion rate of O_2 and thus to the production of hydrogen peroxide. An increase in the Pt loading gives rise to a higher reduction rate, i.e., a higher selectivity toward the production of water.

The electroreduction of dioxygen in 0.5 M H_3PO_4 at polypyrrole film rotating disk electrodes containing nanodispersed colloidal Pt particles was also studied by Chen et al. [13]. In contrast to the results of Holdcroft et al. [11], there is no limitation of the current density by O_2 permeation through thick PPy films (up to 5 μm). The number of electrons involved, n, and the kinetic currents, I_k, are evaluated from Koutecky–Levich plots [14]: n increases from 2 (reduction of O_2 to H_2O_2) to 4 (reduction to H_2O) when the Pt loading increases, while the kinetic current I_k reaches a plateau at high Pt loadings (greater than 25 wt. % of the film).

The use of a PPy film as a convenient matrix for dispersing an oxygen reduction electrocatalyst is questionable since the PPy matrix can react itself with O_2, leading to H_2O_2 by a two-electron transfer reaction [13,15]. Hydrogen peroxide can then oxidize the PPy film, leading to degradation of the polymer and to a decrease of the electrocatalytic activity with time.

Other metallic catalysts, particularly palladium particles, were electrodeposited in polythiophene [16], polypyrrole [17], or polyaniline [18] and investigated for their electrocatalytic activity toward the ORR [16,17]. In particular, Yassar et al. [16] controlled the size of the Pd particles by adjusting the duration of the potential pulses used for Pd electrodeposition: particles sizes from 6–10 nm to 300–600 nm were obtained when the pulse duration was varied from 0.1 to 1000 ms. Increasing the Pd loading leads to an increase of the particle size (e.g., from 10 to 23 nm for Pd loading increasing from 6% to 23%). The specific activity of the Pd/PMeT electrode for the ORR increases from 30 A g^{-1} to 160 A g^{-1} when the particle size decreases from 300 to 6 nm. This is confirmed by a rotating disk electrode study. More recently Scharifker et al. [17] studied the ORR at Pd particles incorporated into a PPy film in 0.5 M H_2SO_4 saturated with O_2, compared to Pd particles deposited on a glassy carbon electrode. They observed an apparent higher rate of O_2 reduction in the first case.

With the aim of using them in an H_2/O_2 PEMFC, Lai et al. prepared composite films of Nafion- and PAni-containing Pt particles as electrocatalysts for the oxygen electroreduction reaction [19]. Films cast from a solution of PAni and Nafion containing chemically reduced Pt displayed good electrocatalytic activity compared to films without PAni.

In a recent paper, Coutanceau et al. made a detailed kinetic study of the ORR on Pt particles dispersed in a polyaniline film with several Pt loadings ranging from a few μg cm^{-2} to a few hundred μg cm^{-2} [20]. The PAni film (0.5-μm thickness) was

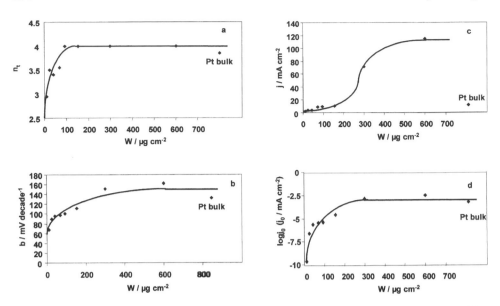

Figure 3 Plots of the total number of electrons n_t (a), of the Tafel slope b (b), of the limiting current density I/A_r, (c) and of the exchange current density I/A_r (d), versus the platinum loading for the reduction of oxygen on a platinum-modified polyaniline–glassy carbon rotating disk electrode (O_2 saturated 0.5 M H_2SO_4; 2 mV s^{-1}; 25 °C; A_r is the true surface area).

electrodeposited by cyclic voltammetry (CV) on a glassy carbon electrode, then Pt particles were introduced in a controlled way by electroreduction at a constant potential [-0.55 V versus mercury sulphate electrode (MSE)] of 10^{-3} M K_2PtCl_6 in 0.5 M H_2SO_4. The platinum loading was varied from 10 to 600 µg cm^{-2} and the platinum deposits were characterized by CV (adsorption–desorption of hydrogen, CO stripping) and by TEM. The ORR kinetics was analyzed by a rotating ring-disk electrode allowing them to separate the different contributions in the overall process: diffusion of molecular oxygen in the electrolyte solution, diffusion inside the PAni film, adsorption process, and electron transfer. From linear Koutecky–Levich plots, the kinetic parameters as functions of the Pt loading were evaluated: total number of exchanged electrons, adsorption limiting current density, Tafel slope, and exchange current density. All these parameters increase with an increase of the Pt loading, reaching the bulk values for loading greater than 200 µg cm^{-2} (Figure 3). At lower Pt loadings ($<$100 µg cm^{-2}) the kinetic is controlled by a Temkin adsorption isotherm, and fewer than four electrons are involved in the ORR, which is confirmed by the detection of hydrogen peroxide at the ring electrode. This last result agrees with what was previously observed by Vork and Barendrecht [12] and by Chen et al. [13].

Other conducting polymers can be investigated as supports for dispersing catalytic metallic particles. Shan and Pickup [21] used a composite of poly(3,4-ethylenedioxythiophene) and poly(styrene-4-sulfonate) (PEDOT/PSS) to disperse Pt particles. They compared the performances of such electrodes with carbon-supported Pt for the electroreduction of oxygen, and they found similar exchange current densities and mechanism. The only difference is the Pt utilization, which is lower in the case of PEDOT/PSS. This seems to be due to electronic isolation of some Pt particles and partial poisoning of these particles.

25.4.2 Hydrogen Oxidation Reaction

The hydrogen oxidation reaction (HOR) was also investigated at noble metals (e.g., Pt and Pd) incorporated in an electron-conducting polymer (e.g., PPy, PAni and PMeT).

The first relevant study is that of Vork et al. [1], who investigated the electro-oxidation of hydrogen dissolved into a 0.5 M H_2SO_4 solution at a glassy carbon (GC) rotating disk electrode modified by a polymer film, mainly polypyrrole, but also polymethylpyrrole (PMP) and polyaniline, in which different amounts (ranging from a few $\mu g\,cm^{-2}$ to $250\,\mu g\,cm^{-2}$) of platinum were electrodeposited at a constant cathodic current density. For platinum loadings greater than $50\,\mu g\,cm^{-2}$, the Pt/PPy/GC electrodes behave similarly to bulk platinum, leading to a limiting apparent current density following the Levich equation at potentials higher than 90 mV/RHE for all rotation speeds. A bare GC electrode and a GC/PPy electrode without platinum particles do not give any hydrogen oxidation current. However, the use of PMP or PAni as a conducting polymer for the preparation of Pt/polymer/GC electrodes containing $0.1\,mg\,Pt\,cm^{-2}$ leads to apparent high overvoltages for hydrogen oxidation (300 mV and 350 mV, respectively), which may be caused by the low electronic conductivity of these two polymers in their reduced state [1]. A similar behavior was obtained with a poly-3-methylthiophene electrode containing $0.1\,mg\,cm^{-2}$ of Pt with an overvoltage of about 200 mV [22].

However, such a behavior was not observed in a more recent work by Laborde et al. [8], where hydrogen oxidation at a PAni film containing $0.1\,mg\,cm^{-2}$ of platinum takes place at low overpotentials, as usual for an electroactive material. In another paper [23] Laborde et al. observed the electro-oxidation of dissolved hydrogen at a copolymer (polypyrrole-polydithiophene) electrode modified with $0.1\,mg\,cm^{-2}$ of platinum, leading to a very small overvoltage.

On the other hand, Chen et al. developed polypyrrole film electrodes containing nanodispersed platinum particles and investigated their catalytic properties for the electro-oxidation of hydrogen [13]. They confirm the remarkable electrocatalytic activity of these PPy/Pt films compared to bulk platinum electrodes, as the film thickness increases to 5 μm.

Using polyaniline as the electron-conducting polymer matrix, Croissant et al. [24] investigated by the RDE technique, the HOR kinetics as a function of the platinum loading [5 to $300\,\mu g\,cm^{-2}$]. It was thus possible to separate the different contributions, e.g., diffusion of molecular hydrogen in the electrolyte, diffusion inside the 0.5-μm PAni film, adsorption step, electron transfer step, in the overall oxidation process. Then the following kinetic parameters were evaluated: total number of exchanged electrons, limiting current of diffusion inside the film, limiting adsorption current density, Tafel slope, and exchange current density. The values obtained increase with the Pt loading until reaching the values for bulk platinum at a Pt loading greater than $100\,\mu g\,cm^{-2}$. The variation of the Tafel slope from 50–$40\,mV\,dec^{-1}$ at low Pt loading to $30\,mV\,dec^{-1}$ at high Pt loading was interpreted by a change in the kinetics control from a slow Heyrowski step to a slow Tafel step [25].

More recently Bouzek et al. investigated the effect of the preparation conditions of Pt-modified polypyrrole films on their electrocatalytic properties for the HOR [26]. Three methods were considered: (1) cathodic deposition of Pt from H_2PtCl_6 in the previously synthesized film, (2) incorporation of colloidal Pt particles during the electropolymerization of polypyrrole; (3) incorporation of $[PtCl_4]^{2-}$ as a counter-ion during the electropolymerization process and its subsequent reduction. Only the first two methods lead to active electrocatalytic films, whereas the last one gives very poor catalysts, maybe because the Pt particles are embedded in the PPy structure and therefore are not accessible to the reactant.

25.4.3 Electro-Oxidation of Small Organic Molecules

The electrocatalytic oxidation of many small organic molecules was carried out at Pt-based catalysts dispersed in an ECP, particularly that of C1 molecules (formic acid, formaldehyde, and methanol).

Electrocatalytic Oxidation of Formic Acid and Formaldehyde

The oxidation of formic acid was one of the first electrocatalytic oxidations at ECP modified by platinum particles, which was studied by Gholamian et al. [27]. They observed that the incorporation of $0.1\,mg\,cm^{-2}$ of Pt into a PAni film enhanced greatly the oxidation rate of formic acid (10 times enhancement). They evaluated the optimum film thickness (around 1 μm) for a maximum enhancement of the oxidation current. They also determined the resistance profile of the PAni film correlatively to the catalytic activity of the Pt particles and found that the maximum electroactivity occurs within the conducting potential window of the polymer. However, this correlation is not definitively established since they do not determine the conductivity of the PAni film in the presence of Pt particles, which must be higher than in the absence of the metallic particles.

Later on many other investigations were carried out on the electro-oxidation of formic acid at Pt and Pt-based catalysts dispersed in a conducting polymer, particularly in our laboratory [8,23,28]. Different conducting polymers, such as PAni [8,28], polythiophene [29], poly-o-toluidine [30], and poly(2-hydroxy-3-amino) phenazine [31], were used as conducting substrates to disperse Pt particles. In all cases, a great enhancement of the electrocatalytic activity was observed, with higher current densities, and above all with a negative shift of the polarization curves (smaller overvoltage). Moreover, the addition of a second metal, such as Ru [28], Sn [28,31], or Pb [29,32] or the modification of the Pt dispersed catalyst by some ad-atoms, such as Pb, Tl, or Bi [31], greatly improves the electrocatalytic behavior of these electrodes, leading to a 200-mV negative shift of the polarization curves and to the quasi-absence of any poisoning effect due to some strongly adsorbed intermediate species (e.g., linearly and bridge-bonded CO) [33].

A very few studies concern the electro-oxidation of formaldehyde at Pt-modified electron-conducting polymers [8,28,33]. It is again observed that dispersion of Pt particles, and above all of Pt-Ru, Pt-Sn, or Pt-Ru-Sn particles in a polyaniline film, greatly enhances the electro-oxidation of formaldehyde (Figure 4). Pt-Sn is again the best electrocatalyst leading to a negative shift of the polarization curves as high as 400 mV, certainly because the presence of Sn prevents the electrode poisoning at low potentials by adsorbed CO [34].

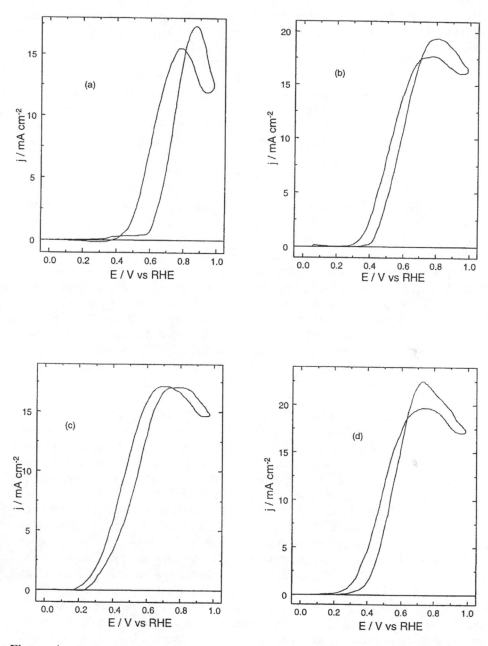

Figure 4 Cyclic voltammograms of PAni/Pt-based electrodes in 0.1 M $HClO_4$ in the presence of 0.1 M HCHO; (a) PAni/Pt; (b) PAni/Pt-Ru; (c) PAni/Pt-Sn; and (d) PAni/Pt-Ru-Sn.

Electro-Oxidation of Methanol

Among the numerous studies concerning the electro-oxidation of small organic molecules, those dealing with methanol are the most common due to the huge potential expected impact from direct methanol fuel cells (DMFC) for various applications [35]. It is well known that one of the key points slowing down the development of DMFC is related to the slow kinetics of the methanol electrocatalytic reaction.

In a DMFC, the oxidation of methanol occurs in an acidic medium to avoid the carbonation of the electrolyte. Unfortunately, under such conditions, only platinum is stable and is able to dissociate the methanol molecule at low potentials. However, this dissociation leads spontaneously to the formation of an adsorbed layer of poisoning species (identified as adsorbed carbon monoxide). This poison can be removed from the electrode surface only at higher potentials when it is oxidized into carbon dioxide.

Two problems are directly related to electrocatalysis, i.e., to increase the overall rate of the reaction of electro-oxidation of methanol together with a reduction of the amount of platinum (and more generally of precious metals) in the electrode catalytic layer, and to improve the activity at low potentials by adding one or more other metals to modify the platinum catalytic properties.

The decrease of the metal loading can be achieved by reducing the size of the particles. There are now various techniques that allow one to obtain nanoparticles highly dispersed in a conducting support necessary to collect the current at the electrode. Even if carbon powder is actually the most convenient support for such a purpose, other conducting matrices can be considered such as the electron-conducting polymers. This alternative was widely studied during the last decade [35,36], mainly for more fundamental studies. In this way, it is possible to prepare not only electrodes with a very low metal loading but also multimetallic electrodes (generally Pt-based) with the objective to act on the kinetics of the reaction as mentioned above.

As it was already mentioned in the experimental section, a convenient procedure to obtain multimetallic electrodes is to electrocodeposit the different metals directly into the polymer-conducting matrix. With this technique it is possible to vary the amount and the composition of the catalyst easily to obtain electrodes with optimum performances. It is thus demonstrated that an optimum amount of platinum can be found for the oxidation of methanol [23,35].

Kost et al. [37] published one of the first papers dealing with the oxidation of methanol at platinum particles dispersed into a polyaniline film. A remarkable stability of the Pt/PAni electrodes was observed during long-term oxidation of methanol (several tens of hours) at a constant potential with small Pt loadings (10 to $30 \, \mu g \, cm^{-2}$), compared to bulk platinum. This behavior was confirmed later on by Ocon-Esteban et al. [38], with different Pt/PAni electrodes containing from $30 \, \mu g \, cm^{-2}$ to $100 \, \mu g \, cm^{-2}$ of Pt, as estimated by a spectrophotometric analysis as suggested by Gholamian et al. [27]. An enhanced electrocatalytic activity (100 times higher for a 100-$\mu g \, cm^{-2}$ platinum loading, compared to bulk platinum) was observed. Moreover, a decrease in the poisoning of the electrode active sites is observed, as confirmed by in-situ infrared reflectance spectroscopy [8,39].

Similar observations concerning the electro-oxidation of methanol were made by Ulmann et al. [40], using platinum microparticles dispersed into polypyrrole films (from 100- to 700-nm thickness) deposited on a gold electrode. The platinum loading was varied from 10 to $300 \, \mu g \, cm^{-2}$, leading to an increase in the current density for platinum loadings up to $150 \, \mu g \, cm^{-2}$, after which the current density reaches a plateau.

PPy films modified by platinum catalyst particles were also considered for electrocatalytic reactions (oxygen reduction and methanol oxidation) by Hepel et al. [41]. The incorporation of a $PtCl_4^{2-}$ anion was performed during the electropolymerization of pyrrole and monitored by the electrochemical quartz crystal microbalance (EQCM) technique, allowing us to evaluate the amount of platinum obtained after reduction of the $PPy/PtCl_4^{2-}$ film.

A recent publication [42] clearly confirms that according to the amount of platinum particles inserted in a polyaniline matrix, the mechanism of the oxidation of methanol can be strongly modified, mainly for very low loadings (around $30 \, \mu g \, cm^{-2}$ of platinum). According to these authors, it seems that the conducting polymers play a significant role in the adsorption of methanol and its further dissociation when the catalyst is highly dispersed.

Polypyrrole was often used as support for platinum particles. Similarly to the case of polyaniline, the activity of such electrodes for the oxidation of methanol depends both on the amount of platinum and on the thickness of the polymer film [43]. In the same study, by using in-situ infrared spectroscopy, it was confirmed that linearly adsorbed CO species are the only detectable species present at the electrode surface. The authors attributed the enhancement of the overall activity observed to the high and uniform dispersion of the metallic particles with, possibly, an effect of the conducting polymer matrix itself. The same conclusions were drawn from another study [44] where the size of the particles obtained by electrodeposition was estimated at 10 nm. In this study, the Pt particles were entrapped into the polymer layer and showed a better activity than particles only deposited on the polymer surface. The authors interpreted their results as a decrease of the poisoning phenomenon in the 3D film in comparison to the only 2D deposit.

However, it is well known that pure platinum is not the best electrocatalyst for methanol oxidation [45]. The addition of at least one second metal is necessary. Pt-Ru catalysts are widely known to give the best performances during the electro-oxidation of methanol. By using a conducting polymer (typically polyaniline or polypyrrole) as a support, it was easy to vary the amount and composition of the metallic electrocatalyst [9,33] to get the optimum performances. Then, to improve the electrocatalytic properties of methanol electrodes, and to reduce the poisoning phenomena observed with platinum, different platinum-based alloys, such as Pt-Ru and Pt-Sn [45], have been considered for a long time. Therefore, such bimetallic systems were also dispersed into electron-conducting polymers. Hable et al. [6,46] were apparently the first authors to disperse Pt-Sn catalyst particles in a polyaniline matrix, with the aim of enhancing the behavior of platinum for the oxidation of methanol. The Pt/Sn ratio was evaluated by x-ray photoelectron spectroscopy, and Pt/Sn ratios of 10:1 were shown to be sufficient to enhance the electrocatalytic oxidation of methanol. Pt was found to be in the Pt(0) state, whereas Sn was in an oxidized form. A significant enhancement was also observed with PAni/Pt-Ru, but from XPS measurements, Ru was observed in two states, Ru(0) and Ru(IV). Similar observations concerning the enhancement of the electrocatalytic activity of Pt-Sn particles inserted in PAni were made by Laborde et al. [47]. Pt-Sn catalysts were also dispersed in poly-3-methylthiophene-conducting polymers to investigate the effect of the catalyst support and Pt loading on the electro-oxidation of methanol [48]. Methanol oxidation currents were appreciable for a platinum loading higher than

$60\,\mu g\,cm^{-2}$, and the presence of a few percent of tin, as analyzed by EDX, decreases the overvoltage by about 150 mV.

Until now, for methanol oxidation the best bimetallic catalyst was found to be Pt-Ru. Several papers deal with the electro-oxidation of methanol at Pt-Ru bimetallic system dispersed in polyaniline [33,46]. From results with bulk alloys, the optimum Pt/Ru ratio of around 6:1 to 4:1 was found [49] and confirmed [50]. The electroactivity of Pt-Ru-modified polyaniline is much better than that displayed by pure Pt particles dispersed into the PAni film. The optimum composition of the Pt-Ru bimetallic system was confirmed from these results [33]. The decrease of the poisoning phenomenon is the consequence of a low coverage in adsorbed CO resulting from the chemisorption of methanol. This was checked by considering the oxidation of CO at the same Pt-Ru/PAni-modified electrode [34], which occurs at low overvoltages (150 mV) in the presence of Ru.

The effect of ruthenium on the oxidation of methanol on Pt is generally interpreted by the ability of ruthenium to increase the coverage in adsorbed OH of the electrode at low potentials. These OH species are necessary to oxidize strongly adsorbed species coming from the methanol dissociation. The optimum composition of Pt-Ru electrocatalysts, i.e., 4:1 to 6:1 atomic ratios, corresponds to a number of Pt sites necessary to accommodate the different methanol adsorbed residues, compared to one Ru site for an optimum coverage in adsorbed OH [50]. However, according to other authors [51], when ruthenium is dispersed in a form of highly reactive Ru oxo centers [polynuclear ruthenium (III,IV) oxide/cyanoruthenates], an improvement of the reaction rate of methanol oxidation with the polypyrrole Pt-Ru modified film is observed. This improvement may be due to the increase in the stability of the conducting polymer film after the incorporation of the polynuclear RuO/CNRu entities.

According to mechanistic considerations, it appears that it seems to be necessary to increase the coverage of adsorbed OH of the electrode surface at low potentials. This can be obtained by the modification of the metallic particles by adding a third metal able to dissociate water, i.e., to form adsorbed OH, at low potentials. Moreover, the catalyst needs to be stable in a strong acidic medium, which is necessary for the DMFC operation.

A large screening was recently done to identify such a third metal, X, to add to platinum and ruthenium [52]. Figure 5 summarizes the behavior of the nine investigated Pt-Ru-X trimetallic electrocatalysts toward methanol oxidation. At low potentials, the Pt-Ru-Mo ternary catalyst gives the highest current densities compared to other ternary electrocatalysts. This catalyst exhibits a current density 10 times greater than Pt-Ru at a potential of 400 mV versus RHE under steady-state conditions (data taken after 5 minutes).

Several of the other tested electrocatalysts give weaker performances than Pt-Ru, regardless of the anode potential, but some of the ternary catalysts such as Pt-Ru-Co, Pt-Ru-Ni, or Pt-Ru-Fe, exhibited good performances but showed, however, poor stability under working conditions, the corrosion of the three metals, Co, Ni, and Fe, occurring in strong acidic media.

These results clearly confirm that modifying a conducting polymer with metallic particles containing Pt, Ru, and Mo leads to enhanced catalytic performances versus the oxidation of methanol in comparison to the behavior of binary catalysts (Pt-Ru).

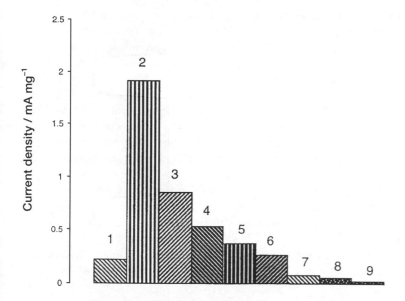

Figure 5   Current densities for the oxidation of 1 M methanol in 0.1 M $HClO_4$ on several PAni/Pt-Ru-X electrodes recorded after 5 minutes of potential hold at 400 mV versus RHE: 1. PAni/Pt-Ru; 2. PAni/Pt-Ru-Mo; 3. PAni/Pt-Ru-Co; 4. PAni/Pt-Ru-W; 5. PAni/Pt-Ru-Fe; 6. PAni/Pt-Ru-Ni; 7. PAni/Pt-Ru-Cu; 8. PAni/Pt-Ru-Sn; and 9. PAni/Pt-Ru-Au.

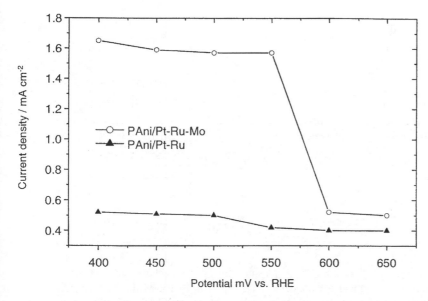

Figure 6   Current densities showing the behavior of two catalysts (Pt-Ru and Pt-Ru-Mo) toward the oxidation of 1 M methanol in 0.1 M $HClO_4$ measured at 400 mV versus RHE after 5 minutes, following different staircase potential to test the stability of the ternary electrode.

An important point observed in these studies is the rather good stability of the ternary Pt-Ru-Mo electrode at relatively high anode potentials. This can be seen by applying a simple potential program to the ternary catalyst in the presence of methanol in order to evaluate the potential at which the electrode is irreversibly altered. It appears that the ternary PAni/Pt-Ru-Mo electrocatalyst is to be stable after 5 minutes of methanol oxidation up to 550 mV versus RHE, while the binary catalyst Pt-Ru seems to be stable for the whole explored potential range (Figure 6). This relative good stability of the ternary catalyst is somewhat unexpected, especially for the third element, molybdenum, which is expected to be completely oxidized to Mo(III) or to Mo(VI) (MoO_3) at the anodic potential and pH encountered in a PEM-DMFC.

The role of molybdenum can be clarified by using information coming from in-situ infrared reflectance spectroscopy. Such a technique allows us to follow the variation in the coverage of the different adsorbed species involved during the oxidation of methanol.

The infrared results are summarized in Figure 7 and in Table 1. These data compare the results obtained during both the oxidation of methanol and the

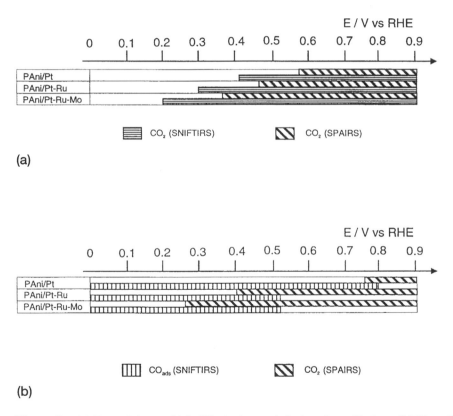

Figure 7 (a) Potentials at which CO_2 is detected during the oxidation of 1 M methanol in 0.5 M $HClO_4$ on three modified PAni electrodes (Pt, Pt-Ru, Pt-Ru-Mo), using both SPAIRS and SNIFTIRS techniques. (b) Potentials at which CO_{ADS} and CO_2 are detected on three modified PAni electrodes (Pt, Pt-Ru, Pt-Ru-Mo) in 0.5 M $HClO_4$, after CO adsorption, using both SPAIRS and SNIFTIRS techniques.

Table 1　Potentials (mV Versus RHE) at which the CO_2 Infrared SPAIRS Band Is Detected for Different Electrodes, PAni/Pt, PAni/Pt-Ru and PAni/Pt-Ru-Mo, During the Oxidation of Methanol and Gaseous CO

	Methanol Oxidation	CO_{ADS} Oxidation
PAni/Pt	550	750
PAni/Pt-Ru	450	400
PAni/Pt-Ru-Mo	350	250

oxidation of CO adsorbed after saturation of the solution by gaseous carbon monoxide. This comparison is instructive as adsorbed CO is the main poisoning species formed during the dissociation of the methanol molecule.

The promoting effect of the third component is also compared with the case of Pt and Pt-Ru catalysts dispersed in PAni. During the oxidation of methanol, the production of carbon dioxide (final product) is observed at a potential as low as 350 mV versus RHE on PAni/Pt-Ru-Mo. Concerning the case of CO adsorption from gaseous CO, formation of CO_2 is observed at 250 mV versus RHE, indicating clearly that Pt-Ru-Mo is less poisoned by CO_{ADS}, in comparison with Pt-Ru and Pt (the formation of CO_2 occurring, respectively, at 400 mV and 750 mV versus RHE).

The potentials at which CO_2 is formed, during the electro-oxidation of methanol or that of CO (from gaseous CO), are summarized in Table 1. These threshold potentials were obtained from infrared data. It clearly appears that the distribution of CO_{ADS} at the surface of the electrocatalyst depends greatly on the source of adsorbed CO, confirming undoubtedly the presence of other adsorbed species in the case of methanol.

The difference in the nature of adsorbed CO (coming from gaseous or during methanol oxidation) is clarified in Figure 8. With PAni/Pt and PAni/Pt-Ru, CO_{ADS} is observed at wavenumbers close to $2050 \sim 2060 \, cm^{-1}$. Moreover, the CO_{ads} band is narrow (10 to $20 \, cm^{-1}$). As potential increases, a slight increase of the average wavenumber is observed, as expected ("Stark effect"). With the ternary catalyst, PAni/Pt-Ru-Mo, CO_{ads} is detected at wavenumbers close to $2015 \sim 2025 \, cm^{-1}$, with a wide band (30 to $50 \, cm^{-1}$). This suggests that linearly bonded CO_{ads} is more weakly adsorbed on the ternary catalyst surface compared with that observed on platinum and on platinum-ruthenium. Moreover, the average wavenumber decreases when the potential increases. These observations prove a lower CO_{ADS} coverage on the ternary catalyst in comparison to the cases of Pt and Pt-Ru catalysts.

In order to test the electrocatalytic activity of the Pt-Ru-Mo ternary electrocatalyst in a direct methanol fuel cell, diffusion electrodes were prepared by electrodeposition of a thin layer of a polyaniline film on a carbon cloth, followed by deposition of $2 \, mg \, cm^{-2}$ of catalyst by electroreduction of a mixture containing K_2PtCl_6, K_2RuCl_5, and $MoCl_5$. Then a membrane-electrode-assembly was realized by hot-pressing this anode on a Nafion 117 membrane, on which is also pressed an oxygen cathode (E-Tek catalyst containing $2 \, mg \, Pt \, per \, cm^2$ dispersed on a carbon powder, e.g., Vulcan XC72). The results obtained, giving the cell voltage and the power density versus the current density (Figure 9), clearly show that the ternary

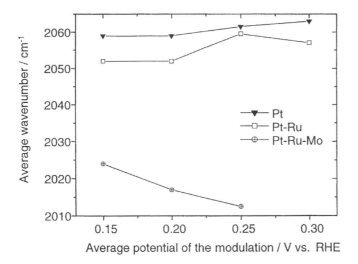

Figure 8 Evolution of the average wavenumber of the CO_{ADS} band as a function of the average potential of the modulation for the three electrodes considered (Pt, Pt-Ru, Pt-Ru-Mo).

catalyst PAni/Pt-Ru-Mo is much better (1.5 times more active) than the binary catalyst PAni/Pt-Ru.

Electrocatalytic Oxidation of CO

Carbon monoxide is a key molecule in the electro-oxidation of C1 compounds and of many alcohols, since it is always produced by the dissociative chemisorption of the molecule, and since it may block the active catalytic sites. Therefore, its electro-oxidation on platinum-based metals dispersed in an electron-conducting polymer, such as PAni, was investigated for a long time in our laboratory [8,28,34].

Pt, Pt-Ru, and Pt-Sn dispersed electrodes were shown to be good electrocatalysts for CO oxidation at low potentials. All these electrodes lead to a large negative shift, compared to smooth platinum, of the polarizations curves, as high as 0.6 V for Pt-Sn/PAni electrodes (Figure 10). In the latter case the active sites are not more blocked at low potentials so that the overall kinetics is not adsorption-limited, but rather becomes diffusion-limited. It was thus possible to analyze quantitatively the reaction mechanism using a rotating disc electrode, consisting of a gold disc substrate covered by a 0.5-μm-thick layer of PAni in which $0.1\,mg\,cm^{-2}$ of Pt-Sn bimetallic catalyst was dispersed [34]. The kinetic parameters were then evaluated, i.e., the charge-transfer coefficient $\alpha n = 0.51$, the exchange current density $j_o = 0.0052\,mA\,cm^{-2}$, and the adsorption limiting current density $j_{ads} = 4.64\,mA\,cm^{-2}$. Moreover, the diffusion coefficient, D_{CO}, of carbon monoxide in the perchloric acid solution was evaluated accurately as $D_{CO} = 2.1 \times 10^{-5}\,cm^2\,s^{-1}$ [34].

Electro-Oxidation of Alcohols and Polyols

The electro-oxidation of ethanol at platinum-based catalysts dispersed in an electron-conducting polymer was relatively sparsely investigated. It seems that

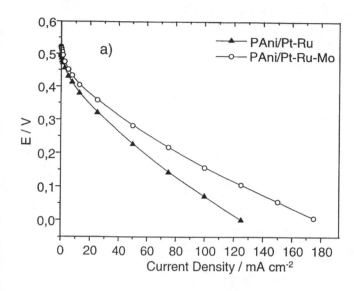

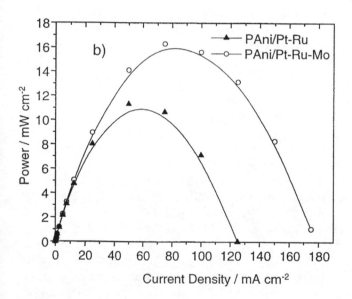

Figure 9 Potential versus current density (a) and power density versus current density (b) curves for two different anodes (PAni/Pt-Ru and PAni/Pt-Ru-Mo) in a direct methanol fuel cell (cell temperature 110 °C; 2 bar O_2 pressure, and $100\,mL\,mn^{-1}$ gas flow; 2 M methanol aqueous solution at $2\,mL\,mn^{-1}$ liquid flow).

Hable and Wrighton were the first to study the electrocatalytic oxidation of ethanol on Pt-Ru and Pt-Sn catalyst particles in PAni [46]. They found that dispersion of Pt, Pt-Ru, and Pt-Sn in PAni greatly enhanced the oxidation current of ethanol, the Pt-Sn electrocatalyst being far superior to the two others, with oxidation current densities reaching $7\,mA\,cm^{-2}$, and with a shift of the polarization curves of at least 300 mV toward the negative potentials.

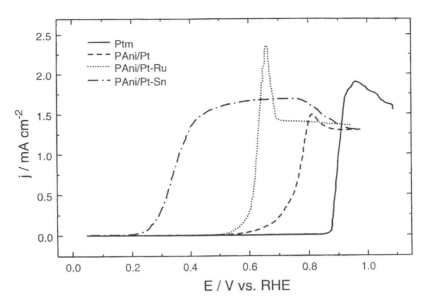

Figure 10 Comparison of the behavior of different platinum-based electrodes dispersed in a PAni film toward the oxidation of CO from a saturated aqueous 0.1 M HClO$_4$ solution (100 μg cm^{-2} metal loading; 5 mV/s sweep rate).

Similar conclusions were reached by Laborde et al. [47], with an optimum atomic ratio Pt/Sn of 3:1 for the dispersed catalyst (Figure 11).

The electro-oxidation of ethylene-glycol (EG) was investigated recently on Pt-doped polyaniline electrodes [53], and on Pt-based catalysts dispersed in polyaniline [54]. In the first case, it was shown that Pt dispersed into PAni gave oxidation current densities comparable to those obtained on bulk platinum, regardless of the film thickness [53]. Conversely, Kelaidopoulou et al. found that Pt dispersed in PAni is a better electrocatalyst than smooth Pt for the electro-oxidation of EG in an acid medium [54]. Furthermore, the modification of Pt by Ru or Sn codeposition greatly enhanced the electro-oxidation currents with quasi-superimposed forward and backward voltammograms (at 5 mV s^{-1}) confirming the quasi-absence of any poisoning effect (by adsorbed CO). However, the underpotential deposition of Tl and Bi ad-atoms inhibits the electro-oxidation of EG, whereas the codeposition of Pb improved the catalytic activity of Pt, but no better than Sn or Ru does it.

Finally, the oxidation of D-glucose at Pt-based electrocatalysts incorporated in polypyrrole [55,56] or in polyaniline [57] was also considered. The first work [55] was carried out in Pt-doped polypyrrole films in a neutral medium (phosphate buffer) in view of biosensor applications. Then the use of Pt-Pd catalysts dispersed in PPy led to higher current densities of glucose oxidation than on pure metal dispersed in PPy. This may be related to the decrease of catalytic poisoning (by adsorbed CO as shown by infrared reflectance spectroscopy [58]), due to the presence of Pd.

The use of PAni instead of PPy to disperse the platinum particles and the increase of the metal loading (from 100 μg cm^{-2} in [55,56] to a few mg cm^{-2} in [57]) greatly increase the oxidation current of D-glucose, which reached about 6 mA cm^{-2}

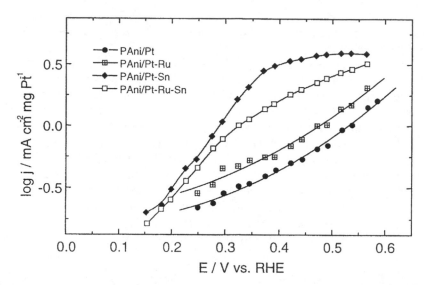

Figure 11 Tafel plots for the electro-oxidation of 0.1 M ethanol in 0.1 M HClO₄ on different Pt-based electrodes dispersed in a 0.5-μm PAni film containing 600 μg cm⁻² of electrocatalysts.

with $2\,mg\,cm^{-2}$ Pt [57]. On the other hand, the modification of Pt particles dispersed in PAni by the underpotential deposition of Tl, Pb, and Bi greatly enhances the electrocatalytic properties of Pt for the electro-oxidation of β-D-glucose, whereas the upd of Cd ad-atoms inhibits them [57].

25.5 CONCLUSIONS AND PERSPECTIVES

From the different examples described here, it clearly appears that electron-conducting polymers are convenient matrices to disperse electrocatalytic materials such as metallic particles or organometallic compounds. Thus, it becomes rather easy to obtain active electrodes with a high dispersion level of the electrocatalysts. One of the main advantages of using such a support is the possibility to design electrodes with chemical bonds between the electrocatalytic material and the supporting matrix. Then tailor-made catalytic electrodes can be conceived for specific reactions, for example, in the application field of organic fine electrochemistry.

The other application of the electron-conducting polymer as a matrix is the replacement of the carbon powder generally used as a support in fuel-cell electrodes. For such a purpose, conducting polymers present the advantage to facilitate the very high dispersion of metallic particles (a few μg cm⁻² is easy to obtain), even in the case of bimetallic particles. Furthermore, using electrodeposition techniques, it is easy to control the metal loading and the atomic composition of bimetallic catalysts. When such dispersion is obtained by electrodeposition, the size of particles is typically in the range of 3–4 nm, larger than particles obtained by other techniques.

Nevertheless, as discussed here, it was demonstrated that fuel-cell electrodes can be obtained with dispersion of plurimetallic particles in an electron-conducting polymer. Actually, the performances of these electrodes remain lower than the state of the art of "classical" electrodes prepared with carbon powder as a support. But, it should be noted that the optimization of fuel-cell electrodes with electron-conducting polymer has not yet been achieved.

The last topic to discuss is the stability of these electron-conducting polymers containing electrocatalytic materials. The long-term stability may be a problem, but it depends on the nature of the polymer used. The most important limitation is probably the temperature limits, mainly for fuel-cell applications. Actually, above 80° to 90°C, the stability of the conducting polymer is not sufficient for such applications. But synthesis of new polymers or copolymers can possibly be considered in the future. However, if it is actually difficult to imagine the use of these conducting polymers in large-scale fuel cells working at intermediate temperatures (100–200°C), numerous applications in small-scale fuel cells (micro-fuel cells) operating under ambient temperature conditions could be considered. In this latter case, electron-conducting polymers may have an unforeseeable future development as convenient matrices to highly dispersed electrocatalysts.

REFERENCES

1. F.T.A. Vork, L.J.J. Janssen, E. Barendrecht, Electrochim. Acta 31, 1569 (1986).
2. L. Coche, J.-C. Moutet, J. Amer. Chem. Soc. 109, 6887 (1987).
3. S. Holdcroft, B.L. Funt, Synth. Met. 28, C121 (1989).
4. M.F.G. Lyons, D.E. McCormack, P.N. Bartlett, J. Electroanal. Chem. 261, 51 (1989).
5. T.N.S. Kumar, A.Q. Contractor, Bull. Electrochem. 6, 333 (1990).
6. C.T. Hable, M.S. Wrighton, Langmuir 7, 1305 (1991).
7. C.S.C. Bose, K. Rajeshwar, J. Electroanal. Chem. 333, 235 (1992).
8. H. Laborde, J.M. Léger, C. Lamy, J. Appl. Electrochem. 24, 219 (1994).
9. C. Lamy, J.-M. Léger, F. Garnier, In *Handbook of Organic Conductive Molecules and Polymers*, H.S. Nalwa (ed.), Vol. 3 (1997) Chap. 10, pp. 471–496.
10. D.E. Stilwell, S.-M. Park, J. Electrochem. Soc. 135, 2491 (1988).
11. S. Holdcroft, B.L. Funt, J. Electroanal. Chem. 240, 89 (1988).
12. F.T.A. Vork, E. Barendrecht, Electrochem. Acta 35, 135 (1990).
13. C.C. Chen, C.S.C. Bose, K. Rajeshwar, J. Electroanal. Chem. 350, 161 (1993).
14. A.J. Bard, L.R. Faulkner, *Electrochemical Methods: Fundamentals and Applications*, John Wiley & Sons, New York (1980).
15. R.C.M. Jakobs, L.J.J. Janssen, E. Barendrecht, Electrochim. Acta 30, 1433 (1985).
16. A. Yassar, J. Roncali, F. Garnier, J. Electroanal. Chem. 255, 53 (1988).
17. A. Leone, W. Marino, B.R. Scharifker, J. Electrochem. Soc. 139, 438 (1992).
18. H.S. Li, M. Josowicz, D.R. Baer, M.H. Engelhard, J. Janata, J. Electrochem. Soc. 142, 798 (1995).
19. E.K.W. Lai, P.D. Beattie, F.P. Orfino, E. Simon, S. Holdcroft, Electrochim. Acta 44, 2559 (1999).
20. C. Coutanceau, M.J. Croissant, T. Napporn, C. Lamy, Electrochim. Acta 46, 579 (2000).
21. J. Shan, P.G. Pickup, Electrochim. Acta 46, 119 (2000).
22. H. Laborde, Ph.D. Thesis, University of Poitiers, November 1992.
23. H. Laborde, J.-M. Léger, C. Lamy, F. Garnier, A. Yassar, J. Appl. Electrochem. 20, 524 (1990).
24. M.J. Croissant, W.T. Napporn, J.-M. Léger, C. Lamy, Electrochim. Acta 43, 2447 (1998).
25. J.O'M. Bockris, S.U.M. Khan, *Surface Electrochemistry*, Plenum Press, New York, 1993, p. 314.

26. K. Bouzek, K.-M. Mangold, K. Jüttner, Electrochim. Acta 46, 661 (2000).
27. M. Gholamian, J. Sundaram, A.Q. Contractor, Langmuir 3, 741 (1987).
28. H. Laborde, J.-M. Léger, C. Lamy, J. Appl. Electrochem. 24, 1019 (1994).
29. R. Schrebler, M.A. del Valle, H. Gomez, C. Veas, R. Cordova, J. Electroanal. Chem. 380, 219 (1995).
30. P. Herrasti, R. Diaz, P. Ocon, New J. Chem. 17, 279 (1993).
31. A. Kelaidopoulou, E. Abelidou, G. Kokkinidis, J. Appl. Electrochem. 29, 1255 (1999).
32. P. Ocon, P. Herrasti, S. Rojas, Polymer 42, 2439 (2001).
33. W.T. Napporn, H. Laborde, J.-M. Léger, C. Lamy, J. Electroanal. Chem. 404, 153 (1996).
34. W.T. Napporn, J.-M. Léger, C. Lamy, J. Electroanal. Chem. 408, 141 (1996).
35. C. Lamy, J.-M. Léger, S. Srinivasan, In *Modern Aspects of Electrochemistry*, J'O.M. Bockris, B.E. Conway, R. White (eds.), Kluwer Academic/Plenum Publishers (New York), Vol. 34 (2001), pp. 53–118.
36. J.-M. Léger, J. Appl. Electrochem. 31, 767 (2001).
37. K. Kost, D. Bartak, B. Kazee, T. Kawana, Anal. Chem. 60, 2379 (1988).
38. P. Ocon-Esteban, J.-M. Léger, C. Lamy, E. Génies, J. Appl. Electrochem. 19, 462 (1989).
39. B. Beden, C. Lamy, In *Spectroelectrochemistry, Theory and Practice*, R.J. Gale (ed.), Chap. 5, Plenum Press, New York (1988), pp. 149–262.
40. M. Ulmann, R. Kostecki, J. Augustynski, D.J. Strike, M. Koudelka-Hep, Chimia 46, 138 (1992).
41. M. Hepel, M. Yi-Chen, U. Stimming, In *Proc. of the First International Symposium on New Materials for Fuel Cell System*, O. Savadogo, P.R. Roberge, T.N. Veziroglu (eds.), Ecole Polytechnique, Montreal (1995), p. 629.
42. A.A. Mikhaylova, E.B. Molodkina, O.A. Khazova, V.S. Bagotsky, J. Electroanal. Chem. 509, 119 (2001).
43. H. Yang, T. Lu, K. Xue, S.G. Sun, G. Lu, S. Chen, J. Electrochem. Soc. 144, 2302 (1997).
44. M. Hepel, J. Electrochem. Soc. 145, 124 (1998).
45. M.M.P. Janssen, J. Moolhuysen, Electrochim. Acta 21, 869 (1976).
46. C.T. Hable, M.S. Wrighton, Langmuir 9, 3284 (1993).
47. H. Laborde, A. Rezzouk, J.-M. Léger, C. Lamy, In *Proc. of the Symposium on Electrode Materials and Processes for Energy Storage and Conversion*, S. Srinivasan, D.D. McDonald, A.C. Khandar (eds.), Vol. 94–23, The Electrochem. Soc., Pennington (1994), p. 275.
48. S. Swathirajan, Y.M. Mikhail, J. Electrochem., Soc. 139, 2105 (1992).
49. H.A. Gasteiger, N. Markovic, P.N. Ross, E.J. Cairns, J. Phys. Chem. 97, 12020 (1993).
50. A. Kabbabi, R. Faure, R. Durand, B. Beden, F. Hahn, J.-M. Léger, C. Lamy, J. Electroanal. Chem. 444, 41 (1998).
51. P.J. Kulesza, M. Matczak, A. Wolkiewicz, B. Grybowska, M. Galkowski, M.A. Malik, A. Wieckowski, Electrochim. Acta 44, 2131 (1999).
52. A. Lima, C. Coutanceau, J.-M. Léger, C. Lamy, J. Appl. Electrochem. 31, 379 (2001).
53. F. Ficicioglu, F. Kadirgan, J. Electroanal. Chem. 451, 95 (1998).
54. A. Kelaidopoulou, E. Abelidou, A. Papoutsis, E.K. Polychroniadis, G. Kokkinidis, J. Appl. Electrochem. 28, 1101 (1998).
55. I. Becerik, F. Kadirgan, J. Electroanal. Chem. 436, 189 (1997).
56. I. Becerik, S. Süzer, F. Kadirgan, J. Electroanal. Chem. 476, 171 (1999).
57. A. Kelaidopoulou, A. Papoutsis, G. Kokkinidis, W.T. Napporn, J.-M. Léger, C. Lamy, J. Appl. Electrochem. 29, 101 (1999).
58. B. Beden, F. Largeaud, K.B. Kokoh, C. Lamy, Electrochim. Acta 41, 701 (1996).

26

Novel Nanostructured Material Based on Transition-Metal Compounds for Electrocatalysis

NICOLÁS ALONSO-VANTE

Université de Poitiers, Poitiers, France

CHAPTER CONTENTS

SUMMARY

The investigation of materials based on transition metal compounds in the nanoscale range for multi-electron charge transfer catalysis is presented. In this respect, this contribution stresses (starting from model systems), the development of nano-structured materials designed from molecular precursors. Ex-situ and in-situ physical-chemical measurements, which allowed an insight into the synthesis and the complex interfacial behavior during catalysis, were performed. These results, taking as a basis the molecular oxygen reduction in acid medium, may confirm that the strategy can be further evaluated and extended to other reactions of technical relevance as to those belonging to fuel cell systems.

26.1 INTRODUCTION AND SCOPE

Up to now most of the research has been mainly focused on the study of platinum and/or platinum-based materials for electrocatalysis [1–35]. These studies have also been aimed at understanding the electrode–electrolyte interfacial behavior, in order to optimize the catalytic properties of such materials. The reason why most of these studies have been devoted to platinum is evident: this material is today the best catalyst, especially for processes occurring at the anode and cathode of low-temperature fuel cells [36–44]. In the abundant literature, one can find reports going from studies on single crystals [11,12,28,30,35,45–51] to highly dispersed powders (particle size 2–4 nm) [7,23,24,37–39,41,42,44,52–56] in which interesting features toward multi-electron charge-transfer reactions have been observed, such as the surface structure sensitivity [12,30,33,34,46,51,54,57–60]. However, other factors, less evident to detect, like the synergetic effects (local electronic modification) and active sites distribution induced by the so-called ensemble effect, can also be operating during electrocatalysis.

All the studies put into evidence the fact that the gap between model systems (single crystals) and the understanding of phenomena occurring on the nanorange scale (dispersed catalysts), particularly on platinum, is becoming smaller. Therefore, highly dispersed systems can also be considered as a model to understand processes in real working conditions in low-temperature fuel cells (FC). In these systems (anode and cathode made of platinum) based on hydrogen (as a fuel) and oxygen (as the oxidant), the limiting factor is the slow kinetics of reduction of the molecular oxygen at the cathode. If the fuel is a liquid such as methanol (in DMFC), the slow kinetics of methanol oxidation at the anode is added to the electrical loss as well. Besides that, depolarization of the cathode occurs due to the permeability of the solid polymer electrolyte, further reducing the electrical output of the system. Aside from developing the anode to increase its reactivity toward methanol oxidation, one must also develop cathodes with high selectivity or tolerance to the presence of methanol.

In order to summarize recent progress in the field of material development for methanol-tolerant electrocatalyst, we wish to present recent findings obtained in our laboratory on ruthenium-based cluster materials for the molecular oxygen reduction reaction (ORR). Tailoring novel materials, in fact, offers the opportunity of arranging individual transition metals to form clusterlike compounds. This has been done using the reactivity of the transition-metal carbonyls in solvents to tailor either novel ruthenium-based chalcogenide [61,62] or metal nanostructured clusterlike materials [63,64]. They were found to be highly selective toward the molecular oxygen reduction in the presence of methanol [65]. Therefore, this chapter will put forward some interesting physical-chemical properties of these novel compounds, for electrocatalysis, which can be modulated, in principle, by a facile chemical synthesis route.

26.2 STATUS OF CLUSTER-BASED MATERIALS FOR MULTI-ELECTRON CHARGE TRANSFER

Metal clusters have been considered as model systems for the study of catalytic reactions [66]. Material bulk properties of clusters, made of a few atoms to a few hundred atoms, are already present in the first atomic layers. The term "cluster" was initially coined to describe a compound in which a group of two or more metal–metal bonds are involved. This is the description for the family of molecular clusters. However, materials based on colloidal metal particles may contain from four to several hundreds of metal atoms in their skeleton. The metal or alloy core is stabilized by solvents and surfactants. Nanoparticles or clusters, e.g., Au_{55} [67] Pd_{561} [68], with dimensions of up to 4 nm can be considered, therefore, as a bridge between molecular clusters [69] and clusters forming metal particles [70]. A review giving account of the synthesis of colloidal particles (with a narrow particle-size distribution) formed by reduction of the metal salts by, e.g., NR_4-hydrotrioorga-noborates, has been written by Bönnemann et al. [71]. The catalytic activities (e.g., cyclooctane hydrogenation process) of these mono- or bimetallic cores, after extraction of the protecting shell (surfactant), are discussed as well. Therefore, the kind of material design summarized in [71] shows the importance of tailoring nanometal powders from organosols for catalysis and electrocatalysis [72,73]. To the best of our knowledge, among the molecular clusters, molecular structures of platinum and platinum-derived cluster carbonyl compounds $[Pt_3(CO)]_n^{2-}$ ($n = 3, 5, 10, \ldots$) [see Figure 1(a)] have been used for the electrocatalytic oxidation of methanol [74,75]. However, the platinum carbonyl cluster complex $Na_2[Pt_3(CO)]_n$, reported in the 1970s by Longoni and Chini [76], was deposited onto surface-modified graphite substrates. This assembly was, thereafter, annealed. The authors [75] reported that Pt_9- or Pt_{15}-complex-derived clusters showed high levels of reactivity for the catalytic oxidation of methanol. Because of the instability of the platinum carbonyl cluster in air, the graphite electrodes were actually covered at the end with Pt nanoparticles resembling, in this way, platinum obtained by means of vacuum deposits. Unfortunately, the authors [75] did not give any detail of the morphology of the platinum-modified electrode surface; therefore, no correlation between particle size and reactivity can be deduced. We have determined that the complex $[Pt_3(CO)]_{10}^{2-}$ decomposes at 410 K under nitrogen, and TEM analysis reveals platinum nanoparticles with a narrow size distribution of about 2 nm [77]; see inset in Figure 1(a). This example shows the interplay between coordination and reactivity. Thus, following this line of reasoning, using molecular cluster compounds as precursors emerges as an attractive alternative since the cluster may predetermine the proper stoichiometry of the metal atoms. Independent of the synthesis route, molecular clusters compounds offer the opportunity to work on their molecular design, as it will be shown below for the development of novel chalcogenide-based materials.

Adams et al. [78,79] have reported a series of synthesis of mixed-metal cluster compounds. One example, $Pt_2Ru_4(CO)_{18}$, is depicted in Figure 1(b). This mixed cluster compound was investigated to study the effect of Pt-Ru nanoparticles developed after the precursor annealing on carbon [80]. In line with the spectroscopic and microscopic measurements, the authors demonstrated that mixed Pt-Ru nanoparticles, with an extremely narrow size distribution (particle size 1.4 nm), reflect an interaction that depends on the nature of the carbon support. Furthermore, as revealed by EXAFS, the Pt-Pt, Pt-Ru, and Ru-Ru coordination distances in the precursor (2.66, 2.64, and 2.84 Å) [79] changed to 2.73, 2.70, and 2.66 Å, respectively, on the mixed-metal nanoparticles supported on carbon black, with an enhanced disorder [80]. Furthermore, some metal segregation could be

a) $[Pt_3(CO)_6]_n^{2-}$ (n = 2, 3, 5)

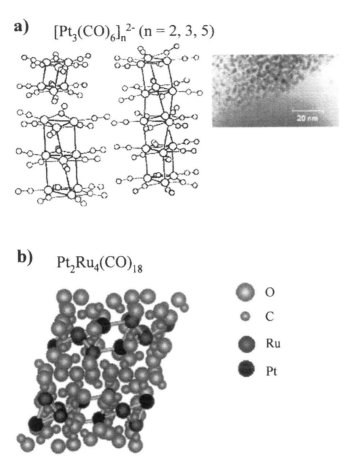

b) $Pt_2Ru_4(CO)_{18}$

O
C
Ru
Pt

Figure 1 Four examples of cluster structures: (a) $Na_2[Pt_3(CO)_6]_n^{2-}$ ($n = 2,3,5$) inset: TEM picture of Pt nanoparticles from $[Pt_3(CO)_6]_{10}^{2-}$ precursor; (b) unit cell of $Pt_2Ru_4(CO)_{18}$ cluster. Coordination distances in a cluster are $R_{Pt-Ru} = 2.70$ Å, $R_{Pt-Pt} = 2.66$ Å, and $R_{Ru-Ru} = 2.16$ Å; (c) unit cell of $Ru_4Se_2(CO)_{11}$ cluster. Coordination distances in a cluster are $R_{Ru-Ru} = 2.79$; 2.77 Å, $R_{Ru-Se} = 2.56$; 2.60 Å. Inset: TEM picture of Ru_xSe_y colloidal nanoparticles from $Ru_3(CO)_{12}$ and Se, via the $Ru_4Se_2(CO)_{11}$ chemical precursor (the stabilizer was octadecanthiol); (d) Chevrel cluster phase: Mo_6Se_8. Coordination distances in a cluster are $R_{Mo-Mo} = 2.68$; 2.83 Å, and $R_{Mo-Se} = 2.598$ Å. For a pseudo-ternary compound: $Mo_4Ru_2Se_8$. Coordination distances in a cluster are $R_{Me-Me} = 2.659$, 2.71 Å; $R_{Me-Se} = 2.623$ Å.

inferred from the measurements (probably due to the top limit of the annealing temperature used); the authors clearly demonstrated that molecular cluster chemical precursors can be an adequate strategy to develop electrodes for, e.g., the DMFC anodes.

The electrocatalytic oxygen reduction in DMFC systems requires the development of highly selective electrodes in the presence of methanol. Present electrolyte membranes based on Nafion® 117 are permeable to methanol (crossover effect), which depolarizes the platinum cathode [81,82]. For this reason, our strategy, some years ago, was to produce electrocatalytic materials from the thermal decomposition of some neutral transition-metal carbonyl compounds in the presence

c) $Ru_4Se_2(CO)_{11}$

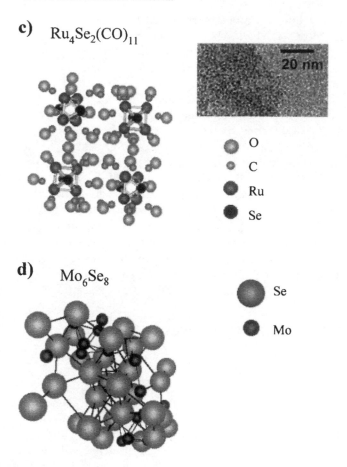

O

C

Ru

Se

d) Mo_6Se_8

Se

Mo

Figure 1 Continued.

of selenium in an organic solvent [61,83]. Very recently it has been recognized that the chemical precursor was not the initial carbonyl complex [$Ru_3(CO)_{12}$] reacting with the chalcogen [64] (see below). It turned out that the main chemical precursor $Ru_4Se_2(CO)_{11}$, formed in situ in the reaction vessel, led to the novel cluster-based metallic compound embedded in selenium: $Ru_xSe_y(x \approx 2, y \approx 1)$. The complex $Ru_4Se_2(CO)_{11}$ is depicted in Figure 1(c). The colloidal catalyst solution derived from such a complex is also monodispersed, as shown in the inset of Figure 1(c). The synthesis of this cluster, via pyrolysis at 185 °C under vacuum of $Ru_3(CO)_{12}$ and PhSeSePh during 19 h, was previously reported by Layer et al. [84]. However, our method of fabricating $Ru_4Se_2(CO)_{11}$ is more facile (see Section 26.3.1). The examples shown above demonstrate that tailoring novel materials is possible via molecular cluster complexes.

The study of the electrocatalytic molecular oxygen reduction [85–88] on a new class of materials—Chevrel cluster compounds: Mo_6X_8 (X = Se, S, or Te) [89]; see Figure 1(d)—led to the development of the chalcogenide clusterlike compound (Ru_xSe_y). Figure 1(d) depicts the structure in which one can recognize the clustering center (octahedral) based on molybdenum atoms surrounded by chalcogens. This cluster unit closely resembles $[Mo_6Cl_8]^{4+}$ [90]. The most active material is the so-called pseudo-ternary cluster, because one or more molybdenum atoms in the unit Mo_6X_8 are replaced by other metals, such as Ru, e.g., $Mo_4Ru_2Se_8$. This substitution modifies the electronic environment of the cluster unit. The electron number is increased from 20 (Mo_6Se_8) to 24 ($Mo_4Ru_2Se_8$), thus reducing the bonding distance by about 15%. This cluster compound reduced molecular oxygen directly to water in acid electrolyte with less than 4% H_2O_2 formation [86]. The systematic study of these materials, for the molecular oxygen reduction, indicated that the presence of two neighboring metals is an important ingredient (synergetic effect). The electroactivity of the base Mo_6 cluster was the least among all the clusters tested [88]. Furthermore, judging by the structure of such materials, we think that a little dependence on the surface orientation can be expected due to a similar number of the statistically exposed clustering units per surface. The advantage of this material structure is that such cluster units (6 atoms) keep a uniform spatial distribution in the chalcogen skeleton, which stabilizes them. Such a precondition is lacking in the nanoparticulate systems. Unfortunately, the Chevrel clusters corrode (mixed potential) and/or react with oxygen from air as revealed by XPS measurements [86].

26.3 THE CHEMICAL ROUTE IN DESIGNING NOVEL MATERIALS

The synthesis of transition-metal chalcogenide-based material (Ru_xSe_y) in mild conditions ($T \leqslant 473\,K$) was performed in two ways: powder [61,83] and colloidal solution [91]. The colloidal material can be easily deposited, e.g., by dipping, on conducting substrates (e.g., glassy carbon, SnO_2:F, etc.). The essential characteristics of these materials, in powder and in the colloidal form, are their high dispersiveness in the nanoscale range and the simplicity of manufacturing them [83,91]. Tailoring the material was also possible using $X = S$, Te, Se [65]. Their electroactivity with respect to the molecular oxygen reduction reaction and their chemical stability are higher than those of Chevrel phase materials [61]. Moreover, in spite of the easy way of synthesis, the route of the chemical reaction has not yet been well established. However, an attempt has already been undertaken to try to detect the intermediates formed during the synthesis and to identify the chemical precursor(s) (see below) of the electrocatalysts chalcogenide-based materials [92]. We can briefly outline some characteristics of this type of compounds: (1) they have a high monodispersiveness; (2) the presence of a chemical stabilizer, during synthesis, does not perturbate the chemical nature of the novel compound, therefore, a similar stoichiometry (Ru_xSe_y, in which $x \approx 2$ and $y \approx 1$) as well as a similar atomic coordination distance can be obtained, as revealed by Rutherford back scattering (RBS) (See Figure 2), and EXAFS [93] measurements (Figure 3) (see Section 26.4). From RBS spectra, the calculated stoichiometry for the powder supported as thin layer onto glassy carbon (GC) was Ru_2Se_1. The corresponding stoichiometry for the ultrathin layer deposited onto the same substrate from colloidal solution (after annealing) corresponds to $Ru_{2.2}Se_1$. One can identify the edge of the carbon (at about $500\,keV$), due to the carbon substrate as well as other "impurities" in/on these layers (oxygen, chlorine), as indicated in Figure 2(a) and (b). On the nanoparticles issued from colloidal solution, no sulphur (from the stabilizer: octadecanthiol) was detected. This shows that one can completely eliminate the surfactant shell.

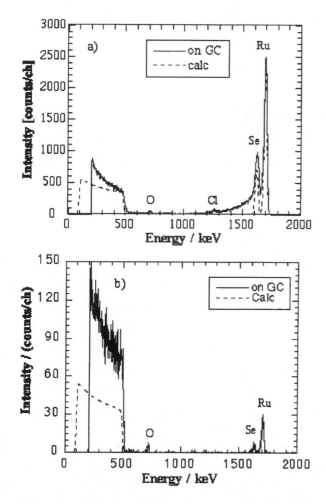

Figure 2 RBS spectra of Ru$_x$Se$_y$, deposited as thin layers onto glassy carbon (GC) substrates, from Ru$_3$(CO)$_{12}$ and Se. (a) Thin layer formed onto the substrate placed in the reaction for powder synthesis; (b) substrate dipped into the colloidal solution, thereafter, annealed at 230 °C (under argon) to eliminate the stabilizer.

26.3.1 The Formation of High-Nuclearity Complexes

This study was performed only during the synthesis of the catalyst (Ru$_x$Se$_y$) in powder form in xylene and/or 1,2 dichlorobenzene solvents. The reaction of tris-ruthenium dodeca-carbonyl [Ru$_3$(CO)$_{12}$], enriched with ^{13}C for NMR analysis, and elemental selenium, dissolved in an organic solvent (e.g., xylene), leads to a polynuclear chemical precursor in the first stages of reaction. After the first 40 min of reaction, new peaks, chemical shift signals [δ or ppm], appear in the ^{13}C-NMR spectrum [Figure 4(a)], namely 198.88, 196.5, and 195.79. The value $\delta = 198.88$ ($\Delta\delta = 0.82$ with respect to the parent complex) is characteristic of Ru$_4$Se$_2$(CO)$_{11}$ [cf. Figure 1(c)]. This finding puts in evidence the fact that selenium must be dissolved in order to participate in the coordination sphere of ruthenium clusters.

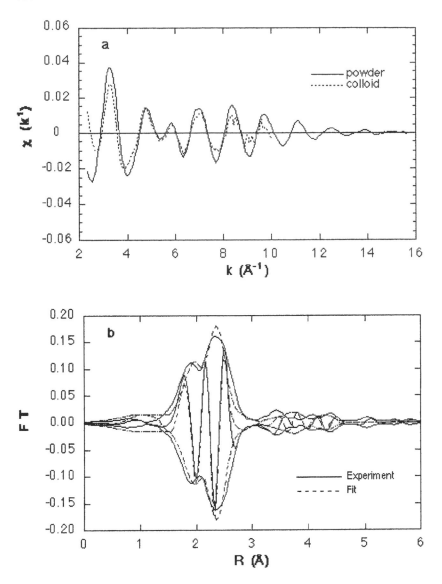

Figure 3 (a) The comparison of the EXAFS interference functions for Ru_xSe_y. In powder form (full line); and as thin layer deposited onto GC substrate, by dipping in the colloidal solution and annealed (230 °C, Ar) (dashed line). (b) The corresponding Fourier transform (k^1 weighed) of raw data from powder [full line in (a)].

The peak intensity with the chemical shift $\delta = 199.7$ corresponds to the parent complex [94]. Furthermore, one observes that between 40 and 80 min the peak intensity of the $Ru_3(CO)_{12}$ complex decreases as a function of the synthesis reaction time [Figure 4(b) and (c)] concomitant to the increase of the peak at 198.88 ppm. This evolution in xylene is apparently slow and accompanied by the formation of other complexes (other minor peak intensities) on both sides of the ^{13}C-NMR

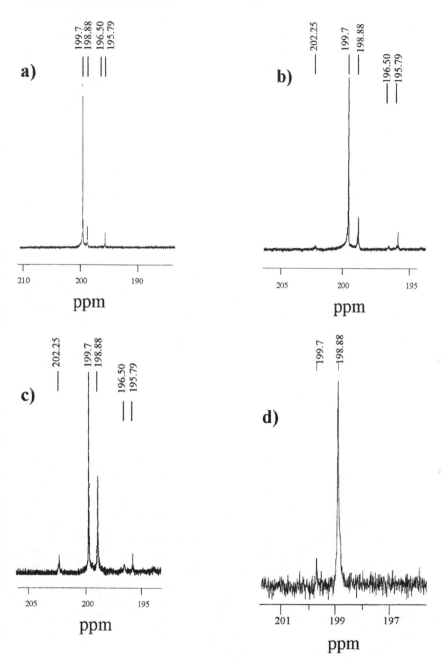

Figure 4 ^{13}C-NMR spectra in the first stages of synthesis of Ru_xSe_y powder in xylene: (a) after 40 min; (b) after 60 min; (c) after 80 min; and (d) in 1,2 dichlorobenzene after 60 min. The substances in xylene were redissolved in CD_2Cl_2.

spectra; Figures 4(b) and (c). The complex interplay of the reaction is evident in these figures. In fact, the use of 1,2 dichlorobenzene allows the synthesis of the same complex in a selective way during the first 60 min (no other cluster complexes were detected); see Figure 4(d). The synthesis performed in this solvent represents a comfortable way of obtaining $Ru_4Se_2(CO)_{11}$ in comparison to the one reported by Layer et al. some years ago [84]. The loss of the carbonyl groups is obtained by keeping the boiling temperature of the solvent in refluxing conditions under argon or under nitrogen during 20 h of reaction. Thus, the intermediate cluster compound $Ru_4Se_2(CO)_{11}$ evolves continually with the reaction time to the novel clusterlike material: Ru_xSe_y. The addition of a chemical stabilizer (octadecanthiol) for the synthesis of a colloidal catalyst solution, at the beginning of the reaction, does not perturbate the chemical route.

26.3.2 Ru_xSe_y Nanoparticles Formed from $Ru_4Se_2(CO)_{11}$ Precursor

The neutral complex $Ru_4Se_2(CO)_{11}$ (formed either in Xylene or dichlorobenzene) as the molecular precursor for Ru_xSe_y was tested. The thermal decomposition of the complex was followed by the thermal differential analysis (TDA) technique under nitrogen atmosphere. Figure 5(a) shows that the mass loss corresponding to CO (35.37%) occurs at a temperature between 140 °C and 225 °C (point 2 in the figure). Higher temperatures (from 360 °C up to 500 °C) lead to a change of phase (peaks corresponding to ruthenium and ruthenium diselenide develop [at point 4 in Figure 5(a)], whose DRX spectra are not shown here). Figure 5(b) depicts the DRX spectrum of the complex (1) as well as the material obtained at 225 °C (2). This latter spectrum is in agreement with the spectrum recorded from the powder synthesized in Xylene after a reaction time of 20 h; see Figure 5(c). The composition of the pyrolized sample at 225 °C was determined by energy-dispersive x-ray (EDX) analysis. In this analysis the incident electron probe beam was located on various regions of the sample and the x-ray fluorescence was detected using a Philips-CM 120 with a filament of LaB_6 microscope. The quantitative analysis, based on the characteristic x-ray fluorescence of Ru and Se [$Ru(K_{\alpha,\beta})$, $Se(K_{\alpha,\beta})$] corrected by Kab factors (obtained from a reference sample: $RuSe_2$) revealed that the stoichiometry of the core is maintained, i.e., 66% at. Ru and 33% at. Se, testifying that the sample has a stoichiometry corresponding to Ru_2Se, and verifying the data obtained some years ago [61]. Now, with respect to the shape of all the DRX spectra recorded on our chalcogenide samples, one can deduce that long-range order does not exist in these materials. The local-order technique, EXAFS, revealed that the chalcogenide-based materials are nanocrystalline; cf. Figure 3. The data in Figure 3, obtained at the Ru-K edge, shows that the short-range order around this element is one of a crystal lattice. Assess, then, the nanocrystalline nature of the novel material. Furthermore, the broad diffraction peak centered at $2\theta = 42.5°$ shows, essentially, the characteristics of a nanodivided ruthenium metal [63,95]. This points out that the active center for electrocatalysis, in these chalcogenide materials, is essentially of metallic nature.

26.4 EVIDENCE OF THE INTERFACIAL DYNAMICS DURING ELECTROCATALYSIS

As mentioned above, EXAFS studies on nanoparticles synthesized in either powder form or nanoparticles from the colloidal solution were performed. The coordination distances for ruthenium–selenium and ruthenium–ruthenium are, respectively, $R_{(Ru-Se)} = 2.43$ Å y $R_{(Ru-Ru)} = 2.64$ Å. The metal–metal coordination distance is of

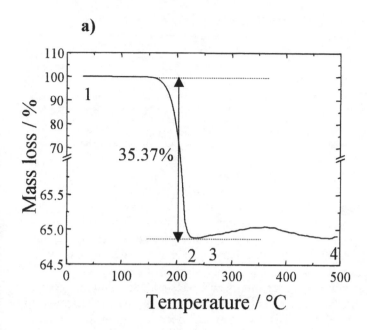

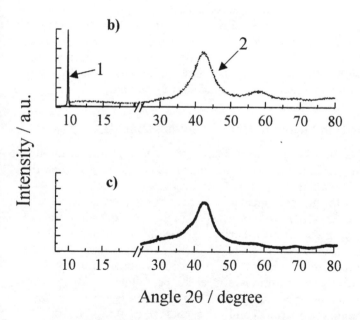

Figure 5 (a) Differential thermal analysis (DTA) of $Ru_4Se_2(CO)_{11}$ cluster compound. The measured mass loss of CO agrees with the theoretical value: 35.39%. (b) DRX diffratograms of (1) $Ru_4Se_2(CO)_{11}$ [see set point 1 in (a)]. Two main diffraction peaks at $2\theta = 9.8°$ (002) and $19.7°$ (004) due to the preferential orientation of complex cluster crystallites are observed; (2) the cluster decomposed at 225 °C [see set point 2 in (a)]; and (c) Ru_xSe_y powder synthesized in xylene. The small peak (spike in the spectrum) at $2\theta = 30°$ corresponds to unreacted selenium. The set points 3 and 4 in (a) indicate the change of phase.

Table 1 Structural Parameters of the Novel Ru_xSe_y Compound

Samples	ΔE_0 (eV)	Ru Ru			Ru Se		
		N	R (Å)	$10^3 \Delta\sigma^2$ (Å²)	N	R (Å)	$10^3 \Delta\sigma^2$, (Å²)
Ru_xSe_y—powder	-1.5 ± 0.5	3.2 ± 0.5	2.64 ± 0.01	3.60 ± 0.50	2.0 ± 1.0	2.43 ± 0.02	8.60 ± 3.00
(a) Ru_xSe_y—coll/GC	0.3 ± 1.0	3.6 ± 0.7	2.63 ± 0.02	6.27 ± 2.00	2.1 ± 1.0	2.43 ± 0.10	12.30 ± 3.00
(b) Ru_xSe_y—coll/CG	6 ± 1.6	4.1 ± 1.2	2.66 ± 0.02	4.9 ± 1.7	2.5 ± 1.5	2.43 ± 0.04	16.8 ± 5.00

(a) Catalyst deposited onto glassy carbon (GC) from colloidal solution. (b) Catalyst deposited onto conducting glass (CG) from colloidal solution. This data fit required the Ru–O coordination. The fitted values are $N = 0.4 \pm 0.1$; $R = 1.64 \pm 0.07$ Å; and $\Delta\sigma^2 = 3.90 \pm 1.4$ Å².

the same order of magnitude as that of well-known cluster-based materials such as the Chevrel phase [88]. More data concerning the structure are reported in Table 1.

Electrochemical studies [via the rotating ring-disk electrode (RRDE) technique] have revealed that the multi-electron charge-transfer process is possible on cluster materials (96% water formation). This is in line with the model proposed by Damjanovic. His model [Figure 6(a)] was discussed by Hsueh et al. [96] and applied to our kinetic measurements [62]. The reaction takes place via k_1 (i.e., 4 electrons). In this model, it is assumed (1) that there is no catalytical decomposition for hydrogen peroxide (2) that adsorption and desorption reactions are fast and in equilibrium, and (3) that the rate constant for the electrochemical oxidation of the same species is negligible. Measurements were performed on massive electrodes [85,86], or with the novel compounds deposited in thin layers onto glassy carbon substrates [61,62]. Using the rotating disk electrode (RDE) technique, with highly dispersed samples, mimicking the electrode used in fuel cells: catalyst powder suspension deposited onto glassy carbon (typically with a charge of $28\,\mu g\,cm^{-2}$), we observe that the process is also via 4 electrons, as shown in Figure 6(b) (Koutecki–Levich analysis).

26.4.1 GIXAS on Ultra-Thin Layers of Chalcogenide Materials

It is interesting to probe, in situ, processes being performed at the electrode–electrolyte interface, using such novel materials, during electrocatalysis with the specific example of the molecular oxygen reduction. In this connection, in-situ electrochemical studies with grazing incidence x-ray absorption spectroscopy (GIXAS) were performed as reported recently [93]. One key question was to understand the process taking place on the surface. x-rays in the grazing incidence mode are used to probe the outermost layers or surface atoms of nanoparticles deposited in ultra-thin layers, even though the surface-to-volume ratio (S/V) of nanoparticles is enhanced. These conditions appear to be fulfilled in our system (Ru_xSe_y dispersed on conducting glass substrate, SnO_2:F). The Fourier transform (FT) of the EXAFS spectra in open-circuit conditions, OCP (E ca. 0.68 V and 0.76 V/RHE in nitrogen and in oxygen-saturated electrolyte solutions, respectively), already indicates dramatic changes (decrease of the prominent peak in oxygen saturated electrolyte); see Figure 7(a). This prominent peak, observed at the lower coordination distance, is mainly attributed to light elements such as oxygen. Indicating that ruthenium surface atoms are interacting with oxygen from water in a nitrogen-saturated electrolyte (formation of metal oxidelike species). This can be

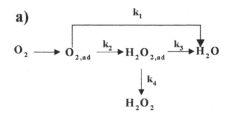

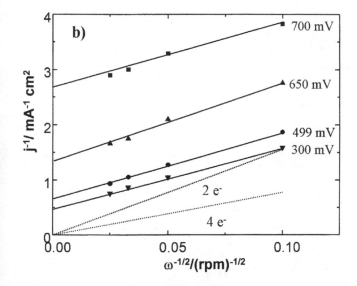

Figure 6 (a) The kinetic model proposed by Damjanovic et al. [96] for the molecular oxygen reduction. (b) Koutecki–Levich analysis for RDE measurements of Ru_xSe_y nanoparticles.

expected since it is known that water has a strong tendency to be adsorbed on ruthenium centers [97]. When oxygen is dissolved in the electrolyte, the magnitude of the peak in the FT spectrum decreases as compared to that in the nitrogen-saturated electrolyte. As observed, the amplitude of the first shell is reduced by a factor of two, and the position of the first shell is slightly shifted to higher R values. The variation of R may indicate that during electrocatalysis the oxygen bound to the electrocatalytic center (the ruthenium) is different in nature. Furthermore, the reduction of the amplitude may indicate a lowering of the coordination number and/ or increase in the Debye–Waller factor. We believe that during electrocatalysis, the oxygen coordination to the metal enhances the disorder of the surface atoms, inducing a more pronounced structure distortion than that of the atoms in the bulk. Indeed, this increased disorder in the local structure of the metal is amazingly similar to the process of amorphous state measured via DRX on naked ruthenium particles exposed to oxygen from air [63,98]. Taking into account these facts, a fitting procedure with an additional oxygen path (due to the prominent peak in the FT results) was employed. The fitting results (averaged values, the coordination number, N, taken from data of powder, was kept constant) for OCP in nitrogen saturated 0.5 M H_2SO_4, were as follows: $R_{(Ru-Se)} = 2.58 \text{ Å}$; $R_{(Ru-Ru)} = 2.78 \text{ Å}$; and

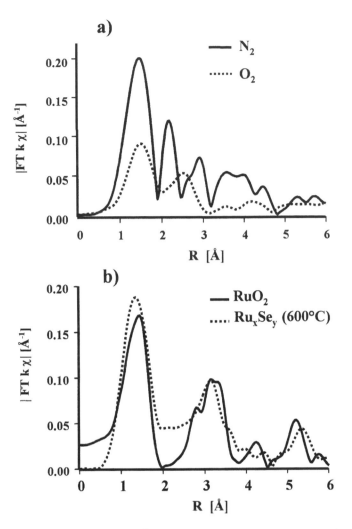

Figure 7 (a) The FT (k^1 weighed) from GIXAS experiments in the open-circuit potential of ultrathin layers deposited onto conducting glass under nitrogen and under oxygen-saturated electrolyte. (b) FT (k^1 weighed) from EXAFS transmission data on RuO_2 and Ru_xSe_y. This latter annealed with nitrogen with <5 ppm per vol of O_2.

$R_{(Ru-O)} - 1.85$ Å; and in oxygen-saturated 0.5 M H_2SO_4, gave the following results: $R_{(Ru-Se)} = 2.66$ Å; $R_{(Ru-Ru)} = 3.08$ Å; and $R_{(Ru-O)} = 2.09$ Å. These values did not change significantly when the potential was varied from open circuit to 0.09 V/RHE [93]. However, to better understand the interaction of oxygen with ruthenium surface atoms, ex-situ measurements performed on differently treated powder samples were necessary; see Figure 7(b). These materials were RuO_2 (Johnson Matthey), and Ru_xSe_y annealed at 600 °C under nitrogen with < 5 ppm per vol of O_2. For RuO_2 the prominent peak centered at R space (1.38 Å) corresponds to an average value of 1.97 Å. This value is, in fact, an average of two Ru–O coordination distances in the first shell. One (1.94 Å) with two oxygen neighbors and the second (1.98 Å) with four oxygen neighbors, according to crystallographic data for the ruthenium dioxide sample [99]. These distances are not easy to distinguish via the EXAFS analysis. The annealing of Ru_xSe_y in the presence of <5 ppm per vol of oxygen produces a change in its structure. It is clear that one phase corresponds to the formation of a ruthenium oxidelike phase, from the position of the first shell. However, the transformation is not complete because higher shells, at $R > 2$ Å, look different from those of the RuO_2. The analysis shows, in fact, that this powder consists of $Ru_xSe_y + Ru_xO_y$ phases [100]. This sample bears a resemblance to the phenomenon observed at the electrochemical interface and supports the idea that a surface complex, more likely a Ru_xO_y species, is responsible for the electrocatalytic reduction of molecular oxygen. In the conditions where oxygen is absent (no electrocalytic currents), the interaction of the oxygen from water is also clear. When comparing these fitting results with those obtained on the Ru_xSe_y powder material, it turns out that the coordination distances on this layer-forming particles (surface) are slightly higher (2–3%) than those obtained for the powder (bulk signal) [101]. This difference can be attributed to the influence of oxygen atoms coordinated to the metal center. As compared to the initial values (under nitrogen, see above) the R_{Ru-O}, R_{Ru-Se}, R_{Ru-Ru} coordination distance variations are 12.6%, 3%, and 13.3%, respectively. The results indicate that important changes occur mainly on the catalytic center, i.e., on ruthenium atoms (see Section 26.4.3). Furthermore, one should note that this change is reversible when changing the atmosphere in the electrolyte, accounting then for interfacial dynamics in electrocatalysis with these clusterlike materials.

26.4.2 EXAFS on Thin Layers of $Ru_x X_y$ Materials

Prior to the development of clusterlike catalyst colloidal solution [91], materials were deposited as thin layers onto conducting supports [glassy carbon, or porous carbon papers (Toray)] [61,62,83]. The nature of chalcogens was varied in order to synthesize Ru_xX_y ($X =$ Se, S, and Te). In the Ru_xSe_y clusters tiny amounts of molybdenum were incorporated (2–4 at %). It was verified by EXAFS that molybdenum atoms were highly disordered in this compound [101], and via XPS this element was found in form of oxides [61]. Its electrochemical behavior was similar to Ru_xSe_y. Furthermore, EXAFS spectra recorded on the Ru–K edge of Ru_xSe_y and $Mo_xRu_ySe_z$ were similar. X-ray fluorescence analysis of Ru_xX_y deposited as thin layers (from xylene solvent) onto porous carbon paper revealed the following composition: $RuSeMo_{0.03}$, $Ru_{1.25}S$, and $RuTe_{0.055}$ [102]. Due to the solubility of elemental chalcogens in xylene, which decreases as follows: S > Se ≫ Te, the tellurium-containing compound was actually based on ruthenium core. It was also possible to carry out in-situ studies in the transmission mode with EXAFS [102]. Figure 8 shows the FT spectra of in-situ measurements at 0.08 V/RHE on Ru_xX_y layers in argon- and oxygen-saturated electrolyte (0.5 M H_2SO_4). In line to what was already observed on powder catalysts [e.g., Ru_xSe_y, Figure 3(b)] one recognizes, in the first coordination sphere, two prominent peaks due to Ru–Ru and Ru–X. The change from argon to oxygen at the same electrode potential induces significant modifications, except for the Ru_xS_y compound. For each system, the applied electrode potential was varied forward (from negative to positive values) and backward. The EXAFS spectra analysis (Debye–Waller factor was kept constant,

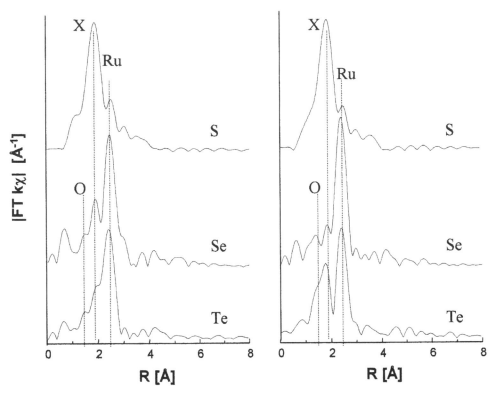

Figure 8 In-situ FT (k^1 weighed) of Ru_xX_y compounds under argon-(left) and under oxygen-(right) saturated electrolyte. The applied electrode potential $= 0.08$ V/RHE.

and coordination numbers were calculated [102]) revealed reversible changes of the coordination distances, R. For instance, R_{Ru-Ru} decreased with the positive potential and was restored when returning to the initial value in all samples. On the other hand, R_{Ru-O} decreases; see Figure 9. This figure summarizes all the coordination data from EXAFS spectra analysis which are displayed as the change of coordination distance ΔR (%) for R_{Ru-Ru}, R_{Ru-O}, and R_{Ru-X}, between argon and oxygen, as a function of the applied electrode potential. The reference point was taken at 0.78 V/RHE. At this potential the corresponding coordination of R_{Ru-Ru}, R_{Ru-O}, and R_{Ru-X}, either in the argon-saturated or oxygen-saturated electrolyte, was taken as reference for such calculations. Under the consideration that catalysis is taking place only on metal cluster centers, the present data put in evidence that structural changes in the catalysts take place. Besides, the surface complex formation species, as discussed above, is also evident from the interaction of ruthenium centers with oxygen. However, at this point, it is worthwhile to note that Ru_xS_y is less electrocatalytic. Although EXAFS reveals some distance changes on the metal core, a variation of the Ru–S coordination distance is negligible [Figure 9(c)]. This might point out that the metal center could be inhibited by the presence of sulphur atoms, and thus hinder an efficient interaction of Ru with oxygen; Figure 9(b). The differences observed between two calculated coordination distances (same symbols bound by a line) in Figure 9 represent measurements on the ongoing positive

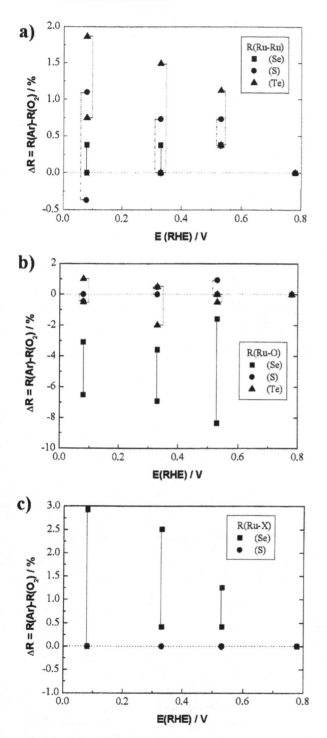

Figure 9 The change of coordination distance between argon- and oxygen-saturated electrolyte, ΔR (%), as a function of the electrode applied potential for $R_{\text{Ru-Ru}}$, $R_{\text{Ru-O}}$, and $R_{\text{Ru-X}}$. $\Delta R = 0$ was taken as a reference at 0.78 V.

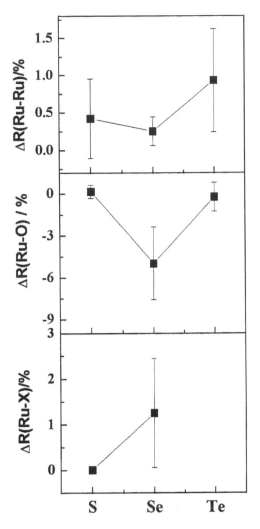

Figure 10 The averaged ΔR (%) for $R_{Ru\text{-}Ru}$, $R_{Ru\text{-}O}$, and $R_{Ru\text{-}X}$ (deduced from Figure 9) as a function of the chalcogen nature (see text).

potential and backwards. $\Delta R = 0$ at 0.78 V/RHE was taken as a reference. In order to rationalize the data, they were averaged for each Ru_xX_y compound. These results are displayed in Figure 10 as a function of the chalcogen nature. Within the limits of calculation and experimental errors, one can recognize that important changes in the structure take place on the selenium-containing compound. Ruthenium nanoparticles are, in principle, catalytically active, in that their surface is not coordinated to oxygen to form RuO_2. Although important variation on the $R_{Ru\text{-}Ru}$ in $RuTe_{0.055}$ takes place, it is difficult to assess, in light of these measurements, the nature of the coordinated oxygen. In summary, we arrive at the conclusion that metal cluster centers, in these novel compounds, are distorted via the adsorption of oxygen. This is in agreement with data generated via GIXAS. The distortion of metal cluster centers is probably a precondition to break the O–O bond, which favors multi-electron charge transfer, generated in the metal clusters.

26.4.3 EXAFS on Chevrel Cluster

The EXAFS spectroscopic analysis of in-situ measurements in transmission mode together with a model system [polycrystalline $Mo_4Ru_2Se_8$; cf. Figure 1(d)] [103], for the reaction of reduction of molecular oxygen in acid medium also revealed several important aspects, namely: (1) the identification of the electrocatalytic center; (2) the distortion of the structure, and absence of oxygen in the electrolyte, (3) the cluster unit works as a reservoir of electrons. The change of the coordination distance, $\Delta R = R(Ar) - R(O)$, deduced in a similar way as described in Section 26.4.2, as a function of the applied electrode potential in argon (Ar) and oxygen (O) reveals the distortion of the structure (Figure 11, and cf. Figure 9). Within the error estimation, the distance variation between Mo–Mo in argon, $R(Ar)$ and Mo–Mo in oxygen $R(O)$ is not significant. However, the coordination distance change, ΔR, for $R_{Ru\text{-}Ru}$, is evident from the electrode potential in which the electrocatalytic process takes place: $0.7\,V > E > 0\,V$ [85,86]. The distance variation for $R_{Ru\text{-}Se}$ follows the same trend as that of $R_{Ru\text{-}Ru}$. The distance $R_{Ru\text{-}Ru}$ decreases by about 1%. Therefore, during the electrocatalytic process for the oxygen reduction, the distortion of the cluster unit also points toward an interfacial dynamics during the charge transfer. Furthermore, EXAFS spectroscopy also provides the information that the catalytic center is located on the ruthenium atoms. This study supports the analysis of the structural changes observed on the novel chalcogenide compounds, as discussed above. However, due to the experimental mode employed, the observed changes were more bulk than surface due to the low S/V ratio of the polycrystalline material.

26.5 PHYSICAL-CHEMICAL ASPECTS OF THE METALLIC NANOSTRUCTURE

In an attempt to understand the nature of the catalytic center, metallic ruthenium nanoparticles were synthesized using the same chemical route described earlier [63,64]. The precursor obtained was named Ru_x and was synthesized using two different solvents [xylene (Xyl) and 1,2 dichlorobenzene (Dcb)]. It turned out that the nature of the solvents influences the way ruthenium atoms are rearranged in order to form clusters. This phenomenon can be visualized in Figure 12(a) and (b). Starting from oxidized nanoparticles [63], the recovery of the metallic precursor by reduction with hydrogen does not occur at the same temperature; see Figure 12(a) and (b). While the nanometallic precursor, synthesized in xylene, is obtained at 95 °C, the precursor nanomaterial prepared in dichlorobenzene is recovered at higher temperatures (183 °C). Under these conditions *hcp* of 2 nm are formed. On the other hand, the intermediate hexagonal (102) peak (at $b = 0.63\,\text{Å}^{-1}$) is significantly diminished on both Ru_x precursors. This testifies toward a very high degree of stacking faults. Debye function analysis (DFA) simulations performed on such materials [95] revealed a bimodal size distribution represented by a small fraction of unfaulted closed-shell *hcp* model clusters ($x = 13$, 57, and 153). On the other hand, larger ones (e.g., $20\,\text{Å} < \delta < 30\,\text{Å}$) are strongly faulted by stacking defects. The respective Debye parameters are $B = 3\,\text{Å}^2$ and $0.15\,\text{Å}^2$. The mean Ru–Ru distances

Alonso-Vante

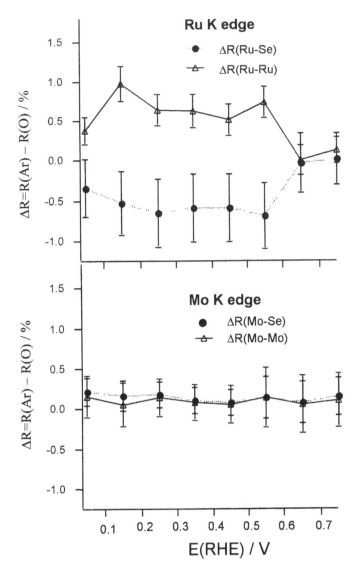

Figure 11 The change of coordination distance between argon and oxygen saturated electrolyte, ΔR (%), as a function of the electrode applied potential for $R_{Ru\text{-}Ru}$, $R_{Ru\text{-}Se}$, $R_{Mo\text{-}Mo}$, $R_{Mo\text{-}Se}$. The points, at $\Delta R = 0$, correspond to the reference.

were found to be appreciably contracted against bulk ruthenium by 3.8% and 0.7% for $Ru_x(Xyl)$ and $Ru_x(Dcb)$. These findings are parallel to the electrochemical behavior, as shown by simple voltammetric measurements for the same equivalent catalyst charge (57 $\mu g\,cm^{-2}$) deposited onto GC substrates; see insets in Figure 12(a) and (b). The electrochemical behavior of Ru_xSe_y bears resemblance to the behavior of Ru_x, both being synthesized in xylene. Moreover, when comparing this behavior with that of Ru_xSe_y prepared in dichlorobenzene, we see that it bears resemblance to Ru_x prepared in the same solvent. Furthermore, the electrocatalytic activity for the

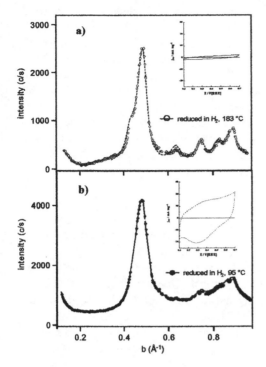

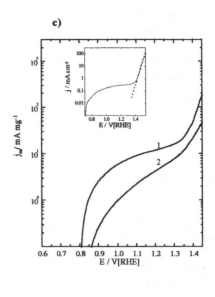

Figure 12 (a) and (b) Diffraction patterns of the ruthenium nanoparticles obtained after hydrogen reduction at the corresponding temperatures. The insets on these figures represent the electrochemical reduced state (cyclic voltammetry at 50 mV/s in 0.5 M H_2SO_4) of such nanoparticles. (c) Tafel plot for the water oxidation on surface-reduced ruthenium nanoparticles. For comparison a massive electrode is also depicted in the inset. The current density measured on nanoparticles is referred to the mass.

reduction of oxygen was higher on xylene-prepared samples than on dichlorobenzene ones. Therefore, based on these observations, it is evident that the catalytic centers of ruthenium clusters must possess a certain degree of disorder, to coordinate the oxygen leading to the active surface complex (Ru_xO_y) responsible for electrocatalysis. Systematic studies are still needed to probe the nature of chalcogen toward the "disordered" metal center synthesized on different solvents. Figure 12(c) also testifies that the electrocatalytic water oxidation is enhanced on the more disordered metallic precursor. However, the naked metallic precursor reacts easily with oxygen. This phenomenon does not take place when such metallic clusters are coordinated with a chalcogen, as, for example, Ru_xSe_y. The chalcogen as ligand, stabilizes the metallic structure against oxidation, as demonstrated in various papers [61,104]. The precondition is that the channels that coordinate molecular oxygen remain open. This is apparently not the case for the sulphur-containing compound.

26.6 SUMMARY AND OUTLOOK

Significant progress has been achieved in the research of materials based on clusterlike chalcogenide in the nanometer-scale range. The organizational effect of molecular precursors to tailor novel materials opens a variety of interesting phenomena for physics and chemistry. Our contribution is an attempt to understand the complex interplay between material design and the physical-chemical aspects that such materials may bring out for electrocatalytical processes. The application of in-situ spectroscopic techniques (e.g., GIXAS, EXAFS) revealed that the adsorption process is accompanied by a distortion of the structure. This seems a precondition to enhance kinetics of multi-electron charge transfer. Ruthenium, as well as other transition metals, presents interesting properties in electrocatalysis when the atoms are arranged in such a way as to fulfill the structure that leads to a novel compound. Ru_xX_y ($X = S$, Se, Te) is just an example among many other compounds that can be tailored by means of the method presented in this chapter.

ACKNOWLEDGMENTS

The author would like to thank his students and co-workers. A special acknowledgement is due to Dr. K. Ellmer (Hahn-Meitner-Institut-Berlin) for his help in RBS measurements, Dr. E. Savinova, and Prof. D. Kochubey (Boreskov Institut of Catalysis, Novosibirsk) for the collaboration in the EXAFS measurements via an INTAS program; Dr. W. Vogel (Fritz-Haber-Institute) for the DRX measurements on ruthenium nanoparticles, and Prof. J.-M. Coustard (Poitiers University) for NMR measurements. Special thanks are also due to the staff of the Hasylab-Hamburg Synchrotron for the beam allocated to perform in-situ measurements.

REFERENCES

1. A. Aramata, M. Masuda, Platinum alloy electrodes bonded to solid polymer electrolyte for enhancement of methanol electro-oxidation and its reaction mechanism. J. Electrochem. Soc. 1991, 138(7), 1949–1957.
2. D. Chu, S. Gilman, Methanol electro-oxidation on unsupported Pt-Ru alloys at different temperatures. J. Electrochem. Soc. 1996, 143(5), 1685–1690.
3. L.D. Burke, J.K. Casey, The role of hydrous oxide species on platinum electrocatalysts in the methanol/air fuel cell. Electrochim. Acta 1992, 37(10), 1817–1829.
4. S.P. Chen, S.G. Sun, T.S. Huang, Structure and electrocatalytic activity of electrode of platinized platinum supported on glassy carbon. Chinese Sci. Bull. 1995, 40(5), 377–381.
5. S.J. Clouser, J.C. Huang, E. Yeager, Temperature dependence of the Tafel slope for oxygen reduction on platinum in concentrated phosphoric acid. J. Appl. Electrochem. 1993, 23, 597–605.
6. B.E. Conway, G. Jerkiewicz, Surface orientation dependence of oxide film growth at platinum single crystals. J. Electroanal. Chem. 1992, 339(1–2), 123–146.
7. F.A. Debruijn, G.B. Marin, J.W. Niemantsverdriet, W. Visscher, J.A.R. Vanveen, Characterization of graphite-supported platinum catalysts by electrochemical methods and XPS. Surface and Interface Analysis 1992, 19(1–12), 537–542.
8. T. Frelink, W. Visscher, A.P. Cox, J.A.R. Vanveen, Ellipsometry and DEMS study of the electrooxidation of methanol at Pt and Ru- and Sn-promoted Pt. Electrochim. Acta 1995, 40(10), 1537–1543.
9. C. Gutierrez, J.A. Caram, Electrooxidation of dissolved CO on a platinum electrode covered with a monolayer of the chemisorbed CO formerly considered to be poison. J. Electroanal. Chem. 1991, 308, 321–325.
10. M. Harada, K. Asakura, N. Toshima, Structural analysis of polymer-protected platinum/rhodium bimetallic clusters using extended X-ray absorption fine structure

spectroscopy—Importance of microclusters for the formation of bimetallic clusters. J. Phys. Chem. 1994, 98(10), 2653–2662.

11. E. Herrero, K. Franaszczuk, A. Wiecskowski, A Voltammetric identification of the surface redox couple effective in methanol oxidation on a ruthenium-covered platinum (110) Electrode. J. Electroanal. Chem. 1993, 361(1–2), 269–273.

12. E. Herrero, K. Franaszczuk, A. Wieckowski, Electrochemistry of methanol at low index crystal planes of platinum: An integrated voltammetric and chronoamperometric study. J. Phys. Chem. 1994, 98(19), 5074–5083.

13. E. Herrero, W. Chrzanowski, A. Wieckowski, Dual path mechanism in methanol electrooxidation on a platinum electrode. J. Phys. Chem. 1995, 99(25), 10423–10424.

14. R. Ianniello, V.M. Schmidt, U. Stimming, J. Stumper, A. Wallau, CO adsorption and oxidation on Pt and Pt-Ru alloys: Dependence on substrate composition. Electrochim. Acta 1994, 39(11–12), 1863–1869.

15. T. Iwasita, W. Vielstich, The electrochemical oxidation of ethanol on platinum. J. Electroanal. Chem. 1988, 257, 319–324.

16. N. Kimizuka, K. Itaya, In situ scanning tunnelling microscopy of underpotential deposition—silver adlayers on Pt(111) in sulfuric acid solutions. Faraday Disc. 1992 (94), 117–126.

17. A.S. Lin, A.D. Kowalak, W.E. O'Grady, Studies of the role of water in the electrocatalysis of methanol oxidation. J. Power Sources 1996, 58(1), 67–72.

18. B.A.L. Lopez De Mishima, H.T. Mishima, G. Castro, Surface studies of Pt-Ru electrodeposits on gold. Electrochim. Acta 1995, 40(15), 2491–2500.

19. N.M. Markovic, H.A. Gasteiger, P.N. Ross, X.D. Jiang, I. Villegas, M.J. Weaver, Electro-oxidation mechanisms of methanol and formic acid on Pt-Ru alloy surfaces. Electrochim. Acta 1995, 40(1), 91–98.

20. N.M. Markovic, A. Widelov, P.N. Ross, O.R. Monteiro, I.G. Brown, Electrooxidation of CO and CO/H_2 mixtures on a Pt-Sn catalyst prepared by an implantation method. Catal. Letters 1997, 43(3–4), 161–166.

21. J. McBreen, S. Mukerjee, In situ x-ray absorption studies of a Pt-Ru electrocatalyst. J. Electrochem. Soc. 1995, 142(10), 3399–3404.

22. J. Munk, P.A. Christensen, A. Hamnett, E. Skou, The electrochemical oxidation of methanol on platinum and platinum plus ruthenium particulate electrodes studied by insitu FTIR spectroscopy and electrochemical mass spectrometry. J. Electroanal. Chem. 1996, 401(1–2), 215–222.

23. J.A. Poirier, G.E. Stoner, Microstructural effects on electronic oxygen reduction activity of nano-grained thin-film platinum in acid media. J. Electrochem. Soc. 1994, 141(2), 425–430.

24. V. Radmilovic, H.A. Gasteiger, P.N. Ross, Structure and chemical composition of a supported Pt-Ru electrocatalyst for methanol oxidation. J. Catal. 1995, 154(1), 98–106.

25. O. Savadogo, P. Beck, Five percent platinum-tungsten oxide-based electrocatalysts for phosphoric acid fuel cell cathodes. J. Electrochem. Soc. 1996, 143(12), 3842–3846.

26. P.J. Slezak, A. Wieckowski, Interfacing surface electrochemistry with solid-state NMR—characterization of surface CO on polycrystalline platinum. J. Magnetic Resonance Series A 1993, 102(2), 166–172.

27. M. Watanabe, Y. Genjima, K. Turumi, Electrochemical oxidation of methanol on platinum electrodes with ruthenium ad-atoms in hot phosphoric acid. Denki Kagaku, 1996, 64(6), 462–463.

28. K. Asakura, J. Lauterbach, H.H. Rotermund, G. Ertl, Spatio-temporal pattern formation during catalytic CO oxidation on a Pt(100) surface modified with submonolayers of Au. Surf. Sci. 1997, 374, 125–141.

29. T. Frelink, W. Visscher, J.A.R. Veen, The third anodic hydrogen peak on platinum; subsurface H_2 adsorption. Electrochim. Acta 1995, 40, 545–549.

30. W. Gorodetskii, A.V. Matveev, P.D. Cobden, B.E. Nieuwenhuys, Study of H_2, O_2, CO adsorption and $CO + O_2$ reaction on Pt(100), Pd(110) monocrystal surfaces. Molecular Catal.: A, 2000, 158, 155–160.

31. Y. Ishikawa, M.-S. Liao, C.R. Cabrera, Oxidation of methanol on platinum, ruthenium and mixed Pt-M metals (M = Ru, Sn): A theoretical study. Surf. Sci. 2000, 463, 66–80.

32. T.D. Jarvi, T.H. Madden, E.M. Stuve, Vacuum and electrochemical behavior of vapor deposited ruthenium on platinum (III). Electrochem. Solid State Lett. 1999, 2(5), 224–227.

33. C.A. Lucas, N.M. Markovic, B.N. Grgur, P.N. Ross, Structural effects during CO adsorption on Pt-bimetallic surfaces—I—The Pt(100) electrode. Surf. Sci. 2000, 448, 65–76.

34. C.A. Lucas, N.M. Markovic, P.N. Ross, Structural effects during CO adsorption on Pt-bimetallic surfaces. II—The Pt(111) electrode. Surf. Sci. 2000, 448, 77–86.

35. M.S. Zei, G. Ertl, On the structural transformation of the reconstructed Pt(100) in electrolyte solutions. Surf. Sci. 1999, 442, 19–26.

36. J. Aragane, H. Urushiba, T. Murahashi, Platinum utilization in a phosphoric acid fuel cell. Denki Kagaku 1995, 63(7), 642–647.

37. A.S. Arico, V. Antonucci, N. Giordano, A.K. Shukla, M.K. Ravikumar, A. Roy, S.R. Barman, D.D. Sarma, Methanol oxidation on carbon-supported platinum tin electrodes in sulfuric acid. J. Power Sources 1994, 50(3), 295–309.

38. A.S. Arico, Z. Poltarzewski, H. Kim, A. Morana, N. Giordano, V. Antonucci, Investigation of a carbon supported quaternary Pt-Ru-Sn-W catalyst for direct methanol fuel cells. J. Power Sources 1995, 55(2), 159–166.

39. P.C. Biswas, Y. Nodasaka, M. Enyo, Electrocatalytic activities of graphite-supported platinum electrodes for methanol electrooxidation. J. Appl. Electrochem. 1996, 26(1), 30–35.

40. P.S. Kauranen, E. Skou, J. Munk, Kinetics of methanol oxidation on carbon-supported Pt and Pt + Ru catalysts. J. Electroanal. Chem. 1996, 404(1), 1–13.

41. Z.D. Wei, H.T. Guo, Z.Y. Tang, Methanol electro-oxidation on platinum and platinum-tin alloy catalysts dispersed on active carbon. J. Power Sources 1996, 58(2), 239–242.

42. P.C. Biswas, Y. Nodaska, M. Enyo, Electrocatalytic activities of graphite-supported platinum electrodes for methanol electrooxidation. J. Appl. Electrochem. 1996, 26, 30–35.

43. K. Lasch, L. Jorissen, J. Garche, The effect of metal oxides as co-catalysts for the electro-oxidation of methanol on platinum-ruthenium. J. Power Sources 1999, 84, 225–230.

44. J.M. Rheaume, B. Muller, M. Schulze, XPS analysis of carbon-supported platinum electrodes and characterization of CO oxidation on PEM fuel cell anodes by electrochemical half cell methods. J. Power Sources 1998, 76, 60–68.

45. A.V. Tripkovic, K.D. Popovic, Oxidation of methanol on platinum single crystal stepped electrodes from [110] zone in acid solution. Electrochim Acta 1996, 41(15), 2385–2394.

46. E. Herrero, K. Franaszczuk, A. Wieckowski, Electrochemistry of methanol at low index crystal planes of platinum: An integrated voltammetric and chronoamperometric study. J. Phys. Chem. 1994, 98, 5074–5083.

47. H. Kim, I. Rabelo de Moraes, G. Tremiliosi-Filho, R. Haasch, A. Wieckowski, Chemical state of ruthenium submonolayers on a Pt(111) electrode. Surf. Sci. 2001, 474, L203–L212.

48. N.M. Markovic, C.A. Lucas, V. Climent, V. Stemenkovic, P.N. Ross, Surface electrochemistry on an epitaxial palladium film on Pt(111): Surface microstructure and hydrogen electrode kinetics. Surf. Sci. 2000, 465, 103–114.

49. M.S. Zei, G. Ertl, On the structural transformation of the reconstructed Pt(100) in electrolyte solutions. Surf. Sci. 1999, 442, 19–26.

50. H.A. Gasteiger, N.M. Markovic, P.N. Ross, H_2 and CO electrooxidation on well-characterized Pt, Ru, and Pt-Ru.1. Rotating disk electrode studies of the pure gases including temperature effects. J. Phys. Chem. 1995, 99(20), 8290–8301.

51. T.D. Jarvi, S. Sriramulu, E.M. Stuve, Reactivity and extent of poisoning during methanol electrooxidation on platinum (100) and (111): A comparative study. Colloids and Surfaces 1998, 134, 145–153.

52. T. Frelink, W. Visscher, J.A.R. Vanveen, Particle size effect of carbon-supported platinum catalysts for the electrooxidation of methanol. J. Electroanal. Chem. 1995, 382(1–2), 65–72.

53. P.W. Jacobs, S.J. Wind, F.H. Ribeiro, G.A. Somorjai, Nanometer size platinum particle arrays: Catalytic and surface chemical properties. Surf. Sci. 1997, 372(1–3), L249–L253.

54. T. Frelink, W. Visscher, J.A.R. Veen, Particle size effect of carbon-supported platinum catalysts for the electrooxidation of methanol. J. Electroanal. Chem. 1995, 382, 65–72.

55. U.A. Paulus, T.J. Schmidt, H.A. Gasteiger, R.J. Behm, Oxygen reduction on a high-surface area Pt/Vulcan carbon catalyst: A thin-film rotating ring-disk electrode study. J. Electroanal. Chem. 2001, 495, 134–145.

56. Y.Y. Tong, C. Belrose, A. Wieckowski, E. Oldfield, First observation of platinum-195 nuclear magnetic resonance in commercial graphite-supported platinum electrodes in an electrochemical environment. J. Amer. Chem. Soc. 1997, 119, 11709–11710.

57. H.A. Gasteiger, N.M. Markovic, P.N. Ross, Structural effects in electrocatalysis: Electrooxidation of carbon monoxide on Pt_3Sn single-crystal alloy surfaces. Catal. Letters 1996, 36(1–2), 1–8.

58. W. Chrzanowski, A. Wieckowski, Surface structure effects in platinum/ruthenium methanol oxidation electrocatalysis. Langmuir 1998, 14, 1967–1970.

59. S.P.E. Smith, H.D. Abruña, Structural effects on the oxidation of HCOOH by bismuth modified Pt(111) electrodes with (110) monatomic steps. Electroanal. Chem. 1999, 467, 43–49.

60. G. Tamizhmani, J.P. Dodelet, D. Guay, Crystallite size effects of carbon-supported platinum on oxygen reduction in liquid acids. J. Electrochem. Soc. 1996, 143, 18–22.

61. O. Solorza-Feria, K. Ellmer, M. Giersig, N. Alonso-Vante, Novel low-temperature synthesis of semiconducting transition metal chalcogenide electrocatalyst for multi-electron charge transfer: Molecular oxygen reduction. Electrochim. Acta 1994, 39(11–12), 1647–1653.

62. N. Alonso-Vante, H. Tributsch, O. Solorza-Feria, Kinetics studies of oxygen reduction in acid medium on novel semiconducting transition metal chalcogenides. Electrochim. Acta 1995, 40(5), 567 576.

63. V. Le Rhun, E. Garnier, S. Pronier, N. Alonso-Vante. Electrocatalysis on nanoscale ruthenium-based material manufactured by carbonyl decomposition. Electrochem. Commun. 2000, 2, 475–479.

64. V. Le Rhun, N. Alonso-Vante, Tailoring of nanodivided electrocatalyst materials based on transition metal. J. New Mater. Electrochem. Syst. 2000, 2, 475–479.

65. N. Alonso-Vante, Electrocatalyse par l'intermédiaire des centres métalliques de composés de métaux de transition. Reduction de l'oxygène moléculaire. J. Chim. Phys. 1996, 93, 702.

66. B.F.G. Johnson, M.A. Gallup, Y.V. Roberts, Molecular clusters as models of metallic catalysts. J. Mol. Catal. 1994, 86(1–3), 51–69.

67. G. Schmid, U. Giebel, W. Huster, A. Schwenck, Large transition metal clusters, 2[1]. Synthesis and properties of $Rh_{55}[P(tert-Bu)_3]_{12}Cl_{20}$. Inorg. Chim. Acta 1984, 85, 97–102.

68. N.M. Vargaftik, I.I. Moiseev, D.I. Kochubey, K.I. Zamaraev, Giant palladium clusters: Synthesis and characterization. Faraday Discuss. Chem. Soc. 1991, 92, 13–29.

69. M. Mingo, D. Wales, Introduction to Cluster Chemistry, Prentice Hall 1990.

70. J.S. Bradley, In Cluster and Colloids, G. Schmid (ed.), VCH Publishers, Inc.; New York, 1994; Chapter 6.

71. H. Bönnemann, G. Braun, W. Brijoux, R. Brinkmann, A. Schulze Tilling, K. Seevogel, K. Siepen, Nanoscale colloidal metals and alloys stabilized by solvents and surfactants. Preparation and use as catalyst precursors. J. Organomet. Chem. 1996, 520, 143–162.

72. H. Bönnemann, P. Britz, W. Vogel, Structure and chemical composition of a surfactant-stabilized Pt_3Sn alloy colloid. Langmuir 1998, 14(23), 6654–6657.

73. T.J. Schmidt, M. Noeske, H.A. Gasteigert, R.J. Behm, P. Britz, H. Bönnemann, PtRu alloy colloids as precursors for fuel cell catalysts: A combined XPS, AFM, HRTEM and RDE study. J. Electrochem. Soc. 1998, 145(3), 925–931.

74. K.-I. Machida, A. Fukuoka, M. Ichikawa, M. Enyo, Preparation of chemically modified electrodes attachement of platinum carbonyl clusters, and their efficient electrocatalytic action in anodic oxidation of methanol. J. Chem. Soc. Chem. Commun. 1987, 1486–1487.

75. K.-I. Machida, A. Fukuoka, M. Ichikawa, M. Enyo, Preparation of platinum cluster-derived electrodes from metal carbonyl complexes and their electrocatalytic properties for anodic oxidation of methanol. J. Electrochem. Soc. 1991, 138(7) 1958–1965.

76. G. Longoni, P. Chini, Synthesis and chemical characterization of platinum carbonyl dianions $[Pt_3(CO)_6]_n^{2-}$ ($n = \approx 10, 6, 5, 4, 3, 2, 1$). A new series of inorganic oligomers. J. Amer. Chem. Soc. 1976, 98, 7225–7231.

77. Unpublished results.

78. R.D. Adams, W. Wengan, Cluster synthesis. 41. New platinum-ruthenium cluster complexes from the reaction of $PhC \equiv CPh$ with $Pt_2Ru_4(CO)_{18}$. Synthesis and structural characterizations of $Pt_2Ru_3(CO)_8(\mu_3-\eta^2-PhC_2Ph)_2(\mu_4-\eta^2-PhC_2Ph)$, $Pt_3Ru_6(CO)_{14}(\mu_3-\eta^2-PhC_2Ph)_3$, $Pt_2Ru_4(CO)_{14}(\mu_3-\eta^2-PhC_2Ph)$ $(\mu_4-\eta^2-PhC_2Ph)$, $Pt_3Ru_6(CO)_{18}(\mu_3-\eta^2-PhC_2Ph)_3$, and $PtRu_2(CO)_6(\mu_3-\eta^2-PhC_2Ph)(dppe)$. Organometallics 1993, 12, 1248–1256.

79. R.D. Adams, G. Chen, W. Wu, The synthesis and structural analysis of $Pt_2Ru_4(CO)_{18}$ and the products obtained from its reactions with 1,2-bis-diphenylphosphino)ethane. J. Cluster Sci. 1993, 4(2), 119–132.

80. C.W. Hills, S. Nashner, A.I. Frenkel, J.R. Shapley, R.G. Nuzzo, Carbon support effects on bimetallic Pt-Ru nanoparticles formed from molecular precursors. Langmuir 1999, 15, 690–700.

81. P.L. Antonucci, A.S. Arico, P. Creti, E. Ramuni, V. Antonucci, Investigation of a direct methanol fuel cell based on a composite Nafion® -silica electrolyte for high temperature operation. Sol. State Ionics 1999, 125, 431–437.

82. K. Scott, W.M. Taama, P. Argyropoulos, Performance of the direct methanol fuel cell with radiation-grafted polymer membranes. J. Membr. Sci. 2000, 171, 119–130.

83. N. Alonso-Vante, M. Giersig, H. Tributsch, Thin layer semiconducting cluster electrocatalysts for oxygen reduction. J. Electrochem. Soc. 1991, 138, 639–640.

84. (a) T.M. Layer, J. Lewis, A. Martin, P.R. Raithby, W.-T. Wong, The chemistry of the selenium-containing cluster $[Ru_4(\mu_4\text{-}Se)_2(CO)_8(\mu\text{-}CO)_3]$: Crystal structures of $[Ru_3(\mu_4\text{-}Se)_2(CO)_7(Ph_2PC_2PPh_2)_3]$, $[Ru_4(\mu_4\text{-}Se)_2(CO)_{10}(SbPh_3)]$ and $[Ru_3(\mu_3\text{-}Se)_2(CO)_9]$. J. Chem. Soc. Dalton Trans. 1992, 3411–3417. (b) B.F.G.L. Johnson, T.M. Layer, P.R. Raithby, W.-T. Wong, Synthesis and characterization of novel phosphido- and selenido substituted ruthenium carbonyl clusters. J. Chem. Soc. Dalton Trans. 1993, 973–980.

85. N. Alonso-Vante, H. Tributsch, Energy conversion catalysis using semiconducting transition metal cluster compounds. Nature 1986, 323, 431–432.

86. N. Alonso-Vante, W. Jaegermann, H. Tributsch, W. Hönle, K. Yvon, Electrocatalysis of oxygen reduction by chalcogenides containing mixed transition metal clusters. J. Amer. Chem. Soc. 1987, 109, 3251–3257.

87. N. Alonso-Vante, B. Schubert, H. Tributsch, A. Perrin, Influence of d-state density and chemistry of transition metal cluster selenides on electrocatalysis. J. Catal. 1988, 112, 384–391.

88. N. Alonso-Vante, B. Schubert, H. Tributsch, Transition metal cluster materials for multi-electron transfer catalysis. Materials Chem. Phys. 1989, 22, 281–307.

89. R. Chevrel, M. Sergent, In *Superconductivity of Ternary Compounds*, O. Fischer, M.B. Maple (eds.), Springer-Verlag, Heidelberg, New York 1982, p 25.

90. N.N. Greenwood, A. Earnshaw, *Chemistry of the Elements*, 2nd ed.; Butterworth-Heinemann, Oxford 1998, p 1022.

91. N. Alonso-Vante, Verfahren zur Herstellung einer inerten Kathode für die selektive Sauerstoffreduktion und Anwendung der hergestellten Kathode DE 196 44 628 C2, October 17, 1996.

92. V. Le Rhun, Ph.D. Thesis, 2001, University of Poitiers, Poitiers, France.

93. N. Alonso-Vante, M. Fieber-Erdmann, P. Borthen, H. Strehblow, E. Holub-Krappe, An in situ grazing incidence X-ray absorption study of ultra thin Ru_xSe_y cluster-like electrocatalyst layers. Electrochim. Acta, 2000, 45, 4227–4236.

94. A. Forster, B.F. Johnson, J. Lewis, T.W. Matheson, B.H. Robinson, W.G. Jackson, ^{13}C nuclear magnetic resonance study of the trinuclear metal carbonyls $M_3(CO)_{12}$ (M = Fe, Ru and Os) and some related compounds. J.C.S. Chem. Comm. 1974, 1042–1044.

95. W. Vogel, V. Le Rhun, E. Garnier, N. Alonso-Vante, Ru clusters synthesized chemically from dissolved carbonyl: In situ study of novel electrocatalyst in the gas phase and in electrochemical environment. J. Phys. Chem B 2001, 105(22), 5238–5243.

96. K.-L. Hsueh, D.-T. Chin, S. Srinivasan, Electrode kinetics of oxygen reduction. A theoretical and experimental analysis of the rotating ring-disk electrode method. J. Electroanal. Chem. 1983, 153, 79–95.

97. E. Ticianelli, J.G. Beery, M.T. Paffett, S. Gottesfeld, An electrochemical, ellipsometric, and surface science investigation of the PtRu bulk alloy surface. J. Electroanal. Chem. 1989, 258, 61–77.

98. Unpublished results: Ru_x nanoparticles synthesized in xylene has the tendency to coordinate easily with oxygen from air (a gentle treatment with hydrogen at RT releases the oxygen [95]). The EXAFS function $\chi(k)$ damped quickly on these Ru_xO_y nanoparticles revealing that high disorder increases (nanoparticles become amorphous).

99. P. Villars, L.D. Calvert, eds. *Pearson's Handbook of Crystallographic Data for Intermetallic Phase*, American Society for Materials: Metals Park, OH 1985.

100. N. Alonso-Vante, M. Fieber-Bordmann, to be submitted to Electrochim. Acta.

101. N. Alonso-Vante, P. Borthen, M. Fieber-Erdmann, A contribution towards the understanding of the catalytic center on novel clustering materials, In *New Materials for Fuel Cells and Modern Battery Systems* II; O. Savadogo, P.R. Roberge, eds. 1997, pp. 654–665.

102. I.V. Malakov, S. Nikitenko, E. Savinova, D. Kochubey, N. Alonso-Vante, In situ EXAFS study to probe active centers of Ru chalcogenide electrocatalysts during oxygen reduction reaction. Phys. Chem. B. 2002, 106(7), 1670–1676.

103. N. Alonso-Vante, M. Fieber-Erdmann, H. Rossner, E. Holub-Krappe, C. Giorgetti, A. Tadjeddine, E. Dartyge, A. Fontaine, R. Frahm, The catalytic centre of transition metal chalcogenides vis-à-vis the oxygen reduction reaction: An in situ electrochemical EXAFS study. J. de Physique IV France 1997, 7, C2-887–C2-889.

104. N. Alonso-Vante, S. Cattarin, M. Musiani, Electrocatalysis of O_2 reduction at polyaniline + molybdenum-doped ruthenium selenide composite electrodes. J. Electroanal. Chem. 2000, 481, 200–207.

Index